D1658764

Hans-Joachim Gabius · Sigrun Gabius (Eds.)

Glycosciences

Status and Perspectives

Published by
Chapman & Hall GmbH, Pappelallee 3, 69469 Weinheim, Germany

Chapman & Hall, 2–6 Boundary Row, London SE1 8HN, UK

Blackie Academic & Professional, Wester Cleddens Road, Bishopbriggs, Glasgow G64 2NZ, UK

Chapman & Hall GmbH, Pappelallee 3, 69469 Weinheim, Germany

Chapman & Hall USA, 115 Fifth Avenue, New York, NY 10003, USA

Chapman & Hall Japan, ITP-Japan, Kyowa Building, 3F, 2-2-1 Hirakawacho, Chiyoda-ku, Tokyo 102, Japan

Chapman & Hall Australia, 102 Dodds Street, South Melbourne, Victoria 3205, Australia

Chapman & Hall India, R. Seshadri, 32 Second Main Road, CIT East, Madras 600 035, India

ISBN 3-8261-0073-5

Hans-Joachim Gabius · Sigrun Gabius (Eds.)

Glycosciences

Status and Perspectives

With 136 Figures, Some in Color

CHAPMAN & HALL
an International Thomson Publishing company ITP®

London · Glasgow · Weinheim · New York · Tokyo · Melbourne · Madras

Prof. Dr. Hans-Joachim Gabius
Institut für Physiologische Chemie
Tierärztliche Fakultät
Ludwig-Maximilians-Universität
Veterinärstraße 13
D-80539 München
Germany

Dr. Sigrun Gabius
Hämatologisch-Onkologische
Schwerpunktpraxis
Sternstraße 12
D-83022 Rosenheim
Germany

Library of Congress and British Library Cataloging-in-Publication Data applied for

Book Production: PRO EDIT GmbH, D-69126 Heidelberg
Typesetting: Mitterweger Werksatz GmbH, D-68723 Plankstadt
Printing: Stürtz AG, Universitätsdruckerei, D-97017 Würzburg
Printed in the Federal Republic of Germany
Printed on acid-free paper

Foreword

Intuition has started to ascribe the potential of a versatile, but elusive code system to the at first sight puzzling array of sugar structures on proteins and lipids, as underscored by a quotation from the preface to an authoritative symposium nearly two decades ago: it states that "many knowledgeable biologists would say, almost reflexly, that complex carbohydrates probably play a pivotal role in determining the specificity of many biological recognition phenomena" (Marchesi *et al.*, 1978). In the following years, the necessity to address comprehensively the enigmatic issue that the oligosaccharide chains of cellular glycoconjugates continued to be referred to as "molecules in search of a function" has been repetitively emphasized (Cook, 1986). The last decade has unmistakably witnessed marked upheavals in our understanding, with a dynamic development which can legitimately be called a permanent revolution. Technical advances in oligosaccharide synthesis, purification and structural analysis, together with the steadily growing knowledge about endogenous binding partners such as lectins, have enabled research activities in the different areas of glycosciences to exert an obvious influence on various disciplines in basic and applied sciences, as depicted in the figure. Due to the amazingly rapid flow of information, the impending danger that guiding concepts and ideas might easily be submerged should not be carelessly ignored. This line of reasoning provided the impetus for us to contact colleagues from various branches of glycosciences and kindly confront them with our request to prepare a special chapter for this collection. First of all, it was intended to capture and transmit the spirit of jointly shared enthusiasm about our field, simultaneously familiarizing the interested reader with

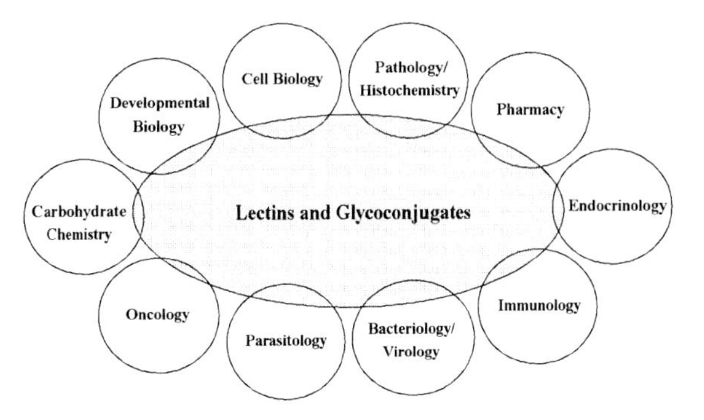

the status of each selected area based on an authoritative, though not necessarily encyclopedic appraisal of the evidence. Conceptual recognition is the essential step for one to be able to follow subsequently the inevitably subjective descriptions of the perspective with an appreciation of the inherently interdisciplinary character of glycosciences. Having been wholeheartedly delighted with the responses, we now have the pleasure of inviting the readers to wander with each author to the edges of the frontiers of our knowledge in glycosciences, to reflect on the current status and then to feel encouraged to go confidently beyond these frontiers. If turning "perspective" into "status" is stimulated by the contributions to this book, authors, publisher and editors will definitely have reached their goal.

References

Cook GMW (1986): Cell surface carbohydrates: molecules in search of a function? In *J. Cell Sci.* Suppl. **4**:45–70.

Marchesi VT, Ginsburg V, Robbins PW *et al.* (1978): Preface. In *Cell Surface Carbohydrates and Biological Recognition*. Progr. Clin. Biol. Res. **23** (Marchesi VT, Ginsburg V, Robbins PW *et al.*, eds) p 13, New York: Alan R. Liss.

H.-J. and S. Gabius (Eds.), Glycosciences

ISBN 3-8261-0073-5

Preface

FRIEDRICH CRAMER: A decade of glycosciences

My first paper was published in 1948 and it was one in the field of glycosciences (Freudenberg and Cramer, 1948). Thus glycosciences opened for me the happy roads to science. It is therefore a great satisfaction for me that the editors have asked me to write this preface to their splendid book on GLYCOSCIENCES, in which so many highly important and relevant topics are treated by the most competent authors. That first paper of mine solved the structure and explained the properties of α-, β-, and γ-cyclodextrin. From then on I was able to develop the story of inclusion compounds and supramolecular chemistry (Cramer, 1952, 1954, 1955; Cramer *et al.*, 1967). Already in this first paper three facts had become apparent.

i: Carbohydrates are capable of forming hydrogen bonds. This fact is obvious, since there are so many hydroxyl groups, which normally make sugars hydrophilic molecules. However, this hydrogen bonding can also be used for internal stabilization of structure and for the interaction with other molecules, e.g. with proteins. This has now been shown in studies of sugar-binding proteins.

ii: Carbohydrates have a fairly rigid structure. For many years this was an assumption, but has now been proved by NMR studies in various ways. In oligosaccharides, as well as in complex carbohydrates, there are a number of internal hydrogen bonds which stabilize certain conformations to a surprising extent. There are no α-helices or β-pleated sheets as in proteins. Nevertheless, complex carbohydrates can be recognized by sugar-binding molecules such as lectins in an unambiguous way. Complex carbohydrate structures on cell surfaces very often play an important role as antigens, whether in bacterial or eukaryotic cells.

iii: Complex carbohydrates may have hydrophobic pockets. This became quite obvious in cyclodextrins (Freudenberg and Cramer, 1948), but it may also play an important role in a variety of carbohydrate interactions. It might well be that certain transport functions, lipid interactions and associated phenomena make use of this partial hydrophobicity of complex carbohydrates.

Chemistry and biochemistry of complex carbohydrates and glycosciences in general have been a neglected field until perhaps one decade ago. This does not mean that this part of science was not highly successful, but that it was a rather heroic area in many respects. Chemistry was just not sufficiently advanced to take up such complex problems. Without the modern techniques of natural product isolation, without NMR, without mass spectroscopy, today one can hardly imagine any success in this field. And yet the basic principles and the basic structures had been determined a long time ago. I still remember when, during my early times in research, some colleagues next door worked on the characterization of blood group-specific substances in the lab of Karl Freudenberg. This was 45 years ago and it was indeed an heroic effort. Thus, it seems hardly surprising that only 12 years ago in a leading textbook one could read the following sentences: *The function of the oligosaccharide side chains in membrane glycolipids and glycoproteins is unclear. It is possible that those in certain transmembrane proteins help to anchor and orient the proteins in the membrane by preventing them from slipping into the cytosol or from tumbling across the bilayer. The carbohydrate also may play a role in stabilizing the folded structure of a glycoprotein. In addition carbohydrate may play a role in guiding a membrane glycoprotein to its appro-*

H.-J. and S. Gabius (Eds.), Glycosciences

ISBN 3-8261-0073-5

priate destination in or on these cells, just as special carbohydrate chains on lysosomal enzymes direct these soluble glycoproteins to lysosomes. However, these cannot be the only function of carbohydrate in membrane glycoproteins... Moreover, functions such as orienting, anchoring, stabilizing and targeting can not account for the carbohydrate in glycolipid molecules nor for the complexity of some of the carbohydrate chains in glycoproteins (Alberts *et al.*, 1983).

Glycoproteins are an essential part of the cellular membrane, and I consider the cellular membrane as a *multipurpose organelle*. It is not simply a sacculus, which separates the inside from the outside. It has an enormous number of functions, it possesses gates, channels, receptors, it provides specific contacts, it is important in cell aggregations and dissociations, it exposes signals and can receive signals, it is responsible for the ordered and regular embryonic growth, and it enables the cell to make highly specific contacts such as synaptic connections in CNS. Most likely all these capacities of the cell reside in the structure and arrangement of the complex carbohydrates on the surface of the cell. Fig. 1 gives a schematic picture of the situation, in which lipids, proteins and complex carbohydrates cooperate through a three-dimensional interaction, the details of which are indeed still largely unknown (from Alberts *et al.*, 1983).

Indeed, one must ask the question: what are these complex carbohydrates good for? The facts

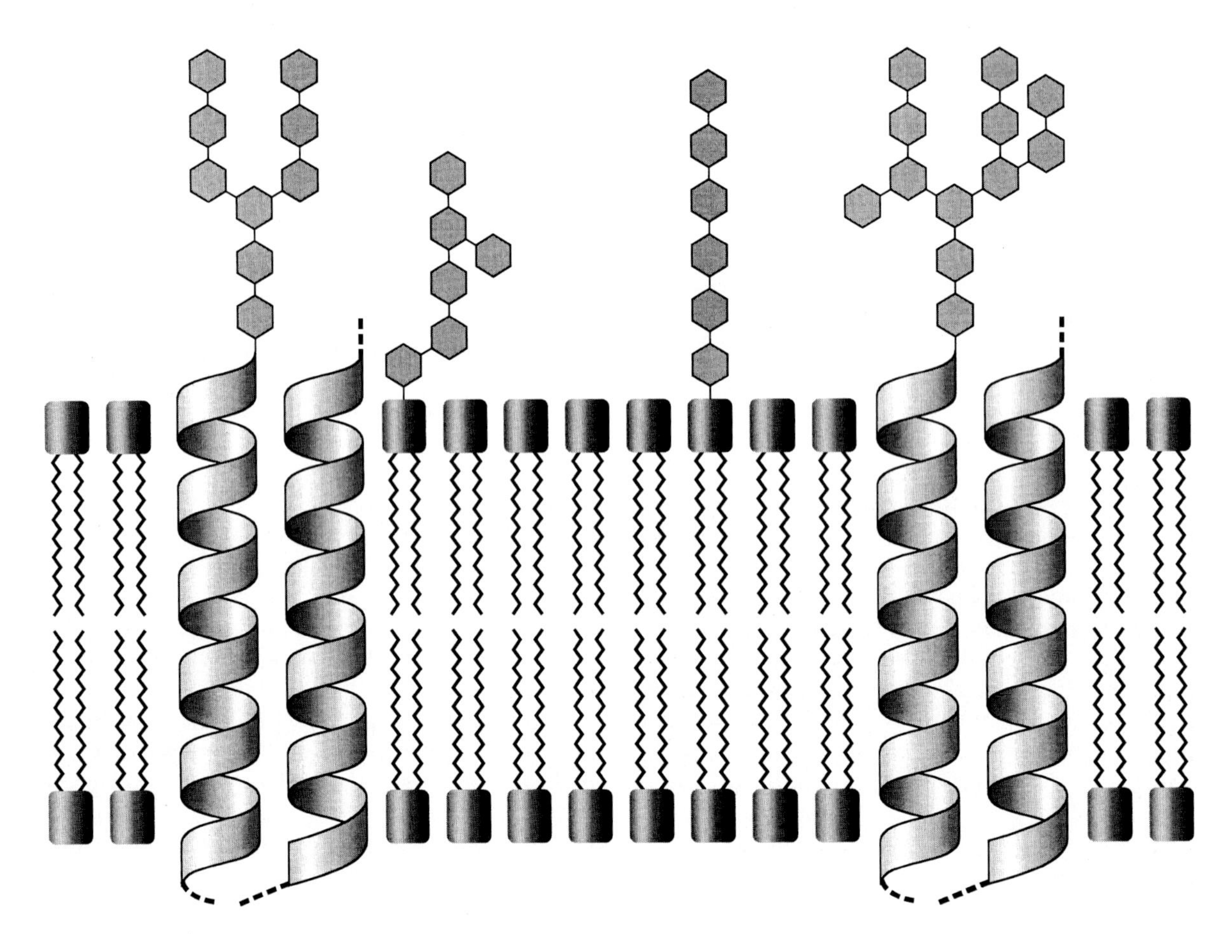

Figure 1. A diagram of the cell code (Glycocalyx) which is made up of the oligosaccharide side chains of glycolipids and integral membrane glycoproteins, and polysaccharide chains of integral proteoglycans. In addition, glycoproteins and adsorbed proteoglycans contribute to the glycocalyx in some cells (Alberts *et al.*, 1983)

that these molecules are synthesized through the action of an enormous number of different and highly specific enzymes, that these enzymes are under strict genetic control, and that the complex carbohydrates are produced with very high precision, all this may well indicate that these glycostructures play a very important and indispensable role in cellular events, and that they are not just there for solubilization purposes. Are they important for *cell adhesion and cell recognition*? If this were the case, there should exist specific receptors for complex carbohydrates, which would allow for such specific recognitions.

About 10 years ago we started the search for such sugar-recognizing molecules (Cramer and Gabius, 1985) and, indeed, found such carbohydrate receptors in many mammalian cells. Carbohydrate-recognizing molecules were known for a very long time in the form of plant lectins. However, the general opinion amongst the scientific community was that such lectins were restricted to the plant kingdom, and nobody really looked into the tissue of higher organisms. As several chapters in this book elegantly describe, tools for lectin detection and their application have enriched our view in this area profoundly. We are now even at a stage where functions can be understood at least in part, as will be discussed in the present volume.

I consider it a great merit of the editors to give – as far as I can see for the first time – a complete overview of modern glycosciences. The topics range from chemical analysis and chemical synthesis through structural determinations to the various kinds of carbohydrate interactions, which have become so important during the last years, and go right into biochemical topics like the glycobiology of host defense mechanisms, glycobiology of signal transduction, or glycobiology of development and fertilization. It is obvious that such a topic also touches problems of medical importance such as mechanisms of infection, tumor biology and histopathology. I believe that after a decade of intensive research in glycosciences, the effort of the editors will be greatly appreciated by the biochemical, biological and medical scientific community.

References

Alberts B, Brays D, Lewis J *et al.* (1983): Molecular Biology of the Cell. p 285, New York: Garland Publ. Inc.
dto., p 284 (with kind permission).

Cramer F (1952): Einschlußverbindungen. In *Angew. Chem.* **64**:437–47.

Cramer F (1954): Einschlußverbindungen. Heidelberg: Springer Verlag. Russian translation, Moskau 1957.

Cramer F (1955): Inclusion compounds. In *Pure Appl. Chem.* **5**: 143–64.

Cramer F, Gabius H-J (1985): New carbohydrate-binding proteins (lectins) in human cancer cells and their possible role in cell differentiation and metastasis. In *Interrelationship among Ageing, Cancer and Differentiation* (Jerusalem Symp. on Quantum Chemistry and Biochemistry, Vol. 18) Pullmann B, Ts'o POP, Schneider EL (eds) pp 187–205, Dordrecht: Reidel.

Cramer F, Saenger W, Spatz HC (1967): Inclusion compounds. XIX.: The formation of inclusion compounds of α-cyclodextrins in aqueous solutions. Thermodynamics and kinetics. In *J. Am. Chem. Soc.* **89**: 14–20.

Freudenberg K, Cramer F (1948): Die Konstitution der Schardinger-Dextrine α, β und γ. In *Z. Naturforsch.* **3b**:464–74.

Contents

Chapter 1 The Information-Storing Potential of the Sugar Code
Roger A. Laine 1

1 Introduction 1
2 Details of Calculation for Isomers of a Trisaccharide 2
2.1 Non-Reducing Oligosaccharides 3
3 Analysis 3
3.1 NMR 3
3.2 Mass Spectrometry 4
4 Synthesis 4
5 Biologically Relevant Oligomer Size . 4
6 Substitutions 4
7 Oligosaccharide Recognition by Proteins 4
8 Evolution of Carbohydrate Code Structures and Receptors 5
9 A "High Level" Biological Code . . . 6
10 Lectins in Biological Recognition . . . 6
11 Biological Mechanisms of Lectins or other Carbohydrate-Binding Proteins . 6
12 Multivalent Effects 7
13 Bacterial/Eucaryotic Interactions . . . 7
14 Pharmaceutical Development 8
15 Glycobiology 8
16 Review of the Formal Calculation of Oligosaccharide Isomers 8
16.1 Linear Oligosaccharides 8
16.2 Branched Oligosaccharides 9
16.3 Formulas 10
17 Oligosaccharide Building Blocks 11
18 Summary 11

Chapter 2 Methods of Glycoconjugate Analysis
Elizabeth F. Hounsell 15

1 Introduction 15
1.1 Monosaccharide Identification 15
1.2 Chemical Reactions 18
2 Biochemical Reagents 18
2.1 Lectin Affinity Interactions 19
2.2 Sequential Exoglycosidase Digestion 20
3 Biophysical Methods 21
3.1 Mass Spectrometry 21
3.2 NMR Spectroscopy 22
4 Glycoprotein Analysis 24
4.1 Proteases and Endoglycosidases 24
4.2 HPLC Purification and Analysis 25
4.3 Capillary Electrophoresis 25

Chapter 3 Strategies for the Chemical Synthesis of Glycoconjugates
Richard R. Schmidt 31

1 Introduction 31
2 General Strategies 31
2.1 Typical Building Blocks for Glycoside Synthesis 31

H.-J. and S. Gabius (Eds.), Glycosciences

ISBN 3-8261-0073-5

2.2 Methodologies in Enzymatic and in Chemical Glycoside Bond Formation – a Comparison . . . 32
2.3 General Aspects of Chemical Glycoside Bond Formation . . . 33
3 The Koenigs-Knorr Procedure and its Variations . . . 34
3.1 The Koenigs-Knorr Procedure . . . 34
3.2 Fluoride as Leaving Group . . . 36
3.3 1,2-Anhydro (1,2-Epoxide) Sugars as Glycosyl Donors . . . 37
3.4 Sulfur as Leaving Group . . . 37
3.5 Nitrogen as Leaving Group . . . 41
4 The Trichloroacetimidate Method, the Phosphite Method, and Other Anomeric Oxygen Activation Procedures . . . 41
4.1 The Trichloroacetimidate Method . . . 41
4.2 The Phosphite Method . . . 48
4.3 Other Anomeric Oxygen Activation Procedures . . . 50
5 Concluding Remarks . . . 51

Chapter 4 Neoglycoconjugates
Reiko T. Lee and Yuan C. Lee . . . 55

1 Introduction . . . 55
2 Advantages of Neoglycoconjugates . . . 55
2.1 Well-Defined Structure . . . 55
2.2 Make as Much as Needed . . . 56
2.3 Glycoside Cluster Effect . . . 56
2.4 Neoglycoconjugates Containing Analogs . . . 56
2.5 Changing Personalities . . . 57
3 Neoglycoproteins . . . 58
3.1 Side Chains of Proteins Useful in Modification . . . 58
3.2 Modification of Primary Amino Groups . . . 58
3.3 Conjugation of Polysaccharides to Proteins . . . 60
3.4 Use of Enzymes . . . 61
3.5 Glycoproteins of Non-Covalent Attachment . . . 63
3.6 Synthetic Glycopeptides . . . 63
4 Neoglycolipids . . . 64
4.1 Neoglycolipids for Thin Layer Chromatographic Separation . . . 64
4.2 Transglycosylation of Ceramide Glycanase . . . 65
5 Glycolipids into Neoglycoproteins and vice versa . . . 65
6 Neoproteoglycans . . . 65
7 Neoglycopolymers . . . 66
8 Derivatives Useful for Preparation of Neoglycoconjugates . . . 68
8.1 Glycosides . . . 68
8.2 Glycosylamines and Glycamines . . . 70
8.3 Attachment of Glycopeptides with a Heterobifunctional Reagent . . . 70
9 Application of Neoglycoconjugates . . . 71
9.1 Probing Carbohydrate-Protein Interactions . . . 71
9.2 Use in Isolation of Carbohydrate-Binding Proteins . . . 72
9.3 Cytochemical Markers . . . 72
9.4 Neoglycoenzymes . . . 73
9.5 Biomedical Applications . . . 73
10 Conclusions and Perspectives . . . 73

5 Glycosyltransferases Involved in *N*- and *O*-Glycan Biosynthesis
Inka Brockhausen and Harry Schachter . . . 79

1 Introduction . . . 79
2 The Roads which Lead to Complex *N*-Glycans . . . 80
3 Glycosyltransferases Involved in the Synthesis of Complex *N*-Glycan Antennae . . . 81
3.1 Galactosyltransferases . . . 81
3.2 Sialyltransferases . . . 84
3.3 *N*-Acetylglucosaminyltransferases . . . 89
3.4 Fucosyltransferases . . . 93
4 Biosynthesis of *O*-Glycans . . . 96

4.1 Initiation of *O*-Glycan Biosynthesis. UDP-GalNAc: Polypeptide α1,3-*N*-Acetylgalactosaminyltransferase (polypeptide GalNAc T; E.C. 2.4.1.41) 97
4.2 Synthesis of *O*-Glycan Core Structures 97
4.3 Elongation of *O*-Glycans 98
4.4 Termination Reactions in the Synthesis of *O*-Glycans. 99
5 Conclusion. 100

Chapter 6 Topology of Glycosylation – a Histochemist's View
MARGIT PAVELKA 115

1 Introduction. 115
2 Topology of *N*-Glycosylation 115
3 Topology of O-Glycosylation 118
4 Summarizing Remarks and Perspectives 118

Chapter 7 Occurrence and Potential Functions of *N*-Glycanases
TADASHI SUZUKI ET AL. 121

1 Introduction. 121
2 *N*-Glycosylation/De-*N*-glycosylation 121
3 Occurrence of PNGases: The Discovery of Animal PNGases . . 122
4 PNGases in Fish 123
4.1 Acid PNGase in Fish Oocyte is Responsible for the Detachment of *N*-Glycan Chains from Glycophosphoproteins 123
4.2 *N*-Glycan Release by Alkaline PNGase during Embryogenesis 123
5 PNGase in Mouse-Derived Cultured Cells: L-929 PNGase 124
6 Neutral, Soluble PNGases are Widely Distributed in Mouse Organs 125
7 Possible Biological Function of Neutral PNGase in "Non-Lysosomal" Degradation. 126
8 Plant PNGases: Possible Regulator Molecules in Cellular Processes 127
9 Concluding Remarks 128

Chapter 8 Glycoproteins: Structure and Function
NATHAN SHARON AND HALINA LIS 133

1 Introduction. 133
2 Structure 134
2.1 Monosaccharide Constituents 135
2.2 Carbohydrate-Peptide Linking Groups 137
2.3 Oligo- and Polysaccharides 139
3 Functions. 146
3.1 Methodology 146
3.2 Modulation of Physicochemical Properties 147
3.3 Modulation of Biological Activity . . . 149
3.4 Cellular Immune Functions 153
3.5 Activities of Free Oligosaccharides . . 154
3.6 Carbohydrates as Recognition Determinants 154

Chapter 9 Glycolipids: Structure and Function

JÜRGEN KOPITZ 163

1 Structure, Classification and Localization of Glycolipids 163
1.1 Bacterial Glycolipids 163
1.2 Plant Glycolipids 164
1.3 Animal Glycolipids 165
2 Elucidation of Glycolipid Structure . . 168
3 Physicochemical Properties and Organization of Glycolipids 169
4 Metabolism and Intracellular Trafficking of Glycolipids 169
5 Physiological Functions of Glycolipids 172
5.1 Bacterial and Plant Glycolipids 172
5.2 Animal Glycolipids 173
6 Functions in Disease Mechanisms . . . 177
6.1 Oncogenic Transformation 177
6.2 Neurodegeneration 178
6.3 Disorders of Glycosphingolipid Catabolism 178
6.4 Attachment Sites for Bacteria or their Toxins 179
6.5 Autoimmune Neuropathies 179
7 Diagnostic and Therapeutic Functions 180
7.1 Cancer 180
7.2 Neurological Disorders 180

Chapter 10 Lectins as Tools for Glycoconjugate Purification and Characterization

RICHARD D. CUMMINGS 191

1 Introduction 191
2 Structures of Plant and Animal Lectins 191
3 Carbohydrate-Binding Specificities of Lectins 192
3.1 Use of Immobilized Lectins in the Chromatography-Based Isolation of Oligosaccharides and Glycoproteins . 193
3.2 Immobilization of Lectins for Affinity Chromatography 193
3.3 Elution of Glycoconjugates Bound to an Immobilized Lectin 196
3.4 Serial Lectin Affinity Chromatography (SLAC) 196
4 Uses of Lectins to Study Phenotypes of Cells 197
5 Uses of Lectins in Solid-Phase Assays for Glycosyltransferases 197
6 Conclusion 198

Chapter 11 Proteoglycans – Structure and Functions

HANS KRESSE 201

1 Introduction 201
2 Glycosaminoglycan Structure 201
3 Hyaluronate 202
3.1 Structure, Biosynthesis and Degradation 202
3.2 Selected Functions 203
4 Proteoglycans 204
4.1 Glycosaminoglycan Chain Assembly 204
4.2 Large Hyaluronate-Binding Matrix Proteoglycans 208
4.3 Small, Leucine-Rich Extracellular Matrix Proteoglycans 209
4.4 Basement Membrane Proteoglycans 211
4.5 Integral Membrane Heparan Sulfate Proteoglycans 212
4.6 Intravesicular Proteoglycans 214
4.7 Optional Proteoglycans 214

Chapter 12 GPI-Anchors: Structure and Functions
VOLKER ECKERT ET AL. 223

1 Introduction 223
1.1 The Discovery of GPI-Anchors: from an Exotic Way to Anchor a Protozoal Surface Antigen to a General Principle among Eukaryotes
1.2 Discovery of Functions beyond the Mere Anchoring of Proteins to a Membrane: the GPI-Anchor as a Molecule with Multiple Functions in Cell Biology 224
1.3 Molecular Biology: from Function to Genes to Therapy? 224
1.4 The Structure of GPI-Anchors: A Structure that has been Conserved throughout Eukaryotic Evolution . . . 225
1.5 Species-Specific and Developmentally Regulated Modification: Additions to a General Theme 225
2 Structural Analysis 225
2.1 An Exercise in Biochemistry and Logical Deduction: a Few Representative Examples 225
2.2 Identification of GPI-Anchored Proteins 226
2.3 Structural Characterization of GPI-Anchors 228
2.4 The Biosynthesis of GPI-Anchors: a Conserved Pathway with Side-tracks and Backup Systems 229
3 The Function of GPI-Anchors 234
3.1 Anchoring of Membrane Proteins; Protozoa vs. Higher Eukaryotes; from a General Principle to Specialized Functions 234
3.2 Transmembrane Signaling 236
3.3 GPI-Anchors in Parasite Pathogenicity: A Small Molecule with Dramatic Effects 236
4 The Molecular Biology of GPI-Anchor Biosynthesis 237
4.1 From Functions to Genes: Mutant Cells Lead the Way 237
4.2 Cloning of Genes Involved in GPI-Anchoring by the Complementation of Defined Defects 238

Chapter 13 The Biology of Sialic Acids: Insights into their Structure, Metabolism and Function in particular during Viral Infection
WERNER REUTTER ET AL. 245

1 Introduction 245
2 Chemical Structure and Expression of Sialic Acids 246
3 Metabolism of Sialic Acids 247
4 Biological Functions of Sialic Acids . . 249
5 Modification of Biological Functions by Neuraminic Acid Analogs 251
6 Summary 254

Chapter 14 The Biology of Sulfated Oligosaccharides
LORA V. HOOPER ET AL. 261

1 Introduction 261
2 Sulfated Oligosaccharides of Luteinizing Hormone 263
3 Sulfated Carbohydrates in Leukocyte Trafficking 269
4 Sulfated Oligosaccharides of Unknown Function 271
5 Conclusion and Future Prospects . . . 272

Chapter 15 Carbohydrate-Carbohydrate Interaction
NICOLAI V. BOVIN 277

1 Introduction 277
2 Historical Aspects 277
3 Experimental Approaches 278
3.1 Cell Aggregation 279
3.2 Aggregation of Liposomes. 279
3.3 Interaction of Liposomes with a Plastic Surface Coated with Glycolipid or Another Glycoconjugate. . . . 280
3.4 Weak-Affinity Chromatography 281
3.5 Equilibrium Dialysis Method is Performed in the Following Way 281
3.6 Enzyme-Linked Immunoassays 281
3.7 Langmuir-Blodgett Technique 282
4 Molecular Nature of Carbohydrate-Carbohydrate Interaction 282
5 Cell Interaction: Static and Dynamic Integration with Other Adhesion Mechanisms 284
6 Polysaccharide-Carbohydrate Interaction 286
7 Conclusions 287

Chapter 16 Carbohydrate-Protein Interaction
HANS-CHRISTIAN SIEBERT ET AL. 291

1 Theoretical Aspects of Carbohydrate-Protein Interaction 291
1.1 Carbohydrate and Protein Flexibility. 291
1.2 Thermodynamic Parameters of Carbohydrate-Protein Interaction . . . 292
2 Modeling of Carbohydrate-Protein Interactions 295
2.1 X-ray Crystallography: Brookhaven Protein Data Bank 295
2.2 Cambridge Structural Database 296
2.3 Knowledge-Based Homology Modeling. 296
2.4 Basic Molecular Features in Carbohydrate-Protein Interactions . . 297
2.5 Estimation of Binding Constants . . . 299
3 NMR Spectroscopy 303
3.1 Free-State Conformation 303
3.2 Transferred NOE-Experiments of Oligosaccharide-Protein Complexes . 304
3.3 Perspectives 306

Chapter 17 Antibody-Oligosaccharide Interactions Determined by Crystallography
DAVID R. BUNDLE 311

1 Crystal Structure Determination. . . . 312
1.1 Sample Purification 312
1.2 Protein Crystals 312
1.3 Collecting X-ray Diffraction Data . . . 313
1.4 Electron Density. 313
1.5 Map Fitting/Model Building 314
1.6 Precision of the Model 315
2 Recently Determined Crystal Structures 315
3 The Structure of Mab Se 155.4 and the Abequose-Based Epitope 319
3.1 Features of a Dodecasaccharide-Fab Complex 320
3.2 Hydrogen-Bonding and the Structured Water Molecule. 322
3.3 Attributing Structural Features to Binding Energy 324
3.4 Oligosaccharide Conformational Change on Binding 324
3.5 A Different Bound Oligosaccharide Conformation in a Single Chain Fv-Trisaccharide Complex 325
3.6 Bound Conformation II in a Fab-Heptasaccharide Complex. 325
4 Role of Water Molecules. 328
5 Solvent-Exposed Hydrogen Bonds . . 329
6 Conformational Change in the Binding Site 329
7 Summary. 329

Chapter 18 Thermodynamic Analysis of Protein-Carbohydrate Interaction
Dipti Gupta and Curtis F. Brewer 333

1 Introduction 333
1.1 Background Studies 334
2 Outline of Experimental Design 335
2.1 Materials 335
2.2 Methodology 336
3 Thermodynamics of Binding of Mono- and Disaccharides 336
4 Thermodynamics of Binding of Trimannoside 10 338
5 Thermodynamics of Binding of Oligosaccharides 8, 11 and 12. 338
6 Binding of Mono- and Dideoxy Derivatives of Trimannoside 10. 339
7 Summary and Perspectives 341

Chapter 19 Analysis of Protein-Carbohydrate Interaction Using Engineered Ligands
Dolores Solis and Teresa Diaz-Mauriño 345

1 Overview 345
2 Analysis of Hydrogen-Bonding and Steric Requirements for Recognition 346
3 Analysis of Hydrogen-Bonding Energetics and Protein Groups Involved in Recognition 348
4 Mapping of Subsites of Anti-Carbohydrate Antibodies 350
5 Probing the Active Site Requirements of Carbohydrate-Binding Enzymes . . 350
6 Probing Conformational Requirements for Recognition. 351

Chapter 20 Application of Site-Directed Mutagenesis to Structure-Function Studies of Carbohydrate-Binding Proteins
Jun Hirabayashi 355

1 General Points on Site-Directed Mutagenesis 355
2 Special Points on Mutagenesis of Carbohydrate-Binding Proteins 357
3 Actual Procedures for Mutagenesis . . 358
3.1 Conventional Procedures 359
3.2 New Procedures 360
3.3 PCR-Aided Mutagenesis. 361
4 Example of Mutagenesis: Human Galectin-1 362
4.1 What is Galectin? 362
4.2 Conserved Amino Acids in the Galectin Family 363
4.3 Primer Design and Mutagenesis 363
4.4 Evaluation of the Results 365

Chapter 21 Bacterial Lectins: Properties, Structure, Effects, Function and Applications
Nechama Gilboa-Garber et al. 369

1 Introduction 369
1.1. Prologue 369
1.2. Bacterial Lectinology – Past and Present Status 369
2 Bacterial Lectin Prevalence, Expression and Detection 371
2.1 State of the Art 371
2.2 Dependence of Lectin Production on Growth Conditions 372
2.3 Lectin Screening Technology 372
3 Bacterial Lectin Properties 373
3.1 Divalent Cation-Binding 373
3.2 Sugar Specificities of the Bacterial Lectins 373

3.3 Physico-Chemical Properties of the Lectins and their Purification 374
4 Bacterial Lectin Size and Structure . . 374
5 Bacterial Lectin Target Molecules . . . 376
5.1 Interactions of Bacterial Lectins with Free Glycosylated Macromolecules 378
5.2 Interactions of Bacterial Lectins/Adhesins with Cell Receptors Including Blood Group Antigens . . . 378
5.3 The Receptor Specificity of Bacterial Lectins Determines Host Cell Selectivity 379
6 Bacterial Lectin Effects on Macromolecules and Cells 381
6.1 Effects on Macromolecules 381
6.2 Effects on Cells 381
7 Bacterial Lectin Functions 383
7.1 Bacterial Self-Protection 383
7.2 Bacterial Cell Organization 384
7.3 Cell Contacts Supplying Nutrition and Physiological Functions 384
7.4 Interactions for Proliferation 384
7.5 Cell Contacts Leading to Bacterial Death 384
8 Involvement of Bacterial Lectins in Pathogenesis 385
8.1 Enzyme/Toxin-Targeting and Trafficking to Host Macromolecules and Cells 385
8.2 The Role of Bacterial Lectins in Adherence of Individual and Coaggregated Bacteria to Host Cells 386
9 Preventive Strategies against Lectin-Mediated Adherence of Pathogenic Bacteria to Host Cells 388
9.1 Prevention of Lectin-Receptor Interactions 388
9.2 Host Cell Receptor Elimination 388
9.3 Prevention of the Bacterial Lectin/Adhesin Production 388
10 Applications of Bacterial Lectins . . . 389

Chapter 22 Glycobiology of Parasites: Role of Carbohydrate-Binding Proteins and their Ligands in the Host-Parasite Interaction

HONORINE D. WARD 399

1 Introduction 399
2 Malaria 400
2.1 Sialic Acid and Sialic Acid-Binding Proteins 401
2.2 Glycosaminoglycan-Binding Proteins 402
2.3 GlcNAc-Binding Proteins 402
3 Chagas' Disease 402
3.1 Sialic Acid and Sialic Acid-Binding Proteins 403
3.2 Heparin-Binding Protein 403
4 Leishmaniasis 404
4.1 Lipophosphoglycan (LPG) and LPG-Binding Proteins 404
4.2 Heparin-Binding Proteins 405
4.3 GlcNAc-Binding Proteins 405
5 Amebiasis 406
5.1 Gal/GalNAc-Binding Lectin 406
5.2 Chitotriose-Binding Proteins 407
6 Giardiasis 407
6.1 Man-6-P-Binding Protein 407
7 Cryptosporidosis 408
7.1 Gal/GalNAc-Binding Protein 408
8 Conclusion 409

Chapter 23 Structure and Function of Plant Lectins

HAROLD RÜDIGER 415

1 Introduction 415
2 Structure 416
2.1 Leguminosae 416
2.2 Non-Leguminous Plants 419
3 Location 421
4 Functions 423
4.1 Internally Directed Activities 423
4.2 Externally Directed Activities 425
5 Conclusion 429

Chapter 24 Lectins and Carbohydrates in Animal Cell Adhesion and Control of Proliferation
Jean-Pierre Zanetta . . . 439

1 Introduction . . . 439
2 The Galectins . . . 439
3 The C-Type Lectins . . . 440
3.1 The Selectins . . . 440
3.2 The NK Sub-Family of C-Type Lectins 441
3.3 The Soluble C-Type Lectins . . . 441
4 Soluble Calcium-Independent Mannose-Binding Lectins . . . 441
4.1 Glycoproteins Ligands of Lectin CSL . . . 442
4.2 Lectin CSL as an Adhesion Molecule. 442
4.3 Lectin CSL as a Mitogen . . . 444
5 Heparin-Binding Growth Factors . . . 446
6 Cytokines . . . 447
6.1 Interleukin 2 (IL-2) . . . 447
6.2 Interleukin 1 (IL-1) and Tumor Necrosis Factor (TNF) . . . 449
7 Conclusions and Perspectives . . . 450
8 Appendix . . . 452

Chapter 25 Galectins in Tumor Cells
David W. Ohannesian and Reuben Lotan . . . 459

1 Introduction . . . 459
2 Galectins in Normal Cells . . . 459
2.1 Galectin-1 . . . 459
2.2 Galectin-3 . . . 460
3 Putative Functions of Galectins . . . 461
4 Galectins in Tumor Cells . . . 462
5 Carbohydrate Specificity of Galectins . . . 463
6 Galectin Ligands . . . 464
6.1 Laminin . . . 464
6.2 Lysosome-Associated Membrane Glycoproteins (LAMPs) . . . 465
6.3 Carcinoembryonic Antigen . . . 465
7 Conclusions . . . 466

Chapter 26 Glycoconjugate-Mediated Drug Targeting
Kevin G. Rice . . . 471

1 Introduction . . . 471
2 Mammalian Lectins that Mediate Drug Delivery . . . 472
3 Design of Glycoconjugate Carriers for Drug Delivery . . . 474
4 Applications of Carbohydrate-Mediated Drug Targeting . . . 475
4.1 Targeting Low Molecular Weight Drug Molecules to Hepatocytes and Macrophages . . . 476
4.2 Targeting Antisense Oligonucleotides . . . 476
4.3 Glyconjugate Targeting of Enzymes. . 477
4.4 Glycoconjugate-Mediated Targeting of DNA . . . 477
4.5 Glycoconjugate Targeting of Liposomes and Lipoproteins . . . 478
5 Conclusions . . . 479

Chapter 27 Glycobiology of Signal Transduction
Antonio Villalobo et al. . . . 485

1 Introduction . . . 485
2 Lectin-Induced Mitogenesis and Immunomodulation . . . 489
3 Lectin-Induced Apoptosis . . . 492
4 Signaling by Membrane-Bound Mammalian Lectins . . . 492
5 Conclusions . . . 493

Chapter 28 Glycobiology of Host Defense Mechanisms
HANS-JOACHIM GABIUS ET AL. 497

1 Introduction 497
2 The Family of Lectins in Host Defense 497
2.1 Acute-Phase Reactants (Pentraxins) 497
2.2 Collectins 498
2.3 Selectins 500
2.4 NK Cell Lectins 500
2.5 Miscellaneous Lectins 501
3 Lectin-Dependent Activation of Defense Mechanisms 501
4 Perspectives 502

Chapter 29 Transgenic Approaches to Glycobiology
HELEN J. HATHAWAY AND BARRY D. SHUR 507

1 Introduction 507
2 Protein Glycosylation 508
3 Approaches to Genetic Manipulation of Oligosaccharide Function 508
3.1 Overexpression of Gene Products . . . 508
3.2 Elimination of Genes 508
3.3 Tissue- and Stage-Specific Alteration of Gene Products 509
4 Genetic Manipulation of Glycosylation Pathways 510
4.1 Altering Terminal Glycosylation 510
4.2 Altering Complex *N*-Linked Glycosylation 511
4.3 Altering Glycosidases 511
5 Altering Carbohydrate Receptors . . . 512
5.1 Selectins 512
5.2 Galectin: L14 512
5.3 β1,4-Galactosyltransferase 512
6 Future Directions 515

Chapter 30 Biomodulation, the Development of a Process-Oriented Approach to Cancer Treatment
PAUL L. MANN ET AL. 519

1 Introduction 519
2 Problem Definition 520
3 A Brief Summary of Other Approaches 521
4 The Biological Response Modifier [BRM] Approach 521
5 Recent Uses of BRM/Biomodulator Terminology 522

Chapter 31 Glycobiology in Xenotransplantation Research
DAVID K.C. COOPER AND RAFAEL ORIOL 531

1 Introduction 531
2 Xenotransplantation – Basic Immunobiology 531
3 Allotransplantation across the ABO Histo-Blood Group Barrier . . . 532
4 Identification of Human Anti-Pig Antibodies as Anti-α-Galactosyl Antibodies 533
5 Identification of Oligosaccharide Epitopes on Pig Vascular Endothelium 536
6 Experimental Studies in Baboons . . . 538
7 Clinical Relevance of Anti-α-Gal Antibody in Xenotransplantation . . . 539
8 The Genetically Engineered α-Gal-Negative Pig 540
8.1 Deletion of the Gene Encoding α1,3-Galactosyltransferase 540
8.2 Increased Expression of Alternate Epitopes 540
9 Other Therapeutic Approaches – Gene Therapy 541
10 Comment 541

Chapter 32 Modern Glycohistochemistry: A Major Contribution to Morphological Investigations

André Danguy et al. . . . 547

1 Introduction . . . 547
2 Why Lectins and Neoglycoproteins are Attractive Reagents in Histology . . . 547
3 Methods . . . 550
3.1 Lectin Cytolabeling . . . 550
3.2 Reverse Lectin Histochemistry . . . 550
3.3 Special Considerations . . . 551
4 Functional Morphological Data . . . 552
4.1 The Kidney . . . 552
4.2 Endocrine Status and the Glycohistochemical Expression . . . 553
4.3 The Integument . . . 553
4.4 Glycan Expression in Skeletal Muscle . . . 556
4.5 Lectin Histochemistry of the Teleost Intestine . . . 557
4.6 Sugar-Binding Sites and Lectin Acceptors in Prokaryotes and Eukaryotic Parasites . . . 557
5 Conclusions and Perspectives . . . 558

Chapter 33 Lectins and Neoglycoproteins in Histopathology

S. Kannan and M. Krishnan Nair . . . 563

1 Introduction . . . 563
1.1 Relevance of Histopathology . . . 563
1.2 Glycoconjugates and Pathology . . . 564
2 Lectins . . . 564
2.1 Methods in Lectin Histochemistry . . . 564
2.2 Exogenous Lectins in Histopathology . . . 565
2.3 Endogenous Lectins in Histopathology . . . 568
3 Neoglycoproteins . . . 569
3.1 Neoglycoprotein Staining Protocols . . . 570
3.2 Neoglycoproteins in Histopathology . . . 570
4 Precautions to be Taken in Lectin Histochemistry . . . 571
5 Perspectives . . . 574
5.1 Modern Roles Played by Histopathology . . . 574
5.2 Perspectives of Lectins and Neoglycoproteins in Histopathology . . . 576

Chapter 34 Glycobiology of Development: Spinal Dysmorphogenesis in Rat Embryos Cultured in a Hyperglycemic Environment

Lori Keszler-Moll et al. . . . 585

1 Introduction . . . 585
2 Problem Identification . . . 586
2.1 The Model, Embryo Culture . . . 587
2.2 The Model, Serum Preparation . . . 587
2.3 Embryo Morphology . . . 587
2.4 The Kinetics of Neural Tube Defect Formation . . . 589
3 Lectin Staining Patterns . . . 590
4 Surface Oligosaccharide and Fusion Mechanism . . . 592

Chapter 35 Glycobiology of Fertilization

FRED SINOWATZ ET AL. 595

1 A Short Review of the Fertilization Pathway 595
2 Structure and Function of the Zona Pellucida 596
2.1 The Glycoprotein Constituents of the Zona Pellucida 596
2.2 Characterization of the Zona Pellucida Oligosaccharides by Lectins 597
2.3 Oligosaccharides of the Porcine Zona Pellucida 597
2.4 Carbohydrates Involved in Sperm-Zona Pellucida Interaction. . . 598
3 Sperm-Associated Zona Pellucida- and Carbohydrate-Binding Proteins 599
3.1 Sperm Membrane C-Type Lectins . . . 600
3.2 Mouse Sperm β1,4-Galactosyl-transferase 601
3.3 Sperm Membrane-Associated Zona Pellucida-Binding Proteins 602
4 Inhibition of Sperm-Egg Interactions by Saccharides and Glycosidases. . . . 605
4.1 Sperm Binding Assays 605
4.2 Inhibition of Sperm-Zona Binding by Saccharides 605
4.3 Do Carbohydrates also Play a Role in the Fusion of the Oocyte Plasma Membrane with the Spermatozoon?. . 606

Chapter 36 Glycobiology of Consciousness

RAYMONDE JOUBERT-CARON ET AL. 611

1 Introduction. 611
2 Brain Lectins 611
3 Neuroimmunomodulation 613
4 Brain Glycoproteins. 615

List of Contributors

ANDRÉ, S.
Institut für Physiologische Chemie
Tierärztliche Fakultät
Ludwig-Maximilians-Universität
Veterinärstr. 13
80539 München
Germany

AVICHEZER, D.
Department of Chemical Immunology
The Weizmann Institute of Science
Rehovot 76100
Israel

BAENZIGER, J. U.
Department of Pathology
Washington University School of Medicine
660 S. Euclid Ave.
St. Louis, MO 63110–1093
USA

BAUM, O.
Freie Universität Berlin, Institut für
Molekularbiologie und Biochemie
Arnimallee 22
14195 Berlin-Dahlem
Germany

BLADIER, D.
UFR Léonard de Vinci
Laboratoire de Biochimie et Technologie des Protéines
74, rue Marcel Cachin
93012 Bobigny Cedex
France

BOVIN, N. V.
Shemyakin and Ovchinnikov Institute of
Bioorganic Chemistry
Russian Academy of Sciences
ul. Miklukho-Maklaya 16/10
117871 GSP-7, V-437 Moscow
Russian Federation

BRAUN, M.
College of Pharmacy
The University of New Mexico
Albuquerque, NM 87131
USA

BREWER, C. F.
Departments of Molecular Pharmacology,
Microbiology and Immunology
Jack and Pearl Resnick Campus
Albert Einstein College of Medicine
1300 Morris Park Ave.
Bronx, NY 10461
USA

BROCKHAUSEN, I.
Biochemistry Department
Research Institute
Hospital for Sick Children
555 University Ave.
Toronto M5G 1X8
Canada

BUNDLE, D. R.
Department of Chemistry
Faculty of Science
University of Alberta
E3–52 Chemistry Bldg.
Edmonton T6G 2G2
Canada

CALVETE, J. J.
Institut für Reproduktionsmedizin
Tierärztliche Hochschule Hannover
Bünteweg 15
30559 Hannover
Germany

H.-J. and S. Gabius (Eds.), Glycosciences

ISBN 3-8261-0073-5

CAMBY, I.
Laboratoire d'Histologie
Faculté de Médecine
Université Libre de Bruxelles
Route de Lennik 808
1070 Bruxelles
Belgium

CARON, M.
UFR Léonard de Vinci
Laboratoire de Biochimie et Technologie des Protéines
74, rue Marcel Cachin
93012 Bobigny Cedex
France

COOPER, D. K. C.
Oklahoma Transplantation Institute
Baptist Medical Center
3300 N.W. Expressway
Oklahoma City, OK 73112–4481
USA

CRAMER, F.
Max-Planck-Institut für experimentelle Medizin
Hermann-Rein-Str. 3
37075 Göttingen
Germany

CUMMINGS, R. D.
Department of Biochemistry and Molecular Biology
University of Oklahoma Health Sciences Center
941 S. L. Young Blvd.
Oklahoma City, OK 73190
USA

DANGUY, A.
Faculté des Sciences
Laboratoire de Biologie Animale et d'Histologie Comparée
Université Libre de Bruxelles
50, avenue F. D. Roosevelt
1050 Bruxelles
Belgium

DÍAZ-MAURIÑO, T.
Instituto de Quimica Física Rocasolano
Consejo Superior de Investigaciones Científicas
Serrano 119, 28006 Madrid
Spain

ECKERT, V.
Institut für Virologie
AG Parasitologie
Philipps-Universität
Robert-Koch-Str. 17
35037 Marburg
Germany

GABIUS, H.-J.
Institut für Physiologische Chemie
Tierärztliche Fakultät
Ludwig-Maximilians-Universität
Veterinärstr. 13
80539 München
Germany

GABIUS, S.
Hämatologisch-Onkologische Schwerpunktpraxis
Sternstr. 12
83022 Rosenheim
Germany

GARBER, N. C.
Department of Life Sciences
Bar-Ilan University
Ramat-Gan 52900
Israel

GARCIA, A.
College of Pharmacy
The University of New Mexico
Health Sciences Center
Albuquerque, NM 87131
USA

GEROLD, P.
Institut für Virologie
AG Parasitologie
Philipps-Universität
Robert-Koch-Str. 17
35037 Marburg
Germany

GILBOA-GARBER, N.
Department of Life Sciences
Bar-Ilan University
Ramat-Gan 52900
Israel

Gilleron, M.
Laboratoire de Pharmacologie et de Toxicologie Fondamentales, Department III
Glycoconjugués et Biomembranes, CNRS
118, route de Narbonne
31062 Toulouse Cedex
France

Gupta, D.
Departments of Molecular Pharmacology, Microbiology and Immunology
Jack and Pearl Resnick Campus
Albert Einstein College of Medicine
1300 Morris Park Ave.
Bronx, NY 10461
USA

Hanosh, J.
College of Pharmacy
The University of New Mexico
Albuquerque, NM 87131
USA

Hathaway, H. J.
Department of Biochemistry and Molecular Biology
The University of Texas
MD Anderson Cancer Center
1515 Holcombe Blvd Cedex
Houston, TX 77030
USA

Hirabayashi J.
Department of Biological Chemistry
Faculty of Pharmaceutical Sciences
Teikyo University
Sagamiko, Kanagawa 199–01
Japan

Hooper, L. V.
Department of Pathology
Washington University School of Medicine
660 S. Euclid Ave.
St. Louis, MO 63110–1093
USA

Horcajadas, J. A.
Instituto de Investigaciones Biomédicas
Consejo Superior de Investigaciones Científicas
Arturo Duperier 4
28029 Madrid
Spain

Hounsell, E. F.
Department of Biochemistry and Molecular Biology
University College of London
Gower Street
London WC1E 6BT
United Kingdom

Inoue, Y.
Department of Biophysics and Biochemistry
Graduate School of Science
University of Tokyo
Hongo-7
Tokyo 113
Japan

Inoue, S.
School of Pharmaceutical Sciences
Showa University
Hatanodai-1
Tokyo 142
Japan

Joubert-Caron, R.
UFR Léonard de Vinci
Laboratoire de Biochimie et Technologie des Protéines
74, rue Marcel Cachin
93012 Bobigny Cedex
France

Kannan, S.
Division of Cancer Research
Regional Cancer Centre
Thiruvananthapuram
695 011, Kerala State
India

Kayser, K.
Abteilung Pathologie, Thoraxklinik
Amalienstr. 5,
69126 Heidelberg
Germany

Kelley, R. O.
College of Pharmacy
The University of New Mexico
Albuquerque, NM 87131
USA

Keszler-Moll, L.
College of Pharmacy
The University of New Mexico
Albuquerque, NM 87131
USA

Kiss, R.
Laboratoire d'Histologie
Faculté de Médecine
Université Libre de Bruxelles
Route de Lennik 808
1050 Bruxelles
Belgium

Kitajima, K.
Department of Biophysics and Biochemistry
Graduate School of Science
University of Tokyo
Hongo-7
Tokyo 113
Japan

Kopitz, J.
Institut für Pathobiochemie und Allg.
Neurochemie
Universität Heidelberg
Im Neuenheimer Feld 220
69120 Heidelberg
Germany

Kresse, H.
Institut für Physiologische Chemie und
Pathobiochemie
Westfälische Wilhelms-Universität Cedex
Waldeyerstr. 15
48129 Münster
Germany

Krishnan Nair, M.
Division of Cancer Research
Regional Cancer Centre
Thiruvananthapuram
695 011, Kerala State
India

Laine, R. A.
Deptartments of Biochemistry and Chemistry
Louisiana State University and The Louisiana
Agricultural Center
Baton Rouge, LA 70803
USA

Lee, R. T.
Department of Biology
Johns Hopkins University
144 Mudd Hall
3400 N. Charles Street
Baltimore, MD 21218–2685
USA

Lee, Y. C.
Department of Biology
Johns Hopkins University
144 Mudd Hall
3400 N. Charles Street
Baltimore, MD 21218–2685
USA

Lis, H.
Department of Membrane Research and
Biophysics
The Weizmann Institute of Science
Rehovot 76100
Israel

Lotan, R.
Department of Tumor Biology
MD Anderson Cancer Center
The University of Texas
1515 Holcombe Blvd., Box 108
Houston, TX 77030
USA

Lutomski, D.
UFR Léonard de Vinci
Laboratoire de Biochimie et Technologie des
Protéines
74, rue Marcel Cachin
93012 Bobigny Cedex
France

Mann, P. L.
College of Pharmacy
The University of New Mexico
Albuquerque, NM 87131
USA

Manzella, S. M.
Department of Pathology
Washington University School of Medicine
660 S. Euclid Ave.
St. Louis, MO 63110–1093
USA

Ohannesian, D. W.
Department of Tumor Biology
MD Anderson Cancer Center
The University of Texas
1515 Holcombe Blvd., Box 108
Houston, TX 77030
USA

Oriol, R.
INSERM U.178
16, Paul Vaillant-Couturier
94807 Villejuif Cedex
France

Pavelka, M.
Institut für Histologie und Embryologie
Universität Innsbruck
Müllerstraße 59
6020 Innsbruck
Austria

Raymond-Stintz, M. A.
College of Pharmacy
The University of New Mexico
Albuquerque, NM 87131
USA

Reuter, G.
Institut für Physiologische Chemie
Tierärztliche Fakultät
Ludwig-Maximilians-Universität
Veterinärstr. 13
80539 München
Germany

Reutter, W.
Institut für Molekularbiologie und Biochemie
Freie Universität Berlin
Arnimallee 22
14195 Berlin-Dahlem
Germany

Rice, K. G.
Division of Medicinal Chemistry and Pharmaceutics
College of Pharmacy
University of Michigan
428 Church Street
Ann Arbor, MI 48109–1065
USA

Rüdiger, H.
Institut für Pharmazie und Lebensmittelchemie
Universität Würzburg
Am Hubland
97074 Würzburg
Germany

Salmon, I.
Service d'Anatomie Pathologique
Hôpital Erasme
Université Libre de Bruxelles
Route de Lennik 808
1050 Bruxelles
Belgium

Schachter, H.
Biochemistry Department
Research Institute
Hospital for Sick Children
555 University Ave.
Toronto M5G 1X8
Canada

Schmidt, R. R.
Institut für Organische Chemie
Universität Konstanz
Universitätsstraße 10
78464 Konstanz
Germany

Schwarz, R. Th.
Institut für Virologie
AG Parasitologie
Philipps-Universität
Robert-Koch-Str. 17
35037 Marburg
Germany

Sharon, N.
Department of Membrane Research and Biophysics
The Weizmann Institute of Science
Rehovot 76100
Israel

Shur, B. D.
Department of Biochemistry and Molecular Biology
MD Anderson Cancer Center
The University of Texas
1515 Holcombe Blvd, Houston, TX 77030
USA

Siebert, H.-C.
Institut für Physiologische Chemie
Tierärztliche Fakultät
Ludwig-Maximilians-Universität
Veterinärstr. 13
80539 München
Germany

Sinowatz, F.
Institut für Tieranatomie
Tierärztliche Fakultät
Ludwig-Maximilians-Universität
Veterinärstr. 13
80539 München
Germany

Solís, D.
Instituto de Quimica Física Rocasolano
Consejo Superior de Investigaciones Científicas
Serrano 119
28006 Madrid
Spain

Stäsche, R.
Institut für Molekularbiologie und Biochemie
Freie Universität Berlin
Arnimallee 22
14195 Berlin-Dahlem
Germany

Stehling, P.
Institut für Molekularbiologie und Biochemie
Freie Universität Berlin
Arnimallee 22
14195 Berlin-Dahlem
Germany

Suzuki, T.
Department of Biophysics and Biochemistry
Graduate School of Science
University of Tokyo, Hongo-7
Tokyo 113
Japan

Töpfer-Petersen, E.
Institut für Reproduktionsmedizin
Tierärztliche Hochschule Hannover
Bünteweg 15
30559 Hannover
Germany

von der Lieth, C.-W.
Deutsches Krebsforschungszentrum
Zentrale Spektroskopie
Im Neuenheimer Feld 280
69120 Heidelberg
Germany

Villalobo, A.
Instituto de Investigaciones Biomédicas
Consejo Superior de Investigaciones Científicas
Arturo Duperier 4
28029 Madrid
Spain

Vliegenthart, J. F. G.
Department of Bio-Organic Chemistry
Bijvoet Center for Biomolecular Research
Utrecht University
P.O. Box 80.075
3508 TB Utrecht
The Netherlands

Ward, H. D.
Division of Geographic Medicine and
Infectious Diseases
Tufts University School of Medicine
750 Washington Street
Boston, MA 02111
USA

Wenk, R.
College of Pharmacy
The University of New Mexico
Albuquerque, NM 87131
USA

Wittmann, J.
Institut für Physiologische Chemie
Tierärztliche Fakultät
Ludwig-Maximilians-Universität
Veterinärstr. 13
80539 München
Germany

Zanetta, J.-P.
Centre de Neurochimie du CNRS
Laboratoire de Neurobiologie Moléculaire des
Interactions Cellulaires
5, rue Blaise Pascal
67084 Strasbourg Cedex
France

1 The Information-Storing Potential of the Sugar Code

ROGER A. LAINE

1 Introduction

Carbohydrates, by their unique multi-linkage monomers and branching structure, contain an evolutionary potential of information content several orders of magnitude higher **in a short sequence** than any other biological oligomer. Therefore a high level of information potential is inherent in biological recognition systems comprised of complex carbohydrate ligands on the one hand which are recognized for targeted activities on the other hand by hapten specific protein receptors, such as lectins.

The potential number of all possible linear and branched isomers of small oligosaccharides has recently been calculated to be much larger than previous estimations (Laine, 1994). Seven structural elements lead to the large number of isomers, including multiple ring sites as points of glycosidic attachment, α/β anomerity, pyranose/furanose configuration and branching structure. For a trisaccharide composed from a set of 3 hexoses, these elements lead to isomer permutations exceeding 38,000. This can be compared with only 27 permutations for 3 amino acids or 3 nucleic acids.

Consider that the trisaccharide could be made up of any of our most commonly found sugars, glucose, mannose, galactose, fructose, *N*-acetylglucosamine, *N*-acetylgalactosamine, fucose, arabinose, xylose, ribose, glucuronic acid, galacturonic acid, mannuronic acid, iduronic acid, and sialic acid, let's say 20 common sugars. The number of possible unsubstituted trisaccharides would be [permutations × anomerics × ring sizes × linkages] or [$20^3 \times 2^3 \times 2^3 \times 12$] (linkage position potential ranges from 9 to 16, taking an average around 12) or >6,000,000 linear structures, plus around 3,000,000 branched structures for a total of 9×10^6 vs 8000 for 20 amino acids (calculation reviewed below).

Among all biological molecules, carbohydrates, *in a short sequence*, can potentially display the largest number of ligand structures to the binding sites of proteins in molecular recognition systems. The 3-dimensional presentation of carbohydrate structures to epitope-specific recognizing proteins comprises a "high level language" biochemical code. In this view, DNA can be looked upon as "machine language", coding for the lectins and the sets of transferases that assemble the sugars. Antibodies are a prime example of binding proteins, being exquisitely sensitive to all of the carbohydrate structural elements. The isomer permutations of small (M_r <1500) carbohydrates are 3–4 orders of magnitude larger than peptides at the trisaccharide level, and 7 orders of magnitude larger at the hexasaccharide level, due to branching. The most common size of binding sites for carbohydrates in proteins are 6 sugars or fewer, but this size allows trillions of possible structures for evolution to exploit.

Most commonly, biologically recognized complex carbohydrates are composed of sets of 3 or fewer epimers of common hexoses or pentoses, 0–2 different amino sugars, 0–1 methyl pentose and 0–1 sialic or uronic acid. In some systems they are often substituted with functional groups such as sulfate, methyl or acetate, usually in a structural motif of fewer than 7 sugars. A simplified set comprised from a set of 6 *D*-hexoses could generate $6^6 \times 2_a^6 \times 2_r^6 \times 4^5$ = [$46{,}656 \times 64 \times 64 \times 1024$] = [195,689,447,424] linear structures (Laine, 1994). Inclusion of branched hexasaccharides raises the number to 1.05×10^{12} (Laine, 1994). The first term represents sequence per-

H.-J. and S. Gabius (Eds.), Glycosciences

ISBN 3-8261-0073-5

mutations, the second, anomeric configuration, the third, ring size and the fourth, linkage position. Individual formulas and a master set of equations with graphs and tables of these results has been published for determination of all possible reducing end isomers for di- to octasaccharides (Laine, 1994). Oligomers higher than dp8, which contain the possibility for numerous branching isomers, generate astronomical numbers, larger than Avagadro's number. For 9-mers there is 1 mole of isomers! These numbers are artificially small, however due to the artifice of using the same size monomer library as the oligomer size. Using a common vocabulary of 20 different sugars to make up the hexasaccharide would give a number 1372 fold larger than either of the above (leading to 2.7×10^{14} linear and 1.44×10^{15} total isomers including branched). In this case the first permutation term would be 20^6 instead of 6^6. A hexapeptide from 20 amino acids would yield 6.4×10^7 structures, 8 orders of magnitude lower.

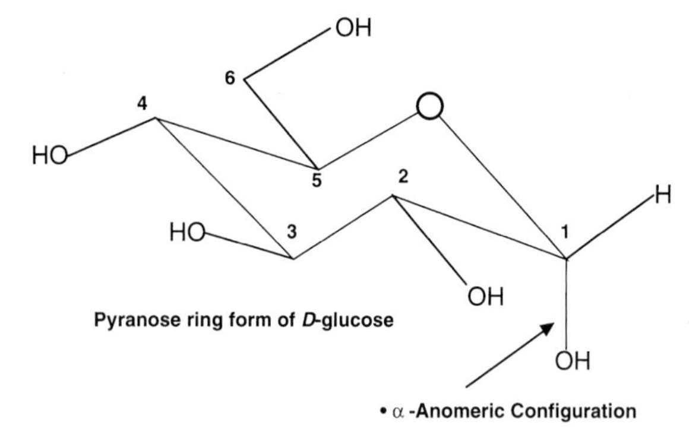

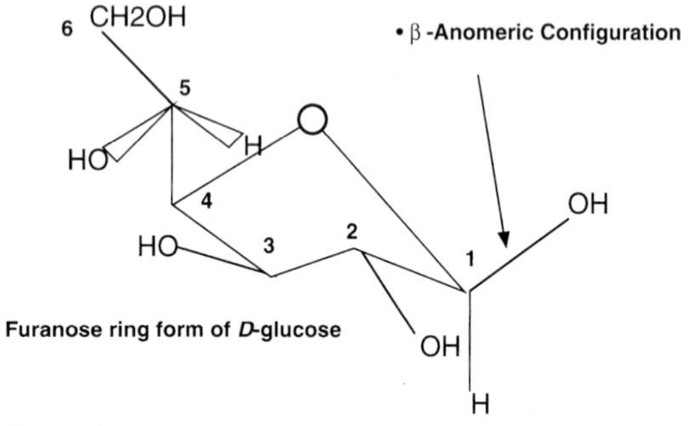

Figure 1

2 Details of Calculation for Isomers of a Trisaccharide

A calculation for the number of possible trisaccharides obtainable from a set of 3 hexoses, uses the formula: Structures = $E^n \times 2^n_r \times 2^n_a \times 4^{n-1}$. The first term E^n represents the permutations from order of sequence including repetitions of the same sugar $3^3 = 27$. In this term, E is the library of sugars (3 in this case), and n is the oligomer size (also 3 in this special case). The total is multiplied by another term for ring size, 2^n_r or $2^3 = 8$ since most sugars can occur in either pyranose or furanose forms. The total is again multiplied by 2^n_a, a term for anomeric configuration: $2^3 = 8$.

The linkage position term 4^{n-1} is relevant for 2 of the 3 sugars, where 4 potential hydroxyls are available for linkage to the previous sugar (hence the n-1) and gives a number of $4^2 = 16$. In furanose forms in a trisaccharide of sequence ABC, sugar A could have been connected through the 5 position of sugar B, for example. This factor is taken into account by the ring size term keeping the total possibilities of linkage positions at 16.

Thus, the correct number for permutations of linear trisaccharides made up from a set of 3 hexoses is $27 \times 8 \times 8 \times 16 = 27{,}648$.

In branched trisaccharides sugars A and B are both glycosides to sugar C by 2,3; 2,4; 2,6; 3,4; 3,6 or 4,6 branching presenting six possibilities. With sugar C as furanose, additional isomers include 2,3; 2,5; 2,6; 3,5; 3,6; or 5,6 for a total of 12 different branched structures. However, the ring size term 2^n_r, takes into account the additional 6 structures engendered if C were furanose.

```
A(1->6)                     B(1->6)
       \                           \
        C(1->R)*    or              C(1->R)*
       /                           /
B(1->3)                     A(1->3)
```

*R = reducing end attachment site (aglycon)

Since each branch can occur in two different ways such as A6,B3 or B6,A3, there are again 12 different ways to branch these three sugars. The permutation term, E^n, however, takes care of the A6,B3 and B6,A3 branching duplex. Therefore, unique branched trisaccharides from a set of 3 hexoses are $27 \times 8 \times 8 \times 6$ = 10,368. The total number of structures from a trisaccharide comprised of 3 hexoses, choosing among a set of only 3 different hexoses, is 27,648 (linear forms) plus 10,368 (branched forms) = 38,016. The formula for isomers of a trisaccharide having a reducing end is thus:

$$E^n \times 2^n_r \times 2^n_a \times 4^{n-1} \text{ (linear forms)}$$
$$+$$
$$E^n \times 2^n_r \times 2^n_a * \times 6^{n-2} \text{ (branched forms)}$$

2.1 Non-Reducing Oligosaccharides

Trisaccharides can assume the trehalose-type disaccharide aldose-1->1-aldose or the raffinose non-reducing aldose-1->2 ketose internal linkage structure, giving a larger number for possible trisaccharides. Longer oligosaccharides can form cyclodextrins. These kinds of permutations would add a large number of oligosaccharides to an isomers calculation. For cyclodextrin hexasaccharides, multiply by 4 the linear permutations number due to a term added by the extra head-to-tail linkage (to 5 possible hydroxyls), making the cyclodextric versions of hexasaccharides alone close to 0.8 trillion. Some of the cyclic "isomers" might be identical, however, depending on the chosen cyclic starting position.

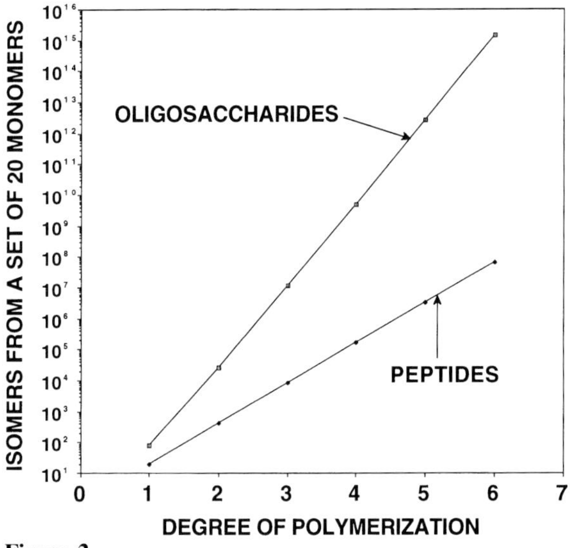

Figure 2

3 Analysis

3.1 NMR

The use of NMR as a single spectroscopic method with the use of a chemical shift library is questionable, even with saccharides as small as 3 sugars. Each trisaccharide from the conservative set using only 3 hexoses would contain 15 ring protons including the anomeric, thus the proton NMR spectrum library would require resolution of $38{,}016 \times 15$ = 570,240 "different" proton environments within 0.5 ppm. This would require a resolution of 10^{-6} ppm, (a **terahertz** instrument) if the line widths were also narrowed concomitantly. It is doubtful that a tenth of this number of lines could be resolved using multi-dimension proton NMR. In fact, the carbon-13 spectrum, thirty times more dispersed, to form a chemical shift library would need to resolve $38{,}016 \times 18$ carbons = 684,288 lines if they all happened to be different. This would require a resolution of 2×10^{-5} ppm, still in the terahertz range. NMR by itself, therefore cannot be used to establish a chemical shift library as a stand-alone identification system for trisaccharides and certainly not for larger oligomers to absolutely identify complete structure by virtue of chemical shift values. NOE and multi-dimension NMR can expand the available data to some degree, but not enough to overcome the chemical shift resolution problem, upon which all of the techniques must ultimately depend. However, much information is available from 1D proton NMR using 100 nanomoles of oligosaccharide.

On the other hand, having 1–5 micromoles of a pure trisaccharide with 15 lines to resolve in 0.5ppm and with "accidental" overlaps minimized, the use of nuclear Overhauser effect (NOE) and 2-D NMR techniques can be used to completely identify the epimers, linkage positions and anomeric configurations. This is useful for confirmation of synthesis, but rarely useful in the case of identification of small amounts of biologically active saccharides where often only nanomole quantities are available.

3.2 Mass Spectrometry

Mass spectrometry cannot be used by itself to identify oligosaccharides. All 10^{12} isomers made of *D*-hexoses, for example, would have the same mass. There may be some fine structure in mass spectra due to linkage position and preferences in cleavage of the rings. Partial fragmentation in collisional activated mass spectrometry might provide the combination of partial degradation and spectral pattern to resolve such parameters as position of linkage (Laine *et al.*, 1988, 1991; Laine, 1989; Yoon and Laine, 1992), but will not be sufficient without other sensitive chemical manipulations (see Hounsell, this volume). The advantage of mass spectrometry for partial structural analysis is its inherent nanomole to picomole sensitivity.

4 Synthesis

Chemical synthesis of a trisaccharide takes 20 man-weeks (Personal Communication: O. Hindsgaul, Edmonton, Alberta; K. Matta, Buffalo, NY; P. Garegg, Stockholm) compared with 3 hours for a tripeptide (automated solid state synthesizer). Part of the difficulty is the isomer problem. If the trisaccharide sought is one out of a possible 38,000 isomers, this is the crux of the synthetic problem and reason for the lack of automated systems.

5 Biologically Relevant Oligomer Size

With few exceptions in glycobiology, hexasaccharides are the upper limit for protein-recognized oligosaccharide sequences (Cisar *et al.*, 1974, 1975; Takeo and Kabat, 1978; Smith-Gill *et al.*, 1984), and repeating units in polysaccharides seldom exceed 6 sugars in size.

6 Substitutions

Often, carbohydrates are substituted with functional groups. Returning to the example of trisaccharides made up from a set of 3 hexoses, each member of 38,000 isomeric structures could be substituted, for example, by one sulfate in any of 10 free positions. Therefore there are more than 380,000 possible singly sulfated trisaccharides composed from a set of 3 hexoses. Also there are 380,000 potential singly-*O*-methylated structures and a similar number of singly acetylated structures, to say nothing of phosphates, carbamoylates, pyruvates, other kinds of derivatives and combinations. There are 44 ways to put 2 sulfates on one trisaccharide. Using 38,000 trisaccharides made of a 3 hexose vocabulary there would be 1.7 million possible structures with 2 sulfates. Using a more reasonable 20 sugar vocabulary for trisaccharides there would be 90,000,000 singly sulfated potential structures, and 4×10^8 potential disulfated trisaccharide isomers. Naturally, the numbers are much higher for oligomers such as hexasaccharides.

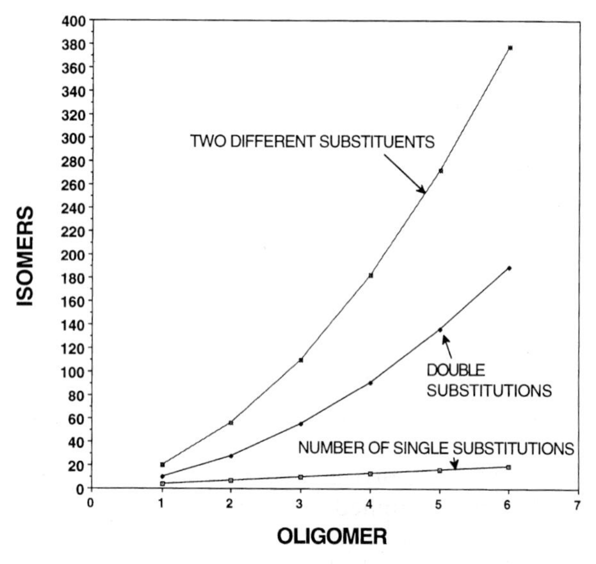

Figure 3

7 Oligosaccharide Recognition by Proteins

The cognate recognition partner for carbohydrate-based biological information is a protein with a specific sugar-binding site, such as a lectin, an antibody, a transferase or a glycosidase. A few investigators also believe that a specific carbohydrate could be recognized by, and bind to, another carbohydrate (see Bovin, this volume). Lectins and antibodies are exquisitely sensitive to carbohydrate molecular structure, with precise recognition for the 7 major saccharide structural motifs:

- epimers (including ***D*** and ***L***)
- sequence of sugars
- anomeric configuration
- ring size
- linkage position
- branching
- charge (COOH, sulfate or NH_3^+, for example)

Other structural motifs that make the biological recognition sites more complex include substitutions such as the following: (see also Sharon and Lis; Hooper *et al.*, this volume)

- phosphate
- phosphonate
- acyl groups (acetates, σ-1 fatty acids in nod factors, mycolic acids in cord factor)
- alkyl ether groups, most commonly *O*-methyl
- pyruvylation
  sulfation
- sulfonation
- carbamoylation and others

The substitution of one methyl group anywhere on the 19 free hydroxyls of each of 10^{12} different hexasaccharides would give 19×10^{12} new structures. For 2 methyl groups, this number would need to be multiplied by a factor of [(H-1)+(H-2)+(H-3)+....+(H-(H-1))], where H is the number of free hydroxyls. For 2 different substituents, this factor would need to be multiplied by 2 to account for reciprocal locations.

8 Evolution of Carbohydrate Code Structures and Receptors

Evolution of receptor/ligand pairs in carbohydrates is probably very slow. This is due to the necessity for development of specific sets of glycosyltransferases to generate the carbohydrate ligand on one hand, and evolution of a binding site in a protein to recognize the new structure on the other. Cell-recognition factors and signalling systems would therefore be expected to be conserved across species and sometimes across genera. The selectins as leukocyte extravasation adhesins in mammals are a good cross-species example (Brandley *et al.*, 1990; Polley *et al.*, 1991; Yuen *et al.*, 1992; Asa *et al.*, 1995; Lasky, 1995). Single point mutations in glycosyltransferase proteins or lectins are not likely to alter target sugar structures or receptor binding sites. In a few known cases a minor amino acid change in a transferase or a lectin brings about recognition of a closely related sugar (Yamamoto *et al.*, 1990; Jordan and Goldstein, 1995). For polypeptide-based carbohydrate recognition, such as in lectins, information is carried in one or more genes. It was an early hypothesis that adhesins could be related to binding sites in glycosyl transferases (Roth *et al.*, 1971). There may be sets of motifs conserved among families of lectins such as the mannose-binding protein and selectins (Bajorath and Aruffo, 1995).

Evolution for biological recognition of just one additional or one altered sugar on an existing structure may require a combination of the following:

1) On the carbohydrate side, mutation of the peptide sequence of an existing glycosyltransferase, or evolution of an entirely novel transferase is needed to produce a new gene product. This new gene product could be a transferase that transfers another sugar, a different sugar, or the same sugar with a different linkage to the precursor structure. The complex carbohydrate ligand is coded into a set of glycosyltransferase genes coding sequentially acting enzymes. Each partially glycosylated precursor is recognized in the binding site of the subsequent glycosyltransferase. This could be called a "sequential binding site pattern". Tri-antennary and tetra-antennary "complex" *N*-linked structures are among the largest non-template-driven precise structures in biology.

2) On the binding protein side, evolution of a new or modified lectin is necessary to express a new binding/recognition site.

9 A "High Level" Biological Code

In all natural polymers, the linear sequence of monomers comprises, in some fashion, a biological code. For carbohydrate polymers, as different from "oligomers", this code may express itself as the chemical properties of the polymer (fibrous, such as cellulose and chitin, soluble gel as starch or agar) which may include specific hydrogen bonds and other elements of self-association. Complex carbohydrate polymers may have subsets of internal oligosaccharides which possess biological activity based on specific proteins that bind small, specific, possibly rare internal sequences. Carbohydrate polymers themselves often contain a complex multifaceted sequence. Specific proteins can bind to relatively short subsets or haptens within longer saccharide sequences, such as in heparin (Riensenfeld *et al.*, 1977; Atha *et al.*, 1987; van Boeckel *et al.*, 1993).

Heparin or heparan sulfate is an especially good example with the following biological activities being ascribed to specific sequences:

- antithrombin III binding pentasaccharide (Lindahl *et al.*, 1979, 1980, 1981; Rosenberg *et al.*, 1981; Casu, 1981, 1994; Choay, 1983; Oscarsson, 1989),
- fibroblast growth factor binding hexasaccharide (Karlsson *et al.*, 1988; Ornitz *et al.*, 1992; Maccarana and Lindahl, 1993; Coltrini *et al.*, 1994; Ishihara *et al.*, 1993, 1994; Schmidt *et al.*, 1995),
- smooth muscle cell growth inhibition (Hoover *et al.*, 1980),
- interaction with platelet factors (Niewiarowski *et al.*, 1979; Maccarana and Lindahl, 1993),
- virus receptors (Lycke *et al.*, 1991),
- anti-neoplastic sequences and
- other binding activities such as superoxide dismutase (Karlsson *et al.*, 1988).

The heparin cofactor II binding hexasaccharide sequence in dermatan sulfate is another example of a biologically active internal sequence in glycosaminoglycans (Tollefsen *et al.*, 1986; Tollefsen, 1992, 1994).

10 Lectins in Biological Recognition

Over the past 10 years, a dramatic number of new biological activities have been ascribed to recognition systems between carbohydrates and proteins.

The number of papers having the word "lectin" in the title or abstract has increased remarkably over the past 20 years. A "Medline" computer search today will show more than 7500 articles with lectin in the title or abstract. There are 6 chapters in this volume with "lectin" in the title, and many of the chapters describe biological recognition systems.

Lectins, enzymes and antibodies can exhibit discriminating binding specificities for the shape, charge, epimers, anomers, linkage positions, ring size, branching and monosaccharide sequence of carbohydrate ligand molecules where the maximum recognized size is usually hexamer or smaller (Cisar *et al.*, 1974, 1975; Takeo and Kabat, 1978; Smith-Gill *et al.*, 1984). Carbohydrate sequences possess unique solution structures which, although dynamic, are shown by NOE NMR and molecular modeling to be populated mainly by minimum energy 3-dimensional conformations (Cumming and Carver, 1987; Poppe *et al.*, 1990; French *et al.*, 1993; Siebert *et al.*, this volume). Oligosaccharide haptens, more rigid than short peptides because of steric crowding, must be envisioned in 3 dimensional space for specific recognition by proteins.

11 Biological Mechanisms of Lectins or other Carbohydrate Binding Proteins

Oligosaccharides can partake in diverse processes:

- signal transduction mechanisms, for example the well known activity of concanavalin A on certain leucocytes, and the activation of neutrophil integrins upon ligandation with P-selectin (Lasky, 1995) or in regulatory molecules (Villalobo *et al.* this volume; Zanetta, this volume),

- as signals for polypeptide location within the cell, such as lysosomal protein markers (Reitman and Kornfeld, 1981; Hooper *et al.*, this volume),
- as ligands for proper protein folding: Glycosylation is apparently important in recognition of proper folding for protein chaperones such as calnexin (Chen *et al.*, 1995; Hebert *et al.*, 1995; Ora and Helenius, 1995),
- in the metazoan, for specific cell surface recognition of one cell by another (Mann *et al.*, this volume),
- in plants, the rhizobium recognition system utilizes a sulfated chitin oligosaccharyl glycolipid (nod factors) with specific acylation and sulfation for species specificity (Lerouge *et al.*, 1990; Roche *et al.*, 1991, 1992; Debelle *et al.*, 1992; Denarie *et al.*, 1992, 1994; Price *et al.*, 1992; Demont *et al.*, 1993, 1994; Denarie and Cullimore, 1993; Horvath *et al.*, 1993; Mergaert *et al.*, 1993; Ardourel *et al.*, 1994; Journet *et al.*, 1994; Lerouge, 1994; Relic *et al.*, 1994; Schwedock *et al.*, 1994; Spaink, 1994; Jabbouri *et al.*, 1995; Price and Carlson, 1995; Stokkermans *et al.*, 1995),
- in plants, pollen tube growth appears to be dependent on specific glycosylation gradients of a resident protein in the pistil (Balanzino *et al.*, 1994),
- in plants, chitin oligosaccharides apparently can stimulate host responses to fungal infections, and chitinases are utilized as plant defense systems (Scheel and Parker, 1990).

The multitude of known plant and animal lectins predicts a host of other very interesting biological activities, yet undiscovered.

12 Multivalent Effects

A collective of low avidity interactions may vastly strengthen intercellular binding (Lee *et al.*, 1990). Specific spacing of carbohydrate moieties within a larger structure may confer several orders of magnitude tighter binding such as in the galactose-binding protein (Rice *et al.*, 1990). Specific charge spacing on sulfated polymers may confer recognition by certain patterns of basic amino acids in cognate proteins. Possible higher complexity might occur where *patterns in sets* of carbohydrates form recognition systems with *sets* of binding proteins. Such systems may play a powerful role in intercellular sociology during development (Feizi, 1985, 1988), in the immune system (Brandley *et al.*, 1990; Polley *et al.*, 1991; Yuen *et al.*, 1992; Asa *et al.*, 1995; Lasky, 1995) and in parasitology (Friedman *et al.*, 1985; Pereira, this volume) and other microbial pathogenesis (Srnka *et al.*, 1992; see Gilboa-Garber *et al.*, this volume).

Numerous reviews and recent papers have been written regarding new discoveries in carbohydrate-based recognition systems as tumor markers (Hoff *et al.*, 1989; Matsushita *et al.*, 1990, 1991; Walz *et al.*, 1990; Irimura *et al.*, 1991; Miller *et al.*, 1992; Chanrasekaran *et al.*, 1995).

13 Bacterial/Eucaryotic Interactions

In another interesting area, microbes simulate mammalian complex carbohydrates to use as immune masks (*E. coli* K5 *N*-acetyl heparosan) (Casu *et al.*, 1994)), *E. coli* K1 colominic acid, *E. coli* K4 fructosyl chondroitin analog, and *Helicobacter pylori* using LeX (Noel *et al.*, 1995). Some bacteria also produce hyaluronic acid. Microbes definitely use lectin-like interactions in pathogenic systems such as enterotoxic *E. coli* (Neeser *et al.*, 1986), which are, interestingly, inhibited by mecomium oligosaccharides (Neeser *et al.*, 1986), the well studied cholera toxin/GM1 ganglioside system for which there are recent studies (Balanzino *et al.*, 1994).

In bacterial/eucaryotic interfaces, apparently there are general systems for recognition of certain microbial cell surface carbohydrate characteristics. For example, CD14 is apparently necessary for recognition of lipopolysaccharides by human endothelial cells (Noel *et al.*, 1995). A long literature attests to the surface binding specificity of certain *E. coli* to Gal(α1–4)Gal sequences in glycosphingolipids of certain individuals susceptible to recurrent urinary tract infections.

In bacteria and other microbes, chemotactic systems are present which recognize simple carbohydrates such as glucose, *N*-acetyl-glucosamine, trehalose and others (Bassler *et al.*, 1989, 1991). These systems predict that a number of proteins with specific sugar-binding sites comprise a sensorium for sugar-based nutrients in the environment (Bassler *et al.*, 1989, 1991). Some of these systems have been preliminarily characterized on a molecular or genetic basis.

14 Pharamaceutical Development

Carbohydrate biochemistry has taken on a new, revolutionary excitement, accompanied by research support in pharmaceutical companies exceeding $500,000,000 since 1988. Much of this has been spent on the "selectin" systems due to the huge market for anti-inflammatory drugs. Several small start-up pharmaceutical companies devote most of their efforts to this field, such as Cytel (San Diego, CA), Glycomed/Alberta Research Council (Alameda, CA), GlycoTech (Rockville, MD), Alpha-Beta (Boston, MA). An outstanding new development in cancer therapy is the use of group B streptococcal polysaccharide to target lectins present only on nascent capillaries as an anti-tumor agent, now in clinical trials (Carbomed Inc., Nashville, TN).

15 Glycobiology

A growing speciality area of biochemistry concerns itself with the biology of protein recognition of specific carbohydrates. This field has been coined "Glycobiology" by Raymond Dwek, which name has also been adopted by the Oxford Dictionary, the new journal "Glycobiology", and has become the name of the North American Scientific Society of some 700+ members (formerly the Society for Complex Carbohydrates).

16 Review of the Formal Calculation of Oligosaccharide Isomers

16.1 Linear Oligosaccharides

The formula for permutations of linear oligosaccharides is as follows:

Formula A: Isomers(S^*) = $E^n \times 2^n_a \times 2^n_r \times (4^{n-1})$ (Laine, 1994).
For a hexasaccharide made up of from a library of 6 hexoses,
$S^* = 46656 \times 64 \times 64 \times 1024 = 195{,}689{,}447{,}424$

Table 1. Linear isomers from a set of 1–6 *D*-hexoses allowing repeating *D*-hexoses

Oligosaccharide size:	Hexose Set	Linear Isomers
Monosaccharide	1	4
Disaccharide	2	256
Trisaccharide	3	27,648
Tetrasaccharide	4	4,194,304
Pentasaccharide	5	819,200,000
Hexasaccharide	6	195,689,447,424
(Hexapeptide)	(6 Amino acids)	46,656

The 825,000,000 mono- to pentasaccharide isomers added together are less than 0.5 % of the total hexasaccharide isomers.

In biology we should consider saccharides made up from a library of 20 common sugars.

Table 2. Linear isomers from a set of 1–6 *D*-hexoses from a set of 20 common sugars

Oligosaccharide size:	Hexose Set	Linear Isomers
Monosaccharide	20	80
Disaccharide	20	25600
Trisaccharide	20	8,192,000
Tetrasaccharide	20	2,621,440,000
Pentasaccharide	20	838,860,800,000
Hexasaccharide	20	268,435,460,000,000
(Hexapeptide)	(20 Amino acids)	64,000,000

16.2 Branched Oligosaccharides (See 1.16.3 for Formulas)

The above numbers are increased by a large number of chemically and biologically possible compounds with branched chains.

The monosaccharide in position "F" is assigned to be the reducing-end throughout, designated as "FR".

Monosaccharide Branches
For the singly branched compounds, examples are as follows:

```
B->C->D->E->FR      B->C->D->E->FR
   |                         |
   A                         A
   I                         II

B->C->D->E->FR      B->C->D->E->FR
         |                         |
         A                         A
        III                        IV
```

293,534,171,136 Isomers (Formula B).

Hexasaccharides with a single disaccharide branch

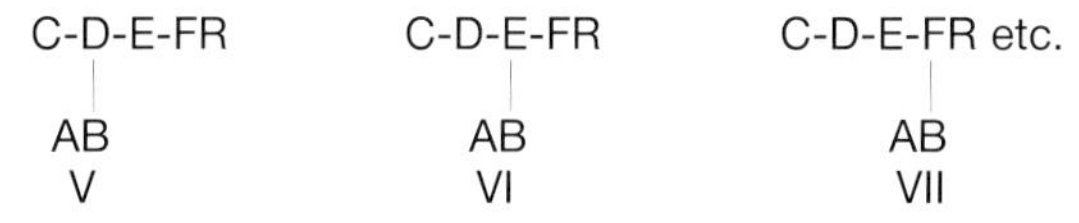

V is the same as II, however, VI and VII are novel arrangements:
146,452,512,768 Isomers (Formula C).

Trisaccharide branches to the core chain

```
D-E-FR     D-E-FR
|              |
ABC           ABC
VIII          IX
```

VIII is the same as III, sugar "D" being the single branch on "E" in the core ABCEF, and IX is the same as VII.

For a hexasaccharide or smaller, no new compounds are generated by considering this pattern (Formula D):

Tetrasaccharide branches

```
E-FR
 |
ABCD
 X
```

X is the same as IV, therefore this pattern produces new compounds only with octasaccharides and higher (Formula E):

Di-branched compounds
Two single branches on two different core monosaccharides give 3 new types of arrangement:

```
C-D-E-FR     C-D-E-FR     C-D-E-FR  etc.
  | |          |   |           | |
  A B          A   B           A B
  XI           XII             XIII
```

330,225,942,528 Isomers (Formula F).

A heptasaccharide is the smallest compound capable of triple single branches as in (Formula G)

```
D-E-F-G-(reducing end)
| | |
A B C
```

Triply branched monosaccharides
For hexasaccharides, two single branches on the same core monosaccharide (Trisubstituted or triple-branched) represent another novel set:

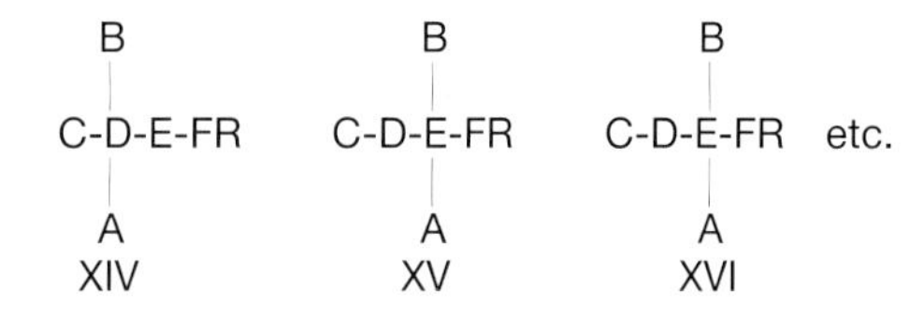

36,691,771,392 (Formula H).

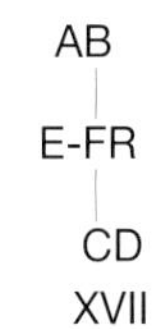

12,230,590,464 (Formula I).

One monosaccharide and one disaccharide branch on different core monosaccharides

```
AB           AB
|             |
D-E-FR       D-E-FR  etc.
  |            |
  C            C
XVIII         XIX
```

XVIII is the same as XIII, while XIX has the core DEF or ABF, a novel arrangement.

A single, itself branched trisaccharide branch, as follows

```
D-E-FR          D-E-FR
  |               |
  C               C
 /\              /\
A B             A B
 XX              XXI
```

27,518,828,544 (Formula J)

```
D
|
E-FR
|        Compound XXII
C
/\
A B
```

4,586,471,424 (Formula K).

Tetra-branched versions are also possible

```
E D
 \/
  F-R   XXIII
 /\
A BC

A B
 \/
E-FR    XXIV
 /\
C D
```

6,115,295,232 (Formula L).

Most biological activities are recognized within a proteinaceous binding site of 6 sugars, (or usually fewer) as exemplified by antibodies, enzymes (lysozyme), heparinoids or lectins (selectins). There are a few examples of proteins requiring higher oligomers for activity, e.g. a few enzyme recognition sites in the *N*-linked anabolic pathway for glycoprotein synthesis which apparently recognize precursors as large as 14 sugars. This calculation covers all possibilities for a *D*-hexasaccharide or smaller where F is the reducing end or is attached to an aglycon.

16.3 Formulas

A: $S^* = E^n \times 2^n_a \times 2^n_r \times (4^{n-1})$

+

B: $S^*=E^n \times 2^n_a \times 2^n_r \times (4^{n-3}) \times [6 \times (n-2)]$

+

C: $S^*=E^n \times 2^n_a \times 2^n_r \times (4^{n-3}) \times [6 \times (n-4)]$

+

D: $S^*=E^n \times 2^n_a \times 2^n_r \times (4^{n-3} \times [6 \times (((n-6)+(Abs.(n-6))/2)]$

+

E: $S^*=E^n \times 2^n_a \times 2^n_r \times (4^{n-3} \times [6 \times (((n-7)+(Abs.(n-7))/2)]$

+

F: $S^*=E^n \times 2^n_a \times 2^n_r \times 4^{n-4} \times [6^2 \times (n-4+n-5+...+n-(n-1))]$

+

G: $S^*=E^n \times 2^n_a \times 2^n_r \times (4^{n-4}) \times [4 \times (n-3)]$

+

H: $S^*=E^n \times 2^n_a \times 2^n_r \times (4^{n-4}) \times [4 \times (n-5)]$

+

J: $S^*=E^n \times 2^n_a \times 2^n_r \times (4^{n-5}) \times [6^2 \times (n-5)]$

+

K: $S^*=E^n \times 2^n_a \times 2^n_r \times (4^{n-6}) \times 6 \times [4 \times (n-5)]$

+

L: $S^*=E^n \times 2^n_a \times 2^n_r \times (4^{n-5})s \times [n-4] \times (4n-5)t$

Totals taken from Formulas A to L for hexasaccharides made up from a library of 6 *D*-hexoses

A	195,689,447,424
B	293,534,171,136
C	146,452,512,768
D	0
E	0
F	330,225,942,528
G	36,691,771,392
H	12,230,590,464
J	27,518,828,544
K	4,586,471,424
L	6,115,295,232

Total: 1,053,045,031,000.

Including the mirror image *L*-sugar forms in this set would increase this number by a factor of 64.

Table 3 shows the number of total reducing oligosaccharides possible from mono- to hexasaccharide:

Table 3. Oligosaccharide isomers from a set of 6 *D*-hexoses, including branched forms

Oligosaccharide size:	Hexose Set	Isomers
Monosaccharide	1	4
Disaccharide	2	256
Trisaccharide	3	38,016
Tetrasaccharide	4	7,602,176
Pentasaccharide	5	2,633,600,000
Hexasaccharide	6	1,053,045,031,000

Using a library of 20 Common Sugars.

Table 4. Oligosaccharide branched and linear isomers from a pool of 20 sugars

Oligosaccharide size:	Hexose Set	Isomers
Monosaccharide	20	80
Disaccharide	20	25,600
Trisaccharide	20	11,264,000
Tetrasaccharide	20	4,751,360,000
Pentasaccharide	20	2,698,064,000,000
Hexasaccharide	20	1,444,506,200,000,000

17 Oligosaccharide Building Blocks

Organisms possess a much larger menu of possibilities than for peptides:

- There exist more than 50 types of amino sugars alone and probably 50 neutral and 50 acidic sugars. Novel monosaccharides are to this day being discovered in plants and microbial cell walls.
- Remember substitutions: Sugars in nature can be substituted with acyl, alkyl, pyruvyl, sulfate, sulfonate, phosphate, phosphonate, and other groups any one of which would raise the possible isomers to a number much greater than the one we have calculated.

18 Summary

There is a very high evolutionary potential in possible epitopes for establishment of a biological recognition "code" consisting of the binding pocket of a specific protein on the one hand and a complex sugar structure on the other. Because proteins can evolve more rapidly than carbohydrates (which must have a substantial enzyme change to add a new sugar), saccharide structures are likely to be very conserved over evolution when compared with proteins whose specificity could change with a single amino acid mutation. Carbohydrate sequences in metazoans will probably have functions that are conserved across genera, such as the selectins and heparinoids in mammals. There is obviously adequate chemistry for much further evolution in carbohydrate-protein or carbohydrate-carbohydrate recognition systems. Calculation of carbohydrate isomers shows the potential for the most complex known chemical code in a short sequence among all biological oligomers.

References

Ardourel M, Demont N, Debelle F *et al.* (1994): *Rhizobium meliloti* lipooligosaccharide nodulation factors: different structural requirements for bacterial entry into target root hair cells and induction of plant symbiotic developmental responses. In *Plant Cell* **6**:1357–74.

Asa D, Raycroft L, Ma L *et al.* (1995): The P-selectin glycoprotein ligand functions as a common human leukocyte ligand for P- and E-selectins. In *J. Biol. Chem.* **270**:11662–70.

Atha DH, Lormeau JC, Petitou M *et al.* (1987): Contribution of 3-*O*- and 6-*O*-sulfated glucosamine residues in the heparin-induced conformational change in antithrombin III. In *Biochemistry* **26**:6454–61.

Bajorath J, Aruffo A (1995): A template for generation and comparison of three-dimensional selectin models. In *Biochem. Biophys. Res. Commun.* **216**:1018–23.

Balanzino LE, Barra JL, Monferran CG *et al.* (1994): Differential interaction of *Escherichia coli* heat-labile toxin and cholera toxin with pig intestinal brush border glycoproteins depending on their ABH and related blood group antigenic determinants. In *Infect. Immun.* **62**:1460–4.

Bassler B, Gibbons P, Roseman S (1989): Chemotaxis to chitin oligosaccharides by *Vibrio furnissii*, a chitinivorous marine bacterium. In *Biochem. Biophys. Res. Commun.* **161**:1172–6.

Bassler BL, Gibbons PJ, Yu C *et al.* (1991): Chitin utilization by marine bacteria. Chemotaxis to chitin oligosaccharides by *Vibrio furnissii*. In *J. Biol. Chem.* **266**:24268–75.

Brandley BK, Swiedler SJ, Robbins PW (1990): Carbohydrate ligands of the LEC cell adhesion molecules. In *Cell* **63**:861–3.

Casu B, Oreste P, Torri G *et al.* (1981): The structure of heparin oligosaccharide fragments with high anti-(factor Xa) activity containing the minimal antithrombin III-binding sequence. Chemical and ^{13}C nuclear-magnetic-resonance studies. In *Biochem. J.* **197**:599–609.

Casu B, Grazioli G, Razi N *et al.* (1994): Heparin-like compounds prepared by chemical modification of capsular polysaccharide from *E. coli* K5. In *Carbohydr. Res.* **263**:271–84.

Chandrasekaran EV, Jain RK, Rhodes JM *et al.* (1995): Expression of blood group Lewis b determinant from Lewis a: association of this novel alpha (1,2)-L-fucosylating activity with the Lewis type alpha (1,3/4)-L-fucosyltransferase. In *Biochemistry* **34**: 4748–56.

Chen W, Helenius J, Braakman I *et al.* (1995): Cotranslational folding and calnexin binding during glycoprotein synthesis. In *Proc. Natl. Acad. Sci. USA* **92**:6229–33.

Choay J, Petitou M, Lormeau JC *et al.* (1983): Structure-activity relationship in heparin: a synthetic pentasaccharide with high affinity for antithrombin III and eliciting high anti-factor Xa activity. In *Biochem. Biophys. Res. Commun.* **116**:492–9.

Cisar J, Kabat EA, Liao J *et al.* (1974): Immunochemical studies on mouse myeloma proteins reactive with dextrans or with fructosans and on human antilevans. In *J. Exp. Med.* **139**:159–79.

Cisar J, Kabat EA, Dorner MM *et al.* (1975): Binding properties of immunoglobulin combining sites specific for terminal or nonterminal antigenic determinants in dextran. In *J. Exp. Med.* **142**:435–59.

Coltrini D, Rusnati M, Zoppetti G *et al.* (1994): Different effects of mucosal, bovine lung and chemically modified heparin on selected biological properties of basic fibroblast growth factor. In *Biochem. J.* **303**:583–90.

Cumming DA, Carver JP (1987): Virtual and solution conformations of oligosaccharides. In *Biochemistry* **26**:6664–76.

Debelle F, Rosenberg C, Denarie J (1992): The *Rhizobium, Bradyrhizobium,* and *Azorhizobium* NodC proteins are homologous to yeast chitin synthases. In *Mol. Plant Microbe Interact.* **5**:443–6.

Demont N, Debelle F, Aurelle H *et al.* (1993): Role of the Rhizobium meliloti nodF and nodE genes in the biosynthesis of lipo-oligosaccharidic nodulation factors. In *J. Biol. Chem.* **268**:20134–42.

Demont N, Ardourel M, Maillet F *et al.* (1994): The Rhizobium meliloti regulatory nodD3 and syrM genes control the synthesis of a particular class of nodulation factors *N*-acylated by (omega-1)-hydroxylated fatty acids. In *EMBO J.* **13**:2139–49.

Denarie J, Cullimore J (1993): Lipo-oligosaccharide nodulation factors: a minireview new class of signaling molecules mediating recognition and morphogenesis. In *Cell* **74**:951–4.

Denarie J, Debelle F, Rosenberg C (1992): Signaling and host range variation in nodulation. In *Annu. Rev. Microbiol.* **46**:497–531.

Denarie J, Truchet G, Prome JC (1994): Lipo-oligosaccharide signalling: the mediation of recognition and nodule organogenesis induction in the legume-Rhizobium symbiosis. In *Biochem. Soc. Symp.* **60**:51–60.

Feizi T (1985): Demonstration by monoclonal antibodies that carbohydrate structures of glycoproteins and glycolipids are onco-developmental antigens. In *Nature* **314**:53–7.

Feizi T (1988): Carbohydrate structures as onco-developmental antigens and components of receptor systems. In *Adv. Exp. Med. Biol.* **228**:317–29.

French AD, Mouhous-Riou N, Perez S (1993): Computer modeling of the tetrasaccharide nystose. In *Carbohydr. Res.* **247**:51–62.

Friedman MJ, Fukuda M, Laine RA (1985): Evidence for a malarial parasite interaction site on the major transmembrane protein of the human erythrocyte. In *Science* **228**:75–7.

Hebert D N, Foellmer B, Helenius A (1995): Glucose trimming and reglucosylation determine glycoprotein association with calnexin in the endoplasmic reticulum. In *Cell* **81**:425–33.

Hoff SD, Matsushita Y, Ota DM *et al.* (1989): Increased expression of sialyl-dimeric LeX antigen in liver metastases of human colorectal carcinoma. In *Cancer Res.* **49**:6883–8.

Hoover RL, Rosenberg R, Haering W *et al.* (1980): Inhibition of rat arterial smooth muscle cell proliferation by heparin. II. In vitro studies. In *Circ. Res.* **47**:578–83.

Horvath B, Heidstra R, Lados M *et al.* (1993): Lipo-oligosaccharides of *Rhizobium* induce infection-related early nodulin gene expression in pea root hairs. In *Plant J.* **4**:727–33.

Irimura T, Matsushita Y, Hoff SD *et al.* (1991): Ectopic expression of mucins in colorectal cancer metastasis. In *Semin. Cancer Biol.* **2**:129–39.

Ishihara M, Shaklee PN, Yang Z *et al.* (1994): Structural features in heparin which modulate specific biological activities mediated by basic fibroblast growth factor. In *Glycobiology* **4**:451–8.

Ishihara, M, Tyrrell, DJ, Stauber GB *et al.* (1993): Preparation of affinity-fractionated, heparin-derived oligosaccharides and their effects on selected biological activities mediated by basic fibroblast growth factor. In *J. Biol. Chem.* **268**:4675–83.

Jabbouri S, Fellay R, Talmont F *et al.* (1995): Involvement of nodS in *N*-methylation and nodU in 6-*O*-carbamoylation of *Rhizobium sp.* NGR234 nod factors. In *J. Biol. Chem.* **270**:22968–73.

Jordan ET, Goldstein IJ (1995): Site-directed mutagenesis studies on the lima bean lectin. Altered carbohydrate-binding specificities result from single amino acid substitutions. In *Eur. J. Biochem.* **230**:958–64.

Journet EP, Pichon M, Dedieu A *et al.* (1994): *Rhizobium meliloti* Nod factors elicit cell-specific transcription of the ENOD12 gene in transgenic alfalfa. In *Plant J.* **6**:241–9.

Karlsson K, Lindahl U, Marklund SL (1988): Binding of human extracellular superoxide dismutase C to sulphated glycosaminoglycans. In *Biochem. J.* **256**:29–33.

Laine RA (1989): Tandem mass spectrometry of oligosaccharides. In *Methods Enzymol.* **179**:157–64.

Laine RA (1994): A calculation of all possible oligosaccharide isomers, both branched and linear yields 1.05×10^{12} structures for a reducing hexasaccharide: The isomer barrier to development of single-method saccharide sequencing or synthesis systems. In *Glycobiology* **4**:1–9.

Laine RA, Pamidimukkala KM, French AL *et al.*. (1988): Linkage position in oligosaccharides by fast atom bombardment ionization, collision-activated dissociation, tandem mass spectrometry and molecular modeling. *L*-Fucosyl$_p$ (β1,X)-*D*-*N*-acetyl-*D*-glucosaminyl$_p$-(β1,3)-*D*-galactosyl$_p$-(β1,0-methyl) where X = 3,4, or 6. In *J. Am. Chem. Soc.* **110**:6931–9.

Laine RA, Yoon E, Mahier TJ *et al.* (1991): Nonreducing terminal linkage position determination in intact and permethylated synthetic oligosaccharides having a penultimate amino sugar: fast atom bombardment ionization, collision-induced dissociation and tandem mass spectrometry. In *Biol. Mass Spectrom.* **20**:505–14.

Lasky LA (1995): Selectin-carbohydrate interactions and the initiation of the inflammatory response. In *Annu. Rev. Biochem.* **64**:113–9.

Lee RT, Ichikawa Y, Allen HJ *et al.* (1990): Binding characteristics of galactoside-binding lectin (galaptin) from human spleen. In *J. Biol. Chem.* 265:7864–71.

Lerouge P (1994): Symbiotic host specificity between leguminous plants and rhizobia is determined by substituted and acylated glucosamine oligosaccharide signals. In *Glycobiology* **4**:127–34.

Lerouge P, Roche P, Faucher C *et al.* (1990): Symbiotic host-specificity of *Rhizobium meliloti* is determined by a sulphated and acylated glucosamine oligosaccharide signal. In *Nature* **344**:781–4.

Lindahl U, Backstrom G, Hook M *et al.* (1979): Structure of the antithrombin-binding site in heparin. In *Proc. Natl. Acad. Sci. USA* **76**:3198–202.

Lindahl U, Backstrom G, Thunberg L *et al.* (1980): Evidence for a 3-*O*-sulfated D-glucosamine residue in the antithrombin-binding sequence of heparin. In *Proc. Natl. Acad. Sci. USA* **77**:6551–5.

Lindahl U, Thunberg L, Backstrom G *et al.* (1981): The antithrombin-binding sequence of heparin. In *Biochem. Soc. Transact.* **9**:499–551.

Lycke E, Johansson M, Svennerholm B *et al.* (1991): Binding of herpes simplex virus to cellular heparan sulphate, an initial step in the adsorption process. In *J. Gen. Virol.* **72**:1131–7.

Maccarana M, Lindahl U (1993): Mode of interaction between platelet factor 4 and heparin. In *Glycobiology* **3**:271–7.

Matsushita Y, Cleary KR, Ota DM *et al.* (1990): Sialyl-dimeric Lewis-X antigen expressed on mucin-like glycoproteins in colorectal cancer metastases. In *Lab. Invest.* **63**:780–91.

Matsushita Y, Hoff SD, Nudelman ED *et al.* (1991): Metastatic behavior and cell surface properties of HT-29 human colon carcinoma variant cells selected for their differential expression of sialyl-dimeric Le(x)-antigen. In *Clin. Exp. Metastasis* **9**:283–99.

Mergaert P, Van Montagu M, Prome JC *et al.* (1993): Three unusual modifications, a D-arabinosyl, an *N*-methyl, and a carbamoyl group, are present on the Nod factors of *Azorhizobium caulinodans* strain ORS571. In *Proc. Natl. Acad. Sci. USA* **90**:1551–5.

Miller KE, Mukhopadhyay C, Cagas P *et al.* (1992): Solution structure of the Lewis x oligosaccharide determined by NMR spectroscopy and molecular dynamics simulations. In *Biochemistry* **31**:6703–9.

Neeser JR, Koellreutter B, Wuersch P (1986): Oligomannoside-type glycopeptides inhibiting adhesion of *Escherichia coli* strains mediated by type 1 pili: preparation of potent inhibitors from plant glycoproteins. In *Infect. Immun.* **52**:428–36.

Niewiarowski , Rucinski, B, James P *et al.* (1979): Platelet antiheparin proteins and antithrombin III interact with different binding sites on heparin molecule. In *FEBS Lett.* **102**:75–8.

Noel RF Jr, Sato TT, Mendez C *et al.* (1995): Activation of human endothelial cells by viable or heat-killed gram-negative bacteria requires soluble CD14. In *Infect. Immun.* **63**:4046–53.

Ora A, Helenius A (1995): Calnexin fails to associate with substrate proteins in glucosidase-deficient cell lines. In *J. Biol.Chem.* **270**:26060–2.

Ornitz DM, Yayon A, Flanagan JG *et al.* (1992): Heparin is required for cell-free binding of basic fibroblast growth factor to a soluble receptor and for mitogenesis in whole cells. In *Mol. Cell. Biol.* **12**:240–7.

Oscarsson LG, Pejler G, Lindahl U (1989): Location of the antithrombin-binding sequence in the heparin chain. In *J. Biol. Chem.* **264**:296–304.

Polley MJ, Phillips ML, Wayner E *et al.* (1991): CD62 and endothelial cell-leukocyte adhesion molecule 1 (ELAM-1) recognize the same carbohydrate ligand, sialyl-Lewisx. In *Proc. Natl. Acad. Sci. USA* **88**:6224–8.

Poppe L, Dabrowski J, von der Lieth CW *et al.* (1990): Three-dimensional structure of the oligosaccharide terminus of globotriaosylceramide and isoglobotriaosylceramide in solution. A rotating-frame NOE study using hydroxyl groups as long-range sensors in conformational analysis by 1H-NMR spectroscopy. In *Eur. J. Biochem.* **189**:313–25.

Price NP, Carlson RW (1995): Rhizobial lipo-oligosaccharide nodulation factors: multidimensional chromatographic analysis of symbiotic signals involved in the development of legume root nodules. In *Glycobiology* **5**:233–42.

Price NP, Relic B, Talmont F *et al.* (1992): Broad-host-range *Rhizobium* species strain NGR234 secretes a family of carbamoylated, and fucosylated, nodulation signals that are *O*-acetylated or sulphated. In *Mol. Microbiol.* **6**:3575–84.

Reitman ML, Kornfeld S (1981): Lysosomal enzyme targeting. *N*-Acetylglucosaminylphosphotransferase selectively phosphorylates native lysosomal enzymes. In *J. Biol. Chem.* **256**:11977–80.

Relic B, Perret X, Estrada-Garcia MT *et al.* (1994): Nod factors of Rhizobium are a key to the legume door. In *Mol. Microbiol.* **13**:171–8.

Rice KG, Weisz OA, Barthel T *et al.* (1990): Defined geometry of binding between triantennary glycopeptide and the asialoglycoprotein receptor of rat heptocytes. In *J. Biol. Chem.* **265**:18429–34.

Riensenfeld J, Hook M, Bjork I *et al.* (1977): Structural requirements for the interaction of heparin with antithrombin III. In *Fed. Proc.* **36**:39–43.

Roche P, Debelle F, Lerouge P *et al.* (1992): The lipo-oligosaccharidic symbiotic signals of *Rhizobium meliloti*. In *Biochem. Soc. Trans.* **20**:288–91.

Roche P, Debelle F, Maillet F *et al.* (1991): Molecular basis of symbiotic host specificity in *Rhizobium meliloti*: nodH and nodPQ genes encode the sulfation of lipo-oligosaccharide signals. In *Cell* **67**:1131–43.

Rosenberg RD, Jordan RE, Favreau LV *et al.* (1979): Highly active heparin species with multiple binding sites for antithrombin. In *Biochem. Biophys. Res. Commun.* **86**:1319–24.

Roth S, McGuire EJ, Roseman S (1971): Evidence for cell-surface glycosyltransferases. Their potential role in cellular recognition. In *J. Cell Biol.* **51**:536–47.

Schmidt A, Skaletz-Rorowski A, Breithardt G *et al.* (1995): Growth status-dependent changes of bFGF compartmentalization and heparan sulfate structure in arterial smooth muscle cells. In *Eur. J. Cell Biol.* **67**:130–5.

Schwedock JS, Liu C, Leyh TS *et al.* (1994): *Rhizobium meliloti* NodP and NodQ form a multifunctional sulfate- activating complex requiring GTP for activity. In *J. Bacteriol.* **176**:7055–64.

Smith-Gill SJ, Rupley JA, Pincus MR *et al.* (1984): Experimental identification of a theoretically predicted "left-sided" binding mode for (*N*-acetylglucosamine)6 in the active site of lysozyme. In *Biochemistry* **23**:993–7.

Spaink HP (1994): The molecular basis of the host specificity of the *Rhizobium* bacteria. In *Antonie Van Leeuwenhoek* **65**:81–98.

Srnka CA, Tiemeyer M, Gilbert JH *et al.* (1992): Cell surface ligands for rotavirus: mouse intestinal glycolipids and synthetic carbohydrate analogs. In *Virology* **190**:794–805.

Stokkermans TJ, Ikeshita S, Cohn J *et al.* (1995): Structural requirements of synthetic and natural product lipo-chitin oligosaccharides for induction of nodule primordia on *Glycine soja*. In *Plant Physiol.* **108**:1587–95.

Takeo K, Kabat EA (1978): Binding constants of dextrans and isomaltose oligosaccharides to dextran-specific myeloma proteins determined by affinity electrophoresis. In *J. Immunol.* **121**:2305–10.

Tollefsen DM (1992): The interaction of glycosaminoglycans with heparin cofactor II: structure and activity of a high-affinity dermatan sulfate hexasaccharide. In *Adv. Exp. Med. Biol.* **313**:167–76.

Tollefsen M (1994): The interaction of glycosaminoglycans with heparin cofactor II. In *Ann. N. Y. Acad. Sci.* **714**:21–31.

Tollefsen DM, Peacock ME, Monafo WJ (1986): Molecular size of dermatan sulfate oligosaccharides required to bind and activate heparin cofactor II. In *J. Biol.Chem.* **261**:8854–8.

van Boeckel CAA, Petitou M (1993): The unique antithrombin III binding domain of heparin: A lead to new synthetic antithrombotics. In *Angew. Chem. Int. Ed.* **32**:1671–90.

Walz G, Aruffo A, Kolanus W *et al.* (1990): Recognition by ELAM-1 of the sialyl-Lex determinant on myeloid and tumor cells. In *Science* **250**:1132–5.

Yamamoto F, Hakomori S-i (1990): Sugar-nucleotide donor specificity of histo-blood group A and B transferases is based on amino acid substitutions. In *J. Biol. Chem.* **265**:19257–62.

Yoon E, Laine RA (1992): Linkage position determination of permethylated neutral novel trisaccharides by collisional induced dissociation and tandem mass spectrometry. In *Biol. Mass Spectrom.* **21**:479–85.

Yuen CT, Lawson AM, Chai W *et al.* (1992): Novel sulfated ligands for the cell adhesion molecule E-selectin revealed by the neoglycolipid technology among *O*-linked oligosaccharides on an ovarian cystadenoma glycoprotein. In *Biochemistry* **31**: 9126–31.

2 Methods of Glycoconjugate Analysis

ELIZABETH F. HOUNSELL

1 Introduction

The monosaccharides which give the name glycose to the glycosciences form a large group of biological building blocks which, unlike the amino acids of proteins, can be linked together in a number of different ways, making their analysis difficult to automate. Several methods are therefore used in concert to distinguish the variety of conformational determinants in order to study their structure/function relationships. The other chapters in this book will attest to the importance of carrying out this analysis as the therapeutic potential of these molecules is gradually being realised. Already close to market are an anticoagulant hexasaccharide (Grootenhuis and van Boeckel, 1991), several inhibitors of microbial infection (Mouricout *et al.*, 1990; von Itzstein *et al.*, 1993; Fischl *et al.*, 1994; Boedeker, 1995; Koketsu *et al.*, 1995) a therapy in neuropathology (Tennant *et al.*, 1995) and an antidiabetic treatment (Bolen and Stolmans, 1995). In addition, the characterisation of carbohydrate antigens and ligands for carbohydrate binding proteins has opened up the potential for immune modulation strategies in inflammation, cancer and xenotransplantation. The majority of cell surface and secreted macromolecules (and some important intracellular proteins; Hart *et al.*, 1989) is glycosylated. Their analysis progresses through the identification of any monosaccharides present, to characterization of oligosaccharide size and monosaccharide sequence, and determination of the linkages between the monosaccharides and those to protein. This is followed by release of oligosaccharides for structural and conformational analysis by NMR spectroscopy and visualization of the data by computer graphics molecular modeling. Most recently the recognition of oligosaccharides by anti-carbohydrate antibodies, enzymes and carbohydrate binding proteins is being studied by microcalorimetry and surface plasmon resonance. In the future X-ray crystallography, at present in its infancy for studies of oligosaccharide-to-protein interaction, will become more routinely used for studies of structure and interactions.

1.1 Monosaccharide Identification

The oligosaccharides of glycoconjugates (glycoproteins, proteoglycans and glycolipids) are composed of several classes of monosaccharides, the most common being the hexopyranosides (hexoses e.g. glucose, galactose, mannose), acetamido sugars (*N*-acetylglucosamine, *N*-acetylgalactosamine), 6-deoxy hexoses (fucose and rhamnose), pentoses (xylose) and the sialic and uronic acids. The presence of hexose can be shown by a phenol-sulfuric acid assay (Smith *et al.*, 1994; by virtue of their having adjacent C1-C2 hydroxyl groups). Acetamido sugars and sialic acids have a UV-absorbing chromophore detected at 190–204 nm. The sialic acids are a family of related sequences, the most common being Neu5Ac (*N*-acetylneuraminic acid, NANA) and Neu5Gc (*N*-glycolylneuraminic acid, NGNA). These can be readily distinguished by release with mild acid (they are the most labile of monosaccharides in an oligosaccharide) and chromatography, particularly high pH anion exchange chromatography (HPAEC) with sensitive electrochemical detection by pulsed amperometry (PAD). The mild acid used is from 0.1 to 0.01 M HCl, the lower concentrations releasing

H.-J. and S. Gabius (Eds.), Glycosciences

ISBN 3-8261-0073-5

sialic acids with intact *O*-acyl groups e.g. acetyl which can occur at the C4, C7, C8 or C9 hydroxyl groups of NeuAc and NeuGc, for example. HPAEC-PAD is also an ideal method for the resolution of sialylated oligosaccharides where separation is achieved not only by number of anionic groups but also by linkage isomer (this and other HPLC and electrophoresis techniques for oligosaccharides are discussed in more detail in subsequent sections).

Although neutral monosaccharide analysis can be carried out by HPAEC-PAD of hydrolysed glycoconjugates or by HPLC of monosaccharide derivatives (Table 1), gas-liquid chromatography (GC) remains a method of choice because of its robustness and ability to separate a large number of very closely related molecules, e.g. the majority of all the α and β, pyranose and furanose methyl glycosides of the common hexoses and pentoses or the partially methylated alditol acetates obtained from a large number of different linkage isomers (Table 1). The latter analysis, called methylation analysis, is usually carried out with on-line mass spectrometric (MS) detection on reliable, relatively inexpensive bench top GC-MS apparatuses. The GC is then also available for accurate monosaccharide quantitation by analysis of the trimethylsilyl ethers of methyl glycosides (Hounsell, 1994a). One additional chromatographic method which has been used for more than a decade (Yamashita *et al.*, 1982), but has recently been packaged into an automatic Glycosequencer (from Oxford Glycosystems) is the use of the gel filtration medium, Biogel P4 (Table 2). This gives carbohydrate composition in terms of glucose equivalents based on hydrodynamic properties and is primarily used in association with specific exoglycosidase digestion to give monosaccharide sequence. High sensitivity fluorescence detection is now the method of choice with 2-amino benzidene (2-AB) derivatives, similar to the pyridylamino group (PA) used for HPLC (Hase *et al.*, 1987). Gel electrophoresis systems (PAGE) have also been commercialized for a similar type of analysis (Bigge *et al.*, 1995; Hu, 1995) using 8-aminonaphthalene-1,3,6-trisulphonic acid (ANTS) or anthranilic acid (2-aminobenzoic acid) derivatives (Tables 1 and 2).

Table 1. Monosaccharide Analysis[a]

HPAEC-PAD	Sialic acids released by hydrolysis (0.1–0.01M HCl) or neuramindases Neutral monosaccharides released by hydrolysis (2M TFA 4 hr 100° C) Acetamido sugars (eg GlcNAc) released as aminosugars (eg $GlcNH_2$) by hydrolysis (4M HCl 2 hrs 100°C) Uronic acids released by methanolysis and subsequent hydrolysis
HPLC	Hydrolysis and derivatisation with 2-aminopyridine (AP)
PAGE	(polyacrylamide gel electrophoresis). Hydrolysis or exoglycosidase digestion and derivatisation with aminonaphthalene – 1,3,6 trisulfonic acid (ANTS) or anthranilic acid
GC	Methanolysis and analysis of trimethylsilyl ethers of methylglycosides eg for galactose. β-*p* α-*p* β-*f* α-*f*

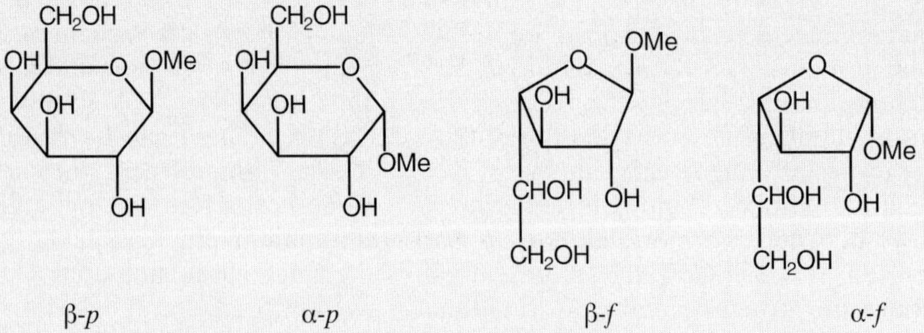

[a] Practical details of the methods are discussed in Hounsell (1994a, b, 1997).

Table 2. Oligosaccharide Sequence and Linkage Analysis[a]

GC-MS	Permethylation CH_2OMe, MeO, O, OMe Methanolysis → Hydrolysis Acetylation; Reduction / Acetylation CH_2OMe, MeO, O, OAc, OMe, OMe CH_2-OAc, CH–OMe, AcO–CH, MeO–CH, CH–OAc, CH_2-OMe
BioGel P4	2-Aminobenzamide labeling of oligosaccharides and analysis on a RAAM™2000 Glycosequencer before and after exoglycosidase digestion
PAGE	ANTS or anthranilic acid (see Table 1) labeling of oligosaccharides before and after exoglycosidase digestion

[a] Practical details of the methods are discussed in Hounsell (1994a, b, 1997).

1.2 Chemical Reactions

Although enzymes can also be used for cleavage of oligosaccharides from glycoproteins (endoglycosidases discussed in Section 4), chemical release is often preferred as this is much less likely to be linkage specific. Hydrazinolysis has been the method of choice for release of *N*-linked chains of glycoproteins since its introduction in the 1960's (Yosizawa *et al.*, 1966; Dimitriev *et al.*, 1975). The methodology has been perfected both on the laboratory bench (Yamashita *et al.*, 1984; Mizuochi, 1993) and in an automated apparatus (Oxford Glycosystems, GlycoPrep). Release of *O*-linked oligosaccharide chains of glycoproteins can also be achieved by a milder set of hydrazinolysis conditions, but this is not yet as reliable. As long as a reducing monosaccharide is not required at the oligosaccharide-to-protein linkage point, these chains can be optimally released by mild alkali-catalysed β-elimination with concomitant reduction to prevent further breakdown of the oligosaccharide chain. For subsequent mass spectrometry or conformational analysis of the protein the reducing agent is excluded as this causes considerable breakdown of the peptide backbone. The glycosaminoglycan chains of proteoglycans are also linked to protein via hydroxylated amino acids, but only linkages to serine have been identified so far. Although β-elimination would release these chains, they tend to be very large and therefore oligosaccharides have classically been obtained by cleavage along the chain at non-*N*-acetylated amino groups using nitrous acid. Alternatively, interchain cleavage can also be achieved by endoglycosidase (lyase) digestion (Section 4).

Glycosphingolipids are the main family of glycolipids where complex oligosaccharide chains attached to ceramide can be released by the chemical method of ozonolysis. Both glycoproteins and glycolipids or their released chains can be analyzed by periodate oxidation which, depending on linkage, cleaves the chain internally or releases monosaccharides from the non-reducing end. This can be used sequentially with intermediate mild hydrolysis steps (Smith degradation) to sequence oligosaccharides. Methylation can be used in association with periodate oxidation (Angel and Nilsson, 1988) for mass spectrometric detection. Specific periodate oxidation conditions can also be designed to only cleave the C6-C9 glycerol side chain of sialic acids or to oxidize the reduced end of oligosaccharide alditols (such as those produced in the alkaline borohydride degradation of *O*-linked glycoproteins). This leads to a reactive aldehyde which can be used to label sialylated oligosaccharides (Haselbeck and Hösel, 1993) or reductively aminate and sequence *O*-linked chains (Stoll and Hounsell, 1988).

2 Biochemical Reagents

Anti-carbohydrate antibodies, lectins and carbohydrate processing enzymes recognize not only monosaccharide type, but also linkage position, anomeric configuration and sequence. They can therefore be used as biochemical reagents in oligosaccharide analysis, but also of course are being studied in their own right in oligosaccharide recognition and function. A large number of plant, marine and microbial lectins have now been identified which have a wide range of specificities. Mammalian carbohydrate-binding proteins (selectins, collectins, galectins and sialoadhesins) will also become available as recombinant proteins. These like monoclonal anti-carbohydrate antibodies (already on the market from the hybridoma technology) tend to recognize a more complex conformational motif than enzymes or lectins and can therefore be used for specific identification of tri- to octa-saccharide sequences. The affinity of binding of these reagents can be increased by positive cooperativity induced by using multivalent oligosaccharides, e.g. those linked to lipids or clustered on proteins (commercially available) or cell surfaces (immunohistochemistry) (Lee and Lee, 1994). Specific glycosyltransferases are also gradually becoming available from biotechnology companies via genetic recombination (Field and Wainwright, 1995). The specificity of the transferases can be used to characterize the oligosaccharide substrate sequence and also, for example, to add radioactive monosaccharides for sensitive detection in cell assays. This latter strategy was employed to identify *O*-GlcNAc as a single monosaccharide on cytoplasmic proteins (Torres and Hart 1984; Greis and Hart 1996). This motif has since been found on a number of important nuclear (Dong *et al.*, 1993) and cytoskeletal (King and Hounsell, 1989) proteins.

The glycosidase enzymes discussed below are more usually used in sequence analysis, but can also be used in reversed synthesis where they have the advantage over glycosyltransferases in using cheap, reducing monosaccharide substrates. However unlike glycosyltransferases, which have strict specificity requirements for acceptor, linkage position and anomeric configuration, the disadvantage of glycosidases in synthesis is that they are not linkage specific; transfer to any hydroxyl group in the acceptor molecule is possible although that on C6 of an aldohexose is usually preferred. For sequential exoglycosidase digestion, however, their ability to distinguish between anomeric configurations and non-reducing monosaccharide type and to provide positional information is invaluable.

2.1 Lectin Affinity Interactions

A series of plant lectins with well defined carbohydrate specificities are now available commercially either as the free enzymes, or immobilized on Sepharose (Yamamoto *et al.*, 1993) or as conjugates for lectin immunohistochemistry and enzyme linked lectin assays (ELLA by comparison with ELISA). Immobilized lectins are particularly used in affinity chromatography to fractionate glycoproteins and glycopeptides by their differences in *N*-linked glycosylation. Released oligosaccharides can be further analyzed by their relative affinities with different lectins usually as their fluorescent or radioactively labeled derivatives for sensitive identification and in concert with exoglycosidase digestion as described next. In general these lectins are treated as if they are specific only for non-reducing monosaccharides of particular linkage position and anomeric configuration. However, as discussed below, their specificity often extends to distant features of the chain imposing additional binding requirements. On the other hand, evidence is accumulating that plant lectins may also be able to recognize peptide mimics.

Information on the specificity of the mammalian lectins is rapidly accumulating due to the interest in finding inhibitors as novel therapeutics. These will also be important reagents in biological studies, but the main work at present is in characterizing their interactions. Our understanding of plant lectins and mammalian lectin interactions with oligosaccharides are being enhanced by studies using NMR and X-ray crystallography of oligosaccharide-protein interactions and other biophysical methods (Siebert *et al.*, this volume; Bundle, this volume). A recent study has looked at the binding of a mammalian lectin (E-selectin) by fluorescence polarisation (Jacob *et al.*, 1995) which obviates the inherent bias in binding and inhibition studies against small molecular weight antagonists. Applications of surface plasmon resonance (BIAcore™) are beginning to be reported, e.g. for plant lectins (Shinohara *et al.*, 1994), which offers significant advances over conventional assays for analysing the affinity and kinetics of weak interactions (Jönsson *et al.*, 1991). The instrument uses the optical phenomenon of surface plasmon resonance to detect binding of macromolecules to ligands immobilised on a dextran matrix within a small flow cell. Interactions are followed in real time providing kinetic data which enables the binding of monomers to be readily distinguished from that of aggregates. A second technique, isothermal titration calorimetry, offers complementary information on binding affinities and both techniques are advised where possible to dispel fears about artifacts. Computer assisted tritration calorimetry has been used to provide direct estimates of the enthalpy of binding of plant lectins as well as the binding constants (Wiseman *et al.*, 1989) and overlapping ligand specificities (Chervenak and Toone, 1995). A related method to the last, that of differential scanning microcalorimetry, has made it easier to obtain accurate, comprehensive thermodynamic data which can elucidate the basis of enthalpy – entropy compensation and hence give information on intramolecular interactions, such as the effect of glycosylation on protein conformation.

From such studies, the fine specificities of lectin binding are now being elucidated. For example, ConA and the lectins from *Dioclea grandiflora* and *Galanthus nivalis* have similar binding properties (to Manα1–6[Manα1–3]Manβ), but the last shares no sequence homology and has a lower affinity for high mannose chains ($Man_9GlcNAc_2$); Con A will also bind high mannose, hybrid and biantennary complex chains. The lectins of pea and lentil (*Pisum sativum* and *Lens culinaris*) share similarities to ConA in carbohydrate binding to biantennary complex chains, but show a definite requirement for core fucose (Fucα1–6GlcNAc) and no affinity for high mannose or hybrid chains. Comparison of their protein structures have been carried out by X-ray crystallography as discussed in section 4.3. *Ricinus communis* lectin binds to hybrid chains, fucosylated biantennary chains and more highly branched oligosaccharides as long as there is a non-reducing terminal Gal residue. When GalNAc replaces Gal the *Wistaria floribunda* lectin binds. The *Dolichos biflorus* lectin also binds to GalNAc but has highest affinity for the blood group A GalNAc linked to Gal in the presence of Fuc. Fucose-specific lectins of plants *Aleuria aurantia*, *Ulex europaeus* and *Lotus tetragonolobus* are complemented in analysis by the lectin of the eel *Anguilla anguilla*. The presence of bisecting GlcNAcβ1–4 gives tight binding to *Griffonia simplicifolia* lectin II and wheat germ agglutinin (WGA). The latter also binds to all sialylated oligosaccharides which can be distinguished by specific binding of the lectin of *Maackia amurensis* seeds to NeuAcα2–6. The T-cell mitogens, poke weed mitogen (*Sambucus nigra* L) and PHA-L (*Phaseolus vulgaris* leukoagglutinin) bind respectively to branched polylactosamine sequences and 2,6 branched tri- or tetraantennary chains. The pokeweed mitogen Pa-4 has a similar amino acid sequence to WGA but does not bind sialic acid even though amino acids of WGA directly involved in primary carbohydrate bindings sites are homologous. Of the Galβ1–3GalNAc binding lectins, that from the peanut (PNA) has a strict specificity for the absence of sialic acid, whereas that from the common mushroom *Agaricus bisporus* can bind in the presence of sialic acid. This may explain the opposite affects of these lectins on human colonic epithelium in vitro where PNA is mitogenic (Campbell *et al.*, 1995), whereas that from mushroom inhibits proliferation (Yu *et al.*, 1993).

The mammalian lectins are at present categorized into four main families based on protein sequence homologies and/or carbohydrate specificity. One group, the selectins, show additional features to those discussed for the plant lectins, such as the combination of protein and oligosaccharide motifs in binding of P-selectin (Sako *et al.*, 1995) and the multivalent requirement of L-selectin for sulphated, sialylated and fucosylated epitopes (Hemmerich *et al.*, 1995). E-selectin has highest affinity for sialyl- or sulpho-Lewis x (Yuen *et al.*, 1992), but there is evidence that this may need to be presented in a specific way on branched tri- or tetraantennary *N*-linked chains. On the other hand, the mannose-binding protein (Gabius *et al.*, this volume) has a more promiscuous ligand requirement, its binding being inhibited by mannose, *N*-acetylglucos-amine and fucose. However, a specific interaction is mediated *in vivo* via the multiple high mannose chains presented on bacterial surfaces which constitutes the pathway of innate immunity. The galectins are a family of related Gal binding lectins that vary in their fine specifity, but for which no specific functions *in vivo* have been proven (Ohannesian and Lotan, this volume). Sialoadhesins (Kelm *et al.*, 1994) are a new family of sialic acid dependent mammalian adhesion molecules, one the first of these to be characterized (CD22) binds some very interesting coreceptors all of which express NeuAcα2–6 in a differentiation dependant manner.

2.2 Sequential Exoglycosidase Digestion

Exoglycosidases will cleave the non-reducing terminal monosaccharides from oligosaccharide chains. This is usually carried out on oligosaccharides released from glycoproteins by either hydrazinolysis or endoglycosidase digestion. Susceptibility to cleavage by exoglycosidases is then assessed by the analysis of the remaining oligosaccharide which can be labeled for sensitive detection throughout successive digestions. Alternatively the released monosaccharide can be analysed. One method employs HPAEC-PAD on a Dionex CarboPac MA1 column to identify without labeling both the released monosaccharide and the remaining oligosaccharide (Weitzhandler *et al.*,

1997). Other methods of oligosaccharide analysis include HPLC of UV absorbing or fluorescent derivatives and Biogel P4 gel filtration chromatography. The latter method has the advantage that release of monosaccharides always results in a decrease in hydrodynamic volume by an amount (given in glucose units) which is indicative of loss of a hexose, deoxyhexose or *N*-acetylhexosamine, whereas in other chromatographic methods retention is much more affected by linkage position, anomeric configuration etc. and therefore the oligosaccharide elution order is not linear with respect to molecular weight.

When assumptions can be made about the oligosaccharide chain under study, sequential enzyme digestions can be carried out to support the proposed structure. In certain circumstances information can also be obtained on mixtures of oligosaccharides. For glycoproteins where the oligosaccharide structure is unknown, an array of enzymes can be used which with computerised data analysis can give the suggested oligosaccharide profile. A reagent array analysis method (RAAM)™ has been commercialized for neutral or sialylated oligosaccharides (Oxford Glycosystems). Although initially set up for the analysis of radioactively labeled oligosaccharides, this technology has been transferred well to fluorescently labeled derivatives (2-AB in this case) which is the probable way of the future. These derivatives also perform well on porous graphitized carbon (PGC or GlycoSep H) columns for both neutral and acidic and mono- or oligosaccharide analysis (Davies and Hounsell, 1995) to give additional structural information and isomer separation.

3 Biophysical Methods

The new biophysical methods discussed above complement the more classical physicochemical methods of nuclear magnetic resonance spectroscopy (NMR) and X-ray crystallography for studies of oligosaccharide-to-protein and oligosaccharide-to-oligosaccharide intractions, which are discussed in separate chapters of this volume. These latter techniques and mass spectrometry are also invaluable in structural analysis as discussed below. Many low molecular weight oligosaccharides and derivatives and polysaccharides have been characterised by X-ray crystallography, but this has so far been unsuccessful for oligosaccharides unless they are complexed to protein. With present methodologies, difficulties are also encountered in the X-ray analysis of glycoproteins. Scanning tunneling and atomic force microscopy have not yet realized their potential for visualisation of glycoprotein structure and interactions (McMaster and Morris, 1993). NMR therefore remains, for the moment, the method of choice for conformational studies of oligosaccharides, glycopeptides and glycoproteins. Fluorescence energy transfer (Imperali and Rickert, 1995) can now be used to study conformational effects in aqueous media at neutral pH at a rapid time scale over which fluctuations occur that can not be detected in the time frame of NMR measurements (resulting in conformational averaging). CD spectroscopy also operates on a faster time scale than NMR, but this has found only limited use in the carbohydrate field, being mostly applicable to peptides with strong spectroscopic signals, such as α-helical motifs.

3.1 Mass Spectrometry (MS)

The original use of MS in the glycosciences was as a discriminatory detection devise on-line to GC for the large number of possible derivatives obtained by methylation analysis (see above) of oligosaccharides. Here, the only disadvantage of MS, that of the relative inability to distinguish within the groups of monosaccharides (e.g. the hexoses mannose, galactose and glucose), is ameliorated by their different retention on GC. In fact detailed analysis of the spectra does show differences between for example the 1,3,5-*O*-acetyl 2,4,6-methyl alditols of the hexoses (Table 2), but these are subtle variations in % abundance of the ions. The detection method of electron impact MS (EIMS) became the method of choice for GC-MS although chemical ionisation MS (CIMS) was also used, particularly of the pertrifluoroacetylated derivatives. These methods have a restriction on the size of molecule that can be ionised and hence detected (<m/z 1000) and therefore

fast atom bombardment (FAB) and liquid secondary ion (LSIMS) ionization methods were introduced in the '80s for oligosaccharide analysis (Dell, 1987; Hounsell, 1994a). The range of methods available and applications has rapidly expanded in the '90s. Matrix-assisted laser desorption ionization (MALDI) has further extended the molecular weight range to several hundred kDa, particularly with time of flight (TOF) analysis. Plasma desorption mass spectrometry (PDMS) enables the production and detection of molecular ions up to about 30kDa and electrospray (ES) MS, by producing a distribution of multiple charged molecular ions, can analyze high molecular weight molecules by exploiting their reduced m/z ratio.

As has been discussed by Whittal *et al.* (1995), it has been shown that underivatized oligosaccharides can be ionized by MALDI, with the limit of detection of 100 fmol (for reference standards in clean samples).

This has been achieved using a matrix such as a mixture of 2,5-dihydroxybenzoic acid (2,5-DHB) and α-cyano-4-hydroxycinnamic acid (4-HCCA) matrix solution at a concentration of 10 mg ml^{-1} in 30 % acetonitrile/water. The matrix solutions are centrifuged to remove undissolved matrix crystals and 0.5 μl aliquot of matrix solution and 0.5–1 μl of analyte solution placed on the sample probe and allowed to dry.

This sensitivity is somewhat higher than that obtained for peptides of similar mass, unlike in FAB and LSIMS where peptides are detected with much greater sensitivity than glycopeptides. As with FAB and LSIMS it has been found that *O*-acylation of neutral oligosaccharides significantly enhances the detection sensitivity and provides more fragmention for sequence determination. The lack of extensive fragment ions can also be ameliorated by combination with MS-MS analysis. These techniques have been used with considerable success for the analysis of released *N*-linked chains and *N*-linked glycoproteins.

For analysis of *O*-linked oligosaccharide chains, characterisation by ESMS of glycopeptides or peptides after oligosaccharide release is now the technique of choice (e.g. for *O*-GlcNAc linked proteins; Greis and Hart, 1997). Mucin type oligosaccharides have also been successfully characterized by LSIMS (Hounsell *et al.*, 1989; Rademaker *et al.*, 1993). Reinhold *et al.* (1995) have carried out the ESMS analysis of a uronic acid containing *O*-linked chain of *Flavobacterium meningosepticum*. Here the known resistance of uronic acid-containing polymers to acid hydrolysis and the instability of a novel amine containing uronic acid (2-*N*-acetylglucuronic acid) greatly complicated the stoichiometric analysis and this was complicated further by the presence of *O*-methyl groups. FAB-MS/ESI has been used to characterize advanced glycation end products (AGEs) of the Maillard reaction (Miyata *et al.*, 1994) implicated in many diseases, but particularly diabetes and (like *O*-linked GlcNAc) in ageing. A variety of proteins are known to undergo this non-enzymatic modification by forming covalent linkages with reducing sugars. The open chain sugar aldehyde groups react with the amino group of proteins to form a labile Schiff base which slowly isomerases to the more stable ketoamine adduct (fructosylamine) via the Amadori rearrangement. Further rearrangements and dehydration result in a series of modifications and protein cross-linking. Classically, these are very difficult to analyze and in the past have probably often missed detection. MS has a great future in picking up such minor (in terms of stoichiometry, molecular mass or occurrence, but not in importance) alterations in protein structure.

3.2 NMR Spectroscopy

If more than 50 µg of a purified oligosaccharide is available, NMR can be carried out to give details of the monosaccharides present and their positional and anomeric linkage. This is achieved by comparison with data in the literature (Hounsell and Wright, 1990) of the chemical shifts of signals arriving from the hydrogen atoms (protons, ^{1}H) in solution in D_2O where the hydroxyl ions have been replaced by DO by exchange. More material is required to carry out ^{1}H-NMR in H_2O (with saturation of the H_2O signal) or to detect natural abundance ^{13}C (approximately 0.1 % of ^{12}C which is not an NMR nucleus). Both ^{13}C and ^{15}N

can be enriched by *in vitro* cell culture in the presence of isotopically labeled intermediates. Analysis can be attempted of two or more milligrams of non enriched low molecular weight (normally less than 20 kDa) glycoproteins where it may be that the more flexible oligosaccharide has sharp signals visible above the broad signals of the more globular protein. For a similar reason it is often possible to carry out NMR studies of intermolecular oligosaccharide-to-protein interactions by concentrating on changes in chemical shifts in the oligosaccharide at different stoichiometric ratios.

The signals in the NMR spectrum can also be analyzed for their *J* splitting (scalar coupling constants) which gives information about the ring conformation of monosaccharides for identification (e.g. mannose from glucose and galactose) and also to provide conformational information about ring structure and glycosidic angles (Φ, ψ, ω). This coupling provides the correlation between signals around the monosaccharide ring which is exploited in 2-dimensional spectroscopy – correlated spectroscopy (COSY) and total correlation spectroscopy (TOCSY). Through space interactions (e.g. between adjacent monosaccharides or those which come into close proximity because of branching) are investigated by measuring the nuclear Overhauser effect (NOE) by 1-dimensional difference or 2-dimensional NOESY and rotating frame NOE spectroscopy (ROESY). Examples and explanations of these analyses are given in Hounsell (1995) and Hounsell and Bailey (1997). Meaningful data can normally be achieved on more than 0.5 mg purified oligosaccharide which can, for example, be incorporated into computer graphics simulation of oligosaccharide conformation and solution behaviour.

NMR data are now beginning to accumulate on intermolecular and intramolecular interactions of oligosaccharides and proteins, as also discussed by Siebert *et al.* in this volume. For the former both the ligands for selectins and antibodies have been studied with particular use being made of transfer NOE or transfer ROE and analysis of relaxation times. For example, Cooke *et al.* (1994) have studied the interaction of E-selectin with sialyl Lewis x (NeuNAcα2–3Galβ1–4[Fucα1–3]GlcNAc) which confirms previous data on the conserved stacking of Fuc on Gal compared to the relative flexibility of the sialic acid and provides new dissociation constant data (kD 1.4$\pm$0.5 mM). Poppe *et al.* (1994) have studied a sialylated determinant of the ganglioside GD1a by ^{13}C, T_1, $T_{1\varrho}$ and ^{1}H-^{13}C NOE in micelles to reveal an additional low energy conformer to that previously found for the sialic acid. X-ray analysis of bacterial poly α2–8sialic acid has been complemented by new NMR studies (Yamasaki and Bacon, 1991; Evans *et al.* 1995) which extend previous conformational studies. The solution conformation of group B polysaccharides of *Neisseria meningitides* were analysed by double quantum filtered (DQF) -COSY and pure absorption 2D NOE NMR at different mixing times. The pyranose ring of the sialic acid in the $^2C_5$ conformation and gauche orientations of H-6 and H-7 and of H-7 and H-8 were obtained from the coupling constants for sialic acid in the polymer ($J_{3a,3e}$ 16, $J_{3a,4}$ 9, $J_{3e,4}$ 5, $J_{4,5}$ 9, $J_{6,7}<3$, $J_{7,8}<3$, $J_{8,9}$ 5, $J_{9,9}$ 11 approximately).

Transfer NOE experiments were used by Bundle *et al.* (1994) to compare the solution and X-ray crystallographic studies of the interactions of antibodies with a *Salmonella* lipopolysaccharide represented by the sequence Galα1–2[Abeα1–3]Manα1. New data are being accumulated on isotopically enriched bacterial lipopolysaccharides studied by heteronuclear NMR. Natural abundance ^{1}H-^{13}C experiments are also providing detailed chemical shift assignments of glycoproteins (Wyss *et al.*, 1995), particularly with the application of pulsed field gradients (de Beer *et al.*, 1994). For the future, major advances lie in the use of pulse field methods (Ruiz-Cabello *et al.*, 1992; Medvedeva *et al.*, 1993; Norwood, 1994; Kay, 1995). For example, gradient enhanced spectroscopy provides the chance to record spectra that are almost free of artifacts and with almost perfect elimination of the water signal and spectra can be recorded up to about pH 8 without losing signals from exchangeable amide proteins. Because of the advantages of this technology, the entire repertoire of pulse sequences used in modern NMR has been and is being rewritten to include pulsed field gradients (and B1 gradient technology).

4 Glycoprotein Analysis

In order to understand the affect of carbohydrates on glycoprotein conformation and function it is important to have a "glycoform analysis", i.e. to show what types of oligosaccharides are present at each consensus *N*-linked site (this is unlikely to be determined by NMR analysis) and the distribution of different *O*-linked and GAG chains. To carry out this analysis, the glycoprotein under investigation is digested with proteases to give peptides and glycopeptides and then treated with endoglycosidases (Table 3) or chemical degradation to release the oligosaccharides. At each step analysis is carried out by MS (and NMR if enough material is available) with purification by high resolution chromatography.

4.1 Proteases and Endoglycosidases

As long as the protein sequence and specificity of the protease is known, the former can be identified by MS of the total digest (around 20 pmole). For ESMS, trypsin is the protease of choice as it cleaves to the carboxyterminus of arginine and lysine amino acids, therefore giving an amino group at both ends of the (glyco)peptide which aids detection. Glycopeptides are then isolated by reversed phase (C18) or porous graphitized carbon (PGC) HPLC and detected by a hexose assay (Section 1). A few steps of amino acid N terminal sequence analysis will identify the glycosylated peptides and the oligosaccharide chains can then be released by PNGase-F and analyzed at picomolar level by either HPAEC-PAD in native form or derivatized with fluorescent labels (see Section 1). *O*-linked chains are mainly still released by alkali catalyzed β-elimination and analyzed by HPLC/HPAEC as their alditols (Davies *et al.*, 1993; Campbell *et al.*, 1994). Other endo-glycosidases besides PNGase F can also be incorporated into this scheme to isolate high mannose chains (Endo H), neutral or sulphated poly *N*-acetyllactosamine repeats (endo-β-galactosidases), glycosaminoglycans (heparinase, heparatinase, chrondroitinases A, B and C) and endo-α-*N*-acetylgalactosaminidase (for *O*-linked chains).

Table 3. Enzymatic Release of Oligosaccharides[a]

Enzyme	Specifity (the arrow signifies the cleavage point)
Endoglycosidase H (Endo-H)	$(Man)_n$GlcNAcβ1–4GlcNAcβ1-Asn (protein) ↑
Peptide N-glycanase F (PNG-ase F)	±NeuAc ±Fuc ±Gal $(GlcNAc)_n$ $(Man)_3$ GlcNAcβ1–4 [±Fucα1–6]GlcNAcβ1-Asn (protein) ↑
Endo-α-N-acetylgalactosaminidase	[a]Galβ1–3GalNAcα1-Ser/Thr ↑
Endo-β-galactosidase	$[\text{-3Gal}\beta1\text{–4GlcNAc}\beta1\text{-}]_n$ ↑ $\pm 6SO_4$
Heparatinase/Heparinase	$[\text{-4GlcNSO}_4\alpha1\text{–4IdoA}\alpha1\text{–4GlcNAc}\alpha1\text{–4GlcA}\beta1\text{-}]_n$ ±3,6 SO_4 ↑ ±2SO_4 ↑ Heparinase Heparatinase I Averaged structure of heparin and major cleavage sites. Heparatinase II cleaves $GlcNSO_4(\pm 6SO_4)$ – GlcA.
Chondroitinase	$[\text{-3GalNAc}\beta1\text{–4GlcA}\beta1\text{-}]_n$ ±4SO_4 ↑ ±6SO_4

[a] This is just one type of GalNAc-Ser/Thr chain.
For further details on O-linked glycosylation structures see Hounsell *et al.* (1996); Sharon and Lis, this volume; Brockhausen and Schachter, this volume.

4.2 HPLC Purification and Analysis

Oligosaccharides offer unique challenges in their chromatographic resolution due to linkage isomerisation and various anionic forms (sialylated, sulfated, phosphorylated). As a generalization neutral oligosaccharides have been analysed or purified by reversed-phase C18 chromatography or normal–phase chromatography and sialylated oligosaccharides by normal-phase or anion exchange chromatography. High pH anion exchange chromatography (HPAEC) is now a method of choice. Gradients have now been perfected in HPAEC for analysis of both neutral and sialylated oligosaccharides in one run and new columns are available for sulfated oligosaccharide analysis. This gives a high sensitivity analysis of different glycoforms and data which can be corroborated by RAAM™ sequencing or HPLC with fluorescence detection. Normal and reversed phase HPLC still have their place for oligosaccharide and glycopeptide separation, but porous graphitized carbon (PGC) columns are an alternative for separation of neutral, sialylated or sulfated oligosaccharides and glycopeptides (Davies and Hounsell, 1996).

Reversed- and normal-phase HPLC columns are classically run in acetonitrile – water gradients the former with a concentration of acetonitrile in water starting out 0 % and increasing, and the latter at 65 % and decreasing. Detection can be achieved by UV-absorbance for neutral oligosaccharides containing acetamido groups at 190–210 nm. Anionic oligosaccharides can be chromatographed by the addition of ion pair reagents to reversed phase and in buffers for normal phase (the latter giving good isomer separation) or more recently on PGC columns run in acetonitrile-aqueous TFA. This last column has the advantage of being able to chromatograph both neutral and acidic oligosaccharides in one run, with isomer separation and in the absence of high salt. Although ion suppression membranes are available for on-line removal of cations, it is difficult to remove sufficient amounts for subsequent high sensitivity MS analysis and may interfere with biological assays. Therefore direct LC-MS analysis is usually carried out with chromatography in acetonitrile-water or TFA gradients. On the other hand, controlled on-line post column addition of metal chlorides has been used for LC/MS/MS analysis of carbohydrate-metal complexes (Kohler and Leary, 1995).

Direct HPLC – NMR is also beginning to be applied in the carbohydrate area. For example, to follow reactions in vitro such as acyl migration, mutarotation etc. Sidelmann *et al.* (1995) have reported that low rates of 1 ml/min is optimal with respect to filling the NMR flow cell (a ^{1}H flow probe of 3 mm i.d. with a volume of 100 μl). Acetonitrile as organic modifier in reverse phase at pH 7.4 was suitable for both HPLC and NMR. Spectra are obtained in stop flow-mode with double presaturation for suppression of acetonitrile and water signals (Wilson *et al.*, 1993).

4.3 Capillary Electrophoresis

The intrinsic high resolving power of electrophoresis makes it particularly suitable for the separation of oligosaccharides. These can be converted *in situ* to charged species by complexation with borate or metal ions or by precolumn derivatization (e.g. ANTS or anthranilic acid discussed above for PAGE analysis). Honda *et al.* (1989) were the first to demonstrate the resolving power of capillary zone electrophoresis (CZE) to 2-AP derivatives separated as anionic borate complexes. Sugar borate complexation was also extended to oligosaccharide 1-phenyl-3-methyl-5-pyrazolone (PMP) derivatives (Honda *et al.*, 1991) which become negatively charged in aqueous basic solutions due to the dissociation of the enolic hydroxy groups. In addition to high resolution high sensitivity analysis of released oligosaccharides, CZE has advantages over HPLC in being able to resolve glycoforms of intact glycoproteins. This was first reported for the separation of transferrins via their di- to hexa-sialylated *N*-linked chains (Kilàr and Hjertén, 1989). CZE of borate complexes was then shown to also be applicable for analysis of neutrally glycosylated proteins, i.e. the various high mannose *N*-linked glycoforms of ribonuclease B (Rudd *et al.*, 1992). The technique has now been used on a range of glycoproteins having both *N*- and *O*-linked glycosylation and has also been applied to glycopeptide mapping (the separation of glycopeptides and peptides formed on protease digestion of a glycoprotein in the analysis of the site specific glycosylation patterns (see above). For the latter analysis Rush *et al.* (1993) have extended CZE of borate complexes (the equivalent of HPAEC in NaOH) to ion-pairing reagents (as discussed above for reversed phase HPLC).

Electrophoresis also has obvious advantages for the analysis of glycosaminoglycans (GAGs) which are highly charged species, both by virtue of the presence of alternate uronic acid residues (except in keratan sulphate) and of high sulphation (except in hyaluronan) of the uronic acids and the alternate hexosamines. The new requirements in this field are a) for the analysis of GAGs which more and more are being found attached to cell membrane glycoproteins and are thus part of the glycoform characterisation discussed here, and b) because specific oligosaccharide motifs of these molecules are being implicated in specific binding or inhibition of proteins important in cell growth and regulation, in addition to their roles in the coagulation pathway which opened this chapter.

References

Angel AS, Nilsson B (1988): Linkage positions in glycoconjugates by periodate oxidation and fast atom bombardment mass spectrometry. In *Methods Enzymol.* **193**:587–607.

Bigge JC, Patel TP, Bruce JA *et al.* (1995): Nonselective and efficient fluorescent labelling of glycans using 2-aminobenzamide and anthranalic acid. In *Anal. Biochem.* **230**:229–38.

Boedeker EC (1995): Gastrointestinal infection. In *Curr. Opinion Gastroenterol.* **11**:49–51.

Bolen M, Stalmans W (1989): The antiglycogenolytic action of 1-deoxynojirimycin results from a specific inhibition of the α-1,6-glucosidase activity of the debranching enzyme. In *Eur. J. Biochem.* **181**: 775–80.

Bundle DR, Baumann H, Brisson J-R *et al.* (1994): Solution structure of a trisaccharide-antibody complex: comparison of NMR measurements with a crystal structure. In *Biochemistry* **33**:5183–92.

Campbell BJ, Davies MJ, Rhodes JM *et al.* (1993): Separation of neutral oligosaccharide alditols from human meconium using high pH anion exchange chromatography. In *J. Chromat.* **622**:137–46.

Campbell JB, Finnie IA, Hounsell EF *et al.* (1995): Direct demonstration of increased expression of Thomsen-Friedenreich (TF) antigen in colonic adenocarcinoma and ulcerative colitis mucin and its concealment in normal mucin. In *J. Clin. Invest.* **95**: 571–6.

Chervenak MA, Toone EJ (1995): Calorimetric analysis of the binding of lectins with overlapping carbohydrate-binding ligand specificities. In *Biochemistry* **34**:5685–95.

Cooke RM, Hale RS, Lister SG *et al.* (1994): Conformation of the sialyl Lewis x ligand changes upon binding to E-selectin. In *Biochemistry* **33**:10591–6.

Davies MJ, Smith KD, Carruthers RA *et al.* (1993): The use of a porous graphitised carbon (PGC) column for the HPLC of oligosaccharides, alditols and glycopeptides with subsequent mass spectrometry analysis. In *J. Chromat.* **646**:317–26.

Davies MJ, Smith KD, Hounsell EF (1994): The release of oligosaccharides from glycoproteins. In Basic Protein and Peptide Protocols *Meth. Mol. Biol.* Vol. 32: (Walker E, ed) pp 129–41, Totowa: Humana Press.

Davies MJ, Hounsell EF (1996): Comparison of separation modes of high performance liquid chromatography for the analysis of glycoprotein- and proteoglycan-derived oligosaccharides. In *J. Chromat.* **720**:227–34.

de Beer T van, Zuylen CWEM, Hård K *et al.* (1994): Rapid and simple approach for the NMR resonance assignment of the carbohydrate chains of an intact glycoprotein. Application of gradient-enhanced natural abundance ^{1}H-^{13}C HSQC and HSQC-TOCSY to the α-subunit of human chorionic gonadotropin. In *FEBS Lett.* **348**:1–6.

Dell A (1987): Mass spectrometry of oligosaccharides. In *Adv. Carbohydr. Chem. Biochem.* **45**:19–72.

Dimitriev BA, Knirel YA, Kochetkov NK (1975): Selective cleavage of glycosidic linkages: studies with the O-specific polysaccharide from *Shigela dysenteriae* type 3. In *Carbohydr. Res.* **40**:365–72.

Dong DLY, Xu ZS, Chevrier MR *et al.* (1993): Glycosylation of mammalian neurofilaments. Localisation of multiple *O*-linked *N*-acetylglucosamine moieties on neurofilament polypeptides L and M. In *J. Biol. Chem.* **268**:16679–87.

Evans SV, Ssigurskjold BW, Jennings HJ *et al.* (1995): Evidence for the extended helical nature of polysaccharide epitopes. The 2.8 Å resolution structure and thermodynamics of ligand binding of an antigen binding fragment specific for α-(2→8)-polysialic acid. In *Biochemistry* **34**:6737–44.

Field MC, Wainwright LJ (1995): Molecular cloning of eukaryotic glycoprotein and glycolipid glycosyltransferases: a survey. In *Glycobiology* **5**:463–72.

Fischl MA, Resnick L, Coombs R *et al.* (1994): The safety and efficacy of combination *N*-butyl-deoxynojirimycin (SC48334) and zidovudine in patients with HIV-1 infection and 200–500 CD4 cells/μl. In *J. Acquir. Defic. Synd.* **7**:139–47.

Greis KD, Hart GW (1997): Analytical methods for the study of *O*-GlcNAc glycoproteins and glycopeptides. In *Meth. Mol. Biol.: Glycoscience Protocols* (Hounsell E, ed) Totowa: Humana Press, in press.

Grootenhuis PDJ, van Boeckel CAA (1991): Constructing a molecular model of the interaction between antithrombin III and a potent heparin analogue. In *J. Am. Chem Soc.* **113**:2743–7.

Haltiwanger RS, Hart GW (1993): Glycosyltransferases as tools in cell biological studies. In *Meth. Mol. Biol.* Vol. **14**: *Glycoprotein Analysis in Biomedicine* (Hounsell EF, ed), pp 175–87, Totowa: Humana Press.

Hart GW, Haltiwanger RS, Holt GD *et al.* (1989): Glycosylation in the nucleus and cytoplasm. In *Annu. Rev. Biochem.* **58**:841–74.

Hase S, Natsuka S, Oku H *et al.* (1987): Identification method for twelve oligomannose-type sugar chains thought to be processing intermediates of glycoproteins. In *Anal. Biochem.* **167**:321–6.

Haselbeck A, Hösel W (1993): Immunological detection of glycoproteins on blots based on labeling with digoxigenin. In *Meth. Mol. Biol.* Vol **14**: *Glycoprotein Analysis in Biomedicine* (Hounsell EF, ed.) pp. 161–73 Totowa: Humana Press.

Hemmerich S, Leffler H, Rosen SD (1995): Structure of the *O*-glycans in GlyCAM-1, an endothelial derived ligand for L-selectin. In *J. Biol. Chem.* **270**:12035–47.

Honda S, Iwase, S, Makino A *et al.* (1989): Simultaneous determination of reducing monosaccharides by capillary zone electrophoresis as the borate complexes of *N*-2-pyridylglycamines. In *Anal. Biochem.* **176**:72–7.

Honda S, Suzuki S, Nose A *et al.* (1991): Capillary zone electrophoresis of reducing mono- and oligosaccharides as the borate complexes of their 3-methyl-1-phenyl-2-pyrazolin-5-one derivatives. In *Carbohydr. Res.* **215**:193–8.

Hounsell EF, Lawson AM, Stoll MS *et al.* (1989): Characterisation by mass spectrometry and 500-MHz proton nuclear magnetic resonance spectroscopy of penta- and hexasaccharide chains of human foetal gastrointestinal mucins (meconium glycoproteins). In *Eur. J. Biochem.* **186**:597–610.

Hounsell EF, Wright DJ (1990): Computer assisted interpretation of ^{1}H-NMR spectra in the structural analysis of oligosaccharides. In *Carbohydr. Res.* **205**:19–29.

Hounsell EF (1994a): Physicochemical analyses of oligosaccharide determinants of glycoproteins. In *Adv. Carbohydr. Chem. Biochem.* **50**:311–50.

Hounsell EF (1994b): Characterisation of the glycosylation status of proteins. In *Mol. Biotech.* **2**:45–60.

Hounsell EF (1995): ^{1}H NMR in the structural and conformational analysis of oligosaccharides and glycoconjugates. In *Progr. Nucl. Res. Spectrosc.* **27**:445–64.

Hounsell EF, Davies MJ, Renouf DV (1996): O-linked protein glycosylation structure and function. In *Glycoconjugate J.* **13**:1–8.

Hounsell EF (ed) (1997): *Glycoscience Protocols*. Totowa: Humana Press, in press.

Hounsell EF, Bailey D (1997): Approaches to the structure determination of oligosaccharides and glycopeptides using NMR. In *Glycopeptides and Related Compounds: Synthesis, Analysis and Applications* (Large DG, Warren CD, eds) New York: M. Dekker Inc., in press.

Hu G (1995): Fluorophore assisted carbohydrate electrophoresis technology and applications. In *J. Chromat.* **705**:89–103.

Hurd RE, John BK (1991): Gradient-enhanced proton-dected heteronuclear multiple-quantum coherence spectroscopy. In *J. Magnetic Res.* **91**:648–53.

Imperiali B, Rickert KW (1995): Conformational implications of asparagine-linked glycosylation. In *Proc. Natl. Acad. Sci. USA* **92**:97–101.

Iwase H, Hotta K (1993): Release of *O*-linked glycoprotein glycans by Endo-α-*N*-Acetylgalactosaminidase. In: *Meth. Mol. Biol.* Vol. **14**: *Glycoprotein Analysis in Biomedicine* (Hounsell EF, ed) pp 151–60, Totowa: Humana Press.

Jacob GS, Kirmaier C, Abbas SZ *et al.* (1995): Binding of sialyl Lewis x to E selectin as measured by flouresence polarization. In *Biochemistry* **34**:1210–7.

Jönsson U, Fagerstam L, Ivarsson B *et al.* (1991): Real-time biospecific interaction analysis using surface plasman resonance and a sensor chip technology. In *Biotechniques* **11**: 620–7.

Kay LE (1995): Field gradient techniques in NMR spectroscopy. In *Curr. Opinion Struct. Biol.* **5**:674–81.

Kelm S, Pelz A, Schauer R *et al.* (1994): Sialoadhesin, myelin-associated glycoprotein and CD22 define a new family of sialic acid-dependent adhesion molecules of the immunoglobulin superfamily. In *Curr. Biol.* **4**:965–72.

Kilàr F, Hjertén S (1989): Separation of the human transferrin isoforms by carrier-free high-performance zone electrophoresis and isoelectric focusing. In *J. Chromatogr.* **480**:351–7.

King IA, Hounsell EF (1989): Cytokeratin 13 contains single *O*-glycosidically linked *N*-acetylglucosamine residues. In *J. Biol. Chem.* **264**:14022–8.

Kohler M, Leary JA (1995): LC/MS/MS of Carbohydrates with postolumn addition of metal chlorides using a triaxial electrospray probe. In *Anal. Chem.* **67**:1500–8.

Koketsu M, Nitoda T, Juneja LR *et al.* (1995): Sialyloligosaccharides from egg yolk as an inhibitor of rotaviral infection. In *J. Agr. Food Chem.* **43**:858–61.

Lee YC, Lee RT (eds) (1994): *Neoglycoconjugates. Preparation and Applications*. Orlando: Academic Press.

McMaster TJ, Morris VJ (1993): Scanning tunneling microscopy of biopolymers. In: *Meth. Mol. Biol.* Vol. **14**: *Glycoprotein Analysis in Biomedicine* (Hounsell EF ed) pp 277–296, Totowa: Humana Press.

Medvedeva S, Simorre JP, Brutscher B *et al.* (1993): Extensive ^{1}H NMR resonance assignment of proteins using natural abundance gradient-enhanced ^{13}C-^{1}H correlation spectroscopy. In *FEBS Lett.* **333**:251–6.

Miyata T, Inagi R, Wada Y *et al.* (1994): Glycation of human β_2-microglobulin in patients with hemodialysis-associated amyloidosis: Identification of the glycated sites. In *Biochemistry* **33**:12215–21.

Mizuochi T (1993): Microscale sequencing of *N*-linked oligosaccharides of glycoproteins using hydrazinolysis, Bio-Gel P-4, and sequential exoglycosidase digestion. In: *Meth. Mol. Biol.* Vol **14**: *Glycoprotein Analysis in Biomedicine* (Hounsell EF ed) pp 55–68, Totowa: Humana Press.

Mouricout M, Petit JM, Carias JR *et al.* (1990): Glycoprotein glycans that inhibit adhesion of *Escherichia coli* mediated by K99 fimbriae: Treatment of experimental colibacillosis. In *Infect. Immun.* **58**:98–106.

Norwood TH (1994): Magnetic field gradients in NMR: friend or foe? In *Chem. Soc. Rev.* **23**:59–66.

Poppe L, van Halbeek H, Acquotti D *et al.* (1994): Carbohydrate dynamics at a micellar surface GD1a headgroup transformations revealed by NMR spectroscopy. In *Biophys. J.* **66**:1642–52.

Rademaker GJ, Haverkamp J, Thomas-Oates J (1993): Determination of glycosylation sites in *O*-linked glycopeptides: A sensitive mass spectrometric protocol. In *Organic Mass Spectrom.* **28**:1536–41.

Reinhold BB, Hauer CR, Plummer TH *et al.* (1995): Detailed structural analysis of a novel, specific *O*-linked glycan from the prokaryote *Flavobacterium meningosepticum*. In *J. Biol. Chem.* **270**:13197–203.

Rudd PM, Scragg IG, Coghill E *et al.* (1992): Separation and analysis of the glycoform populations of ribonuclease B using capillary electrophoresis. In *Glycoconjugate J.* **9**:86–91.

Rush RS, Derby PL, Strickland TW *et al.* (1993): Peptide mapping and evaluation of glycopeptide microheterogeneity derived from endoproteinase digestion of erythropoietin by affinity high-performance capillary electrophoresis. In *Anal. Chem.* **63**: 1834–42.

Ruiz-Cabello J, Vuister GW, Moonen CTW *et al.* (1992): Gradient-enhanced heteronuclear correlation spectroscopy. Theory and experimental aspects. In *J. Magnetic Res.* **100**:282–302.

Sako D, Comess KM, Barone KM *et al.* (1995): A sulfated peptide segment at the amino terminus of PSGL-1 is critical for P-selectin binding. In *Cell* **83**:323–31.

Shinohara Y, Kim F, Shimizu M *et al.* (1994): Kinetic measurement of the interaction between an oligosaccharide and lectins by a biosensor based on surface plasmon resonance. In *Eur. J. Biochem.* **223**:189–94.

Sidelmann UG, Gavaghan C, Carless HAJ *et al.* (1995): Identification of the positional isomers of 2-fluorobenzoic acid 1-*O*-acyl glucuronide by directly coupled HPLC-NMR. In *Anal. Chem.* **67**:3401–4.

Smith KD, Davies MJ, Hounsell EF (1994): Structural profiling of oligosaccharides of glycoproteins. In: *Meth. Mol. Biol.* Vol. **32**: *Basic Proteins and Peptide Protocols* (Walker J, ed) pp 143–55, Totowa: Humana Press.

Stoll MS, Hounsell EF (1988): Selective purification of reduced oligosaccharides using a phenylboronic acid bond elute column: potential application in HPLC, mass spectrometry, reductive amination procedures and antigenic/serum analysis. In *Biomed. Chromatogr.* **2**:249–53.

Teneberg S, Willemsen PTJ, de Graaf FK *et al.* (1994): Characterization of gangliosides of epithelial cells of calf small intestine, with special reference to receptor-active sequences for enteropathogenic *Escherichia coli* K99. In *J. Biochem.* **116**:560–75.

Tennant GA, Lovat LB, Pepys MB (1995): Serum amyloid P component prevents proteolysis of the amylod fibrils of Altzheimer disease and systemic amyloidosis. In *Proc. Nat. Acad. Sci. USA* **92**:4299–303.

Torres CR, Hart GW (1984): Topography and polypeptide distributionof terminal *N*-acetylglucosamine residues on the surface of intact lymphocytes. In *J. Biol. Chem.* **259**:3308–17.

von Itzstein M, Wu W-Y, Kok GB *et al.* (1993): Rational design of potent sialidase-based inhibitors of influenza virus replication. In *Nature* **363**:418–23.

Weitzhandler M, Rohrer J, Thayer JR *et al.* (1997): HPAE-PAD analysis of monosaccharides released by exoglycosidase digestion using the CarboPac MA1 column. In: *Meth. Mol. Biol.* Vol. **30**: *Glycoscience Protocols* (Hounsell EF ed) pp 1–10, Totowa: Humana Press.

Whittal RM, Palcic MM, Hindsgaul O *et al.* (1995): Direct analysis of enzymatic reactions of oligosaccharides in human serum using matrix-assisted laser desorption ionization mass spectrometry. In *Anal. Chem.* **67**:3509–14.

Wilson I, Nicholson JK, Hofmann M *et al.* (1993): Investigation of the human metabolism of antipyrine using coupled liquid chromatography and nuclear resonance spectroscopy of urine. In *J. Chromat.* **617**:323–8.

Wiseman T, Williston S, Brandts JF *et al.* (1989): Rapid measurement of binding constants and heats of binding using a new titration calorimeter. In *Anal. Biochem.* **179**:131–7.

Wyss DF, Choi JS, Wagner G (1995): Composition and sequence specific resonance assignments of the heterogeneous *N*-linked glycan in the 13.6kDa adhesion domain of human CD2 as determined by NMR on the intact glycoprotein. In *Biochemistry* **34**:1622–34.

Yamamoto K, Tsuji T, Osawa T (1993): Analysis of asparagine-linked oligosaccharides by sequential lectin-affinity chromatography. In: *Meth. Mol. Biol. Vol.* **14**: *Glycoprotein Analysis in Biomedicine* (Hounsell EF ed) pp 17–34, Totowa: Humana Press.

Yamasaki R, Bacon B (1991): Three-dimensional structural analysis of the group B polysaccharide of *Neiseria meningitidis* 6275 by two-dimensional NMR: the polysaccharide is suggested to exist in helical conformations in solution. In *Biochemistry* **30**:851–7.

Yamashita K, Mizuochi T, Kobata A (1982): Analysis of oligosaccharides by gel filtration. In *Methods Enzymol.* **83**:105–26.

Yamashita K, Okura T, Tachibana Y *et al.* (1984): Comparative study of the oligosaccharides released from baby hamster kidney cells and their polyomer transformants by hydrazinolysis. In *J. Biol. Chem.* **259**:10834–40.

Yosizawa Z, Sato T, Schmid K (1966): Hydrazinolysis of α_1-acid glycoprotein. In *Biochem. Biophys. Acta* **121**:417–20.

Yu L, Fernig DG, Smith JA *et al.* (1993): Reversible inhibition of epithelial cell lines by *Agaricus bisporus* (edible mushroom) lectin. In *Cancer Res.* **53**:4627–32.

Yuen CT, Lawson AM, Chai W *et al.* (1992) Novel sulfated ligands for the cell adhesion molecule E-selectin revealed by the neoglycolipid technology among O-linked oligosaccharides on an ovarian cystadenoma glycoprotein. In *Biochemistry* **31**:9126–31.

3 Strategies for the Chemical Synthesis of Glycoconjugates

RICHARD R. SCHMIDT

1 Introduction

Glycoside synthesis is a very common reaction in nature, providing a great variety of oligosaccharides and glycoconjugates, as for instance, glycolipids (especially glycosphingolipids), glycopeptides and glycoproteins, glycopeptidolipids, glycosylphosphates, glycophospholipids, and glycophosphopeptidolipids. Their manifold functions, more and more recognized only recently (Schmidt, 1986; Hakomori, 1990; Ledeen, 1992; Paulsen, 1990; Jung and Schmidt, 1991; Kunz, 1993; Sharon and Lis, this volume; Kopitz, this volume), are based on a great structural diversity of the oligosaccharide portion, which is inherent in the variability in glycoside bond formation (Schmidt, 1986), thus rendering oligosaccharides ideal as carriers of biological information and specificity.

This knowledge is mainly based on an enormous extension and improvement of the methodologies for structural analysis (Vliegenthart *et al.*, 1983; Hounsell, this volume) and for chemical synthesis (Schmidt, 1986, 1989, 1994; Paulsen, 1990; Jung and Schmidt, 1991; Banoub, 1992; Kunz, 1993; Toshima and Tatsata, 1993; Barresi and Hindsgaul, 1995; Khan *et al.*, 1996; Whitfield and Douglas, 1996), as will be outlined below. The synthesis of glycosides, oligosaccharides, and glycoconjugates is characterized by a much larger number of possibilities for coupling than that of other biopolymers, such as peptides or proteins and ribo- or deoxyribo-nucleotides. In contrast to peptides and nucleotides, in which the informational content is determined solely by the number and sequence of different monomer units, the informational content of oligosaccharides is determined additionally by the site of coupling (regioselectivity), the configuration of the glycosidic linkage (anomeric selectivity/diastereoselectivity, α or β), and the occurrence of branching (dimensional selectivity). Thus, oligomers made up of carbohydrates can carry considerably more information per atomic unit than any other biopolymers, such as proteins and nucleotides, as has been clearly demonstrated (Schmidt, 1986; Laine, this volume). A simple example may illustrate this phenomenon. Three alanine or adenosine monophosphate residues yield only one trimer; however, when three glucopyranose residues are combined by two glycosidic linkages (not counting possible trehalose formation and glucofuranose combinations) anomeric selectivity (two possibilities), connection to 2-,3-,4-, and 6-position (four possibilities), and branching [(1–2)/(1–3)-,(1–2)/(1–4)-,(1–2)/(1–6)-,(1–3)/(1–4)-,(1–3)/(1–6)-,(1–4)/(1–6)-connections: six possibilities] totals 176 different trisaccharides. Obviously, the combination of a few different sugar residues yields an astronomical number of combinatorial structural variants (see also Laine, this volume).

2 General Strategies

2.1 Typical Building Blocks for Glycoside Synthesis

Nature has provided, through biosynthetic means, all the important sugar residues which are generally employed as basic building blocks for oligosaccharide synthesis. However, this availability in terms of quantity and cost varies dramatically, as exhibited for some sugars in Table 1.

H.-J. and S. Gabius (Eds.), Glycosciences
© Chapman & Hall, Weinheim, 1997
ISBN 3-8261-0073-5

Table 1. Approximate catalogue prizes for common sugars and for the derived natural glycosyl donors

Sugar	US$/g[a]	Glycosyl donor	US$/g[a]
D-Glucose (Glc)	0.010	UDP-Glc	150.000
D-Galactose (Gal)	0.100	UDP-Gal	5.000
D-Mannose (Man)	1.000	GDP-Man	4.000
D-Glucosamine (GlcN)	0.200	UDP-GlcNAc	600.000
D-Galactosamine (GalN)	35.000	UPD-GalNAc	15.000
L-Fucose (Fuc)	10.000	GDP-Fuc	150.000
N-Acetyl-neuraminic Acid (Neu5Ac)	400.000	CMP-Neu5Ac	25.000
Lactose	0.020	–	–
N-Acetyllactosamine	6.000	–	–

[a] Based on the price for the biggest catalogue quantity

Yet, due to the high functionality and the presence of several stereocenters, chemical synthesis of common sugars found in glycoconjugates is generally not rewarding. Therefore, basically all chemical syntheses of glycoconjugates start from the natural precursors listed in Table 1.

A great advantage in the ligation of sugar residues would be the use of preexisting di- and trisaccharide building blocks, which can be obtained from natural sources. However, apart from the readily available lactose, even disaccharides are very rare and practically unobtainable in reasonable amounts. Not even the lactosamine residue, which frequently occurs as a constituent of glycoconjugates, is available at an affordable price (Table 1). Thus, chemical (in vitro) glycoconjugate synthesis generally necessitates regio- and diastereo-controlled formation of all glycosidic linkages – similar to the natural approach. Therefore, the question remains: are there any basic differences between chemical and enzymatic methods for glycoside bond formation, or are the underlying chemical principles essentially the same?

2.2 Methodologies in Enzymatic and in Chemical Glycoside Bond Formation – a Comparison

Enzymatic glycosylation is generally based on specific glycosyltransferases which employ nucleoside diphosphate or, in some cases, nucleoside monophosphate sugars as glycosyl donors; the nucleoside di- or monophosphate residue is the leaving group and sugars, saccharides or any other aglycones are the glycosyl acceptors (Wong, 1996) (Scheme 1). Thus, the energy for this condensation reaction is contained in the activated glycosyl donor, which is obtained from sugars and nucleoside triphosphates by independent (and often quite lengthy) enzymatic processes. Cleavage of the nucleoside di- or monophosphate residue from the activated sugars is the driving force for irreversible glycoside bond formation; in this process the glycosyltransferase (a direct gene product, which requires a complex machinery and a large amount of energy for its construction) provides the desired regio- (positional) and diastereoselectivity (anomeric stereoselectivity). Obviously, this methodology starts from the reducing end of a sugar or from a lipid, a peptide, or a phosphate residue as acceptor, and eventually leads via subsequent glycosylation steps to highly glycosylated glycoconjugates. This almost universally adopted strategy is based on quite labile and very expensive glycosyl donors (Table 1), if they are available at all. Therefore, enzymatic glycosylation is often attempted via enzymatic *in situ* preparation of the glycosyl donor, which generally requires a chemically synthesized, energy-rich phosphate donor (acyl phosphate, phosphoenol pyruvate, etc.) as shown in Scheme 1 for *N*-acetyl lactosamine synthesis.

Glycosidases have also been extensively investigated for their use in glycoside bond formation (Thiem and Sauerbrei, 1991; Nilsson, 1996). However, the reversibility, lower regio- and dia-

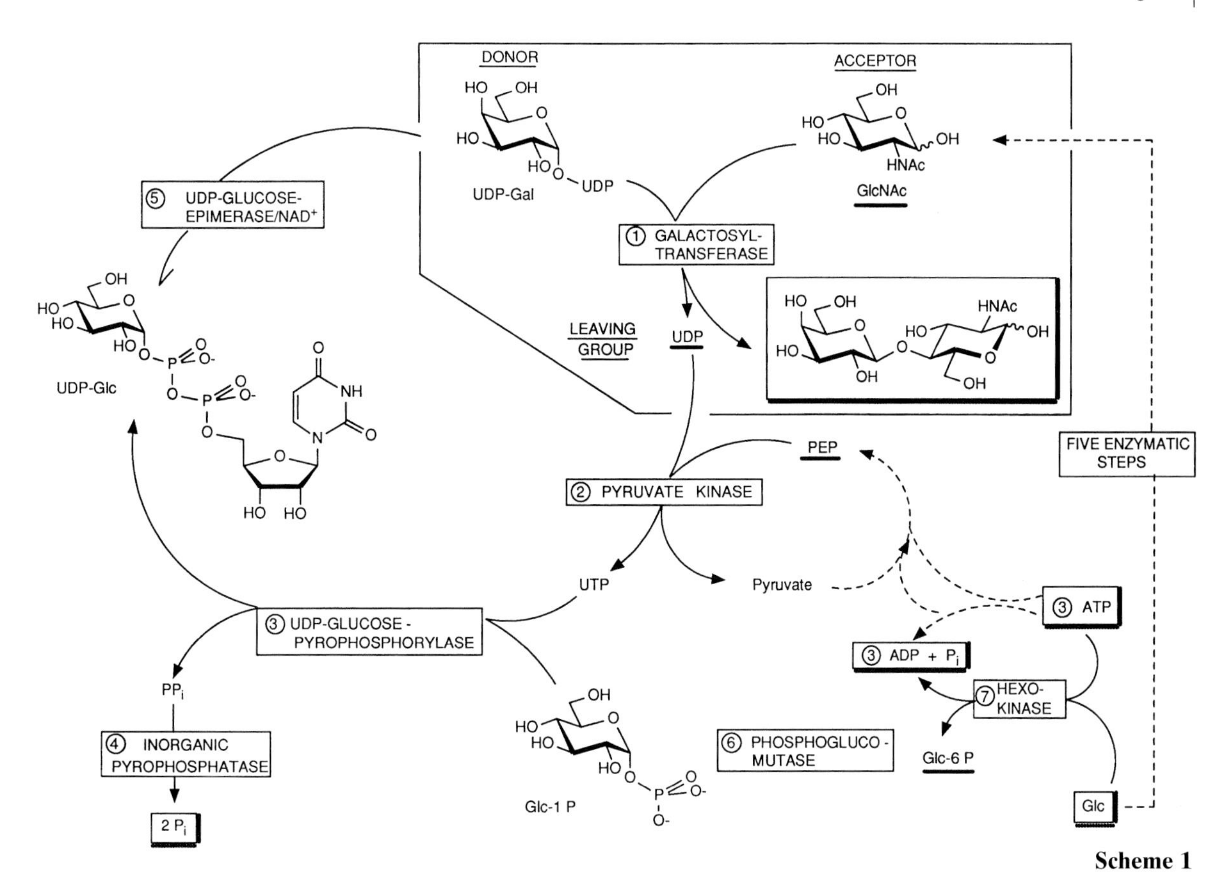

Scheme 1

stereoselectivity, and the lower product yields compared to glycosyltransferases precluded the wide use of glycosidases in the demanding glycoconjugate synthesis.

The limited availability of glycosyltransferases, the complex generation of expensive glycosyl donors, and the various pitfalls in carrying out these enzymatic reactions have led in the last twenty years to an extensive development of chemical glycosylation procedures; these methods possess the inherent advantage of also making available non-natural compounds.

2.3 General Aspects of Chemical Glycoside Bond Formation

A versatile and generally applicable method for diastereo-controlled glycoside bond formation – closely related to the natural process – seems to necessitate a two-step procedure with the following requirements (Schmidt, 1986). The first step, activation of the anomeric center generating the glycosyl donor requires convenient formation of a sterically uniform glycosyl donor, having, according to choice, either α- and β-configuration, and possessing thermal stability at least up to room temperature; eventually chromatographic purification should be possible. The second step, the glycosyl transfer to the acceptor providing the glycoside, requires catalysis of the glycoside transfer by simple means, irreversibility of the reaction, no effect on other glycosidic linkages, high chemical yield, and high α- or β-selectivity (anomeric control) via a diastereo-controlled reaction at the anomeric center.

The different approaches to solving this problem are compiled in Scheme 2. Obviously, the type and generation of the glycosyl donor is also a decisive step in the chemical methodologies for

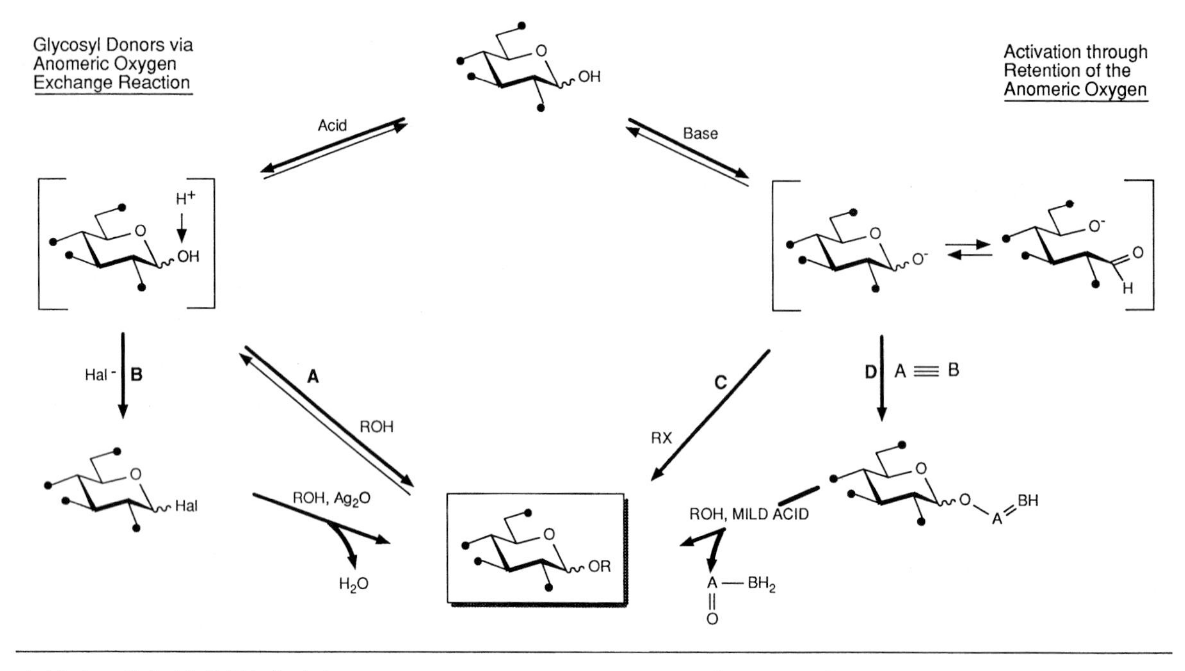

A Fischer - Helferich (Acid Activation)
B Koenigs - Knorr : Br, Cl, (I) Activation
F - Activation
S - Activation

C Anomeric O-Alkylation : Base Activation
D Trichloroacetimidate (Imidate) Activation
– OSO_2R – Activation
– $OPO(OR)_2$, $OP(OR)_2$ – Activation

Scheme 2

glycoside bond formation. Two basic principles have been employed for this purpose, which are used to categorize the methodologies developed thus far:

(i) anomeric oxygen exchange reactions containing the Fischer-Helferich procedure and the Koenigs-Knorr procedure and its many variations, for instance, fluoride and sulfur activation.

(ii) anomeric oxygen retaining reactions containing the anomeric *O*-alkylation and trichloroacetimidate, phosphate (similar to nature), phosphite, sulfate, and sulfite activation.

Because of the irreversibility required for the construction of complex glycoconjugates, the Fischer-Helferich procedure is not applicable for solving this task. This procedure has its merits in the synthesis of simple alkyl glycosides. Also anomeric *O*-alkylation, though of great advantage in glycoside synthesis, because protective groups are often not required, has not gained wide use in glycoconjugate synthesis. The frequently observed low reactivity of activated secondary hydroxylic groups and the need for inversion of configuration at this center limits the general applicability of this procedure; it is an excellent and very efficient method for glycoside bond formation to primary hydroxy groups.

Thus, the Koenigs-Knorr procedure and its variations and the trichloroacetimidate procedure, and other methods based on retention of the anomeric oxygen during activation, are the methods of choice. Important aspects of these procedures will be discussed below.

3 The Koenigs-Knorr Procedure and its Variations

3.1 The Koenigs-Knorr Procedure

In the classical Koenigs-Knorr procedure (Koenigs and Knorr, 1901; Schmidt, 1989; Paulsen, 1990) and the subsequently developed efficient variations, the activation step is achieved through the formation of glycosyl halides (halogen = chlorine and bromine). Glycosyl transfer occurs in the second step with the help of heavy metal salts (preferably silver and mercury salts, which are highly halophilic). This method has been critically reviewed, giving the following general picture.

3.1.1 Glycosyl Halide Formation

The exchange of the anomeric hydroxy group for chlorine or bromine is carried out with typical halogenating agents while all other hydroxy groups are *O*-protected; because of the "thermodynamic anomeric effect", the product with the halogen atom in axial position (α-anomer of the most common aldohexopyranoses) is mainly obtained.

3.1.2 Glycosyl Halide Stability and Reactivity

The stability and the reactivity of glycosyl halides is highly dependent on the halogen, the sugar residue and the protective groups, thus offering a wide selection of glycosyl donors to solve specific problems in glycosylation reactions. For instance, the thermal stability of glycosyl halides increases from bromide to chloride and is also increased by substituting electron-donating protective groups for electron-withdrawing protective groups. However, reactivity generally increases in reverse order compared with stability. Additionally, protective groups at different positions exert "remote electronic effects" (van Boeckel and Beetz, 1985) thus augmenting the methodological repertoire in glycoside synthesis. Yet, due to instability and sensitivity toward hydrolysis, glycosyl halides are frequently generated in situ and used without further purification, which has an impact on chemical yields and the ease of anomeric control in glycosylation reactions.

3.1.3 Promoters for the Glycosylation

For the glycosylation reaction at least equivalent amounts of a "promoter" (often falsely termed "catalyst") is required (Scheme 2 and 3). Because of their halophilicity, various silver or mercury salts are applied as promoters. Important aspects are differences in reactivity and the generation of equivalent amounts of acid and/or water in the course of glycoside bond formation. For the most common promoter systems the following order of reactivity has been generally confirmed: $AgOTf/Ag_2CO_3$ > $AgClO_4$ > $Hg(CN)_2/HgBr_2$ > $Hg(CN)_2$ (Paulsen, 1990). The formation of acid (HX, Scheme 3) can be circumvented by the addition of Ag_2CO_3 in combination with catalytic amounts of AgOTf (silver trifluoromethanesulfonate) or $AgClO_4$ and the use of a base, for instance lutidine or collidine; water is usually removed with Drierite or molecular sieves.

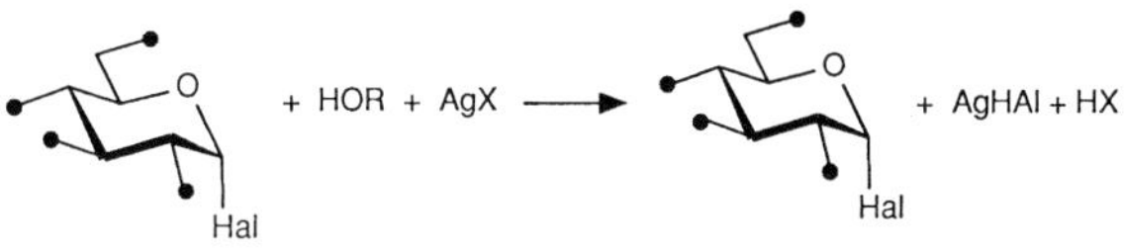

+ HOR + AgX → + AgHal + HX

$Ag_2CO_3 + 2\,HX \rightarrow 2\,AgX + H_2O + CO_2\uparrow$

$HX + B \rightarrow HB^{\oplus}X^{\ominus}$

X = OTf, ClO_4; B = Base

Scheme 3

3.1.4 Reaction Conditions

Solvents of low donicity are commonly used in glycoside synthesis, for instance dichloromethane, cyclohexane, or petroleum ether. These solvents favor S_N2-type reactions. Solvents of higher donicity, for instance ether, acetonitrile, etc., each produce a typical change in the reaction course, due to their varying participation in the stabilization of reaction intermediates. These effects will be discussed below. Due to the low thermal stability of many glycosyl halides, reaction temperatures above room temperature are usually not applied. Taking advantage of the stereo-controlled S_N2-type reaction for anomeric stereocontrol requires the use of reaction temperatures as low as possible. However, solubility and reactivity are commonly the decisive factors for reaction temperature selection.

3.1.5 Structural Types of Glycopyranosides – α/β-Selectivity

Due to the importance of the hydroxy group next to the anomeric position, four different structural situations are generally differentiated (Table 2). The ease of formation of these different anomeric linkages is mainly dependent on the strength of the anomeric effect, which is stronger in the α-manno-type than in the α-gluco- (and the α-galacto) type sugars, and on possible neighboring group participation of protective groups at the functional group at C-2. These effects explain the experimentally found ease of formation of α-manno-type glycosidic linkages, because they are favored by the thermodynamic anomeric effect and by neighboring group participation. β-Gluco-type structures are readily accesible via neighboring group participation. However, the formation of α-gluco-type structures depends strongly on thermodynamic control or on S_N2-type reactions when the "in situ-anomerization" procedure of axial to equatorial glycosyl halides (introduced by Lemieux and Haymi, 1965) is employed. Obviously, the formation of β-manno-type structures, most commonly found in the branching position of *N*-glycoproteins, is supported neither by the thermodynamic anomeric effect nor by neighboring group participation. Therefore, only a kinetically controlled S_N2-type reaction with the help of the readily available α-mannopyranosyl halides could solve the problem. This was indeed found by the use of insoluble promoters (as reported by Paulsen *et al.*, 1981; Garegg and Ossowski, 1983).

Table 2. Structural types differential in hexopyranosides

1,2-trans-type		1,2-cis-type	
α-manno-type	β-gluco-type	α-gluco-type	β-manno-type

Typical examples: gluco-type: D-Glc, D-GlcNAc, D-GlcUA, MurAc, D-Qui

D-Gal, D-GalNAc, L-Fuc

manno-type: D-Man, D-ManNAc, L-Rha

3.1.6 Examples

The Koenigs-Knorr procedure has been quite successfully employed for the synthesis of all prominent glycoconjugates containing all types of glycosidic linkages. A very recent example (Nakahara *et al.*, 1996) is shown in Scheme 4, exhibiting the synthesis of a complex *N*-glycopeptide structure which is based on a building block strategy, and which demonstrates concomitant formation of up to three glycosidic linkages.

3.1.7 Conclusion

In the Koenigs-Knorr procedure, the reactivity of the glycosyl donor can be varied over a wide range by the choice of the halogen, the protective group pattern, the promoter, the solvent, and the temperature. Therefore, the glycosyl donor reactivity can be adjusted to the glycosyl acceptor reactivity, which can be also varied (see below). The anomeric stereocontrol can be attained by S_N2-type reaction, by neighboring group participation, or with the help of the thermodynamic anomeric effect. The application of this methodology has led to many excellent results. However, severe, partly inherent disadvantages of the Koenigs-Knorr procedure could not be overcome. These are the low thermal stability and the sensitivity to hydrolysis of many glycosyl halides, and above all the use of heavy metal salt promoters in at least equimolar amounts (often up to four equivalents). Therefore improved methods, which would also be applicable to large scale preparations, are in great demand.

3.2 Fluoride as Leaving Group

Closely related to the Koenigs-Knorr procedure with bromide and chloride as leaving groups, fluorine has been introduced by Micheel and Borrmann (1960) (Scheme 2, X = F) and has recently been investigated for glycoside synthesis (Schmidt, 1989; Paulsen, 1990; Kunz, 1993; Toshima and Tatsuta, 1993). The higher thermal stability and the higher stability towards many reagents is due to the higher strength of the carbon-fluorine bond compared with other carbon-halogen bonds. However, this concomitantly lowers the glycosyl donor properties of glycosyl fluorides. Therefore, specifically fluorophilic promoters were sought, which would also be required in at least equimolar amounts. Recently, besides the inexpensive $BF_3 \cdot OEt_2$, the mainly more expensive $SnCl_2/AgClO_4$, tin trifluoromethanesulfonate [$Sn(OTf)_2$], cyclopentadienyl-zirconium bistrifluoromethanesulfonate [$CpZr(OTf)_2$], and cyclopentadienyl-hafnium bistrifluoromethanesulfonate [$CpHf(OTf)_2$] systems were successfully employed as promoters. However, relatively few applications of glycosyl fluorides in complex oligosaccharide synthesis have been reported so far (Ito *et al.*, 1988). This shows the limited value of glycosyl fluorides as glycosyl donors. The major advantage seems to be the access to base-catalyzed modifications of unprotected hydroxy groups in the presence of anomeric fluorine atoms (Thiem and Wiesner, 1988). However, better alternatives that take account of this point, which has an impact on the strategy of glycoconjugate synthesis, are now available (see below).

3.3 1,2-Anhydro (1,2-Epoxide) Sugars as Glycosyl Donors

Cyclic ethers possessing ring strain, such as epoxides, are generally quite good alkylating agents which can be activated by acid catalysts. Even stronger alkylating properties are observed for alkoxy-epoxides, structural moieties found in 1,2-anhydro sugars. Therefore, these compounds should be good glycosyl donors under acid catalysis (Schmidt, 1986). These long-known compounds have been recently reinvestigated, and the glycosylation methodology has also been greatly improved and successfully employed in glycoconjugate synthesis (Bilodeau and Danishefsky, 1996). These syntheses generally start from the glycals as precursors for the individual sugars; therefore, stereocontrol is required not only at the anomeric position (C-1) but also at C-2 in product formation; thus far, mainly 1,2-trans-glycosidic bonds have been generated with this method.

3.4 Sulfur as Leaving Group

Thioglycosides have attracted considerable interest because they offer the above described temporary protection of the anomeric center, and, in addition, various sulfur-specific thiophilic activation reactions for generating glycosyl donor properties are available (Schmidt, 1989; Toshima and Tatsuta, 1993; Norberg, 1996). However, again all these thiophilic promoters are required in at least equimolar amounts. Particularly obvious is the selective activation of thioglycosides by thiophilic metal salts of mercury, copper, silver, and even lead. Due to the high toxicity of most of these salts other reagents became more fashionable. Besides metal salts, halonium ions (especially iodonium, bromonium, and chloronium ions) are also highly thiophilic. Thus, with bromine and chlorine, thioglycosides can be readily transformd into halogenoses, enabling the application of the Koenigs-Knorr methodology for glycoside bond formation. If the counterion of the halonium ion is a poor nucleophile [as for instance in *N*-iodo- or *N*-bromosuccinimide (NIS or NBS)], then direct reaction with alcohols is preferred; additional use of some trifluoromethanesulfonic acid (TfOH) as promoter became one of the most frequently employed activation systems for thioglycosides. This is exhibited in Scheme 5 for the successful synthesis of the important sialylgalactosyl globoside ganglioside of the globo series (Ishida *et al.*, 1996) which had previously been synthesized via the trichloroacetimidate method (Lassaletta and Schmidt, 1995). The generation of sulfonium ions with the help of methyl trifluoromethanesulfonate (methyl triflate) has also been exploited. However, the health hazard of this compound and the formation of methylation products in other side reactions are disadvantages of this method. However, dimethyl(methylthio)-sulfonium triflate (DMTST), a methyl triflate adduct of dimethyl disulfide, is also very thiophilic, less toxic, and gives rise to good glycosylation yields. Therefore, selenonium type compounds (for instance *N*-phthaloyl-phenylselenamide (PhSeNPhth), phenylselenyl trifluoromethanesulfate (PhSeOTf) have also been successfully used as promoters for thioglycosides. Besides, methyl, ethyl, and phenyl thio-

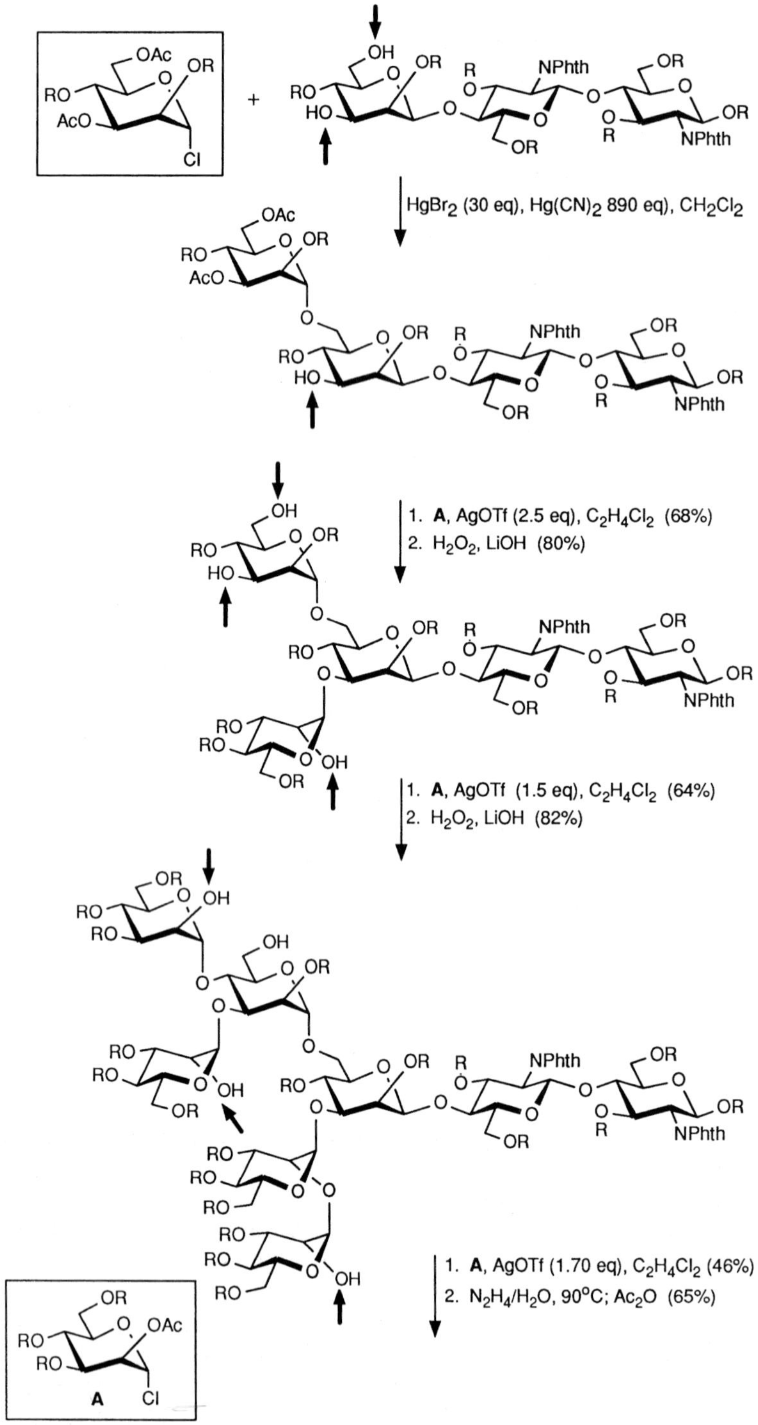

$HgBr_2$ (30 eq), $Hg(CN)_2$ 890 eq), CH_2Cl_2

1. **A**, AgOTf (2.5 eq), $C_2H_4Cl_2$ (68%)
2. H_2O_2, LiOH (80%)

1. **A**, AgOTf (1.5 eq), $C_2H_4Cl_2$ (64%)
2. H_2O_2, LiOH (82%)

1. **A**, AgOTf (1.70 eq), $C_2H_4Cl_2$ (46%)
2. N_2H_4/H_2O, 90°C; Ac_2O (65%)

A

Scheme 4

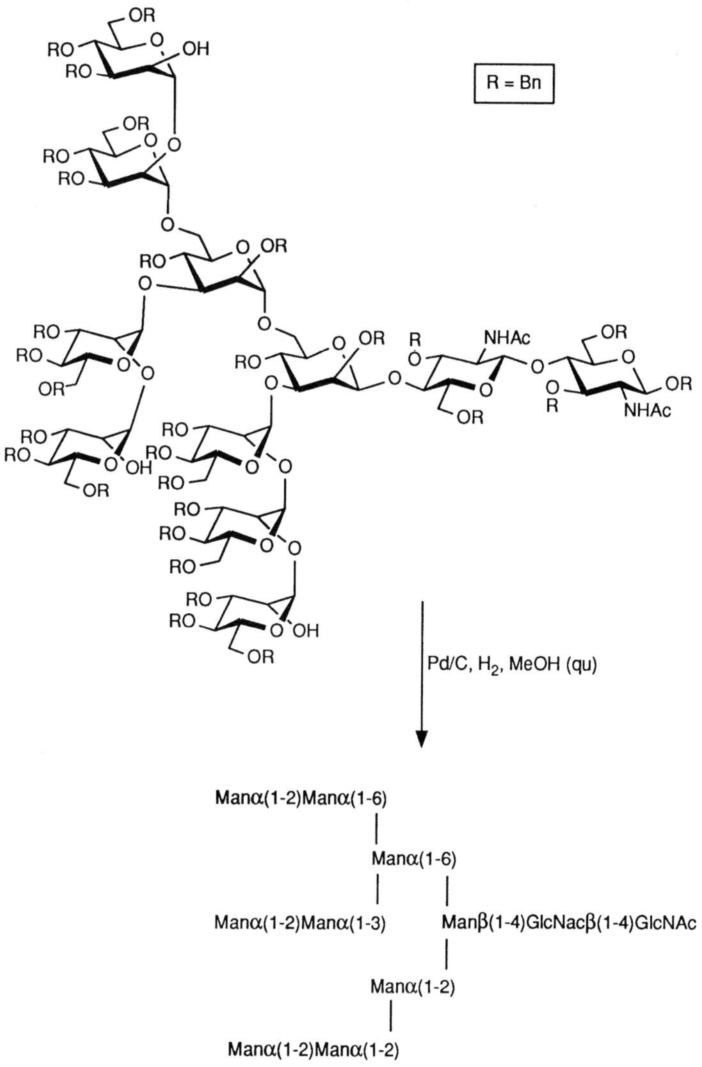

R = Bn

Pd/C, H_2, MeOH (qu)

Manα(1-2)Manα(1-6)
Manα(1-6)
Manα(1-2)Manα(1-3)
Manβ(1-4)GlcNacβ(1-4)GlcNAc
Manα(1-2)
Manα(1-2)Manα(1-2)

Scheme 4 (continued)

glycosides, glycosyl xanthates have also been used with similar activation systems, but with less success in terms of glycosylation yields.

The recent popularity of the use of thioglycosides as glycosyl donors is based on their convenient generation, their stability and access to further modifictions, and their activation with commercially available promoter systems. Disadvantages of thioglycosides are again the requirement of at least equimolar amounts of the promoters, which can also lead to undesired side reactions (for instance, to transglycosylation reactions), their limited reactivity as glycosyl donors and their negative interference with the often required palladium- or platinum-catalyzed hydrogenolysis for partial or complete deprotection.

Recently, glycosyl sulfoxides have also been introduced als glycosyl donors (Norberg, 1996); they are readily obtained from the thioglycosides via mono-oxygenation (giving generally diastereomeric mixtures). Their activation is performed by strong acids or acid anhydrides (for instance, trimethyl trifluoromethanesulfonate (TMSOTf), TMSOTf/triethyl phosphite, trifluoromethanesulfonic anhydride (Tf_2O). Because of the few reports so far available, a general statement on

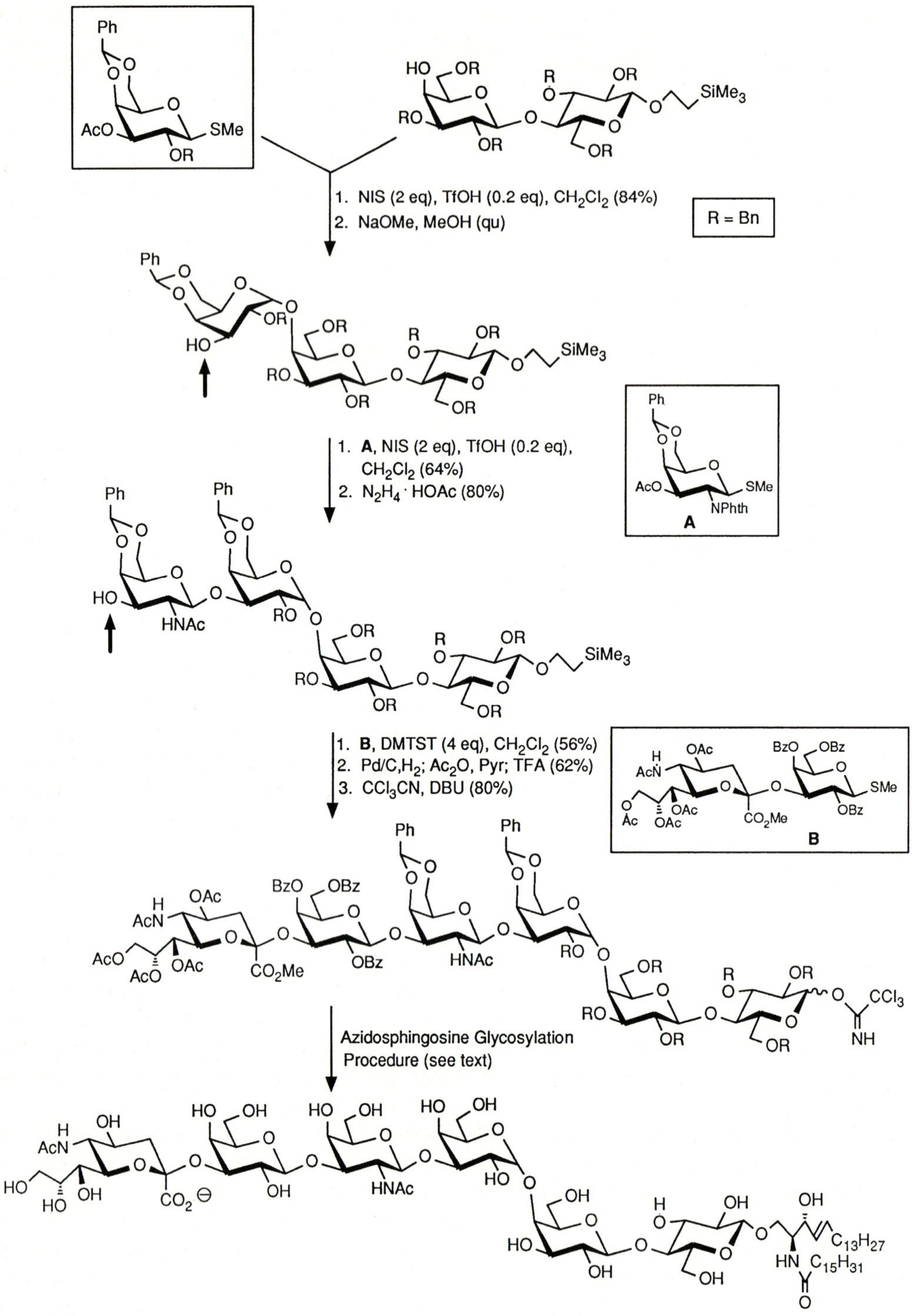

1. NIS (2 eq), TfOH (0.2 eq), CH_2Cl_2 (84%)
2. NaOMe, MeOH (qu)

R = Bn

1. **A**, NIS (2 eq), TfOH (0.2 eq), CH_2Cl_2 (64%)
2. $N_2H_4 \cdot HOAc$ (80%)

A

1. **B**, DMTST (4 eq), CH_2Cl_2 (56%)
2. Pd/C,H_2; Ac_2O, Pyr; TFA (62%)
3. CCl_3CN, DBU (80%)

B

Azidosphingosine Glycosylation Procedure (see text)

Scheme 5

their scope cannot be made. However, the requirement of quite large quantities of promoter for generating leaving group character with the sulfoxide moiety will presumably not result in any superior glycosyl donor properties.

3.5 Nitrogen as Leaving Group

Spiro-diazirines including the anomeric carbon are an interesting new type of potentially useful glycosyl donors (Uhlmann *et al.*, 1994). However, the preparation of the glycosyl donors requires several steps and is still complicated on a large scale; also the results in glycosylation reactions are not yet satisfactory.

4 The Trichloroacetimidate Method, the Phosphite Method, and Other Anomeric Oxygen Activation Procedures

4.1 The Trichloroacetimidate Method

4.1.1 Introduction and Earlier Attempts at Anomeric Oxygen Activation

The requirements for glycoside synthesis discussed above are obviously not met by any of the methods described. However, it seems that the general strategy for glycoside synthesis is appropriate, namely that the first step consists of the generation of a glycosyl donor, and that the second step yields a catalyzed, sterically uniform, irreversible glycosyl transfer to the acceptor.

Obviously, apart from the acid-supported anomeric oxygen exchange for a leaving group (Scheme 2, path Ⓑ) discussed above, a base-promoted transformation of the anomeric oxygen into a good leaving group (Scheme 2, path Ⓓ) can be employed as an alternative method for generating a glycosyl donor. Therefore, it is not surprising that anomeric oxygen acylation, sulfonylation, and silylation, reaction with the Mitsunobu reagent, and 1,2-orthoester formation were employed (Schmidt, 1986, 1989). However, none of these methods gained wide application in complex glycoconjugate synthesis. In some cases the glycosyl donor properties are too weak (acylation, silylation, orthoester formation), and/or the stability of the glycosyl donors is too low and anomeric stereocontrol in the glycosylation step is poor (sulfonylation). Recent modifications of the orthoester method have led to interesting improvements in this regard (Backinovsky *et al.*, 1980); however, the effort required to make of these glycosyl donors limits the general applicability.

4.1.2 Trichloroacetimidate Formation

Obviously, for a stereo-controlled transformation of the anomeric oxygen into a good leaving group, the anomerisation of the anomeric hydroxy group or the anomeric oxide oxygen in the presence of base has to be taken into account. Thus, in a reversible activation process, and with the help of kinetic and thermodynamic reaction control, both anomers of the glycosyl donor should be accessible (Scheme 2). These considerations will be met when pyranoses and furanoses are able to undergo base-catalyzed addition directly and in a stereo-controlled manner to suitable triple bond systems A ≡ B (or to compounds containing suitable cumulative double bond systems A = B = C; Scheme 2, path Ⓓ). The intermediates thus obtained in the first step should, by appropriate choice of the centers A and B (or A, B, and C), be stable under basic and neutral conditions and have good glycosyl donor properties in the presence of acid. The water liberated in the glycoside bond formation is transferred in two steps to the activating species A ≡ B (or A = B = C), thus providing the driving force for the reaction.

From the various A ≡ B (and A = B = C) systems tested, the highly electron deficient trichloroacetonitrile gave in all respects by far the best results (Schmidt, 1986, 1989, 1994, 1996). A detailed study of trichloroacetonitrile addition to 2,3,4,6-tetra-*O*-benzyl *D*-glucose showed (Scheme 6) that from the 1-oxide the equatorial β-trichloroacetimidate is formed preferentially or even exclusively in a very rapid and reversible addition reaction. However, this product anomerizes in a slow, base-catalyzed reaction (via retroreaction, 1-oxide anomerization, and then renewed addition) practically completely to the α-trichloroacetimidate, with the electron-withdrawing anomeric trichloroacetimidoyloxy-substituent in the axial position as favored by the thermodynamic anomeric effect (Schmidt, 1986). The higher nucleophilicity of the equatorial β-oxide can be attributed to a steric effect in combination with a kinetically effective stereoelectronic effect, resulting from repulsions of lone electron pairs or from dipole effects; this effect was termed "kinetic anomeric effect" (Schmidt and Michel, 1984; Schmidt *et al.*, 1984). Thus, with different bases, for instance K_2CO_3 or NaH and 1,8-diazabicyclo[5.4.0]undec-7-ene (DBU), both anomers can be isolated in pure form and in high yield via kinetic and thermodynamic reaction control. Both anomers are thermally stable and can be easily stored.

The diastereoselective *O*-glycosyl trichloroacetimidate formation is applicable to all aldoses and practically to all important *O*-protected hexopyranoses (Glc, Gal, Man, Qui, Fuc, Rha, GlcN, GalN), hexofuranoses, pentopyranoses, and pentofuranoses, as well as to uronic acids (GlcUA, GalUA), to deoxy sugars (Qui, Fuc, Rha, 2-Deoxy-Glc etc.), and to other important sugar derivatives (e.g. muramic acid); also some partially *O*-protected sugars have been successfully transformed into *O*-glycosyl-trichloroacetimidates. The most common bases for the catalysis of the trichloroacetimidate formation are NaH, DBU, K_2CO_3, and Cs_2CO_3 in dichloromethane as solvent, thus commonly providing stable compounds in a stereo-controlled manner. With this method, the requirements for the first step, namely convenient and diastereoselective formation of stable glycosyl donors, are almost ideally fulfilled.

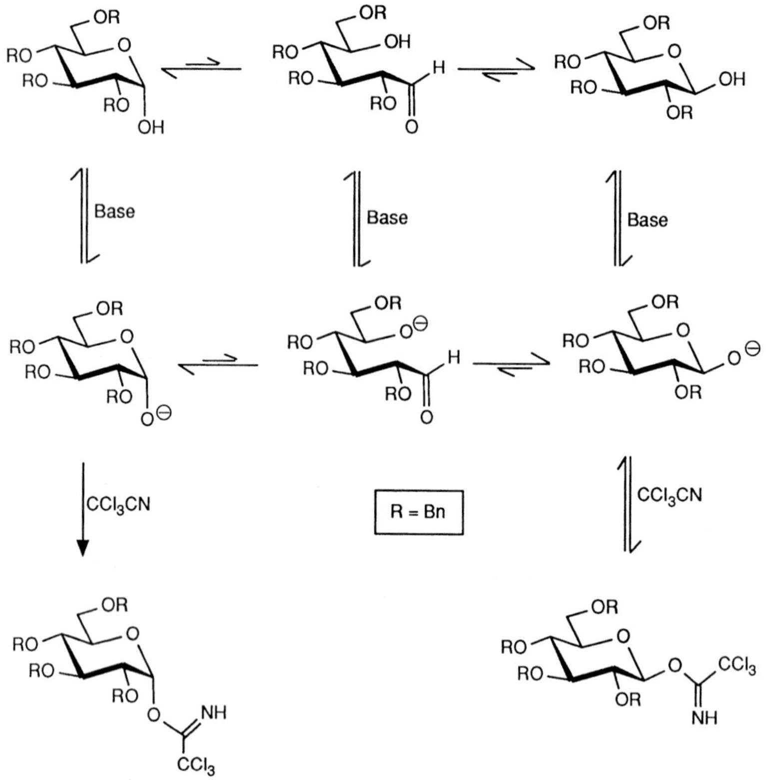

Scheme 6

4.1.3 The Glycosylation Step

The significance of the *O*-glycosyl trichloroacetimidates as glycosyl donors is derived from their high glycosylation potential under acid catalysis, as confirmed in various laboratories with various glycosyl donors and acceptors (Barresi *et al.*, 1995).

Brønsted Acids as Acceptors.
The direct glycosylation of Brønsted acids is a particularly advantageous property of *O*-glycosyl trichloroacetimidates. For instance, carboxylic acids and phosphorous acid and derivatives generally react with inversion of configuration at the anomeric center; thus, glycosyl acetates and glycosyl phosphates, respectively, are readily available with high diastereo-control and without the addition of any catalyst. This opens a convenient route to nucleoside di- and -monophosphate sugars and to glycophospholipids (Schmidt, 1992) which are of interest as intermediates in biological glycosyl transfer and as constituents of cell membranes, respectively. A typical example is shown in Scheme 7 with the synthesis of UDP-sulfoquinovose, which was found on this basis to be the natural sulfoquinovosyl donor (Heinz *et al.*, 1989).

Alcohols and Sugar Hydroxy Groups as Glycosyl Acceptors.
Alcohols or sugars as components for reaction as glycosyl acceptors generally require the presence of an acidic catalyst for the activation of the glycosyl donor (Schmidt, 1986, 1989, 1992, 1994, 1996); thus, mild acid treatment leads irreversibly to formation of glycosides. The water liberated upon glycoside bond formation is transferred in two separate steps (namely glycosyl donor and glycoside bond formation) to the activating species (i.e. trichloroacetonitrile, CCl_3CN) with the formation of a stable, nonbasic water adduct (i.e. trichloroacetamide, CCl_3-$CONH_2$ = O=A-BH_2 in Scheme 2, Path D), thus providing the driving force for the glycosylation reaction. Due to the very low basicity of the liberated trichloroacetamide, the (Lewis) acid required for the activation (of the basic *O*-glycosyl trichloroacetimidates) is released after the glycosyl transfer and is available for further activation of unreactive glycosyl donors, thus leading to a truly catalytic process. Therefore, this methodology very much resembles enzymatic glycoside bond formation, as exhibited in Scheme 8 for the synthesis of a lactosamine intermediate (Lay and Schmidt, 1995).

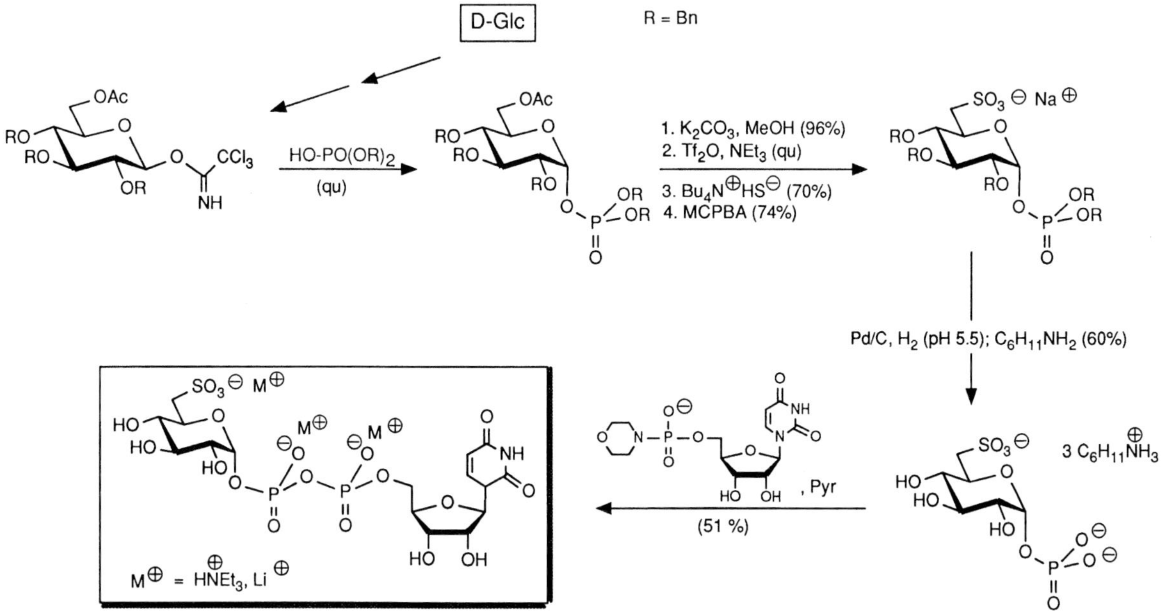

Scheme 7

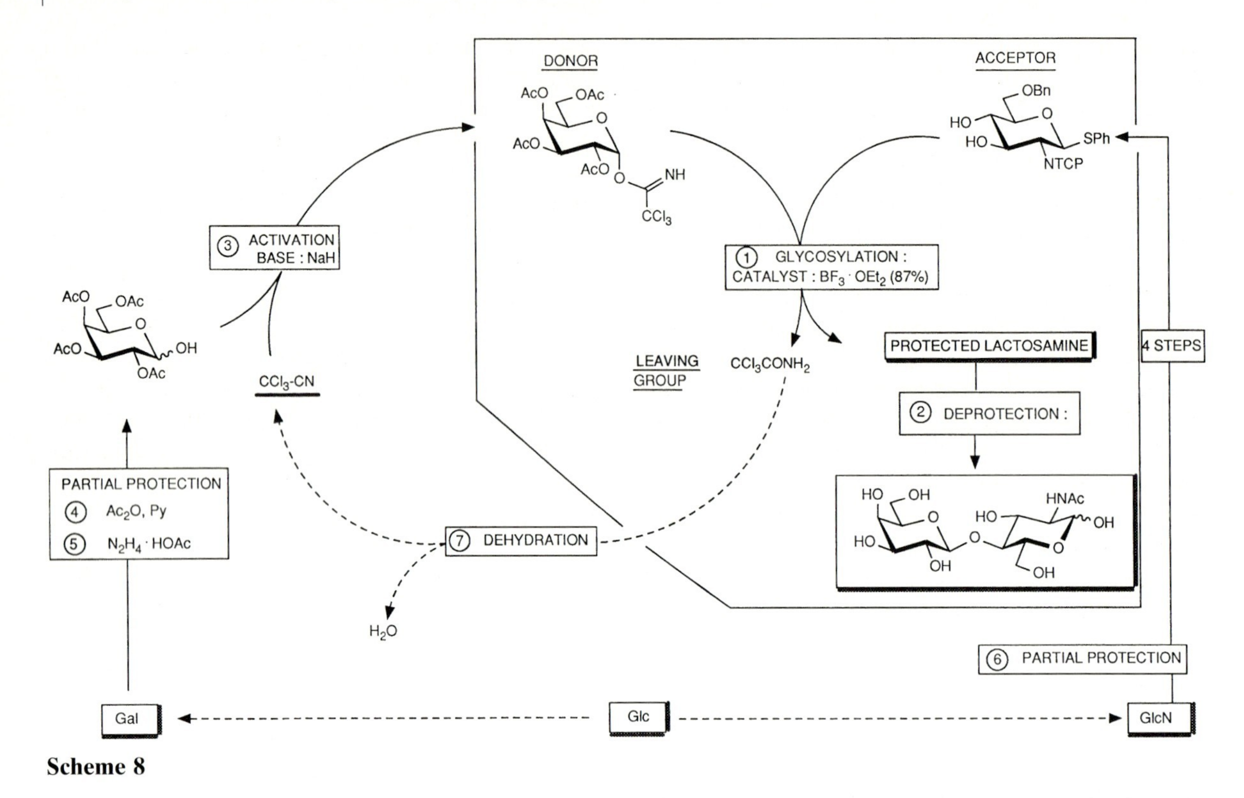

Scheme 8

4.1.4 Reaction Conditions

The general importance of *O*-glycosyl trichloroacetimidates lies in their ability to act as strong glycosyl donors, often in the presence of only 0.005–0.02 equivalents of an acid catalyst. The most frequently employed catalysts are trimethylsilyl trifluoromethanesulfonate (TMSOTf), boron trifluoride-etherate ($BF_3{\cdot}OEt_2$), stannous trifluoromethanesulfonate [$Sn(OTf)_2$], silver trifluoromethanesulfonate (AgOTf), and zinc chloride-etherate ($ZnCl_2{\cdot}OEt_2$). The most frequently used solvents are dichloromethane (CH_2Cl_2), hexane (C_6H_{14}) and mixtures of these two solvents, ether (Et_2O), and acetonitrile (CH_3CN) or propionitrile (CH_3CH_2CN). The reaction temperatures generally range from room temperature to as low as -80°C.

4.1.5 Anomeric Stereocontrol (α/β-Selectivity)

The choice of an *O*-α- or *O*-β-glycosyl trichloroacetimidate and/or the choice of the reaction conditions is crucial for the anomeric stereocontrol. However, again – as discussed already for the Koenigs-Knorr procedure – the relative position of the hydroxy group next to the anomeric center exerts a strong influence on the α/β-selectivity.

In general, neighboring group participation of 2-*O*-acyl or 2-*N*-acyl protecting groups is a dominating effect in anomeric stereocontrol independent of the configuration of the trichloroacetimidate; thus, α-manno- and β-gluco-type (or 1,2-trans-)glycosides (see Scheme 2) are readily generated. When non-participating protective groups are selected, in nonpolar solvents (C_6H_{14}, CH_2Cl_2, C_6H_{14}/CH_2Cl_2) and supported by low temperatures and BF_3OEt_2 as catalyst, S_N2-type reactions can be quite frequently carried out; thus, α-trichloroacetimidates yield β-glycosides and β-trichloroacetimidates yield α-glycosides.

This has been demonstrated especially for α-gluco- and β-gluco-type (1,2-cis- and 1,2-trans-) glycosides, respectively. Stronger acid catalysts, as for instance TMSOTf or trifluoromethanesulfonic acid (TfOH), especially at higher temperatures (room temperature) and in more polar solvents (CH_2Cl_2 and Et_2O which exerts also a specific effect; see below) support formation of the thermodynamically more stable product independent of the configuration of the trichloroacetimidate; thus, under S_N1-type conditions α-manno- and α-gluco-type glycosides, respectively, are obtained. Under these reaction conditions the *O*-glycosyl trichloroacetimidate is transformed into a possibly tight or solvent separated carbenium ion having half-chair conformation; because of stereoelectronic reasons, this is preferentially attacked from the axial (α-)face, thus leading, for kinetic reasons, to the α-product with chair conformation, which is also favored for thermodynamic reasons.

Of particular interest is the influence of solvents under S_N1-type conditions, as has been well studied for ethers (Schmidt, 1986) and recently also for nitriles (Schmidt, 1994). The participation of ethers results, due to the reverse anomeric effect (Lemieux and Koto, 1974), in the preferential generation of equatorial *O*-glycosyl oxonium ions which favor formation of the thermodynamically more stable axial (α-)products. The dramatic effect of nitriles as participating solvents in glycoside synthesis ("nitrile effect") has only recently been discovered by us (Schmidt *et al.*, 1990), leading to a more complex picture of the solvent influence on anomeric stereocontrol. When the excellent trichloroacetamide leaving group is released at low temperatures (-40°C for CH_3CN and -80°C for CH_3CH_2CN) with TMSOTf as catalyst, the carbenium ion having half-chair conformation is attacked by nitrile molecules preferentially from the axial (α-)face to afford under kinetic control α-nitrilium-nitrile conjugates, leading, due to their alkylating properties, to the equatorial (β-)product. Therefore, in nitriles the β-product can become the exclusive product, as found for *O*-benzyl protected *O*-(2-azido-2-deoxyglycopyranosyl) trichloroacetimidates as donors (Toepfer *et al.*, 1994) and also for various other cases (Schmidt, 1996) (β-gluco-type glycosides). With this procedure some success in the difficult β-manno-type glycoside bond formation has also been obtained (Schmidt *et al.*, 1990). However, for the success of the "nitrile effect" the carbenium ion stability, the glycosyl donor properties of the intermediate nitrilium-nitrile conjugates, and the relative rate of the transformation of the kinetically preferred α-nitrilium-nitrile conjugate into the thermodynamically more stable β-nitrilium-nitrile conjugate (reverse anomeric effect) seem to play decisive roles in the anomeric stereocontrol. Yet these factors cannot be fully rationalized at this time. Based on the experience described above, high anomeric stereocontrol can generally be accomplished with the trichloroacetimidate method when the reactions are carried out in homogeneous phase. For solid phase glycosylation reactions the development of an equally well-established methodology would be of great current interest.

4.1.6 Reactivity of *O*-Glycosyl Trichloroacetimidates

The relative reactivities of different glycosyl donors has been compared several times. A typical example exhibiting the high reactivity of *O*-glycosyl trichloroacetimidates was reported by van Boom and collaborators (see Scheme 9) in their endeavor to synthesize an important bacterial cell wall constituent (Smid *et al.*, 1992). Even at -30°C in the presence of TMSOTf as catalyst, the *O*-glycosyl trichloroacetimidate affords higher yields of the desired disaccharide than any of the corresponding thioglycosides at room temperature, which in addition require more than equivalent amounts of promoter.

The high reactivity of *O*-glycosyl trichloroacetimidates once activated may also lead to side reactions, for instance with counterions, or even to decomposition when the acceptor is not close to the reaction center (concentration effect) and/or when the acceptor is sterically hindered (steric effect). Besides stabilization of labile glycosyl donors by electron-withdrawing protective groups, high concentrations, or often more effectively application of the "inverse procedure" (i.e. addition of the donor to an acceptor-catalyst

A-H, Et2O/C2H4Cl2, NIS/TfOH, RT (64 %)
A-H, CH2Cl2, TMSOTf, -30°C (75 %)
(a) A-H, Et2O/C2H4Cl2, NIS/TfOH, RT (16 %)
(b) A-H, CH2Cl2, Cu(OTf)2, RT (23 %)

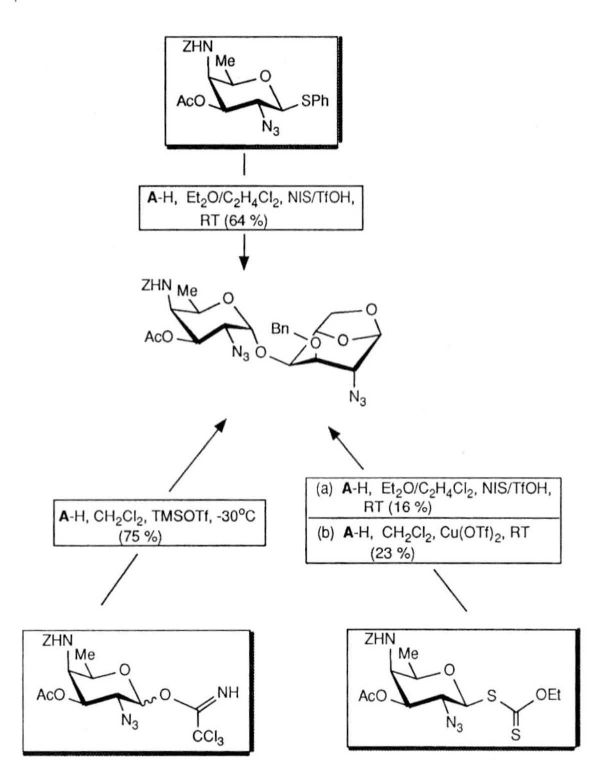

Scheme 9

solution) can also lead to highly improved product yields (Schmidt and Toepfer, 1991). In this case, cluster formation between the acceptor and the catalyst presumably takes place, thus leading to intramolecular glycoside bond formation with the incoming glycosyl donor.

4.1.7 Examples

The trichloroacetimidate method, like the Koenigs-Knorr procedure, has been quite successfully employed for the synthesis of all prominent classes of glycoconjugates containing practically all types of glycosidic linkages and various protective group patterns. A literature search revealed that this method was the most widely used method in 1994 (Barresi *et al.*, 1995). A very few examples are shown in Scheme 10 in which the synthesis is shown of sialyl Lewis X (sia Le[x]) which consists of nine sugar residues and of the β-linked ceramide moiety (Hummel and Schmidt, 1996).

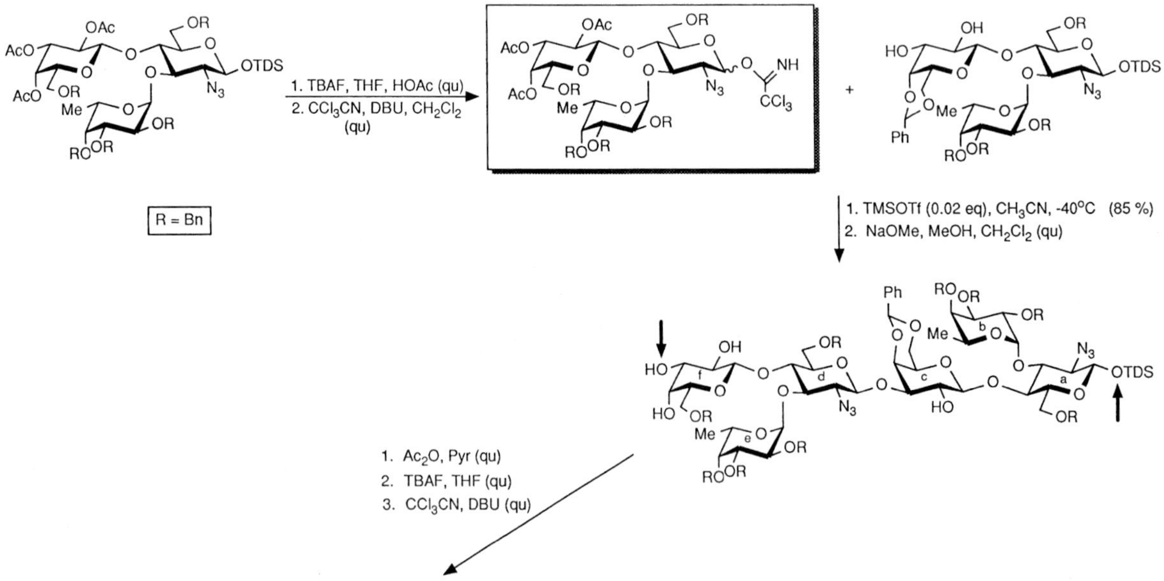

R = Bn
1. TBAF, THF, HOAc (qu)
2. CCl3CN, DBU, CH2Cl2 (qu)
1. TMSOTf (0.02 eq), CH3CN, -40°C (85 %)
2. NaOMe, MeOH, CH2Cl2 (qu)
1. Ac2O, Pyr (qu)
2. TBAF, THF (qu)
3. CCl3CN, DBU (qu)

Scheme 10

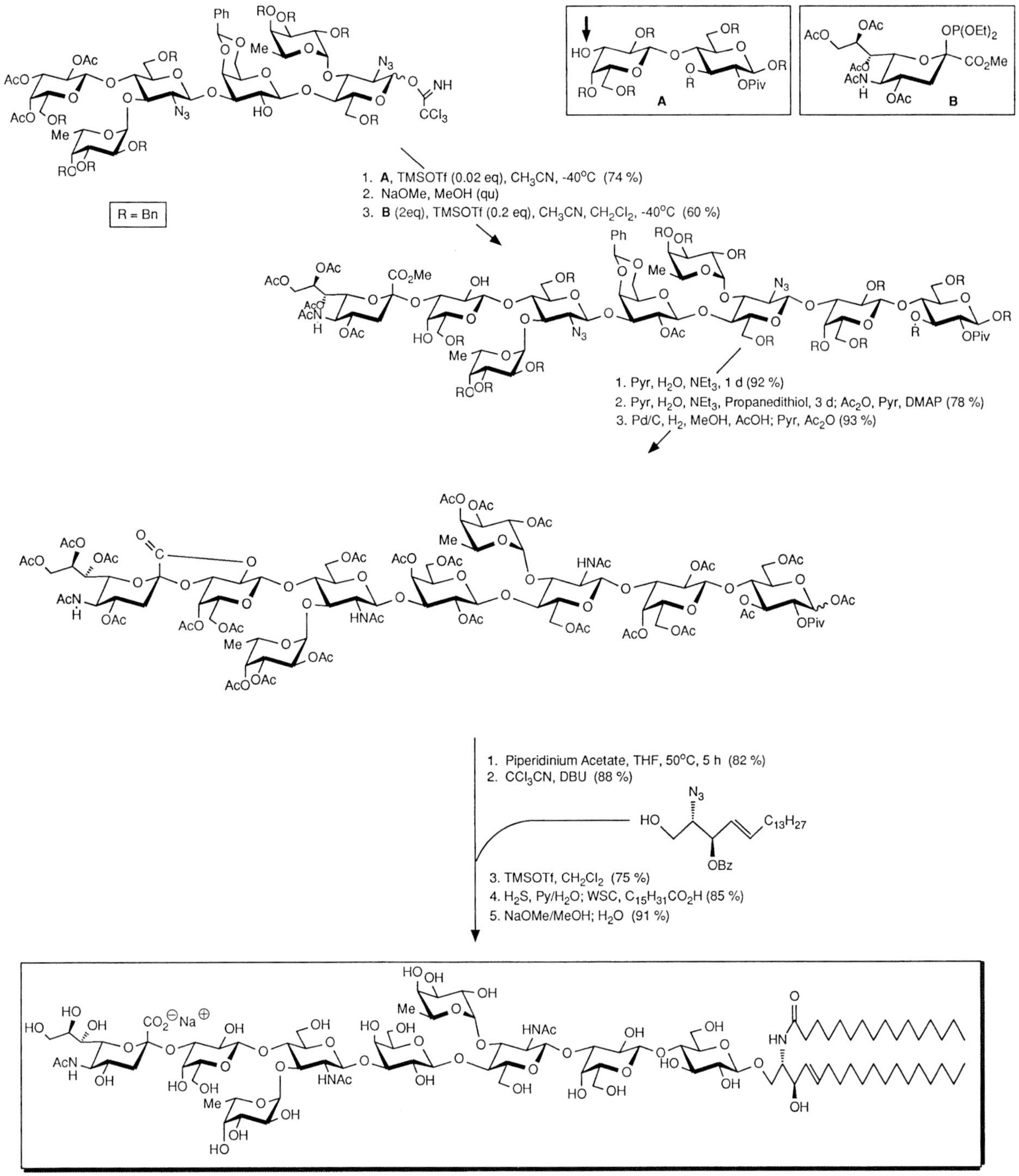

R = Bn

1. **A**, TMSOTf (0.02 eq), CH_3CN, -40°C (74 %)
2. NaOMe, MeOH (qu)
3. **B** (2eq), TMSOTf (0.2 eq), CH_3CN, CH_2Cl_2, -40°C (60 %)

1. Pyr, H_2O, NEt_3, 1 d (92 %)
2. Pyr, H_2O, NEt_3, Propanedithiol, 3 d; Ac_2O, Pyr, DMAP (78 %)
3. Pd/C, H_2, MeOH, AcOH; Pyr, Ac_2O (93 %)

1. Piperidinium Acetate, THF, 50°C, 5 h (82 %)
2. CCl_3CN, DBU (88 %)
3. TMSOTf, CH_2Cl_2 (75 %)
4. H_2S, Py/H_2O; WSC, $C_{15}H_{31}CO_2H$ (85 %)
5. NaOMe/MeOH; H_2O (91 %)

Scheme 10 (continued)

4.1.8 Other Nucleophiles as Glycosyl Acceptors

Besides oxygen, other nucleophiles have also been successfully employed as acceptors. For instance, *N*-, *C*-, *S*- and *P*-glycosyl derivatives have been obtained from corresponding nucleophiles and *O*-glycosyl trichloroacetimidates in the presence of acid catalysts (Schmidt, 1989). Of particular interest are nucleoside (*N*-glycoside) and *C*-glycoside formations, which have been performed quite successfully. The importance of thioglycosides as structural analogues in biological investigations has recently led to an entirely thio-linked sialyl Lewis X synthesis which is exhibited in Scheme 11. Also glycosyl phosphonates – interesting structural analogues of glycosyl phosphates – have been directly obtained from *O*-glycosyl trichloroacetimidates and phosphite esters as acceptors (Eisele *et al.*, 1996).

4.1.9 Conclusion

The trichloroacetimidate method meets to a large extent the requirements put forward for efficient glycosylation methods (see p. 5). This can be clearly deduced from the many successful applications of this method. These results can be summarized as follows.

Typical features of the activation step are: (i) convenient base-catalyzed trichloroacetimidate formation; (ii) often stereo-controlled access to either α- or β-compounds depending on the choice of the base; (iii) thermal stability of *O*-glycosyl α- and β-trichloroacetimidates up to room temperature; (iv) silica gel chromatography for further purification if required; (v) compatibility to some extent with partial protection.

Typical features of the glycosylation step are: (i) catalysis by (Lewis) acids under mild conditions; (ii) irreversible reaction; (iii) liberated water is chemically bound to the leaving group; (iv) other glycosidic bonds are not affected; (v) regioselectivity with partially protected acceptors; (vi) usually high chemical yield, (vii) anomeric stereocontrol is generally good to excellent, when the influence of the donor configuration, of neighboring group participation, of solvent, of temperature, and of kinetic and thermodynamic effects are considered.

Thus, *O*-glycosyl trichloroacetimidates exhibit outstanding glycosyl donor properties in terms of ease of formation, stability, reactivity, and general applicability, thus offering a very competitive and most economical alternative to the Koenigs-Knorr procedure. Their resemblance in various respects to the nucleoside diphosphate sugars, employed by nature as glycosyl donors, may be the reason for the great success of the trichloroacetimidate method.

4.2 The Phosphite Method

For glycosyl donors requiring excellent leaving groups for their activation (sugar uronates of aldoses, common aldoses, and to some extent deoxy aldoses) *O*-glycosyl trichloroacetimidates have been shown to be highly efficient. For sugars requiring relatively low activation to generate glycosyl donor properties (ketoses, 3-deoxy-2-glyculosonates: 3-deoxy-2-octulosonic acid (KDO), Neu5Ac) other methods have proved to be more successful (Schmidt, 1994).

In the search for a catalytic procedure based on a convenient anomeric oxygen activation, consideration of various leaving groups led us to phosphite moieties and their derivatives (Martin *et al.*, 1992). Thus, the readily available *O*-acetylated neuraminic acid derivative N (Scheme 12) is transformed into sialyl phosphite D as donor which with alcohols ROH (A) as acceptors will eventually provide the desired target molecules T. The catalyst C (for instance, TMSOTf), due to the basicity of the leaving group, will attack the phosphite moiety and release a phosphonate residue which is neither strongly acidic nor basic; therefore, the acidic catalyst is released to activate the next sialyl phosphite molecule, thus initiating a catalytic cycle. Provided the assumptions regarding the "nitrile effect" are also applicable to sialyl donors, then using nitriles as solvents and under kinetic control, the equatorial product, i.e. the desired α-glycoside, should be obtained. Indeed, this was observed in many sialylation reactions based on this methodology (Schmidt,

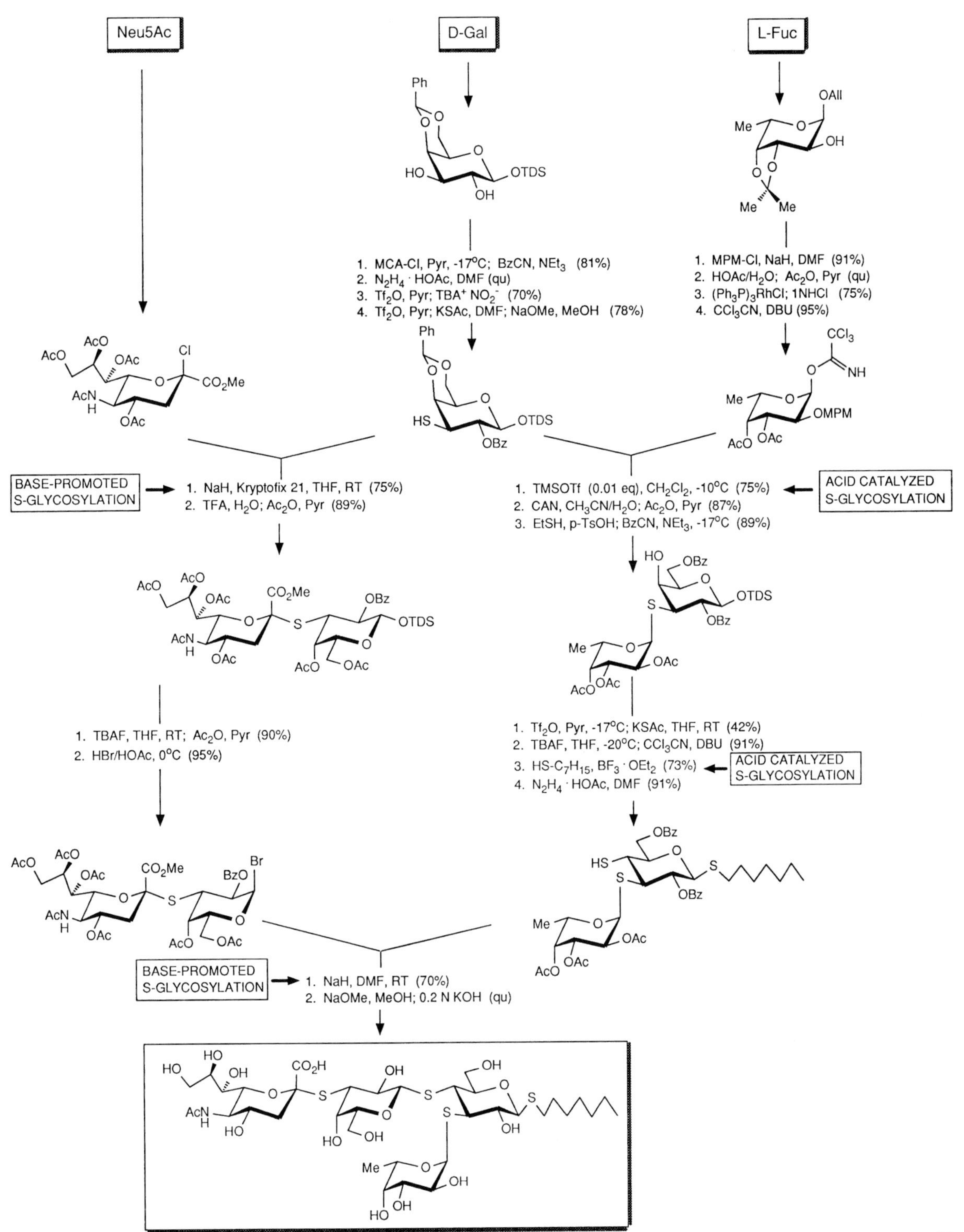
Neu5Ac
D-Gal
L-Fuc
1. MCA-Cl, Pyr, -17°C; BzCN, NEt3 (81%)
2. N2H4 · HOAc, DMF (qu)
3. Tf2O, Pyr; TBA+ NO2- (70%)
4. Tf2O, Pyr; KSAc, DMF; NaOMe, MeOH (78%)
1. MPM-Cl, NaH, DMF (91%)
2. HOAc/H2O; Ac2O, Pyr (qu)
3. (Ph3P)3RhCl; 1NHCl (75%)
4. CCl3CN, DBU (95%)
BASE-PROMOTED S-GLYCOSYLATION
1. NaH, Kryptofix 21, THF, RT (75%)
2. TFA, H2O; Ac2O, Pyr (89%)
1. TMSOTf (0.01 eq), CH2Cl2, -10°C (75%)
2. CAN, CH3CN/H2O; Ac2O, Pyr (87%)
3. EtSH, p-TsOH; BzCN, NEt3, -17°C (89%)
ACID CATALYZED S-GLYCOSYLATION
1. TBAF, THF, RT; Ac2O, Pyr (90%)
2. HBr/HOAc, 0°C (95%)
1. Tf2O, Pyr, -17°C; KSAc, THF, RT (42%)
2. TBAF, THF, -20°C; CCl3CN, DBU (91%)
3. HS-C7H15, BF3 · OEt2 (73%)
4. N2H4 · HOAc, DMF (91%)
ACID CATALYZED S-GLYCOSYLATION
BASE-PROMOTED S-GLYCOSYLATION
1. NaH, DMF, RT (70%)
2. NaOMe, MeOH; 0.2 N KOH (qu)

Scheme 11

1994; Lassaletta and Schmidt, 1995; Hummel and Schmidt, 1996; Wong, 1996). A typical example is shown in the synthesis of the biologically important ganglioside sialyl Lewis X (Scheme 10). Therefore, sialyl phosphites became important sialyl donors due to the ease of their formation and their convenient activation by catalytic amounts of TMSOTf or similar catalysts.

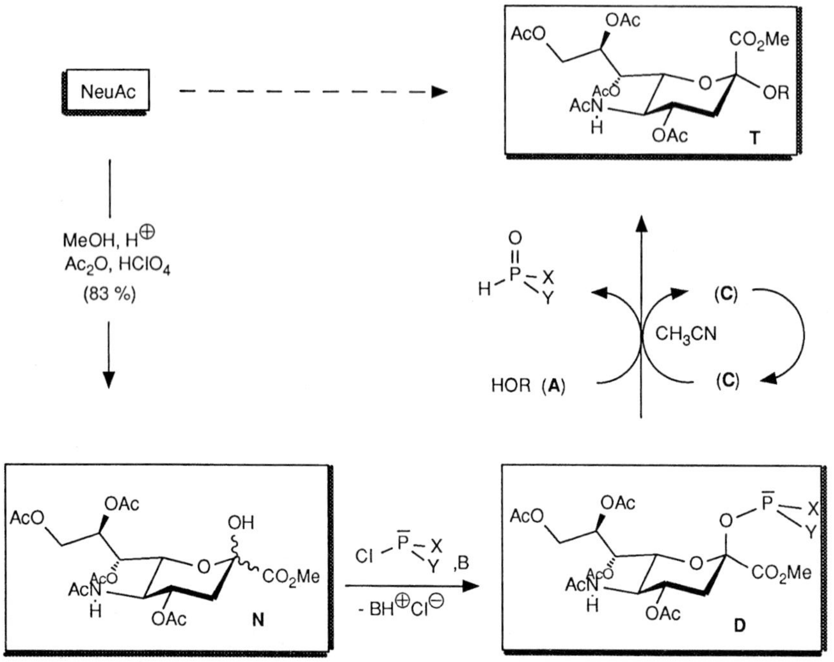

Scheme 12

Meanwhile, phosphites were also successfully employed as leaving groups with other sugars. Particularly good results were obtained for ketoses and high reactivities were also found for deoxy sugars. Because of the lower glycosyl donor properties of *O*-glycosyl phosphites compared with *O*-glycosyl trichloroacetimidates, the phosphite method nicely complements the trichloroacetimidate method (Müller *et al.*, 1994).

A particularly advantageous property of sialyl phosphites is their ready access to phosphite/phosphate exchange reactions with phosphorous acid derivatives as acceptors. On this basis a simple synthesis of nucleoside monophosphate neuraminic acid derivatives could be developed, which includes an efficient synthesis of the natural sialyl donor CMP-Neu5NAc (Martin and Schmidt, 1993).

4.3 Other Anomeric Oxygen Activation Procedures

4.3.1 Glycosyl Phosphates

As discussed above, nature generally employs glycosyl phosphate derivatives as glycosyl donors. However, for chemical glycosylations the use of glycosyl phosphates exhibits no advantages over existing methodologies. The phosphate group possesses moderate leaving group character (comparable to fluoride) and low basicity; therefore, quite strong acid catalysts are required for the activation (Schmidt, 1992). Additionally, the phosphorous acid released increases the acidity of the reaction mixture during the reaction course, which may eventually result in uncontrolled side reactions.

4.3.2 Glycosyl Carboxylates

Quite a few glycosyl carboxylates have been used in glycosylation reactions. For instance, p-nitrobenzoates seemed to offer the required leaving group character under acid catalysis (Schmidt, 1986). However, the drawback found for glycosyl phosphates are essentially also pertinent to glycosyl carboxylates.

4.3.3 4-Pentenyl Glycosides

The 4-pentenyloxy group as a leaving group can also be included among the anomeric oxygen activation procedures, though Fischer-Helferich glycosylation (of 4-penten-1-ol) and not anomeric *O*-alkylation is the common procedure for their preparation (generally yields α/β-mixtures). The activation of the 4-pentenyloxy moiety for glycosylation reactions requires at least equimolar amounts of a promoter; typically iodonium dicollidine perchlorate (IDCP), *N*-iodosuccinimide (NIS)/TfOH, or NIS/triethylsilyl trifluoromethanesulfonate (TESOTf) are employed. The number of applications of this method is still rather limited; the reports are essentially from one laboratory (Madsen *et al.*, 1996). However, in terms of glycosyl donor property, promoter requirement, glycosylation yield and anomeric stereocontrol there seems to be no advantage over the classical Koenigs-Knorr procedure or the trichloroacetimidate method. Yet, similar to thioglycosides, 4-pentenyl glycosides are accessible to protective group manipulations, because they also offer temporary protection for the anomeric center.

5 Concluding Remarks

Obviously, various methods for the generation of the glycosidic bond by chemical means are available. For the synthesis of complex glycoconjugates the use of glycosyl halides, thioglycosides, and *O*-glycosyl trichloroacetimidates has proved to be particularly successful. The high glycosyl donor properties of trichloroacetimidates and the requirement of only catalytic amounts of an acid catalyst for their activation led to wide application of this method and its usefulness in large scale preparations. Glycosyl phosphites exhibit similar fetures, and these are the glycosyl donors of choice for ketoses and especially for the attachment of *N*-acetylneuraminic acid.

References

Backinovskii LV, Tsvetkov YE, Balan NF *et al.* (1980): Synthesis of 1,2-*trans*-disaccharides via sugar thio orthoesters. In *Carbohydr. Res.* **85:** 209–21.

Banoub J (1992): Synythesis of oligosaccharides of 2-amino-2-deoxy sugars. In *Chem. Rev.* **92:** 1167–95.

Barresi F, Hindsgaul O (1995): Glycosylation methods in oligosaccharide synthesis. In *Modern Synthetic Methods* (Ernst B, Leumann C, eds) pp 283–330, Basel: Verlag Helvetica Chimica Acta.

Bilodeau MT, Danishefsky SJ (1996): Coupling of glycals: A new strategy for the rapid assembly of oligosaccharides. In *Modern Methods in Carbohydrate Synthesis* (Khan SH, O'Neill RA, eds) pp 171–93; and references therein; Amsterdam: Harwood Academic Publishers.

Eisele T, Toepfer A, Kretzschmar G *et al.* (1996): Synthesis of a thio-linked analogue of sialyl Lewis X. In *Tetrahedron Lett.* **37:** 1389–92.

Garegg PJ, Ossowski P (1983): Silver zeolite as promoter in glycoside synthesis. The synthesis of β-D-mannopyranosides. In *Acta Chem. Scand., Ser. B* **37:** 249–50.

Hakomori S (1990): Bifunctional role of glycosphingolipids. Modulators for transmembrane signaling and mediators for cellular interactions. In *J. Biol. Chem.* **265:** 18713–16; and references therein.

Heinz E, Schmidt H, Hoch M *et al.* (1989): Synthesis of different nucleoside 5'-diphospho sulfoquinovoses and their use for studies on sulfolipidbiosynthesis in chloroplasts. In *Eur. J. Biochem.* **184:** 445–53.

Hummel G, Schmidt RR (1996): Efficient synthesis of sialyl Lewis x (sLe^x). Manuscript to be submitted.

Ishida, H, Miyawaki R, Kiso M *et al.* (1996): First total synthesis of sialyl globopentaosyl ceramide ($V^3Neu^5AcGb_5Cer$) and its positional isomer ($V^6Neu^5AcGb_5Cer$). In *J. Carbohydr. Chem.* **15:** 163–82.

Ito S, Sato S, Mori M *et al.* (1988): An efficient approach to the synthesis of lacto-*N*-triosylceramide and related substances. In *J. Carbohydr. Chem.* **7:** 359–76; and references therein.

Jung K-H, Schmidt RR (1991): Structure and synthesis of biologically active glycopeptides and glycolipids. In *Curr. Opinion Struct. Biol.* **1:** 721–31.

Khan SH, O'Neill RA + many contributors (1996): *Modern Methods in Carbohydrate Synthesis* Amsterdam: Harwood Academic Publishers.

Koenigs W, Knorr E (1901): Über einige Derivate des Traubenzuckers und der Galactose. In *Ber. Dtsch. Chem. Ges.* **34:** 957–60.

Kunz H (1993): Glycopeptides of biological interest: a challenge for chemical synthesis. In *Pure Appl. Chem.* **65:** 1223–36; and references therein.

Lassaletta JM, Schmidt RR (1995): Versatile approach to the Synthesis of globoside glycosphingolipids – Synthesis of Sialyl-galactosyl-globoside. In *Tetrahedron Lett.* **36:** 4209–12.

Lay L, Schmidt RR (1996): Convenient synthesis of lactosamine intermediates. Manuscript to be submitted.

Ledeen RW, Wu G (1992): Ganglioside function in the neuron. In *Trends Glycosci. Glycotechnol.* **4:** 174–87.

Lemieux RU, Haymi JI (1965): The mechanism of the anomerization of the tetra-*O*-acetyl-*D*-glucopyranosyl chlorides. In *Can. J. Chem.* **43:** 2162–73.

Lemieux RU, Koto S (1974): Conformational properties of glycosidic linkages. In *Tetrahedron* **30:** 1933–44.

Madsen R, Fraser-Reid B (1996): n-Pentenyl glycosides in oligosaccharide synthesis. In *Modern Methods in Carbohydrate Synthesis* (Khan SH, O'Neill RA eds) pp 155–70; and references therein; Amsterdam: Harwood Academic Publishers.

Martin TJ, Schmidt RR (1992): Efficient sialylation with phosphite as leaving group. In *Tetrahedron Lett.* **33:** 5161–4.

Martin TJ, Schmidt RR (1993): Convenient chemical synthesis of CMP-*N*-acetylneuraminate (CMP-Neu5NAc). In *Tetrahedron Lett.* **34:** 1765–8.

Micheel F, Borrmann D (1960): Ein neues Verfahren zur Synthese höherer Disaccharide. *In Chem. Ber.* **93:** 1143–7.

Müller T, Schneider R, Schmidt RR (1994): Utility of glycosyl phosphites as glycosyl donors – Fructofuranosyl and 2-deoxyhexopyranosyl phosphites in glycoside bond formation. In *Tetrahedron Lett.* **35:** 4763–6.

Nakahara Y, Shibayama S, Nakahara Y *et al.* (1996): Rationally designed syntheses of high-mannose and complex-type undecasaccharides. In *Carbohydr. Res.* **280:** 67–84.

Nilsson KGI (1996): Synthesis with glycosidases. In *Modern Methods in Carbohydrate Synthesis* (Khan SH, O'Neill RA eds) pp 518–47; and references therein; Amsterdam: Harwood Academic Publishers.

Norberg T (1996): Glycosylation properties and reactivity of thioglycosides, sulfoxides, and other S-glycosides: Current scope and future prospects. In *Modern Methods in Carbohydrate Synthesis* (Khan SH, O'Neill RA, eds) pp 82–106; and references therein.; Amsterdam: Harwood Academic Publishers.

Paulsen H (1990): Synthesen, Konformationen und Röntgenstrukturanalysen von Saccharidketten der Core-Regionen von Glycoproteinen. In *Angew. Chem.* **102:** 851–67; or: Syntheses, conformations, an X-ray-analyses of saccharide chains of glycoprotein core regions. In *Angew. Chem. Int. Ed. Engl.* **29:** 823–39; and references therein.

Paulsen H, Lockhoff O (1981): Neue effektive β-Glycosidsynthese für Mannose-Glycoside. Synthese von Mannose-haltigen Oligosacchariden. *In Chem. Ber.* **114:** 3102–14.

Schmidt RR (1986): Neue Methoden zur Synthese von Glycosiden und Oligosacchariden – Gibt es Alternativen zur Koenigs-Knorr-Methode? In *Angew. Chem.* **98:** 213–36; or: New methods for the synthesis of glycosides and oligosaccharides – are there alternatives to the Koenigs-Knorr method? In *Angew. Chem. Int. Ed. Engl.* **25:** 212–35; and references therein.

Schmidt RR (1989): Recent developments in the synthesis of glycoconjugates. In *Pure Appl. Chem.* **61:** 1257–70.

Schmidt RR (1992): New aspects of glycosylation reactions. In *Carbohydrates – Synthetic Methods and Applications in Medicinal Chemistry* (Ogura H, Hasegawa A, Suami T, eds) pp 68–88; and references therein. Tokyo: Kodanasha Ltd.

Schmidt RR (1994): Chemical synthesis of sialylated glycoconjugates. In *Synthetic Oligosaccharides – Indispensible Probes for Life Sciences* (Kovac P) pp 276–96, Washington DC: ACS Symposium Series, 560.

Schmidt RR (1996): The anomeric *O*-alkylation and the trichloroacetimidate method – Versatile strategies for glycoside bond formation. In *Modern Methods in Carbohydrate Synthesis* (Khan SH, O'Neill RA, eds) pp 20–54; and references therein; Amsterdam: Harwood Academic Publishers.

Schmidt RR, Michel J (1984): Direct *O*-glycosyl trichloroacetimidate formation. Nucleophilicity of the anomeric oxygen atom. In *Tetrahedron Lett.* **25:** 821–4.

Schmidt RR, Toepfer A (1991): Glycosylation with highly reactive glycosyl donors: efficiency of the inverse procedure. In *Tetrahedron Lett.* **32:** 3353–6; and references therein.

Schmidt RR, Michel J, Roos M (1984): Direkte Synthese von α- und β-*O*-Glycosylimidaten. Bedeutung des thermodynamischen und kinetischen anomeren Effektes. In *Liebigs Ann. Chem:* 1343–57.

Schmidt RR, Behrendt M, Toepfer A (1990): Nitriles as solvents in glycosylation reactions: highly selective β-glycoside synthesis. In *Synlett:* 694–6; and references therein.

Smid P, Joerning WPA, van Duuren AMG *et al.* (1992): Stereoselective synthesis of a dimer containing an α-linked 2-acetamido-4-amino-2,4,6-trideoxy-D-galactopyranose (SUGp) unit. In *J. Carbohydr. Chem.* **11:** 849–56.

Thiem J, Wiesner M (1988): Alkylation of glycosyl fluorides. In *Synthesis:* 124–7.

Thiem J, Sauerbrei B (1991): Chemoenzymatische Synthesen von Sialyloligosacchariden mit immoblisierter Sialidase. In *Angew. Chem.* **103:** 1521–3; or Chemoenzymatic syntheses of sialooligosaccharides with immobilized sialidase. In *Angew. Chem. Int. Ed. Engl.* **30:** 1503–5.

Toepfer A, Kinzy W, Schmidt RR (1994): Efficient synthesis of the Lewis antigen x (Le^x) family. In *Liebigs Ann. Chem.*: 449–64.

Toshima K, Tatsuta K (1993): Recent progress in *O*-glycosylation methods and its application to natural products synthesis. *In Chem. Rev.* 93: 1503–31.

Uhlmann P, Vasella A (1994): Glycosidation of benzyl β-D- and β-L-ripopyranosides. Further evidence for the effect of stereoelectronic control on the regioselectivity of glycosidation. In *Helv. Chim. Acta* **77:** 1175–92; and references therein.

Van Boeckel CAA, Beetz T (1985): Substituent effects in carbohydrate chemistry. Part II. Coupling reactions involving gluco- and galactopyranosyl bromides promoted by insoluble silver salts. In *Recl. Trav. Chim. Pays-Bas* **104:** 171–3; and references therein.

Vliegenthart JFG, Dorland L, van Halbeck H (1983): High resolution ^{1}H-nuclear magnetic resonance spectroscopy as a tool in the structural analysis of carbohydrates related to glycoproteins. In *Adv. Carbohydr. Chem. Biochem.* **41:** 209–374.

Whitfield DM, Douglas SP (1996): Glycosylation reactions – present status, future directions. In *Glycoconjugate J.* **13:** 5–17.

Wong C-H (1996): Practical synthesis of oligosaccharides based on glycosyltransferases and glycosyl phosphites. In *Modern Methods in Carbohydrate Synthesis* (Khan SH, O'Neill RA eds) pp 467–91; and references therein, Amsterdam: Harwood Academic Publishers.

4 Neoglycoconjugates

Reiko T. Lee and Yuan C. Lee

1 Introduction

What are "neoglycoconjugates"? First we must define what "glycoconjugates" are. Glycoconjugates are those **hybrid** biochemicals which contain carbohydrates chemically bonded to peptides (glycopeptides or glycoproteins), lipids (glycolipids), or others. Nucleic acids themselves are glycoconjugates, albeit of a very special kind. Carbohydrate components contained in these natural glycoconjugates usually have complex and non-homogeneous structures. For example, even when there is a single site of carbohydrate attachment on a protein, more than one kind of oligosaccharide structures are usually present, so that the resulting glycoprotein is not made of a homogeneous molecular species. Such molecular species having different carbohydrate structures on the same protein backbone is referred to as "glycoforms" (Rademacher *et al.*, 1988). Supposing that the **glyco** part of a glycoprotein is implicated in its biological activity, the different glycoforms of this glycoprotein may provide subtle modulation of a biological activity. However, the heterogeneity of glycans in such glycoconjugates makes any in-depth investigation of the biological role of the carbohydrate moiety much more difficult. It would be desirable to use glycoconjugates that contain a single carbohydrate species of well-defined structure for such investigations. When the role of individual oligosaccharide chain is understood, we may be able to tackle the collective property of glycoforms as an ensemble.

About twenty years ago, we first coined the term "neoglycoproteins" to describe proteins chemically modified with carbohydrates of defined structre (Stowell and Lee, 1980). Subsequently, the usage of the prefix "neo" has spread to carbohydrate-modified lipids (neoglycolipids), carbohydrate-modified polymers (neoglycopolymers), and eventually "neoglycoconjugates" including all of these carbohydrate hybrids appeared (Magnusson, 1986). There have been several reviews on the topics of neoglycoproteins (Stowell and Lee, 1980; Aplin and Wriston, 1981; Lee and Lee, 1982) and neoglycoconjugates (Magnusson, 1986; Lee and Lee, 1991; Lee and Lee, 1992). Readers are advised to consult these articles as well as a recent book on neoglycoconjugates (Lee and Lee, 1994a) and collected volumes on methodologies (Lee and Lee, 1994b, c).

2 Advantages of Neoglycoconjugates

2.1 Well-Defined Structure

The foremost of the advantages of neoglycoconjugates is that they contain carbohydrates of known sturctures and assured purity. As briefly mentioned above, workers in the areas of glycoproteins, glycolipids, and proteoglycans are fully aware that it is rare to encounter a glycoconjugate containing a single, unique carbohydrate structure.

The problem of glycoforms exists even for **genetically engineered** glycoproteins produced in a single cell line transfected with a single DNA (Sasaki *et al.*, 1988; Spellman, 1990). A famous enzymologist Efraim Racker once said, "One cannot do clean experiments with dirty enzymes." When heterogeneous oligosaccharides are used in glycobiology research, the results will be of course ambiguous.

H.-J. and S. Gabius (Eds.), Glycosciences

ISBN 3-8261-0073-5

Glycolipids have an additional source of hereogeneity. Whereas the glycoproteins mostly contain different oligosaccharide chains on the same peptide backbone, glycolipids can vary also in their fatty acid structures. Natural glycolipids, even those with a single species of carbohydrate chain, often exhibit multiple bands or peaks because of the variations in fatty acids. In such cases, a neoglycolipid containing a simple, well-defined lipid structure may simplify the data interpretation.

2.2 Make as Much as Needed

Isolation, purification, and subsequent complete structural characterization of bioactive oligosaccharides obtained from glycoconjugates are extremely laborious even with modern, sophisticated techniques. However, it has become increasingly clear that the carbohydrate structures involved in biological activities are often limited to a terminal sugar residue or a short oligosaccharide sequence, and thus the total synthetic reproduction of the natural oligosaccharide structures is usually not required for the construction of effective neoglycoconjugates. For example, for mammalian hepatic carbohydrate receptors/lectins (will be henceforth referred to as hepatic lectins), Gal/GalNAc-containing neoglycoproteins (Krantz *et al.*, 1976) are as effective or more effective ligands than any natural glycoprotein ligands. Since a large quantity of Gal/GalNAc-neoglycoproteins (e.g., Gal-BSA, bovine serum albumin containing Gal residues) can be produced with relative ease, a number of informative experiments, which may be impossible or impractical to perform with natural glycoproteins, can be carried out with such neoglycoproteins. Sepharose modified with a high-affinity neoglycoprotein also makes a superior affinity medium.

2.3 Glycoside Cluster Effect

Carbohydrates, unlike peptides or nucleotides, can have branched sturctures. The recognition of carbohydrates in biological systems often involves such multi-branched structures, and the binding affinity is often strongly influenced by the branching pattern as well as the number of such branches. One of the great assets of neoglycoconjugates is that the multivalency can easily be attained even beyond the levels provided by natural glycoconjugates. As mentioned above, neoglycoproteins heavily modified with Gal or GalNAc have higher affinity for mammalian hepatic lectins than the most strongly bound natural glycoproteins. This strong enhancement of affinity by multivalency is termed "glycoside cluster effect."

There are many examples of the glycoside cluster effect documented in literature. The most spectacular example is that of mammalian hepatic lectins. These lectins bind certain mono-, bi- and tri-valent Gal-terminated oligosaccharides with a dramatically increasing affinity in the ratio of approximately 1:1,000:1,000,000 (Lee *et al.*, 1983; Lee, 1992), a numerical increase in the valence resulting in a geometrical increase in affinity. The reason for this tremendous cluster effect resides in the rigid organization of multiple receptor sites in a sterical arrangement complementary to the ligand oligosaccharide structure, so that the terminal Gal residues fit effortlessly into multiple receptor binding sites. The glycoside cluster effect can be achieved with simple synthetic compounds as well. We have prepard a number of small, multi-valent (with respect to sugar component) ligands that contain only the required terminal sugars (Gal or GalNAc for the mammalian and GlcNAc for the avian hepatic lectins) using Asp or Glu as the branching point. By varying the length of the arm connecting sugars to Asp (or Glu), we showed that the structures that allow their sugar residues to straddle across multiple binding sites will produce a large affinity enhancement (Lee *et al.*, 1984; Lee and Lee, 1987). The basis for glycoside cluster effect is discussed in detail in a recent review (Lee and Lee, 1994d). Different carbohydrate-protein interactions give varying degrees of glycoside cluster effect.

2.4 Neoglycoconjugates Containing Analogs

Binding of a carbohydrate ligand by a carbohydrate-receptor or lectin involves both hydrogen bonding and hydrophobic interactions.

Quite often not all the functional groups of a carbohydrate ligand are needed for binding. Neoglycoproteins containing an analog of the natural ligand with a specific structural alteration (e.g., removal of a single hydroxyl group) can provide valuable information on the binding mode of the receptor/lectin. For instance, the mammalian hepatic lectins bind neoglycoproteins containing 6-*O*-methylated Gal residues as well as the Gal counterparts. By preparing many different kinds of analog neoglycoproteins, we discovered that this receptor allows a large substituent at the 6-position of Gal, but does not allow a negatively charged group at this position (Lee *et al.*, 1982). Such informations were invaluable in the later designs of the affinity labeling reagent (Lee and Lee, 1986; Rice and Lee, 1990) for these receptors as well as attachment of fluorescent probes for conformational analysis of a triantennary glycopeptide (Rice and Lee, 1993).

2.5 Changing Personalities

Sometimes neoglycoconjugates are prepared for the purpose of changing physical properties. If the hydrophobic nature of glycolipids is undesirable, they can be transformed into neoglycoproteins more soluble in aqueous solutions (Chatterjee *et al.*, 1985; Tiemeyer *et al.*, 1989). This approach was instrumental in the discovery of a ganglioside receptors in the central nervous system (Tiemeyer *et al.*, 1990). Similarly, neoglycoproteins containing active oligosaccharide fragments (antigenic determinants) of a glycolipid from *Mycobacterium leprae* (See Fig. 1) were proven extremely useful for diagnosis of the diseases caused by *Mycobacteria* (Gaylord and Brennan, 1987).

Transformation in the opposite direction (from glycoproteins to neoglycolipids) can also be profitable. A mixture of neoglycolipids prepared from oligosaccharides of glycoproteins can be conveniently separated by thin layer chromatography (Tang *et al.*, 1985), which offers a sensitive method for detecting potential ligands of lectins. This methodology will be described in more detail in Section 4.1. Hydrophobicity of lipid group can be utilized in another way as shown by Kimata's group (Kimata, 1993). This will be described in detail in Section 6.

In the following sections, we will examine each of the different types of neoglycoconjugates more closely.

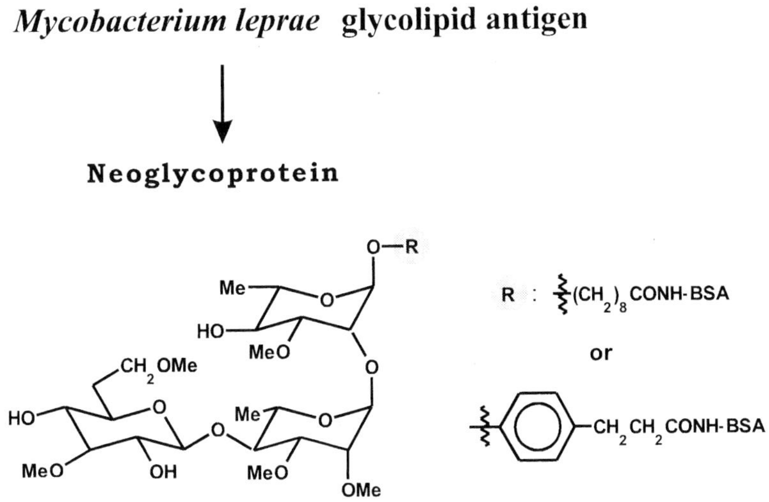

Figure 1. Neoglycoproteins containing Mycobacterium antigenic oligosaccharide

3 Neoglycoproteins

3.1 Side Chains of Proteins Useful in Modification

Proteins are quite suitable for modification with carbohydrate derivatives, because of their good solubility in aqueous solutions and the availability of many different types of side chains. Neoglycoproteins can also be prepared from bioactive proteins such as enzymes, immunoglobulins, or growth factors. This would produce a sort of "double agents" having both the carbohydrate-mediated function as well as the original biological activity of the proteins. Neoglycoproteins can be made even from natural glycoproteins. For example, *Aspergillus* α-amylase, which contains a single oligosaccharide chain of the oligomannose type, can be modified with galactose derivatives (Krantz *et al.*, 1976).

Table I lists the side chains of proteins and the linkage types useful for the formation of neoglycoproteins. The most frequently used functional group for the modification of proteins is the ε-amino group of lysyl side chains. This is because a lysyl side chain is quite flexible and most of them are solvent accessible, so that a large number of sugar residues can be attached with ease. In the next section, we describe only the sugar attachment via the primary amino groups (amino termini and ε-amino groups).

Table 1. Linkage between carbohydrates and proteins

Side chain functional groups		Functional group of carbohydrate derivatives	(Type of reaction)
amino	(Lys, *N*-terminal)	carboxyl	(amidation)
		carbonyl	(alkylation)
		imidate	(amidination)
		isothio-cyanate	(thioureidation)
phenol	(Tyr)	diazo	(diazo coupling)
thiol	(Cys)	maleimido	(Michael addition)
carboxyl	(Asp, Glu, *C*-terminal)	amino	(amidation)

3.2 Modification of Primary Amino Groups

Most of the modifications described below were first explored on proteins, but they can easily be adapted to other carriers that contain amino groups.

3.2.1 Reductive amination

Reductive amination (Gray, 1974) of the carbonyl group of reducing di- or oligosaccharides is one of the gentlest ways to attach the reducing oligosaccharides directly to proteins or other amino-compounds (Lee and Lee, 1992). It is widely used by biochemists and biologists because of the simplicity of its operation. The reaction is typically carried out at near neutrality and at room temperature. The reaction between the carbonyl function of sugar and an amino group of protein initially produces a Schiff's base which is then reduced to produce a permanent covalent bond. The modified amino groups (becoming secondary or tertiary amines) remain positively charged under physiological conditions, so that little perturbation of the tertiary structure of proteins is observed (Lee and Lee, 1980). This is an extremely important factor to consider when bioactive proteins are to be modified. The original method of reductive amination for neoglycoprotein preparation uses sodium cyanoborohydride as the reducing agent (Gray, 1974), but pyridine borane and other amino borane complexes (Cabacungan *et al.*, 1982) are attractive alternatives.

With a ready availability of endo-β-hexosaminidases or glycoamidases (Takahashi and Muramatsu, 1992), the preparation of reducing oligosaccharides from *N*-glycosides of glycoproteins is becoming very practical. Furthermore, with the advent of instrumentation for automated hydrazinolysis, oligosaccharides can be liberated from both *N*- and *O*-glycosides of glycoproteins, so that the attainment of pure reducing oligosaccharides is limited only by the efficiency of subsequent purification.

A shortcoming of the reductive amination is that it inevitably converts the reducing terminus into an acyclic form. Therefore, it is important to know *a priori* that the biological function of the oligosaccharide is not affected by the modification of its reducing end. Another drawback of reductive amination using reducing di- or oligosaccharides is its slow reaction rate. This is caused by the extremely low concentration of the acyclic form of reducing sugars which is the reacting species. Indeed, the use of glycosides possessing an ω-aldehydo group in the aglycon, such as shown in Fig. 2, can facilitate a rapid reductive amination (Lee and Lee, 1980). In another approach to provide an ω-aldehydo group, a polyethylene glycol-based aglycon possessing a cyclic acetal has been proposed (Verez-Bencomo *et al.*, 1991).

In the case of a reducing oligosaccharide obtained from natural sources, one can reduce it to a sugar alcohol, then selectively oxidize it by periodate to generate a new aldehydo group. The condition of periodate oxidation can be controlled in such a way that only the vicinal glycols of the **reduced** sugar (acyclic) is oxidized to generate aldehydo group(s) without affecting the vicinal glycols on the sugar rings. The newly generated aldehydo group remains mostly acyclic because the oxidized terminal sugar chain is usually too short for cyclization. This approach is especially useful for the preparation of neoglycoproteins from *O*-glycosides of glycoproteins which are usually liberated by β-elimination concurrent with reduction to yield oligosaccharide alditols.

3.2.2 Amidination

As shown in Fig. 3, an imidoester-containing thioglycoside, obtained by treating the corresponding cyanomethyl thioglycoside with sodium methoxide, will react with a protein amino group to form an amidino group, which, like the parent primary amino group, is positively charged under physiological conditions. The cyanomethyl thioglycosides of mono- or disaccharides can be prepared fairly easily; however, the preparation becomes more cumbersome for larger oligosaccharides. The advantages of the thioglycosidic linkage are its resistance to hydrolytic enzymes

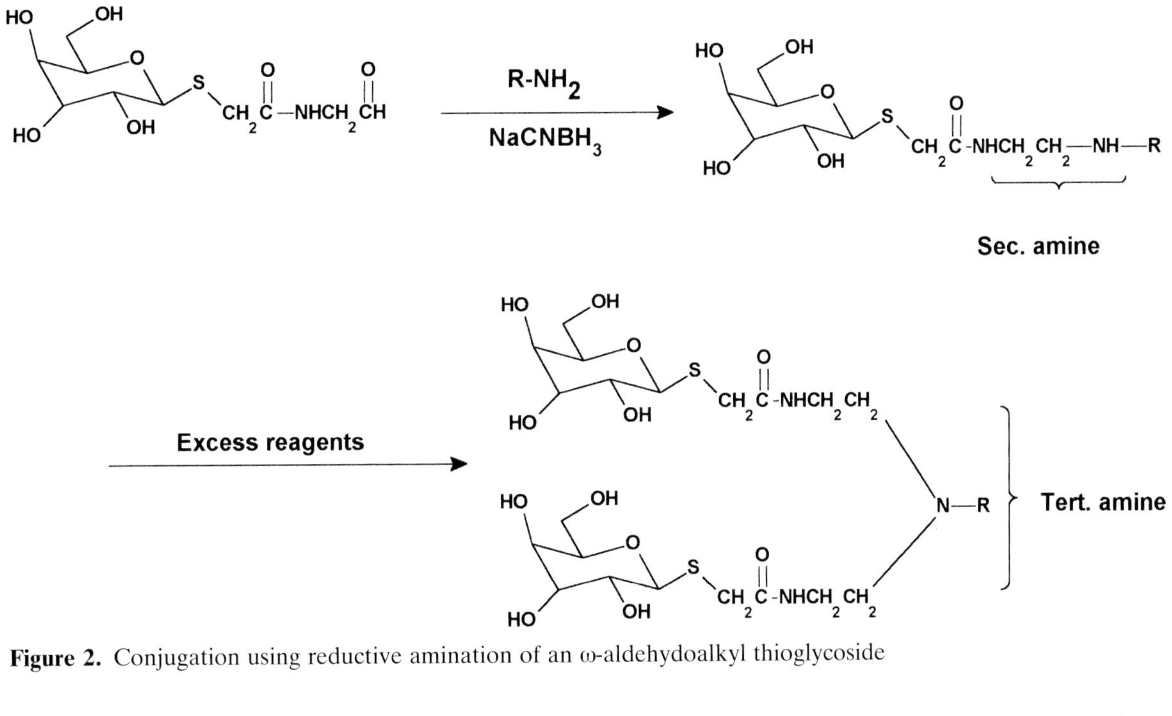

HO OH O S $CH_2C(=O)-NHCH_2CH(=O)$ HO OH

R-NH_2 / $NaCNBH_3$

HO OH O S $CH_2C(=O)$-$NHCH_2CH_2$—NH—R HO OH

Sec. amine

Excess reagents

HO OH O S $CH_2C(=O)$-$NHCH_2CH_2$ HO OH

N—R Tert. amine

HO OH O S $CH_2C(=O)$-$NHCH_2CH_2$ HO OH

Figure 2. Conjugation using reductive amination of an ω-aldehydoalkyl thioglycoside

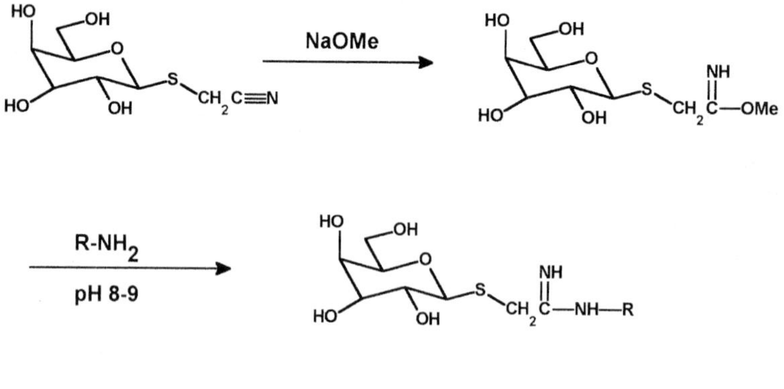

R: Protein

Figure 3. Conjugation of a sugar imidate to protein

and the ease of specific chemical hydrolysis by mercuric ion (Krantz and Lee, 1976). The thioglycosidic linkage was also employed in the earlier example of the ω-aldehydo glycosides (Fig. 2).

3.2.3 Acylation

Acylation of amino groups of proteins is also an easy reaction, but it changes positively charged amino groups into neutral amide groups. Such a change may alter physical and biological properties of the proteins modified, so that the resultant neoglycoproteins may not be suitable for the purpose intended.

Attaching carbohydrate groups by acylation can be done with glycosides containing activated carboxyl groups, such as active esters (Lee and Lee, 1992; Andersson and Oscarson, 1993). Or, one can attach an unactivated carboxylic acid to the amino groups of a protein by using a suitable water-soluble carbodiimide or other coupling agents (Andersson and Oscarson, 1993). A carboxyl function can be created by oxidizing the reducing end of di- or oligosaccharides with bromine or other mild oxidizing agents to form aldonic acids (or their lactones). Synthetic glycosides having an ω-ester group, such as 7-methoxycarbonylhexyl aglycon, can be converted to the corresponding azido group for coupling to the protein amino groups (Lemieux *et al.*, 1975). The ester function can also be hydrolyzed to generate the carboxylic acid to be coupled by the carbodiimide method.

Another useful method of attaching sugars *via* amino groups is to perform thiocarbamylation of the protein amino groups. Commonly available *p*-aminophenyl glycoside is converted to isothiocyanate by reacting the anilino group with thiophosgene. The resulting isothiocyanate reacts readily with ε-amino groups (McBroom *et al.*, 1972). An undesirable feature of this method, in addition to neutralizing the positively charged amino groups, is the introduction of an aromatic ring with each sugar attachment, dramatically increasing hydrophobicity, often causing precipitation of the protein.

3.3 Conjugation of Polysaccharides to Proteins

The method of selectively oxidizing the reducing terminal sugar with periodate to generate an aldehydo function has been applied successfully to polysaccharides, such as hyaluronic acid. The oxidized hyaluronic acid was first conjugated to 1,6-diaminohexane by reductive amination, and the resulting conjugate was then attached to proteins (Raja *et al.*, 1984). An interesting way of incorporating aldehydic functions to a polysaccharide is to react the latter with chloroacetaldehyde dimethyl acetal. After unmasking of the acetal, the modified polysaccharide is readily conjugated to proteins by reductive amination (Bogwald *et al.*, 1986). Since the initial reaction of chloroacetaldehyde dimethyl acetal with polysaccharide lacks selectivity, this method suffers from randomness of the modification sites on the polysaccharides.

Certain polysaccharides (e.g., heparin and chitosan) possess free or potentially free amino groups. Treatment of such polysaccharides with nitrous acid yields 2,5-anhydro-D-mannose residues with a concomitant fragmentation of the polysaccharide due to cleavage of the affected glycosidic bonds. Oligosaccharides containing this active aldehydo group at the reducing terminus can be used for conjugation to proteins or amino compounds by reductive amination. This was the approach used to prepare low-mass heparin derivatives possessing anticoagulant activities (Malsch *et al.*, 1994).

Polysaccharides from certain microorganisms have been effectively converted into vaccines by partial hydrolysis, and reductive amination with appropriate protein carriers (Jennings and Sood, 1994).

3.4 Use of Enzymes

Total chemical synthesis of oligosaccharides, even as small as a trisaccharide, is quite tedious. The use of appropriate enzymes to modify mono-, di- or oligosaccharide is often a profitable approach for the preparation of more complex sugar structures. These reactions can be carried out on free sugars or sugars already conjugated to proteins or other carriers. The strength of enzymes in constructing oligosaccharides is their exquisite specificity. For example, when a glycosyltransferases is used for glycosylation, the nature of the acceptor sugar, the position of the attachment site, as well as the anomeric configuration are quite specific. There are also enzymes which allow transfer of the whole preexisting glycan moieties.

3.4.1 Use of glycosyltransferases

With the exception of transglutaminase to be described below, a direct enzymatic attachment of a sugar unit to peptide side chain has not been reported. However, a stepwise addition of glycosyl unit to preformed neoglycoproteins are quite feasible with glycosyltransferases.

Paulson's group (Paulson *et al.*, 1984) used α-2,3- and α-2,6-specific sialyltransferases to resialylate red blood cells that had been previously desialylated enzymatically. These reconstructed cell surface glycoconjugates containing only a single specific type of sialyl linkages enabled them to probe specificities of viral binding. In another example of reconstruction of glycoconjugates, Berman *et al.* (Berman *et al.*, 1985; Berman *et al.*, 1986; Berman and Lis, 1987), used a galactosyltransferase to incorporate ^{13}C-labeled Gal onto GlcNAc for binding studies by CMR. Chemical synthesis of such ^{13}C-labeled oligosaccharide chains would be extremely laborious.

Hill and coworkers added Gal and L-Fuc sequentially to GlcNAc-BSA (which can be prepared by a simple procedure (Lee *et al.*, 1976)) using a galactosyltransferase and one of two different L-Fuc-transferases to produce two isomeric L-Fuc-containing neoglycoproteins for their studies of fucose-binding proteins (Lehrman *et al.*, 1986a; Lehrman *et al.*, 1986b) (Fig. 4). A similar enzymatic approach of successive galactosylation and sialylation has also been applied to a polyacrylamide polymer containing GlcNAc pendants (Nishimura *et al.*, 1994). This type of chemoenzymatic approach, also employed by others (Sabesan and Paulson, 1986; Sabesan *et al.*, 1992), is expected to grow in popularity with expanding availability of genetically engineered glycosyltransferases.

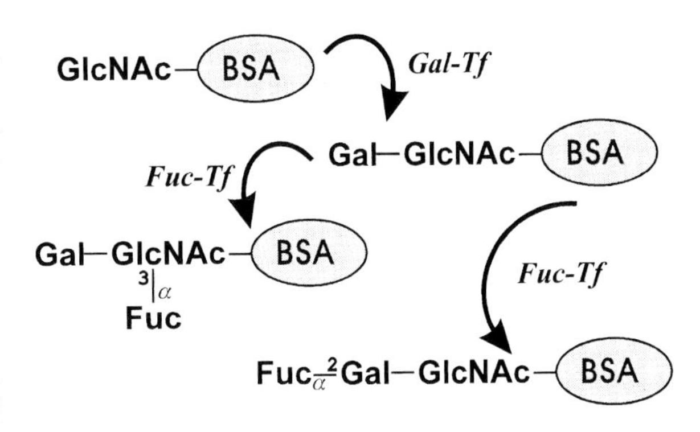

Figure 4. Modification of neoglycoproteins with glycosyltransferases

Modification of the sugar residues on neoglycoproteins or natural glycoproteins needs not be with the natural sugar residues. Many glycosyltransferases can be tricked to use sugar nucleotides containing unnatural sugars as donor substrates. For example, a fucosyltransferase can transfer a derivative of GDP-fucose having the blood group B trisaccharide structure linked via a spacer arm (Srivastava *et al.*, 1992) (Fig. 5). Modification of cell surface with such a fucose-derivative ("sneaky-B") resulted in a change in serological properties of the cells. *N*-Acetylneuraminic acid (Neu5Ac) modified with a fluorescent or photoactivatable group at carbon-9 can be activated into the corresponding CMP-derivative (with CMP-Neu5Ac synthetase) which is transferred onto appropriate oligosaccharide acceptors by several of the sialyltransferases (Gross *et al.*, 1989).

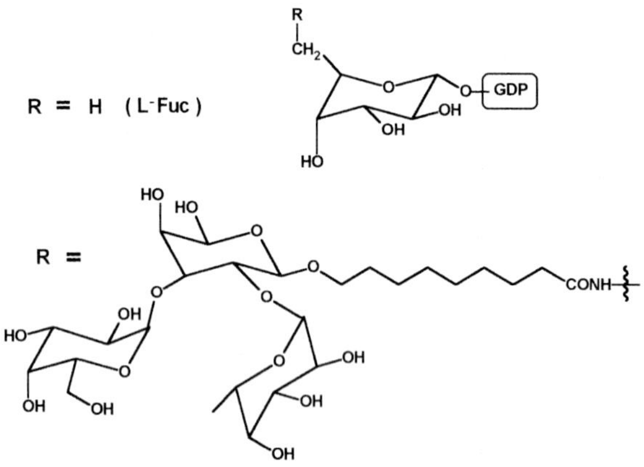

Figure 5. GDP-derivative of a fucose modified to carry a "B-trisaccharide" at C6 to be used by fucosyltransferase

3.4.2 Use of glycosidases in transglycosylation

Hydrolytic activity of glycosidases can be reversed to some extent to obtain di- or oligosaccharides (Ichikawa *et al.*, 1992). Due to a heavily favored equilibrium toward hydrolysis, the true reversal of hydrolysis is not a very profitable process. However, the yield can be greatly increased, if one harnesses the ability of glycosidases to transfer a glycosyl unit from certain suitable glycosides (e.g., phenyl glycoside) to an acceptor monosaccharide.

The main advantages of glycosidases in comparison to glycosyltransferases are lower costs for enzymes as well as substrates (no need to use sugar nucleotides). Glycosidase-catalyzed oligosaccharide formations are usually specific with respect to the anomeric configuration but lack stringent positional specificity.

3.4.3 Use of transglycosylases in transglycosylation

Some glycosidases manifest surprisingly high transglycosylase activities. A good example is *Trypanosoma cruzi* trans-sialylase, which is capable of transferring an α-2,3-linked Neu5Ac residue to a galactosyl residue (See Fig. 6). This enzyme is interesting in that its requirement for the substrate structure is quite strict as in the case of sialyltransferases, both the donor substrate and

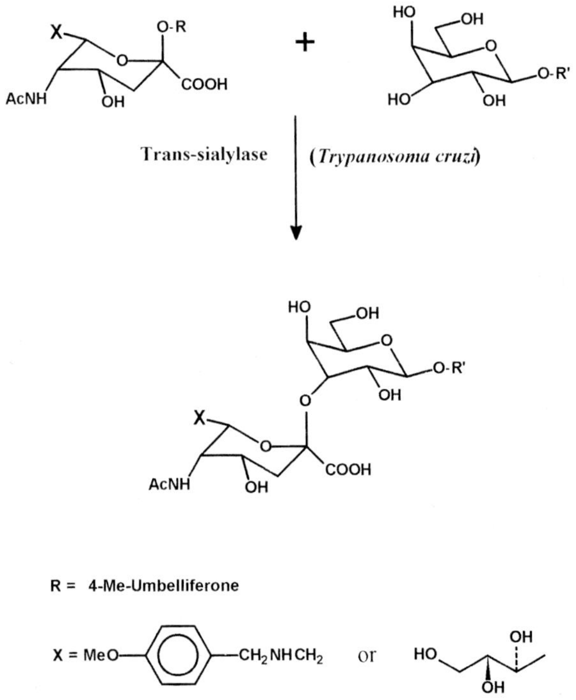

Figure 6. Transfer of Neu5Ac and derivatized Neu5Ac by *Trypanosoma cruzi* trans-sialylation

the product being of α-2,3-linkage to a Gal residue. The advantage is that the donor substrate is a ketoside of Neu5NAc rather than CMP-Neu5NAc (which is more laborious to prepare) which is required by the sialyl transferases. Interestingly, 4-methylumbelliferyl or *p*-nitrophenyl ketoside of Neu5NAc (4-MU-Neu5NAc and PNP-Neu5NAc, respectively), common chromogenic substrates for sialidase activity measurement, can also serve as glycosyl donors. Since chemical synthesis of these ketosides is relatively simple, an isotopically labeled Neu5NAc donor (e.g., PNP-Neu5NAc labeld with ^{13}C at *N*-Ac group) can easily be prepared, which is then used to incorporate the tagged Neu5NAc into oligosaccharide chains. Recently, we showed that 4-MU-Neu5NAc derivatives modified at the exocyclic side chain can serve as donor substrates (Lee KB and Lee YC, 1994) (Fig. 6).

The transglycosidase activities of endo-β-*N*-acetyl-glucosaminidase from *Arthrobacter* and *Mucor* will be described later in the section of glycopeptides (Section 3.6).

3.4.4 Use of transglutaminase

Yan and Wold (Yan and Wold, 1984) developed an ingenious method of attaching glycopeptides by the action of transglutaminase. For example, the γ-$CONH_2$ groups of Gln in β-casein can be modified with the α-amino group of Asn-oligosaccharides from ovalbumin by this enzyme. Of course, all the amino groups in β-casein must be masked first so that cross-linking of the proteins would not occur. In the example cited, conjugation of four Gln sites with Asn-oligosaccharides was achieved.

3.5 Glycoproteins of Non-Covalent Attachment

The biotin-avidin technology is well established in biological sciences and biotechnology. The interaction between biotin and avidin/streptavidin is so strong that it can almost be considered as a covalent bond. In a unique application of this interaction, Wold's group (Chen and Wold, 1984) let biotinylated 6-aminoalkyl glycosides bind to avidin or streptavidin, which has four biotin-binding sites. Typically three biotinylated glycosides or glycopeptides can be bound readily, but the fourth insertion is more difficult. This type of neoglycoproteins, though limited in the number of oligosaccharide being attached, the sites of attachment are geometrically well defined (Shao *et al.*, 1994).

In a recent development, an oligosaccharide mixture was derivatized with 2,6-diaminopyridine-derivatized biotin. This derivatization allowed excellent monitoring and separation of the oligosaccharides by reverse-phase HPLC (Rothenberg *et al.*, 1993). The separated oligosaccharide derivatives can be used for construction of neoglycoproteins *via* the biotin-avidin interaction, or used in a number of other ways.

A similar non-covalent attachment based on the enzyme-coenzyme complexing has also been examined. For example, aspartate aminotransferase complexed with phosphopyridoxylated Asn-$GlcNAc_2Man_5$ (Chen and Wold, 1984) has been used in the study of binding by alveolar macrophages. Unfortunately, this type of complex is not stable and its use may be of limited scope.

3.6 Synthetic Glycopeptides

Synthesis of glycopeptides is more difficult than the conventional peptide synthesis. This is because the protection-deprotection reactions usable in the peptide synthesis can cause serious problems to the protective groups commonly used for carbohydrates. Recent technical innovations have solved many of these problems, and glycopeptide synthesis is actively pursued (Meldal, 1994). The current preference is to use 9-fluorenylmethyloxycarbonyl (Fmoc) group for *N*-protection and pentafluorophenyl (PFP) group as the activating group for the carboxylic acid. Usually, Asn-GlcNAc or Ser/Thr-GalNAc as a building block is incorporated into the peptide using solid-phase synthesis, and a sugar chain is built onto the monosaccharide stub by the aid of enzymes.

Some endo-β-hexosaminidases show a great promise in the *en bloc* transfer of oligosaccharides rather than one single monosaccharide at a time done by glycosyltransferases. For example, endo-A from *Arthrobacter* (Fan *et al.*, 1995; Takegawa *et al.*, 1995) and endo-M from *Mucor* (Kadowaki *et al.*, 1991; Yamamoto *et al.*, 1993) can use as donor the high-mannose type and the complex type oligosaccharides, respectively, and transfer them to acceptors, which can be a free mono- or oligosaccharide or a peptide containing terminal GlcNAc residues. We have successfully raised the yield of transglycosylation catalyzed by endo-A to more than 90 % by inclusion of some organic solvent in the reaction medium (Fan *et al.*, 1995).

4 Neoglycolipids

4.1 Neoglycolipids for Thin Layer Chromatographic Separation

As mentioned earlier (Section 3.2.1), *O*-glycosides in glycoproteins are most frequently liberated as reduced oligosaccharides by alkaline β-elimination under reducing conditions. Such reduced oligosaccharides can be selectively oxidized with periodate to generate terminal aldehydic group(s) (Fig. 7). Conjugation of this aldehydic group with distearylphosphatidyl ethanolamine by reductive amination yields neoglycolipids (Tang *et al.*, 1985) amenable to thin layer chromatographic separation. When a mixture of oligosaccharides is converted to neoglycolipids, separated by thin layer chromatography and then probed by the overlaying technique analogous to that originally developed by Magnani *et al.* (Mag-

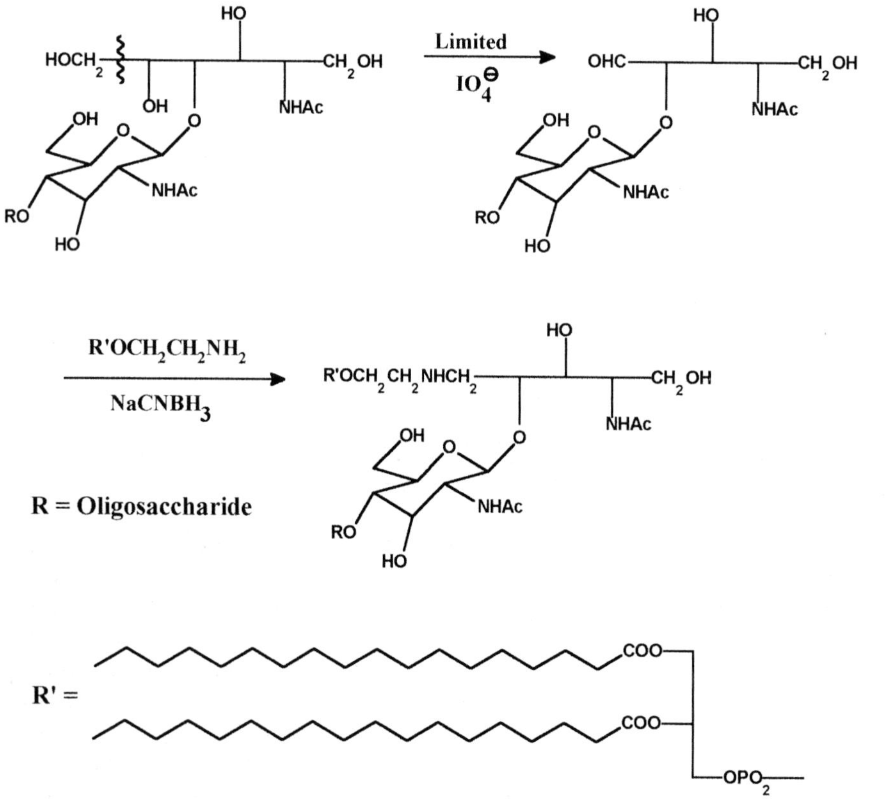

Figure 7. Preparation of neoglycolipids via periodate oxidation and reductive amination

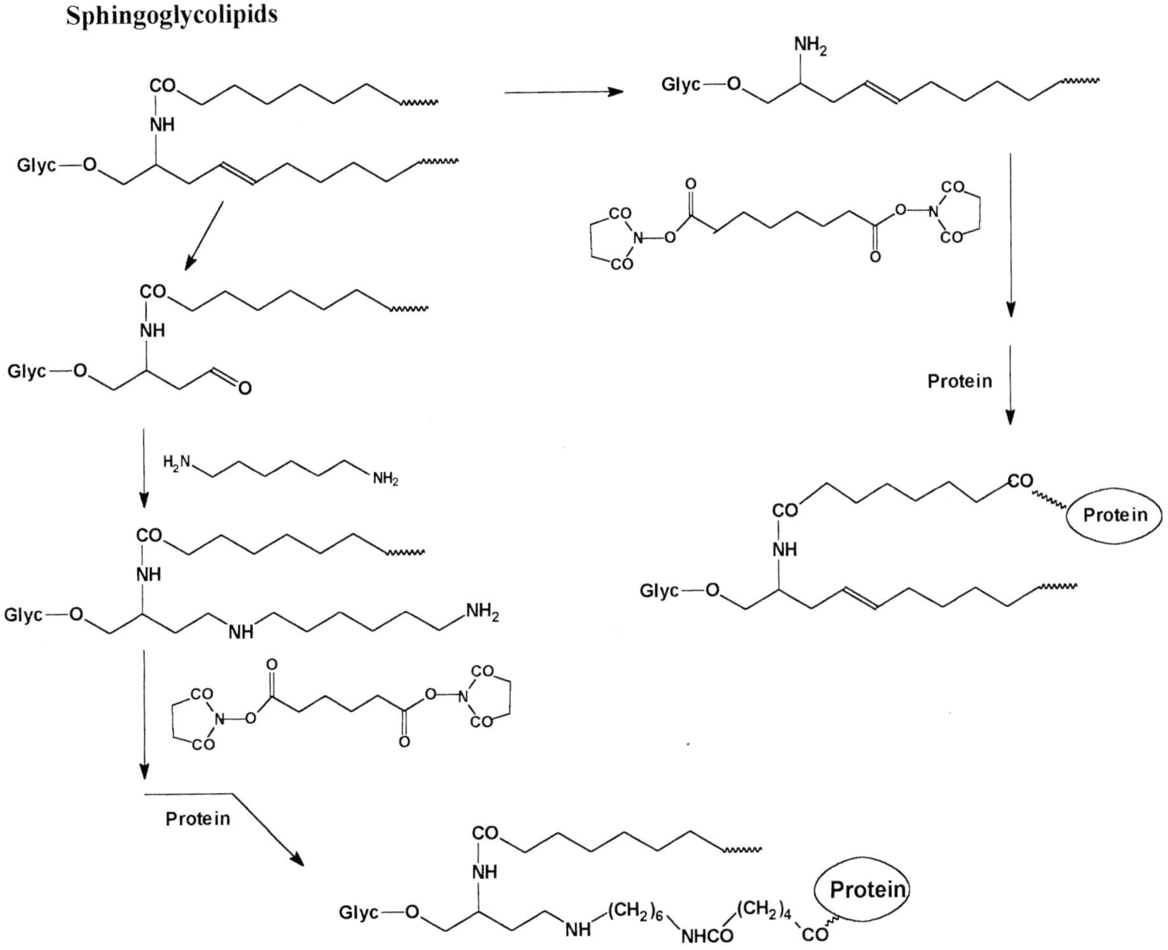

Sphingoglycolipids

Figure 8. Conversion of sphingolipids into neoganglioproteins

nani, 1987), binding specificity of a carbohydrate-binding protein can be examined quickly. Usefulness of this methodology has been demonstrated in the examination of the binding specificity of serum mannose-binding protein (Childs *et al.*, 1990) and E-selectin (Yuen *et al.*, 1992). The added advantage of these neoglycolipids is that they are also suitable for mass spectrometric examination (Stoll *et al.*, 1990).

4.2 Transglycosylation of Ceramide Glycanase

A ceramide glycanase from leech, which hydrolyzes the linkage between the ceramide and the glycan chain, has been found to have high transglycosidase activity. The enzyme can transfer the glycan of G_{M1} for example to simple long-chain (C>6) 1-alkanols as well as to ω-functionalized 1-alkanols. The ω-function (e.g., *t*-BOC or trifuloroacetyl protected amino groups) after deprotection can then be used to conjugate to proteins (Li *et al.*, 1991; Li and Li, 1994).

5 Glycolipids into Neoglycoproteins and Vice Versa

The method described in Section 4.1 demonstrates the advantages of using neoglycolipids. However, the reverse transformation of glycolipids into neoglycoproteins sometimes offers different kinds of advantage, such as improved solubility and higher ligand density (See also Section 2.5.).

The oligosaccharide chain of a ganglioside can be incorporated into a protein *via* lysoganglioside in the following fashion. Lysogangliosides, produced by alkaline hydrolysis of amide bonds in the gangliosides, can be treated in such a way as to selectively re-*N*-acetylate the amino sugars to restore the original glycan structure, leaving the amino group of sphingosine available for modification (e.g., acylation). The use of an ω-functionalized acylating reagent will extend the linking arm between the glycan and protein. The spacer added improves the efficiency of the conjugation and the accessibility of the glycans on the newly formed neoglycoprotein. A convenient way of accomplishing this is to modify a lysoganglioside with a bifunctional reagent (Yasuda *et al.*, 1987), such as bis(sulfosuccinimidyl)suberate (Tiemeyer *et al.*, 1989) (Fig.8).

Alternatively, oxidation of the double bond in the ceramide portion leads to a glycosyl derivative containing a terminal aldehydo group that can be used for coupling to proteins *via* an extention arm (e.g., 1,6-diaminohexane) (Fig. 8). These neoglycoproteins are sometimes called **neoganglioproteins**, and have been instrumental in demonstrating the presence of a membrane receptor for ganglioside in association with the central nervous system myelin (Tiemeyer *et al.*, 1990). Likewise, Pacuszka and Fishman (1990) used neoganglioproteins to prove that the interaction of cholera toxin with G_{M1} requires both the oligosaccharide chain of G_{M1} and the lipid moiety.

6 Neoproteoglycans

Kimata and coworkers (Kimata, 1993) oxidized the terminal xylose of chondroitin sulfate into xylonic acid and conjugated it to phosphatidyl ethanolamine by amide formation (Fig. 9). A term **neoproteoglycans** was coined to designate this type of neoglycoconjugates, although it should perhaps be called a type of neoglycolipid. Neoproteoglycans can be coated on the polystyrene well surface for the studies of cell adhesion. A clear demonstration of the role of chondroitin sulfates as inhibitors of cell adhesion was achieved by using this type of neoproteoglycans.

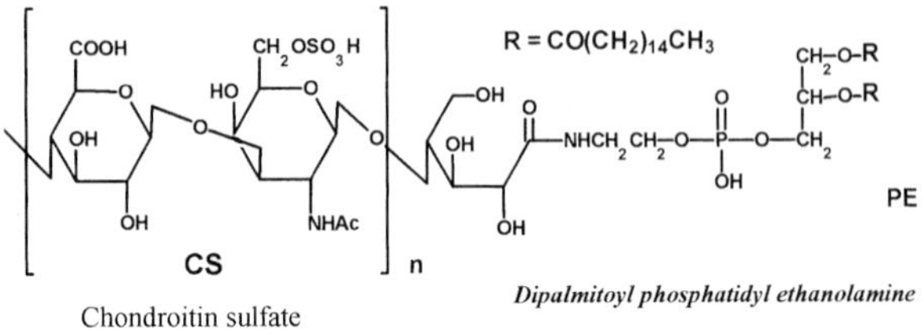

Figure 9. Neoproteoglycan formation from chondroitin sulfate

7 Neoglycopolymers

Nature offers infinite varieties of polysaccharides, some of them are linear but most are branched. Often, biological interactions of proteins with polysaccharides are through the side chains of branched polysaccharides. This is perhaps the reason that most of the synthetic glycopolymers recently reported are made of non-carbohydrate backbones with oligosaccharide pendants. It can be assumed that most, if not all, of the carbohydrate segments in such a polymer behave identically.

There are many advantages for preparing polymers containing side chains of structurally defined carbohydrates. (i) Polymerization is the simplest method to increase the valency, so that the glycoside cluster effect (see Section 2.3 above) can be attained easily; (ii) The backbone material can be inert (e.g., polyacrylamide) and thus the problem of instability or undesirable interactions inherent in natural polysaccharides can be eliminated; (iii) It is often possible to control the size of polymer and the density of the pendant carbohydrate groups by design; (iv) In some applications, neoglycopolymers facilitate isolation of bound proteins.

The most often used backbone material is polyacrylamide (Schnaar, 1984; Nishimura *et al.*,

1994; Schnaar, 1994). The basic scheme for preparing such polymers is illustrated in Fig. 10. Although ω-unsaturated glycosides (e.g., allyl and *n*-pentenyl glycosides) can be incorporated into polyacrylamide (Nishimura *et al.*, 1990), the efficiency of incorporation is much lower than that of ω-acrylamido-derivatives.

Another popular backbone material is polystyrene (Kobayashi *et al.*, 1985; Kobayashi *et al.*, 1986). The neoglycopolymers built on polystyrene have an advantage of greater adhesiveness towards the polystyrene-based multi-well plates. This fact was taken advantage of in the hepatocyte culturing systems (Pless *et al.*, 1983; Kobayashi *et al.*, 1986). Thus, the lactose-bearing polystyrene (PVLA) derivatives adhere tightly to the plastic surface of the microtiter wells, and the terminal galactose residues allow tight adhesion of hepatocytes to these plates. Cell life is prolonged and some cell functions are improved when cultured on such a surface.

Glycosyl groups are most usually attached to backbones *via* a spacer arm. The length of the spacer arm can be quite important in manifestation of biological activities. When acrylamide polymers containing 6-aminohexyl galactosides separated from the backbone by 0, 1 and 2 units of 6-aminohexanoic acid (AHA) were compared, hepatocytes appear to adhere with greater efficiency to the gels with AHA spacer arms than to the gels without them (Pless *et al.*, 1983; Schnaar, 1984). During isolation of certain lectins, it was found that the length of spacer arm caused different affinity for different isolectins (Gabius, 1990).

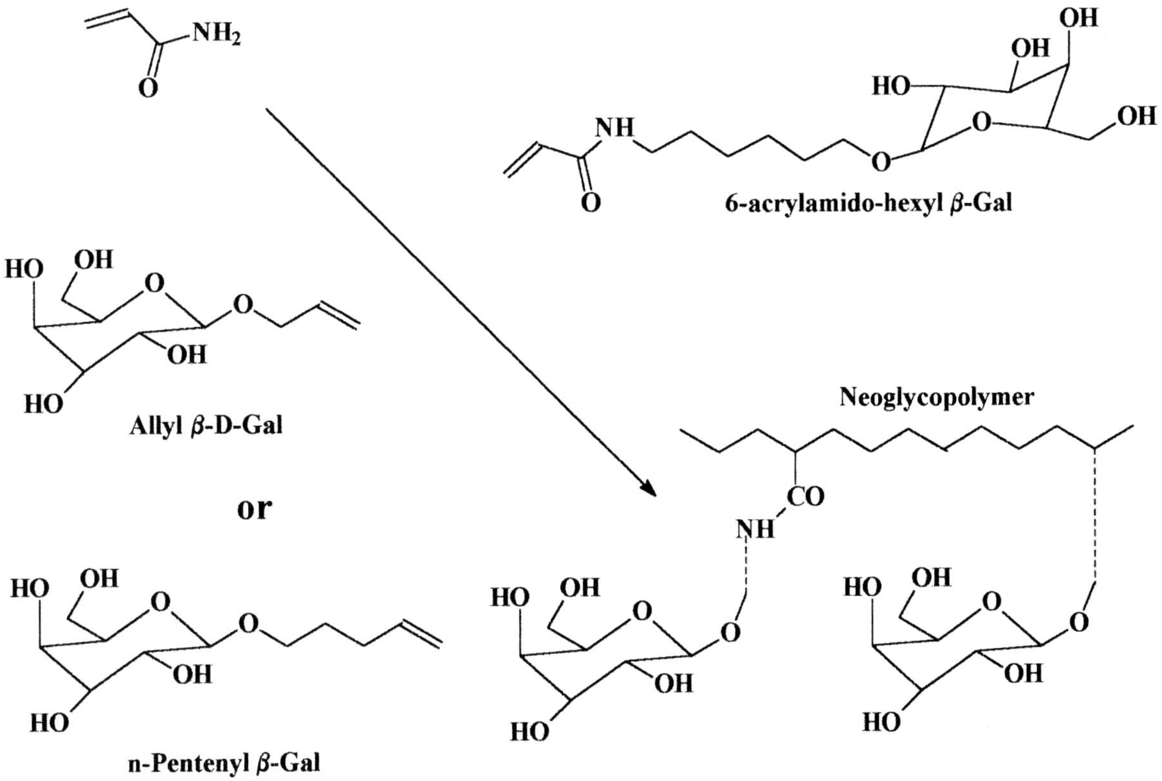

Figure 10. Use of glycosides with ω-unsaturated aglycon in neoglyopolymer preparation

8 Derivatives Useful for Preparation of Neoglycoconjugates

Most of the preparation of neoglycoconjugates begins with the preparation of a carbohydrate derivative suitable for conjugation. This can be from a natural source or totally synthetic or the combination of the two. The nature of target neoglycoconjugates to be prepared dictates the form of such derivatives.

8.1 Glycosides

Glycosides are the most versatile form of derivatives for making mono- or oligosaccharides suitable for conjugation. Glycosides can be *O*-, *N*- or *S*-linked, but the aglycon must be functionalized at the terminal position to enable conjugation or polymerization.

8.1.1 ω-Aminoalkyl glycosides (Chipowsky and Lee, 1973; Weigel *et al.*, 1979).

The terminal amino group in an aglycon can be used to react directly with carboxyl groups of proteins or other carriers. Alternatively, it can be readily transformed into other useful functionalities. For instance, succinylation will produce a carboxyl function and acryloylation will convert it into a readily polymerizable unit (See Section 7). Several ω-amino alkanols are commercially available. A recent report describes polyethyleneglycol glycosides whose ω-terminal has been transformed into an amino group (Bertozzi and Bednarski, 1991, see Fig. 12 for structures.)

8.1.2 ω-Unsaturated glycosides

ω-Unsaturated glycosides are versatile derivatives of carbohydrates, useful in the preparation of neoglycopolymers (See Section 7) and other conjugates. Allyl and *n*-pentenyl glycosides, which can be used as co-monomers as illustrated in Fig. 10, are easily prepared by glycosylating (e.g., using 1-bromo-derivatives) the respective alcohols (Mootoo *et al.*, 1989). *n*-Pentenyl glycosides (Mootoo *et al.*, 1989) were originally developed for activation of anomeric carbon, and have been utilized successfully in synthesis of some complex oligosaccharides. Nishimura *et al.* (1990) used synthetic glycosides with terminal-unsaturated aglycons of C3, C4, C5, and C11 in length for polymerization. Gel layers prepared with longer aglycons were found to be better for rat hepatocyte culture (Kobayashi and Akaike, 1990).

However, for higher efficiency of polymerization, glycosides with ω-acrylamido group is preferable. As mentioned above, such glycosides are most conveniently prepared by acryloylation of ω-aminoalkyl thioglycosides (Schnaar *et al.*, 1993). A convenient way of synthesizing thioglycosides with a terminal acrylamido group is to react one of the two double bonds in *bis*-methylene- diacrylamide with 1-thiosugars via Michael addition (Lee *et al.*, 1979). Conversely, one can generate an ω-amino aglycon from ω-unsaturated aglycon. For instance, allyl glycosides were modified with cysteamine by Michael addition to acquire a terminal amino group (Lee and Lee, 1974).

8.1.3 Dibromoalkyl glycosides

Dibromoalkyl (e.g. dibromoisopropyl, DBI) glycosides are versatile compounds for preparation of a wide variety of neoglycoconjugates (Magnusson, 1986; Magnusson *et al.*, 1990). For example, a DBI glycoside (Fig. 11) can react with thiols to yield *bis*-sulfide. Alternatively, aminolysis of the bromo groups yields a diamino compound which can be further modified. The strength of the DBI glycoside is that it can generate two-pronged neoglycolipids which more closely resemble natural glycolipids than those described in Section 4. Such two-pronged aglycons are also found to be useful for immobilization of the glycosides onto silica gel surface (Magnusson *et al.*, 1990).

8.1.4 Glycosides with ω-aldehydo or ω-carboxyl group

Thioglycosides containing terminal acetal (Fig. 2) have been used effectively for neoglycoprotein

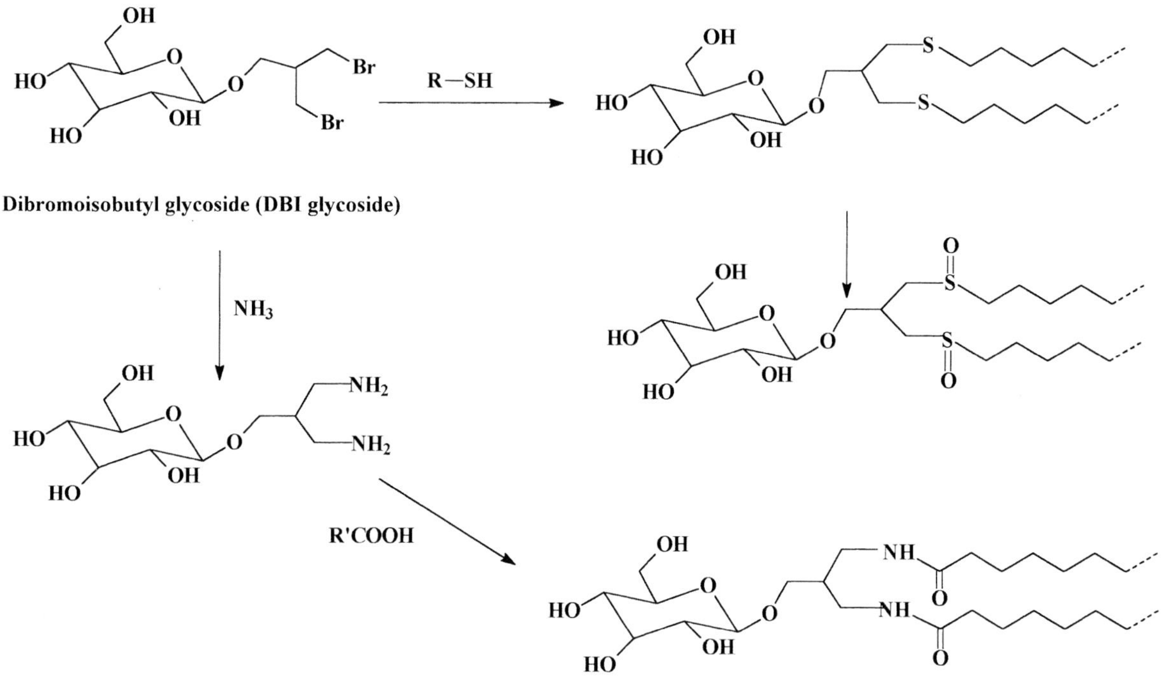

Dibromoisobutyl glycoside (DBI glycoside)

Figure 11. Versatile transformation of dibromoisobutyl glycoside

preparation (Lee and Lee, 1980). In contrast to imidate derivative (Krantz *et al.*, 1976), which allows attachment of one thioglycoside to each reactive amino group, an aldehydo-derivative can potentially attach two glycosides to each amino group. This would result in a higher sugar density of the resultant neoglycoprotein.

Glycosides containing an ω-carboxyl aglycon are widely used for generating synthetic antigens (Lemieux *et al.*, 1975). The optimum length of the aglycon chain was determined to be 8-carbon. Polyethyleneglycol glycosides containing an ω-cyclic acetal (Verez-Bencomo *et al.*, 1991) and polyethyleneglycol glycosides having a terminal carboxyl group (Andersson *et al.*, 1993) have been described recently. The structures of these glycosides are shown in Fig. 12.

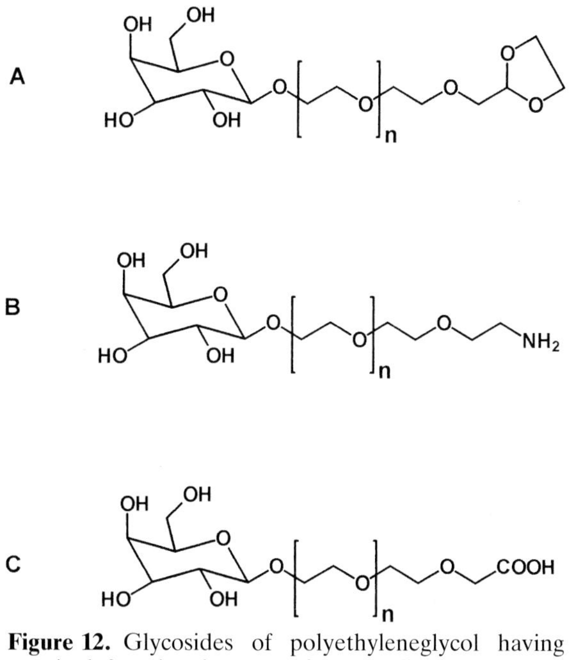

Figure 12. Glycosides of polyethyleneglycol having terminal functional group of: A. cyclic acetal group (Verez-Bencomo *et al.*, 1991); B. an amino group (Bertozzi and Bednarski, 1991); C. a carboxyl group (Andersson *et al.*, 1993)

8.2 Glycosylamines and Glycamines

8.2.1 Glycosylamine derivatives

There has been a considerable effort to chemically synthesize "natural" glycopeptides, i.e. peptides containing *N*-glycosides via β-amide of asparagine (Asn-Glyc) (Garg and Jeanloz, 1985; Meldal, 1994). The classic method of preparing an Asn-linked oligosaccharide is to start with glycosyl halide, which is converted to glycosyl azide. Hydrogenolysis of the azido group yields glycosylamine, which is coupled to β-COOH of suitably protected aspartic acid derivative to yield Asn-Glyc.

In a simpler method (Kallin *et al.*, 1986), reducing oligosaccharides are kept in saturated ammonium bicarbonate solution for several days, with daily supplemental addition of solid ammonium bicarbonate, and after removal of excess ammonium bicarbonate, the glycosylamine formed is purified on a column of cation exchange resin. Although glycosylamines are unstable as such, acylation with Asp-derivatives to form Asn-oligosaccharides will stabilize them.

In one of the newer schemes (Manger *et al.*, 1992), glycosylamine is first chloroacetylated, which allows the resultant aglycon to be extended further. In another recent scheme (Tamura *et al.*, 1994), glycosylamines from an oligosaccharide mixture are tyrosinated, and then separated by RP-HPLC. The addition of a tyrosyl group improves separation on RP-HPLC columns, and permits iodination as well (Fig. 13).

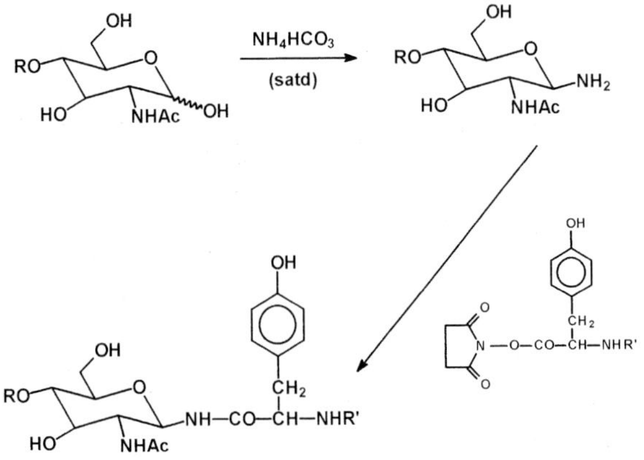

Figure 13. Formation of glycosylamine from a reducing sugar and its derivatization

8.2.2 Glycamine derivatives

When a reducing sugar is reductively aminated, the sugar structure is fixed in a linear form, and the product is called glycamine. This is in contrast to a glycosylamine which retains the cyclic sugar structure. Taking advantage of the fact that an anilino group reacts more readily in reductive amination, Kallin *et al.* (Kallin *et al.*, 1986; Kallin, 1994) used 4-trifluoroacetamido-aniline as the amino component (Fig. 14 see page 71). After the reductive amination, the trifluoroacetyl group is removed to expose a new aryl amino group, which can then be diazotized or converted to isothiocyanate for conjugation to proteins. The efficiency of sugar derivatization with this compound is quite high, though the introduction of aromatic ring is not desirable for some experiments.

Recently, we developed an efficient method of preparing glycamine derivatives using benzylamine (Yoshida and Lee, 1994). This method allows efficient formation of glycamine even from minute quantities of reducing oligosaccharide.

8.3 Attachment of Glycopeptides with a Heterobifunctional Reagent

If a glycopeptide with structurally well-defined oligosaccharide chain can be obtained, a direct conjugation of such a glycopeptide can alleviate the effort of releasing and further modifying the reducing oligosaccharide for conjugation. Moreover, either the amino or the carboxyl terminus of the glycopeptide can be utilized for conjugation.

We developed a heterobifunctional reagent (Lee *et al.*, 1989) which contains an acyl hydrazide at one end for coupling to the amino group of a glycopeptide and an aldehydo group (masked as acetal) on the other end to be used for conjugation to proteins (Fig. 15 see page 72). This reagent can conjugate glycopeptides to proteins with a high efficiency (ca. 90 % conjugation when the protein amino group is in a large excess) even at submicromole levels.

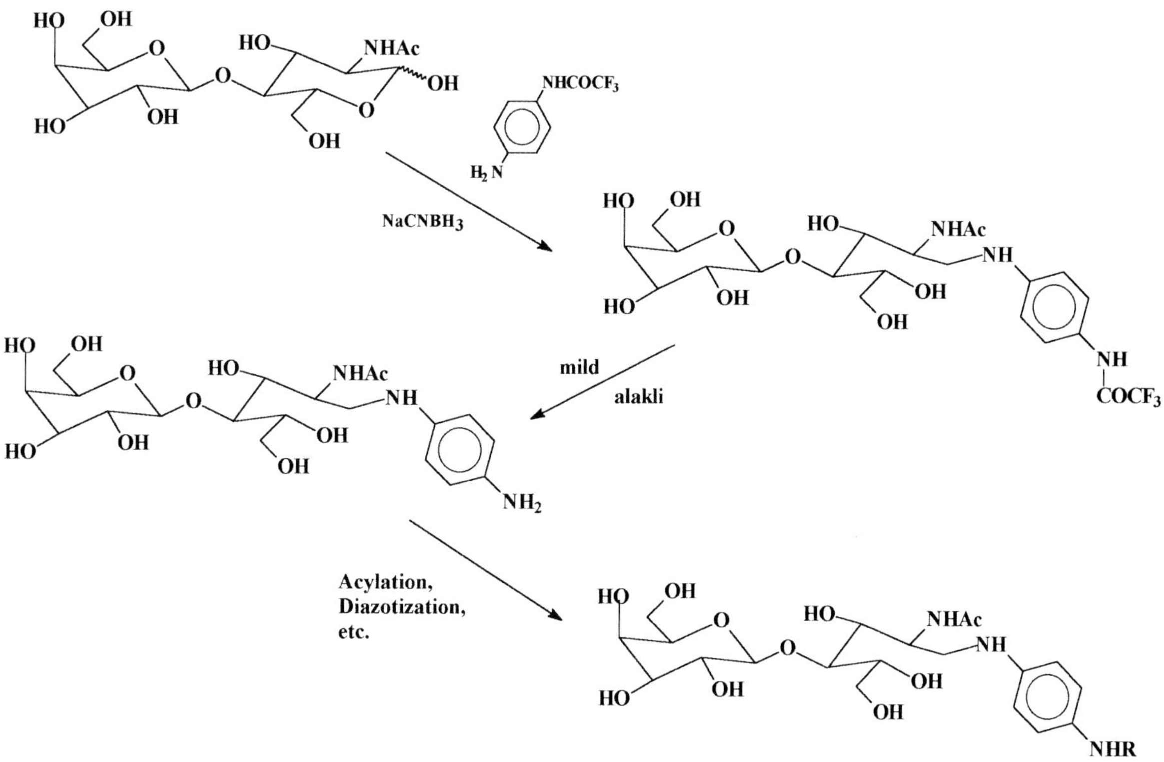

Figure 14. Derivatization of oligosaccharide via glycamine formation

9 Applications of Neoglycoconjugates

The purpose of preparing neoglycoconjugates is to use them in many aspects of glycobiology/ glycosciences in such ways as to be less ambiguous and more expedient and efficient than attainable with the natural glycoconjugates. Neoglycoconjugates are indeed being used in many different fields. Here we cite a few important examples.

9.1 Probing Carbohydrate-Protein Interactions

Neoglycoproteins with different sugars have proven extremely valuable in deciphering the binding specificity of hepatic lectins (Lee *et al.*, 1991; Lee, 1992). Neoglycoproteins can be labeled easily with radioactive iodine, which greatly facilitates the binding assays. When radiolabeled BSA derivatives modified with Gal, GlcNAc, Man etc were allowed to bind various hepatic lectins, the lectin from mammalian hepatocytes bound only Gal-BSA, and the lectin from chicken hypatocytes bound only GlcNAc-BSA (Lee, 1992). At high sugar densities on BSA (>30 mol/mol), both lectins bound the respective BSA-derivatives with $K_d < \text{nM}$, whereas the irrelevant sugar-BSA derivatives were usually not inhibitory at μM levels. Thus, the binding specificities of these hepatic lectins were unequivocally determined (Lee, 1992).

These experiments also clearly demonstrated two other important aspects of these lectins: (i) Only the terminal sugar residue is recognized; (ii) A linear increase in the sugar density brings about near logarithmic increase in the binding potency. The latter phenomenon (glycoside cluster effect) is a very important aspect of many carbohydrate-protein interactions.

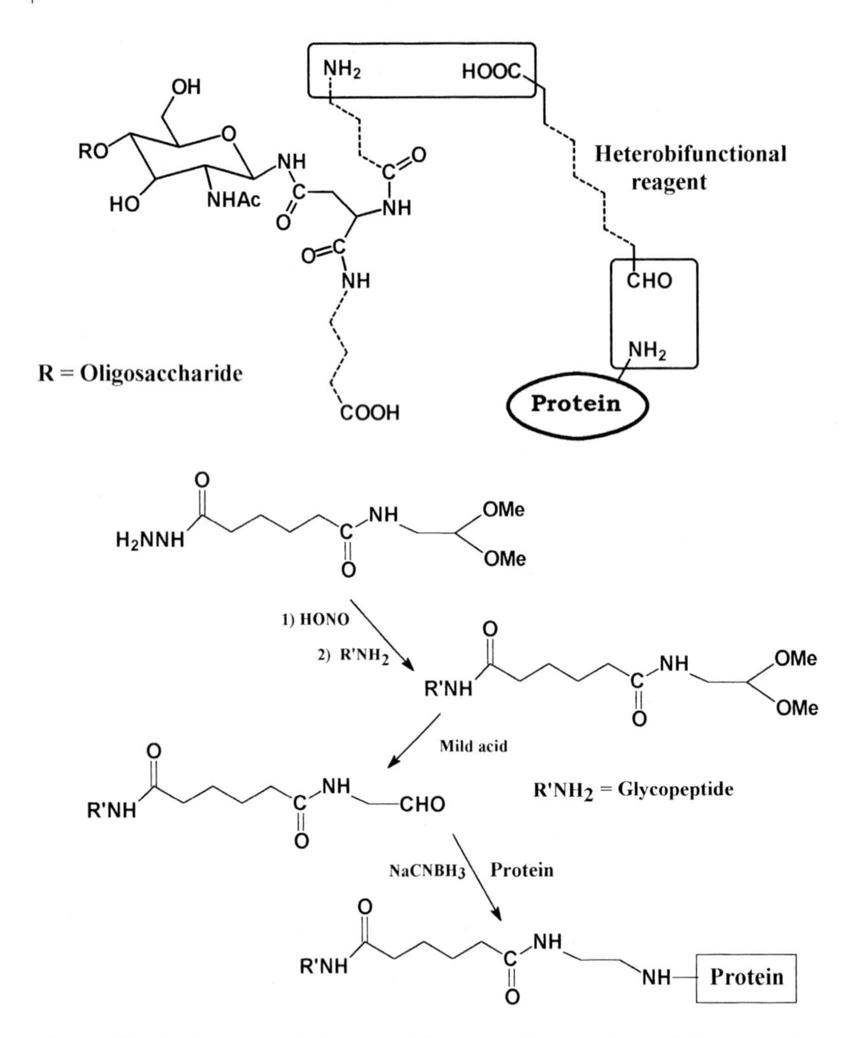

Figure 15. Conjugation of glycopeptide to protein via a heterobifunctional reagent

The neoglycolipids derived from reduced oligosaccharides and limited periodate oxidation (see section 4.1) have proven their merits in elucidating carbohydrate-binding specificities of a number of lectins (Feizi, 1991; Yuen *et al.*, 1992).

9.2 Use in Isolation of Carbohydrate-Binding Proteins

Unlike natural glycoproteins or oligosaccharides thereof, relatively large quantities of neoglycoconjugates of mono- and disaccharides can be easily prepared. Thus the use of neoglycoconjugates for isolation of carbohydrate-binding proteins is quite attractive. For example, the hepatic lectins mentioned above were isolated by affinity chromatography using Sepharose to which appropriate neoglycoproteins had been immobilized.

9.3 Cytochemical markers

The presence of carbohydrate-binding proteins in tissue sections can be readily detected with neoglycoproteins marked with fluorescent probes. In such a process, the protein moiety not only functions as the carrier of carbohydrate and fluorescent marker, but also provides the basis for a better survival of the probe during cytochemical procedures because of its molecular size. The aspect of polyvalency is also advantageous, because a tight binding is a prerequisite for successful cytochemical detection. The use of neoglycoenzymes (see below) in similar applications can increase the detection sensitivity even further. Changes occurring during malignancy have been probed and documented by such methods (Gabius *et al.*, 1989).

The flow cytometric assay can also greatly benefit from the use of neoglycoproteins (Midoux *et al.*, 1987) or other suitable neoglycoconjugates. A fluorescently labeled neoglycoconjugate with a strong binding affinity is required for successful cell sorting based on carbohycrate binding.

9.4 Neoglycoenzymes

Enzymes modified with carbohydrates (neoglycoenzymes) can be used in cytochemistry as described above or in biochemical detection of lectins in solid-phase assays. For example, bacterial β-galactosidase modified with *p*-aminophenyl α-D-mannopyranoside via amide linkage was useful in determination of Con. A immobilized on plastic microtiter plates. Lactose-modified β-galactosidase was effective in histochemical detection of galactoside-specific lectins (Gabius *et al.*, 1989).

9.5 Biomedical Applications

Many vaccines against microorganisms utilize the outer coat of these organisms which often contains a large amount of carbohydrates. A number of effective vaccines in use are conjugates of bacterial saccharides attached to carrier proteins (Jennings and Sood, 1994). Conjugation to proteins is especially important in vaccines for infants, since oligo- and polysaccharides alone are poor immunogens in general, and especially for infants.

Gal-modified human serum albumin, further modified with technetium, has been useful for the *in vivo* monitoring of liver receptor function, which in turn can be correlated to the desease status (Kudo *et al.*, 1994). Neoglycoproteins have been valuable in tumor diagnosis (Gabius *et al.*, 1994). A more intriguing application is to achieve targeted delivery of DNA via a carbohydrate-mediated entry into specific cells (Lee-Young *et al.*, 1994). Carbohydrate-mediated targeting of drugs in general has been reviewed recently (Ouchi and Ohya, 1994).

10 Conclusions and Perspectives

It should be clear from the foregoing presentation that neoglycoconjugates have a wide range of applicability. As the synthetic methodology advances, more sophisticated neoglycoconjugates will be prepared and used. In such an endeavor, chemoenzymatic methods are predicted to be instrumental. Progress in molecular biology of glycosyltransferases accelerates the usage of this class of enzymes in construction of neoglycoconjugates.

Acknowledgment

This article was written while YCL was a Scholar-in-residence at the Fogarty International center for Advanced Study in the Health Scienses, National Institutes of Health, Bethesda, MD, USA.

References

Andersson M, Oscarson S (1993): Synthesis of glycoconjugates by covalent coupling of *O*-glycopyranosyl-*N*-hydroxysuccinimide derivatives of lactose to proteins and lipids and polymerization of their parent acryloyl derivatives into acrylamide polymers. In *Bioconjugate Chem.* **4**:246–9.

Andersson M, Oscarson S, Öberg L (1993): Synthesis of oligosaccharides with oligoethyleneglycol spacers and their conversion into glycoconjugates using N,N,N', N"-tetramethyl(succinimido)uronium tetrafluoroborate as coupling reagent. In *Glycoconjugate J.* **10**:197–201.

Aplin JD, Wriston JC (1981): Preparation, properties, and applications of carbohydrates of proteins and lipids. In *CRC Crit. Rev. Biochem.* **10**:259–306.

Berman E, Brown J, Lis H *et al.* (1985): Binding of [1-^{13}C]galactose-labeled *N*-acetyllactosamine to *Erythrina cristagalli* agglutinin as studied by ^{13}C-NMR. In *Eur. J. Biochem.* **152**:447–51.

Berman E, Lis H (1987): ^{13}C-NMR study of the binding of [^{13}C]galactose-labeled *N*-acetyllactosamine and [1-^{13}C]galactose-enriched hen ovalbumin to soybean agglutinin. In *Biochim. Biophys. Acta* **924**:403–7.

Berman E, Lis H, James TL (1986): Binding of [1-^{13}C]galactose-enriched hen ovalbumin to *Erythrina cristagalli* agglutinin as studied by ^{13}C-NMR spectroscopy. In *Eur. J. Biochem.* **161**:589–94.

Bertozzi CR, Bednarski MD (1991): The synthesis of heterobifunctional linkers for the conjugation of ligands to molecular probes. In *J. Org. Chem.* **56**:4326–9.

Bogwald J, Seljelid R, Hoffman J (1986): Coupling of polysaccharides activated by means of chloroacetaldehyde dimethyl acetal to amines or proteins by reductive amination. In *Carbohydr. Res.* **148**:101–7.

Cabacungan JC, Ahmed AI, Feeney RE (1982): Amine boranes as alternative reducing agents for reductive alkylation of proteins. In *Anal. Biochem.* **124**:272–8.

Chatterjee D, Douglas JT, Cho S-N *et al.* (1985): Synthesis of neoglycoproteins containing the 3,6-di-*O*-methyl-β-D-glucopyranosyl epitope and their use in serodiagnosis of leprosy. In *Glycoconjugate J.* **2**:187–208.

Chen VJ, Wold F (1984): Neoglycoproteins: Preparation of non-covalent glycoproteins through high-affinity protein-(glycosyl) ligand complexes. In *Biochemistry* **23**:3306–11.

Childs RA, Feizi T, Yuen C-T *et al.* (1990): Differential recognition of core and terminal portions of oligosaccharide ligands by carbohydrate-recognition domains of two mannose-binding proteins. In *J. Biol. Chem.* **265**:20770–7.

Chipowsky S, Lee YC (1973): Synthesis of 1-thioaldosides having an amino group at the aglycon terminal. In *Carbohydr. Res.* **31**:339–46.

Fan JQ, Takegawa K, Iwahara S *et al.* (1995): Enhanced transglycosylation activity of *Arthrobacter protophormiae* endo-β-*N*-acetylglucosaminidase in media containing organic solvents. In *J. Biol. Chem.* **270**:17723–9.

Feizi T (1991): Carbohydrate differentiation antigens: probable ligands for cell adhesion molecules. In *Trends Biochem. Sci.* **16**:84–6.

Gabius H-J (1990): Influence of type of linkage and spacer on the interaction of β-galactoside-binding proteins with immobilized affinity ligands. In *Anal. Biochem.* **189**:91–4.

Gabius H-J, Brinck U, Kayser K *et al.* (1994): Neoglycoproteins: Use in tumor diagnosis. In *Neoglycoconjugates: Preparation and applications* (Lee YC, Lee RT, eds) pp 404–24, San Diego, CA: Academic Press.

Gabius S, Hellmann K-P, Hellmann T *et al.* (1989): Neoglycoenzymes: A versatile tool for lectin detection in solid-phase assays and glycohistochemistry. In *Anal. Biochem.* **182**:447–51.

Garg HG, Jeanloz RW (1985): Synthetic *N*- and *O*-glycosyl derivatives of L-asparagine, L-serine, and L-threonine. In *Adv. Carbohydr. Chem. Biochem.* **43**:135–201.

Gaylord H, Brennan PJ (1987): Leprosy and the leprosy bacillus: Recent developments in characterization of antigens and immunology of the disease. In *Annu. Rev. Microbiol.* **41**:645–75.

Gray GR (1974): The direct coupling oligosaccharides to proteins and derivatized gels. In *Arch. Biochem. Biophys.* **163**:426–8.

Gross HJ, Rose U, Karuse JM *et al.* (1989): Transfer of synthetic sialic acid analogues to *N*- and *O*-linked glycoprotein glycans using four different mammalian sialyltransferases. In *Biochemistry* **28**:7386–92.

Ichikawa Y, Look GC, Wong C-H (1992): Enzyme-catalyzed oligosaccharide synthesis. In *Anal. Biochem.* **202**:215–38.

Jennings HJ, Sood RK (1994): Synthetic glycoconjugates as human vaccine. In *Neoglycoconjugates: Preparation and Applications* (Lee YC, Lee RT, eds) pp 324–71, San Diego, CA: Academic Press.

Kadowaki S, Yamamoto K, Fujisaki M *et al.* (1991): Microbial endo-β-*N*-acetylglucosaminidases acting on complex-type sugar chains of glycoproteins. In *J. Biochem.* **110**:17–21.

Kallin E (1994): Use of aminoalditols and glycosylamines in neoglycoconjugate synthesis. In *Neoglycoproteins: Preparation and Applications* (Lee YC, Lee RT, eds) pp 199–223, San Diego, CA: Academic Press.

Kallin E, Loenn H, Norberg T (1986): New derivatization and separation procedure for reducing oligosaccharides. In *Glycoconjugate J.* **3**:311–9.

Kimata K (1993): Proteoglycans as regulators of cell adhesion and cell growth. In *Experimental Medicine (Japanese)* **11**:138–45.

Kobayashi A, Akaike T, Kobayashi K *et al.* (1986): Enhanced adhesion and survival efficienty of liver cells in culture dishes coated with a lactose-carrying styrene homopolymer. In *Macromol. Chem., Rapid Commun.* **7**:645–50.

Kobayashi K, Akaike T (1990): Hepatocyte adhesion using carbohydrate polymer as substrate material and expression of sugar chain high density effect. In *Trends Glycosci. Glycotechnol.* **2**:26–33.

Kobayashi K, Sumitomo H, Ina Y (1985): Synthesis and functions of polystyrene derivatives having pendant oligosaccharides. In *Polymer J.* **17**:567–75.

Krantz MJ, Holtzman NA, Stowell CP *et al.* (1976): Attachment of thioglycosides to proteins: Enhancement of liver membrane binding. In *Biochemistry* **15**:39633–8.

Krantz MJ, Lee YC (1976): Quantitative hydrolysis of thioglycosides. In *Anal. Biochem.* **71**:318–21.

Kudo M, Vera DR, Stadalnik RC (1994): Hepatic receptor imaging using radiolabeled asialoglycoprotein analogs. In *Neoglycoconjugates: Preparation and applications* (Lee YC, Lee RT, eds) pp 373–402, San Diego, CA: Academic Press.

Lee KB, Lee YC (1994a): Transfer of modified sialic acids by *Trypanosoma cruzi* trans-sialidase for attachment of functional groups to oligosaccharides. In *Anal. Biochem.* **216**:358–64.

Lee RT, Cascio S, Lee YC (1979): A simple method for the preparation of polyacrylamide gels containing thioglycoside ligands. In *Carbohydr. Res.* **95**: 260–9.

Lee RT, Lee YC (1974): Synthesis of 3-(2-aminoethylthio)propyl glycosides. In *Carbohydr. Res.* **37**:191–201.

Lee RT, Lee YC (1980): Preparation and some biochemical properties of neoglycoproteins produced by reductive amination of thioglycosides containing an ω-aldehydoaglycon. In *Biochemistry* **19**:156–63.

Lee RT, Lee YC (1986): Preparation of a high-affinity photolabeling reagent for the Gal/GalNAc lectin of mammalian liver: Demonstration of galactose-combining sites on each subunit of rabbit hepatic lectin. In *Biochemistry* **25**:6835–41.

Lee RT, Lee YC (1987): Preparation of cluster glycosides of *N*-acetylgalactosamine that have subnanomolar binding constants towards the mammalian hepatic Gal/GalNAc-specific receptor. In *Glycoconjugate J.* **4**:317–28.

Lee RT, Lin P, Lee YC (1984): New synthetic cluster ligands for galactose/*N*-acetylgalactosamine-specific lectin of mammalian liver. In *Biochemistry* **23**:4255–61.

Lee RT, Myers RW, Lee YC (1982): Further studies on the binding characteristics of rabbit liver galactose/*N*-acetylgalactosamine-specific lectin. In *Biochemistry* **24**:6292–8.

Lee RT, Wong T-C, Lee R *et al.* (1989): Efficient coupling of glycopeptides to proteins with a heterobifunctional reagent. In *Biochemistry* **28**:1856–61.

Lee YC (1992): Biochemistry of carbohydrate-protein interactions. In *FASEB J.* **6**:3193–200.

Lee YC, Lee RT (1982): Neoglycoproteins as probes for binding and cellular uptake of glycoconjugates. In *The Glycoconjugates,* **IV** (Horowitz MI, eds) pp 57–83, New York, NY: Academic Press.

Lee YC, Lee RT (1991): Neoglycoconjugates: Fundamentals and recent progress. In *Lectins and Cancer* (Gabius H-J, Gabius S, eds) pp 53–70, Berlin: Springer Verlag.

Lee YC, Lee RT (1992): Synthetic glycoconjugates. In *Glycoconjugates* (Allen HJ, Kisailus EC, eds) pp 121–65, New York: Marcel Dekker, Inc.

Lee YC, Lee RT (1994a): Neoglycoconjugates: Preparation and Applications. San Diego, CA: Academic Press.

Lee YC, Lee RT (1994b): Neoglycoconjugates, Part A. Synthesis. In *Methods in Enzymol.* **242** (Abelson JN, Simon MI, eds) San Diego, CA: Academic Press.

Lee YC, Lee RT (1994c): Neoglycoconjugates. Part B. Biomedical Applications. In *Methods in Enzymol.* **247** (Abelson JN, Simon MI, eds) San Diego, CA: Academic Press.

Lee YC, Lee RT (1994d): Enhanced biochemical affinities of multivalent neoglycoconjugates. In *Neoglycoconjugates: Preparation and Applications* (Lee YC, Lee RT, eds) pp 23–50, San Diego, CA: Academic Press.

Lee YC, Lee RT, Rice K *et al.* (1991): Topography of binding sites of animal lectins: Ligands' view. In *Pure & Appl. Chem.* **63**:499–506.

Lee YC, Stowell CP, Krantz MJ (1976): 2-Imino-2-methoxyethyl 1-thioglycosides: New reagents for attaching sugars to proteins. In *Biochemistry* **15**:3956–63.

Lee YC, Townsend RR, Hardy MR *et al.* (1983): Binding of synthetic oligosaccharides to the hepatic Gal/GalNAc lectin. In *J. Biol. Chem.* **258**:199–202.

Lee-Young AW, Wu GY, Wu CH (1994): Delivery of polynucleotides to hepatocytes. In *Neoglycoconjugates: Preparation and Applications* (Lee YC, Lee RT, ed.) pp 511–37, San Diego, CA: Academic Press.

Lehrman M, Haltiwanger RS, Hill RL (1986a): The binding of fucose-containing glycoproteins by hepatic lectins. The binding specificity of the rat liver fucose lectin. In *J. Biol. Chem.* **261**:7426–32.

Lehrman M, Pizzo SV, Imber MJ *et al.* (1986b): The binding of fucose-containing glycoproteins by hepatic lectins. Re-examination of the clearance from blood and the binding to membrane receptors and pure lectins. In *J. Biol. Chem.* **261**:7412–8.

Lemieux RU, Bundle DR, Baker DA (1975): Properties of a "synthetic" antigen related to the human blood-group Lewis a. In *J. Am. Chem. Soc.* **97**:4076–83.

Li Y-T, Carter BZ, Rao BN *et al.* (1991): Synthesis of neoglycoconjugates using oligosaccharide transferring activity of ceramide glycanase. In *J. Biol. Chem.* **266**:10723–6.

Li Y-T, Li S-C (1994): Synthesis of neoglycoconjugates using oligosaccharide-transferring activity of ceramide glycanase. In *Neoglycoconjugates: Preparation and Applications* (Lee YC, Lee RT, eds) pp 250–60, San Diego, CA: Academic Press.

Magnani J (1987): Immunostaining free oligosaccharides directly on thin-layer chromatography. In *Methods Enzymol.* **138**:208–12.

Magnusson G (1986): Synthesis of neoglycoconjugates. In *Protein-Carbohydrate Interactions in Biological Systems* (Lark DA, ed) pp 215–28, London, UK: Academic Press.

Magnusson G, Ahlfors S, Dahmen J *et al.* (1990): Prespacer glycosides in glycoconjugate chemistry. Dibromoisobutyl glycosides for the synthesis of neoglycolipids, neoglycoproteins, neoglycoparticles, and soluble glycosides. In *J. Org. Chem.* **55**:3932–46.

Malsch R, Guerrini M, Torri G *et al.* (1994): Synthesis of a N'-alkylamine anticoagulant active low-molecular-mass heparin for radioactive and fluorescent labeling. In *Anal. Biochem.* **217**:255–64.

Manger ID, Wong SYC, Rademacher TW *et al.* (1992): Synthesis of 1-*N*-glycyl β-oligosaccharide derivatives. Reactivity of *Lens culinaris* lectin with a fluorescent labeled streptavidin pseudoglycoprotein and immobilized neoglycolipid. In *Biochemistry* **31**:10733–40.

McBroom CR, Samanen CH, Goldstein IJ (1972): Carbohydrate antigens: Coupling of carbohydrates to proteins by diazonium and phenylisothiocyanate. In *Methods Enzymol.* **28**:212–9.

Meldal M (1994): Glycopeptide Synthesis. In *Neoglycoconjugates: Preparation and Applications* (Lee YC, Lee RT, eds) pp 144–96, San Diego, CA: Academic Press.

Midoux P, Roche A-C, Monsigny M (1987): Quantitation of the binding, uptake, and degradation of fluoresceinylated neoglycoproteins by flow cytometry. In *Cytometry* **8**:327–34.

Mootoo DR, Konradsson P, Fraser-Reid B (1989): *n*-Pentenyl glycosides facilitate a stereoselective synthesis of the pentasaccharide core of the protein membrane ancor found in *Trypanosoma brucei*. In *J. Am. Chem. Soc.* **111**:8540–2.

Nishimura S, Matsuoka K, Kurita K (1990): Synthetic glycoconjugates: Simple and potential glycoprotein models containing pendant *N*-acetyl-D-glucosamine and *N,N'*-diacetylchitobiose. In *Macromolecules* **23**:4182–4.

Nishimura S-I, Lee KB, Matsuoka K *et al.* (1994): Chemoenzymic preparation of a glycoconjugate polymer having a sialyloligosaccharide: Neu5Acα(2–3)Galβ(1–4)GlcNAc. In *Biochem. Biophys. Res. Commun.* **199**:249–52.

Ouchi T, Ohya Y (1994): Drug delivery systems using carbohydrate recognition. In *Neoglycoconjugates: Preparation and Applications* (Lee YC, Lee RT, eds) pp 464–98, San Diego, CA: Academic Press.

Pacuszka T, Fishman PH (1990): Generation of cell surface neogangliproteins. G_{M1}-neoganglioproteins are non-functional receptors for cholera toxin. In *J. Biol. Chem.* **265**:7673–8.

Paulson JC, Rogers GN, Carroll SM *et al.* (1984): Selection of influenza virus variants based on sialyloligosaccharide receptor specificity. In *Pure & Appl. Chem.* **56**:797–805.

Pless DD, Lee YC, Roseman S *et al.* (1983): Specific cell adhesion to immobilized glycoproteins demonstrated using new reagents for protein and glycoprotein immobilization. In *J. Biol. Chem.* **258**:2340–9.

Rademacher TW, Parekh RB, Dwek RA (1988): Glycobiology. In *Annu. Rev. Biochem.* **57**:785–838.

Raja RH, LeBoeuf RD, Stone GW *et al.* (1984): Preparation of alkylamine and ^{125}I-radiolabeled derivatives of hyaluronic acid uniquely modified at the reducing end. In *Anal. Biochem.* **139**:168–77.

Rice KG, Lee YC (1990): Modification of triantennary glycopeptide into probes for the asialoglycoprotein receptor of hepatocytes. In *J. Biol. Chem.* **265**:18423–8.

Rice KG, Lee YC (1993): Oligosaccharide valency and conformations in determining binding to the asialoglycoprotein receptor of rat hepatocytes. In *Adv. Enzymol.* **66**:41–83.

Rothenberg BE, Hayes BK, Toomre D *et al.* (1993): Biotinylated diaminopyridine: An approach to tagging oligosaccharides and exploring their biology. In *Proc. Natl. Acad. Sci. USA* **90**:11939–43.

Sabesan S, Duus J, Neirs S *et al.* (1992): Cluster sialoside inhibitors for influenza virus: Synthesis, NMR, and biological studies. In *J. Am. Chem. Soc.* **114**:8363–75.

Sabesan S, Paulson JC (1986): Combined chemical and enzymatic synthesis of sialyloligosaccharides and characterization by 500-MHz ^{1}H and ^{13}C NMR spectroscopy. In *Biochemistry* **108**:2068–80.

Sasaki H, Ochi N, Dell A *et al.* (1988): Site-specific glycosylation of human recombinant erythropoietin: Analysis of glycopeptides or peptides at each glycosylation site by fast atom bombardment mass spectrometry. In *Biochemistry* **27**:8618–26.

Schnaar R (1984): Immobilized glycoconjugates for cell adhesion studies. In *Anal. Biochem.* **143**:1–13.

Schnaar RL (1994): Immobilized glycoconjugates for cell recognition studies. In *Neoglycoconjugates: Preparation and Applications* (Lee YC, Lee RT, eds) pp 425–43, San Diego, CA: Academic Press.

Schnaar RL, Weigel PH, Roseman S *et al.* (1993): Immobilization of carbohydrates on poly(acrylamide) gels: I. Poly(acrylamide) gels copolymerized with active esters. In *Methods in Carbohydr. Chem.* **IX** (BeMiller JN, Whitler RL, eds) pp 181–6, New York, NY: John Wiley & Sons, Inc.

Shao M-C, Sokolik CW, Wold F (1994): Non-covalent neoglycoproteins. In *Neoglycoproteins: Preparation and Applications* (Lee YC, Lee RT, eds) pp 224–49, San Diego, CA: Academic Press.

Spellman MW (1990): Carbohydrate characterization of reconmbinant glycoproteins of pharmaceutical interest. In *Anal. Chem.* **62**:1714–22.

Srivastava G, Kaur KJ, Hindsgaul O *et al.* (1992): Enzymatic transfer of a preassembled trisaccharide antigen to cell surfaces using a fucosyltransferase. In *J. Biol. Chem.* **267**:22356–61.

Stoll MS, Hounsell EF, Lawson AM *et al.* (1990): Microscale sequencing of *O*-linked oligosaccharides using mild periodate oxidation of alditols, coupling to phospholipid and TLC-MS analysis of the resulting neoglycolipids. In *Eur. J. Biochem.* **189**: 499–507.

Stowell CP, Lee YC (1980): Neoglycoproteins. The preparation and application of synthetic glycoproteins. In *Adv. Carbohydr. Chem. Biochem.* **37**:225–81.

Takahashi N, Muramatsu T (1992): Handbook of endoglycosidases and glycoamidases. Boca Raton, FL: CRC Press.

Takegawa K, Tabuchi M, Yamaguchi S *et al.* (1995): Synthesis of neoglycoproteins using oligosaccharide-transfer activity with endo-β-*N*-acetylglucosaminidase. In *J. Biol. Chem.* **270**:3094–9.

Tamura T, Wadhwa MS, Rice KG (1994): Reducing-end modification of *N*-linked oligosaccharides with tyrosine. In *Anal. Biochem.* **216**:335–44.

Tang PW, Gooi HC, Hardy M *et al.* (1985): Novel approach to the study of the antigenecities and receptor functions of carbohydrate chains of glycoproteins. In *Biochem. Biophys. Res. Commun.* **132**:474–80.

Tiemeyer M, Swank-Hill P, Schnaar RL (1990): A membrane receptor for gangliosides is associated with central nervous system myelin. In *J. Biol. Chem.* **265**:11990–9.

Tiemeyer M, Yasuda Y, Schnaar RL (1989): Ganglioside receptors on rat brain membranes. In *J. Biol. Chem.* **264**:1671–81.

Verez-Bencomo V, Campos-Valdes M, Marino-Albernas JR *et al.* (1991): Glycosides of 8-hydroxy-3,6-dioxaoctanal. A synthesis of a new spacer for synthetic oligosaccharides. In *Carbohydr. Res.* **271**:263–7.

Weigel PH, Naoi M, Roseman S *et al.* (1979): Preparation of 6-aminohexyl D-aldopyranosides. In *Carbohydr. Res.* **70**:83–91.

Yamamoto K, Watanabe J, Kadowaki S *et al.* (1993): Transglycosylation activity of *Mucor hiemalis* endo-β-*N*-acetyl-glucosaminidase. In *Seikagaku* **65**:1014.

Yan S-CB, Wold F (1984): Neoglycoproteins: In vitro introduction of glycosyl units at glutamines in β-casein using transglutaminase. In *Biochemistry* **23**:3759–65.

Yasuda Y, Tiemeyer M, Blackburn CC *et al.* (1987): Neuronal recognition of gangliosides: Evidence for a brain ganglioside receptor. In *New Trends in Ganglioside Research: Neurochemical and Neuroregenerative Aspects* (Ledeen R, Tettamanti G, Yu R, Yates H, eds) pp 229–43, Heidelberg-Berlin: Springer Verlag.

Yoshida T, Lee YC (1994): Glycamine formation via reductive amination of oligosaccharides with benzylamine: efficient coupling of oligosaccharides to protein. In *Carbohydr. Res.* **251**:249–54.

Yuen C-T, Lawson AM, Chai W *et al.* (1992): Novel sulfated ligands for the cell adhesion molecule E-selectin revealed by the neoglycolipid technology among *O*-linked oligosaccharides on an ovarian cystadenoma glycoprotein. In *Biochemistry* **31**:9126–31.

5 Glycosyltransferases Involved in *N*- and *O*-Glycan Biosynthesis

INKA BROCKHAUSEN AND HARRY SCHACHTER

1 Introduction

Proteins, nucleic acids and glycoconjugates (glycoproteins and glycolipids) are large molecules essential to all living cells. Whereas nucleic acids and proteins are linear molecules in which the building blocks are joined together by identical bonds (phosphodiester and amide bonds respectively), the glycan moieties of glycoconjugates are usually branched and the monosaccharide building blocks may be joined to one another in many different linkages. This has important implications for biosynthesis. In the case of proteins and nucleic acids every new molecule is copied from a pre-existing template molecule acting as a mould. Glycans, like automobiles, cannot be copied from a mould and are manufactured on an assembly line in which individual components are incorporated sequentially. The glycan assembly line is the cell's endomembrane system – the endoplasmic reticulum and Golgi apparatus (Kornfeld and Kornfeld, 1985). The workers are glycosidases and glycosyltransferases chained to this assembly line. They act sequentially on the growing oligosaccharide as it moves along the lumen of the endomembrane system.

This chapter will discuss some of the glycosyltransferases which make protein-bound *N*-glycans (Asn-GlcNAc *N*-glycosidic linkage) (Beyer *et al.*, 1981; Snider, 1984; Kornfeld and Kornfeld, 1985; Schachter, 1986, 1995; Hemming, 1995; Verbert, 1995) and *O*-glycans (Ser/Thr-GalNAc *O*-glycosidic linkage) (Beyer *et al.*, 1981; Sadler, 1984; Schachter and Brockhausen, 1989, 1992; Brockhausen, 1995). The reactions catalyzed by these enzymes are:

Donor (X-sugar) + acceptor (R-OH or R-NH) $\rightarrow$ R-O(or N)-sugar + X

where sugar is either a mono- or oligosaccharide, R can be a free saccharide, a saccharide linked to an aglycone, a protein or a lipid, and X can be a nucleotide, dolichol-phosphate, or dolichol-pyrophosphate. Hundreds of different glycosidic linkages have been described and for every linkage there are usually one or more specific glycosyltransferases; a few exceptions to this rule have been reported in which a single enzyme makes more than one linkage, e.g., the Lewis blood group-dependent α1,3/4-fucosyltransferase. A recent review (Field and Wainwright, 1995) reported the cloning of 28 distinct glycosyltransferases from a variety of vertebrate species and the list is growing at a rapid pace.

The glycosyltransferases are intrinsic membrane proteins firmly bound to the endomembrane assembly line with their catalytic domains within the lumen. Detergent treatment is required for solubilization and full expression of enzymatic activity *in vitro*. Most of these enzymes require divalent cation which probably serves to bind the negatively charged nucleotide-sugar to the protein. Exogenous cation is not required for the assay of fucosyltransferases, sialyltransferases and several β1,6-*N*-acetylglucosaminyltransferases. Glycosyltransferases usually show precise substrate specificities. However, under non-physiological conditions *in vitro*, some enzymes can catalyze interesting promiscuous reactions at relatively low rates. For example, the human blood group B α1,3-galactosyltransferase can transfer GalNAc instead of Gal to its acceptor substrate *in vitro* and thereby make the human blood group A epitope (Greenwell *et al.*, 1986). If this reaction were to occur *in vivo*, a severe blood incompatibility would result.

H.-J. and S. Gabius (Eds.), Glycosciences

ISBN 3-8261-0073-5

2 The Roads Which Lead to Complex *N*-Glycans

The synthesis of *N*-glycans begins with the transfer within the lumen of the endoplasmic reticulum of $Glc_3Man_9GlcNAc_2$ from dolichol-pyrophosphate-$Glc_3Man_9GlcNAc_2$ to an asparagine residue of the nascent polypeptide chain (Hemming, 1995; Verbert, 1995) within Asn-X-Ser/Thr sequences. Oligosaccharide processing by glucosidases I and II, endoplasmic reticulum mannosidase and Golgi mannosidase I forms $Man_5GlcNAc_2$-Asn-X which is the entry point for the formation of hybrid and complex *N*-glycans due to the action of UDP-GlcNAc:Man(α1–3)R [GlcNAc to Man(α1–3)] β1,2-*N*-acetylglucosaminyltransferase I (GnT I) (Figure 1).

This enzyme step is essential for the subsequent action of several enzymes in the processing pathway, i.e., α3/6-mannosidase II (Figure 1), GnT II, III and IV (Figure 2), and the α1,6-fucosyltransferase which adds fucose in (α1–6) linkage to the Asn-linked GlcNAc (Schachter, 1986; Schachter, 1991b).

GnT II to VI (Figure 2) act on the product of α-mannosidase II to initiate the various complex *N*-glycan "antennae". The most common antenna is sialyl(α2–6)Gal(β1–4)GlcNAc-. Sialic acid can also be incorporated in α2,3 linkage to Gal, the Gal(β1–4) residue may be replaced by a Gal(β1–3) residue or by a GalNAc(β1–4) resi-

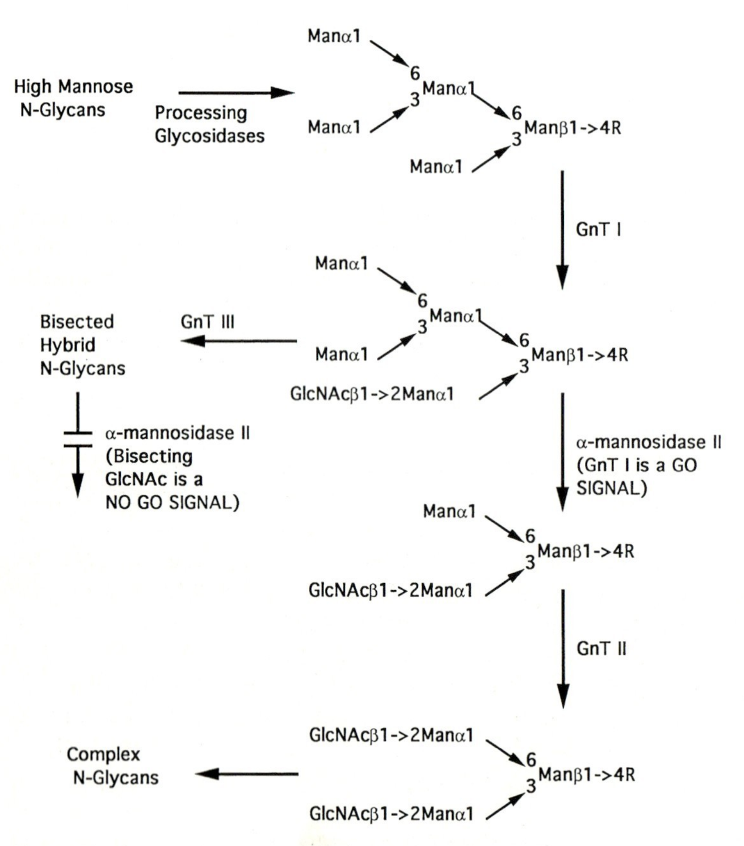

Figure 1. Conversion of high mannose *N*-glycans to hybrid and complex *N*-glycans GnT I, UDP-GlcNAc:Man(α1–3)R [GlcNAc to Man(α1–3)] β1,2-*N*-acetylglucosaminyltransferase I. GnT II, UDP-GlcNAc: Man(α1–6)R [GlcNAc to Man(α1–6)] β1,2-*N*-acetyl-glucosaminyltransferase II

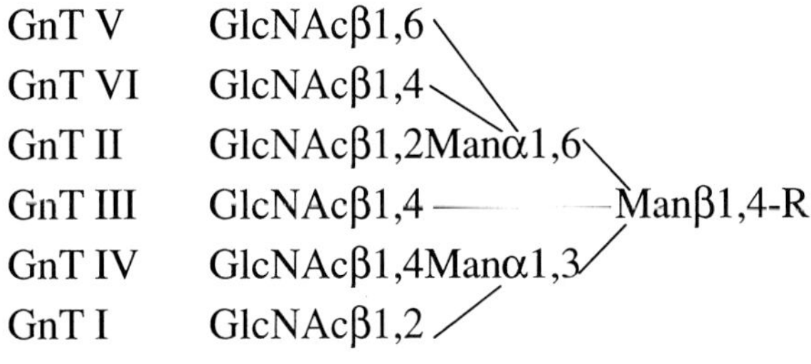

Figure 2. GnT I to VI incorporate GlcNAc residues into the Man(α1–6)[Man(α1–3)]Manβ-R *N*-glycan core

due, the terminal sialic acid residue may be replaced by a Gal(α1–3) residue, Gal and GlcNAc residues may be fucosylated, poly-*N*-acetyllactosamine, sulphate or phosphate groups may be present and other modifications have been reported (Cummings, 1992).

If GnT III acts on the product of GnT I before α-mannosidase II to form the bisected hybrid structure (Figure 1), the pathway is committed to hybrid structures because α-mannosidase II cannot act on bisected oligosaccharides (Harpaz and Schachter, 1980). The reverse order of action leads to complex *N*-glycans. The relative abundance of GnT III and α-mannosidase II in a particular tissue therefore controls the pathway towards hybrid or complex *N*-glycans. The route taken at such a divergent branch point is dictated primarily by the relative activities of glycosyltransferases which compete for a common substrate. The insertion of a bisecting GlcNAc by GnT III prevents the actions of GnT II, IV and V, α-mannosidase II and core α1,6-fucosyltransferase and is an example of a glycosyl residue acting as a STOP signal whereas the action of GnT I is a GO signal. Competition and STOP and GO signals are "substrate-level controls" of the biosynthetic pathways as opposed to control at the transcriptional, translational or post-translational levels.

3 Glycosyltransferases Involved in the Synthesis of Complex *N*-Glycan Antennae

A series of glycosyltransferases elongate the *N*-glycan antennae and add terminal sugars, sulfate groups and blood group and tissue antigens. The kinetic properties, substrate specificity, purification, cell biology and molecular biology of the *N*-glycan and ABO blood group glycosyltransferases have been thoroughly reviewed (Kornfeld and Kornfeld, 1985; Rademacher *et al.*, 1988; Paulson and Colley, 1989; Schachter, 1991a, 1995; Hart, 1992; Joziasse, 1992; Shaper and Shaper, 1992; Kleene and Berger, 1993; van den Eijnden and Joziasse, 1993; Brockhausen 1993, 1995). The review by Kleene and Berger (Kleene and Berger, 1993) is especially useful because it has detailed summary tables on many aspects of the glycosyltransferase literature. The present review will be limited to relatively recent data on some of these enzymes.

3.1 Galactosyltransferases

3.1.1 UDP-Gal:GlcNAc-R β1,4-Galactosyltransferase (E.C. 2.4.1.38/90)

UDP-Gal:GlcNAc-R β1,4-Galactosyltransferase (β4GalT) (Strous, 1986; Shur, 1991; Shaper and Shaper, 1992) is ubiquitous and acts on *N*-glycans, *O*-glycans and glycolipids. The enzyme requires divalent metal ions *in vitro* and exhibits an ordered mechanism with sequential binding to Mn^{2+}, UDP-Gal and GlcNAc substrate (Morrison and Ebner, 1971). β4GalT is involved in the synthesis of poly *N*-acetyllactosamine chains which form the basis for the attachment of terminal carbohydrate antigens. The enzyme alters its substrate specificity when it complexes with α-lactalbumin to form lactose synthetase which transfers Gal to Glc to make the milk sugar lactose. Although β4GalT can transfer GalNAc from UDP-GalNAc to GlcNAc at a relatively low rate (Palcic and Hindsgaul, 1991), the addition of α-lactalbumin greatly stimulates this activity (Do *et al.*, 1995) which may explain the presence of the GalNAc(β1–4)GlcNAc moiety on bovine milk glycoproteins.

Domain structure of β4GalT

Bovine β4GalT cDNA was isolated in 1986, the first glycosyltransferase to be cloned (Narimatsu *et al.*, 1986; Shaper *et al.*, 1986; D'Agostaro *et al.*, 1989). Human (Masri *et al.*, 1988; Mengle-Gaw *et al.*, 1991), murine (Nakazawa *et al.*, 1988; Shaper *et al.*, 1988; Hollis *et al.*, 1989) and chicken (Ghosh *et al.*, 1992; Shaper *et al.*, 1995) genes have also been cloned [consult (Field and Wainwright, 1995) for EMBL/GenBank accession numbers]. The human β4GalT gene is located on chromosome 9p13–21 (Duncan *et al.*, 1986).

β4GalT is a type II integral membrane protein (N_{in}/C_{out} orientation) with a short amino-terminal cytoplasmic domain (11–24 residues), a 20-residue non-cleavable signal/anchor transmembrane domain and a long (386–402 residues) intra-lumenal carboxy-terminal catalytic domain. This domain structure (Figure 3) has been found in all the Golgi-localized glycosyltransferases for which genes have been cloned to date (Paulson and Colley, 1989; Schachter, 1994, 1995). The amino-terminal, transmembrane and stem domains are not required for catalytic activity but are essential for accurate targeting and anchoring

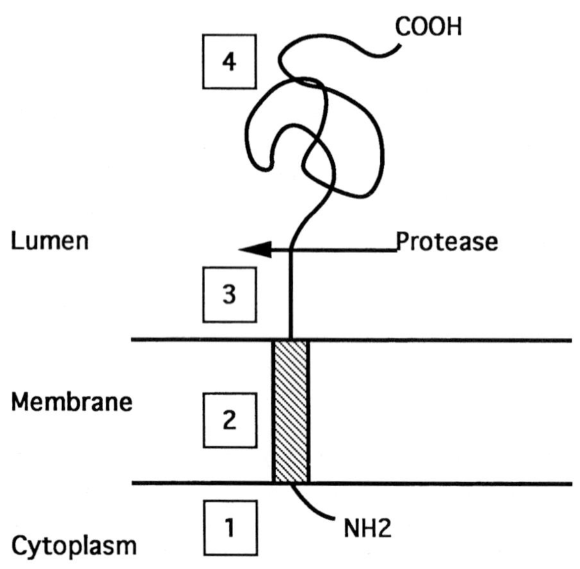

Figure 3. Domain structure of the glycosyltransferases. All glycosyltransferases cloned to date have an amino acid sequence compatible with a type II transmembrane glycoprotein (N_{in}/C_{out}). (1) Short amino-terminal cytoplasmic domain. (2) Trans-membrane non-cleavable signal/anchor domain. (3) Stem or neck region which acts as a tether to hold the catalytic domain in the lumen and can be cleaved by proteases to release a catalytically active soluble enzyme. (4) Carboxy-terminal catalytic domain.

of the enzyme to a specific region of the Golgi membrane. Comparison of the amino acid sequences of bovine, human and murine β4GalT shows over 90 % sequence similarity in the transmembrane anchor and C-terminal catalytic domains but higher variability occurs in the stem region. Tyr and Trp residues have been shown to be important for the UDP-Gal binding site by site specific mutagenesis (Ats *et al.*, 1992). A β4GalT probe has been used to clone the gene for a snail *N*-acetylglucosaminyltransferase (Bakker *et al.*, 1994) but no similarities to other proteins have as yet been detected.

Control of β4GalT gene transcription

Bovine, human and murine β4GalT cDNAs have two in-frame ATG codons at the 5'-end. Northern analysis showed the presence of two sets of transcripts (3.9 and 4.1 kb) (Shaper *et al.*, 1988; Hollis *et al.*, 1989). The first transcription initiation site was upstream of the first ATG codon and the second site was between the two ATG codons indicating that at least two promoters control the gene. Translation from the two in-frame ATG codons predicts proteins differing in length by 13 amino acids. *In vitro* translation experiments showed that only a single protein product of the predicted size was obtained from both the short and long mRNAs respectively proving that the long transcript initiated translation only at the upstream Met codon. When *in vitro* translation was carried out in the presence of dog pancreas microsomes, both the long and short proteins increased in size by about 3 kDa. Endoglycosidase H treatment removed most of this extra material indicating that both proteins had been glycosylated. It was concluded that both proteins contain a non-cleavable transmembrane segment and are oriented as type II integral membrane proteins (Figure 3) (Russo *et al.*, 1990).

The long and short forms of β4GalT are expressed in a tissue-specific manner in the mouse and provide a mechanism for regulation of enzyme levels (Harduin-Lepers *et al.*, 1993). Tissues which express relatively low levels of β4GalT (such as brain) are under the control of the upstream promoter and contain only the long transcript. Most somatic mouse tissues express intermediate levels of enzyme and are under the control of both promoters. Tissues which express β4GalT at very high levels (lactating mammary gland) have 10 times more short transcript than long. The upstream promoter appears to be a typical constitutive "house-keeping" promoter. It lacks classical CAAT and TATA boxes and has six upstream GC boxes (*cis*-acting positive regulatory elements) under the control of the Sp1 transcription factor. Between the two promoters three more GC boxes are located, at least two mammary gland-specific positive regulatory elements and a negative *cis*-acting regulatory element. Lactation is believed to turn on the synthesis of mammary gland specific transcription factors which bind to the positive regulatory elements and deactivate the negative control thereby activating the tissue-specific downstream promoter and causing a marked increase in production of the short form of β4GalT. The presence of a germ cell-specific promoter which regulates expression of β4GalT in haploid round spermatids has been reported (Shaper *et al.*, 1990, 1994; Harduin-Lepers *et al.*, 1992). Shur's group has provided evidence that β4GalT on the mouse sperm surface mediates fertilization by binding to terminal GlcNAc residues on the egg zona pellucida glycoprotein ZP3 (Shur, 1991; Miller *et al.*, 1992).

Targeting to the Golgi apparatus

The domains responsible for targeting β4GalT to the trans-cisternae of the Golgi apparatus (Strous, 1986) have been investigated with chimeric cDNAs encoding hybrid proteins in which various domains of β4GalT are connected to reporter proteins not normally retained in the Golgi apparatus (Nilsson *et al.*, 1991; Aoki *et al.*, 1992; Teasdale *et al.*, 1992; Gleeson *et al.*, 1994). Following either stable or transient expression of these hybrid constructs in mammalian cells, the intracellular destination of the reporter protein is determined by immunofluorescence and immunoelectron microscopy. The transmembrane domain is essential for accurate Golgi retention while sequences outside the transmembrane domain play accessory roles for β4GalT (Teasdale *et al.*, 1994) and for other glycosyltransferases (Munro, 1991; Dahdal and Colley, 1993; Burke *et al.*, 1994).

There is a divergence of opinion on the functional significance of the long and short transcripts of β4GalT. It has been suggested that the long form may be preferentially targeted to the cell surface where it plays a role in cell-cell recognition (Lopez and Shur, 1988; Lopez *et al.*, 1989, 1991; Evans *et al.*, 1993) while other workers have concluded that both the long and short forms are targeted to the Golgi apparatus (Nilsson *et al.*, 1991; Aoki *et al.*, 1992; Russo *et al.*, 1992; Shaper and Shaper, 1992; Teasdale *et al.*, 1992; Harduin-Lepers *et al.*, 1993; Masibay *et al.*, 1993;). Escape of the long form, but not the short form, from the Golgi apparatus to the plasma membrane may be due to the abnormally high β4GalT expression levels in transient expression systems (Dinter and Berger, 1995).

A possible mechanism for the specific retention of proteins within the Golgi apparatus is the presence of a unique retention signal on every protein and a unique Golgi membrane receptor for every such signal. However, since Golgi retention is not saturable even at the very high levels of β4GalT expression of transient transfection experiments, a more likely mechanism is retention due to homo-oligomerization of the glycosyltransferase or hetero-oligomer formation with other Golgi proteins. The large aggregate may be unable to enter budding vesicles either because of its size or due to interaction with the Golgi membrane lipid bilayer (Nilsson *et al.*, 1993). Evidence for such hetero-oligomers or "kin oligomers" has recently been obtained (Nilsson *et al.*, 1993, 1994; Slusarewicz *et al.*, 1994). Kin recognition seems to be mediated primarily by the lumenal domain rather than the transmembrane domain (Teasdale *et al.*, 1994; Munro, 1995b) but it has been reported that oligomer formation and Golgi retention of β4GalT are both abolished by mutations in the transmembrane domain (Yamaguchi and Fukuda, 1995). The transmembrane domains of plasma membrane-targeted proteins are broader and more hydrophobic than those of Golgi-targeted proteins and the length of the transmembrane domain has been shown to be critical for accurate Golgi localization (Masibay *et al.*, 1993; Munro, 1995a,b) suggesting that sorting may be mediated by interaction with lipid microdomains of different thicknesses.

3.1.2 UDP-Gal:Gal(β1–4)GlcNAc-R (Gal to Gal) α1,3-Galactosyltransferase (α3GalT, E.C. 2.4.1.124/151)

Gal(α1–3) is an uncharged alternative to sialic acid at the non-reducing terminus of *N*-glycan antennae. The Gal(α1–3)Gal(β1–4)GlcNAc epitope is not expressed in fish, amphibians, reptiles, birds, Old World monkeys, apes and humans but has a wide distribution in non-primate mammals, lemurs and New World monkeys (Galili *et al.*, 1988; Galili, 1989, 1992; Galili and Swanson, 1991). Since 1 % of circulating IgG in human serum is a natural antibody against this epitope, severe rejection reactions occur when tissues containing this epitope are used for transplantation purposes (e.g., pig heart) (Galili, 1993).

Genes for α3GalT (EMBL/GenBank accession numbers in brackets) have been cloned from cow (J04989 (Joziasse *et al.*, 1989)), pig (L36535, L36152 (Sandrin *et al.*, 1995; Strahan *et al.*, 1995a,b)), non-human primates (M73307, M73311, M72426, M73308, M73304, M73306, M73309, M73305, M73310, S71333 (Galili and Swanson, 1991; Henion *et al.*, 1994)) and mouse (M26925, M85153 (Larsen *et al.*, 1989; Joziasse *et al.*, 1992)); two non-functional pseudogenes have been isolated from human DNA (J05421, M65082, M60263 (Larsen *et al.*, 1990b; Joziasse *et al.*, 1991; Yamamoto *et al.*, 1991)). The two human pseudogenes are located on chromosomes 12q14-q15 and 9q33-q34 respectively (Shaper *et al.*, 1992). Although α3GalT has a domain structure typical of the glycosyltransferases (Figure 3), the only sequence similarity is an approximately 55 % identity of the human α3GalT pseudogene to the human blood group A α1,3-*N*-acetylgalactosaminyltransferase and blood group B α3GalT (Yamamoto and Hakomori, 1990). The gene locus encoding the allelic blood group A and B transferase genes is on human chromosome 9q34 suggesting that the blood group genes and the human α3GalT pseudogene are derived from the same ancestral gene by gene duplication and subsequent divergence.

The murine α3GalT gene consists of 9 exons spanning at least 35 kb (Joziasse *et al.*, 1992) (Figure 3). A single exon encodes the cytoplasmic and transmembrane domains and part of the stem region while the translation termination codon and all of the relatively long 3'-untranslated region are on another exon (Joziasse, 1992). The hypothesis that the Gal(α1–3)Gal epitope on the mouse egg zona pellucida glycoprotein ZP3 is required for fertilization (Bleil and Wassarman, 1988; Cheng *et al.*, 1994) is supported by the fact that the mouse α3GalT gene is expressed in female but not in male germ cells (Joziasse *et al.*, 1992; Johnston *et al.*, 1995) but has been disproved by the finding that female mice lacking the α3GalT gene are fertile (Thall *et al.*, 1995).

Truncation of New World monkeys α3GalT by removal of 89 amino acids from the amino-terminal region (the entire cytoplasmic and transmembrane domains and 67 amino acids from the stem region) had no effect on catalytic activity (Henion *et al.*, 1994).

3.1.3 UDP-Gal:GlcNAc-R β1,3-Galactosyltransferase (β3GalT)

Another elongating enzyme, β3GalT, has been characterized and purified from pig trachea (Sheares *et al.*, 1982; Sheares and Carlson, 1983) and the activity has been found in a number of other cell types (Bailly *et al.*, 1986; Messer and Nicholas, 1991). Both β3GalT and β4GalT appear to be subject to developmental regulation in the mammary tissue of the tammar wallaby. β3GalT is probably involved in the elongation of *O*-glycans since it acts effectively on *O*-glycan core 3, GlcNAcβ1–3GalNAc- (Sheares and Carlson, 1983).

3.2 Sialyltransferases

Sialic acid residues may be part of important antigenic determinants (e.g. the cancer-associated sialyl-Tn antigen) and ligands for cell-cell interactions (e.g. the selectin ligand sialyl-Le^x) or for virus binding to cell surfaces. *N*-glycan synthesis involves primarily two of these enzymes, CMP-sialic acid:Gal(β1–4)GlcNAc-R α2,6-sialyltransferase (E.C. 2.4.99.1) and CMP-sialic acid: Gal(β1–3/4)GlcNAc-R α2,3-sialyltransferase (E.C. 2.4.99.6). Abbreviations for the α2,6- and α2,3-sialyltransferases are respectively ST6N or ST6Gal I and ST3N or ST3Gal III (Table 1).

Table 1. Sialyltransferases involved in *N*- and *O*-glycan synthesis

Abbreviations		Enzyme Product	Tissue	EC No.	Comments	Acc.No.	References
ST3Gal I	ST3O I ST3GalA.1 SiaT-4a	**Sia**(α2–3)- Gal(β1–3)GalNAc-	Pig SM gland	2.4.99.4	*O*-glycans > glycolipids	M97753 M98463	(Gillespie *et al.*, 1992)
			Mouse brain			X73523	(Lee *et al.*, 1993)
			Chick embryo			X80503	(Kurosawa *et al.*, 1995)
			Human placenta			L29555	(Kitagawa and Paulson, 1994b)
			Human submax. gland			L13972	(Chang *et al.*, 1995)
ST3Gal II	ST3O II SAT-4 ST3GalA.2 SiaT-4b	**Sia**(α2–3)- Gal(β1–3)GalNAc-	Rat brain Mouse brain		glycolipids > *O*-glycans	X76988 X76989	(Lee *et al.*, 1994)
ST3Gal III	ST3N I	**Sia**(α2–3)- Gal(β1–3/4)GlcNAc-	Human placenta Mouse brain Rat liver	2.4.99.6	Gal(β1–3) GlcNAc > Gal(β1–4) GlcNAc	L23768 D28941 M97754, M98462	(Kitagawa and Paulson, 1993) (Sasaki *et al.*, 1994c) (Wen *et al.*, 1992a)
	ST3N II	**Sia**(α2–3)- Gal(β1–3/4)-GlcNAc-	Human melanoma		Gal(β1–4) GlcNAc > Gal(β1–3) GlcNAc	X74570	(Sasaki *et al.*, 1993)
ST3 Gal IV	STZ SAT-3 SiaT-4c	**Sia**(α2–3)- Gal(β1–3)GalNAc- **Sia**(α2–3)- Gal(β1–4)GlcNAc-	Human placenta Human		Does not act on Gal(β1–3) GlcNAc-; acts on *N*- and *O*-glycans, glycolipids	L23767 L29553	(Kitagawa and Paulson, 1994a)
ST6Gal I	ST6N	**Sia**(α2–6)- Gal(β1–4)GlcNAc-	Human placenta	2.4.99.1		S55689, S55693, S55697–9	(Wang *et al.*, 1993)
			Human			X17247	(Grundmann *et al.*, 1990)
			Human			X62822	(Bast *et al.*, 1992)
			Human			X54363	(Stamenkovic *et al.*, 1990)
			Human			L11720	(Aasheim *et al.*, 1993)
			Rat			M73985–7	(O'Hanlon *et al.*, 1989) (Wang *et al.*, 1990)
			Rat liver			M18769	(Weinstein *et al.*, 1987)
			Rat kidney			M83142–4	(Wen *et al.*, 1992b)
			Mouse liver			D16106	(Hamamoto *et al.*, 1993)
			Chick embryo			X75558	(Kurosawa *et al.*, 1994b)

Table 1. (continued)

Abbreviations		Enzyme Product	Tissue	EC No.	Comments	Acc.No.	References
ST6Gal NAc I	ST6O I	**Sia**(α2–6)GalNAc-	Chick embryo	2.4.99.3	Acts on R-GalNAc-where R can be H-, Gal(β1–3)- or Sia(α2–3) Gal(β1–3)-	X74946	(Kurosawa *et al.*, 1994a)
ST6GalNAc II	ST6O III	**Sia**(α2–6)-[Gal(β1–3)]GalNAc-	Chick testis		Acts on R-GalNAc-where R can be Gal(β1–3)- or Sia(α2–3) Gal(β1–3)-	X77775	(Kurosawa *et al.*, 1994c)
ST6GalNAc III	ST6O II	Sia(α2–3)Gal(β1–3)-**Sia**(α2–6)]GalNAc-	Fetal calf liver Pig submax. gland Sheep submax. gland	2.4.99.7	Acts only on Sia(α2–3)-Gal(β1–3) GalNAc-	Not cloned	(Bergh and van den Eijnden, 1983)
ST8Sia I	SAT-2	GD$_3$	Human melanoma Human melanoma Human Mouse brain	2.4.99.8	GD$_3$ synthase; no activity towards glycoproteins	X77922 D26360 L32867 X84235	(Sasaki *et al.*, 1994b) (Nara *et al.*, 1994) (Haraguchi *et al.*, 1994) (Yoshida *et al.*, 1995a)
ST8Sia II	STX	[**Sia**(α2–8)]$_n$Sia(α2–3)-Gal(β1–3/4)GlcNAc-	Human fetal brain Human Mouse brain Rat brain		Fetal and neonatal brain; acts on Sia(α2–3)-; is a polysialic acid synthase; no action on glycolipids	L29556 U33551 X83562 L13445	(Kitagawa, Paulson, 1994b) (Scheidegger *et al.*, 1995) (Kojima *et al.*, 1995a) (Livingston, Paulson, 1993)
ST8Sia III		**Sia**(α2–8)Sia(α2–3)-Gal(β1–4)GlcNAc-	Mouse brain		*N*-glycans > glycolipids; in brain and testis	X80502	(Yoshida *et al.*, 1995a)
ST8Sia IV		[**Sia**(α2–8)]$_n$Sia(α2–3)-Gal(β1–3/4)GlcNAc-	Hamster fibroblasts Mouse lung Human fetal brain		Acts on Sia(α2–3)-; is a polysialic acid synthase	Not listed X86000 L41680	(Eckhardt *et al.*, 1995) (Yoshida *et al.*, 1995b) (Nakayama *et al.*, 1995)

Abbreviations: ST, sialyltransferase; Sia, sialic acid; the first column shows the most recent nomenclature proposed by J.C.Paulson and S.Tsuji based on the chronological order of published sequences; the second column lists other nomenclatures previously used by J.C.Paulson, S.Tsuji, J.T.Y.Lau and S.Basu; Acc.No. is the EMBL/GenBank data base accession number.

3.2.1 CMP-sialic acid:Gal(β1–4)GlcNAc-R α2,6-sialyltransferase (ST6N, E.C. 2.4.99.1)

CMP-sialic acid:Gal(β1–4)GlcNAc-R α2,6-sialyltransferase (ST6N) catalyzes the incorporation of sialic acid in (α2–6) linkage to the Gal(β1–4)GlcNAc termini of *N*-glycans, with a marked preference for the Man(α1–3) arm in biantennary substrates (Joziasse *et al.*, 1987; Kurosawa *et al.*, 1994b). Studies with modified Gal(β1–4)GlcNAc-R substrates have shown that the 2-acetamido group of GlcNAc and the 6-hydroxy group of Gal are essential for activity but Gal can be replaced by sugars such as Man, Glc, GlcNAc and GalNAc (van Pelt *et al.*, 1989; Hokke *et al.*, 1993; Wlasichuk *et al.*, 1993). Cell surface sialic acid variations have been shown to occur with cell transformation and tumour progression (Dennis, 1992). In general, cells with reduced sialylation show attenuated metastatic potential although over-expression of ST6N also causes reduction in metastatic phenotype (Takano *et al.*, 1994). On the other hand, transformation of rat fibroblasts with the oncogene *c-Ha-ras* induced a 10-fold increase in ST6N due to elevated mRNA levels and protein expression (Le Marer *et al.*, 1992; Delannoy *et al.*, 1993) whereas other oncogenes did not have this effect; the ST6N levels of *ras*-transformed fibroblasts correlated with increased invasiveness (Le Marer and Stéhelin, 1995). Human cancer has also been associated with high ST6N levels (Dall'Olio *et al.*, 1989, 1992).

Control of α2,6-sialyltransferase gene transcription

Analysis of rat (Weinstein *et al.*, 1987) and human (Lance *et al.*, 1989; Grundmann *et al.*, 1990; Bast *et al.*, 1992) cDNAs encoding ST6N indicate that the enzyme has a typical glycosyltransferase domain structure (Figure 3). The rat ST6N gene spans at least 80 kb of genomic DNA and contains at least 11 exons. There are at least three promoters responsible for the production of five or more messages from a single gene sequence (O'Hanlon *et al.*, 1989; Paulson *et al.*, 1989; Wang *et al.*, 1990; Wen *et al.*, 1992b). Promoter P_L produces a liver-specific 4.3 kb mRNA, promoter P_C is constitutive in several tissues and produces a 4.7 kb mRNA and promoter P_K is restricted to the kidney and produces several 3.6 kb mRNAs missing the 5'-half of the coding sequence. The proteins encoded by the 3.6 kb messages are not catalytically active (O'Hanlon and Lau, 1992). There is good evidence that the human gene also has at least three promoters (Aasheim *et al.*, 1993; Wang *et al.*, 1993).

Cell type-specific transcription of genes is controlled by transcription factors which themselves have a restricted pattern of expression. Promoter P_L was demonstrated to be about 50-fold more active in a hepatoma cell line (HepG2) known to express ST6N than in a cell line (Chinese hamster ovary) which does not express this enzyme (Svensson *et al.*, 1990). The P_L promoter sequence contains consensus binding sites for liver-restricted transcription factors (Svensson *et al.*, 1992) thereby accounting for the high level expression of ST6N in rat liver.

Targeting to the Golgi apparatus

Rat ST6N has been localized to the *trans*-Golgi cisternae and the *trans*-Golgi network of hepatocytes, hepatoma cells and intestinal goblet cells but may be more diffusely distributed in the Golgi apparatus of other cell types (Colley *et al.*, 1989). The cytoplasmic, transmembrane and stem domains of ST6N are not required for catalytic activity. The transmembrane domain is, however, essential for proper Golgi targeting and both the stem and cytoplasmic domains may also play a role which is as yet not understood (Munro, 1991; Wong *et al.*, 1992; Dahdal and Colley, 1993).

The sialylmotif

Comparison of the amino acid sequences of ST6N, CMP-sialic acid: Gal (β1–3)GalNAc-R α2,3-sialyltransferase (ST3O) and CMP-sialic acid:Gal(β1–3/4)GlcNAc-R α2,3-sialyltransferase I (ST3N I) reveals a region of 55 amino acids with extensive homology (40 % identity; 58 % conservation) in the middle of the catalytic domain (Wen *et al.*, 1992a). There is a second 23 residue region of similarity near the COOH-terminus (Drickamer, 1993). This so-called "sialylmotif" has been used to clone chicken ST6N (Kurosawa *et al.*, 1994b) and several other sialyltransferases (Table 1). Site-directed mutagenesis experiments have suggested that the sialylmotif participates in the binding of CMP-sialic acid to the enzyme (Datta and Paulson, 1995).

3.2.2 CMP-sialic acid:Gal(β1–3/4)GlcNAc-R α2,3-sialyltransferase I (ST3N I, E.C. 2.4.99.6)

Recombinant rat liver (Wen *et al.*, 1992a) and human placental (Kitagawa and Paulson, 1993) CMP-sialic acid:Gal(β1–3/4)GlcNAc-R α2,3-sialyltransferase I (ST3N I) catalyze the incorporation of sialic acid in (α2–3) linkage to both the Gal(β1–3)GlcNAc and Gal(β1–4)GlcNAc termini of *N*-glycans with a marked preference for the type 1 chain, Gal(β1–3)GlcNAc-, thereby indicating that this enzyme is different from the previously reported human placental enzyme which showed preferential activity towards type 2 chains, Gal(β1–4)GlcNAc- (Nemansky and van den Eijnden, 1993). The amino acid sequences deduced from rat (Wen *et al.*, 1992a) and human (Kitagawa and Paulson, 1993) cDNAs encoding ST3N I indicated a domain structure typical of the glycosyltransferases (Figure 3). Expression of a truncated form of ST3N I lacking the cytoplasmic and transmembrane domains showed that these domains are not required for enzyme activity but are necessary for targeting to the Golgi apparatus.

3.2.3 CMP-sialic acid:Gal(β1–3/4)GlcNAc-R α2,3-sialyltransferase II (ST3N II)

Selection with a cytotoxic lectin and an expression cloning approach were used to clone a human melanoma cDNA encoding CMP-sialic acid:Gal(β1–3/4)GlcNAc-R α2,3-sialyltransferase II (ST3N II) which differs from the previously cloned ST3N I (above) in showing a 3:1 preferential activity towards type 2 chains, Gal(β1–4) GlcNAc, relative to type 1 chains, Gal(β1–3) GlcNAc (Sasaki *et al.*, 1993). Human ST3N I and II have only 34 % homology but both enzymes have the sialylmotif typical of all sialyltransferases cloned to date. The relationship of ST3N II to a similar human placental enzyme (Nemansky and van den Eijnden, 1993) is as yet unknown.

3.2.4 CMP-sialic acid:Gal(β1–4)GlcNAc-R/Gal(β1–3)GalNAc-R α2,3-sialyltransferase (STZ, SAT-3, ST3Gal IV)

The sialylmotif was used to clone a novel CMP-sialic acid:Gal(β1–4)GlcNAc-R/Gal(β1–3)Gal NAc-R α2,3-sialyltransferase (STZ, SAT-3, ST3Gal IV) (Table 1) which transfers sialic acid in α2–3-linkage to the terminal Gal of Gal(β1–4)GlcNAc or Gal(β1–3)GalNAc of oligosaccharide, glycoprotein and glycolipid acceptors (Kitagawa and Paulson, 1994a). The human enzyme could not act on Gal(β1–3)GlcNAc termini (Kitagawa and Paulson, 1994b) and was therefore different from ST3N I and II described above. The enzyme may be the glycolipid α2,3-sialyltransferase SAT-3 (Basu, 1991).

3.2.5 Brain specific CMP-sialic acid: [sialyl(α2–8)]$_n$sialyl(α2–3)Gal(β1–3/4) GlcNAc-R α2,8-sialyltransferase (STX, ST8Sia II)

There are at least four α2–8-sialyltransferases (Table 1) which attach sialic acid in (α2–8) linkage to an (α2–3) or (α2–8) linked sialic acid. Three of these (ST8Sia II, III and IV) attach sialic acid to *N*-glycan antennae.

The sialylmotif was used to clone cDNAs encoding a novel CMP-sialic acid:[sialyl(α2–8)]$_n$sialyl(α2–3)Gal(β1–3/4)GlcNAc-R α2,8-sialyltransferase (STX, ST8Sia II) which is expressed in the brains of fetal and newborn rats (Livingston *et al.*, 1992; Livingston and Paulson, 1993) and mice (Kojima *et al.*, 1995a,b) but not in adult brain or any other adult or newborn tissues. The expression of ST8Sia II is therefore both tissue-specific and developmentally regulated. The enzyme was inactive towards non-sialylated substrates and gangliosides but transferred sialic acid in (α2–8) linkage to the terminal sialic acid of sialyl(α2–3)Gal(β1–3/4)GlcNAc-glycoproteins (Kojima *et al.*, 1995a) and was also able to extend these chains with polysialic acid [sialyl(α2–8)]$_n$ (Kojima *et al.*, 1995b; Scheidegger *et al.*, 1995).

3.2.6 CMP-sialic acid:sialyl(α2–3) Gal(β1–4)GlcNAc-R α2,8-sialyltransferase (ST8Sia III)

The sialylmotif was used to clone two other mouse genes encoding α2,8-sialyltransferases which act on *N*-glycan antennae (ST8Sia III and IV, Table 1) (Yoshida *et al.*, 1995a,b). ST8Sia III (Yoshida *et al.*, 1995a) attaches sialic acid in (α2–8) linkage to the terminal sialic acid of both protein- and lipid-linked sialyl(α2–3)Gal(β1–4) GlcNAc whereas ST8Sia I acts only on sialylated glycolipids and ST8Sia II and IV both act only on sialylated glycoproteins. ST8Sia III is expressed only in brain and testis whereas ST8Sia IV is strongly expressed in lung, heart and spleen but only weakly in brain.

3.2.7 CMP-sialic acid:[sialyl(α2–8)]$_n$sialyl (α2–3)Gal(β1–3/4)GlcNAc-R α2,8-sialyltransferase (ST8Sia IV)

Mouse ST8Sia IV (Yoshida *et al.*, 1995b) exhibits relatively low amino acid sequence identities with ST8Sia I (15 %), II (56 %) and III (26 %) but shows 99 % identity with recently cloned hamster (Eckhardt *et al.*, 1995) and human fetal brain (Nakayama *et al.*, 1995) polysialyltransferases. ST8Sia IV can initiate polysialic acid synthesis by attaching an (α2–8)-linked sialic acid to an (α2–3)-linked sialic acid and can also synthesize polysialic acid *in vitro*. Polysialic acid is a developmentally regulated product of post-translational modification of the neural cell adhesion molecule (*N*-CAM) and may act as a regulator of *N*-CAM mediated cell-cell adhesion (Troy, 1992). *N*-CAM with polysialic acid is abundant in embryonic brain but adult *N*-CAM lacks this glycan structure. Hamster polysialyltransferase induced polysialic acid synthesis in all *N*-CAM-expressing cell lines tested (Eckhardt *et al.*, 1995). The human enzyme (Nakayama *et al.*, 1995) can attach polysialic acid to *N*-CAM and is strongly expressed in fetal brain but weakly in adult brain although it is expressed in other adult tissues. The roles of ST8Sia II, III and IV in nervous system development remain to be determined.

3.3 *N*-Acetylglucosaminyltransferases

Six *N*-acetylglucosaminyltransferases (GnT I to VI) are involved in the synthesis of complex *N*-glycans (Figure 2, Table 2) (Schachter, 1986, 1991b, 1995). The molecular biology of the *N*-acetylglucosaminyltransferases has been reviewed recently (Taniguchi and Ihara, 1995).

3.3.1 UDP-GlcNAc:Man(α1–3)R [GlcNAc to Man(α1–3)] β1,2-*N*-acetylglucosaminyltransferase I (GnT I, EC 2.4.1.101)

GnT I controls the synthesis of hybrid and complex *N*-glycans (Figure 1) by initiating the synthesis of the first antenna. The enzyme has been purified from several species and was thoroughly characterized (reviewed in Schachter 1995). The genes for rabbit, human, mouse, rat, chicken and frog GnT I have been cloned (Table 2) and there are at least 3 *Caenorhabditis elegans* genes listed in the EMBL/GenBank data base which are homologous to mammalian GnT I. The amino acid sequences of the mammalian enzymes are over 90 % identical but there is no sequence homology to any other known glycosyltransferase. GnT I has the type II integral membrane protein domain structure typical of all cloned Golgi glycosyltransferases (Figure 3). The transmembrane segment of GnT I is essential for retention in medial-Golgi cisternae but the other domains also play a role (Burke *et al.*, 1992, 1994). The protein expressed in ldlD mutant CHO cells is not *N*-glycosylated because there are no Asn-X-Ser(Thr) sequons but it is *O*-glycosylated (Nilsson *et al.*, 1994; Hoe *et al.*, 1995).

Part of the 5'-untranslated region and all of the coding and 3'-untranslated regions of the human and mouse GnT I genes are on a single 2.5 kb

Table 2. *N*-acetylglucosaminyltransferases involved in *N*- and *O*-glycan synthesis

Abbrev.	Enzyme Product	Tissue	EC No.	Comments	Acc.No.	References
GnT I	**GlcNAc**(β1–2)- Man(α1–3)Manβ	Rabbit Human Human Mouse Mouse Mouse Rat Chicken Frog	2.4.1.101	First step towards hybrid and complex *N*-glycans	M57301 M61829 M55621 X77487–8 M73491 L07037 D16302 Not listed Not listed	(Sarkar *et al.*, 1991) (Hull *et al.*, 1991) (Kumar *et al.*, 1990) (Yang *et al.*, 1994a) (Pownall *et al.*, 1992) (Kumar *et al.*, 1992) (Fukada *et al.*, 1994) (Narasimhan *et al.*, 1993) (Mucha *et al.*, 1995)
GnT II	**GlcNAc**(β1–2)-Man(α1–6)Manβ	Human Rat Frog	2.4.1.143	Synthesis of biantennary *N*-glycans	U15128, L36537 U21662 Not listed	(Tan *et al.*, 1995a) (D'Agostaro *et al.*, 1995) (Mucha *et al.*, 1995)
GnT III	**GlcNAc**(β1–4)-Manβ	Human Rat	2.4.1.144	Synthesis of bisected *N*-glycans	D13789 D10852	(Ihara *et al.*, 1993) (Nishikawa *et al.*, 1992)
GnT IV	**GlcNAc**(β1–4)- [GlcNAc(β1–2)]Man(α1–3)Manβ	Hen oviduct	2.4.1.145	Initiation of antenna on Man(α1–3) arm	Not cloned	(Gleeson and Schachter, 1983)
GnT V	**GlcNAc**(β1–6)- [GlcNAc(β1–2)]Man(α1–6)Manβ	Human Rat	2.4.1.155	Initiation of antenna on Man(α1–6) arm	D17716 L14284	(Saito *et al.*, 1994) (Shoreibah *et al.*, 1993)
GnT VI	**GlcNAc**(β1–4)- [GlcNAc(β1–2)][GlcNAc(β1–6)] Man(α1–6)Manβ	Chicken		Initiation of antenna on Man(α1–6) arm	Not cloned	(Brockhausen *et al.*, 1989)
C2β6GnT	**GlcNAc**(β1–6)- [Gal(β1–3)]GalNAc-R	Human Human Mouse	2.4.1.102	Synthesis of core 2 *O*-glycans	M97347 L41415, L41390 U19265	(Bierhuizen and Fukuda, 1992) (Bierhuizen *et al.*, 1995) (Warren *et al.*, 1995)
C3β3GnT	**GlcNAc**(β1–3)GalNAc-R	Pig, rat	2.4.1.147	Core 3 *O*-glycans	Not cloned	(Brockhausen *et al.*, 1985)
C4β6GnT	**GlcNAc**(β1–6)- [GlcNAc(β1–3)]GalNAc-R	Pig, rat	2.4.1.148	Core 4 *O*-glycans	Not cloned	(Brockhausen *et al.*, 1985)
iβ3GnT	**GlcNAc**(β1–3)Galβ-	Human	2.4.1.149	Blood group i epitope	Not cloned	(Schachter and Brockhausen, 1992)
Iβ6GnT I	**GlcNAc**(β1–6)- [GlcNAc(β1–3)]Galβ-	Human Human	2.4.1.150	Blood group I epitope	L41605–7 L19659, Z19550	(Bierhuizen *et al.*, 1995) (Bierhuizen *et al.*, 1993)
Iβ6GnT II	**GlcNAc**(β1–6)- [Gal(β1–4>)GlcNAc(β1–3)]Galβ-	Human Rat		Blood group I epitope due to internal branching	Not cloned	(Leppänen *et al.*, 1991) (Gu *et al.*, 1992)
Elongation β3GnT	**GlcNAc**(β1–3)- Gal(β1–3)(R1)GalNAc-R2	Pig	2.4.1.146	Elongates core 1 and 2 *O*-glycans	Not cloned	(Brockhausen *et al.*, 1983) (Brockhausen *et al.*, 1984)

Abbreviations: GnT, *N*-acetylglucosaminyltransferase; Acc.No. is the EMBL/GenBank data base accession number

exon (Hull *et al.*, 1991; Kumar *et al.*, 1992; Pownall *et al.*, 1992). The remaining 5'-untranslated sequence of the gene is on a different exon (or exons) at least 5 kb upstream. The human GnT I gene has been localized to chromosome 5q35. There is only a single copy of the gene in the haploid human and mouse genomic DNA.

Mouse embryos lacking a functional GnT I gene die prenatally at about 10 days of gestation (Ioffe and Stanley, 1994; Metzler *et al.*, 1994). Among the anatomical defects noted was a failure of normal neural tube development. These findings indicate that complex *N*-glycans are essential for normal embryogenesis and development, particularly of the nervous system.

3.3.2 UDP-GlcNAc:Man(α1–6)R [GlcNAc to Man(α1–6)] β1,2-*N*-acetylglucosaminyltransferase II (GnT II, E.C. 2.4.1.143)

GnT II initiates the first antenna on the Man (α1–6) arm of the *N*-glycan core and is therefore essential for normal complex *N*-glycan formation (Schachter, 1995). The human and rat GnT II genes have been cloned (Table 2). The enzyme has a typical glycosyltransferase domain structure (Figure 3). The open reading frame and 3'-untranslated region of the human and rat genes are on a single exon. There is no sequence homology to any previously cloned glycosyltransferase including GnT I. The human GnT II gene is on chromosome 14q21.

Fibroblasts from two children with Carbohydrate-Deficient Glycoprotein Syndrome (CDGS) Type II have been shown to lack GnT II activity (Jaeken *et al.*, 1994) due to point mutations in the GnT II gene (Charuk *et al.*, 1995; Tan *et al.*, 1995b). CDGS is a group of autosomal recessive diseases with multisystemic abnormalities including a severe disturbance of nervous system development (Jaeken *et al.*, 1993).

3.3.3 UDP-GlcNAc:R_1-Man(α1–6)[GlcNAc (β1–2)Man(α1–3)]Man(β1–4)R_2 [GlcNAc to Man(β1–4)] β1,4-*N*-acetylglucosaminyltransferase III (GnT III, E.C. 2.4.1.144)

GnT III incorporates a bisecting GlcNAc residue into the *N*-glycan core. It is of interest because it causes a STOP signal, i.e., several enzymes involved in *N*-glycan synthesis cannot act on bisected substrates as discussed above (Figure 1). The enzyme has been shown to be elevated in various types of rat hepatoma (Narasimhan *et al.*, 1988; Ishibashi *et al.*, 1989; Nishikawa *et al.*, 1990; Miyoshi *et al.*, 1993) and human leukemia (Yoshimura *et al.*, 1995a,b). The genes for human and rat GnT III have been cloned (Table 2). The enzyme has a typical glycosyltransferase domain structure (Figure 3). There is no apparent sequence similarity to any known glycosyltransferase. Analysis of the human GnT III gene shows that the entire coding region is on a single exon on chromosome 22q.13.1.

3.3.4 UDP-GlcNAc:R_1Man(α1–3)R_2 [GlcNAc to Man(α1–3)] β1,4-*N*-acetylglucosaminyltransferase IV (GnT IV, E.C. 2.4.1.145)

GnT IV adds a GlcNAc in (β1–4) linkage to the Man(α1–3)Manβ- arm of the *N*-glycan core (Gleeson and Schachter, 1983). The enzyme has not been purified nor has the gene been cloned. Simple oligosaccharides of the form GlcNAc(β1–2)Manα-R can be converted to GlcNAc(β1–4)[GlcNAc(β1–2)]Manα-R by hen oviduct extracts, possibly due to the action of GnT IV (Brockhausen *et al.*, 1992b).

3.3.5 UDP-GlcNAc:R_1Man(α1–6)R_2 [GlcNAc to Man(α1–6)] β1,6-*N*-acetylglucosaminyltransferase V (GnT V, E.C. 2.4.1.155)

GnT V adds a GlcNAc in (β1–6) linkage to the Man(α1–6)Manβ arm of the *N*-glycan core. One of the most common alterations in transformed or malignant cells is the presence of larger *N*-glycans due primarily to a combination of increased GlcNAc branching, sialylation and poly-*N*-acetyllactosamine content (Smets and van Beek, 1984; Dennis *et al.*, 1987; Dennis and Laferte, 1988; Dennis, 1992). GnT V plays a major role in these effects. Polyoma virus, Rous sarcoma virus- and T24 H-ras-transformed cells were shown to have significantly increased GnT V activity (Yamashita *et al.*, 1985; Arango and Pierce, 1988; Dennis *et al.*, 1989; Crawley *et al.*, 1990; Palcic *et al.*, 1990; Lu and Chaney, 1993). Transforming growth factor β was found to up-regulate expression of GnT V in mouse melanoma cells (Miyoshi *et al.*, 1995). Poly-*N*-acetyllactosamine chains have been shown to carry cancer-associated antigens and the initiation of these chains is favored on the antenna initiated by GnT V (Cummings and Kornfeld, 1984; van den Eijnden *et al.*, 1988; Dennis, 1992). Transfection of the GnT III gene into a highly metastatic mouse melanoma cell line resulted in decreased (β1–6) branching of *N*-glycans without altering GnT V enzyme levels (Yoshimura *et al.*, 1995c) due to the fact that GnT V cannot act on bisected *N*-glycans (Schachter, 1986); the GnT III-transfected cells showed a marked reduction in metastatic potential. Transfection of the GnT V gene into premalignant epithelial cells resulted in relaxation of growth controls and reduced substratum adhesion (Demetriou *et al.*, 1995). Evidence has been obtained for post-translational activation of GnT V by phosphorylation (Ju *et al.*, 1995). The genes for human and rat GnT V have been cloned (Table 2) and the human gene has been mapped to chromosome 2q21. The human GnT V gene has 17 exons spanning 155 kb (Saito *et al.*, 1995).

3.3.6 UDP-GlcNAc:$R_1(R_2)$Man(α1–6)R_3 [GlcNAc to Man(α1–6)] β1,4-*N*-acetylglucosaminyltransferase VI (GnT VI)

GnT VI adds a GlcNAc residue in (β1–4) linkage to the Man(α1–6)Manβ arm of the *N*-glycan core (Brockhausen *et al.*, 1989). The minimum structural requirement for GnT VI substrate is the trisaccharide GlcNAcβ1–6[GlcNAc(β1–2)]Manα-, i.e., the enzyme requires the prior actions of GnT I, II and V. Unlike *N*-glycan GnT I to V, GnT VI can act on both bisected and non-bisected substrates. The enzyme has been demonstrated in avian but not in mammalian tissues. The enzyme has not been purified nor has the gene been cloned.

3.3.7 UDP-GlcNAc- GalR β1,3-*N*-acetylglucosaminyltransferase synthesizing poly *N*-acetyllactosamine chains (i β3GnT)

i β3GnT synthesizes the GlcNAcβ1–3Gal linkages and therefore the poly *N*-acetyllactosamine chains (repeating Galβ1–4 GlcNAcβ1–3 units, i antigen) of *N*- and *O*-glycans and glycolipids. i β3GnT is ubiquitous and has been described in many different cell types and in human serum. The gene coding for the enzyme has not yet been cloned. Similar activities synthesizing the GlcNAcβ1–3Galβ linkage can be distinguished from the i β3GnT by their *in vitro* properties and tissue distribution (Brockhausen *et al.*, 1983, 1984; Stults and Macher, 1993). This suggests that there is a family of β3GnT.

3.3.8 UDP-GlcNAc [+/-Galβ1–4] GlcNAcβ1–3 Gal-R (GlcNAc to Gal) β1,6-*N*-acetylglucosaminyltransferase (I β6GnTI, I β6GnTII)

The i antigen which is present at an early developmental stage in erythrocytes is replaced by the branched I antigen, Galβ1–4 GlcNAcβ1–6 (Galβ1–4 GlcNAcβ1–3) Gal- in a developmentally regulated fashion (Feizi, 1985). The I antigen is found on many glycoproteins and mucins. A number of different activities that can be distinguished by their substrate specificities synthesize the GlcNAcβ1–6 branch by alternate pathways (Brockhausen, 1995). One of these activities, I β6GnT I from pig gastric mucosa, adds a GlcNAcβ1–6 branch to the Gal residue of GlcNAcβ1–3Gal- terminal structures (Piller *et al.*, 1984; Brockhausen *et al.*, 1986). In contrast, Iβ6GnTII from human serum and rat intestine acts on Galβ1–4 GlcNAcβ1–3Gal- substrates by adding a GlcNAcβ1–6 branch to the internal Gal residue (Leppänen *et al.*, 1991; Gu *et al.*, 1992). Activities have also been described that synthesize linear GlcNAcβ1–6 Gal linkages (Koenderman *et al.*, 1987). Although none of these enzymes has been purified to homogeneity, the cDNA encoding one of these enzymes has been cloned from human teratocarcinoma cells by expression cloning using an antibody against the I antigen (Bierhuizen *et al.*, 1993). The cDNA sequence revealed a protein with homology in the putative catalytic domain to core 2 β6GnT but not to other glycosyltransferases. The Iβ6GnT gene was localized to chromosome 9 q21 (Bierhuizen *et al.*, 1993).

3.4 Fucosyltransferases

Fucose is found in *N*- and *O*-glycans and glycolipids attached to Gal or GlcNAc residues in α-linkage to carbons 2, 3, 4 or 6. Several fucosylated structures form human blood group antigenic epitopes (A, B, H, Le^a, Le^b, Le^x and Le^y). Fucosylated oligosaccharides are often expressed in a regulated manner in development, differentiation and progression of metastasis (Gooi *et al.*, 1981; Hakomori, 1981, 1989; Feizi, 1988, 1990). Fucosyltransferases are expressed in a tissue-specific fashion (Brockhausen, 1995).

3.4.1 Human Blood Group H and Se GDP-Fuc:Galβ-R α1,2-Fucosyltransferases (α2FucT)

There are at least two human GDP-Fuc: Galβ-R α1,2-fucosyltransferases (α2FucT), encoded respectively by the H (E.C. 2.4.1.69) and Se (secretory) loci (Le Pendu *et al.*, 1985; Sarnesto *et al.*, 1992a; Kelly *et al.*, 1994). The H locus is expressed mainly in tissues derived from mesoderm (hematopoietic tissues, plasma) or ectoderm while Se locus expression is restricted to tissues derived from endoderm (e.g. epithelial cells lining salivary glands, stomach and intestine) (Oriol, 1990). The H and Se loci are closely linked on human chromosome 19q13.3 separated by only about 35 kb (Mollicone *et al.*, 1988; Rouquier *et al.*, 1995). The genes for both enzymes have been cloned (Table 3) and encode proteins with the typical glycosyltransferase domain structure (Figure 3). The non-secretor phenotype occurs in about 20 % of individuals and at least some of these are due to homozygosity for a nonsense allele at the Se locus (Kelly *et al.*, 1995). Neither the H nor Se gene is essential for normal human survival but inactivation of these genes (such as occurs in the rare Bombay and para-Bombay phenotypes) can cause serious problems if blood transfusion is required (Kelly *et al.*, 1994).

3.4.2 Human Blood Group Lewis GDP-Fuc:Gal(β1–4/3)GlcNAc (Fuc to GlcNAc) α1,3/4-Fucosyltransferase III (FucT III, E.C. 2.4.1.65)

There are at least five distinct human GDP-Fuc: Gal(β1–4)GlcNAc (Fuc to GlcNAc) α1,3-fucosyltransferases (α3FucT) (Tetteroo *et al.*, 1987; Mollicone *et al.*, 1990, 1992; Couillin *et al.*, 1991; Macher *et al.*, 1991) named FucT III to VII (Table 3). FucT III (the Lewis type enzyme) has the broadest α3FucT substrate specificity since it can incorporate Fuc in either (α1–3) linkage to Gal(β1–4)GlcNAc (to make Le^x) or (α1–4) linkage to Gal(β1–3)GlcNAc (to make Le^a) even when the Gal residue in these structures is substituted by a fucose (to make Le^y or Le^b respectively) or sialic acid (to make Sialyl Le^x or Sialyl Le^a respectively). The gene for human FucT III has been cloned (Table 3) and contains an open reading frame on a single exon mapping to human chromosome 19. The protein encoded by this gene shows the domain structure typical of all glycosyltransferases cloned to date (Figure 3). FucT III, V and VI share approximately 85–90 % amino acid sequence identity while FucT IV shares only about 60 % identity with the other α3FucTs. The α3FucT family shows no sequence similarities to any other glycosyltransferases.

Table 3. Fucosyltransferases involved in *N*- and *O*-glycan synthesis

Abbrev.	Enzyme Product	Tissue	EC No.	Comments	Acc.No.	References
Hα2FucT	**Fuc**(α1–2)GalβR	Human Rat Rabbit	2.4.1.69	Blood group H epitope	M35531 L26009–10 X80225	(Larsen *et al.*, 1990a) (Piau *et al.*, 1994) (Hitoshi *et al.*, 1995)
Seα2FucT	**Fuc**(α1–2)GalβR	Human Rabbit		Secretory gene	U17894–5 X80226	(Kelly *et al.*, 1995) (Hitoshi *et al.*, 1995)
FucT III	**Fuc**(α1–3/4)-[Gal(β1–4/3)]GlcNAc-	Human	2.4.1.65	Broad specificity, makes Le^a, Le^b, Le^x, Le^y,	X53578 U27326–8 S52874,	(Kukowska-Latallo *et al.*, 1990) (Cameron *et al.*, 1995) (Nishihara *et al.*, 1993)
		Cow		SiaLe^a, SiaLe^x; Lewis type	S52967–9 X87810	(Oulmouden *et al.*, 1995)
FucT IV	**Fuc**(α1–3)-[Gal(β1–4)]GlcNAc-	Human		Narrow specificity, makes only Le^x: myeloid type	M65030 S65161 M58596–7	(Lowe *et al.*, 1991a) (Kumar *et al.*, 1991) (Goelz *et al.*, 1990)
		Mouse			U33457–8	(Gersten *et al.*, 1995)
FucT V	**Fuc**(α1–3)-[Gal(β1–4)]GlcNAc-	Human		Makes Le^x, SiaLe^x	M81485 U23729–30	(Weston *et al.*, 1992a) (Cameron *et al.*, 1995)
FucT VI	**Fuc**(α1–3)-[Gal(β1–4)]GlcNAc-	Human		Makes Le^x, SiaLe^x; plasma type	L01698 M98825 U27331–7 S52874, S52967–9	(Weston *et al.*, 1992b) (Koszdin and Bowen, 1992) (Cameron *et al.*, 1995) (Nishihara *et al.*, 1993)
FucT VII	**Fuc**(α1–3)-[Gal(β1–4)]GlcNAc-	Human		Makes SiaLe^x but not Le^x; candidate for making E-selectin ligand on leukocyte cell surface	X78031 U08112, U11282	(Sasaki *et al.*, 1994a) (Natsuka *et al.*, 1994)
core α6FucT	**Fuc**(α1–6)-[R]GlcNAc-Asn-X	Mammals Insects	2.4.1.68	Makes core Fuc(α1–6) GlcNAc-Asn-X	Not cloned	(Longmore and Schachter, 1982) (Staudacher *et al.*, 1992b)
core α3FucT	**Fuc**(α1–3)-[R]GlcNAc-Asn-X	Plants Insects		Makes core Fuc(α1–3) GlcNAc-Asn-X	Not cloned	(Johnson and Chrispeels, 1987) (Staudacher *et al.*, 1991)

Abbreviations: FucT, Fucosyltransferases; Le, Lewis; Sia, sialic acid; Acc.No. is the EMBL/GenBank data base accession number.

3.4.3 Human GDP-Fuc:Gal(β1–4)GlcNAc (Fuc to GlcNAc) α1,3-Fucosyltransferase IV (FucT IV, myeloid type)

Human FucT IV-VII differ from FucT III in that they cannot form the Fuc(α1–4)GlcNAc linkage. FucT IV, found in myeloid tissues, has the narrowest acceptor specificity in this group since it is not effective in the synthesis of the sialyl(α2–3) Gal(β1–4)[Fuc(α1–3)]GlcNAc moiety (sialyl Le^x). The origin of the sialyl Le^x structure on neutrophils has aroused a great deal of recent interest because sialyl Le^x is the most likely ligand for an endothelial cell receptor called endothelial-leukocyte adhesion molecule-1 (ELAM-1; E-selectin) which is essential for a normal inflammatory response involving inflammation-activated homing of leukocytes to endothelial cells (Brandley *et al.*, 1990; Lowe *et al.*, 1990; Phillips *et al.*, 1990; Walz *et al.*, 1990; Kumar *et al.*, 1991; Lowe *et al.*, 1991a,b; Macher *et al.*, 1991; Polley *et al.*, 1991). Although FucT IV is a myeloid enzyme, it is probably not responsible for synthesis of the E-selectin ligand since it is relatively ineffective in sialyl Le^x synthesis; more likely, it is FucT VII which carries out the synthesis of neutrophil sialyl Le^x (Natsuka *et al.*, 1994). The gene for FucT IV has been cloned (Table 3) and maps to chromosome 11q21 (Reguigne *et al.*, 1994). The open reading frame is on a single exon and encodes a protein with a domain structure typical for the glycosyltransferases (Figure 3).

3.4.4 Human GDP-Fuc:Gal(β1–4)GlcNAc (Fuc to GlcNAc) α1,3-Fucosyltransferases V and VI (FucT V, VI)

A probe prepared from the cDNA encoding human FucT III was used to screen a human genomic DNA library under conditions of low stringency and two genes were isolated (Table 3), each with an open reading frame on a single exon, encoding two proteins (FucT V and VI) with domain structures typical of glycosyltransferases (Figure 3) (Weston *et al.*, 1992a,b). Although Fuc T V and VI differ quantitatively in substrate specificity, both can synthesize Le^x and sialyl-Le^x but not Le^a nor sialyl-Le^a. Comparison of FucT V and VI substrate specificities with those of the plasma-type (Sarnesto *et al.*, 1992b) and lung carcinoma-type (Macher *et al.*, 1991) enzymes indicates that FucT VI is probably the plasma type enzyme but that neither enzyme type corresponds to FucT V. Study of an Indonesian population which lacks plasma α3FucT activity proves that FucT VI is the plasma-type enzyme (Mollicone *et al.*, 1994). The genes for FucT III and VI are closely linked on human chromosome 19 (Nishihara *et al.*, 1993).

3.4.5 Human GDP-Fuc:Gal(β1–4)GlcNAc (Fuc to GlcNAc) α1,3-fucosyltransferase VII (FucT VII)

The gene for a human leukocyte α3FucT has been cloned (FucT VII) which may be responsible for synthesis of the leukocyte cell surface E-selectin ligand (sialyl Le^x) required for inflammation-activated homing of leukocytes to endothelial cells (Natsuka *et al.*, 1994). FucT VII differs from the other α3FucTs in that it can make sialyl Le^x but not Le^x, Le^a nor sialyl Le^a. The gene for FucT VII is localized to chromosome 9 and shares about 39 % amino acid sequence identity with FucT III (a prototype of chromosome 19-localized FucT III, V and VI) and about 38 % with chromosome-localized FucT IV.

3.4.6 GDP-Fuc:β-*N*-acetylglucosaminide (Fuc to Asn-linked GlcNAc) α1,3- and α1,6-fucosyltransferases

Fucose can be transferred to the asparagine-linked *N*-acetylglucosamine of the *N*-glycan core either in (α1–6) linkage (mammals, insects) or (α1–3) linkage (plants, insects) by GDP-Fuc: β-*N*-acetylglucosaminide (Fuc to Asn-linked GlcNAc) α1,6-Fucosyltransferase (core α6FucT, E.C. 2.4.1.68) or α1,3-Fucosyltransferase (core α3FucT) respectively (Table 3). GlcNAc (β1–2) Man(α1–3)Manβ- on the *N*-glycan core, due to prior action of GnT I, is essential for both FucT (Longmore and Schachter, 1982; Shao *et al.*, 1994) and FucT (Johnson and Chrispeels, 1987; Staudacher *et al.*, 1991; Altmann *et al.*, 1993).

Human α6FucT (Voynow *et al.*, 1991) and mung bean α3FucT (Staudacher *et al.*, 1995) have been purified but neither enzyme has been cloned. Insects have been shown to contain both FucT and FucT which can act in concert to place two fucose residues on the same Asn-linked GlcNAc residue (Staudacher *et al.*, 1991, 1992a; Kubelka *et al.*, 1993, 1995; März *et al.*, 1995).

4 Biosynthesis of *O*-Glycans

Most glycoproteins contain *N*- and *O*-glycans while mucins carry mainly *O*-glycans. Mucins, the major glycoproteins of the mucus gel, are highly *O*-glycosylated and serve an important function in protecting the underlying mucosa. The biosynthesis of *O*-glycans starts mainly in the cis Golgi and therefore does not involve dolichol derivatives which are restricted to the ER. There is no processing by glycosidases; instead, sugars are transferred individually from nucleotide sugars to synthesize eight *O*-glycan core structures which may be further elongated and terminated to build hundreds of different *O*-glycan chains. Core 1, Galβ1–3 GalNAc-, and core 2, GlcNAcβ1–6 (Galβ1–3) GalNAc-, are found in most mucins and glycoproteins. Core 3, Glc NAcβ1–3 GalNAc-, and core 4, GlcNAcβ1–6 (GlcNAcβ1–3) GalNAc-, are common in mucins. The other core structures (Brockhausen, 1995) are less common. Many of the enzymes elongating and terminating *N*-glycans also act on *O*-glycans. These include the enzymes synthesizing linear and branched poly *N*-acetyllactosamine chains, α2-, α3- and α4-Fuc-transferases, the enzymes synthesizing the ABO blood groups and others. The enzymes which synthesize the common core structures 1 to 4, and elongate core 1 or 2, and the α3- and α6-sialyltransferases (ST3O and ST6O) are specific for the synthesis of *O*-glycans (Figure 4) and will be discussed below.

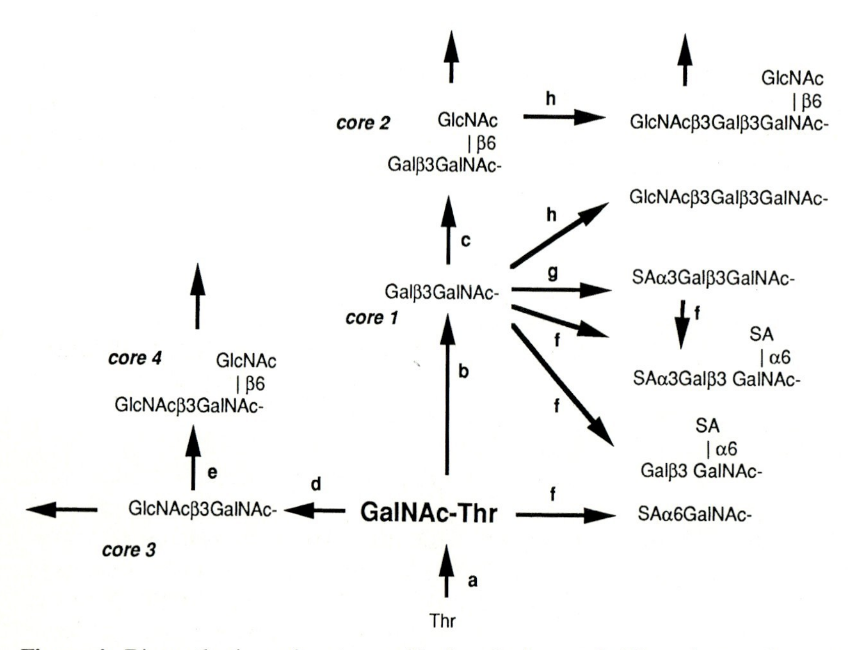

Figure 4. Biosynthetic pathways specific for *O*-glycans 2. The scheme shows the biosynthesis of the common *O*-glycan core structures 1 to 4 and some of the elongation and sialylation reactions that are specific for *O*-glycans. Numbers indicate the enzymes catalyzing the reactions. Path a, polypeptide α-GalNAc-transferase; path b, core 1 β3-Gal-transferase; path c, core 2 β6-GlcNAc-transferase; path d, core 3 β3-GlcNAc-transferase; path e, core 4 β6-GlcNAc-transferase; path f, α6-sialyltransferase; path g, α3-sialyltransferase; path h, elongation β3-GlcNAc-transferase

4.1 Initiation of *O*-Glycan Biosynthesis. UDP-GalNAc: Polypeptide α1,3-*N*-Acetylgalactosaminyltransferase (Polypeptide GalNAcT; EC 2.4.1.41)

The first step in the synthesis of all *O*-glycans is catalyzed by polypeptide GalNAcT. The enzyme is probably expressed in all mammalian cells and acts mainly in the cis Golgi, thus after the addition of *N*-glycans (Deschuyteneer *et al.*, 1988; Piller *et al.*, 1989). *In vitro*, the enzyme efficiently transfers GalNAc only to Thr but not to Ser, and requires Mn^{2+}. The enzyme has been purified from porcine submaxillary glands (Wang *et al.*, 1993) and other tissues. The cDNAs encoding several members of this glycosyltransferase family with slightly different substrate specificities have been cloned from bovine and human sources (Hagen *et al.*, 1993; Homa *et al.*, 1993; Sorensen *et al.*, 1995; White *et al.*, 1995). The protein has the typical type II membrane protein domain structure but no significant homology to other glycosyltransferases.

The peptide structure as well as the existing glycosylation of the substrate play an important role in the activity of the enzyme. However, attempts to define a specific acceptor consensus sequence have been largely unsuccessful (reviewed in Brockhausen *et al.*, 1995). Pro is usually found near *O*-glycosylation sites and may serve to expose Ser/Thr residues. Clearly, the three-dimensional structure of the substrate plays an important role in determining *O*-glycosylation.

4.2 Synthesis of *O*-Glycan Core Structures

4.2.1 UDP-Gal: GalNAc-R β1,3-Galtransferase (core 1 β3GalT, E.C. 2.4.1.122)

Core 1 β3GalT is present in most mammalian cells and synthesizes *O*-glycan core 1. The enzyme has not yet been purified to homogeneity or cloned and requires divalent cation such as Mn^{2+} or Cd^{2+}, and is inhibited by Zn^{2+} *in vitro* (Cheng and Bona, 1982). Rat liver core 1 β3GalT recognizes the 2-*N*-acetyl group and the 3- and 4- but not the 6-hydroxyl of GalNAc and acts on GalNAc-R where R may be a peptide or a synthetic group (Brockhausen *et al.*, 1990, 1992a). The activity is strongly influenced by the composition, length and sequence of the peptide backbone of the substrate and the attachment position and number of sugar residues present (Brockhausen *et al.*, 1990; Granovsky *et al.*, 1994). These studies suggest that the synthesis of core structures at individual *O*-glycosylation sites is regulated by the peptide structure and glycosylation near these sites.

Core 1 β3GalT is deficient in erythrocytes (Tn-erythrocytes) from patients with permanent mixed-field polyagglutinability (Cartron *et al.*, 1978; Thurnher *et al.*, 1992). Although the defect is stable in clones of affected lymphocytes (Thurnher *et al.*, 1992), the condition is reversible with certain differentiation agents (Thurnher *et al.*, 1993). Other cell types that show premature termination of *O*-glycan synthesis and a reduction in core 1 β3GalT activity are the human T-lymphoblastoid cell line Jurkat (Thurnher *et al.*, 1993) and a human colon cancer cell line LSC (Brockhausen *et al.*, 1995b). In these cell types core 1 β3GalT can not be reactivated by differentiating agents although the activity appears to be regulated during differentiation of human colonic adenocarcinoma Caco-2 cells (Brockhausen *et al.*, 1991a).

4.2.2 UDP-GlcNAc: GalNAc-R β1,3-GlcNAc-transferase (core 3 β3GlcNAcT, E.C.2.4.1.147)

Core 3 is synthesized by core 3 β3GlcNAcT, which catalyzes the transfer of GlcNAc in β1–3 linkage to GalNAcα-R in the presence of Mn^{2+}, where R may be peptide or a synthetic group (Brockhausen *et al.*, 1985). Core 3 β3GlcNAcT is expressed in mucin secreting tissues in a tissue specific fashion. The activity is reduced in colon cancer tissue (King *et al.*, 1994; Yang *et al.*, 1994b) and appears to be absent from cultured cancer cells (Vavasseur *et al.*, 1993, 1995). The regulation of this enzyme remains to be elucidated after purification and cloning.

4.2.3 UDP-GlcNAc: Galβ1–3GalNAc-R (GlcNAc to GalNAc) β1,6-GlcNAc-transferase (core 2 β6GnT, EC 2.4.1.102) and UDP-GlcNAc: GlcNAcβ1–3GalNAc-R (GlcNAc to GalNAc) β1,6-GlcNAc-transferase (core 4 β6GnT, EC 2.4.1.148)

Core 2 β6GnT activity synthesizing *O*-glycan core 2, GlcNAcβ1–6 (Galβ1–3) GalNAc-, from core 1 occurs in most tissues but the activity appears to be regulated in a tissue specific and differentiation specific fashion (Williams *et al.*, 1980; Wingert and Cheng, 1984). Substitution of core 1 will inhibit core 2 formation (Williams *et al.*, 1980; Brockhausen *et al.*, 1986). The enzyme in mucin secreting tissue is accompanied by activities synthesizing core 4, GlcNAcβ1–6 (GlcNAcβ1–3) GalNAc-, from core 3, and the GlcNAcβ1–6 (GlcNAcβ1–3) Gal branch (part of the I antigen) from the linear GlcNAcβ1–3 Gal structure (part of the i antigen) (Brockhausen *et al.*, 1986; Kuhns *et al.*, 1993). This activity has been named core 2 β6GnT M and has been purified from bovine trachea (Ropp *et al.*, 1991). The cDNA encoding human core 2 β6GnT L which only catalyzes the conversion of core 1 to core 2 has been cloned from human myelocytic HL60 cells by expression cloning using an antibody against a sialylated core 2 structure attached to leukosialin to select expressing clones (Bierhuizen and Fukuda, 1992). The gene for the L enzyme was localized to chromosome 9 q21 (Bierhuizen *et al.*, 1993), near the locus for the blood group A and B transferases and the I β6GnT.

The L enzyme has an absolute requirement for the 4- and 6-hydroxyls of GalNAc and the 6-hydroxyl of the Gal residue of Galβ1–3 GalNAcα-R substrate. It is inhibited by a derivative with a 6-deoxy group on the GalNAc, or a p-nitrophenyl group as the aglycon, upon UV irradiation (Toki *et al.*, 1994). The protein predicted from the cDNA sequence has the typical type II membrane protein domain structure but no apparent homology to other transferases with the exception of Iβ6GnT. The two potential *N*-linked glycosylation sites at the amino terminus have to be occupied to produce fully active recombinant enzyme secreted from insect cells (Toki *et al.*, in preparation).

Core 2 β6GnT L is increased upon activation of lymphocytes with anti CD3 antibody and interleukin-2 (Piller *et al.*, 1988) or by differentiation of mouse F9 teratocarcinoma cells (Heffernan *et al.*, 1993) or CaCo-2 human adenocarcinoma cells (Brockhausen *et al.*, 1991a). The enzyme activity is also elevated in lymphocytes and platelets from children with the Wiscott-Aldrich immunodeficiency syndrome (Higgins *et al.*, 1991, Piller *et al.*, 1991) and in leukocytes from patients with various leukaemias such as chronic (CML) and acute (AML) myelogenous leukaemias (Brockhausen *et al.*, 1991b) and in lymphocytes from patients with acute lymphocytic leukaemia (ALL) and chronic lymphocytic leukaemia (CLL) (Saitoh *et al.*, 1991). Mucin derived from breast cancer cell lines were shown to contain more core 1 and less core 2 structures (Hull *et al.*, 1989). This correlated with the loss of core 2β6GnT L activity and of mRNA expression in several breast cancer cell lines (Brockhausen *et al.*, 1995a).

Core 4 is synthesized from core 3 by core 4 β6GnT. Core 4 cannot be formed, however, after galactosylation of core 3 to form Galβ1–4GlcNAcβ1–3GalNAc- (I. Brockhausen, unpublished). The activity synthesizing core 4 is reduced relative to core 2 β6GnT L in a number of model cancer cells. This suggests either a decrease of the M enzyme or a switch from the M to the L enzyme in cancer (Vavasseur *et al.*, 1993; Yang *et al.*, 1994b).

4.3 Elongation of *O*-Glycans

4.3.1 UDP-GlcNAc: Galβ1–3 (R1–6) GalNAc-R (GlcNAc to Gal) β1,3-GlcNAc-transferase (elongation β3GnT, E.C.2.4.1.146)

Pig gastric mucosa and other tissue and cell types contain an elongation enzyme (Brockhausen *et al.*, 1983, 1984) which catalyzes the elongation at the Gal residues of core 1 and 2 structures by synthesizing a GlcNAcβ1–3Gal linkage. The GlcNAc residues of cores 2, 3 and 4 are likely to be elongated by β4GalT and i β3GnT to form poly *N*-acetyllactosamine chains. Elongation β3GnT shows a strict *in vitro* requirement for divalent cation and its substrate specificity and tissue distribution differ from those of the i β3GnT and core 3 β3GnT. The elongation β3GnT appears to be turned off in certain cell types such as in leukocytes from CML patients.

4.4 Termination Reactions in the Synthesis of *O*-Glycans

Many of the terminal *O*-glycan structures are recognized as blood group or tissue antigens, for example the Lewis, ABO, Cad and S^d blood groups. Terminal sugars include SA, Fuc, Gal, GalNAc and GlcNAc, usually in α-linkage. Internal or terminal Gal and GlcNAc residues may be sulfated by specific sulfotransferases (Carter *et al.*, 1988; Goso and Hotta 1993; Kuhns *et al.*, 1995). Sialic acid residues on *O*-glycans appear in a developmentally regulated and tissue-specific fashion. Sialylated *O*-glycans may be part of several cancer-associated antigens and play an important role in the cell surface properties of cells. The sialyltransferases that sialylate GalNAc in α2–6 linkage and the Gal residue of core 1 and core 2 in α2–3 linkage appear to be specific for *O*-glycans. It is not clear how many of the sialyltransferases described above acting on *N*-glycans are also involved in *O*-glycan synthesis. The blood group AB transferases, and α2-, α3- and α4-Fuc-transferases acting on *O*-glycans are the same as those described above for *N*-glycans.

4.4.1 CMP-sialic acid: R_1-GalNAc-R α6-sialyltransferase I (ST6O I, ST6GalNAc I, EC 2.4.99.3) and CMP-sialic acid: R_1-GalNAc-R α6-sialyltransferase II (ST6O III, ST6GalNAc II)

Sialic acidα2–6 GalNAc- chains are often found in cancer mucins as the sialyl-Tn antigen; the occurrence of sialyl-Tn has been associated with poor prognosis (Itzkowitz *et al.*, 1990; Yonezawa *et al.*, 1992). ST6GalNAc I synthesizing this linkage uses GalNAcα-peptide as a substrate and terminates *O*-glycan chain growth since sialic acidα2–6GalNAc- is not a substrate for any of the known glycosyltransferases. The enzyme may also act on core 1 and on sialic acidα2–3 Galβ1–3 GalNAc-R. Sialic acidα2–6 (Galβ1–3) GalNAc- may be a substrate for α2FucT and blood group A and B determinants may subsequently be attached. The enzyme has been purified from porcine submaxillary glands (Sadler *et al.*, 1979b). cDNA cloning from chicken cDNA libraries using the sialyl-motif revealed sequences for at least two different sialyltransferase species, ST6GalNAc I and II (Table 1), which differ in substrate specificity (Kurosawa *et al.*, 1994a,c), although both require peptide in the substrate.

4.4.2 CMP-sialic acid: sialic acidα2–3 Galβ1–3 GalNAc-R (sialic acid to GalNAc) α6-sialyltransferase II (ST6O II, ST6GalNAc III, EC 2.4.99.7)

In contrast to ST6GalNAc I and II, ST6GalNAc III does not act on neutral *O*-glycan core 1 substrates but requires the previous attachment of an α2–3 linked sialic acid residue to Gal. The enzyme appears to be common in non-mucin secreting tissues such as liver. The enzyme acts *in vitro* on sialic acid α2–3 Galβ1–3 GalNAc-R substrates where R may be peptide or a synthetic group (Bergh *et al.*, 1983; Bergh and van den Eijnden, 1983).

4.4.3 CMP-sialic acid: Galβ1–3 GalNAc-R α3-sialyltransferase (ST3O I, ST3Gal I, EC 2.4.99.4)

ST3O I transfers sialic acid in α2–3 linkage to the Gal residue of *O*-glycan core 1 and 2 substrates (Rearick *et al.*, 1989; Kuhns *et al.*, 1993). The enzyme has been purified from pig submaxillary gland (Sadler *et al.*, 1979a) but it is not clear how many of the presently cloned sialyltransferases are responsible for the α3-sialylation of mucins (Gillespie *et al.*, 1992; Lee *et al.*, 1993, 1994; Kitagawa and Paulson, 1994a,b; Chang and Lau, 1995; Kurosawa *et al.*, 1995). Two species of the enzyme have now been defined, ST3O I and II with ST3O I probably being responsible for the synthesis of most *O*-glycans, since ST3O II preferably acts on glycolipids (Lee *et al.*, 1994).

The kinetic analysis by Rearick *et al.* (1979) suggests a random equilibrium mechanism. Substrate specificity studies indicate that the enzyme from placenta and from AML cells (Kuhns *et al.*, 1993) has an absolute requirement only for the 3-hydroxyl of the Gal residue and can be competitively inhibited by the 3-deoxy-Gal-derivative of Galβ1–3 GalNAcα-benzyl substrate (Kuhns *et al.*, 1993) although the corresponding 3-deoxy-Gal substrate analogue is not an inhibitor for the α3-sialyltransferase acting on *N*-glycans (Hindsgaul *et al.*, 1991). Reboul *et al.* (1992) postulated that the ST3O may be regulated by a phosphorylation/dephosphorylation mechanism.

The sialic acid α2–3 Galβ1–3GalNAc- structure synthesized by ST3O is a receptor for virus binding since Sendai virus-mediated erythrocyte hemagglutination and infection of bovine kidney cells can be prevented by removing sialic acid, and can be restored with ST3O I whereas other sialyltransferases are ineffective (Markwell and Paulson, 1980). The expression of ST3O appears to be regulated during growth and differentiation. The activity is increased in AML and CML and has been correlated with hypersialylation of leukocyte cell surfaces (Baker *et al.*, 1987). The enzyme is increased several-fold in human colon cancer compared to normal tissue (Yang *et al.*, 1994b) and in several human breast cancer cell lines compared to normal mammary cell lines (Brockhausen *et al.*, 1995a). This may cause a premature termination of *O*-glycan processing and shorter, more sialylated *O*-glycans (Fukuda *et al.*, 1986).

Ha-ras oncogene transfection into FR3T3 rat fibroblasts caused a decreased expression of the ST3O and an increased expression of the ST6N (Delannoy *et al.*, 1993). However, since there was no concomitant increase in *Peanut* lectin binding to cell surfaces, the expression of sialylated *O*-glycans may not be affected. In situ hybridization experiments (Gillespie *et al.*, 1993) using digoxigenin-labeled RNA probes, showed that the enzyme is regulated during maturation of thymocytes.

5 Conclusions

With the cloning of glycosyltransferase genes it became clear that a number of glycosyltransferase families exist; the members of a family have homology and similar specificities. Glycosyltransferase genes fall into two categories (Joziasse, 1992): (i) single-exonic and (ii) multi-exonic. The entire coding regions for FucT III to VI, GnT I, GnT II and core 2 GnT are within a single exon whereas the coding regions for ST6N, β4GalT and α3GalT are distributed over 5 or more exons. The genes for the latter enzymes span 35 kb or more of genomic DNA and there are one or more non-coding exons several kb upstream of the coding region. Many of these genes have a single long (2–3 kb) 3'-terminal exon which carries the entire relatively long 3'-untranslated region, the translational stop site and part of the 3'-terminal coding region. The tissue- and time-specific expression of glycosyltransferase genes is due at least in part to differential activation of multiple promoters, e.g., β4GalT (Shaper *et al.*, 1990, 1994; Harduin-Lepers *et al.*, 1992, 1993), ST6N (Paulson *et al.*, 1989; Svensson *et al.*, 1990; Wang *et al.*, 1990; Wen *et al.*, 1992b; Wang *et al.*, 1993), GnT I (Yang *et al.*, 1994a), GnT III (Koyama *et al.*, 1995) and GnT V (Perng *et al.*, 1994; Saito *et al.*, 1995).

There are at least four glycosyltransferase gene families (Joziasse, 1992; Kleene and Berger, 1993; van den Eijnden and Joziasse, 1993): (i) α3FucT (Table 3), (ii) α3GalT, (iii) β6GnT (Table 2) and (iv) sialyltransferases (Table 1). The α3GalT family includes the human blood group B α3GalT and human blood group A α-1,3-*N*-acetylgalactosaminyltransferase. The β6GnT family has at least two members (core 2 GnT and IGnT I, Table 2) localized to human chromosome 9q21. The members of a family share varying degrees of sequence similarity. The fact that homologous regions for the β6GnT and sialyltransferase families are distributed over more than one exon indicates that evolution occurred through gene duplication followed by intron insertion rather than by exon shuffling (Bierhuizen *et al.*, 1995).

References

Aasheim HC, Aas-Eng DA, Deggerdal A *et al.* (1993): Cell-specific expression of human β-galactoside α2,6-sialyltransferase transcripts differing in the 5' untranslated region. In *Eur. J. Biochem.* **213**:467–75.

Altmann F, Kornfeld G, Dalik T *et al.* (1993): Processing of asparagine-linked oligosaccharides in insect cells – *N*-acetylglucosaminyltransferase I and II activities in cultured Lepidopteran cells. In *Glycobiology* **3**:619–25.

Aoki D, Lee N, Yamaguchi N *et al.* (1992): Golgi retention of a trans-Golgi membrane protein, galactosyltransferase, requires cysteine and histidine residues within the membrane-anchoring domain. In *Proc. Natl. Acad. Sci. USA* **89**:4319–23.

Arango J, Pierce M (1988): Comparison of *N*-acetylglucosaminyltransferase V activities in Rous Sarcoma-transformed baby hamster kidney (RS-BHK) and BHK cells. In *J. Cell. Biochem.* **37**:225–31.

Ats S, Lehmann J, Petry S (1992): A spacer-modified disaccharide as photoaffinity reagent for the acceptor-binding area of bovine β (1–4)-(1–4)-D-galactosyltransferase: Comparison of its acceptor properties with those of other 2-acetamido-2-deoxy-β-D-glucopyranosides. In *Carbohydr. Res.* **233**:125–39.

Bakker H, Agterberg M, Vantetering A *et al.* (1994): A *Lymnaea stagnalis* gene, with sequence similarity to that of mammalian β 1,4-galactosyltransferases, encodes a novel UDP-GlcNAc:GlcNAc β-R β 1,4-*N*-acetylglucosaminyltransferase. In *J. Biol. Chem.* **269**:30326–33.

Bast BJEG, Zhou LJ, Freeman GJ *et al.* (1992): The HB-6, CDw75, and CD76 differentiation antigens are unique cell-surface carbohydrate determinants generated by the β-galactoside α2,6-sialyltransferase. In *J. Cell Biol.* **116**:423–35.

Basu SC (1991): The serendipity of ganglioside biosynthesis: pathway to CARS and HY-CARS glycosyltransferases. In *Glycobiology* **1**:469–75.

Bergh MLE, van den Eijnden DH (1983): Aglycon specificity of fetal calf liver and ovine and porcine submaxillary gland α-*N*-acetylgalactosaminide α2,6-sialyltransferase. In *Eur. J. Biochem.* **136**:113–8.

Bergh MLE, Hooghwinkel GJM, van den Eijnden DH (1983): Biosynthesis of the *O*-glycosidically linked oligosaccharide chains of fetuin. Indication for an α-*N*-acetylgalactosaminide α2,6 sialyltransferase with a narrow acceptor specificity in fetal calf liver. In *J. Biol. Chem.* **258**:7430–6.

Beyer TA, Sadler JE, Rearick JI *et al.* (1981): Glycosyltransferases and their use in assessing oligosaccharide structure and structure-function relationships. In: Adv. Enzymol. 52 (Meister A, ed) pp 23–175, New York: John Wiley and Sons.

Bierhuizen MFA, Fukuda M (1992): Expression cloning of a cDNA encoding UDP-GlcNAc:Galβ1–3-GalNAc-R (GlcNAc to GalNAc) β1–6GlcNAc transferase by gene transfer into CHO cells expressing polyoma large tumor antigen. In *Proc. Natl. Acad. Sci. USA* **89**:9326–30.

Bierhuizen M, Mattei MG, Fukuda M (1993): Expression of the developmental I antigen by a cloned human cDNA encoding a member of a β-1,6-*N*-acetylglucosaminyltransferase gene family. In *Genes Dev.* **7**:468–78.

Bierhuizen MFA, Maemura K, Kudo S *et al.* (1995): Genomic organization of core 2 and I branching β-1,6-*N*-acetylglucosaminyltransferases. Implication for evolution of the β-1,6-*N*-acetylglucosaminyltransferase gene family. In *Glycobiology* **5**:417–25.

Bleil JD, Wassarman PM (1988): Galactose at the non-reducing terminus of *O*-linked oligosaccharides of mouse egg zona pellucida glycoprotein ZP3 is essential for the glycoprotein's sperm receptor activity. In *Proc. Natl. Acad. Sci. USA* **85**:6778–82.

Brandley BK, Swiedler SJ, Robbins PW (1990): Carbohydrate ligands of the LEC cell adhesion molecules. In *Cell* **63**:861–3.

Brockhausen I (1993): Clinical aspects of glycoprotein biosynthesis. In *Crit. Rev. Clin. Lab. Sci.* **30**:65–151.

Brockhausen I (1995): Biosynthesis of *O*-glycans of the *N*-acetylgalactosamine-α-Ser/Thr linkage type. In *Glycoproteins* 29a (Montreuil J, Vliegenthart JFG, Schachter H, eds) pp 201–59, Amsterdam, The Netherlands: Elsevier.

Brockhausen I, Williams D, Matta KL *et al.* (1983): Mucin synthesis. III. UDP-GlcNAc:Galβ1–3 (GlcNAcβ1–6) GalNAc-R (GlcNAc to Gal) β3-*N*-acetylglucosaminyltransferase, an enzyme in porcine gastric mucosa involved in the elongation of mucin-type oligosaccharides. In *Can. J. Biochem. Cell Biol.* **61**:1322–33.

Brockhausen I, Orr J, Schachter H (1984): Mucin synthesis. V. The action of pig gastric mucosal UDP-GlcNAc:Galβ1–3(R_1)GalNAc-R_2 (GlcNAc to Gal) β3-*N*-acetylglucosaminyltransferase on high molecular weight substrates. In *Can. J. Biochem. Cell Biol.* **62**:1081–90.

Brockhausen I, Matta KL, Orr J *et al.* (1985): Mucin synthesis. VI. UDP-GlcNAc:GalNAc-R β3-*N*-acetylglucosaminyltransferase and UDP-GlcNAc: GlcNAcβ1–3GalNAc-R (GlcNAc to GalNAc) β6-*N*-acetylglucosaminyltransferase from pig and rat colon mucosa. In *Biochemistry* **24**:1866–74.

Brockhausen I, Matta KL, Orr J *et al.* (1986): Mucin synthesis. VII. Conversion of R_1-β1–3Gal-R_2 to R_1-β1–3(GlcNAcβ1–6)Gal-R_2 and of R_1-β1–3GalNAc-

R_2 to R_1-β1–3(GlcNAcβ1–6)GalNAc-R_2 by a β6-*N*-acetylglucosaminyltransferase in pig gastric mucosa. In *Eur. J. Biochem.* **157**:463–74.

Brockhausen I, Hull E, Hindsgaul O *et al.* (1989): Control of glycoprotein synthesis. Detection and characterization of a novel branching enzyme from hen oviduct, UDP-*N*-acetylglucosamine: GlcNAcβ1–6 (GlcNAcβ1–2)Manα-R (GlcNAc to Man) β-4-*N*-acetylglucosaminyltransferase VI. In *J. Biol. Chem.* **264**:11211–21.

Brockhausen I, Möller G, Merz G *et al.* (1990): Control of glycoprotein synthesis: The peptide portion of synthetic *O*-glycopeptide substrates influences the activity of *O*-glycan core 1 uridine 5'-diphosphogalactose: *N*-acetylgalactosamine-α-R β3-galactosyltransferase. In *Biochemistry* **29**:10206–12.

Brockhausen I, Romero P, Herscovics A (1991a): Glycosyltransferase changes during differentiation of CaCo-2 human adenocarcinoma cells. In *Cancer Res.* **51**:3136–42.

Brockhausen I, Kuhns W, Schachter H *et al.* (1991b): Biosynthesis of *O*-glycans in leukocytes from normal donors and from patients with leukemia: increase in *O*-glycan core 2 UDP-GlcNAc: Galβ3GalNAcα-R (GlcNAc to GalNAc) β(1–6)-*N*-acetylglucosaminyltransferase in leukemic cells. In *Cancer Res.* **51**:1257–63.

Brockhausen I, Möller G, Pollex-Krüger A *et al.* (1992a): Control of *O*-glycan synthesis: specificity and inhibition of *O*-glycan core 1 UDP-galactose:*N*-acetylgalactosamine-α-R β3-galactosyltransferase from rat liver. In *Biochem. Cell Biol.* **70**:99–108.

Brockhausen I, Möller G, Yang JM *et al.* (1992b): Control of glycoprotein synthesis. Characterization of (1,4)-*N*-acetyl-β-D-glucosaminyltransferases acting on the α-D-(1,3)- and α-D-(1,6)-l inked arms of *N*-linked oligosaccharides. In *Carbohydr. Res.* **236**:281–99.

Brockhausen I, Yang J, Burchell J *et al.* (1995a): Mechanism underlying aberrant glycosylation of the MUC1 mucin in breast cancer cells. In *Eur. J. Biochem.* **233**:607–17.

Brockhausen I, Dickinson N, Ogata S *et al.* (1995b): Enzymatic basis for the high sialyl-Tn expression in a colon cancer cell line. In *Glycoconjugate J.* **12**: 566.

Burke J, Pettitt JM, Schachter H *et al.* (1992): The transmembrane and flanking sequences of β1,2-*N*-acetylglucosaminyltransferase I specify medial-Golgi localization. In *J. Biol. Chem.* **267**:24433–40.

Burke J, Pettitt JM, Humphris D *et al.* (1994): Medial-Golgi retention of *N*-acetylglucosaminyltransferase I – Contribution from all domains of the enzyme. In *J. Biol. Chem.* **269**:12049–59.

Cameron HS, Szczepaniak D, Weston BW (1995): Expression of human chromosome 19p α(1,3)-fucosyltransferase genes in normal tissues – Alternative splicing, polyadenylation, and isoforms. In *J. Biol. Chem.* **270**:20112–22.

Carter SR, Slomiany A, Gwozdzinski K *et al.* (1988): Enzymatic sulfation of mucus glycoprotein in gastric mucosa. Effect of ethanol. In *J. Biol. Chem.* **263**:11977–84.

Cartron J, Andrev J, Cartron J *et al.* (1978): Demonstration of T-transferase deficiency in Tn-polyagglutinated blood samples. In *Eur. J. Biochem.* **92**:111–9.

Chang ML, Eddy RL, Shows TB *et al.* (1995): Three genes that encode human β-galactoside α 2,3-sialyltransferases. Structural analysis and chromosomal mapping studies. In *Glycobiology* **5**:319–25.

Charuk JHM, Tan J, Bernardini M *et al.* (1995): Carbohydrate-deficient glycoprotein syndrome type II – An autosomal recessive *N*-acetylglucosaminyltransferase II deficiency different from typical hereditary erythroblastic multinuclearity, with a positive acidified-serum lysis test (HEMPAS). In *Eur. J. Biochem.* **230**:797–805.

Cheng A, Le T, Palacios M *et al.* (1994): Sperm-egg recognition in the mouse: characterization of sp56, a sperm protein having specific affinity for ZP3. In *J. Cell Biol.* **125**:867–78.

Cheng P-W, Bona SJ (1982): Mucin biosynthesis. Characterization of UDP-galactose:α-*N*-acetylgalactosaminide β3-galactosyltransferase from human tracheal epithelium. In *J. Biol. Chem.* **257**:6251–8.

Colley KJ, Lee EU, Adler B *et al.* (1989): Conversion of a Golgi apparatus sialyltransferase to a secretory protein by replacement of the NH2-terminal signal anchor with a signal peptide. In *J. Biol. Chem.* **264**:17619–22.

Couillin P, Mollicone R, Grisard MC *et al.* (1991): Chromosome 11q localization of one of the three expected genes for the human α-3-fucosyltransferases, by somatic hybridization. In *Cytogenet. Cell Genet.* **56**:108–11.

Crawley SC, Hindsgaul O, Alton G *et al.* (1990): An enzyme-linked immunosorbent assay for *N*-acetylglucosaminyltransferase-V. In *Anal. Biochem.* **185**:112–7.

Cummings RD (1992): Synthesis of asparagine-linked oligosaccharides: pathways, genetics and metabolic regulation. In *Glycoconjugates. Composition, Structure and Function* (Allen HJ, Kisailus EC, eds) pp 333–60, New York, NY: Marcel Dekker, Inc.

Cummings RD, Kornfeld S (1984): The distribution of repeating Galβ1–4GlcNAcβ1–3 sequences in asparagine-linked oligosaccharides of the mouse lymphoma cell line BW5147 and PHA® 2.1. In *J. Biol. Chem.* **259**:6253–60.

D'Agostaro G, Bendiak B, Tropak M (1989): Cloning of cDNA encoding the membrane-bound form of bovine β 1,4-galactosyltransferase. In *Eur. J. Biochem.* **183**:211–7.

D'Agostaro GAF, Zingoni A, Moritz RL *et al.* (1995): Molecular cloning and expression of cDNA encoding the rat UDP-*N*-acetylglucosamine:α-6-D-mannoside β-1,2-*N*-acetylglucosaminyltransferase II. In *J. Biol. Chem.* **270**:15211–21.

Dahdal RY, Colley KJ (1993): Specific sequences in the signal anchor of the β-galactoside α-2,6-sialyltransferase are not essential for Golgi localization – membrane flanking sequences may specify Golgi retention. In *J. Biol. Chem.* **268**:26310–9.

Dall'Olio F, Malagolini N, Di Stefano G *et al.* (1989): Increased CMP-NeuAc:Galβ1,4GlcNAc-R α2,6-sialyltransferase activity in human colorectal cancer tissues. In *Int. J. Cancer* **44**:434–9.

Dall'Olio F, Malagolini N, Serafini-Cessi F (1992): Enhanced CMP-NeuAc:Galβ1,4GlcNAc-R α2,6-sialyltransferase activity of human colon cancer xenografts in athymic nude mice and of xenograft-derived cell lines. In *Int. J. Cancer* **50**:325–30.

Datta AK, Paulson JC (1995): The sialyltransferase "sialylmotif" participates in binding the donor substrate CMP-NeuAc. In *J. Biol. Chem.* **270**:1497–500.

Delannoy P, Pelczar H, Vandamme V *et al.* (1993): Sialyltransferase activity in FR3T3 cells transformed with ras oncogene. In *Glycoconjugate J.* **10**:91–8.

Demetriou M, Nabi IR, Coppolino M *et al.* (1995): Reduced contact-inhibition and substratum adhesion in epithelial cells expressing GlcNAc-transferase V. In *J. Cell Biol.* **130**:383–92.

Dennis JW (1992): Changes in glycosylation associated with malignant transformation and tumor progression. In *Cell Surface Carbohydrates and Cell Development* (Fukuda M, ed) pp 161–94, Boca Raton: CRC Press.

Dennis JW, Laferte S (1988): Asn-linked oligosaccharides and the metastatic phenotype. In *Altered Glycosylation in Tumor Cells 79* (Reading CL, Hakomori S-I, Marcus DM, eds) pp 257–67, New York, NY: Alan R. Liss Inc.

Dennis JW, Laferte S, Waghorne C *et al.* (1987): β1–6 branching of Asn-linked oligosaccharides is directly associated with metastasis. In *Science* **236**:582–5.

Dennis JW, Kosh K, Bryce D-M *et al.* (1989): Oncogenes conferring metastatic potential induce increased branching of Asn-linked oligosaccharides in rat2 fibroblasts. In *Oncogene* **4**:853–60.

Deschuyteneer M, Eckhardt A, Roth J *et al.* (1988): The subcellular localization of apomucin and nonreducing terminal *N*-acetylgalactosamine in porcine submaxillary glands. In *J. Biol. Chem.* **263**:2452–9.

Dinter A, Berger EG (1995): Targeting of transfected β1,4galactosyltransferase to the Golgi apparatus and the cell surface in COS-1 cells: role of the *N*-terminal 13 amino acids. In *Glycoconjugate J.* **12**:448.

Do KY, Do SI, Cummings RD (1995): α-lactalbumin induces bovine milk β 1,4-galactosyltransferase to utilize UDP-GalNAc. In *J. Biol. Chem.* **270**: 18447–51.

Drickamer K (1993): A conserved disulphide bond in sialyltransferases. In *Glycobiology* **3**:2–3.

Duncan AMV, McCorquodale MM, Morgan C *et al.* (1986): Chromosomal localization of the gene for a human galactosyltransferase (GT-1). In *Biochem. Biophys. Res. Commun.* **141**:1185–8.

Eckhardt M, Mühlenhoff M, Bethe A *et al.* (1995): Molecular characterization of eukaryotic polysialyltransferase-1. In *Nature* **373**:715–8.

Evans SC, Lopez LC, Shur BD (1993): Dominant negative mutation in cell surface β1,4-galactosyltransferase inhibits cell-cell and cell-matrix interactions. In *J. Cell Biol.* **120**:1045–57.

Feizi T (1985): Demonstration by monoclonal antibodies that carbohydrate structures of glycoproteins and glycolipids are onco-developmental antigens. *Nature* **314**:53–7.

Feizi T (1988): Oligosaccharides in molecular recognition. In *Biochem. Soc. Transact.* **16**:930–4.

Feizi T (1990): Observations and thoughts on oligosaccharide involvement in malignancy. In *Environ. Health Perspect.* **88**:231–2.

Field MC, Wainwright LJ (1995): Molecular cloning of eukaryotic glycoprotein and glycolipid glycosyltransferases: A survey. In *Glycobiology* **5**:463–72.

Fukuda M, Carlsson SR, Klock JC *et al.* (1986): Structures of *O*-linked oligosaccharides isolated from normal granulocytes, chronic myelogenous leukemia cells, and acute myelogenous leukemia cells. *J. Biol. Chem.* **261**:12796–806.

Fukada T, Iida K, Kioka N *et al.* (1994): Cloning of a cDNA encoding *N*-acetylglucosaminyltransferase I from rat liver and analysis of its expression in rat tissues. In *Biosci. Biotechnol. Biochem.* **58**:200–1.

Galili U (1989): Abnormal expression of α-galactosyl epitopes in man. A trigger for autoimmune processes. In *Lancet* **2**:358–61.

Galili U (1992): The natural anti-Gal antibody: evolution and autoimmunity in man. In *Molecular Immunobiology of Self-Reactivity* (Bona CA, Kaushik AK, eds) pp 355–73, New York: Marcel Dekker, Inc.

Galili U (1993): Interaction of the natural anti-Gal antibody with α-galactosyl epitopes – A major obstacle for xenotransplantation in humans. In *Immunol. Today* **14**:480–2.

Galili U, Swanson K (1991): Gene sequences suggest inactivation of α-1,3-galactosyltransferase in catarrhines after the divergence of apes from monkeys. In *Proc. Natl. Acad. Sci. USA* **88**:7401–4.

Galili U, Shohet SB, Kobrin E *et al.* (1988): Man, apes, and old world monkeys differ from other mammals in the expression of α-galactosyl epitopes on nucleated cells. In *J. Biol. Chem.* **263**:17755–62.

Gersten KM, Natsuka S, Trinchera M *et al.* (1995): Molecular cloning, expression, chromosomal assignment, and tissue-specific expression of a murine α-(1,3)-fucosyltransferase locus corresponding to the human ELAM-1 ligand fucosyl transferase. In *J. Biol. Chem.* **270**:25047–56.

Ghosh S, Sankar BS, Basu S (1992): Isolation of a cDNA clone for β1–4 galactosyltransferase from embryonic chicken brain and comparison to its mammalian homologs. In *Biochem. Biophys. Res. Commun.* **189**:1215–22.

Gillespie W, Kelm S, Paulson JC (1992): Cloning and expression of the Galβ1,3GalNAc α2,3-sialyltransferase. In *J. Biol. Chem.* **267**:21004–10.

Gillespie W, Paulson J, Kelm S *et al.* (1993): Regulation of α2,3-sialyltransferase expression correlates with conversion of peanut agglutinin $(PNA)^+$ to PNA^- phenotype in developing thymocytes. In *J. Biol. Chem.* **268**: 3801–4.

Gleeson PA, Schachter H (1983): Control of glycoprotein synthesis. VIII. UDP-GlcNAc:GnGn (GlcNAc to Manα1–3) β4-*N*-acetylglucosaminyltransferase IV, an enzyme in hen oviduct which adds GlcNAc in β1–4 linkage to the α1–3-linked Man residue of the trimannosyl core of *N*-glycosyl oligosaccharides to form a triantennary structure. In *J. Biol. Chem.* **258**:6162–73.

Gleeson PA, Teasdale RD, Burke J (1994): Targeting of proteins to the Golgi apparatus. In *Glycoconjugate J.* **11**:381–94.

Goelz SE, Hession C, Goff D *et al.* (1990): ELFT: a gene that directs the expression of an ELAM-1 ligand. In *Cell* **63**:1349–56.

Gooi HC, Feizi T, Kapadia A *et al.* (1981): Stage-specific embryonic antigen involves α1–3 fucosylated type 2 blood group chains. In *Nature* **292**:156–8.

Goso Y, Hotta K (1993): Characterization of a rat corpus sulfotransferase for the 6-*O*-sulfation of β-D-*N*-acetylglucosamine residues on oligosaccharides. In *Glycoconjugate J.* **10**:226.

Granovsky M, Bielfeldt T, Peters S *et al.* (1994): *O*-glycan core 1 UDP-Gal: GalNAc β3-galactosyltransferase is controlled by the amino acid sequence and glycosylation of glycopeptide substrates. In *Eur. J. Biochem.* **221**:1039–46.

Greenwell P, Yates AD, Watkins WM (1986): UDP-*N*-acetyl-D-galactosamine as a donor substrate for the glycosyltransferase encoded by the B gene at the human blood group ABO locus. In *Carbohydr. Res.* **149**:149–70.

Grundmann U, Nerlich C, Rein T *et al.* (1990): Complete cDNA sequence encoding human β-galactoside α-2,6-sialyltransferase. In *Nucleic Acids Res.* **18**: 667.

Gu J, Nishikawa A, Fujii S *et al.* (1992): Biosynthesis of blood group I and i antigens in rat tissues. Identification of a novel β1,6-*N*-acetylglucosaminyltransferase. In *J. Biol. Chem.* **267**:2994–9.

Hagen FK, Van Wuyckhuyse B, Tabak LA (1993): Purification, cloning, and expression of a bovine UDP-GalNAc: polypeptide *N*-acetylgalactosaminyltransferase. In *J. Biol. Chem.* **268**:18960–5.

Hakomori S (1981): Glycosphingolipids in cellular interaction, differentiation and oncogenesis. In *Annu. Rev. Biochem.* **50**:733–64.

Hakomori SI (1989): Aberrant glycosylation in tumors and tumor-associated carbohydrate antigens. In *Adv. Cancer Res.* **52**:257–332.

Hamamoto T, Kawasaki M, Kurosawa N *et al.* (1993): Two step single primer mediated polymerase chain reaction. Application to cloning of putative mouse β-galactoside α-2,6-sialyltransferase cDNA. In *Bioorganic & Medicinal Chem.* **1**:141–5.

Haraguchi M, Yamashiro S, Yamamoto A *et al.* (1994): Isolation of GD_3 synthase gene by expression cloning of GM_3 α-2,8-sialyltransferase cDNA using anti-GD_2 monoclonal antibody. In *Proc. Natl. Acad. Sci. USA* **91**:10455–9.

Harduin-Lepers A, Shaper NL, Mahoney JA *et al.* (1992): Murine β1,4-galactosyltransferase: round spermatid transcripts are characterized by an extended 5'-untranslated region. In *Glycobiology* **2**:361–8.

Harduin-Lepers A, Shaper JH, Shaper NL (1993): Characterization of two cis-regulatory regions in the murine β1,4-galactosyltransferase gene: evidence for a negative regulatory element that controls initiation at the proximal site. In *J. Biol. Chem.* **268**: 14348–59.

Harpaz N, Schachter H (1980): Control of glycoprotein synthesis. V. Processing of asparagine-linked oligosaccharides by one or more rat liver Golgi α-D-mannosidases dependent on the prior action of UDP-*N*-acetylglucosamine:α-D-mannoside β-2-*N*-acetylglucosaminyltransferase I. In *J. Biol. Chem.* **255**:4894–902.

Hart GW (1992): Glycosylation. In *Curr. Opinion Cell Biol.* **4**:1017–23.

Heffernan M, Lotan R, Amos B *et al.* (1993): Branching β1–6*N*-acetylglucosaminetransferases and polylactosamine expression in mouse F9 teratocarcinoma cells and differentiated counterparts. In *J. Biol. Chem.* **268**:1242–51.

Hemming FW (1995): The coenzymic role of phosphodolichols. In *Glycoproteins 29a* (Montreuil J, Vliegenthart JFG, Schachter H, eds) pp 127–43, Amsterdam, The Netherlands: Elsevier.

Henion TR, Macher BA, Anaraki F *et al.* (1994): Defining the minimal size of catalytically active primate α1,3 galactosyltransferase: Structure-function studies on the recombinant truncated enzyme. In *Glycobiology* **4**:193–201.

Higgins EA, Siminovitch KA, Zhuang D *et al.* (1991): Aberrant *O*-linked oligosaccharide biosynthesis in lymphocytes and platelets with the Wiscott-Aldrich Syndrome. In *J. Biol. Chem.* **266**:6280–90.

Hindsgaul O, Kaur KJ, Srivastava G *et al.* (1991): Evaluation of deoxygenated oligosaccharide acceptor analogs as specific inhibitors of glycosyltransferases. In *J. Biol. Chem.* **226**:17858–62.

Hitoshi S, Kusunoki S, Kanazawa I *et al.* (1995): Molecular cloning and expression of two types of rabbit β-galactoside α1,2-fucosyltransferase. In *J. Biol. Chem.* **270**:8844–50.

Hoe MH, Slusarewicz P, Misteli T *et al.* (1995): Evidence for recycling of the resident medial/trans Golgi enzyme, *N*-acetylglucosaminyltransferase I, in ldlD cells. In *J. Biol. Chem.* **270**:25057–63.

Hokke CH, Van der Ven JGM, Kamerling JP *et al.* (1993): Action of rat liver Galβ1–4GlcNAc α(2–6)-sialyltransferase on Manβ1–4GlcNAcβ-OMe, GalNAcβ1–4GlcNAcβ-OMe, Glcβ1–4GlcNAcβ-OMe and GlcNAcβ1–4GlcNAcβ-OMe as synthetic substrates. In *Glycoconjugate J.* **10**:82–90.

Hollis GF, Douglas JG, Shaper NL *et al.* (1989): Genomic structure of murine β-1,4-galactosyltransferase. In *Biochem. Biophys. Res. Commun.* **162**:1069–75.

Homa F, Hollander T, Lehman D *et al.* (1993): Isolation and expression of a cDNA clone encoding a bovine UDP-GalNAc:polypeptide *N*-acetylgalactosaminyltransferase. In *J. Biol. Chem.* **268**:12609–16.

Hull E, Sarkar M, Spruijt MPN *et al.* (1991): Organization and localization to chromosome 5 of the human UDP-*N*-acetylglucosamine:α-3-D-mannoside β-1,2-*N*-acetylglucosaminyltransferase I gene. In *Biochem. Biophys. Res. Commun.* **176**:608–15.

Hull SR, Bright A, Carraway KL *et al.* (1989): Oligosaccharide differences in the DF3 sialomucin antigen from normal human milk and the BT-20 human breast carcinoma cell line. In *Cancer Commun.* **1**:261–7.

Ihara Y, Nishikawa A, Tohma T *et al.* (1993): cDNA cloning, expression, and chromosomal localization of human *N*-acetylglucosaminyltransferase III (GnT-III). In *J. Biochem.* **113**:692–8.

Ioffe E, Stanley P (1994): Mice lacking *N*-acetylglucosaminyltransferase I activity die at midgestation, revealing an essential role for complex or hybrid *N*-linked carbohydrates. In *Proc. Natl. Acad. Sci. USA* **91**:728–32.

Ishibashi K, Nishikawa A, Hayashi N *et al.* (1989): *N*-acetylglucosaminyltransferase III in human serum, and liver and hepatoma tissues: Increased activity in liver cirrhosis and hepatoma patients. In *Clin. Chim. Acta* **185**:325–32.

Itzkowitz SH, Dahiya R, Byrd JC *et al.* (1990): Blood group antigen synthesis and degradation in normal and cancerous colonic tissues. In *Gastroenterology* **99**:431–42.

Jaeken J, Carchon H, Stibler H (1993): The carbohydrate-deficient glycoprotein syndromes – Pre-Golgi and Golgi disorders? In *Glycobiology* **3**:423–8.

Jaeken J, Schachter H, Carchon H *et al.* (1994): Carbohydrate deficient glycoprotein syndrome type II: A deficiency in Golgi localised *N*-acetylglucosaminyltransferase II. In *Arch. Dis. Child.* **71**:123–7.

Johnson KD, Chrispeels MJ (1987): Substrate specificities of *N*-acetylglucosaminyl-, fucosyl-, and xylosyltransferases that modify glycoproteins in the Golgi apparatus of bean cotyledons. In *Plant Physiol.* **84**:1301–8.

Johnston DS, Shaper JH, Shaper NL *et al.* (1995): The gene encoding murine α1,3-galactosyltransferase is expressed in female germ cells but not in male germ cells. In *Dev. Biol.* **171**:224–32.

Joziasse DH (1992): Mammalian glycosyltransferases: genomic organization and protein structure. In *Glycobiology* **2**:271–7.

Joziasse DH, Schiphorst WECM, van den Eijnden DH *et al.* (1987): Branch specificity of bovine colostrum CMP-sialic acid:Galβ1,4GlcNAc-R α2,6-sialyltransferase. Sialylation of bi-, tri-, and tetraantennary oligosaccharides and glycopeptides of the *N*-acetyllactosamine type. In *J. Biol. Chem.* **262**:2025–33.

Joziasse DH, Shaper JH, Van den Eijnden DH *et al.* (1989): Bovine α1,3-galactosyltransferase: Isolation and characterization of a cDNA clone. Identification of homologous sequences in human genomic DNA. In *J. Biol. Chem.* **264**:14290–7.

Joziasse DH, Shaper JH, Jabs EW *et al.* (1991): Characterization of an α1,3-galactosyltransferase homologue on human chromosome 12 that is organized as a processed pseudogene. In *J. Biol. Chem.* **266**: 6991–8.

Joziasse DH, Shaper NL, Kim D *et al.* (1992): Murine α1,3-galactosyltransferase. A single gene locus specifies four isoforms of the enzyme by alternative splicing. In *J. Biol. Chem.* **267**:5534–41.

Ju T-Z, Chen H-L, Gu J-X *et al.* (1995): Regulation of *N*-acetylglucosaminyltransferase V by protein kinases. In *Glycoconjugate J.* **12**:767–72.

Kelly RJ, Ernst LK, Larsen RD *et al.* (1994): Molecular basis for H blood group deficiency in Bombay (*O*-h) and para-Bombay individuals. In *Proc. Natl. Acad. Sci. USA* **91**:5843–7.

Kelly RJ, Rouquier S, Giorgi D *et al.* (1995): Sequence and expression of a candidate for the human secretor blood group α(1,2)fucosyltransferase gene (FUT2) – Homozygosity for an enzyme-inactivating nonsense mutation commonly correlates with the non-secretor phenotype. In *J. Biol. Chem.* **270**:4640–9.

King MJ, Chan A, Roe R *et al.* (1994): Two different glycosyltransferase defects that result in GalNAcα-*O*-peptide (Tn) expression. In *Glycobiology* **4**:267–9.

Kitagawa H, Paulson JC (1993): Cloning and expression of human Galβ1,3(4)GlcNAc α2,3-sialyltransferase. In *Biochem. Biophys. Res. Commun.* **194**:375–82.

Kitagawa H, Paulson JC (1994a): Cloning of a novel α2,3-sialyltransferase that sialylates glycoprotein and glycolipid carbohydrate groups. In *J. Biol. Chem.* **269**:1394–401.

Kitagawa H, Paulson JC (1994b): Differential expression of five sialyltransferase genes in human tissues. In *J. Biol. Chem.* **269**:17872–8.

Kleene R, Berger EG (1993): The molecular and cell biology of glycosyltransferases. In *Biochim. Biophys. Acta* **1154**:283–325.

Koenderman AH, Koppen PL, van den Eijnden DH (1987): Biosynthesis of polylactosaminoglycans. In *Eur. J. Biochem.* 166:199–208.

Kojima N, Yoshida Y, Kurosawa N *et al.* (1995a): Enzymatic activity of a developmentally regulated member of the sialyltransferase family (STX): Evidence for α2,8-sialyltransferase activity toward *N*-linked oligosaccharides. In *FEBS Lett.* **360**:1–4.

Kojima N, Yoshida Y, Tsuji S (1995b): A developmentally regulated member of the sialyltransferase family (ST8Sia II, STX) is a polysialic acid synthase. In *FEBS Lett.* **373**:119–22.

Kornfeld R, Kornfeld S (1985): Assembly of asparagine-linked oligosaccharides. In *Annu. Rev. Biochem.* **54**:631–64.

Koszdin KL, Bowen BR (1992): The cloning and expression of a human α1,3 fucosyltransferase capable of forming the E-selectin ligand. In *Biochem. Biophys. Res. Commun.* **187**:152–7.

Koyama N, Miyoshi E, Ihara Y *et al.* (1995): Structure of human *N*-acetylglucosaminyltransferase III gene and its transcripts: multiple promoters and splicing patterns. In *Glycoconjugate J.* **12**:475.

Kubelka V, Altmann F, Staudacher E *et al.* (1993): Primary structures of the *N*-linked carbohydrate chains from honeybee venom phospholipase A2. In *Eur. J. Biochem.* **213**:1193–204.

Kubelka V, Altmann F, Marz L (1995): The asparagine-linked carbohydrate of honeybee venom hyaluronidase. In *Glycoconjugate J.* **12**:77–83.

Kuhns W, Rutz V, Paulsen H *et al.* (1993): Processing *O*-glycan core 1, Galβ1–3GalNAcα-R. Specificities of core 2 UDP-GlcNAc:Galβ1–3GalNAc-R β6-*N*-acetylglucosaminyltransferase and CMP-SA: Galβ1–3GalNAc-R α3-sialyltransferase. In *Glycoconjugate J.* **10**:381–94.

Kuhns W, Jain R, Matta KL *et al.* (1995): Characterization of sulfotransferase activity from rat colonic mucosa synthesizing sulfated *O*-glycan core 1, 3-sulfate-Galβ1–3GalNAcα-R. In *Glycobiology* **5**:689–97.

Kukowska-Latallo JF, Larsen RD, Nair RP *et al.* (1990): A cloned human cDNA determines expression of a mouse stage-specific embryonic antigen and the Lewis blood group α(1,3/1,4)fucosyltransferase. In *Genes Dev.* **4**:1288–303.

Kumar R, Yang J, Larsen RD *et al.* (1990): Cloning and expression of *N*-acetylglucosaminyltransferase I, the medial Golgi transferase that initiates complex *N*-linked carbohydrate formation. In *Proc. Natl. Acad. Sci. USA* **87**:9948–52.

Kumar R, Potvin B, Muller WA *et al.* (1991): Cloning of a human α(1,3)-fucosyltransferase gene that encodes ELFT but does not confer ELAM-1 recognition on Chinese hamster ovary cell transfectants. In *J. Biol. Chem.* **266**:21777–83.

Kumar R, Yang J, Eddy RL *et al.* (1992): Cloning and expression of the murine gene and chromosomal location of the human gene encoding *N*-acetylglucosaminyltransferase I. In *Glycobiology* **2**:383–93.

Kurosawa N, Hamamoto T, Lee YC *et al.* (1994a): Molecular cloning and expression of GalNAcα2,6-sialyltransferase. In *J. Biol. Chem.* **269**:1402–9.

Kurosawa N, Kawasaki M, Hamamoto T *et al.* (1994b): Molecular cloning and expression of chick embryo Galβ1,4 GlcNAcα2,6-sialyltransferase – Comparison with the mammalian enzyme. In *Eur. J. Biochem.* **219**:375–81.

Kurosawa N, Kojima N, Inoue M *et al.* (1994c): Cloning and expression of Galβ1,3GalNAc-specific GalNAcα2,6-sialyltransferase. In *J. Biol. Chem.* **269**:19048–53.

Kurosawa N, Hamamoto T, Inoue M *et al.* (1995): Molecular cloning and expression of chick Galβ1,3GalNAcα2,3-sialyltransferase. In *Biochim. Biophys. Acta* **1244**:216–22.

Lance P, Lau KM, Lau J (1989): Isolation and characterization of a partial cDNA for a human sialyltransferase. In *Biochem. Biophys. Res. Commun.* **164**:225–32.

Larsen RD, Rajan VP, Ruff MM *et al.* (1989): Isolation of a cDNA encoding a murine UDPgalactose:β-D-galactosyl-1,4-*N*-acetyl-D-glucosaminide α1,3-galactosyltransferase: Expression cloning by gene transfer. In *Proc. Natl. Acad. Sci. USA* **86**:8227–31.

Larsen RD, Ernst LK, Nair RP *et al.* (1990a): Molecular cloning, sequence, and expression of a human GDP-L-fucose:β-D-galactoside 2-α-L-fucosyltransferase cDNA that can form the H blood group antigen. In *Proc. Natl. Acad. Sci. USA* **87**:6674–8.

Larsen RD, Rivera-Marrero CA, Ernst LK *et al.* (1990b): Frame-shift and nonsense mutations in a human genomic sequence homologous to a murine UDP-Gal:β-D-Gal(1,4)-D-GlcNAcα1,3-galactosyltransferase cDNA. In *J. Biol. Chem.* **265**:7055–61.

Le Marer N, Laudet V, Svensson EC *et al.* (1992): The c-Ha-ras oncogene induces increased expression of β-galactoside α2,6-sialyltransferase in rat fibroblast (FR3T3) cells. In *Glycobiology* **2**:49–56.

Le Marer N, Stéhelin D (1995): High α2,6-sialylation of *N*-acetyllactosamine sequences in ras-transformed rat fibroblasts correlates with high invasive potential. In *Glycobiology* **5**:219–26.

Le Pendu J, Cartron JP, Lemieux RU *et al.* (1985): The presence of at least two different H-blood-group-related β-D-Gal α-2-L-fucosyltransferases in human serum and the genetics of blood group H substances. In *Am. J. Human Genet.* **37**:749–60.

Lee YC, Kurosawa N, Hamamoto T *et al.* (1993): Molecular cloning and expression of Galβ1,3GalNAcα2,3-sialytransferase from mouse brain. In *Eur. J. Biochem.* **216**:377–85.

Lee YC, Kojima N, Wada E *et al.* (1994): Cloning and expression of cDNA for a new type of Galβ1,3GalNAcα2,3-sialyltransferase. In *J. Biol. Chem.* **269**:10028–33.

Leppänen A, Penttilä L, Niemelä R *et al.* (1991): Human serum contains a novel β1,6-*N*-acetylglucosaminyltransferase activity that is involved in midchain branching of oligo(*N*-acetyllactosaminoglycans). In *Biochemistry* **30**:9287–96.

Livingston BD, Paulson JC (1993): Polymerase chain reaction cloning of a developmentally regulated member of the sialyltransferase gene family. In *J. Biol. Chem.* **268**:11504–7.

Livingston B, Kitagawa H, Wen D *et al.* (1992): Identification of a sialylmotif in the sialyltransferase gene family. In *Glycobiology* **2**:489.

Longmore GD, Schachter H (1982): Control of glycoprotein synthesis. VI. Product identification and substrate specificity studies of the GDP-L-fucose:2-acetamido-2-deoxy-β-D-glucoside (Fuc to Asn-linked GlcNAc) 6-α-L-fucosyltransferase in a Golgi-rich fraction form porcine liver. In *Carbohydr. Res.* **100**:365–92.

Lopez LC, Shur BD (1988): Comparison of two independent cDNA clones reported to encode β1,4-galactosyltransferase. In *Biochem. Biophys. Res. Commun.* **156**:1223–9.

Lopez LC, Maillet CM, Oleszkowicz K *et al.* (1989): Cell surface and Golgi pools of β1,4-galactosyltransferase are differentially regulated during embryonal carcinoma cell differentiation. In *Mol. Cell. Biol.* **9**:2370–7.

Lopez LC, Youakim A, Evans SC *et al.* (1991): Evidence for a molecular distinction between Golgi and cell surface forms of β1,4-galactosyltransferase. In *J. Biol. Chem.* **266**:15984–91.

Lowe JB, Stoolman LM, Nair RP *et al.* (1990): ELAM-1-dependent cell adhesion to vascular endothelium determined by a transfected human fucosyltransferase cDNA. In *Cell* **63**:475–84.

Lowe JB, Kukowska-Latallo JF, Nair RP *et al.* (1991a): Molecular cloning of a human fucosyltransferase gene that determines expression of the Lewis x and VIM-2 epitopes but not ELAM-1-dependent cell adhesion. In *J. Biol. Chem.* **266**:17467–77.

Lowe JB, Stoolman LM, Nair RP *et al.* (1991b): A transfected human fucosyltransferase cDNA determines biosynthesis of oligosaccharide ligand(s) for endothelial-leukocyte adhesion molecule I. In *Biochem. Soc. Transact.* **19**:649–54.

Lu Y, Chaney W (1993): Induction of *N*-acetylglucosaminyltransferase V by elevated expression of activated or proto-Ha-ras oncogenes. In *Mol. Cell. Biochem.* **122**:85–92.

Macher BA, Holmes EH, Swiedler SJ *et al.* (1991): Human α1–3fucosyltransferases. In *Glycobiology* **1**:577–84.

Markwell MAK, Paulson JC (1980): Sendai virus utilizes specific sialyloligosaccharides as host cell receptor determinants. In *Proc. Natl. Acad. Sci. USA* **77**:5693–97.

März L, Altmann F, Staudacher E *et al.* (1995): Protein glycosylation in insects. In *Glycoproteins* 29a (Montreuil J, Vliegenthart JFG, Schachter H, eds) pp 543–63, Amsterdam: Elsevier.

Masibay AS, Balaji PV, Boeggeman EE *et al.* (1993): Mutational analysis of the Golgi retention signal of bovine β1,4-galactosyltransferase. In *J. Biol. Chem.* **268**:9908–16.

Masri KA, Appert HE, Fukuda MN (1988): Identification of the full-length coding sequence for human galactosyltransferas (β-*N*-acetylglucosaminide: β1,4-galactosyltransferase). In *Biochem. Biophys. Res. Commun.* **157**:657–63.

Mengle-Gaw L, McCoy-Haman MF, Tiemeier DC (1991): Genomic structure and expression of human β1,4-galactosyltransferase. In *Biochem. Biophys. Res. Commun.* **176**:1269–76.

Messer M, Nicholas KR (1991): Biosynthesis of marsupial milk oligosaccharides: characterization and developmental changes of two galactosyltransferases in lactating mammary gland of the tammar wallaby, *Marcropeus eugenii*. In *Biochim. Biophys. Acta* **1077**:79–85.

Metzler M, Gertz A, Sarkar M *et al.* (1994): Complex asparagine-linked oligosaccharides are required for morphogenic events during post-implantation development. In *EMBO J.* **13**:2056–65.

Miller DJ, Macek MB, Shur BD (1992): Complementarity between sperm surface β1,4-galactosyltransferase and egg-coat ZP3 mediates sperm-egg binding. In *Nature* **357**:589–93.

Miyoshi E, Nishikawa A, Ihara Y *et al.* (1993): *N*-acetylglucosaminyltransferase III and V messenger RNA levels in LEC rats during hepatocarcinogenesis. In *Cancer Res.* **53**:3899–902.

Miyoshi E, Nishikawa A, Ihara Y *et al.* (1995): Transforming growth factor-β up-regulates expression of the *N*-acetylglycosaminyltransferase V gene in mouse melanoma cells. In *J. Biol. Chem.* **270**:6216–20.

Mollicone R, Dalix AM, Jacobsson A *et al.* (1988): Red cell H-deficient, salivary ABH secretor phenotype of Reunion Island. Genetic control of the expression of H antigen in the skin. In *Glycoconjugate J.* **5**:499–512.

Mollicone R, Gibaud A, François A *et al.* (1990): Acceptor specificity and tissue distribution of three human α-3-fucosyltransferases. In *Eur. J. Biochem.* **191**:169–76.

Mollicone R, Candelier JJ, Mennesson B *et al.* (1992): Five specificity patterns of 1,3-α-L-fucosyltransferase activity defined by use of synthetic oligosaccharide acceptors. Differential expression of the enzymes during human embryonic development and in adult tissues. In *Carbohydr. Res.* **228**:265–76.

Mollicone R, Reguigne I, Fletcher A *et al.* (1994): Molecular basis for plasma α1,3-fucosyltransferase gene deficiency (Fut6). In *J. Biol. Chem.* **269**:12662–71.

Morrison JF, Ebner KE (1971): Studies on galactosyltransferase. Kinetic investigations with *N*-acetylglucosamine as the galactosyl group acceptor. In *J. Biol. Chem.* **246**:3977–84.

Mucha J, Kappel S, Schachter H *et al.* (1995): Molecular cloning and characterization of cDNAs coding for *N*-acetylglucosaminyltransferases I and II from Xenopus laevis ovary. In *Glycoconjugate J.* **12**: 473.

Mulder H, Schachter H, de Jong-Brink M *et al.* (1991): Identification of a novel UDP-Gal:Gal NAcβ1–4GlcNAc-Rβ-3-galactosyltransferase in he connective tissue of the snail Lymnaea stagnalis. In Eur. J. Biochem. **201**:459–65.

Munro S (1991): Sequences within and adjacent to the transmembrane segment of α2,6-sialyltransferase specify Golgi retention. In *EMBO J.* **10**:3577–88.

Munro S (1995a): A comparison of the transmembrane domains of Golgi and plasma membrane proteins. In *Biochem. Soc. Transact.* **23**:527–30.

Munro S (1995b): An investigation of the role of transmembrane domains in Golgi protein retention. In *EMBO J.* **14**:4695–704.

Nakayama J, Fukuda MN, Fredette B *et al.* (1995): Expression cloning of a human polysialyltransferase that forms the polysialylated neural cell adhesion molecule present in embryonic brain. In *Proc. Natl. Acad. Sci. USA* **92**:7031–5.

Nakazawa K, Ando T, Kimura T *et al.* (1988): Cloning and sequencing of a full-length cDNA of mouse *N*-acetylglucosamine β1–4-galactosyltransferase. In *J. Biochem.* **104**:165–8.

Nara K, Watanabe Y, Maruyama K *et al.* (1994): Expression cloning of a CMP-NeuAc:NeuAc α2–3Galβ1–4Glcβ1–1'-Cerα2,8-sialyltransferase (GD3 synthase) from human melanoma cells. In *Proc. Natl. Acad. Sci. USA* **91**:7952–6.

Narasimhan S, Schachter H, Rajalakshmi S (1988): Expression of *N*-acetylglucosaminyltransferase III in hepatic nodules during rat liver carcinogenesis promoted by orotic acid. In *J. Biol. Chem.* **263**:1273–81.

Narasimhan S, Yuen R, Fode CJ *et al.* (1993): Cloning of cDNA encoding chicken β1,2-*N*-acetylglucosaminyltransferase I. In *Glycobiology* **3**:531.

Narimatsu H, Sinha S, Brew K *et al.* (1986): Cloning and sequencing of cDNA of bovine *N*-acetylglucosamine β1–4-galactosyltransferase. In *Proc. Natl. Acad. Sci. USA* **83**:4720–4.

Natsuka S, Gersten KM, Zenita K *et al.* (1994): Molecular cloning of a cDNA encoding a novel human leukocyte α1,3-fucosyltransferase capable of synthesizing the sialyl Lewis x determinant. In *J. Biol. Chem.* **269**:16789–94.

Nemansky M, van den Eijnden DH (1993): Enzymatic characterization of CMP-NeuAc:Galβ1–4GlcNAc-Rα2–3-sialyltransferase from human placenta. In *Glycoconjugate J.* **10**:99–108.

Nilsson T, Lucocq JM, Mackay D *et al.* (1991): The membrane spanning domain of β1,4-galactosyltransferase specifies trans Golgi localization. In *EMBO J.* **10**:3567–75.

Nilsson T, Slusarewicz P, Hoe MH *et al.* (1993): Kin recognition: a model for the retention of Golgi enzymes. In *FEBS Lett.* **330**:1–4.

Nilsson T, Hoe MH, Slusarewicz P *et al.* (1994): Kin recognition between medial Golgi enzymes in HeLa cells. In *EMBO J.* **13**:562–74.

Nishihara S, Nakazato M, Kudo T *et al.* (1993): Human α1,3 fucosyltransferase (FucT-VI) gene is located at only 13 Kb 3' to the Lewis type fucosyltransferase (FucT-III) gene on chromosome 19. In *Biochem. Biophys. Res. Commun.* **190**:42–6.

Nishikawa A, Gu J, Fujii S *et al.* (1990): Determination of *N*-acetylglucosaminyltransferases III, IV and V in normal and hepatoma tissues of rats. In *Biochim. Biophys. Acta* **1035**:313–8.

Nishikawa A, Ihara Y, Hatakeyama M *et al.* (1992): Purification, cDNA cloning, and expression of UDP-*N*-acetylglucosamine:β-D-mannoside β1,4*N*-acetylglucosaminyltransferase III from rat kidney. In *J. Biol. Chem.* **267**:18199–204.

O'Hanlon TP, Lau JTY (1992): Analysis of kidney mRNAs expressed from the rat β-galactoside α2,6-sialyltransferase gene. In *Glycobiology* **2**:257–66.

O'Hanlon TP, Lau KM, Wang XC *et al.* (1989): Tissue-specific expression of β-galactoside α2,6-sialyltransferase. In *J. Biol. Chem.* **264**:17389–94.

Oriol R (1990): Genetic control of the fucosylation of ABH precursor chains. Evidence for new epistatic interactions in different cells and tissues. In *J. Immunogenet.* **17**:235–45.

Oulmouden A, Wierinckx A, Petit JM *et al.* (1995): Molecular cloning and expression of bovine α1,3/4-fucosyltransferases. In *Unpublished EMBL/GenBank listing.*

Palcic MM, Hindsgaul O (1991): Flexibility in the donor substrate specificity of β1,4-galactosyltransferase: application in the synthesis of complex carbohydrates. In *Glycobiology* **1**:205–9.

Palcic MM, Ripka J, Kaur KJ *et al.* (1990): Regulation of *N*-acetylglucosaminyltransferase V activity. Kinetic comparisons of parental, Rous sarcoma virus-transformed BHK, and L-phytohemagglutinin-resistant BHK cells using synthetic substrates and an inhibitory substrate analog. In *J. Biol. Chem.* **265**:6759–69.

Paulson JC, Colley KJ (1989): Glycosyltransferases. Structure, localization, and control of cell type-specific glycosylation. In *J. Biol. Chem.* **264**: 17615–8.

Paulson JC, Weinstein J, Schauer A (1989): Tissue-specific expression of sialyltransferases. In *J. Biol. Chem.* **264**:10931–4.

Perng GS, Shoreibah M, Margitich I *et al.* (1994): Expression of *N*-acetylglucosaminyltransferase V mRNA in mammalian tissues and cell lines. In *Glycobiology* **4**:867–71.

Phillips ML, Nudelman E, Gaeta F *et al.* (1990): ELAM-1 mediates cell adhesion by recognition of a carbohydrate ligand, sialyl-Le^x. In *Science* **250**: 1130–2.

Piau JP, Labarriere N, Dabouis G *et al.* (1994): Evidence for two distinct α1,2-fucosyltransferase genes differentially expressed throughout the rat colon. In *Biochem. J.* **300**:623–6.

Piller F, Cartron J-P, Maranduba A *et al.* (1984): Biosynthesis of blood group I antigens. Identification of a UDP-GlcNAc: GlcNAcβ1–3Gal (-R)β1–6 (GlcNAc to Gal) *N*-acetylglucosaminyltransferase in hog gastric mucosa. In *J. Biol. Chem.* **259**:13385–90.

Piller F, Piller V, Fox R *et al.* (1988): Human T-lymphocyte activation is associated with changes in *O*-glycan biosynthesis. In *J. Biol. Chem.* **263**: 15146–50.

Piller V, Piller F, Klier FG *et al.* (1989): Glycosylation of leukosialin in K562 cells. Evidence for initiation and elongation in early Golgi compartments. In *Eur. J. Biochem.* **183:**23–135.

Piller F, Le Deist F, Weinberg KI *et al.* (1991): Altered *O*-glycan synthesis in lymphocytes from patients with Wiscott-Aldrich syndrome. In *J. Exp. Med.* **173**:1501–10.

Polley MJ, Phillips ML, Wayner E *et al.* (1991): CD62 and endothelial cell-leukocyte adhesion molecule 1 (ELAM-1) recognize the same carbohydrate ligand, sialyl-Lewis x. In *Proc. Natl. Acad. Sci. USA* **88**:6224–8.

Pownall S, Kozak CA, Schappert K *et al.* (1992): Molecular cloning and characterization of the mouse UDP-*N*-acetylglucosamine: α-3-D-mannoside β1,2-*N*-acetylglucosaminyltransferase I gene. In *Genomics* **12**:699–704.

Rademacher TW, Parekh RB, Dwek RA (1988): Glycobiology. In *Annu. Rev. Biochem.* **57**:785–838.

Rearick JI, Sadler JE, Paulson JC *et al.* (1979): Enzymatic characterization of β-D-galactoside α2–3 sialyltransferase from porcine submaxillary gland. In *J. Biol. Chem.* **254**:4444–51.

Reboul P, George P, Geoffrey J *et al* (1992): Study of *O*-glycan sialylation in C6 cultured glioma cells: regulation of a β-galactoside α2,3 sialyltransferase activity by Ca^{2+} / calmodulin antagonists and phosphatase inhibitors. In *Biochem. Biophys. Res. Com.* **186**:1575–81.

Reguigne I, James MR, Richard CW *et al.* (1994): The gene encoding myeloid α-3-fucosyltransferase(Fut4) is located between D11S388 and D11S919 on 11Q21. In *Cytogenet. Cell Genet.* **66**:104–6.

Ropp PA, Little MR, Cheng P-W (1991): Mucin biosynthesis: Purification and characterization of a mucin β6-*N*-acetylglucosaminyltransferase. In *J. Biol. Chem.* **266**:23863–71.

Rouquier S, Lowe JB, Kelly RJ *et al.* (1995): Molecular cloning of a human genomic region containing the H blood group α1,2-fucosyltransferase gene and two H locus-related DNA restriction fragments – Isolation of a candidate for the human secretor blood group locus. In *J. Biol. Chem.* **270**:4632–9.

Russo RN, Shaper NL, Taatjes DJ *et al.* (1992): β1,4-Galactosyltransferase: a short NH_2-terminal fragment that includes the cytoplasmic and transmembrane domain is sufficient for Golgi retention. In *J. Biol. Chem.* **267**:9241–7.

Russo RN, Shaper NL, Shaper JH (1990): Bovine β1,4-galactosyltransferase: Two sets of mRNA transcripts encode two forms of the protein with different amino-terminal domains. *In vitro* translation experiments demonstrate that both the short and the long forms of the enzyme are type II membrane-bound glycoproteins. In *J. Biol. Chem.* **265**:3324–31.

Sadler JE (1984): Biosynthesis of glycoproteins: Formation of *O*-linkedoligosaccharides. In *Biology of Carbohydrates* 2 (Ginsburg V, Robbins PW, eds) pp 199–288, New York, NY: John Wiley and Sons.

Sadler JE, Rearick JI, Paulson JC *et al.* (1979a): Purification to homogeneity of a β-galactoside α2,3-sialyltransferase and partial purification of an α-*N*-acetylgalactosaminide α2,6-sialyltransferase from porcine submaxillary glands. In *J. Biol. Chem.* **254**:4434–43.

Sadler JE, Rearick JI and Hill RL (1979b): Purification to homogeneity and enzymatic characterization of an α-*N*-acetylgalactosaminide α2–6-sialyltransferase from porcine submaxillary glandsIn *J. Biol. Chem.* **254**:5934–41.

Saito H, Nishikawa A, Gu JG *et al.* (1994): CDNA cloning and chromosomal mapping of human *N*-acetylglucosaminyltransferase V. In *Biochem. Biophys. Res. Commun.* **198**:318–27.

Saito H, Gu JG, Nishikawa A *et al.* (1995): Organization of the human *N*-acetylglucosaminyltransferase V gene. In *Eur. J. Biochem.* **233**:18–26.

Saitoh O, Piller F, Fox R *et al.* (1991): T-lymphocytic leukemia expresses complex, branched *O*-linked oligosaccharides on a major sialoglycoprotein, leukosialin. In *Blood* **77**:1491–9.

Sandrin MS, Dabkowski PL, Henning MM *et al.* (1995): Characterization of cDNA clones for porcine α1,3-galactosyltransferase: The enzyme generating the Galα1,3Gal epitope. In *Unpublished EMBL/GenBank data.*

Sarkar M, Hull E, Nishikawa Y *et al.* (1991): Molecular cloning and expression of cDNA encoding the enzyme that controls conversion of high-mannose to hybrid and complex *N*-glycans: UDP-*N*-acetylglucosamine:α-3-D-mannoside β1,2-*N*-acetylglucosaminyltransferase I. In *Proc. Natl. Acad. Sci. USA* **88**:234–8.

Sarnesto A, Köhlin T, Hindsgaul O *et al.* (1992a): Purification of the secretor-type β-galactoside α1,2-fucosyltransferase from human serum. In *J. Biol. Chem.* **267**:2737–44.

Sarnesto A, Köhlin T, Hindsgaul O *et al.* (1992b): Purification of the β-*N*-acetylglucosaminide α1,3-fucosyltransferase from human serum. In *J. Biol. Chem.* **267**:2745–52.

Sasaki K, Watanabe E, Kawashima K *et al.* (1993): Expression cloning of a novel Gal β1–3/1–4G1cNAcα2,3-sialyltransferase using lectin resistance selection. In *J. Biol. Chem.* **268**:22782–7.

Sasaki K, Kurata K, Funayama K *et al.* (1994a): Expression cloning of a novel α1,3-fucosyltransferase that is involved in biosynthesis of the sialyl Lewis x carbohydrate determinants in leukocytes. In *J. Biol. Chem.* **269**:14730–7.

Sasaki K, Kurata K, Kojima N *et al.* (1994b): Expression cloning of a GM_3-specific α2,8-sialyltransferase (GD_3 synthase). In *J. Biol. Chem.* **269**:15950–6.

Sasaki K, Watanabe E, Kawashima K *et al.* (1994c): Unpublished, EMBL/GenBank data.

Schachter H (1986): Biosynthetic controls that determine the branching and microheterogeneity of protein-bound oligosaccharides. In *Biochem. Cell Biol.* **64**:163–81.

Schachter H (1991a): Enzymes associated with glycosylation. In *Curr. Opinion Struct. Biol.* **1**:755–65.

Schachter H (1991b): The "yellow brick road" to branched complex *N*-glycans. In *Glycobiology* **1**:453–61.

Schachter H (1994): Molecular cloning of glycosyltransferase genes. In *Molecular Glycobiology* (Fukuda M, Hindsgaul O, eds) pp 88–162, Oxford: Oxford University Press.

Schachter H (1995): Glycosyltransferases involved in the synthesis of *N*-glycan antennae. In *Glycoproteins* 29a (Montreuil J, Vliegenthart JFG, Schachter H, eds) pp 153–99, Amsterdam: Elsevier.

Schachter H, Brockhausen I (1989): The biosynthesis of branched *O*-glycans. In *Mucus and Related Topics*, 43 (Chantler E, Ratcliffe NA, eds) pp 1–26, Cambridge: Society for Experimental Biology.

Schachter H, Brockhausen I (1992): The biosynthesis of serine(threonine)-*N*-acetylgalactosamine-linked carbohydrate moieties. In *Glycoconjugates. Composition, Structure and Function* (Allen HJ, Kisailus EC, eds) pp 263–332, New York, NY: Marcel Dekker, Inc.

Scheidegger EP, Sternberg LR, Roth J *et al.* (1995): A human STX cDNA confers polysialic acid expression in mammalian cells. In *J. Biol. Chem.* **270**:22685–8.

Shao MC, Sokolik CW, Wold F (1994): Specificity studies of the GDP-[l]-fucose – 2-acetamido-2-deoxy-β-D-glucoside (fuc-asn-linked GlcNAc) 6-α-fucosyltransferase from rat-liver golgi membranes. In *Carbohydr. Res.* **251**:163–73.

Shaper JH, Shaper NL (1992): Enzymes associated with glycosylation. In *Curr. Opinion Struct. Biol.* **2**:701–9.

Shaper NL, Shaper JH, Meuth JL *et al.* (1986): Bovine galactosyltransferase: a clone identified by direct immunological screening of a cDNA expression library. In *Proc. Natl. Acad. Sci. USA.* **83**:1573–7.

Shaper NL, Hollis GF, Douglas JG *et al.* (1988): Characterization of the full length cDNA for murine β1,4-galactosyltransferase. Novel features at the 5'-end predict two translational start sites at two in-frame AUGs. In *J. Biol. Chem.* **263**:10420–8.

Shaper NL, Wright WW, Shaper JH (1990): Murine β1,4-galactosyltransferase: Both the amounts and structure of the mRNA are regulated during spermatogenesis. In *Proc. Natl. Acad. Sci. USA* **87**: 791–5.

Shaper NL, Lin S, Joziasse DH *et al.* (1992): Assignment of two human α1,3-galactosyltransferase gene sequences (GGTA1 and GGTA1P) to chromosomes 9q33-q34 and 12q14-q15. In *Genomics* **12**:613–5.

Shaper NL, Harduinlepers A, Shaper JH (1994): Male germ cell expression of murine β4-galactosyltransferase – A 796-base pair genomic region, containing two cAMP-responsive element (CRE)-like elements, mediates male germ cell-specific expression in transgenic mice. In *J. Biol. Chem.* **269**:25165–71.

Shaper JH, Joziasse DH, Meurer JA *et al.* (1995): The chicken genome contains two functional non-allelic β1,4-galactosyltransferase genes. In *Glycoconjugate J.* **12**:477.

Sheares BT, Carlson DM (1983): Characterization of UDP-galactose:2-acetamido-2-deoxy-D-glucose 3β-galactosyltransferase from pig trachea. In *J. Biol. Chem.* **258**:9893–8.

Sheares BT, Lau JYT, Carlson DM (1982): Biosynthesis of galactosyl-β1,3-*N*-acetylglucosamine. In *J. Biol. Chem.* **257**:599–602.

Shoreibah M, Perng GS, Adler B *et al.* (1993): Isolation, characterization, and expression of a cDNA encoding *N*-acetylglucosaminyltransferase V. In *J. Biol. Chem.* **268**:15381–5.

Shur BD (1991): Cell surface β1,4-galactosyltransferase: twenty years later. In *Glycobiology* **1**:563–75.

Slusarewicz P, Nilsson T, Hui N *et al.* (1994): Isolation of a matrix that binds medial Golgi enzymes. In *J. Cell Biol.* **124**:405–13.

Smets LA, Van Beek WP (1984): Carbohydrates of the tumor cell surface. In *Biochim. Biophys. Acta* **738**:237–49.

Snider MD (1984): Biosynthesis of glycoproteins: Formation of *N*-linked oligosaccharides. In *Biology of Carbohydrates* 2 (Ginsburg V, Robbins PW, eds) pp 163–98, New York, NY: John Wiley and Sons.

Sorensen T, White T, Wandall HH *et al.* (1995): UDP-*N*-acetyl-α-D-galactosamine: polypeptide *N*-acetylgalactosaminyltransferase. In *J. Biol. Chem.* **270**: 24166–73.

Stamenkovic I, Asheim HC, Deggerdal A *et al.* (1990): The B cell antigen CD75 is a cell surface sialyltransferase. In *J. Exp. Med.* **172**:641–3.

Staudacher E, Altmann F, Glössl J *et al.* (1991): GDP-fucose: β-*N*-acetylglucosamine (Fuc to (fucα1, 6GlcNAc)-Asn-peptide) α1,3-fucosyltransferase activity in honeybee (*Apis mellifica*) venom glands. The difucosylation of asparagine-bound *N*-acetylglucosamine. In *Eur. J. Biochem.* **199**:745–51.

Staudacher E, Altmann F, März L *et al.* (1992a): α1,6(α1,3)-difucosylation of the asparagine-bound *N*-acetylglucosamine in honeybee venom phospholipase A2. In *Glycoconjugate J.* **9**:82–5.

Staudacher E, Kubelka V, März L (1992b): Distinct *N*-glycan fucosylation potentials of three lepidopteran cell lines. In *Eur. J. Biochem.* **207**:987–93.

Staudacher E, Dalik T, Wawra P *et al.* (1995): Functional purification and characterization of a GDP-fucose:β-*N*-acetylglucosamine (fuc to Asn linked GlcNAc) α1,3-fucosyltransferase from mung beans. In *Glycoconjugate J.* **12**:780–6.

Strahan KM, Gu F, Andersson L *et al.* (1995a): Pig α1, 3-galastosyltransferase: Sequence of a full-length cDNA clone, chromosomal localisation of the corresponding gene, and inhibition of expression in cultured pig endothelial cells. In *Transplant. Proc.* **27**:245–6.

Strahan KM, Gu F, Preece AF *et al.* (1995b): cDNA sequence and chromosome localization of pig α1,3-galactosyltransferase. In *Immunogenetics* **41**:101–5.

Strous GJ (1986): Golgi and secreted galactosyltransferase. In *CRC Crit. Rev. Biochem.* **21**:119–51.

Stults CLM, Macher BA (1993): β1,3-*N*-acetylglucosaminyltransferase in human leukocytes: properties and role in regulating neolacto glycosphingolipid biosynthesis. In *Arch. Biochem. Biophys.* **303**:125–133.

Svensson EC, Soreghan B, Paulson JC (1990): Organization of the β-galactoside α2,6-sialyltransferase gene. Evidence for the transcriptional regulation of terminal glycosylation. In *J. Biol. Chem.* **265**: 20863–8.

Svensson EC, Conley PB, Paulson JC (1992): Regulated expression of α2,6-sialyltransferase by the liver-enriched transcription factors HNF-1, DBP, and LAP. In *J. Biol. Chem.* **267**:3466–72.

Takano R, Muchmore E, Dennis JW (1994): Sialylation and malignant potential in tumour cell glycosylation mutants. In *Glycobiology* **4**:665–74.

Tan J, D'Agostaro GAF, Bendiak B *et al.* (1995a): The human UDP-*N*-acetylglucosamine: α-6-D-mannoside-β1,2-*N*-acetylglucosaminyltransferase II gene (MGAT2) – Cloning of genomic DNA, localization to chromosome 14q21, expression in insect cells and purification of the recombinant protein. In *Eur. J. Biochem.* **231**:317–28.

Tan J, Dunn J, Jaeken J *et al.* (1995b): Carbohydrate-deficient glycoprotein syndrome type II. An autosomal recessive disease due to point mutations in the coding region of the *N*-acetylglucosaminyltransferase II gene. In *Glycoconjugate J.* **12**:478.

Taniguchi N, Ihara Y (1995): Recent progress in the molecular biology of the cloned *N*-acetylglucosaminyltransferases. In *Glycoconjugate J.* **12**:733–8.

Teasdale RD, D'Agostaro G, Gleeson PA (1992): The signal for Golgi retention of bovine β1,4-galactosyltransferase is in the transmembrane domain. In *J. Biol. Chem.* **267**:4084–96.

Teasdale RD, Matheson F, Gleeson PA (1994): Post-translational modifications distinguish cell surface from Golgi-retained β1,4 galactosyltransferase molecules. Golgi localization involves active retention. In *Glycobiology* **4**:917–28.

Tetteroo PAT, de Heij HT, van den Eijnden DH *et al.* (1987): A GDP-fucose:[Galβ1,4]GlcNAc α1,3-fucosyltransferase activity is correlated with the presence of human chromosome 11 and the expression of the Le^x, Le^y, and sialyl-Le^x antigens in human-mouse cell hybrids. In *J. Biol. Chem.* **262**:15984–9.

Thall AD, Maly P, Lowe JB (1995): Oocyte Galα1,3Gal epitopes implicated in sperm adhesion to the zona pellucida glycoprotein ZP3 are not required for fertilization in the mouse. In *J. Biol. Chem.* **270**: 21437–40.

Thurnher M, Clausen H, Fierz W *et al.* (1992): T-cell clones with normal or defective *O*-galactosylation from a patient with permanent mixed-field polyagglutinability. In *Eur. J. Immunol.* **22**:1835–42.

Thurnher M, Rusconi S, Berger EJ (1993): Persistant repression of a functional allele can be responsible for galactosyltransferase deficiency in Tn syndrome. In *J. Clin. Invest.* **91**:2103–10.

Toki D, Granovsky M, Reck F *et al.* (1994): Inhibition of UDP-GlcNAc: Galβ1,3GalNAc (GlcNAc to GalNAc) β6-GlcNAc-transferase from acute myelogenous leukemia cells by nitrophenyl substrate derivatives. In *Biochem. Biophys. Res. Comm.* **198**:417–23.

Toki D, Sarkar M, Yip B *et al.* (1996) Role of *N*-linked glycosylation of recombinant human core 2 β6-*N*-acetylglucosaminyltransferase expressed in Sf9 insect cells. Submitted.

Troy FA, II (1992): Polysialylation: From bacteria to brains. In *Glycobiology* **2**:5–23.

van den Eijnden DH, Joziasse DH (1993): Enzymes associated with glycosylation. In *Curr. Opinion Struct. Biol.* **3**:711–21.

van den Eijnden DH, Koenderman AHL, Schiphorst WECM (1988): Biosynthesis of blood group i-active polylactosaminoglycans. Partial purification and properties of an UDP-GlcNAc:*N*-acetyllactosaminide β1,3-*N*-acetylglucosaminyltransferase from Novikoff tumor cell ascites fluid. In *J. Biol. Chem.* **263**:12461–71.

van Pelt J, Dorland L, Duran M *et al.* (1989): Transfer of sialic acid in α2,6 linkage to mannose in Manβ1,4GlcNAc and Manβ1,4GlcNAcβ1,4GlcNAc by the action of Galβ1,4GlcNAc α2,6-sialyltransferase. In *FEBS Lett.* **256**:179–84.

Vavasseur F, Dole K, Yang J *et al.* (1994): *O*-glycan biosynthesis in human colorectal adenoma cells during progression to cancer. In *Eur. J. Biochem.* **222**: 415–24.

Vavasseur F, Yang J, Dole K *et al.* (1995): Characterization of UDP-GlcNAc: GalNAc β3-*N*-acetylglusaminyl-transferase activity from colonic tissues. Loss of the activity in human cancer cell lines. In *Glycobiology* **5**:351–7.

Verbert A (1995): From $Glc_3Man_9GlcNAc_2$-protein to $Man_5GlcNAc_2$-protein: transfer "en bloc" and processing. In *Glycoproteins* 28a (Montreuil J, Vliegenthart JFG, Schachter H, eds) pp 145–52, Amsterdam: Elsevier.

Voynow JA, Kaiser RS, Scanlin TF *et al.* (1991): Purification and characterization of GDP-L-fucose-*N*-acetyl β-D-glucosaminide α1,6 fucosyltransferase from cultured human skin fibroblasts. In *J. Biol. Chem.* **266**:21572–7.

Walz G, Aruffo A, Kolanus W *et al.* (1990): Recognition by ELAM-1 of the sialyl-Le^x determinant on myeloid and tumor cells. In *Science* **250**:1132–5.

Wang X-C, O'Hanlon TP, Young RF *et al.* (1990): Rat β-galactoside α2,6-sialyltransferase genomic organization: alternate promoters direct the synthesis of liver and kidney transcripts. In *Glycobiology* **1**:25–31.

Wang XC, Vertino A, Eddy RL *et al.* (1993): Chromosome mapping and organization of the human β-galactoside α2,6-sialyltransferase gene. Differential and cell-type specific usage of upstream exon sequences in B-lymphoblastoid cells. In *J. Biol. Chem.* **268**:4355–61.

Wang Y, Abernethy JL, Eckhardt AE *et al.* (1992): Purification and characterization of a UDP-GalNAc:polypeptide *N*-acetylgalactosaminyltransferase specific for glycosylation of threonine residues. In *J. Biol. Chem.* **267**:12709–26.

Warren CE, Smookler DS, Dennis JW (1995): UDP-GlcNAc:Galβ1,3GalNAc-R (GlcNAc to GalNAc) β1,6GlcNAc-transferase (core2-GnT) mouse cDNA sequence. *Unpublished, GenBank data.*

Weinstein J, Lee EU, McEntee K *et al.* (1987): Primary structure of β-galactoside α2,6-sialyltransferase. Conversion of membrane-bound enzyme to soluble forms by cleavage of the NH_2-terminal signal abchor. In *J. Biol. Chem.* **262**:17735–43.

Wen DX, Livingston BD, Medzihradszky KF *et al.* (1992a): Primary structure of Galβ1,3(4)GlcNAc α2,3-sialyltransferase determined by mass spectrometry sequence analysis and molecular cloning. In *J. Biol. Chem.* **267**:21011–9.

Wen DX, Svensson EC, Paulson JC (1992b): Tissue-specific alternative splicing of the β-galactoside α2,6-sialyltransferase gene. In *J. Biol. Chem.* **267**: 2512–8.

Weston BW, Nair RP, Larsen RD *et al.* (1992a): Isolation of a novel human α1,3-fucosyltransferase gene and molecular comparison to the human Lewis blood group α1,3/1,4-fucosyltransferase gene. Syntenic, homologous, nonallelic genes encoding enzymes with distinct acceptor substrate specificities. In *J. Biol. Chem.* **267**:4152–60.

Weston BW, Smith PL, Kelly RJ *et al.* (1992b): Molecular cloning of a fourth member of a human α1,3-

fucosyltransferase gene family. Multiple homologous sequences that determine expression of the Lewis x, sialyl Lewis x, and difucosyl sialyl Lewis x epitopes. In *J. Biol. Chem.* **267**:24575–84.

White T, Bennett EP, Takio K, Sorensen T *et al.* (1995): Purification and cDNA cloning of a human UDP-*N*-acetyl-α-D-galactosamine:polypeptide *N*-acetylgalactosaminyltransferase. In *J. Biol. Chem.* **270**: 24156–65.

Wlasichuk KB, Kashem MA, Nikrad PV *et al.* (1993): Determination of the specificities of rat liver Galβ1,4GlcNAcα2,6-sialyltransferase and Galβ1,3/4GlcNAcα2,3-sialyltransferase using synthetic modified acceptors. In *J. Biol. Chem.* **268**:13971–7.

Wong SH, Low SH, Hong W (1992): The 17-residue transmembrane domain of β-galactoside α2,6-sialyltransferase is sufficient for Golgi retention. In *J. Cell Biol.* **117**:245–58.

Yamaguchi N, Fukuda MN (1995): Golgi retention mechanism of β1,4-galactosyltransferase – Membrane-spanning domain-dependent homodimerization and association with α- and β-tubulins. In *J. Biol. Chem.* **270**:12170–6.

Yamamoto F, Hakomori S (1990): Sugar-nucleotide donor specificity of histo-blood group A and B transferases is based on amino acid substitutions. In *J. Biol. Chem.* **265**:19257–62.

Yamamoto F, McNeill PD, Hakomori S (1991): Identification in human genomic DNA of the sequence homologous but not identical to either the histo-blood group ABH genes or α1,3-galactosyltransferase pseudogene. In *Biochem. Biophys. Res. Commun.* **175**:986–94.

Yamashita K, Tachibana Y, Ohkura T *et al.* (1985): Enzymatic basis for the structural changes of asparagine-linked sugar chains of membrane glycoproteins of baby hamster kidney cells induced by polyoma transformation. In *J. Biol. Chem.* **260**: 3963–9.

Yang J, Bhaumik M, Liu Y *et al.* (1994a): Regulation of *N*-linked glycosylation. Neuronal cell-specific expression of a 5' extended transcript from the gene encoding *N*-acetylglucosaminyltransferase I. In *Glycobiology* **4**:703–12.

Yang JM, Byrd J, Siddiki B *et al.* (1994b): Alterations of *O*-glycan biosynthesis in human colon cancer tissues. In *Glycobiology* **4**:873–84.

Yonezawa S, Tachikawa T, Shin S *et al.* (1992): Sialosyl-Tn antigen: its distribution in normal human tissues and expression in adenocarcinomas. In *Am. J. Clin. Pathol.* 98:167–74.

Yoshida Y, Kojima N, Kurosawa N *et al.* (1995a): Molecular cloning of Sia α2,3Galβ1,4GlcNAcα2,8-sialyltransferase from mouse brain. In *J. Biol. Chem.* **270**:14628–33.

Yoshida Y, Kojima N, Tsuji S (1995b): Molecular cloning and characterization of a third type of *N*-glycan α2,8-sialyltransferase from mouse lung. In *J. Biochem.* **118**:658–64.

Yoshimura M, Ihara Y, Taniguchi N (1995a): Changes of β1,4-*N*-acetylglucosaminyltransferase III (GnT-III) in patients with leukaemia. In *Glycoconjugate J.* **12**:234–40.

Yoshimura M, Nishikawa A, Ihara Y *et al.* (1995b): High expression of UDP-*N*-acetylglucosamine: β-D mannoside β1,4-*N*-acetylglucosaminyltransferase III (GnT-III) in chronic myelogenous leukemia in blast crisis. In *Int. J. Cancer* **60**:443–9.

Yoshimura M, Nishikawa A, Ihara Y *et al.* (1995c): Suppression of lung metastasis of B16 mouse melanoma by *N*-acetylglucosaminyltransferase III gene transfection. In *Proc. Natl. Acad. Sci. USA* **92**:8754–8.

6 Topology of Glycosylation – a Histochemist's View

MARGIT PAVELKA

1 Introduction

The preceding chapter by Brockhausen and Schachter has expertly explained the enzymological aspects of glycosylation. In addition to clarifying the questions of enzyme structure and mechanisms which produce the carbohydrate chains of complex carbohydrates, the spatial organization of the enzymes involved needs to be deciphered to understand the ways in which the carbohydrate code elements, whose complexity is reviewed in the chapters of Laine and of Sharon and Lis, are generated. Cytochemical work in the past ten to twenty years has contributed considerably to our knowledge of the subcellular organization of glycosylation. Not only were biochemical results confirmed; immunocytochemical and lectinocytochemical patterns helped to characterize the main organelles where glycosylation takes place. Novel techniques at the electronmicroscopic level (e.g. Bernhard and Avrameas, 1971; Newman *et al.*, 1983; Roth, 1983; Slot and Geuze, 1983; Ellinger and Pavelka, 1985; Pavelka and Ellinger, 1989a; Bennett and Wild, 1991) provided approaches for the precise subcellular localization of various glycoconjugates and of an array of enzymes of the glycosylation machinery, although we are still some way from a complete description. Indeed, electron microscopic immuno- and lectinocytochemistry as well as radioautography are continually enhancing our level of understanding of the, at first, puzzling and complicated series of steps during the biosynthesis of *N*- and *O*-linked glycans; subcompartments of the main glycosylation organelles could be defined. Furthermore, cytochemical studies presented the glycosylation organelles as highly dynamic structures. It became clear that their architecture varies from one cell type to another, changing during development and depending upon the functional state of the cells. Various aspects of the subcellular distribution of glycosylation sites and their variability are summarized in detail in recent reviews (Farquhar and Palade, 1981; Farquhar, 1985; Griffiths and Simons, 1986; Pavelka, 1987; Roth, 1987, 1991, 1993; Griffiths *et al.*, 1989; Geuze and Morré, 1991; Taatjes and Roth, 1991; Mellman and Simons, 1992; Farquhar *et al.*, 1993; Pavelka and Ellinger, 1994; Rothman, 1994; Fiedler and Simons, 1995).

2 Topology of *N*-Glycosylation

The biosynthesis of *N*-glycosidically linked glycans of glycoproteins is a multistep process (for reviews see Kornfeld and Kornfeld, 1985; Schachter, 1995 and the preceding chapter by Brockhausen and Schachter), which starts in the endoplasmic reticulum (ER) with the insertion of high-mannose oligosaccharides, while the polypeptide is still growing. It proceeds post-translationally in the ER and in subcompartments of the Golgi apparatus; here, step by step conversion of high-mannose oligosaccharides into complex- and hybrid-type glycans takes place. The organelles and compartments which are involved in the subsequent glycosylation steps are summarized in Figure 1.

Endoplasmic reticulum (ER)

Dolichol-pyrophosphate-linked oligosaccharides with terminal glucose-residues – $glc_3man_9glcNAc_2$ – are preformed. In the lumen of the ER they are transferred en bloc to distinct asparagine residues

H.-J. and S. Gabius (Eds.), Glycosciences

ISBN 3-8261-0073-5

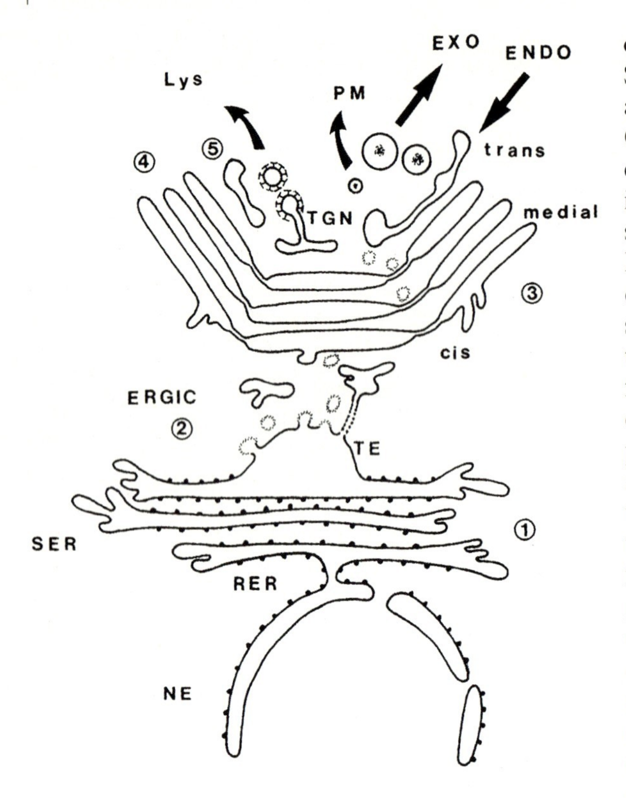

Figure 1. The main organelles of the glycosylation machinery and their subcompartments, namely (1) – (5), in this drawing.
In the endoplasmic reticulum (1), initial *N*-glycosylation takes place; the newly synthesized glycans are modified co-translationally as well as post-translationally and exported out of the ER at specialized regions, the transitional elements. The glycoconjugates travel across (2), ERGIC (an intermediate sorting compartment between ER and the Golgi apparatus), to the Golgi stacks. At the *cis* Golgi side (3), *O*-glycosylation is initiated and *N*-glycosylation continues. In the Golgi cisternae, the high-mannose glycans produced in the ER are converted step by step into complex- and hybrid-type glycans. In medial Golgi cisternae (4), *N*-acetylglucosaminyltransferase I has been localized. In the *trans* Golgi cisternae and *trans* Golgi network (5), *N*- and *O*-linked oligosaccharide chains are completed; β1,4 galactosyltransferase and α2,6 sialyltransferase both have been localized in this Golgi region.
NE – nuclear envelope; RER – rough endoplasmic reticulum; SER – smooth endoplasmic reticulum; TE – transitional elements of the endoplasmic reticulum; ERGIC – endoplasmic reticulum Golgi apparatus intermediate compartment; cis – *cis* Golgi cisternae; medial – medial Golgi cisternae; trans – *trans* Golgi cisternae; TGN – *trans* Golgi network; Lys – lysosomal system; PM – plasma membrane; EXO – exocytosis; ENDO – endocytosis

of the nascent polypeptide chains (Fig. 1–1). Still co-translationally, the oligosaccharide chains are subjected to trimming by ER-glucosidases. Glucosidase I, which removes one of the 3 glucoses of the *N*-linked oligosaccharide chains, initiates entry into the deglucosylation/reglucosylation cycle (for a review see Hammond and Helenius, 1995). Following the action of glucosidase II, the processed glycoprotein will not necessarily exit from the ER region. As a quality control step, conformations that are not properly folded can be recognized by UDP-glucose:glycoprotein glucosyltransferase, a soluble protein of the ER (Parodi, 1996). Reglucosylation allows the ensuing interaction with calnexin/calreticulin, retaining the re-processed glycoprotein in the ER for the required correction. After correct folding a protein is no longer recognized by the glucosyltransferase and is released from the deglucosylation/reglucosylation cycle by glucosidase II. By high-resolution immunogold labeling, glucosidase II has been localized to the lumen of the nuclear envelope and ER in pig liver hepatocytes (Lucocq *et al.*, 1986); although transitional elements of the ER and associated smooth membranes were reactive, the Golgi apparatus with its stacked cisternae was clearly devoid of label, thus revealing glucosidase II as an enzyme residing in the ER and not in the Golgi apparatus. Multiple cytochemical studies with different cell types have shown the dominance of high-mannose glycans in the nuclear envelope and in the lumen of the ER by using mannose-/glucose-reactive lectins such as concanavalin A or the pea and lentil lectins (e.g. Bernhard and Avrameas, 1971; Tartakoff and Vassalli, 1983; Pavelka and Ellinger, 1985, 1989b; Roth, 1987).

Further trimming by cleavage of mannose residues occurs, at least in part, post-translationally, before the regularly folded and possibly oligomerized proteins are exported out of the ER. On their route from the ER to the Golgi apparatus and to further destinations, the newly synthesized glycoproteins bearing high-mannose oligosaccharide chains traverse an intermediate compartment (ER Golgi intermediate compartment – ERGIC, Fig. 1–2; Hauri and Schweizer, 1992). This appears to be mainly a salvage compartment, from where luminal and membrane ER constituents are separated from the bulk of others destined for various cellular compartments and are returned to the ER. Sorting systems such as the KDEL-system (Lewis and Pelham, 1992; Tang *et al.*, 1993; Griffiths *et al.*, 1994) for luminal proteins, and the di-lysine-motif for membrane proteins are well characterized (for review Hammond and Helenius, 1995; Pelham, 1995). It is still a question whether the terminal mannoses of high-mannose glycans are involved in maintaining the correct route as well. Interestingly, ERGIC 53, which is one of the main ERGIC constituents and recycles to the ER, has been proved to be a mannose-binding lectin (Arar *et al.*, 1995).

Conspicuously, labeling for galactose residues was observed in limited, dilated regions of the rough ER of rat duodenal goblet cells (Ellinger and Pavelka, 1992). Since galactose residues are known to become inserted into glycans in the Golgi stacks and not in the ER, the ER-localized galactose presumably belongs to resident ER-glycoproteins, which are retrieved from the Golgi apparatus (for review Pelham, 1995). Calreticulin, which appears to gain access to *trans* Golgi cisternae and becomes galactosylated before returning to the ER (Peter *et al.*, 1992), may account for the galactose-specific reactions.

Golgi apparatus

N-glycosylation continues after glycoconjugates have been exported out of the ER and then been transported across ERGIC to cisternae of the Golgi apparatus. Golgi cisternae are the sites where the *N*-linked high-mannose oligosaccharide chains synthesized in the ER become converted into complex-type and hybrid-type glycans. Some of the Golgi enzymes involved have been localized by immunoelectron microscopy. These are *N*-acetylglucosaminyltransferase I, which has been demonstrated in cisternae of the medial part of the Golgi stacks (Fig. 1–4; Dunphy *et al.*, 1985), β1,4-galactosyltransferase and α2,6 sialyltransferase, the localization of which predominates in the *trans and transmost* Golgi cisternae (Fig. 1–5; Roth and Berger, 1982; Roth *et al.*, 1986; Berger *et al.*, 1993; Rabouille *et al.*, 1995). Such patterns indicate a subcompartmentation of *N*-glycosylation proceeding from the *cis* to the *trans* Golgi side. This has been confirmed by radioautography: the insertion of galactose and sialic acid could be localized to the *trans* Golgi region (for review see Bennett and Wild, 1991). Furthermore, lectinocytochemical patterns mainly confirm the results mentioned above. Mannose residues as visualized by concanavalin A and the pea and and lentil lectins dominate at the *cis* Golgi side and, on the other hand, galactose and sialic acid are concentrated at the *trans* side of the stacks (Tartakoff and Vassalli, 1983; Roth *et al.*, 1984; Pavelka and Ellinger, 1985, 1993; Hedman *et al.*, 1986; for reviews see Pavelka and Ellinger, 1991, 1994; Farquhar *et al.*, 1993; Roth, 1993). Galactose residues were mainly localized by means of binding of *Ricinus communis I* agglutinin, *Erythrina cristagalli* lectin and *Allomyrina dichotoma* lectin; *Limax flavus* lectin and the elderberry lectin were used for demonstrating sialic acid. The changes in the reactivities for the different lectins from the *cis* to the *trans* Golgi side may correspond to the conversion of high mannose glycans into complex-type ones. However, the patterns are highly variable. Different cell types residing side by side, such as duodenal goblet cells and resorptive cells, showed very different distributions of Golgi transferases and their products, and reaction patterns changed during differentiation (e.g. Roth *et al.*, 1986; Pavelka and Ellinger, 1987). Similarly, for Golgi mannosidase II, highly variable localizations have been reported (Novikoff *et al.*, 1983; Velasco *et al.*, 1993; Rabouille *et al.*, 1995). In a recent study with HeLa-cells, β1,2-*N*-acetylglucosaminyltransferase I has been co-localized with α1,3–1,6-mannosidase II in medial cisternae and, similarly, β1,4-galactosyltransferase showed co-localization with α2,6-sialyltransferase in the *trans* Golgi network (Rabouille *et al.*, 1995). There was remarkable co-localization of all four enzymes in one *trans* cisterna of the stacks. From these cytochemical results and from transport studies performed previously (Nilsson *et al.*, 1994), it has been postulated that a process connected with kin recognition may be responsible for the correct targeting of Golgi enzymes to their right "home" cisternae; a functional implication of this suggested interaction for the establishment of the observable stack-organization of the Golgi cisternae seems likely.

3 Topology of O-Glycosylation

O-linked oligosaccharides exhibit a much greater heterogeneity in their structure than *N*-linked ones (for review Roth, 1987,1991; Brockhausen and Schachter, this volume; Sharon and Lis, this volume). Cytochemical results mainly concern *O*-linked glycans with a serine/threonine-*N*-acetylgalactosaminyl-core linkage. The biosynthesis of these *O*-linked glycans appears to be a late post-translational event, which takes place by a number of consecutive "classical" glycosyl transfer reactions. The first step in the synthesis of *O*-glycosidically linked glycans is the addition of *N*-acetylgalactosamine (galNAc) to threonine or serine. By immunogold electron microscopy, UDP-*N*-acetyl-D-galactosamine:polypeptide-*N*-acetylgalactosamininyltransferase has recently been localized in porcine and bovine submaxillary gland (Roth *et al.*, 1994): specific gold labeling was found over the *cis* cisternae of the Golgi apparatus and vesicular structures close to the *cis* Golgi side (Fig. 1–3), but was absent from the endoplasmic reticulum. This indicates that the galNAc-transferase-mediated initial step of *O*-glycosylation occurs at the *cis* side of the Golgi apparatus and not in the ER. Possibly, the intermediate compartment in between the ER and the Golgi apparatus is involved in the initial *O*-glycosylation as well, but the ER is clearly excluded. This is in accordance with lectinocytochemical studies using *Helix pomatia* lectin and *Vicia villosa* isolectin B4, the binding reactions of which mirror the immunolocalization of the transferase (Roth, 1984; Roth *et al.*, 1994). The initial insertion of galNAc is followed by the consecutive step-by-step addition of further sugar residues, such as galactose, fucose and sialic acid. By using HRP-labeled peanut lectin, which specifically recognizes galβ1,3-galNAc-serine/threonine, the insertion of galactose has been attributed to the medial Golgi cisternae in goblet cells of the gastrointestinal tract (Fig. 1–4; Sato and Spicer, 1982). These patterns and several other data available indicate that at least certain steps during the synthesis of *O*-linked glycans take place in distinct subcompartments of the Golgi apparatus.

4 Summarizing Remarks and Perspectives

Cytochemical patterns have helped us to gain more insight into the organization of the organelles involved in glycosylation. In the ER as well as in the Golgi apparatus, subcompartments could be defined in which certain steps during *N*- and *O*-glycosylation take place. In the Golgi stacks, the general *cis*-to-*trans* orientation of consecutive glycosylation steps could be confirmed; we learned about the great differences between different cell types in the appearance of glycosylation compartments and realized the enormous dynamics of the ER and Golgi architecture. We got an impression as to how the consecutive steps of glycosylation could be organized, but we are a long way from detailed knowledge. Further studies will have to concentrate even more intensely on the function – architecture relationship of the glycosylation compartments, for example by influencing traffic out of the ER and across the Golgi stack by brefeldin A (for a review see Klausner *et al.*, 1992), and will have to include glycosylation and reglycosylation of recycling molecules as well (for a review see Tauber *et al.*, 1993). There is no doubt that the topology of glycosylation will remain an attractive topic of cytochemists' work with unmistakable relevance for other areas of glycosciences.

References

Arar C, Carpentier V, Le Caer JP *et al.* (1995): ERGIC-53, a membrane protein of the endoplasmic reticulum – Golgi intermediate compartment, is identical to MR60, an intracellular mannose-specific lectin of myelomonocytic cells. In *J. Biol. Chem.* **270**:3551–3.

Bennett G, Wild G (1991): Traffic through the Golgi apparatus as studied by radioautography. In *J. Electr. Microsc. Techn.* **17**:132–49.

Berger EG, Grimm K, Bächi T *et al.* (1993): Double immunofluorescent staining of α2,6 sialyltransferase and β 1,4 galactosyltransferase in monensin-treated cells: Evidence for different Golgi compartments? In *J. Cell. Biochem.* **52**:275–88.

Bernhard W, Avrameas S (1971): Ultrastructural visualization of cellular carbohydrate components by means of concanavalin A. In *Exp. Cell Res.* **64**: 232–6.

Dunphy WG, Brands R, Rothman JE (1985): Attachment of terminal *N*-acetylglucosamine to asparagine-linked oligosaccharides occurs in central cisternae of the Golgi stack. In *Cell* **40**:463–72.

Ellinger A, Pavelka M (1985): Post-embedding localization of glycoconjugates by means of lectins on thin sections of tissues embedded in LR White. In *Histochem. J.* **17**:1321–36.

Ellinger A, Pavelka M (1992): Subdomains of the rough endoplasmic reticulum in colon goblet cells of the rat: Lectin-cytochemical characterization. In *J. Histochem. Cytochem.* **40**:919–30.

Farquhar MG, Palade GE (1981): The Golgi apparatus (complex)-(1954–1981)-from artifact to center stage. In *J. Cell Biol.* **91**:77s-103s.

Farquhar MG (1985): Progress in unraveling pathways of the Golgi traffic. In *Annu. Rev. Cell Biol.* **1**:447–88.

Farquhar MG, Hendricks L, Noda T *et al.* (1993): Cell organelles. Contributions of cytochemistry and immunocytochemistry to our understanding of the organization and function of the Golgi apparatus. In *Electron Microscopic Cytochemistry and Immunocytochemistry in Biomedicine* (Ogawa K and Barka T, eds) pp 441–79, Boca Raton: CRC Press.

Fiedler K, Simons K (1995): The role of *N*-glycans in the secretory pathway. In *Cell* **81**:309–12.

Geuze HJ, Morré DJ (1991): *Trans* Golgi reticulum. In *J. Electron Microsc. Techn.* **17**:24–34.

Griffiths G, Simons K (1986): The *trans* Golgi reticulum: sorting at the exit site of the Golgi complex. In *Science* **234**:438–43.

Griffiths G, Fuller SD, Back F *et al.* (1989): The dynamic nature of the Golgi complex. In *J. Cell Biol.* **108**:277–97.

Griffiths G, Ericsson M, Krijnse-Locker J *et al.* (1994): Localization of the Lys, Asp, Glu, Leu tetrapeptide receptor to the Golgi complex and the intermediate compartment in mammalian cells. In *J. Cell Biol.* **127**:1557–74.

Hammond C, Helenius A (1995): Quality control in the secretory pathway. In *Curr. Opinion Cell Biol.* **7**:523–9.

Hauri HP, Schweizer A (1992): The endoplasmic reticulum – Golgi intermediate compartment. In *Curr. Opinion Cell Biol.* **4**:600–8.

Hedman K, Pastan I, Willingham MC (1986): The organelles of the trans domain of the cell. Ultrastructural localization of sialoglycoconjugates using *Limax flavus* agglutinin. In *J. Histochem. Cytochem.* **34**: 1069–77.

Klausner RD, Donaldson JG, Lippincott-Schwartz J (1992): Brefeldin A: insights into the control of membrane traffic and organelle structure. In *J. Cell Biol.* **116**: 1071–80.

Kornfeld R, Kornfeld S (1985): Assembly of asparagine-linked oligosaccharides. In *Annu. Rev. Biochem.* **54**:631–64.

Lewis MJ, Pelham HRB (1992): Ligand – induced redistribution of a human KDEL receptor from the Golgi complex to the endoplasmic reticulum. In *Cell* **68**:353–64.

Lucocq JM, Brada D, Roth J (1986): Immunolocalization of the oligosaccharide trimming enzyme glucosidase II. In *J. Cell Biol.* **102**:2137–46.

Mellman I, Simons K (1992): The Golgi complex: *In vitro* veritas? In *Cell* **68**:829–40.

Newman GR, Jasani B, Williams ED (1983): A simple postembedding system for the rapid demontration of tissue antigens under the electron microscope. In *Histochem. J.* **15**:543–55.

Nilsson T, Hoe MH, Slusarewicz P *et al.* (1994): Kin recognition between medial Golgi enzymes in HeLa cells. In *EMBO J.* **13**:562–74.

Novikoff PM, Tulsiani DRP, Touster O *et al.* (1983): Immunocytochemical localization of α-D-mannosidase II in the Golgi apparatus of the rat liver. In *Proc. Natl. Acad. Sci USA* **80**:4364–8.

Parodi AJ (1996): The UDP-glc:glycoprotein glucosyltransferase and the quality control of glycoprotein folding in the endoplasmic reticulum. In *Trends Glycosci. Glycotechnol.* **8**:1–12.

Pavelka M (1987): Functional morphology of the Golgi apparatus. In *Adv. Anat. Embryol. Cell Biol.* **106**: 1–94.

Pavelka M, Ellinger A (1985): Localization of binding sites for concanavalin A, *Ricinus communis* I and *Helix pomatia* lectin in the Golgi apparatus of rat small intestinal absorptive cells. In *J. Histochem. Cytochem.* **33**:905–14.

Pavelka M, Ellinger A (1987): The Golgi apparatus in the acinar cells of the developing embryonic pancreas. II. Localization of lectin binding sites. In *Am. J. Anat.* **178**:224–30.

Pavelka M, Ellinger A (1989a): Pre-embedding labeling techniques applicable to intracellular binding sites. In *Electron Microscopy of Subcellular Dynamics* (Plattner H, ed) pp 199–218, Boca Raton: CRC Press.

Pavelka M, Ellinger A (1989b): Affinity-cytochemical differentiation of glycoconjugates of small intestinal absorptive cells using *Pisum sativum* and *Lens culinaris* lectins. In *J. Histochem. Cytochem.* **37**: 877–84.

Pavelka M, Ellinger A (1991): Cytochemical characteristics of the Golgi apparatus. In *J. Electr. Microsc. Techn.* **17**:35–50.

Pavelka M, Ellinger A (1993): Early and late transformations occurring at organelles of the Golgi area under the influence of brefeldin A: An ultrastructural and lectin cytochemical study. In *J. Histochem. Cytochem.* **41**:1031–42.

Pavelka M, Ellinger A (1994): Functional morphology of the Golgi region: A lectino-electron-microscopic exploration. In *Advances in Molecular and Cell Biology 8* (Bittar EE, Chen LB eds) pp 63–85, Greenwich, London: JAI Press Inc.

Pelham HRB (1995): Sorting and retrieval between the endoplasmic reticulum and Golgi apparatus. In *Curr. Opinion Cell Biol.* **7**:530–5.

Peter F, Nguyen van P, Söling HD (1992): Different sorting of KDEL proteins in rat liver. In *J. Biol. Chem.* **267**:10631–7.

Rabouille C, Hui N, Hunte F *et al.* (1995): Mapping the distribution of Golgi enzymes involved in the construction of complex oligosaccharides. In *J. Cell Sci.* **108**:1617–27.

Roth J (1983): Application of lectin-gold complexes for electron microscopic localization of glycoconjugates on thin sections. In *J. Histochem. Cytochem.* **31**: 987–99.

Roth J (1984): Cytochemical localization of terminal *N*-acetyl-D-galactosamine residues in cellular compartments of intestinal goblet cells: implications for the topology of *O*-glycosylation. In *J. Cell Biol.* **98**: 399–406.

Roth J (1987): Subcellular organization of glycosylation in mammalian cells. In *Biochim. Biophys. Acta* **906**:405–36.

Roth J (1991): Localization of glycosylation sites in the Golgi apparatus using immunolabeling and cytochemistry. In *J. Electr. Microsc. Techn.* **17**:121–31.

Roth J (1993): Cellular sialoglycoconjugates: a histochemical perspective. In *Histochem. J.* **25**:687–710.

Roth J, Berger EG (1982): Immunocytochemical localization of galactosyltransferase in Hela cells: Codistribution with thiamine pyrophosphatase in trans-Golgi cisternae. In *J. Cell Biol.* **93**:223–9.

Roth J, Lucocq JM, Charest PM (1984): Light and electron microscopic demonstration of sialic acid residues with the lectin from *Limax flavus*: A cytochemical affinity technique with the use of fetuin-gold complexes. In *J. Histochem. Cytochem.* **32**: 1167–76.

Roth J, Taatjes DJ, Weinstein J *et al.* (1986): Differential subcompartmentation of terminal glycosylation in the Golgi apparatus of intestinal absorptive and goblet cells. In *J. Biol. Chem.* **261**:14307–12.

Roth J, Wang Y, Eckhardt AE *et al.* (1994): Subcellular localization of the UDP-*N*-acetyl-D-galactosamine: polypeptide *N*-acetylgalactosaminyltransferase-mediated *O*-glycosylation reaction in the submaxillary gland. In *Proc. Natl. Acad. Sci. USA* **91**:8935–9.

Rothman J (1994): Mechanisms of intracellular protein transport. In *Nature* **372**:55–63.

Sato S, Spicer SS (1982): Ultrastructural visualization of galactosyl residues in various alimentary tract epithelial cells with the peanut lectin-horseradish peroxidase procedure. In *Histochemistry* **73**:607–24.

Schachter H (1995): Biosynthesis. 2c. Glycosyltransferases involved in the synthesis of *N*-glycan antennae. In *Glycoproteins* (Montreuil J, Schachter H, Vliegenhart JFG, eds) pp 153–99, Amsterdam: Elsevier Science B.V.

Slot JW, Geuze HJ (1983): Immunoelectron microscopic exploration of the Golgi complex. In *J. Histochem. Cytochem.* **31**:1049–56.

Taatjes DJ, Roth J (1991): Glycosylation in intestinal epithelium. In *Int. Rev. Cytol.* **126**:135–93.

Tauber R, Volz B, Kreisel W *et al.* (1993): Reprocessing of membrane glycoproteins. In *Glyco- and Cell Biology. Biosynthesis, Transport and Function of Glycoconjugates* (Wieland F, Reutter W, eds) pp 119–30, Berlin, Heidelberg, New York: Springer Verlag.

Tang BL, Wong SH, Qi LX *et al.* (1993): Molecular cloning, characterization, subcellular localization and dynamics of p 23, the mammalian KDEL receptor. In *J. Cell Biol.* **120**:325–38.

Tartakoff AM, Vasalli P (1983): Lectin-binding sites as markers of Golgi subcompartments: proximal to distal maturation of oligosaccharides. In *J. Cell Biol.* **97**:1243–8.

Velasco A, Hendricks L, Moremen KW *et al.* (1993): Cell type-dependent variations in the subcellular distribution of α-mannosidase I and II. In *J. Cell Biol.* **122**:39–51.

7 Occurrence and Potential Functions of *N*-Glycanases

Tadashi Suzuki, Ken Kitajima, Sadako Inoue, and Yasuo Inoue

1 Introduction

N-glycosylation is now acknowledged as an important post-translational modification reaction of proteins in eukaryotic cells, and plays some modulatory roles in protein stabilization, affinity of receptor-ligand interaction, and susceptibility to proteases (Varki, 1993). On the other hand, no attention has been paid to the possible functional importance of "de-*N*-glycosylation" of glycoproteins, irrespective of the fact that the removal of carbohydrate moieties from certain glycoproteins is known to cause structural and physicochemical changes in their core proteins, thereby leading to significant effects on the functional properties of such glycoproteins. Thus, de-*N*-glycosylation could be recognized as a post-translational remodification of proteins. In this chapter, we have endeavored to highlight several lines of evidence showing the biological significance of de-*N*-glycosylation, through which we put forward a new concept of an "*N*-glycosylation/de-*N*-glycosylation system" as a basic mechanism for protein modification.

2 *N*-Glycosylation/ De-*N*-Glycosylation

N-Glycosylation occurs cotranslationally on proteins in the lumen of the endoplasmic reticulum (ER) by the action of the oligosaccharyl transferase complex, after which the structure of the glycan moieties undergoes modification or processing to high mannose-, hybrid-, and complex-type glycan forms by a number of glycosidases and glycosyltransferases, as the proteins migrate through the intracellular transport pathway to the Golgi, lysosome, plasma membrane or outside cells (see Pavelka, this volume). Contrary to the well-established mechanism for *N*-glycosylation and subsequent processing of *N*-glycan chains (Kornfeld and Kornfeld, 1985), the fate of the *N*-glycan chain on glycoproteins, especially the detachment of the *N*-glycan chain from glycoproteins, mostly remains veiled except for the case with lysosomal degradation, where *N*-glycan chains are known to be detached from glycoproteins through glycoasparaginasis (Brassart *et al.*, 1987).

Our hypothetical scheme for the *N*-glycosylation/de-*N*-glycosylation system is depicted in Figure 1. *N*-glycanase (peptide-N^4-[*N*-acetyl-β-D-glucosaminyl] asparagine amidase; EC 3.5.1.52, PNGase as a trivial name) is a key enzyme responsible for de-*N*-glycosylation of *N*-glycosylated proteins. PNGase is an enzyme catalyzing cleavage of the amide bond between the proximal GlcNAc and the linker amino acid Asn residues on glycoprotein molecules. This reaction not only removes a glycan chain but also introduces a negative charge into proteins by converting the glycosylated asparagine into an aspartic acid residue, resulting in alteration of various physiological and/or physicochemical properties of proteins. Both *N*-glycosylation and de-*N*-glycosylation are expected to affect the functional properties of the core proteins, and the *N*-glycosylation/de-*N*-glycosylation system is thus believed to make it possible to exert a dynamic control of functional changes of proteins.

H.-J. and S. Gabius (Eds.), Glycosciences

ISBN 3-8261-0073-5

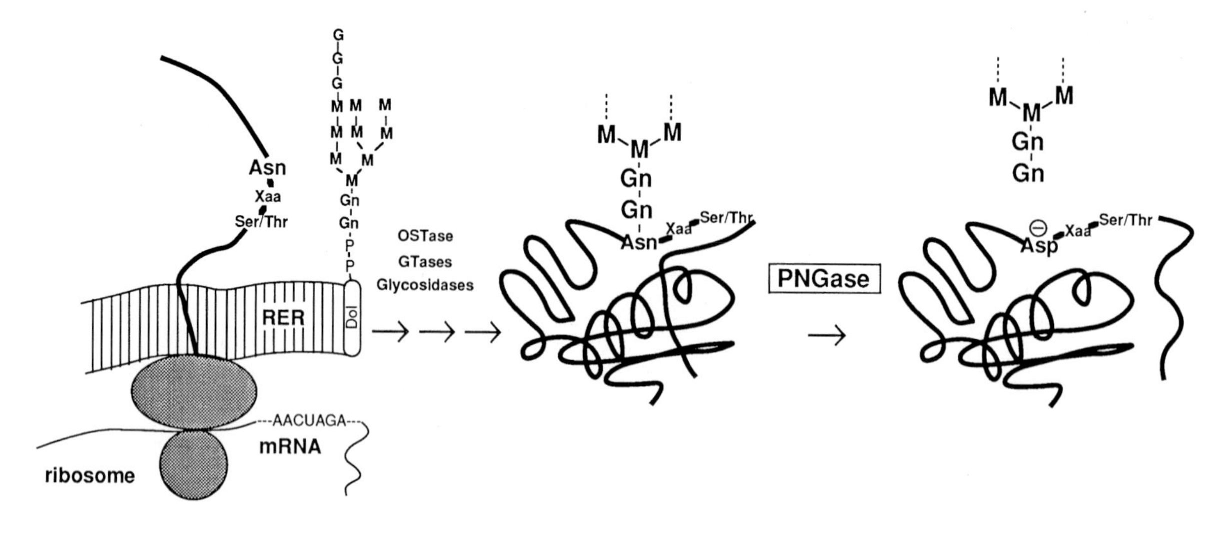

N-glycosylation
co- and post-translational modification

de-N-glycosylation
post-translational remodification

Figure 1. Schematic representation of the *N*-glycosylation/de-*N*-glycosylation system. This figure is reproduced from Suzuki *et al.* (1994c) with the permission of Oxford University Press. The *N*-glycosylation of protein occurs co-translationally by transferring a pre-assembled oligosaccharide ($Glc_3Man_9GlcNAc_2$) from a dolicholpyrophosphoryl carrier to the nascent peptide chain, in the lumen of the rough endoplasmic reticulum, by oligosaccharyl transferase (Kornfeld and Kornfeld, 1985). Subsequently, the transferred *N*-linked chain is processed further by various glycosyltransferases and glycosidases localized in the ER and Golgi during the intracellular transport, to give rise to a complete glycoprotein. These co- and post-translational events are well established to have various important biological roles (Varki, 1993). Some of these completed *N*-glycosylated glycoproteins presumably undergo site-specific de-*N*-glycosylation, catalyzed by PNGase, and glycosylated asparagine residue(s) are converted into aspartic acid residue(s). This post-translational remodification process may have critical roles in modulating physicochemical and/or physiological properties of target glycoproteins in cellular system of eukaryotic cells.
OSTase : oligosaccharyl transferase; GTase : glycosyltransferase; Dol : dolichol

3 Occurrence of PNGases: The Discovery of Animal PNGases

PNGase had been found to occur widely in plant cells (Takahashi, 1977; Plummer *et al.*, 1987) and in a bacterium, *Flavobacterium meningosepticum* (Plummer *et al.*, 1984), but there had been no report of the occurrence of PNGase in animal cells until we demonstrated for the first time the presence of the enzyme activity in the early embryos of Medaka fish, *Oryzias latipes* (Seko *et al.*, 1991a). Our initial search for PNGase in animals came from the finding of the accumulation of free oligosaccharides in the oocytes and early embryos of several fish species (Inoue *et al.*, 1989; Ishii *et al.*, 1989; Seko *et al.*, 1989; Inoue, 1990; Seko *et al.*, 1991b; Iwasaki *et al.*, 1992). The striking structural feature of these oligosaccharides was retention of *N,N'*-diacetylchitobiosyl structure at their reducing termini, which led us to propose that there must be PNGase in fish oocytes and embryos, and that PNGase-catalyzed de-*N*-glycosylation might be involved in some essential events for oogenesis or embryogenesis. Since the first discovery of animal PNGase in fish, PNGase activities have been identified in various animal cells (Seko *et al.*, 1991a; Inoue *et al.*, 1993; Suzuki *et al.*, 1993, 1994a-c, 1995a,b; Kitajima *et al.*, 1995) including mammals and birds. Occurrence of PNGase in nature is shown in Table 1.

Table 1. Reported occurrence of PNGase in nature

	Source	References
Animal	Medaka (*Oryzias latipes*)	Seko *et al.*, 1991a Inoue *et al.*, 1993
	Cultured cells from human and mouse	Suzuki *et al.*, 1993, 1994a
	Mouse	Kitajima *et al.*, 1995
	Chicken	Suzuki *et al.*, 1995a
Plant	Almond emulsin	Takahashi, 1977 Taga *et al.*, 1985
	Jack bean	Sugiyama *et al.*, 1983 Yet and Wold, 1988
	Various plant seeds	Plummer *et al.*, 1987
	Culture cells from white campion	Lhernould *et al.*, 1992, 1995
	Radish	Berger *et al.*, 1995
Bacteria	*Flavobacterium meningosepticum*	Plummer *et al.*, 1984 Tarentino *et al.*, 1990

Table 2. Free oligosaccharides found in oocytes and embryos of several fish species

Fish (Species)	Cells	Progenitor Protein
Freshwater Trout (*Plecoglossus altivelis*) Dace (*Tribolodon hakonensis*) Medaka (*Oryzias latipes*)	Unfertilized Eggs	Glycophospho-protein
Flounder (*Paralichthys olivaceus*) Medaka (*Oryzias latipes*)	Fertilized Eggs	Hyosophorin

4 PNGases in Fish

Free oligosaccharides found in fish ovary and embryo were divided into two groups according to their progenitor glycoproteins (Table 2). Now we have identified two distinct PNGases, namely acid PNGase and alkaline PNGase, depending on their optimal pH for enzyme activity, in Medaka oocyte and embryos, respectively (Inoue *et al.*, 1993; Seko *et al.*, unpublished).

4.1 Acid PNGase in Fish Oocyte is Responsible for the Detachment of *N*-Glycan Chains from Glycophosphoproteins

Glycophosphoproteins (GPP) are abundant in fish oocytes and are believed to be one of the discrete molecular forms of phosvitin derived from vitellogenin. The PNGase-catalyzed detachment of the glycan chains from GPP during oogenesis in fish was thought to be catalyzed by acid PNGase and to be important in vitellogenesis (Inoue, 1990; Suzuki *et al.*, 1994c). One plausible explanation for the significance of de-*N*-glycosylation was that it might be related to the recycling of vitellogenin receptor on egg surface. Incorporation of vitellogenin into oocytes is known to be a typical receptor-mediated pinocytosis (Opresko *et al.*, 1980; Busson-Mabillot, 1984), and dissociation of the vitellogenin ligand from its receptor is considered to be a prerequisite for the recycling of the receptor molecules incorporated into the cell. De-*N*-glycosylation would be required to facilitate dissociation of the vitellogenin-receptor complex within the cell (Inoue, 1990; Suzuki *et al.*, 1994c). An alternative explanation was that de-*N*-glycosylation may possibly be a rate-limiting step necessary for the subsequent proteolytic processing of vitellogenin.

4.2 *N*-Glycan Release by Alkaline PNGase during Embryogenesis

Another target substrate for PNGase identified in fish is hyosophorin, a glycopolyprotein localized in the cortical alveoli of unfertilized eggs; the apo-protein is made up of tandem repetition of the identical oligopeptide sequence (Kitajima *et al.*, 1989). Upon fertilization, cortical alveolar exocytosis occurs and H-hyosophorin is discharged from the vesicles to the perivitelline space (Inoue and Inoue, 1986; Inoue *et al.*, 1987) concomitantly with depolymerization of H-hyosophorin into the repeating unit, L-hyosophorin, by the action of a unique protease, hyosophorinase (Kitajima and Inoue, 1988). During early embryogenesis, a fraction of the L-hyosophorin molecules is further processed by the action of alkaline PNGase, which was found in the blastodisc of Medaka embryos, to give rise to de-*N*-glycosylated peptide and a free glycan chain (Seko *et al.*, 1989, 1991a,b; Inoue *et al.*, 1993; Seko *et al.*, unpublished). Accumulation of hyosophorin-derived free glycan was stage-specific (Seko *et al.* 1991b), indicating that de-*N*-glycosylation of L-hyosophorin in early embryos could be an essential event required for the conversion of L-hyosophorin into functional peptide or oligosaccharide. We have observed that polyclonal antibodies against the deglycosylated peptide formed from L-hyosophorin by PNGase catalysis, but not against L-hyosophorin itself or free glycan from hyosophorin, was able to block cell migration during gastrulation, suggesting that the deglycosylated peptide from L-hyosophorin has hormone-like activity essential for the subsequent normal development of the embryos (Seko *et al.*, unpublished).

5 PNGase in Mouse-Derived Cultured Cells: L-929 PNGase

Following the first demonstration of PNGase activities in an animal, we began to search for PNGase activity in several mammalian-derived cultured cells to investigate how widespread the occurrence of this class of enzymes was. We found PNGase activities, albeit at neutral pH, in all cell lines from mouse and human examined (Suzuki *et al.*, 1993). We were also successful in purifying to homogeneity the PNGase from C3H mouse-derived fibroblast cells, L-929, designated L-929 PNGase, and characterized its enzymatic properties (Suzuki *et al.*, 1994a).

L-929 PNGase was shown to be distinguished from the known PNGases, one (PNGase A) from almond emulsin and the other (PNGase F) from *Flavobacterium meningosepticum* (Suzuki *et al.*, 1994a,b; 1995b) in having several distinct properties (Table 3). The L-929 enzyme showed the maximal activity at pH 7.0, and this enzyme was unable to hydrolyze glycoasparagines (Suzuki *et al.*, 1994a). These results collectively suggest that the newly found enzyme activity is clearly distinct from lysosomal glycoasparaginase activity, which is known to occur commonly in vertebrates, because the latter enzyme requires the substrates to have both the α-COOH and α-NH_2 groups of their *N*-glycosylated asparagine residue unsubstituted (Tarentino and Maley, 1969; Kaartinen *et al.*, 1991). L-929 PNGase is a soluble enzyme, and this enzyme activity was not detected in the culture medium of L-929 cells, suggesting that L-929 PNGase is localized within the cells (Suzuki *et al.* 1994a). It is noteworthy that L-929 PNGase has higher affinity to its substrate glycopeptides (e.g., K_m = 114 μM for fetuin-derived glycopentapeptide) than the other known PNGases (Table 3) and that PNGase-catalysis is strongly inhibited by the free oligosaccharide products but not by the free peptides formed. Thus, it is suggested that L-929 PNGase can bind to a certain type of carbohydrate chain and can have a dual functional role as an enzyme and a lectin-like molecule. Recently, we have examined the carbohydrate-binding activity using yeast mannan (Suzuki *et al.*, 1994b). L-929 PNGase was, in contrast to the other known PNGases, found to bind strongly to yeast mannan-coupled Sepharose 4B in a carbohydrate-dependent manner (Suzuki *et al.*, 1994b).

Very recently, we have demonstrated directly the binding of the enzyme to oligosaccharides, and revealed the mechanism of carbohydrate-binding and inhibition of the L-929 PNGase by the oligosaccharides, based on kinetic and binding experiments (Suzuki *et al.*, 1995b). L-929 PNGase bound strongly to oligosaccharides having the triomannosido-di-*N*-acetylchitobiosyl ($Man_3GlcNAc_2$) structure (K_d, ~10 μM), and this binding was inhibited by triomannose (Man_3), Manα1→3(Manα1→6)Man, but not by *N,N'*-diacetylchitobiose ($GlcNAc_2$) (Suzuki *et al.*, 1995b). Scatchard analysis suggested that approximately two binding sites for these oligosaccharides exist on a homodimeric form of the 105 K molecules. The minimum structures that inhibit the PNGase activity were Man_3 and $GlcNAc_2$. Kinetic studies showed that the mechanism of inhibition by the oligosaccharides and Man_3 fits

Table 3. Comparison of properties of L-929 PNGase with those of PNGase A from almond and PNGase F from F. meningosepticum*1

	L-929 PNGase	PNGase A	PNGase F
Molecular Weight (K)	212	66.8	34.8
Subunit	dimeric	monomeric	monomeric
pH Optimum	7.0	4.5	8.5
Km (mM)*2	0.114	1.46	0.525
Action on GlcNAc-peptide	no	yes	no
Action on Glycoasparagine	no	no	no
Inhibitory Effect of Fuc attached $\alpha 1 \rightarrow 3$ to the proximal GlcNAc	yes	no	yes
Inhibitory Effect of Fuc attached $\alpha 1 \rightarrow 6$ to the proximal GlcNAc	yes	no	no
Requirement of -SH Group(s) for Enzyme Activity	yes	no	no
Binding Activity to Yeast Mannan-Sepharose 4B Column	yes	no	no
IC_{50} of *N,N'*-diacetylchitobiose (mM)*3	6.9	1.5	3.9
Inhibitory Effect of Man_3 on the Enzyme Activity	yes	no	no

*1 See Suzuki *et al.*, 1994a-c; 1995b and references therein.
*2 Values for fetuin glycopeptide I [Leu-Asn($NeuAc_3Gal_3GlcNAc_5Man_3$)-Asp-Ser-Arg].
*3 IC_{50}, concentration required for 50 % inhibition of enzyme activity (Suzuki *et al.*, 1995b).

well with the assumption that there are two inhibitor binding sites on the enzyme, while inhibition by $GlcNAc_2$ followed the competitive mode (Suzuki *et al.*, 1995b). Interestingly, alkylation of -SH group(s) in L-929 PNGase resulted in loss of the enzymatic activity, but complete retention of the carbohydrate-binding ability, indicating that carbohydrate-binding sites can be discriminated from the catalytic sites, although the spatial relationship between these sites on the enzyme is unknown (Suzuki *et al.*, 1995b). Unlike PNGase A or PNGase F, L-929 PNGase has a relatively high molecular weight and exists in its native state as a dimeric form (Mr, 212,000) of the intact subunits (Mr, 105,000) (Table 3), suggesting that this enzyme, like most of animal lectins (Drickamer and Taylor, 1993), may possibly be a member of a "mosaic protein" family, which is composed of distinct functional domains. Thus, the observed carbohydrate-recognition (lectin-like) activity of L-929 PNGase may be associated not only with regulation of the enzyme activity, but also with receptor and carrier functions for glycoconjugates in certain intracellular processes.

6 Neutral Soluble PNGases Are Widely Distributed in Mouse Organs

We have developed a new assay method to detect PNGase activity in a crude enzyme fraction by a combination of paper chromatographic and paper electrophoretic analyses of reaction products, and using this method a wide distribution of PNGases among mouse organs has been demonstrated. Soluble fractions and membranous or particulate fractions of all organs tested were found to contain PNGase activity, although the level of the enzyme activity was different among organs (Kitajima *et al.*, 1995). Total PNGase activity for soluble fractions of brain and liver was shown to be relatively high. The soluble and membranous fractions of brain and thymus, and the soluble fraction of spleen and heart, showed relatively higher specific activity (Kitajima *et al.*, 1995).

Soluble PNGases were partially purified from brain, liver, kidney, and spleen of the mouse by TSK butyl-Toyopearl 650 M hydrophobicity chromatography, and characterized for enzymatic properties. The soluble enzymes were found to share the following properties with L-929

PNGase: (*a*) high hydrophobicity; (*b*) sensitivity to metal cations such as Zn^{2+}, Cu^{2+}, and Fe^{3+}; (*c*) neutral pH for optimal enzyme activity; (*d*) inactivity on glycoasparagine substrate; and (*e*) requirement of sulfhydryl group(s) for enzyme activity (Kitajima *et al.*, 1995). Since the PNGase activity was not detected in mouse serum, PNGases found in mouse organs were suggested to be localized intracellularly, as was the case with L-929 PNGase.

7 Possible Biological Function of Neutral PNGase in "Non-Lysosomal" Degradation

Since soluble and neutral PNGases have been shown to occur ubiquitously within various types of animal cells, it would be interesting to consider that these enzymes might have some essential roles in yet undefined cellular processes. One possible biological function of these enzymes is the modulation of the biological activity of the target substrate, as outlined in a recent review (Suzuki *et al.*, 1995a) (Table 4). Another possibility is that they might participate in some "non-lysosomal" degradation pathways. Soluble and neutral PNGase is suggested to be localized in the cytosol or in intracellular vesicle(s), as judged from the data based on the centrifugation experiments.

Cytosol, where ubiquitination is a degradation signal (Hershko and Ciechanover, 1992), is known to be the major site for intracellular "non-lysosomal" protein degradation. If this type of PNGase is a *bona fide* cytosolic enzyme, *N*-glycosylated glycoproteins, possible substrates for PNGases, should occur in cytosol. This is controversial in view of the current understanding that they do not have the same topological localization as cytosolic proteins (Hirschberg and Snider, 1987; Abeijon and Hirschberg, 1992), or otherwise, the PNGase would have to leave the cytosol on certain occasions to encounter its physiological substrates.

In the former case, it is worth noting here that several examples of *N*-glycosylated glycoproteins in cytosol, or membranous proteins with cytosol-oriented *N*-glycan, have been reported (Haeuw *et al.*, 1991; Pedemonte *et al.*, 1990; Suzuki *et al.*, 1994c, 1995a). Very recently, the dramatic cellular redistribution of heat-shock *N*-glycosylated glycoproteins was reported. In response to heat stress, glycosylated calreticulin, the major heat-induced glycosylated protein in CHO cells (Jethmalani *et al.*, 1994), was shown to disappear from

Table 4. Potential biological functions of PNGase as a modulator of bioactivity*

Potential Function	Examples
Modulation of Receptor-Ligand Interaction	De-*N*-glycosylation of glycophosphoproteins (GPP) appears to be necessary for the dissociation of the ligand vitellogenin from the receptor.
Formation of Bioactive Molecule	De-*N*-glycosylation of L-hyosophorin at gastrulation stage of *O. latipes* is an essential molecular event for the normal development. "Unconjugated *N*-glycans" (UNGs), a part of which is assumed to be derived from glycoproteins by the action of PNGase, are hormone factors in both growth and development.
Regulation of Proteolytic Processing	De-*N*-glycosylation of some plant lectins such as Con A might possibly precede proteolysis for formation of mature molecules.
Generation of Structural Polymorphism	De-*N*-glycosylation of prolactin, a protein hormone known to have various molecular forms, produces a novel molecular form ("deglycosylated form"), through which it could generate a variety of functions within the multiple target organs.

* See Suzuki *et al.*, 1995a for reference.

the endoplasmic reticulum (ER) fraction and appear rapidly in the cytosolic fraction (Henle *et al.*, 1995).

In the latter case, it would be interesting to point out a story of "ectocytosis". Certain lectins, such as galectin and annexin, are known to alter their topological localization by secretion from cytosol to an extracellular site (Christmas *et al.*, 1991; Hirabayashi, 1994). This extracellular transport was termed "ectocytosis" (Cooper and Barondes, 1990). Such a translocation mechanism would enable PNGases in cytosol to encounter their substrates in extracellular compartments.

Another possibility is that this type of PNGase is localized within certain non-lysosomal vesicles. ER, where the folding and assembly of integral membrane, lumenal, and secretory proteins occur, has recently been recognized as the site of "non-lysosomal" degradation in various cellular system (Klausner and Sitia, 1990; Stafford and Bonifatino, 1991). The rapid degradation observed in ER seems to be involved in the disposal of newly synthesized abnormal or unassembled proteins (Klausner and Sitia, 1990; Stafford and Bonifatino, 1991). The detailed mechanism of degradation is still veiled, but PNGases may possibly be involved in such degradation processes as a "quality control" of *N*-glycosylated glycoproteins in this compartment. Thus, abnormally glycosylated or malfolded glycoproteins can be corrected to normal ones or degraded into unglycosylated ones, to be catabolized *via* the PNGase-catalyzed deglycosylation pathway.

In this regard, a recent elegant model for the proof-reading of *N*-glycosylated glycoprotein in ER is most interesting. The monoglucosylated form of *N*-glycosyl glycoprotein is known to be an intermediate in the processing of the *N*-glycan chain following *N*-glycosylation of newly synthesized polypeptides. According to the proposed model, this monoglucosylated form of glycoproteins is inspected for the denaturating state and refolded if necessary by calnexin, a membrane-bound molecular chaperone that resides in ER (Hammond *et al.*, 1994; Helenius, 1994; Hebert *et al.*, 1995). Recently, this monoglucosylated form was also shown to be synthesized by UDP-Glc: glycoprotein glucosyltransferase from glycoprotein which was completely deglucosylated by the action of glucosidase I and II, but malfolded (Sousa *et al.*, 1992; Parker *et al.*, 1995). Thus, this glucosyltransferase can be recognized as determining if the glycoprotein could leave the ER or should be repaired by refolding with the aid of calnexin. It seems reasonable to assume here that persistently malfolded proteins which fail to acquire the normal folding after several rounds of reglucosylation by the glucosyltransferase would have to be degraded by mechanisms as yet unknown. The soluble PNGase that we found in mouse organs and cultured cells may be one of the enzyme protein candidates, that are associated with such a degradation process. In this context, it is noteworthy that L-929 PNGase can hydrolyze ovalbumin-derived glycopeptide bearing high-mannose type glycans more than 10-fold faster than those having hybrid-type glycans with an equivalent peptide structure (Suzuki *et al.* 1994a).

8 Plant PNGases: Possible Regulator Molecules in Cellular Processes

The biological significance of PNGase has also been demonstrated in a plant system. In 1992, deglycosylation of an *N*-linked glycan chain in the pro-region of Concanavalin A (Con A), a lectin from jack bean, was shown to be essential for expression of lectin activity (Min *et al.*, 1992; Sheldon and Bowles, 1992). Deglycosylation of Con A precursor can also be interpreted as the deglycosylation-regulated maturation of Con A, since the removal of the glycan chain could alter the accessibility to a processing protease (Suzuki *et al.*, 1994a). A similar model was proposed to explain the possible biological roles of deglycosylation in maturation of *Gramineae* lectins, where deglycosylation of proprotein was rate-limiting and preceded proteolytic processing (Wilkins *et al.*, 1990), although the enzyme activity responsible for the detachment of the glycan has not yet been identified. Interestingly, several examples have been reported for plant lectins with a potential *N*-glycosylation site in the processing region of proproteins (Suzuki *et al.*,

1995a), indicating that deglycosylation may possibly be a more common basic reaction in the processing of at least certain plant lectins than is recognized at present.

Recently, it was reported that free oligosaccharides, called "unconjugated *N*-glycans" (UNGs) (Priem *et al.*, 1994), were isolated from the extracellular medium of a plant-cell suspension of white campion, *Silene alba* (Priem *et al.*, 1990a; Lhernould *et al.*, 1992) as well as from the tomato fruit pericarp (Priem *et al.*, 1993). Since the free glycan concentration increased on carbon deprivation, it was hypothesized that the UNGs may possibly arise from carbon starvation (Lhernould *et al.*, 1992). Indeed, PNGase activity increased as soon as nutritive sugars were totally assimilated by the cells (Lhernould *et al.*, 1994). Since the production of UNGs only occurs after complete disappearance of sugar nutrients, while PNGase activity was found throughout the culture cycle, one can assume that PNGase and its substrates are not always in the same compartment, and that their proximity could occur when drastic change happens inside the cells, e.g. in the case of autophagy (Lhernould *et al.*, 1994). PNGase Se, that could be responsible for the formation of a portion of UNGs in white campion cells, was characterized (Lhernould *et al.*, 1995).

UNGs are capable of influencing both growth (Priem *et al.*, 1990b) and senescence (Priem and Gross, 1992; Yunovitz and Gross, 1994a,b) in plants. These effects were dose-dependent, suggesting that these free glycans can be a kind of plant hormonal factor. In tomato fruit, the amount of UNGs in the total oligosaccharidic extract was shown to increase with maturation (Priem *et al.*, 1992, 1994). The deglycosylating enzyme such as PNGase may thus be responsible for generating such hormonal factors in plants. The target physiological glycoprotein(s) of the enzyme in plant cells remains to be determined.

Most recently, PNGase activities during germination and postgerminative development in *Raphanus sativus* (radish) were monitored, which revealed that PNGase activity was found in dry seeds and that its level was constant during germination as well as postgermination (Berger *et al.*, 1995). The authors stated that the enzyme activity might be responsible for the release of complex-type (Xyl-containing) UNGs, although the target glycoproteins are also yet to be unidentified.

9 Concluding Remarks

Ever since we discovered direct evidence for the accumulation of intracellular free glycan chains in unfertilized fish eggs, and during early stages of embryogenesis of certain fish species, we have postulated that "non-lysosomal de-*N*-glycosylation" of glycoproteins by PNGases is a more widespread phenomenon than recognized, and hypothesized *N*-glycosylation/de-*N*-glycosylation as a system which constitutes a basic biological processes. As clearly described in this article, we can now demonstrate that deglycosylation is widely distributed and is crucial for at least certain cellular processes. The next logical step to elucidate the exact role of PNGases may be to identify their localization and target glycoproteins *in vivo*. Studies aiming in this direction are currently underway in this laboratory.

Finally, it should be noted that the biological significance of the deglycosylation reaction is by no means limited to that caused by the action of PNGase but should be extended to that caused by other "proximal glycanases" responsible for the detachment of intact glycan from glycoconjugates (e.g. endo-α-*N*-acetylgalactosaminidase, ceramide glycanase etc.) (Suzuki *et al.*, 1994c), and that caused by endo-β-*N*-acetylglucosaminidase, another deglycosylating enzyme that acts upon *N*-linked glycan chains (Suzuki *et al.*, 1995a), although there have so far been few reports to clarify the biological function of such enzymes, other than involvement in the degradative path. We hope that this article provides a new aspect of the function of glycan chains for those who are interested in the field of glycosciences.

References

Abeijon C, Hirschberg CB (1992): Topography of glycosylation reactions in the endoplasmic reticulum. In *Trends Biochem. Sci.* **17**:32–6.

Berger S, Menudier A, Julien R *et al.* (1995): Endo-*N*-acetyl-β-D-glucosaminidase and peptide-N^4-(*N*-

acetyl-glucosaminyl) asparagine amidase activities during germination of *Raphanus sativus*. In *Phytochemistry* **39**:481–7.

Brassart D, Baussant T, Wieruszeski J-M *et al.* (1987) Catabolism of *N*-glycosylglycoprotein glycans: evidence for a degradation pathway of sialylglycoasparagines resulting from the combined action of the lysosomal aspartylglucosaminidase and endo-*N*-acetyl-β-D-glucosaminidase. A 400-MHz ^{1}H-NMR study. In *Eur. J. Biochem.* **169**:131–6.

Busson-Mabillot S (1984): Endosomes transfer yolk proteins to lysosomes in the vitellogenic oocyte of the trout. In *Biol. Cell* **51**:53–66.

Christmas P, Callaway J, Fallon J *et al.* (1991): Selective secretion of annexin 1, a protein without a signal sequence, by the human prostate grand. In *J. Biol. Chem.* **266**:2499–507.

Cooper DNW, Barondes SH (1990): Evidence for export of a muscle lectin from cytosol to extracellular matrix and for a novel secretory mechanism. In *J. Cell Biol.* **110**:1681–91.

Drickamer K, Taylor ME (1993): Biology of animal lectins. In *Annu. Rev. Cell Biol.* **9**:237–64.

Haeuw J-F, Michalski J-C, Strecker G *et al.* (1991) Cytosolic glycosidases: do they exist? In *Glycobiology* **1**:487–92.

Hammond C, Braakman I, Helenius A (1994): Role of *N*-linked oligosaccharide recognition, glucose trimming, and calnexin in glycoprotein folding and quality control. In *Proc. Natl. Acad. Sci. USA* **91**:913–17.

Hebert DN, Foellmer B, Helenius A (1995): Glucose trimming and reglucosylation determine glycoprotein association with calnexin in the endoplasmic reticulum. In *Cell* **81**:425–33.

Helenius A (1994) How *N*-linked oligosaccharides affect glycoprotein folding in the endoplasmic reticulum. In *Mol. Biol. Cell* **5**:253–65.

Henle KJ, Jethamalani SM, Nagle WA (1995): Heat stress-induced protein glycosylation in mammalian cells. In *Trends Glycosci. Glycotechnol.* **7**:191–204.

Hershko A, Ciechanover A (1992): The ubiquitin system for protein degradation. In *Annu. Rev. Biochem.* **61**:761–807.

Hirabayashi J (1993) A general comparison of two major families of animal lectins. In *Trends Glycosci. Glycotechnol.* **5**:251–70.

Hirschberg CB, Snider MD (1987): Topography of glycosylation in the rough endoplasmic reticulum and Golgi apparatus. In *Annu. Rev. Biochem.* **56**: 63–87.

Inoue S, Inoue Y (1986): Fertilization (activation)-induced 200- to 9-kDa depolymerization of polysialoglycoprotein, a distinct component of cortical alveoli of rainbow trout eggs. In *J. Biol. Chem.* **261**:5256–61.

Inoue S, Kitajima K, Inoue Y *et al.* (1987): Localization of polysialoglycoprotein as a major glycoprotein component in cortical alveoli of the unfertilized eggs of *Salmo gairdneri*. In *Dev. Biol.* **123**:442–54.

Inoue S, Iwasaki M, Ishii K *et al.* (1989): Isolation and structures of glycoprotein-derived free sialooligosaccharides from the unfertilized eggs of *Tribolodon hakonensis*, a dace. Intracellular accumulation of a novel class of biantennary disialooligosaccharides. In *J. Biol. Chem.* **264**:18520–6.

Inoue S (1990): De-*N*-glycosylation of glycoproteins in animal cells: evidence for occurrence and significance. In *Trends Glycosci. Glycotechnol.* **2**:225–34.

Inoue S, Iwasaki M, Seko A *et al.* (1993): Identification and developmentally regulated expression of acid and alkaline Peptide-N^4(*N*-acetyl-β-glucosaminyl) asparagine amidases (PNGase) in *Oryzias latipes* embryos: the first demonstration of the occurrence of an enzyme responsible for de-*N*-glycosylation in animal origin. In *Glycoconjugate J.* **10**: 223 (abstract).

Ishii K, Iwasaki M, Inoue S *et al.* (1989): Free sialooligosacharides found in the unfertilized eggs of a freshwater trout, *Plecoglossus altivelis*. A large storage pool of complex-type bi-, tri-, and tetraantennary sialooligosaccharides. In *J. Biol. Chem.* **264**: 1623–30.

Iwasaki M, Seko A, Kitajima K *et al.* (1992): Fish egg glycophosphoproteins have species-specific *N*-linked glycan units previously found in a storage pool of free glycan chains. In *J. Biol. Chem.* **267**:24287–96.

Jethmalani SM, Henle KJ, Kaushal GP (1994): Heat shock-induced prompt glycosylation. Identification of P-SG67 as calreticulin. In *J. Biol. Chem.* **269**:23603–9.

Kaartinen V, Williams JC, Tomich J *et al.* (1991): Glycoasparaginase from human leukocytes. Inactivation and covalent modification with diazo-oxonorvaline. In *J. Biol. Chem.* **266**:5860–9.

Kitajima K, Inoue S (1988): A proteinase associated with cortices of rainbow trout eggs and involved in fertilization-induced depolymerization of polysialoglycoproteins. In *Dev. Biol.* **129**:270–4.

Kitajima K, Inoue S, Inoue Y (1989): Isolation and characterization of a novel type of sialoglycoproteins (hyosophorin) from the eggs of Medaka, *Oryzias latipes*: Nonapeptide with a large *N*-linked glycan chain as a tandem repeat unit. In *Dev. Biol.* **132**:544–53.

Kitajima K, Suzuki T, Kouchi Z *et al.* (1995): Identification and distribution of peptide:*N*-glycanase in mouse organs. In *Arch. Biochem. Biophys.* **319**:393–401.

Kornfeld R, Kornfeld S (1985) Assembly of asparagine-linked oligosaccharides. In *Annu. Rev. Biochem.* **54**:631–64.

Klausner RD, Sitia R (1990): Protein degradation in the endoplasmic reticulum. In *Cell* **62**:611–4.

Lhernould S, Karamanos Y, Bourgerie S *et al.* (1992): Peptide-N^4-(*N*-acetylglucosaminyl)asparagine amidase (PNGase) activity could explain the occurrence of extracellular xylomannosides in a plant cell suspension. In *Glycoconjugate J.* **9**: 191–7.

Lhernould S, Karamanos Y, Priem B *et al.* (1994): Carbon starvation increases endoglycosidase activities and production of "Unconjugated *N*-glycans" in *Silene alba* cell-suspension cultures. In *Plant Physiol.* **106**:779–84.

Lhernould S, Karamanos Y, Lerouge P *et al.* (1995): Characterization of the peptide-N^4-(*N*-acetylglucosaminyl)asparagine amidase (PNGase Se) from *Silene alba* cells. In *Glycoconjugate J.* **12**:94–8.

Min W, Dunn AJ, Jones DH (1992): Non-glycosylated recombinant pro-concanavalin A is active without polypeptide cleavage. In *EMBO J.* **11**:1303–7.

Opresko L, Wiley HS, Wallace RA (1980): Differential postendocytotic compartmentation in *Xenopus* oocytes is mediated by a specifically bound ligand. In *Cell* **22**:47–57.

Parker CG, Fessler LI, Nelson RE *et al.* (1995): *Drosophila* UDP-glucose:glycoprotein glucosyltransferase: sequence and characterization of an enzyme that distinguishes between denatured and native proteins. In *EMBO J.* **14**:1294–303.

Pedemonte CH, Sachs G, Kaplan JH (1990) An intrinsic membrane glycoprotein with cytosolically oriented *N*-linked sugars. In *Proc. Natl. Acad. Sci. USA* **87**:9789–93.

Plummer TH Jr, Elder JH, Alexander S *et al.* (1984): Demonstration of peptide:*N*-glycosidase F activity in endo-β-*N*-acetylglucosaminidase F preparations. In *J. Biol. Chem.* **259**:10700–4.

Plummer TH Jr, Phelan AW, Tarentino AL (1987): Detection and quantification of peptide-N^4-(*N*-acetyl-β-glucosaminyl)asparagine amidases. In *Eur. J. Biochem.* **163**:167–73.

Priem B, Solo Kwan J, Wieruszeski J-M *et al.* (1990a): Isolation and Characterization of free glycans of the oligomannoside type from the extracellular medium of a plant cell suspension. In *Glycoconjugate J.* **7**:121–32.

Priem B, Morvan H, Hafez AMA *et al.* (1990b): Influence of plant glycan of the oligomannoside type on the growth of flax plantlets. In *C. R. Acad. Sci. III* **311**: 411–6.

Priem B, Gross KC (1992): Mannosyl- and xylosyl-containing glycans promote tomato (*Lycopersicon esculentum* Mill.) fruit ripening. In *Plant Physiol.* **98**:399–401.

Priem B, Gitti R, Bush CA *et al.* (1993): Structure of ten free *N*-glycans in ripening tomato fruit. Arabinose is a constituent of a plant *N*-glycan. In *Plant Physiol.* **102**:445–58.

Priem B, Morvan H, Gross KC (1994): Unconjugated *N*-glycans as a new class of plant oligosaccharins. In *Biochem. Soc. Transact.* **22**:398–402.

Seko A, Kitajima K, Iwasaki M *et al.* (1989): Structural studies of fertilization-associated carbohydrate-rich glycoproteins (hyosophorin) isolated from the fertilized and unfertilized eggs of flounder, *Paralichthys olivaceus*. Presence of a novel penta-antennary *N*-linked glycan chain in the tandem repeating glycopeptide unit of hyosophorin. In *J. Biol. Chem.* **264**: 15922–9.

Seko A, Kitajima K, Inoue Y *et al.* (1991a): Peptide:*N*-glycosidase activity found in the early embryos of *Oryzias latipes* (Medaka fish). The first demonstration of the occurrence of peptide:*N*-glycosidase in animal cells and its implication for the presence of a de-*N*-glycosylation system in living organisms. In *J. Biol. Chem.* **266**:22110–4.

Seko A, Kitajima K, Inoue S *et al.* (1991b): Identification of free glycan chain liberated by de-*N*-glycosylation of the cortical alveolar glycoprotein (hyosophorin) during early embryogenesis of the Medaka fish, *Oryzias latipes*. In *Biochem. Biophys. Res. Commun.* **180**:1165–71.

Sheldon PS, Bowles DJ (1992): The glycoprotein precursor of concanavalin A is converted to an active lectin by deglycosylation. In *EMBO J.* **11**:1297–301.

Sousa MC, Ferrero-Garcia MA, Parodi AJ (1992): Recognition of the oligosaccharide and protein moieties of glycoproteins by UDP-Glc:glycoprotein glucosyltransferase. *Biochemistry* **31**:97–105.

Stafford FJ, Bonifacino JS (1991): A permeabilized cell system identifies the endoplasmic reticulum as a site of protein degradation. In *J. Cell Biol.* **115**: 1225–36.

Sugiyama K, Ishihara H, Tejima S *et al.* (1983): Demonstration of a new glycopeptidase, from Jack-bean meal, acting on aspartylglucosamine linkages. In *Biochem. Biophys. Res. Commun.* **112**:155–60.

Suzuki T, Seko A, Kitajima K *et al.* (1993): Identification of peptide:*N*-glycanase activity in mammalian-derived cultured cells. In *Biochem. Biophys. Res. Commun.* **194**:1124–30.

Suzuki T, Seko A, Kitajima K *et al.* (1994a): Purification and enzymatic properties of peptide:*N*-glycanase (PNGase) from C3H mouse-derived L-929 fibroblast cells. A possible widespread occurrence of post-translational remodification of proteins by *N*-deglycosylation. In *J. Biol. Chem.* **269**:17611–8.

Suzuki T, Kitajima K, Inoue S *et al.* (1994b): Does an animal peptide:*N*-glycanase have the dual role as an enzyme and a carbohydrate-binding protein? In *Glycoconjugate J.* **11**: 469–76.

Suzuki T, Kitajima K, Inoue S *et al.* (1994c): Occurrence and biological roles of "proximal glycanases" in animal cells. In *Glycobiology* **4**:777–89.

Suzuki T, Kitajima K, Inoue S *et al.* (1995a): *N*-glycosylation/deglycosylation as a mechanism for the post-translational modification/remodification of proteins. In *Glycoconjugate J.* **12**:183–93.

Suzuki T, Kitajima K, Inoue Y *et al.* (1995b): Carbohydrate-binding property of peptide:*N*-glycanase from mouse fibroblast L-929 cells as evaluated by inhibition and binding experiments using various oligosaccharides. In *J. Biol. Chem.* **270**: 15181–6.

Taga EM, Waheed A, Van Etten RL (1985) Structural and chemical characterization of a homogenous peptide *N*-glycosidase from almond. In *Biochemistry* **23**:815–22.

Takahashi N (1977): Demonstration of a new amidase acting on glycopeptides. In *Biochem. Biophys. Res. Commun.* **76**:1194–201.

Tarentino AL, Maley F (1969): The purification and properties of a β-aspartyl *N*-acetylglucosylamine amidohydrolase from hen oviduct. In *Arch. Biochem. Biophys.* **130**:295–303.

Tarentino AL, Quinones G, Trumble A *et al.* (1990) Molecular cloning and amino acid sequence of peptide-N^4-(*N*-acetyl-β-D-glucosaminyl)asparagine amidase from *Flavobacterium meningosepticum*. In *J. Biol. Chem.* **265**: 6961–6.

Varki A (1993): Biological roles of oligosaccharides: all of the theories are correct. In *Glycobiology* **3**:97–130.

Wilkins TA, Bednarek SY, Raikhel NV (1990): Role of propeptide glycan in post-translational processing and transport of barley lectin to vacuoles in transgenic tobacco. In *Plant Cell* **2**:301–13.

Yet M-G, Wold F (1988): Purification and characterization of two glycopeptide hydrolases from Jack beans. In *J. Biol. Chem.* **263**:118–22.

Yunovitz H, Gross KC (1994a): Delay of tomato fruit ripening by an oligosaccharide *N*-glycan. Interactions with IAA, galactose and lectins. In *Physiol. Planta* **90**:152–6.

Yunovitz H, Gross KC (1994b): Effect of tunicamycin on metabolism of unconjugated *N*-glycans in relation to regulation of tomato fruit ripening. In *Phytochemistry* **37**:663–8.

8 Glycoproteins: Structure and Function[1]

Nathan Sharon and Halina Lis

1 Introduction

Proteins containing covalently bound carbohydrate, named glycoproteins, are ubiquitous in nature[1]. They are the most common and most versatile products of post-translational modification of proteins in all living organisms, from archaebacteria to humans, although they are rare in eubacteria. In fact, most proteins are glycoproteins. They include all kinds of biologically active substance, such as enzymes, antibodies, hormones, cytokines and receptors, as well as structural proteins, such as collagens. Glycoproteins are found inside cells, both in the cytoplasm and the subcellular organelles, in cell membranes and in extracellular fluids. Blood serum is a particularly rich source of these compounds: of the almost 100 proteins identified in this fluid, almost all are glycosylated. One of the very few exceptions is serum albumin; a genetic variant of the protein which is glycosylated has however been found (Brennan *et al.*, 1990). Another rich source of glycoproteins is hen's egg white, from which ovalbumin was originally isolated and crystallized at the turn of the century; it was proven to be a glycoprotein in 1938 – the first example of a well characterized compound of this class. By now, many hundreds of glycoproteins have been identified, and their number is growing fast. Glycoproteins are also produced by genetic engineering techniques, for research and clinical use. The latter include erythropoietin, which is the number one product of the modern biotechnological industry, as well as enzymes (e.g. β-glucocerebrosidase), interferons, colony stimulating factors and blood clotting factors.

Glycoproteins vary in carbohydrate content, from less than 1 %, as in some of the collagens, to over 99 %, as in glycogen. They also differ greatly in the number of their carbohydrate units, which ranges from one to two hundred, and in the size and structure of the units. These occasionally consist of a monosaccharide or disaccharide, but more frequently are in the form of oligosaccharides or polysaccharides (up to 200 monosaccharide units in size), linear or branched, generally referred to as glycans. In a glycoprotein containing multiple carbohydrate units, they are attached to the polypeptide backbone at distinct positions by the same or by different kinds of chemical bonds. Only very seldom does the carbohydrate crosslink two or three polypeptides, as recently found in inter-α-trypsin inhibitor from human plasma (Morelle *et al.*, 1994). Identical glycans are often found in glycoproteins from diverse sources. On the other hand, different glycoproteins from the same cell often carry different glycans. Moreover, the structure of these glycans changes during growth, differentiation, development, and disease, including malignant transformation, which suggests that they are required for the normal functioning of the organism.

A striking feature of glycoproteins is the heterogeneity of the glycans, which contrasts with the uniformity of the polypeptide chain of the particular protein to which they are linked. In other

[1] This subject has been extensively reviewed in several recent books (Allen and Kisailus, 1992; Fukuda and Hindsgaul, 1994; Montreuil *et al.*, 1995) and major review articles (Lis and Sharon, 1993; Varki, 1993; Dwek, 1995), where references not included in this chapter can be found. Here, mainly selected references are given. For the nomenclature of glycoproteins, see IUPAC-IUB, 1986.

H.-J. and S. Gabius (Eds.), Glycosciences
© Chapman & Hall, Weinheim, 1997
ISBN 3-8261-0073-5

words, individual molecules of a given glycoprotein usually carry different saccharides at the same attachment site in the polypeptide chain. One of the extremely rare exceptions is soybean agglutinin, a glycoprotein that possesses a single uniform oligosaccharide per subunit (Ashford *et al.*, 1991). Another such exception is amyloid P component, a universal constituent of the abnormal tissue deposits in amyloidosis, the single glycan of which is homogeneous (Pepys *et al.*, 1994). As mentioned, however, heterogeneity of the carbohydrate is the rule, even in glycoproteins with a single oligosaccharide chain. It may be minor, as when there are slight changes in the structure of the glycan, involving very few of its monosaccharide constituents. These lead to what has been referred to as "microheterogeneity", a term coined when the phenomenon was discovered some 30 years ago in studies of hen ovalbumin, in which more than a dozen different structures have been identified at its single glycosylation site. A more recent example is human IgG, with an average number of approximately 3 glycans per molecule and more than 30 different oligosaccharide structures (Dwek, 1995). Despite this complex nature, the ratio of the different structures is relatively constant from individual to individual. A similar heterogeneity is found upon analysis of monoclonal IgG, demonstrating that the large number of different structures associated with this glycoprotein is not the result of polyclonality. Major changes in glycosylation include substantial structural differences in the carbohydrate units, such as increased branching, the complete loss of one or more arms of the glycan, or the deletion or addition of one or more carbohydrate units in the molecule as a whole.

Whether the change is minor or major, it results in the formation of discrete molecular subsets of the glycosylated protein, often referred to as glycoforms, the number of which can be staggering. In ovalbumin, the number of glycoforms is close to 20, while in IgG their number could be as high as 900. The various glycoforms may have different physical and biochemical properties, which, in turn, may lead to functional diversity of the particular glycoprotein. The earlier view that heterogeneity of the carbohydrates of glycoproteins is random, mainly due to the lack of fidelity in their synthesis, now seems unlikely, since under constant physiological conditions the relative proportions of such glycoforms appear to be reproducible and highly regulated.

The fact that the carbohydrate units of glycoproteins have been conserved in evolution, and the growing awareness of the widespread occurrence and structural diversity of glycoproteins, coupled with the finding that oligosaccharide structures of glycoproteins change under different physiological conditions, serve as continuing stimuli in the search for their biological role(s). It is now clear that there is no single unifying function for the carbohydrates present in glycoproteins (Varki, 1993). Carbohydrates may modify the physicochemical and biological properties of proteins to which they are attached by altering the charge, stability, susceptibility to proteases, or quaternary structure. Further, the large size of many glycans relative to the peptide moiety may allow them to cover functionally important areas of the proteins and to modulate their interactions with other molecules. Currently, much interest is focused on the function of carbohydrates as recognition determinants in a variety of physiological and pathological processes. It should be noted that such determinants are frequently found also on glycolipids, since identical carbohydrate structures are often present on both classes of glycoconjugate. This is particularly relevant when dealing with the functions of cell surface sugars, many of which are constituents of both glycoproteins and glycolipids.

2 Structure

The development of highly refined preparative and analytical techniques for the study of glycoproteins, even those available in microgram quantities or less (Hounsell, this volume), has resulted during the last decade in a vast increase in the number and types of well characterized compounds of this class. New monosaccharide constituents and carbohydrate-peptide linking groups have been discovered. In addition, the detailed structures of well over a thousand glycoprotein oligosaccharides have been elucidated, and the partial structures of numerous other such oligo- and polysaccharides are also known. These structures are exceedingly diverse, and although we have a fair knowledge of the pathways of their synthesis, we are still far from understanding the principles that guide their formation.

2.1 Monosaccharide Constituents

The monosaccharides found in glycoproteins, including the proteoglycans, can be roughly divided into two groups: commonly occurring (Table 2.1) and rare (Table 2.2). Of those belonging to the first group, galactose, mannose, and *N*-acetylglucosamine are present in numerous glycoproteins, derived from all kinds of organism, whereas glucose is confined to a small number of animal glycoproteins (mainly the collagens) and is found only occasionally in glycoproteins of other organisms. *N*-Acetylgalactosamine, and the uronic acids, occur predominantly in animals but rarely, if at all, in plants. Many plant glycoproteins contain fucose, a component of numerous animal glycoproteins, and xylose, which in animals is confined to the proteoglycans. Other plant glycoproteins contain arabinose, not found in animal glycoproteins. The sialic acids – a family consisting of *N*-acetylneuraminic acid and close to 40 of its derivatives – are, with notable exceptions (e.g *Drosophila melanogaster* and certain bacteria), confined to higher invertebrates and vertebrates (Varki, 1992; Schauer *et al.*, 1995). While *N*-acetylneuraminic acid occurs widely, other sialic acids have a limited distribution (Table 2.3).

The sialic acids differ not only in the substituent on the amino group (acetyl or glycoloyl), but also in the number (up to three), position (4, 7, 8 and 9) and nature (acetyl, lactoyl and methyl) of substituents on the hydroxyl groups of neuraminic acid. A recent addition is 4,6-anhydro-*N*-acetylneuraminic acid found in edible bird's nests. *N*-Glycoloylneuraminic acid (Neu5Gc), once believed to be confined to pigs and horses, has now been found in most animals, except adult humans and birds. Using immunological methods, tiny amounts of this sialic acid have, however, been detected in antigenic glycoproteins of

Table 2.1. Common monosaccharide constituents of glycoproteins

Class	Compound[a]	Abbr.	Comments
Hexoses	*D*-Galactose	Gal	
	D-Glucose	Glc	Mainly in collagens
	D-Mannose	Man	
Deoxyhexoses	*L*-Fucose	Fuc	
Pentoses	*L*-Arabinofuranose	Ara*f*	In plant glycoproteins
	D-Xylose	Xyl	In proteoglycans and plant glycoproteins
Hexosamines	*N*-Acetyl-*D*-galactosamine	GalNAc	
	N-Acetyl-*D*-glucosamine	GlcNAc	
Uronic acids	*D*-Glucuronic acid	GlcA	In proteoglycans
	L-Iduronic acid	IdoA	In proteoglycans
Sialic acids	*N*-Acetylneuraminic acid[b]	Neu5Ac	Mainly in higher vertebrates and invertebrates

[a] For the sake of simplicity, the optical configuration (*D*- or *L*-) of the monosaccharides is not mentioned in text. The ring form is pyranose, unless otherwise stated.
[b] Other sialic acids are also found in glycoproteins (see Table 2.3)

Table 2.2. Rare monosaccharide constituents of glycoproteins

Monosaccharide[a]	Source
2-Acetamido-4-amino-2,4,6-trideoxyglucose	*Clostridium symbiosum*
6-Deoxyaltrose	Salmonid fish eggs
2,3-Diacetamido-2,3-dideoxymannuronic acid	*Bacillus stearothermopilus*
3,6-Dideoxymannose (tyvelose)	*Trichinella spiralis*
Fuc2Me	Nematode
Galactofuranose	Bacteria; trypanosoma; fungi
Gal3Me	Snail
Gal4Me	Nematode
Gal6Me	Algae
Gal3SO_3-	Thyroglobulin; mucins in cystic fibrosis
GalNAc4SO_3-	Pituitary glycohormones
Glc3Me	*Methanothermus fervidus*
GlcNAc3Me	*Clostridium thermocellum*
GlcNAc6SO_3-	Thyroglobulin
Gulose	Algae
2-Keto-3-deoxy-nonulosonic acid (Kdn)	Salmonid fish eggs
Man3Me	Snail
Man4SO_3^-	Ovalbumin
Man6SO_3^-	Ovalbumin; slime mold
Man6PO_3^-Me	Slime mold
ManNAc	*Clostridium symbiosum*
Rhamnose	Eubacteria

[a] All compounds are of the *D*-configuration, except for fucose and rhamnose that are *L*-sugars

Table 2.3. Sialic acids found in glycoproteins

Sialic acid	Occurrence
Neu5Gc	Most animals[a]
Neu5,7Ac_2	Higher animals and certain bacteria[c]
Neu5,9Ac_2	Higher animals and certain bacteria[b]
Neu4,5Ac_2	Ungulates[b]
Neu4,5$Ac_2$9Lactyl	Horses[c]
Neu5Ac9Lactyl	Higher animals
Neu5Gc9Lactyl	Porcine submandibullary gland
4,8-Anhydro-Neu5Ac	Edible bird's nest
Neu5Gc8Me9Ac	Starfish[c]
Neu5Gc7,9$Ac_2$8Me	Starfish[c]
Kdn[d]	Fish, mammals
Kdn9Ac	Fish

[a] Neu5Gc and its derivatives are not found in adult humans (except in some cancers) nor in birds
[b] The same derivatives of Neu5Gc also occur in these organisms
[c] Not known whether in glycoproteins or glycolipids
[d] Kdn is classified as a sialic acid, although not strictly belonging to this class of compound

some human tumors. The existence of these antigens has raised the possibility that the gene responsible for the synthesis of *N*-glycoloylneuraminic acid (by hydroxylation of CMP-Neu5Ac to CMP-Neu5Gc) is suppressed under normal conditions in humans, but may be induced in the course of oncogenesis. A related monosaccharide, classified with the sialic acids (although not an *N*- or *O*-substituted neuraminic acid in the strict sense), is 3-deoxy-*D*-glycero-*D*-galactononulosonic acid in which the 5-amino group of neuraminic acid, has been replaced by a hydroxyl group. This compound also goes under the name of 2-keto-3-deoxy-nonulosonic acid, KDN or Kdn. It is quite abundant in glycoproteins of salmonid fish and in batracians and is also widespread in mammalian tissues (Troy, 1995).

Close to 30 other monosaccharides occur each in a small number of glycoproteins (Table 2.2). Thus, besides fucose, the only 6-deoxyhexose known until the late 1980's as a glycoprotein constituent, 6-deoxyaltrose, has been found in glycoproteins of salmonid fish eggs, and 6-deoxymannose (rhamnose) and 2,4-diamino-2,4,6-trideoxyglucose in those of bacteria. Another rare 6-deoxyhexose is tyvelose (3,6-di-deoxymannose), believed in the past to be confined to bacterial lipopolysaccharides, and now found also in a glycoprotein from the larvae of the parasite *Trichinella spiralis* (Wisniewski *et al.*, 1993; Reason *et al.*, 1994). As a rule, the monosaccharides of glycoproteins are in the pyranose form. However furanose forms of monosaccharides, usually ignored by biochemists (even though present in the nucleic acids), do occur in glycoproteins. Arabinofuranose has been found in *O*-glycans of plant glycoproteins (e.g. tomato and potato lectins) and galactofuranose in bacterial and protozoan ones (Lederkremer and Colli, 1995).

In animals in particular, several of the monosaccharides are *O*- and/or *N*-sulfated. Such derivatives are characteristic constituents of the proteoglycans (Kresse, this volume), but are also present in other glycoproteins. For instance, $GalNAc4SO_4$ is a component of certain pituitary hormones and of viral glycoproteins. Galactose sulfated at 3-*O* is found in thyroglobulins from various sources and in respiratory mucins of cystic fibrosis patients; thyroglobulin contains also 6-sulfated *N*-acetylglucosamine. Mannose, sulfated at positions 4- or 6, occurs in ovalbumin and in the slime mold (*Dictyostelium discoideum*), respectively. Another monosaccharide derivative of interest is mannose-6-phosphate, present in many of the lysosomal enzymes, where it serves as a marker for their targeting to the lysosomes (Neufeld, 1992).

2.2 Carbohydrate-Peptide Linking Groups

Attachment of carbohydrates to proteins occurs by three major types of linkage: (a) *N*-glycosidic, between the reducing terminal sugar and the amide group of asparagine (*N*-glycans), (b) *O*-glycosidic, between the sugar and a hydroxyl group of an amino acid, most commonly serine and threonine (*O*-glycans) (Table 2.4), and (c), *via* ethanolamine phosphate, between the C-terminal residue of the protein and an oligosaccharide attached to phosphatidylinositol, generally known as glycosylphosphatidylinositol (GPI or GlPtdIn) anchor. Only the first two types of linkage are dealt with in this chapter; for a discussion of glycoproteins containing the third type of linkage, the reader is referred to the chapter by R.T. Schwarz in this volume.

2.2.1 *N*-Linkages

The most common *N*-linking group, discovered in 1963, is *N*-acetylglucosaminyl-asparagine (GlcNAc-Asn). The asparagine is part of the consensus sequence (or sequon) Asn-X-Ser/Thr, where X may be any amino acid, with the exception of proline; not all such asparagine residues are, however, glycosylated. During the last decade, several new carbohydrate-asparagine linking groups have been discovered, mainly in bacterial glycoproteins, in which *N*-acetylglucosamine is replaced by α- or β-glucose, by β-*N*-acetylgalactosamine or rhamnose. The β-glucosylasparagine linkage unit has recently been identified also in the mammalian protein laminin, an extracellular basement membrane component (Schreiner *et al.*, 1994). Whenever carefully studied, the asparagine in the novel linkages, too, is part of the consensus sequence mentioned above. Perhaps the only exception is nephritogenoside, in which glucose is bound in an α-linkage to the amino terminal sequence Asn-Pro-Leu.

2.2.2 *O*-Linkages

A large variety of *O*-glycosidic linkages between the carbohydrate and the polypeptide are found in glycoproteins (Table 2.4). These include the well known GalNAc-Ser/Thr linking group, found in numerous animal glycoproteins, Gal-Hyl restricted to the collagens, Xyl-Ser found almost exclusively in the proteoglycans and Ara-Hyp in certain plant glycoproteins. A widely occurring *O*-glycosidic linkage, first identified about a decade ago, is that between *N*-acetylglucosamine and the hydroxyl group of serine/threonine (Haltiwanger *et al.*, 1992). Quite unusually, no other sugars are attached to this *N*-acetylglucosamine. The *O*-GlcNAc linking group is present predominantly in intracellular proteins

Table 2.4. Carbohydrate-peptide linking groups

Type	Monosaccharide	Amino acid	Occurrence
N-glycoside	β-GalNAc	Asn	Archaebacteria
	β-Glc	Asn	Animals[a]
	β-Glc	Asn	Archaebacteria
	β-GlcNAc	Asn	Common
	Rhamnose	Asn	Eubacteria
O-glycoside	Ara*f*	Hyp	Plants
	α-Fuc	Ser,Thr	Animals
	α-Gal	Hyp	Plants, eubacteria
	α-Gal	Ser	Plants, eubacteria
	β-Gal	Hyl	Animals[b]
	β-Gal	Tyr	Eubacteria
	α-GalNAc	Ser, Thr	Common
	α-Glc	Tyr	Animals[c]
	β-Glc	Ser, Thr	Eubacteria, animals
	β-GlcNAc	Ser, Thr	Animals
	α-Man	Ser, Thr	Yeasts, animals
	β-Xyl	Ser	Animals[d]

[a] To date found only in laminin
[b] Found only in collagens
[c] Only in glycogen
[d] Confined to proteoglycans

exposed to the nuclear and cytoplasmic compartments. Among these are components of the nuclear pore, the membrane and cytoskeleton of the endoplasmic reticulum, numerous chromatin proteins, including transcription factors, as well as viral proteins. Fucose bound to serine or threonine in the consensus sequence Gly-Gly-Ser/Thr has been found in the epidermal growth factor domain of several blood coagulation and fibrinolytic proteins (Harris and Spellman, 1993). These domains also contain glucose β-linked to serine. Tyrosine has been added to the list of *O*-linked amino acids with the unequivocal identification of Glcα-Tyr in glycogenin, the primer for glycogen synthesis (Alonso *et al.*, 1995), and of Galβ-Tyr in glycoproteins of the crystalline surface layers (termed the S-layer) of eubacteria (Bock *et al.*, 1994; Messner *et al.*, 1995). The Galα-Ser linkage is not confined to plant glycoproteins (e.g. extensin and potato lectin), as originally thought; it is also present in the glycoproteins from the cellulosome, an extracellular complex of different cellulases, produced by the cellulolytic bacteria *Clostridium thermocellum* and *Bacteroides cellulosolvens.* Glucose bound in a β-linkage to serine has been found in the bovine blood clotting factor IX and a 2,4-diacetamido-2,4,6-trideoxyhexose in a glycoprotein from *Neisseria meningitidis* (Stimson *et al.*, 1995).

2.2.3 Other

Recently, a new type of linkage between a carbohydrate and a protein has been discovered in human RNase U_s, in which α-mannose is attached *C*-glycosidically to the indole ring of a tryptophan (Hofsteenge *et al.*, 1994; de Beer *et al.*, 1995). Another unusual carbohydrate-peptide linking group is that of GlcNAc-1-P attached by a diester bond to serine, found in proteinase I from *D. discoideum*.

A different kind of protein modification, involving the formation of a carbohydrate-peptide bond, is ADP-ribosylation. It leads to the covalent linkage of the 3-hydroxyl of ribose with glutamic (or aspartic) acid, arginine or cysteine, and serves to attach monomeric or polymeric ADP-ribosyl residues to various proteins in different cellular compartments. Discovered about

20 years ago, ADP-ribosylation is now known to be widespread and leads to modulation of diverse processes such as DNA repair, differentiation, transmembrane signaling, carcinogenesis, and bacterial nitrogen fixation. ADP-ribosylation of low-molecular-mass GTP-binding proteins by the toxins of e.g. *Vibrio cholerae, Bordetella pertussis* and enterotoxigenic *E. coli* is a key step in the pathology of cholera, whooping cough and so-called travellers' diarrhea, respectively (Burnette, 1994).

2.3 Oligo- and Polysaccharides

2.3.1 *N*-linked

The tremendous diversity of asparagine-linked oligosaccharides derives from variations in the number and structure of branches and other substituents attached to the pentasaccharide core, Manα3(Manα6)Manβ4GlcNAcβ4GlcNAc, common to all members of this class. On the basis of the structure of the branches, *N*-glycans are traditionally classified into three subgroups – oligomannose, complex and hybrid, to which a fourth, xylose- and mannose containing has recently been added (Vliegenthart and Montreuil, 1995) (Table 2.5). Oligomannose glycans contain as a rule 2 to 6 α-mannose residues in their branches. However, elongated oligomannose chains, with

Table 2.5. Representative structures of the different types of *N*-linked glycans

Oligomannose

Man(α1−2)Man(α1−6)\
          Man(α1−6)\
Man(α1−2)Man(α1−3)/   Man(β1−4)GlcNAc(β1−4)GlcNAcβ
          /
Man(α1−2)Man(α1−2)Man(α1−3)

Complex

          Fuc(α1−6)$_{0,1}$
Neu5Ac(α2−6)Gal(β1−4)GlcNAc(β1−2)Man(α1−6)\ |
          [GlcNAc(β1−4)]$_{0,1}$Man(β1−4)GlcNAc(β1−4)GlcNAcβ
Neu5Ac(α2−6)Gal(β1−4)GlcNAc(β1−2)Man(α1−3)/

Hybrid

Man(α1−6)\
          Man(α1−6)\
Man(α1−3)/   \
          [GlcNAc(β1−4)]$_{0,1}$Man(β1−4)GlcNAc(β1−4)GlcNAcβ
Gal(β1−4)GlcNAc(β1−2)Man(α1−3)/

Xylose-containing

          Fuc(α1−3)$_{0,1}$
Man(α1−6)\ |
          Man(β1−4)GlcNAc(β1−4)GlcNAcβ
Man(α1−3)/ |
          Xyl(β1−2)$_{0,1}$

Table 2.6. Branching patterns in complex *N*-glycans[a]

Fuc(α1–6)$_{0,1}$
R¹.Man(α1–6)\ |
Man(β1–4)GlcNAc(β1–4)GlcNAcβ
R²-Man(α1–3)/

R¹	R²
Di-branched (biantennary)	
GlcNAc(β1–2)	GlcNAc(β1–2)
Tri-branched (triantennary)	
GlcNAc(β1–2)	GlcNAc(β1–4)\ GlcNAc(β1–2)/
or	
GlcNAc(β1–6)\ GlcNAc(β1–2)/	GlcNAc(β1–2)
Tetra-branched (tetraantennary)	
GlcNAc(β1–6)\ GlcNAc(β1–2)/	GlcNAc(β1–4)\ GlcNAc(β1–2)/
Penta-branched (pentaantennary)	
GlcNAc(β1–6)\ GlcNAc(β1–4)– GlcNAc(β1–2)/	GlcNAc(β1–4)\ GlcNAc(β1–2)/

[a] The *N*-acetylglucosamine in the branches is usually substituted by β4-linked galactose, to which other mono- or oligosaccharides may be attached. Examples of commonly occurring structures are given in Table 2.7

up to 100–200 mannose residues, are produced by yeasts. In glycans of the complex (or *N*-acetyllactosamine) type, up to five units of the disaccharide Galβ4GlcNAc (*N*-acetyllactosamine, LacNAc or NAL in brief), frequently substituted with sialic acid, are β-linked to the two α-mannose residues of the pentasaccharide core (Table 2.6). The structure of complex type glycans is further diversified, for example by the attachment of fucose linked to the pentasaccharide core and to the outer branches and of an *N*-acetylglucosamine residue linked β1–4 to the β-linked mannose of the core ("bisecting" GlcNAc) (Tables 2.7 and 2.8), and sometimes by the addition of Galα3 to the galactose of *N*-acetyllactosamine. A novel structure present on complex type chains (although rare in higher animals) is the disaccharide GalNAcβ4GlcNAc (LacdiNAc), in which *N*-acetylgalactosamine replaces the galactose in *N*-acetyllactosamine.

The third subgroup, called the hybrid type, has the characteristic features of both complex type and oligomannose glycans. One or two α-mannose residues are linked to the Manα1,6 arm of the core, as found in the oligomannose type, and a complex type outer chain is attached to the Manα1,3 arm of the core.

Table 2.7. Commonly occurring side chains in complex *N*-glycans

Gal(β1–4)GlcNAcβ

Neu5Ac(α2–3/6)Gal(β1–4)GlcNAcβ

Fuc(α1–3)
|
Gal(β1–4)GlcNAcβ

Fuc(α1–3)
|
Neu5Ac(α2–3/6)Gal(β1–4)GlcNAcβ

[Gal(β1–4)GlcNAc(β1–3)]$_n$Gal(β1–4)GlcNAcβ

Gal(β1–4)GlcNAc(β1–6) Gal(β1–4)GlcNAc(β1–6)
\ \
[Gal(β1–4)GlcNAc(β1–3)]$_n$Gal(β1–4)GlcNAcβ

Gal(β1–3)GlcNAcβ

Fuc(α1–4)
|
NeuAc(α2–3)Gal(β1–3)GlcNAcβ

Table 2.8. Some uncommon *N*-glycan structures

A. Linked to the pentasaccharide core[a]

Fuc(α1−6)$_{0,1}$
R_1-Man(α1−6) \|
\ Man(β1−4)GlcNAc(β1−4)GlcNAcβ
/
R_2-Man(α1−3)

Structure	R	Source	Ref[b]
GalNAc(β1−4)GlcNAc(β1−2) | Neu5Ac(α2−3/6)[c]	R_1 and R_2	Bowes melanoma tPA; lactotransferrin; glycodelin; snake venom batroxobin	(1)
GalNAc(β1−4)GlcNAc(β1−2) | Fuc(α1−3)	R_1 and R_2	Glycodelin	(1)
Neu5Ac(α2−6)Gal(β1−3)GlcNAc(β1−4) | Neu5Ac(α2−3)	R_1 and R_2	Fetuin	
Gal(β1−4)GlcNAc(β1−2) | |6 Neu5Ac(α2−6) SO_4	R_2	Thyroglobulin	
Neu5Ac(α2−3)Gal(β1−4)GlcNAc(β1−2) |6 SO_4	R_1	Recombinant tPA	
Gal(β1−4)GlcNAc(β1−6/4) | | Fuc(α1−2) Fuc(α1−3)	R_1/R_2.	Human γ-seminoprotein	
Tyv(1−3)GalNAc(β1−4)GlcNAc | Fuc(α1−3)$_{0,1}$	$R_1$and R_2	*Trichinella spiralis*	(2)
GlcNAc(β1−2) |6 2-Aminoethylphosphonate	R_1 and R_2	*Locusta migratoria*	(3)
GalNAc(β1−4)GlcNAc(β1−2)	R_1 and R_2	Human kallidinogenase	
Gal3Me(β1−3)GalNAc(β1−4)GlcNAc(β1−2) |	R_1 and R_2	Hemocyanin of *Lymnaea stagnalis*	
Gal*f* Man(α1−2)Man(α1−6) Gal*f* Man(α1−2)Man(α1−3)	R_1	*Leptomonas samueli*	
Glc(α1−3)Man(α1−2)Man(α1−6)\ Man(α1−3)/	R_1	Hen egg-yolk antibody	
Glc(α1−3)Glc(α1−3)Man(α1−2)Man(α1−2)	R_2	Starfish egg	

Table 2.8 Continued

B. Modifications of the pentasaccharide core		
Fuc(α1–6) | -4GlcNAc-Asn | Fuc(α1–3)	Honeybee venom phospholipase and hyaluronidase	(4)
Man(α1–6)\\ Man(β1–4)GlcNAc(β1–4)GlcNAc-Asn	interleukin-6	
Man(α1–6)\\ Man(β1–4)GlcNAc(β1–4)GlcNAc-Asn Man3Me(α1–3)/	Hemocyanin of *Lymnaea stagnalis*	
Fuc(α1–6) Man(α1–6)\\ | Man(β1–4)GlcNAc(β1–4)GlcNAc-Asn Man(α1–3)/ | GlcNAc(α/β1–6)	Lectin resistant mutant of CHO cells	(5)

[a] Only branches containing unusual structures are shown.
[b] Only references not appearing in Lis and Sharon (1993) are given: (1) Dell *et al.*, 1995; (2) Reason *et al.*, 1994; (3) Hard *et al.*, 1993; (4) Kubelka *et al.*, 1995; (5) Raju *et al.*, 1995.

The fourth subgroup consists of oligosaccharides with a xylose residue attached in a β1,2 linkage to the β-linked mannose of the core, and often also fucose linked α1,3 (and not α1,6 as in the complex or hybrid subgroups) to the innermost *N*-acetylglucosamine of the core. Originally identified in plants, the same structure is present in neural tissue of *Drosophila* and other insects, while the β1,2 linked xylose (but not the α1,3 linked fucose) is found in molluscan hemocyanins.

Branches that contain sulfated sugars are generally not sialylated. An unusual structure, containing both sulfate and sialic acid on the same branch, Neu5Acα2,6Galβ4Glc(SO_4)NAcβ2 is present in thyroglobulin, while in recombinant tPA, expressed in mouse epithelial cells, both subsituents are attached to the same galactose residue. A difucosylated *N*-glycan, with both fucose residues linked to the asparagine-bound *N*-acetylglucosamine, i.e. Fucα6(Fucα3)GlcNAc, was isolated from honey bee venom glycoproteins (Kubelka *et al.*, 1995). Di- and trisaccharides of α2,5-linked Neu5Gc8Me, in which the glycosidic linkage is through the hydroxyl of the *N*-glycoloyl moiety, have been isolated from the starfish *Asterias rubens*; its origin – whether glycoprotein or glycolipid – is, however, not known. These and some other uncommon structures are listed in Table 2.8.

The branches of many *N*-oligosaccharides in animal cells and, as shown recently, also in a glycoprotein from the protozoan *Trypanosoma brucei*, contain poly-*N*-acetyllactosamine, $[Gal\beta4\ GlcNAc\beta3]_n$, where n may be as high as 50 (Fukuda, 1995). Due to the branch specificity of β1,3-*N*-acetylglucosaminyltransferase – the enzyme that initiates the formation of poly-*N*-acetyllactosamine chains and catalyses their elongation – such chains are predominant on the α6-linked mannose of the core pentasaccharide. These chains, with their repeating disaccharide and frequent substitution of their galactose residues by GlcNAcβ6, Galβ4GlcNAcβ6 or Fucα3/4, can form a tremendous number of different structures. Among others, they serve as a backbone for the ABH(O), I/i and Lewis (Le^a and Le^b) blood group antigenic determinants on human erythrocyte membrane glycoproteins (and glycolipids), as well as the Le^x and sialyl Le^x ($SiaLe^x$) determinants on granulocytes and other cells (Table 2.9).

Table 2.9. Antigenic determinants

Structure of determinants	Name of antigen
*Type 1**	
Fuc(α1–2)Gal(β1–3)GlcNAc	Blood group H(O), type 1
GalNAc(α1–3)[Fuc(α1–2)]Gal(β1–3)GlcNAc	Blood group A, type 1***
Gal(α1–3)[Fuc(α1–2)]Gal(β1–3)GlcNAc	Blood group B, type 1***
Gal(β1–3)[Fuc(α1–4)]GlcNAc	Blood group Lewis^a (Le^a)
Neu5Ac(α2–3)Gal(β1–3)[Fuc(α1–4)]GlcNAc	Sialyl Lewis^a (SiaLe^a)
Fuc(α1–2)Gal(β1–3)[Fuc(α1–4)]GlcNAc	Blood group Lewis^b (Le^b)
*Type 2***	
Fuc(α1–2)Gal(β1–4)GlcNAc	Blood group H(O), type 2
GalNAc(α1–3)[Fuc(α1–2)]Gal(β1–4)GlcNAc	Blood group A, type 2
Gal(α1–3)[Fuc(α1–2)]Gal(β1–4)GlcNAc	Blood group B, type 2
Gal(β1–4)[Fuc(α1–3)]GlcNAc	Lewis^x (Le^x)
Neu5Ac(α2–3)Gal(β1–4)[Fuc(α1–3)]GlcNAc	Sialyl Lewis^x (SiaLe^x)
Polyl-N-acetyllactosaminoglycan	
[Gal(β1–4)GlcNAc(β1–3)]$_n$	Blood group i
Gal(β1–4)GlcNAc(β1–6)\ Gal(β1–4)GlcNAc Gal(β1–4)GlcNAc(β1–3)/	Blood group I

* Containing the sequence Gal(β1–3)GlcNAc
** Containing the sequence Gal(β1–4)GlcNAc
*** These determinants often contain an additional, α4-linked fucose on the GlcNAc, forming the ALe^b and BLe^b structures

The expression of certain poly-*N*-acetyllactosamines is strictly controlled, both spatially and temporally. In fetal erythrocytes, poly-*N*-acetyllactosamine is linear and tetraantennary and it expresses blood type i activity, while in adult cells it is branched and biantennary, and acquires blood type I activity. A large, fucosylated, highly branched poly-*N*-acetyllactolactosamine glycan (called embryoglycan) is abundant in early mouse embryos; with development its amount progressivly decreases.

Long chain acidic oligosaccharides of the glycosaminoglycan type, linked to asparagine via *N*-acetylgalactosamine or galactose, are found in halobacteria, while in eubacteria the oligosaccharides are linked to the asparagine via rhamnose (Sumper and Wieland, 1995). Examples of these unusual glycans are presented in Table 2.10.

Another type of polymer, confined largely to glycoproteins of neural origin of different animals and of fish eggs, are the polysialic acids (Troy, 1995). They may be *N*-linked to the protein backbone via the pentasaccharide core, or *O*-linked (see below). In neural glycoproteins, where polysialic acid is part of a class of adhesion molecules known as *N*-CAMs, it is made up exclusively of up to 55 α2,8- linked *N*-acetylneuraminic acid residues. The size of the polysialic acid in these molecules varies with tissue type and is developmentally regulated. A similar polysialic acid was also detected in kidney cells in early differentiation stages and in Wilms' tumor, a highly malignant kidney tumor of childhood. In contrast, polysialic acids from fish egg glycoproteins exhibit an extraordinary variation in chemical structure, due to the presence of both *N*-acetyl- and *N*-glycoloylneuraminic acid, of different linkages (α2,8, α2,9 and both in the same polymer) and diverse substituents on their hydroxyls. A polymer related to the polysialic acids is polyKdn, identified in glycoproteins of a wide range of animal cells and tissues (Inoue, 1993).

Table 2.10. Structures of some bacterial glycans

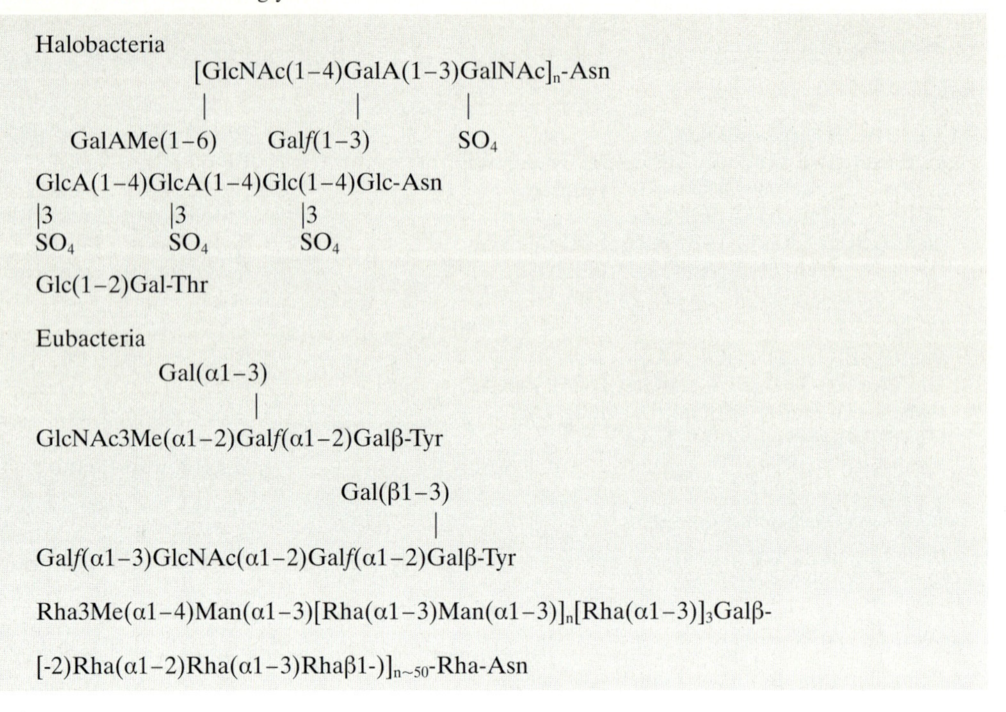

Halobacteria

[GlcNAc(1–4)GalA(1–3)GalNAc]$_n$-Asn
with GalAMe(1–6) attached to GlcNAc, Gal*f*(1–3) attached to GalA, and SO_4 attached to GalNAc

GlcA(1–4)GlcA(1–4)Glc(1–4)Glc-Asn
with SO_4 attached at position 3 of each of the first three residues

Glc(1–2)Gal-Thr

Eubacteria

GlcNAc3Me(α1–2)Gal*f*(α1–2)Galβ-Tyr
with Gal(α1–3) attached to Gal*f*

Gal*f*(α1–3)GlcNAc(α1–2)Gal*f*(α1–2)Galβ-Tyr
with Gal(β1–3) attached to the second Gal*f*

Rha3Me(α1–4)Man(α1–3)[Rha(α1–3)Man(α1–3)]$_n$[Rha(α1–3)]$_3$Galβ-

[-2)Rha(α1–2)Rha(α1–3)Rhaβ1-)]$_{n\sim 50}$-Rha-Asn

2.3.2 *O*-linked

Saccharides, bound to the proteins via the GalNAc-Ser/Thr linkage, represent a large and ubiquitous family of *O*-glycans (Schachter and Brockhausen, 1992). They are extremely heterogeneous, but can be divided into subgroups according to the structure of the core region, closest to the protein (Table 2.11).

O-Glycans vary in size from monomers to polymers that consist typically of up to 20 monosaccharide constituents. Monomeric and dimeric units are not common, one example being *N*-acetylgalactosamine and Neu5Acα2,6GalNAc in glycophorin of individuals of the Tn phenotype and on human cancer cells. Neu5Acα2,6GalNAc is also a constituent of ovine submaxillary mucin, each molecule of which is composed of some 600 amino acid residues, with one third substituted by such disaccharides. As its name implies, this glycoprotein is a member of the mucins, a group of glycoproteins found in epithelial secretions and as membrane constituents. Most of the mucins are, however, of higher molecular size (M_r up to $4–6\times10^6$), with numerous oligosaccharides larger than these found in ovine submaxillary mucin, and they exhibit a striking structural heterogene-

Table 2.11. Core structures of *O*-glycans

Core	Structure	Occurrence
1	Gal(β1–3)GalNAcα-	Mucins and other glycoproteins
2	GlcNAc(β1–6) | Gal(β1–3)GalNAcα-	Mucins and other glycoproteins
3	GlcNAc(β1–3)GalNAcα-	Mucins
4	GlcNAc(β1–6) \\ GlcNAc(β1–3)GalNAcα-	Mucins
5	GalNAc(α1–3)GalNAcα-	Glycoproteins
6	GlcNAc(β1–6)GalNAcα-	Mucins and other glycoproteins[a]

[a] To date found only in humans

Table 2.12. Some uncommon *O*-glycan structures

Structure	Source	Ref[a]
A. Linked directly to serine or threonine		
Xyl(α1–3)Xyl(α1–3)Glcβ[b]	Blood clotting factors, EGF[c]	(1)
Neu5Ac(α2–6)Gal(β1–4)GlcNAc(β1–3)Fucα	Blood clotting factors, EGF	(1)
Glcα 1-P-Man	Phosphoglucomutase	(2)
Gal(α1–3)GalNAc	Embryonal carcinoma cells	
GlcNAc(α1–3)Gal	*Neisseria gonorrhoeae*	(3)
Neu5Ac(α2–3)Gal(β1–-3)GalNAc | Neu5Ac(α2–8)Neu5Ac(α2–6)	Glycophorin	
Gal4Me(β1–-3)GalNAc | Fuc2Me(α1–2)	Nematode	
Neu5Ac(α2–6) | Gal(β1–3)GalNAc |6 SO_4	Mucin of patient with cystic fibrosis	
SO_4 |6 GlcNAc(β1–6) | Neu5Ac(α2–6)Gal(β1–3)GalNAc	Mucin of patient with cystic fibrosis	
SO_4 |6 Neu5Ac(α2–3)Gal(β1–4)GlcNAc(β1–6)GalNAc | | Fuc(α1–3) | Neu5Ac(α2–3)Gal(β1–3)	Mucin of patient with cystic fibrosis	(4)
B. Peripheral oligosaccharides		
Gal(β1–4)GlcNAc(β1–3)Gal(β1–4)-R[d] | | | Fuc(α1–2) | Fuc(α1–2) Fuc(α1–3)	Mucin of patient with bronchiectasis	
Fuc(α1–2)$_{0,1}$Fuc(α1–3)GalNac(β1–4)GlcNAc(β1–3)Gal(α1–3)R[e] | Fuc(α1–2)$_{0,1}$Fuc(α1–2)Fuc(α1–3)	Glycocalyx of *Schistosoma mansoni*	(5)

[a] Only references not quoted in Sharon and Lis (1993) are given: (1) Harris and Spellman, 1993; (2) Dey *et al.*, 1994; (3) Parge *et al.*, 1993; (4) Lo-Guidice *et al.*, 1994; (5) Khoo *et al.*, 1995

[b] Found only as serine linked

[c] EGF – epidermal growth factor

[d] R = GlcNAc(β1–6)[GlcNAc(β1–3)]GalNAcα-Ser/Thr or GlcNAc(β1–6)[Fuc(α1–2)Gal(β1–3)]GalNAcα-Ser/Thr

[e] R=GalNAc(β1–4)[Fuc(α1–3)]$_{0,1}$GlcNAc(β1–3)Gal(β1–3)GalNAcα-Ser/Thr or GalNAc(β1–4)[Fuc(α1–3)]$_{0,1}$GlcNAc(β1–6)GalNAcα-Ser/Thr

ity. *N*-Acetylgalactosamine, the typical constituent of these oligosaccharides, is accompanied by galactose, *N*-acetylglucosamine, fucose, various sialic acids and sulfate. One example is the sulfated trisaccharide Gal($3SO_4$)β4GlcNAcβ6GalNAc found in mucins isolated from sputum of patients with cystic fibrosis. Tracheobronchial glycoproteins from such patients also contain oligosaccharides with both sulfate and sialic acid, sometimes on the same branch, e.g. SO_4-Galβ3(Neu5Acα2,6)GalNAc (Table 2.12). Poly-*N*-acetyllactosamine chains carrying several fucose residues, e.g. Fucα2Galβ4(Fucα3)GlcNAcβ3(Fucα2) Galβ4-R, were isolated from respiratory mucins of a patient suffering from bronchiectasis. Some mucins bear blood type as well as Le^x and $SiaLe^x$ determinants. Glycoproteins from the eggs of rainbow trout contain *O*-linked polysialic acid, Galβ3[$(Neu5Gc\alpha2,8)_n$Neu5Gc α2,6]GalNAc, in which *N*-glycoloylneuraminic acid accounts for about 60 % of the weight.

Linear, short chains of α2-linked mannoses, e.g. ManαMan and $[Man\alpha3]_{0,1}$Manα2Manα2-Man, attached to the protein by a ManαSer/Thr linkage, occur in yeasts, while in plants chains of 1 to 4 β-arabinofuranosyl residues, β-linked to hydroxyproline, are present. The disaccharide GlcNAcα3Gal, *O*-linked to serine, has recently been identified in *Neisseria gonorrhoeae* (Parge *et al.*, 1995), a finding unprecedented in prokaryotic proteins.

2.3.3 Free oligosaccharides

Several unusual structures, related to those present in glycoproteins, have been found in free oligosaccharides. Thus, Neu5Acα2,6Manβ4GlcNAc has been isolated from the urine of a patient with β-mannosidosis, a genetic disease of glycoprotein catabolism. This trisaccharide is probably the product of the enzymatic transfer of *N*-acetylneuraminic acid to Manβ4GlcNAc that accumulates in such patients. In bovine colostrum, the trisaccharide with the structure Neu5Ac α2,6Galβ4GlcNAc-6-P is present, but the biosynthetic origin of this compound is not known. A novel arabinosyl-containing oligosaccharide, Manα6(Ara*f*α2)Manβ4GlcNAcβ4(Fucα3)GlcNAc, has been isolated from ripening tomatoes (Priem *et al.*, 1993). It is possible that structures of the types described occur in glycoproteins.

3 Functions

3.1 Methodology

Diverse approaches are being employed in the quest to unravel the role(s) of the carbohydrate units of glycoproteins. They include modification of glycans by glycosidases and transferases, the use of inhibitors of glycosylation or glycoprotein processing, and of cell mutants with known defects in glycosylation and, more recently, techniques of molecular genetics. Recombinant glycoproteins can be expressed in different cells and organisms, resulting in different patterns of glycosylation. An extreme example is that of bacteria (e.g. *E. coli*), which produce completely non-glycosylated proteins. Most insects lack the ability to sialylate glycoproteins, while Chinese hamster ovary (CHO) cells do not transfer sialic acid in α2,6-linkage, nor can they attach galactose in α1,3-linkage to glycoprotein glycans. Such α-linked galactose is not normally found in humans and New World primates, but is present in glycoproteins (and glycolipids) of other animals, including pigs and mice. Most useful are lectin resistant mammalian cell mutants lacking particular glycosyltransferases (Stanley and Ioffe, 1995). Site-directed mutagenesis can be applied to modify glycosylation sites so that they will no longer serve as acceptors for glycan attachment. With *N*-glycoproteins, where the carbohydrate is linked to the Asn-Xaa-Ser/Thr sequon, modification of either the first or third amino acid will abolish glycosylation at this site. In glycoproteins with more than one carbohydrate unit, whether *N*- or *O*-linked (or containing both types of linkage), the sites can be systematically eliminated in various combinations, to form a panel of mutants in which the roles of each carbohydrate chain can be assessed. Recently, many glycosyltransferases have been cloned (Schachter and Brockhausen, this volume). The availability of the glycosyltransferase genes, in conjunction with the transgenic and gene disruption techniques, provides powerful strategies for investigation of the functions for carbohydrates, especially in intact organisms (Stanley and Ioffe, 1995; see also Shur and Hathaway, this volume).

Each of the above approaches suffers, however, from certain limitations. Thus, inhibitors of glycosylation act indiscriminately, and lack of glycosylation of other proteins may indirectly modulate the expression or function of the glycoprotein under investigation. Similarly, in transgenic animals, deletion or insertion of a glycosyltransferase gene, for example, may modify many glycoproteins, since glycans are secondary gene products. Effects observed after elimination of a glycosylation site by mutagenesis may be due to differences in protein folding caused by the change in amino acid sequence and not to the absence of the glycan in the particular site.

Most of our knowledge of the functions of carbohydrates in glycoproteins comes from investigations of *N*-glycans. Studies of the functions of *O*-glycans have been hampered, not least because of the scarcity of specific inhibitors of *O*-glycosylation. Such glycans are also less amenable to genetic manipulation since, unlike *N*-glycosylation, the requirements for *O*-glycosylation are not obvious from the primary structure of the acceptor protein. The availability of glycosyltransferase genes will however certainly change this situation.

The overall conclusion from the numerous studies is that for many glycoproteins the carbohydrate, whether *N*- or *O*-linked, is dispensable, since it appears to have at most trivial effects on their properties. At the other extreme there are glycoproteins the properties of which depend critically on their glycans. Moreover, the same glycan may mediate different functions at different locations on the same glycoprotein, or in different cells, tissues or stages of the development of an organism. Each glycoprotein must therefore be evaluated individually to determine the contribution of the carbohydrate to its properties and functions.

3.2 Modulation of Physicochemical Properties

3.2.1 Nascent glycoproteins

Perhaps the most important functions of *N*-glycosylation are to aid in folding of the nascent polypeptide chain in the endoplasmic reticulum and to stabilize the conformation of the mature glycoprotein. In the absence of such glycosylation, some (glyco)proteins aggregate and/or are degraded, and as a result, are not secreted from the cells in which they are synthesized. Other glycoproteins are less affected and are secreted, but have compromised biological activities, while some appear to be totally unaffected.

One way in which *N*-linked oligosaccharide chains affect protein folding is by mediating the interaction of the newly-synthesized peptide with calnexin, a chaperone with apparent selectivity for *N*-glycoproteins (Helenius, 1994; Williams, 1995). Calnexin is a non-glycosylated membrane protein of 65.4 kDa molecular weight; it is located in the membrane of the endoplasmic reticulum, has a large, negatively charged carboxy-terminal cytoplasmic tail and an external domain containing three internal repeats of a conserved hexapeptide sequence. Experiments with several glycoproteins showed that calnexin binds transiently to the newly synthesized glycoproteins that have partially trimmed, monoglucosylated oligosaccharides; the binding coincides with protein folding and oligomer assembly. The proteins remain bound to calnexin for different periods of time, depending on the rates at which they achieve conformational maturation. During this time, the single terminal glucose is rapidly turned over in a deglucosylation-reglucosylation cycle. Once a glycoprotein is folded or assembled into oligomeric form, it is no longer reglucosylated and therefore does not react any more with calnexin and is free to leave the endoplasmic reticulum.

Studies reported very recently have shown that *N*-glycans act as apical sorting signals in epithelial cells (Scheiffele *et al.*, 1995). In these experiments, secretion of a non-glycosylated rat growth hormone from Madin-Darby canine (MDCK) cells was compared with mutants of the same protein that carry one or two engineered *N*-glycosylation sites. Whereas the non-glycosylated hormone was secreted from both the apical and basolateral sides of the cells, the glycosylated mutant proteins were secreted from the apical side only. Moreover, the introduction of two glycosylation sites led to increased apical delivery as compared to monoglycosylation. Examination of the literature revealed that almost all proteins reported to be apically secreted contain *N*-glycan(s), and in several cases the presence of such glycans was shown to be critical for apical delivery. On the other hand, non-glycosylated secretory proteins are usually secreted both apically and basolaterally.

Not only is glycosylation at a particular site important in directing protein folding, assembly and trafficking, but the precise structure of the glycan may also be critical. A clear case is that of human chorionic gonadotropin, a glycoprotein hormone composed of two subunits, for which it was shown that an abnormally glycosylated α-subunit is unable to associate with the β-subunit to form the mature hormone.

3.2.2 Mature glycoproteins

Quite frequently, the glycans modify the physical properties of the protein to which they are attached (Table 3.1). The influence is more pronounced the higher the carbohydrate content of the glycoprotein. The negative charges of sialic acid residues and sulfate groups change the overall charge of glycoproteins, increase their solubility and modify their conformation. These effects are of particular importance for the function of the highly glycosylated mucins, whose protein backbone is often densely substituted by sialic acid-containing oligosaccharides. Such mucins assume rigid, rod-like structures that may reach a length of several hundred nanometers. Mucin regions are often found on cell surface receptors and it is thought that the role of the resulting molecular rigidity is to extend the functional domains away from the cell surface.

Table 3.1. Carbohydrates modify the properties of the proteins to which they are linked

OFTEN
Affect solubility, charge and viscosity
Control folding and subunit assembly
Stabilize protein conformation
Protect against proteolysis
Affect the lifetime in circulation
Change the immunological properties
Modify the transmission of signals by cellular receptors
▼ Modify the activity of enzymes and hormones
RARELY

Since the surface area of the glycans is quite significant when compared to that of the peptide moiety, they may, in addition, influence other properties of proteins, such as heat stability and susceptibility to proteolysis. For instance, a comparison of two different β1,3/1,4 glucanases from *Bacillus* species that have been expressed in *E. coli* and in *Saccharomyces cerevisiae* has shown that the heavily glycosylated enzymes (with a carbohydrate content of about 45 %) secreted from the yeasts were considerably more heat stable than their unglycosylated counterparts synthesized by *E. coli* (Olsen and Thomsen, 1991). Such findings should be of special interest to industries producing commercially used enzymes, for which stability is a key requirement.

The carbohydrate may change markedly the quartenary structure of a protein to which it is attached, as demonstrated in the X-ray crystallographic study of the *Erythrina corallodendron* lectin, a member of the large family of legume lectins (Shaanan *et al.*, 1991). The heptasaccharide, linked at Asn-17 of each of the two subunits of this lectin, prevents the formation of the characteristic dimer observed in other members of the legume lectin family (e.g. concanavalin A and pea lectin). As a result, these subunits adopt a completely different quaternary structure.

Glycosylation is a major feature of the extracellular portion of human CD2, a leukocyte antigen on T lymphocytes that mediates cellular adhesion. The NH_2-terminal, membrane-distal domain of the protein, which mediates the adhesion function of the molecule, carries a single oligomannose carbohydrate chain. Nuclear magnetic resonance studies have shown that the *N*-acetylglucosamine residues of the core interact with the polypeptide chain, mainly through hydrogen bonds and van der Waals contacts, stabilizing an exposed cluster of five surface lysines (Wyss *et al.*, 1995). Elimination of this glycan led to partial or full unfolding of the protein and loss of its cellular adhesion functions.

3.3 Modulation of Biological Activity

The ability of carbohydrates to modulate the activities of biologically functional proteins, occasionally even in an all-or-nothing manner, has been established unequivocally during the last decade in a limited number of cases only. For most glycoproteins, however, such an effect of carbohydrates has not been demonstrated.

3.3.1 Antibodies

As mentioned previously (see Section 1), IgG contains close to 3 glycans per molecule, of which two are at a conserved glycosylation site, Asn-297 in the Fc fragment of the immunoglobulin, one per chain. The fact that these glycans are highly conserved suggests that they may have an important structural or functional role. Indeed, carbohydrate-free IgG, obtained by removal of the carbohydrate by site-directed mutagenesis, by glycosidase digestion of mature IgG, or by treating IgG secreting cells with tunicamycin (an inhibitor of *N*-glycosylation), lost its ability to bind to Fc receptors on macrophages (Dwek, 1995). It also exhibited a threefold lower affinity for the complement component $C1_q$ than untreated antibody. Although the antigen-binding properties of IgG were not affected, the antigen-antibody complexes formed from carbohydrate-deficient antibodies failed to be eliminated from the circulation. Similar effects were observed when IgG was expressed in lectin-resistant mutants of CHO cells that produce only oligomannose *N*-glycans (Wright and Morrison 1994). The immunoglobulin was properly assembled and secreted, but was significantly deficient in several effector functions, thus stressing the importance of the precise structure of the glycan for biological activity.

The serum IgG of patients with rheumatoid arthritis contains the same set of over 30 different oligosaccharides found in healthy individuals, but in very different proportions. There is a significant increase in structures devoid of galactose and terminating in *N*-acetylglucosamine (Dwek, 1995). Women during pregnancy show a low concentration of this form and an increase to normal level post-partum. This correlates well with the course of the disease, which enters remission in most women during pregnancy and recurs post-partum. The lower levels of galactose appear to be confined to IgG, since other serum glycoproteins examined, e.g. transferrin, are normally glycosylated. It has been reported that human B cells contain a galactosyltransferase specific for the *N*-oligosaccharides of IgG and that this enzyme is much less effective in patients with rheumatoid arthritis due to its lower affinity for UDP-Gal.

X-Ray crystallographic studies of the Fc fragment indicate that galactose residues present on the α1,6 branch of the IgG oligosaccharide can interact with residues Phe-243, Pro-244 and Thr-260 on the surface of the protein. It has been postulated (Dwek, 1995) that an increase in the level of glycoforms lacking terminal galactose on this branch could thus affect the interaction of the carbohydrate with the protein; this, in turn, could lead to changes in the conformation of the Fc moiety of IgG and to exposure of new antigenic determinants that may elicit an immune response in the patient, with possible relevance to rheumatoid arthritis. In addition, the sites on the protein originally occupied by the galactose of the α1,6 branch may interact with galactose still present on other IgG molecules, resulting in the appearance in the patient's serum of complexes typical of the disease, without an actual autoimmune response.

Another consequence of changes in the conformation of Fc in galactose-poor IgG is the exposure of terminal *N*-acetylglucosamine residues on the surface of the molecule. As a result, the modified immunoglobulin binds to the mannose binding protein, a serum lectin specific for mannose and *N*-acetylglucosamine (Malhotra *et al.*, 1995). The binding leads to activation of the classical complement pathway and this activation could be a major source of inflammation.

3.3.2 Antigens

The first documented case illustrating the importance of carbohydrates as immunodeterminants was that of ABH(O) human blood type determinants, α-linked *N*-acetylgalactosamine in A type, α-linked galactose in B type and α-linked fucose in H(O) type. The A and B blood type-specific oligosaccharides (antigens) are most abundant in glycoproteins of intestinal and gastric mucosa, lungs and salivary glands. Significant amounts are also found in other tissues, for instance kidney, bladder, and bone marrow glycoproteins. Indeed, much of the early structural work on the human blood group antigens was carried out on glycoproteins isolated from ovarian cysts, although the receptors are also present on glycolipids (Kopitz, this volume). Molecules carrying the antigens include membrane enzymes, membrane structural proteins and receptors [e.g., epidermal growth factor (EGF) receptor in A431 cells]. Besides the ABH(O) blood type determinants, glycoprotein glycans carry many other antigenic determinants, such as Le^x, $SiaLe^x$ and blood type i/I (Table 2.9). Other strongly immunogenic carbohydrate determinants are the β1,2 xylose and α1,3 fucose attached to the *N*-linked pentasaccharide core, found in plant glycoproteins. The latter monosaccharide, present on glycoproteins of bee venom, is responsible to a large extent for the allergenic properties of the venom (Tretter *et al.*, 1993).

The effect of a glycan on the antigenicity of a glycoprotein can be indirect, resulting from its impact on protein folding. In its absence, altered folding may either eliminate epitopes present on the native glycoprotein or create new ones. Thus, two (out of 11) conformational epitopes of vesicular stomatitis virus glycoprotein disappeared when the glycoprotein was produced in the presence of tunicamycin (Grigera *et al.*, 1991). These epitopes were, however, unaffected if the *N*-glycans of normally formed virus glycoprotein were removed enzymatically. It is thus possible that *N*-glycans may direct the folding of the glycoprotein into its native conformation and that once this conformation has been attained, it can be maintained in their absence. In human CD2, on the other hand, elimination of the glycan located in the NH_2-terminal domain of the molecule abolished the interaction of the antigen with three anti-CD2 monoclonal antibodies directed against distinct epitopes of the native molecule (Wyss *et al.*, 1995; see also 3.2.2).

A case of the masking of epitopes by carbohydrates emerged in the course of a clinical trial of granulocyte-macrophage colony stimulating factor (GM-CSF) (Gribben *et al.*, 1990). From a total of 16 patients that have been treated with recombinant GM-CSF synthesized in yeasts, four developed serum antibodies directed against native GM-CSF. The antibodies reacted with sites on the native protein backbone that are normally protected by *O*-glycosylation, but are exposed in the recombinant GM-CSF produced in yeasts and *E. coli*. Masking of antigenic epitopes by carbohydrates may be independent of oligosaccharide size. A single *N*-acetylglucosamine at Asn-149 was sufficient to prevent recognition of a peptide epitope of influenza virus hemagglutinin by its antibody (Munk *et al.*, 1992).

The interaction between an antigen and its antibody can be influenced as well by the presence of carbohydrate on the latter. Moreover, the effect on antigen binding, whether enhancement or inhibition, depends on the position of the carbohydrate. Comparison of a number of anti-α-1,6 dextran monoclonal antibodies revealed that those in which Asn-58 of the variable region of the heavy chain was glycosylated bound dextran with a 15-fold higher apparent K_a than those that lack glycosylation at this site. Introduction, by site-directed mutagenesis, of a glycosylation site at Asn-54 in the variable region of a non-glycosylated anti-dextran antibody blocked antigen binding, while glycosylation at Asn-60 increased the affinity of the antigen-antibody interaction about 5-fold (Wright *et al.*, 1991). The presence of an occupied *N*-glycosylation site in the heavy chain variable region of an antibody against the glycolipid galactosylgloboside abrogated the ability of the antibody to bind antigen (Marcus *et al.*, 1992). It has been suggested that the presence of carbohydrate affects the conformation of the combining site. However, the possibility that the carbohydrate blocks access to the binding site has not been excluded.

3.3.3 Enzymes

Tissue plasminogen activator (tPA) is one of the very few enzymes the activity of which is modulated by its carbohydrate (Dwek, 1995). This serine protease converts plasminogen into plasmin and thereby induces clot lysis (fibrinolysis). Two major molecular species of tPA occur naturally, type I and type II, that differ only in the number of *N*-linked glycans – three and two, respectively. Plasminogen itself also exists in two forms, with either one or more oligosaccharide units attached. The proteolytic activity of tPA is greatly enhanced by the formation of a ternary complex with its substrate and fibrin. The rate of formation and turnover of the complex spans a range that is dependent on the glycosylation site occupancy of both tPA and plasminogen (Mori *et al.*, 1995). At the extremes, this activity for type II tPA with type 2 plasminogen (possessing one *O*-linked glycan) is 2–3 times the value for type I tPA and type 1 plasminogen with one *O*-linked and one *N*-linked glycan, respectively.

A second enzyme is bovine pancreatic RNase, which occurs both in unglycosylated (RNase A) and glycosylated (RNase B) forms; the latter is a collection of isoforms, in which nine different oligomannose chains (from Man_5 to Man_9) are attached to the same polypeptide at the single *N*-glycosylation site (Asn34). RNase A and RNase B have always been reported as having the same enzymatic activity and were frequently quoted as a proof, provided by nature, that such activity is not affected by the presence of carbohydrates in the molecule. With the aid of a novel sensitive assay using double stranded RNA substrate it was shown that RNase A was more than three times as active as RNase B (Rudd *et al.*, 1994). The individual glycoforms RNase-Man_5, RNase-Man_1 and RNase-Man_0, prepared by exoglycosidase treatment of naturally occurring RNase B and separated by capillary electrophoresis, exhibited intermediate activity (Table 3.2). These differences were attributed to an overall increase in dynamic stability of the molecule with glycosylation (as demonstrated by measurements of proton exchange rates of the various RNase forms) and to steric effects, and supported by molecular modeling (Woods *et al.*, 1994).

Table 3.2. Enzymatic activity of RNase A and glycoforms of RNase B[a]

RNase	Carbohydrate	Relative activity
A	None	1.0
B[b]	$Man_0[GlcNAc]_2$	0.62
B[b]	$Man_1[GlcNAc]_2$	0.45
B[b]	$Man_5[GlcNAc]_2$	0.28
B	$Man_{5-9}[GlcNAc]_2$	0.26

[a] Based on Rudd *et al.*, 1994
[b] Modified by enzymatic removal of some mannose residues

Even a minor change in the glycosylation of an enzyme may have a dramatic effect on its acivity, as shown recently in studies of toxin B of *Clostridium difficile,* one of the two major virulence factors of this organism. The toxin was known to act by inhibiting a low molecular mass GTPase, involved in the regulation of the actin cytoskeleton. It has now been shown that the toxin functions as a glucosyltransferase that catalyzes the incorporation of one mole of glucose per mole of GTPase at threonine in position 37; this modification by a single sugar unit renders the enzyme inactive (Just *et al.*, 1995). Several other low-molecular-mass GTP-binding proteins, involved in the regulation of the actin skeleton, are similarly glucosylated by *C. difficile* toxins (Aktories and Just, 1995).

3.3.4 Hormones

The role of *N*-linked oligosaccharides in glycoprotein hormones is exerted at several levels. Intracellularly, they are involved in subunit folding, heterodimer assembly and secretion; extracellularly, the oligosaccharides modulate hormonal clearance and signal transduction. Diversity in the *N*-linked glycans can affect the bioavailability and bioactivity of the hormones. At least one role for the *O*-linked oligosaccharides in the β-subunit of chorionic gonadotropin is to increase the circulatory half-life of the hormone where there is a demand for sustained levels during pregnancy. For details, see Hooper *et al.*, in this volume, and Bielinska and Boime (1995).

3.3.5 Lectins

Most lectins from plants or animals are glycoproteins and, whenever tested, their activity was not affected by modification or absence of their glycans. For instance, recombinant *E. corallodendron* lectin expressed in *E. coli* has the same hemagglutinating activity and carbohydrate specificity as the native, glycosylated protein (Arango *et al.*, 1992). An exception seems to be ricin, the two-chain toxic lectin from *Ricinus communis*, the B-chain of which has been reported to lose its carbohydrate-binding activity when produced in *E. coli* (Richardson *et al.*, 1991). An unusual case of a different kind is that of concanavalin A, the lectin from jack bean. The mature lectin is not glycosylated, but during biosynthesis in the plant an inactive, glycosylated precursor appears transiently. Transformation of the precursor into mature lectin involves, besides deglycosylation, a complex series of polypeptide cleavages and rearrangements. However, *in vitro* enzymatic deglycosylation is sufficient to render the precursor active (Sheldon and Bowles, 1992). Also, the non-glycosylated precursor expressed in *E. coli* was active without further processing (Min *et al.*, 1992). These and other findings led to the conclusion that glycosylation of the pro-lectin is essential for its intracellular trafficking in the plant, possibly by preventing interactions with glycoproteins on its way from the endoplasmic reticulum to its final destination in the vacuoles. Wheat germ agglutinin, another lectin which is non-glycosylated in its mature form, is transiently glycosylated during biosynthesis in the plant; no information on the effect of the carbohydrate on the precursor is available.

3.3.6 Neural cell adhesion molecules

The neural adhesion molecules L1 and NCAM interact with each other within the membrane of one cell to form a functional complex with enhanced affinity for L1 on another cell. This interaction is based on the binding of oligomannose glycans on L1 to a lectin-like region on NCAM (Horstkorte *et al.*, 1993). Such glycans on the surface of neuronal cells may be involved in important cellular functions, as shown for neurite outgrowth *in vitro*: the latter was strongly inhibited when glycoconjugates containing such glycans were added to the culture medium. Neurite outgrowth was also inhibited by a synthetic peptide comprising part of the lectin-like region of NCAM. These and other results suggest that the L1/NCAM complex is important for the regulation of neural cell behavior.

3.3.7 Receptors

As with other classes of biologically active glycoproteins, the role of glycosylation in receptor function is highly variable. Like most membrane glycoproteins, they need, as a rule, to be *N*-glycosylated to attain a conformation necessary for transport to the cell surface. Loss of one (out of three) glycosylation sites of the human transferrin receptor by site-directed mutagenesis was sufficient to prevent the mutated receptor from leaving the endoplasmic reticulum, where it underwent specific cleavage and subsequent degradation (Hoe and Hunt, 1992). Similar mutational studies of the β-adrenergic receptor revealed that glycosylation is important for correct intracellular trafficking of the receptor (Rands *et al.*, 1990). In contrast, no such requirement was observed for cell surface localization of the m2 muscarinic acetylcholine receptor (van Koppen and Nathanson, 1990).

The effect of receptor glycosylation on its ability to combine with its ligand is also variable. For instance, glycosylation plays no role in the binding characteristics of the β-adrenergic receptor and of the acetylcholine receptor mentioned above, whereas it is required for binding by the basic fibroblast growth factor receptor (a heparan sulfate proteoglycan). Of the six potential glycosylation sites on the rat luteinizing hormone receptor, two were shown to carry carbohydrate chains essential for hormone binding activity (Zhang *et al.*, 1995). Although binding of fibronectin to its receptor (which is a member of the integrin family) is based on interactions of the α- and β-receptor subunits with defined peptide sequences of the ligand, the strength of association between them is modulated by the glycosylation status of the receptor. As recently shown, de-*N*-glycosylation of the receptor induces dissociation (or altered association) of the subunits, with concomitant loss of affinity for fibronectin (Zheng *et al.*, 1994).

Different glycoforms of the low-molecular weight mannose-6-phosphate (Man-6-P) receptor (see Section 3.6.1) differ in their affinity for the ligand: a more highly glycosylated form with a high content of sialic acid had a lower affinity than the form lacking poly-*N*-acetyllactosamine units and most of the sialic acid. Quite surprisingly, the non-glycosylated receptor, produced in the presence of tunicamycin, was reported to have unchanged binding properties (Wendland *et al.*, 1991). The same change in glycosylation may selectively modify the binding properties of closely related receptors, as shown for the insulin and insulin-like growth factor receptors from CHO cell glycosylation mutants.

Carbohydrates on receptors may affect the functional coupling of the latter to effector systems such as adenylate cyclase [*via* guanine nucleotide-binding proteins (G-proteins)] and tyrosine kinase, essential for the transmission of signals from the ligand to the cell. Thus, insulin receptor in which all four potential *N*-glycosylation sites of the β-subunit had been eliminated by site-directed mutagenesis had similar affinity for its ligand to the wild type receptor but had lost its transmembrane signaling ability, as evidenced by lack of stimulation of glucose transport and glycogen synthesis by the hormone (Leconte *et al.*, 1992).

3.4 Cellular Immune Functions

Carbohydrates on cell surfaces often modulate cellular immune functions. One of the earliest demonstrations came from studies with lymphocytes, showing that binding of lectins to their surface carbohydrates had a mitogenic effect on the cells. Treatment of the cells with periodate under conditions that oxidize specifically the side chains at C-7, C-8 and C-9 of sialic acid had a similar effect, as did treatment with galactose oxidase, an enzyme that generates cell surface aldehydes on C-6 of terminal galactose and *N*-acetyl-galactosamine residues. Also, treatment of antigen-presenting cells with the enzyme greatly enhanced the immunogenicity of viral, bacterial and protozoal vaccines in mice (Zheng *et al.*, 1992). All these effects could be due to transient Schiff base formation between aldehyde groups on one type of the interacting cells and amino groups on the other type, which may be required for activation of T lymphocytes by antigen-presenting cells. It is not clear, however, whether sugar residues on cell surface glycoproteins or glycolipids (or both) were responsible for this effect.

3.4.1 Natural killer cells

Natural killer (NK) cells, a population of lymphocytes capable of lysing target tumor cells, are likely candidates for primary defence of the body against cancer cells. Tumor cells vary greatly in their susceptibility to NK-induced lysis. A number of observations have implicated carbohydrates on the surface of the target cells in this selectivity (McCoy and Chambers, 1991). For instance, mutant CHO cell lines, synthesizing exclusively oligomannose and hybrid type *N*-glycans were more susceptible to lysis by NK cells than their parent cells that synthesize complex oligosaccharide chains. Furthermore, oligomannose glycopeptides were efficient inhibitors of such lysis. Similarly, K-562 cells, a standard target for human NK cells, exhibited increased sensitivity to the latter cells when grown in the presence of an inhibitor of α-mannosidase I (thus synthesizing exclusively oligomannose glycans) (Ahrens, 1993). In contrast, insertion of glycophorin A into the membrane of K562 cells increased their resistance to NK cells; the increase was related to the number of glycophorin A molecules inserted (El Ouagari *et al.*, 1995). This protective effect was shown to be dependent on the presence of the single complex *N*-glycan of glycophorin A. It thus appears that the two types of sugar chains – oligomannose and complex – can modulate the sensitivity of target cells to NK cell activity in different directions.

Recent studies have implicated a group of molecules containing extracellular C-type lectin domains on NK cells as potential mediators of carbohydrate-dependent NK-target cell interactions (Yokoyama, 1995). Such interactions appear to be crucial both for target cell recognition and for delivery of stimulatory and inhibitory signals linked to the NK cytolytic activity. Two functionally distinct subsets of lectin-like molecules on NK cells have been identified. Among the first subset is NKR-P1 in rats, which transmits activating signals into NK cells, while the second subset, represented by mouse Ly-49 proteins, is responsible for the transmission of negative signals that can cause total inactivation of NK cells. Information about the carbohydrate specificity of Ly-49 proteins is still limited, but they appear to recognize sulfated sugars and/or fucose (Daniels *et al.*, 1994; Brennan *et al.*, 1995). Detailed studies of NKR-P1 have shown (Bezouska *et al.*, 1994) that this protein is specific for *N*-acetygalactosamine and *N*-acetylglucosamine, preferably in a terminal nonreducing position in oligosaccharides such as GlcNAcβ3Galβ4Glc. However the best ligands were acidic, highly sulfated, proteoglycan-derived oligosaccharides. A striking correlation was found between the concentration of carbohydrate needed for inhibition of binding of NK cells to target cells and inhibition of killing of the latter, clearly implicating NKR-P1-carbohydrate interactions in events leading to the killing process.

3.5 Activities of Free Oligosaccharides

Diverse activities are exhibited by free oligosaccharides, either derived from glycoproteins or from other sources (Table 3.3). This was originally demonstrated with heparin oligosaccharides that, like the parent molecule, act as anticoagulants. Subsequently, it was found that a heparin/heparan sulfate dodecasaccharide activates cell-bound fibroblast growth factor, similarly to full-size heparin (or heparan sulfate proteoglycans, such as syndecan; for further information on proteoglycans, please see chapter of Kresse, this volume). The saccharide binds both to the growth factor and its receptor, and the formation of such a trimeric complex appears to be a prerequisite for signal transduction. Oligosaccharides of the type found in *N*-glycans of plant glycoproteins at appropriate concentrations promote or delay tomato ripening (Yunovitz and Gross, 1993). Other oligosaccharides that act on plants in different ways have been described, some examples of which are listed in Table 3.3.

Table 3.3. Biologically active oligosaccharides

Structure	Activity
In animals	
Heparin-derived	Anticoagulant
Heparan sulfate-derived	Growth factor activators
In plants	
Oligoglucosides	Induce disease resistance
Pectin-derived	Anti-auxins
Fuc-Xyl-*N*-glycans	Delay tomato ripening
Xyloglucan-derived	Inhibit or promote elongation of pea stem segments
Lipo-chito-oligosaccharides	Nodulation of legumes by Rhizobia

3.6 Carbohydrates as Recognition Determinants

There is increasing evidence for the concept, formulated over 20 years ago, that carbohydrates act as recognition determinants in a variety of physiological and pathological processes (Sharon and Lis, 1993). This concept evolved with the realization that carbohydrates have an enormous potential for encoding biological information. The messages encoded in the structures of complex carbohydrates are deciphered through interactions with complementary sites on carbohydrate-binding proteins, chiefly lectins (Sharon and Lis, 1995). Processes in which the participation of carbohydrate-lectin interactions was clearly demonstrated include intracellular trafficking of enzymes, clearance of glycoproteins from the circulatory system (Table 3.4) and a wide range of cell-cell interactions (Table 3.5). Particularly exciting is the demonstration that binding of carbohydrates on the surface of leukocytes by a class of animal lectins designated selectins, controls leukocyte traffic by mediating adhesion of these

Table 3.4. Clearance and targeting of glycoproteins

Glycoprotein	Receptor	
	Specificity	Location
Asialoglyco-proteins	Galactose	Liver (hepatocytes)
Diverse	Fucose	Liver (Kupffer cells)
Hormones	$GalNAc4SO_4$	Liver (Kupffer cells, endothelial cells)
Lysosomal enzymes	Man-6-phosphate	Ubiquitous
Diverse	Mannose	Macrophages, liver (endothelial cells)

Table 3.5. Carbohydrates and lectins in cell-cell recognition

Process	Sugars on	Lectins on
Infection	Host cells	Microorganisms
Defense	Phagocytes	Microorganisms
	Microorganisms	Phagocytes
Fertilization	Eggs	(Sperm)[a]
Leukocyte traffic	Leukocytes	Endothelial cells
	Endothelial cells	Lymphocytes
Metastasis	Target organs	Malignant cells
	Malignant cells	(Target organs)[a]

[a] Presumed, no experimental evidence available

cells to restricted portions of the endothelium and their recruitment to inflammatory sites. An earlier finding attracting increasing attention is that carbohydrates mediate the adhesion of bacteria to epithelial cells. Intensive efforts are now being made, in both academia and industry, to develop agents for anti-adhesion therapy of inflammation and infectious diseases.

3.6.1 Clearance (traffic) markers

The rapid removal of desialylated glycoproteins from rabbit serum via the hepatic asialoglycoprotein receptor (hepatic binding protein or lectin), a phenomenon discovered in the late 1960's, is the prototype of the saccharide-based recognition system. Such a general mechanism would allow cells to use the same oligosaccharide markers on many different proteins without the need to "code" this in the DNA of the proteins. However, the role in nature proposed for this clearance system has not yet been proven beyond doubt. There are several other systems in which the traffic of glycoproteins is controlled by their carbohydrate constituents (Table 3.4). A prominent example is the intracellular routing of lysosomal enzymes to their compartment, which is mediated by the recognition between Man-6-P that is part of the oligomannose unit(s) of such enzymes, and the Man-6-P receptors. Two such receptors have been described, one cation-independent and of high molecular weight (220 kDa), the other cation-dependent and of low molecular weight (48 kDa). A defect in the synthesis of the Man-6-P marker recognized by the receptors results in I-cell disease (also called mucolipidosis II or MLII), an inherited lysosomal storage disease, characterized by a lack in the lysosomes of those enzymes, among them β-glucocerebrosidase, *L*-iduronidase, α-*N*-acetylglucosaminidase, that normally carry the marker (Kornfeld, 1992; Neufeld, 1992). It is caused by a deficiency of GlcNAc-phosphotransferase, the first enzyme in the pathway of mannose phosphorylation, and is thus a processing disease, the first of its kind to be identified. Therefore, even though the disease is transmitted by a single gene, some 20 enzymes are affected. The enzymes lacking the recognition marker do not reach their destination (the lysosomes); consequently, they are secreted into the extracellular milieu, which is one of the biochemical abnormalities of the affected cells. The specificity of the GlcNAc-phosphotransferase for certain lysosomal enzymes is based on its ability to recognize a specific lysine residue and a particular tertiary domain of the acceptor glycoprotein (Baenziger, 1994). Another carbohydrate-specified clearance system, discussed elsewhere in this volume (Hooper *et al.*), is that of the sulfated glycoprotein hormones (see also Table 3.4).

The presence of well-defined carbohydrate binding proteins on cell surfaces is being exploited for drug targeting to specific organs. Gaucher disease is caused by a deficiency of the enzyme β-glucocerebrosidase, resulting in accumulation of glucocerebroside in Kupffer and endothelial (non-parenchymal) cells of the liver. These cells contain on their surface a mannose- (and *N*-acetylglucosamine) specific lectin. To target the β-glucocerebrosidase into the above cells, the complex and hybrid sugar chains of the enzyme were trimmed down with the aid of exoglycosidases to expose the mannose residues of the pentasaccharide core (Brady *et al.*, 1994). In this way the administered glucocerebrosidase is effectively delivered to the deficient cells where the enzyme is needed to degrade the accumulated glucocerobroside.

3.6.2 Fertilization

The interaction between oligosaccharides on the extracellular coat (zona pellucida) of the egg and a carbohydrate-binding protein on the sperm appears to be involved in the highly precise gamete (egg and sperm) recognition required for species-specific fertilization in mammals (Litscher and Wassarman, 1993). In the mouse, there is substantial evidence to suggest that the ability of the zona pellucida to bind sperm is conferred by terminal α-linked galactose residues on *O*-glycans of one of three zona pellucida glycoproteins, designated ZP3. Recent findings, however, cast some doubt on the role of this sugar in sperm-egg interactions. Mice, in which the gene for the α1,3-galactosyltransferase, responsible for the synthesis and expression of the Galα3Gal epitope, has been inactivated, were fully fertile (Thall *et al.*, 1995). Other studies suggested *N*-acetylglucosamine on ZP3 as being involved in the binding of sperm. This saccharide serves as an acceptor for the β-1,4-galactosyltransferase, an enzyme shown to be an integral sperm plasma membrane component and to recognize selectively *O*-linked oligosaccharides on ZP3 (Miller *et al.*, 1992).

3.6.3 Infection

The oligosaccharide repertoire on the host-cell surface is among the key genetic susceptibility factors in viral and microbial infection and in toxin action. A number of viral, mycoplasmal, bacterial and protozoan pathogens use specific carbohydrate structures of glycoproteins (or glycolipids) on host cells as attachment sites in the initial stages of infection (Ofek and Doyle, 1994; Sharon and Lis, 1996). In a few cases, the glycoproteins in question have been identified (Sharon and Ofek, 1995). Recent examples include the receptor for *Streptococcus sanguis* OMZ9 on human buccal epithelial cells (Neeser *et al.*, 1995), for enterotoxigenic *E. coli* on rabbit intestinal cells (Wennerås *et al.*, 1995) and for influenza C virus in a canine kidney cell line (Zimmer *et al.*, 1995).

Experiments in intact animals have indeed proved that it is possible to prevent bacterial infection by blocking the attachment of the responsible organism with an appropriate sugar (Table 3.6). These findings have provided an impetus for the development of carbohydrate-based anti-adhesion drugs to combat infections. Should a bacterium mutate so that it no longer recognizes the anti-adhesive carbohydrates, it will also fail to bind to its cell surface receptors, and therefore lose the ability to cause infection. Moreover, since such drugs do not kill the pathogens, they will not exert selection pressure and their use will not result in the development of resistance.

Even if a particular carbohydrate has been established as an inhibitor for a disease-causing microorganism in an animal (or in humans), it must be determined whether the use of this carbohydrate, or its analogues, will not interfere with other processes in the body. One such process is lectinophagocytosis, well documented for the mannose specific *E. coli* (Ofek *et al.*, 1995). This mode of phagocytosis may result from binding of the bacteria to phagocytes, e.g. macrophages or neutrophils, which is followed by activation of the phagocytes and uptake and killing of the bacteria. Lectinophagocytosis may occur *in vivo* and may provide protection against infection of nonimmune hosts by bacteria or in sites that are poor in opsonins. The latter include lungs, renal medulla, the cerebrospinal fluid and the peritoneal cavity, especially during peritoneal dialysis.

Table 3.6. Inhibitors of sugar-specific adhesion preventing infection *in vivo*

Organism	Animal, site	Inhibitor
Escherichia coli type 1	Mice, UT	MeαMan
	Mice, GIT	Mannose
	Mice, UT	Anti-Man antibody
Klebsiella pneumoniae type 1	Rats, UT	MeαMan
Shigella flexnerii type 1	Guinea pigs, eye	Mannose
Escherichia coli type P	Mice	Globotetraose
	Monkeys	Galα4GalβOMe
Escherichia coli K99	Calves. GIT	Glycopeptides of serum glycoproteins
Pseudomonas aeruginosa	Human, ear	Gal+Man+Neu5Ac

UT, urinary tract; GIT, gastrointestinal tract

In another mode of lectinophagocytosis, a wide range of microorganisms (bacteria, fungi and protozoa) that express mannose on their surface bind to the mannose-specific lectin present on the surface of macrophages. This binding, too, may lead to the uptake of the microorganisms by the phagocytic cell and occasionally also their killing. A particularly interesting example of such a microorganism is the pathogenic fungus, *Pneumocystis carinii*, a major case of death among AIDS patients.

Glycoprotein glycans play also a role in the infection by human immunodeficiency virus (HIV), the causative agent of AIDS. This has been demonstrated for gp120, the predominant envelope glycoprotein of the virus. gp120 contains 20–25 potential glycosylation sites, to which a diverse range of more than 100 oligomannose, hybrid, bi-tri- and tetraantennary structures are attached. It interacts with the membrane glycoprotein CD4 of T lymphocytes, a key step in the HIV infection. Correct glycosylation appears to be a prerequisite for virus infectivity, since inhibitors of glycosylation that affect *N*-glycan processing were shown to be potent inhibitors of infection *in vitro*. Originally it was believed that the decrease in infectivity was due to the effect of incorrect gp120 glycosylation on CD4 binding and thus viral entry into the cell. Recent results suggest that glycosylation is mainly necessary for a post-binding event, such as the fusion of the viral and cellular membranes (Fischer *et al.*, 1995). Of the various inhibitors tested, the most dramatic anti-viral effects observed have been with *N*-butyldeoxynojirimycin. Second-generation derivatives of this inhibtor, aimed at preventing various side effects, as well as other glycosylation inhibitors, are now undergoing clinical tests (Jacobs, 1995).

3.6.4 Leukocyte traffic

Research carried out mainly since 1990 has demonstrated that adhesive interactions mediated by surface carbohydrates and surface lectins play a crucial role in leukocyte trafficking to sites of inflammation and hemostasis, and in the migration (homing) of lymphocytes to specific lymphoid organs (Zanetta, this volume). In these processes, the carbohydrates serve as "area codes" which are interpreted by E-selectin, P-selectin and L-selectin, members of a family of endogenous lectins (McEver *et al.*, 1995; Nelson *et al.*, 1995). The selectins are highly asymmetric membrane-bound proteins. Their extracellular part consists of an amino-terminal carbohydrate recognition domain (CDR), an epidermal growth factor-like domain, and several short repeating units related to complement-binding protein. They bind specifically to $SiaLe^x$ and its positional isomer $SiaLe^a$ (Table 2.9), with both fucose and sialic acid required for binding; sialic acid can be replaced by another negatively charged group such as sulfate. These proteins recognize the carbohydrate ligands only when the latter are present on particular glycoproteins, such as cell surface mucins, pointing to the role of the carrier molecule and carbohydrate presentation in the recognition of the latter by lectins.

The selectins provide the best paradigm for the role of sugar-lectin interactions in biological recognition. In broad outline, they all mediate, although with some differences, the adhesion of circulating leukocytes to endothelial cells of blood vessels, leading to the exit of the former cells from the circulation. The extravasation is necessary for the migration of leukocytes into tissues, such as occurs under normal recirculation of lymphocytes between different lymphoid organs, or in recruitment of leukocytes to sites of inflammation. L-selectin, also known as "homing receptor", is found on all leukocytes. It is predominantly involved in the recirculation of lymphocytes, directing them specifically to peripheral lymph nodes. In contrast to the homing receptor, the two other selectins are expressed mainly on endothelial cells, and only when these cells are activated by inflammatory mediators, mainly cytokines (e.g. interleukin-2 and tumor necrosis factor). The latter are released from tissue leukocytes in response to e.g wounding, infection or ischemia, and induce the expression of P-selectin on the endothelial surface within minutes and of E-selectin within 3–4 hours.

Studies in animals have provided direct evidence for the role of selectins in the control of leukocyte traffic. For instance, in P-selectin-deficient mice, generated by targeted gene disruption, the recruitment of neutrophils to the inflamed peritoneal cavity was significantly delayed (Mayadas *et al.*, 1993). The clinical importance of selectin-carbohydrate interactions in acute inflammatory responses in humans is illustrated by the finding that the neutrophils of two patients with recurrent bacterial infections had a deficiency in $SiaLe^x$ and were unable to bind to E-selectin (Etzioni *et al.*, 1995). The specific biochemical lesion responsible for these defects has not yet been established, but is believed to be a reflection of a general deficiency in fucose metabolism in these patients. This assumption is strengthened by the fact that both unrelated patients belong to the extremely rare Bombay blood type, so named since the first cases of persons lacking the ABH(O) blood types were discovered in Bombay, India. Such persons are known to lack fucosyltransferase, a key enzyme in the synthesis of blood type ABH(O) determinants on glycoproteins and glycolipids. The above findings imply that the inability to synthesize $SiaLe^x$ prevents the neutrophils from migrating to the sites of infection and suggest that inhibitors of the selectins may be potent anti-inflammatory agents. Prevention of adverse inflammatory reactions by inhibition of leukocyte extravasation has become a major aim of many pharmacological industries. It has indeed been demonstrated that oligosaccharides recognized by the selectins exert protective effects against experimentally induced lung injury in experimental animals. These approaches are now being evaluated for treatment of human disease.

References

Ahrens PB (1993): Role of target cell glycoproteins in sensitivity to natural killer cell lysis. In *J. Biol. Chem.* **268**:385–91.

Aktories K, Just I (1995): Monoglucosylation of low-molecular-mass GTP-binding Rho proteins by clostridial cytotoxins. In *Trends Cell Biol.* **5**:441–3.

Allen HJ, Kisailus EC (eds) (1992): *Glycoconjugates. Composition, Structure and Function.* 685 pp. New York Basel Hong Kong: Marcel Dekker, Inc.

Alonso MD, Lomako J, Lomako WM *et al.* (1995): A new look at the biogenesis of glycogen. In *FASEB J.* **9**:1126–37.

Arango R, Adar R, Rozenblatt S *et al.* (1992): Expression of *Erythrina corallodendron* lectin in *Escherichia coli.* In *Eur. J. Biochem.* **205**:575–81.

Ashford DA, Dwek RA, Rademacher TW *et al.* (1991): The glycosylation of glycoprotein lectins: intra- and inter-genus variation in *N*-linked oligosaccharide expression. In *Carbohydr. Res.* **213**:215–27.

Baenziger JU (1994): Protein-specific glycosyltransferases: how and why they do it. In *FASEB J.* **8:** 1019–25.

Bezouska K, Yuen C-T, O'Brien J *et al.* (1994): Oligosaccharide ligands for NKR-P1 protein activate NK cells and cytotoxicity. In *Nature* **372**:150–7.

Bielinska M, Boime I (1995): The glycoprotein hormone family: structure and function of the carbohydrate chains. In *Glycoproteins* (Montreuil J, Vliegenthart JFG, Schachter H, eds) pp 565–88, Amsterdam: Elsevier.

Bock K, Schuster-Kolbe J, Altman E *et al.* (1994): Primary structure of the *O*-glycosidically linked glycan chain of the crystalline surface layer glycoprotein of *Thermoanaerobacter thermohydrosulfuricus* L 111–69: galactosyl tyrosine as a novel linkage unit. In *J. Biol. Chem.* **269**:7137–44.

Brady RO, Murray GJ and Barton NW (1994): Modifying exogenous glucocerebrosidase for effective replacement therapy in Gaucher disease. In *J. Inherited. Metab. Dis.* **17**:510–9.

Brennan J, Takei F, Wong S *et al.* (1995): Carbohydrate recognition by a natural killer cell receptor, Ly-49C. In *J. Biol. Chem.* **270**:9691–4.

Brennan SO, Myles T, Peach RJ *et al.* (1990): Albumin Redhill (-1 Arg, 320 Ala-Thr): A glycoprotein variant of human serum albumin whose precursor has an abberant signal peptidase cleavage site. In *Proc.* Natl. Acad. *Sci. USA* **87**:26–30.

Burnette WN (1994): AB_5 ADP-ribosylating toxins: comparative anatomy and physiology. In *Structure* **2**:151–8.

Daniels BF, Nakamura MC, Rosen SD *et al.* (1994): Ly-49A, a receptor for H-$2D^d$, has a functional carbohydrate recognition domain. In *Immunology* **1**:785–92.

de Beer T, Vliegenthart JFG, Loffler A *et al.* (1995): The hexapyranosyl residue that is C-glycosidically linked to the side chain of tryptophan-7 in human RNase U_s is α-mannopyranose. In *Biochemistry* **34**: 11785–9.

Dell A, Morris HR, Easton RL *et al.* (1995): Structural analysis of the oligosaccharides derived from glycodelin, a human glycoprotein with potent immunosuppressive and contraceptive activities. In *J. Biol. Chem.* **270**:24116–26.

Dey NB, Bounelis P, Fritz TA *et al.* (1994): The glycosylation of phosphoglucomutase is modulated by carbon source and heat schock in *Saccharomyces cerevisiae*. In *J. Biol. Chem.* **269**:27143–8.

Dwek RA (1995): Glycobiology: towards understanding the function of sugars. In *Biochem. Soc. Transact.* **23**:1–25.

El Ouagari K, Teissie J, Benoist H (1995): Glycophorin A protects K562 cells from natural killer cell attack. Role of oligosaccharides. *In J. Biol. Chem.* **270**:26970–5.

Etzioni A, Phillips LM, Paulson JC *et al.* (1995): Leukocyte adhesion deficiency (LAD) II. In *Cell Adhesion and Human Disease* (Marsh J, Goode JA, eds) (Ciba Foundation Symposium 189) pp 51–62, Chichester: Wiley.

Fischer PB, Collin M, Karlsson GB *et al.* (1995): The α-glucosidase inhibitor *N*-butyldeoxynojirimycin inhibits human immunodeficiency virus entry at the post-CD4 binding level. In *J. Virol.* **69**:5791–7.

Fukuda M (1994): Cell surface carbohydrates: cell-type specific expression. In *Molecular Glycobiology* (Fukuda M, Hindsgaul O, eds) pp 1–52, Oxford: University Press.

Fukuda M, Hindsgaul O, eds (1994): *Molecular Glycobiology*, 261 pp, Oxford: University Press.

Gribben JG, Devereux S, Thomas NSB *et al.* (1990): Development of antibodies to unprotected glycosylation sites on recombinant GM-CSF. In *Lancet* **335**:434–7.

Grigera PR, Mathieu ME, Wagner RR (1991): In *Virology* **180**:1–9.

Grinnell BW, Walls JD, Gerlitz B (1991): Glycosylation of human protein C affects its secretion, processing, functional activities and activation by thrombin. In *J. Biol. Chem.* **226**:9778–85.

Haltiwanger RS, Kelly WG, Roquemore EP *et al.* (1992): Glycosylation of nuclear and cytoplasmic proteins is ubiquitous and abundant. In *Biochem. Soc. Trans.* **20**:264–9.

Hard K, van Dorn JM, Thomas-Oates JE *et al.* (1993): Structure of the Asn-linked oligosaccharides of apolipophorin III from the insect *Locusta migratoria*. Carbohydrate-linked 2-aminoethylphosphonate as a constituent of a glycoprotein. In *Biochemistry* **32**:766–75.

Harris RJ, Spellman MW (1993): *O*-linked fucose and other post-translational modifications unique to EGF molecules. In *Glycobiology* **3:**219–24**.**

Hayes BK, Hart GW (1994): Novel forms of protein glycosylation. In *Curr. Opinion Struct. Biol.* **4**: 692–6.

Helenius A (1994): How *N*-linked oligosaccharides affect glycoprotein folding in the endoplasmic reticulum. In *Mol. Biol. Cell* **5**: 253–6.

Hoe MH, Hunter RC (1992): Loss of one asparagine-linked oligosaccharide from human transferrin receptors results in specificicleavage and association with the endoplasmic reticulum. In *J. Biol. Chem.* **267**:4916–23.

Hofsteenge J, Muller DR, de Beer T *et al.* (1994): New type of linkage between a carbohydrate and a protein: C-glycosylation of a specific tryptophan residue in human RNase U_s. In *Biochemistry* **33**:13524–30.

Horstkorte R, Schachter M, Magyar JP *et al.* (1993): The fourth immunoglobulin-like domain of NCAM contains a carbohydrate recognition domain for oligomannosidic glycans implicated in association with L1 and neurite outgrowth. In *J. Cell Biol.* **121**:1409–21.

Inoue Y (1993): Glycobiology of fish egg polysialoglycoproteins (PSPG) and deaminated neuraminic-rich glycoproteins (KD*N*-gp). In *Polysialic acid* (Roth J, Rutishauser U, Troy II FA, eds) pp 183–9 Basel: Birkhauser.

IUPAC-IUB Joint Commision on Biochemical Nomenclature (JCBN) (1986): Nomenclature of gly-

coproteins, glycopeptides and peptidoglycans. In *Eur. J. Biochem.* **159**:1–6.
Jacobs GS (1995): Glycosylation inhibitors in biology and medicine. In *Curr. Opinion Struct. Biol.* **5**:605–11.
Just I, Selzer J, Wilm M *et al.* (1995): Glucosylation of Rho proteins of *Clostridium difficile* toxin B. In *Nature* **375**:500–3.
Khoo K-H, Sarda S, Xu X *et al.* (1995): A unique multifucosylated -3GalNAcβ1–4GlcNAcβ1–3Galα1-motif constitutes the repeating unit of the complex *O*-glycans derived from the cercarial glycocalyx of *Schistosoma mansoni*. In *J. Biol.* Chem. **270**: 17114–23.
Kobata A (1992): Structures and functions of the sugar chains of glycoproteins. In *Eur. J.* Biochem. **209**:483–501.
Kornfeld S (1992): Structure and function of the mannose 6-phosphate insulin-like growth factor II receptors. In *Annu. Rev. Biochem.* **61**:307–30.
Kubelka V, Altmann F, Marz L (1995): The asparagine-linked carbohydrate of honeybee venom hyaluronidase. In *Glycoconjugate J.* **12**:77–83.
Leconte I, Azuan C, Debant A *et al.* (1992): *N*-linked oligosaccharide chains of the insulin receptor β-subunit are essential for transmembrane signaling. In *J. Biol. Chem.* **267**: 17415–23.
Lederkremer de RM, Colli W (1995): Galactofuranose-containing glycoconjugates in trypanosomatids. In *Glycobiology* **5**: 547–52.
Lis H, Sharon N (1993): Proteins glycosylation. Structural and functional aspects. In *Eur. J. Biochem.* **218**:1–27.
Litscher ES, Wassarman PM (1993): Carbohydrate-mediated adhesion of eggs and sperm during mammalian fertilization. In *Trends Glycosci. Glycotechnol.* **5**:369–88.
Lochnit G, Geyer R (1995): Carbohydrate structure analysis of batroxobin, a thrombin-like protease from *Bothrops moojeni* venom. In *Eur. J. Biochem.* **228**:805–16.
Lo-Guidice J-M, Wieruszeski J-M, Lemoine J *et al.* (1994): Sialylation and sulfation of the carbohydrate chains in respiratory mucins from a patient with cystic fibrosis. In *J. Biol. Chem.* **269**:18794–813.
Malhotra R, Wormald R, Rudd PM *et al.* (1995): Glycosylation changes of IgG associated with rheumatoid arthritis can activate complement via the mannose binding protein. In *Nature Medicine* **1**:237–41.
Marcus DM, Yu-Lee L-Y, Dinh Q *et al.* (1992): Glycosylation of an antibody (Ab) framework region may inhibit antigen binding. In *J. Cell Biochem.* **16D**:156.
Mayadas TN, Johnson RC, Rayburn H *et al.* (1993): Leukocyte rolling and extravasation are severely compromised in P-selectin deficient mice. In *Cell* **74**:541–4.
McCoy JP, Chambers WJ (1991): Carbohydrates in the function of natural killer cells. In *Glycobiology* **1**:321–8.
McEver RP, Moore KL, Cummings RD (1995): Leukocyte trafficking mediated by selectin-carbohydrate interactions. In *J. Biol. Chem.* **270**:11025–8.
Messner P, Christian R, Neununger C *et al.* (1995): Similarity of "core" structures in two different glycans of tyrosine-linked eubacterial S-layers. In *J. Bacteriol.* **177**:2188–93.
Miller DJ, Macek MB, Shur BD (1992): Complementarity between sperm surface β-1,4-galactosyltransferase and egg coat ZP3 mediates sperm-egg binding. In *Nature* **357**:589–93.
Min W, Dunn AJ, Jones DH (1992): Non-glycosylated recombinant pro-concanavalin A is active without polypeptide cleavage. In *EMBO J.* **11**:1303–7.
Montreuil J, Vliegenthart JFG, Schachter H (eds) (1995): *Glycoproteins*. Amsterdam: Elsevier.
Morelle W, Capon C, Balduyck M *et al.* (1994): Chondroitin sulphate covalently cross-links the three polypeptide chains of inter-α-trypsin inhibitor. In *Eur. J. Biochem.* **221**:881–8.
Mori, K, Dwek RA, Downing AK et al (1995): The activation of type 1 and type 2 plasminogen by type I and type II tissue plasminogen activator. In *J. Biol. Chem.* **270**:3261–7.
Munk K, Pritzer E, Kretzschmar E *et al.* (1992): Carbohydrate masking of an antigenic epitope of influenza virus haemagglutinin independent of oligosaccharide size. In *Glycobiology* **2**:233–40.
Nelson RM, Venot A, Bevilacqua MP *et al.* (1995): Interactions in vascular biology. In *Annu. Rev. Cell Dev. Biol.* **11**:601–13.
Neeser J-R, Grafstrom RC, Woltz A *et al.* (1995): A 23 kDa glycoprotein bearing NeuNAcα2–3Galβ1–3GalNAc *O*-linked carbohydrate chain acts as a receptor for *Streptococcus sanguis* OMZ9 on human buccal epithelial cells. In *Glycobiology* **5**:97–104.
Neufeld EF (1992): Lysosomal storage diseases. In *Annu. Rev. Biochem.* **60**:257–80.
Ofek I, Doyle RJ (1994) *Bacterial Adhesion to Cells and Tissues*. London: Chapman and Hall.
Ofek I, Goldhar J, Keysari Y *et al.* (1995): Nonopsonic phagocytosis of microorganisms. In *Annu. Rev. Microbiol.* **49**:239–70.
Olsen O, Thomsen KK (1991): Improvement of bacterial β-gluconase thermostability by glycosylation. In *J. Gen. Microbiol.* **137**:579–85.
Parge HE, Forest KT, Hickey MJ *et al.* (1995): Structure of the fibre-forming protein pilin at 2.6 Å resolution. In *Nature* **378**:32–8.
Pepys MB, Rademacher TW, Amatayakul-Chantler S *et al.* (1995): Human serum amyloid P component is an invariant constituent of amyloid deposits and has a

uniquely homogenous glycostructure. In *Proc. Natl. Acad. Sci. USA* **91**:5602–6.
Priem B, Gitti R, Bush CA *et al.* (1993): Structure of ten free *N*-glycans in ripening tomato fruit. Arabinose is a constituent of a plant *N*-glycan. In *Plant Physiol.* **102**:445–58.
Puanglarp N, Oxley D, Currie GJ *et al.* (1995): Structure of the *N*-linked oligosaccharides from tridacnin, a lectin found in the haemolymph of the giant clam *Hippopus hippopus*. In *Eur. J. Biochem.* **232**:873–80.
Rands E, Candelore MR, Cheung AH *et al.* (1990): Mutational analysis of β-adrenergic receptor glycosylation. In *J. Biol. Chem.* **265**:10759–64.
Raju, TS, Ray MK, Stanley P (1995): LEC18, a dominant Chinese hamster ovary glycosylation mutatnt synthesizes *N*-linked carbohydrates with a novel core structure. In *J. Biol. Chem.* **270**:30294–302.
Reason AJ, Ellis LA, Appleton JA *et al.* (1994): Novel tyvelose-containing tri- and tetra-antennary *N*-glycans in the immunodominant antigens of the intracellular parasite *Trichinella spiralis*. In *Biochemistry* **34**:593–603.
Richardson PT, Hussain K, Woodland HR *et al.* (1991): The effects of *N*-glycosylation on the lectin activity of recombinant ricin B chain. In *Carbohydr. Res.* **213**:19–25.
Rudd PM, Joao HC, Coghill E *et al.* (1994): Glycoforms modify the dynamic stabiliy and functional activity of an enzyme. In *Biochemistry* **33**:17–22.
Schachter H, Brockhausen I (1992): The biosynthesis of serine (threonine)-*N*-acetylgalactosamine-linked carbohydrate moieties. In *Glycoconjugates. Composition, Structure and Function.* (Allen HJ, Kisailus, EC, eds) pp 263–332, New York Basel Hong Kong: Marcel Dekker, Inc.
Schauer R, Kelm S, Reuter G *et al.* (1995): Biochemistry and role of sialic acids. In *Biology of the sialic acids.* (Rosenberg A, ed) pp 7–67, New York and London: Plenum Press.
Scheiffele P, Peranen J, Simons K (1995): *N*-glycans as apical sorting signals in epithelial cells. In *Nature* **378**:96–8.
Schreiner R, Schnabel E, Wieland F (1994): Novel *N*-glycosylation in eukaryotes: laminin contains the linkage unit β-glycosylasparagine. In *J. Cell Biol.* **124**:1071–81.
Shaanan B, Lis H, Sharon N (1991): Structure of a legume lectin with an ordered *N*-linked carbohydrate in complex with lactose. In *Science* **254**:862–6.
Sharon N, Lis H (1993): Carbohydrates in cell recognition. In *Sci. Am.* **268(1)**:82–9.
Sharon N, Lis H (1995): Lectins – proteins with a sweet tooth: functions in cell recognition. In *Essays Biochem.* **30**:59–75.
Sharon N, Lis H (1996): Microbial lectins and their glycoprotein receptors. In *Glycoproteins* (Montreuil J, Vliegenthart JFG, Schachter H, eds) Vol. 2, Amsterdam: Elsevier, in press
Sharon N, Ofek I (1995): Identification of receptors for bacterial lectins by blotting techniques. In *Meth. Enzymol.* **253**:91–8.
Sheldon PS, Bowles DJ (1992):The glycoprotein precursor of concanavalin A is converted to an active lectin by deglycosylation. In *EMBO J.* **11**: 1297–301.
Stanley P and Ioffe E (1995): Glycosyltransferase mutants: key new insights in glycobiology. In *FASEB* J. **9**:1436–44.
Stimson E, Makepeace K, Dell A *et al.* (1995): Meningococcal pilin: a glycoprotein substituted with digalactosyl 2,4-diacetamido-2,4,6-trideoxy hexose. In *Mol. Microbiol.* **17**:1201–14.
Sumper M, Wieland FT (1995): Bacterial glycoproteins. In *Glycoproteins* (Montreuil J, Vliegenthart JFG, Schachter H, eds) pp 455–73, Amsterdam: Elsevier.
Thall AD, Maly P, Lowe JB (1995): Oocyte Galα1,3Gal epitopes implicated in sperm adhesion to zona pellucida glycoprotein ZP3 are not required for fertilization in the mouse. In *J. Biol. Chem.* **270**:21437–40.
Tretter V, Altmann F, Kubelka V *et al.* (1993) Fucose α 1,3-linked to the core region of glycoprotein *N*-glycans creates an important epitope for IgE from honeybee venom allergic individuals. In *Arch. Allergy. Immunol.* **102**:259–66.
Troy II FA (1995): Sialobiology and the polysialic acid glycotype: occurence, structure, function, synthesis and glycopathology. In *The Biology of the Sialic Acids* (Rosenberg A, ed) pp 95–144, New York and London: Plenum Press.
van Koppen CL, Nathanson NM (1990): Site directed mutagenesis of the m2 muscarinic acetylcholine receptor. Analysis of the role of *N*-glycosylation in receptor expression and function. In *J. Biol. Chem.* **265**:20887–92.
Varki A (1992): Diversity in the sialic acids. In *Glycobiology* **2**:25–40.
Varki A (1993): Biological roles of oligosaccharides: all of the theories are correct. In *Glycobiology* **3**:97–130.
Vliegenthart JFG, Montreuil J (1995): Primary structure of glycoprotein glycans. In *Glycoproteins* (Montreuil J, Vliegenthart JFG, Schachter H, eds) Vol. 1, pp 13–28, Amsterdam: Elsevier.
Wendland M, Waheed A, Schmidt B *et al.* (1991): Glycosylation of the M_r 46,000 mannose-6-phosphate receptor. Effect on ligand binding, stability and conformation. In *J. Biol. Chem.* **266**:4598–604.
Wennerås C, Neeser J-R, Svennerholm A-M (1995): Binding of the fibrillar CS3 adhesin of entrotoxigenic *Escherichia coli* to rabbit intestinal glycoproteins is

competitively prevented by GalNAcβ1–4Gal-containing glycoconjugates. In *Infect. Immun.* **63**:640–6.
Williams DB (1995): Calnexin leads glycoproteins into the fold. In *Glycoconjugate J.* **12**: iii-iv.
Wisniewski N, McNeil M, Grieve RB *et al.* (1993): Characterization of novel fucosyl- and tyvelosyl-containing glycoconjugates from *Trichinella spiralis* muscle stage larvae. In *Mol. Biochem. Parasitol.* **61**:25–36.
Woods RJ, Edge CJ, Dwek RA (1994): Protein surface oligosaccharides and protein function. In *Nature Struct. Biol.* **1**:499–500.
Wright A, Morrison SL (1994): Effect of altered C_H2-associated carbohydrate structure on the functional properties and *in vivo* fate of chimeric mouse-human immunoglobulin G1. In *J. Exp. Med.* **180**:1087–96.
Wright A, Tao M, Kabat EA *et al.* (1991): Antibody variable region glycosylation: position effects on antigen binding and carbohydrate structure. In *EMBO J.* **10**:2717–23.
Wyss DF, Choi JS, Li J *et al.* (1995): Conformation and function of the *N*-linked glycan in the adhesion domain of human CD2. In *Science* **269**:1273–8.
Yokoyama WM (1995): Natural killer cell receptors. In *Curr. Opin. Immunol.* **7**:110–20.
Yunovitz H, Gross KC (1993): Delay of tomato fruit ripening by an oligosaccharide *N*-glycan. Interactions with IAA, galactose and lectins. In *Physiol. Plant.* **90**:152–6.
Zhang R, Huiqing C, Naheed F *et al.* (1995): Functional glycosylation sites of the rat luteinizing hormone receptor required for ligand binding. In *J. Biol. Chem.* **270**:21722–8.
Zheng B, Brett SJ, Tite JP *et al.* (1992): Galactose oxidation in the design of immunogenic vaccines. In *Science* **256**:1560–3.
Zheng M, Fang H, Hakomori S (1994): Functional role of *N*-glycosylation in α5β1 integrin receptor. De-*N*-glycosylation induces dissociation or altred association of α5 and β1 sububits and concomitant loss of fibronectin binding activity. In *J. Biol. Chem.* **269**: 12325–31.
Zimmer G, Klenk HD, Herrler G (1995): Identification of a 40 kDa cell surface sialoglycoprotein with the characteristics of a major influenza C virus receptor in a Madin-Darby canine kidney cell line. In *J. Biol. Chem.* **270**:17815–22.

9 Glycolipids: Structure and Function

Jürgen Kopitz

1 Structure, Classification and Localization of Glycolipids

The glycolipids are a structurally heterogeneous group of membrane components found in all species ranging from bacteria to man. The term 'glycolipid' designates any compound containing one or more monosaccharide residues linked by a glycosyl linkage to a lipid moiety (IUPAC-IUB Commission on Biochemical Nomenclature, 1977). Since their discovery more than 100 years ago a large number of glycolipid structures have been cataloged and the qualitative and quantitative glycolipid content has been determined for many cell types. Most bacterial and plant glycolipids are glycerol-containing compounds (glycoglycerolipids), whereas in animals and man glycosphingolipids are predominant.

1.1 Bacterial Glycolipids

Glycolipid structures of bacteria are enormously varied and only relatively few species have been investigated in detail (Shaw, 1974; Luzzati *et al.*, 1987; Kates, 1992; Koga *et al.*, 1993). Mycobacterial plasma membranes, containing numerous highly immunogenic glycolipids, have been most extensively studied, due to their pathological

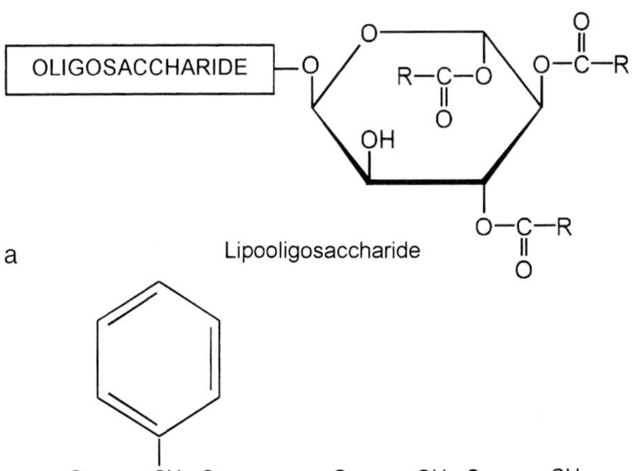

Figure 1a–c. Structures of mycobacterial glycolipids. R–C(=O)– = Long chain fatty acyl

H.-J. and S. Gabius (Eds.), Glycosciences

ISBN 3-8261-0073-5

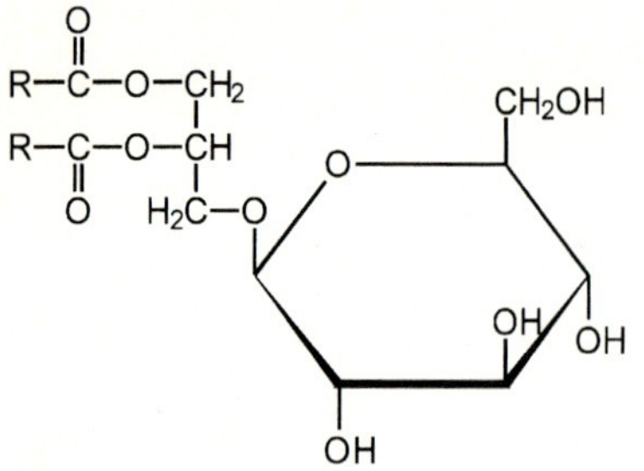

Figure 2. Structure of monoglucosyldiacylglycerol

importance. Three major classes of glycolipids have been identified in mycobacteria: lipooligosaccharides, glycopeptidolipids and phenolic glycolipids (Fig. 1a – c) (Brennan, 1981, 1989; Chatterjee *et al.*, 1989; Puzo, 1990; McNeil and Brennan, 1991).

Mono- to penta-glycosyldiacylglycerols have been found in bacterial membranes (Fig. 2). A wide variety of sugars, including glucose, galactose, mannose, rhamnose, trehalose, glucuronic acid and glucosamine, with numerous modifications (e.g. methylation, alkylation or acetylation), are observed (Goldfine, 1982).

All archaebacterial membrane lipids are unique in consisting entirely of derivatives of a C_{20}-C_{20} diacylglycerol ether and its dimer, from which abundant glycolipids are derived. An example of these unusual glycolipids is the macrocyclic tetraether lipids (Fig. 3), which are of special interest, due to their exceptional physical properties, which stabilise the bacterial membrane in harsh habitats (De Rosa *et al.*, 1986).

The outer membrane of Gram-negative bacteria contains an extremely complex class of glycolipids, which are neither diacylglycerol- nor sphingoid-based. They consist of a basic unit ('Lipid A': Diglucosamine-phosphate with amide linked and ester-bound β-hydroxy-myristic acid). A glycosyl head group is linked to Lipid A, which is composed of approximately 40 sugar residues (Fig. 4) (Cullis and de Kruijff, 1979). It goes without saying that there is again an extreme heterogeneity in the lipopolysaccharide structures.

1.2 Plant Glycolipids

In plants, glycolipids based on diacylglycerol, particulary mono- and digalactosyldiacylglycerol, are major components of the stacked thylakoid membranes of chloroplasts (Fig. 5a). Sulfoquinovosyldiacylglycerol is a charged glycolipid of thylakoids, possessing a glycosyl headgroup which is sulfated in the 6-position (Fig. 5b) (Nishihara *et al.*, 1980).

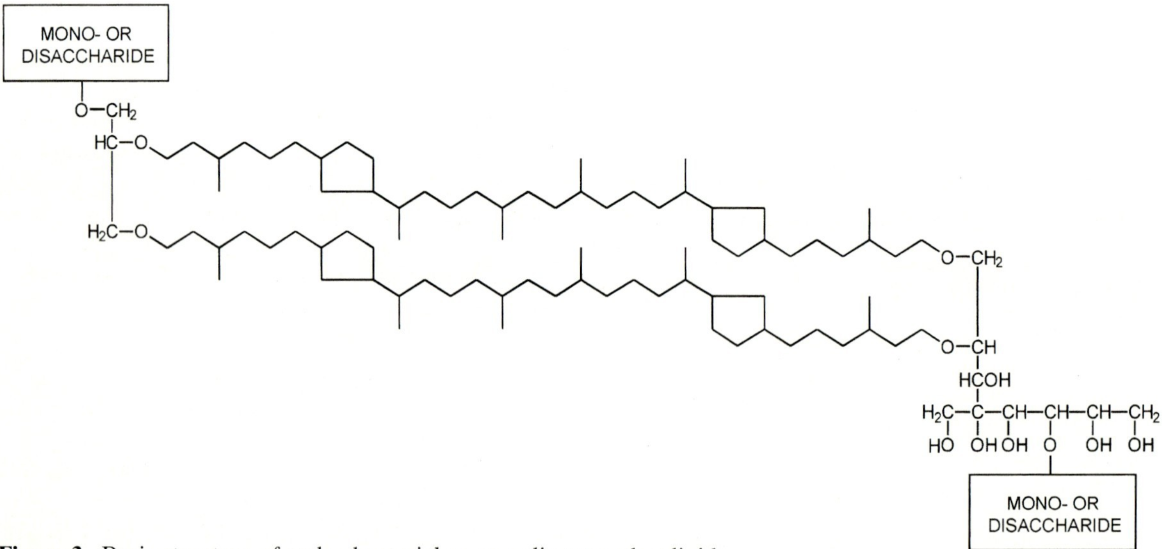

Figure 3. Basic structure of archaebacterial marocyclic tetraether lipids

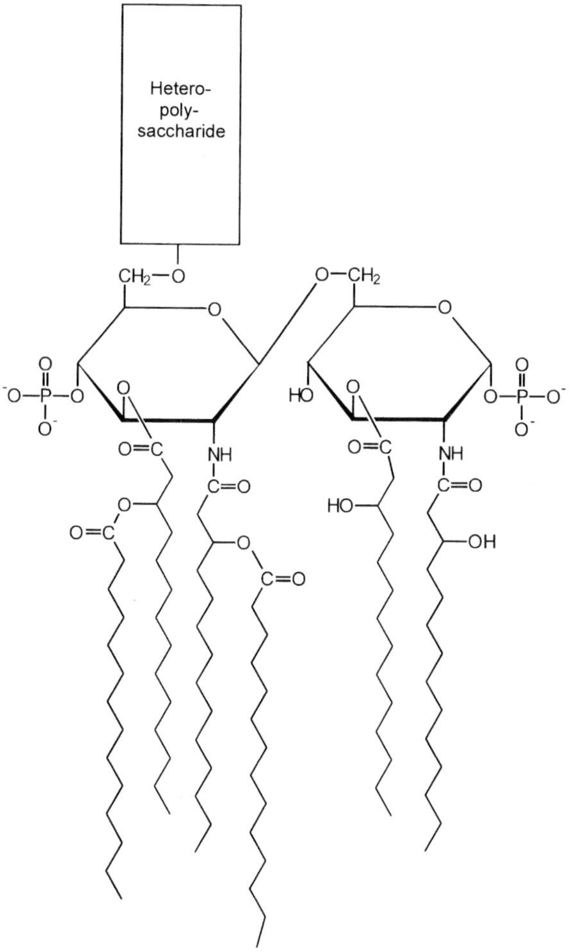

Figure 4. Basic structure of bacterial lipopolysaccharides

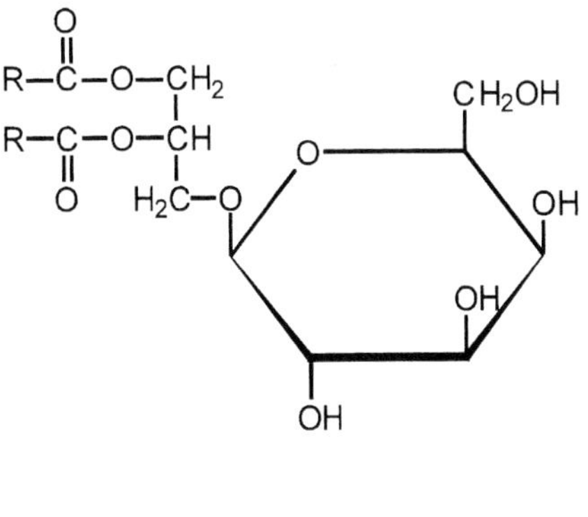

a Monogalactosyldiacylglycerol

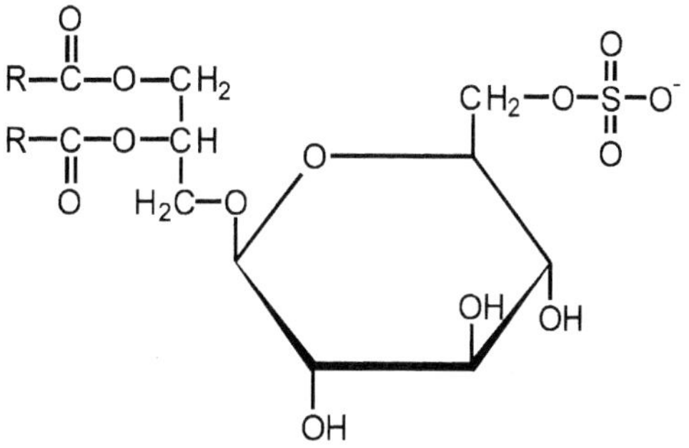

b Sulfoquinovosyldiacylglycerol

Figure 5a, b. Structures of plant glycolipids

1.3 Animal Glycolipids

In most animal glycolipids the long chain aminodiol sphingosine (4-sphingenine) (Fig. 6a) serves as the lipid backbone; such glycolipids are therefore termed glycosphingolipids. To the aminogroup of sphingosine, fatty acids of varying lengths are attached via amide bonds, forming the glycolipid's hydrophobic tail, called ceramide (Fig. 6b). The carbohydrate units are bound to the primary hydroxyl of ceramide via glycosidic linkages. Since the systematic names of the oligosaccharides linked to the ceramide are so cumbersome, suitable trivial names were created for frequently observed parent oligosaccharides (IUPAC-IUB Commission on Biochemical Nomenclature (CBN), 1977). General principles in constructing these names are as follows:

The number of monosaccharide units is indicated by the suffixes 'biose', 'triaose', 'tetraose' etc. According to their structure and biogenic relationship, they are grouped in series, like globo, muco, lacto, ganglio and gala, which are used as prefixes in the trivial names. Differences in linkage, such as 1→4 rather than 1→3, in otherwise identical sequences, are indicated by 'iso' or 'neo' as additional prefix. Table 1 summarizes a set of symbols which have been devised for a simple representation of complex glycosphingolipids (IUPAC-IUB Commission on Biochemical Nomenclature (CBN), 1977).

Glycolipids are found primarily in the plasma membrane of all vertebrate tissues, and they are

Table 1. Trivial names and abbreviations of glycosphingolipids

Structure	Trivial name	Symbol
Galα1→4Galβ1→4GlcCer	Globotriaosylceramide	$GbOse_3Cer$
GalNAcβ1→3Galα1→4Galβ1→4GlcCer	Globotetraosylceramide	$GbOse_4Cer$
Galα1→3Galβ1→4GlcCer	Isoglobotriaosylceramide	$iGbOse_3Cer$
GalNAcβ1→3Galα1→3Galβ1→4GlcCer	Isoglobotetraosylceramide	$iGbOse_4Cer$
Galβ1→4Galβ1→4GlcCer	Mucotriaosylceramide	$McOse_3Cer$
Galβ1→3Galβ1→4Galβ1→4GlcCer	Mucotetraosylceramide	$McOse_4Cer$
GlcNAcβ1→3Galβ1→4GlcCer	Lactotriaosylceramide	$LcOse_3Cer$
Galβ1→3GlcNAcβ1→3Galβ1→4GlcCer	Lactotetraosylceramide	$LcOse_4Cer$
Galβ1→4GlcNAcβ1→3Galβ1→4GlcCer	Neolactotetraosylceramide	$nLcOse_4Cer$
GalNAcβ1→4Galβ1→4GlcCer	Gangliotriaosylceramide	$GgOse_3Cer$
Galβ1→3GalNAcβ1→4Galβ1→4GlcCer	Gangliotetraosylceramide	$GgOse_4Cer$
Galα1→4GalCer	Galabiosylceramide	$GaOse_2Cer$
Galα1→4Galα1→4GalCer	Galatriaosylceramide	$GaOse_3Cer$

particulary abundant in the nervous system. The major glycolipid of mammalian brain is galactosylceramide (galactocerebroside) (Fig. 6c), constituting about 16 % of total lipid in adult brain (Agranoff and Hajra, 1994). Lactosylceramide, and neutral tri- and tetraglycosylceramides are abundant in all tissues (Macher and Sweeley, 1978). Glycosphingoplipids with complex branched oligosaccharide structures received special interest, due to their being blood group antigens, and their modifications acting as human cancer antigens (Hakomori, 1984a, 1986).

Sulfation of the saccharide part turns neutral glycosphingolipids into charged sulfatides. The major mammalian sulfatide consists of a monogalactosylceramide in which the 3 position of the galactosyl headgroup is sulfated (Fig. 6d). This compound occurs in high concentrations in myelin and as a derivative (sulfogalactosyl 1-*O*-alkyl-2-*O*-acylglycerol) in the plasma membrane of mammalian spermatozoa, comprising >4 moles percent of total lipid (Vos *et al.*, 1994).

Gangliosides are a group of acidic glycosphingolipids characterized by the presence of one or more sialic acids (Fig. 6e). The sialic acid of the gangliosides may be *N*-acetylneuraminic acid or the less common *N*-glycolylneuraminic acid. Additionally, the sialic acid can be chemically modified, e.g. by deacetylation or by *O*-acetylation. Naming of glycosphingolipids as represented before is also applied to gangliosides, which are named *N*-acetyl- (or *N*-glycoloyl) neuraminosyl-(X)osylceramide, where (X) stands for the root name of the neutral oligosaccharide to which the sialosyl residue is attached. The location of the sialosyl residue is indicated by a Roman numeral designating the position of the monosaccharide residue in the parent oligosaccharide to which the sialosyl residue is attached, with an Arabic numeral superscript indicating the position within that residue to which the sialic acid is attached (IUPAC-IUB Commission on Biochemical Nomenclature (CBN), 1977). Frequently, the Svennerholm (Svennerholm, 1994a) designation for gangliosides is preferred; examples are also shown in Table 2: G = ganglioside; M = monosialo, D = disialo, T = trisialo, Q = quatrasialo etc.; the length of the neutral sugar chain is designated by a subscript number following the formula $5 - n$, where n equals the number of neutral sugars in the ganglioside. The position of the sialic acid(s) can be characterized further by a subscript lower case a, b or c, depending on the biosynthetic pathway for the ganglioside (see also Fig. 7).

Gangliosides have been found in every vertebrate tissue studied thus far, mainly localized in the plasma membrane, although they are also present at lower concentrations in the endoplasmatic reticulum, the Golgi apparatus, the lysoso-

a Sphingosine

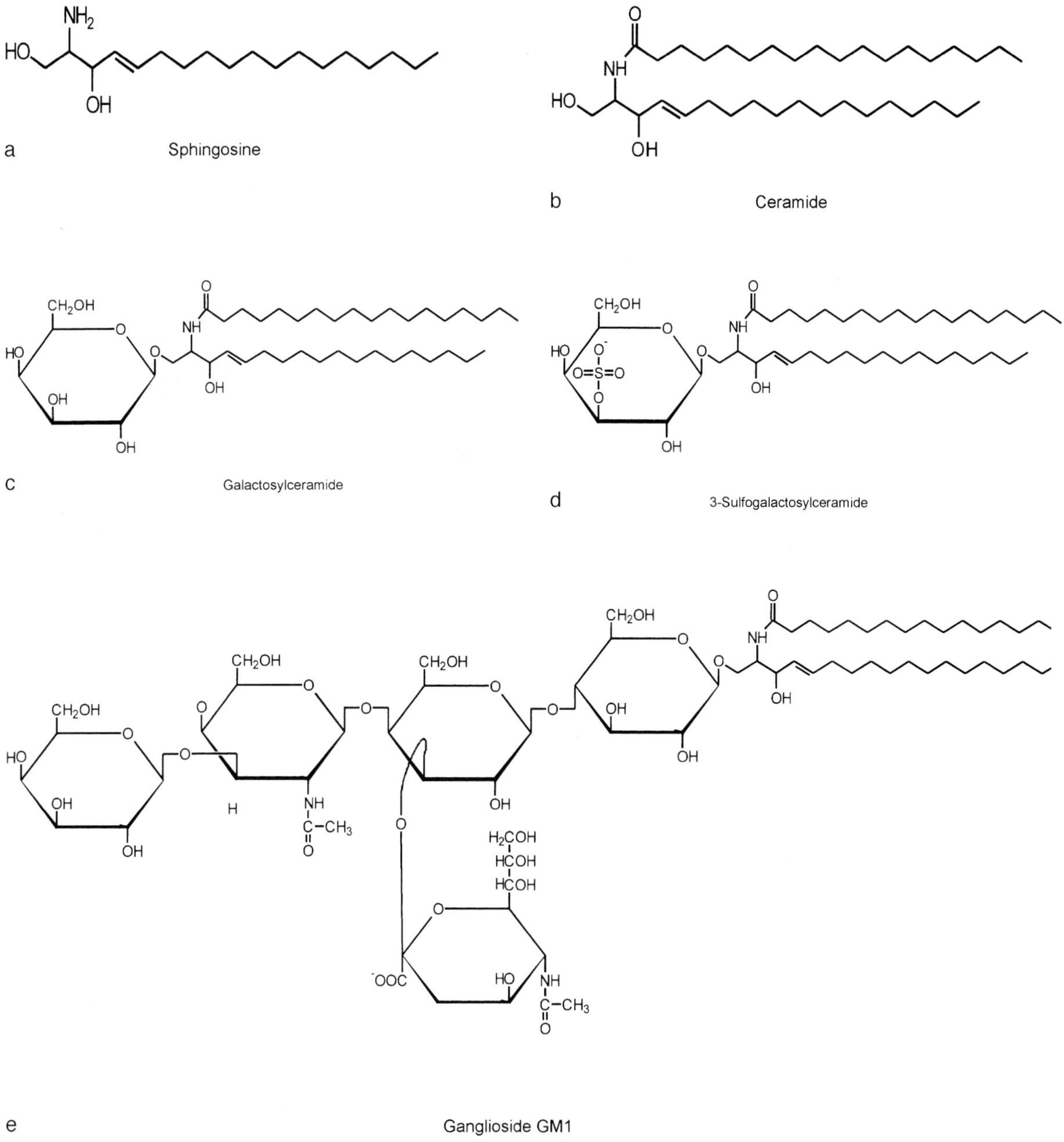

b Ceramide

c Galactosylceramide

d 3-Sulfogalactosylceramide

e Ganglioside GM1

Figure 6a–e. Components and structures of glycosphingolipids

Table 2. Examples for the Svennerholm designation of gangliosides

Svennerholm designation	Abbreviated structure
GM3	II^3NeuAc-LacCer
GM2	II^3NeuAc-$GgOse_3$Cer
GM1	II^3NeuAc-$GgOse_4$Cer
GD3	$II^3(NeuAc)_2$-LacCer
GD2	$II^3(NeuAc)_2$-$GgOse_3$Cer
GD1a	IV^3NeuAc,II^3NeuAc-$GgOse_4$Cer
GD1b	$II^2(NeuAc)_2$-$GgOse_4$Cer
GD1c	$II^3(NeuAc)_2$-$GgOse_4$Cer
GT1a	$IV^3(NeuAc)_2$,II^3NeuAc-$GgOse_4$Cer
GT1b	IV^3NeuAc,$II^3(NeuAc)_2$-$GgOse_4$Cer
GT1c	$II^3(NeuAc)_3$-$GgOse_4$Cer
GQ1b	$IV^3(NeuAc)_2$,$II^3(NeuAc)_2$-$GgOse_4$

mes and the nuclear membrane (Yu and Saito, 1989; Zeller and Marchase, 1992; Wu *et al.*, 1995). Again, there is great heterogenity in the oligosaccharide structure of gangliosides. Thus, more than 100 different gangliosides have been isolated from various sources (Yu and Saito, 1989; Miller-Podraza *et al.*, 1992), about half of them in the nervous system. Quantitatively, gangliosides are only minor or trace components of the plasma membrane in most cell types, with the exception of neural cells, where they are found in high concentrations, thereby constituting an important part of the glycocalyx network and being crucial in determining the properties and functions of these cells.

Glycosphingolipids in invertebrates, particulary insects, were found to be quite similar to the structures in vertebrates; however, two principal differences were observed. Mannose replaced galactose in the parent oligosaccharide chain, and glucuronic acid took the place of sialic acids, which insects are unable to synthesize. Such glucuronic acid-containing insect glycosphingolipids have been given the generic name arthrosides, with implied synonymy to the gangliosides (Wiegandt, 1992; Dennis and Wiegandt, 1993).

2 Elucidation of Glycolipid Structure

Basics of structural analysis of glycoconjugates are covered by other sections of this book. Therefore only a short outline, with special reference to glycolipids, will be given here. Elucidation of the oligosaccharide structure of glycolipids requires, first of all, analysis of sugar composition and sequence, followed by determination of glycosidic substitution sites and stereochemistry of glycosidic bonds. Analysis of composition of oligosaccharide chains is commonly based on gas-liquid chromatographic assay of individual sugars in the form of methyl glycosides or alditol derivatives. Classical procedures for determination of carbohydrate sequence applied the sequential hydrolysis of sugars from the glycolipid, and analysis of the liberated sugars or of the unhydrolysed remnant. Enzymatic analysis, applying specific glycosidases for sequential reaction, greatly improved sequencing procedures for oligosaccharide chains of glycolipids. Mass spectrometry is often given preference over other methods, its principal advantage being the small quantities required to establish oligosaccharide sequence in glycolipids. Standard methods for determining glycosidic substitution pattern of oligosaccharide chains in glycolipids have been classical chemical methods such as periodate oxidation and permethylation in combination with gas-liquid chromatographic analysis of the resulting derivatives. Nuclear magnetic resonance (NMR) spectroscopy is also finding increasing use for elucidation of glycosidic substitution, the appealing feature of NMR methods being the lack of destruction of the sample. The stereochemical specificity of glycosylexohydrolases has been widely used for determining the anomeric conformation of glycolipids. In addition, NMR spectroscopy has proven to be a versatile tool to characterize the conformation of glycolipid molecules.

Citation of a few examples of the manifold uses of these methods in structural analysis of these methods will highlight their great importance in glycolipid research (Clausen *et al.*, 1986; Yu *et al.*, 1986; Blaszczyk *et al.*, 1987; Levery and Hakomori, 1987; Dabrowski *et al.*, 1988; Kushi *et al.*, 1988; Riviere *et al.*, 1988; Ando *et al.*, 1989; Daffe and Servin, 1989; Jardine *et al.*, 1989; Levery *et al.*, 1989, 1992; Smith *et al.*, 1989a; Suzuki *et al.*, 1989; Poppe *et al.*, 1990; Scarsdale *et al.*, 1990; Stoll *et al.*, 1990; Costello *et al.*, 1994; Hartmann *et al.*, 1994).

3 Physicochemical Properties and Organization of Glycolipids

Besides the determination of glycolipid structures and glycolipid content of different cell types, measurement of the physical behavior of glycolipids in model membranes was used as an approach to understanding the function of glycolipids (Curatolo, 1987a). The physical properties of glycolipids were studied using various structural and spectroscopic means of characterization, including differential scanning calorimetry, X-ray diffraction, NMR spectroscopy, and electron microscopy. Due to the enormous structural diversity, only rough generalizations of their physical behavior are possible. Since the major difference between glycolipids and phospholipids lies in the extensive hydrogen bonding capacity of the glycolipids, their phase behavior is dominated by this hydrogen bonding capability. Two major consequences arise from hydrogen bonding: the acyl chain order-disorder transition temperature is generally quite high, and many glycolipids exhibit metastable polymorphism in the gel state, related to hydration and dehydration of the glycosyl headgroups. The high transition temperature provides structural integrity to membranes containing glycolipids and causes glycolipids to form clusters (microdomains) in the membrane plane. The ability of glycolipids to affect gross membrane morphology, in particular to form membranes with non-spherical morphology, might prove to be a prerequisite for glycolipid function (Thompson and Tillack, 1985; Curatolo, 1987b; Smith *et al.*, 1989b; Koynova and Caffrey, 1994). Negatively charged glycolipids, such as gangliosides, might carry out unique functions in biological membranes due to their unusual properties, such as the formation of a micellar phase (Sonnino *et al.*, 1994).

Detailed information on phase behavior of sphingolipids and glycoglycerolipids can be found in recent reviews from Koynova and Caffrey (1994, 1995). They summarized and analyzed data obtained from LIPIDAT, a computerized database providing access to the wealth of information scattered throughout the literature concerning the polymorphic and mesomorphic phase behavior of synthetic and biologically derived polar lipids (Caffrey, 1993).

4 Metabolism and Intracellular Trafficking of Glycolipids

Initial steps of sphingolipid biosynthesis are catalyzed by membrane-bound enzymes at the cytosolic face of the endoplasmic reticulum, and involve condensation of serine and palmitoyl-CoA producing sphiganine. *N*-acetylation and introduction of a 4-trans double bond yields ceramide. Since glycosylation of ceramide takes place in the Golgi apparatus, transport from ER to Golgi is required, but there is still a controversy whether the means of transportation is vesicular membrane flow or nonvesicular transport. Glycosphingolipid biosynthesis is completed by membrane-bound glycosyltransferases, which are restricted to the luminal face of the Golgi apparatus, and catalyse the sequential addition of monosaccharides or sialic acids from nucleotide sugar donors (van Echten and Sandhoff, 1993) (Fig. 7). Variety in the oligosaccharide composition of glycolipids is achieved primarily by the specificity of the enzymes for their donor and acceptor substrates (Paulson and Colley, 1989). Exact topology of these glycosylation reactions is still far from understood. Elongation of the oligosaccharide chains seems to take place in successive compartments in a sequence which has analogies with the well known mechanism for processing of proteins in the ER. However, in contrast to protein glycosylation, glycolipids from all stages of the biosynthetic pathway can escape from further glycosylation, providing possibilities for greater variability of the glycosylation process and consequently of their appearance on the cell surface (Zeller and Marchase, 1992).

3-Sulfogalactosylceramide, the major sulfatide, originates from the transfer of sulfate from 3'-

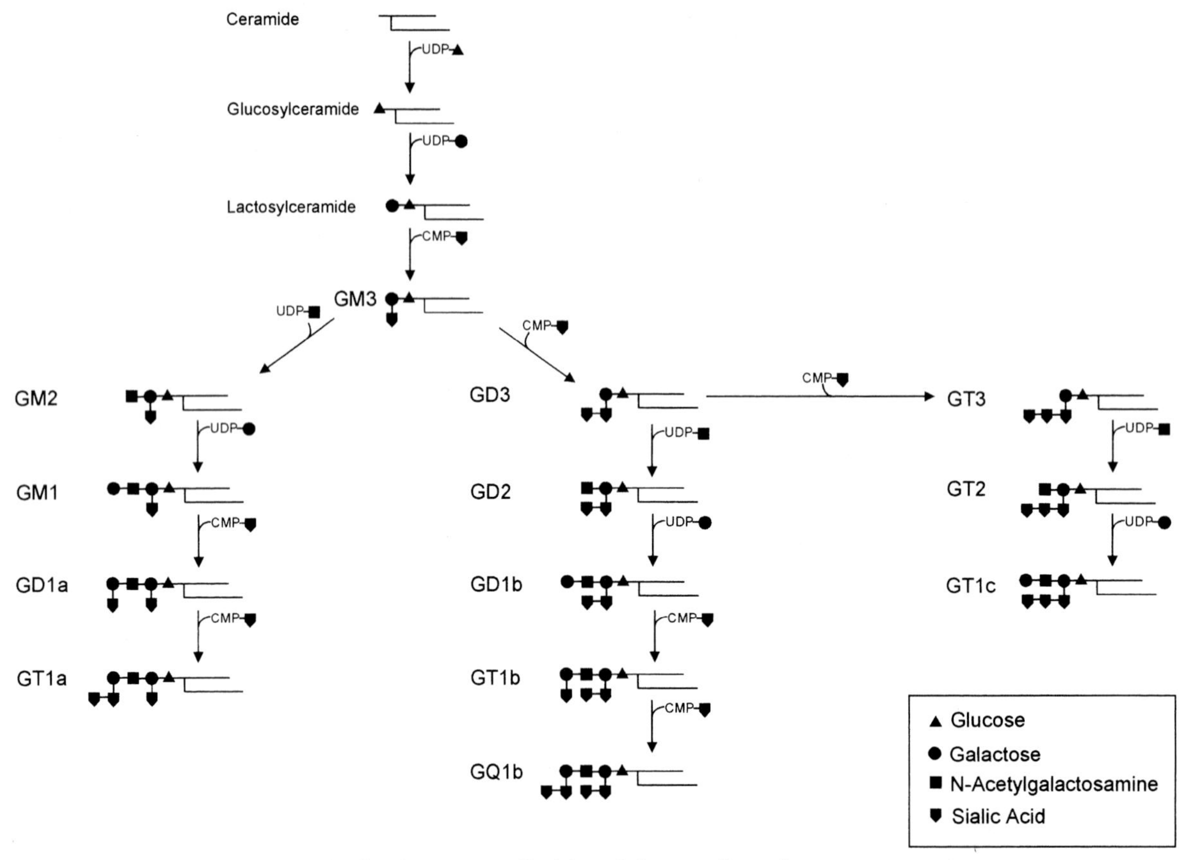

Figure 7. Pathways for biosynthesis of some gangliosides of the ganglio series

phosphoadenosine 5'-phosphosulfate (PAPS) to galactocerebroside, catalyzed by the enzyme cerebroside sulfotransferase, residing at the luminal side of Golgi stacks (Dorne *et al.*, 1990).

Mechanisms for the regulation of of glycolipid biosynthesis pathways are still poorly understood. Control of the expression of relevant glycosyltransferases seems to be a major principle to control glycosylation patterns. In addition, activity of certain glycosyltransferases can be controlled by post-translational modifications, such as cAMP-dependent phosphorylation / dephosphorylation (van Echten and Sandhoff, 1993; Yu, 1994; Gu *et al.*, 1995). Transfer of glycosphingolipids from their site of synthesis to the cell surface requires approximately 20 min, and it appears that it is accomplished exclusively by vesicular transport (Allan and Kallen, 1993).

Glycolipid degradation proceeds in stepwise fashion with liberation of individual sugars. Most of the enzymes of the catabolic pathway reside in the lysosomal compartment, into which glycolipids are transported together with other components and fragments of the plasma membrane, mainly by endocytotic membrane flow. Ganglioside desialylation may also occur in extralysosomal compartments, particularly in the plasma membrane and in cytosol (Miyagi and Tsuiki, 1985; Riboni *et al.*, 1991; Kopitz *et al.*, 1994, 1996). Lysosomal enzymes involved in glycoconjugate degradation act as exoglycosidases with high specificity for the sugars recognized and the anomeric linkage to be split, the nature of the aglycon usually being less important. Degradation of complex carbohydrates thus requires the sequential action of various glycosidases and, for

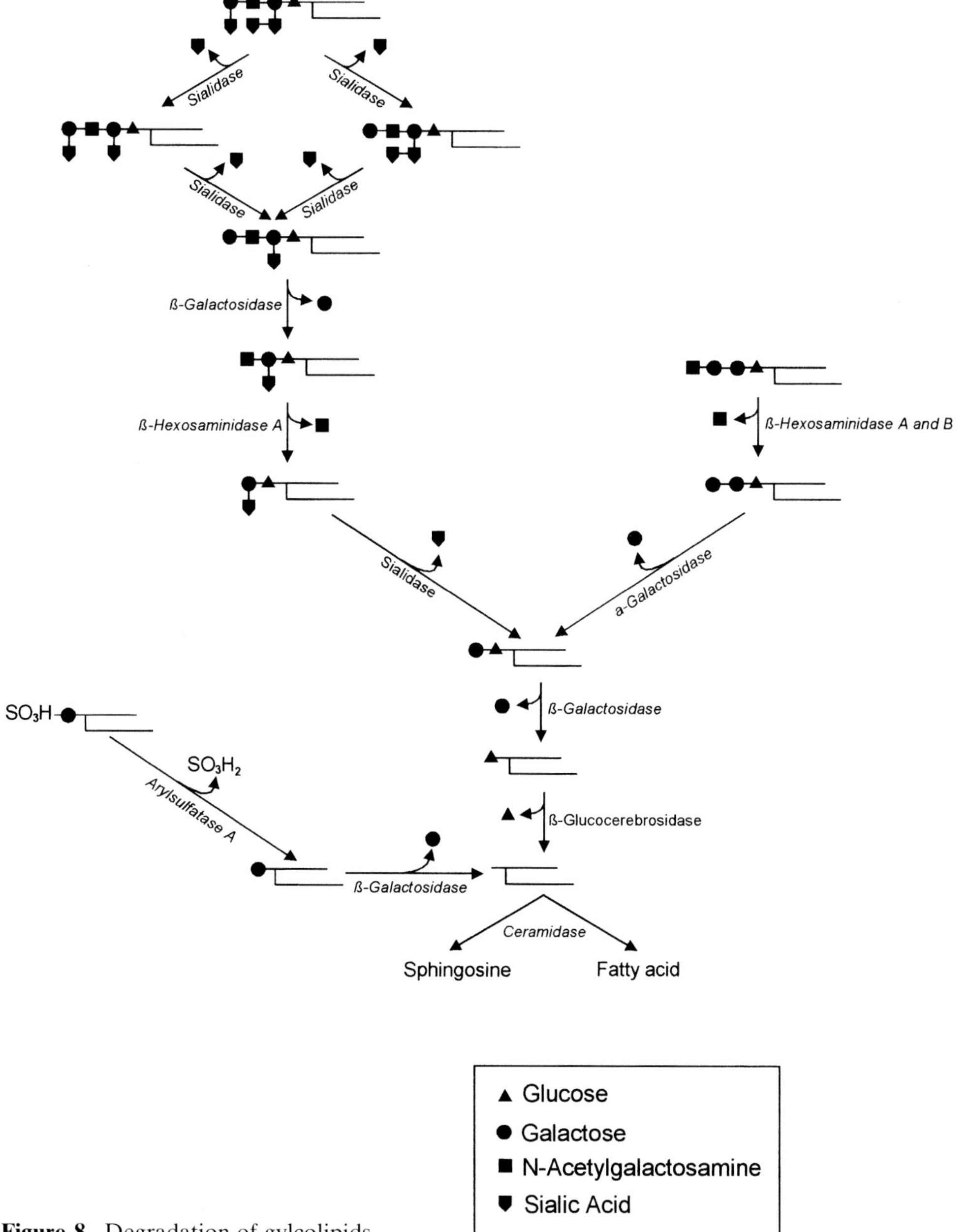

Figure 8. Degradation of gylcolipids

sulfated compounds, also sulfatases (Fig. 8) (Conzelmann and Sandhoff, 1987).

Some soluble lysosomal hydolases involved in glycolipid degradation cannot attack membrane-bound lipid substrates directly; instead, glycolipids have to be solubilized by specific nonenzymatic 'activator proteins'. In particular, the lysosomal degradation of sphingolipids with short hydrophilic headgroups depends on non-enzymatic sphingolipid activator proteins (named saposins or SAPs). Thus, hydrolysis of certain glycolipids, such as ganglioside GM2 by hexosaminidase A, requires complexing of the lipid with an GM2 activator protein and specific interaction of the complex with the glycosidase. The activator proteins sap-A, sap-B, sap-C and sap-D are small heat-stable glycoproteins derived from a common precursor, prosaposin or SAP precursor. *In vitro*, they activate the hydrolysis of a variety of glycolipids by different enzymes, but their *in vivo* function is still poorly understood. Sap-B also plays a crucial role in the desulfation of sulfatides by arylsulfatase A (Kishimoto *et al.*, 1992; Sandhoff *et al.*, 1995).

5 Physiological Functions of Glycolipids

5.1 Bacterial and Plant Glycolipids

The most probable function for glycolipids in bacterial membranes is based on their capability to undergo interlipid hydrogen bonding via the glycosyl headgroups. Especially in species without cell walls, such as mycoplasma, the plasma membrane must provide resistance to osmotic stress, and thus needs extra stabilization (Curatolo, 1987b). Also archaebacteria, in contrast to eubacteria, generally do not possess peptidoglycan-based cell walls. The plasma membranes of these organisms, which are unusually stable to lysis by osmotic shock, detergents, heat and low pH, consist of highly glycosylated proteins and unique membrane lipids. A glycolipid class based on a macrocyclic tetraether lipid structure is found in all archaebacterial membranes, suggesting that these lipids have functions that are common to all archaebacteria. Thus, the unique structure of these glycolipids indicates that the archaebacterial membrane may consist primarily of a bipolar monolayer of two-headed lipids which span the entire membrane. As an example of glycolipid function in archaebacteria, such structure strongly stabilizes the membrane at high temperature. Additionally, the alkyl ether bonds, in contrast with the usual acyl ester structure, impart stability to the lipids over a wide range of pH (Kates, 1992). Besides the membrane stabilizing effects of glycolipids, they may also contribute to specific cellular functions by influencing membrane bound proteins, ion transport, proton transport and conductance, energy transduction, and permeability to nutrients (McElhaney, 1982). Perhaps detailed understanding of these highly unusual glycolipids will help to reveal the function of more familiar lipids in eukaryotes.

Investigations of glycolipid functions in plants focused on chloroplasts due to the unique photosynthetic properties of these organelles and to the strikingly unusual lipid composition of the thylakoid membrane. The dominant species is the uncharged monogalactosyldiacylglycerol with 57 % by weight, followed by digalactosyldiacylglycerol which represents 27 % of the total lipid (Dorne *et al.*, 1990). As the thylakoids are the predominant membranes in plant leaf cells, representing 60–80 % of the total cellular membranes in higher plant mesophyll cells, mono- and diacylgalactosylglycerol are probably the most abundant glycolipids in nature. Comparison of physical properties of the major thylakoid lipids with phospholipids, the predominant lipid species in animals, bacteria and non-chloroplastic compartments in plants, reveals two striking differences: (1) lack of charge on more than 80 % of chloroplast lipids, and (2) monogalactosyldiacylglycerol does not form bilayers spontanously when dispersed in purified form, but rather adopts hexagonal II configurations (inverted micelles or tubes). Therefore, structural roles, such as the formation of the highly curved edges of the thylakoid sacs, are likely. Also, specific physiological functions have been suggested: maintainance of ion permeability and electrical properties of the thylakoid membrane, and insertion and stabilization of the large photosystem protein complexes in the thylakoid membranes. Also, for the plant sulfolipid sulfoquinovosyldiacylglycerol, a vital role in photosynthesis, particulary in electron transport, was hypothesized (Sprague, 1987; Webb and Green, 1991; Nussberger *et al.*, 1993; Fuks and Homble, 1994).

Although we are beginning to understand the importance of specific glycolipids in thylakoid membranes, the complexity of structure and function of this system remains a challenging field for investigations of glycolipid function. Future work will have to find plausible physical explanations why galactolipids are the dominant components of the thylakoids, and particularly why an uncharged and non-bilayer lipid should make up such a large fraction of the total lipid. Another major task will be elucidation of specific lipid-protein interactions.

5.2 Animal Glycolipids

The physicochemical properties and the localization of glycosphingolipids in the outer half of the lipid bilayer suggested that they contribute to the structural rigidity of the surface leaflet (Curatolo, 1987a). Until the 1970s there were only few hints as to what the more specific functions of glycosphingolipids may be, but during the past two decades their involvement in important cellular functions has been established, making glycolipid research an attractive and important subject of glycochemistry and cell biology (Zeller and Marchase, 1992; Hakomori and Igarashi, 1993; Nagai and Tsuji, 1994; Vos *et al.*, 1994).

5.2.1 Cell Adhesion, Cell-Cell Interaction and Recognition

Glycosphingolipids, particulary gangliosides and sulfolipids bearing a charged oligosaccharide chain, possess a high binding potential, and consequently a variety of proteins interacting with the glycolipid's oligosaccharide chain have been described. This is usually screened by the adherence of radiolabeled proteins to plastic wells or silica gel, coated with the glycolipid. In many cases the results of such binding experiments cannot be extrapolated to the situation *in vivo* because the lipid is exposed to the protein quite differently from when it is in its natural position in the plasma membrane. But further experiments using intact cells demonstrated, for various extracellular adhesive matrix proteins (such as laminin) and adhesion receptors (such as P- and L-selectin), a high attachment-promoting affinity towards certain glycolipids, indicating their involvement in cell adhesion (Evans, 1992; Hughes, 1992; Levy *et al.*, 1994; Vos *et al.*, 1994). In neuronal cells (and many others), attachment to the extracellular matrix components strongly supports differentiation, emphasising the importance of such interactions (Matsushima and Bogenmann, 1992; Adams and Watt, 1993; Taira *et al.*, 1993). For oligodendrocytes there is increasing evidence that sulfomonogalactosylceramide (and its precursor monogalactosylceramide) mediate recognition or adhesion to neurons or extracellular matrix, thereby playing a role in synthesis, formation, wrapping and maintenance of myelin membranes (Vos *et al.*, 1994). Gangliosides are involved in the formation of selective intraneuronal connections, as shown, for example, by work using explants of fetal mouse spinal cord-dorsal root ganglia, where gangliosides induced the formation of selective sensory afferent projection patterns within the cord (Baker, 1988). Glycolipid-selectin interactions are likely to be the basis of leukocyte tethering and rolling on activated endothelium under physiologic flow conditions (Alon *et al.*, 1995). Sulfogalactolipids on mammalian sperm cells interact with oocyte plasma membrane components during binding and the fusion process (covered by Sinowatz *et al.*, this volume).

A modulatory influence of gangliosides on cell adhesion was initially discovered upon analysis of the effects of gangliosides on binding of cells to fibronectin. A ganglioside-binding domain was detected on fibronectin, and binding of ganglioside appeared to act synergistically with binding to the fibronectin receptor, as shown for fibroblasts, which could adhere to fibronectin in the absence of gangliosides, although organization of fibronectin into fibrillar arrays required the presence of gangliosides. Recent studies in other cell lines supported the idea of a possible modulation of integrin functions by gangliosides (Zheng *et al.*, 1993; Merzak *et al.*, 1995).

The observation of dramatic changes in glycosphingolipid composition during early embryogenesis, and their close correlation with cell recognition, suggested that certain glycolipids act as receptor sites in cell-cell recognition. Classic examples of glycolipid-bound oligosaccharides involved in embryonic development are among the so called stage-specific embryonic antigens (SSEAs), whose expression is both quantitatively and qualitatively regulated during development. SSEA-3 (Gal $\beta1\rightarrow3$ GalNAc $\beta1\rightarrow3$ Gal $\beta1\rightarrow4$ Gal $\beta1\rightarrow4$ Glc $\beta1\rightarrow1$ Cer) and SSEA-4 (NeuAc α2→3 Gal β1→3 GalNAc β1→3 Gal β1→4 Gal β1→4 Glc β1→1Cer) are synthesized during oogenesis and are present in oocytes, zygotes and blastulas until they disappear at the eight cell stage. SSEA-1 (Gal β1→4 [Fuc α1→3] GlcNAc-R) is expressed on both glycolipids and glycoproteins, and initially appears at the eight-cell stage (Feizi, 1985; Fenderson *et al.*, 1990). Specific modifications of glycosphingolipids, like 9-*O*-acetyl ganglioside GD3 or sulfoglucuronyl glycolipids, are also functionally involved in embryogenesis, in particular in the development of the nervous system (Schlosshauer *et al.*, 1988; Jungalwala, 1994). The concept that glycolipids may mediate cellular interactions, was supported by experiments where glycosphingolipids or corresponding oligosaccharide sequences blocked cell-cell recognition events (Hakomori, 1990; Fenderson *et al.*, 1990).

Cellular interactions via cell surface glycosphingolipids require their specific binding to complementary receptors on apposing membranes, but at present there are few hints on the identity of these receptors (Schnaar, 1991; Yu and Saito, 1992; Hattori *et al.*, 1995). Interestingly, besides carbohydrate-protein interactions, the possibility of binding of a cell surface carbohydrate to a complementary carbohydrate structure at the surface of another cell was demonstrated in studies of glycosphingolipid-to-glycosphingolipid interaction (Fenderson *et al.*, 1989). This kind of interaction would provide the basis for a specific cell recognition system independent of the fibronectin / integrin or surface lectin systems, occurring earlier during a cell recognition event (Hakomori, 1990; Bovin, this volume).

A promising strategy for detection and characterization of novel receptors for glycosphingolipids is the use of biotinylated or radioiodinated synthetic glycosphingolipid-based ligands (neoganglioproteins, consisting of a biotinylated or radioiodinated carrier protein covalently derivatized with gangliosides) to characterize the structural and tissue specificity of binding activities, and to purify the responsible receptors (Gabius *et al.*, 1990; Schnaar *et al.*, 1994). This methodology has already been successfully applied in the detection of a membrane receptor for gangliosides associated with central nervous system myelin (Tiemeyer *et al.*, 1990).

5.2.2 Signal Transduction

The striking alterations in glycolipid patterns observed during the cell cycle, density-dependent growth inhibition, differentiation, and oncogenic transformation indicate that sphingolipid metabolism may play a role in the regulation of cell growth and differentiation. Consistent with this concept are observations that addition of sphingolipids to the culture medium influenced cell growth and differentiation of fibroblasts, lymphocytes, neural cells and others. Exogenously added glycosphingolipids are readily incorporated into the plasma membrane, where they affect activities of growth factor-associated protein kinases and ion flux (Hakomori, 1990; Tettamanti and Riboni, 1993; Bremer, 1994; Taga *et al.*, 1995). Thus, glycolipids appear to be capable of modulating transmembrane signal transduction. One of the best studied examples of such a mechanism is cell growth modulation through ganglioside-mediated modification of EGF receptor activity. Ganglioside GM3 can regulate the intrinsic tyrosine kinase activity of the EGF receptor. This modulation is not associated with alterations of hormone binding to the receptor. GM3 blocks tyrosyl kinase activity by direct binding to the receptor, thereby interrupting the transmembrane signalling process and causing growth inhibition (Song *et al.*, 1993; Zhou *et al.*, 1994).

The neuritogenic and neuronotrophic effects of gangliosides *in vitro* and *in vivo* are well documented, but their mechanism of action is still puzzling (Ledeen, 1989; Tettamanti and Riboni, 1993). Ganglioside (particularly GM1) effects were found to be similar to those of well-known trophic substances, such as nerve growth factor. Since it was subsequently demonstrated that GM1 treatment potentiates NGF-induced trophic effects in cell culture and animal models, a synergistic interaction between GM1 and NGF was hypothesized. Recent studies suggested that GM1 may share some part of the NGF signal transduction pathway, but comprehensive analysis of the neurochemical correlates is still lacking (Cuello *et al.*, 1989; Garofalo *et al.*, 1993; Garofalo and Cuello, 1995). In a neuroblastoma cell line GM1 was directly and tightly associated with

Trk, the high-affinity tyrosine kinase type receptor for NGF, and addition of GM1 to the culture medium strongly enhanced neurite outgrowth and neurofilament expression in cells, elicited by a low dose of NGF that alone was insufficient to induce neuronal differentiation. In these cells GM1 seems to operate as an endogenous activator of NGF receptor function, since there was also a >3-fold increase in NGF-induced autophosphorylation of Trk as compared with NGF alone, and *in vitro* studies demonstrated that GM1 could directly enhance NGF-activated autophosphorylation of immunoprecipitated Trk (Mutoh *et al.*, 1995). The ganglioside GQ1b exhibited NGF-like activities itself in two human neuroblastoma cell lines, and evidence for a glycoreceptor specifically recognizing the carbohydrate structure of GQ1b was presented. Further studies suggested that the neuritogenic activity of GQ1b is induced, by the action of the GQ1b sugar-specific glycoreceptor of the cell surface membrane and a unique cell surface-localised protein kinase (ectoprotein kinase), to phosphorylate particular cell surface proteins (Nagai, 1995).

During analysis of the responses of neuronal cells to treatment with gangliosides, stimulation of Ca^{2+} flux followed by modulation of Ca^{2+}-dependent enzyme activities, such as protein kinases, was observed. GM1-induced neuritogenesis was found to be diminished in neuroblastoma cells in the absence of extracellular calcium (Spoerri *et al.*, 1990; Wu *et al.*, 1990; Guerold *et al.*, 1992; Wu and Ledeen, 1994). Such effects of gangliosides are likely to be mediated by their interaction with ion channels, such as L-type voltage-dependent calcium channels (Carlson *et al.*, 1994). Ganglioside-modulated calcium influx from extracellular sources and its influence on differentiation and proliferation has also been reported for many other cell types, including neurons, astrocytes, lymphocytes, thymocytes and fibroblasts (Buckley *et al.*, 1995). Considering the central role of calcium in regulation of a myriad of critical physiological processes, it can be expected that our current knowledge of the physiological significance of ganglioside / calcium flux interaction is only the tip of the iceberg.

Another mechanism by which sphingolipids may act in signal transduction emerged when sphingosine, a metabolite of membrane sphingolipids, was found to inhibit potently and specifically protein kinase C activity, and that exogenous sphingosine inhibited various protein kinase C-dependent processes *in vivo* (Olivera and Spiegel, 1992). Subsequently, many other sphingolipid-derived compounds (sphingosine-1-phosphate, *N*-monoethylsphingosine, *N,N*-dimethylsphingosine, ceramide, ceramide-1-phosphate, lyso-gangliosides and de-*N*-acetyl-gangliosides) were shown to exert powerful regulatory effects on enzymes, which are fundamental in the control of cellular metabolism (Merrill, 1991; Zhang *et al.*, 1991; Hakomori and Igarashi, 1993; Olivera and Spiegel, 1993; Tettamanti and Riboni, 1993). Free sphingosine, ceramide and its derivatives originate from the breakdown of sphingomyelin and glycosphingolipid (Riboni *et al.*, 1992, 1994), indicating that controlled glycolipid degradation might contribute to the generation of lipid second messengers.

5.2.3 Synaptic transmission

Neuronal activities, especially electroresponsiveness, synaptic transmission of information, and storage of information, are considered to be dependent on a controlled exchange of calcium between the extracellular space (mM Ca^{2+}) and the neuroplasm (μM Ca^{2+}). Rahmann combined detailed physicochemical data on ganglioside/calcium interactions with physiological, biochemical and electron microscopic investigations of the effects of gangliosides and calcium on neuronal cells. On the basis of these experiments he presented a comprehensive hypothesis in which calcium/ganglioside interactions play a crucial role in neuronal transmission and storage of information (Rahmann *et al.*, 1982, 1992; Thomas and Brewer, 1990; Rahmann, 1995).

5.2.4 Antigenic Determinants and Modifiers of Immune Responses

Glycolipids are implicated in a variety of immunological phenomena, in particular the expression of glycolipid antigens on cell surfaces and an involvement in immunoregulation. Carbohydrate structures of the cell surface are important determinants in self-nonself recognition. These antigenic epitopes are found in glycoproteins or glycosaminoglycans, or they exist as oligosaccharide chains of glycosphingolipids. Blood group antigens (ABO, Lewis, Hh and Ii) are probably the best characterized glycolipid-bound antigenic determinants (Table 3) due to their importance for transfusion medicine, organ transplantation, and infectious and autoimmune diseases. About 600 discrete antigens have been defined, and to appreciate their significance one must realize that the name blood-group antigen is misleading; these antigens are found not only in blood, but also in many kinds of tissue. In particular, they are present in high concentrations on the surface of all kinds of epithelial cells, which constitute the mucous linings of many organs. Intriguingly, it has been reported that many of these glycolipid structures are altered in content or display in tumor cells, or not found in normal cells, making these antigens valuable for tumor detection and treatment (Hakomori, 1984a, 1986; Clausen and Hakomori, 1989; Green, 1989; Marsh, 1990; Holgersson *et al.*, 1992).

Besides their action as antigenic determinants, glycosphingolipids are involved in mechanisms leading to proliferation, differentiation and activation of lymphocytes. There is increasing evidence that almost the full repertoire of actions already described for glycolipids is used, providing a suitable model to sum up their importance for cellular functions in animals:

1. Glycolipid patterns change dramatically during hematopoiesis and gangliosides act as dif-

Table 3. Structures of some blood group A antigens

```
GalNAcα1→3Galβ1→4GlcNAcβ1→3Galβ1→4Glc→Cer                                         Aª
             2
             ↑
         Fucα1

GalNAcα1→3Galβ1→4GlcNAcβ1→3Galβ1→4GlcNAcβ1→3Galβ1→4Glc→Cer                        Aᵇ
             2
             ↑
         Fucα1

         Fucα1
             ↓
             2
GalNAcα1→3Galβ1→4GlcNAcβ1→3Galβ1→4GlcNAcβ1→3Galβ1→4Glc→Cer                        Aᶜ
                               6
                               ↑
         Fucα1→2Galβ1→4GlcNAcβ1
                 3
                 ↑
          GalNAcα1

                  Fucα1
                      ↓
                      2
         GalNAcα1→3Galβ1→4GlcNAcβ1→6Galβ1→4GlcNAcβ1→3Galβ1→4Glc→Cer              Aᵈ
                                        3
                                        ↑
Fucα1→2Galβ1→4GlcNAcβ1→3Galβ1→4GlcNAcβ1
        3
        ↑
 GalNAcα1
```

ferentiation inducers for hematopoietic cells (Pique *et al.*, 1991; Nohara *et al.*, 1993, 1994; Saito, 1993; Taga *et al.*, 1995).

2. Glycolipid-based leucocyte differentiation antigens mediate cellular interactions via cell adhesion molecules (Tiemeyer *et al.*, 1991; Foxall *et al.*, 1992, Mangeney *et al.*, 1993).
3. Glycolipids interact with the T-cell receptor, thereby influencing T-cell activation (Morrison *et al.*, 1990, 1993; Yuasa *et al.*, 1990).
4. Gangliosides can influence Ca^{2+}-influx from extracellular sources (Gouy *et al.*, 1994; Kato *et al.*, 1994).
5. Sphingolipid breakdown products act in second messenger mechanisms (Ballou, 1992).

5.2.5 Perspectives

Evidence for important functions of glycolipids comes from various experimental systems. Although in many cases our knowledge of the physiological significance of the observed phenomena is still fragmentary, it is clear that glycolipids are much more than just structural components of the plasma membrane. They can modulate cellular interactions and transmembrane signaling, thereby probably influencing basic cellular functions. It can be expected that further understanding of glycolipid functions will make valuable contributions to glycobiology and related fields such as neurobiology, immunology and developmental biology.

6 Functions in Disease Mechanisms

Accumulating evidence for the functions of glycolipids in the cellular processes described above also points to their possible involvement in the pathogenesis of diseases where an impairment of these functions is observed. Consequently, results of basic research on glycolipid functions inspired experiments proving the involvement of glycolipids in disease mechanisms. On the other hand, comparison of the glycolipid composition of plasma membranes in diseased cells or tissues with their normal counterparts not only provides valuable information on pathogenic mechanisms, but also promotes understanding of the normal functions of glycolipids. Therefore, the following examples of their role in various diseases should also be considered as further indication of their physiological importance, and as hints of their normal cellular activities.

6.1 Oncogenic Transformation

Oncologists became aware of glycolipids by the basic observation that alterations in carbohydrate structures of sphingolipids are associated with the cancerous state. Pioneering experiments demonstrated that tumor cells, whether transformed by viruses or by chemicals, produced different profiles and structures of cell surface glycolipids from those of non-transformed progenitor cells. Many of the tumor-associated glycosphingolipids are of the lacto and neolacto series, and these epitopes are also found on glycoproteins. Ganglioside antigens (e.g. GD2 or GD3) of the ganglio series are preferentially detected in tumors of neuroectodermal origin, have binding epitopes that are not found in glycoproteins. Since essentially all tumor cells display altered glycosphingolipid patterns, these compounds are not only suitable targets in the diagnosis and treatment of human tumors, but they might also contribute to the biological evolution of tumors themselves (Brady and Fishman, 1974; Hakomori, 1984b; Ladisch, 1987; Campanella, 1992; Nilsson, 1992; Fredman, 1993, 1994). Considering the involvement of glycolipids in cell to cell or cell to matrix interactions, it is likely that aberrant glycolipids may be instrumental in failure of functional cell contact or cell communication, thereby causing invasive and infiltrative properties of tumor cells. Additionally, changes in growth factor receptor associated glycolipids might promote tumor growth. Although strong evidence for such events remains to be furnished, further understanding of the roles of glycolipids in cellular functions are likely to provide one point of departure for detection of molecular mechanisms causing development, growth and metastasis of tumors.

6.2 Neurodegeneration

Immunochemical and biochemical studies demonstrated that changes in glycolipids of neuronal membranes are related to degenerative changes characteristic of Alzheimer's disease. Diminution of total brain gangliosides is a primary event in brains from early onset Alzheimer cases (Svennerholm, 1994b). Additionally, shifts in the ganglioside pattern are observed; e.g. in the temporal and frontal cortex of Alzheimer brains the more complex gangliosides (GM1, GD1a, GD1b, and GT1b) are decreased, whereas simple gangliosides (GM2, GM3) are relatively elevated (Kracun *et al.*, 1991, 1992a). Furthermore, direct association of gangliosides with sites of Alzheimer degeneration (senile plaques and neurofibrillary tangles) were detected with ganglioside-specific monoclonal antibodies, and a contribution of these gangliosides to the formation of senile plaques or neurofibrillary tangles was suggested (Emory *et al.*, 1987; Kingsley *et al.*, 1988; Majocha *et al.*, 1989; Sparkman *et al.*, 1991; Nishinaka *et al.*, 1993).

Whether gangliosides are involved in other neurodegenerative diseases is unclear, but gangliosides display neuroprotective and neuroregenerative effects in diseases like Parkinson's disease or ischemic stroke (Nobile *et al.*, 1994). For the normal aging process of brains in individuals without any sign of neurological or psychiatric disorders there is also no direct evidence for an involvement of glycolipids. But the normal aging process is accompanied by characteristic region-specific alterations in quantitity and quality of brain gangliosides (Ohsawa and Shumiya, 1991; Svennerholm *et al.*, 1991; Kracun *et al.*, 1992b; Walsh and Opello, 1992). Ganglioside GM1 administration enhanced cholinergic parameters in the brains of senescent rats (Hadjiconstantinou *et al.*, 1992).

6.3 Disorders of Glycosphingolipid Catabolism

Disorders of glycosphingolipid catabolism (sphingolipidoses) result from a genetic deficiency of a single lysosomal enzyme, which causes accumulation of undegraded substrate. The biochemical study of these genetic diseases provided the basis for detailed understanding of catabolic pathways for lysosomal glycolipid degradation (Neufeld, 1991; Gieselmann, 1995). A summary of enzymes involved in glycosphingolipid degradation and the disorder resulting from deficiency of the corresponding enzyme is given in Table 4. The majority of these disorders become manifest in infants and children, presenting severe neurological and somatic symptoms. The discovery of disorders caused by the deficiency of sphingolipid activator proteins provided direct evidence of their important role in glycolipid degradation and will promote further understanding of their specificity and mechanism of action (Sandhoff *et al.*, 1995).

Although some of these disorders are well known also as adult forms, some other sphingolipidoses were previously thought to be confined to children, and it is only recently that adult forms have been recognized. These late-onset forms show fewer somatic signs, but present a psychiatric disease. Symptoms range from behavioral problems to frank psychosis. Since such psychiatric patients are still rarely tested for sphingolipidoses it is unclear how many psychiatric patients of unknown etiology suffer from impaired glycosphingolipid degradation (Rapola, 1994).

Table 4. Disorders of glycosphingolipid degradation

Enzyme Deficiency	Disorder
Arylsulfatase A	Metachromatic Leucodystrophy (Kolodony and Fluharty, 1995)
Ceramidase	Farber Disease (Moser, 1995)
β-Galactosidase	GM1 Gangliosidosis (Suzuki *et al.*, 1995a)
α-Galactosidase	Fabry Disease (Desnick *et al.*, 1995)
Galactosylceramidase	Krabbe Disease (Suzuki *et al.*, 1995b)
β-Glucocerebrosidase	Gaucher Disease (Beutler and Grabowski, 1995)
β-Hexosaminidase A	Tay-Sachs Disease (Gravel *et al.*, 1995)
β-Hexosaminidases A and B	Sandhoff Disease (Gravel *et al.*, 1995)

6.4 Attachment Sites for Bacteria or their Toxins

Attachment of bacteria to host cell surfaces is the initial event in the process of infection and is considered an important virulence factor. Often a high selectivity of bacterial infections towards their target (animal species or tissue) is observed, indicating a specific molecular recognition. Although knowledge of the underlying mechanisms is still extremely limited, the molecular basis of host-pathogen interaction is well-defined in relatively few case so far. Substances involved are mainly specific proteins on the bacterial site, whereas the dominant fraction of receptor substances of the host are oligosaccharide structures, in many cases characterized as glycosphingolipids (Bock *et al.*, 1988; Karlsson, 1989; Boren *et al.*, 1993; Lingwood, 1993; Falk *et al.*, 1994). Therefore, the distinct expression of separate glycolipids in different animals and tissue is of great relevance for the development of infections. Furthermore, detailed receptor knowledge will probably provide new ways of diagnosing, treating, and preventing important infections.

The classical and most well characterized receptor function of a glycolipid is that of ganglioside GM1 as the receptor for cholera toxin. This toxin, which mediates its effects on cells by activating adenylate cyclase, binds with high affinity and specificity to ganglioside GM1. Toxin-resistant cells which lack GM1 can be sensitized to cholera toxin by treating them with GM1. Cholera toxin specifically protects GM1 from cell surface labeling procedures and only GM1 is recovered when toxin-receptor complexes are isolated by immunoadsorption. Use of these criteria identified GM1 as the only natural, functional receptor for cholera toxin (Fishman, 1982). Although the binding determinants for cholera toxin reside in the oligosaccharide moiety, more recent studies indicated that the ceramide is also important for biological activity of the toxin. Thus, various neogangliolipid analogs of GM1 with an altered lipid moiety were tested for their ability to sensitise GM1-deficient cells to cholera toxin. These changes in the lipid moiety either enhanced or decreased the biological response to cholera toxin. Interestingly, when the GM1-oligosaccharide was attached to proteins, they only behaved as non-functional receptors (Fishman *et al.*, 1993). The importance of the lipid moiety observed here should be kept in mind when elucidating functions of oligosaccharide structures that may occur naturally on both glycolipids and glycoproteins.

6.5 Autoimmune Neuropathies

In recent years, considerable interest has developed in the role of antibodies against gangliosides and other glycosphingolipids in the pathogenesis of certain neurological disorders, including peripheral neuropathies and the Guillain-Barré syndrome. Ganglioside antibodies were found in sera from patients suffering from different forms of neuropathies that involve motor neurons or motor nerve fibres, but in most cases it is unclear whether these antibodies are a primary manifestation or a consequence of the disease (Marcus, 1990; Asbury, 1994; Marcus and Weng, 1994). Since in some cases immunosuppression had beneficial effects, an autoimmune reaction towards membrane components of neurons might be involved in the pathogenic mechanism, gangliosides being a candidate as the molecular target (Serratrice, 1993; Steck and Kappos, 1994; Griffin, 1994; Garcia *et al.*, 1995).

Guillain-Barré syndrome is the most common cause of acute neuromuscular paralysis. Sera from patients exhibiting this syndrome frequently have anti-GM1 antibody, probably acting as antineural antibodies in the autoimmune pathogenesis of the disease. 'Molecular mimicry' by infectious agents of surface glycolipids has been postulated to trigger loss of self-tolerance towards neuronal membrane components (Hughes and Rees, 1994).

7 Diagnostic and Therapeutic Functions

7.1 Cancer

Tumor-associated glycosphingolipids have attracted a great deal of interest with regard to the potential role of these antigens as a target for antibodies in the diagnosis and therapy of cancer. The advent of hybridoma technology has provided a great variety and virtually unlimited amounts of specific high affinity monoclonal antibodies to tumor-associated glycosphingolipids. In this context it has to be stressed that glycolipids designated 'tumor-associated' are not necessarily exposed exclusively on tumor cells. Rather, criteria for glycosphingolipids to be considered as tumor-associated are: (1) they are not detectable with immunological methods in their progenitor cells or in the organs where the tumor is located, or (2) they are expressed in fetal but not adult tissue, that is, they are oncofetal antigens, or (3) they are found in a significantly increased proportion or quantity in the tumor as compared to progenitor cells or surrounding tissue (Fredman, 1993).

The use of glycosphingolipids as antigens for production and characterization of monoclonal antibodies to tumor oligosaccharides is advantageous, because they can be purified to homogeneity, whereas glycoproteins show microheterogeneity in their carbohydrate moieties. The antibodies can be used in solid phase immunoassays (ELISA), immuno-thin-layer chromatography (an adaptation of the Western blotting method to lipid analysis on silica plates) or in immunohistology, to screen for tumor-associated antigens (Reaman *et al.*, 1990; Sell, 1990; Sariola *et al.*, 1991; Suresh, 1991; Leoni *et al.*, 1992; Vangsted and Zeuthen, 1993; Li *et al.*, 1994; Wikstrand *et al.*, 1994). Since some of these antigens are shed into circulating blood, their immunological detection is a useful aid in early diagnosis (Valentino *et al.*, 1990; Nakamura *et al.*, 1991; Valentino and Ladisch, 1992; Merritt *et al.*, 1994; Vangsted *et al.*, 1994). The observation of premalignant changes of glycolipids in experimental models suggested that some of them might support premalignant diagnosis (Holmes and Hakomori, 1983; Dahiya *et al.*, 1987 and 1988).

An attractive perspective is the use of antibodies towards tumor-associated antigens in immunotherapy of cancer. Binding of these antibodies to tumor cell membranes can attract and activate immune effector cells, such as T-lymphocytes, macrophages, and monocytes, thereby inducing cytotoxic attack on the tumor cell (Cheung *et al.*, 1994; Handgretinger *et al.*, 1995; Minasian *et al.*, 1995). Additionally, immunotoxins composed of monoclonal antibodies to tumor cell glycolipids have been developed for the treatment of malignancies. Experiments in animals suggest that the establishment of appropriate anti-glycolipid immunotoxins may lead to effective immunotargeting therapy of cancer (Mujoo *et al.*, 1991; Gottstein *et al.*, 1994; Usui and Hakomori, 1994).

7.2 Neurological Disorders

The biological effects of exogenously administered glycolipids in cell cultures and animals suggested that the neurotrophic and neuritogenic properties of some glycolipids might be beneficial in the treatment of neurological disorders. Therefore, glycolipids (particulary gangliosides) were tested for their effects on various neurological diseases in experimental animal models and in clinical trials.

Monogangliosides were shown to attenuate the degenerative process of the rat optic nerve after crush injury (Yoles *et al.*, 1992; Zalish *et al.*, 1993). Chronic treatment with GM1 was effective in advancing functional recovery in rats following spinal cord transection (Bose *et al.*, 1986). Evidence for promotion by GM1 of recovery after experimental brain injury in animals was also presented (Feeney and Sutton, 1987; Karpiak *et al.*, 1987; Antal *et al.*, 1992). Preliminary reports from clinical trials suggest that treatment with gangliosides in humans might be also effective in spinal cord injury (Papadopoulos, 1992; Geisler *et al.*, 1993; Walker and Harris, 1993).

Glutamate, an excitatory amino acid in the CNS, plays an important role in neuron to neuron signaling by binding to specific receptors. Overactivation of these receptors by excess glutamate as observed during brain hypoxia-ischemia, causes neuronal cell death (excitotoxic cell damage) (Choi, 1988). Antagonism of glutamate neurotoxity by ganglioside GM1 or its synthetic derivatives might be of therapeutic benefit, since in cell cultures and animals these compounds were incorporated into neuronal membranes, and reduced excitotoxic cell damage (Favaron *et al.*, 1988; Cahn *et al.*, 1989; Lipartiti *et al.*, 1991, 1992; Costa *et al.*, 1993; Saulino and Schengrund, 1993). Clinical trials suggest that GM1 and derivatives can improve recovery in patients who experienced ischemic stroke (Argentino *et al.*, 1989; Lenzi *et al.*, 1994; Svennerholm, 1994c).

Treatment with MPTP (1-methyl-4-phenyl-1,2,3,6-tetrahydropyridine) damages dopaminergic cells of substantia nigra in animals, which is considered to be an experimental model for Parkinson's disease. It has been shown that administered gangliosides enhanced recovery from MPTP-induced lesions of substantia nigra in these models (Gupta *et al.*, 1990; Schneider, 1992; Herrero *et al.*, 1993; Kastner *et al.*, 1994). Furthermore, gangliosides counteracted the MPTP-induced inhibition of proliferation in cultured cells (Saulino and Schengrund, 1993). In a pilot study with a small number of Parkinson's disease patients, improvement of functional measures was present in most patients after GM1 treatment (Schneider *et al.*, 1995). Since no suitable animal models exist for Alzheimer's disease, therapeutic strategies using gangliosides could not be tested in laboratory experiments. The few data from clinical studies of ganglioside treatment in Alzheimer patients are controversial. Promising results were obtained by continuous intraventricular infusion of GM1, which had a significant beneficial effect in early onset Alzheimer's disease (Stern, 1992; Flicker *et al.*, 1994; Gottfries, 1994; Svennerholm, 1994c).

Although there is already widespread use of gangliosides in clinical neurology, their possitive effects need to be confirmed in larger controlled trials (Nobile *et al.*, 1994). Nevertheless, the important functions of glycolipids in cellular processes point to their potential pharmacological action in various diseases.

References

Adams JC, Watt FM (1993): Regulation of development and differentiation by the extracellular matrix. In *Development* **117**:1183–98.

Agranoff BW, and Hajra AK (1994): Lipids. In *Basic Neurochemistry: Molecular, Cellular, and Medical Aspects* (Siegel GJ ed) pp 97–116, New York: Raven Press, Ltd.

Allan D, Kallen KJ (1993): Transport of lipids to the plasma membrane in animal cells. In *Prog. Lipid Res.* **32**:195–219.

Alon R, Feizi T, Yuen CT *et al.* (1995): Glycolipid ligands for selectins support leukocyte tethering and rolling under physiologic flow conditions. In *J. Immunol.* **154**:5356–66.

Ando S, Yu RK, Scarsdale JN *et al.* (1989): High resolution proton NMR studies of gangliosides. Structure of two types of GD3 lactones and their reactivity with monoclonal antibody R24. In *J. Biol. Chem.* **264**:3478–83.

Antal J, d'Amore A, Nerozzi D *et al.* (1992): An EEG analysis of drug effects after mild head injury in mice. In *Life Sci.* **51**:185–93.

Argentino C, Sacchetti ML, Toni D *et al.* (1989): GM1 ganglioside therapy in acute ischemic stroke. Italian Acute Stroke Study. In *Stroke* **20**:1143–9.

Asbury AK (1994): Gangliosides and peripheral neuropathies: an overview. In *Prog. Brain Res.* **101**:279–87.

Baker RE (1988): Gangliosides as cell adhesion factors in the formation of selective connections within the nervous system. In *Prog. Brain Res.* **73**:491–508.

Ballou LR (1992): Sphingolipids and cell function. In *Immunol. Today* **13**:339–41.

Beutler E, Grabowski GA (1995): Gaucher disease. In *The Metabolic and Molecular Bases of Inherited Disease,* 7th edition (Scriver CR, Beaudet AL, Sly WS, Valle D, eds) pp 2641–70, New York: McGraw-Hill.

Blaszczyk TM, Thurin J, Hindsgaul O *et al.* (1987): Y and blood group B type 2 glycolipid antigens accumulate in a human gastric carcinoma cell line as detected by monoclonal antibody. Isolation and characterization by mass spectrometry and NMR spectroscopy. In *J. Biol. Chem.* **262**:372–9.

Bock K, Karlsson KA, Stromberg N *et al.* (1988): Interaction of viruses, bacteria and bacterial toxins with host cell surface glycolipids. Aspects on receptor identification and dissection of binding epitopes. In *Adv. Exp. Med. Biol.* **228**:153–86.

Boren T, Falk P, Roth KA *et al.* (1993): Attachment of Helicobacter pylori to human gastric epithelium mediated by blood group antigens. In *Science* **262**:1892–5.

Bose B, Osterholm JL, Kalia M (1986): Ganglioside-induced regeneration and reestablishment of axonal continuity in spinal cord-transected rats. In *Neurosci. Lett.* **63**:165–9.

Brady RO, Fishman PH (1974): Biosynthesis of glycolipids in virus-transformed cells. In *Biochim. Biophys. Acta* **355**:121–48.

Bremer EG (1994): Glycosphingolipids as effectors of growth and differentiation. In *Cell Lipids* **1**:387–411.

Brennan PJ (1981): Structures of the typing antigens of atypical mycobacteria: a brief review of present knowledge. In *Rev. Infect. Dis.* **3**:905–13.

Brennan PJ (1989): Structure of mycobacteria: recent developments in defining cell wall carbohydrates and proteins. In *Rev. Infect. Dis.* **11 Suppl 2**:420–30.

Buckley NE, Sy Y, Milstien S *et al.* (1995): The role of calcium influx in cellular proliferation induced by interaction of endogenous ganglioside GM1 with the B subunit of cholera toxin. In *Biochim. Biophys. Acta* **1256**:275–83.

Caffrey, M (1993) LIPIDAT: A database of thermodynamic data and associated information on lipid mesomorphic and polymorphic transitions. Boca Raton: CRC Press.

Cahn R, Borzeix MG, Aldinio C *et al.* (1989): Influence of monosialoganglioside inner ester on neurologic recovery after global cerebral ischemia in monkeys. In *Stroke* **20**:652–6.

Campanella R (1992): Membrane lipids modifications in human gliomas of different degree of malignancy. In *J. Neurosurg. Sci.* **36**:11–25.

Carlson RO, Masco D, Brooker G *et al.* (1994): Endogenous ganglioside GM1 modulates L-type calcium channel activity in N18 neuroblastoma cells. In *J. Neurosci.* **14**:2272–81.

Chatterjee D, Hunter SW, McNeil M *et al.* (1989): Structure and function of mycobacterial glycolipids and glycopeptidolipids. In *Acta Leprol.* **7 Suppl 1**:81–4.

Cheung NK, Kushner BH, Yeh SJ *et al.* (1994): 3F8 monoclonal antibody treatment of patients with stage IV neuroblastoma: a phase II study. In *Prog. Clin. Biol. Res.* **385**:319–28.

Choi DW (1988): Glutamate toxicity and diseases of the nervous system. In *Neuron* **1**:623–34.

Clausen H, Hakomori S (1989): ABH and related histo-blood group antigens; immunochemical differences in carrier isotypes and their distribution. *Vox Sang.* **56**:1–20.

Clausen H, Levery SB, Kannagi R *et al.* (1986): Novel blood group H glycolipid antigens exclusively expressed in blood group A and AB erythrocytes (type 3 chain H). I. Isolation and chemical characterization. In *J. Biol. Chem.* **261**:1380–7.

Conzelmann E, Sandhoff K (1987): Glycolipid and glycoprotein degradation. In *Adv. Enzymol.* **60**: 89–216.

Costa E, Armstrong D, Guidotti A *et al.* (1993): Ganglioside GM1 and its semisynthetic lysogangliosides reduce glutamate neurotoxicity by a novel mechanism. In *Adv. Exp. Med. Biol.* **341**:129–41.

Costello CE, Juhasz P, Perreault H (1994): New mass spectral approaches to ganglioside structure determinations. In *Prog. Brain Res.* **101**:45–61.

Cuello AC, Garofalo L, Kenigsberg RL *et al.* (1989): Ganglioside potentiate in vivo and in vitro effects of nerve growth factor on central cholinergic neurons. In *Proc. Natl. Acad. Sci. USA* **86**:2056–60.

Cullis PR, de Kruijff B (1979): Lipid polymorphism and the functional roles of lipids in biological membranes. In *Biochim. Biophys. Acta* **559**:399–420.

Curatolo W (1987a): The physical properties of glycolipids. In *Biochim. Biophys. Acta* **906**:111–36.

Curatolo W (1987b): Glycolipid function. In *Biochim. Biophys. Acta* **906**:137–60.

Dabrowski J, Trauner K, Koike K *et al.* (1988): Complete 1H-NMR spectral assignments for globotriaosyl-Z- and isoglobotriaosyl-E-ceramide. In *Chem. Phys. Lipids* **49**:31–7.

Daffe M, Servin P (1989): Scalar, dipolar-correlated and J-resolved 2D-NMR spectroscopy of the specific phenolic mycoside of Mycobacterium tuberculosis. In *Eur. J. Biochem.* **185**:157–62.

Dahiya R, Dudeja PK, Brasitus TA (1987): Premalignant alterations in the glycosphingolipid composition of colonic epithelial cells of rats treated with 1,2-dimethylhydrazine. In *Cancer Res.* **47**:1031–5.

Dahiya R, Dudeja PK, Brasitus TA (1988): Premalignant alterations in the glycosphingolipids of small intestinal mucosa of rats treated with 1,2-dimethylhydrazine. In *Lipids* **23**:445–51.

Dennis RD, Wiegandt H (1993): Glycosphingolipids of the invertebrata as exemplified by a cestode platyhelminth, *Taenia crassiceps*, and a dipteran insect, *Calliphora vicina*. In *Adv. Lipid Res.* **26**:321–51.

De Rosa M, Gambacorta A, Gliozzi A (1986): Structure, biosynthesis, and physicochemical properties of archaebacterial lipids. In *Microbiol. Rev.* **50**:70–80.

Desnick RJ, Ioannou YA, Eng CM (1995): α-Galactosidase A deficiency: Fabry disease. In *The Metabolic and Molecular Bases of Inherited Disease*, 7th edition (Scriver CR, Beaudet AL, Sly WS, Valle D, eds) pp 2741-84, New York: McGraw-Hill.

Dorne AJ, Joyard J, Douce R (1990): Do thylacoids really contain phosphatidylcholine? In *Proc. Natl. Acad. Sci. USA* **87**:71–4.

Emory CR, Ala TA, Frey WH (1987): Ganglioside monoclonal antibody (A2B5) labels Alzheimer's neurofibrillary tangles. In *Neurology* **37**:768–72.

Evans CW (1992): Mini-Review – Cell adhesion and metastasis. In *Cell Biol. Int. Rep.* **16**:1–10.

Falk P, Boren T, Normark S (1994): Characterization of microbial host receptors. In *Meth. Enzymol.* **236**:353–74.

Favaron M, Manev H, Alho H *et al.* (1988): Gangliosides prevent glutamate and kainate neurotoxicity in primary neuronal cultures of neonatal rat cerebellum and cortex. In *Proc. Natl. Acad. Sci. USA* **85**:7351–5.

Feeney DM, Sutton RL (1987): Pharmacotherapy for recovery of function after brain injury. In *Crit. Rev. Neurobiol.* **3**:135–97.

Feizi T (1985): Demonstration by monoclonal antibodies that carbohydrate structures of glycoproteins and glycolipids are onco-developmental antigens. In *Nature* **314**:53–7.

Fenderson BA, Eddy EM, Hakomori S (1989): A role of carbohydrate-carbohydrate interaction in the process of specific cell recognition during emryogenesis and organogenesis. In *Biochem. Biophys. Res. Commun.* **158**:913–20.

Fenderson BA, Eddy EM, Hakomori S (1990): Glycoconjugate expression during embryogenesis and its biological significance. In *Bioessays* **12**:173–9.

Fishman PH (1982): Role of membrane gangliosides in the binding and action of bacterial toxins. In *J. Membr. Biol.* **69**:85–97.

Fishman PH, Pacuszka T, Orlandi PA (1993): Gangliosides as receptors for bacterial enterotoxins. In *Adv. Lipid Res.* **25**:165–87.

Flicker C, Ferris SH, Kalkstein D *et al.* (1994): A double-blind, placebo-controlled crossover study of ganglioside GM1 treatment for Alzheimer's disease. In *Am. J. Psychiatry* **151**:126–9.

Foxall C, Watson SR, Dowbenko D *et al.* (1992): The three members of the selectin receptor family recognize a common carbohydrate epitope, the sialyl Lewis(x) oligosaccharide. In *J. Cell Biol.* **117**:895–902.

Fredman P (1993): Glycosphingolipid tumor antigens. In *Adv. Lipid Res.* **25**:213–34.

Fredman P (1994): Gangliosides associated with primary brain tumors and their expression in cell lines established from these tumors. In *Prog. Brain Res.* **101**:225–40.

Fuks B, Homble F (1994): Permeability and electrical properties of planar lipid membranes from thylacoid lipids. In *Biophys. J.* **66**:1404–14.

Gabius S, Kayser K, Hellmann KP *et al.* (1990): Carrier-immobilized derivatized lysoganglioside GM1 is a ligand for specific binding sites in various human tumor cell types and peripheral blood lymphocytes and monocytes. In *Biochem. Biophys. Res. Commun.* **169**:239–44.

Garcia GC, Garcia A, Rubio G (1995): Presence and isotype of anti-ganglioside antibodies in healthy persons, motor neuron disease, peripheral neuropathy, and other diseases of the nervous system. In *J. Neuroimmunol.* **56**:27–33.

Garofalo L, Cuello AC (1995): Pharmacological characterization of nerve growth factor and/or monosialoganglioside GM1 effects on cholinergic markers in the adult lesioned brain. In *J. Pharmacol. Exp. Therap.* **272**:527–45.

Garofalo L, Ribeirodasilva A, Cuello AC (1993): Potentiation of nerve growth factor-induced alterations in cholinergic fibre length and presynaptic terminal size in cortex of lesioned rats by the monosialoganglioside GM1. In *Neuroscience* **57**:21–40.

Geisler FH, Dorsey FC, Coleman WP (1993): Past and current clinical studies with GM-1 ganglioside in acute spinal cord injury. In *Ann. Emerg. Med.* **22**:1041–7.

Gieselmann V (1995): Lysosomal storage diseases. In *Biochim. Biophys. Acta* **1270**:103–36.

Goldfine H (1982): Lipids of prokaryotes – Structure and distribution. In *Curr. Top. Membr. Transp.* **17**:1–43

Gottfries CG (1994): Therapy options in Alzheimer's disease. In *Br. J. Clin. Pract.* **48**:327–30.

Gottstein C, Schon G, Tawadros S *et al.* (1994): Anti-disialoganglioside ricin A-chain immunotoxins show potent antitumor effects in vitro and in a disseminated human neuroblastoma severe combined immunodeficiency mouse model. In *Cancer Res.* **54**:6186–93.

Gouy H, Deterre P, Debre P *et al.* (1994): Cell calcium signaling via GM1 cell surface gangliosides in the human Jurkat T cell line. In *J. Immunol.* **152**:3271–81.

Gravel RA, Clarke JT, Kaback MM *et al.* (1995): The GM2 gangliosidoses. In *The Metabolic and Molecular Bases of Inherited Disease,* 7th edition (Scriver CR, Beaudet AL, Sly WS, Valle D, eds) pp 2839–79, New York: McGraw-Hill.

Green C (1989): The ABO, Lewis and related blood group antigens; a review of structure and biosynthesis. *FEMS Microbiol. Immunol. Lett.* **1**:321–30.

Griffin JW (1994): Antiglycolipid antibodies and peripheral neuropathies: links to pathogenesis. In *Prog. Brain Res.* **101**:313–23.

Gu XB, Preuss U, Gu TJ *et al.* (1995): Regulation of sialyltransferase activities by phosphorylation and dephosphorylation. In *J. Neurochem.* **64**:2295–302.

Guerold B, Massarelli R, Forster V *et al.* (1992): Exogenous gangliosides modulate calcium fluxes in cultured neuronal cells. In *J. Neurosci. Res.* **32**: 110–5.

Gupta M, Schwarz J, Chen XL *et al.* (1990): Gangliosides prevent MPTP toxicity in mice–an immunocytochemical study. In *Brain Res.* **527**:330–4.

Hadjiconstantinou M, Karadsheh NS, Rattan AK *et al.* (1992): GM1 ganglioside enhances cholinergic parameters in the brain of senescent rats. In *Neuroscience* **46**:681–6.

Hakomori S (1984a): Philip Levine award lecture: blood group glycolipid antigens and their modifications as human cancer antigens. In *Am. J. Clin. Pathol.* **82**:635–48.

Hakomori S (1984b): Tumor-associated carbohydrate antigens. In *Annu. Rev. Immunol.* **2**:103–26.

Hakomori S (1986): Glycosphingolipids. In *Sci. Am.* **254/5**:32–41.

Hakomori S (1990): Bifunctional role of glycosphingolipids. Modulators for transmembrane signaling and mediators for cellular interactions. In *J. Biol. Chem.* **265**:18713–6.

Hakomori S, Igarashi Y (1993): Gangliosides and glycosphingolipids as modulators of cell growth, adhesion, and transmembrane signaling. In *Adv. Lipid Res.* **25**:147–62.

Handgretinger R, Anderson K, Lang P *et al.* (1995): A phase I study of human/mouse chimeric antiganglioside GD2 antibody ch14.18 in patients with neuroblastoma. In *Eur. J. Cancer* **31A**:261–7.

Hartmann S, Minnikin DE, Mallet AI *et al.* (1994): Fast atom bombardment mass spectrometry of mycobacterial phenolic glycolipids. In *Biol. Mass. Spectrom.* **23**:362–8.

Hattori M, Horiuchi R, Hosaka K *et al.* (1995): Sialyllactose-mediated cell interaction during granulosa cell differentiation. Identification of its binding proteins. In *J. Biol. Chem.* **270**:7858–63.

Herrero MT, Perez Otano I, Oset C *et al.* (1993): GM-1 ganglioside promotes the recovery of surviving midbrain dopaminergic neurons in MPTP-treated monkeys. In *Neuroscience* **56**:965–72.

Holgersson J, Breimer ME, Samuelsson BE (1992): Basic biochemistry of cell surface carbohydrates and aspects of the tissue distribution of histo-blood group ABH and related glycosphingolipids. *APMIS Suppl.* **27**:18–27.

Holmes EH, Hakomori S (1983): Enzymatic basis for changes in fucoganglioside during chemical carcinogenesis: induction of a specific α-fucosyltransferase and status of an α-galactosyltransferase in precancerous rat liver and hepatoma. In *J. Biol. Chem.* **258**:3706–13.

Hughes RA, Rees JH (1994): Guillain-Barre' syndrome. In *Curr. Opinion Neurol.* **7**:386–92.

Hughes RC (1992): Role of glycosylation in cell interaction with extracellular matrix. In *Biochem. Soc. Transact.* **20**:279–84.

IUPAC-IUB Commission on Biochemical Nomenclature (CBN) (1977): The nomenclature of lipids. In *Lipids* **12**:455–68.

Jardine I, Scanlan G, McNeil M *et al.* (1989): Plasma desorption mass spectrometric analysis of mycobacterial glycolipids. In *Anal. Chem.* **61**:416–22.

Jungalwala FB (1994): Expression and biological functions of sulfoglucuronyl glycolipids (SGGLs) in the nervous system–a review. In *Neurochem. Res.* **19**:945–57.

Karlsson KA (1989): Animal glycosphingolipids as membrane attachment sites for bacteria. In *Annu. Rev. Biochem.* **58**:309–50.

Karpiak SE, Li YS, Mahadik SP (1987): Ganglioside treatment: reduction of CNS injury and facilitation of functional recovery. In *Brain Inj.* **1**:161–70.

Kastner A, Herrero MT, Hirsch EC *et al.* (1994): Decreased tyrosine hydroxylase content in the dopaminergic neurons of MPTP-intoxicated monkeys: effect of levodopa and GM1 ganglioside therapy. In *Ann. Neurol.* **36**:206–14.

Kates M (1992): Archaebacterial lipids: structure, biosynthesis and function. In *Biochem. Soc. Symp.* **58**:51–72.

Kato E, Akiyoshi K, Furuno T *et al.* (1994): Interaction between ganglioside-containing liposome and rat T-lymphocyte: confocal fluorescence microscopic study. In *Biochem. Biophys. Res. Commun.* **203**:1750–5.

Kingsley BS, Gaskin F, Fu SM (1988): Human antibodies to neurofibrillary tangles and astrocytes in Alzheimer's disease. In *J. Neuroimmunol.* **19**:89–99.

Kishimoto Y, Hiraiwa M, O'Brien JS (1992): Saposins: structure, function, distribution, and molecular genetics. In *J. Lipid Res.* **33**:1255–67.

Koga Y, Nishihara M, Morii H *et al.* (1993): Ether polar lipids of methanogenic bacteria: structures, comparative aspects, and biosyntheses. In *Microbiol. Rev.* **57**:164–82.

Kolodony EH, Fluharty AL (1995): Metachromatic leucodystrophy and multiple sulfatase deficiency: sulfatide lipidosis. In *The Metabolic and Molecular Bases of Inherited Disease,* 7th edition (Scriver, CR, Beaudet AL, Sly WS, Valle D, eds) pp 2693–739, New York: McGraw-Hill.

Kopitz J, von Reitzenstein C, Mühl C *et al.* (1994): Role of plasma membrane ganglioside sialidase of human neuroblastoma cells in growth control and differentiation. In *Biochem. Biophys. Res. Commun.* **199**:1188–93.

Kopitz J, von Reitzenstein C, Sinz K *et al.* (1996): Selective ganglioside desialylation in the plasma membrane of human neuroblastoma cells. In *Glycobiology* **6**: 367–76.

Koynova R, Caffrey M (1994): Phases and phase transitions of the glycoglycerolipids. In *Chem. Phys. Lipids* **69**:181–207.

Koynova R, Caffrey M (1995): Phases and phase transitions of the sphingolipids. In *Biochim. Biophys. Acta* **1255**:213–36.

Kracun I, Rosner H, Drnovsek V *et al.* (1991): Human brain gangliosides in development, aging and disease. In *Int. J. Dev. Biol.* **35**:289–95.

Kracun I, Kalanj S, Talanhranilovic J *et al.* (1992a): Cortical distribution of gangliosides in Alzheimer's disease. In *Neurochem. Int.* **20**:433–8.

Kracun I, Rosner H, Drnovsek V *et al.* (1992b): Gangliosides in the human brain development and aging. In *Neurochem. Int.* **20**:421–31.

Kushi Y, Rokukawa C, Handa S (1988): Direct analysis of glycolipids on thin-layer plates by matrix-assisted secondary ion mass spectrometry: application for glycolipid storage disorders. In *Anal. Biochem.* **175**:167–76.

Ladisch S (1987): Tumor cell gangliosides. In *Adv. Pediatr.* **34**:45–58.

Ledeen RW (1989): Biosynthesis, metabolism, and biological effects of gangliosides. In *Neurobiology of Glycoconjugates*, 1st edition (Margolis RU, Margolis RK, eds.) pp. 43–84, New York: Plenum Press.

Lenzi GL, Grigoletto F, Gent M *et al.* (1994): Early treatment of stroke with monosialoganglioside GM-1. Efficacy and safety results of the Early Stroke Trial. In *Stroke* **25**:1552–8.

Leoni F, Colnaghi MI, Canevari S *et al.* (1992): Glycolipids carrying Le(y) are preferentially expressed on small-cell lung cancer cells as detected by the monoclonal antibody MLuC1. In *Int. J. Cancer* **51**:225–31.

Levery SB, Hakomori S (1987): Microscale methylation analysis of glycolipids using capillary gas chromatography-chemical ionization mass fragmentography with selected ion monitoring. In *Meth. Enzymol.* **138**:13–25.

Levery SB, Holmes EH, Harris DD *et al.* (1992): 1H NMR studies of a biosynthetic lacto-ganglio hybrid glycosphingolipid: confirmation of structure, interpretation of "anomalous" chemical shifts, and evidence for interresidue amide-amide hydrogen bonding. In *Biochemistry* **31**:1069–80.

Levery SB, Nudelman ED, Salyan ME *et al.* (1989): Novel tri-and tetrasialosylpoly-*N*-acetyllactosaminyl gangliosides of human placenta: structure determination of pentadeca- and eicosaglycosylceramides by methylation analysis, fast atom bombardment mass spectrometry, and 1H NMR spectroscopy. In *Biochemistry* **28**:7772–81.

Levy DE, Tang PC, Musser JH (1994): Cell adhesion and carbohydrates. In *Ann. Rep. Med. Chem.* **29**:215–24.

Li J, Pearl DK, Pfeiffer SE *et al.* (1994): Patterns of reactivity with anti-glycolipid antibodies in human primary brain tumors. In *J. Neurosci. Res.* **39**:148–58.

Lingwood CA (1993): Verotoxins and their glycolipid receptors. In *Adv. Lipid Res.* **25**:189–211.

Lipartiti M, Lazzaro A, Manev H (1992): Ganglioside derivative LIGA20 reduces NMDA neurotoxicity in neonatal rat brain. In *Neuroreport* **3**:919–21.

Lipartiti M, Lazzaro A, Zanoni R *et al.* (1991): Monosialoganglioside GM1 reduces NMDA neurotoxicity in neonatal rat brain. In *Exp. Neurol.* **113**:301–5.

Luzzati V, Gambacorta A, De Rosa M *et al.* (1987): Polar lipids of thermophilic prokaryotic organisms: chemical and physical structure. In *Annu. Rev. Biophys. Biophys. Chem.* **16**:25–47.

Macher BA, Sweeley CC (1978): Glycosphingolipids: structure, biological source, and properties. In *Meth. Enzymol.* **50**:236–51.

Majocha RE, Jungalwala FB, Rodenrys A *et al.* (1989): Monoclonal antibody to embryonic CNS antigen A2B5 provides evidence for the involvement of membrane components at sites of Alzheimer degeneration and detects sulfatides as well as gangliosides. In *J. Neurochem.* **53**:953–61.

Mangeney M, Lingwood CA, Taga S *et al.* (1993): Apoptosis induced in Burkitt's lymphoma cells via Gb3/CD77, a glycolipid antigen. In *Cancer Res.* **53**:5314–9.

Marcus DM (1990): Measurement and clinical importance of antibodies to glycosphingolipids. In *Ann. Neurol.* **27 Suppl**:53–5.

Marcus DM, Weng N (1994): The structure of human anti-ganglioside antibodies. In *Prog. Brain Res.* **101**:289–93.

Matsushima HBogenmann E (1992): Modulation of neuroblastoma cell differentiation by the extracellular matrix. In *Int. J. Cancer* **51**:727–32.

Marsh WL (1990): Biological roles of blood group antigens. *Yale J. Biol. Med.* **63**:455–60.

McNeil MR, Brennan PJ (1991): Structure, function and biogenesis of the cell envelope of mycobacteria in relation to bacterial physiology, pathogenesis and drug resistance; some thoughts and possibilities arising from recent structural information. In *Res. Microbiol.* **142**:451–63.

McElhaney, RN (1982): Effects of membrane lipids on transport and enzymic activities. In *Curr. Top. Membr. Transp.* 17:317–40.

Merrill AHJ (1991): Cell regulation by sphingosine and more complex sphingolipids. In *J. Bioenerg. Biomembr.* **23**:83–104.

Merritt WD, De Minassian V, Reaman GH (1994): Increased GD3 ganglioside in plasma of children with T-cell acute lymphoblastic leukemia. In *Leukemia* **8**:816–2.

Merzak A, Koochekpour S, Pilkington GJ (1995): Adhesion of human glioma cell lines to fibronectin, laminin, vitronectin and collagen is modulated by gangliosides in vitro. In *Cell Adh. Commun.* **3**:27–43.

Miller Podraza H, Mansson JE, Svennerholm L (1992): Isolation of complex gangliosides from human brain. In *Biochim. Biophys. Acta* **1124**:45–51.

Minasian LM, Yao TJ, Steffens TA *et al.* (1995): A phase I study of anti-GD3 ganglioside monoclonal antibody R24 and recombinant human macrophage-colony stimulating factor in patients with metastatic melanoma. In *Cancer* **75**:2251–7.

Miyagi T, Tsuiki S (1985): Purification and characterization of cytosolic sialidase from rat liver. In *J. Biol. Chem.* **260**:6710–16.

Morrison WJ, Offner H, Vandenbark AA (1990): Transmembrane signalling associated with ganglioside-induced CD4 modulation. In *Immunopharmacology* **20**:135–41.

Morrison WJ, Young K, Offner H *et al.* (1993): Ganglioside (GM1) distinguishes the effects of CD4 on signal transduction through the TCR/CD3 complex in human lymphocytes. In *Cell. Mol. Biol. Res.* **39**:159–65.

Moser WM (1995): Ceramidase deficiency: Farber lipogranulomatosis. In *The Metabolic and Molecular Bases of Inherited Disease,* 7th edition (Scriver, CR, Beaudet AL, Sly WS, Valle D, eds) pp 2589–99, New York: McGraw-Hill.

Mujoo K, Reisfeld RA, Cheung L *et al.* (1991): A potent and specific immunotoxin for tumor cells expressing disialoganglioside GD2. In *Cancer Immunol. Immunother.* **34**:198–204.

Mutoh T, Tokuda A, Miyadai T *et al.* (1995): Ganglioside GM1 binds to the Trk protein and regulates receptor function. In *Proc. Natl. Acad. Sci. USA* **92**:5087–91.

Nagai Y (1995): Functional roles of gangliosides in bio-signaling. In *Behav. Brain Res.* **66**:99–104.

Nagai Y, Tsuji S (1994): Significance of ganglioside-mediated glycosignal transduction in neuronal differentiation and development. In *Prog. Brain Res.* **101**:119–26.

Nakamura O, Iwamori M, Matsutani M *et al.* (1991): Ganglioside GD3 shedding by human gliomas. In *Acta Neurochir. Wien.* **109**:34–6.

Neufeld EF (1991): Lysosomal storage diseases. In *Annu. Rev. Biochem.* **60**:257–80.

Nilsson O (1992): Carbohydrate antigens in human lung carcinomas. In *APMIS Suppl.* **27**:149–61.

Nishihara M, Yokota K, Kito M (1980): Lipid molecular species composition of thylakoid membranes. In *Biochim. Biophys. Acta* **617**:12–19.

Nishinaka T, Iwata D, Shimada S *et al.* (1993): Anti-ganglioside GD1a monoclonal antibody recognizes senile plaques in the brains of patients with Alzheimer-type dementia. In *Neurosci. Res.* **17**:171–6.

Nobile E, Carpo M, Scarlato G (1994): Gangliosides. Their role in clinical neurology. In *Drugs* **47**:576–85.

Nohara K, Nakauchi H, Spiegel S (1994): Glycosphingolipids of rat T cells. Predominance of asialo-GM1 and GD1c. In *Biochemistry* **33**:4661–6.

Nohara K, Sano T, Shiraishi F (1993): An activation-associated ganglioside in rat thymocytes. In *J. Biol. Chem.* **268**:24997–5000.

Nussberger S, Dörr K, Wang D *et al.* (1993): Lipid-protein interactions in crystals of plant light-harvesting complex. In *J. Mol. Biol.* **234**:347–56.

Ohsawa T, Shumiya S (1991): Age-related alteration of brain gangliosides in senescence-accelerated mouse (SAM)-P/8. In *Mech. Ageing Dev.* **59**:263–74.

Olivera A, Spiegel S (1992): Ganglioside GM1 and sphingolipid breakdown products in cellular proliferation and signal transduction pathways. In *Glycoconjugate J.* **9**:110–7.

Olivera A, Spiegel S (1993): Sphingosine-1-phosphate as second messenger in cell proliferation induced by PDGF and FCS mitogens. In *Nature* **365**:557–60.

Papadopoulos SM (1992): Spinal cord injury. In *Curr. Opinion Neurol. Neurosurg.* **5**:554–7.

Paulson JC, Colley KJ (1989): Glycosyltransferases. Structure, localization, and control of cell type-specific glycosylation. In *J. Biol. Chem.* **264**:17615–8.

Pique C, Mahe Y, Scamps C *et al.* (1991): Analysis of phenotypic and functional changes during ganglioside-induced inhibition of human T cell proliferation. In *Mol. Immunol.* **28**:1163–70.

Poppe L, Dabrowski J, von der Lieth CW *et al.* (1990): Three-dimensional structure of the oligosaccharide terminus of globotriaosylceramide and isoglobotriaosylceramide in solution. A rotating-frame NOE study using hydroxyl groups as long-range sensors in conformational analysis by 1H-NMR spectroscopy. In *Eur. J. Biochem.* **189**:313–25.

Puzo G (1990): The carbohydrate- and lipid-containing cell wall of mycobacteria, phenolic glycolipids: structure and immunological properties. In *Crit. Rev. Microbiol.* **17**:305–27.

Rahmann H (1995): Brain gangliosides and memory formation. In *Behav. Brain Res.* **66**:105–16.

Rahmann H, Probst W, Muhleisen M (1982): Gangliosides and synaptic transmission. In *Jpn. J. Exp. Med.* **52**:275–86.

Rahmann H, Schifferer F, Beitinger H (1992): Calcium-ganglioside interactions and synaptic plasticity effect of calcium on specific ganglioside/peptide (Valinomycin, Gramicidin-A)-complexes in mixed monolayers and bilayers. In *Neurochem. Int.* **20**:323–38.

Rapola J (1994): Lysosomal storage diseases in adults. In *Pathol. Res. Pract.* **190**:759–66.

Reaman GH, Taylor BJ, Merritt WD (1990): Anti-GD3 monoclonal antibody analysis of childhood T-cell acute lymphoblastic leukemia: detection of a target antigen for antibody-mediated cytolysis. In *Cancer Res.* **50**:202–5.

Riboni L, Bassi R, Sonnino S *et al.* (1992): Formation of free sphingosine and ceramide from exogenous ganglioside GM1 by cerebellar granule cells in culture. In *FEBS Lett.* **300**:188–92.

Riboni L, Prinetti A, Bassi R *et al.* (1991): Cerebellar granule cells in culture exhibit a ganglioside-sialidase presumably linked to the plasma membrane. In *FEBS Lett.* **287**:42–6.

Riboni L, Prinetti A, Bassi R *et al.* (1994): Formation of bioactive sphingoid molecules from exogenous sphingomyelin in primary cultures of neurons and astrocytes. In *FEBS Lett.* **352**:323–6.

Riviere M, Fournie JJ, Vercellone A *et al.* (1988): Particular matrix for fast atom bombardment mass spectrometric analysis of phenolic glycolipid antigens isolated from pathogen mycobacteria. In *Biomed. Environ. Mass. Spectrom.* **16**:275–8.

Saito M (1993): Bioactive gangliosides: differentiation inducers for hematopoietic cells and their mechanism(s) of actions. In *Adv. Lipid Res.* **25**:303–27.

Sandhoff K. , Harzer K, Fürst W (1995): Sphingolipid activator proteins. In *The Metabolic and Molecular Bases of Inherited Disease,* 7th edition (Scriver, CR, Beaudet AL, Sly WS, Valle D, eds) pp 2427–41, New York: McGraw-Hill.

Sariola H, Terava H, Rapola J *et al.* (1991): Cell-surface ganglioside GD2 in the immunohistochemical detection and differential diagnosis of neuroblastoma. In *Am. J. Clin. Pathol.* **96**:248–52.

Saulino MF, Schengrund CL (1993): Effects of specific gangliosides on the in vitro proliferation of MPTP-susceptible cells. In *J. Neurochem.* **61**:1277–83.

Scarsdale JN, Prestegard JH, Yu RK (1990): NMR and computational studies of interactions between remote residues in gangliosides. In *Biochemistry* **29**:9843–55.

Schlosshauer B, Blum AS, Mendez-Otero R *et al.* (1988): Developmental regulation of ganglioside antigens recognized by the JONES antibody. In *J. Neurosci.* **8**:580–92.

Schnaar RL (1991): Glycosphingolipids in cell surface recognition. In *Glycobiology* **1**:477–85.

Schnaar RL, Mahoney JA, Swank Hill P *et al.* (1994): Receptors for gangliosides and related glycosphingolipids on central and peripheral nervous system cell membranes. In *Prog. Brain Res.* **101**:185–97.

Schneider JS (1992): MPTP-induced parkinsonism: acceleration of biochemical and behavioral recovery by GM1 ganglioside treatment. In *J. Neurosci. Res.* **31**:112–9.

Schneider JS, Roeltgen DP, Rothblat DS *et al.* (1995): GM1 ganglioside treatment of Parkinson's disease: An open pilot study of safety and efficacy. In *Neurology* **45**:1149–54.

Sell S (1990): Cancer-associated carbohydrates identified by monoclonal antibodies. In *Hum. Pathol.* **21**:1003–19.

Serratrice G (1993): Motor neuropathies and antiglycolipid antibodies. In *Clin. Exp. Neurol.* **30**:25–32.

Shaw N (1974): Lipid composition as a guide to the classification of bacteria. In *Adv. Appl. Microbiol.* **17**:63–108.

Smith IC, Auger M, Jarrell HC (1989a): Structure and dynamics of the glycolipid components of membrane receptors: 2H NMR provides a route to in vivo observation. In *Ann. N. Y. Acad. Sci.* **568**:44–51.

Smith IC, Baenziger J, Auger M *et al.* (1989b): Deuterium NMR as a monitor of organization and dynamics at the surface of membranes: the glycolipids. In *Prog. Clin. Biol. Res.* **292**:13–22.

Song WX, Welti R, Hafner-Strauss S *et al.* (1993): Synthesis and characterization of *N*-parinaroyl analogs of ganglioside-G(M3) and de-*N*-acetyl GM3 – Interactions with the EGF receptor kinase. In *Biochemistry* **32**:8602–7.

Sonnino S, Cantu L, Corti M *et al.* (1994): Aggregative properties of gangliosides in solution. In *Chem. Phys. Lipids* **71**:21–45.

Sparkman DR, Goux WJ, Jones CM *et al.* (1991): Alzheimer disease paired helical filament core structures contain glycolipid. In *Biochem. Biophys. Res. Commun.* **181**:771–9.

Spoerri PE, Dozier AK, Roisen FJ (1990): Calcium regulation of neuronal differentiation: the role of calcium in GM1-mediated neuritogenesis. In *Dev. Brain Res.* **56**:177–88.

Sprague SG (1987): Structural and functional consequences of galactolipids on thylakoid membrane organization. In *J. Bioenerg. Biomembr.* **19**: 691–703.

Steck AJ, Kappos L (1994): Gangliosides and autoimmune neuropathies: classification and clinical aspects of autoimmune neuropathies. In *J. Neurol. Neurosurg. Psychiatry* **57 Suppl**:26–8.

Stern GM (1992): New drug interventions in Alzheimer's disease. In *Curr. Opinion Neurol. Neurosurg.* **5**:100–3.

Stoll MS, Hounsell EF, Lawson AM *et al.* (1990): Microscale sequencing of *O*-linked oligosaccharides using mild periodate oxidation of alditols, coupling to phospholipid and TLC-MS analysis of the resulting neoglycolipids. In *Eur. J. Biochem.* **189**:499–507.

Suresh MR (1991): Immunoassays for cancer-associated carbohydrate antigens. In *Semin. Cancer Biol.* **2**:367–77.

Suzuki K, Suzuki Y, Suzuki K (1995b): Galactosylceramide lipidosis: globoid cell leukodystrophy (Krabbe disease). In *The Metabolic and Molecular Bases of Inherited Disease,* 7th edition (Scriver, CR, Beaudet AL, Sly WS, Valle D, eds) pp 2671–92, New York: McGraw-Hill.

Suzuki M, Sekine M, Yamakawa T *et al.* (1989): High-performance liquid chromatography-mass spectrometry of glycosphingolipids: I. Structural characterization of molecular species of GlcCer and IV3 beta Gal-Gb4Cer. In *J. Biochem.* **105**:829–33.

Suzuki Y, Sakuraba H, Oshima A (1995a): β-Galactosidase deficiency (β-galactosidosis): GM1 gangliosidosis and Morquio B disease. In *The Metabolic and Molecular Bases of Inherited Disease,* 7th edition (Scriver CR, Beaudet AL, Sly WS, Valle D, eds) pp 2785–823, New York: McGraw-Hill.

Svennerholm L (1994a): Designation and schematic structure of gangliosides and allied glycosphingolipids. In *Prog. Brain Res.* **101**:11–14.

Svennerholm L (1994b): Ganglioside loss is a primary event in Alzheimer disease type I. In *Prog. Brain Res.* **101**:391–404.

Svennerholm L (1994c): Gangliosides–a new therapeutic agent against stroke and Alzheimer's disease. In *Life Sci.* **55**:2125–34.

Svennerholm L, Bostrom K, Helander CG *et al.* (1991): Membrane lipids in the aging human brain. In *J. Neurochem.* **56**:2051–9.

Taga S, Tetaud C, Mangeney M *et al.* (1995): Sequential changes in glycolipid expression during human B cell differentiation: Enzymatic bases. In *Biochim. Biophys. Acta* **1254**:56–65.

Taira E, Takaha N, Miki N (1993): Extracellular matrix proteins with neurite promoting activity and their receptors. In *Neurosci. Res.* **17**:1–8.

Tettamanti G, Riboni L (1993): Gangliosides and modulation of the function of neural cells. In *Adv. Lipid Res.* **25**:235–67.

Thomas PD, Brewer GJ (1990): Gangliosides and synaptic transmission. In *Biochim. Biophys. Acta* **1031**:277–89.

Thompson TE, Tillack TW (1985): Organization of glycosphingolipids in bilayers and plasma membranes of mammalian cells. In *Annu. Rev. Biophys. Chem.* **14**:361–86.

Tiemeyer M, Swank-Hill P, Schnaar RL (1990): A membrane receptor for gangliosides is associated with central nervous system myelin. In *J. Biol. Chem.* **265**:11990–9.

Tiemeyer M, Swiedler SJ, Ishihara M *et al.* (1991): Carbohydrate ligands for endothelial-leukocyte adhesion molecule. In *Proc. Natl. Acad. Sci. USA* **88**:1138–42.

Usui H, Hakomori S (1994): Evaluation of ricin A chain-containing immunotoxins directed against glycolipid and glycoprotein on mouse lymphoma cells. In *Acta Med. Okayama* **48**:305–9.

Valentino L, Moss T, Olson E *et al.* (1990): Shed tumor gangliosides and progression of human neuroblastoma. In *Blood* **75**:1564–7.

Valentino LA, Ladisch S (1992): Localization of shed human tumor gangliosides: association with serum lipoproteins. In *Cancer Res.* **52**:810–4.

van Echten G, Sandhoff K (1993): Ganglioside metabolism. Enzymology, topology, and regulation. In *J. Biol. Chem.* **268**:5341–4.

Vangsted A, Drivsholm L, Andersen E *et al.* (1994): New serum markers for small-cell lung cancer. I. The ganglioside fucosyl-GM1. In *Cancer Detect. Prev.* **18**:221–9.

Vangsted AJ, Zeuthen J (1993): Monoclonal antibodies for diagnosis and potential therapy of small cell lung cancer–the ganglioside antigen fucosyl-GM1. In *Acta Oncol.* **32**:845–51.

Vos JP, Lopes Cardozo M, Gadella BM (1994): Metabolic and functional aspects of sulfogalactolipids. In *Biochim. Biophys. Acta* **1211**:125–49.

Walker JB, Harris M (1993): GM-1 ganglioside administration combined with physical therapy restores ambulation in humans with chronic spinal cord injury. In *Neurosci. Lett.* **161**:174–8.

Walsh TJ, Opello KD (1992): Neuroplasticity, the aging brain, and Alzheimer's disease. In *Neurotoxicology* **13**:101–10.

Webb MS, Green BR (1991): Biochemical and biophysical properties of thylacoid acyl lipids. In *Biochim. Biophys. Acta* **1060**:133–58.

Wiegandt H (1992): Insect glycolipids. In *Biochim. Biophys. Acta* **1123**:117–26.

Wikstrand CJ, Fredman P, Svennerholm L *et al.* (1994): Detection of glioma-associated gangliosides GM2, GD2, GD3, 3'-isoLM1 3',6'-isoLD1 in central nervous system tumors in vitro and in vivo using epitope-defined monoclonal antibodies. In *Prog. Brain Res.* **101**:213–23.

Wu G, Ledeen RW (1994): Gangliosides as modulators of neuronal calcium. In *Prog. Brain Res.* **101**:101–12.

Wu G, Lu ZH, Ledeen RW (1995): Induced and spontanous neuritogenesis are associated with enhanced expression of ganglioside GM1 in the nuclear membrane. In *J. Neurosci.* **15**:3739–3746.

Wu G, Vaswani KK, Lu ZH *et al.* (1990): Gangliosides stimulate calcium flux in Neuro-2A cells and require exogenous calcium for neuritogenesis. In *J. Neurochem.* **55**:484–91.

Yoles E, Zalish M, Lavie V *et al.* (1992): GM1 reduces injury-induced metabolic deficits and degeneration in the rat optic nerve. In *Invest. Ophthalmol. Vis. Sci.* **33**:3586–91.

Yu RK (1994): Development regulation of ganglioside metabolism. In *Prog. Brain Res.* **101**:31–44.

Yu RK, Koerner TA, Scarsdale JN *et al.* (1986): Elucidation of glycolipid structure by proton nuclear magnetic resonance spectroscopy. In *Chem. Phys. Lipids* **42**:27–48.

Yu RK, Saito M (1989): Structure and localization of gangliosides. In *Neurobiology of Glycoconjugates*,1st edition (Margolis RU, Margolis RK, eds) pp 1–42, New York: Plenum Press.

Yu RK, Saito M (1992): Gangliosides and neuronal-astrocytic interactions. In *Prog. Brain Res.* **94**:333–8.

Yuasa H, Scheinberg DA, Houghton AN (1990): Gangliosides of T lymphocytes: evidence for a role in T-cell activation. In *Tissue Antigens* **36**:47–56.

Zalish M, Lavie V, Duvdevani R *et al.* (1993): Gangliosides attenuate axonal loss after optic nerve injury. In *Retina* **13**:145–7.

Zeller CB, Marchase RB (1992): Gangliosides as modulators of cell function. In *Am. J. Physiol.* **262**:C1341–55.

Zhang H, Desai NN, Olivera A *et al.* (1991): Sphingosine-1-phosphate, a novel lipid, involved in cellular proliferation. In *J. Cell Biol.* **114**:155–67.

Zheng MZ, Fang H, Tsuruoka T *et al.* (1993): Regulatory role of GM3 ganglioside in $\alpha_5\beta_1$ integrin receptor for fibronectin-mediated adhesion of FUA169 cells. In *J. Biol. Chem.* **268**:2217–22.

Zhou Q, Hakomori S, Kitamura K *et al.* (1994): GM3 directly inhibits tyrosine phosphorylation and de-*N*-acetyl-GM3 directly enhances serine phosphorylation of epidermal growth factor receptor, independently of receptor-receptor interaction. In *J. Biol. Chem.* **269**:1959–65.

10 Lectins as Tools for Glycoconjugate Purification and Characterization

Richard D. Cummings

1 Introduction

Glycoconjugates from different organisms display tremendous variations in structure and function. This vast array of glycans presents a formidable problem for investigators wanting to study and characterize their expression and structures. In many cases the set of oligosaccharides derived from the total glycoprotein fraction of cells may vary only within the range of 300–3,000 daltons, but this set may contain hundreds of structurally different compounds (Laine, this volume). Separation of the oligosaccharides within such a mixture by traditional physico-chemical techniques, for example using gel filtration, can take intense labor and expense and still may fail to distinguish isomeric and near isomeric compounds (Hounsell, this volume). The carbohydrate chemist who could wave a magic wand to rectify this problem might create a great panel of specific monoclonal antibodies, each of which could recognize a unique structure and could be used to readily identify and isolate specific oligosaccharides within a complex mixture. Fortunately, nature has already waved the magic wand to some extent and provided us with a vast number of carbohydrate-binding proteins called lectins. A lectin is a carbohydrate-binding protein that is not an antibody and lacks enzymatic activity toward carbohydrates.

Lectins were discovered more than a hundred years ago in castor beans, and since then hundreds of different lectins have been isolated and characterized from plants, animals and microorganisms. The available evidence suggests that lectins are produced by all living things. Many of these lectins are known to be important for promoting cellular adhesion and acting as receptors for other glycoconjugates. Most lectins have high-affinity interactions with specific carbohydrate determinants, usually larger than a monosaccharide, and thus lectins can be used to study, characterize, and isolate glycoconjugates on the basis of specific structural features, instead of the size and/or charge of the glycan. In addition, lectins can be used in a manner akin to antibodies to probe carbohydrate structures on cell surfaces, in matrices, or on intact glycoproteins, and to study the biosynthesis of glycoconjugates.

This chapter will present the use of lectins as tools for studying the expression and characterization of complex glycoconjugates. There have been many excellent reviews on one or more aspects of this subject in recent years and the reader is encouraged to consult them for additional information (Finne and Krusius, 1982; Merkle and Cummings, 1987; Osawa and Tsuji, 1987; Gabius and Gabius, 1993; Kobata and Yamashita, 1993; Cummings, 1994; West and Goldring, 1994; Yamamoto *et al.*, 1995).

2 Structures of Plant and Animal Lectins

Lectins derived from plants are usually multimeric and soluble proteins (some lectins are themselves glycoproteins) with multiple binding sites for carbohydrates (see Lis and Sharon, 1986; Rüdiger, this volume). The multivalent nature of plant lectins allows them to agglutinate cells to which they bind, and many lectins in plants were identified historically as agglutinins. In contrast, the physical nature of animal lectins is more varied and they can be placed into three groups: (1) intrinsic membrane-bound glycoproteins; (2)

H.-J. and S. Gabius (Eds.), Glycosciences

ISBN 3-8261-0073-5

soluble monomeric and monovalent proteins and glycoproteins; and (3) soluble, multimeric and multivalent proteins and glycoproteins.

Many plant lectins have been characterized from leguminous plants, such as peas and beans, and many of these have shared primary structures and represent the class of lectins called leguminous plant lectins. These include pea lectin (*Pisum sativum*) from the common garden pea, concanavalin A (*Canavalia ensiformis*) from the jack bean, *Ricinus communis* agglutinin from the castor bean, and *Phaseolus vulgaris* agglutinin (L_4-phytohemagglutinin; L_4-PHA) from the red kidney bean. The so-called non-leguminous plant lectins comprise a predictably large list. For example, lectins from the nightshade family of plants, including tomato lectin (*Lycopersicon esculentum*), potato lectin (*Solanum tuberosum*) and jimsonweed (*Datura stramonium*) agglutinin are closely related in terms of structure and carbohydrate-binding specificity and they have no relationship to the leguminous plant lectins in their primary structures.

Animal lectins are also classified according to their sequence homologies (Drickamer and Taylor, 1993; Powell and Varki, 1995; Gabius *et al.*, this volume; Ohannesian and Lotan, this volume; Zanetta, this volume). The C-type lectin family is a group of lectins with a common sequence motif that require Ca^{2+} for binding activity. The galectin family is a group of lectins with a common sequence motif and do not require Ca^{2+} for binding activity. Members of the I-type family have immunoglobulin-(Ig)-like domains and the P-family binds to phosphorylated mannosyl residues. Several other "families" of animal lectins are known and the list is growing rapidly. Except for a few exceptions, however, most lectins used in studies of animal cell glycoconjugates are plant-derived. This is due primarily to the expense of preparing animal lectins versus plant lectins and the commercial availability of the latter. It is hoped that animal lectins will become available in the future, since the specificity of their interactions are often quite different from those of plant lectins

3 Carbohydrate-Binding Specificities of Lectins

The plant lectins were originally shown to bind sugars by the ability of certain monosaccharides or their derivatives to inhibit lectin-induced agglutination of cells. Thus, plant lectins were historically classified on the basis of their so-called monosaccharide-binding specificity. For example, concanavalin A (Con A), *Ricinus communis* agglutinin (RCA) and *Lotus tetragonolobus* agglutinin (LTA) were classified as mannose-binding, galactose-binding, and fucose-binding lectins, respectively. Since animal lectins were also found to be inhibited in their binding by monosaccharides, they were also accordingly classified. As examples, galectin-1 from bovine heart and spleen was classified as a galactose-binding lectin and the hepatocyte-derived asialoglycoprotein receptor was termed a galactose/*N*-acetylgalactosamine-binding lectin (see review by Zhou and Cummings, 1992).

These simple designations, though convenient in some ways, belie the high degree of specificity and range of affinity exhibited by lectins toward complex carbohydrates. A detailed presentation of some of the complex determinants in glycans recognized by lectins is illustrated in Fig. 1. In each case the boxed portion represents the necessary determinant recognized by the lectin. Not only are lectins able to distinguish certain sugars but they also recognize the anomeric linkage of the sugars. For example, RCA-I binds well to glycans containing terminal β-1,4 galactosyl residues, but it does not bind to those containing α-linked galactosyl residues and binds poorly to those containing β-1,3 linked galactosyl residues.

It should be obvious that each of the lectins presented in Fig. 1 recognizes a different determinant and can be used to discriminate among all the possible glycans presented. However, it should be noted that the boxed determinant is necessary, but not sufficient for interaction in many cases. Additional modifications of the glycan may inhibit lectin recognition. For example, lentil and pea lectins bind to bi- and triantennary *N*-glycans containing the structures shown in Fig. 1 (see page 194–195) and a core α-1,6-linked fucose residue. The core fucose is required for high-affinity binding. However, tetraantennary *N*-glycans are not bound by these lectins even when the core fucose is present.

3.1 Use of Immobilized Lectins in the Chromatography-Based Isolation of Oligosaccharides and Glycoproteins

In introducing the concept of lectin affinity chromatography it is important to consider historically the two different approaches taken to define and understand the carbohydrate-binding specificity of a lectin. In one approach using lectin inhibition assays, one measures the ability of a panel of defined monosaccharides and oligosaccharides to inhibit lectin-induced agglutination. An order of increasing affinity can be established and this may help to discern the carbohydrate determinants being recognized by the lectin. This approach is limited by the availability of defined reagents and their purity. In many cases one may only find the "best" inhibitor among those tested, but still be far from understanding the complex determinants required for high-affinity binding of oligosaccharides to a lectin. Nevertheless, this approach allows the identification of a protein as a lectin and usually identifies a simple and commercially available monosaccharide or disaccharide (termed haptens) to promote dissociation of a lectin bound to glycoconjugates.

In the second approach using affinity chromatography, a lectin is immobilized on a support and a mixture of oligosaccharides, derived from the cell to which the lectin is known to bind, is applied to the lectin column. Many glycans do not bind, but those that do bind may be eluted with high concentrations of an inhibitory hapten. The structural features shared by the bound glycans can then reveal the important carbohydrate determinants required for high-affinity binding by the lectin. This approach may be hampered by the lack of material and the difficulty in defining the structures of the bound glycans, which may differ in many ways, except for a shared determinant required for high-affinity lectin binding.

Although both these approaches are useful, the appeal of the affinity chromatography approach is its practical value. Not only does that approach allow one to define the determinants required for high-affinity binding to a lectin, but the immobilized lectin is demonstrably useful in isolating the subset of glycans containing a specific structural feature. It is self-evident that a combination of many lectins, differing in their precise requirements for high-affinity binding to glycans, can provide a high degree of fractionation of oligosaccharides into sets that share common structural features. The combination of many lectins in such a chromatographic scheme, which is called serial lectin affinity chromatography, has been enormously successful in the analysis of complex mixtures of oligosaccharides and is discussed in more detail below.

The oligosaccharides of animal cell glycoproteins are usually linked in either amide linkage to Asn (*N*-glycans) or in *O*-glycosidic linkage to Ser/Thr (*O*-glycans). The *N*-glycans are subclassified as either high mannose-type, hybrid-type or complex-type depending on the mannose content (Fig. 1). The *O*-glycans are further subclassified as mucin-type containing R-GalNAc-Ser/Thr, glycosaminoglycan-type containing R-Xyl-Ser, *O*-linked GlcNAc-type containing only GlcNAc-Ser/Thr, collagen-type containing Gal-hydroxylysine, or non-traditional containing a variety of possible structures such as R-Glc-Ser/Thr, Glc_n-Glc-Tyr, R-Fuc-Ser/Thr-, etc. All of these *N*- and *O*-glycan types differ dramatically in their structures and biosynthesis (Brockhausen and Schachter, this volume). Some lectins are useful for binding one or the other sets of glycans and some are useful for studying determinants shared between different glycans (Fig. 1).

3.2 Immobilization of Lectins for Affinity Chromatography

Lectins are usually covalently coupled to insoluble supports such as Sepharose 4B, AffiGel, or Emphaze beads. Some studies have shown the usefulness of immobilizing lectins in other types of matrices for high performance liquid chromatography (Green and Baenziger, 1989). Most studies with lectin affinity chromatography, however, use gravity flow columns (Sueyoshi *et al.*, 1994; Fukushima *et al.*, 1995; Goulut-Chassaing

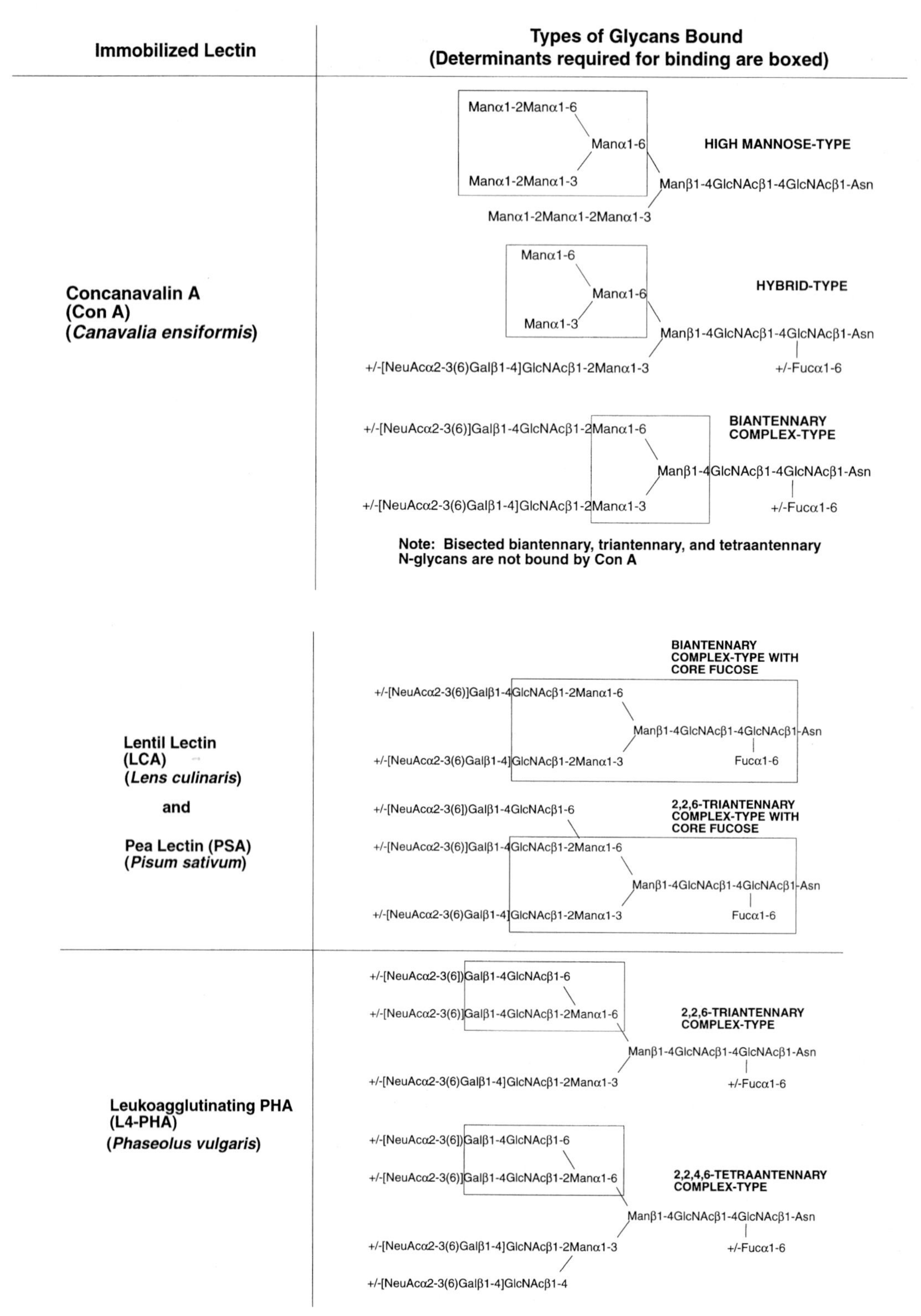

Figure 1. Glycan Recognition by Lectins

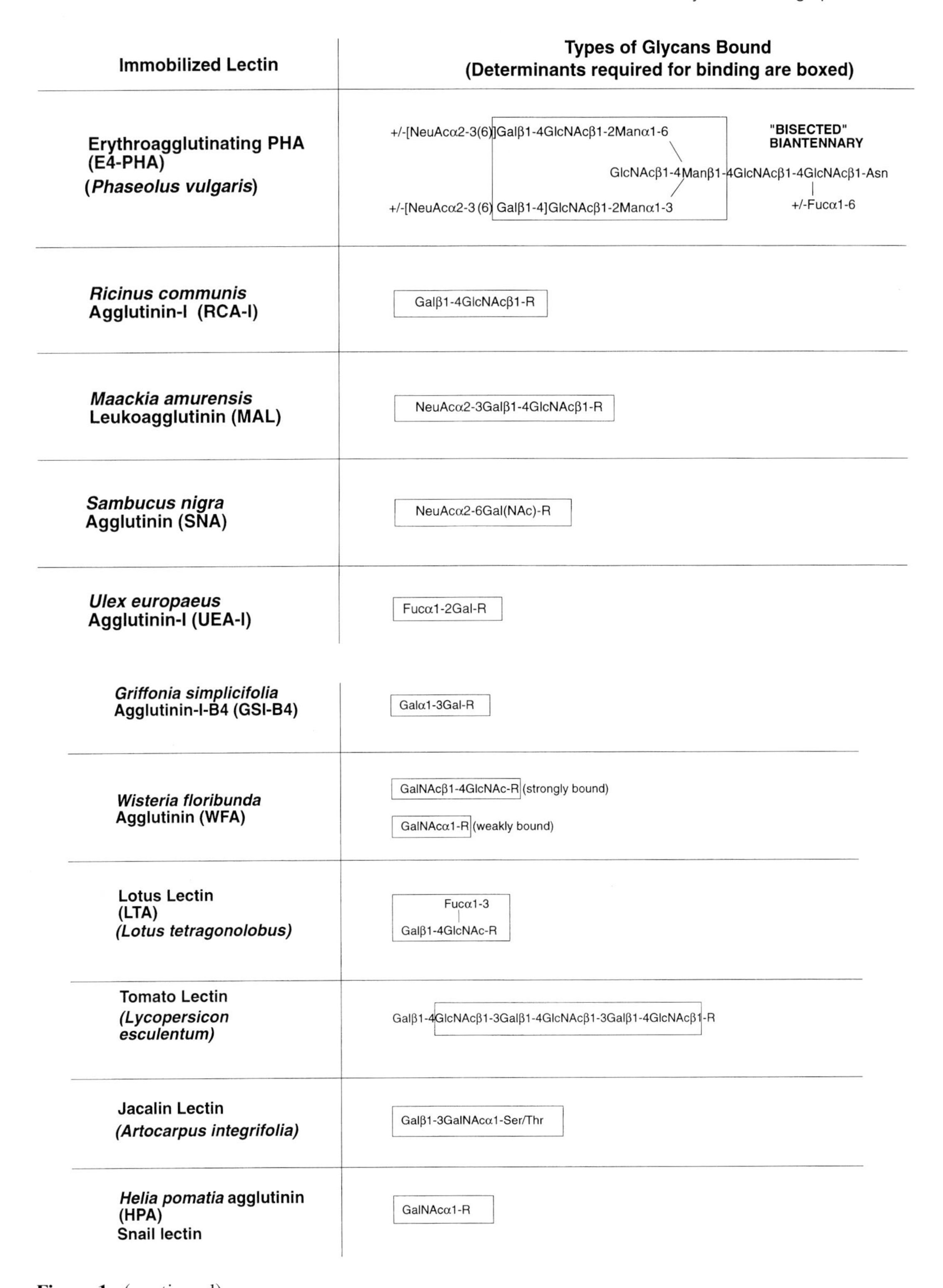

Figure 1. (continued)

and Bourrilon, 1995). Several companies sell lectins already derivatized to Sepharose or agarose. However, it is important to note that the density of coupling of each lectin is a major factor in determining the usefulness of the lectin for chromatographic purposes. Usually high density supports (>5 mg/ml) are required for affinity chromatography of free glycans or glycopeptides, whereas lower density supports may be useful for intact glycoproteins. This is because the affinity of plant lectins for free glycans is in the range of 10^{-4} to 10^{-6}, whereas the avidity of lectin binding to an intact glycoprotein containing multiple glycans recognized by the lectin may be several times higher.

3.3 Elution of Glycoconjugates Bound to an Immobilized Lectin

Although lectins can distinguish complex determinants within a glycan, most lectins can be inhibited in their binding by high concentrations of rather simple sugars, which in some cases are not present in the bound glycan. Two examples are the following. L_4-PHA recognizes a galactose-containing pentasaccharide structure within a complex-type *N*-glycan (Fig. 1) (Cummings and Kornfeld, 1982a; Hammarström *et al.*, 1982), but free GalNAc at a concentration of 20–40 mM can usually elute bound glycans from the lectin. Lactose will elute glycans bound by either *Sambucus nigra* agglutinin (SNA) or *Maackia amurensis* leukoagglutinin (MAL), although both lectins recognize sialic acid-containing determinants in glycans (Shibuya *et al.* 1987; Wang and Cummings, 1988). Other reviews on the detailed techniques of lectin affinity chromatography describe haptens and their concentrations necessary for elution of glycans bound by immobilized lectins. In some cases glycans are "retarded" in their elution from an immobilized lectin with which they interact and no hapten elution is required (Merkle and Cummings, 1987). Such elution patterns are often seen when either the density of immobilized lectin is low, the column geometry is not optimal, or the flow rates are too fast. Nevertheless, the degree of separation and purification of the glycans is often equivalent to that found when glycans are bound much tighter by the lectin. This hapten-independent elution is actually an advantage in many cases, since the chromatography is faster, less immobilized lectin is required, and it is not necessary in subsequent steps to separate the glycans from unwanted hapten sugars.

Lectin affinity chromatography is especially advantageous for isolating free glycans or glycopeptides. It is also a practical method for isolating intact glycoproteins from whole cells or tissues extracts, and facilitates the isolation a specific set of glycoproteins from cells containing a shared determinant. Two examples of this technique are the following. Glycoproteins containing *O*-linked α-GalNAc in a terminal position can be efficiently isolated on immobilized *Helix pomatia* agglutinin (HPA) (Do and Cummings, 1992). Both wheat germ (*Triticum vulgaris*) agglutinin and *Ricinus communis* agglutinin (RCA-I) can be used to identify *O*-linked GlcNAc residues on glycoproteins, either before, or after, addition of a terminal β1,4-linked galactosyl residue to the *O*-linked GlcNAc (Roquemore *et al.*, 1994; Hayes *et al.*, 1995). However, great care should be taken in the chromatography of whole cell extracts or impure glycoproteins on immobilized lectins, since the samples may be contaminated with proteases and glycosidases that could destroy both the lectin and the glycans. In addition, detergents used to solubilize cell-derived glycoproteins may interfere with lectin recognition.

3.4 Serial Lectin Affinity Chromatography (SLAC)

The combination of many lectins to serially fractionate complex mixtures of free glycans or glycopeptides is called serial lectin affinity chromatography or SLAC (Cummings and Kornfeld, 1982b). This approach has been successful in identifying and isolating both *N*- and *O*-glycans for subsequent structural analysis. A typical approach in SLAC is to apply a mixture of *N*-glycans to Con A-Sepharose and differentially separate the biantennary *N*-glycans from high mannose-type glycans by elution with with 10 mM α-methylglucoside and 100 mM α-methylmannoside, respectively. The separation of *N*-glycans on Con A-Sepharose was one of the first demonstrated uses of immobilized lectins to isolate specific glycans from a complex mixture (Kornfeld and Ferris, 1975; Ogata *et al.*, 1975; Krusius *et al.*, 1976). Most *O*-glycans and complex-type tri-/tetraantennary *N*-glycans are not bound by Con A. The bound and unbound glycans are then desalted by passage over a column of BioGel P-2 or Sephadex G-25 and the glycans are subsequently fractionated further by passage over a column of immobilized lentil or pea lectin. The glycans bound and unbound by these lectins can then be desalted and applied to columns of other immobilized lectins, as originally described by Cummings and Kornfeld (1982b). SLAC provides fractionation of mixtures based on the structures of the glycans and not their charge or size and can be expanded to include a vast number of lectins with different specificities (Merkle and Cummings, 1987; Osawa and Tsuji, 1987). When SLAC is combined with other more conventional chromatographic procedures, such as high performance liquid chromatography, ion-exchange column chromatography, paper chromatography and paper electrophoresis, one can achieve a high degree of purification for definitive structural characterization. These procedures are outlined by Hounsell (this volume).

4 Uses of Lectins to Study Phenotypes of Cells

In addition to their utility in isolating and characterizing glycoconjugates, lectins can be used to study the phenotypes (or glycotypes) of animal cells. In fact, plant lectins were originally found useful in blood typing because of their ability to discriminate ABO blood groups (Lis and Sharon, 1986). Studies on the glycotypes of cells are usually conducted in either one of two ways. In one approach, a lectin is simply mixed with a suspension of cells in the presence or absence of hapten sugar and the specific agglutination of the cells is observed. For example, human erythrocytes are not agglutinated by either L_4-PHA or MAL, because these cells lack the determinants recognized by these lectins. However, both of these lectins are potent agglutinins of lymphocytes.

In the second approach, which is more quantitative, a biotinylated derivative of a lectin is incubated with cells and fluorescently-labeled streptavidin is added and the fluorescent labeling of the cells is examined by flow cytometry (see for example, Cho *et al.*, 1996). A simpler approach is to directly use a fluorescently labeled form of the lectin without the use of biotin-avidin. The specificity of binding can be demonstrated by the use of hapten sugars to block binding. Such methods also allow populations of cells to be sorted based on their differential display of determinants recognized by the lectin. Fluorescently labeled lectins or peroxidase-labeled lectins can also be used in histochemical studies of cells and tissues to discern expression patterns for glycoconjugates on cell surfaces and in intracellular compartments, such as granules, lysosomes, or Golgi (Pavelka, this volume). WGA is often used in this way as a marker of the Golgi, since this organelle in most cells is the site of the major reaction with this lectin.

5 Uses of Lectins in Solid-Phase Assays for Glycosyltransferases

The exquisite specificity of lectins have also made them useful for assaying glycosyltransferases. In this approach an acceptor for the glycosyltransferase is immobilized in the wells of a microtiter plate and into the well is added a source of enzyme, sugar nucleotide and divalent cations. Carrier-immobilized acceptor structures (neoglycoconjugates) are valuable tools to establish a matrix with accessible reactant (Bovin and Gabius, 1995; Lee and Lee, this volume). The enzyme transfers sugar to the acceptor generating a new "glycodeterminant" recognizable by a specific lectin. After washing the wells, a derivatized form of a lectin is added, excess lectin is removed and the bound lectin is determined by alkaline phosphatase or bioluminescent protein reagents. This approach has been used successfully to assay β1,4-galactosyltransferase, α2,3-sialyltransferase, α2,6-sialyltransferase and α1,3-fucosyltransferase

(Zatta *et al.*, 1991; Mattox *et al.*, 1992; Yan *et al.*, 1994; Yeh and Cummings, 1996). An excellent variation of this approach is to use a soluble derivatized acceptor that is glycosylated by the enzyme in solution-phase. Both the acceptor and product are then captured by an antibody and the product may then be probed with a derivatized lectin (Mengeling *et al.*, 1991).

6 Conclusion

In this chapter the use of lectins as tools to isolate and characterize animal glycans has been described. Lectins provide a simple and direct method to purify and characterize glycoconjugates from complex mixtures and provide a means for a researcher to quickly identify relevant or interesting glycans for study. In addition, they supplement studies with antibodies to carbohydrates, since many lectins recognize determinants not known to be recognized by available antibodies, such as sialic acid linkages and internal branch positions of *N*-glycans. Recently, kits containing panels of derivatized lectins have become available and used in lectin blotting (akin to Western blotting) of glycoproteins after transfer to nitrocellulose filters (Haselbeck and Hosel, 1993). As the fine specificities of an increasing number lectins are mapped (Debray *et al.*, 1994; Du *et al.*, 1994; Misquith *et al.*, 1994) and new lectins discovered, the utility of lectins will undoubtedly expand. It is also anticipated that animal lectins will become more available for isolating specific glycoproteins, as has been demonstrated for the bovine mannose-6-phosphate receptor (Varki and Kornfeld, 1983) and bovine galectin-1 (Merkle and Cummings, 1988).

References

Bovin NV, Gabius HJ (1995): Polymer-immobilized carbohydrate ligands: versatile chemical tools for biochemistry and medical sciences. In *Chem. Soc. Rev.* **24**:413–21.

Cho SK, Yeh J-C, Cho M *et al.* (1996): Transcriptional regulation of α1,3-galactosyltransferase in embryonal carcinoma cells by retinoic acid: masking of Lewis x antigens by α-galactosylation. In *J. Biol. Chem.* **271**:3238–46.

Cummings RD (1993): Structural characterization of *N*-glycans obtained from metabolically radiolabeled glycoproteins. In *Glycobiology: A Practical Approach* (Fukuda M, Kobata A, eds) pp 243–90, Oxford: IRL Press.

Cummings RD (1994): Use of lectins in analysis of glycoconjugates. In *Meth. Enzymol.* **230**:66–85.

Cummings RD, Kornfeld S (1982a): Characterization of the structural determinants required for the high-affinity interaction of asparagine-linked oligosaccharides with immobilized *Phaseolus vulgaris* leukoagglutinating and erythroagglutinating lectins. In *J. Biol. Chem.* **257:**11230–4.

Cummings RD, Kornfeld S (1982b): Fractionation of asparagine-linked oligosaccharides by serial lectin-agarose affinity chromatography. A rapid, sensitive, and specific technique. In *J. Biol. Chem.* **257**: 11235–40.

Debray H, Montreuil J, Franz H (1994): Fine sugar specificity of the mistletoe (*Viscum album L.*) lectin I. In *Glycoconjugate J.* **11**:550–7.

Do S-I, Cummings RD (1992): The hamster transferrin receptor contains Ser/Thr-linked oligosaccharides: use of a lectin-resistant CHO cell line to identify glycoproteins containing these linkages. In *J. Biochem. Biophys. Meth.* **24**:153–65.

Drickamer K, Taylor ME (1993): Biology of animal lectins. In *Annu. Rev. Cell Biol.* **9**:237–64.

Du MH, Spohr U, Lemieux RU (1994): The recognition of three different epitopes for the H-type 2 human blood group determinant by lectins of *Ulex europaeus*, *Galactia tenuiflora* and *Psophocarpus tetragonolobus* (winged bean). In *Glycoconjugate J.* **11**:443–61.

Finne J, Krusius T (1982): Preparation and fractionation of glycopeptides. In *Meth. Enzymol.* **83**:269–77.

Fukushima K, Ohkura T, Kanai M *et al.* (1995): Carbohydrate structures of a normal counterpart of the carcinoembryonic antigen produced by colon epithelial cells of normal adults. In *Glycobiology* **5:**105–15.

Gabius S, Gabius H-J (1993) (eds): *Lectins and Glycobiology*, Heidelberg: Springer Verlag.

Goulut-Chassaing C, Bourrillon R (1995): Structural differences between complex-type Asn-linked glycan chains of glycoproteins in rat hepatocytes and Zajdela hepatoma cells. In *Biochim. Biophys. Acta* **1244:**30–40.

Green ED, Baenziger JU (1989): Characterization of oligosaccharides by lectin affinity high-performance liquid chromatography. In *Trends Biochem. Sci.* **14:**168–72.

Hammarström S, Hammarström ML, Sundblad G *et al.* (1982): Mitogenic leukoagglutinin from *Phaseo-*

lus vulgaris binds to a pentasaccharide unit in *N*-acetyllactosamine-type glycoprotein glycans. In *Proc. Natl. Acad. Sci. USA* **79**:1611–5.

Haselbeck A, Hosel W (1993): Immunological detection of glycoproteins on blots based on labeling with digoxigenin. In *Meth. Mol. Biol.* **14**:161–73.

Hayes BK, Greis KD, Hart GW (1995): Specific isolation of *O*-linked *N*-acetylglucosamine glycopeptides from complex mixtures. In *Anal. Biochem.* **228**:115–22.

Kobata A, Yamashita K (1993): Fractionation of oligosaccharides by serial affinity chromatography with use of immobilized lectin columns. In *Glycobiology: A Practical Approach* (Fukuda M, Kobata A, eds), pp 103–26, Oxford: IRL Press.

Kornfeld R, Ferris C (1975): Interaction of immunoglobulin glycopeptides with concanavalin A. In *J. Biol. Chem.* **250**:2614–9.

Krusius T, Finne J, Rauvala H (1976): The structural basis of the different affinities of two types of acidic *N*-glycosidic glycopeptides for concanavalin A-Sepharose. In *FEBS Lett.* **72**:117–20.

Lis H, Sharon N (1986): Lectins as molecules and as tools. In *Annu. Rev. Biochem.* **55**:35–67.

Mattox S, Walrath K, Ceiler D *et al.* (1992): A solid-phase assay for the activity of CMPNeu Ac: Galβ1,4GlcNAc-R α-2,6-sialyltransferase. In *Anal. Biochem.* **206**:430–6.

Mengeling BJ, Smith PL, Stults NL *et al.* (1991): A microplate assay for analysis of solution-phase glycosyltransferase reactions: determination of kinetic constants. In *Anal. Biochem.* **199:**286–92.

Merkle R, Cummings RD (1987): Lectin affinity chromatography of glycopeptides. In *Meth. Enzymol.* **138**:232–59.

Merkle R, Cummings RD (1988): Asparagine-linked oligosaccharides containing poly-*N*-acetyllactosamine chains are preferentially bound by immobilized calf heart agglutinin. In *J. Biol. Chem.* **263**:16143–9.

Misquith S, Rani PG, Surolia A (1994): Carbohydrate-binding specificity of the B-cell maturation mitogen from *Artocarpus integrifolia* seeds. In *J. Biol. Chem.* **269**, 30393–401.

Ogata S, Muramatsu T, Kobata A (1975): Fractionation of glycopeptides by affinity column chromatography on concanavalin A-Sepharose. In *J. Biochem.* **78**:687–96.

Osawa T, Tsuji T (1987): Fractionation and structural assessment of oligosaccharides and glycopeptides by use of immobilized lectins. In *Annu. Rev. Biochem.* 56:21–42.

Powell LD, Varki A (1995): I-type lectins. In *J. Biol. Chem.* **270**:14243–6.

Roquimore E, Chou T-H, Hart GW (1995): Detection of *O*-linked *N*-acetylglucosamine (*O*-GlcNAc) on cytoplasmic and nuclear proteins. In *Meth. Enzymol.* **230**:443–60.

Shibuya N, Goldstein IJ, Broekaert WF *et al.* (1987): The elderberry (*Sambucus nigra* L.) bark lectin recognizes the Neu5Ac(α2–6)Gal/GalNAc sequence. In *J. Biol. Chem.* **262**:1596–601.

Sueyoshi S, Sawada R, Fukuda M (1994): Carbohydrate structures of recombinant soluble lamp-1 and leukosialin. In *Bioorg. Med. Chem.* **2:**1331–8.

Varki A, Kornfeld S (1983): The spectrum of anionic oligosaccharides released by endo-β-*N*-acetylglucosaminidase H from glycoproteins. Structural studies and interactions with the phosphomannosyl receptor. In *J. Biol. Chem.* **258**:2808–18.

Wang W-C, Cummings RD (1988): The immobilized leukoagglutinin from the seeds of *Maackia amurensis* binds with high-affinity to complex-type Asn-linked oligosaccharides containing terminal sialic acid linked α2–3 to penultimate galactose residues. In *J. Biol. Chem.* **263**:4576–85.

West I, Goldring O (1994): Lectin affinity chromatography. In *Mol. Biotech.* **2**:147–55.

Yamamoto K, Tsuji T, Osawa T (1995): Analysis of asparagine-linked oligosaccharides by sequential lectin affinity chromatography. In *Mol. Biotech*. **3:**25–36.

Yan L, Smith DF, Cummings RD (1994): Determination of GDPFuc:Galβ1,4GlcNAc-R (Fuc to GlcNAc) α1,3-fucosyltransferase activity by a solid-phase method. In *Anal. Biochem.* **223**:111–8.

Yeh Y-C, Cummings RD (1996): Absorbance and light-based solid-phase assays for CMPNeuAc: Galβ1,4GlcNAc-R α2,3-sialyltransferase. In *Anal. Biochem.* **236** in press.

Zatta PF, Smith DF, Cormier MJ *et al.* (1991): A solid-phase assay for β1,4-galactosyltransferase activity in human serum using recombinant aequorin. In *Anal. Biochem.* **194**:185–91.

Zhou Q, Cummings RD (1992): Animal lectins: a distinct group of carbohydrate binding proteins involved in cell adhesion, molecular recognition and development. In *Cell Surface Carbohydrates and Cell Development* (Fukuda M, ed) pp 99–126, Boca Raton FL: CRC Press.

11 Proteoglycans – Structure and Functions

HANS KRESSE

1 Introduction

The definition of a proteoglycan – a protein that is covalently linked with at least one glycosaminoglycan chain – is fulfilled by a wide variety of diverse macromolecules. Proteoglycans, like other glycoproteins, do not possess a unifying functional feature nor are they unique members of subcellular or extracellular compartments. It is now recognized that proteoglycans could best be divided into distinct families which share structural properties of their protein backbone and hence have several functions in common (see for example Evered and Whelan, 1986; Ruoslahti, 1988; Kjellén and Lindahl, 1991; Ruoslahti and Yamaguchi, 1991; for reviews see Hardingham and Fosang, 1992; Jollès 1993). In spite of the proposal that proteoglycans could best be classified according to their core protein structures, important functions of the proteoglycans are also fulfilled by their glycosaminoglycan chains. This is the reason for considering hyaluronate, also named hyaluronan, as an "honorary proteoglycan" although it is synthesized in a protein-free form (see Evered and Whelan, 1989; Laurent and Fraser, 1992; Knudson and Knudson, 1993 for reviews). For these reasons hyaluronate and the most important families of proteoglycans are covered together in this review.

2 Glycosaminoglycan Structure

As outlined by Scott (1993), the basic structure of glycosaminoglycans can best be outlined by defining them as unbranched polymers of anionic derivatives of either glucose-glucose or galactose-glucose-derived disaccharides. Table 1 gives an overview of the basic structures of the glycosaminoglycans. It is seen that hyaluronate is a polyanion because of its repeating disaccharide structure [*N*-acetylglucosamine-β1,4-glucuronic acid β1,3]$_n$. Hyaluronate does not carry sulfate ester groups. Heparan sulfate and heparin differ from

Table 1. Glycosaminoglycan Composition

Building principle	Basic disaccharide structure	Modified and/or substituted monosaccharides	Common names
[Glcβ1-4Glcβ1-3]$_n$	GlcNAcβ1-4GlcAβ1-3	not found	Hyaluronate (Hyaluronan)
[Glcα1-4Glcβ1-4]$_n$	GlcNAcα1-4GlcAβ1-4	GlcNS, GlcNS6S, GlcNAc6S, GlcNS3S6S (GlcA2S), L-IdoA, L-Ido A2S	Heparan sulfate, Heparin
[Galβ1-4Glcβ1-3]$_n$	GalNAcβ1-4GlcAβ1-3	GalNAc4S, GalNAc6S, GalNAc4S6S	Chondroitin sulfate (A,C)
		L-IdoA, L-IdoA2S	Dermatan sulfate
[Galβ1-4Glcβ1-3]$_n$	Galβ1-4GlcNAc	Gal6S, GlcNAc6S	Keratan sulfate

H.-J. and S. Gabius (Eds.), Glycosciences

ISBN 3-8261-0073-5

hyaluronate by the type of linkage between the monomeric constituents and by the wide variety of possible substitutions with ester sulfate and sulfamate groups. When the glucuronic acid moiety becomes epimerized, a C-atom 5 in an *L*-iduronosyl residue in β-linkage is then present. Heparin is distinguished from heparan sulfate by a higher *N*-sulfate/*N*-acetyl ratio and by a larger number of 3-sulfated glucosamine-*N*-sulfate residues.

The repeating disaccharide units of the chondroitin sulfates are composed of *N*-acetyl-galactosamine and glucuronic acid, both monosaccharides being linked with each other β-glycosidically. Unsulfated sequences with this structure are named chondroitin. Chondroitin-4-sulfate contains a GalNac-4-sulfate and chondroitin-6-sulfate a 6-sulfated amino sugar moiety. In contrast to the glucosaminoglycans, the galactosaminoglycan with a C5-epimerized glucuronic acid residue received a separate name, dermatan sulfate. It will be discussed below that glycosaminoglycan chains containing solely *L*-iduronic acid as the hexuronic acid moiety do not exist. There is still no unambiguous convention what the proportion of iduronic acid residues must be to name a molecule dermatan sulfate. In the past, chondroitin sulfate B was used in place of dermatan sulfate.

There is an additional glycosaminoglycan that does not contain a hexuronic acid moiety. Its basic structure consists of β-glycosidically linked galactose and *N*-acetylglucosamine residues. Preferentially the amino sugar, but also galactose, can carry sulfate ester groups on C6.

3 Hyaluronate

3.1 Structure, Biosynthesis and Degradation

Hyaluronate is a linear polymer built from repeating disaccharide units of the structure $[O\text{-}(\beta\text{-D-GlcA})\text{-}(1-3)\text{-}\beta\text{-d-GlcNAc})\text{-}(1-4)]_n$. *In vivo* it is usually a high molecular mass component ($10^6 - 10^7$ Da) of the extracellular matrix and of biological fluids. The contour length of a chain of 4×10^6 Da is about 10μm (Fessler and Fessler, 1966). Physico-chemical studies on the conformation of hyaluronate in solution led to the proposal of an extended helical conformation with a greater flexibility of the 1–4 linkage compared to the 1–3 linkage. In one conformation there are approximately 2.4 residues per turn of the helix, compared with approximately 4.4 residues per turn in the alternative structure (Holmbeck *et al.*, 1994). It is proposed that the secondary structure of hyaluronate leads to a clustering of hydrophobic groups which could form a surface for chain-chain-interactions. A network of aggregated hyaluronate chains has indeed been observed electron microscopically (Scott *et al.*, 1991). The stiffened helical configuration gives the molecule an expanded coil structure in solution, containing about 1 l of immobilized water within the coil structure generated by 1 g of hyaluronate (Laurent, 1970).

Hyaluronate biosynthesis differs in two important aspects from that of other glycosaminoglycans. Hyaluronate synthase is located within the plasma membrane and polymerization occurs on its cytosolic side, whereas it is proposed that the growing chain is transported through a "pore" to the pericellular space. Polymerization of all other glycosaminoglycans occurs in intracellular membrane compartments. The second fundamental difference concerns the chain elongation mechanism, which takes place uniquely at the reducing end. The reducing end is transferred successively to UDP-glucuronic acid and UDP-*N*-acetylglucosamine, which presumably bind to two different sites of the synthase. Thus, a sugar nucleotide is always present at the reducing end of the nascent hyaluronate chain, as in the case of a nascent polypeptide chain at the ribosome which is covalently linked with a tRNA molecule. The factor(s) leading to the termination of chain growth are not yet known (Prehm, 1983, 1989). The gene for hyaluronate synthase from group A streptococci has recently been cloned (DeAngelis *et al.*, 1993; DeAngelis and Weigel, 1994). The eukaryotic hyaluronate synthase is obviously related to the enzyme from streptococci because of the recognition of the human enzyme by antibodies against the streptococcal synthase (Klewes *et al.*, 1993). Eukaryotic hyaluronate synthase could be activated by phosphorylation (Klewes and Prehm, 1994) which may be mediated by growth factors like EGF, PDGF, TGF-β, IGF-I and others (Heldin *et al.*, 1989; Westergren-Thorsson *et al.*, 1990; Sampson *et al.*, 1992).

The degradation of hyaluronate is initiated by hyaluronidases, which in mammals are hyaluronate and chondroitin sulfate-degrading endo-hexosaminidases (see Kresse and Glössl, 1987; Frazer and Laurent, 1989; Rodén *et al.*, 1989 for reviews). The lysosomal-type hyaluronidases are largely inactive at physiological pH, and it had therefore been proposed that cellular hyaluronate catabolism requires receptor-mediated endocytosis of the polysaccharide. Some hyaluronate receptors belong to the CD44 family of lymphocyte homing receptors. Members of this receptor family are responsible for the uptake of hyaluronate by chondrocytes (Hua *et al.*, 1993). Hyaluronate could also be removed from the extracellular matrix through drainage into the lymphatic system and catabolism in regional lymph nodes, or after uptake by liver endothelial cells (Smedsrød *et al.*, 1984; McGary *et al.*, 1989). The receptor has recently been purified from liver endothelial cells and shown to be identical with the interstitial cell adhesion protein ICAM-1 (Forsberg and Gustafson, 1991; Gustafson *et al.*, 1995). The importance of the different hyaluronate receptors is not yet clear. ICAM-1 seems to play a special role during inflammation. However, the majority of ICAM-1 molecules are not accessible for hyaluronate, and the events which make the binding sites available remain to be investigated.

3.2 Selected Functions

It is widely accepted that hyaluronic acid serves as a lubricant in joints and separates tissue surfaces that slide along each other, for example in the fibrils of skeletal muscle. The lubricating function depends on the special rheological properties and the visco-elastic behavior and shear dependence of hyaluronate solutions (Gibbs *et al.*, 1968).

Hyaluronate has also been shown to affect cell behavior, including cell-cell adhesion and cell migration, thereby fulfilling an essential function during embryogenesis. In particular, it is proposed that hyaluronate is an important factor for cell locomotion. Hyaluronate accumulates as cells begin to move during morphogenesis, and it disappears as cell movement ceases (Toole, 1991). Turley and coworkers have characterized a hyaluronate receptor in the cell membrane that mediates hyaluronate-induced cell locomotion. The receptor has therefore been named RHAMM for receptor for hyaluronate-mediated mobility (Hardwick *et al.*, 1992; Yang *et al.*, 1993). It seems possible that binding of hyaluronate to RHAMM is an important factor for regulating the chain length of newly synthesized hyaluronate. Very recently it was shown that RHAMM could transform fibroblasts. When the expression of RHAMM was suppressed, transformed cells were completely non-tumorigenic and non-metastatic. Furthermore, RHAMM was found to act downstream of *ras*. RHAMM participated in the signaling within focal adhesions and mediated signaling via the focal adhesion kinase $pp125^{FAK}$ (Hall *et al.*, 1995).

As stated above, CD44 is one of the hyaluronate receptors. It had been shown recently that the expression of CD44 isoforms that bind hyaluronate augments the rapidity of tumor formation by melanoma cells *in vivo*, whereas the expression of a CD44 mutant which does not mediate cell attachment to hyaluronate fails to do so. These observations, too, suggest that tumor growth depends on the ability of tumor cells to mediate cell attachment to hyaluronate (Bartolazzi *et al.*, 1994).

Hyaluronate might also play a regulatory role in inflammation. Hyaluronate was shown to stimulate phagocytosis and to activate granulocytes. In rheumatoid arthritis it had been shown that the complement component C1q and IgG bind covalently to synovial hyaluronate. These unphysiological linkages may create new antigenic sites and elicit antibodies, which in turn may be responsible for the chronic phase of rheumatoid arthritis (Prehm, 1995).

An analysis of hyaluronate in synovial tissue of arthritic joints indicated that hyaluronate became markedly degraded as a consequence of the disease. Hyaluronate degradation was accompanied by massive oxygen radical production. Radical scavengers protected hyaluronate from degradation. These data demonstrate an enzyme-independent pathway of hyaluronate degradation under pathological conditions (Schenck *et al.*, 1995). Interestingly, in rheumatoid arthritis there is also an upregulation of CD44 in synovial tissue, indicating again the role of hyaluronate and its receptors in the inflammatory disease (Haynes *et al.*, 1991).

4 Proteoglycans

4.1 Glycosaminoglycan Chain Assembly

4.1.1 Chain initiation

Chondroitin sulfate, dermatan sulfate, heparan sulfate and heparin are all bound *O*-glycosidically to a serine residue of the core protein via a characteristic tetrasaccharide linkage region. The serine is most often followed by a glycine residue. The linkage region consists of the sequence GlcA-β1,3-Gal-β1,3-Gal-β1,4-Xyl-β1-*O*-Ser. Keratan sulfate on the contrary is either linked *N*-glycosidically to an asparagine residue of the core protein by a typical linkage for biantennary, hybrid, *N*-linked oligosaccharides found in many glycoproteins (Keller *et al.*, 1981; Nilsson *et al.*, 1983; Oeben *et al.*, 1987). Sialic acid can cap one branch of the oligosaccharide leaving keratan sulfate extending the other branch. In their unsulfated polysaccharide form, i e. as lactosaminoglycans, these molecules are a common glycoprotein species (Hounsell, 1989). In other proteoglycans, keratan sulfate chains are *O*-glycosidically bound to a distinct core protein region. This linkage closely resembles the structure of mucin-type oligosaccharides, and either serine or threonine residues can be utilized. The non-reducing terminal of the keratan sulfate chain can also be capped by an α2,3-linked neuraminic acid residue (Huckerby *et al.*, 1995).

An unusual protein-glycosaminoglycan-protein structure has been found in bikunin proteins which are members of the pancreatic trypsin inhibitor (Kunitz) family. The bikunin proteins are multi-chain plasma proteins composed of bikunin and one or two distinct but homologous heavy chains. These polypeptides are assembled by a chondroitin-4-sulfated chain originating from Ser-10 of bikunin and one or two heavy chains covalently bound by an ester bond between the α-carbon of the C-terminal aspartate residue and carbon-6 of an internal *N*-acetylgalactosamine (Enghild *et al.*, 1991, 1993).

The fact that quite different glycosaminoglycan chains (heparan sulfate and heparin as well as chondroitin sulfate and dermatan sulfate) have a common linkage tetrasaccharide to particular serine sites of the core proteins raises the question as to the determinants for the assembly of a specific glycosaminoglycan chain in core proteins containing Ser-Gly sequences. One of the core proteins named serglycin (see below) can either be substituted with heparin chains, while the same cells also have the latent capacity to synthesize oversulfated chondroitin sulfate-type glycosaminoglycans. In other species, chondroitin sulfate and dermatan sulfate have been observed as substituents of serglycin (see Stevens *et al.*, 1988 and Kresse *et al.*, 1993 for reviews). In the cartilage proteoglycan aggrecan, the Ser-Gly-rich domain of the core protein is used for chondroitin sulfate chain attachment.

When a proteoglycan carries only a single or a few glycosaminoglycan chains the glycosaminoglycan attachment sites most often have the following sequence: Asp/Glu-Xaa-Ser-Gly-Xaa-Gly (Bourdon *et al.*, 1987; Mann *et al.*, 1990). However, the postulated Ser-Gly-Xaa-Gly-sequence is not present in all proteoglycan core proteins and chick decorin contains a Gly-Ser instead of a Ser-Gly (Li *et al.*, 1992). This sequence reversal is also present in the attachment site of the glycosaminoglycan chain in type IX collagen-proteoglycan (McCormick *et al.*, 1987).

From these data it is obvious that there is no short amino acid consensus sequence which can unambiguously be characterized as necessary and sufficient for glycosaminoglycan chain attachment. With regard to the type of glycosaminoglycan chain being synthesized, Shworak *et al.* (1994) demonstrated that in the proteoglycan syndecan-4 (also named ryudocan or amphiglycan) all three functional attachment sites are capable of bearing either heparan sulfate or chondroitin sulfate. Each site functioned in a relatively independent fashion. In comparison with other proteoglycans it was concluded that the functional promiscuity of the glycosaminoglycan sites of syndecan-4 must be a property for which the amino acid composition of the core protein is responsible. In another hybrid proteoglycan containing chondroitin sulfate as well as heparan sulfate chains, named betaglycan (see below), it was found that one site supported synthesis of only chondroitin sulfate and the other one synthesis of both chondroitin sulfate and heparan sulfate (Zhang and Esko, 1994).

4.1.2 Linkage region biosynthesis

The linkage region which, with the exception of keratan sulfate, connects the repeating disaccharide units of all glycosaminoglycans with the protein moiety, is synthesized by the sequential action of specific transferases using nucleotide activated monosaccharides as substrates. Since a xylosylated core protein without repetitive glycosaminoglycan disaccharide units could be isolated from fibroblasts and chondrocytes, and since intermediates of the chain-elongation process have only very short half-lives (Hoppe *et al.*, 1985), it is hypothesized that the components of the linkage region become attached to the core protein in a compartment proximal to the trans-Golgi cisternae where the glycosaminoglycan chains are polymerized. This conclusion is supported by the observation that xylosyltransferase is localized in chondrocytes in the rough endoplasmic reticulum (Hoffman *et al.*, 1984). In rat hepatocytes, however, strong evidence was obtained that xylosyl transfer to heparan sulfate proteoglycan core protein occurred in the Golgi apparatus and not in the endoplasmic reticulum (Nuwayhid *et al.*, 1986).

None of the enzymes involved in the biosynthesis of the linkage region has been purified to homogeneity nor have any of the enzymes been molecularly cloned. The xylosyl transferase can relatively easily be solubilized and forms complexes spontaneously with galactosyltransferase I and galactosyltransferase II, which appear to be separate gene products. A review on the chain initiation is given by Rodén *et al.* (1985).

In addition to xylosylated core proteins, galactosyltransferase I also recognizes artificial acceptors. *p*-Nitrophenyl-xylosylpyranoside and methylumbelliferyl xylosylpyranoside have been widely used as competitors for proteoglycan biosynthesis. Depending on the concentration of the xylosides, dermatan sulfate and chondroitin sulfate polymers can be obtained. Using specific aglycones such as estradiol, heparan sulfate chains can additionally be induced (Lugemwa and Esko, 1991; Fritz *et al.*, 1994a).

The linkage region of native proteoglycans as well as of artificially initiated chains can be phosphorylated at C-atom2 of the xylose unit (Oegema *et al.*, 1984; Fransson *et al.*, 1985; Glössl *et al.*, 1986; Greve and Kresse, 1988). Sulfate ester groups can also be present at the galactose moiety of the linkage region (Sugahara *et al.*, 1992a,b).

We have observed a single patient who represented a progeroid variant with signs of the Ehlers-Danlos syndrome (Kresse *et al.*, 1987). The patient's disorder could be traced back to a mutant galactosyltransferase I protein (Quentin *et al.*, 1990). The partial inactivity of this enzyme resulted in the secretion of mature and of unglycanated forms of the small proteoglycans decorin and biglycan. Although evidence for the existence of more than one galactosyltransferase I is lacking (Esko *et al.*, 1987), the patient's skin fibroblasts produced normally behaving heparan sulfate proteoglycans and large proteoglycans. Since the patient could produce a considerable proportion of mature decorin and biglycan, it seems reasonable to postulate sufficient residual activity for apparently normal production of proteoglycans carrying several glycosaminoglycan chains.

4.1.3 Glycosaminoglycan chain polymerization

In the biosynthesis of heparin and heparan sulfate, the first specific step is the attachment of an α1,4-glycosidically-linked *N*-acetylglucosamine residue to the linkage region tetrasaccharide. This step is catalyzed by a specific enzyme which is not involved in the subsequent polymerization reaction (Fritz *et al.*, 1994b). The subsequent alternating transfer of *N*-acetylglucosamine and glucuronic acid residues to the non-reducing termini of the chain occurs by a single 70-kDa protein. The dual catalytic activities of the protein presumably require two different catalytic sites (Linde *et al.*, 1993). The existence of a single enzyme with dual transferase activities explains the previous findings of a single mutation in Chinese hamster ovary cells which affected both *N*-acetylglucosaminyltransferase and glucuronosyltransferase activities (Lidholt *et al.*, 1992).

During the course of the polysaccharide chain polymerization, the nascent precursor becomes progressively modified. These modifications comprise a series of successive reactions that occur in a strictly regulated order, because the first modification reactions are prerequisites for the action of the following enzymes. It is important to realize that these modifications occur in a blockwise manner, but they are incomplete, thereby creating a variety of specific sequences that may be responsible for distinct biological functions.

The first modification reaction consists of the enzymatic removal of *N*-acetyl groups and the subsequent formation of sulfamate groups in the presence of 3'-phosphoadenylylsulfate. A single 110 kDa protein catalyzes deacetylation as well as sulfation (Pettersson *et al.*, 1991; Wei *et al.*, 1993; Eriksson *et al.*, 1994; Orellana *et al.*, 1994).

The next modification reaction consists of a C5-epimerization of glucuronic acid to *L*-iduronic acid. In a Chinese hamster ovary cell mutant defective in *N*-deacetylase/*N*-sulfotransferase, the content of *L*-iduronic acid residues was reduced as much as the content of sulfamate groups (Dame *et al.*, 1991). The enzyme, which has recently been purified (Campbell *et al.*, 1994), has been shown to require *N*-sulfation of at least one amino sugar adjacent to the glucuronic acid residue which is to be epimerized. The subsequent sulfation at C-2 of the iduronosyl residue prevents "back epimerization" to glucuronic acid.

A glucosamine 6-sulfotransferase has recently been purified which catalyzed 6-sulfation of glucosamine *N*-sulfate residues adjacent to iduronic acid, but not adjacent to glucuronic acid. *N*-acetylglucosamine residues were not sulfated (Habuchi *et al.*, 1995). Nevertheless, 6-sulfated *N*-acetylglucosamine residues can be found in heparin and heparan sulfate preparations, and it remains therefore an open question whether or not different 6-sulfotransferases do exist. There is a further sulfotransferase forming ester groups on the C3 of the amino sugar. This enzyme requires a specific sequence in sulfate distribution in the substrate, thereby performing the last step in the biosynthesis of the antithrombin binding site of heparin (Razi and Lindahl, 1995).

A summary of the enzymatic processes involved in heparan sulfate/heparin biosynthesis is given in Figure 1.

The properties of the enzymes required for chondroitin and dermatan sulfate biosynthesis are less well known. There is evidence that the first *N*-acetylgalactosamine residue which is β-glycosidically linked to the glucuronic acid moiety of the linkage region is transferred by an enzyme other than the one involved in chain-elongation (Rohrmann *et al.*, 1985). Whether the subsequent chain polymerization by glucuronosyl and *N*-acetylgalactosaminyl transferases occurs by a single enzyme as in heparin/heparan sulfate or by two separate proteins is not known. Since chondroitin sulfate bears sulfate ester groups at position 6 or 4 of *N*-acetylgalactosamine residues, the existence of two different sulfotransferases has been proposed. An *N*-acetylgalactosaminyl 6-sulfotransferase involved in chondroitin sulfate biosynthesis has recently been cloned (Fokuta *et al.*, 1995). The sequence predicts a transmembrane domain, as was to be expected from the localization of chondroitin sulfate biosynthesis in trans-Golgi cisternae. Interestingly, the enzyme was also found to catalyze the sulfation of keratan sulfate. Molecular details of an *N*-acetylgalactosaminyl 4-sulfotransferase are not available in the literature. In mammalian tissues the ratio of 6-sulfation and 4-sulfation varies with development (Habuchi *et al.*, 1986), malignant change (Adany *et al.*, 1990), and susceptibility to arteriosclerosis (Edwards and Wagner, 1988).

As in the case of heparin/heparan sulfate biosynthesis, epimerization of D-GlcA to L-IdoA occurs at the polymer level, requiring probably a separate enzyme. Epimerization was shown to be tightly connected to 4-sulfation of the parent polysaccharide chain (Malmström, 1984). It

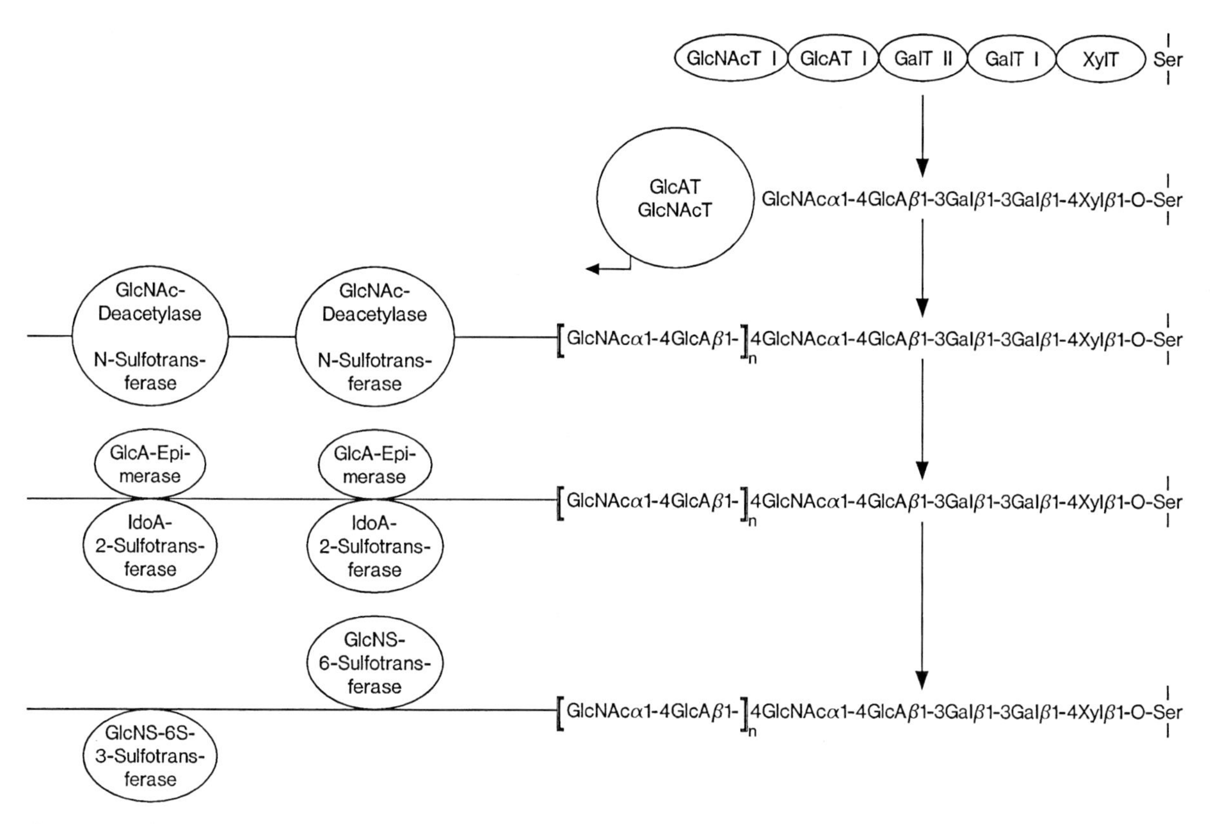

Figure 1. Scheme of heparan sulfate biosynthesis. The abbreviations are:
GalT, galactosyltransferase; GlcA-Epimerase, glucuronate 5-epimerase; GlcAT, glucuronosyl-transferase; GlcNAc-Deacetylase, *N*-acetylglucosamine acetylesterase; GlcNAcT, *N*-acetyl-glucosaminyltransferase; GlcNS-6-Sulfotransferase, glucosamine-*N*-sulfate 6-sulfotransferase;
GlcNS-6S-3-Sulfotransferase, 6-sulfoglucosamine-*N*-sulfate 3-sulfotransferase; IdoA-2-Sulfo-transferase, *L*-iduronate 2-sulfotransferase; *N*-sulfotransferase, glucosamine *N*-sulfotransferase; XylT, xylosyltransferase.

appears that epimerization is directly followed by 4-sulfation of the *N*-acetylgalactosamine residue immediately proximal to the epimerized uronic acid. Iduronic acid residues may also become sulfated in position 2.

From these data it follows that dermatan sulfate chains are mainly composed of the following four disaccharides: GlcA-GalNAc4S, IdoAGalNAc4S, GlcA-GalNAc6S, IdoA2SGalNAc4S. However, dermatan sulfate is also characterized by a block structure, and this block structure differs in dermatan sulfate species from different sources. In skin dermatan sulfate for example, GlcAGalNAc6S was the first disaccharide at the reducing end of the chain and iduronic acid-rich disaccharides were most often found in positions 4–7 and around position 15. In the sclera, the first disaccharide was epimerized and was followed by two glucuronic acid-containing disaccharides. The reasons for the different locations and for the wide diversity of the relative amount of GlcA and IdoA is unclear. The protein core does not play any major direct role in iduronic acid formation, although it is probably of importance in sorting the various cores to different compartments along the secretory route. An important role may be exerted by the linkage of sulfation and epimerization. In the presence of the ionophore monensin, which disturbs the transport in the medial part of the Golgi apparatus, only a minor influence on 6-sulfation, but a dramatic one on epimerization and 4-sulfation, was observed (Hoppe *et al.*, 1985). One could, therefore, speculate that glycosaminoglycan chain polymerization and 6-sulfation occur in an earlier compartment than epimerization and 4-sulfation.

4.2 Large Hyaluronate-Binding Matrix Proteoglycans

There is a limited number of unique protein domains which during evolution have been utilized in the formation of large multidomain proteins (Engel, 1991). The large hyaluronate-binding proteoglycans of the extracellular matrix represent such a well characterized family of proteins, in which characteristic motifs are assembled to allow the proteins to fulfill specific biological functions in their proteoglycan form.

4.2.1 Family members

Among the family of large hyaluronate-binding proteoglycans, aggrecan, a proteoglycan from cartilage, has received the greatest attention and represents a textbook prototype of a proteoglycan. Its core protein of about 225–250 kDa has been molecularly cloned (Doege *et al.*, 1987, 1991; Chandrasekaran and Tanzer, 1992). The gene of human aggrecan has been assigned to 15q25 (Just *et al.*, 1993; Korenberg *et al.*, 1993). Electron microscopical studies showed that aggrecan contains a globular domain (G1-domain) at the *N*-terminus, which serves as a hyaluronate-binding region and interacts with a protein of similar structure that can also interact with hyaluronate. The G1-domain is separated by an extended region of about 12 kDa followed by a second globular domain, G2. These three domains contain glycoprotein-type oligosaccharides which are *N*- or *O*-glycosidically linked. Some of the *N*-glycosidically linked oligosaccharides may be processed to keratan sulfate chains. The G2-domain is followed by a region which carries 20–30 *O*-glycosidically linked keratan sulfate chains. Then two regions follow which are substituted with about 100 chondroitin sulfate and a few additional keratan sulfate chains. The C-terminal G3-domain is a composite of several structural motifs, and contains an epidermal growth factor-like sequence which is variably expressed in human aggrecan mRNA, a lectin-like and a complement-regulatory domain; the latter may also be lacking due to alternative splicing.

Both the G1-domain and the link protein bind to a decasaccharide unit of hyaluronate via homologous tandem repeat loops resembling immunoglobulin folds. A single hyaluronate molecule of 10^6 Da may form a supramolecular aggregate with 100 aggrecan and link protein molecules, raising the molecular mass of the complex to about 200×10^6 Da (see Hardingham and Fosang, 1992; Upholt *et al.*, 1993 for reviews).

In other tissues in cartilage, i. e. skin, bone, aorta and nervous tissue, a proteoglycan homologous to aggrecan was detected. Since its amino acid sequence, deduced from cloned cDNA, suggested the ability to perform versatile functions, it received the name versican (Zimmermann and Ruoslahti, 1989). Versican differs from aggrecan in the absence of the G2-domain and by the presence of many fewer glycosaminoglycan attachment sites. Common features of the proteoglycan are the hyaluronate-binding domain at the amino terminus and the epidermal growth factor-like, the lectin-like, and the complement regulatory protein elements in the carboxy-terminal portion of the molecule. Several splice variants differ in the number of glycosaminoglycan attachment domains and in their tissue expression pattern (Dours-Zimmermann and Zimmermann, 1994). A homologous proteoglycan (PG-M) was first described in chicken but PG-M has also been detected in mammals, although it appears to be the product of the versican gene. There exist alternatively spliced forms without a chondroitin sulfate attachment region (Zako *et al.*, 1995). The versican gene has been localized to 5q12–5q14 (Iozzo *et al.*, 1992). The presence of 60–70 kDa-hyaluronic acid-binding proteins named hyaluronectin in brain and other tissues is probably the result of proteolysis of versican in the tissue or during protein purification (Margolis and Margolis, 1993).

From nervous tissue, a chondroitin sulfate proteoglycan has been isolated which contained three chondroitin sulfate chains and a core glycoprotein of 245 kDa in 7-day rat brains and of 150 kDa postnatally. Molecular cloning indicated that both forms are derived from the same gene product, designated neurocan. Neurocan, although a protein of only 136 kDa, contains a similar hyaluronate binding site to aggrecan and versican and it also contains two EGF-like, a lectin-like and a complement regulatory-like domain at the C-terminus (Rauch *et al.*, 1992).

There is a further family member which has only a very short glycosaminoglycan attachment region and was therefore named brevican (Yamada *et al.*, 1994).

4.2.2 Functions

The main function of aggrecan is certainly a physicochemical one. Aggrecan is present at very high concentrations within the cartilage matrix in an underhydrated state. In fact, only about 25 % of its water-binding capacity is available due to the tight meshwork of collagen fibrils. Therefore, the proteoglycans exert considerable pressure, assuring resilience of the tissue to external compressive forces. At the same time, however, they control diffusion and transport of ions through the cartilage matrix (Maroudas *et al.*, 1991). Interestingly, aggrecan has also been found in bovine deep flexor tendons in regions subjected to compression (Vogel *et al.*, 1994). In rheumatoid arthritis and osteoarthritis there is increased formation of aggrecan fragments, generated primarily by an "aggrecanase" which cleaves the interglobular domain between G1 and G2 (see Hardingham and Fosang, 1995 and Lohmander, 1995, for most recent reviews).

The functional importance of aggrecan is best elucidated in mouse cartilage matrix deficiency. This model disease is caused by a 7bp deletion in the aggrecan gene (Watanabe *et al.*, 1994) and results in disproportionate dwarfism, cleft palate and osteochondrodysplasia.

The functions of versican and neurocan are much less well understood. Versican was shown to have antiadhesive effects in cultured cells, and could therefore play a role in growth control (Yamagata *et al.*, 1993; Yamagata and Kimata 1994). It has been suggested that versican participates in the formation of the elastic network of the skin, but the consequences of the colocalization of versican and microfibrillar proteins are not yet known (Zimmermann *et al.*, 1994).

An interesting functional aspect of neurocan resulted from studies of the neural cell adhesion molecules *N*-CAM and Ng-CAM that are expressed on neurons, and which have been implicated in various aspects of cell adhesion, cell migration and neurite fasciculation. Low concentrations of neurocan inhibited the homophilic interaction between these adhesion molecules. It appeared, therefore, that the extracellular chondroitin sulfate proteoglycans may act as repulsive molecules which modulate cell-cell and cell-matrix interactions, providing a mechanism for diminishing adhesive forces (Margolis and Margolis, 1993).

4.3 Small, Leucine-Rich Extracellular Matrix Proteoglycans

4.3.1 Family members

Small leucine-rich proteoglycans of the extracellular matrix are characterized by long arrays of leucine-rich repeat motifs of about 24 amino acids in length. The structural motifs have been detected in many species, from yeast to man (Hashimoto *et al.*, 1988). Recently, the first complete high resolution X-ray study of a cytosolic protein made of leucine-rich repeats (ribonuclease inhibitor) has been reported (Kobe and Deisenhofer, 1993). A novel tertiary structure was elucidated consisting of a set of β-turns which is enclosed in an external circle of α-helices resulting in a horseshoe shape. As shown below, the leucine-rich proteoglycans contain 10–12 such leucine-rich repeats. Secondary structure predictions according to the Chou and Fasman-method suggest, however, that one of the leucine-rich repeats in the middle of the molecule is unlikely to fit into the shape of a horseshoe. It is nevertheless reasonable to assume that the small proteoglycans exhibit bipartite features of a horseshoe and that these structures are the prerequisites for a variety of protein-protein interactions. One member of the family of small proteoglycans, decorin, had previously been considered to contain a globular core protein. A re-analysis of rotary-shadowed decorin suggested indeed its horseshoe shape (Scott and Cummings, 1995).

Protein and cDNA sequencing indicated the existence of at least five members of the family of leucine-rich proteoglycans of the extracellular matrix: biglycan (Fisher *et al.*, 1989; Neame *et al.*, 1989), decorin (Krusius and Ruoslahti, 1986; Day *et al.*, 1987), fibromodulin (Oldberg *et al.*, 1989; Antonsson *et al.*, 1993), lumican (Blochberger *et al.*, 1992) and proteoglycan-Lb (Shinomura and Kimata, 1992). Biglycan and decorin are interstitial chondroitin/dermatan sulfate proteoglycans. Biglycan received its name because it is most often but not always substituted with two glycosaminoglycan chains. Decorin, which carries a single polysaccharide chain only, decorates the surface of interstitial collagen fibrils (Scott, 1988). Fibromodulin, which is a keratan sulfate proteoglycan in some species (Plaas *et al.*, 1990) but is glycosaminoglycan-free in others, affects collagen fibril formation. Lumican is a major constituent of the corneal stroma, and may play a significant role in the acquisition of corneal transparency by regulating collagen fibril diameter and interfibrillar spacing (Rada *et al.*, 1993).

The core proteins of biglycan and decorin are characterized by a putative propeptide which is cleaved before the proteoglycan is secreted (Yeo *et al.*, 1995). The mature core proteins can be characterized as follows: The glycosaminoglycan attachment sites are located near the *N*-terminus in a hypervariable region. This site is followed by a cysteine-rich region, and the central portion of the core protein contains the characteristic leucine-rich repeats where the sequence of Leu-Xaa-Xaa- Leu-Xaa-Leu-Xaa-Xaa-Asn-Xaa-[Leu-Ile]-[Ser, Thr]-Xaa-[Val, Ile] is followed by a more variable sequence of 7–10 amino acids. In the C-terminal portion there is again a conserved disulfide loop. In humans, the gene for biglycan has been mapped to Xq27-q28 (McBride *et al.*, 1990; Traupe *et al.*, 1992), and that for decorin to 12q21-q22 (Pulkkinen *et al.*, 1992), where the gene for lumican was also found (Chakravarti *et al.*, 1995). Normally, biglycan contains two glycosaminoglycan chains which are linked with serine 5 and serine 10 of the mature core protein. However, neither of these sites is necessarily used all the time. In decorin the single glycosaminoglycan chain is linked to serine 4 of the mature core protein. Since the glycosaminoglycan chain composition is not under direct genetic control, a tremendous variability in the fine structure of the polysaccharide of both proteoglycans has been reported. However, within a given tissue the composition of the glycosaminoglycan chains is similar (Cheng *et al.*, 1994).

There is a developmentally regulated program of expression of biglycan and decorin, at least in bone and cartilage. In bone, biglycan levels were high in fetal cells and in cells derived from pubescent donors. Decorin was produced maximally in cells from adolescent donors (Fedarko *et al.*, 1992), suggesting that biglycan may have a growth-promoting function. Several reports on the influence of cytokines on expression of small proteoglycans agree that TGF-β is a potent reagent in upregulating biglycan. Divergent results, however, were found when the influence of the cytokine on decorin expression was investigated. Marked upregulation, marginal effects or even downregulation were observed (see Kresse *et al.*, 1993, for a review). These differences in TGF-β responsiveness are remarkable as there is a TGF-β-negative element in the 5' untranslated region of decorin cDNA (Danielson *et al.*, 1993). Glucocorticoids, interleukin-1, cell density and other factors were shown to be potent regulators of decorin expression (Kähäri *et al.*, 1995).

4.3.2 Functions of small leucine-rich proteoglycans

Small leucine-rich proteoglycans have at least two functions. They are involved in matrix assembly and in the control of cell adhesion and cell proliferation.

Biglycan, decorin, fibromodulin and lumican all associate with type I and/or type II collagen via their core proteins, the interaction of biglycan being of lower affinity (Scott, 1988, 1992; Vogel *et al.*, 1984; Schönherr *et al.*, 1995a). Decorin appears regularly and orthogonally arrayed at the d-band of the gap region of type I collagen fibrils, which is probably also the location of biglycan. Fibromodulin has been shown to be located at a different site (Hedbom and Heinegård, 1993). In bone, decorin was not found at the d-band, which is the first place where apatite crystals are deposited along collagen fibrils. Hence, it was assumed that decorin-collagen interactions in soft connective tissues are responsible for preventing mineralization. As decorin binds to the surface of collagen fibrils, the lateral assembly of individual triple helical collagen molecules is delayed, and the final diameter of the fibrils become thinner (Vogel *et al.*, 1987). It was shown recently that decorin contains several domains which interact with reconstituted collagen fibrils and that the binding site to preformed fibrils is not necessarily the same site which is involved in the regulation of the kinetics of fibril formation (Schönherr *et al.*, 1995b).

While the importance of the core protein for collagen binding is firmly established, the role of the glycosaminoglycan chain is less clear. It has been proposed that they are responsible for regulating the distance between adjacent collagen fibrils. Dermatan sulfate chains are able to self-associate (Fransson *et al.*, 1982) thereby determining the distance between adjacent collagen fibrils (Scott, 1993). Recent data, however, show that the protein moiety is also engaged in dermatan sulfate-decorin interactions (Bittner *et al.*, in press).

Functions of small leucine-rich proteoglycans in the control of cell proliferation have been deduced from a variety of observations (see Ruoslahti and Yamaguchi, 1991, for review). Decorin and biglycan have antiadhesive properties, at least for fibroblasts, to a variety of substrates (Winnemöller *et al.*, 1991; Bidanset *et al.*, 1992). The observation that over-expression of decorin in Chinese hamster ovary cells inhibited cell proliferation (Yamaguchi and Ruoslahti, 1988) because of a decorin-mediated block of transforming growth factor-β activity (Yamaguchi *et al.* 1990) was received with great attention. This growth inhibition, however, has not been observed by other investigators (Bittner *et al.*, in press), and it has been shown that *in vitro* only selected functions of TGF-β are inhibited upon complex formation with decorin (Hausser *et al.*, 1994). Nevertheless, decorin immobilized in the extracellular matrix by binding to collagen fibrils may still be able to interact with the cytokine, and may therefore indeed be useful as a therapeutic agent in fibrotic diseases caused by overproduction of TGF-β (Border *et al.*, 1992).

A further interesting observation concerns the increased expression of decorin in the stroma of human colon carcinoma (Adany *et al.*, 1991). This overexpression could be traced back to hypomethylation of the decorin gene promotor (Adany and Iozzo, 1991). It has recently been shown that overexpression of decorin in the tumor cells, in contrast to overexpression in the stroma cells surrounding the tumor, suppresses the malignant phenotype (Santra *et al.*, 1995). These data suggest that decorin is an antiproliferative molecule and a possible tumor suppressor, synthesized by the host stroma cells to counteract and block tumor cell growth *in vivo*.

As mentioned in section 3.1.2, a patient with an impaired capacity to substitute decorin and biglycan with glycosaminoglycan chains had progeroid appearance, growth failure, developmental delay, lax skin and bone and teeth anomalies (Kresse *et al.*, 1987). These findings support the proposal of diverse functions for members of the small proteoglycan family.

4.4 Basement Membrane Proteoglycans

4.4.1 Prototype member

Although basement membranes contain several proteoglycans, only one component has been characterized in detail. It was given the name perlecan because of the rotary shadowing images showing 5–7 globular domains with small connecting rods like pearls on a string (Paulsson *et al.*, 1987). Perlecan is widely distributed in the animal kingdom from the nematode *C. elegans* to humans. In man, the cDNA encodes for a 267 kDa protein (Kallunki and Tryggvason, 1992; Murdoch *et al.*, 1992). The perlecan gene is in the telomeric region of human chromosome 1 and contains with 94 exons, one of the largest human genes (Dodge *et al.*, 1991; Cohen *et al.*, 1993).

Perlecan exhibits an interesting domain structure. At the *N*-terminal end there is a unique region rich in acid amino acids which, in basement membranes, carry three heparan sulfate chains. This domain is followed by four cysteine-rich repeats homologous to similar regions of the LDL-receptor. In the third domain, there is homology with the short arm of the α-chain of the basement membrane protein laminin. This region is followed by a domain which harbors the largest collection of IgG repeats so far described, and which is also homologous to *N*-CAM. At the C-terminal end there are three globular and four epidermal growth factor-like motifs, which are homologous to laminin α-chains.

Perlecan is not exclusively a component of basement membranes. It is present in the pericellular matrix of a number of organs, and perlecan message has been found in fibroblasts and chondrocytes. In cartilage, however, perlecan is substituted with both chondroitin and heparan sulfate side chains (SundarRaj *et al.*, 1995).

There are several other proteoglycans which can be found in basement membranes. A chondroitin sulfate proteoglycan isolated from Reichert's membrane and from kidney consists of a 150 kDa core protein and about 20 chondroitin sulfate chains (McCarthy *et al.*, 1989). A heterodimeric large dermatan sulfate proteoglycan has also been found in some but not all basement membranes (Schittny *et al.*, 1995). These proteins, however, have not yet been sequenced, and it is not known whether they belong to the perlecan family or not.

4.4.2 Functions

There are several important biological functions that can be attributed to basement membrane proteoglycans. They were found to be essential for branching morphogenesis, probably because of their matrix protein-binding and cell-binding properties (see Timpl, 1993, for a review). Binding partners of perlecan are laminin and type IV collagen. Nidogen, as a further component of basement membranes, is able to mediate the formation of a complex between perlecan and laminin, exclusively by protein-protein interactions, i.e. in a heparan sulfate-independent fashion (Battaglia *et al.*, 1992). The cell-binding properties of perlecan are mediated by β1-integrins. Cell binding could not be attributed to the core protein or the heparan sulfate chains alone and, therefore, apparently requires the cooperation of both structures.

With regard to the filtration control of basement membranes, it is accepted that heparan sulfate with its polyanionic binding sites forms a charge-selective filter. Proteinuria, for example, is associated with abnormalities of perlecan biosynthesis (Farquhar, 1991). Further functions that could be attributed to the heparan sulfate chains of perlecan are discussed in the section on plasma membrane proteoglycans.

4.5 Integral Membrane Heparan Sulfate Proteoglycans

4.5.1 Families and their members

Heparan sulfate proteoglycans are regular components of plasma membranes and are perhaps absent only in the plasma membrane of erythrocytes. These plasma membrane proteoglycans can be grouped into several distinct families (see Bernfield *et al.*, 1992 and David, 1993, for reviews). All of them have in common that the heparan sulfate chains point towards the extracellular space. The largest family is that of the syndecans, which are all small cysteine-free type I membrane proteins. The original member, syndecan-1, consists of a transmembrane core protein of about 31 kDa which is linked with up to 3 heparan sulfate and 1 or 2 chondroitin sulfate chains (Saunders *et al.*, 1989). Syndecan-1 is a major proteoglycan in epithelial cells but only a minor one in fibroblasts. Syndecan-2, which is also called fibroglycan, has a core protein of about 20 kDa, is apparently linked exclusively with heparan sulfate chains, and is a predominant component in fibroblasts (Marynen *et al.*, 1989). Syndecan-3 or *N*-syndecan has the largest core protein (41 kDa) of any of the members of this family, and is found predominantly but not exclusively in neural tissues (Carey *et al.*, 1992; Gould *et al.*, 1992). The last member, syndecan-4, also named amphiglycan and ryudocan, occurs nearly ubiquitously and has the smallest core protein (19.7 kDa) (David *et al.*, 1992; Kochima *et al.*, 1992). Syndecan-4 is selectively enriched in focal adhesions (Woods and Couchman, 1994).

All members of the syndecan family carry conserved tyrosine residues in the cytosolic tail of their core proteins. Emerging data suggest that the expression of the individual syndecan family members is highly regulated. Transcriptional control and post-translational regulatory mechanisms have been observed (Kim *et al.*, 1994).

Another heparan sulfate proteoglycan is connected with the cell surface via a covalent linkage to glycosyl phosphatidylinositol and is, therefore, in contact only with the extracytosolic leaflet of the plasma membrane. This proteoglycan has been named glypican (David *et al.*, 1990) and there probably exist other family members to be characterized in detail. Glypican consists of a core protein of about 62 kDa which is substituted with heparan sulfate chains close to the cell surface. It is expressed by many different cell lines and is a major proteoglycan of the brain. Glypican appears to be transported by specific routes after its insertion into the plasma membrane. It can be internalized, recycled via the Golgi apparatus where heparan sulfate chains can be added, and finally can be re-deposited at the cell surface (Fransson *et al.*, 1995).

There are other transmembrane proteoglycans which are not structurally related to the syndecans or to glypican. Betaglycan has been described as a non-signaling TGF-β-receptor of type III. It may either carry heparan sulfate and/or chondroitin sulfate chains or it may be glycosaminoglycan-free (Andres *et al.*, 1989). Furthermore, CD44, the hyaluronate receptor mentioned above, may either be a proteoglycan or not, and can be linked with chondroitin sulfate or, in its epidermal splice variant, with heparan sulfate (Kugelman *et al.*, 1992).

4.5.2 Functions of integral membrane heparan sulfate proteoglycans

So far, specific functions of the core protein moieties of the integral membrane heparan sulfate proteoglycans concerned intracellular trafficking and anchoring, as well as specific interactions with TGF-β and hyaluronate in the cases of betaglycan and CD44, respectively. It was first shown by Lindahl and coworkers that heparan sulfate-derived oligosaccharides interact specifically and selectively with proteins. The classic example represents the interaction of a heparin/heparan sulfate-derived pentasaccharide of the sequence GlcNS/NAc6Sα1–4GlcAβ1-4GlcNS3S6Sα1–4L-Ido2Sα1–4GlcNS6S with antithrombin III (see Kjellén and Lindahl, 1991, for a review). A wealth of information has now been accumulated about various ligands of heparan sulfate, showing that these interactions are most often highly specific and depend on defined sugar sequences within the glycosaminoglycan chains. Selected examples are given in Table 2. References can be found in the reviews mentioned above. The incomplete list of high affinity binding molecules makes it obvious that membrane-associated heparan sulfate is of central importance for growth control, cell-cell recog-

Table 2. Examples of proteins interacting with heparan sulfate/heparin

Protease inhibitors: antithrombin III, heparin cofactor II, plasminogen activator inhibitor-3, protease nexin I.
Growth factors: fibroplast growth factors, hepatocyte growth factor/scatter factor, vascular endothelial growth factor, heparin-binding epithelial growth factor.
Chemokines and related proteins: MIP-Iβ, interleukin-8, interleukin-10, interferon-γ
Adhesion molecules: L-selectin, PECAM(CD31).
Matrix components: fibronectin, thrombospondin, laminin.
Receptors: FGF receptor, LDL receptor-related protein, decondrin endocytosis receptor.
Enzymes: lipoprotein lipase, superoxide dismutase.
Infectious agents: herpes simplex virus, HIV-1, malaria sporozoites, trypanozomas.

nition and cell matrix adhesion. It is, however, likely that the different heparan sulfate proteoglycans can at least in part substitute for each other in these functions. There is no unambiguous example of a specific saccharide sequence unique to a single member of the heparan sulfate proteoglycan families. Nevertheless, it may turn out that the enrichment in syndecan-4 in focal adhesions, or the preferential transfer of glypican to the apical site of polarized cells gives rise to a unique function. The general importance of heparan sulfate is best demonstrated by the findings of Esko *et al.* (1988) that Chinese hamster ovary cell mutants which have a reduced capacity for heparan sulfate biosynthesis are tumorigenic in nude mice, whereas wild-type cells are not.

4.6 Intravesicular Proteoglycans

4.6.1 Prototype family

Intravesicular proteoglycans have been isolated from many hematopoietic cells, e.g. from mast cells, basophilic and promyelocytic leukemia cells, eosinophils, natural killer cells, platelets and lymphocytes (see Stevens *et al.*, 1988 and Kresse *et al.*, 1993 for reviews). They all have a core protein with an extended sequence of alternating serine and glycine residues, which can be heavily substituted with chondroitin sulfate and/or heparin chains. Because of these structural features, these proteoglycans are named serglycins. The primary translation products of the single serglycin gene undergoes an extreme variety of post-translational events. In rat and mouse serosal mast cells, serglycins mature to proteoglycans of 750–1000 kDa, whereas mucosal mast cells contain 150 kDa and bone marrow mast cells 200 kDa proteoglycans. This indicates that the number and/or the length of the glycosaminoglycan chains may vary considerably. As mentioned above, at least some types are able to substitute the core protein with different glycosaminoglycan chains. The same mast cell type, for example, may produce heparin and oversulfated chondroitin sulfate-type proteoglycans simultaneously, but the two glycosaminoglycan types are bound to different core protein molecules.

Some populations of synaptic vesicles and constitutive secretory vesicles, in contrast to those containing regulated secretory proteins, contain heparan sulfate proteoglycans (Tooze and Huttner, 1990). Their protein structures, however, have not yet been determined.

4.6.2 Functions

Serglycins are thought to serve specific functions while they are stored in secretory vesicles, as well as after their release into the extracellular space. A couple of basic proteases are complexed to the proteoglycans, resulting in enzyme inactivation or modulation of proteolytic activity. The polyanionic properties of serglycins may serve to maintain electrical neutrality, to concentrate secretory material efficiently, and to reduce the osmotic pressure in histamine-rich vacuoles. In natural killer cells, the perforins are inactivated by complex formation with heparin at the acidic pH of the secretory granule. After discharge of the vesicle the complex dissociates at neutral pH, thereby allowing target cell lysis. In other instances, the complexes may remain stable and the proteoglycan may serve to delay diffusion and to retain the complex in the extracellular matrix.

4.7 Optional Proteoglycans

With the progress made on proteoglycan research during recent years it became evident that a considerable number of proteins could exist in a glycosaminoglycan-containing and a glycosaminoglycan-free form. The factors governing the attachment of a glycosaminoglycan chain to such proteins, also named "part-time" proteoglycans, are not known, and often it is not firmly established what the functional consequences of the glycosaminoglycan chain attachment might be. A few examples are given below.

It has already been mentioned that the hyaluronate receptor CD44 may carry a chondroitin sulfate chain. A functionally related example of an optional proteoglycan is the invariant chain (Ii), which associates with major histocompatibility complex class II molecules, and has been shown to mediate several functions in class II-restricted antigen presentation. A small proportion of Ii is expressed at the cell surface in association with newly synthesized class II molecules, and this cell surface form is specifically the chondroitin sulfate proteoglycan. CD44 binds directly to the proteoglycan but not to the chondroitin sulfate-free invariant chain, suggesting that the proteoglycan variant functions as an accessory molecule at the cell surface, which facilitates T-cell interaction with antigen-presenting cells (Naujokas *et al.*, 1993).

Another optional proteoglycan which is a constituent of plasma membrane is thrombomodulin. Thrombomodulin is an endothelial cell surface anticoagulant which exists in a chondroitin sulfate-containing and a chondroitin sulfate-free form. The proteoglycan form is more effective in the inhibition of thrombin-clotting activity, and it also accelerates the inhibition of thrombin by antithrombin-III (Lin *et al.*, 1994).

A further example of an optional chondroitin sulfate proteoglycan is a macrophage growth factor, the colony-stimulating factor CSF-1 (Price *et al.*, 1992; Suzu *et al.*, 1992a). The presence of a chondroitin sulfate chain enables the cytokine to interact with type V collagen (Suzu *et al.*, 1992b). Furthermore, the glycosaminoglycan chain makes the molecule partially inactive, and full activity occurs only after glycosaminoglycan removal and partial proteolytic processing (Partenheimer *et al.*, 1995, in press).

A further exciting example of an optional chondroitin sulfate proeoglycan is the Alzheimer amyloid precursor protein, which is called appican. It has recently been found that the splicing out of exon 15 generates a glycosaminoglycan chain attachment site close to the amyloid peptide sequence within the amyloid precursor protein (Pangalos *et al.*, 1995). Appican is not produced by neurons but by astrocytes (Shioi *et al.*, 1995), which are found associated with neuritic plaques and participate in the formation of brain scars. Following neuronal injury, appican may well be involved in the development of pathological structures. The observation that chondroitin sulfate may serve as a survival factor for neurons (Junghans *et al.*, 1995) points to a further role of brain proteoglycans.

Finally, some collagen types may also, in some instances, be linked with chondroitin sulfate chains. Interestingly, all the examples of collagen proteoglycans described so far belong to the family of fibril-associated collagens with interrupted triple helices, i.e. to types IX-, XII-, and XIV-collagen (Vaughan *et al.*, 1985; Koch *et al.*, 1992). These collagens bind to the surface of fibrillar collagen bundles, and it may be, therefore, that the glycosaminoglycan chain of these collagens serves a similar function to the glycosaminoglycan chains of the small interstitial leucine-rich proteoglycans, which also bind to collagen fibrils.

Acknowledgment

The expert secretarial help of Friederike Eickholt is gratefully acknowledged. Work from the author's laboratory reported in this review was supported financially by the Deutsche Forschungsgemeinschaft (SFB 310, Teilprojekt B2; Kr 304/3–2) and by the Fonds der Chemischen Industrie.

References

Adany R, Heimer R, Caterson B *et al.* (1990): Altered expression of chondroitin sulfate proteoglycan in the stroma of human colon carcinoma. In *J. Biol. Chem.* **265**: 11389–96.

Adany R, Iozzo RV (1991): Hypomethylation of the decorin proteoglycan gene in human colon cancer. In *Biochem J.* **276**: 301–6.

Andres JL, Stanley K, Cheifetz S *et al.* (1989): Membrane-anchored and soluble forms of betaglycan, a polymorphic proteoglycan that binds transforming growth factor-β. In *J. Cell Biol.* **109**: 3137–45.

Antonsson P, Heinegård D, Oldberg Å (1993): Structure and deduced amino acid sequence of the human fibromodulin gene. In *Biochim. Biophys. Acta* **1174**: 204–6.

Bame KJ, Lidholt K, Lindahl U *et al.* (1991): Biosynthesis of heparan sulfate. Coordination of polymer-modification reactions in a Chinese hamster ovary cell mutant defective in *N*-sulfotransferase. In *J. Biol. Chem.* **266**: 10287–93.

Bartolazzi A, Peach R, Aruffo A *et al.* (1994): Interaction between CD44 and hyaluronate is directly implicated in the regulation of tumor developement. In *J. Exp. Med.* **180**: 53–66.

Battaglia C, Mayer U, Aumailley M *et al.* (1992): Basement membrane heparan sulfate proteoglycan binds to laminin by its heparan sulfate chains and to nidogen by sites in the protein core. In *Eur. J. Biochem.* **208**: 359–66.

Bernfield M, Kokenyesi R, Kato M *et al.* (1992): Biology of the syndecans: A family of transmembrane heparan sulfate proteoglycans. In *Annu. Rev. Cell Biol.* **8**: 365–93.

Bidanset DJ, LeBaron R, Rosenberg L *et al.* (1992): Regulation of cell substrate adhesion: effects of small galactosaminoglycan-containing proteoglycans. In *J. Cell Biol.* **118**: 1523–31.

Bittner K, Liszio C, Blumberg P *et al.* (1996): Modulation of collagen gel contraction by decorin. In *Biochem. J.*, in press.

Blochberger TC, Vergnes J-P, Hempel J *et al.* (1992): cDNA to chick lumican (corneal keratan sulfate proteoglycan) reveals homology to the small interstitial proteoglycan gene family and expression in muscle and intestine. In *J. Biol. Chem.* **267**: 347–352.

Border WA, Noble NA, Yamamoto T *et al.* (1992): Natural inhibitors transforming growth factor-β protects against scarring in experimental kidney disease. In *Nature* **360**: 361–4.

Bourdon MA, Krusius T, Campbell S *et al.* (1987): Identification and synthesis of a recognition signal for the attachment of glycosaminoglycans to proteins. In *Proc. Natl. Acad. Sci. USA* **84**: 3194–8.

Campbell P, Hannesson HH, Sandbäck D *et al.* (1994): Biosynthesis of heparin/heparan sulfate. Purification of the d-glucuronyl C5-epimerase from bovine liver. In *J. Biol. Chem.* **269**: 26953-8.

Carey DJ, Evans DM, Stahl RC *et al.* (1992): Molecular cloning and characterization of *N*- syndecan, a novel transmembrane heparan sulfate proteoglycan. In *J. Cell Biol.* **117**: 191–201.

Chakravarti S, Stallings RL, SundarRaj N*et al.* (1995): Primary structure of human lumican (keratan sulfate proteoglycan) and localization of the gene (LUM) to chromosome 12q21.3-q22. In *Genomics* **27**: 481–8.

Chandrasekaran L, Tanzer ML (1992): Molecular cloning of chicken aggrecan: Structural analysis. In *Biochem. J.* **288**: 903–10.

Cheng F, Heinegård D, Malmström A *et al.* (1994): Patterns of uronosyl epimerization and 4-/6-*O*-sulphation in chondroitin/dermatan sulphate from decorin and biglycan of various bovine tissues. In *Glycobiology* **4**: 685–96.

Cohen IR, Grässel S, Murdoch AD *et al.* (1993): Structural characterization of the complete human perlecan gene and its promoter. In *Proc. Natl. Acad. Sci. USA* **90**: 10404–8.

Danielson KG, Fazzio A, Cohen I *et al.* (1993): The human decorin gene: Intron-exon organization, discovery of two alternatively spliced exons in the 5' untranslated region, and mapping the gene to 12q23. In *Genomics* **15**: 146–60.

David G (1993): Integral membrane heparan sulfate proteoglycans. In *FASEB J.* **7**: 1023–30.

David G, Lories V, Decock B *et al.* (1990): Molecular cloning of a phosphatidylinositol-anchored membrane heparan sulfate proteoglycan from human lung fibroblasts. In *J. Cell Biol.* **111**: 3165- 76.

David G, van der Schueren B, Marynen P *et al.* (1992): Molecular cloning of amphiglycan, a novel integral membrane heparan sulfate proteoglycan expressed by epithelial and fibroblastic cells. In *J. Cell Biol.* **118**: 961–9.

Day AA, McQuillan CI, Termine JD *et al.* (1987): Molecular cloning and sequence analysis of the cDNA for small proteoglycan II of bovine bone. In *Biochem J.* **248**: 801–5.

DeAngelis PL, Papaconstantinou J, Weigel PH (1993): Molecular cloning, identification, and sequence of the hyaluronan synthase gene from group A *Streptococcus pyogenes. J. Biol. Chem.* **268**: 19181–4.

DeAngelis PL, Weigel PH (1994): Immunochemical confirmation of the primary structure of streptococcal hyaluronan synthase and synthesis of high molecular weight product by the recombinant enzyme. In *Biochemistry* **33**: 9033–9.

Dodge GR, Kovalszky I, Chu M-L *et al.* (1991): Heparan sulfate proteoglycan of human colon: partial molecular cloning, cellular expression and mapping of the gene (HSPGII) to the short arm of human chromosome 1. In *Genomics* **10**: 673–80.

Doege K, Sasaki M, Horigan E *et al.* (1987): Complete primary structure of the rat cartilage proteoglycan core protein deduced from cDNA clones. In *J. Biol. Chem.* **262**: 17757–67.

Doege KJ, Sasaki M, Kimura T *et al.* (1991): Complete coding sequence and deduced primary structure of the human cartilage large aggregating proteoglycan, aggrecan. Human specific repeats, and additional alternatively spliced forms. In *J. Biol. Chem.* **266**: 894–902.

Dours-Zimmermann MT, Zimmermann DR (1994): A novel glycosaminoglycan attachment domain identified in two alternative splice variants of human versican. In *J. Biol. Chem.* **269**: 32992–8.

Edwards IJ, Wagner WD (1988): Distinct synthetic and structural characteristics of proteoglycans produced by cultured artery smooth muscle cells of atherosclerosis-susceptible pigeons. In *J. Biol. Chem.* **263**: 9612–20.

Engel J (1991): Common structural motifs in proteins of the extracellular matrix. In *Curr. Opinion Cell Biol.* **3**: 779–85.

Enghild JJ, Salvesen G, Hefta SA *et al.* (1991): Chondroitin 4-sulfate covalently cross-links the chains of the human blood protein pre-α-inhibitor. In *J. Biol. Chem.* **266**: 747–51.

Enghild JJ, Salvesen G, Thøgersen IB *et al.* (1993): Presence of the protein-glycosaminoglycan- protein covalent cross-link in the inter-α-inhibitor-related proteinase inhibitor heavy chain 2/bikunin. In *J. Biol. Chem.* **268**: 8711–6.

Eriksson I, Sandbäck D, Ek B *et al.* (1994): cDNA cloning and sequencing of mouse mastocytoma glucosaminyl *N*-deacetylase/*N*-sulfotransferase, an enzyme involved in the biosynthesis of heparin. In *J. Biol. Chem.* **269**: 10438–43.

Evered D, Whelan J (eds) (1986): Functions of the Proteoglycans. CIBA Foundation Symposium, Vol. **124**, Chichester: Wiley.

Esko JD, Weinke JL, Taylor WH *et al.* (1987): Inhibition of chondroitin and heparan sulfate biosynthesis in Chinese hamster ovary cell mutants defective in galactosyltransferase I. In *J. Biol. Chem.* **262**: 12189–95.

Esko JD, Rostand KS, Weinke JL (1988): Tumor formation dependent on proteoglycan biosynthesis. In *Science* **241**: 1092–5.

Farquhar MG (1991): The glomerular basement membrane: a selective macromolecular filter. In *Cell Biology of Extracellular Matrix* (Hay ED, ed.) pp 365–418, New York: Plenum Press.

Fedarko NS, Vetter UK, Weinstein S *et al.* (1992): Age-related changes in hyaluronan, proteoglycan, collagen, and osteonectin synthesis by human bone cells. In *J. Cell Physiol.* **151**: 215–27.

Fessler JH, Fessler LI (1966): Electron microscopic visualization of the polysaccharide hyaluronic acid. In *Proc. Natl. Acad. Sci. USA* **56**: 141–7.

Fisher LW, Termine JD, Young MF (1989): Deduced protein sequence of bone small proteoglycan I (biglycan) shows homology with proteoglycan II (decorin) and several nonconnective tissue proteins in a variety of species. In *J. Biol. Chem.* **264**: 4571–6.

Fokuta M, Uchimura K, Nakashima K *et al.* (1995): Molecular cloning and expression of chick chondrocyte chondroitin 6-sulfotransferase. In *J. Biol. Chem.* **270**: 18575–80.

Forsberg N, Gustafson S (1991): Characterization and purification of the hyaluronan-receptor on liver endothelial cells. In *Biochim. Biophys. Acta* **1078**: 12–8.

Fransson L-Å, Cöster L, Malmström A *et al.* (1982): Self-association of scleral proteodermatan sulfate. In. *J. Biol. Chem.* **257**: 6333–8.

Fransson L-Å, Silverberg I, Carlstedt I (1985): Structure of the heparan sulfate-protein linkage region. Demonstration of the sequence galactosyl-galactosyl-xylose-2-phosphate. In *J. Biol. Chem.* **260**: 14722–6.

Fransson LÅ, Edgren G, Havsmark B *et al.* (1995): Recycling of a glycosylphosphatidyl-inositol- anchored heparan sulphate proteoglycan (glypican) in skin fibroblasts. In *Glycobiology* **5**: 407–15.

Frazer JE, Laurent TC (1989): Turnover and metabolism of hyaluronan. In: *The Biology of Hyaluronan*. Ciba Foundation Symposium **143** (Evered D, Whelan J, eds) pp 41–59, Chichester: Wiley.

Fritz TA, Lugemwa FN, Sarkar AK *et al.* (1994a): Biosynthesis of heparan sulfate on β-D-xylosides depends on aglycone structure. In *J. Biol. Chem.* **269**: 300–7.

Fritz TA, Gabb MM, Wei G *et al.* (1994b): Two *N*-acetylglucosaminyltransferases catalyze the biosynthesis of heparan sulfate. In *J. Biol. Chem.* **269**: 28809–14.

Funderburgh JL, Funderburgh ML, Mann MM *et al.* (1991): Arterial lumican. In *J. Biol. Chem.* **266**: 24773–7.

Gibbs DA, Merril EW, Smith KA *et al.* (1968): Rheology of hyaluronic acid. In *Biopolymers* **6**: 777–91.

Glössl J, Hoppe W, Kresse H (1986): Post-translational phosphorylation of proteodermatan sulfate. In *J. Biol. Chem.* **261**: 1920–3.

Gould SE, Upholt WB, Kosher RA (1992): Syndecan 3: a member of the syndecan family of membrane-intercalated proteoglycans that is expressed in high amounts at the onset of chicken limb cartilage differentiation. In *Proc. Natl. Acad. Sci. USA* **89**: 3271–5.

Greve H, Kresse H (1988): Secretion of unphosphorylated and phosphorylated xyloside-induced glycosaminoglycan chains. In *Glycoconjugate J.* **5**: 175–83.

Gustafson S, Björkman T, Forsberg N *et al.* (1995): Accessible hyaluronan receptors identical to ICAM-1 in mouse mast-cell tumours. In *Glycoconjugate J.* **12**: 350–5.

Habuchi H, Kimata K, Suzuki S (1986): Changes in proteoglycan composition during development of rat skin. In *J. Biol. Chem.* **261**: 1031–40.

Habuchi H, Habuchi O, Kimata K (1995): Purification and characterization of heparan sulfate 6- sulfotransferase from the culture medium of Chinese hamster ovary cells. In *J. Biol. Chem.* **270**: 4172–9.

Hall CL, Yang B, Yang X *et al.* (1995): Overexpression of the hyaluronan receptor RHAMM is transforming and is also required for H-*ras* transformation. In *Cell* **82**: 19–28.

Hardingham TE, Fosang AJ (1992): Proteoglycans: Many forms and many functions. In *FASEB J.* **6**: 861–71.

Hardingham TE, Fosang AJ (1995): The structure of aggrecan and its turnover in cartilage. In *J. Rheumatol.* **22** suppl. 43: 86–90.

Hardwick C, Hoare K, Owens R *et al.* (1992): Molecular cloning of a novel hyaluronan receptor that mediates tumor cell motility. In *J. Cell Biol.* **117**: 1343–50.

Hashimoto C, Hudson KL, Anderson KV (1988): The Toll gene of Drosopohila, required for dorsal-ventral embryonic polarity, appears to encode a transmembrane protein. In *Cell* **52**: 269-79.

Hausser H, Gröning A, Hasilik A *et al.* (1994): Selective inactivity of TGF-β/decorin complexes. In *FEBS Lett.* **353**: 243–5.

Haynes BF, Hale LP, Patton KL *et al.* (1991): Measurement of an adhesion molecule as an indicator of inflammatory disease activity. In *Arthritis Rheum.* **34**: 1434–43.

Hedbom E, Heinegård D (1993): Binding of fibromodulin and decorin to separate sites on fibrillar collagen. In *J. Biol. Chem.* **268**: 27307–12.

Heldin P, Laurent TC, Heldin C-H (1989): Effect of growth factors on hyaluronan synthesis in cultured human fibroblasts. In *Biochem. J.* **258**: 919–22.

Hoffmann H-P, Schwartz NB, Rodén L *et al.* (1984): Location of xylosyltransferase in cisternae of the rough endoplasmic reticulum of embryonic cartilage cells. In *Connect. Tissue Res.* **12**: 151- 63.

Holmbeck SMA, Peptillo PA, Lerner LA (1994): The solution conformation of hyaluronan: a combined NMR and molecular dynamics study. In *Biochemistry* **39**: 14246–55.

Hoppe W, Glössl J, Kresse H (1985): Influence of monensin on biosynthesis, processing and secretion of proteodermatan sulfate by skin fibroblasts. In *Eur. J. Biochem.* **152** : 91–7.

Hounsell EF (1989): Structural and conformational analysis of keratan sulphate oligosaccharides and related carbohydrate structures. In: *Keratan Sulphate* (Greiling H, Scott JE, eds) pp 12–15. London: The Biochemical Society.

Hua Q, Knudson CB, Knudson W (1989): Internalization of hyaluronan by chondrocytes occurs via receptor-mediated endocytosis. In *J. Cell Sci.* **106**: 365–75.

Huckerby TN, Dickenson JM, Brown GM *et al.* (1995): Keratenase digestion of keratan sulphates: characterization of large oligosaccharides from the *N*-acetyllactosamine repeat sequence and from the non-reducing terminal chain caps. In *Biochim. Biophys. Acta* **1244**: 17–29.

Iozzo RV, Naso MF, Cannizzaro LA *et al.* (1992): Mapping of the versican proteoglycan gene (CSPG2) to the long arm of human chromosome 5 (5q12–5q14). In *Genomics* **14**: 845–51.

Jollès P (ed) (1993): Proteoglycans. In *Experientia* **49**: 365–470.

Junghans U, Koops A, Westmeyer A *et al.* (1995): Purification of a meningeal cell-derived chondroitin sulfate proteoglycan with neurotrophic activity for brain neurons and its identification as biglycan. In *Eur. J. Neurosci,* in press.

Just W, Klett C, Vetter U *et al.* (1993): Assignment of the human aggrecan gene AGC1 to 15q25 → q26.2 by in situ hybridization. In *Hum. Genet.* **92**: 516–8.

Kähäri V-M, Häkkinen L, Westermarck J *et al.* (1995):Differential regulation of decorin and biglycan gene expression by dexamethasone and retinoic acid in cultured human skin fibroblasts. In *J. Invest. Dermatol.* **104**: 503–8.

Kallunki P, Tryggvason K (1992): Human basement membrane heparan sulfate proteoglycan core protein: A 467-kD protein containing multiple domains ressembling elements of the low density lipoprotein receptor, laminin, neural cell adhesion molecules, and epidermal growth factor. In *J. Cell Biol.* **116**: 559–71.

Keller R, Stein T, Stuhlsatz HW *et al.* (1981): Studies on the characterization of the linkage-region between polysaccharide chain and core protein in bovine corneal proteokeratan sulfate. In *Hoppe-Seyler's Z. Physiol. Chem.* **362**: 327–36.

Kim CW, Goldberger OA, Gallo RL *et al.* (1994): Members of the syndecan family of heparan sulfate proteoglycans are expressed in distict cell-, tissue-, and development-specific patterns. In *Mol. Biol. Cell* **5**: 797–805.

Kjellén L, Lindahl U (1991): Proteoglycans: Structures and interactions. In *Annu Rev. Biochem.* **60**: 443–75.

Klewes L, Turley EA, Prehm P (1993): The hyaluronan synthase from a eukaryotic cell line. In *Biochem. J.* **290**: 791–5.

Klewes L, Prehm P (1994): Intracellular signal transduction for serum activation of the hyaluronan synthase in eukaryotic cell lines. In *J. Cell. Physiol.* **160**: 539–44.

Kobe B, Deisenhofer J (1993): Crystal structure of porcine robonuclease inhibitor, a proein with leucine-rich repeats. In *Nature* **366**: 751–6.

Koch M, Bernasconi C, Chiquet M (1992): A major oligomeric fibroblast proteoglycan identified as a novel large form of type XII collagen. In *Eur. J. Biochem.* **207**: 847–56.

Kojima T, Shworak NW, Rosenberg RD (1992): Molecular cloning and expression of two distinct cDNA-encoding heparan sulfate proteoglycan core proteins from a rat endothelial cell line. In *J. Biol. Chem.* **267**: 4870–7.

Korenberg JR, Chen XN, Doege K *et al.* (1993): Assignment of the human aggrecan gene (AGC1) to 15q26 using fluorescence in situ hybridization analysis. In *Genomics* **16**: 546–8.

Kresse H, Glössl J (1987): Glycosaminoglycan degradation. In *Adv. Enzymol. 60* (Meister A, ed.) pp 217–311, New York: J. Wiley.

Kresse H, Rosthøj S, Quentin E *et al.* (1987): Glycosaminoglycan-free small proteoglycan core protein secreted by fibroblasts from a patient with a syndrome ressembling progeroid. In *Am. J. Hum. Genet.* **41**: 436–53.

Krusius T, Ruoslahti E (1986): Primary structure of extracellular matrix proteoglycan core protein deduced from cloned cDNA. *Proc. Natl. Acad. Sci. USA* **83**: 7683–7.

Kugelman LC, Ganguly S, Haggerty JG *et al.* (1992): The core protein of epican, a heparan sulfate proteoglycan on keratinocytes, is an alternative form of CD44. In *J. Invest. Dermatol.* **99**: 381–5

Laurent TC (1970): Structure of hyaluronate acid. In: *Chemistry and Molecular Biology of the Intercellular Matrix* (Balazs EA ed) pp 703–32, London: Academic Press.

Li W, Vergnes J-P, Cornuet PK *et al.* (1992): cDNA clone to chick corneal chondroitin/ dermatan sulfate proteoglycan reveals identity to decorin. In *Arch. Biochem. Biophys.* **296**: 190–7.

Lidholt K, Weinke JL, Kiser CS *et al.* (1992): A single mutation affects both *N*-acetylglucos-aminyltransferase and glucuronosyltransferase activities in a Chinese hamster ovary cell mutant defective in heparan sulfate biosynthesis. In *Proc. Natl. Acad. Sci. USA* **89**: 2267–71.

Lin J-H, McLean K, Morser J *et al.* (1994): Modulation of glycosaminoglycan addition in naturally expressed and recombinant human thrombomodulin. In *J. Biol. Chem.* **269**: 25021–30.

Lind T, Lindahl U, Lidholt K (1993): Biosynthesis of heparin/heparan sulfate. Identification of a 70-kDa protein catalyzing both the D-glucuronosyl- and the *N*-acetyl-D-glucosaminyltransferase reactions. In *J. Biol. Chem.* **268**: 20705–8.

Lohmander LS (1995): The release of aggrecan fragments into synovial fluid after joint injury and in osteoarthritis. In *J. Rheumatol.* **22** suppl 43: 75–7.

Lugemwa FN, Esko JD (1991): Estradiol-β-D-xyloside, an efficient primer for heparan sulfate biosynthesis. In *J. Biol. Chem.* **266**: 6674–7.

Malmström A (1984): Biosynthesis of dermatan sulfate. II. Substrate specificity of C-5 uronosyl epimerase. In *J. Biol. Chem.* **259**: 161–5.

Mann DM, Yamaguchi Y, Bourdon MA *et al.* (1990): Analysis of glycosaminoglycan substitution in decorin by site-directed mutagenesis. In *J. Biol. Chem.* **265**: 5317–23.

Margolis RK, Margolis RU (1993): Nervous tissue proteoglycans. In *Experientia* **49**: 429–46.

Maroudas A, Wachtel E, Grushko G *et al.* (1991): The effect of osmotic and mechanical pressures on water partitioning in articular cartilage. In *Biochim. Biophys. Acta* **1093**: 285–94.

Marynen P, Zhang J, Cassiman J-J *et al.* (1989): Partial primary structure of the 48- and 90- kilodalton core proteins of cell surface-associated heparan sulfate proteoglycans of lung fibroblasts. In *J. Biol. Chem.* **264**: 7017–24.

McBride OW, Fisher LW, Young MF (1990): Localization of PGI (biglycan, BGN) and PGII (decorin, DCN, PG-40) genes on human chromosomes Xq13-qter and 12q, respectively. In *Genomics* **6**: 219–25.

McCarthy KJ, Accavitti MA, Couchman JR (1989): Immunological characterization of a basement membrane-specific chondroitin sulfate proteoglycan. In *J. Cell Biol.* **109**: 3187–98.

McCormick D, van der Rest M, Goodship J *et al.* (1987): Structure of the glycosaminoglycan domain in the type IX collagen-proteoglycan. In *Proc. Natl. Acad. Sci. USA* **84**: 4044–8.

McGary CT, Raja RH, Weigel PH (1989): Endocytosis of hyaluronic acid by rat liver endothelial cells. Evidence of receptor recycling. In *Biochem. J.* **257**: 875–84.

Murdoch AD, Dodge GR, Cohen I *et al.* (1992): Primary structure of the human heparan sulfate proteoglycan from basement membrane (HSPG2/Perlecan). In *J. Biol. Chem.* **267**: 8544–57.

Naujokas MF, Morin M, Anderson MS *et al.* (1993): The chondroitin sulfate form of invariant chain can enhance stimulation of T cell responses through interaction with CD44. In *Cell* **74**: 257-68.

Neame PJ, Choi H, Rosenberg LC (1989): The primary structure of the small, leucine-rich proteoglycan (PGI) from bovine articular cartilage. In *J. Biol. Chem.* **264**: 8653–61.

Nilsson B, Nakazawa K, Hassell JR *et al.* (1983): Structure of oligosaccharides and the linkage region between keratan sulfate and the core protein on proteoglycans from monkey cornea. In *J. Biol. Chem.* **258**: 6056–63.

Nuwayhid N, Glaser JH, Johnson JC *et al.* (1986): Xylosylation and glucuronosylation reactions in rat liver Golgi apparatus and endoplasmic reticulum. In *J. Biol. Chem.* **261**: 12936–41.

Oeben M, Keller R, Stuhlsatz HW *et al.* (1987): Constant and variable domains of different disaccharide structure in corneal keratan sulphate chains. In *Biochem. J.* **248**: 85–93.

Oegema TR, Jun., Kraft EL, Jourdian GW *et al.* (1984): Phosphorylation of chondroitin sulfate in proteoglycans from the Swarm rat chondrosarcoma. In *J. Biol. Chem.* **259**: 1720–6.

Oldberg Å, Antonsson P, Lindblom K *et al.* (1989): A collagen-binding 59-kd protein (fibromodulin) is structurally related to the small interstitial proteoglycans PG-S1 and PG-S2 (decorin). In *EMBO J.* **8**: 2601–4.

Orellana A, Hirschberg CB, Wei Z *et al.* (1994): Molecular cloning and expression of a glycosaminoglycan *N*-acetylglucosaminyl *N*-deacetylase/*N*-sulfotransferase from a heparin-producing cell line. In *J. Biol. Chem.* **269**: 2270–6.

Pangalos MN, Efthimiopoulos S, Shioi J *et al.* (1995): The chondroitin sulfate attachment site of appican is formed by spicing out exon 15 of the amyloid precursor gene. In *J. Biol. Chem.* **270**: 10388–91.

Partenheimer A, Schwarz K, Wrocklage C *et al.* (1995): Proteoglacan form of colony-stimulating factor-1 (proteoglycan-100): Stimulation of activity by glycosaminoglycan removal and proteolytic processing. In *J. Immunol.* in press.

Paulsson M, Yurchenko PD, Ruben GC *et al.* (1987): Structure of low density heparan sulfate proteoglycan isolated from a mouse tumor basement membrane. In *J. Mol. Biol.* **197**: 297–313.

Pettersson I, Kusche M, Unger E *et al.* (1991): Biosynthesis of heparin. Purification of a 110-kDa mouse mastocytoma protein required for both glucosaminyl *N*-deacetylation and *N*-sulfation. In *J. Biol. Chem.* **266**: 8044–9.

Plaas AHK, Neame PJ, Nivens CM *et al.* (1990): Identification of the keratan sulfate attachment sites on bovine fibromodulin. In *J. Biol. Chem.* **265**: 20634–41.

Prehm P (1983): Synthesis of hyaluronate in differentiated teratocarcinoma cells. In *Biochem J.* **211**: 181–9 and 191–8.

Prehm P (1989): Identification and regulation of the eukaryotic hyaluronate synthase. In: *The Biology of Hyaluronan.* Ciba Foundation Symposium **143** (Evered D, Whelan J, eds) pp 21–40, Chichester: Wiley Press.

Prehm P (1995): Synovial hyaluronate in rheumatoid arthritis binds C1q and is covalently bound to antibodies: a model for chronicity. In *Annals Rheum. Dis.* **54**: 408–12

Price LKH, Choi, HU, Rosenberg L *et al.* (1992): The predominant form of a secreted colony- stimulating factor-1 is a proteoglycan. In *J. Biol. Chem.* **267**: 2190–9.

Pulkkinen L, Alitalo T, Krusius T *et al.* (1992): Expression of decorin in human tissues and cell lines and defined chromosomal assignment of the gene locus (DCN). In *Cytogenet. Cell Genet.* **60**: 107–11.

Quentin E, Gladen A, Rodén L *et al.* (1990): A genetic defect in the biosynthesis of dermatan sulfate proteoglycan: galactosyltransferase I-deficiency in fibroblasts from a patient with a progeroid syndrome. In *Proc. Natl. Acad. Sci. USA* **87**: 1342–6.

Rada JA, Cornuet PK, Hassell JR (1993): Regulation of corneal collagen fibrillogenesis *in vitro* by corneal proteoglycan (lumican and decorin) core proteins. In *Exp. Eye Res.* **56**: 635–48.

Rauch U, Karthikeyan L, Maurel P *et al.* (1992): Cloning and primary structure of neurocan, a developmentally regulated, aggregating chondroitin sulfate proteoglycan of brain. In *J. Biol. Chem.* **267**: 19536–47.

Razi N, Lindahl U (1995): Biosynthesis of heparin/heparan sulfate. The D-glucosaminyl 3-*O*-sulfotransferase reaction; target and inhibitor saccharides. In *J. Biol. Chem.* **270**, in press.

Rodén L, Campbell P, Fraser RE *et al.* (1989): Enzymatic pathways of hyaluronan catabolism in the biology of hyaluronan. In: *The Biology of Hyaluronan.* Ciba Foundation Symposium **143** (Evered D and Whelan J, eds) pp 61–86, Chichester: Wiley Press.

Rodén L, Koerner T, Olson C *et al.* (1985): Mechanisms of chain initiation in the biosynthesis of connective tissue polysaccharides. In *Fed. Proceedings* **44**: 373–89.

Rohrmann K, Niemann R, Buddecke E (1985): Two *N*-acetylgalactosaminyltransferase are involved in the biosynthesis of chondroitin sulfate. In *Eur. J. Biochem.* **148**: 463–9.

Ruoslahti E (1988): Structure and biology of proteoglycans. In *Annu. Rev. Cell Biol.* **4**: 229–55.

Ruoslahti E, Yamaguchi Y (1991): Proteoglycans as modulators of growth factor activities. In *Cell* **64**: 867–9.

Sampson PM, Rochester CL, Freundlich B *et al.* (1992): Cytokine regulation of human lung fibroblast hyaluronan (hyaluronic acid) production. In *J. Clin. Invest.* **90**: 1492–503.

Santra M, Skorski T, Calabretta B *et al.* (1995): *De novo* decorin gene expression suppresses the malignant phenotype in human colon cancer cells. In *Proc. Natl. Acad. Sci. USA* **92**: 7016–20.

Saunders S, Jalkanen M, O'Farrell S *et al.* (1989): Molecular cloning of syndecan, an integral membrane proteoglycan. In *J. Cell Biol.* **108**: 1547–56.

Schenck P, Schneider S, Miehlke R *et al.* (1995): Synthesis and degradation of hyaluronate by synovia from patients with rheumatoid arthritis. In *J. Rheumatol.* **22**: 400–5.

Schittny JC, Kresse H, Burri PH (1995): Immunostaining of a heterodimeric dermatan sulphate proteoglycan is correlated with smooth muscles and some basement membranes. In *Histochemistry* **103**: 271–79.

Schönherr E, Witsch-Prehm P, Harrach B *et al.* (1995a): Interaction of biglycan with type I collagen. In *J. Biol. Chem.* **270**: 2776–83.

Schönherr E, Hausser H, Beavan L *et al.* (1995b): Decorin-type I collagen interaction. In *J. Biol. Chem.* **270**: 8877–83.

Scott JE (1988): Proteoglycan-fibrillar collagen interactions. In *Biochem. J.* **252**: 313–23.

Scott JE (1992): Supramolecular organization of extracellular matrix glycosaminoglycans, *in vitro* and in the tissues. In *FASEB J.* **6**: 2639–45.

Scott JE (1993): The nomenclature of glycosaminoglycans and proteoglycans. In *Glycoconjugate J.* **10**: 419–21.

Scott JE, Cummings C (1995): The proteins of proteodermatan and proteokeratan sulphates (decoron and fibromodulon/lumicon) are horseshoe shaped resembling ribonuclease inhibitor. In *Biochem. Soc. Transact.* **23**: 515S.

Scott JE, Cummings C, Brass A *et al.* (1991): Secondary and tertiary structures of hyaluronan in aqueous solution, investigated by rotary shadowing-electron microscopy and computer simulation. In *Biochem. J.* **274**: 699–705.

Shinomura T, Kimata K (1992): Proteoglycan-Lb, a small dermatan sulfate proteoglycan expressed in embryonic chick epiphyseal cartilage, is structurally related to osteoinductive factor. In *J. Biol. Chem.* **267**: 1265–70.

Shioi J, Pangalos MN, Ripellino JA *et al.* (1995): The Alzheimer amyloid precursor proteoglycan (appican) is present in brain and is produced by astrocytes but not by neurons in primar neural cultures. In *J. Biol. Chem.* **270**: 11839–44.

Shworak NW, Shirakawa M, Mulligan RC *et al.* (1994): Characterization of ryudocan glycosaminoglycan acceptor sites. In *J. Biol. Chem.* **269**: 21204–14.

Smedsrød B, Pertoft H, Eriksson S *et al.* (1989): Studies *in vitro* on the uptake and degradation of sodium hyaluronate in rat liver endothelial cells. In *Biochem. J.* **223**: 617–26.

Stevens RL, Kamada MM, Serafin WE (1988): Structure and function of the family of proteoglycans that reside in the secretory granules of natural killer cells and other effector cells of the immune response. In *Curr. Topics Microbiol. Immun.* **140**: 93–108.

Sugahara K, Mizono N, Okumura Y *et al.* (1992a): The phosphorylated and/or sulfated structure of the carbohydrate-protein linkage region isolated from chondroitin sulfate in the hybrid proteoglycans of Engelbreth-Holm-Swarm mouse tumor. In *Eur. J. Biochem.* **204**: 401–6.

Sugahara K, Yamada S, Yoshida K *et al.* (1992b): A novel sulfated structure in the carbohydrate- protein linkage region isolated from porcine intestinal heparin. In *J. Biol. Chem.* **267**: 1528–33.

SundarRaj N, Fite D, Ledbetter S *et al.* (1995): Perlecan is a component of cartilage matrix and promotes chondrocyte attachment. In *J. Cell Sci.* **108**: 2663–72.

Suzu S, Ohtsuki T, Yanai N *et al.* (1992a): Identification of a high molecular weight macrophage colony-stimulating factor as a glycosaminoglycan-containing species. In *J. Biol. Chem.* **267**: 4345–8.

Suzu S, Ohtsuki T, Makashima M *et al.* (1992b): Biological activity of a proteoglycan form of macrophage colony-stimulating factor and its binding to type V collagen. In *J. Biol. Chem.* **267**: 16812–5.

Timpl R (1993): Proteoglycans of basement membranes. In *Experientia* **49**: 417–28.

Toole BP (1991): Proteoglycans and hyaluronan in morphogenesis and differentiation. In Cell Biology of Extracellular Matrix (Hay D, ed) pp 305–39, New York: Plenum Press.

Tooze SA, Huttner WB (1990): Cell-free protein sorting to the regulated and constitutive secretory pathways. In *Cell* **60**: 837–47.

Traupe H, van den Ouweland AMW, van Oost BA *et al.* (1992): Fine mapping of the human biglycan (BGN) gene within the Xq28 region employing a hybrid cell panel. In *Genomics* **13**: 481-3.

Upholt WB, Chandrasekaran L, Tanzer ML (1993): Molecular cloning and analysis of the protein modules of aggrecan. In *Experientia* **49**: 384–92.

Vaughan L, Winterhalter K, Bruckner P (1985): Proteoglycan Lt from chicken embryo sternum identified as type IX collagen. In *J. Biol. Chem.* **260**: 4758–63.

Vogel KG, Paulsson M, Heinegård D (1984): Specific inhibition of type I and type II collagen fibrillogenesis by the small proteoglycan of tendon. In *Biochem. J.* **223**: 587–97

Vogel KG, Sandy JD, Pogány G *et al.* (1994): Aggrecan in bovine tendon. In *Matrix Biology* **14**: 171–9.

Vogel KG, Koob TJ, Fisher LW (1987): Characterization and interactions of a fragment of the core protein of the small proteoglycan (PGII) from bovine tendon. In *Biochem. Biophys. Res. Commun.* **148**: 658–63.

Watanabe H, Kimata K, Line S *et al.* (1994): Mouse cartilage matrix deficiency (cmd) caused by a 7 bp delition in the aggrecan gene. In *Nature Genet.* **7**: 154–7.

Wei Z, Swiedler SJ, Ishihara M *et al.* (1993): A single protein catalyzes both *N*-deacetylation and *N*-sulfation during the biosynthesis of heparan sulfate. In *Proc. Natl. Acad. Sci. USA* **90**: 3885–8.

Westergren-Thorsson G, Särnstrand B, Fransson L-Å *et al.* (1990): TGF-β enhances the production of hyaluronan in lung but not in skin fibroblasts. In *Exp. Cell Res.* **186**: 192–5.

Winnemöller M, Schmidt G, Kresse H (1991): Influence of decorin on fibroblast adhesion to fibronectin. In *Eur. J. Cell Biol.* **54**: 10–7.

Woods A, Couchman JR (1994): Syndecan 4 heparan sulfate proteoglycan is a selectively enriched and widespread focal adhesion component. In *Mol. Biol. Cell* **5**: 183–92

Yamada H, Watanabe K, Shimonaka M *et al.* (1994): Molecular cloning of brevican, a novel brain proteoglycan of the aggrecan/versican family. In *J. Biol. Chem.* **269**: 10119–26.

Yamagata M, Saga S, Kato M *et al.* (1993). Selective distributions of proteoglycans and their ligands in pericellular matrix of cultured fibroblasts. Implications for their roles in cell-substratum adhesion. In *J. Cell Sci.* **106**: 55–65.

Yamagata M, Kimata K (1994): Repression of a malignant cell-substratum adhesion phenotype by inhibiting the production of the anti-adhesive proteoglycan PG-M/versican. In *J. Cell Sci.* **107**: 2581–90.

Yamaguchi Y, Ruoslahti E (1988): Expression of human proteoglycan in Chinese hamster ovary cells inhibits cell proliferation. In *Nature* **336**: 244–6.

Yamaguchi Y, Mann DM, Ruoslahti E (1990): Negative regulation of transforming growth factor-β by the proteoglycan decorin. In *Nature* **346**: 281–4.

Yang B, Zhang L, Turley EA (1993): Identification of two hyaluronan-binding domains in the hyaluronan receptor RHAMM. In *J. Biol. Chem.* **268**: 8617–23.

Yeo T-K, Torok MA, Kraus HL *et al.* (1995): Distribution of biglycan and its propeptide form in rat and bovine aortic tissue. In *J. Vasc. Res.* **32**: 175–82.

Zako M, Shinomura T, Ujitta M *et al.* (1995): Expression of PG-M (V3), an alternatively spliced form of PG-M without a chondroitin sulfate attachment region in mouse and human tissues. In *J. Biol. Chem.* **270**: 3914–8.

Zhang L, Esko JD (1994): Amino acid determinants that drive heparan sulfate assembly in a proteoglycan. In *J. Biol.Chem.* **269**: 19295–9.

Zimmermann DR, Ruoslahti E (1989): Multiple domains of the large fibroblast proteoglycan, versican. In *EMBO J.* **8**: 2975–81.

Zimmermann DR, Dours-Zimmermann MT, Schubert M *et al.* (1994): Versican is expressed in the proliferating zone in the epidermis and in association with the elastic network of the dermis. In *J. Cell Biol.* **124**: 817–25.

12 GPI-Anchors: Structure and Functions

VOLKER ECKERT, PETER GEROLD, AND RALPH THOMAS SCHWARZ

1 Introduction

1.1 The Discovery of GPI-Anchors: From an Exotic Way to Anchor a Protozoal Surface Antigen to a General Principle Among Eukaryotes

Glycosyl-phosphatidylinositol (GPI) membrane anchors were first discovered in the parasitic protozoan *Trypanosoma brucei brucei* by Low, Holder, Cardoso and Boothroyd, who showed that the carboxy terminus of the main surface protein of this parasite, the variant surface glycoprotein (VSG), was associated with ethanolamine and a carbohydrate structure associated with a hydrophobic component (Boothroyd *et al.*, 1981; Holder and Cross, 1981; Holder, 1983) and was efficiently released into the medium as a water soluble form by a trypanosome phosphatidylinositol-specific phospholipase C (PI-PLC) (Cardoso-de-Almeida and Turner, 1983). Together with earlier data of Low and co-workers, who showed that PI-PLC was able to release selectively certain membrane proteins such as alkaline phosphatase, acetylcholinesterase and 5'-nucleotidase (reviewed by Low, 1987), it was concluded that these proteins were linked to the membrane associated phosphatidylinositol via ethanolamine and a carbohydrate bridge, and the term glycosyl-phosphatidylinositol anchor was defined (Ferguson *et al.*, 1985). Subsequently Ferguson and co-workers elucidated the *T. b. brucei* VSG GPI-anchor structure (Ferguson *et al.*, 1988), which consists of ethanolaminephosphate6Manα1–2Manα1–6Manα1–4GlcN1–6inositolphosphate-lipid, as shown in Figure 1. At the same time, the GPI attachment to main surface proteins of several protozoan parasites such as *Toxoplasma gondii, Plasmodium falciparum* and *Leishmania spp.* as well as to surface glycoproteins of higher eukaryotes (rat brain Thy-1, torpedo-fish and human erythrocyte acetylcholinesterase) was demonstrated. Although the information from higher eukaryotes continued to increase, GPI-anchoring of surface proteins was often regarded as an exotic way of anchoring membrane proteins used predominantly by protozoans, in contrast to the "classical" transmembrane domains. Since then, well over 100 proteins which are GPI-anchored have been described (reviewed in: Cross, 1990; Englund, 1993; McConville and Ferguson, 1993; Field and Menon, 1994), demonstrating that this principle is widely distributed among eukaryotes (with the exception of

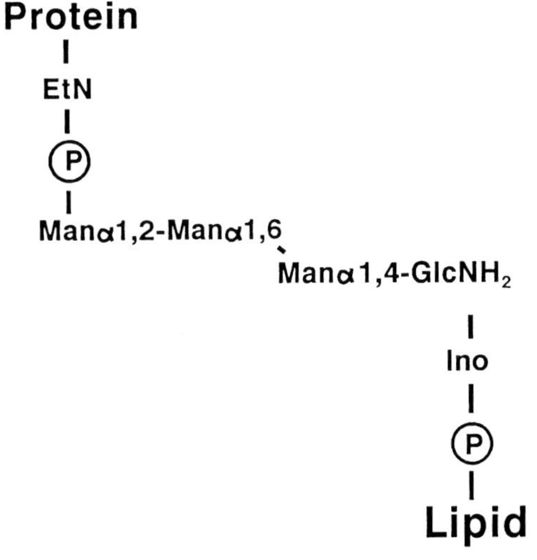

Figure 1. The minimal structure of a GPI-anchor. P, phosphodiester bridge; Ino, inositol; $GlcNH_2$, glucosamine; Man, mannose; EtN, ethanolamine.

H.-J. and S. Gabius (Eds.), Glycosciences

ISBN 3-8261-0073-5

plants, from which such data are still lacking), with GPI-anchoring seeming to be a more general principle among protozoans, whereas higher eukaryotes use this principle predominantly for certain proteins with specialized functions (for review see Anderson, 1994).

1.2 Discovery of Functions Beyond the Mere Anchoring of Proteins to a Membrane: the GPI Anchor as a Molecule with Multiple Functions in Cell Biology

GPI-anchored proteins, due to their association with the membrane, exhibit different properties from proteins with a transmembrane domain. One of the obvious differences is, of course, the lack of a transmembrane and a cytoplasmic domain, and the consequent inability to interact directly with intracellular components. This may, especially in the case of protozoans, be an advantage when it is necessary to protect the integrity of the cell against the extracellular environment. This necessity may be reflected by the dense surface coat of most of these organisms, which in many species consists of GPI-anchored proteins. The prototype for such a surface protein is the VSG of African trypanosomes. It covers the entire cell surface, with a calculated copy number of 10 million per cell, and is considered the parasite's primary defense against the host's immune system. This protection is achieved by the dense packing of the protein, only allowing small molecules <20 kDa such as nutrients to pass, and thus shielding the cell from, for example, lysis by the complement pathway. Protection against antibody-mediated lysis is achieved by switching between different VSG types, and thus undergoing antigenic variation, as reviewed in detail by Cross (1990).

Insulation may also be the reason for GPI-anchoring for different hydrolases such as alkaline phosphatase, acetylcholinesterase and 5'-phosphatase on the surface of some mammalian cells. Other cells of higher eukaryotes present a different situation. In contrast to protozoans, most surface proteins are anchored via a "classical" transmembrane domain with the exception of some polarized cells, where the apical membranes contain a higher concentration of GPI-anchored proteins than other cells. In cells where GPI-anchored proteins represent only a minor fraction of the surface proteins, they exhibit specialised functions such as intracellular targeting, podocytosis and signal transduction, which will be discussed below.

1.3 Molecular Biology: from Function to Genes to Therapy?

Since it became obvious that GPI-anchors can perform a multitude of functions in different systems, and that their core structure has been conserved throughout eukaryotic evolution, much attention has been focused on identifying genes involved in GPI-biosynthesis. The first data came from mutant cell lines that were unable to produce GPI-anchored surface proteins. These mutations were grouped in several complementation classes by genetic analysis, and the block in biosynthesis could be identified biochemically by analysing the precursor glycolipids that accumulate in these cells (Hyman, 1988). This allowed cloning of several mammalian genes by complementation. Another powerful system to analyze the genetics of GPI-anchor biosynthesis is yeast, since yeast molecular genetics is well established and GPI-anchors are essential for cell viability. Several groups could identify genes spanning most of the biosynthetic pathway by using temperature-sensitive mutants, and some genes have been cloned already by the same complementation approach as with mammalian cell lines. Cloning and expression of these genes in various organisms, for example, protozoan parasites such as *T. b. brucei* or *P. falciparum,* allows direct comparison at the sequence as well as at the functional level. Although GPI-anchor biosynthesis has been conserved throughout eukaryotic evolution, biochemical analysis of corresponding enzymes between such distantly related organisms as mammals and protozoans may reveal subtle differences in their biochemical parameters, which could be used in a therapeutic approach towards protozoal or fungal infections. First indications of such differences have been shown for the final step in the biosynthesis of GPI-anchored proteins, the transamidase reaction (see below). These data, together with the finding that the human disease paroxysmal nocturnal hemoglobinuria (PNH) is caused by a defect in GPI-anchor biosynthesis (Selvaraj *et al.*, 1988; Carothers *et al.*, 1990; Schubert *et al.*, 1990), and that proteins with such high medical relevance as the prion protein are GPI-anchored (Cashman *et al.*, 1990), show that genetic and functional analysis is now gaining more and more importance in conjunction with the "classical" structural analysis.

1.4 The Structure of GPI-Anchors: A Structure that Has Been Conserved throughout Eukaryotic Evolution

The structural analysis that was pioneered by Michael A. J. Ferguson and co-workers in solving the *T. b. brucei* VSG-GPI-anchor structure revealed a structure containing a phospholipid that was substituted with inositol, a non-acetylated glucosamine (a highly unusual carbohydrate used in biological systems), followed by three mannoses and a terminal ethanolamine-phosphate to which the protein is attached via its amino function (Ferguson *et al.*, 1988; Homans *et al.*, 1988). The nature of the core glycan was finally determined by two-dimensional proton NMR studies (Homans *et al.*, 1988). Studies on the biosynthesis of GPI-anchors were performed by radioactive labeling in parasite lysates, followed by a combination of chemical and enzymatic cleavages in conjunction with analysis of the compounds by HPLC, TLC and other classical biochemical methods (Menon *et al.*, 1988; Doering *et al.*, 1989; 1990; Masterson *et al.*, 1989; 1990; Mayor *et al.*, 1990a; 1990b). A comprehensive and personal overview of these initial and crucial experiments is given in a transcript of the 1991 Colworth Medical Lecture given by M.A.J. Ferguson (1992). Based on these initial experiments, strategies for the structural analysis of GPI-anchors have been developed which consist of a combination of chemical, enzymatic and spectroscopic methods and will be discussed later. The conserved structural features shown in Figure 1 were then confirmed by the analysis of GPI anchors from various other sources, and are proven to be correct for all GPI-anchor structures determined so far.

1.5 Species-Specific and Developmentally Regulated Modifications: Additions to a General Theme

The conserved structure of a GPI-anchor, as shown in Figure 1, is usually modified, and thus gives rise to a plethora of structures characteristic for a given species as well as for certain stages. A complete description of these structures is beyond the scope of this article and is found in various reviews (Ferguson, 1992; Englund, 1993; McConville and Ferguson, 1993; Ferguson *et al.*, 1994; Field and Menon, 1994). The most prominent of such modifications are additional carbohydrates attached to the trimannosyl core; additional ethanolamine-phosphates (a modification that seems to be specific for higher eukaryotes) or an additional fatty acid covalently linked to the inositol. Figure 2 gives an overview over some of these modifications. An example for stage specific modifications is again the protozoan *T. b. brucei,* which exhibits different sugar side chains and lipid moieties on the GPI-anchors of the insect stage (PARP-protein), and the form that lives in the warm blooded host (VSG-protein) (for review see McConville and Ferguson, 1993; Ferguson *et al.*, 1994). These different structures may be a reflection of the strikingly different environmental conditions although a definitive answer is still missing.

2 Structural Analysis

2.1 An Exercise in Biochemistry and Logical Deduction: a Few Representative Examples

As mentioned in the preceding section, the basic structure of all GPI-anchors is conserved and can thus guide biochemical analysis. In addition to this, sets of chemical and enzymatic cleavage reactions are available, which give rise to charac-

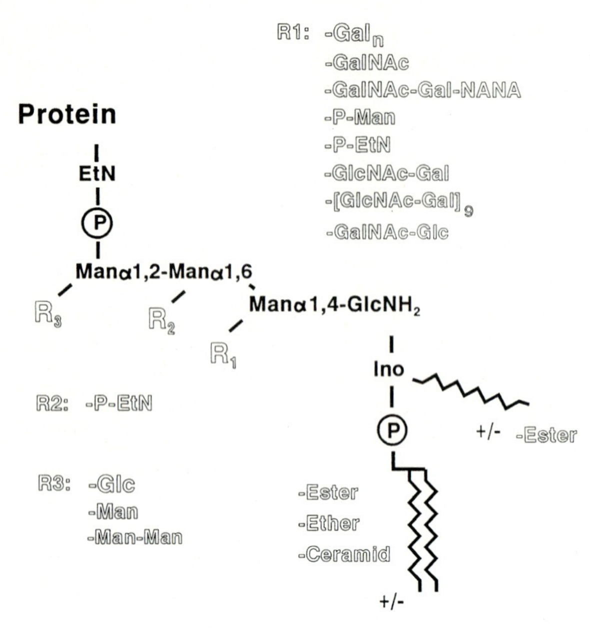

Figure 2. Modifications of protein-bound GPI-anchors. The minimal structure of a GPI-anchor can be modified by a variety of hydrophilic and hydrophobic modifications, some of which are shown in this figure. Gal, galactose; GalNAc, *N*-acetyl-galactosamine; NANA, *N*-acetyl-neuraminic acid; Man-P, mannose-phosphate; Et*N*-P, ethanolamine-phosphate; GlcNAc, *N*-acetyl-glucosamine, Glc, glucose; P, phosphodiester bridge; Ino, inositol; $GlcNH_2$, glucosamine.

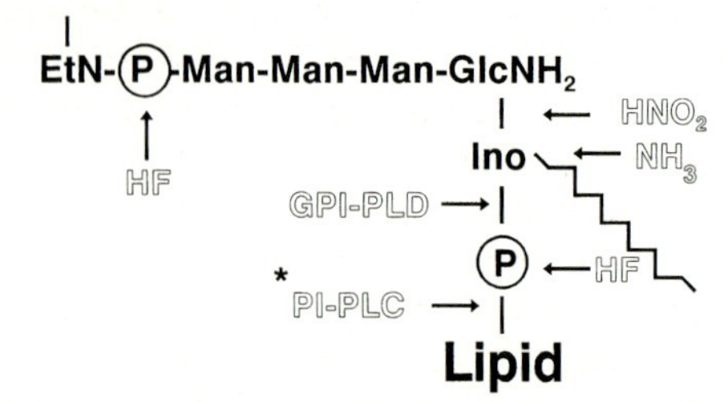

Figure 3. Potential cleavage sites of a GPI-anchor. The cleavage reactions indicated in this figure are those most commonly used for the identification and structural analysis of GPI-anchors. HNO_2, nitrous acid deamination; NH_3, mild alkaline saponification; HF, aqueous hydrofluoric acid dephosphorylation; GPI-PLD, glycosyl-phosphatidylinositol-specific phospholipase D; PI-PLC*, glycosyl-phosphatidylinositol-specific phospholipase C cleaves only if the inositol is not modified; P, phosphodiester bridge; Ino, inositol; $GlcNH_2$, glucosamine; Man, mannose, EtN, ethanolamine.

teristic reaction products that can be detected by biochemical methods. Some of the most commonly used reactions are summarised in Figure 3. In the following paragraph some of these reactions, the analysis of the reaction products, and the deductions that lead to a structural proposal are described. The first step, before analysing the structure, is to confirm that the protein in question indeed is GPI-anchored.

2.2 Identification of GPI-Anchored Proteins

Since the addition of a glycolipid tail to the C-terminus of a membrane protein has significant implications for its biological function, the often necessary identification of such a GPI-anchor does not have to rely solely upon direct structural determination, but on the fact that the amphipathic and hydrophilic forms of a protein exhibit different chemical properties. In addition, a number of techniques have been established to identify GPI-anchored proteins even if these proteins are scarce (for review see Field and Menon, 1992; Ferguson, 1993; Menon, 1994).

The most powerful method to identify a GPI-anchor on a given protein is the use of a bacterial phosphatidylinositol-specific phospholipase C (PI-PLC; from *B. cereus*, *B. thuringiensis*, *S. aureus*, or *C. novyi*) or a glycosyl-phosphatidylinositol specific phospholipase C (GPI-PLC; from *T. b. brucei*, or rat liver). Most of these enzymes are commercially available and can be used on intact cells or in cell lysates. After treatment with these enzymes, the protein should lose its amphipathic character, which is reflected in a change in solubility. However, GPI-anchors carrying a modified inositol ring are insensitive to cleavage by (G)PI-PLC. Alternative methods to identify a GPI-anchor and/or to confirm results obtained by (G)PI-PLC treatment are the use of a mammalian GPI-specific phospholipase D (GPI-PLD), or deamination by nitrous acid. Both treatments cleave GPI-anchors from only a few solubilized proteins since GPI-anchored proteins on intact cells are only marginally susceptible to GPI-PLD or nitrous acid deamination.

Whether a protein is sensitive to these GPI-specific treatments can be determined by a shift in mobility of the protein in SDS-polyacrylamide gels. However, this gel shift assay does not work for all GPI-proteins, since the amphipathic and the hydrophilic forms of some proteins migrate identically in conventional SDS-polyacrylamide gels. However, when using non-denaturating polyacrylamide gel electrophoresis in the presence of detergents (for example, Triton X-100), the interaction of the GPI-anchor lipid moiety with the detergent dramatically changes the mobility of the GPI-anchored protein (for details see Hooper, 1992).

The release of a GPI-anchor by (G)PI-PLC can also be assessed by the identification of the so-called cross-reacting determinant by mono- or polyclonal antibodies. These antibodies recognize the cyclic phosphate at the inositol ring in a Western blot analysis. This cross-reacting determinant is cryptic in the amphipathic form of the protein, and is only formed when the protein is converted to its hydrophilic form by (G)PI-PLC treatment (for details see Hooper, 1992).

GPI-anchored proteins can also be identified by temperature-induced phase separation in Triton X-114. This technique discriminates between GPI-anchored proteins and integral membrane proteins versus cytosolic proteins. It is based on the ability of the non-ionic detergent Triton X-114 to partition into two distinct phases at 30 °C: a detergent-rich phase, with the major part of the membrane proteins, and an aqueous phase, containing predominantly non-membrane proteins (for details see Ko and Thompson, 1995).

Another technique which makes use of the amphipathic character of GPI-anchored proteins is the reconstitution of proteins into artificial lipid vesicles or liposomes. Integral membrane proteins as well as GPI-anchored proteins reconstitute in the lipid vesicles and can be purified as proteoliposomes by centrifugation through a sucrose cushion (for details see Hooper, 1992).

Based on the amphipathic character of GPI-anchored proteins, hydrophobic interaction chromatography on phenyl- or octyl-Sepharose can also be a powerful tool to identify and purify GPI-anchored proteins, and to discriminate between their amphipathic and hydrophilic forms. The amphipathic form of the protein interacts via the hydrophobic domain of a GPI-anchor with the hydrophobic gel matrix and can be eluted specifically with increasing amounts of organic solvents (for details see Ferguson, 1993).

Another approach to identify a protein-bound GPI-anchor is the use of metabolic radioactive labeling techniques. As GPI-anchors exhibit highly conserved structural features containing ethanolamine and inositol, GPI-anchor components can be labeled with radioactive precursors in cell culture, in crude cell lysates or with purified membrane fractions. The incorporation of ethanolamine and inositol is highly diagnostic for the presence of a GPI-anchor. Labeling with radioactive glucosamine and mannose, although they are predominant components of GPI-anchors, is not sufficient to postulate a GPI-anchor, because a lot of membrane proteins are modified by *N*- and *O*-glycans, consisting in part of these two sugars. The use of the antibiotic tunicamycin, which inhibits the synthesis of lipid-linked precursors of the *N*-glycosylation, is recommended when labeling GPI-anchors with these sugar precursors (for details see Field and Menon, 1992; Hooper, 1992).

The release of a hydrophobic fragment by (G)PI-PLC, GPI-PLD and nitrous acid deamination from proteins labeled with fatty acids is also a strong indication for the presence of a GPI-anchor, and can easily be asssessed by phase partition between an aqueous and an organic phase. The most reliable ways to identify a GPI-anchor with certainty are sensitivity of a protein to purified (G)PI-PLC or GPI-PLD, and labeling the protein anchor with radioactive ethanolamine and inositol. Techniques based only on the hydrophobicity of the GPI-anchor are not very diagnostic, since integral membrane proteins, carrying a stretch of hydrophobic amino acids as a membrane anchor, might behave similarly to a GPI-anchored protein.

Each of these above mentioned techniques provides good evidence for the presence of a GPI-anchor, but usually a combination of several methods is required to confirm the presence of a GPI-anchor unequivocally.

2.3 Structural Characterization of GPI-Anchors

All GPI-anchors described so far show some highly conserved structural features (see above) but vary by numerous modifications of their hydrophobic and hydrophilic moieties. Naturally the strategies and methods to analyze GPI-anchors depend on the amount available. Complete structural analysis usually relies on two-dimensional ^{1}H-NMR studies (Ferguson *et al.*, 1988; Homans *et al.*, 1988) but such analyses require more than 100 nmol of highly purified protein. A lot of structural data can be obtained with relatively small amounts of starting material (1–50 nmol) using techniques like gas-liquid-chromatography/mass spectrometry (e.g. compositional analysis of peracetylated hydrolysis products of the glycan part and hydroxyester fatty acid analysis; Ferguson, 1993), fast-atom-bombardment or electrospray/mass spectrometry (e.g. determination of the mass of methylated glycan fragments or of the lipid part; Dell *et al.*, 1993; Ferguson, 1993) and liquid chromatography (e.g. comparison of defined fragments with standards of known structure; Ferguson, 1993). Another technique to obtain information about the mass of the intact GPI-anchor or a defined fragment from only 10–100 pmol is the use of matrix-assisted-laser-desorption/ionisation time-of-flight mass-spectrometry (Gusev *et al.*, 1995). These sophisticated analyses of intact GPI-anchors or defined parts of GPI-anchor peptides are only useful on purified material. Besides hydrophobic interaction chromatography, there are some convenient techniques to purify GPI-anchors, such as iatrobead chromatography, anion-exchange chromatography on DEAE-cellulose, thin-layer chromatography and size-exclusion chromatography on LH-20.

These physico-chemical techniques, although highly informative, require either a relatively high amount of purified material and/or highly sophisticated and expensive equipment, which may not be readily available to most laboratories. We will therefore focus on more conventional biochemical techniques, which can be established more easily.

The most sensitive techniques to analyze GPI-anchors rely on metabolically labeled material. Using radioactive precursors and simple GPI-anchor purification techniques, such as hydrophobic-interaction-chromatography of protease-generated GPI-anchor peptides (Ferguson, 1993), GPI-anchor components can be identified easily by subjecting this purified material to specific chemical and enzymatic reactions followed by chromatographic analysis.

During the course of identifying and defining GPI-anchor structures, initial information is available due to the sensitivity towards specific enzymatic treatments. The use of (G)PI-PLC, GPI-PLD and sphingomyelinase provides structural information about the hydrophobic moiety. A GPI-anchor which is insensitive towards (G)PI-PLC, but can be cleaved by GPI-PLD, implies a modified inositol structure. A GPI-anchor additionally sensitive to sphingomyelinase is likely to be ceramide based. So far, diacyl-glycerol, alkyl/acyl-glycerol, monoacyl-glycerol and ceramide have been described as hydrophobic parts of GPI-anchors (for review see Englund, 1993; McConville and Ferguson 1993; Field and Menon, 1994). Detailed information about the hydrophobic moiety, for example, the fatty acids of GPI-anchors, can be obtained by metabolic labeling with various radioactive fatty acids, followed by their release from the GPI-peptides using alkaline hydrolysis, specific for ester-linked fatty acids or ceramide-based lipids (Field and Menon, 1992). The intact hydrophobic moiety of a GPI-anchor can be released from the protein or GPI-anchor peptide by enzymatic treatments using (G)PI-PLC, GPI-PLD and sphingomyelinase (Field and Menon, 1992). Besides these enzymatic treatments, the hydrophobic part can be released by chemical treatments such as dephosphorylation with aqueous hydrofluoric acid, deamination with nitrous acid, and acetolysis (for review see Field and Menon, 1992; Ferguson, 1993; Menon, 1994). All these treatments generate different hydrophobic structures (e.g. for a diacyl-glycerol based GPI-anchor, (G)PI-PLC produces diacyl-glycerol; GPI-PLD produces phosphatidc acid; dephophorylation produces diacyl-glycerol; deamination gives phosphatidyl-inositol; and acetolysis produces acetyl-diacyl-glycerol). The hydrophobic moieties released can be analyzed by high-performance liquid chromatography or thin-layer chromatography, along with commercially available standards.

Structural information about the hydrophilic part of a GPI-anchor relies on generating defined hydrophilic fragments such as neutral core-glycans, generated by dephosphorylation, deamination, and reduction of sugar-labeled material (Field and Menon, 1992; Ferguson, 1993; Menon, 1994). The intensively desalted core-glycans can be analyzed by size-exclusion chromatography on Bio-Gel P4 (Bio-Rad) or by high-pH-anion-exchange chromatography on a Dionex system (Dionex Corporation, Sunnyvale, CA). Both systems are widely used to analyze GPI-anchor core-glycans and tables with the elution position of various core-glycans are indicated in different reviews (e.g. Ferguson, 1992; Field and Menon, 1992; Ferguson, 1993; Menon, 1994). The elution positions of the glycans are indicated in relation to partly hydrolyzed dextran as glucose units (Bio-Gel P4) or dextran-units (Dionex system). The dextran is included in the radioactive sample as an internal standard, and is detected colorimetrically after oxidation with orcinol/sulfuric acid (Bio-Gel P4) or by pulsed amperometric detection (Dionex system). The use of internal standards with known structures makes it possible to identify core-glycans by their elution behavior in comparison to the elution position of such standards.

A potent method to analyze and purify neutral core-glycans is the use of non-inert HPLC-systems supplied with amino-propyl columns. The major drawback of this method is that they are not commonly used, so that the amount of structural data based on this method is very limited.

Additional information about the structure of a GPI-anchor core-glycan can be obtained using exoglycosidases, which are available with various specificities. For example, the presence of the evolutionarily conserved trimannosyl core-glycan can be confirmed by elution positions of the core-glycan on Bio-Gel P4 and Dionex-HPAEC and by its sensitivity towards α-mannosidases. The treatment with α1,2-mannosidase from *Aspergillus saitoi* (Oxford Glycosystems), which releases the last mannose from a trimannosyl core-glycan, is especially diagnostic for GPI-anchor core-glycans.

The structural characterization of a GPI-anchor core-glycan has to make use of a combination of the techniques described. Even the use of the very powerful high-pH-anion-exchange chromatography (Dionex) does not necessarily reveal the identity of a core-glycan just by simple co-elution. The identity must always be confirmed using exoglycosidases and/or a second analytical system. Bio-Gel P4 size-exclusion chromatography is easier to set up, but is less specific than analysis by a Dionex system. Therefore, analysis of core-glycans on Bio-Gel P4 always relies on the use of exoglycosidases.

The structural analysis of GPI-anchors is based on a set of different techniques, dependent on the amount of material accessible for the analysis. In most cases a combination of the above mentioned techniques will give a detailed insight in the structure of a unknown GPI-anchor.

2.4 The Biosynthesis of GPI-Anchors: a Conserved Pathway with Sidetracks and Backup Systems

The first analysis of GPI-anchor biosynthesis, like the structural analysis, was undertaken in *T. b. brucei,* by establishing a cell-free system which allowed the detailed analysis of each step. Since the core structure of all GPI-anchors is highly conserved, it was assumed that the biosynthetic pathway for this molecule might be equally conserved. This assumption was confirmed by data from various organisms ranging from other protozoans to higher eukaryotes (for review see Ferguson, 1992; Englund, 1993; Ferguson, *et al.*, 1994; Field and Menon, 1994; Stevens, 1995; Takeda and Kinoshita, 1995). However, the biosynthetic pathways, as might be expected, differ with respect to specific modifications which are added after core structure assembly.

2.4.1 Assembly of the conserved core structure

The first detailed studies on the pathway for GPI-anchor biosynthesis came from the trypanosome system (Menon *et al.*, 1988, 1990a,b; Doering *et al.*, 1989, 1990; Masterson *et al.*, 1989, 1990; Mayor *et al.*, 1990a,b) using washed trypanosome membranes or cell-lysates and tritiated sugar-nucleotides. Using UDP-*N*-acetyl-[^{3}H]glucosamine and GDP-[^{3}H]mannose, biosynthetic intermediates were readily detectable. Structural analysis of these GPI-anchor biosynthesis intermediates led to the identification of non-protein bound GPIs which were built by sequential transfer of sugars and ethanolamine-phosphate. Additional information on the GPI-anchor biosynthetic pathway of bloodstream forms of *T. b. brucei* came from the use of specific inhibitors, such as antibiotics (e.g. amphomycin) and sugar analogues (e.g. mannosamine), which interfere with defined steps of this pathway. The GPI-anchor biosynthesis of bloodstream forms of *T. b. brucei* can be summarized as follows: The first step is the transfer of *N*-acetylglucosamine from UDP-*N*-acetylglucosamine to phosphatidylinositol. The product, *N*-acetylglucosamine-phosphatidylinositol, is then deacetylated to form glucosamine-phosphatidylinositol (Doering *et al.*, 1989). This reaction creates the non-acetylated glucosamine, which is highly specific for GPIs. Studies relying on mutants of GPI-biosynthesis in mammalian cells indicate that three genes are involved in the synthesis of glucosamine-phosphatidylinositol (Stevens and Raetz, 1991). The addition of the mannoses, which build the evolutionarily conserved trimannosyl core-glycan, involves the transfer of activated mannose from GDP-mannose on to a hydrophobic mannose donor, dolichol-phosphate-mannose (Schwarz *et al.*, 1989; Menon *et al.*, 1990b). Dolichol-phosphate-mannose is the donor for all three mannoses of the trimannosyl core-glycan, as shown by studies using inhibitors of dolichol-phosphate-mannose synthesis (e.g. amphomycin) and by the analysis of cell lines which were deficient in dolichol-phosphate-mannose synthesis (Conzelmann *et al.*, 1986; Fatemi and Tartakoff, 1986). Ethanolamine-phosphate is then added to the third mannose (counted from the reducing end of the core-glycan) by the hydrophobic donor phosphatidylethanolamine (Menon and Stevens, 1993; Menon *et al.*, 1993), thus completing the biosynthesis of the conserved core glycan.

GPIs carrying an ethanolamine-phosphate on the third mannose are potential GPI-anchor precursors, because they can be transferred on to a newly synthesized protein by replacing a C-terminal GPI-anchor signal sequence by the GPI-anchor precursor in a transamidase-like reaction (see below).

Studies on GPI-anchor biosynthesis in other systems (e.g. insect-forms of *T. b. brucei*, *T. gondii*, *P. falciparum*, yeast, and mammalian cell lines) reveal similarities among different organisms (for review see Englund, 1993; McConville and Ferguson, 1993; Field and Menon, 1994; Stevens, 1995; Takeda and Kinoshita, 1995). The GPI-anchor biosynthesis in all organisms investigated so far involves the same donors (UDP-*N*-acetyl-glucosamine, dolichol-phosphate-mannose and ethanolamine-phosphate) producing a GPI-anchor precursor having at least one ethanolamine-phosphate and the trimannosyl core-glycan. The same holds true for the transfer of *N*-acetylglucosamine on to the inositol-phosphate-lipid, the subsequent deacetylation and transfer of mannoses from the donor dolichol-phosphate-mannose, which is also common to all organisms investigated so far. The genetics underlying this biosynthetic pathway is described below.

2.4.2 Completion of the full GPI-anchor structure: (It's a zoo out there)

As mentioned above, the biosynthetic pathways can differ strikingly between different organisms after completion of the core glycan. This not only includes additions to the carbohydrate backbone but also the hydrophobic moiety, and the point when certain modifications are introduced, e.g. before or after transfer of the GPI-moiety to the protein. We will therefore mention only a few examples briefly, since detailed descriptions of these sometimes highly complex biosynthetic pathways can be found in several recent review articles (e.g. Englund, 1993; McConville and Ferguson, 1993; Field and Menon, 1994; Stevens, 1995; Takeda and Kinoshita, 1995).

There are some striking differences concerning the hydrophobic fragments and additional decorations on GPI-anchors. Ester-linked fatty acids, ether-linked fatty acids and ceramide are described for the hydrophobic fragments of GPIs (for review see Englund, 1993; McConville and Ferguson, 1993; Ferguson *et al.*, 1994, Field and Menon, 1994). In *T. b. brucei* and *Trypanosoma congolense* the fatty acids at GPI-anchor precursors are replaced by myristic acid in a very elaborate process called fatty acid-remodeling (Masterson *et al.*, 1990).

A different type of remodeling reaction seems to be involved in the GPI-anchor biosynthesis in yeast. In yeast there is evidence that the GPI-anchor biosynthetic intermediates, precursors, and newly synthesized GPI-anchored proteins have predominantly acyl-based GPIs (Conzelmann *et al.*, 1992; Costello and Orlean, 1992), whereas most mature GPI-anchored proteins contain ceramide-inositol-phosphate as the hydrophobic moiety. This suggests that yeast may perform a novel form of lipid remodeling involving the exchange of glycerolipid for ceramide (Conzelmann *et al.*, 1992).

Mammalian cells and the malaria parasite *P. falciparum* add an ester-linked fatty acid to the inositol ring of the early biosynthetic intermediates prior to the addition of the first mannose residue of the trimannosyl core-glycan. Therefore all mannosylated GPI biosynthesis intermediates of these organisms are insensitive to (G)PI-PLC treatment. Interestingly, many protein-bound GPI-anchors of mammalian cells are (G)PI-PLC-sensitive. They must therefore have lost their acylation at the inositol on their way to the cell surface. Recently it has been reported that *T. b. brucei* is also capable of synthesizing mannosylated GPI-anchor biosynthesis intermediates having an acylated inositol (Güther and Ferguson, 1995). These results imply that *T. b. brucei,* like *P. falciparum* and mammalian cells, synthesizes acylated forms of GPIs. In the trypanosomal system they are in equilibrium with the non-acylated forms and are present only in very small amounts (Güther and Ferguson, 1995).

A characteristic feature of evolutionarily higher eukaryotes such as mammals is the addition of at least one extra ethanolamine-phosphate on to the growing mannosyl-chain of GPI-anchor biosynthetic intermediates. The first additional ethanolamine-phosphate is added to the mannose next to the glucosamine, prior the addition of the second mannose. These additional ethanolamine-phosphates are a unique feature of GPIs in higher eukaryotes.

Modifications other than ethanolamine-phosphates were also described for GPI-anchor precursors of other organisms: *P. falciparum* and *T. gondii* synthesize GPI-anchor precursors with trimannosyl core-glycans modified by sugar side-chains prior to their addition to the protein. In some strains of *T. b. brucei* and *T. congolense,* the carbohydrate moiety of GPI-anchors is modified on their way to the cell surface, thus generating protein-bound GPI-anchors different from those originally transferred to protein (for review see McConville and Ferguson, 1993).

Another very interesting aspect of GPI-anchor biosynthesis is the topology of the reaction products and enzymes involved. As all donors for the synthesis of GPI-anchor precursors, such as phosphatidylinositol, UDP-*N*-acetylglucosamine, dolichol-phosphate-mannose and phosphatidylethanolamine are available on both leaflets of the endoplasmic reticulum, it was for a long time uncertain where the GPI-biosynthesis takes place. Recently, evidence has been presented that biosynthesis of GPI-anchor precursors occurs on the cytoplasmic leaflet of the endoplasmic reticulum (Mensa-Willmot *et al.*, 1993; Vidugiriene and Menon, 1993, 1994; Menon and Vidugiriene, 1994). This implies the existence of an enzyme which helps the preformed GPI-anchor precursors to cross the endoplasmic reticulum (ER) membrane to the luminal leaflet where the transfer on to the protein will occur. This enzyme, the so-called "flipase", still needs to be identified. Evidence for an enzymatic activity mediating this translocation from the cytosolic to the luminal side of the ER-membrane has recently been published for dolichol-phosphate-mannose (Rush and Waechter, 1995), supporting the existence of such an activity for GPI-anchor precursors. At the moment several genes for enzymes involved in GPI-biosynthesis have been cloned (see below) but none has been completely purified yet. Altogether, GPI-anchor biosynthesis seems to have some highly conserved reactions with a lot of species-specific modifications.

2.4.3 GPI-transamidase

The crucial step in the biosynthesis of GPI-anchored proteins is obviously the transfer of the preassembled GPI-anchor to the corresponding proteins. This reaction, which occurs at the luminal side of the ER, proceeds rapidly as soon as the protein is appropriately inserted and positioned into the ER membrane. This translocation of the nascent protein chain across the ER membrane is a requirement for subsequent GPI-anchoring (Caras, 1991). The reaction itself is a transamidase-type process, where the original carboxy terminus is cleaved off and replaced by the GPI-anchor via the amino function of the terminal ethanolamine-phosphate (reviewed in Field and Menon, 1994; Udenfriend and Kodukula, 1995). The crucial question that underlies this process is how the transamidase-complex recognizes the appropriate proteins.

Proteins, to be GPI-anchored, have to fulfill certain criteria to assure correct processing. At first the protein in question has to be translocated into the ER, which is mediated via an amino-terminal stretch of hydrophobic amino acids that functions as a signal sequence which is cleaved off by the so-called signal peptidase after entering the ER. After translation is complete, it must be established that the protein is anchored in the ER-membrane for recognition by the GPI-transamidase. This is achieved by a C-terminal stretch of hydrophobic amino acids, which acts as a transmembrane domain. This anchoring of the newly synthesized protein in the ER membrane is absolutely necessary for subsequent GPI-anchor transfer, since proteins from which the original hydrophobic C-termini have been deleted are usually secreted into the medium. After correct insertion into the ER membrane is achieved, the original C-terminus is cleaved and the new carboxy-terminal amino acid is attached to the readily assembled GPI-anchor. All these data about the transamidase reaction are summarized in a recent review by Udenfriend and Kodukula (1995). Figure 4 shows the stucture and functional sites of a typical protein destined to be GPI-anchored.

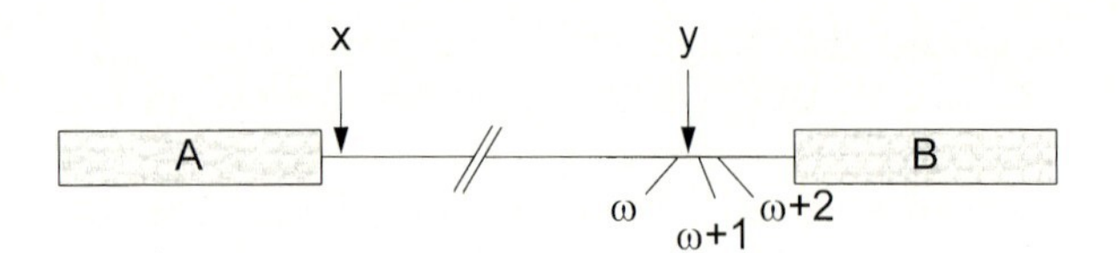

Figure 4. Overall structure and functional sites of a potentially GPI-anchored protein. A, *N*-terminal signal peptide; B, C-terminal transmembrane domain; X, cleavage site for the signal peptidase; Y, cleavage/recognition site for the GPI-transamidase; ω, ω+1, ω+2, amino acid positions involved in the recognition and cleavage process with w being the amino acid to which the GPI-anchor will be attached.

2.4.4 Determination of the cleavage/attachment site

The identification of the attachment sites came from a variety of lines of evidence, and was pioneered by the groups of Sidney Udenfriend and Ingrid W. Caras. One line of evidence came from experiments designed to determine the actual cleavage/attachment site by the preparation and purification of protease-generated GPI-peptides. These peptides were then sequenced by conventional sequencing methods to determine the carboxy-terminal amino acid. These data showed that all GPI-peptides analyzed contained a small amino acid as the residue the GPI-anchor was attached to. This turned out to be the amino-terminal amino acid of three small amino acids found in all cDNAs sequenced so far. Since this is a very elaborate and tedious process, only about 20 attachment sites have been determined experimentally (Gerber *et al.*, 1992). More evidence about the nature of the transamidase recognition site was provided by comparison of cDNA sequences of over 125 cloned GPI-proteins (Udenfriend and Kodukula, 1995). As mentioned, all carboxy-terminal sequences fulfilled the criteria for transmembrane domains, ranging in size from 17–31 residues, with the amino acid composition differing considerably from protein to protein. There was no motif-like feature, other than the hydrophobicity and the lack of a cytoplasmic tail, that these sequences had in common, indicating that anchoring in the ER membrane is the main function of these hydrophobic tails. Another feature noticed in these analyses was the presence of a stretch of three small amino acids 10–12 residues amino-terminal to the hydrophobic domain, with the most amino-terminal amino acid being the residue for attachment of the GPI-anchor. This attachment site was then termed ω-site, with the following positions being ω+1, ω+2 etc (Gerber *et al.*, 1992). This agreement between the DNA sequence data and the experimentally determined attachment site now allows predictions to be made about the GPI-anchoring of a given protein and its cleavage site, by analysis of the DNA sequence. These deductions are now widely used, since the actual experimental determination of the ω-site, as mentioned already, is a very laborious process; from more than 125 GPI-anchored proteins known so far, the ω-site has been determined experimentally for only about 20 (for review see Englund, 1993; Field and Menon, 1994; Udenfriend and Kodukula, 1995). Based on these data, the cleavage/attachment site has been defined, and the so-called ω/ω+1/ω+2 rule was postulated, with ω being the amino acid the GPI-anchor becomes attached to (Gerber *et al.*, 1992; Kodukula *et al.*, 1993). As already mentioned, this site is defined by three consecutive small amino acids, each of the three positions being characterized by the amino acids that are tolerated to allow efficient recognition by the transamidase. The residues found at the ω-site are aspartic acid, asparagine, glycine, alanine, serine, and cysteine. The ω+1 site allows these six amino acids plus glutamic acid and threonine, making this site the least conserved. The most conserved position is the ω+2 site, at which only glycine, alanine, and serine are permitted except for threonine in DAF (decay accelerating factor). To verify these findings each of the sites in a model protein was analyzed by site directed mutagenesis or saturation mutagenesis of all three sites simultaneously, confirming the results obtained by sequence comparison (Micanovic *et al.*, 1990; Moran *et al.*, 1991; Moran and Caras, 1991a, b, 1992). This ω/ω+1/ω+2 rule in now widely used to predict GPI-anchoring and the attachment site from sequence data. Similar experiments showed that the amino acids adjacent to the cleavage/attachment site have no effect on GPI anchoring. Since these rules for GPI-anchoring of proteins were established, it is now possible to convert a non-anchored protein to a perfectly GPI-anchored form properly targeted to the plasma membrane, not only by replacing the whole carboxy-terminus but merely by introducing an appropriate cleavage/attachment site at the correct position upstream a carboxy-terminal hydrophobic domain (reviewed in Udenfriend and Kodukula, 1995).

Although these rules apply to all GPI-anchored proteins so far, two striking exceptions have been described which may indicate significant differences between higher eukaryotes and protozoans, especially parasites. Moran and Caras showed that two protozoan surface antigens, the *T. b. brucei* VSG (the prime example for a GPI-anchored protein) and the *P. falciparum* circumsporozoite protein (where GPI-anchoring has only been deduced from the cDNA sequence), were not GPI-anchored when expressed in mammalian (COS) cells (Moran and Caras, 1993). The proteins themselves were expressed in significant amounts but were retained in the ER, thus showing no or extremely inefficient GPI-anchoring. By exchanging their carboxy-termini with the corresponding DAF-sequence, correct targeting to the cell membrane could be restored. By further analysis, the authors could show that it indeed was the cleavage/attachment site of the protozoal proteins that was responsible for the aberrant processing in COS-cells. These experiments suggest that although the functional sequences at the protein level fit the established rules, there might be subtle differences in the processing machinery between distantly related taxa such as protozoans and parasites, which might be a useful approach for developing new antiparasitic therapeutics via selective inhibition of GPI-anchoring.

Finally, this rather simple recognition signal gives cells the flexibility to express certain proteins in different forms, which can be either GPI-anchored, secreted, or anchored via a classical transmembrane domain simply by synthesizing these proteins with different carboxy-termini. This is achieved either by differential splicing (Caras *et al.*, 1987) or by expression of similar but distinct genes (for reviews see Field and Menon, 1994; Udenfriend and Kodukula, 1995).

3 The Function of GPI-Anchors

3.1 Anchoring of Membrane Proteins: Protozoa vs. Higher Eukaryotes; from a General Principle to Specialized Functions

The most obvious function of GPI-anchors is of course the stable, oriented association of a given protein with a membrane, usually the outer leaflet of the cell membrane, with the protein being oriented towards the extracellular environment. This orientation may indicate a general principle why certain proteins are GPI-anchored, rather than anchored via transmembrane domains. One of the main advantages may be the physical isolation of surface proteins, which have extracellular functions, from the cytoplasm, thus protecting the inside of the cell. As mentioned previously, GPI-anchoring of proteins seems to be a more general phenomenon in protozoa, some of which are parasitic, and are the cause of very important human diseases such as African sleeping sickness (*Trypanosoma* spp.) leishmaniasis (*Leishmania spp.*) malaria (*Plasmodium spp.*) and toxoplasmosis (*T. gondii*). It is therefore important to gain detailed knowledge about such a predominant phenomenon as the anchoring of membrane proteins via GPI-anchors, especially since parasite surface structures such as glycoconjugates are the prime target for the immune system and thus are crucial for parasite survival. The prime example for GPI-anchoring in protozoans is again the parasite *T. b. brucei*. The surface of the bloodstream form of this organism is entirely covered with the GPI-anchored variant surface protein (VSG). It is estimated that this surface coat contains up to 10^7 molecules, making the VSG the most abundant protein in *T. b. brucei* cells. Structural predictions about the orientation and geometry of VSG show that anchoring via GPI allows extremely dense packing which could not be achieved by a transmembrane protein, thus providing perfect insulation against a hostile environment. This holds true, not only for the bloodstream VSG, but also for the insect-stage specific procyclic acidic repetitive protein (PARP or procyclin) (for review see McConville and Ferguson, 1993; Ferguson *et al.*, 1994). In both stages the GPI anchors are highly substituted with carbohydrate side chains. By three dimensional modeling it was concluded that these side chains contribute to the shielding of the cell from the environment by forming an additional glycocalyx immediately adjacent to the outer membrane (for review see Ferguson, 1992; Ferguson *et al.*, 1994). This phenomenon – forming a surface coat with lipid-linked carbohydrate structures which are not protein-linked – occurs in other kinetoplastids such as *Leishmania* which produce GPI-related glycosylinositol phospholipids (GIPLS) or the rather complex lipophosphoglycans (LPGs) (McConville and Ferguson, 1993). Their relatedness to GPI-anchors is shown by the common structural motif Manα1–4GlcN1-inositol-phosphate. These structures also occur on the cell surface at a high density, forming a densely packed glycocalyx, which is viewed as a protective coat like the VSG in *Trypanosoma brucei*. This interpretation of a clearly protective function of GPI-linked structures by forming a dense surface coat, is straightforward in these kinetoplastid parasites.

Another example of such surface structures is the sporozoite stage of malaria parasites, *Plasmodium spp*. The stage which is injected into the warm blooded host by the mosquito also carries a dense protein surface coat made of the circumsporozoite protein. The circumsporozoite protein is thought to be GPI-anchored, although this has not been shown directly. This assumption was made by analysis of the cDNA sequence, which shows a typical cleavage attachment site for GPI-anchored proteins. Other stages of this parasite, for example, the invasive stage of the erythrocytic cycle (merozoites), present a different situation. Although most surface proteins are GPI-anchored, the cell surface is not covered by a rather uniform surface coat as in the sporozoite stage or in trypanosomes and leishmania. The same picture is found in *T. gondii*, like *P. falciparum* an apicomplexan parasite, where several GPI-anchored proteins have been identified in the tachyzoite stage (Tomavo *et al.*, 1989). Whether these proteins play also a largely protective role or are necessary for other parasite functions, such as invasion, membrane stability or intracellular survival, or if this reflects the general tendency of lower eukaryotes to anchor membrane proteins via GPI-anchors, remains to be elucidated.

This phenomenon of a dense surface coat that consists of glycolipid-anchored proteins has recently been described in the free living ciliate *Paramecium primaurelia* (Capdeville *et al.*, 1993). The surface antigens (SAGs) of this protozoan organism undergo elaborate antigenic variation by mutual intergenic and interallelic exclusion of a multigene family. The switching between individual SAGs can be controlled experimentally, which makes this organism an excellent model for studying antigenic variation. By biosynthetic labeling, followed by chemical and biochemical analysis of the protein-bound glycolipids, it could be shown that, as in protozoan parasites, the proteins of this dense surface coat are indeed anchored via a GPI-like structure which has been identified as ceramide based glycosylinositol-phospholipids (Azzouz *et al.*, 1995).

Another single-celled organism that has recently gained importance in GPI research is the yeast *Saccharomyces cerevisiae*. It is an organism that can be grown and easily manipulated genetically. Several GPI-anchored proteins have been identified which, as expected, are inserted into the outer leaflet of the cell membrane, whereas one protein, the α-agglutinin, is cleaved off its GPI-anchor after reaching the cell wall and subsequently becomes covalently linked to the β-glucans of the cell wall (Lu *et al.*, 1994). GPI biosynthesis and correct anchoring of proteins is essential in this organism since defects in this pathway lead to a lethal phenotype (Leidich *et al.*, 1994). In addition, incorrect anchoring and processing leads to destabilisation of the cell wall, which also gives rise to a selectable phenotype, making yeast an excellent system for the genetic analysis of GPI-anchoring (see below). These data on yeast also support the notion that GPI-anchoring might indeed be essential for lower eukaryotes, although this has not been shown for the protozoans, for example by selective inactivation of genes via gene targeting.

As mentioned earlier, higher eukaryotes use GPI-anchoring of surface proteins only for proteins with specialized functions, and in contrast to protozoans and yeast, defects in GPI-anchor biosynthesis are not lethal for mammalian cells in culture.

One of the striking consequences of GPI-anchoring is seen in polarized epithelial cells, where the plasma membrane is divided into an apical and a basolateral domain, divided by a ring of tight junctions. In most of these cells, GPI-anchored proteins are found almost exclusively on the apical surface which indicates that in these cells the GPI-anchor can be viewed as an efficient apical targeting signal (for review see Anderson, 1994; Field and Menon, 1994). This was shown by exchanging the C-terminus of basolaterally targeted proteins with the C-terminus of a GPI-anchored, apically targeted protein, which resulted in apical expression of the chimeric proteins (Brown *et al.*, 1989; Lisanti *et al.*, 1989). These cell surfaces face the "outside" of a tissue, which may represent a strikingly different, even harsher environment than the one facing the basolateral surface, thus needing a more basic insulation function of GPI-proteins than might be the case in protozoan systems. The phenomenon of targeted secretion or insertion into distinct membrane areas is also found in neurons, where Thy-1 is expressed exclusively on axonal membranes (for review see Anderson, 1994).

One explanation for this highly ordered intracellular transport might be the observation that GPI-anchored proteins become sequestered into specialized transport vesicles during their passage through the Golgi apparatus or shortly thereafter (Brown and Rose, 1992). It is proposed that these vesicles contain clusters of GPI-anchored proteins, glycosphingolipids and accessory proteins (Simons and Wandinger-Ness, 1990). The sequestration into specialised vesicles may also explain the often inhomogeneous distribution of GPI-anchored proteins on mammalian cell membranes, in contrast to protozoan cells where the surface distribution of these proteins is mainly uniform. It was postulated that GPI-proteins in higher eukaryotes are sequestered in specialized membrane microdomains (Simons and Wandinger-Ness, 1990). These microdomains persist at the cell surface and can contain a mixture of GPI-anchored proteins. This sequestration may also explain the exclusion of these proteins from the clathrin-dependent endocytosis pathway and thus their usually low turnover rates (Rothberg, 1995).

3.2 Transmembrane Signaling

One of the most interesting and controversial aspects of GPI-function is their ability to participate in signaling mechanisms, or to function directly as second messengers. Most information on GPI-mediated signaling phenomena has been obtained using mammalian lymphocytes (for review see Robinson, 1991) from which many GPI-anchored proteins have been described (e.g. CD-59, CD-55 (DAF), Thy-1, CD48, Qa-2). Studies to investigate the function of the GPI-anchors of the surface antigens Qa-2 and H-2D^b made use of transgenic mice expressing either the GPI-anchored form or the transmembrane form of these proteins (Robinson *et al.*, 1989). It was shown that the GPI-anchor on these T-cell antigens is essential for mitogenesis of T-cells. Similar approaches, using transmembrane and GPI-anchored forms of decay-accelerating factor prepared by genetic engineering, and other GPI-anchored lymphocyte surface antigens, were undertaken to show that only the GPI-anchored forms of these proteins were associated with specific protein tyrosine kinases in coprecipitation studies (Stefanova *et al.*, 1991; Shenoy-Scaria *et al.*, 1992). This association of GPI-anchored lymphocyte antigens with different tyrosine kinases (especially of the family of *src*-kinases) has been confirmed by several authors (for review see Gaulton and Pratt, 1994). The requirement of GPI-anchored forms of proteins for lymphocyte activation, and their association with tyrosine kinases, imply that GPI-anchors are directly involved in signal transduction processes on the lymphocyte surface, leading to the activation of signaling pathways within the cells. These processes might be involved in the pathology of infection with some parasitic protozoa, which release large amounts of protein-bound and free GPIs during infection (see below). The structural requirements of GPIs to interfere with the intracellular signaling cascades have not been elucidated so far. There are only limited data about the effects of non-protein-bound GPIs on cell activation. Nothing is known about how a signal can be transferred from the GPI-anchored protein or a non-protein-bound GPI localized in the outer leaflet, to a tyrosine kinase at the cytosolic site of the plasma membrane. GPIs might play important roles in the activation of lymphocytes by pathways synergistic or alternative to classical receptor-mediated cell activation.

In addition to the activation of tyrosine kinases, GPIs could serve as second messengers in the plasma membrane. As they are structurally related to more common second messengers such as inositol-phosphate, diacyl-glycerol, phosphatidic acid and ceramide, GPIs or their cleavage products, hydrolyzed following receptor ligation, play a role in cellular signaling and hormone action (for reviews see Saltiel *et al.*, 1988; Saltiel, 1991; Gaulton and Pratt, 1994). Free and protein-released GPIs are reported to mimic the effects of three hormone-like peptides: interleukin-2, nerve growth factor and insulin (for review see Gaulton and Pratt, 1994).

The best investigated function of a GPI-fragment is the insulin-mimetic activity of the inositol-glycan released from intact GPIs by GPI-PLD and (G)PI-PLC (for review see Kilgour, 1993). It has been shown that the inositol-glycans lead to the stimulation of glucose oxidation and lipogenesis, as well as the inhibition of lipid lysis when exposed to intact rat adipocytes (for review see Kilgour, 1993). The structure of mediators of insulin-like activity may differ. Besides inositol-mediators of mammalian origin, these effects have also been described for GPIs from parasitic protozoa (Low and Saltiel, 1988; Schofield and Hackett, 1993; Schofield *et al.*, 1996; Tachado *et al.*, 1996).

3.3 GPI-Anchors in Parasite Pathogenicity: A Small Molecule with Dramatic Effects

GPIs are ubiquitous among eukaryotes investigated so far, but they are highly abundant in protozoa. Among the parasitic protozoa investigated for GPIs are some of the most dangerous species for mankind and domestic lifestock. Some of these parasitic protozoa (e.g. *P. falciparum, T.*

brucei, T. congolense) share a striking similarity in their pathology: they induce the release of high amounts of tumor necrosis factor α (TNF-α) by their hosts. Investigations on human *malaria tropica* (severe falciparum malaria) and different mouse models of severe malaria confirm the deadly role of high TNF-α levels in malaria pathology (for review see Clark *et al.*, 1991; Jakobsen *et al.*, 1995). The increase of TNF-α levels during malaria infection is induced by a parasite toxin released during synchronous rupture of infected erythrocytes (Kwiatkowski, 1989). This parasite toxin has been shown to be malarial GPIs (Schofield and Hackett, 1993; Schofield *et al.*, 1993; 1994; Schofield *et al.*, 1996; Tachado *et al.*, 1996). Beside the induction of TNF-α release, it has been shown that malarial GPIs (free and protein-bound) are involved in the upregulation of endothelial cell surface markers (Schofield *et al.*, 1996), which are receptors for *P. falciparum*-infected erythrocytes, and of nitric oxide synthase (Tachado *et al.*, 1996), which synthesises nitric oxide, another mediator involved in cell activation. The effects of parasite GPIs are shown to be an activation of different cell types (e.g. macrophages and endothelial cells) by stimulation of protein kinase C and various protein tyrosine kinases (Schofield *et al.*, 1994). These data show that malarial GPIs induce signaling pathways similar to GPI-anchors of lymphocyte antigens, but that perhaps they activate these pathways without the normal control mechanisms. Thus malarial GPIs seem to represent a potent toxin involved in several mechanisms of malarial pathology. A similar result was obtained for the activation of TNF-α and interleukin-1 release induced by GPIs of *T. b. brucei* (Tachado and Schofield, 1994). The identification of GPIs as potential mediators of cytokine release might suggest a new aim of vaccination. A vaccine capable of neutralizing GPIs could lead, besides other effects, to an inhibition of TNF-α release and therefore help in treating severe malaria. Such a vaccine should be called an "anti-disease vaccine" (Playfair *et al.*, 1990) because it is not directed against the parasite but against its toxic agents.

4 The Molecular Biology of GPI-Anchor Biosynthesis

4.1 From Functions to Genes: Mutant Cells Lead the Way

As described above, a lot of knowledge about structure, biosynthesis and function of GPI anchors has been obtained, revealing some highly interesting aspects of GPI-anchor biology. It is therefore of equal importance to gain insight into the genetics of this biosynthetic pathway.

The first information that led to the identification of genes involved in GPI-anchor biosynthesis came from mutant murine thymoma cell lines, that were unable to express and synthesize GPI-anchored Thy1 on the cell surface. These mutants were grouped into complementation classes by somatic cell fusion analysis (Hyman, 1988). The cell lines in which the lack of Thy1-expression could be attributed to deficient GPI-anchor biosynthesis or GPI-anchoring of Thy1 were called A, B, C, E, F, H and K. Similar cell lines have been isolated from other mammalian sources and allocated to these complementation classes (reviewed in Kinoshita and Takeda, 1994). The defects in these cell lines were subsequently characterized biochemically by identifying the GPI-biosynthetic intermediates that accumulate in these cells. Cell lines of classes A, C and H are unable to synthesize the first intermediate, GlcNAc-phosphatidylinositol (Stevens and Raetz, 1991; Sugiyama *et al.*, 1991; Hirose *et al.*, 1992), indicating that at least three independent genes are necessary for the initial step of GPI-biosynthesis. Class E cells are deficient in dolichol-phophate-mannose biosynthesis, and thus are unable to add mannose to Glc*N*-phosphatidylinositol (DeGasperi *et al.*, 1990; Urakaze *et al.*, 1992). Cell lines belonging to class B accumulate intermediates containing only two mannoses (Puoti *et al.*, 1991; Sugiyama *et al.*, 1991), while class F cells are deficient in adding the terminal ethanolamine-phosphate, thus accumulating Man_3-Glc*N*-phosphatidylinositol (Sugiyama *et al.*, 1991), and class K cells are deficient in the transfer of the GPI-anchor to protein (Mohney *et al.*, 1994). Descriptions of these cell lines are found in recent reviews by Kinoshita

(Kinoshita and Takeda, 1994; Takeda and Kinoshita, 1995) and references therein. By establishing cell lines from paroxysmal nocturnal hemoglobinuria (PNH) patients, it could be demonstrated that these cells belong to class A and carry inactivating somatic mutations in the PIG-A (phosphatidylinositolglycan A) gene (Takeda *et al.*, 1993; Bessler *et al.*, 1994). These cell lines provide an indispensable tool for the cloning of the genes of interest for which no sequence information is available: a selectable phenotype.

The second organism that has been used to clone and identify genes involved in GPI-anchor biosynthesis is the budding yeast *S. cerevisiae*. The main advantages of working with this single celled organism are that it is genetically and biochemically well characterized and thus has become one of the most important systems in eukaryotic molecular biology. What makes yeast especially useful for the cloning of genes for GPI-anchor biosynthesis is the fact that GPI-anchoring is essential in this species, and a block in this pathway leads to a lethal phenotype (Leidich *et al.*, 1994, 1995), thus providing a tool to select for temperature-sensitive mutants, for example (Leidich *et al.*, 1995).

4.2 Cloning of Genes Involved in GPI-Anchoring by the Complementation of Defined Defects

One of the main obstacles in cloning genes for GPI-anchor biosynthesis was the fact that no peptide sequence information was available for the initial studies, since none of the enzymes involved in GPI-anchor biosynthesis has been purified so far. Therefore it was impossible to design DNA probes for the cloning of such genes by hybridization, or to produce antibodies for the detection of expression-positive clones.

The strategy used was cloning by complementation, which uses mutant cell lines or yeast strains. A prerequisite for this strategy to work is the presence of a selectable phenotype, as emphasized above. The cells are first transformed with an expression cDNA library, and transformed cells are then screened for reversal of the mutant phenotype. In theory, the exogenous DNA, if carrying the gene in question, should be able to restore the defect, and thus enable the recipient cells to produce a wild-type phenotype. This restored phenotype can be detected, the plasmid DNA isolated and the coding sequence can then be determined by DNA sequencing.

In the case of the mammalian cells, the restoration of Thy1 surface expression in conjunction with biochemical analysis of GPI-anchor intermediates was used as screening parameter to identify mammalian genes for GPI-anchor biosynthesis. These experiments were largely done by T. Kinoshita and co-workers, and led to the cloning of the genes for complementation classes A, F, and H (Inoue *et al.*, 1993; Kamitani *et al.*, 1993; Miyata *et al.*, 1993). The genes were termed PIG (phosphatidylinositolglycan) A, F and H, respectively.

The second organism used in cloning GPI-anchor genes is yeast. Since GPI-anchoring in yeast is essential, and thus a complete block in biosynthesis would be lethal, conditionally lethal yeast mutants have to be used. In general, temperature-sensitive (ts) mutants are used for this purpose. Orlean and co-workers have cloned the dolichol-phosphate-mannose synthase from yeast using such a yeast strain (Orlean *et al.*, 1988), and have recently described three ts-yeast strains which are deficient in the first step in GPI-anchor biosynthesis. They obtained these strains by classical mutagenesis, followed by screening for reduced ^{3}H-inositol incorporation at the non-permissive temperature. Three strains (gpi1/2/3) could be characterized, and the corresponding genes (GPI1/2/3) were cloned by complementation, thus supporting the notion that at least three genes are involved in the synthesis of GlcNAc-phosphatidylinositol (Leidich *et al.*, 1994, 1995). Howard Riezman and co-workers cloned a yeast gene (SPT14) involved in this step by comparison of sequence homologies with the mammalian PIG-A; SPT14 turned out to be identical to the GPI3 gene cloned by Orlean's group (Schönbächler *et al.*, 1995). Another gene that was cloned by using a temperature-sensitive mutant (gene GAA1) was shown to be involved in the transfer of the GPI-anchor to protein (Hamburger *et al.*, 1995).

Recently, Conzelmann and coworkers reported the identification of six complementation classes involved in the biosynthesis of GPI-anchors in yeast, spanning most of the biosynthetic pathway for the conserved core glycolipid (Benghezal *et al.*, 1995). This group also used the approach of mutagenising yeast and subsequently screening for cells deficient in surface expression of GPI-anchored proteins, as well as making use of reduced cell wall stability in GPI-deficient yeast.

The approaches discussed so far made use of homologous complementation to identify genes of interest. This strategy might also be extended by using, for example, yeast mutants and transfecting them with cDNA libraries from other organisms. Since GPI-anchor biosynthesis is an evolutionarily highly conserved pathway, it might be expected that non-homologous DNA could also compensate functionally for a given defect in this pathway. This approach of heterologous complementation has been used successfully by Orlean and co-workers, compensating a mammalian cell line deficient in the dolichol-phosphate-mannose synthase gene by the corresponding yeast gene (Beck *et al.*, 1990; DeGasperi *et al.*, 1990). In our group, heterologous complementation has been used to clone the dolichol-phosphate-mannose synthase from the parasitic protozoan *T. b. brucei,* using the temperature-sensitive yeast strain DPM 1–6 of Peter Orlean (Mazhari *et al.*, 1996). By expanding the approach of complementation to heterologous strategies, it might be expected that, by using the existing yeast mutants and mammalian cell lines, genes for the biosynthesis of GPI-anchors from various organisms could be cloned, thus providing information for a detailed understanding of GPI-anchor biosynthesis and function.

Another aspect of GPI-molecular biology has already been mentioned. Since the attachment/cleavage site was defined, this information can be used to convert almost any given protein to a GPI-anchored form and *vice versa* (Moran and Caras, 1991a, b), even proteins of prokaryotic origin (Soole *et al.*, 1995), which will not only be expressed at the surface of the recipient cells but also correctly targeted to the apical membrane of polarized cells, thus providing a tool for functional studies.

In conclusion, it can be said that GPI-research has evolved from the biochemical/analytical field to a subject also involving genetics, cell biology, molecular biology and medical research, progressing to a truly interdisciplinary, fast-growing area, with very interesting prospects for the future. In short: there's more to come.

References

Anderson RG (1994): Functional specialisation of the glycosylphosphatidylinositol membrane anchor. In *Sem. Immunol.* **6**:89–95.

Azzouz N, Striepen B, Gerold P *et al.* (1995): Glycosylinositol-phosphoceramide in the free-living protozoan *Paramecium primaurelia*: modification of core glycans by mannosyl phosphate. In *EMBO J.* **14**:4422–33.

Beck PJ, Orlean P, Albright C *et al.* (1990): The *Saccharomyces cerevisiae* DPM1 gene incoding dolichol-phosphate-mannose synthase is able to complement glycosylation-defective mammalian cell line. In *Mol. Cell. Biol.* **10**:4612–22.

Benghezal M, Lipke PN, Conzelmann A (1995): Identification of six complementation classes involved in the biosynthesis of glycosylphosphatidylinositol anchors in *Saccharomyces cerevisiae*. In *J. Cell Biol.* **130**:1333–44.

Bessler M, Mason PJ, Hillmen P *et al.* (1994): Paroxysmal nocturnal hemoglobinuria (PHN) is caused by somatic mutations in the PIG-A gene. In *EMBO J.* **13**:110–7.

Boothroyd JD, Paynter CA, Cross GA *et al.* (1981): Variant surface glycoproteins of *Trypanosoma brucei* are synthezised with cleavable hydrophobic sequences at the carboxy and amino termini. In *Nucl. Acids Res.* **9**:4735–43.

Brown DA, Rose JK (1992): Sorting of GPI-anchored proteins to glycolipid enriched membrane subdomains during transport to the cell surface. In *Cell* **68**:533–42.

Brown DA, Crise B, Rose JK (1989): Mechanism of membrane anchoring affects polarized expression of two proteins in MDCK cells. In *Science* **245**:1499–503.

Capdeville Y, Charret R, Antony A *et al.* (1993): Ciliary and plasma membrane proteins in *Paramecium*: description, localisation, and intracellular transit. In *Advances in Cell and Molecular Biology of Membranes* (Plattner H, ed) pp 181–226, Hampton Hill: JAI Press.

Caras IW (1991): An internally positioned signal can directed attachment of a glycophospholipid membrane anchor. In *J. Cell Biol.* **113**:77–85.

Caras IW, Davitz MA, Rhee L *et al.* (1987): Cloning of decay-accelerating factor suggests novel use of splicing to generate two proteins. In *Nature* **325**:545–9.

Cardoso-de-Almeida ML, Turner MT (1983): The membrane form of variant surface glycoproteins of *Trypanosoma brucei*. In *Nature* **302**:349–52.

Carothers DJ, Harza SV, Anderson SB *et al.* (1990): Synthesis of aberrant decay-accelerating factor proteins by affected paroxysmal nocturnal hemoglubinuria leukocytes. In *J. Clin. Invest.* **85**:47–55.

Cashman NR, Loertscher R, Nalbantoglu J *et al.* (1990): Cellular isoform of the scrapie agent protein participates in lymphocyte activation. In *Cell* **61**:185–92.

Clark IA, Rockett KA, Crowden WB (1991): Proposed link between cytokines, nitric oxide and human cerebral malaria. In *Parasitol. Today* **7**:205–7.

Conzelmann A, Spiazzi A, Hyman R *et al.* (1986): Anchoring of membrane proteins via phosphatidylinositol is deficient in two classes of Thy-1 negative mutant lymphoma cells. In *EMBO J.* **5**:3291–6.

Conzelmann A, Puoti A, Lester RL *et al.* (1992): Two different types of lipid moieties are present in glycophosphoinositol-anchored membrane proteins of *Saccharomyces cerevisiae*. In *EMBO J.* **11**:457–66.

Costello LC, Orlean P (1992): Inositol acylation of a potential glycosyl phosphoinositol anchor precursor from yeast requires acyl coenzyme A. In *J. Biol. Chem.* **267**:8599–603.

Cross GAM (1990): Cellular and genetic aspects of antigenic variation in trypanosomes. In *Annu. Rev. Immunol.* **8**:83–110.

DeGasperi R, Thomas LJ, Sugiyama E *et al.* (1990): Correction of a defect in mammalian GPI anchor biosynthesis by a transfected yeast gene. In *Science* **250**:988–91.

Dell A, Khoo KH, Panico M *et al.* (1993): FAB-MS and ES-MS of glycoproteins. In *Glycobiology: A Practical Approach* (Fukuda M, Kobata A, eds) pp 187–222, Oxford: IRL-Press.

Doering TL, Masterson WJ, Englund PT *et al.* (1989): Biosynthesis of the glycosyl-phosphatidylinositol membrane anchor of the trypanosoma variant surface glycoprotein. Origin of the non-acetylated glucosamine. In *J. Biol. Chem.* **264**:11168–73.

Doering TL, Masterson WJ, Hart GW *et al.* (1990): Biosynthesis of glycosyl phosphatidylinositol membrane anchors. In *J. Biol. Chem.* **265**:611–4.

Englund PT (1993): The structure and biosynthesis of glycosyl phosphatidylinositol protein anchors. In *Annu. Rev. Biochem.* **62**:121–38.

Fatemi SH, Tartakoff AM (1986): Hydrophilic anchor-deficient Thy-1 is secreted by a class E mutant T lymphoma. In *Cell* **46**:653–7.

Ferguson MA (1992): Glycosyl-phosphatidylinositol membrane anchors: The tale of a tail. In *Biochem. Soc. Transact.* **20**:243–56

Ferguson MA (1993): GPI membrane anchors: isolation and analysis. In *Glycobiology: A Practical Approach* (Fukuda M, Kobata A, eds) pp 349–84, Oxford: IRL-Press.

Ferguson MA, Low MG, Cross GA (1985): Glycosyl-sn-1,2-dimyristylphosphatidylinositol is covalently linked to *Trypanosoma brucei* variant surface glycoprotein. In *J. Biol. Chem.* **260**:14547–55.

Ferguson MA, Homans SW, Dwek RA *et al.* (1988): Glycosyl-phosphatidylinositol moiety that anchors *Trypanosoma brucei* variant surface glycoprotein to the membrane. In *Science* **239**:753–9.

Ferguson MA, Brimacombe JS, Cottaz S *et al.* (1994): Glycosyl-phosphatidylinositol molecules of the parasite and the host. In *Parasitol.* **108**:S45–54.

Field MC, Menon MA (1992): Biosynthesis of glycosyl-phosphatidylinositol membrane protein anchors. In *Lipid Modification of Proteins: A Practical Approach* (Hooper NM, Tuner AJ, eds) pp 155–90, Oxford: IRL-Press.

Field MC, Menon AK (1994): Glycolipid anchoring of cell surface proteins. In *Lipid Modifications of Proteins* (Schlesinger M, ed) pp 83–133, Boca Raton: CRC Press.

Gaulton GN, Pratt JC (1994): Glycosylated phosphatidylinositol molecules as second messengers. In *Sem. Immunol.* **6**:97–104.

Gerber LD, Kodukula K, Udenfriend S (1992): Phosphatidylinositol glycan (PI-G) anchored membrane proteins. Amino acid requirements adjacent to the site of cleavage and PI-G attachment in the COOH-terminal signal peptide. In *J. Biol. Chem.* **267**:12168–73.

Gusev AJ, Wilkinson WR, Procter A *et al.* (1995): Improvement of signal reproducibility and matrix/comatrix effects in MALDI analysis. In *Anal. Chem.* **34**:1034–41.

Güther ML, Ferguson MA (1995): The role of inositol acylation and inositol deacylation in GPI biosynthesis in *Trypanosoma brucei*. In *EMBO J.* **14**:3080–93.

Hamburger D, Egerton M, Riezman H (1995): Yeast Gaa1p is required for attachment of a completed GPI anchor onto proteins. *J. Cell Biol.* **129**:629–39.

Hirose S, Mohney RP, Mutka SC (1992): Derivation and characterization of glycoinositol-phospholipid anchor-defective human K562 cell clones. In *J. Biol. Chem.* **267**:5272–8.

Holder AA (1983): Carbohydrate is linked through ethanolamine to the C-terminal amino acid of *Trypanosoma brucei* variant surface glycoprotein. In *Biochem. J.* **209**:261–2.

Holder AA, Cross GAM (1981): Glycopeptides from variant surface glycoproteins of *Trypanosoma brucei*: C-terminal location of antigenically cross-reacting carbohydrate moieties. In *Mol. Biochem. Parasitol.* **2**:135–50.

Homans SW, Ferguson MA, Dwek RA *et al.* (1988): Complete structure of the glycosyl phosphatidylinositol membrane anchor of rat brain Thy-1 glycoprotein. In *Nature* **333**:269–72.

Hooper NM (1992): Identification of a glycosylphosphatidylinositol anchor on membrane proteins. In *Lipid Modification of Proteins: A Practical Approach* (Hooper NM, Tuner AJ, eds) pp 89–116, Oxford: IRL-Press.

Hyman R (1988): Somatic genetic analysis of the expression of cell surface molecules. In *Trends Genet.* **4**:5–8.

Inoue N, Kinoshita T, Orii T *et al.* (1993): Cloning of a human gene, PIG-F, a component of glycosylphosphatidylinositol anchor biosynthesis, by a novel expression cloning strategy. In *J. Biol. Chem.* **268**:6882–5.

Jakobsen PH, Bate CA, Taverne J *et al.* (1995): Malaria: toxins, cytokines and disease. In *Parasite Immunol.* **17**:223–31.

Kamitani T, Chang HM, Rollins C *et al.* (1993): Correction of the class H defect in glycosylphosphatidylinositol anchor biosynthesis in Ltk-cells by a human cDNA clone. In *J. Biol. Chem.* **268**:20733–6.

Kilgour E (1993): A role for inositol-glycan mediators and G-proteins in insulin action. In *Cell. Signal.* **5**:97–105.

Kinoshita T, Takeda J (1994): GPI-anchor synthesis. In *Parasitol. Today* **10**:139–43.

Ko YG, Thompson GA (1995): Purification of glycosylphosphatidylinositol-anchored proteins by modified trition X-114 partitioning and preparative gel electrophoresis. In *Anal. Biochem.* **224**:166–72.

Kodukula K, Gerber LD, Amthauer R *et al.* (1993): Biosynthesis of glycosylphosphatidylinositol (GPI)-anchored membrane proteins in intact cells: specific amino acid requirements adjacent to the site of cleavage and GPI attachment. In *J. Cell Biol.* **120**:657–64.

Kwiatkowski D (1989): Febrile temperatrures can synchronize the growth of *Plasmodium falciparum in vitro*. In *J. Exp. Med.* **169**:357–61.

Leidich SD, Drapp DA, Orlean P (1994): A conditionally lethal yeast mutant blocked at the first step in glycosyl phosphatidylinositol anchor synthesis. In *J. Biol. Chem.* **269**:10193–6.

Leidich SD, Kostova Z, Latek RR *et al.* (1995): Temperature-sensitive yeast GPI anchoring mutants gpi2 and gpi3 are defective in the synthesis of *N*-acetylglucosaminyl phosphatidylinositol. Cloning of the GPI2 gene. In *J. Biol. Chem.* **270**:13029–35.

Lisanti MP, Caras I, Davitz MA *et al.* (1989): A glycophospholipid membrane anchor acts as an apical targeting signal in polarized epithelial cells. In *J. Cell Biol.* **109**:2145–56.

Low MG (1987): Biochemistry of the glycosylphosphatidylinositol membrane protein anchors. In *Biochem. J.* **244**:1–13.

Low MG, Saltiel AR (1988): Structural and functional roles of glycosyl-phosphatidylinositol in membranes. In *Science* **239**:268–75.

Lu CF, Kurjan J, Lipke PN (1994): A pathway for cell wall anchorage of *Saccharomyces cerevisiae* α-agglutinin. In *Mol. Cell. Biol.* **14**:4825–33.

Masterson WJ, Doering TL, Hart GW *et al.* (1989): A novel pathway for glycan assembly: biosynthesis of the glycosyl-phosphatidylinositol anchor of the trypanosome variant surface glycoprotein. In *Cell* **56**:793–800.

Masterson WJ, Raper J, Doering TL *et al.* (1990): Fatty acid remodeling: a novel reaction sequence in the biosynthesis of trypanosome glycosyl phosphatidylinositol membrane anchors. In *Cell* **62**:73–80.

Mayor S, Menon AK, Cross GA *et al.* (1990a): Glycolipid precursors for the membrane anchor of *Trypanosoma brucei* variant surface glycoproteins. I. Glycan structure of the phosphatidylinositol-specific phospholipase C sensitive and resistant glycolipids. In *J. Biol. Chem.* **265**:6164–73.

Mayor S, Menon AK, Cross GA (1990b): Glycolipid precursors for the membrane anchor of *Trypanosoma brucei* variant surface glycoproteins. II. Lipid structures of phosphatidylinositol-specific phospholipase C sensitive and resistant glycolipids. In *J. Biol. Chem.* **265**:6174–81

Mazhari R, Eckert V, Blank M *et al.* (1996): Cloning and functional expression of glycosyltransferases from parasitic protozoa by heterologous complementation in yeast: The DOL-P-Man synthase from *Trypanosoma brucei brucei*. In *Biochem. J.* in press.

McConville MJ, Ferguson MA (1993): The structure, biosynthesis and function of glycosylated phosphatidylinositols in the parasitic protozoa and higher eukaryotes. In *Biochem. J.* **294**:305–24.

Menon AK (1994): Structural analysis of glycosylphosphatidylinositol anchors. In *Meth. Enzymol.* **230:**418–42.

Menon AK, Mayor S, Ferguson MA *et al.* (1988): Candidate glycophospholipid precursor for the glycosylphosphatidylinositol membrane anchor of *Trypanosoma brucei* variant surface glycoproteins. In *J. Biol. Chem.* **263**:1970–7.

Menon AK, Schwarz RT, Mayor S *et al.* (1990a): Cell-free synthesis of glycosyl-phosphatidylinositol precursors for the glycolipid membrane anchor of *Trypanosoma brucei* variant surface glycoproteins. Structural characterization of putative biosynthetic intermediates. In *J. Biol. Chem.* **265**:9033–42.

Menon AK, Mayor S, Schwarz RT (1990b): Biosynthesis of glycosyl-phosphatidylinositol lipids in *Trypanosoma brucei*: involvement of mannosylphosphoryldolichol as the mannose donor. In *EMBO J.* **9**:4249–58.

Menon AK, Eppinger M, Mayor S *et al.* (1993): Phosphatidylethanolamine is the donor of the terminal phosphoethanolamine group in trypanosome glycosylphosphatidylinositols. In *EMBO J.* **12**:1907–14.

Menon AK, Stevens VL (1993): Phosphatidylethanolamine is the donor of the ethanolamine residue linking a glycosylphosphatidylinositol anchor to protein. In *J. Biol. Chem.* **267**:15277–80.

Menon AK, Vidugiriene J (1994): Topology of GPI biosynthesis in the endoplasmic reticulum. In *Braz. J. Med. Biol. Res.* **27**:167–75.

Mensa-Wilmot K, LeBowitz JH, Chang KP *et al.* (1993): A glycosylphosphatidylinositol (GPI)-negative phenotyp produced in *Leishmania major* by GPI-phospholipase C from *Trypanosoma brucei*: topology of two GPI pathways. In *J. Cell Biol.* **124**:935–47.

Micanovic R, Gerber LD, Berger J *et al.* (1990): Selectivity of the cleavage/attachment site of phosphatidylinositol-glycan-anchored membrane proteins determined by site-specific mutagenesis at Asp-484 of placental alkaline phosphatase. In *Proc. Natl. Acad. Sci. USA* **87**:157–61.

Miyata T, Takeda J, Iida Y *et al.* (1993): The cloning of PIG-A, a component in the early step of GPI-anchor biosynthesis. In *Science* **259**:1318–20.

Mohney RP, Knez JJ, Ravi L *et al.* (1994): Glycoinositol phospholipid anchor-defective K562 mutants with biochemical lesions distinct from those in Thy-1-murine lymphoma mutants. In *J. Biol. Chem.* **269**:6536–42.

Moran P, Caras IW (1991a): Fusion of sequence elements from non-anchored proteins to generate a fully functional signal for glycophosphatidylinositol membrane anchor attachment. In *J. Cell Biol.* **115**: 1595–1600.

Moran P, Caras IW (1991b): A nonfunctional sequence converted to a signal for glycophosphatidylinositol membrane anchor attachment. In *J. Cell Biol.* **115**:329–36.

Moran P, Caras IW (1992): Proteins containing an uncleaved signal for glycophosphatidylinositol membrane anchor attachment are retained in a post-ER compartment. In *J. Cell Biol.* **119**:763–72.

Moran P, Caras IW (1993): Requirements for glycosylphosphatidylinositol attachment are similar but not identical in mammalian cells and parasitic protozoa. In *J. Cell Biol.* **125**:333–43.

Moran P, Raab H, Kohr WJ *et al.* (1991): Glycophospholipid membrane anchor attachment. Molecular analysis of the cleavage/attachment site. In *J. Biol. Chem.* **266**:1250–7.

Orlean P, Albright C, Robbins PW (1988): Cloning and sequencing of the yeast gene for dolichol phosphate mannose synthase, an essential protein. In *J. Biol. Chem.* **263**:17499–507.

Playfair JH, Taverne J, Bate CA *et al.* (1990): The malaria vaccine: anti-parasite or anti-disease? In *Immunol. Today* **11**:25–7.

Puoti A, Desponds C, Fankhauser C *et al.* (1991): Characterization of glycophospholipid intermediate in the biosynthesis of glycophosphatidylinositol anchors accumulating in the Thy-1-negative lymphoma line SIA-b. In *J. Biol. Chem.* **266**:21051–9.

Robinson, PJ (1991): Signal transduction by GPI-anchored membrane proteins. In *Cell. Biol. Intern. Rep.* **15**:761–7.

Robinson PJ, Millrain M, Antoniou J *et al.* (1989): A glycophospholipid anchor is required for Qa-2-mediated T-cell activation. In *Nature* **342**:85–7.

Rothberg KG (1995): Caveolar targeting of glycosylphosphatidylinositol-anchored proteins. In *Meth. Enzymol.* **250**:669–79.

Rush JS, Waechter CJ (1994): Transmembrane movement of a water-soluble analogue of mannosylphosphoryldolichol is mediated by an endoplasmic reticulum protein. In *J. Cell Biol.* **130**:529–36.

Saltiel AR (1991): The role of glycosylphosphoinositides in hormone action. In *J. Bioenerg. Biomembr.* **23**:29–41.

Saltiel AR, Osterman DG, Darnell JC *et al.* (1988): The function of glycosyl phosphoinositides in hormone action. In *Philos. Trans. R. Soc. Lond. Biol.* **320**:345–58.

Schönbächler M, Horvath A, Fassler J *et al.* (1995): The yeast spt14 gene is homologous to the human PIG-A gene and is required for GPI anchor synthesis. In *EMBO J.* **14**:1637–45.

Schofield L, Hackett F (1993): Signal transduction in host cells by a glycosylphosphatidylinositol toxin of malaria parasites. In *J. Exp. Med.* **177**:145–53.

Schofield L, Vivas L, Hackett F *et al.* (1993): Neutralizing monoclonal antibodies to glycosylphosphatidylinositol, the dominant TNF-alpha-inducing toxin of *Plasmodium falciparum*: prospects for the immunotherapy of severe malaria. In *Ann. Trop. Med. Parasitol.* **87**:617–26.

Schofield L, Gerold P, Schwarz RT *et al.* (1994): Signal transduction in host cells mediated by glycosylphosphatidylinositols of the parasitic protozoa, or why do the parasitic protozoa have so many GPI molecules? In *Braz. J. Med. Biol. Res.* **27**:249–54.

Schofield L, Novakovic S, Gerold P *et al.* (1996): Glycosylphosphatidylinositol toxin of *Plasmodium falciparum* up-regulates intercellular adhesion molecule-1, vascular cell adhesion molecule-1, and E-selectin expression in vascular endothelial cells and increases leukocyte kinase-dependent signal transduction. In *J. Immunol.* in press.

Schubert J, Uzeichowski P, Delany P *et al.* (1990): The PIG-anchoring defect NK-lymphocytes of PNH-patients. In *Blood* **76**:1181–7.

Schwarz RT, Mayor S, Menon AK (1989): Biosynthesis of the glycolipid membrane anchor of *Trypanosoma brucei* variant surface glycoproteins: Involvement of dolichol-phosphate-mannose. In *Biochem. Soc. Transact.* **17**:746.

Selvaraj P, Rosse WF, Silber R *et al.* (1988): The major Fc receptor in blood has a phosphatidylinositol anchor and is deficient in paroxysmal nocturnal haemoglobinuria. In *Nature* **333**:565–7.

Shenoy-Scaria AM, Kwong J, Fujita T *et al.* (1992): Signal transduction through decay-accelerating factor. Interaction of glycosyl-phosphatidylinositol anchor and protein tyrosine kinases p56lck and p59fyn 1. In *J. Immunol.* **149**:3535–41.

Simons K, Wandinger-Ness A (1990): Polarized sorting in epithelia. In *Cell* **62**:207–10.

Soole KL, Jepson MA, Hazlewood GP *et al.* (1995): Epithelial sorting of a glycosyl-phosphatidylinositol-anchored bacterial protein expressed in polarized renal MDCK and intestinal Caco-2 cells. In *J. Cell Sci.* **108**:369–77.

Stefanova I, Horejsi V, Ansotegui IJ *et al.* (1991) GPI-anchored cell-surface molecules complexed to protein tyrosine kinases. In *Science* **254**:1016–9.

Stevens VL (1995): Biosynthesis of glycosylphosphatidylinositol membrane anchors. In *Biochem. J.* **310**:361–70.

Stevens VL, Raetz CR (1991): Defective glycosylphosphatidylinositol biosynthesis in extracts of three Thy-1 negative lymphoma cell mutants. In *J. Biol. Chem.* **266**:10039–42.

Sugiyama E, DeGasperi R, Urakaze M *et al.* (1991): Identification of defects in glycosylphosphatidylinositol anchor biosynthesis in the Thy-1 expression mutants. In *J. Biol. Chem.* **266**:12119–22.

Tachado SD, Schofield L (1994): Glycosylphosphatidylinositol toxin of *Trypanosoma brucei* regulates IL-1 alpha and TNF-alpha expression in macrophages by protein tyrosine kinase mediated signal transduction. In *Biochem. Biophys. Res. Commun.* **205**:984–91.

Tachado SD, Gerold P, McConville MJ *et al.* (1996): Glycosylphosphatidylinositol toxin of *Plasmodium falciparum* induces nitric oxide synthase expression in macrophages and vascular endothelial cells by a protein tyrosine kinase-dependent and protein kinase C-dependent signaling pathway. In *J. Immunol.* in press.

Takeda J, Kinoshita T (1995): GPI-anchor biosynthesis. In *Trends Biochem. Sci.* **20**:367–71.

Takeda J, Miyata T, Kawagoe K *et al.* (1993): Deficiency of the GPI-anchor caused by a somatic mutation of the *PIG-A* gene in paroxysmal nocturnal hemoglobinuria. In *Cell* **73**:703–11.

Tomavo S, Schwarz RT, Dubremetz JF (1989): Evidence for glycosyl-phosphatidylinositol anchoring of *Toxoplasma gondii* major surface antigens. In *Mol. Cell. Biol.* **9**:4576–80.

Udenfriend S, Kodukula K (1995): How glycosylphosphatidylinositol-anchored membrane proteins are made. In *Annu. Rev. Biochem.* **64**:563–91.

Urakaze M, Kamitani T, DeGasperi R *et al.* (1992): Identification of a missing link in glycosylphosphatidylinositol anchor biosynthesis in mammalian cells: In *J. Biol. Chem.* **267**:6459–62.

Vidugiriene J, Menon AK (1993): Early lipid intermediates in glycosyl-phosphatidylinositol anchor assembly are synthesized in the ER and located in the cytoplasmic leaflet of the ER-membrane bilayer. In *J. Cell Biol.* **121**:987–96.

Vidugiriene J, Menon AK (1994): The GPI anchor of cell-surface proteins is synthesized on the cytoplasmic face of the endoplasmic reticulum. In *J. Cell Biol.* **127**:333–41.

13 The Biology of Sialic Acids: Insights into their Structure, Metabolism and Function in particular during Viral Infection

WERNER REUTTER, ROGER STÄSCHE, PEER STEHLING, AND OLIVER BAUM

Abbreviations

CMP-*N*-acetyl-D-neuraminic acid	CMP-Neu5Ac
2-Deoxy-2(propanoyl-amido)-D-mannose	ManNProp
2-Deoxy-2(butanoyl-amido)-D-mannose	ManNBut
2-Keto-3-deoxy-nonulosonic acid	KDN
D-Fructose-6-phosphate	Frc-6-P
D-Glucosamine	GlcN
D-Glucosamine-6-phosphate	GlcN-6-P
D-Glucose	Glc
D-Glucose-6-phosphate	Glc-6-P
Human polyoma virus BK	BKV
B-lymphotropic papovavirus	LPV
N-Acetyl-D-glucosamine	GlcNAc
N-Acetyl-D-glucosamine-6-phosphate	GlcNAc-6-P
N-Acetyl-D-glucosamine-1-phosphate	GlcNAc-1-P
N-Acetyl-D-mannosamine	ManNAc
N-Acetyl-D-mannosamine-6-phosphate	ManNAc-6-P
N-Acetyl-neuraminic acid	Neu5Ac
N-Acetyl-neuraminic acid-9-phosphate	Neu5Ac-9-P
N-Glycolyl-neuraminic acid	Neu5Gc
Polysialic acid	PSA
D-Phosphoenolpyruvate	PEP
UDP-*N*-acetyl-D-glucosamine	UDP-GlcNAc
UDP-D-galactose	UDP-Gal
UDP-D-glucose	UDP-Glc

1 Introduction

Sialic acids are a family of monosaccharides which are expressed as important components within the oligosaccharide residues of complex glycoconjugates. Sialylated carbohydrates are detectable in all higher animals and also in a few prokaryotic organisms. The wide occurrence of sialic acids in nature indicates their biological importance. Members of this group of sugars are abundant constituents of plasma membranes and also of most secreted proteins. Biological molecules containing sialic acids in the peripheral position of their oligosaccharide chains are glycolipids and glycoproteins, as well as components of mucus.

In extending the chapters on glycoproteins and glycolipids by Sharon and Lis and by Kopitz in this volume, we focus on this class of carbohydrates from different perspectives. We present a brief overview of current knowledge of the chemical structure and metabolism of sialic acids, together with some biological functions, in particular with respect to viral infection. In addition, we focus our report on recent investigations aimed at elucidating the biological role of sialic acids by the use of sugar analogs. For further information on these topics the reviews by Corfield and Schauer, 1982a, b; Reutter *et al.*, 1982; Schauer, 1985, 1991; Varki, 1992, 1993; and Schauer *et al.*, 1995 are recommended.

H.-J. and S. Gabius (Eds.), Glycosciences

ISBN 3-8261-0073-5

2 Chemical Structure and Expression of Sialic Acids

Most monosaccharides detected at the non-reducible end of the carbohydrate residues of complex glycoconjugates are sialic acids. This term embraces all derivatives of neuraminic acid which share an elementary 9-carbon carboxylated skeleton (Fig. 1). Since the chemical backbone can be modified by various substituents, the sialic acids are the structurally most varied of any group of natural monosaccharides.

The most prominent member of this family of sugars is *N*-acetylneuraminic acid (Neu5Ac). Replacement of the 5-amino group by a hydroxyl group results in the conversion of Neu5Ac into 2-keto-3-deoxy-nonulosonic acid (KDN), originally found in rainbow trout eggs (Nadano *et al.*, 1986). Another common modification of Neu5Ac is the hydroxylation of the C5-*N*-acetyl group resulting in *N*-glycolyl-neuraminic acid (Neu5Gc) (Schauer *et al.*, 1968). Neuraminic acid molecules with various other modifications (e.g. *O*-acetylation) are also widely distributed on the surface of animal cells (Fig. 1). Whereas the *O*-acetylation at C4 is found only in few species, many cells carry C7 and C9 acetylated hydroxyl residues of sialic acids in their glycoconjugates (Hutchins *et al.*, 1988; Schauer, 1991). Functionally, the most important compound within this group is the ganglioside 9-*O*-acetyl-GD_3 as the differentiation antigen CDw60 (Kniep *et al.*, 1992; Sjoberg *et al.*, 1992). Atypical modifications such as *N*-*O*-diacetyl-neuraminic acid occur in the flagellate *Crithidia fasciculata* (do Valle Matta *et al.*, 1995). The different members of the family of sialic acids are listed by Varki (1992).

Sialic acids occur in all living organisms with the exception of certain bacteria (Corfield and Schauer, 1982a; Rodriguez *et al.*, 1995). The prokaryotes might have acquired the ability to express sialic acids by horizontal gene transfer (Higa and Varki, 1988). Early reports suggested that the expression pattern of each modified sialic acid displayed a relatively strict species, tissue and developmental specificity. However, due to the development of highly sensitive detection methods it has been demonstrated that, on the contrary, many of them show an almost ubiquitous distribution. Since the expression of some sialic acids is markedly regulated in a variety of species during development, it is clear that spe-

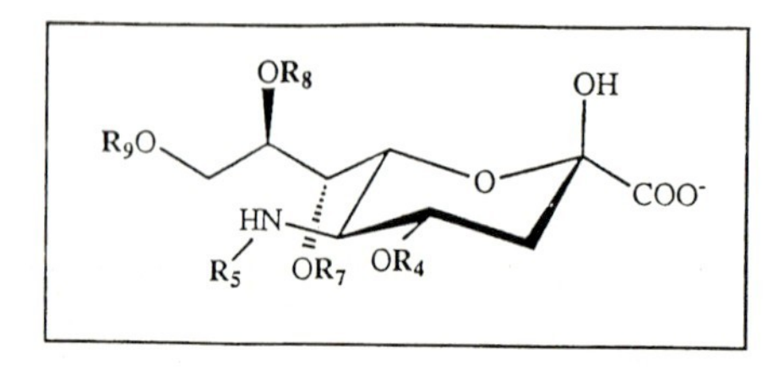

R_5	$R_{4,7,8,9}$
-CO-CH_3 *(N-acetylneuraminic acid)* -CO-CH_2OH *(N-glycolylneuraminic acid)* -H *(neuraminic acid; only stable in glycosidic bonds)*	-H *(4,7,8,9)* -CO-CH_3 *(4,7,8,9)* -CO-CHOH-CH_3 *(9)* -CH_3 *(8)* -SO_3H *(8)* -PO_3H_2 *(9)*

Figure 1. Chemical structure of neuraminic acids. All natural derivatives of neuraminic acid share an elementary 9-carbon carboxylated structure. However, several positions ($R_{4,5,7,8,9}$) can be substituted by different chemical residues as indicated in the table. More than 40 naturally occurring sialic acids have been identified so far.

cific enzymatic mechanisms exist for the generation of these sialic acids, and that the resulting modifications might play an important role in organogenesis and -stasis.

3 Metabolism of Sialic Acids

The cellular metabolism of Neu5Ac generally includes three distinct processes. First, the molecule is synthesized *de novo* from various carbohydrates by a series of enzymatic reactions, and activated to yield the corresponding nucleotide carbohydrate. Second, the activated amino sugar is added stepwise to the oligosaccharide side chain of glycoconjugates. Third, Neu5Ac can be removed again from these glycoconjugates and degraded into molecular fragments. Two of these three processes (biosynthesis with activation, and the degradation of Neu5Ac) will be briefly described here. Additionally, an overview of the metabolism of the other sialic acids will also be given.

D-glucose (Glc) most frequently serves as the starter molecule for the biosynthesis of all monosaccharides and amino sugars, including Neu5Ac and its activated form. In summary, Glc is metabolized to Glc*N*-6-P in three consecutive reactions: the ATP-dependent activation of Glc to *D*-glucose-6-phosphate (Glc-6-P), isomerization of Glc-6-P to *D*-fructose-6-phosphate (Frc-6-P), and the irreversible transamination with the amino-donor *L*-glutamine, to produce *D*-glucosamine-6-phosphate (Glc*N*-6-P), as indicated in Fig. 2.

Other carbohydrates such as *D*-glucosamine (GlcN) or *N*-acetyl-*D*-hexosamines, which are

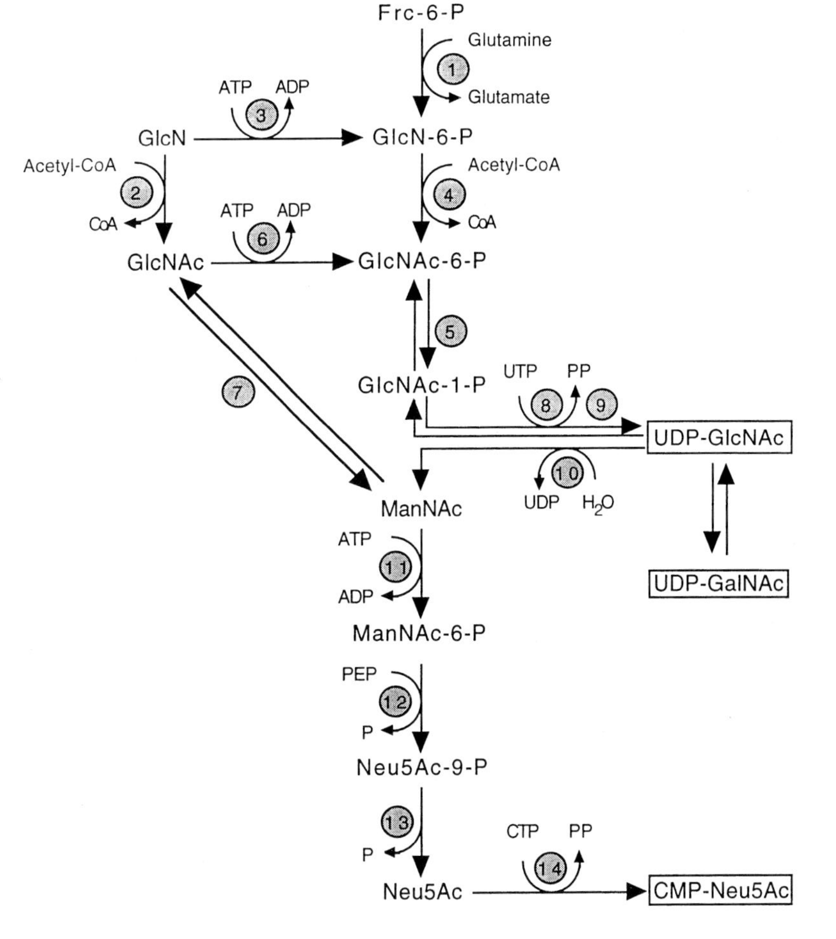

Figure 2. Overview of the biosynthesis of Neu5Ac and its activated nucleotide sugar. The biosynthesis of Neu5Ac starts either with fructose-6-phosphate or with glucosamine-phosphate, which is derived from cellular degradation. Apart from the synthesis of the sialic acid and its activated form, other nucleotide sugars are also generated as precursors for the biosynthesis of glycoconjugates. **1:** glucosamine-6-phosphate-synthase; **2:** glucosamine-*N*-acetyltransferase; **3:** hexokinase/glucokinase; **4:** glucosamine-6-phosphate-*N*-acetyl-transferase; **5:** *N*-acetylglucosamine-6-phosphate-mutase; **6:** *N*-acetyl-glucosamine-kinase; **7:** *N*-acetyl-glucosamine-2-epimerase; **8:** UDP-*N*-acetyl-glucosamine-pyrophosphorylase; **9:** pyrophosphatase; **10:** UDP-*N*-acetylglucosamine-2-epimerase; **11:** *N*-acetylmannosamine-kinase; **12:** *N*-acetylneuraminate-9-phosphate-synthase; **13:** *N*-acetyl-neuraminate-9-phosphate-phosphatase and **14:** CMP-acetyl-neuraminate-synthase.

derived from intracellular degradation of glycoconjugates or from nutrition, are potent precursors of Neu5Ac. GlcN can be introduced by three pathways into the biosynthesis of Neu5Ac; all three pathways produce ManNAc. First, and most relevant in cells, it is possible that this amino sugar is phosphorylated to *D*-glucosamine-6-phosphate (Glc*N*-6-P) followed by acetyl-CoA dependent acetylation to *N*-acetyl-*D*-glucosamine-6-phosphate (GlcNAc-6-P). Following isomerization, the resulting *N*-acetyl-*D*-glucosamine-1-phosphate (GlcNAc-1-P) is activated to UDP-GlcNAc. UDP is then released in a two-stage process, and GlcNAc is simultaneously epimerized into ManNAc. Alternatively, GlcN might enter the Neu5Ac *de novo* synthesis pathway by direct acetylation to GlcNAc, which is then either phosphorylated to GlcNAc-6-P (followed by metabolism to ManAc as described above) or in some cases directly epimerized to ManNAc. ManNAc is phosphorylated by ATP to *N*-acetyl-*D*-mannosamine-6-phosphate (Man NAc-6-P), which cannot be directly generated from GlcNAc-6-P in animal tissues (van Rinsum, 1984). ManNAc-6-P reacts at position C1 with *D*-phosphoenolpyruvate (PEP) to form *N*-acetyl-*D*-neuraminic acid-9-phosphate (Neu5Ac-9-P) which is finally dephosphorylated to Neu 5Ac.

The attachment of any sialic acid to nascent glycoconjugates is catalyzed by specific sialyltransferases. These enzymes display binding specificity (α2,3- α2,6- and α2,8-linkages) and their expression is subject to species-, organ- and development-dependent control mechanisms. However, prior to conjugation with the glycoconjugate, sialic acids must be activated to nucleotide sugars. This activation occurs by the reaction of Neu5Ac with CTP to form CMP-Neu5Ac and two molecules of inorganic phosphate. This process is localized in the nucleus, probably because of the high concentration of CTP (Coates *et al.*, 1980). After its biosynthesis, CMP-Neu5Ac is transported into the Golgi complex.

All enzymes involved in the biosynthesis of sialic acids are localized in the cytoplasm, with the exception of CMP-Neu5Ac-synthase, which is localized in the nucleus (Corfield and Schauer, 1982b; van Rinsum, 1984). The key enzymes of Neu5Ac biosynthesis are UDP-GlcNAc-2-epimerase and ManNAc-kinase. The UDP-GlcNAc-2-epimerase is allosterically inhibited in a feed-back mechanism by CMP-Neu5Ac, whereas its expression is reduced in hepatoma cell lines (Harms *et al.*, 1973).

As described above, important biological modifications of sialic acids include hydroxylation, acetylation, methylation and phosphorylation of the 9-carbon carboxylated backbone. All of these derivatives are synthesized from Neu5Ac as the precursor molecule. Neu5Gc and the corresponding nucleotide sugar (CMP-Neu5Gc) can be generated either by the direct reduction of the *N*-acetyl residue, or by *de novo* synthesis from ManNGc, which is derived from the degradation of Neu5Gc-containing glycoconjugates. *O*-Acetylation of sialic acids occurs after transfer to the glycoconjugate (Taatjes *et al.*, 1988). *O*-Acetyl residues are added by 7(9)-*O*-acetyltransferase to the sialic acids at position C7, and then migrate rapidly and non-enzymatically from this position to C9 (Kamerling *et al.*, 1987; Schauer, 1991). The occurrence of de-*N*-acetylation has only been described for gangliosides containing neuraminic acid derivatives. The relevant enzyme, designated deacetylase, is probably located in the Golgi complex. This enzyme takes part in the deacetylation/reacetylation cycle which is responsible for the accelerated turnover of the acetyl residue of the sialic acid compared with the rest of the molecule (Manzi *et al.*, 1990).

The release of terminal sialic acids from glycoproteins, glycolipids, and polysaccharides is catalyzed by the superfamily of sialidases (or neuraminidases), a class of glycosyl hydrolases found mainly in higher eukaryotes and in some, mostly pathogenic, viruses, bacteria and protozoans. The functions of sialidases are as yet only poorly understood (Vimr, 1994).

O-Acetyl residues are removed from substituted sialic acids by special sialate *O*-acetylesterases (e.g. sialate 9-*O*-acetylesterases). Sulfate groups are removed by glycosulfatases and phosphate groups are removed by special phosphatases. *N*-Acetylneuraminate lyase catalyzes the cleavage of Neu5Ac to form pyruvate and ManNAc, which can be degraded further or remetabolized to sialic acids.

4 Biological Functions of Sialic Acids

Besides the physicochemical functions of the negatively charged backbone, which enable the glycocalyx to bind water and ions, sialic acids also fulfill diverse biochemical functions. Since they often occupy the peripheral residues of glycoconjugates and are therefore exposed at the cell surface, sialic acids are the most accessible molecules for other cells or extracellular components. Additionally, sialic acids display great structural versatility, so that they are well suited as molecular determinants of specific biological functions. However, it has not so far been possible to identify any standard governing the function of a defined carbohydrate residue, and, in particular, of Neu5Ac and its derivatives. In some cases, sialic acids clearly make no apparent contribution to any molecular process, whereas in others they seem to be involved in the modulation of crucial cellular processes such as development and growth. Thus, the biological function of a sialic acid depends strongly on the biological context in which it is expressed.

Four main biological functions can be attributed to sialic acids: (1) modulation of cellular adhesion, aggregation or agglutination processes and receptor functions, (2) influence on conformation, solubility, viscosity and charge of glycoproteins, (3) protection of glycoconjugates and cells from degradation and (4) modulation of enzyme activity. These four functions of sialic acids will be briefly reviewed here.

Sialic acids contribute to the total negative charge of cell surfaces which *per se* might affect the weak intrinsic interactions (adhesion and repulsion) between cells and/or the surrounding extracellular matrix (Shimamura *et al.*, 1994). Therefore, it is reasonable that the removal of sialic acids by neuraminidase treatment decreases cellular adhesion in many cell lines. However, cell-cell as well as cell-matrix adhesion are mostly mediated by specific adhesion receptors. Many efforts have been made to characterize these cellular adhesion molecules. In many cases it has been shown that carbohydrates (especially sialic acid) are involved in the recognition by cellular adhesion molecules of their respective counterparts. The most prominent family of specific adhesion receptors is the selectins, which mediate the interaction of leukocytes with endothelial cells (Stoolman, 1989). At least one member of this protein family (E-selectin) specifically recognizes the terminal tetrasaccharide structures sialyl-Lewis^x (Neu5Ac α2–3 Gal β1–4 (Fuc α1–3) GlcNAc-) and sialyl-Lewis^a (Neu5Ac α2–3 Gal β1–3 (Fuc α1–4) GlcNAc-) expressed on glycoproteins (Bevilacqua *et al.*, 1991).

Neural cell adhesion molecule (*N*-CAM), which belongs to the Ig-superfamily, another family of adhesion receptors, participates in cell-cell binding by a homophilic mechanism. In embryonic cells, *N*-CAM carries long chains of polysialic acid (PSA), which allow a weak binding between cells (Horstkorte *et al.*, 1993; Rougon, 1993; Yang *et al.*, 1994). However, the length of the expressed PSA is gradually reduced during ontogenic development, so that in the adult state *N*-CAM gives rise to strong intercellular adhesion (Roth *et al.*, 1992; Seki and Arai, 1993). Further members of this family which actually bind sialic acids are sialoadhesin, CD22, CD33 and myelin-associated glycoprotein (Powell and Varki, 1995; Schauer *et al.*, 1995).

Sialic acids also seem to be involved in some malignant and infectious processes. In general, cancer is often accompanied by an oncofetal cellular phenotype including ectopic glycosylation, especially sialylation. Therefore, there appears to be a correlation between cancer and degree of cell sialylation (Reglero *et al.*, 1993; Kageshita *et al.*, 1995). The expression of *O*-acetylated structures on cancer cells also has been observed (Cheresh *et al.*, 1984). However, the altered glycosylation pattern, including highly branched and sialylated *N*-linked oligosaccharides expressed in increased amounts on some malignant cell lines, also seems to enhance the metastatic potential of tumor cells (Dennis and Laferte, 1987; Pilatte *et al.*, 1993). Specific types of gangliosides expressed by tumor cells, particularly those with an increased sialic acid content, have suppressive effects on cell proliferation in the immune response, and are important for the immunosuppressive effects of cancer (Marcus, 1984).

Terminal sialic acids of glycoconjugates also form part of many highly specific receptors for viruses, bacteria, parasites and toxins (Karlsson, 1995). Carrier-immobilized ganglioside epitopes have been instrumental in detecting binding sites for such oligosaccharides in human cells as well (Gabius *et al.*, 1990; Lee and Lee, 1994). In these cases it is obvious that the efficiency of sialic acids as a recognition signal depends crucially on the binding type and substitution of the carbohydrate.

Hemagglutinins are the surface receptors by which many viruses recognize terminal sialic acids on their host cells, leading to adsorption to the plasma membrane, the first step in the viral infectious process. The influenza A virus strain binds to Neu5Ac α2–6 Gal-residues, whereas the B strain of the same influenza virus shows specificity for Neu5Ac α2–3 Gal-residues on glycoconjugates. Even a single modification of the Neu5Ac (9-*O*-acetylation) prevents recognition by these two influenza strains (Schauer, 1985). However, precisely this type of modified sialic acid is preferred by the influenza C strain for binding to cell surfaces (Rogers *et al.*, 1986; Zimmer *et al.*, 1994). The binding ability of viruses is also determined by the linkage type between the sialic acids and the neighboring galactose residues, as shown for the adsorption of viruses to erythrocytes and subsequent agglutination (Weis *et al.*, 1988).

Gangliosides (sialic acid-containing glycolipids) serve as receptors for several bacterial toxins, for example, cholera toxin (Schengrund and Ringler, 1989; Corfield, 1992), botulinum toxin (Schengrund *et al.*, 1991), or tetanus toxin (Schiavo *et al.*, 1991).

Parasitic protozoans of the species *Trypanosoma cruzi* also recognize terminal sialic acids on host glycoconjugates. After binding, they transfer host sialic acids to their own surface by the coordinated activity of a neuraminidase and a sialyltransferase. This trans-sialylation activity protects the parasites from the host immune response and complement system (Colli, 1993; Tomlinson *et al.*, 1994).

Enzymatic desialylation causes conformational changes of single glycoconjugates, plasma membrane fractions and whole cells. A recent example is given by altered surface exposure of histidine, tryptophan and tyrosine residues after desialylation of human serum amyloid P component (Siebert *et al.*, 1995). Thus, it is reasonable to conclude that sialic acids have a stabilizing effect on the tertiary structure of glycoconjugates and their immediate environment. This hypothesis is confirmed by many other observations. The sialic acid residues of fibrinogen influence fibrin assembly (Dang *et al.*, 1989), and play a functional role in the stabilization of fibrin clot formation (Okuda *et al.*, 1995). Sialylation of the somatostatin receptor is required for high affinity binding of the hormone. Desialylation of the receptor led to a conformational change, a significantly weaker binding of the hormone and, therefore, to a reduced biological effect of somatostatin (Rens-Domiano and Reisine, 1991).

Unlike the normal sialylated molecule, desialylated plasminogen tends to cleave fibrin spontaneously and to bind to the cell surface (Stack *et al.*, 1992), whereas more highly sialylated forms of plasminogen are activated more readily by tissue-type plasminogen activator (Pirie-Shepherd *et al.*, 1995). However, it is not always possible to draw general conclusions on the stabilizing effect of sialic acids. After desialylation, erythropoietin, a hormone which stimulates the formation of red blood cells, showed increased biological activity *in vitro* but decreased biological activity *in vivo* (Wasley *et al.*, 1991).

In many cases it has been observed that sialic acids might protect proteins from degradation. This effect can be realized directly by steric hindrance of the proteolytic action (Sjoberg *et al.*, 1994) or indirectly by changing the biophysical-chemical environment. Olden *et al.* (1982) showed that sialic acids protect the acetylcholine receptor against proteolytic degradation. Old erythrocytes lose substantial amounts of their surface-exposed sialic acids, and are therefore internalized and degraded in the liver (Schlepper-Schäfer *et al.*, 1980). However, the sialic acid dependence of homeostasis seems to operate via a common mechanism, because serum glycoconjugates (Ashwell and Harford, 1982) and antigen-antibody complexes (Day *et al.*, 1980) are recognized in a similar way. The rates of degradation are markedly influenced by the structure of the sialic acid. The half-lives of C7-, C8- or C9-*O*-acetylated sialic acids are markedly longer than that of unmodified Neu5Ac, because the former are removed more slowly by sialidases (Schauer, 1991).

The major component of gastric mucus is a glycoprotein with many *O*-linked sialylated oligosaccharide chains forming a highly viscous gel by adsorption of water. Because the sialic acids also bind differing quantities of bicarbonate ions, the mucous gel establishes a pH-gradient from 1 to 2 at the lumen to pH 6 to 7 near to the cell surface. Therefore, the stomach is protected from self-digestion by secreted HCl (Bhaskar *et al.*, 1992; Raju and Davidson, 1994).

Sialic acids also influence the antigenicity of cellular components (Schauer, 1985; Pilatte *et al.*, 1993). Removal of the sialic acids from the blood group substances on the cell surface led to the loss of antigenicity, thereby demonstrating their role as components of antigenic determinants. Addition of sialic acids to core *O*-linked oligosaccharides (T and Tn antigens) abolished the autoimmune reactivity (Itzkowitz *et al.*, 1991). Finally, it has been demonstrated that sialic acids or sialyloligosaccharides expressed on the surface of bacteria might block both the generation and the reactivity of antimicrobial antibodies, thereby protecting the microorganism from the host immune response (Markham *et al.*, 1982).

As shown recently (Emig *et al.*, 1995), sialic acids probably have a functional significance in the active nuclear protein transport from the cytosol into the nucleus through nucleoporin p62. *N*-Acetyl residues of sialic acid moieties are particularly important for the inhibition of DNA polymerase α activity by ganglioside GM_3 (Simbulan *et al.*, 1994). The sialylation of glycolipids is also reduced in cystic fibrosis cells, leading to the accumulation of asialo-gangliosides (Barasch *et al.*, 1991).

Most of the investigations reported above do not distinguish between different members of the sialic acid family. Therefore, only a little information (see next paragraph) is available on the different biological properties of different sialic acids, e.g. the modifications of the Neu5Ac responsible for the recognition by the different influenza strains as mentioned above. In addition, an increase in *O*-acetylation of sialic acids resulted in a decrease of degradation, leading to an extension of the half-life of glycoconjugates (Kiehne and Schauer, 1992). Therefore, this modification might protect the colonic mucosa from bacterial degradation (Schauer, 1991). Reviews on the function of *O*-acetylated sialic acids, whose *O*-acetyl group is a contact point for a natural immunoglobulin G from human serum (Ahmed and Gabius, 1989; Zeng and Gabius, 1992; Siebert *et al.*, in preparation) are given in the papers of Suzuki (1990) and Schauer (1991). This example reinforces the potential relevance of distinct sugar epitopes for medical considerations, e.g. in tumor diagnosis and therapy (Gabius, 1988; Gabius and Gabius, 1991, 1993; Gabius *et al.*, 1995).

Despite the wide range of functions involving sialic acids, the importance of this sugar has not yet been clearly delineated. A promising approach to elucidate the significance of sialic acids will become possible with the availability of "knock-out" mice, the overexpression of sialic acids in transgenic mice, or the inhibition of the biosynthesis or modulation of sialic acid structure by sugar analogs.

5 Modification of Biological Functions by Neuraminic Acid Analogs

Analogs may be used *in vivo* in two ways to gain a greater insight into the biological functions of Neu5Ac: first, by the application of chemically synthesized neuraminic acid analogs as competitive inhibitors of sialidases and, second, by the introduction of chemically synthesized precursors with a modified *N*-acyl side chain into the glycoconjugates. Almost all synthetic sialic acid analogs tried so far show only small structural differences from Neu5Ac, although, for example, their polarity was strongly increased. Chemical derivatization has been performed mostly on the *N*-acyl side chain or one of the *O*-acetyl side chains within the C9-carbohydrate backbone. As indicated above, hydroxylation of the acetamido group of Neu5Ac in particular seems to be functionally significant, since in most of the different physiological neuraminic acids this group has a relatively constant structure. Therefore, the *N*-acyl side chain is of major interest for derivatization.

Most of the work with neuraminic acid analogs was done as a result of searching for effective inhibitors for influenza virus infection, because Neu5Ac plays an important role, both during attachment to the host cell and during viral replication (Hirst, 1941). One of the key viral proteins of the influenza strains A and B is sialidase, which presumably helps by eluting newly synthesized virions from infected cells (Palese and Schulman, 1974; Lin and Air, 1993). The enzyme cleaves α2–3 and α2–6 bound neuraminic acids from the preceding sugar residue, mainly galactose. Therefore, blockage of the enzyme activity by a sialic acid analog should hinder the establishment and progression of influenza virus infection. 2-Deoxy-2,3-dehydro-

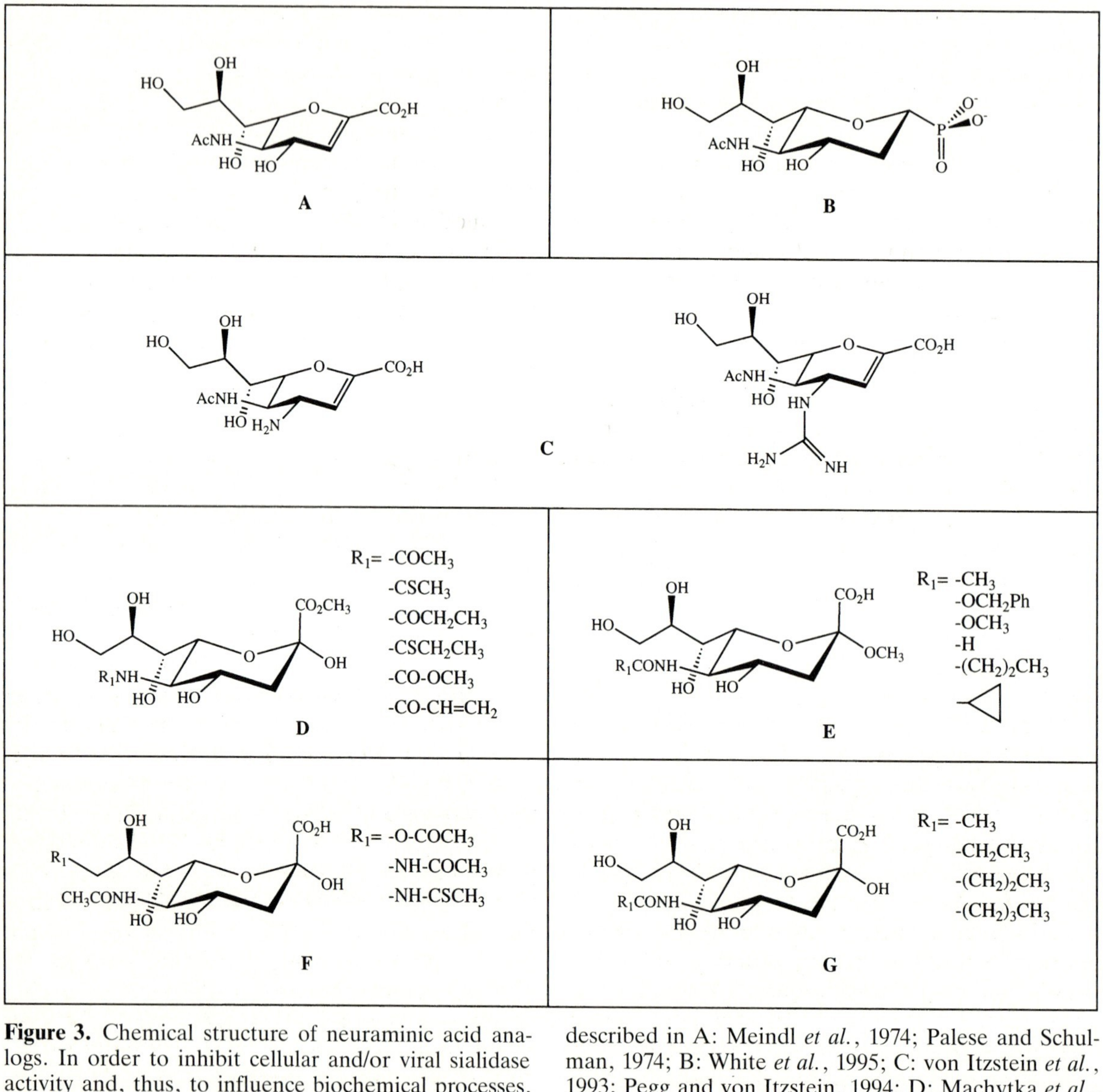

Figure 3. Chemical structure of neuraminic acid analogs. In order to inhibit cellular and/or viral sialidase activity and, thus, to influence biochemical processes, several analogs of neuraminic acid have been synthesized and investigated in different systems. The synthesis of molecules and the experiments performed are described in A: Meindl *et al.*, 1974; Palese and Schulman, 1974; B: White *et al.*, 1995; C: von Itzstein *et al.*, 1993; Pegg and von Itzstein, 1994; D: Machytka *et al.*, 1993; E: Sparks *et al.*, 1993; F: Brossmer *et al.*, 1993 and G: Kayser *et al.*, 1992a; b.

Neu5Ac (Fig. 3A) was identified as a moderate (and non-selective) inhibitor of sialidase (Meindl *et al.*, 1974; Palese and Schulman, 1974), but unfortunately it had no effect on the infection in animal models. Another approach to the inhibition of influenza virus sialidase activity was devised by White *et al.* (1995). This group synthesized a sialic acid-derived phosphonate analog in which the monobasic carboxyl group was replaced by a dibasic phosphonyl group (Fig. 3B). This substitution strengthened the ionic interaction between the carbohydrate and the amino acids in the active site of sialidase, thereby inhibiting sialidase activity about ten-fold, compared with Neu5Ac in the same concentration. With the elucidation of the crystal structure of influenza virus sialidase and its active site, it became possible to design new neuraminic acid analogs intended to compete with the physiological substrate for the active site (von Itzstein *et al.*, 1993). Both 4-amino- and 4-guanidino-2-deoxy-2,3-dehydroneuraminic acid (Fig. 3C) proved to be specific high-affinity inhibitors of influenza virus sialidase; in addition, they strongly blocked the replication of influenza virus strains A and B (Pegg and von Itzstein, 1994; von Itzstein *et al.*, 1994).

A second key protein of the influenza virus strains A and B is hemagglutinin. This viral surface protein binds to sialic acid residues of glycoconjugates expressed on the plasma membrane of the host system. Therefore, it is reasonable that a particle presenting sialic acid analogs should bind tightly to the active site of hemagglutinin and block the adsorption of the influenza virus to the host cell, preventing the viral infection. This approach was employed by Machytka *et al.* (1993), who synthesized several methyl-α-glycoside analogs of Neu5Ac differing in their *N*-acyl side chain moiety (Fig. 3D). They demonstrated that the multivalent binding to optimally spaced surface carbohydrate epitopes significantly increased the binding affinity of influenza virions to their cellular receptors. A comparable strategy was adopted by Sparks *et al.* (1993), who synthesized N5-analogues of Neu5Ac (Fig. 3E) which inhibited the influenza virus-cell adhesion process mediated by hemagglutinin.

In contrast, the influenza virus strain C possesses only one cell surface glycoprotein expressing both hemagglutinin and esterase (=sialidase) activity (HE). Therefore, none of the Neu5Ac analog inhibitors effective against influenza virus A and B infection have any influence on influenza virus C. However, as mentioned above, the attachment of this virus strain to the cell surface is accomplished by the ability of HE to recognize host receptors with *N*-acetyl-9-*O*-acetyl-neuraminic acid as a crucial determinant (Rogers *et al.*, 1986; Herrler and Klenk, 1987). Consequently, Brossmer *et al.* (1993) were able to show that analogs of *N*-acetyl-9-*O*-acetyl-neuraminic acid (Fig. 3F) inhibit the binding of influenza virus C to the host cell-system.

Apart from the construction of anti-viral inhibitors, sialic acid analogs have only been used for the modulation of a few other specific cellular processes. It has been demonstrated that the de-acetylation of Neu5Ac in the ganglioside GM_3 results in a molecule that serves as a strong promoter of epidermal growth factor receptor kinase and as a stimulator of cell growth (Hanai *et al.*, 1988). However, this analog only showed an effect when it was attached to lactosylceramide.

In the experiments described above, sialic acid analogs were administered as chemically synthesized derivatives of Neu5Ac to the biological model system, and were expected to be introduced in this form into the glycoconjugate. A different approach to the incorporation of biologically effective sialic acid analogs into glycoconjugates was made by our group (Kayser *et al.*, 1993). As mentioned above, both hexosamines, GlcNAc and ManNAc, are precursors in the biosynthetic pathway to Neu5Ac. However, the enzymes involved in this transformation are rather unspecific with respect to the *N*-acyl side chain of the hexosamines (Zeitler *et al.*, 1992). Thus, in lymphoma cells we could show that *N*-2-analogs (Fig. 3G) of GlcNAc (Kayser *et al.*, 1992a) and ManNAc (Kayser *et al.*, 1992b) are incorporated and metabolized to their respective Neu5Ac analogs, although the uptake of ManNAc derivatives was higher than that of GlcNAc derivatives (Kayser *et al.*, 1992b). Furthermore, we also demonstrated that almost 50% of the naturally occurring Neu5Ac of membrane-bound glycoproteins had been substituted by the non-physiological *N*-acylneuraminic acid (Keppler *et al.*, 1995).

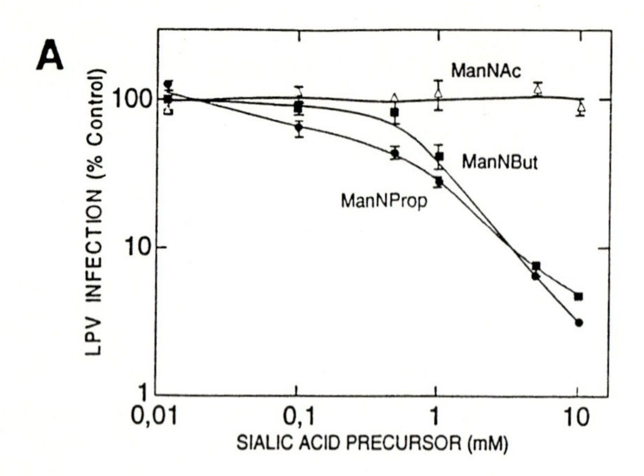

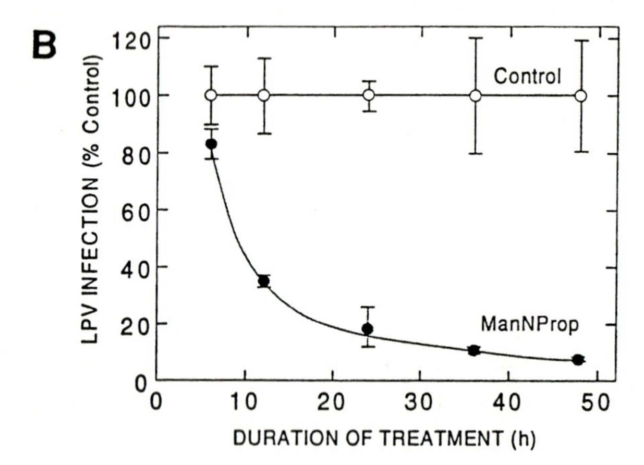

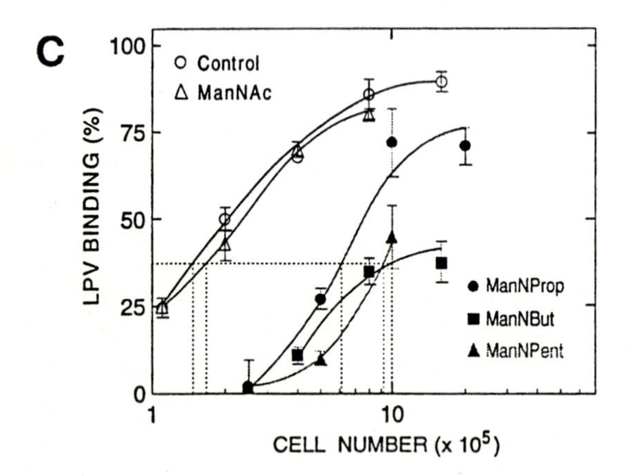

Figure 4. Reduction of LPV infection (a and b) and binding (c) in BJA-B cells containing *N*-substituted sialic acids. (a): BJA-B cells were incubated with different concentrations of sugar analogs as indicated for 48 h, and subsequently infected with LPV. The viral antigen was measured 48 h after infection by a specific ELISA. (b) After ManNProp pretreatment (10 mM) for 6–48 h, cells were infected with LPV and analyzed as described above. (c) The LPV binding capacity of BJA-B cells pretreated with different *N*-acyl-*D*-mannosamines (5 mM, 48 h) was tested in an indirect virus binding assay. The cell numbers required to bind 37.5 % of the administered LPV were used to determine virus binding capacities of pretreated cells relative to control cells. For details of the experiments see Keppler *et al.*, 1995.

Replacement of Neu5Ac by its derivatives on the cell surface of some lymphoma (BJA-B and Vero) cell lines caused a marked enhancement and/or abolition of the susceptibility of these cells to infection by B-lymphotropic papovavirus (LPV) and human polyoma virus BK (BKV), both of which exploit distinct sialylated cell surface receptors for the infection process (Fig. 4). When *N*-pentanoylmannosamine was used, the uptake of sialic acid-dependent virus was greatly decreased (Fig. 5). In contrast, in control experiments with a sialic acid-dependent virus-cell receptor system (Simian virus 40 on Vero cells), virus adsorption was unaffected by biosynthetic modulation of cell surface glycoconjugates (Keppler *et al.*, 1995).

Recent research shows that it is, in principle, possible to clarify special aspects of cell biology by the *in vivo* application of modified sialic acid. This strategy offers a new means of influencing specifically sialic acid-dependent ligand-receptor interactions.

6 Summary

N-Acylneuraminic (sialic) acids are found mainly in vertebrates, but also in viruses, bacteria, plants, protozoa and other metazoa. Our understanding of the biological functions of this family of molecules, which are chemically characterized as 9-carbon carboxylated sugars with various substitutions at C4-, C5-, C7-, C8-, and C9-, is still

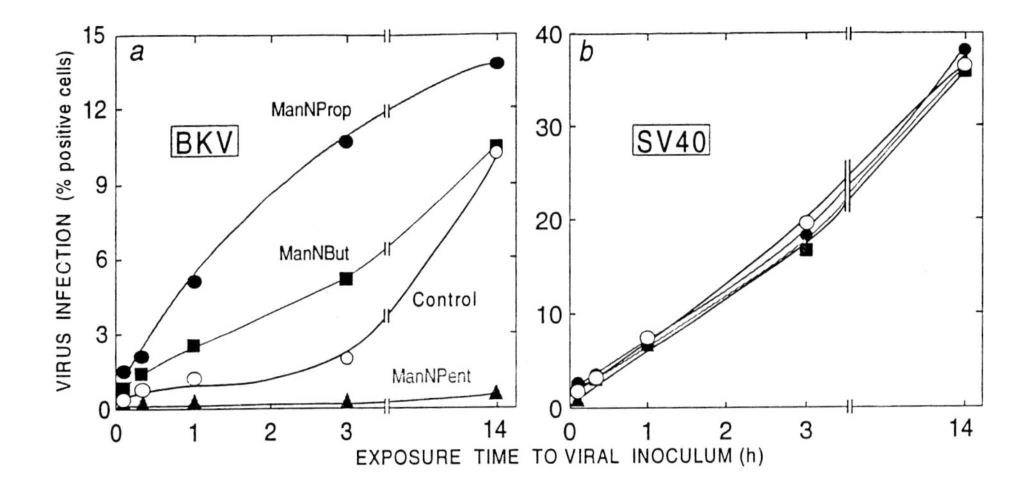

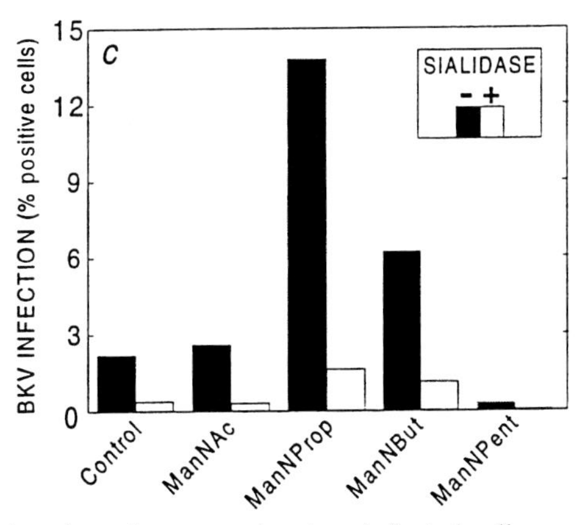

Figure 5. Sialic acid-dependent BKV infection (a and c) but not sialic acid independent SV40 infection (b) is affected in Vero cells precultivated in the presence of *N*-substituted *D*-mannosamines (5 mM, 48 h). BKV (a) or SV40 (b) were allowed to attach to pretreated Vero cells. Unbound viruses were washed away and, subsequently, virus-infected cells were determined by immunofluorescence. The increased BKV susceptibility of ManNProp- and ManNBut-pretreated Vero cells was sensitive to sialidase treatment (c). For details of the experiments see Keppler *et al.*, 1995.

growing. Hitherto, four main functions have been attributed to sialic acids: (1) contribution to physicochemical properties (conformation, solubility, viscosity, charge), (2) mediation of recognition processes (adhesion, receptor function), (3) regulation of biological stability, and (4) modulation of the activity of some enzymes.

In this report evidence is presented for the participation of sialic acid in the recognition and binding of different viruses by host cells. The virus-host cell interaction may be influenced by at least two different means. First, analogs of sialic acid inhibit uptake of the virus by competitive inhibition of virus-specific sialidases. Using this approach, the most promising results have been obtained with the chemically synthesized analog 2,3-dideoxy-4-guanidino-neuraminic acid, which inhibits strongly infection by influenza A and B virus. Another analog, *N*-acetyl-9-*O*-acetyl-neuraminic acid, prevents the binding of influenza C virus to host cells.

In a second approach, the structure of sialic acid was modulated by *in vivo* application of a chemically modified precursor of sialic acid. This new procedure ("*in vivo* modulation" of sialic acid) was successful when *N*-propanoyl-, *N*-butanoyl-or *N*-pentanoyl-*D*-mannosamine were used. It was shown that these unphysiological sugar analogs are metabolized to yield the corresponding *N*-acylneuraminic acid. Thus, the infection of lymphoma-derived cells with human polyoma or B-lymphomatropic papovaviruses was drastically reduced after pretreatment with these amino sugar analogs. These results demonstrate the importance of the side chains of sialic acids for infection with sialic acid-dependent viruses.

Acknowledgement

This work was supported by the BMBF and the Fonds der Chemischen Industrie, Frankfurt am Main, Germany.

References

Ahmed H, Gabius HJ (1989): Purification and properties of a Ca^{2+}-independent sialic acid-binding lectin from human placenta with preferential affinity to *O*-acetylsialic acids. In *J. Biol. Chem.* **264**:18673–8.

Ashwell G, Harford J (1982): Carbohydrate-specific receptors of the liver. In *Annu. Rev. Biochem.* **51**:531–54.

Barasch J, Kiss B, Prince A *et al.* (1991): Defective acidification of intracellular organelles in cystic fibrosis. In *Nature* **352**:70–3.

Bevilacqua M, Butcher E, Furie B *et al.* (1991): Selectins: A family of adhesion molecules. In *Cell* **67**:233–6.

Bhaskar KR, Garik P, Turner BS *et al.* (1992): Viscous fingering of HCl through gastric mucin. In *Nature* **360**:458–61.

Brossmer R, Isecke R, Herrler G (1993): A sialic acid analogue acting as a receptor determinant for binding but not for infection by influenza C virus. In *FEBS Lett.* **323**:96–8.

Cheresh DA, Reisfeld RA, Varki A (1984): *O*-Acetylation of disialoganglioside GD_3 by human melanoma cells creates an unique antigenic determinant. In *Science* **225**:844–6.

Coates SW, Gurney T, Sommers LW *et al.* (1980): Subcellular localization of sugar nucleotide synthetases. In *J. Biol. Chem.* **255**:9225–9.

Colli W (1993): Trans-sialidase: a unique enzyme activity discovered in the protozoan *Trypanosoma cruzi.* In *FASEB J.* **7**:1257–64.

Corfield AP, Schauer R (1982a): Occurrence of sialic acids. In *Sialic Acids, Chemistry, Metabolism and Function* (Cell Biol. Monographs **10**) Schauer R (ed) pp 5–50, Wien-New York: Springer.

Corfield AP, Schauer R (1982b): Metabolism of sialic acids. In *Sialic Acids, Chemistry, Metabolism and Function* (Cell Biol. Monographs **10**) Schauer R (ed) pp 173–94, Wien-New York: Springer.

Corfield T (1992): Bacterial sialidases-roles in pathogenicity and nutrition. In *Glycobiology* **2**:509–21.

Dang CV, Shin CK, Bell WR *et al.* (1989): Fibrinogen sialic acid residues are low affinity calcium-binding sites that influence fibrin assembly. In *J. Biol. Chem.* **264**:15104–8.

Day JF, Thornburg RW, Thorpe SR *et al.* (1980): Carbohydrate-mediated clearance of antibody-antigen complexes from the circulation. In *J. Biol. Chem.* **255**:2360–5.

Dennis JW, Laferte S (1987): Tumor cell surface carbohydrate and the metastatic phenotype. In *Cancer Metastasis Rev.* **5**:185–204.

do Valle Matta MA, Alekstich V, Angluster J *et al.* (1995): Occurrence of *N*-acetyl- and *N-O*-diacetyl-neuraminic acid derivatives in wild and mutant *Crithidia fasciculata.* In *Parasitol. Res.* **81**: 426–33.

Emig S, Schmalz D, Shakibaei M *et al.* (1995): The nuclear pore complex protein p62 is one of several sialic acid-containing proteins of the nuclear envelope. In *J. Biol. Chem.* **270**:13787–93.

Gabius H-J (1988): Tumor lectinology: At the intersection of carbohydrate chemistry, biochemistry, cell biology and oncology. In *Angew. Chem. Int. Ed. Engl.* **27**:1267–76.

Gabius H-J, Gabius S (eds) (1991): Lectins and cancer. Heidelberg-New York: Springer.

Gabius H-J, Gabius S (eds) (1993): Lectins and glycobiology. Heidelberg-New York: Springer.

Gabius S, Kayser K, Hellmann KP *et al.* (1990): Immobilisized derivatized lysoganglioside GM_1 is a ligand for specific binding sites in various human tumor cell types and peripheral blood lymphocytes and monocytes. In *Biochem. Biophys. Res. Comm.* **169**:239–44.

Gabius H-J, Kayser K, Gabius S (1995): Protein-Zucker-Erkennung. Grundlagen und medizinische Anwendung am Beispiel der Tumorlektinologie. In *Naturwissenschaften* **82**:533–43.

Hanai N, Dohi T, Nores G *et al.* (1988): A novel ganglioside, de-*N*-acetyl-GM_3, acting as a strong promotor for epidermal growth factor receptor kinase and as a stimulator for cell growth. In *J. Biol. Chem.* **263**:6296–301.

Harms E, Kreisel W, Morris HP *et al.* (1973): Biosynthesis of *N*-acetylneuraminic acid in Morris hepatomas. In *Eur. J. Biochem.* **32**:254–62.

Herrler G, Klenk H-D (1987): The surface receptor is a major determinant of the cell tropism of influenza C virus. In *Virology* **159**:102–8.

Higa HH, Varki A (1988): Acetyl-coenzyme A: polysialic acid *O*-acetyl transferase from K1-positive *Escherichia coli*: the enzyme responsible for the *O*-acetyl plus phenotype and for *O*-acetyl form variation. In *J. Biol. Chem.* **263**:8872–8.

Hirst GK (1941): The agglutination of red blood cells by allontoic fluid of chicken embryos infected with influenza virus. In *Science* **94**:22–3.

Horstkorte R, Schachner M, Magyar JP *et al.* (1993): The fourth immunoglobulin-like domain of *N*-CAM contains a carbohydrate recognition domain for oligomannosidic glycans implicated in association with L1 and neurite outgrowth. In *J. Cell Biol.* **121**:1409–21.

Hutchins JT, Reading CL, Giavazzi R *et al.* (1988): Distribution of mono-, di- and tri-*O*-acetylated sialic acids in normal and neoplastic colon. In *Cancer Res.* **48**:483–9.

Itzkowitz S, Kjeldsen T, Friera A *et al.* (1991): Expression of Tn and T antigens in human pancreas. In *Gastroenterology* **100**:1691–700.

Kageshita T, Hirai S, Kimura T *et al.* (1995): Association between sialyl Lewis[a] expression and tumor progression in melanoma. In *Cancer Res.* **55**:1748–51.

Kamerling JP, Schauer R, Shukla AK *et al.* (1987): Migration of *O*-acetyl groups in *N,O*-acetylneuraminic acids. In *Eur. J. Biochem.* **162**: 601–7.

Karlsson K-A (1995): Microbial recognition of target-cell glycoconjugates. In *Curr. Opinion Struct. Biol.* **5**:622–35.

Kayser H, Geilen CC, Paul C *et al.* (1992a): Incorporation of *N*-acyl-2-amino-2-deoxy-hexoses into glycosphingolipids of the pheochromocytoma cell line PC 12. In *FEBS Lett.* **301**:137–40.

Kayser H, Zeitler R, Kannicht C *et al.* (1992b): Biosynthesis of a non-physiological sialic acid in different rat organs, using *N*-propanoyl-D-hexosamines as precursors. In *J. Biol. Chem.* **267**:16934–8.

Kayser H, Ats C, Lehmann J *et al.* (1993): New amino sugar analogues are incorporated at different rates into glycoproteins of mouse organs. In *Experientia* **49**:885–7.

Keppler OT, Stehling P, Herrmann M *et al.* (1995): Biosynthetic modulation of sialic acid-dependent virus-receptor interactions of two primate polyoma viruses. In *J. Biol. Chem.* **270**:1308–14.

Kiehne K, Schauer R (1992): The influence of α- and β-galactose residues and sialic acid *O*-acetyl groups of rat erythrocytes on the interaction with peritoneal macrophages. In *Biol. Chem. Hoppe-Seyler* **373**:1117–23.

Kniep B, Peter-Katalinic J, Flegel W *et al.* (1992): CDw60 antibodies bind to acetylated forms of ganglioside GD_3. In *Biochem. Biophys. Res. Commun.* **187**:1343–9.

Lee YC, Lee RT (eds) (1994): Neoglycoconjugates. Preparation and applications. San Diego: Academic Press.

Lin C, Air GM (1993): Selection and characterization of a neuraminidase minus mutant of influenza virus and its rescue by cloned neuraminidase genes. In *Virology* **193**:1–5.

Machytka D, Kharitonenkov I, Isecke R *et al.* (1993): Methyl-α-glycoside of *N*-thioacetyl-D-neuraminic acid: a potential inhibitor of influenza A virus. A 1H NMR study. In *FEBS Lett.* **334**:117–20.

Manzi AE, Sjoberg ER, Diaz S *et al.* (1990): Biosynthesis and turnover of *O*-acetyl and *N*-acetyl groups in the gangliosides of human melanoma cells. In *J. Biol. Chem.* **265**:13091–103.

Marcus DM (1984): A review of the immunogenic and immunomodulatory properties of glycosphingolipids. In *Mol. Immunol.* **21**:1083–91.

Markham RB, Weller AN, Schiffman G *et al.* (1982): The presence of sialic acid on two related bacterial polysaccharides determines the site of the primary immune response and the effect of complement depletion on the response in mice. In *J. Immunol.* **128**:2731–3.

Meindl P, Bodo G, Palese P *et al.* (1974): Inhibition of neuraminidase activity by derivatives of 2-deoxy-2,3-dehydro-*N*-acetylneuraminic acid. In *Virology* **58**:457–63.

Nadano D, Iwasaki M, Endo S *et al.* (1986): A naturally occurring deaminated neuraminic acid 3-deoxy-D-glycero-D-galacto-nonulosonic acid (KDN). Its unique occurrence at the nonreducing ends of oligosialyl chains in polysialoglycoprotein of rainbow trout eggs. In *J. Biol. Chem.* **262**:11550–7.

Okuda M, Yamanaka A, Akihama S (1995): The effects of pH on the generation of turbidity and elasticity associated with fibrinogen-fibrin conversion by thrombin are remarkably influenced by sialic acid in fibrinogen. In *Biol. Pharm. Bull.* **18**:203–7.

Olden K, Parent JB, White SL (1982): Carbohydrate moieties of glycoproteins. A re-evaluation of their function. In *Biochim. Biophys. Acta* **650**:206–32.

Palese P, Schulman JL (1974): Inhibition of influenza and parainfluenza virus replication in tissue culture by 2-deoxy-2,3-dehydro-*N*-trifluoroacetylneuraminic acid. In *Virology* **59**:490–8.

Pegg MS, von Itzstein M (1994): Slow-binding inhibition of sialidase from influenza virus. In *Biochem Mol. Biol. Int.* **32**:851–8.

Pilatte Y, Bignon J, Lambre CR (1993): Sialic acids as important molecules in the regulation of the immune system: pathophysiological implications of sialidases in immunity. In *Glycobiology* **3**:201–18.

Pirie-Shepherd S, Jett EA, Andon NL *et al.* (1995): Sialic acid content of plasminogen 2 glycoforms as a regulator of fibrinolytic activity. Isolation, carbohydrate analysis, and kinetic characterization of six glycoforms of plasminogen. In *J. Biol. Chem.* **270**:5877–81.

Powell LP, Varki A (1995): I-type lectins. In *J. Biol. Chem.* **270**:14243–6.

Raju TS, Davidson EA (1994): Role of sialic acid on the viscosity of canine tracheal mucin glycoprotein. In *Biochem. Biophys. Res. Commun.* **205**:402–9.

Reglero A, Rodriguez AL, Luengo JM (1993): Polysialic acids. In *Int. J. Biochem.* **25**:1517–27.

Rens-Domiano S, Reisine T (1991): Structural analysis and functional role of the carbohydrate component of somatostatin receptors. In *J. Biol. Chem.* **266**:20094–102.

Reutter W, Köttgen E, Bauer C *et al.* (1982): Biological significance of sialic acids. In *Sialic Acids, Chemistry, Metabolism and Function* (Cell Biol. Monographs **10**) Schauer R (ed) pp 263–306, Wien-New York: Springer.

Rodriguez AL, Ferrero MA, Reglero A (1995): *N*-Acetyl-D-neuraminic acid synthesis in *Escherichia coli* K1 occurs through condensation of *N*-acetyl-D-mannosamine and pyruvate. In *Biochem. J.* **308**: 501–5.

Rogers GN, Herrler G, Klenk H-D *et al.* (1986): Influenza C virus uses 9-*O*-acetyl-*N*-neuraminic acid as a high affinity receptor determinant for attachment to cells. In *J. Biol. Chem.* **261**:5947–51.

Roth J, Rutishauser U, Troy F (1992): *Polysialic Acids*, Basel: Birkhäuser Verlag.

Rougon G (1993): Structure, metabolism and cell biology of polysialic acids. In *Eur. J. Cell Biol.* **61**:197–207.

Schauer R (1985): Sialic acids and their role as biological masks. In *Trends Biochem. Sci.* **10**:357–60.

Schauer R (1991): Biosynthesis and function of *N*- and *O*-substituted sialic acids. In *Glycobiology* **1**: 449–52.

Schauer R, Schoop HJ, Faillard H (1968): Zur Biosynthese der Glycolylgruppe der *N*-Glycolyl-neuraminsäure. In *Hoppe Seyler's Z. Physiol. Chem.* **349**:645–52.

Schauer R, Kelm S, Reuter G *et al.* (1995): Biochemistry and role of sialic acids. In *Biology of Sialic Acids* (Rosenberg A, ed) pp 7–67, New York: Plenum Press.

Schengrund CL, DasGupta BR, Ringler NJ (1991): Binding of botulinum and tetanus neurotoxins to ganglioside GT_{1b} and derivatives thereof. In *J. Neurochem.* **57**:1024–37.

Schengrund CL, Ringler NJ (1989): Binding of *Vibrio cholerae* toxin and the heat-labile enterotoxin of *Escherichia coli* to GM_1, derivatives of GM_1 and nonlipid oligosaccharid polyvalent ligands. In *J. Biol. Chem.* **264**:13233–7.

Schiavo G, Demel R, Montecucco C (1991): On the role of polysialoglycosphingolipids as tetanus toxin receptors. A study with lipid monolayers. In *Eur. J. Biochem.* **199**:705–11.

Schlepper-Schäfer J, Kolb-Bachofen V, Kolb H (1980): Analysis of lectin-dependent recognition of desialylated erythrocytes by Kupffer cells. In *Biochem. J.* **186**:827–31.

Seki T, Arai Y (1993): Distribution and possible roles of the highly polysialylated neural cell adhesion molecule (NCAM-H) in the developing and adult central nervous system. In *Neurosci. Res.* **17**:265–90.

Shimamura M, Shibuya N, Ito M *et al.* (1994): Repulsive contribution of surface sialic acid residues to cell adhesion to substratum. In *Biochem. Mol. Biol. Int.* **33**:871–8.

Siebert H-C, André S, Reuter G *et al.* (1995): Effect of enzymatic desialylation of human serum amyloid P component on surface exposure of laser photo CIDNP (chemically induced dynamic nuclear polarization)-reactive histidine, tryptophan and tyrosine-residues. In *FEBS Lett.* **371**:13–6.

Simbulan CM, Taki T, Tamiya-Koizumi K *et al.* (1994): Sulfate- and sialic acid-containing glycolipids inhibit DNA polymerase α activity. In *Biochim. Biophys. Acta* **1205**:68–74.

Sjoberg ER, Manzi AE, Khoo KH *et al.* (1992): Structural and immunological characterization of *O*-acetylated GD_2. Evidence that GD_2 is an acceptor for ganglioside *O*-acetyltransferase in human melanoma cells. In *J. Biol. Chem.* **267**:16200–11.

Sjoberg ER, Powell LD, Klein A *et al.* (1994): Natural ligands of the B cell adhesion molecule CD22β can be masked by 9-*O*-acetylation of sialic acids. In *J. Cell Biol.* **126**:549–62.

Sparks MA, Williams KW, Whitesides GM (1993): Neuraminidase-resistant hemagglutination inhibitors: acrylamide copolymers containing a C-glycoside of *N*-acetylneuraminic acid. In *J. Med. Chem.* **36**:778–83.

Stack MS, Pizzo SV, Gonzalez-Gronow M (1992): Effect of desialylation on the biological properties of human plasminogen. In *Biochem. J.* **284**:81–6.

Stoolman LM (1989): Adhesion molecules controlling lymphocyte migration. In *Cell* **56**:907–10.

Suzuki Y (1990): Biological role of *O*-acetylated sialic acid. In *Trends Glycosci. Glycotechnol.* **2**:112–5.

Taatjes DJ, Roth J, Weinstein J et al (1988): Post-Golgi apparatus localization and regional expression in rat intestinal sialyltransferase detected by immunoelectronmicroscopy with polypeptide epitope-purified antibody. In *J. Biol. Chem.* **263**:6302–9.

Tomlinson S, Pontes de Carvalho LC, Vandekerckhove F *et al.* (1994): Role of sialic acid in the resistance of *Trypanosoma cruzi* trypomastigotes to complement. In *J. Immunol.* **153**:3141–7.

van Rinsum J (1984): Some aspects of the *N*-acetylneuraminic acid biosynthesis in rat tissues. In *Vrije Universiteit te Amsterdam (thesis)*.

Varki A (1992): Diversity in the sialic acids. In *Glycobiology* **2**:25–40.

Varki A (1993): Biological roles of oligosaccharides: all of the theories are correct. In *Glycobiology* **3**:97–130.

Vimr ER (1994): Microbial sialidases: does bigger always mean better? In *Trends Microbiol.* **2**:271–7.

von Itzstein M, Wu W-Y, Kok GB *et al.* (1993): Rational design of potent sialidase-based inhibitors of influenza virus replication. In *Nature* **363**:418–23.

von Itzstein M, Wu W-Y, Jin B (1994): The synthesis of 2,3-didehydro-2,4-dideoxy-4-guanidinyl-*N*-acetylneuraminic acid: a potent influenza virus sialidase inhibitor. In *Carbohydr. Res.* **259**:301–5.

Wasley LC, Timony G, Murtha P *et al.* (1991): The importance of *N*- and *O*-linked oligosaccharides for the biosynthesis and *in vitro* and *in vivo* biologic activities of erythropoietin. In *Blood* **77**:2624–32.

Weis W, Brown JH, Cusack S *et al.* (1988): Structure of the influenza virus haemagglutinin complexed with its receptor, sialic acid. In *Nature* **333**:426–31.

White CL, Janakiraman MN, Laver WG *et al.* (1995): A sialic acid-derived phosphonate analog inhibits different strains of influenza virus neuraminidase with different efficiencies. In *J. Mol. Biol.* **245**:623–34.

Yang P, Major D, Rutishauser U (1994): Role of charge and hydration in effects of polysialic acid on molecular interactions on and between cell membranes. In *J. Biol. Chem.* **269**:23039–44.

Zeitler R, Giannis A, Danneschewski S *et al.* (1992): Inhibition of *N*-acetylglucosamine kinase and *N*-acetylmannosamine kinase by 3-*O*-methyl-*N*-acetyl-D-glucosamine *in vitro*. In *Eur. J. Biochem.* **204**:1165–8.

Zeng F-Y, Gabius H-J (1992): Sialic acid-binding proteins: Characterization, biological functions and applications. In *Z. Naturforschung* **47c**:641–53.

Zimmer G, Suguri T, Reuter G *et al.* (1994): Modification of sialic acids by 9-*O*-acetylation is detected in human leucocytes using the lectin property of influenza C virus. In *Glycobiology* **4**:343–9.

14 The Biology of Sulfated Oligosaccharides

LORA V. HOOPER, STEPHEN M. MANZELLA, AND JACQUES U. BAENZIGER

1 Introduction

A characteristic feature of the carbohydrates attached to proteins and lipids is the almost limitless number of distinct structures which can be produced by assembling different combinations of monosaccharides into oligosaccharides. The number of combinatorial possibilities is enormous because the number of different monosaccharides is large, monosaccharides can be assembled in many different types of linkages, and one or more branch points may be present on a single oligosaccharide. Modifications such as sulfation, phosphorylation, methylation, and acetylation provide additional structural complexity. The enormous structural diversity of oligosaccharides makes them ideal information-bearing molecules. However, the search for biological functions for carbohydrate moieties has been confounded by the very same structural diversity and by the fact that the oligosaccharide structures found on individual glycoproteins or glycolipids are generally highly heterogeneous. Rarely is a single structure found even at a single glycosylation site on a glycoprotein. In spite of these challenges, a number of oligosaccharides have now been shown to have definite biological functions. These biological functions are generally associated with the presence of one or more oligosaccharides with unique structural features such as sulfate or phosphate moieties that form part of a determinant which is recognized by a specific receptor or receptors.

The number of proteins which have been shown to bear *N*- or *O*-glycosidically linked oligosaccharides containing sulfate is rapidly growing. Although sulfated monosaccharides were known to be present on proteoglycans, mucins, and glycolipids, their presence on a wide range of glycoproteins produced by many different cells and tissues has only more recently been appreciated. Sulfate is incorporated into carbohydrate moieties from organisms extending from bacteria to man. Oligosaccharides containing sulfate vary with respect to the monosaccharides which are modified (e.g. galactose (Gal), *N*-acetylgalactosamine (GalNAc), *N*-acetylglucosamine (GlcNAc), mannose (Man) etc., the hydroxyl to which the sulfate is linked (e.g. 3-OH, 4-OH, etc.), and the structure of the underlying oligosaccharide moiety. The diversity of sulfated structures remains to be fully defined but is likely to be large.

Sulfate is transferred to oligosaccharide acceptors by sulfotransferases which utilize 3'-phosphoadenosine-5'-phosphosulfate (PAPS) as the high energy sulfate donor. Sulfate addition is a late synthetic event which occurs within the lumen of the Golgi (Green *et al.*, 1984; Schwarz *et al.*, 1984; Niehrs *et al.*, 1994). The large number of distinct sulfated structures already known and the presence of these structures on only a limited number of the glycoproteins produced by a given cell or tissue indicate that the sulfotransferases which mediate addition of sulfate to glycoprotein oligosaccharides are both numerous and highly specific. Examples of glycoproteins bearing sulfated monosaccharides are summarized in Table 1 along with information about the structures of the sulfated oligosaccharides. Based on these examples and the sulfotransferases which have thus far been characterized, one can conclude that sulfotransferases generally must either recognize unique features within the structures of their oligosaccharide targets and/or within the peptide to which the oligosaccharides are attached.

H.-J. and S. Gabius (Eds.), Glycosciences

ISBN 3-8261-0073-5

Table 1. Examples of sulfated monosaccharides identified on glycoproteins

Sulfated monosaccharide	Glycoproteins known to contain this monosaccharide
Gal-3-SO_4 SO_4-3-Galβ1,4GlcNAcβ-R	Thyroglobulin: *N*-linked glycans (Kamerling *et al.*, 1988; Spiro and Bhoyroo, 1988; De Waard *et al.*, 1991)
SO_4-3-Galβ1,4[Fucα1,3]GlcNAcβ-R	Ovarian cystadenoma glycoproteins (Yuen *et al.*, 1992)
GalNAc-4-SO_4 SO_4-4-GalNAcβ1,4GlcNAcβ-R	LH (Baenziger and Green, 1991; Stockell Hartree and Renwick, 1992), TSH (Baenziger and Green, 1991; Stockell Hartree and Renwick, 1992), carbonic anhydrase VI (Hooper *et al.*, 1995a), Tamm-Horsefall (Hård *et al.*, 1992), urokinase (Bergwerff *et al.*, 1995) :*N*-linked glycans. Proopiomelanocortin (Skelton *et al.*, 1992; Siciliano *et al.*, 1993; Siciliano *et al.*, 1994): *N*- and *O*-linked glycans.
GlcNAc-6-SO_4 Galβ1,4[SO_4-6]GlcNAcβ-R	Thyroglobulin (Kamerling *et al.*, 1988; Spiro and Bhoyroo, 1988; De Waard *et al.*, 1991), HIV gp120 (Shilatifard *et al.*, 1993): *N*-linked glycans. GlyCAM-1 (Hemmerich *et al.*, 1995): *O*-linked glycans
Man-4-SO_4	Ovalbumin (Yamashita *et al.*, 1983): *N*-linked glycans
Man-6-SO_4	*Dictyostelium discoideum* glycoproteins: *N*-linked glycans (Freeze and Wolgast, 1986)
GlcUAc-3-SO_4 SO_4-3-GlcUAcβ1,4Gal	P_o glycoprotein (Burger *et al.*, 1990): *N*-linked glycans.

Methods to identify and analyze sulfated oligosaccharides have rapidly improved over the past decade as more structures have been examined. In most instances, glycoproteins bearing sulfated oligosaccharides have initially been identified on the basis of metabolic incorporation of $[^{35}S]SO_4$. The high specific activity of $[^{35}S]SO_4$ makes it possible to identify even rare glycoproteins bearing sulfate; however, this level of sensitivity has also posed a problem for those interested in the functional significance of sulfation. In many cases it has been found that only a minor fraction of the oligosaccharides on a particular glycoprotein or at a specific glycosylation site are modified with sulfate even though they can readily be labeled by metabolic incorporation of $[^{35}S]SO_4$ (Hortin *et al.*, 1986). The biological significance of sulfation in such instances remains unclear since it is not yet known if this reflects poor fidelity on the part of one or more of the transferases involved in the synthesis of these structures or a specific form of modification.

Enzymatic removal of sulfate moieties from monosaccharides or oligosaccharides also requires the action of highly specific sulfatases. In contrast to phosphate which is sensitive to release by alkaline phosphatase regardless of its linkage or the underlying oligosaccharide structure, an equivalent sulfatase of broad specificity has not been identified. Those sulfatases which have been identified and characterized are highly specific for both the monosaccharide and the hydroxyl to which the sulfate is attached (Freeman *et al.*, 1987; Gibson *et al.*, 1987; Freeman and Hopwood, 1989; Bielicki *et al.*, 1990; Bielicki and Hopwood, 1991). This raises the possibility that the removal of sulfate moieties from glycoprotein oligosaccharides may be highly regulated and occur under specific circumstances. The ability to both synthesize sulfated oligosaccharides and to selectively remove sulfate moieties from these structures under specific circumstances may contribute to the biological functions of these structures.

Strikingly, unique sulfated glycoconjugates have been shown to play specific biological roles in several systems, ranging from symbiotic interactions between plants and nitrogen fixing bacteria to the homing of lymphocytes to lymph nodes. We will focus on the roles that sulfated glycoconjugates play in biological phenomena and consider the origin of specificity both in the synthesis of sulfated sugar chains and in the recognition events which mediate biological outcomes. Rather than considering all of the systems in which sulfation may have an important biological role, we will describe in detail two systems in which sulfated carbohydrates have been shown to play specific roles in well-defined biological processes control of the circulatory half life of lutropin and homing of lymphocytes to lymphnodes. Biological function in these two examples is correlated with unique and specific structures which occur on a distinct and limited subset of glycoproteins. The structures of the sulfated oligosaccharides involved in these two processes are shown in Figures 1 and 2, respectively. In these examples, there is evidence for a great deal of specificity in the synthetic machinery responsible for the production of the sulfated structures. Furthermore, the unique sulfated oligosaccharides produced in both cases participate in a specific molecular recognition event; i.e. binding to a carbohydrate receptor or lectin, which is responsible for the biological outcome. It is important to note that many of the principles which emerge from our consideration of the sulfated oligosaccharides on glycoproteins will likely apply equally to other sulfated glycoconjugates such as proteoglycans and glycolipids, as well as to carbohydrate structures which bear other unique modifications.

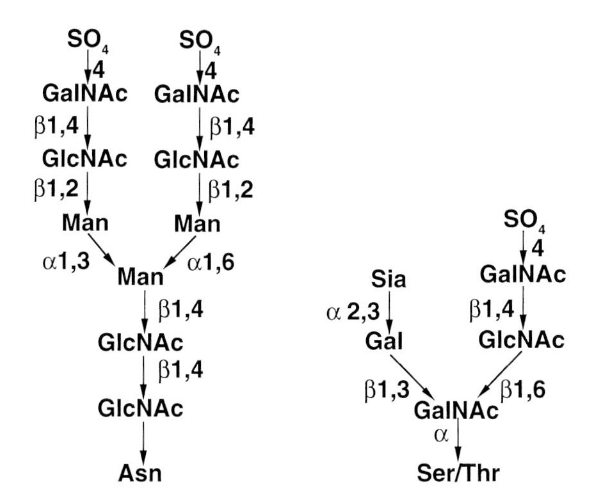

Figure 1. Examples of *N*- and *O*-linked oligosaccharides terminating with β1,4-linked GalNAc-4-SO_4. *N*-linked oligosaccharides terminating with β1,4-linked GalNAc-4-SO_4 have been described on the glycoprotein hormones LH and TSH (Baenziger and Green, 1988; 1991; Stockell Hartree and Renwick, 1992) proopiomelanocortin (POMC) (Siciliano *et al.*, 1993; Skelton *et al.*, 1992), urokinase (Bergwerff *et al.*, 1995) and carbonic anhydrase VI (Hooper *et al.*, 1995a). POMC bears *O*-linked structures which also terminate with β1,4-linked GalNAc-4-SO_4 (Siciliano *et al.*, 1994). The *N*-linked structures can display considerable microheterogeneity.

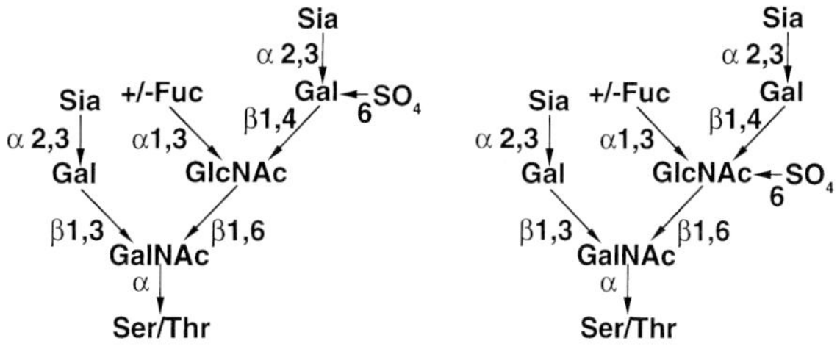

Figure 2. Structures of the sulfated *O*-linked oligosaccharides present on GlyCAM-1. The *O*-linked oligosaccharides on GlyCAM-1 contain either Gal-6-SO_4 or GlcNAc-6-SO_4 α2,3-linked sialic acid and α1,3-linked fucose (Hemmerich *et al.*, 1995). These structures can account for the dependence of L-selectin recognition on sulfate, fucose, and sialic acid (Rosen, 1993a, b).

2 Sulfated Oligosaccharides of Luteinizing Hormone

One of the first biological systems in which sulfated carbohydrates were shown to play a crucial role was that involving luteinizing hormone (LH). LH is essential for the regulation of a number of physiological processes involved in reproduction, including follicular maturation, ovulation, and the secretion of estradiol and progesterone. LH is synthesized by gonadotrophs in the anterior lobe of the pituitary and stored in secretory granules prior to release into the bloodstream under stimulation by gonadotropin releasing factor (GnRH) (Pierce and Parsons, 1981). The hormone exerts its physiological effects by binding to and stimulating the lutropin/chorionic gonadotropin (LH/CG) receptor in the ovary. A unique aspect of the biology of LH is its pulsatile pattern of appearance in the bloodstream. It is thought that this episodic rise and fall of LH levels is necessary for the expression of its bioactivity (Veldhuis *et al.*, 1984, 1987; Crowley and Hofler, 1985; Evans *et al.*, 1992). The LH/CG receptor is a G protein-coupled receptor which activates adenylate cyclase upon binding of LH or CG. Like other members of this family of receptors, the LH/CG receptor is desensitized upon ligand binding (Wang *et al.*, 1991; Lefkowitz, 1993; Segaloff and Ascoli, 1993). The rise and fall of LH levels in the circulation thus may be crucial for maintaining maximal stimulation of the LH/CG receptor in the ovary, since the decrease in circulating LH levels between pulses allows desensitized receptor to be replaced by fully active, unoccupied receptor ready for another round of stimulation. There are a number of factors which contribute to producing the pulsatile rise and fall in LH levels in the bloodstream the frequency of stimulation of LH release from the pituitary by GnRH, the release of LH from granules in secretory bursts, and the short circulatory half-life of LH.

Even though the short circulatory half life of LH had been appreciated for a number of years (Ascoli *et al.*, 1975), the basis for the rapid clearance of LH from the circulation was not understood prior to elucidation of the role played by its *N*-linked oligosaccharide moieties (Fiete *et al.*, 1991; Baenziger *et al.*, 1992). LH is a member of a closely related family of glycoproteins, the glycoprotein hormones, which includes LH, thyrotropin (TSH), follitropin (FSH), and chorionic gonadotropin (CG) (Pierce and Parsons, 1981). LH and TSH bear unique *N*-linked oligosaccharide moieties terminating with the sequence SO_4-4GalNAcβ1,4GlcNAc instead of Sialic acidα2,6/3Galβ1,4GlcNAc (Fig.1), which is found on the *N*-linked oligosaccharides of the vast majority of secreted glycoproteins present in the circulation including those on FSH and CG (Baenziger and Green, 1988, 1991; Stockell Hartree and Renwick, 1992). A biological function for the sulfated oligosaccharides on LH was suggested by clearance studies in rats showing that native LH, bearing oligosaccharides which terminate with GalNAc-4-SO_4, is removed from the circulation 4–5 fold more rapidly than recombinant LH bearing oligosaccharides which terminate with Sialic acid-Gal. Since the only difference between native and recombinant bovine LH is the terminal sequence of their *N*-linked oligosaccharides, these experiments indicated that the shorter circulatory half life is a result of the presence of sulfated rather than sialylated oligosaccharides on LH (Baenziger *et al.*, 1992).

The rapid clearance from the circulation of LH bearing sulfated oligosaccharides has a significant impact on its bioactivity *in vivo* (Baenziger *et al.*, 1992). The shortened circulatory half life following release into the blood reduces the potency of LH *in vivo*, but is essential for obtaining the pulsatile rise and fall in LH levels in the blood. As was noted above, episodic stimulation of the LH/CG receptor serves to prevent desensitization during the preovulatory surge of LH in the blood.

The reduced half life in the blood stream resulting from the presence of sulfated *N*-linked oligosaccharides on LH suggested the existence of a receptor specific for terminal GalNAc-4-SO_4. This receptor would rapidly remove LH and other glycoproteins bearing oligosaccharides with the terminal sequence SO_4-4GalNAcβ1,4GlcNAc from the bloodstream. A receptor with the predicted properties was identified in the endothelial cells and Kupffer cells of the liver (Fiete *et al.*, 1991). The receptor is specific for GalNAc-4-SO_4-bearing structures since the identical structure terminating with GalNAc-3-SO_4 is not recognized; i.e. the SO_4 must be attached to a specific hydroxyl on the GalNAc in order to be recognized by this receptor. The GalNAc-4-SO_4 receptor is plentiful, with more than 500,000 binding sites detected at the surface of hepatic endothelial cells, and has an apparent K_D of 1.6×10^{-7} M (Fiete *et al.*, 1991), thus accounting for the rapid removal of LH from the circulation. This suggests the model of LH function illustrated in Figure 3. LH, bearing sulfated oligosaccharides, is stored in secretory granules in the pituitary and is secreted in bursts due to episodic stimulation by GnRH. LH which has been released into the blood binds to the LH/CG receptor in the ovary while excess LH is rapidly cleared from the bloodstream as the blood passes through the liver where it is bound by the GalNAc-4-SO_4 receptor on hepatic endothelial cells.

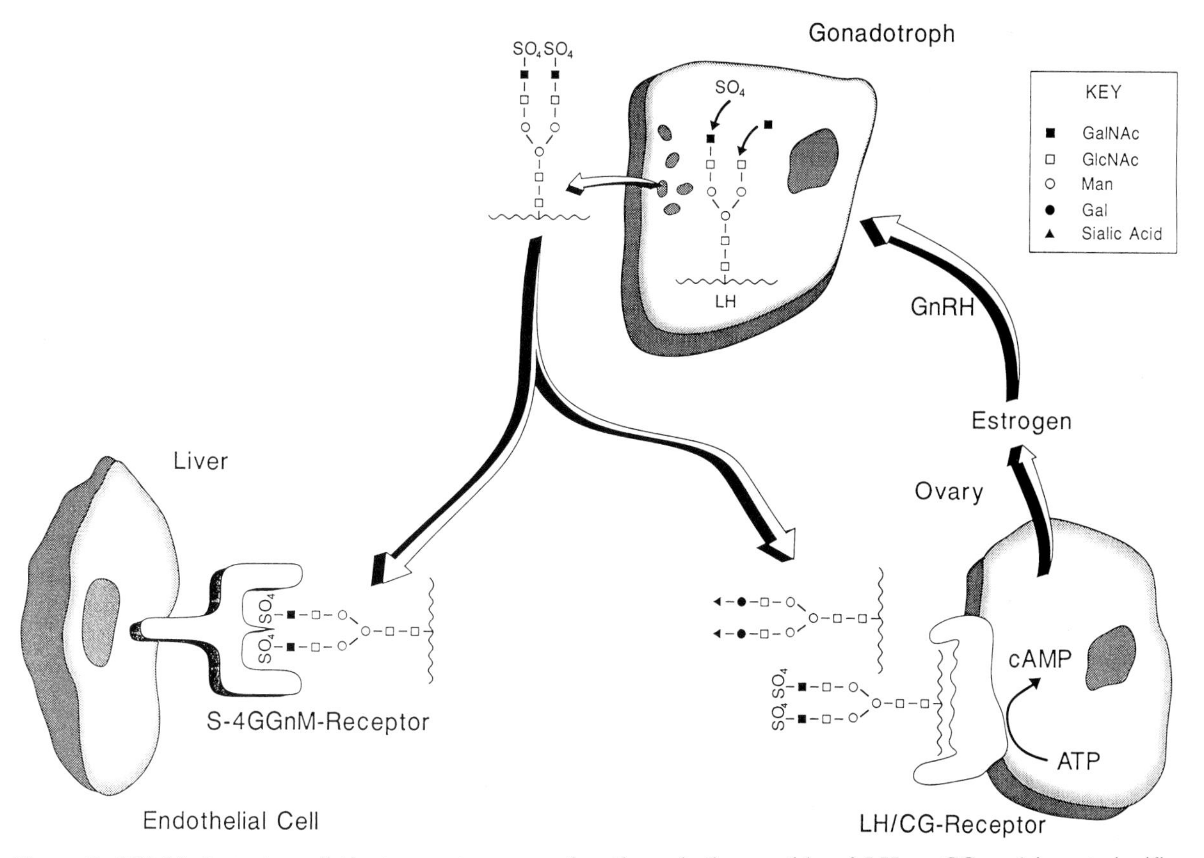

Figure 3. LH binds to two distinct receptors one of which recognizes carbohydrate and the other peptide. LH is synthesized in gonadotrophs found the anterior pituitary. Following addition of GalNAc and sulfate to its *N*-linked oligosaccharides, LH is stored in granules from which it is released upon stimulation with GnRH. LH binds to the LH/CG receptor in the ovary and stimulates the production of cyclic AMP. Binding is through the peptide of LH or CG and is not significantly affected by terminal glycosylation. Excess LH is rapidly removed from the blood by the GalNAc-4-SO_4 receptor present in hepatic endothelial cells. Removal of excess LH from the circulation allows any LH/CG receptor which has been desensitized following LH binding to be replaced by fully active receptor.

The importance of the sulfated oligosaccharides on LH for the ovulatory cycle is further evidenced by the absence of sulfated oligosaccharides on chorionic gonadotropin (CG). CG, which binds to the same receptor as LH, is produced by placental trophoblasts during the early stages of pregnancy in humans and horses. The β-subunit of human (h) CG has a different sequence than the β-subunit of hLH and arises from a separate gene, whereas in the β-subunits of equine (e) CG and eLH are identical and arise from the same gene (Moore *et al.*, 1980; Pierce and Parsons, 1981; Fiddes and Talmagde, 1984; Murphy and Martinuk, 1991; Sherman *et al.*, 1992). Equine LH, synthesized in the pituitary, bears oligosaccharides terminating with SO_4-4GalNAcβ1,4GlcNAc (Smith *et al.*, 1993) whereas eCG, synthesized in the placenta, bears oligosaccharides terminating with sialic acidα2,3Galβ1,4GlcNAc (Christakos and Bahl, 1979; Damm *et al.*, 1990; Matsui *et al.*, 1991). eLH, but not eCG, is recognized by the hepatic endothelial cell GalNAc-4-SO_4 receptor and as a result eLH has a shorter circulatory half life than eCG (Smith *et al.*, 1993). The difference in terminal glycosylation and thus circulatory half life appears to be the major difference between the hormone of the ovulatory cycle and that of early pregnancy since both bind to the LH/CG receptor in the ovary. This suggests episodic stimulation of the LH/CG receptor during the ovulatory cycle is essential whereas tonic stimulation is required during the early stages of pregnancy.

N-Linked oligosaccharides terminating with GalNAc-4-SO_4 have thus far been found on only a limited number of glycoproteins. What is the basis for addition of GalNAc-4-SO_4 to the *N*-linked oligosaccharides of LH and TSH and its absence on the *N*-linked oligosaccharides of FSH and other glycoproteins synthesized in the anterior pituitary? LH, TSH, and FSH are closely related hormones which share a common subunit and have highly homologous β-subunits (Pierce and Parsons, 1981). The α- and β-subunits each bear *N*-linked oligosaccharides and combine to form the respective dimeric hormones prior to exiting the endoplasmic reticulum (Corless *et al.*, 1987; Keene *et al.*, 1989). The synthesis of oligosaccharides terminating with GalNAc-4-SO_4 is accomplished by two highly specific transferases expressed in the pituitary a β1,4 *N*-acetylgalactosaminyltransferase (β1,4-GalNAc-transferase) and an *N*-acetylgalactosamine-4-sulfotransferase (GalNAc-4-sulfotransferase). Notably, Gal and GalNAc are added in β1,4-linkage to the same precursor core oligosaccharide, $GlcNAc_2Man_3GlcNAc_2Asn$ (Fig. 4). This indicates that features encoded within the peptide as well as the oligosaccharide acceptor must be recognized by the β1,4-GalNAc-transferase in order to account for addition of β1,4-linked GalNAc to the identical oligosaccharide acceptor on a selective subset of glycoproteins.

We identified a specific peptide motif recognized by the β1,4-GalNAc-transferase in LH and other glycoproteins bearing terminal GalNAc-4-SO_4 (Smith and Baenziger, 1990). In the presence of the peptide determinant the apparent K_m for transfer of GalNAc to $GlcNAc_2Man_3GlcNAc_2$-Asn by the β1,4-GalNAc-transferase is in the range of 5–10 mM whereas in the absence of the peptide recognition determinant the apparent K_m for transfer to the same acceptor is 1–2 mM. In contrast, β1,4-galactosyltransferase (β1,4 Gal-transferase) does not display peptide specificity, transferring Gal to terminal GlcNAc on the acceptor $GlcNAc_2Man_3GlcNAc_2Asn$ with an apparent K_m of 1–2 μM for all proteins (Smith and Baenziger, 1990). Thus, in cells which express both the β1,4-GalNAc-transferase and the β1,4 Gal-transferase, glycoproteins containing the recognition determinant are modified with β1,4-linked GalNAc while those which do not contain the determinant are modified with β1,4-linked Gal. We have located peptide recognition determinants for the β1,4-GalNAc-transferase within the common a subunit and the β-subunits of CG and LH. This determinant is absent from the β-subunit of FSH due to an amino terminal truncation in the sequence of FSH-β as compared to CG and LH β-subunits (Smith and Baenziger, 1992). Combination of the FSH-β subunit with the α-subunit masks the recognition determinant within the α-subunit (Smith and Baenziger, 1990) resulting in addition of β1,4-linked Gal rather than GalNAc.

LH oligosaccharides modified with β1,4-linked GalNAc are substrates for a GalNAc-4-sulfotransferase which is also expressed in the pituitary and transfers sulfate from the donor PAPS to terminal β1,4-linked GalNAc. This GalNAc-4-sulfotransferase is specific for oligosaccharides bearing terminal GalNAc, and will not recognize carbohydrates terminating with Gal or GlcNAc (Skelton *et al.*, 1991; Hooper *et al.*, 1995b). In addition, this transferase adds sulfate exclusively to the 4-OH of the terminal GalNAc. In contrast to the β1,4-GalNAc-transferase, there is no evidence for protein specificity on the part of the GalNAc-4-sulfotransferase, which has an apparent K_m of 30 mM for the terminal disaccharide GalNAcβ1,4GlcNAc (Hooper *et al.*, 1995b). Thus, the presence of terminal GalNAc-4-SO_4 on LH oligosaccharides, but not on oligosaccharides of the majority of other pituitary glycoproteins, derives from the fact that the β1,4-GalNAc-transferase recognizes a peptide motif present in LH and the GalNAc-4-sulfotransferase specifically transfers sulfate to terminal β1,4-linked GalNAc. In contrast, there is no evidence of protein specificity for the transferases responsible for the addition of either sialic acid or galactose to *N*-linked oligosaccharides.

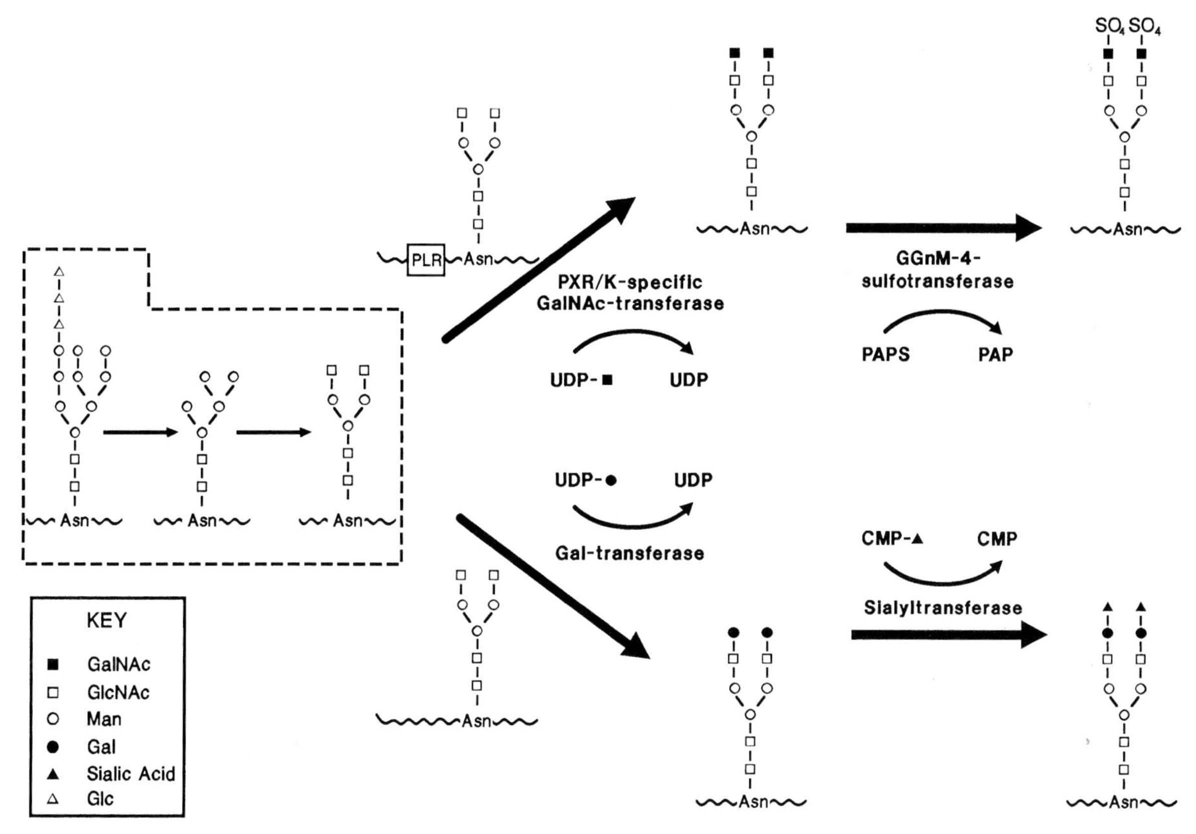

Figure 4. Peptide specific recognition by the GalNAc-transferase accounts for the presence of GalNAc-4-SO_4 on the *N*-linked oligosaccharides of LH but not other pituitary glycoproteins. The oligosaccharides enclosed by the broken line are synthetic intermediates found on many glycoproteins. Both β1,4-Gal-transferase and β1,4-GalNAc-transferase are able to add Gal and GalNAc to the acceptor $GlcNAc_2Man_3GlcNAc_2Asn$. In the presence of the peptide recognition determinant, PLRSKK, found on the α-subunit the apparent K_m for GalNAc addition is reduced from 1–2 mM to 5–10 μM. In contrast the apparent K_m of 1–2 mM for Gal addition is not influenced by the peptide. As a result GalNAc is added to the oligosaccharides of glycoproteins containing the recognition determinant while other glycoproteins are modified with Gal (Smith and Baenziger, 1988, 1990). The GalNAc-4-sulfotransferase (GGnM-4-sulfotransferase) only adds sulfate to the terminal sequence GalNAcβ1,4GlcNAcβ-R (Skelton *et al.*, 1991; Hooper *et al.*, 1995b).

Features of the peptide recognition determinant utilized by the β1,4-GalNAc-transferase have been extensively characterized. In the case of the glycoprotein hormone a subunit, the information required for recognition is contained within a 23 amino acid glycopeptide fragment (Fig. 5) which has the same apparent K_m for GalNAc addition as the intact, native α-subunit (Smith and Baenziger, 1992). We have used site directed mutagenesis to identify amino acids within this region of the α-subunit which affect recognition by the β1,4-GalNAc-transferase *in vitro* (Mengeling *et al.*, 1995). The basic residues Arg^{42}, Lys^{44}, and Lys^{45} (Fig. 5) are essential for recognition, whereas the Pro^{40} is not essential even though the motif ProXaaArg/Lys is present in many of the glycoproteins believed to contain a recognition determinant for the β1,4-GalNAc-transferase (Skelton *et al.*, 1992; Smith *et al.*, 1992). The rate of GalNAc transfer is enhanced by basic or aromatic residues at position 43 and reduced by acidic residues. The residues critical for recognition are found within two turns of an α helix which projects from the surface of the α-sub-

AYPT$\underline{P}^{40}\underline{L}^{41}\underline{R}^{42}S\underline{K}^{44}\underline{K}^{45}$TMLVQKN^{52}VTSE

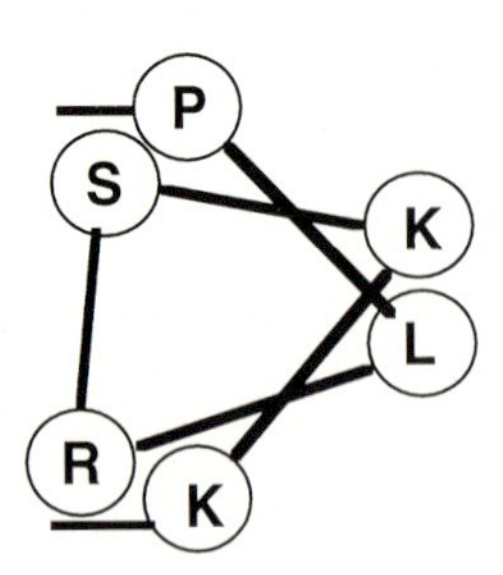

Figure 5. Sequence of the 21 amino acid peptide from the glycoprotein hormone α-subunit which contains all the information required for recognition by the glycoprotein hormone specific β1,4-GalNAc-transferase. The apparent K_m for transfer of GalNAc to the *N*-linked oligosaccharide located at Asn^{52} of this 21 amino acid glycopeptide from the glycoprotein hormone α-subunit is the same as that for transfer to the intact, native hormone (Smith and Baenziger, 1992). Mutagenesis studies have revealed that residues which have a critical impact on recognition by the β1,4-GalNAc-transferase include Leu^{41}, Arg^{42}, Lys^{44}, and Lys^{45}. Substitution of Leu^{41} with an aromatic or basic amino acid increases the rate of transfer while acid residues reduce the rate of GalNAc transfer (Mengeling *et al.*, 1995). The critical residues are found within two turns of an α-helix present on the native α-subunit, placing them in a cluster on one side of the helix (Mengeling *et al.*, 1995). The peptide recognition determinant used by the β1,4-GalNAc-transferase appears to consist of a cluster of closely placed basic amino acids. The residues critical for recognition as well as the sulfated oligosaccharides have been conserved in the glycoprotein hormones from all classes of vertebrates (Manzella *et al.*, 1995).

unit (Lapthorn *et al.*, 1994; Wu *et al.*, 1994; Mengeling *et al.*, 1995). Basic residues are also critical for recognition of the β-subunit determinant; however, in the case of the β-subunit these do not appear to be part of an α-helical structure. The peptide recognition determinant for the β1,4-GalNAc-transferase appears to consist of a cluster of basic amino acids which are able to attain a conformation allowing interaction with the transferase. A single recognition determinant can serve to mediate GalNAc addition to multiple oligosaccharides. It should be noted that within the native glycoprotein it is the spatial relationship of the peptide determinant to oligosaccharides which are modified rather than proximity within the linear amino acid sequence which dictates which oligosaccharides will be modified.

The amino acids essential for recognition by the β1,4-GalNAc-transferase are located within a region of the a subunit which has been conserved in all classes of vertebrates from fish to mammals. As a result, all classes of vertebrates have a subunits which have the potential to be recognized by the β1,4-GalNAc-transferase. Furthermore, β1,4-GalNAc-transferase and GalNAc-4-sulfotransferase activities with the expected specificity are detected in the pituitaries of all classes of vertebrates, and *N*-linked oligosaccharides on glycoprotein hormones from all classes of vertebrates terminate with GalNAc-4-SO_4 (Manzella *et al.*, 1995). This region of the a subunit is highly conserved because it plays a role in activation of the LH/CG receptor and binding to the β-subunit during dimer formation, as well as recognition by the β1,4-GalNAc-transferase. This is one of the first instances in which a unique form of terminal modification of an oligosaccharide has been found to be conserved during evolution of a protein over many classes of vertebrates. Conservation of the structure of the unique sulfated oligosaccharides on the glycoprotein hormones must be viewed as a manifestation of their biological importance. If regulation of circulatory half life is the critical function served by GalNAc-4-SO_4 bearing oligosaccharides then the GalNAc-4-SO_4 receptor should also be present within the livers of all vertebrate classes; however, this remains to be established.

Expression of the β1,4-GalNAc-transferase and of the GalNAc-4-sulfotransferase is hormonally and developmentally regulated (Dharmesh and Baenziger, 1993). LH first appears in the circulation around day 21 following birth in the rat. β1,4-GalNAc-transferase and GalNAc-4-sulfotransferase activities are also first detected in the pituitary at this time. In the mature female, LH levels rise and fall in the pituitary and the circulation in response to the amount of estrogen being produced by the ovary. Similarly, β1,4-GalNAc-transferase and GalNAc-4-sulfotransferase activities increase in the pituitary in parallel to LH levels as estrogen falls and decrease as estrogen levels rise. Thus, in the gonadotroph, expression of β1,4-GalNAc-transferase and GalNAc-4-

sulfotransferase are regulated co-ordinately with their substrate, LH, which is a major product of this cell (Dharmesh and Baenziger, 1993). This may be essential to assure that the oligosaccharides on LH, but not those on FSH or other glycoproteins produced in the gonadotroph, are efficiently modified with terminal GalNAc-4-SO_4. This in turn assures that LH produced at all times during the ovulatory cycle will be recognized by the hepatic GalNAc-4-SO_4 receptor and have a short half life following release into the blood even during the preovulatory surge in LH levels.

The function of GalNAc-4-SO_4-bearing oligosaccharides almost certainly extends beyond the clearance function demonstrated for LH. This unique structure is also present on thyroid stimulating hormone (TSH), also a member of the pituitary glycoprotein hormone family, as well as pro-opiomelanocortin (Skelton *et al.*, 1992; Siciliano *et al.*, 1993, 1994), a pituitary hormone precursor unrelated to either LH or TSH. POMC is unique, providing the first example of an *O*-linked oligosaccharide containing β1,4-linked GalNAc-4-SO_4 (Siciliano *et al.*, 1993) (Fig. 1). In the case of proopiomelanocortin, a function for terminal GalNAc-4-SO_4 in proteolytic processing has been proposed (Skelton *et al.*, 1992; Siciliano *et al.*, 1994). The GalNAc-transferase and GalNAc-4-sulfotransferase, in addition to being expressed in the pituitary, are also expressed in a limited number of other tissues, including submaxillary gland, lacrimal gland, and kidney, suggesting the existence of endogenous glycoproteins in these tissues which bear oligosaccharides terminating with GalNAc-4-SO_4 (Dharmesh *et al.*, 1993). One such glycoprotein from submaxillary gland has been identified as carbonic anhydrase VI, a salivary gland-specific form of carbonic anhydrase secreted into the saliva (Hooper *et al.*, 1995a). Since carbonic anhydrase VI presumably never enters the bloodstream, it is not likely to encounter the GalNAc-4-SO_4-receptor in the liver, suggesting that the presence of GalNAc-4-SO_4 on its *N*-linked oligosaccharides must serve a function other than rapid clearance from the circulation. Urokinase, which is synthesized in the kidney, has also been shown to be extensively modified with terminal GalNAc-4-SO_4 (Bergwerff *et al.*, 1995). Since urokinase is a potent activator of plasminogen (Dano *et al.*, 1985) its rapid clearance from the circulation would help assure that plasminogen is activated locally and not systemically. The presence of these unique sulfated oligosaccharides on an increasing number of unrelated glycoproteins raises the intriguing possibility that these structures may serve distinct biological functions depending upon when and where these structures are produced and to which receptor(s) they will be exposed.

3 Sulfated Carbohydrates in Leukocyte Trafficking

Leukocytes monitor the body for infection by continuously recirculating between blood, tissue, and lymph nodes. The targeting of subsets of leukocytes to sites of inflammation and to lymphoid organs is accomplished by specific interactions between the circulating cells and vascular endothelium and platelets. Members of the selectin receptor family are important participants in these adhesive interactions. The selectin family is comprised of three members, designated E-, L-, and P-selectin, with E-and P-selectin present on activated endothelium and platelets and L-selectin expressed on leukocytes. These closely related receptors share a number of structural features, including domains characteristic of C-type (calcium-binding) lectins positioned as a unit at the *N*-terminus. The lectin domains present in E-, P-, and L-selectin exhibit 60% to 70% sequence similarity to each other. Consistent with the presence of these lectin domains, E-, P-, and L-selectin have each been shown to bind carbohydrate, and a number of approaches have been used to identify their oligosaccharide ligands. Each of the selectins is capable of recognizing structures terminating with sialyl Lewis^x (Siaα2,3Galβ1,4[Fucα1,3]GlcNAc) or related structures (reviewed in Varki, 1992). However, this tetrasaccharide structure is found on a number of cell types and secreted proteins other than those recognized by the selectins. Furthermore, most studies to date have demonstrated low-affinity binding to these structures *in vitro* (Zhou *et al.*, 1991; Foxall *et al.*, 1992; Cooke *et al.*, 1994;

Jacob *et al.*, 1995), suggesting that sialyl Lewis^x (sLe^x) is necessary but not sufficient to account for selectin-ligand interactions *in vivo*. This raises the question as to the basis for specificity as well as avidity in selectin-carbohydrate interactions. Several possible mechanisms could be envisioned to answer this question, including a role for protein-protein interactions between selectins and their *in vivo* ligands, multivalency of oligosaccharide ligands, or specific modifications of a sLe^x "backbone".

The oligosaccharide structures on the natural ligands for L-selectin are the best characterized among the selectin ligands. L-selectin, widely expressed on leukocytes, is involved both in leukocyte recruitment to sites of inflammation and in homing of lymphocytes to lymph nodes where they contact sequestered antigens displayed by antigen presenting cells (reviewed in Rosen, 1993b). In this case, the answer to the question of avidity and selectivity probably involves at least in part the presence of one or more specifically-placed sulfate residues on the sLe^x tetrasaccharide, turning a relatively common structural motif into a unique ligand.

The carbohydrate ligand for L-selectin was first shown to be present on two lymph node proteins of 50 and 90 kD, identified as having the ability to bind to a soluble L-selectin-IgG chimera (Imai *et al.*, 1991). These glycoproteins were shown to be sulfated by their incorporation of [^{35}S]SO_4 in metabolic labeling experiments. Furthermore, both sialylation (Imai *et al.*, 1991) and sulfation (Imai *et al.*, 1993) were found to be required for binding to L-selectin. The 50 kD protein has been cloned and identified as a novel glycoprotein with clusters of *O*-linked saccharide chains (Lasky *et al.*, 1992). This mucin-like molecule, termed GlyCAM-1 (for glycosylated cell adhesion molecule-1), has a highly specific pattern of tissue expression in which it is specifically associated with high endothelial venules, consistent with its proposed role in lymphocyte homing. The 90 kD protein was identified by amino acid sequencing as CD34, another mucin-like glycoprotein expressed on hematopoetic stem cells and endothelium (Baumhueter *et al.*, 1993). Like GlyCAM-1, CD34 contains clusters of *O*-linked oligosaccharides and is associated with endothelium in peripheral lymph nodes. In agreement with its proposed role in L-selectin mediated lymphocyte homing, the sulfated glycoform of CD34 appears to be restricted to lymph node high endothelial venules.

The detailed structure of the carbohydrate moiety on GlyCAM-1 necessary for recognition by L-selectin has been elucidated (Hemmerich *et al.*, 1995). Like many other carbohydrates shown to bind selectins *in vitro*, the *O*-linked glycans of GlyCAM-1 were found to terminate with sLe^x. However, the structures on GlyCAM-1 are distinct in that they are 6-*O*-sulfated on either the underlying Gal or GlcNAc (Hemmerich *et al.*, 1994, 1995; Hemmerich and Rosen, 1994) (Fig. 2). This structure therefore accounts for all of the L-selectin binding requirements, including sialylation, sulfation, and probably fucosylation (Imai *et al.*, 1991, 1993). Furthermore, this novel 6-*O*-sulfated sialyl Lewis^x structure is incorporated into the core 2 structure (Galβ1,3 [GlcNAcβ1,6]GalNAc), a common structural precursor for *O*-linked carbohydrates on many glycoproteins. Thus the oligosaccharide ligand for L-selectin as expressed on GlyCAM-1 consists of a unique sulfated outer sequence and a common underlying core sequence. It remains to be determined whether this structure is also expressed on CD34.

Although the enzymes responsible for the synthesis of this unique sulfated sLe^x oligosaccharide have not yet been isolated, elucidation of this structure gives some indication of what to expect regarding synthesis of the sulfated ligand for L-selectin. We would predict that synthesis of this structure is restricted to GlyCAM-1, CD34, and perhaps a limited number of other endothelium-associated proteins in lymph node. This implies that synthesis of this sulfated structure must include at least one step involving recognition of the underlying protein. Furthermore, the core 2 structure is not unique; it is a common component of *O*-linked glycans present on a large number of glycoproteins from many different tissues and cell types. Synthesis of the basic sLe^x tetrasaccharide is also common to a number of glycoproteins in a number of cell types and tissues, and is not known to be protein-specific. Thus, it would seem most likely that the sulfotransferase(s) involved in the transfer of sulfate to the 6'-OH of either Gal or GlcNAc has a requirement for some feature of the underlying protein in addition to requiring structures terminating with Siaα2,3Galβ1,4GlcNAc. Confirmation of these predictions awaits isolation and characterization of the sulfotransferase(s) responsible for the 6'-*O*-sulfation of sialyl Lewis^x on GlyCAM-1.

The extent to which the presence of 6'-sulfated sLe^x accounts for the observed properties of L-selectin remains to be fully addressed. Although L-selectin exhibits a relatively weak interaction with sLe^x (K_d>1 mM) (Nelson *et al.*, 1993), it may be that the affinity of L-selectin for sLe^x-bearing structures increases dramatically by the addition of specifically-placed sulfate residues. Sulfation of Lewis$^{x/a}$ oligosaccharides may also play a role in biologically relevant E- and P-selectin interactions. SLe^x is expressed both by the myeloid cell line HL-60 and by CHO cells which have been transfected with an α1,3/4-fucosyltransferase. Even so, P-selectin interacts with a higher affinity and saturably with HL-60 cells, indicating that additional modifications may be required for P-selectin recognition of sLe^x on myeloid cells (Zhou *et al.*, 1991). A membrane glycoprotein which can account for the specific recognition of myeloid cells by P-selectin, P-selectin glycoprotein ligand-1 (PSGL-1), was recently identified by expression cloning from an HL-60 cDNA library (Sako *et al.*, 1993). PSGL-1 is a mucin-like glycoprotein which, like GlyCAM-1 and CD34, bears clusters of core 2 type *O*-linked oligosaccharides terminating with sLe^x (Stockert *et al.*, 1980). PSGL-1 also has a consensus sequence for tyrosine sulfation near its *N*-terminus (Sako *et al.*, 1993). Sulfation of one or more of the 3 Tyr residues which are potential sites for sulfate addition is essential for recognition of PSGL-1 by P-selectin but not by E-selectin (Pouyani and Seed, 1995; Sako *et al.*, 1995; Wilkins *et al.*, 1995). Thus, sulfation is essential for the specific recognition of PSGL-1 by P-selectin, but in contrast to L-selectin recognition of GlyCAM-1 or CD34 the essential sulfate is located on tyrosine rather than the target oligosaccharide itself. The epidermal growth factor-like domain of P-selectin has been shown to play a prominent role in P-selectin recognition of myeloid cells and may account for binding to the region with Tyr-SO_4 on PSGL-1 (Kansas *et al.*, 1994). Since E-, P-, and L-selectin each have an EGF-like domain immediately following their C-type lectin domain it is possible that each of these selectins will interact with a sulfate located either on a monosaccharide or on a Tyr. In this regard it is interesting that Lewis blood group oligosaccharides modified with sulfate residues can support avid binding of E-selectin. In vitro studies show that 3'-sulfated Le^a (SO_4-3-Galβ1,3[Fucα1,4]GlcNAcβ1,3Gal) is a 15 to 45-fold more potent inhibitor of E-selectin binding to sialyl Le$^{x/a}$ than sialyl Le^x itself (Yuen *et al.*, 1994). Thus, sulfation may also play a role in E-selectin binding under some circumstances. The biological significance of these observations for E-selectin will not become clear until its natural ligands are isolated and their detailed oligosaccharide structures are determined.

4 Sulfated Oligosaccharides of Unknown Function

The number of glycoproteins identified as bearing one or more sulfated oligosaccharides is rapidly increasing. Although many of these represent unique structures, their biological significance has not yet been established. Sulfated oligosaccharides have also been encountered recombinant glycoproteins; however, the cells expressing recombinant glycoproteins often have a different repertoire of transferases than is present in the cells which produce the glycoprotein *in vivo*. As a result, the sulfated structures on recombinant glycoproteins may not be present on the native form of the protein or be of biological significance for the native protein. Nonetheless, the presence of sulfated oligosaccharides on recombinant glycoproteins indicates that the host cell most likely synthesizes these same structures on one or more of its endogenous glycoproteins. Developing an understanding the importance of these sulfated structures for endogenous glycoproteins will require their identification and characterization.

Human and bovine thyroglobulin have been shown to bear *N*-linked oligosaccharides with terminal β1,4-linked Gal-3-SO_4 and/or internal GlcNAc-6-SO_4 (Kamerling *et al.*, 1988; Spiro and Bhoyroo, 1988; De Waard *et al.*, 1991). A Gal-3-sulfotransferase which requires only the terminal sequence Galβ1,4GlcNAc for transfer is present in extracts of thyroid (Kato and Spiro, 1989). Since the acceptor oligosaccharide is a common structure whereas structures terminating with Gal-3-SO_4 are rare, it is likely that the Gal-3-sulfotransferase will be found to display some form of peptide as well as oligosaccharide specificity. The GlcNAc-6-sulfotransferase has not to date been characterized; however, oligosaccharides with internal GlcNAc-6-SO_4 moieties have been reported on a number of glycoproteins including the human immunodeficiency virus (HIV) envelope glycoprotein gp120 when expressed in Molt-3 cells (Shilatifard *et al.*, 1993) and as yet unidentified glycoproteins from a number of different mammalian cell lines (Roux *et al.*, 1988; Sundblad *et al.*, 1988a, b). Even though no biological function for the sulfated structures on these glycoproteins has thus far been established, the presence of Gal-3-SO_4 termini on thyroglobulin from at least two different vertebrate species suggests the possibility that these structures will play a role in the transport and/or processing of thyroglobulin.

Glucuronic acid-3-SO_4 (GlcUA-3-SO_4) is a part of the epitope defined as HNK-1. In glycolipids this epitope consists of the structure SO_4-3-GlcUAβ1,3Galβ1,4GlcNAcβ-R (Chou *et al.*, 1985). Although the same epitope has been found on the oligosaccharides of myelin glycoproteins such as P_o, the structures of the oligosaccharides bearing this epitope on glycoproteins have not yet been chemically defined (Field *et al.*, 1992). Glucuronyltransferase activities which are specific for Galβ1,4GlcNAcβ-R acceptors on glycoprotein and glycolipid oligosaccharides, respectively, have been identified and separated (Kawashima *et al.*, 1992; Oka *et al.*, 1992), and a glucuronyl-3-sulfotransferase has been identified (Chou and Jungalwala, 1993). It seems likely that the glucuronyltransferase will ultimately be found to be protein specific, recognizing features encoded within the peptide as well as the oligosaccharide acceptor and producing a substrate for the glucuronyl-3-sulfotransferase on only a selected group of glycoproteins. Such structures may be of considerable importance since they are thought to play a role in neural cell adhesion and neurite outgrowth (Needham and Schnaar, 1991, 1993; Hall *et al.*, 1993). Again one would predict the existence of a receptor for these structures if this is the case. Clearly the potential for playing an important role is present for these structures even though we do not yet fully understand biological function of these structures at the molecular level.

5 Conclusions and Future Prospects

It is already apparent that many distinct sulfated oligosaccharide structures are synthesized and utilized for biological roles. Sulfated oligosaccharides are well suited for this because in most instances the incorporation of sulfate, even onto a structural precursor common to other non-sulfated oligosaccharides, results in the production of a unique structure which can contribute to recognition by a receptor. The existence of sulfated oligosaccharides which differ in the linkage of the sulfate, the monosaccharide which is sulfated, or the structure of the underlying oligosaccharide indicates that there must be a large number of highly specific sulfotransferases to account for the synthesis of these distinct structures. In most instances the unique sulfated structures are based on structural precursors which are common to other non-sulfated oligosaccharides. The presence of such sulfated structures on specific glycoproteins indicates that one or more steps in their synthesis must involve protein-specific transferases, as has been shown to be the case for synthesis of sulfated oligosaccharides on LH and phosphorylated oligosaccharides on lysosomal hydrolases. An understanding of sulfotransferases at the molecular level awaits their purification and cloning. This will reveal the extent to which sulfotransferases are structurally related and will provide tools to examine the regulation of their expression.

The demonstration that P-selectin recognition of PSGL-1 involves recognition of both oligosaccharide and a peptide component in the form of Tyr-SO_4 is particularly notable, because it indicates that the context of the oligosaccharides being recognized is also likely to be critical in many instances. It is clear that to develop an understanding of the biological function of sulfated oligosaccharides will in most instances require that each component of these complex systems be identified and ultimately characterized *in vivo*. Although this may seem a daunting challenge, the rewards of success will no doubt be considerable.

References

Ascoli M, Liddle RA, Puett D (1975): The metabolism of luteinizing hormone. Plasma clearance, urinary excretion, and tissue uptake. In *Mol. Cell. Endocrinol.* **3**:21–36.

Baenziger JU, Green ED (1988): Pituitary glycoprotein hormone oligosaccharides structure synthesis and function of the asparagine-linked oligosaccharides on lutropin follitropin and thyrotropin. In *Biochim. Biophys. Acta* **947**:287–306.

Baenziger JU, Green ED (1991): Structure synthesis and function of the asparagine-linked oligosaccharides on pituitary glycoprotein hormones. In *Biology of Carbohydrates* Volume 3, (Ginsberg V, Robbins PW, eds) pp 1–46, London: JAI Press Ltd.

Baenziger JU, Kumar S, Brodbeck RM *et al.* (1992): Circulatory half-life but not interaction with the lutropin/chorionic gonadotropin receptor is modulated by sulfation of bovine lutropin oligosaccharides. In *Proc. Natl. Acad. Sci.* USA **89**:334–8.

Baumhueter S, Singer MS, Henzel W *et al.* (1993): Binding of L-selectin to the vascular sialomucin CD34. In *Science* **262**:436–8.

Bergwerff AA, Van Oostrum J, Kamerling JP *et al.* (1995): The major *N*-linked carbohydrate chains from human urokinase. The occurrence of 4-*O*-sulfated (α2–6-sialylated or (α1–3)-fucosylated *N*-acetylgalactosamine(β1–4)-4-*N*-acetylglucosamine elements. In *Eur. J. Biochem.* **228**:1009–19.

Bielicki J, Hopwood JJ (1991): Human liver *N*-acetylgalactosamine 6-sulphatase. Purification and characterization. In *Biochem. J.* **279**:515–20.

Bielicki J, Freeman C, Clements PR *et al.* (1990): Human liver iduronate-2-sulphatase. Purification characterization and catalytic properties. In *Biochem. J.* **271**:75–86.

Burger D, Simon M, Perruisseau G *et al.* (1990): The epitope(s) recognized by HNK-1 antibody and IgM paraprotein in neuropathy is present on several *N*-linked oligosaccharide structures on human P0 and myelin-associated glycoprotein. In *J. Neurochem.* **54**:1569–75.

Chou DKH, Jungalwala FB (1993): Characterization and developmental expression of a novel sulfotransferase for the biosynthesis of sulfoglucuronyl glycolipids in the nervous system. In *J. Biol. Chem.* **268**:330–6.

Chou KH, Ilyas AA, Evans JE *et al.* (1985): Structure of a glycolipid reacting with monoclonal IgM in neuropathy and with HNK-1. In *Biochem. Biophys. Res. Commun.* **128**:383–8.

Christakos S, Bahl OP (1979): Pregnant mare serum gonadotropin purification and physiochemical biological and immunological characterization. In *J. Biol. Chem.* **254**:4253–61.

Cooke RM, Hale RS, Lister SG *et al.* (1994): The conformation of the sialyl Lewis X ligand changes upon binding to E-selectin. In *Biochemistry* **33**:10591–6.

Corless CL, Matzuk MM, Ramabhadran TV *et al.* (1987): Gonadotropin beta subunits determine the rate of assembly and the oligosaccharide processing of hormone dimer in transfected cells. In *J. Cell Biol.* **104**:1173–81.

Crowley WF, Hofler JG (1985): In The Episodic Secretion of Hormones (Crowley WF, Hofler JG, eds) pp 121–235, New York: Wiley.

Damm JBL, Hård K, Kamerling JP *et al.* (1990): Structure determination of the major *N*- and *O*-linked carbohydrate chains of the β subunit from equine chorionic gonadotropin. In *Eur. J. Biochem.* **189**:175–83.

Dano K, Andreasen PA, Grondahl-Hansen J *et al.* (1985): Plasminogen activators tissue degradation and cancer. In *Adv. Cancer Res.* **44**:139–266.

De Waard P, Koorevaar A, Kamerling JP *et al.* (1991): Structure determination by 1H NMR spectroscopy of (sulfated) sialylated *N*-linked carbohydrate chains released from porcine thyroglobulin by peptide-N4-(*N*-acetyl-β-glucosaminyl)asparagine amidase-F. In *J. Biol. Chem.* **266**:4237–43.

Dharmesh SM, Baenziger JU (1993): Estrogen modulates expression of the glycosyltransferases that synthesize sulfated oligosaccharides on lutropin. In *Proc. Natl. Acad. Sci.* USA 90:11127–31.

Dharmesh SM, Skelton TP, Baenziger JU (1993): Coordinate and restricted expression of the ProXaaArg/Lys-specific GalNAc-transferase and the GalNAcβ1,4GlcNAcβ1,2Manα-4-sulfotransferase. In *J. Biol. Chem.* **268**:17096–102.

Evans WS, Sollenberger MJ, Booth RAJ *et al.* (1992): Contemporary aspects of discrete peak-detection algorithms. II. The paradigm of the luteinizing hormone pulse signal in women. In *Endocr. Rev.* **13**:81–104.

Fiddes JC, Talmadge K (1984): Structure expression and evolution of the genes for the human glycoprotein hormones. In *Recent Prog. Horm. Res.* **40**:43–79.

Field MC, Wing DR, Dwek RA *et al.* (1992): Detection of multisulphated *N*-linked glycans in the L2/HNK-1 carbohydrate epitope expressing neural adhesion molecule P0. In *J. Neurochem.* **58**:993–1000.

Fiete D, Srivastava V, Hindsgaul O *et al.* (1991): A hepatic reticuloendothelial cell receptor specific for SO_4-4-GalNAcβ1,4GlcNAcβ1,2Manα that mediates rapid clearance of lutropin. In Cell **67**:1103–10.

Foxall C, Watson SR, Dowbenko D *et al.* (1992): The three members of the selectin receptor family recognize a common carbohydrate epitope the sialyl Lewis x oligosaccharide. In *J. Cell Biol.* **117**: 895–902.

Freeman C, Clements PR, Hopwood JJ (1987): Human liver *N*-acetylglucosamine-6-sulphate sulfatase. Purification and characterization. In *Biochem. J.* **246**:347–54.

Freeman C, Hopwood JJ (1989): Human liver glucuronate-2-suphatase. Purification, characterization and catalytic properties. In *Biochem. J.* **259**:209–16.

Freeze HH, Wolgast D (1986): Structural analysis of *N*-linked oligosaccharides from glycoproteins secreted by *Dictyostelium discoideum*. Identification of mannose-6-sulfate. In *J. Biol. Chem.* **261**:127–34.

Gibson GJ, Saccone GT, Brooks DA *et al.* (1987): Human liver *N*-acetylgalactosamine-4-sulphate sulfatase. Purification monoclonal antibody production and native and subunit Mr values. In *Biochem. J.* **248**:755–64.

Green ED, Gruenebaum J, Bielinska M *et al.* (1984): Sulfation of lutropin oligosaccharides with a cell-free system. In *Proc. Natl. Acad. Sci.* USA **81**:5320–4.

Hall H, Liu L, Schachner M *et al.* (1993): The L2/HNK-1 carbohydrate mediates adhesion of neural cells to laminin. In *Eur. J. Neurosci.* **5**:34–42.

Hård K, Van Zadelhoff G, Moonen P *et al.* (1992): The Asn-linked carbohydrate chains of human Tamm-Horsfall glycoprotein of one male–Novel sulfated and novel *N*-acetylgalactosamine-containing *N*-linked carbohydrate chains. In Eur. *J. Biochem.* **209**:895–915.

Hemmerich S, Rosen SD (1994): 6'-Sulfated sialyl Lewis x is a major capping group of GlyCAM-1. In *Biochemistry* **33**:4830–5.

Hemmerich S, Bertozzi CR, Leffler H *et al.* (1994): Identification of the sulfated monosaccharides of GlyCAM-1 an endothelial-derived ligand for L-selectin. In *Biochemistry* **33**:4820–9.

Hemmerich S, Leffler H, Rosen SD (1995): Structure of the *O*-glycans in GlyCAM-1 an endothelial-derived ligand for L-selectin. In *J. Biol. Chem.* **270**:12035–47.

Hooper LV, Beranek MC, Manzella SM *et al.* (1995a): Differential expression of GalNAc-4-sulfotransferase and GalNAc-transferase results in distinct glycoforms of carbonic anhydrase VI in parotid and submaxillary glands. In *J. Biol. Chem.* **270**:5985–93.

Hooper LV, Hindsgaul O, Baenziger JU (1995b): Purification and characterization of the GalNAc-4-sulfotransferase responsible for sulfation of GalNAcβ1,4GlcNAc-bearing oligosaccharides. In *J. Biol. Chem.* **270**:16327–32.

Hortin G, Green ED, Baenziger JU *et al.* (1986): Sulfation of proteins secreted by the human hepatoma derived cell line HEP-G2 sulfation of *N*-linked oligosaccharides attached the α2HS-glycoprotein. In *Biochem. J.* **235**:407–14.

Imai Y, Singer MS, Fennie C *et al.* (1991): Identification of a carbohydrate-based endothelial ligand for a lymphocyte homing receptor. In *J. Cell Biol.* **113**: 1213–21.

Imai Y, Lasky LA, Rosen SD (1993): Sulphation requirement for GlyCAM-1 an endothelial ligand for L-selectin. In *Nature* **361**:555–7.

Jacob GS, Kirmaier C, Abbas SZ *et al.* (1995): Binding of sialyl Lewis x to E-selectin as measured by fluorescence polarization. In *Biochemistry* **34**:1210–7.

Kamerling JP, Rijkse I, Maas AAM *et al.* (1988): Sulfated *N*-linked carbohydrate chains in porcine thyroglobulin. In *FEBS Lett.* **241**:246–50.

Kansas GS, Saunders KB, Ley K *et al.* (1994): A role for the epidermal growth factor-like domain of P-selectin in ligand recognition and cell adhesion. In *J. Cell Biol.* **124**:609–18.

Kato Y, Spiro RG (1989): Characterization of a thyroid sulfotransferase responsible for the 3-*O*-sulfation of terminal β-D-galactosyl residues in *N*-linked carbohydrate units. In *J. Biol. Chem.* **264**:3364–71.

Kawashima C, Terayama K, Ii M *et al.* (1992): Characterization of a glucuronyltransferase neolactotetraosylceramide glucuronyltransferase from rat brain. In *Glycoconjugate J.* **9**:307–14.

Keene JL, Matzuk MM, Otani T *et al.* (1989): Expression of biologically active human follitropin in chinese hamster ovary cells. In *J. Biol. Chem.* **264**: 4769–75.

Lapthorn AJ, Harris DC, Littlejohn A *et al.* (1994): Crystal structure of human chorionic gonadotropin. In *Nature* **369**:455–61.

Lasky LA, Singer MS, Dowbenko D *et al.* (1992): An endothelial ligand for L-selectin is a novel mucin-like molecule. In *Cell* **69**:927–38.

Lefkowitz RJ (1993): G protein-coupled receptor kinases. In *Cell* **74**:409–12.

Manzella SM, Dharmesh SM, Beranek MC *et al.* (1995): Evolutionary conservation of the sulfated oligosaccharides on vertebrate glycoprotein hormones that control circulatory half-life. In *J. Biol. Chem.* **270**:21665–71.

Matsui T, Sugino H, Miura M *et al.* (1991): β-Subunits of equine chorionic gonadotropin and lutenizing hormone with an identical amino acid sequence have different asparagine-linked oligosaccharide chains. In *Biochem. Biophys. Res. Commun.* **174**:940–5.

Mengeling BJ, Manzella SM, Baenziger JU (1995): A cluster of basic amino acids within an α-helix is essential for α-subunit recognition by the glycoprotein hormone *N*-acetylgalactosaminyltransferase. In *Proc. Natl. Acad. Sci.* USA **92**:502–6.

Moore WTJ, Burleigh BD, Ward DN (1980): Chorionic gonadotropins comparative studies and comments on relationships to other glycoprotein hormones. In *Chorionic Gonadotropin* (Segal SJ, ed) pp 89–125, New York: Plenum Press.

Murphy BD, Martinuk SD (1991): Equine chorionic gonadotropin. In *Endocr. Rev.* **12**:27–44.

Needham LK, Schnaar RL (1991): Adhesion of primary Schwann cells to HNK-1 reactive glycosphingolipids. Cellular specificity. In *Ann. NY Acad. Sci.* **633**:553–5.

Needham LK, Schnaar RL (1993): Carbohydrate recognition in the peripheral nervous system. A calcium-dependent membrane binding site for HNK-1-reactive glycolipids potentially involved in Schwann cell adhesion. In *J. Cell Biol.* **121**:397–408.

Nelson RM, Dolich S, Aruffo A *et al.* (1993): Higher affinity oligosaccharide ligands for E-selectin. In *J. Clin. Invest.* **91**:1157–66.

Niehrs C, Beisswanger R, Huttner WB (1994): Protein tyrosine sulfation 1993–An update. In *Chem. Biol. Interact.* **92**:257–71.

Oka S, Terayama K, Kawashima C *et al.* (1992): A novel glucuronyltransferase in nervous system presumably associated with the biosynthesis of HNK-1 carbohydrate epitope on glycoproteins. In *J. Biol. Chem.* **267**:22711–4.

Pierce JG, Parsons TF (1981): Glycoprotein hormones structure and function. In *Annu. Rev. Biochem.* **50**:465–95.

Pouyani T, Seed B (1995): PSGL-1 recognition of P-selectin is controlled by a tyrosine sulfation consensus at the PSGL-1 amino terminus. In *Cell* **83**: 333–43.

Rosen SD (1993a): Ligands for L-selectin: Where and how many. In *Res. Immunol.* **144**: 699–703.

Rosen SD (1993b): L-selectin and its biological ligands. In *Histochemistry* **100**:185–91.

Roux L, Holojda S, Sundblad G *et al.* (1988): Sulfated *N*-linked oligosaccharides in mammalian cells. I. Complex-type chains with sialic acids and *O*-sulfate esters. In *J. Biol. Chem.* **263**:8879–89.

Sako D, Chan X-J, Barone KM *et al.* (1993): Expression cloning of a functional glycoprotein ligand for P-selectin. In *Cell* **75**:1179–86.

Sako D, Comess KM, Barone KM *et al.* (1995): A sulfated peptide segment at the amino terminus of PSGL-1 is critical for P-selectin binding. In *Cell* **83**:323–31.

Schwarz JK, Capasso JM, Hirschberg CB (1984): Translocation of adenosine 3'-phosphate 5'-phosphosulfate into rat liver Golgi vesicles. In *J. Biol. Chem.* **259**:3554–9.

Segaloff DL, Ascoli M (1993): The lutropin/choriogonadotropin receptor...4 years later. In *Endocr. Rev.* **14**:324–47.

Sherman GB, Wolfe MW, Farmerie TA *et al.* (1992): A single gene encodes the β-subunits of equine luteinizing hormone and chorionic gonadotropin. In *Mol. Endocrinol.* **6**:951–9.

Shilatifard A, Merkle RK, Helland DE *et al.* (1993): Complex-type *N*-linked oligosaccharides of gp120 from human immunodeficiency virus type 1 contain sulfated *N*-acetylglucosamine. In *J. Virol.* **67**: 943–52.

Siciliano RA, Morris HR, McDowell RA *et al.* (1993): The Lewis x epitope is a major non-reducing structure in the sulphated *N*-glycans attached to Asn-65 of bovine pro-opiomelanocortin. In *Glycobiology* **3**: 225–39.

Siciliano RA, Morris HR, Bennett HPJ *et al.* (1994): *O*-glycosylation mimics *N*-glycosylation in the 16-kDa fragment of bovine pro-opiomelanocortin. The major *O*-glycan attached to Thr-45 carries SO_4-4GalNAcβ1–4GlcNAcβ1- which is the archetypal non-reducing epitope in the *N*-glycans of pituitary glycohormones. In *J. Biol. Chem.* **269**:910–20.

Skelton TP, Hooper LV, Srivastava V *et al.* (1991): Characterization of a sulfotransferase responsible for the 4-*O*-sulfation of terminal β-*N*-acetyl-D-galactosamine on asparagine-linked oligosaccharides of glycoprotein hormones. In *J. Biol. Chem.* **266**: 17142–50.

Skelton TP, Kumar S, Smith PL *et al.* (1992): Proopiomelanocortin synthesized by corticotrophs bears asparagine-linked oligosaccharides terminating with SO_4-4GalNAcβ1,4GlcNAcβ1,2Manα. In *J. Biol. Chem.* **267**:12998–3006.

Smith PL, Baenziger JU (1988): A pituitary *N*-acetylgalactosamine transferase that specifically recognizes glycoprotein hormones. In *Science* **242**:930–3.

Smith PL , Baenziger JU (1990): Recognition by the glycoprotein hormone-specific *N*-acetylgalactosaminetransferase is independent of hormone native conformation. In *Proc. Natl. Acad. Sci.* USA **87**: 7275–9.

Smith PL, Baenziger JU (1992): Molecular basis of recognition by the glycoprotein hormone-specific *N*-acetylgalactosamine-transferase. In *Proc. Natl. Acad. Sci.* USA **89**:329–33.

Smith PL, Skelton TP, Fiete D *et al.* (1992): The asparagine-linked oligosaccharides on tissue factor pathway inhibitor terminate with SO_4-4Gal NAcβ1,4GlcNAcβ1, 2Manα. In *J. Biol. Chem.* **267**: 19140–6.

Smith PL, Bousfield GS, Kumar S *et al.* (1993): Equine lutropin and chorionic gonadotropin bear oligosaccharides terminating with SO_4-4-GalNAc and sialic acidα2,3Gal, respectively. In *J. Biol. Chem.* **268**: 795–802.

Spiro RG, Bhoyroo VD (1988): Occurrence of sulfate in the asparagine-linked Complex carbohydrate units of thyroglobulin – Identification and localization of galactose 3-sulfate and *N*-acetylglucosamine 6-sulfate residues in the human and calf proteins. In *J. Biol. Chem.* **263**:14351–8.

Stockell Hartree A, Renwick AGC (1992): Molecular structures of glycoprotein hormones and functions of their carbohydrate components. In *Biochem. J.* **287**:665–79.

Stockert RJ, Gartner U, Morell AG *et al.* (1980): Effects of receptor-specific antibody on the uptake of desialylated glycoproteins in the isolated perfused rat liver. In *J. Biol. Chem.* **255**:3830–41.

Sundblad G, Holojda S, Roux L *et al.* (1988a): Sulfated *N*-linked oligosaccharides in mammalian cells. ii. Identification of glycosaminoglycan-like chains attached to complex-type glycans. In *J. Biol. Chem.* **263**:8890–6.

Sundblad G, Kajiji S, Quaranta V *et al.* (1988b): Sulfated *N*-linked oligosaccharides in mammalian cells. iii. Characterization of a pancreatic carcinoma cell surface glycoprotein with *N*- and *O*-sulfate esters on asparagine-linked glycans. In *J. Biol. Chem.* **263**: 8897–903.

Varki A (1992): Selectins and other mammalian sialic acid-binding lectins. In *Curr. Opinion Cell Biol.* **4:** 257–66.

Veldhuis JD, Beitins IZ, Johnson ML *et al.* (1984): Biologically active luteinizing hormone is secreted in episodic pulsations that vary in relation to stage of the menstrual cycle. In *J. Clin. Endocrinol. Metab.* **58:** 1050.

Veldhuis JD, Carlson ML, Johnson ML (1987): The pituitary gland secretes in bursts. Appraising the nature of glandular secretory impulses by simultaneous multiple-parameter deconvolution of plasma hormone concentrations. In *Proc. Natl. Acad. Sci.* USA **84**:7686–90.

Wang H, Segaloff DL, Ascoli M (1991): Lutropin/choriogonadotropin down-regulates its receptor by both receptor-mediated endocytosis and a cAMP-dependent reduction in receptor mRNA. In *J. Biol. Chem.* **266**:780–5.

Wilkins PP, Moore KL, McEver RP *et al.* (1995): Tyrosine sulfation of P-selectin glycoprotein ligand-1 is required for high affinity binding to P-selectin. In *J. Biol. Chem.* **270**:22677–80.

Wu H, Lustbader JW, Liu Y *et al.* (1994): Structure of human chorionic gonadotropin at 2.6 Å resolution from MAD analysis of the selenomethionyl protein. In *Structure* **2**:545–58.

Yamashita K, Ueda I, Kobata A (1983): Sulfated asparagine-linked sugar chains of hen egg albumin. In *J. Biol. Chem.* **258**:14144–7.

Yuen C-T, Lawson AM, Chai W *et al.* (1992): Novel sulfated ligands for the cell adhesion molecule E-selectin revealed by the neoglycolipid technology among *O*-linked oligosaccharides on an ovarian cystadenoma glycoprotein. In *Biochemistry* **31:** 9126–31.

Yuen C-T, Bezouska K, O'Brien J *et al.* (1994): Sulfated blood group Lewis a. A superior oligosaccharide ligand for human E-selectin. In *J. Biol. Chem.* **269**:1595–8.

Zhou Q, Moore KL, Smith DF *et al.* (1991): The selectin GMP-140 binds to sialylated fucosylated lactosaminoglycans on both myeloid and nonmyeloid cells. In *J. Cell Biol.* **115**:557–64.

15 Carbohydrate-Carbohydrate Interaction

NICOLAI V. BOVIN

1 Introduction

Homotypic interactions between biopolymers such as nucleic acid-nucleic acid, protein-protein and lipid-lipid interactions are the basis for biorecognition and cell functioning, as they mediate such processes as transcription, signal transduction and cell membrane formation. Until recently there were practically no data on the existence of biologically significant carbohydrate-carbohydrate interactions, and only after the publication of a number of articles by S. Hakomori and co-workers (first publication, Eggens *et al.*, 1989), concerning the interaction of glycolipids at the molecular and cellular level, did it become evident that this type of interaction exists and that it can have a general, fundamental character for cell biology. In this review the presently known examples of specific carbohydrate-carbohydrate interaction are discussed, paying attention to the biologically significant processes of molecular recognition; in our opinion, it seems reasonable to exclude from the range of subjects discussed polysaccharide-polysaccharide interactions in the solid phase (as, for example, during the formation of cellulose crystals), and the interaction of oppositely charged molecules, as the attraction in this case is exclusively electrostatic.

2 Historical Aspects

The hypothesis concerning the possibility of a specific carbohydrate-carbohydrate interaction was formulated for the first time in 1982 by T. Endo and co-workers (Endo *et al.*, 1982), who found that glycophorin (the heavily glycosylated major protein of human erythrocytes) introduced into a liposome membrane did not interact with lectin if Forssman glycolipid or globoside had been inserted into the same membrane. In other words, these authors found a specific interaction between two types of biopolymers (glycoprotein and glycolipid) in the composition of the same membrane, i.e. the "side-by-side" type (Figure 1a).

All the following papers, including Brewer and Thomas (1984), in which the hypothesis on gangliosides as the mediators of artificial membrane adhesion was formulated, and the key article of I. Eggens and co-workers (1989) dealt principally with other systems, namely cell or artificial membranes of two different types interacting in a ,head-to-head" arrangement, i.e. two surfaces were involved in the interaction (Figure 1b). This work is summarized in several reviews (Hakomori, 1990, 1993; 1994a, b; Fenderson, 1993). The work on glycosphingolipid-glycosphingolipid interaction has an intriguing history. It had been known since 1981 that the Le^x determinant was expressed to a large extent during the early stages of mice embryo development (compaction, 8–16 cell stage) (Gooi *et al.*, 1981; Hakomori *et al.*, 1981) and that the appearance of Le^x glycolipid correlates in time exactly with the compaction stage, which can be inhibited by the free oligosaccharide Le^x. Although it was suggested that membrane Le^x was recognized by some complementary molecule (lectin) on the neighboring cells, it was not possible to find the protein partner. Only eight years later it became clear that the partner of Le^x on the neighboring cells was Le^x itself, and that the interaction was of the carbohydrate-carbohydrate type (Eggens *et al.*, 1989). It seems that the authors of this outstanding work had to overcome two main obstacles, the first of them being the stereotype according to which a

H.-J. and S. Gabius (Eds.), Glycosciences

ISBN 3-8261-0073-5

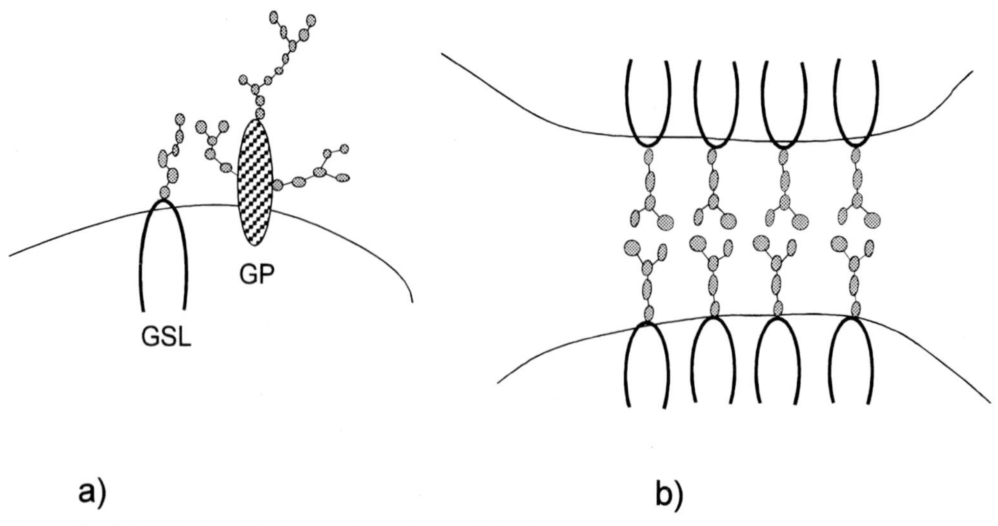

Figure 1. *(a)* Side-by-side interaction of glycolipid (GSL) with glycoprotein (GP) on cell membrane; *(b)* head-to-head contact of glycolipids belonging to different cells.

carbohydrate must interact only with a protein partner, and the second being the low value of the effect itself, as it was clearly manifested only during the contact of a large number of Le^x copies tightly arranged on the membrane (see details below). As soon as the prevailing paradigm was broken, other examples of carbohydrate-carbohydrate recognition were found, including those not of glycolipid-glycolipid type (Stewart and Boggs, 1993; Siuzdak *et al.*, 1993; Gupta *et al.*, 1994; Mikhalchik *et al.*, 1994; Bovin *et al.*, 1995).

A paper on the interaction of the carbohydrate chains of hyaluronate and chondroitin sulfate was published in 1980 (Turley and Roth, 1980), and several reports on the interaction of polysaccharides and glycosaminoglycans from sponges were published in the eighties and nineties (Misevic *et al.*, 1982; Misevic and Burger, 1993; Dammer *et al.*, 1995), though, as has been mentioned above, these results are not within the scope of this review. It is only interesting to note that the specific carbohydrate-mediated agglutination of sponge cells has been explained for a long time by the action of lectins, and only in a recent paper (Misevic and Burger, 1993) has the carbohydrate-carbohydrate mechanism of this aggregation been confirmed; in fact, the story is quite similar to that with Le^x glycolipid on mice embryonic cells.

3 Experimental Approaches

Carbohydrate-carbohydrate interaction has a distinct cooperative character. To observe this effect special approaches are required, giving the possibility of observing simultaneous or concerted recognition of several dozens or even hundreds of molecules which have a suitable spatial arrangement. The interaction of single molecules is so weak that it cannot be detected by the usual methods. Thus, a unsuccessful attempt to register Le^x/Le^x interaction by the NMR method has been described (Wormald *et al.*, 1991). The only method that permitted detection of the complex Le^x-Ca^{2+}-Le^x was ion spray mass spectrometry (Siuzdak *et al.*, 1993). However, the experimental conditions in mass spectrometry, in which the interaction is in vacuum, are quite different from the conditions of interaction characteristic for biological recognition, which occur in solution. The following approaches have been used for the study of multivalent carbohydrate-carbohydrate interaction.

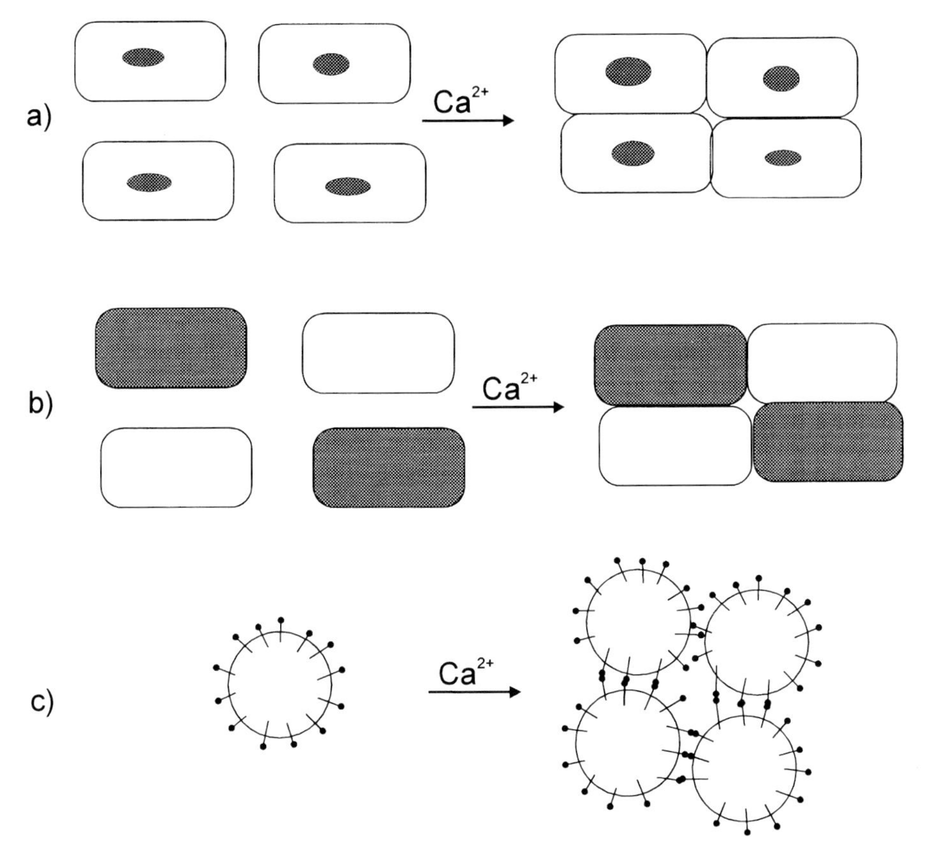

Figure 2. *(a)* Homotypic and *(b)* heterotypic cell aggregation; *(c)* aggregation of liposomes.

3.1 Cell Aggregation (Figure 2a, b)

A high concentration of Le^x glycolipid on the cell membrane of mice teratocarcinoma F9 cells allowed observation of the carbohydrate-carbohydrate interaction using light microscopy. The indispensable condition for carbohydrate-carbohydrate interaction is the presence of Ca^{2+} or other bivalent cations (see the details below). In the absence of Ca^{2+}, F9 exists almost exclusively as single cells but, after the addition of Ca^{2+} to the medium, a considerable and visible cell aggregation takes place. The process is reversible and can be inhibited by addition of either EDTA, that eliminates Ca^{2+}, or Le^x oligosaccharide; structurally related oligosaccharides do not manifest any significant inhibitory effect (Eggens *et al.*, 1989). Specific aggregation of two cell types (Figure 2b), namely mice lymphoma cells and mice melanoma cells, caused the establishment of another type of carbohydrate-carbohydrate interaction between glycolipids GM_3 and Gg_3 (see Table 1 for the structures of saccharides) (Kojima and Hakomori, 1991 a). Lymphoma AA12 cells express a large amount of glycolipid Gg_3, whereas melanoma cell line B16 expresses ganglioside GM_3. The aggregation of these cells takes place due to the interaction Gg_3/GM_3 in presence of Ca^{2+}. The aggregation can be inhibited by oligosaccharide fragments of Gg_3 and GM_3 (e.g. 5 mM 3'-sialyllactose), monoclonal antibodies to Gg_3 and GM_3, and by EDTA.

3.2 Aggregation of Liposomes (Figure 2c)

Specific aggregation of cells offers the chance of finding new examples of carbohydrate-carbo-

Table 1. Oligosaccharides participating in carbohydrate-carbohydrate interaction

Structure	Designation
Fucα1–3 \\ GlcNAc Galβ1–4 /	Le^x
Fucα1–4 \\ GlcNAc Fucα1–2Galβ1–3 /	Le^b
Fucα1–3 \\ GlcNAc Fucα1–2Galβ1–4 /	Le^y
Fucα1–2Galβ1–3GlcNAc	H (type 1)
Fucα1–2Galβ1–4GlcNAc	H (type 2)
Galβ1–4GlcNAc	LacNAc
Neu5Acα2–3(GalNAcβ1–4)Galβ1–4Glc	GM_2
Neu5Acα2–3Galβ1–4Glc	GM_3
KDNα2–3Galβ1–4Glc	(KDN)GM_3
GalNAcβ1–3Galα1–4Galβ1–4Glc	Gb_4
Galα1–4Galβ1–4Glc	Gb_3
GalNAcβ1–4Galβ1–4Glc	Gg_3
Galβ1–4Glc	Lac
Galβ1–3GalNAcβ1–4Galβ1–4Glc	Gg_4

hydrate interaction, although this method is not convenient just for the study of this phenomenon, due to the extremely high diversity of the cell surface molecules. The liposome model is, in principle, very similar, and it is suitable in practice for this study. The glycolipid(s) under study is introduced into the liposome membrane, and the aggregation of the liposomes in the presence of different substances is followed. Liposome aggregation can be evaluated quantitatively with the help of a spectrophotometer or aggregometer (Eggens *et al.*, 1989). This approach gives extensive possibilities for the investigation of fine details of carbohydrate-carbohydrate interaction. For example, it is possible to study the influence of the oligosaccharide density on the membrane surface on the value of binding to the carbohydrate partner, or the contribution of the lipid part to binding (by the variations in the size and nature of the lipid "tail") (Stewart and Boggs, 1993). More exact and reliable in interpretation (in comparison to the first method) are the data on the inhibition of aggregation by oligosaccharides, antibodies, EDTA and glycosidases obtained by this method. An approach which is analogous in principle is based on the coating of plastic beads (3 μm diameter) with glycolipid (Kojima *et al.*, 1994).

3.3 Interaction of Liposomes with a Plastic Surface Coated with Glycolipid or Another Glycoconjugate (Figure 3)

This is one of the most convenient and informative approaches. The wells of 96-well polystyrene plates are coated with glycolipid (about 1 nmol/well), this forming a regular monolayer imitating the membrane of a cell or liposome. After removing unbound glycolipid by washing, liposomes into which the second partner of the carbohydrate-carbohydrate interaction is integrated are added to the wells. Radioactive label is additionally inserted into the liposomes by any suitable method. After incubation for several hours, unbound liposomes are removed, whereas the bound ones are extracted, followed by detection

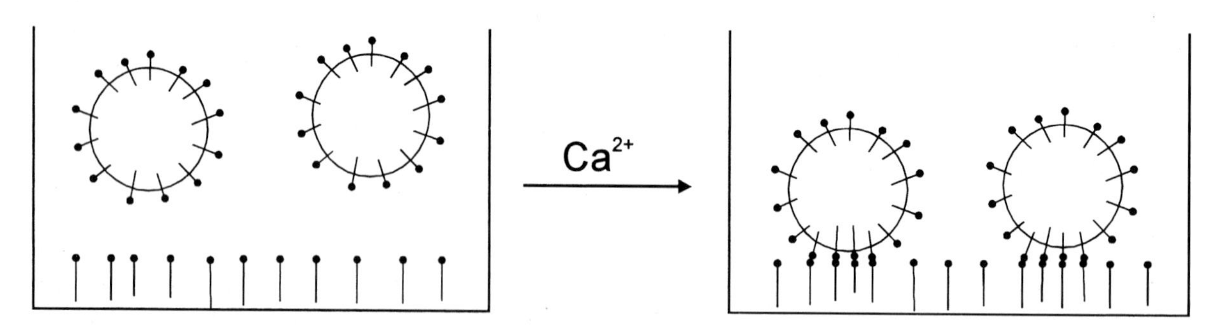

Figure 3. Attachment of Le^x-liposomes to Le^x-coated plastic surface.

of the amount of radioactivity in the extract, which is proportional to the value of the interaction under test. As in the previous approaches, the binding can be inhibited by oligosaccharides, EDTA, etc. (Eggens *et al.*, 1989).

3.4 Weak-Affinity Chromatography (Eggens *et al.*, 1989; Zopf and Ohlson, 1990; Kojima *et al.*, 1994)

Weak interactions can be measured with the help of so-called weak-affinity chromatography (Zopf and Ohlson, 1990). The essence of this methodology is that one of the components of the interaction studied (such as glycoside, glycoprotein, oligosaccharide or even cell extract) is immobilized on the column (preferably a high performance one), whereas the second partner is passed through the column and the delay time is determined relative to the known undelayed control. The mathematical theory of this approach is well developed and generally allows calculation of the constants of weak interaction, to which type the carbohydrate-carbohydrate interaction belongs. This approach has one undoubted advantage, because it allows measurement of the dynamic interaction, and carbohydrate-carbohydrate mediated cell adhesion is mostly manifested only during dynamic cell interaction (see below).

3.5 Equilibrium Dialysis Method is Performed in the Following Way (Eggens *et al.*, 1989)

Liposomes bearing the first of the interacting partners are placed into the dialysis chamber together with the radioactively labeled oligosaccharide, which is the second partner. Due to the specific interaction, the elution of oligosaccharide from the cell is delayed compared to the control (an oligosaccharide of the same size but incapable of interaction). It seems that this approach is the only one of the methods mentioned that gives the possibility of studying the involvement of a monovalent partner in the carbohydrate-carbohydrate interaction.

3.6 Enzyme-Linked Immunoassays

All the methods described above are primarily designed for the study of glycolipid-glycolipid interaction. To study the carbohydrate-carbohydrate interactions, in which polysaccharides are involved, and to model glycoprotein-glycolipid and glycoprotein-glycoprotein interactions, solid phase ELISAs have been examined using synthetic glycoconjugates (neoglycoconjugates) as the major components (Figure 4). Polysaccharide-carbohydrate interaction (Bovin *et al.*, 1995) was studied using zymosan from yeast (a water-insoluble complex of glucan with mannan and a small amount of protein) as the first partner and water soluble polyacrylamide-based neoglycoconjugates (Sug-PAA, Bovin *et al.*, 1993; Bovin and Gabius, 1995) as the second partner. The zymosan suspension was incubated with the solution of biotinylated Sug-PAA, followed by sedimentation into the wells of the 96-well polystyrene plate. After removal of the unbound neoglycoconjugate by washing, the material bound to zymosan could be detected quantitatively using an enzymatic reaction with the streptavidin-peroxidase conjugate (Figure 4a). A biotinylated natural glycoprotein can be used for the reaction with polysaccharide instead of a neoglycoconjugate. The interaction of zymosan with Sug-PAA-biotin is reversible and can be inhibited by low molecular weight saccharide, glycoprotein, soluble polysaccharide, or EDTA. There is a simplified version of this approach: an analog of Sug-PAA-biotin bearing a fluorescent label, Sug-PAA-flu, is used; after the incubation of zymosan with Sug-PAA-flu, the zymosan-neoglycoconjugate complex is destroyed and the label is quantified with the help of a fluorimeter (Bovin *et al.*, 1995).

Two completely artificial model systems have also been tested. The interaction of Le^x-PAA applied to polystyrene with soluble Le^x-PAA-biotin (Figure 4b, the model of glycoprotein-glycoprotein interaction) was studied in the first system. A synthetic analog of Le^x-glycolipid was coated on polystyrene in the second system and its interaction with Le^x-PAA-biotin was measured (Figure 4c, the model of glycolipid-glycoprotein interaction).

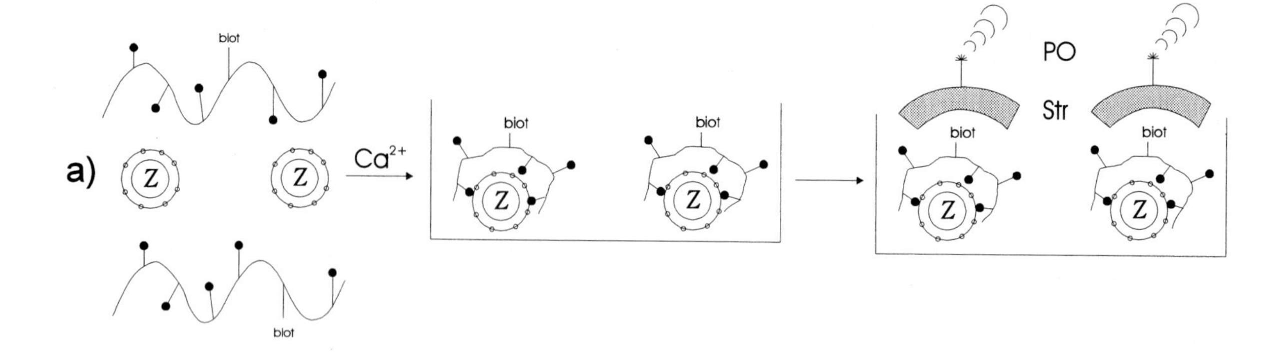

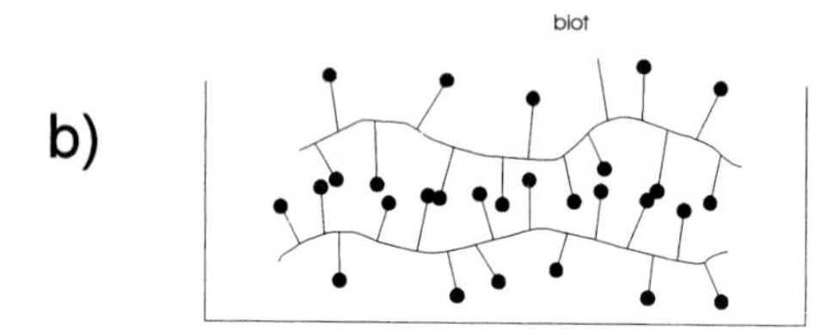

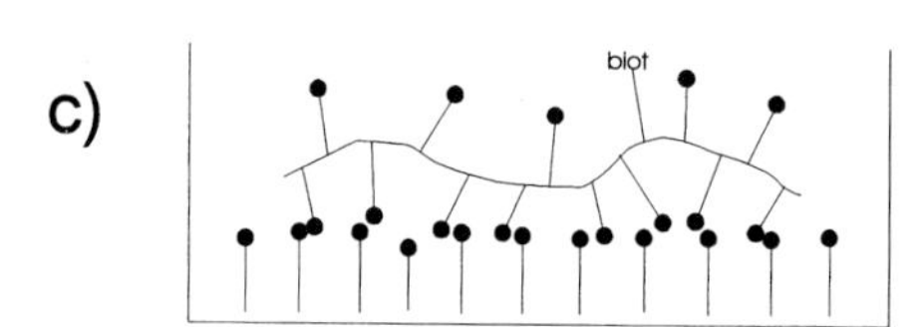

Figure 4. *(a)* Interaction of zymosan (Z) with biotinylated neoglycoconjugate, followed by visualization of the sedimented complex by streptavidin (Str)-peroxidase (PO); *(b)* binding of adsorbed neoglycoconjugate with a second neoglycoconjugate in solution, as a model of glycoprotein-glycoprotein interaction; *(c)* binding of adsorbed neoglycolipid with polyvalent neoglycoconjugate as a model of glycolipid-glycoprotein interaction.

3.7 Langmuir-Blodgett Technique

This technique, developed in colloid chemistry, enables measurement of the force between two solid surfaces in water as a function of their separation, with an accuracy of 0.1 Å. Trisaccharide Le^x was synthesized and connected to a long alkyl triple tail. Two surfaces were covered with neoglycolipid and the forces acting between them were measured using Langmuir balances in the presence and absence of calcium ions. When nucleosides (A and T) were coated on the surfaces, a positive interaction was observed (Pincet *et al.*, 1994) but no interaction was found in the case of Le^x neoglycolipid (Le Bouar *et al.*, 1994). It is interesting that in both cases no effect was observed when synthetic trisaccharide Le^x was used for measurement of Le^x-Le^x interaction. Perhaps the recognition of natural Le^x glycolipids requires not only the trisaccharide determinant Fucα1–3(Galβ1–4)GlcNAc but also the inner core Galβ1–4Glc.

4 Molecular Nature of the Carbohydrate-Carbohydrate Interaction

The following examples of carbohydrate-carbohydrate interaction are known at present:

1. Le^x – Le^x — Hakomori, 1994a
2. GM_3 – Gg_3 — Hakomori, 1994a
3. (KDN)GM_3 – Gg_3 — Yu Song *et al.*, 1994
4. Le^y – H(type 1) — Fenderson, 1993
5. Le^y – H(type 2) — Fenderson, 1993
6. GM_2 – GM_3 — Hakomori, 1995
7. GM_3 – Lac — Hakomori, 1994a
8. H – H — Hakomori, 1994a
9. Gal-3 – HSO_3Gal — Stewart and Boggs, 1993
10. Le^x – H(type 2) — Hakomori, 1994a
11. GM_3 – Gb_4 — Hakomori, 1994a
12. GM_3 – Gb_3 — Hakomori *et al.*, 1991
13. Glucan – Man — Bovin *et al.*, 1995

Moreover, it is necessary to note the interaction of glycophorin with glycolipids (Endo *et al.*, 1982) (see above) and the interaction of two glycoproteins – laminin and gp120/140 (a glycoprotein of mice melanoma cell line B16-F10) with each other (Chammas *et al.*, 1991), although the nature of the interacting carbohydrate epitopes has not been established in the last example. In most cases, the carrier of oligosaccharides is a glycolipid, this being quite understandable because glycolipids form flat surfaces (such as membranes, liposomes or similar assemblies) with high oligosaccharide density, which is an ideal construction for the manifestation of a weak carbohydrate-carbohydrate interaction. Generally, the work of S. Hakomori and co-workers has demonstrated that the higher the oligosaccharide density on the membrane (the closer the glycolipids are situated on the membrane), the stronger the carbohydrate-carbohydrate interaction, indicating the dependence of the interaction on cooperativity. Obviously, it is not possible to find conditions for glycoproteins as favorable as those for glycolipids tightly arranged on a flat membrane, as it is difficult to imagine a better situation than the contact of two congruent surfaces (e.g. two planes). Even in the case of the mucin type of glycoprotein glycosylation (when oligosaccharides are attached to several neighboring residues of serine or threonine), oligosaccharides, though possibly not separated by significant distances, are placed at a considerable angle to each other. This arrangement does not help to establish multivalent recognition of another glycoprotein or glycolipid assembly. This fact can explain the low number of examples of carbohydrate-carbohydrate interaction in which glycoproteins are involved. However, one example of such an unambiguously proven carbohydrate-carbohydrate interaction is known (Kojima *et al.*, 1994), namely the interaction of embryoglycan from F9 cells (a highly multivalent glycoprotein, the major carrier of Le^x epitopes in this cell line) with other Le^x-containing glycoconjugates. It seems that in natural conditions one would only detect the glycoprotein-mediated interaction when several heavily glycosylated proteins form highly ordered assemblies with a high density of regularly arranged identical carbohydrate epitopes on the cell surface. One more example of a non-glycolipid arrangement of carbohydrate epitopes which can be involved in carbohydrate-carbohydrate interaction is a polysaccharide. A high degree of repetition of identical epitopes is characteristic of polysaccharides, while chain spiralization can be of such type that all these epitopes are directed to one side. If several polysaccharide chains are closely located in parallel, a surface can be established similar in principle to that established by glycolipids on a membrane. Such a situation can be realized in nature on the polysaccharide walls of bacteria and yeast, and the only example of the carbohydrate-carbohydrate interaction of this type is the interaction of zymosan (yeast cell wall fragment) with mannose-containing glycoconjugates (Mikhalchik *et al.*, 1994; Bovin *et al.*, 1995) (see the details below).

According to affinity values, Hakomori (Hakomori, 1994a) described strong (e.g. Le^x-Le^x, GM_3-Gg_3), moderate (Le^y-H, GM_3-Lac), weak (Le^x-H, GM_3-Gb_4) and repulsive interactions (Le^y-Le^y, GM_3-GM_3). This sequence reflects only relative affinities obtained during the comparison of the partners in the same experimental conditions (identical test systems); absolute values of the affinity constants are not given in any of the cited reports. Taking into consideration the obligate multivalency of binding (surfaces but not single molecules interact in typical cases), one can imagine the principal problems during the K_a value determination. Even when it is possible to determine the actual number of molecules participating in the interaction, the approximation of K^n_a of the multimeric interaction to K^1_a of the singular interaction loses its physical sense. When comparing carbohydrate-carbohydrate interactions with, for example, protein-protein or protein-carbohydrate ones, the first type of interaction certainly belongs to a "weak field", i.e. they represent the weakest type of specific biological recognition. Thus, the first characteristic feature of the carbohydrate-carbohydrate interaction is extremely low affinity; the interaction at the monovalent level is effectively undetectable. The second feature is the fact that in spite of extremely low affinity the interaction is nevertheless a specific one. Comparing the pairs listed above, one can see that the addition or removal of one monosaccharide residue can basically change not only the value (GM_3-Gg_3 → GM_3-Gb_4) but the interaction sign (GM_3-Lac → GM_3-GM_3). Moreover, an analogous role can be played not only by a monosaccharide residue but by a single hydroxyl group (GM_3(Ac) →GM_3(Gc)). Especially revealing are the experiments on the inhibition of multivalent interaction (e.g. of liposome aggregation) by monovalent oligosaccharides (Eggens *et al.*, 1989), when only Le^x oligosaccharide manifests a significant inhibitory effect, while the isomeric oligosaccharide H (see Table 1) is practically inactive. The experiments on inhibition specificity show unambiguously that monomers really do take part in the interaction, since it would not be possible to interpret the inhibition phenomenon in any other way. One more characteristic feature of the carbohydrate-carbohydrate interaction is its dependence on Ca^{2+} (Eggens *et al.*, 1989), or, to be more exact, its dependence on the bivalent ions calcium, magnesium and manganese. Although, in some cases, the addition of EDTA to the reaction medium does not abolish the interaction completely, Ca^{2+} is still necessary in typical cases. It is well known that calcium ions are capable of forming co-ordination bonds with hydroxyl groups of carbohydrates. This interaction is notably specific, as shown in an example of isomeric trisaccharides, which bound Ca^{2+} in a significantly different way (Fura and Leary, 1993). It seems that bivalent cations play the role of a bridge binding two oligosaccharides simultaneously (e.g. Le^x and Le^x). Such binding saturates the co-ordination capabilities of calcium. If oligosaccharides themselves are capable of additional interaction between each other (independently of the ion), a positive co-operative effect takes place, the metal-dependent and metal-independent components of the interaction enhancing each other. Minimization of Le^x oligosaccharide energy with the help of the HSEA method (GESA program) has been performed (Kojima *et al.*, 1994), followed by the minimization of a complex of two oligosaccharides, and evaluation of the most probable co-ordination sites with calcium (two binding sites have been found). According to the authors' model one side of the Le^x oligosaccharide molecule is significantly more hydrophobic than the other one; geometrically, hydrophobic sites of the two molecules are complementary and can form a complex in which the bound water is excluded, leading to energetic gain (due to the entropy factor). The structure of the dimer permits attachment of one or two calcium ions bridging the primary complex (Fig. 5a), and making it more stable without distortion (bivalent cation co-ordinates with two hydroxyls of each oligosaccharide). According to the model of Suizdak *et al.* (1993), the formation of the complex between one oligosaccharide and one ion takes place first, followed by the attachment of the second oligosaccharide. Taking into consideration the tendency of calcium to co-ordinate with saccharides, this sequence of events (Fig. 5b) seems to be most probable. It is interesting that, if oligosaccharides are replaced by glycolipids in the complex, the lipid "tails" are directed to the opposite sides (Fig. 5c), this topography correlating ideally with the suggested arrangement in the interacting membranes. The real composition of the complex can be shown only after the decoding of X-ray structure, if a derivative of Le^x in presence of Ca^{2+} ion could be crystallized.

5 Cell Interaction: Static and Dynamic Integration with Other Adhesion Mechanisms

Several general mechanisms of animal cell recognition and adhesion are known: carbohydrate-carbohydrate, protein-carbohydrate (selectin-mediated) and protein-protein (integrin-mediated) ones. Binding affinity increases from the first mechanism to the third one, whereas the velocity increases in the opposite direction. It is supposed by Kojima *et al.* (1994) that the sequence of events finally leading to the close cell interaction is the following: very fast though labile primary binding takes place at first, due to relatively weak carbohydrate-carbohydrate and selectin-carbohydrate cell contacts. Weak multipoint interaction enables a single cell to roll along a row of other cells ("rolling") searching the most complement-

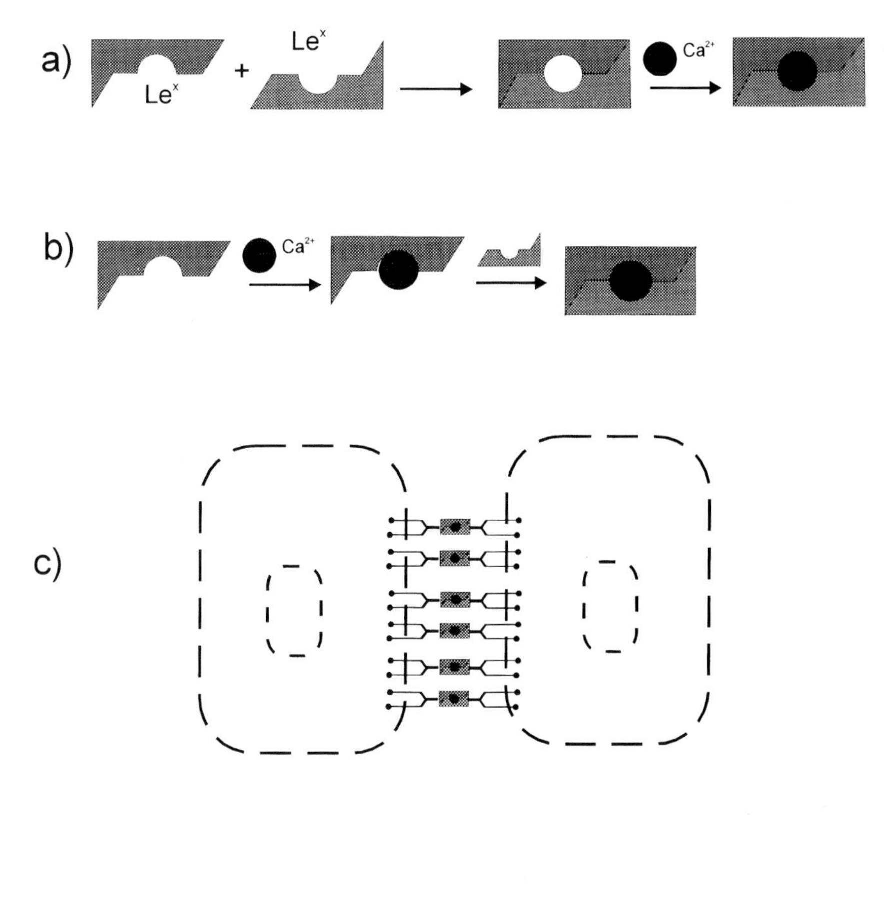

Figure 5. The mechanism of interaction of Le^x with Ca^{2+}: *(a)* Le^x interacts with Le^x followed by the ion attachment; *(b)* the ion attaches to the first molecule of Le^x followed by complex formation with the second molecule of Le^x; *(c)* the topography of Le^x glycolipid interactions is such that the hydrophobic lipid "tails" are directed strictly in opposite directions.

ary partner, or a pair of cells to "palpate" the surfaces of each other searching for the receptor-counterreceptor sites on the membrane with the highest affinity. It is obvious that the primary binding should not be too strong, to permit easy breakage of contacts with a wrong or sub-optimal partner, so that contact with a new partner can be repeated until the optimal contact is established. An analogy can be found in clothes fasteners. Weak interactions resemble a burdock fastener where the exactness of contact is not necessary, and the fastener can be clasped and unclasped rapidly and without effort, whereas the strong cell adhesions of the receptor type resemble a fastened button, where exact penetration is necessary, requiring a long time although the fastening is relatively tight.

Relatively weak carbohydrate-mediated interactions are the ideal ones for primary recognition. Moreover, they prevail in protein-protein processes in dynamic systems (Kojima *et al.*, 1992), for example during the interactions of leukocytes or tumor cells (during metastasis) with endothelial cells in the blood stream. This has been shown by elegant experiments *in vitro*, when one cell type has been forced to move along another one or along a surface coated with individual (purified) adhesion molecules.

Adhesion of B16 cells, based on the interaction GM_3/Gg_3 or GM_3/LacCer (see "Experimental Approaches" and "Molecular Nature of Carbohydrate-Carbohydrate Interaction"), is quite different in static and dynamic modes (Hakomori, 1994b). In a dynamic system, cell suspension was pumped through a narrow passage. Its glass surface was coated with adhesion molecules. The conditions were selected for maximal resemblance to the microvascular environment in which accumulation of tumor cells takes place. In contrast to the static system, the degree of adhesion in the dynamic system depended heavily on time: integrin-dependent adhesion (requiring a longer incubation time than carbohydrate-carbohydrate-mediated adhesion in static conditions) became minimal in dynamic conditions, whereas GM_3-Gg_3-dependent adhesion remained unchanged. Carbohydrate-carbohydrate interaction also prevailed in lectin-carbohydrate interaction in dynamic conditions, whereas the opposite was observed in static conditions. The experiments on the adhesion of B16 cells in dynamic conditions (including inhibition with oligosaccharides and antibodies) permitted the conclusion (Hakomori, 1994b) that attachment of melanoma cells to endothelium *in vivo* was based just on GM_3-LacCer interaction, and that it was the first stage of metastatic deposition. This process launches a cascade of reactions leading to the activation of endothelial cells, expression of selectins and Ig-family receptors, and an increase in adhesion and migration of tumor cells (Kojima *et al.*, 1992). Clearly, synergism is characteristic for cell adhesion mediated by carbohydrate-carbohydrate and protein-protein recognition, since the joint effect is considerably larger than the simple sum of the effects caused by these two processes taken separately (Kojima and Hakomori, 1991b). It should be noted especially that the carbohydrate-carbohydrate mechanism of primary adhesion of metastasizing cells does not require the activation of endothelial cells, in contrast to the selectin-mediated mechanism. The carbohydrate-carbohydrate mechanism (e.g. the Le^x-mediated one) can help tumor cells to form microemboli, further activating endothelial cells. Thus, the following series of steps can be envisaged: (1) tumor cells attach to non-activated endothelial cells or undergo self-aggregation due to carbohydrate-carbohydrate interaction; (2) after activation by tumor cells, endothelial cells and platelets express on their surface E- and P-selectins; (3) molecules containing $SiaLe^a$ and $SiaLe^x$ (selectin ligands) located on the surface of tumor cells form clusters; (4) the process of adhesion is completed by high affinity binding of integrins with Ig-family receptors. The inhibition of carbohydrate-carbohydrate interaction seems to be a possible future approach for blocking the metastatic process (Hakomori, 1994a).

It seems that carbohydrate-carbohydrate interactions of the repulsive type (such as GM_3-GM_3 and Le^y-Le^y) could also be the basis for some cellular social processes during development and morphogenesis, and it may be a crucial mechanism for understanding cleft formation, demarcation, etc. (Kojima and Hakomori, 1991b).

6 Polysaccharide-Carbohydrate Interaction

Carbohydrate-carbohydrate interaction, where a neutral polysaccharide takes part, fills a special place in the range of the examples mentioned above and requires special examination. This type of interaction was established during the study of respiration "explosion" of macrophages under the action of zymosan (Mikhalchik *et al.*, 1994). It was found that the conjugate of mannose with polyacrylamide Man-PAA bound both macrophages and zymosan. Systematic study of the binding of zymosan and its components with glycoconjugates using different approaches (see "Experimental Approaches") showed that the binding was based on Ca^{2+}-dependent carbohydrate-carbohydrate interaction. As zymosan is a complex of mannan, glucan and protein, it has been primarily necessary to clarify whether the protein part of zymosan is a lectin. To do so, zymosan has been treated with proteolytic enzymes, alkali and periodate. Strong alkali and enzymes did not influence the ability of the complex to bind with Man-PAA, whereas it was considerably decreased by periodate treatment. Thus, the carbohydrate rather than the protein part of zymosan bound to glycoconjugate. Further study showed that the active part of the polysaccharide complex was represented by glucan (Bovin *et al.*, 1995), a branched polysaccharide where Glcβ1–3Glc was the major bond and Glcβ1–6Glc was

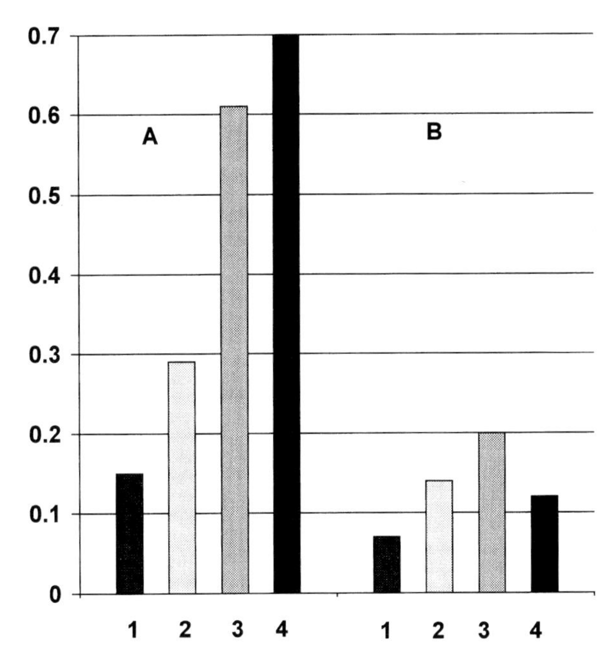

Figure 6. Binding of neoglycoconjugates Sug-PAA-biotin with zymosan in ELISA: (1) Glcβ; (2) Fucα; (3) GlcNAcβ; (4) Manα. (A) in the presence of Ca^{2+}; (B) in the absence of Ca^{2+}.

the minor bond. The binding of zymosan with Man-PAA was inhibited by the addition of EDTA (though not completely) and by mannose and α-methylmannoside (in a dose-dependent manner). The interaction was also inhibited by mannose-rich carbohydrate chains from glycoproteins, and by the glycoproteins bearing these chains. It should be noted that the binding of zymosan with Man-PAA was not as specific as the interaction Le^x-Le^x or GM_3-Gg_3. It can be seen from Figure 6 that the conjugates of other monosaccharides, e.g. Fuc and GlcNAc, also bound to zymosan, although to a lesser extent, whereas the binding could be inhibited by EDTA and mannose (but not by glucose). The binding is reversible: when treating the zymosan/Man-PAA complex with EDTA or Man the glycoconjugate is released from the complex. It is interesting that not only saccharides but also simple polyhydroxyl compounds, e.g. $(HOCH_2)_3C$- (as Tris-PAA-biotin) possess the capability to bind with zymosan. It is possible that the observed limited specificity is related to the irregularity of glucan composition (unlike glycolipids, which are highly regulated on the membrane surface). Different oligosaccharide fragments (or epitopes) of the polysaccharide chain can interact selectively with different polyhydroxyl partners. Possibly, the observed interaction, having low affinity and limited specificity, is a substitute for the specific interaction of polysaccharide with a natural but more complex carbohydrate partner such as glycoprotein or a cluster of glycolipids. Interactions such as those of polysaccharides from the cell wall of bacteria or yeast with mannose-rich proteins of other cells can mediate the primary cellular interactions which are analogous to the glycolipid-mediated recognition of animal cells. It cannot be excluded that the driving force in the formation of polysaccharide complexes such as zymosan, or other fragments of cell wall, is the carbohydrate-carbohydrate interaction (of glucan-mannan type). The possibility of a similar interaction between mannan and dextran has been shown by Kang *et al.* (1995). It is intriguing to speculate that the physiological action of some glucans, such as the proposed anti-tumor and anti-AIDS action, is related to the interaction of glucan with mannose-rich chains of the glycoproteins from tumor cells or viral gp120.

7 Conclusions

In spite of its simplicity, clearness and fruitfulness, the classic lock and key model does not encompass many real mechanisms of biological recognition. This statement relates mostly to the recognition between cells when arrays of molecules situated on large areas of the membrane surface make contact, rather than single molecules. High specificity of recognition can be realized in different ways:

(1) exact complementarity of atomic groups of the two interacting molecules by the lock and key principle, which is characteristic for proteins;
(2) cooperative interaction of two surfaces through a large number of copies of identical molecules (or the repeating motifs of a macromolecule) in the presence of calcium ions, which is characteristic for carbohydrates;

(3) an intermediate way, when a subunit or clustered domains of a membrane protein interacts with a multivalent or clustered ligand, which is characteristic for lectin-carbohydrate recognition. To use an analogy to the dualism of quantum physics, the intercellular interaction appears to encompass both types of processes – strictly determined ("classic") ones, and random ones, delocalized in space ("relativistic"). Carbohydrate-mediated processes tend to be the second type of interaction, in which the relativism is connected both with the mechanism of interaction and with the indistinctiveness of the carbohydrate chain structure, for which structural heterogeneity (especially of glycoproteins) and conformational multiplicity are characteristic.

It is obvious that it is considerably more difficult to study and interpret unambiguously weak (and not always specific, in the usual sense) processes than the habitual protein-ligand interactions. Not surprisingly, weak biological processes have been discovered only recently, and the number of established carbohydrate-carbohydrate interactions is smaller than that of the known protein-protein interactions. It is not surprising that the very fact of the existence of biologically significant carbohydrate-carbohydrate recognition produces some scepticism among glycobiologists. Two factors can be taken to indicate the validity of the concept: the number of publications is increasing, and the results are being reproduced in different laboratories. It is thus reasonable to assume that the study of carbohydrate-carbohydrate interaction has the potential to become a "hot" topic of glycobiology in the very near future.

Acknowledgments

The author acknowledges I.M. Belyanchikov for help in the manuscript arrangement.

References

Acebes JL, Lorences EP, Revilla G *et al.* (1993): Pine xyloglucan. Occurrence, localization and interaction with cellulose. In *Physiol. Plant.* **89**:417–22.

Bovin NV, Gabius HJ (1995): Polymer-immobilized carbohydrate ligands: versatile chemical tools for biochemistry and medical sciences. In *Chem. Soc. Rev.* **24**:419–23.

Bovin NV, Korchagina EY, Zemlyanukhina TV *et al.* (1993): Synthesis of polymeric neoglycoconjugates based on *N*-substituted polyacrylamides. In *Glycoconjugate J.* **10**:142–51.

Bovin NV, Shiyan SD, Mikhalchik EV (1995): New type of carbohydrate-carbohydrate interaction. In *Glycoconjugate J.* **12**:427.

Brewer GJ, Thomas PD (1984): Role of gangliosides in adhesion and conductance changes in large spherical model membranes. In *Biochim. Biophys. Acta,* **776**:279–87.

Chammas R, Veiga SS, Line S *et al.* (1991): Asn-linked oligosaccharide-dependent interaction between laminin and gp120/140. An α6/β1 integrin. In *J. Biol. Chem.* **266**:3349–55.

Coteron JM, Vicent C, Bosso C *et al.* (1993): Glycophanes, cyclodextrin-cyclophane hybrid receptors for apolar binding in aqueous solutions. A stereoselective carbohydrate-carbohydrate interaction in water. In *J. Am. Chem. Soc.* **115**:10066–76.

Dammer U, Popescu O, Wagner P *et al.* (1995): Binding strength between cell adhesion proteoglycans measured by atomic force microscopy. In *Science* **267**:1173–5.

Eggens I, Fenderson B, Toyokuni T *et al.* (1989): Specific interaction between Le^x and Le^x determinants. A possible basis for cell recognition in preimplantation embryos and in embrional carcinoma cells. In *J. Biol. Chem.* **264**:9476–84.

Endo T, Nojima S, Inoue K (1982): Intermolecular interaction between glycolipids and glycophorin on liposomal membranes. In *J. Biochem.* **92**:1883–90.

Fenderson B (1993): Saccharides involved in implantation. In *Trends Glycosci. Glycotechnol.* **5**:271–85.

Fura A, Leary JA (1993): Differentiation of Ca^{2+}- and Mg^{2+}-coordinated branched trisaccharide isomers: an electrospray ionization and tandem mass spectrometry study. In *Analyt. Chem.* **65**:2805–11.

Gooi HC, Feizi T, Kapadia A *et al.* (1981): Stage-specific embryonic antigen involves α1–3 fucosylated type 2 blood group chains. In *Nature* **292**:156–8.

Gupta D, Arango R, Sharon N *et al.* (1994): Differences in the cross-linking activities of native and recombinant *Erythrina corallodendron* lectin with asialofetuin. Evidence for carbohydrate-carbohydrate interactions in lectin-glycoprotein complexes. In *Biochemistry* **33**:2503–8.

Hakomori S (1990): Bifunctional role of glycosphingolipids. Modulators for transmembrane signalling and mediators for cellular interactions. In *J. Biol. Chem.* **265**:18713–6.

Hakomori S (1993): Structure and function of sphingolipids in transmembrane signalling and cell-cell interactions. In *Biochem. Soc. Trans.* **21**:583–95.

Hakomori S (1994a): Role of carbohydrates in cell adhesion and recognition. In *Complex Carbohydrates in Drug Research* (Bock K, Clausen H, eds) pp 337–49, Copenhagen: Munksgaard.

Hakomori S (1994 b): Role of gangliosides in tumor progression. In *Biological Function of Gangliosides* (Svennerholm L, ed) Progr. Brain Res. Vol. 101, pp 241–50, Amsterdam: Elsevier.

Hakomori S (1995): Karl Meyer Award Lecture. In XIII International Symposium on Glycoconjugates, Seattle.

Hakomori S, Nudelman E, Levery SB *et al.* (1981): The hapten structure of a developmentally regulated glycolipid antigen (SSEA-1) isolated from human erythrocytes and adenocarcinoma: a preliminary note. In *Biochem. Biophys. Res. Commun.* **100**:1578–86.

Hakomori S, Igarashi Y, Kojima N *et al.* (1991): Functional role of cell surface carbohydrates in ontogenesis and oncogenesis. In *Glycoconjugate J.* **8**:178.

Kang EC, Akioshi K, Sunamoto J (1995): Partition of polysaccharide-coated liposomes in aqueous two-phase systems. In *Int. J. Biol. Macromol.* **16**: 348–53.

Kojima N, Fenderson BA, Stroud MR *et al.* (1994): Further studies on cell adhesion based on Le^x-Le^x interaction, with new approaches: embryoglycan aggregation of F9 teratocarcinoma cells, and adhesion of various tumor cells based on Le^x expression. In *Glycoconjugate J.* **11**: 238–48.

Kojima N, Hakomori S (1989): Specific interaction between gangliotetraosylceramide (Gg_3) and sialosyllactosylceramide (GM_3) as a basis for specific cellular recognition between lymphoma and melanoma cells. In *J. Biol. Chem.* **264**:20159–62.

Kojima N, Hakomori S (1991a): Cell adhesion, spreading, and motility of GM_3-expressing cells based on glycolipid-glycolipid interaction. In *J. Biol. Chem.* **266**:17552–8.

Kojima N, Hakomori S (1991b): Synergistic effect of two cell recognition system: glycosphingolipid-glycosphingolipid interaction and integrin receptor interaction with pericellular matrix protein. In *Glycobiology* **1**:623–30.

Kojima N, Shiota M, Sadahira Y *et al.* (1992): Cell adhesion in a dynamic flow system as compared to static system. Glycosphingolipid-glycosphingolipid interaction in the dynamic system predominates over lectin- or integrin-based mechanisms in adhesion of B16 melanoma cells to non-activated endothelial cells. In *J. Biol. Chem.* **267**:17264–70.

Le Bouar T, Perez E, Zhang YM *et al.* (1995): Direct measurement of interaction Le^x-Le^x. In Philippe Laudat Conference "Carbohydrate mediated cell-cell interactions in inflammation and metastasis". Aix-les-Bains, France.

Mikhalchik EV, Korkina LG, Shiyan SD *et al.* (1994): Effect of neoglycoconjugates on *in vitro* luminol-dependent chemiluminescence of rat peritoneal macrophages. In *Biol. Membrany* **11**:581–7.

Misevic GN, Burger MM (1993): Carbohydrate-carbohydrate interactions of a novel acidic glycan can mediate sponge cell adhesion. In *J. Biol. Chem.* **268**:4922–9.

Misevic GN, Jumblatt JE, Burger MM (1982): Cell binding fragments from a sponge proteoglycan-like aggregation factor. In *J. Biol. Chem.* **257**:6931–6.

Pincet F, Perez E, Bryant G *et al.* (1994): Long-range attraction between nucleosides with short-range specificity: direct measurements. In *Physical Rev. Lett.* **73**:2780–3.

Siuzdak G, Ichikawa Y, Caulfield TJ *et al.* (1993): Evidence of Ca^{2+}-dependent carbohydrate association through ion spray mass spectrometry. In *J. Am. Chem. Soc.* **115**:2877–81.

Stewart RJ, Boggs JM (1993): A carbohydrate-carbohydrate interaction between galactosyl-ceramide-containing liposomes and cerebroside sulfate-containing liposomes: dependence on the glycolipid ceramide composition. In *Biochemistry* **32**: 10666–74.

Turley EA, Roth S (1980): Interactions between the carbohydrate chains of hyaluronate and chondroitin sulphate. In *Nature* **283**:268–71.

Wormald MR, Edge CJ, Dwek RA (1991): The solution conformation of the Le^x group. In *Biochem. Biophys. Res. Commun.* **180**:1214–21.

Yu Song, Kitajima K, Muto Y *et al.* (1994): KDN-gangliosides. Carbohydrate-carbohydrate interaction as a basic molecular mechanism for sperm-egg association. In *XVIIth International Carbohydrate Symposium, Ottawa, Canada. Abstracts.* Abstr. NoC1.44, p 431.

Zopf D, Ohlson S (1990): Weak-affinity chromatography. In *Nature* **346**:87–8.

16 Carbohydrate-Protein Interaction

HANS-CHRISTIAN SIEBERT, CLAUS-WILHELM VON DER LIETH, MARTINE GILLERON, GERD REUTER, JOSEF WITTMANN, JOHANNES F.G. VLIEGENTHART AND HANS-JOACHIM GABIUS

1 Theoretical Aspects of Carbohydrate-Protein Interaction

Monosaccharides are small, rigid, polar molecules that can participate in hydrogen bonding and hydrophobic interactions. Their assembly into oligo- or polysaccharides gives rise to an immense variety of different structures that may be suitable for transfer of information. As discussed in preceding chapters, the interaction of the carbohydrate part of cellular glycoconjugates with proteins is assumed to mediate various physiological functions. Thus, the underlying recognition process must display a high level of accuracy to transmit the stored information. Several chapters in the second part of this book will deal with the biological consequences of the recognition process, whereas this and the following contributions will try to answer the question of how the carbohydrate-protein interaction can be explained at the molecular level. Principally, several questions have to be addressed to understand the driving forces that are responsible for this recognition process: how flexible are saccharides? Can they adopt a large number of different conformations or is their flexibility restricted to only a few possibilities? Is the flexibility the same for all monosaccharide residues within a complex carbohydrate chain? Will the apparent hydrophilicity of saccharides exclude non-polar interactions with protein surfaces? Is there any influence of the solvent on this interaction? Can we develop appropriate models to explain the actual high specificity of carbohydrate-protein interactions? Do we have tools to test the validity of these theoretical models?

1.1 Carbohydrate and Protein Flexibility

Pyranose rings that almost exclusively form the carbohydrate chains of cellular glycoconjugates normally adopt a well defined chair conformation, either $^{1}C_4$ or $^{4}C_1$, depending on the nature of the individual sugar, without flipping from one form to the other. The energy barrier between the two different ring conformations for D-glucopyranose is 42 kJ/mol (Joshi and Rao, 1979), exceeding the value of 12.5 kJ/mol which can be overcome by thermal motion at the normal body temperature. Flexibility within such a monosaccharide is therefore restricted to rotational freedom of all exocyclic parts, i.e. CH_2OH and OH groups. Thus, the glycosidic linkages of hexopyranose residues, also being exocyclic bonds, confer flexibility on the chain structure. This aspect contributes markedly to the overall spatial variability for oligo- or polysaccharides. The different conformations that can be adopted by rotation of substituents around these bonds are defined by the angles ω, Φ, and ψ that are formed, for example, by H5'-C5'-C6'-O6', H1'-C1'-O1'-C4, and C1'-O1'-C4-H4, respectively, corresponding to rotations around the C5'-C6', C1'-O1' and O1'-C4-linkages, as illustrated in Figure 1 for lactose.

Due to the rigidity of each monosaccharide ring, and the inherent flexibility of the exocyclic bonds, different models can be devised theoretically to understand carbohydrate-protein interaction, by assuming either a low or a high level of flexibility of the carbohydrate chain. In the first model, saccharide chains are highly flexible, thereby permitting a considerable number of conformations of a given saccharide unit with different levels of free energies (local minima)

H.-J. and S. Gabius (Eds.), Glycosciences

ISBN 3-8261-0073-5

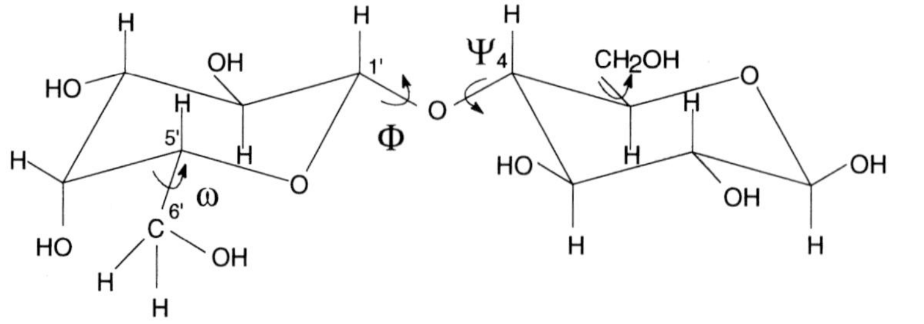

Figure 1. Possible flexibility of saccharide structures by rotation of exocyclic bonds. Dihedral angles Φ, ψ are formed by a vector connecting the first two atoms and a plane determined by the second, third and fourth atom. Dihedral angles Φ and ψ are defined as follows: Φ: H1-C1-*O*-CX and ψ : C1-*O*-CX-HX. Φ is considered to be 0°, when the C1-C2- or the H1-C1-bond eclipse the *O*-CX-bond. Φ corresponds to Φ_H, ψ corresponds to ψ_H according to the IUPAC conventions. ψ is considered to be 0°, when the C1-*O*-bond eclipses the CX-HX-bond. Angles Φ and ψ have values between –180° and 180°, and the positive sense of rotation about a given chemical bond corresponds to clockwise motion of the group of several atoms pendant on the atom defining the remote end of that bond, when the pendant atoms are viewed along the bond (Brant and Christ, 1990).

(Homans, 1993a; Imberty *et al.*, 1993; Perez *et al.*, 1994). One out of this pool is ideally suited to bind to the protein receptor and is thus selected. The inherent flexibility, together with the relatively low activation energy needed for interconversion of the different forms, ensures that the ligand population is not deprived of the receptor-binding conformer. The level of affinity to the receptor and its specificity should correlate with the number of functional groups on different monosaccharide residues within one oligosaccharide that are involved in recognition.

This model can be logically extended to allow spatial adaptation if even the most suitable conformer does not fit completely to the binding site of the protein. Upon interaction with the protein the carbohydrate's conformation may be modified by provoking an "induced fit". It can be postulated that the energy needed for receptor-mediated distortion of the saccharide structure would be supplied by a gain in energy upon binding (Imberty *et al.*, 1993). In the second model, the carbohydrate structure is considered to be rigid, yielding only one conformer without the possibility of an induced fit to the protein structure. Under these circumstances, only the protein can exhibit conformational flexibility to accommodate the rigid saccharide chain (Carver *et al.*, 1989; Carver, 1993).

Since it is clearly the aim to describe biological systems as completely as possible, the natural environment of the binding partners has to be considered. As a consequence, it is an obvious drawback of both models at this stage that the nature of the solvent and its specific interaction with the sugar and the protein are neglected.

1.2 Thermodynamic Parameters of Carbohydrate-Protein Interaction

Before considering how to obtain the experimental evidence needed to decide between these models, some general considerations should be made of the parameters that might change during the recognition process. An interaction process between a carbohydrate ligand and a protein receptor will affect common thermodynamic parameters such as enthalpy and entropy, from which the free energy can be calculated according to

$$(1)\quad \Delta G = \Delta H - T\Delta S$$

This change in free energy © must be negative to let a chemical reaction proceed, as is known from basic thermodynamics. A transformation of equation (1) to

$$(2)\quad -\Delta H = -\Delta G - T\Delta S$$

suggests plotting the values for -ΔH against -TΔS. As can be deduced from equation (2), a straight line with a slope of 1 and -ΔG as intercept should result (Carver *et al.*, 1989; Sigurskjold and Bundle, 1992).

Performing this computation for one protein receptor and one appropriate carbohydrate ligand with thermodynamic parameters measured at different temperatures (Sigurskjold and Bundle, 1992), or for different receptors with corresponding ligands at one temperature (Carver *et al.*, 1989; Lemieux *et al.*, 1991; Toone, 1994), a constant value for ΔG at about – 21 kJ/mol is obtained, suggesting that the total driving force of all analyzed carbohydrate-protein interactions is quantitatively the same. Deviations from the theoretically expected slope of 1 to slightly higher values are observed for extended ligand structures. This result indicates an increase in affinity of the more complex ligands relative to smaller ones (Carver *et al.*, 1989).

From the simple mathematical model of equation (2) it is also obvious that if $-\Delta G$ is constant $-\Delta H$ is proportional to $-T\Delta S$. In other words, any change in $-\Delta H$ must apparently be coupled to a corresponding change in $-T\Delta S$. Thus, a binding reaction that occurs despite of an unfavorable loss in entropy is driven by an advantageous loss in enthalpy. In the converse situation, an unfavorable enthalpy change may well be compensated by the entropic factor.

On the basis of a limited number of examples that do not, at present, allow one to draw final conclusions, it was shown that these thermodynamic values differ slightly when lectins and antibodies are compared. A slope of less than one and a more negative intercept were seen in the case of certain antibodies (Toone, 1994). However, most values measured so far are sufficiently close to each other that similar models of interaction can be assumed for lectins (Reeke and Becker, 1988) and for carbohydrate-specific antibodies (Thurin, 1988). These principal considerations can also be applied to describe the recognitive interplay between antibodies and peptide antigens (Braden and Poljak, 1995).

After this rather general and formalistic discussion of the thermodynamic parameters that are expected to change during carbohydrate-protein interaction, we can proceed to apply our consideration to the molecular level. All systems tend to increase their entropy by a decrease in order. A high level of molecular flexibility and thus a large number of possible conformers of protein or carbohydrate will necessarily establish a rather large amount of entropy if no kinetic barrier for the interconversion of the different conformers exists. Binding of a carbohydrate ligand to a protein receptor immediately reduces conformational flexibility of both partners and consequently results in a loss of entropy. Hence, it is hardly conceivable that specific recognition can be achieved with two completely flexible molecules, unless pronounced enthalpic forces occur that compensate for the drastic loss of spatial flexibility and thus of entropy. Since the entropy of the whole system is considered, it is at present not important whether it is primarily the protein or the carbohydrate part that is reduced in flexibility.

From the considerations made above, it is clear that the expected loss in flexibility (i.e. entropy) upon binding of carbohydrate and protein is overcome by a corresponding loss in enthalpy, rendering the reaction more exothermic. The greater the loss of flexibility in this interaction, the more reaction heat must be formed. This obvious reasoning leads to an inspection of the types of molecular interactions between carbohydrate and protein that influence the extent of change in enthalpy. Strong ionic interactions will not be found very often in glycosciences, since saccharides are uncharged molecules, with the exception of uronic acids, sialic acids, carbohydrate sulfates or phosphates, and ketodeoxynonulosonic acid (KDN), as well as 3-deoxyoctulosonic acid (KDO) in the bacterial endotoxins. Polar interactions such as hydrogen bonding will usually be observed instead. The carbohydrates' hydroxyl functions that are present at almost every carbon atom, and the amino functions that are mostly acylated, are ideally suited to form bridges to the side chains of all charged or hydrophilic amino acids. Among others, aspartate, asparagine, glutamate, glutamine, threonine and serine residues of lectins or antibodies have been recognized as hydrogen bonding partners of OH groups of saccharides (Quiocho, 1986, 1989; Delbaere *et al.*, 1990). If we assume a well defined carbohydrate structure with the functional groups in a distinct conformational arrangement, depending on the nature of the monosaccharides and the type of glycosidic linkages, the protein receptor must have a set of appropriate (i.e. complementary) amino acids in its binding pocket to allow accommodation of the ligand by this hydrophilic interaction mechanism.

A close look at the panel of carbohydrates present in glycoconjugates suggests, however, that the carbohydrate-protein interaction may not be based exclusively on hydrophilic forces. L-Fucose, for example, carries a 6-deoxy function, which means that it has a reduced capacity for hydrophilic interaction (Lemieux, 1989; Lemieux *et al.*, 1991). This monosaccharide residue is a common part of biologically important structures such as Lewis blood group epitopes. Similarly, abequose (a 3,6-dideoxyhexopyranose), as part of a *Salmonella* antigen, has been found to interact with tryptophan, histidine and phenylalanine residues of a monoclonal antibody (Sigurskjold *et al.*, 1991). Besides this peculiarity of deoxy sugars, one has to take into account that each C-atom not only carries an OH-function but also a hydrogen atom, which forms a C-hydrogen bond of low polarity. Depending on their orientation within a given monosaccharide, an area with hydrophobic character may be formed that can even expand over several residues within a carbohydrate chain. As prerequisites for establishment of this structural feature, the other monosaccharides should also display hydrophobic areas, and the angles Φ and ψ that define the glycosidic linkages must permit such an arrangement, as seen, for example, for the Lewis b type human blood group determinant and its recognition by *Griffonia simplicifolia* lectin (Lemieux, 1989; Delbaere *et al.*, 1990), or the H-type determinant that interacts with the *Psophocarpus tetragonolobus* lectin (Acharya *et al.*, 1990; Lemieux *et al.*, 1994). In contrast to a superficial view of carbohydrates as entirely polar substances, a more detailed inspection reveals that hydrophobic as well as van der Waals interactions between carbohydrate and protein can add to the mechanism of interaction, and thereby contribute to the observed specificity of this process (Lemieux, 1989; Quiocho, 1989). On the basis of the overall properties of carbohydrates one could, however, hardly imagine binding to a protein without involvement of hydrophilic interactions. It is thus reasonable to assume that the additional possibility of hydrophobic contacts should permit fine tuning of the recognition. Up to this point, any contribution of the solvent has not been taken into account. We therefore expand our considerations to discuss the possible roles of the solvent, which is water for all biological systems. Despite the presence of hydrophobic areas on protein and carbohydrate, the characteristics of both types of compounds will be dominated by hydrophilic properties as long as they exhibit biological functions as soluble molecules. As a consequence, they carry water molecules attached to them by hydrogen bonding. This hydration sphere can even extend to form what may be called a cave of highly ordered water molecules. Due to the strong polar interactions between, for example, the saccharides' hydroxyl functions and the water molecules, there will be a gain in enthalpy but a concomitant loss in entropy for both: for the carbohydrate, forming a larger, ordered and thus less flexible structure than without water, and for the water molecules, which would have a higher degree of flexibility in bulk water than when associated with the saccharide. The formation of such a solvation sphere for the ligand, which is equally conceivable at least for parts of the binding site of the receptor, will strongly be influenced by their individual functionalities. The hydrophobic areas of carbohydrates, for example, are hydrated to a significantly lesser extent than the hydrophilic areas. What would now be expected to happen if a hydrated saccharide approaches a hydrated protein? On the basis of the required specificity of interaction transmitted by the sum of hydrophilic, hydrophobic and van der Waals contacts, a great number, if not the majority, of water molecules in the vicinity of the interacting areas have to be removed from the hydration sphere because they hide those contact sides; with the water in place only very loose unspecific interactions would be possible. Water repulsion during the establishment of specific binding between carbohydrate and protein should result in a favorable change in enthalpy. Concomitantly, the entropy of this complex may decrease, because its mobility may be lower than that of the solvated individual binding partners. For the water molecules, however, being released from the ordered hydration cave to the less ordered bulk water, a significant gain in entropy can be expected (Lemieux, 1989; Lemieux *et al.*, 1991; Sigurskjold and Bundle, 1992; Imberty *et al.*, 1993; Beierbeck *et al.*, 1994). A further important role of the solvent may be that a distinct, spatially fixed water molecule can be used as a molecular bridge between a receptor and a ligand (Lemieux, 1989; Lemieux *et al.*, 1991; Imberty *et al.*, 1993).

In summary, all models considering exclusively the direct interaction between carbohydrate and protein will give a loss in entropy due to reduced flexibility and a compensatory loss in enthalpy upon binding, to yield an unchanged value of ΔG for the interaction. Entropically driven carbohydrate-protein interactions can easily be imagined only by involvement of the solvent. All these models and factors which might possibly influence carbohydrate-protein interaction seem to be reasonable at a theoretical level. We now have to consider what computational and experimental approaches can be designed to test which of these models is valid, or if a satisfactory explanation requires integration of several of the models developed above.

2 Modeling of Carbohydrate-Protein Interactions

Up to the present, X-ray crystallography of proteins is the major tool for studying carbohydrate-protein interactions at the molecular level, as discussed in detail in the accompanying chapter. Most of our experimental information in this field is derived from corresponding studies forming a database for application of modern computational techniques.

Such computational methods can be used for: (1) refinement and detailed analysis of 3D-structures of carbohydrate-protein complexes experimentally determined by X-ray studies, because protein crystallography is unable to locate hydrogen atoms. This drawback is overcome by computer-assisted modeling techniques. Thus, the formation of hydrogen bonds, hydrogen bond networks and their energetic contributions for binding processes can be evaluated; (2) detailed analysis of the thermodynamic parameters of carbohydrate-protein interactions at the molecular level, if the binding site for the ligand and the mode of binding are known (Imberty *et al.*, 1994; Bundle *et al.*, 1995); (3) modeling of the docking of carbohydrates into appropriate binding sites of the protein (Loris *et al.*, 1994) with possible changes of the oligosaccharide structures upon binding; the procedure includes the evaluation of geometrical and energetic consequences of the complex formation; (4) constructing reasonable 3D-structures in families of related proteins on the basis of the knowledge-based homology modeling technique, when no experimental 3D-data are available (Imberty and Pérez, 1994); and (5) exploring the presence of binding sites within proteins, when the location of these sites and the mode of binding are not known (Nyholm *et al.*, 1995).

When 3D-structures are available from X-ray crystallography and/or NMR measurements of the sugar receptor (e.g. a lectin), as well as for the ligand of a carbohydrate-protein complex, these data can be used as starting conformations for a computational approach to define the inherent characteristics of the carbohydrate-protein interaction. Since the number of experimentally solved structures of carbohydrate-protein complexes is increasing steadily, knowledge-based approaches to construct 3D-structures of homologous proteins, where 3D-structures have not yet been determined, will be of increasing importance.

2.1 X-Ray Crystallography: Brookhaven Protein Data Bank (PDB) (Bernstein, 1977)

The 3D-structures of more than one thousand different proteins of various classes are collected in the Brookhaven protein data bank. Less than one hundred entries deal with proteins that can bind carbohydrates. In this context, the structural data of an animal galactose-binding lectin (galectin-1; for details, see chapter by Ohannesian and Lotan, this volume) is one example of interest (Fig. 2). Since a resolution of at least 2.5 Å of the X-ray structures is necessary to identify the loca-

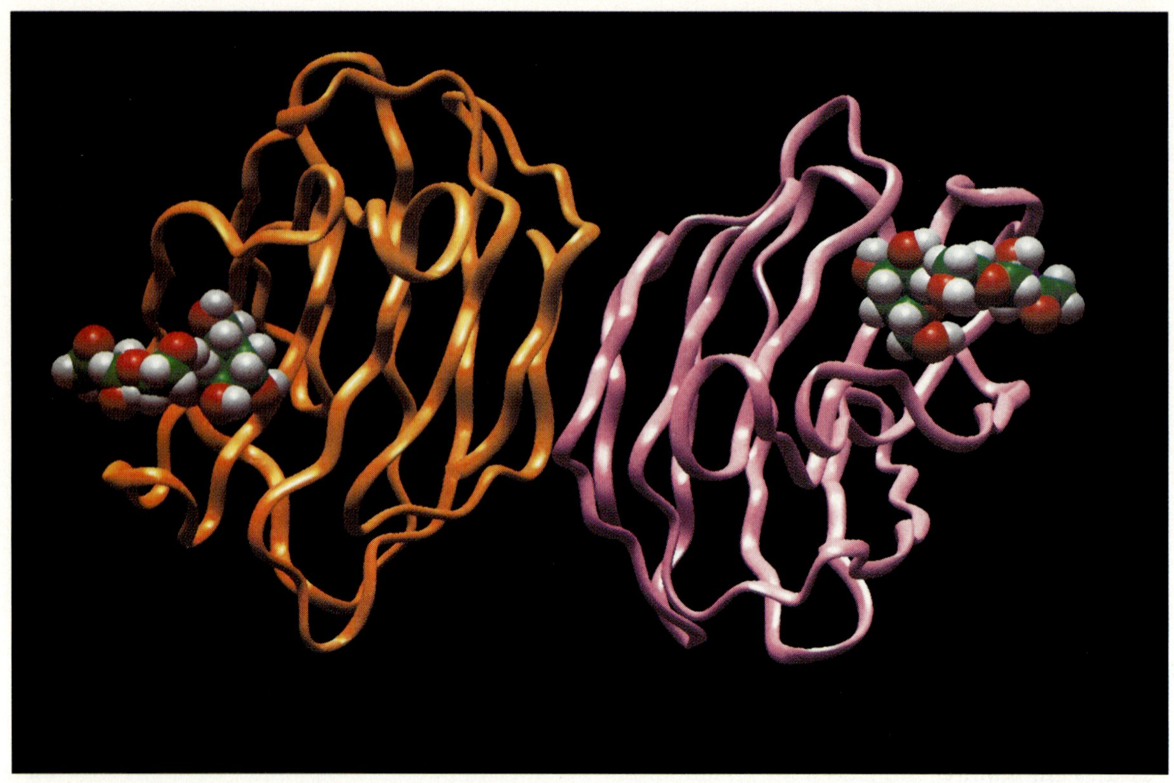

Figure 2. Graphical representation of a 14 kDa bovine spleen S-lectin (galectin-1) complexed with the disaccharide *N*-acetyllactosamine (entry 1SLT taken from the Brookhaven protein data bank) (Liao *et al.*, 1994). The backbone of the lectin is displayed as a ribbon. The bound sugar moieties are visualized as space filling (CPK) models. Two monomers associate to form an extended β-sandwich. Each monomer binds one *N*-acetyllactosamine molecule.

tion and the mode of binding of saccharides with sufficient accuracy, the total number of high quality 3D-structures of carbohydrate-protein complexes is rather limited. Although the number of available high resolution X-ray structures has grown in the recent years (more than 1000 new entries were added to the Brookhaven protein data bank in 1994), the present lack of a reliable 3D-structure can hamper the use of modeling studies.

2.2 Cambridge Structural Database (Allen *et al.*, 1991)

The 3D-structures of crystallized carbohydrates are stored in the Cambridge structural database, where all structures, solved by X-ray diffraction techniques, of small organic molecules have to be entered before publication. The current release contains less than 1000 entries for mono- and oligosaccharides, reflecting the difficulties in crystallizing complex carbohydrates. Thus, for many carbohydrates of biological interest no or only a few X-ray crystal structures are accessible. Therefore, various NMR techniques in combination with computational approaches are often needed to elucidate in detail the conformation of oligosaccharides in solution.

2.3 Knowledge-Based Homology Modeling

Starting from an entry into the protein data bank, the knowledge-based homology modeling technique can be used logically to extend the information to related proteins. When, for example, no experimental 3D-structural data of a lectin are available, further investigation can proceed on the basis of information about a homologous agglutinin. The feasibility of this approach thus requires access to at least one experimentally solved 3D-structure of a receptor, whose sequence allows an extended alignment to the target protein of the modeling study. If the sequence similarity is sufficiently high (40% seems to be the limit in many cases), one has a good chance of deducing a realistic picture of the 3D-structure of the lectin, as well as of its interaction with the non-covalently bound oligosaccharide. A convenient procedure for knowledge-based protein model building will take the following steps (Sutcliffe *et al.*, 1987a,b) into account: (1) the amino acid sequence of the protein under investigation and of the homologous receptors, whose 3D-structures are known, have to be compiled; (2) an alignment of the selected proteins has to be performed in order to find the so-called structurally conserved regions; (3) an average structure of the conserved region(s) is determined and serves as a framework to construct a model of the new structure; (4) the backbone of the conserved region(s) is generated and the amino acid side-chains are added (Sutcliffe *et al.*, 1987a); (5) the various conserved region(s) are connected by variable linkers that are provided by a library, which stores experimentally known loop fragments of different lengths and sequences. The optimal loop which fulfills the requirements of the model protein is found automatically; and (6) the geometry of the model protein is optimized using force-field calculations.

In principle, this strategy seems to be attractive. However, uncertainties remain in many cases about the location of insertions and deletions of amino acids between different sequences (Sali *et al.*, 1990). These ambiguities lower the quality of the modeling of carbohydrate-protein complexes. It should not be overlooked that it has recently been shown that the carbohydrate-binding sites may be located in the regions of greatest variability (Young and Oomen, 1992).

The value of such modeling studies strongly depends on the level of information about the conformation of the binding partners, and the nature of the binding process itself. In cases where sufficient information is available, educated guesses for the location of the site of contact and the mode of binding can be made. For example, the interaction between antithrombin III and a sulfated pentasaccharide with heparin-like activity has been predicted on the basis of documented structural relationships of a series of pentasaccharide analogs, of the knowledge about the amino acids essential for heparin-binding, and of the crystal structure of a homologous protein (Grootenhuis and van Boeckel, 1991). For a better understanding of this fundamental process, we now return to the elaboration of our thermodynamic description of the recognition.

2.4 Basic Molecular Features in Carbohydrate-Protein Interactions

Force-field calculations consider the main molecular interactions that govern the recognition process. As already briefly discussed in the context of thermodynamic parameters in this chapter, these include stabilization through a network of hydrogen bonds with involvement of water molecules, hydrophobic and van der Waals interactions, as well as the stacking of aromatic amino acid side-chains with the saccharide. An instructive example of the integration of such diverse compounds into the overall scheme of interaction is given in Figure 3. It shows a schematic 2D-representation of the main molecular interactions between *N*-acetyllactosamine and a β-galactoside-binding protein (galectin-1) from bovine spleen, using the program LIGPLOT (Wallace *et al.*, 1995). Schematic 2D-representations of carbohydrate-lectin interactions have the advantage that the most important features, especially hydrogen bonds, can often be seen more clearly than in complex 3D-graphical representations.

Hydrogen bonds play a major role in the formation of carbohydrate-protein complexes. They exhibit a strong directionality and provide stability in the order of a few kJ per hydrogen bond formed. Thus, their strength is sufficiently low to allow dissociation of the ligand from the protein. Hydroxyl groups of the sugar can serve as hydrogen bond donors and acceptors. More-

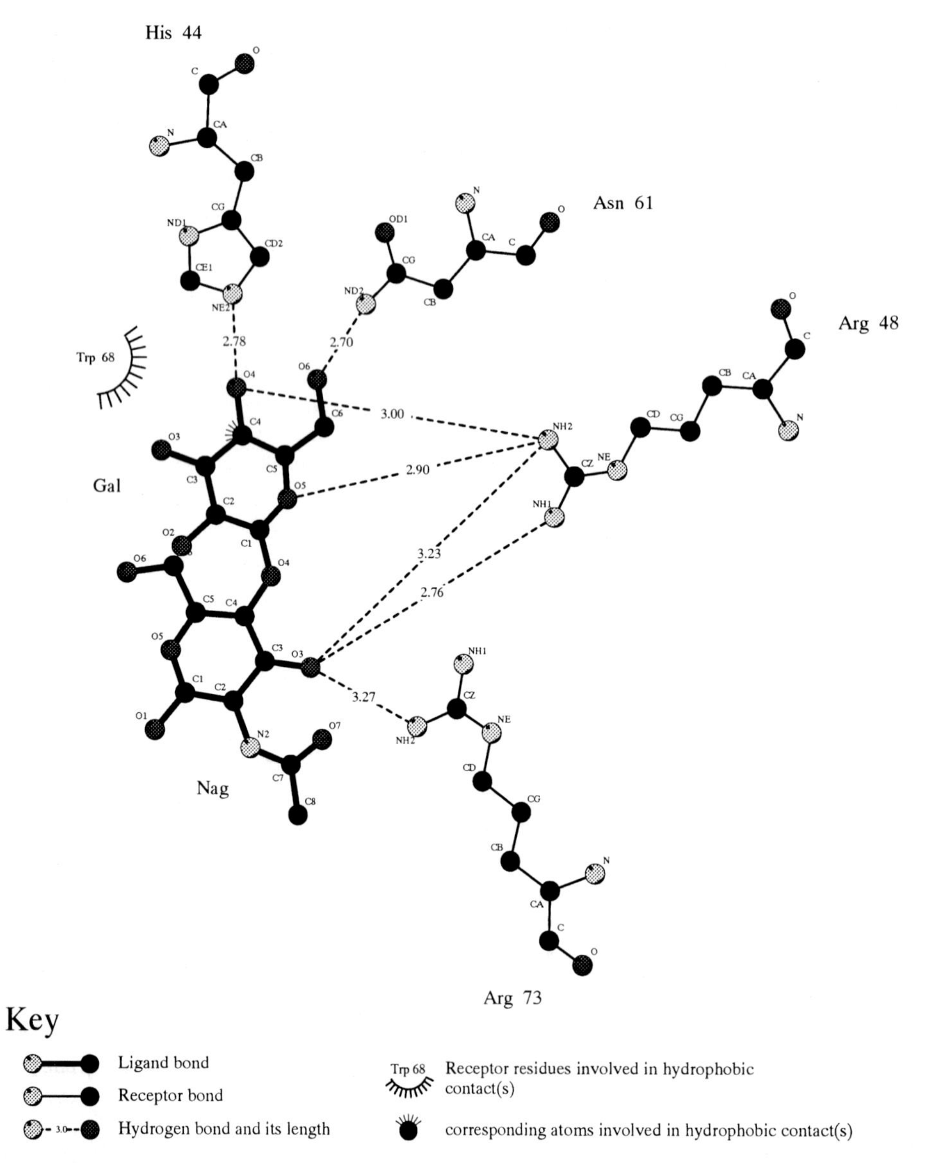

Figure 3. Schematic 2D-representation of the main molecular features stabilizing the complex of *N*-acetyllactosamine and bovine spleen S-lectin (galectin-1, entry 1SLT taken from Brookhaven protein data bank) (Liao *et al.*, 1994). The plot was automatically generated with the program LIGPLOT (Wallace *et al.*, 1995). The interactions shown by LIGPLOT are those mediated by hydrogen bonds and by hydrophobic contacts. Hydrogen bonds are indicated by dashed lines with specification of their lengths. Hydrophobic contacts are indicated schematically: residues from the protein involved in these contacts are represented by an arc with spikes radiating towards the ligand atoms, which they contact.

over, the hydroxyl groups can rotate freely and can thus easily adopt the energetically most favorable linear orientation as donors with acceptor groups on the protein surface. Suitable partners for the interaction might be not only amino acid side-chains, but also water molecules tightly associated to the protein's surface, as already indicated. The readily accessible parts of the oligosaccharide, which are not in direct contact with the protein surface, tend to form hydrogen bonds to water molecules of the solvent. Some of these water molecules can interact with the protein surface, creating hydrogen bond networks between the carbohydrate and the surface (Bourne *et al.*, 1993). In the case of a complex between isolectin I from *Lathyrus ochrus* and a diantennary octasaccharide of the *N*-acetyllactosamine-type, 21 hydrogen bonds were detected, fourteen being located directly at the interface between the protein and the saccharide, while the remaining seven involve bridging water molecules (Bourne *et al.*, 1993).

Non-polar contacts, which do not show such a high degree of directionality, appear to be almost as important as polar interactions. They provide specificity for the carbohydrate binding site and can account for about half of the total energy needed to establish the complex (Imberty *et al.*, 1993). Additional stability is conferred by a partial or nearly complete stacking of aromatic residues with pyranose ring structures. In the case of the *Lathyrus ochrus* isolectin I, the initial complex is probably formed at two contact points at each side of two closely related binding sites. Between these two attachment points, which are mainly formed by hydrogen bonds, the oligosaccharide folds around a non-polar region created by a phenylalanine moiety, which contributes considerably to the stability of the complex (Bourne *et al.*, 1991). Furthermore, van der Waals contacts may well enhance the extent of interactions. In this context it is important to note that the precise quantification of these forces is difficult to derive from experimental structures. For an elegant and detailed description of the current status of X-ray analysis of proteins and carbohydrate-protein complexes the reader is referred to the accompanying chapter.

2.5 Estimation of Binding Constants

The accurate prediction of binding constants for carbohydrate-protein complexes of known 3D-structures is a key step towards a profound understanding and modeling of molecular associations. Many efforts are currently devoted to enhancing our knowledge of the processes by which a certain carbohydrate specifically and reversibly binds to a lectin in solution. However, up till now, such approaches face a number of limitations. In the following paragraphs, we will outline the basic assumptions, which are generally used to achieve a semiquantitative estimation of binding constants, extending the description given in the first part of this chapter.

2.5.1 Qualitative modeling of binding constants

The free energy of binding (ΔG) can be calculated on the basis of measurement of the binding and/or dissociation constant K according to equation (3).

$$(3)\quad \Delta G = -RT\ln K$$

As already outlined in section 1.2, this free energy of binding comprises an enthalpic and an entropic contribution. Both terms need to be subdivided, as graphically emphasized in Figure 4. Theoretically, the free energy of binding for molecular associations in solutions can be expressed by several terms (Williams *et al.*, 1991; Searle *et al.*, 1992). These terms are assumed to be additive, as indicated in equation (4), and should either be experimentally accessible or inferable from known physical parameters.

$$(4)\quad \Delta G = \Delta H_{binding} + \Delta H_{conform} - T(\Delta S_{rot+trans} + \Delta S_{flex} + \Delta S_{solv})$$

The meaning of the individual terms can be explained as follows:

$\Delta H_{binding}$ favorable van der Waals, electrostatic and hydrogen bond contributions to the interaction between the carbohydrate ligand and the protein;

$\Delta H_{conform}$ unfavorable change in conformational enthalpy to reach the bound state;

$\Delta S_{rot+trans}$ unfavorable loss of translational and rotational degrees of freedom upon binding;

ΔS_{flex} unfavorable change of entropy associated with the freezing of torsional angles in the bound state; and

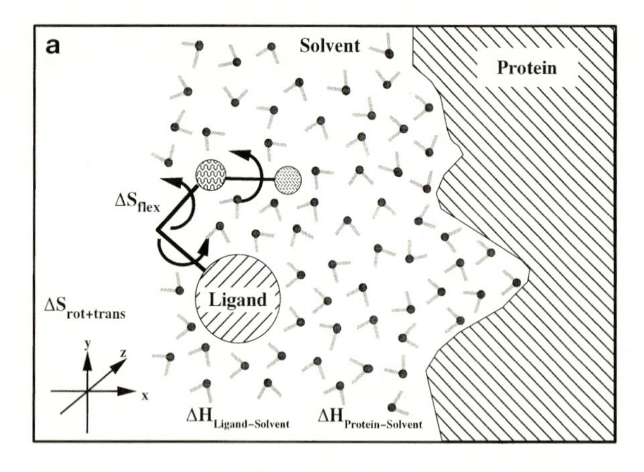

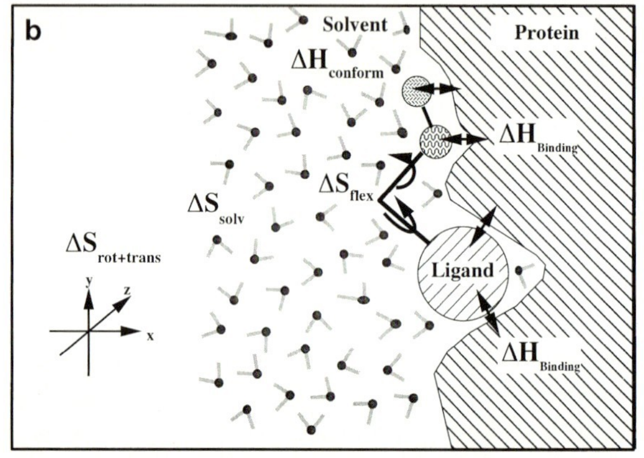

Figure 4: Schematic representation of the factors which have to be evaluated for a detailed description of the basic steps involved in molecular association. Fig. 4a shows the isolated molecules before the formation of the complex. Fig. 4b represents the formed complex.

ΔS_{solv} favorable or unfavorable change of entropy associated with the removal of solvent molecules from the binding site of the protein and the ligand during formation of the complex.

Any molecular association inherently has an unfavorable entropy term $T\Delta S_{rot+trans}$, since each of the two components has three degrees of translational and three degrees of rotational freedom prior to association. These twelve degrees of freedom are reduced to only six when the complex is formed. Normally, this process requires the freezing of internal movements of the protein receptor as well as the carbohydrate ligand. This gives rise to a second unfavorable entropy term, namely ΔS_{flex}. Both binding partners often have to adopt a certain conformation within the complex, which is energetically less favorable than the free-state conformation. This third energetically unfavorable term $\Delta H_{conform}$ is indispensable to enable both molecules to exhibit the conformation that is required for association. In other words, $\Delta H_{conform}$ is the strain energy in the complex associated with the introduction of unfavorable bond lengths and bond angles.

The three terms $T\Delta S_{rot+trans}$, $T\Delta S_{flex}$ and $\Delta H_{conform}$ need to be counterbalanced to yield an association process. The main favorable enthalpic contribution is derived from the term $\Delta H_{binding}$, which comprises the contributions of van der Waals, electrostatic, polar and hydrogen bond interactions. The last term to be considered, namely ΔS_{solv}, is associated with the removal of solvent molecules from the protein's binding site as well as from the ligand during the formation of the complex. In aqueous solution this term is often called the hydrophobic effect. Depending on the features of the individual complex, this process causes a positive or negative change of entropy.

2.5.2 Quantification of the free energy of binding

Since the term $\Delta H_{binding}$ summarizes van der Waals, electrostatic and hydrogen bond contributions, its amount can be estimated using standard force-field calculations (Kollman, 1994). The main driving forces for these non-covalent interactions are the electrostatic and van der Waals contributions. They are generic terms in molecular mechanics and molecular dynamics calculations. The amount of $\Delta H_{conform}$ can likewise be obtained with good accuracy. For isolated oligosaccharides various methods have been developed to explore the conformational space accessible for these molecules. In addition to systematic conformational searches, which yield results of which an example is given in Figure 5, molecular dynamics and Monte Carlo simulation are commonly employed (for a recent review, see von der Lieth *et al.*, 1996). It has to be mentioned

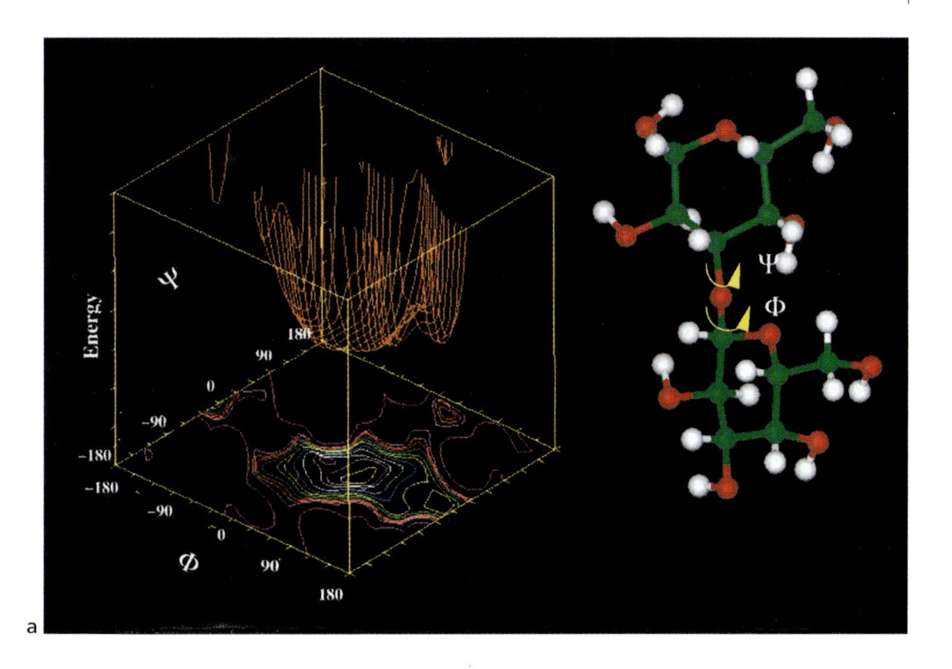

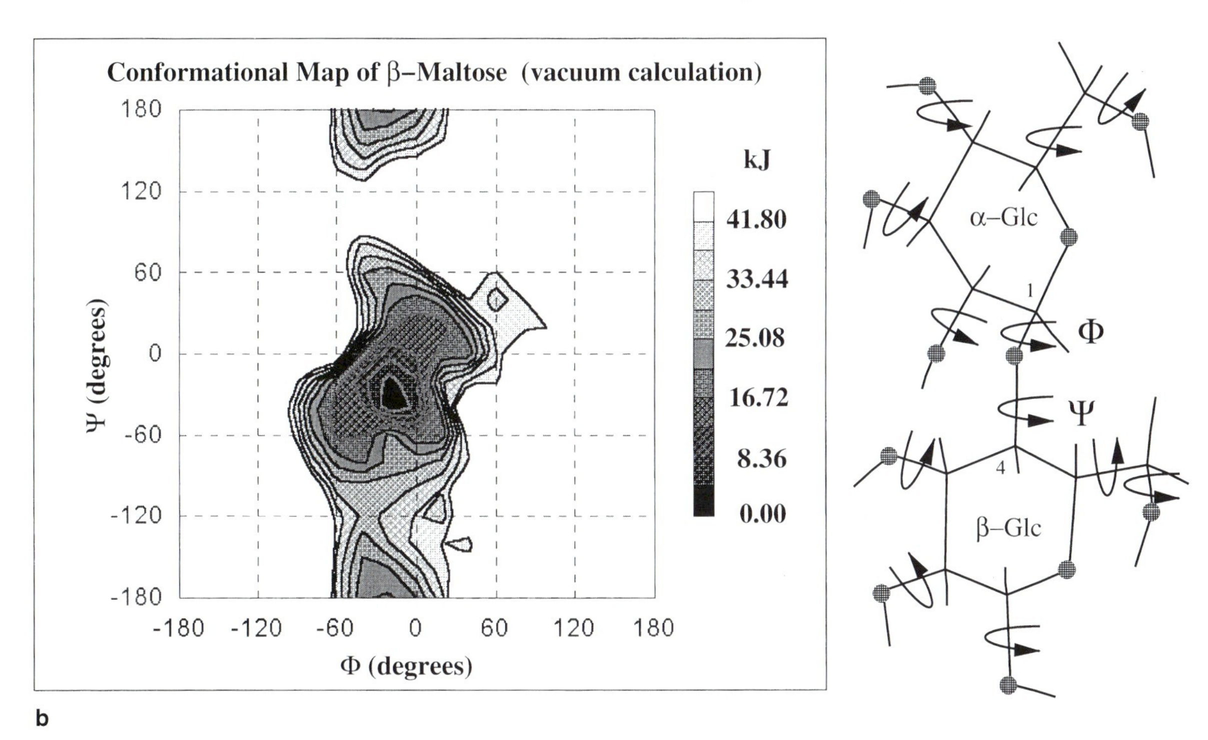

Figure 5. An example of the results of systematic conformational search methods, namely the energy profile on the basis of a step-wise evaluation of the conformational space around the glycosidic linkage of a disaccharide (a) and a comprehensive conformational map of α1–2 glycosidic bond of a trisaccharide (b). The influence of all 15 rotatable bonds on the conformational map of this linkage has been evaluated systematically. For this purpose the program RAMM (Kozár *et al.*, 1990) was applied.

that association with the receptor will not necessarily involve the energetically most favored conformation of the oligosaccharide in solution. The main conformational changes upon binding result from rotations around glycosidic bonds (Imberty *et al.*, 1993). Torsional angles of the glycosidic linkage in experimentally determined oligosaccharide-protein complexes appear to correspond to local low energy regions predicted from force-field calculations. The energetic implications of conformational changes of the proteins are normally neglected in this approach, because their potential energy surfaces are even more shallow than those of carbohydrates. This simplified consideration is justified, unless an induced fit (i.e. a considerable conformational change of the protein) occurs upon binding to enhance the association further. In these cases, thorough evaluation of the conformation of the protein's binding site has to be performed. Unfortunately, this may not be practicable due to the presumed enormous extent of conformational freedom which can exist in proteins. The individual energetic loss of the translational and rotational degrees of freedom $T\Delta S_{rot+trans}$ upon binding is nearly impossible to estimate in biological systems including solvent molecules. Assuming as a simplification that these contributions for the interaction of the protein and the small ligand are rather similar, they can be summarized and then determined. The unfavorable entropy term ΔS_{flex}, associated with the freezing of torsion angles in the bound state, may be estimated, if the number of conformations and their population densities in the free state of the oligosaccharide relative to the bound state are known. In principle, force-field calculations provide tools to obtain these quantities as the so-called conformational or adiabatic maps (French and Dowed, 1993; von der Lieth *et al.*, 1996). Since the demanding task of a complete conformational analysis for all degrees of internal freedom has to be performed, a simple estimate is often applied at present, where ΔS_{flex} is quantified by a certain constant energy value per rotatable bond in a molecule. The entropy change ΔS_{solv} is caused by the removal of solvent molecules from the binding site and the ligand during formation of the complex. It presents the most difficult problem for calculation, for which an approximation has been proposed (Williams *et al.*, 1991; Searle *et al.*, 1992). On the basis of experiments and theoretical calculations it is assumed that the hydrophobic effect can stabilize a complex by about 0.124 kJ/mol for every $Å^2$ of hydrocarbons removed from water accessibility upon formation of the complex. When the spatial extensions of the components are known in the free and bound state, this area can be calculated, gaining access to the last term for the computer-assisted calculations.

2.5.3 General considerations

Although the computer-assisted model calculations are already remarkably elaborate, it is still impossible in many cases to obtain accurate values of all enthalpic and entropic terms involved in the process of molecular association. Hence, the quantitative description encompasses some uncertainties. As outlined above, we presently have to resort to using approximations, namely in the quantification of the entropic term governed by the release of solvent molecules upon binding. Keeping these drawbacks in mind, it becomes evident that the predictive power to estimate binding constants with the outlined approach will provide only a limited level of accuracy. Recently, a simple but very fast empirical scoring function has been described that yields binding constants of protein-ligand complexes with known 3D-structure, based on the principles and assumptions discussed above (Böhm, 1994). This procedure has been calibrated with 45 known protein-ligand complexes with various kinds of ligands, and can be improved as the number of accessible structures grows and our understanding of the driving forces for molecular association increases. While X-ray crystallography provides parameters of the molecules in a lattice, NMR spectroscopy can provide experimental insights into the mode of binding in solution. It also offers possibilities of investigating the conformational behavior of ligands in the bound as well as in the free state. This capacity makes NMR spectroscopy a valuable tool, in combination with computational approaches, to disclose conformational aspects before and after complex formation, as shown for example in the case of an animal galectin and a polyclonal human immunoglobulin G fraction (Siebert *et al.*, 1996a, b).

3 NMR Spectroscopy

3.1 Free-State Conformation

NMR spectroscopic methods have a firmly established place in the structural analysis of biomolecules in solution (Ernst *et al.*, 1985; Wüthrich, 1986; Bax, 1994; Hull, 1994). While computer calculations such as molecular dynamics simulations describe the conformation on the basis of a set of approximations, experimental information about the actual conformation and dynamic aspects of these molecules is obtained from different types of spectra (Kaptein *et al.*, 1988; Karplus and Pestko, 1990). Structural analysis of oligo- and polysaccharides has been carried out successfully with numerous one-, two- and three-dimensional NMR methods (Vuister *et al.*, 1989; Hård and Vliegenthart, 1993; Kamerling and Vliegenthart, 1993; Siebert *et al.*, 1993; Dabrowski, 1994; Homans, 1995). The most valuable information about conformations obtainable from NMR experiments is based on the measurement of the **N**uclear **O**verhauser **E**ffect (NOE). It originates from dipole-dipole interactions between protons less than 3.5 Å apart for carbohydrates. The NOE intensity is proportional to the minus sixth power of the distance r_{ij} between the involved protons i and j (Noggle and Schirmer, 1971; Neuhaus and Williamson, 1989), thus yielding information about the oligosaccharide conformations, when distinct protons of two monosaccharide residues of such a chain are observed (Homans, 1990, 1993a, b, 1995; van Halbeek, 1994). However, in flexible molecules like oligosaccharides, the NOE-derived distances must be considered as time-averaged data not necessarily leading to true conformations (Jardetzky, 1980; Cumming and Carver, 1987; Carver *et al.*, 1989).

The distance mapping procedure (Acquotti *et al.*, 1990; Poppe *et al.*, 1990; Siebert *et al.*, 1992) takes this source of ambiguity into account. It assumes the presence of at least two true interproton distances for each proton-proton interaction inducing a NOE, one larger and one smaller than the distance calculated from the measured NOE. Since an interproton distance of more than 3.5 Å does not give a detectable NOE in practice, credible conformers with larger interproton distances can be neglected. Thus, only those conformations need to be considered in which at least one interproton distance with $r_{ij} < 3.5$ Å can be found with confidence. The lower reasonable limit is normally derived from the van der Waals radii of about 2.0 Å. For the conformational analysis of glycosidic linkages that are determined by the rotation angles, as indicated in Figure 1, it is convenient to use the Ramachandran-type representation. The distance constraints for each pair of atoms are expressed in the form of their Φ, ψ coordinates. When a distinct distance is set for a proton-proton pair, iterative variation of the two angles yields a series of theoretically possible Φ, ψ values. Consequently, a single pair of contours drawn for the lower and upper limits of r_{ij} for one NOE contact still encloses a large number of Φ, ψ combinations and thus does not allow one to define one particular conformation. If, however, this region overlaps with another area defined by a second NOE contact of protons of the same two saccharide residues, the number of reasonable glycosidic angles Φ and ψ is appreciably reduced (Siebert *et al.*, 1996a). As an example, the results of the distance mapping for the free Galβ1–4Glc disaccharide is given in Figure 6, revealing a remarkable area of overlap. All included Φ, ψ combinations are in principle in agreement with the NMR data. The probability that a specific, well-defined conformation exists increases as the number of NOE contacts with overlapping contour regions becomes larger and the size of the area of overlap decreases. Unfortunately, observable NOE contacts in oligosaccharides are often restricted to interactions between the protons linked to the two carbons involved in the glycosidic linkage. To address this problem, distance mapping for oligosaccharides based on ROESY (**R**otating Framc Nuclear **O**verhauser **E**ffect **S**pectroscop**Y**) (Bothner-By *et al.*, 1984; Bax and Davis, 1985) has been developed, which uses hydroxyl and amide protons as long range sensors (Dabrowski and Poppe, 1989). The

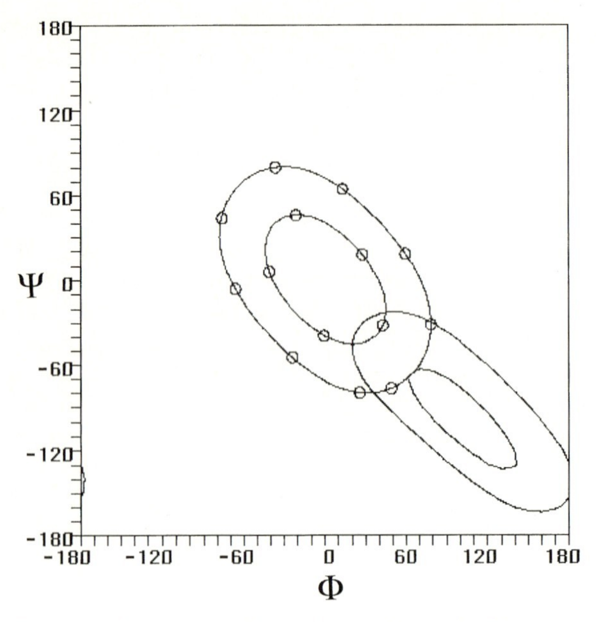

Figure 6. Distance mapping of free Galβ1–4Glc. The contour lines of the two inter-residual contacts between H1 of galactose and H4 of glucose with distance limits between 2.2 and 2.6 „ (○-○) and both H6 protons of galactose and OH3 of glucose with distance limits of 2.2 and 3.2 Å (–) are shown.

number of distance constraints obtained by this method improves the value of the distance mapping procedure. This approach has been applied successfully to detect energetically possible conformations that could not be recognized by C-H proton NOE contacts alone, as shown for the Galβ1–3Glcβ1-OMe disaccharide (Dabrowski *et al.*, 1995). The concomitant consideration of the energy profile of the derivable conformations, as shown, for example, in Figure 5, clearly enables one to assign different levels of probability to the array of structures. Often the free-state conformations, which for example in the case of a number of biologically relevant disaccharides have been compiled in a data bank (Imberty *et al.*, 1990, 1991, 1995), can fit easily in the receptor's binding pocket (Imberty and Perez, 1994; Imberty *et al.*, 1994). Similarly, in the well studied examples of the conformations of the sialic acid α2–3 or α2–6 linkages to galactose (Breg *et al.*, 1989; Poppe *et al.*, 1992; Siebert *et al.*, 1992), one of the possible free-state conformations is actually found in a lectin/hemagglutinin-bound state (Wright, 1990, 1992; Sauter *et al.*, 1992).

3.2 Transferred NOE-Experiments of Oligosaccharide-Protein Complexes

TRansferred **NOE** (TRNOE) experiments are specifically designed to analyze the interaction of a ligand with its receptor, including conformational aspects (Albrand *et al.*, 1979; Clore and Gronenborn, 1983; Ni, 1994). The TRNOE experiments are based on the fact that ligand protons in high molecular mass complexes tumble more slowly in solution than those of rapidly reorienting small ligands in the free state (Otting, 1993). This slow tumbling allows a fast magnetization transfer between different ligand protons in the bound state. The corresponding NOE signals observed for the small ligand, which moves rapidly between bound and free state, are almost entirely dominated by the NOEs of the bound state. TRNOE experiments usually employ a large excess of ligand, and, consequently, the average line width is small, because it is dominated by the free ligand. In this way, the ligand signals can readily be resolved, even when the molecular mass of the complex would lead to extremely broad signals typical for high molecular mass complexes, as illustrated in Figure 7 for a plant lectin-lactose complex. Therefore, it is possible to observe NOEs developed in the bound state at the resonance of a distinct ligand

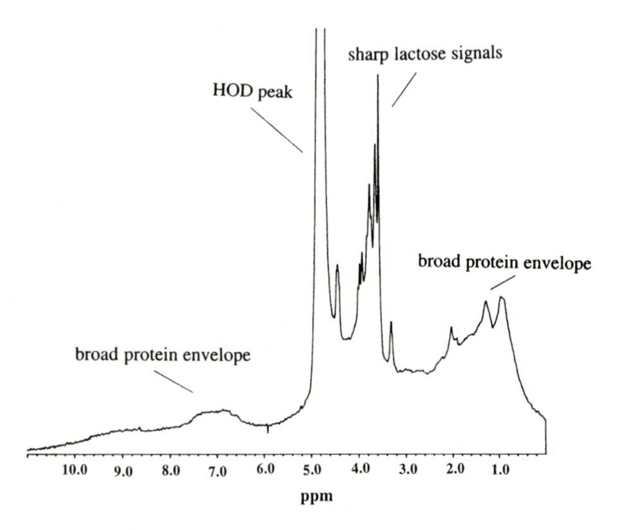

Figure 7. 1D-NMR spectrum of the complex between mistletoe lectin (*Viscum album* agglutinin) and lactose at a molar ratio of ligand:lectin of 10:1.

proton with negligible effects of line broadening and interference by the lectin signals. In this way, the TRNOE provides information on complexes in solution, complementing X-ray data of oligosaccharide-protein complexes (Shaanan *et al.*, 1992; Bundle *et al.*, 1994; Cygler *et al.*; 1994; Cambillau, 1995).

In principle, any lectin can be used for TRNOE experiments, if the binding parameters meet the prerequisites, which are described in the theory of TRNOE (Clore and Gronenborn, 1983). This type of experiment has proven to be very effective in this field (Bevilacqua *et al.*, 1990, 1992; Glaudemans *et al.*, 1990; Bundle *et al.*, 1994; Siebert *et al.*, 1996a). Three conditions, which ensure a meaningful study, have to be fulfilled in the TRNOE-experiments: (1) the negative NOE signals of the bound ligand have to be predominant in relation to the positive NOE signals of the free ligand, which depends on the molar ratios of lectin to ligand; (2) the effects of spin diffusion, by which the magnetization may be transferred to the protein instead of a saccharide residue, thus possibly leading to erroneous results, have to be minimized (Kalk and Berendsen, 1975; Neuhaus and Williamson, 1989); and (3) exclusion of signals originating from protein-protein interactions must be possible by recording a reference spectrum. This reference spectrum can be obtained by irradiation at resonance frequencies, when only broad protein signals are observed.

The value of such measurements in combination with computer-assisted simulations of the free-state conformation is illustrated for a complex of a high-affinity carbohydrate ligand with a galactoside-binding plant lectin, namely TRNOE experiments of Galβ1–4Glc in solution with the galactose-specific mistletoe lectin (*Viscum album* agglutinin, VAA). The molar ratio of carbohydrate ligand and lectin receptor was adjusted to a 10 to 20-fold molar excess of the disaccharide in relation to the agglutinin. A strong negative NOE of the bound ligand was detectable. The negative peaks of a TRNOE spectrum of a VAA-lactose complex, from which the reference spectrum had been subtracted, demonstrate that the bound state dominates the NOEs under these conditions. The inter-residual NOE contact between the galactose proton 1 (Gal H1) and the glucose proton 4 (Glc H4) is indicated by the presence of the Glc H4 signal, which occurs after irradiation at the resonance frequency of Gal H1 (Fig. 8). However, an unambiguous NOE-based decision on the Φ and ψ values of the glycosidic linkage is not possible. It is nonetheless instructive to note the results of the computer-assisted calculations of Φ, ψ combinations in relation to the energy level. In the example of lactose, the global free-state energy minimum of the ligand, as depicted in a stereo plot in Figure 9, is consistent with the measurable TRNOE distance in the complex.

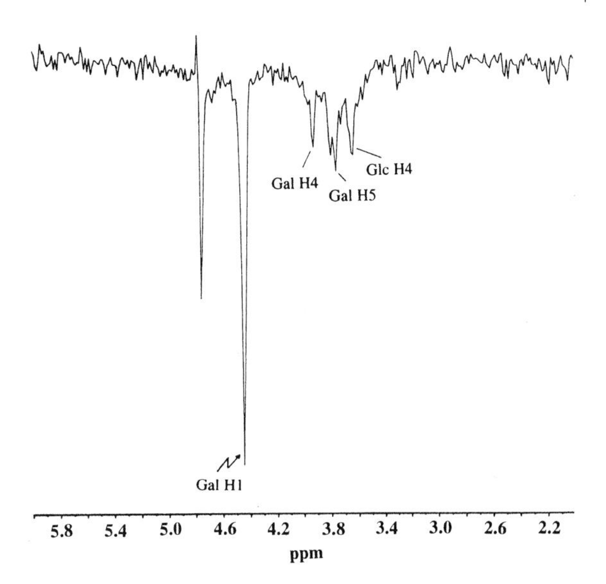

Figure 8. TRNOE difference spectrum of mistletoe lectin-lactose complex, irradiated at the resonance frequency of Gal H1. The molar ratio of ligand:lectin was 10:1. The spectrum was recorded at 300 K.

Although X-ray structures are not yet available for a great number of lectins, modeling of the unknown structure of a selected receptor on the basis of X-ray data from a related protein can be applied, as already discussed in a preceding paragraph of this chapter. Combination of conformational data for ligands obtained by, for example, TRNOE measurements, with X-ray structures of carbohydrate receptors, will open a wide field for optimally tailoring carbohydrate-protein complexes that may be used for drug design, for example. Modeling studies will not only allow a fit for a carbohydrate ligand, but will also allow

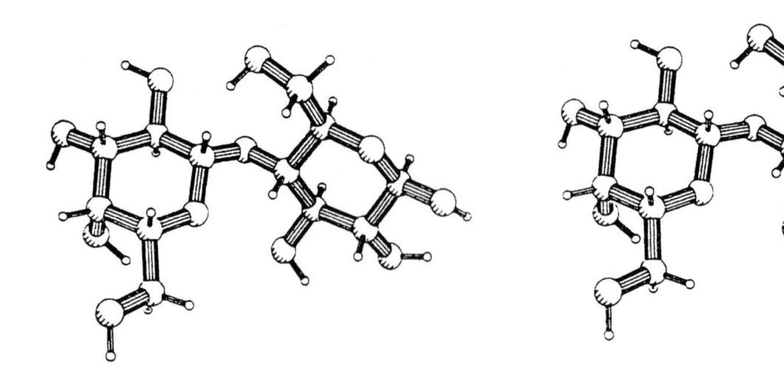

Figure 9. Stereo plot of Galβ1–4Glc in an energy minimum conformation in free solution. The conformer has torsion angles Φ and ψ of (55°/0°).

one to assign certain amino acid residues to the surface of a protein. This calculated parameter is experimentally determined by a sophisticated NMR technique, namely laser photo CIDNP (**C**hemically **I**nduced **D**ynamic **N**uclear **P**olarization). By this method it is possible to detect the side chains of tyrosine, tryptophan and histidine moieties that are accessible on the protein surface (Kaptein, 1982; Hore and Broadhurst, 1993). Observation of the corresponding signals in the presence and absence of an appropriate carbohydrate ligand may reveal an involvement of these amino acids in ligand binding (Siebert *et al.*, in preparation). Moreover, the possible influence of modifications of the glyco-part of glycoconjugates or of the protein part of the lectin (e.g. by site-directed mutagenesis, as explained in the chapter by Hirabayashi in this volume) on the surface accessibility of the given amino acid side chains can be determined (Hård *et al.*, 1992; Siebert *et al.*, 1995). Finally, results on sugar receptors in solution can validate the relevance of corresponding data from crystals.

3.3 Perspectives

The level of sophistication of computer hard- and software and NMR field strengths, as well as of the efficiency of complex algorithms and experimental measurement techniques, is increasing rapidly. This will bring further insights into the rapidly growing field of glycosciences, because structure and function are complementary aspects of life. X-ray and NMR structures of a limited number of saccharide-binding proteins are already known. However, many more structures related to carbohydrate-protein interactions have still to be solved, encouraging and inviting efforts of students and newcomers in this area. Crystallization of carbohydrate-lectin complexes and their X-ray analysis is often very time-consuming, and can only provide structural information for a molecule whose dynamic behavior is frozen in the crystal structure. A proper combination of X-ray and NMR data with computational methods such as molecular modeling will lead to a thorough description of molecular details of carbohydrate-protein complexes. Even a reasonable prediction of the impact of the exchange of a single amino acid on the overall conformation and binding specificity can be envisioned, allowing serious consideration of custom-made tailoring of pharmaceutically relevant glycodrugs.

References

Acharya S, Patanjali SR, Sajjan SU *et al.* (1990): Thermodynamic analysis of ligand binding to winged bean (*Psophocarpus tetragonolobus*) acidic agglutinin reveals its specificity for terminally monofucosylated H-reactive sugars. In *J. Biol. Chem.* **265**:11586–94.

Acquotti D, Poppe L, Dabrowski J *et al.* (1990): Three-dimensional structure of the oligosaccharide chain of GM_1 ganglioside revealed by a distance-mapping procedure: a rotating and laboratory frame nuclear Overhauser effect investigation of native glycolipid in dimethyl sulfoxide and in water-dodecylphosphocholine solutions. In *J. Am. Chem. Soc.* **112**:7772–8.

Albrand JP, Birdsall B, Feeney J *et al.* (1979): The use of transferred nuclear Overhauser effects in the study of the conformations of small molecules bound to proteins. In *Int. J. Biol. Macromol.* **1**:37–41.

Allen F, Davies J, Galloy J *et al.* (1991): The development of version 3 and 4 of the Cambridge structural database system. In *J. Chem. Inf. Comp. Sci.* **31**:187–204.

Bax A (1994): Multidimensional nuclear magnetic resonance methods for protein studies. In *Curr. Opinion Struct. Biol.* **4**:738–44.

Bax A, Davis DG (1985): Practical aspects of two-dimensional transverse NOE spectroscopy. In *J. Magn. Reson.* **63**:207–13.

Beierbeck H, Delbaere LTJ, Vandonselaar M *et al.* (1994): Molecular recognition. XIV. Monte Carlo simulation of the hydration of the combining site of a lectin. In *Can. J. Chem.* **72**:463–70.

Bernstein, F (1977): The protein data bank: a computer-based archival file for macromolecular structures. In *J. Mol. Biol.* **112**:535–42.

Bevilacqua VL, Thomson DS, Prestegard JH (1990): Conformation of methyl-β-lactoside bound to ricin B-chain: interpretation of transferred nuclear Overhauser effects facilitated by spin simulation and selective deuteration. In *Biochemistry* **29:**5529–37.

Bevilacqua VL, Kim Y, Prestegard JH (1992): Conformation of β-methylmelibiose bound to the ricin B-chain as determined from transferred nuclear Overhauser effects. In *Biochemistry* **31**:9339–49.

Böhm H (1994): The development of a simple empirical scoring function to estimate the binding constant of protein-ligand complex of known three-dimensional structure. In *J. Comput. Aided Mol. Design* **8**:243–56.

Bothner-By AA, Stephens RL, Lee J-M *et al.* (1984): Structure determination of a tetrasaccharide: transient nuclear Overhauser effects in rotating frame. In *J. Am. Chem. Soc.* **106**:811–3.

Bourne Y, Rougé P, Cambillau C (1991): X-ray structure of a biantennary octasaccharide- lectin complex refined at 2.3-Å resolution. In *J. Biol. Chem.* **267**:197–203.

Bourne Y, van Tilbeurgh H, Cambillau C (1993): Protein-carbohydrate interactions. In *Curr. Opinion Struct. Biol.* **3**:681–6.

Braden BC, Poljak RJ (1995): Structural features of the reactions between antibodies and protein antigens. In *FASEB J.* **9**:9–16.

Brant DA, Christ MD (1990): Realistic conformational modeling of carbohydrates: applications and limitations in the context of carbohydrate-high polymers. In *Computer Modeling of Carbohydrate Molecules* (French AD, Brady JW, eds) pp 42–68, Washington: ACS.

Breg J, Kroon-Batenburg LMJ, Strecker G *et al.* (1989): Conformational analysis of the sialylα(2–3/6)*N*-acetyllactosamine structural element, occurring in glycoproteins, by two-dimensional NOE ^{1}H-NMR spectroscopy in combination with energy calculations by hard-sphere exo-anomeric and molecular mechanics force-field with hydrogen bonding potential. In *Eur. J. Biochem.* **178**:727–39.

Bundle DR, Baumann H, Brisson J-R *et al.* (1994): Solution structure of a trisaccharide- antibody complex: comparison of NMR measurements with a crystal structure. In *Biochemistry* **33**:5183–92.

Cambillau C (1995): The structural features of protein-carbohydrate interactions revealed by X-ray crystallography. In *Glycoproteins* (Montreuil J, Vliegenthart JFG, Schachter H, eds) pp 67–86, Amsterdam-New York: Elsevier

Carver JP (1993): Oligosaccharides: how can flexible molecules act as signals? In *Pure Appl. Chem.* **65**:763–70.

Carver JP, Michnick SW, Imberty A *et al.* (1989): Oligosaccharide-protein interactions: a three-dimensional view. In *Carbohydrate Recognition in Cellular Function* (Ciba Found. Symp. 145) pp 6–26, Chichester: Wiley.

Clore GM, Gronenborn AM (1983): Theory of time-dependent transferred nuclear Overhauser effect: applications to structural analysis of ligand-protein complexes in solution. In *J. Magn. Reson.* **53**: 423–42.

Cumming DA, Carver JP (1987): Virtual and solution conformations of oligosaccharides. In *Biochemistry* **26**:6664–76.

Cygler M, Rose DR, Bundle DR (1994): Recognition of a cell-surface oligosaccharide of pathogenic salmonella by an antibody fab fragment. In *Science* **253**:442–5.

Dabrowski J, Poppe L (1989): Hydroxyl and amido groups as long-range sensors in conformational analysis by nuclear Overhauser effect: a source of experimental evidence for conformational flexibility of oligosaccharides. In *J. Am. Chem. Soc.* **111**:1510–1.

Dabrowski J (1994): Two-dimensional and related NMR methods in structural analyses of oligosaccharides and polysaccharides. In *Two-dimensional NMR Spectroscopy* (Croasmun WR, Carlson NK, eds) pp 741–83, Weinheim: VCH-Publishers.

Dabrowski J, Kozár T, Grosskurth H *et al.* (1995): Conformational mobility of oligosaccharides: experimental evidence for the existence of an "anti" conformer of the Galβ1–3Glcβ1-OMe disaccharide. In *J. Am. Chem. Soc.* **117**:5534–9.

Delbaere LTJ, Vandonselaar M, Prasad L *et al.* (1990): Molecular recognition of a human blood group determinant by a plant lectin. In *Can. J. Chem.* **68**:1116–21.

Ernst RR, Bodenhausen G, Wokaun A (1985): Principles of Nuclear Magnetic Resonance in One and Two Dimensions. Oxford: Clarendon Press.

French AD, Dowd MK (1993): Exploration of disaccharide conformations by molecular mechanics. In *J. Mol. Struct.* **286**:183–201.

Glaudemans CPJ, Lerner L, Daves Jr GD *et al.* (1990): Significant conformational changes in an antigenic carbohydrate epitope upon binding to a monoclonal antibody. In *Biochemistry* **29**:10906–11.

Grootenhuis P, van Boeckel C (1991): Constructing a molecular model of the interaction between antithrombin III and a potent heparin analogue. In *J. Am. Chem. Soc.* **113**:2743–7.

Hård K, Kamerling JP, Vliegenthart JFG (1992): Application of laser photo-CIDNP for an intact glycoprotein in solution. In *Carbohydr. Res.* **236**:321–6.

Hård K, Vliegenthart JFG (1993): Nuclear magnetic resonance spectroscopy of glycoprotein-derived carbohydrate chains. In *Glycobiology. A Practical Approach* (Fukuda M, Kobata A, eds) pp 223–42, Oxford-New York-Tokyo: Oxford University Press.

Homans SW (1990): Oligosaccharide conformations: application of NMR and energy calculations. In *Progress in NMR Spectroscopy* **22**:55–81.

Homans SW (1993a): Conformation and dynamics of oligosaccharides in solution. In *Glycobiology* **3**:551–5.

Homans SW (1993b): Characterization of the extent of internal motions in oligosaccharides. In *Biochemistry* **32**:12715–24.

Homans SW (1995): Three dimensional structure of oligosaccharides explored by NMR and computer calculations. In *Glycoproteins* (Montreuil J, Vliegenthart JFG, Schachter H, eds) pp 67–86, Amsterdam-New York: Elsevier.

Hore PJ, Broadhurst RW (1993): Photo-CIDNP of biopolymers. In *Progress in NMR Spectroscopy* **25**:345–402.

Hull WE (1994): Experimental aspects of two-dimensional NMR. In *Two-dimensional NMR Spectroscopy* (Croasmun WR, Carlson NK, eds) pp 67–424, Weinheim: VCH-Publishers.

Imberty A, Pérez S (1994): Molecular modeling of protein-carbohydrate interactions. Understanding the specificities of two legume lectins towards oligosaccharides. In *Glycobiology* **4**:351–66.

Imberty A, Gerber S, Tran V *et al.* (1990): Data bank of three-dimensional structures of disaccharides, a tool to build 3-D-structures of oligosaccharides. Part I. Oligomannose type *N*-glycans. In *Glycoconjugate J.* **7**:27–54.

Imberty A, Delage M-M, Bourne Y *et al.* (1991): Data bank of three-dimensional structures of disaccharides. Part II. *N*-Acetyllactosaminic type *N*-glycans. Comparison with the crystal structure of a biantennary octasaccharide. In *Glycoconjugate J.* **8**:456–83.

Imberty A, Bourne Y, Cambillau C *et al.* (1993): Oligosaccharide conformation in protein/carbohydrate complexes. In *Adv. Biophys. Chem.* **3**:71–117.

Imberty A, Casset F, Gegg C *et al.* (1994): Molecular modeling of the *Dolichos biflorus* seed lectin and its specific interactions with carbohydrates: α-D-*N*-acetylgalactosamine, Forssman disaccharide and blood group A trisaccharide. In *Glycoconjugate J* . **11**:400-13.

Imberty A, Mikros E, Koca J *et al.* (1995): Computer simulation of histo-blood group oligosaccharides: energy maps of all constituting disaccharides and potential energy surfaces of 14 ABH and Lewis carbohydrate antigens. In *Glycoconjugate J.* **12**:331- 49.

Jardetzky O (1980): On the nature of molecular conformations inferred from high-resolution NMR. In *Biochim. Biophys. Acta* **621**:227–32.

Joshi NV, Rao VSR (1979): Flexibility of the pyranose ring in α- and in β-D-glucoses. In *Biopolymers* **18**:2993–3004.

Kalk A, Berendsen HJC (1975): Proton magnetic relaxation and spin diffusion in proteins. In *J. Magn. Reson.* **24**:343–66.

Kamerling JP, Vliegenthart JFG (1993) High-resolution ^{1}H-nuclear magnetic resonance spectroscopy of oligosaccharide-alditols released from mucin-type *O*-glycoproteins. In *Carbohydrates and Nucleic Acids, Biological Magnetic Resonance* **10** (Berliner LJ, Reuben J, eds) pp 1–194, New York- London: Plenum Press.

Kaptein R (1982) Photo CIDNP studies of proteins. In *Biological Magnetic Resonance* (Berliner LJ, ed) Vol. 4, pp 145–91, New York-London: Plenum Press.

Kaptein R, Boelens R, Scheek RM *et al.* (1988): Protein structures from NMR. In *Biochemistry* **27**:5389–95.

Karplus M, Petsko GA (1990): Molecular dynamics simulations in biology. In *Nature* **347**:631–9.

Kollman P (1994): Theory of macromolecule-ligand interactions. In *Curr. Opinion Struct. Biol.* **4**:240–5.

Kozár T, Petrak F, Galova S *et al.* (1990): RAMM – a new procedure for theoretical conformational analysis of carbohydrates. In *Carbohydr. Res.* **204**: 27–36.

Lemieux RU (1989): The origin of the specificity in the recognition of oligosaccharides by proteins. In *Chem. Soc. Rev.* **18**:347–74.

Lemieux RU, Delbaere LTJ, Beierbeck H *et al.* (1991): Involvement of water in host-guest interactions. In *Host-Guest Molecular Interactions: from Chemistry to Biology* (Ciba Found. Symp. 158) pp 231–48, Chichester: Wiley.

Lemieux RU, Du M-H, Spohr U *et al.* (1994): Molecular recognition XIII. The binding of H- type 2 human blood group determinant by a winged bean (*Psophocarpus tetragonolobus*) acidic lectin. In *Can. J. Chem.* **72**:158–63.

Liao D, Kapadia G, Ahmed H *et al.* (1994): Structure of S-lectin, a developmentally regulated vertebrate β-galactoside-binding protein. In *Proc. Natl. Acad. Sci. USA* **91**:1428-32.

Loris RF, Casset BJ, Pletincks J *et al.* (1994): The monosaccharide binding site of lentil lectin: an X-ray and molecular modeling study. In *Glycoconjugate J.* **11**:507–17.

Neuhaus D, Williamson MP (1989): The Nuclear Overhauser Effect in Structural and Conformational Analysis. New York: VCH Publishers.

Ni F (1994): Recent developments in transferred NOE methods. In *Progress in NMR Spectroscopy* **26**:517–606.

Noggle JH, Schirmer RE (1971): The Nuclear Overhauser Effect – Chemical Applications. New York: Academic Press.

Nyholm P-G, Brunton J, Lingwood CA (1995): Modeling of interactions of verotoxin-1 (VT1) with its glycolipid receptor, globotriaosylceramide (Gb_3). In *Int. J. Biol. Macromol.* **17**:199–204.

Otting G (1993): Experimental NMR techniques for studies of protein-ligand interactions. In *Curr. Opinion Struct. Biol.* **3**:760–8.

Perez S, Imberty A, Carver JP (1994): Molecular modeling: an essential component in the structure determination of oligosaccharides and polysaccharides. In *Adv. Comput. Biol.* **1**: 147–202.

Poppe L, von der Lieth C-W, Dabrowski J (1990): Conformation of the glycolipid globoside head group in various solvents and in the micelle-bound state. In *J. Am. Chem. Soc.* **112**:7762–71.

Poppe L, Stuike-Prill R, Meyer B *et al.* (1992): The solution conformation of sialyl-α(2–6)- lactose studied by modern NMR techniques and Monte Carlo simulations. In *J. Biomol. NMR* **2**:109–36.

Quiocho FA (1986): Carbohydrate-binding proteins: tertiary structures and protein-sugar interactions. In *Annu. Rev. Biochem.* **55**:287–315.

Quiocho FA (1989): Protein-carbohydrate interactions: basic molecular features. In *Pure Appl. Chem.* **61**:1293–306.

Reeke GN, Becker JW (1988): Carbohydrate-binding sites of plant lectins. In *Curr. Top. Microbiol. Immunol.* **139**:35–58.

Sali A, Overington J, Johnson M *et al.* (1990): From comparisons of protein sequences and structures to protein modeling and design. In *Trends Biochem. Sci.* **15**:235–40.

Sauter NK, Hanson JE, Glick GD *et al.* (1992) Binding of influenza virus hemagglutinin to analogs of its cell-surface receptor, sialic acid: analysis by proton nuclear magnetic resonance spectroscopy and X-ray crystallography. In *Biochemistry* **31**: 9609–21.

Searle M, Williams D, Gerhard U (1992): Partitioning of free energy contributions in the estimation of binding constants: residual motions and consequences for amide-amide hydrogen bond strength. In *J. Am. Chem. Soc.* **114**:10697–704.

Shaanan B, Gronenborn AM, Cohen GH *et al.* (1992): Combining experimental information from crystal and solution studies: joint X-ray and NMR refinement. In *Science* **257**:961–4.

Siebert HC, Reuter G, Schauer R *et al.* (1992): Solution conformation of GM_3 gangliosides containing different sialic acid residues as revealed by NOE-based distance-mapping, and molecular mechanics and molecular dynamics calculations. In *Biochemistry* **31**:6962–71.

Siebert HC, Kaptein R, Vliegenthart JFG (1993): Study of oligosaccharide-lectin interaction by various nuclear magnetic resonance (NMR) techniques and computational methods. In *Lectins and Glycobiology* (Gabius H-J, Gabius S, eds) pp 105–16, Heidelberg – New York: Springer Publ. Co.

Siebert HC, André S, Reuter G *et al.* (1995): Effect of enzymatic desialylation of human serum amyloid P component on surface exposure of laser photo CIDNP (chemically induced dynamic nuclear polarization) – reactive histidine, tryptophan and tyrosine residues. In *FEBS Lett.* **371**:13–6.

Siebert HC, Guilleron M, Kaltner H *et al.* (1996a): NMR-based, molecular dynamics- and random walk molecular mechanics-supported study of conformational aspects of a carbohydrate ligand (Gal β1,2Galβ1-R) for an animal lectin in the free and the bound state. In *Biochem. Biophys. Res. Commun.* **219**:205–12.

Siebert HC, von der Lieth C-W, Dong X *et al.* (1996b): Molecular dynamics-derived conformation and intramolecular interaction analysis of the *N*-acetyl-9-*O*-acetylneuraminic acid-containing ganglioside GD1a and NMR-based analysis of its binding to a human polyclonal immunoglobulin G fraction with selectivity for *O*-acetylated sialic acids. In *Glycobiology*, in press.

Sigurskjold BW, Bundle DR (1992): Thermodynamics of oligosaccharide binding to a monoclonal antibody specific for *Salmonella O*-antigen point to hydrophobic interactions in the binding site. In *J. Biol. Chem.* **267**:8371–6.

Sigurskjold BW, Altman E, Bundle DR (1991): Sensitive titration microcalorimetric study of the binding of *Salmonella O*-antigenic oligosaccharides by a monoclonal antibody. In *Eur. J. Biochem.* **197**: 239–46.

Sutcliffe MJ, Hayes FRF, Blundell TL (1987a): Knowledge based modelling of homologous proteins, part II: rules for the conformations of substituted sidechains. In *Prot. Eng.* **1**:385–92.

Sutcliffe MJ, Haneef I, Carney D *et al.* (1987b): Knowledge based modelling of homologous proteins, part I: three-dimensional frameworks derived from simultaneous superposition of multiple structures. In *Prot. Eng.* **1**:377–84.

Thurin J (1988): Binding sites of monoclonal anti-carbohydrate antibodies. In *Curr. Top. Microbiol. Immunol.* **139**:59–79.

Toone EJ (1994): Structure and energetics of protein-carbohydrate complexes. In *Curr. Opinion Struct. Biol.* **4**:719–28.

van Halbeek H (1994): NMR development in structural studies of carbohydrates and their complexes. In *Curr. Opinion Struct. Biol.* **4**:697–709.

von der Lieth C-W, Kozár T, Hull WE (1996): A (critical) survey of modeling protocols to explore the conformational space of oligosaccharides. In *J. Mol. Struct.*, in press.

Vuister GW, de Waard P, Boelens R *et al.* (1989): The use of 3D-NMR in structural studies of oligosaccharides. In *J. Am. Chem. Soc.* **111**:772–4.

Wallace AC, Laskowski RA, Thornton JM (1995): LIGPLOT: a program to generate schematic diagrams of protein-ligand interactions. In *Prot. Eng.* **8**:127–34.

Williams D, Cox J, Doig A *et al.* (1991): Toward the semiquantitative estimation of binding constants. Guides for peptide-peptide binding in aqueous solution. In *J. Am. Chem. Soc.* **113**:7020–30.

Wright CS (1990): 2.2 Å resolution structure analysis of two refined *N*-acetylneuraminyl- lactose-wheat germ agglutinin isolectin complexes. In *J. Mol. Biol.* **215**:635–51.

Wright CS (1992): Crystal structure of a wheat germ agglutinin/glycophorin-sialoglycopeptide receptor complex. In *J. Biol. Chem.* **267**:14345–52.

Wüthrich K (1986): NMR of proteins and nucleic acids. New York: Wiley.

Young NM, Oomen RP (1992): Analysis of sequence variation among legume lectins: a ring of hypervariable residues forms the perimeter of the carbohydrate-binding site. In *J. Mol. Biol.* **228**: 924–34.

17 Antibody-Oligosaccharide Interactions Determined by Crystallography

David R. Bundle

Abbreviations

IgG	Immunoglobulin, antibody molecule weight of 150 kDa
ΔG	Gibbs free energy
Heptasaccharides A and B	-These heptasaccharides were generated by mild acid hydrolysis of two repeating units of the polysaccharie antigen.
ScFv	-Single chain antibody molecule in which the heavy and light chain are covalently attached via a polypeptide linker that connects the C terminus of one chain to the N terminus of the other.
Fab and Fc	-The 50 kDa fragments liberated from a whole immunoglobulin when it is subjected by proteolytic cleavage by papain.
NOE	-Nuclear Overhauser Effect
abe	-3,6-Dideoxy-D-*xylo*hexopyranose
Gal	Galactose
Man	Mannose

The interactions of cell surface oligosaccharides with antibodies and lectins are important events that allow a host to ward off infection and challenge transplantation (Cooper and Oriol, this volume; Gabius *et al.*, this volume; Gilboa-Garber *et al.*, this volume), or, in the case of mammalian lectins, guide correct trafficking of cells and macromolecules (Hooper *et al.*, this volume; Ohannesian and Lotan, this volume; Rice, this volume; Sharon and Lis, this volume; Villalobo *et al.*, this volume; Ward, this volume). The initial molecular event that triggers these essential processes is the recognition and binding of oligosaccharide epitopes (the smallest structural element that fills the protein's binding site). Continuing the description of protein-carbohydrate recognition, given in the preceding chapter (Siebert *et al.*, this volume), this chapter presents a brief overview of the crystal structure of an antibody-oligosaccharide complex and tabulates recently published structures of oligosaccharides complexed with antibodies or lectins.

The carbohydrate-binding sites of antibodies and lectins exhibit similar features of protein-carbohydrate interaction with affinity constants in the high millimolar to low micromolar range. Antibodies, however, possess sites that are generally deeper than those found in lectins (Cygler *et al.*, 1991; Rose *et al.*, 1993; Vyas *et al.*, 1993; Bundle *et al.*, 1994; Jeffrey *et al.*, 1995). However, the size of the binding sites and the number of monosaccharide residues required to fill them are generally similar at three to four hexose units and within these, the crucial recognition element can be still smaller. Antibodies with their inherent ability to adapt to an antigen-driven immune response may on occasion form binding sites with interactions that extend over a larger carbohydrate antigen surface, although this situation seems to represent the exception to the general case (Evans *et al.*, 1995). Chemical synthesis of complex oligosaccharides (Schmidt, this volume) is an important aid to structural studies since it provides not only free ligand for complex formation but also effective affinity matrices for functional protein purification (Shibata *et al.*, 1982; Spohr *et al.*, 1985). As the availability of 5–10 mg quantities of highly pure protein is a major hurdle on the path to obtaining crystals of oligosaccharide-

H.-J. and S. Gabius (Eds.), Glycosciences

ISBN 3-8261-0073-5

protein complexes, this capability can provide a crucial advantage.

1 Crystal Structure Determination

Protein crystallography provides a snapshot of the complex between these two biopolymers. Since the method establishes the coordinates of protein/ligand non-hydrogen atoms, a solved crystal structure of medium to high resolution remains the gold standard against which other methods such as NMR (Siebert *et al.*, this volume) are compared. Although it is quite common to view three-dimensional models of proteins either on graphics workstations or as stereoplots in journals, it should be borne in mind that protein crystallography provides more detail than is revealed by the graphic image. A summary of the steps involved in the determination of a protein crystal structure highlights the basis upon which the final graphics model rests as well as the requirements for solving a structrue. It also points out the importance of an appreciation of the information that is contained in a published structure. A recent text is especially useful to non-specialists (Rhodes, 1993).

1.1 Sample Purification

Structurally the simplest immunoglobulin is an IgG. Each molecule has 2 binding sites positioned at the *N*-terminus of the polypeptide chains. The molecule possesses a two-fold axis of symmetry and is built from 2 heavy chains (H) and 2 light chains (L). Domains each containing about 100 amino acids provide the structural elements from which the whole molecule is built. The amino acid sequences of the constant domains are highly conserved whilst the two variable domains of the H and L chains, V_H and V_L, show greater variability and very high variability in three peptide segments, referred to as the hypervariable loops or CDRs. These 6 loops, 3 from the V_H and 3 from the V_L domains, come together in space to create the binding pocket or groove (Nisonoff, 1982). Estimates of the size of the binding site, the volume circumscribed by these loops, in terms of the carbohydrate antigen have ranged from 2 up to 8 hexose residues (Kabat, 1966, 1976). Digestion of whole IgG with the proteolytic enzyme papain provides a univalent, 50,000 Dalton antigen-binding fragment, Fab, which is the smallest and most readily prepared source of antibody for crystallographic study of the binding site.

Since a crystal is composed of ordered and regular arrays of molecules, reliable and reproducible production of crystals is facilitated by high sample purity. Lectins and antibodies may generally be purified by affinity chromatography and provided the intrinsic affinity of the binding site is at least 10^4 M^{-1}, this method is feasible (Spohr *et al.*, 1985). For IgG this permits the separation of active Fab from denatured or degraded sample and from the Fc portion that lacks an antigen-binding site. The Fc fragment usually has an acidic isoelectric point (pI) and the Fab a neutral to basic pI, thereby permitting the use of ion exchange as an effective method of purification. Protein A or protein G columns may also be used to remove Fc.

1.2 Protein Crystals

Protein crystals are grown by slow controlled precipitation from aqueous solution. The conditions are selected so that the protein is not denatured. Salts such as ammonium sulphate are used to precipitate proteins by a process called "salting out". Polyethylene glycol is also widely used because it is effective as a precipitant, and usually causes minimal denaturation. One of the most frequently used approaches for obtaining crystals is to bring the level of precipitant in a fairly concentrated solution of protein in aqueous buffer (5–10 mg/ml) to a point just below that required to cause precipitation. Water is allowed to evaporate slowly from this solution thereby effecting slow concentration. When precipitation occurs from the supersaturated solution crystals may form and the success of this process depends on many factors including protein concentration, level of protein purity, temperature, pH and ionic strength. Extensive screening of a variety of conditions is generally made in order to identify conditions that can reliably yield good quality crystals that diffract x-rays well. The hanging drop method is widely used and consists of an inverted coverslip from which is suspended 5–20 μl of protein solution containing precipitant at 50 % of the concentration required to precipitate the protein. The coverslip seals a small chamber that contains a larger volume of precipitant at a concentration close to that needed to bring about precipitation. On standing vapour diffusion slowly concentrates the protein solution in the hanging drop. Under favorable conditions several small crystals are obtained and these are used to seed further hanging drops in a manner that gives rise to large single crystals.

Diffraction studies require crystals that are at least 0.5 mm in their shortest dimension, although modern methods of data collection may allow the use of smaller crystals.

Antibody molecules because of their large size and inherent flexibility must first be cleaved to Fab fragments prior to crystallization. The proteolytic cleavage can produce ragged carboxy-terminal ends in the peptide chain that results in multiple forms of the fragment each with a characteristic isoelectric point. Alternatively, a two domain fragment, the Fv, can be obtained by cloning and expression (Anand *et al.*, 1991). It has been observed that isoelectric focusing and separation of isoforms prior to crystallization can yield bigger, more ordered crystals (Bott *et al.*, 1982; Boodhoo *et al.*, 1988) and eventually a higher resolution structure. It should be noted that a protein crystal contains a very large percentage of water molecules some of which are ordered and form an essential element of the crystal lattice and others (much larger in number) are disordered.

1.3 Collecting X-Ray Diffraction Data

A single crystal sealed in a thin walled capillary is mounted in the path of a narrow beam of x-rays, situated between the source of x-rays and an x-ray detector. The simplest form of detector is an x-ray film that displays dark spots where x-rays are diffracted and impinge on the film. The intensity of the spot is proportional to the number of x-rays striking the emulsion. The spots are referred to as reflections. In order to compute an image of the molecule causing diffraction the intensity and direction of each diffracted beam must be determined. Data collection is simplified by devices such as area detectors which replace x-ray film by electronic detection of diffracted x-rays and directly digitize the direction and intensity of reflections.

Single atoms or molecules scatter only a small portion of the x-ray beam and the success of the method relies upon the ordered array of atoms in a crystal to create a detectable diffraction pattern. The smallest and simplest volume element that is representative of the whole crystal is called the unit cell. Numerous copies of the unit cell are arrayed in a single crystal and produce a diffraction pattern. Diffraction data are collected for all unique orientations of the crystal. The resulting list of intensities for each reflection is the raw material for determining the structure of molecules in the crystal.

1.4 Electron Density

Crystallography provides an image of the electron clouds of the atoms in the unit cell of a crystal. To deduce a structure, an electron density map must be obtained from the diffraction data, and eventually this map is converted to an image of the molecule. However, this physical model is only developed after several iterative stages in the refinement of the electron density map.

Diffraction data lists the intensities and direction of diffracted x-rays. Fourier transformation is used to convert a Fourier series description of the x-ray reflections into a Fourier series description of the electron density. However, to convert diffraction data into electron density another piece of essential information is required, the phase of the reflection. This must be determined independently and several methods are used. The heavy atom method (isomorphous replacement) involves crystallization of the protein as a heavy atom derivative (usually Hg, Pt or Au), which must be isomorphic with the native crystal. Collection of another diffraction data set for the heavy atom derivative may permit the computation of the phase of the reflections. Often more than one heavy atom derivative may be necessary to achieve this goal. Alternatively, molecular replacement, the use of solved structures of related proteins, may be used to compute phasing models. For instance, antibody molecules share a highly conserved structural motif and often a known antibody structure may be used to derive phases by placing the known antibody in the unit cell of the new antibody. This approach is particularly useful when comparing the structure of protein-ligand complexes, when the structure of the native protein is known.

The electron density map that first results from a successful outcome of the process described above is usually not immediately useful for matching against the known protein sequence. Iterative corrections to the phases and original diffraction data eventually produce an electron density map that allows a correlation with most amino acid main chain and side chain elements of the protein. As the model becomes more detailed, the computed phases improve and the iterative fitting and recalculation of electron density continuously improve. At all points during the refinement the atomic coordinates from least squares refinement in the current molecular model are used to compare the convergence of the diffraction data calculated for the model with the original diffraction data. Expected bond lengths and valence angles are almost always imposed on the current model to guarantee that the refinement process does not produce a model with improper stereochemistry.

1.5 Map Fitting/Model Building

When the electron density map becomes sufficiently clear, the protein chain can be traced through it. In fact, it is rarely possible to trace the complete peptide chain since the quality of the map varies from place to place, and likely even exhibits broken density in some sections. However, certain features will be discernible, such as α helices and sheets of β structure. Also hydrophobic amino acids with bulky side chains (Trp, Phe, Tyr) are often readily identified. Using computer graphics to display the electron density a stick model of the known sequence can be fitted to sections of the electron density map. New phases are calculated from the partial molecular model. These and subsequent iterations improve the electron density map allowing more protein sequence to be fitted to a continually improving map. Thus by making the model structurally realistic the resolution of the model is steadily improved. Regions of electron density that cannot be aligned with amino acid sequence are often built with polyalanine, since an alanine simulates the first β-carbon of the side chain. Knowledge of protein structural motifs such as pleated sheets can be used to make sensible guesses as to the positions of amino acid residues and atoms.

Molecular dynamics may be used to advantage to relax advanced structural models in an attempt to arrive at a structure that more closely matches the measured reflection intensities. In refinement by simulated annealing the model is allowed to move at high temperature to lift it out of local minima and then it is cooled slowly to find its preferred conformation. During this process the computer searches for the conformation of lowest energy with the assigned energy partially dependent on agreement with diffraction data. XPLOR is a widely used software package of refinement programs that is used for simulated annealing.

In the last stages of structure determination constraints and restraints are relaxed and agreement with experimental intensities are given high priority. At this stage ordered water molecules may be discernible and these are added to the model and occupancy may be allowed to vary since not all water sites in each unit cell may be occupied.

The primary measure of convergence between the experimental data and the model is the residual index or R-factor. A desirable target R-factor for a protein model with data to 2.5Å is 0.2. Structural parameters such as the bond lengths and angles are compared to an accepted set of values and the root-mean-square (rms) deviations from these values are computed. In a well-refined structure observed rms deviations on bond lengths should be no larger than 0.02Å and on bond angles no larger than 4°.

1.6 Precision of the Model

Although a crystal structure may be cited as "solved to a resolution of 2Å", this does not mean that objects that are less than 2Å apart remain unresolved. The resolution in this case refers to the model that takes into account diffraction from sets of parallel planes spaced as closely as 2Å. Depending on the R-factor and the rms errors, good data can yield atom positions that are precise to within 0.2–0.1 of the stated resolution (for a 2Å resolution structure with R-factor of 0.2 and rms error of 0.15 Å this corresponds to an average uncertainty of atom positions about one fifth the length of a carbon-carbon bond). Another useful check of a structure is Ramachandran plots that examine the torsional angles Φ and ψ that define backbone conformational angles. In practice the pair of Φ,ψ angles are severely restricted by steric repulsion. With the exception of glycine residues that lack a side chain group all Φ,ψ pairs should fall within allowed regions. Failing this, an adequate explanation should be provided in terms of structural constraints that overcome the energetic costs of the unusual conformation.

Vibration and disorder contribute to the uncertainty in atom positions. Thermal motion corresponds to vibration of an atom about its position of rest. Disorder refers to the fact that groups of atoms may not occupy identical positions in each unit cell. The temperature factor B_j should reflect thermal motion while the occupancy n_j reflects disorder. In practice it is difficult to separate these two sources of uncertainty. Plots of mean values for temperature factors against residue number are quite informative as to which portions of a structure show flexibility or motion.

Other limitations of crystallographic models that may arise include disordered regions, where portions of the protein are never found in the electron density maps. This presumably arises, because the region is highly disordered or in motion. For this reason it is not uncommon to find *N*- or C-terminal residues missing from a model. Possible explanations for unexplained electron density for which there is no apparent content include ions and reagents used in protein manipulations and purification.

When reading a protein structure paper and using the corresponding protein data bank coordinate file it is advisable to critically evaluate the various aspects discussed above (Brändén and Jones, 1990; Rhodes, 1993).

2 Recently Determined Crystal Structures

The number of solved crystal structures of oligosaccharide-protein complexes has increased rapidly in recent years. A list of structures, the coordinates of which may be found in the Protein Data Base is given (Table 1). Although many structures are solved without bound sugar, these are intrinsically less interesting since no information is available on the binding site interactions and the conformation of the bound state. This information is of potentially crucial importance to understanding protein function and vital to structure function relationships. In general the bound sugar conformation is significant since oligosaccharides are flexible molecules that sample many low-energy conformational states in solution (Siebert *et al.*, this volume). Therefore, it is tempting to model ligands into the protein site, if its structure is known (Oomen *et al.*, 1991; Rose *et al.*, 1993; Evans *et al.*, 1995). Unfortunately, it is difficult to model such ligand-protein interactions, because there is no generally accepted force-field to deal with protein-sugar interactions, no way to discriminate between the many accessible sugar conformations and no guide on initial placement of the ligand.

Two examples of the manual docking approach have been published for protein antibody complexes, a *Brucella* polysaccharide antigen-antibody complex (Oomen *et al.*, 1991; Rose *et al.*, 1993) and the group B meningococcal capsular antigen-antibody complex (Evans *et al.*, 1995). The Fabs of both antibodies were solved at intermediate resolution and oligosaccharides were manually positioned in the Fab binding site in a manner consistent with ligand-binding data. The advantage of these systems is that the location of antibody-binding sites is well-known but for new proteins such as novel lectins or sugar-binding bacterial toxins the typically shallow

Table 1. Solved Protein – Carbohydrate Crystal Structures Reported in the Protein Data Bank (PDB)

PDB Code	Name of system	Sugar sequence	Journal reference	Resolution of the solved structure (Å)	R value	Lectin (L)/ Antibody (A)	# of sugars in ligand	# of observed sugars	Structure and co-ordinates available from PDB
1MFA	Single chain Fv (Murine SE155–4)	α-D-Gal-(1–2)-[α-D-Abe-(1–3)]-α-D-Man-OMe	Zdanov *et al.* (1994)	1.7	0.166	A	3	3	Y
1MFB	FAB Fragment (Murine SE155–4)	α-D-Gal-(1–2)-α-D-Man-(1–4)-α-L-Rha-(1–3)-α-D-Gal-(1–2)-[α-D-Abe-(1–3)]-α-D-Man-(1–4)-α-L-Rha	Unpublished data	2.1	0.160	A	7	7	Y
1MFC	FAB Fragment (Murine SE155–4)	α-D-Gal-(1–2)-[α-D-Abe-(1–3)]-α-D-Man-(1–4)-α-L-Rha-(1–3)-α-D-Gal-(1–2)-α-D-Man-(1–4)-α-L-Rha	Unpublished data	2.1	0.163	A	7	4	Y
1MFD	FAB Fragment (Murine SE155–4)	α-D-Gal-(1–2)-[α-D-Abe-(1–3)]-α-D-Man-OMe	Bundle *et al.* (1994)	2.1	0.183	A	3	3	Y
1MFE	FAB Fragment (Murine SE155–4)	$(\text{-}\{α\text{-D-Gal-(1–2)-[α-D-Abe(1–3)]-α-D-Man-(1–4)-α-L-Rha-(1–3)}\}\text{-})_3$	Cygler *et al.* (1991)	2.0	0.185	A	12	3	Y
1CLZ	IGG FAB Fragment (Murine BR96)	nLe^Y	Jeffrey *et al.* (1995)	2.78	0.197	A	4	4	(6/96)
1CLY	IGG FAB Frag. (Human Chimera BR96)	nLe^Y	Jeffrey *et al.* (1995)	2.5	0.238	A	4	4	(6/96)
1LEM	Lentil Lectin *(Lens culinaris)*	α-D-Glucopyranose	Loris (1994)*et al.*	3.0	0.206	L	1	1	Y
1LES	"	Sucrose	Casset *et al.* (1995)	1.9	0.188	L	2	2	Y
1SLT	Galectin (Bovine Spleen)	GlcNAc-Gal	Liao *et al.* (1994)	1.9	0.167	L	2	2	Y
1LEC	"	"	"	"	0.187	"	"	"	"
1RIN	Pea Lectin *(Pisum sativum)*	Methyl 3,6-di-*O*-(α-D-mannopyranosyl)-α-D-mannopyranoside	Rini *et al.* (1993)	2.6	0.183	L	3	1	Y

Table 1. (continued)

PDB Code	Name of system	Sugar sequence	Journal reference	Resolution of the solved structure (Å)	R value	Lectin (L)/ Antibody (A)	# of sugars in ligand	# of observed sugars	Structure and co-ordinates available from PDB
1SLA	Galectin-1 (Bovine heart galectin)	β-D-Gal-(1–4)-β-D-GlcNAc-(1–2)-α-D-Man-(1–6)-[β-D-Gal-(1–4)-β-D-GlcNAc-(1–2)-α-D-Man-(1–6)-]-β-D-Man-(1–4)-β-D-GlcNAc	Bourne *et al.* (1994)	2.45 (H form)	0.20	L	8	8	Y
1SLC	"	"	"	2.15 (T Form)	0.194	"	"	"	"
1SLB	"	"	"	2.3 (M form)	0.177	"	"	"	"
1HLC	S-Lac lectin (Human L-14-II)	Lactose	Lobsanov *et al.* (1993)	2.9	0.177	L	2	2	Y
1LOC	Isolectin-I (Legume, *Lathyrus ochrus*)	*N*-Carboxy-*N*-methyl-muramyl-dipeptide D-ALA-D-IGLN	Bourne *et al.* (1994)	2.0	0.189	L	1	1	Y
1LOD	"	Muramic acid	"	2.05	0.197	L	1	1	Y
1LOA	"	Methyl α-D-glucopyranoside	Bourne *et al.* (1990)	2.2	0.179	L	1	1	Y
1LOB	"	Methyl α-D-mannopyranoside	Bourne *et al.* (1990)	2.0	0.182	L	1	1	Y
1LTE	Lectin (Coral tree, *Erythrina corallodendron*)	Lactose	Shaanan *et al.* (1991)	2.0	0.190	L	2	2 (Gal fully, and Glc barely)	Y
2MSB	Mannose-binding protein A (Lectin domain)	{α-D-Man-(1–6)-[α-D-Man-(1–3)]-α-D-Man-(1–6)}-[α-D-Man-(1–2)-α-D-Man-(1–3)]-β-D-Man-(1–4)-β-D-GlcNAc-(1–4)-β-D-GlcNAc-(1-N)-Asn	Weis *et al.* (1992)	1.7	0.174	L	7	5 (plus three additional atoms	Y

Table 1. (continued)

PDB Code	Name of system	Sugar sequence	Journal reference	Resolution of the solved structure (Å)	R value	Lectin (L)/ Antibody (A)	# of sugars in ligand	# of observed sugars	Structure and co-ordinates available from PDB
1LGB	Legume Isolectin II (*Lathyrus ochrus*)	Lactotransferrin glycopeptide	Bourne *et al.* *To be published* (*Also*:J. Mol. Biol. (1992) **227**:938)	3.3	0.210	L	9	?	Y
1LGC	"	Lactotransferrin glycopeptide	"	2.8	0.185	L	13	?	Y
1WGC	Wheat Germ Agglutinin (Isolectin 1, *Triticum vulgaris*)	*N*-Acetyl-neuraminyl-lactose	Wright *et al.* (1990)	2.2	0.172	L	3	3	Y
1MSA	Mannose-specific agglutinin (*Galanthus nivalis*)	Methyl α-D-mannopyranoside	Hester *et al.* (1995)	2.29	0.177	L	1	1	Y
5CNA	Concanavalin A (Jack bean, *Canavalia ensiformis*)	Methyl α-D-mannopyranoside	Naismith *et al.* (1994)	2.0	0.199	L	1	1	Y
1SBA	Soybean agglutinin	{β-D-Gal-(1–4)-β-D-GlcNAc-(1–2)}-[β-D-Gal-(1–4)-β-D-GlcNAc-(1–6)]-GalOR	Dessen *et al.* (1995)	2.6	0.201	L	5	4+	Y

sugar-binding site may not be easily identified (Stein *et al.*, 1994). An algorithm has been proposed to investigate sugar-protein interactions and group potential binding site hits (Imberty and Perez, 1994). At this point neither the quantitative approach nor the manual fitting approach seem particularly satisfactory since the overall binding energy is low, the relative strengths of potential hydrogen bonds are difficult to assess, the potential roles of water molecules remain unknown and the positioning of the ligand should have a stringent confidence level of less than ± 0.5 Å, if the primary interactions are to yield meaningful free energy differences. Taken together, these requirements are highly demanding and preclude the confident docking of ligands by manual or computer docking approaches, and highlight the importance of solved structures of oligosaccharide complexes. Structurally, the most well-documented antibody-carbohydrate system is monoclonal antibody (Mab) Se 155.4 and its corresponding bacterial antigen (Cygler *et al.*, 1991; Bundle *et al.*, 1994; Zdanov *et al.*, 1994). As indicated in Table 2, several crystal structures of this binding site have been solved with a variety of oligosaccharides. The counterpart in lectin crystal structures is the *Griffonia simplicifolia* lectin complexed with the Lewis[b] blood group antigen, a comparably well defined lectin-carbohydrate structure that is at least as well or more thoroughly defined in terms of binding data and resolution of protein-sugar interactions (Delbaere *et al.*, 1990, 1993).

Table 2.

Complex	Electron density for pyranose residues	Conformation of bound ligand
Fab + (Gal[Abe]ManRha)$_3$	3	A[a]: Abe → Gal H-bond
Fab + Gal[Abe]Man→OMe	3	A[b]
Fab + GalManRhaGal[Abe] ManRha	7	B: Abe → Gal H-bond bridged by water
Single chain F$_v$ + Gal[Abe]Man → OMe	3	B[c]

a Cygler *et al.* (1991)
b Bundle *et al.* (1994)
c Zdanov *et al.* (1994)

3 The Structure of Mab Se 155.4 and the Abequose-Based Epitope

One of the primary difficulties in seeking a crystal structure of an antibody-oligosaccharide complex is the fact that most anti-carbohydrate antibodies belong to the IgM class. Carbohydrates readily induce IgM antibodies, but these high molecular weight pentameric molecules are virtually impossible to efficiently fragment into Fab. The binding sites also exhibit very low intrinsic affinities. Serendipity or substantial efforts are required to secure an antibody of the IgG class, the best source of Fab. The IgG1 monoclonal antibody Se 155.4 was produced in response to immunization of BALB/c mice with whole cells of killed *Salmonella* bacteria. The mice were boosted by an injection with protein-oligosaccharide conjugate and immunization was continued with the whole cell vaccine (Bundle *et al.*, 1994). The mice received 8 injections over a 6 month period prior to generating the Mabs by the hybridoma technique. Even with such long immunization schedules it is often difficult to induce a secondary response. The antibody was purified from ascitic fluid by affinity chromatography and exhibited a conveniently pH dependent binding profile, such that it could be eluted from an affinity column prepared from the polysaccharide antigen at pH 4.5. Proteolytic cleavage by papain gave excellent yields of Fab that was also easily recovered and eluted from the same affinity column. Sequencing of the Fc portion of the heavy chain showed ragged C-terminal ends. The sequence of the λ1 light chain showed that 8 somatic mutations of the germ line Vλ1/Jλ1 sequence had taken place during immunization, an indication that a secondary immune response had taken place. Although the crude Fab preparation from the affinity column gave crystals, their quality and size was improved by isolating protein with a single pI, either by ion exchange chromatography or by quantitative isoelectric focusing (Bott *et al.*, 1982; Boodhoo *et al.*, 1988; Bundle *et al.*, 1994).

3.1 Features of a Dodecasaccharide-Fab Complex

Several structures of the Se 155.4 Fab complexed with various oligosaccharides have been solved (Table 2). The first crystal structure was obtained for a complex with a dodecasaccharide ligand (Cygler *et al.*, 1991). The *O*-antigens have a biological repeating unit structure shown in Figure 1a and phage enzyme degradation of this polysaccharide cleaves the Rha-Gal linkage to give oligomeric fragments with a degree of polymerization, n = 2–5. Since the polysaccharide precipitated the antibody, it was reasoned that the most likely epitope consisted of the internal repeating sequence of the polysaccharide and this was confirmed by the reaction of Se 155.4 with the synthetic conjugates, tetrasaccharide **1** and trisaccharide **2** (Figure 1b). Thus, it was inferred that the antibody accommodated at least the trisaccharide. This inference was later supported by the crystal structure and by calorimetry measurements on a panel of modified oligosaccharides ranging in size from the monosaccharide of abequose up to the tetrasaccharide **1** (Bundle *et al.*, 1994). When the crystal structure of the dodecasaccharide complex was examined, clear electron density for the oligosaccharide was only observable for three hexose residues (Figure 2a). The density at the base of a binding pocket could be clearly assigned to the dideoxyhexose. It should be appreciated that electron density does not necessarily readily distinguish between various orientation of a pyranose ring. Thus, depending on the stereochemistry of a sugar several orientations of a hexopyranose may be realistically fitted to the electron density. However, for the case in point the replacements of the hydroxymethyl group at C-5 by a methyl group and a hydroxyl group at C-3 by a hydrogen atom allows the electron density attributable to the abequose residue to be readily identified. Fitting of three hexoses (Abe, Man, Gal) to the electron density produced the branched trisaccharide **2**, a conclusion that nicely accounted for the observation that the tetrasaccharide **1** was not a better inhibitor than **2**.

This crystal structure and that of the native Fab without bound sugar revealed several important details concerning the type of amino acids in the binding site, the size of the site, bound sugar conformation, the role of immobilized water molecules and protein movement on sugar-binding:

- aromatic amino acids were the dominant antigen contact residues that lined the binding site.
- the binding site was filled by a trisaccharide and one bound water molecule (Figure 2b).
- the dideoxyhexose was completely buried in the binding site with both of its hydroxyl groups involved in hydrogen-bonding networks, one of which involved the bound water molecule.
- a saccharide-saccharide hydrogen bond results from a relatively small but notable change in the conformation of the bound antigen.
- mannose and galactose were each partially exposed to bulk water and lay across the surface of the V_H and V_L domains.
- the binding site corresponded to a cavity or pocket rather than a groove type site, with the result that antigen runs across the V_H and V_L domains rather than parallel to the V_H:V_L interface.
- only very small changes of the amino acid side chain positions occurred when the site was filled by antigen.
- although a dodecasaccharide was co-crystallized with Fab, only 3 of the 12 hexoses residues were located by electron density, implying a disordered and mobile state for the other 9 residues.

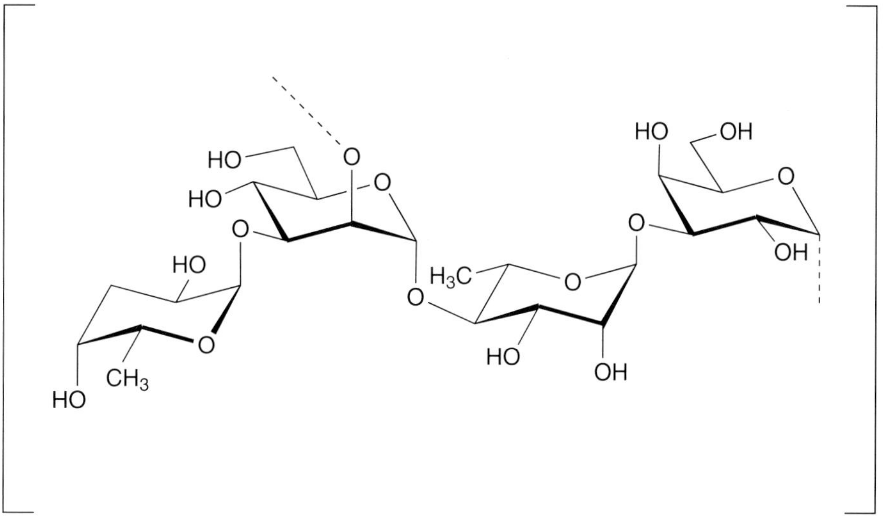

Figure 1a. The biological repeating unit of the *Salmonella* serogroup B *O*-antigen. Phage degradation breaks the Rha-Gal bond to yield oligosaccharides with the sequence Gal[Abe]ManRha.

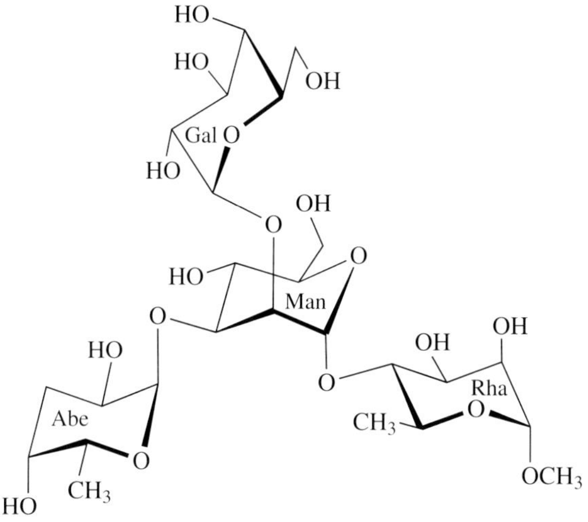

Figure 1b. Tetrasaccharide methyl glycoside **1** corresponds to the chemical repeating unit liberated by phage enzyme. Trisaccharide methyl glycoside **2**, the structural element that fills the Se 155.4 binding site.

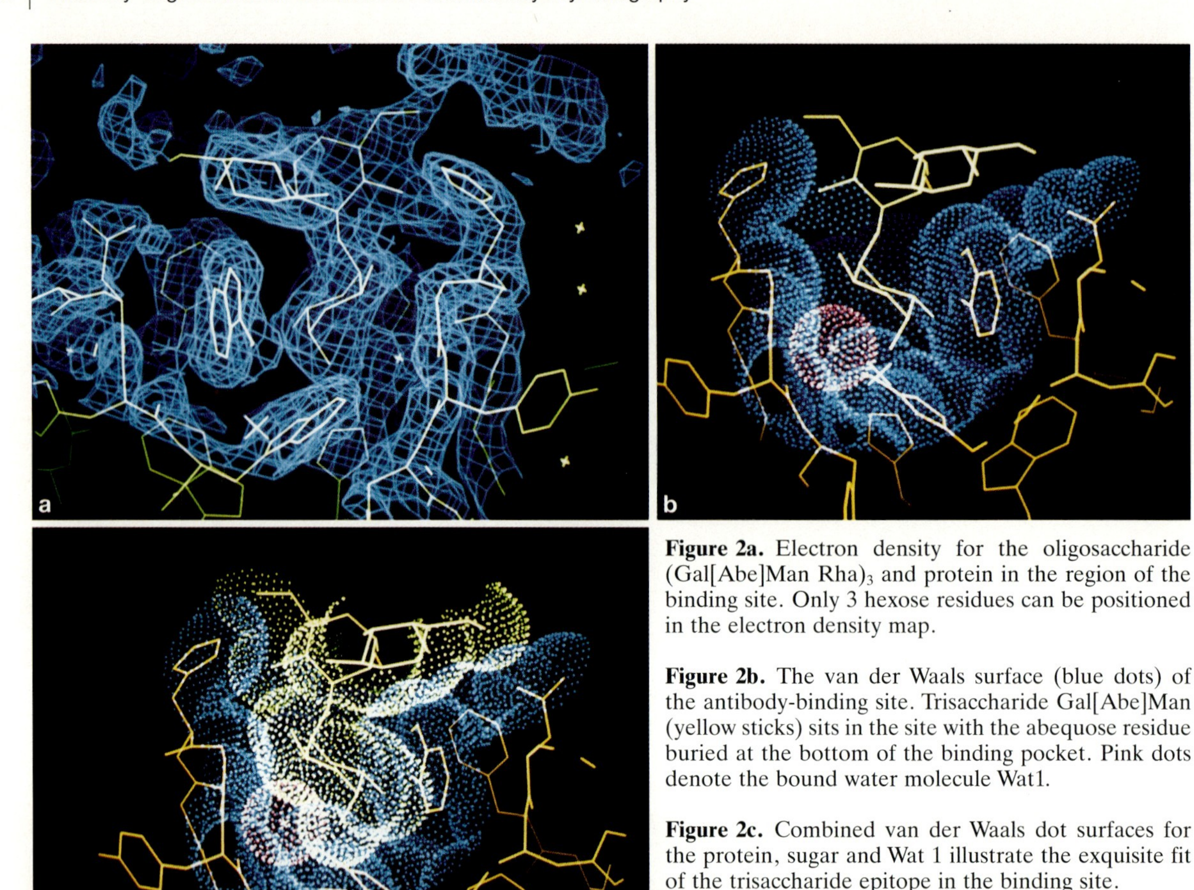

Figure 2a. Electron density for the oligosaccharide (Gal[Abe]Man Rha)$_3$ and protein in the region of the binding site. Only 3 hexose residues can be positioned in the electron density map.

Figure 2b. The van der Waals surface (blue dots) of the antibody-binding site. Trisaccharide Gal[Abe]Man (yellow sticks) sits in the site with the abequose residue buried at the bottom of the binding pocket. Pink dots denote the bound water molecule Wat1.

Figure 2c. Combined van der Waals dot surfaces for the protein, sugar and Wat 1 illustrate the exquisite fit of the trisaccharide epitope in the binding site.

3.2 Hydrogen-Bonding and the Structured Water Molecule

The antigen presents a relatively hydrophobic surface to the protein compared to that of typical hexoses. The antigen is known as one with an immunodominant 3,6-dideoxyhexose residue (Bundle, 1990). This sugar is buried in the deepest part of the antibody site. A water molecule (Wat 1) that is observed in the unliganded native Fab structure remains at the bottom of the pocket (Figure 2b) in every one of the five liganded structures that have been solved to date (Table 2). The role of this water molecule in coordinating hydrogen bonds between the sugar and the protein accounts for its ubiquitous presence in the structure. This water molecule may also be seen to contour the surface at the bottom of the binding site, filling what would otherwise be a partial void and as an extension of the protein surface ensures a tight fit and optimal complementarity between the sugar and protein (Figure 2c).

The hydrogen-bonding map for the dodecasaccharide-Fab complex (Figure 3) demonstrates the importance of conjugated networks

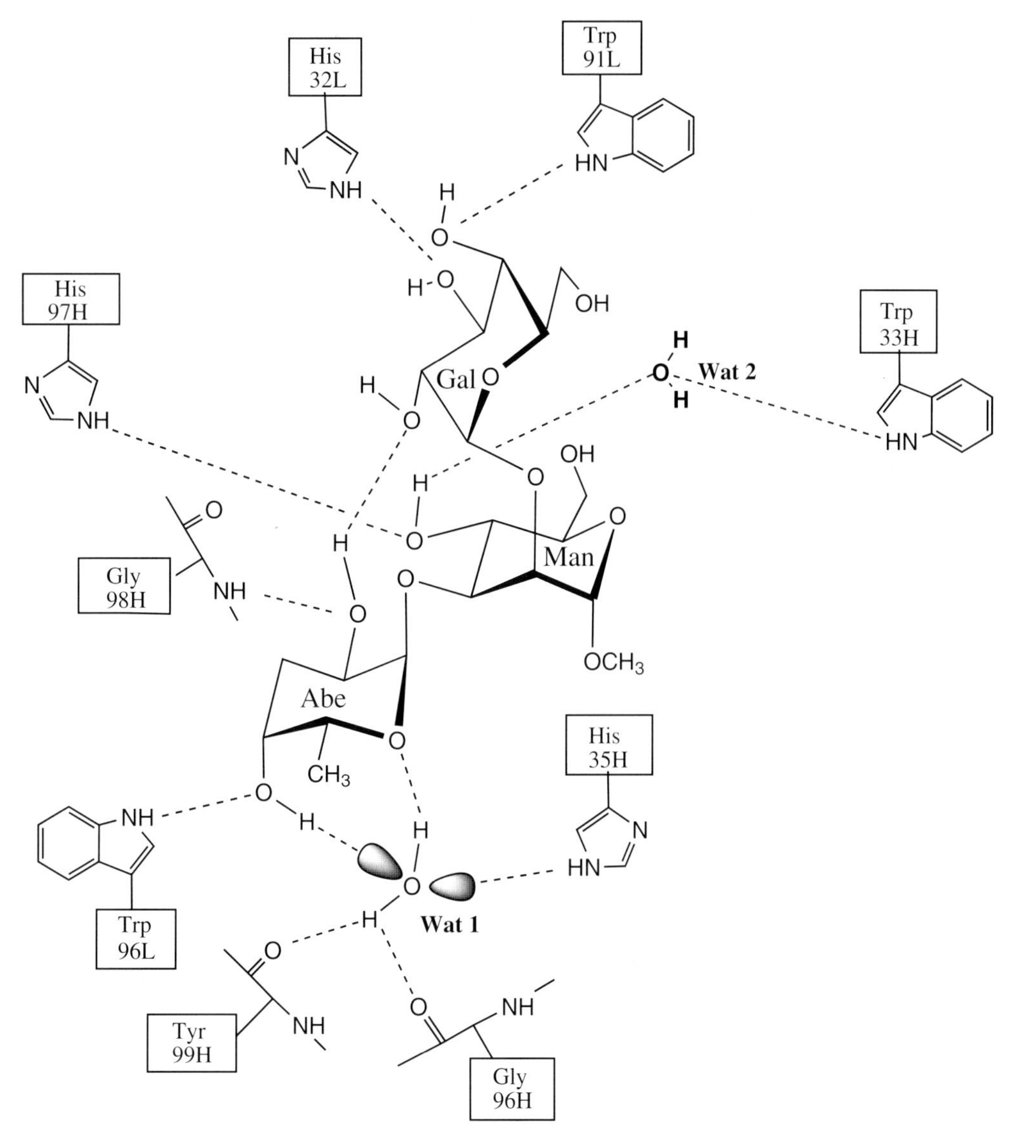

Figure 3. The hydrogen-bonding map for Fab-dodecasaccharide and Fab-trisaccharide complexes. The bound conformation I places Abe *O*-2 and Gal *O*-2 in proximity principally through rotation of the Gal-Man torsional angle Φ.

of hydrogen bonds of the type first reported by Quiocho (1986, 1989): *N*-H...*O*-H...A, where A is a hydrogen bond acceptor such as another hydroxyl group, an oxygen atom of a water molecule or an amino acid side chain. The bound water molecule (Wat 1) is potentially anchored at the base of the binding site by three residues His 35H, Gly 96H and Tyr 99H. Both of the latter residues originate in the third hypervariable loop of the heavy chain and immobilize the water molecule, probably by a bifurcated hydrogen bond to main chain carbonyl groups. His 35H therefore donates a hydrogen bond to the water molecule. This water molecule is then able to accept a hydrogen bond from the Abe *O*-4 hydroxyl group when it is hydrogen-bonded to Trp 96L. Finally, water is able to maximize its hydrogen-bonding potential by contact formation with the ring oxygen atom, *O*-5 of abequose acting as an acceptor. The remaining C-2 hydroxyl group of abequose accepts a hydrogen bond from the amide nitrogen of Gly 98H, another residue of the H-chain third hypervariable loop. The only close acceptor for the hydrogen of *O*-2 Abe is in fact the Gal C-2 hydroxyl group. This oxygen atom is only sufficiently close to Abe *O*-2 for hydrogen-bonding, if the Gal residue undergoes a small but significant torsional rotation at the Gal-Man anomeric linkage (Table 3). Mannose *O*-4 accepts a His 97H hydrogen bond and a water molecule (Wat 2) accepts the hydrogen. Two additional hydrogen bonds can be postulated in this structure. They are His 32L with Gal *O*-3 and Trp 91L with Gal *O*-4. Neither the Gal nor the Man *O*-6 atoms are involved in hydrogen bonds within the unit cell.

3.3 Attributing Structural Features to Binding Energy

Several questions immediately arise concerning important interactions between sugar and protein. Which hydrogen bonds are essential to the stability of the complex? What is the relative importance of the intramolecular saccharide-saccharide hydrogen bond? To what extent is this interaction essential to binding and does protein induce the conformational change that makes this hydrogen bond possible? The answers to these questions cannot be found in this structure without reference to other data. Calorimetry measurements, also discussed by Gupta and Brewer (this volume), and EIA inhibition data provided some of the necessary information (Bundle *et al.*, 1994). The buried hydrogen bonds involving Wat 1 and the abequose appear to be crucial to the interaction, since none of the isomeric 3,6-dideoxyhexoses nor any of their monodeoxygenated analogues bind the antibody. Whereas these changes produce inactive antigens, abolition of the partially solvent-exposed Man *O*-4 hydrogen bond results in 1 kcal mol^{-1} lower binding energy, while absence of hydrogen-bonding potential from individual sites within the galactose unit have a much smaller effect (less than 0.5 kcal mol^{-1}). These data suggest that not only are the solvent-exposed hydrogen bonds involving Gal *O*-3 and *O*-4 of minor importance but that the saccharide hydrogen bond to Gal *O*-2 is not essential to an active antigenic determinant. However, movement of the galactose residue about the Gal to Man glycosidic linkage does remove a possible unfavorable saccharide-protein interaction. For this reason it was originally concluded that the Abe *O*-2/Gal *O*-2 hydrogen bond results from a protein-induced conformational change. This was supported by the observation of the same intramolecular hydrogen bond in a second structure between trisaccharide **2** and Fab (Table 2) (Bundle *et al.*, 1994).

3.4 Oligosaccharide Conformational Change on Binding

When the interaction between Fab and **2** was investigated by transferred NOE experiments as outlined in principle by Siebert *et al.* (this volume), it was noted that the distance constraints for the trisaccharide **2** were consistent with two families of conformers (Table 3) (Bundle *et al.*, 1994). One clustered around the observed conformation I seen in the Fab oligosaccharide complexes (Cygler *et al.*, 1991; Bundle *et al.*, 1994), the other set clustered around oligosaccharide conformations that were close to those of the minimum-energy conformation. At this point it was concluded that the second set of conforma-

Table 3

Bound conformation	Glycosidic torsional angles Abe → Man ϕ, υ	Gal 1 → Man ϕ, ψ
Conformation A[a] (Fab + trisaccharide)	-45°, -15°	-10°, -30°
Conformation B[b] (scFv + trisaccharide)	-45°, -15°	-35°, +23°
Global minimum energy	-51°, -13°	-49°, -21°
NMR transferred NOE (sol'n)	-40°, -20°	-10°, -25°

a Bundle *et al.* (1994)
b Zdanov *et al.* (1994)

tions was not represented in the crystal structure and that a wider range of galactose conformations could be sampled in solution than in the crystal.

3.5 A Different Bound Oligosaccharide Conformation in a Single Chain Fv-Trisaccharide Complex

Crystallization of **2** with an *E. coli* expressed single chain antibody Fv fragment showed that despite covalent attachment of the V_L to the V_H domain, the position of the binding site residues were virtually identical to their position in the Fab (Zdanov *et al.*, 1994). Furthermore, the conformation of trisaccharide **2** (conformation II) was quite distinct from conformation I seen in earlier structures of Fab (Table 3). The hydrogen-bonding map for this structure now showed a third structured water molecule (Wat 3) had bridged the Abe *O*-2/Gal *O*-2 via hydrogen bonds (Figure 4 see page 326). In order to accomplish this one may either consider the relaxation of the constrained Gal-Man glycosidic conformation by insertion of a water molecule in conformation I, or more likely the acceptance of the lower energy conformation II with a bound water molecule. This water molecule is able to satisfy the requirement that a buried hydroxyl group (Abe *O*-2) acting as a hydrogen bond acceptor must in turn find a suitable acceptor for its hydrogen atom made relatively more acidic when its parent oxygen atom accepts a hydrogen bond. Examination of the Φ,ψ torsional angles of the two inter-residue glycosidic linkages of **2** revealed that the Abe-Man linkage adopted relatively unchanged angles in both bound conformers I and II, whereas the Gal-Man linkage showed significant conformational change in both the Φ and ψ angles (Table 3). Thus, the most solvent-exposed residue Gal is able to adopt 2 conformations in the bound state. Needless to say, the hydrogen-bonding pattern involving the galactose residue shows the largest change between the two structures. Conformation II has only one direct protein-sugar hydrogen bond from Trp 91L to Gal *O*-2 (bound to Gal *O*-4 in conformation I). The hydrogen of Gal *O*-2 is presumable hydrogen-bonded to the bridging water molecule Wat 3 and a completely new feature of this scheme is the hydrogen-bonding of Wat 3 to Wat 4 which in turn is secured by a hydrogen bond to Man *O*-6. There are now no hydrogen bonds to Gal *O*-4 and His 32L formerly directly bound to Gal *O*-3.

3.6 Bound Conformation II in a Fab-Heptasaccharide Complex

Bound conformation II was also found in a fourth crystal structure formed when Fab crystallized with heptasaccharide B (Table 2). This heptasaccharide was designed to be capable of monogamous binding (Figure 5a, see page 327) since it contained a single abequose residue (Baumann *et al.*, 1993). The obligatory binding mode for the related heptasaccharide A (Baumann *et al.*, 1993) was predicted by molecular modeling to clash with protein at the central Gal residue (Figure 5b). While heptasaccharide A has a steric clash with the protein when it adopts low-energy conformations, heptasaccharide B orients its extended chain to the other side of the site and in its low-energy conformations no steric clashes with protein result. In fact, the extended chain makes favorable contacts beyond the immediate binding site as well as many hydrogen bonds to the protein via water molecules. Calorimetry data show that the enthalpy of binding is identical for both heptasaccharide A and B and the observed difference in ΔG can be attributed to an entropy penalty imposed on binding heptasaccharide A (Bundle *et al.*, 1994). The crystal structure of heptasaccharide B complexed with Fab shows all seven monosaccharide residues can be located by electron density, while a fifth crystal structure of heptasaccharide A with Fab shows density for the 3 binding site residues plus the immediately adjacent rhamnose residue. The 3 saccharide units Gal, Man, Rha that must move away from low-energy conformations to avoid a steric clash with protein (Figure 5b) presumably adopt a disordered state, with the result that electron density for them is lacking.

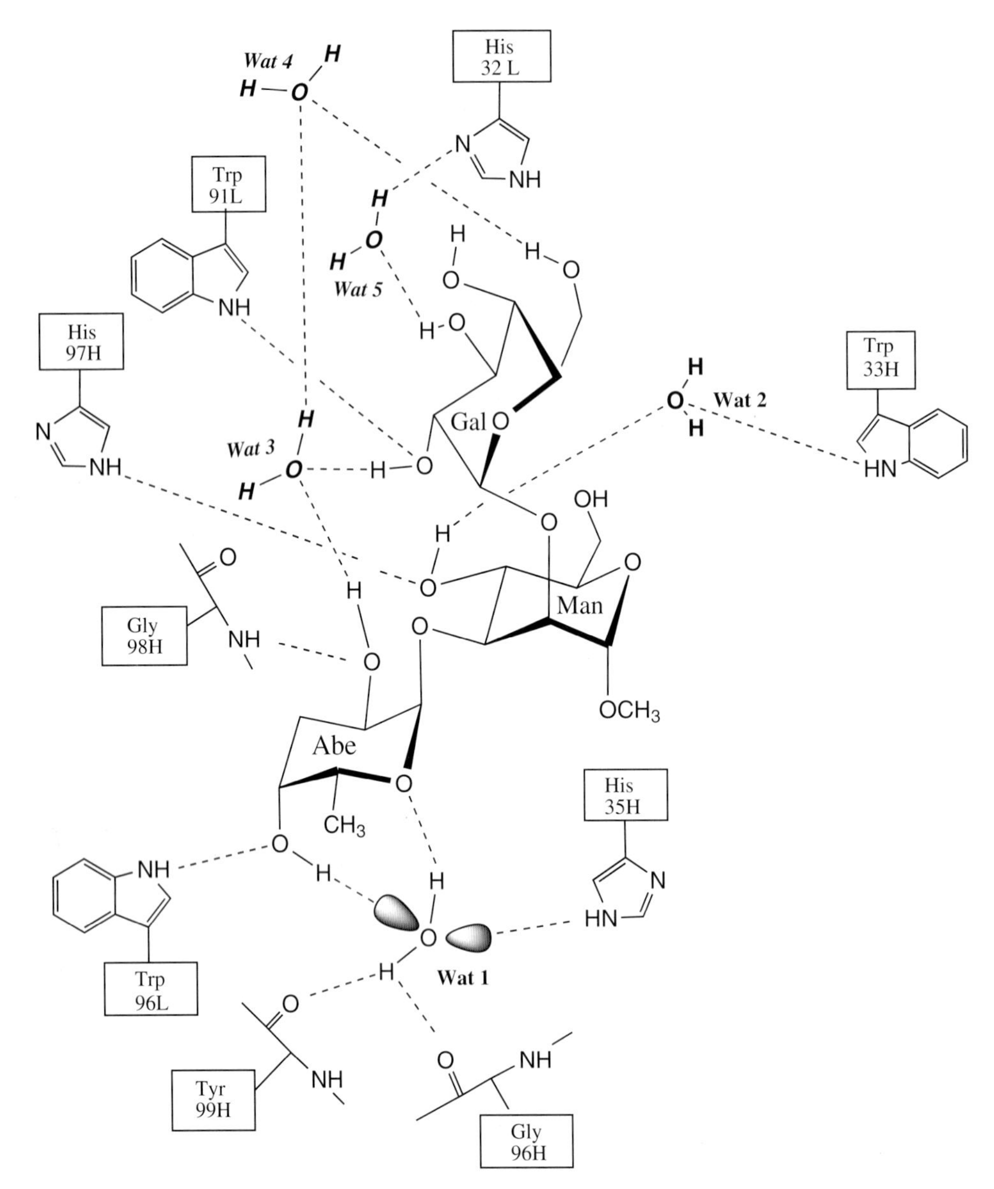

Figure 4. Hydrogen-bonding map for the scFv-trisaccharide **2** complex. Additional ordered water molecules Wat 3-Wat 5 give rise to a different bound conformation II. Wat 3 allows the Gal-Man Φ torsional angle to adopt low energy values. The top of the diagram represents the most solvent-exposed region of the bound epitope.

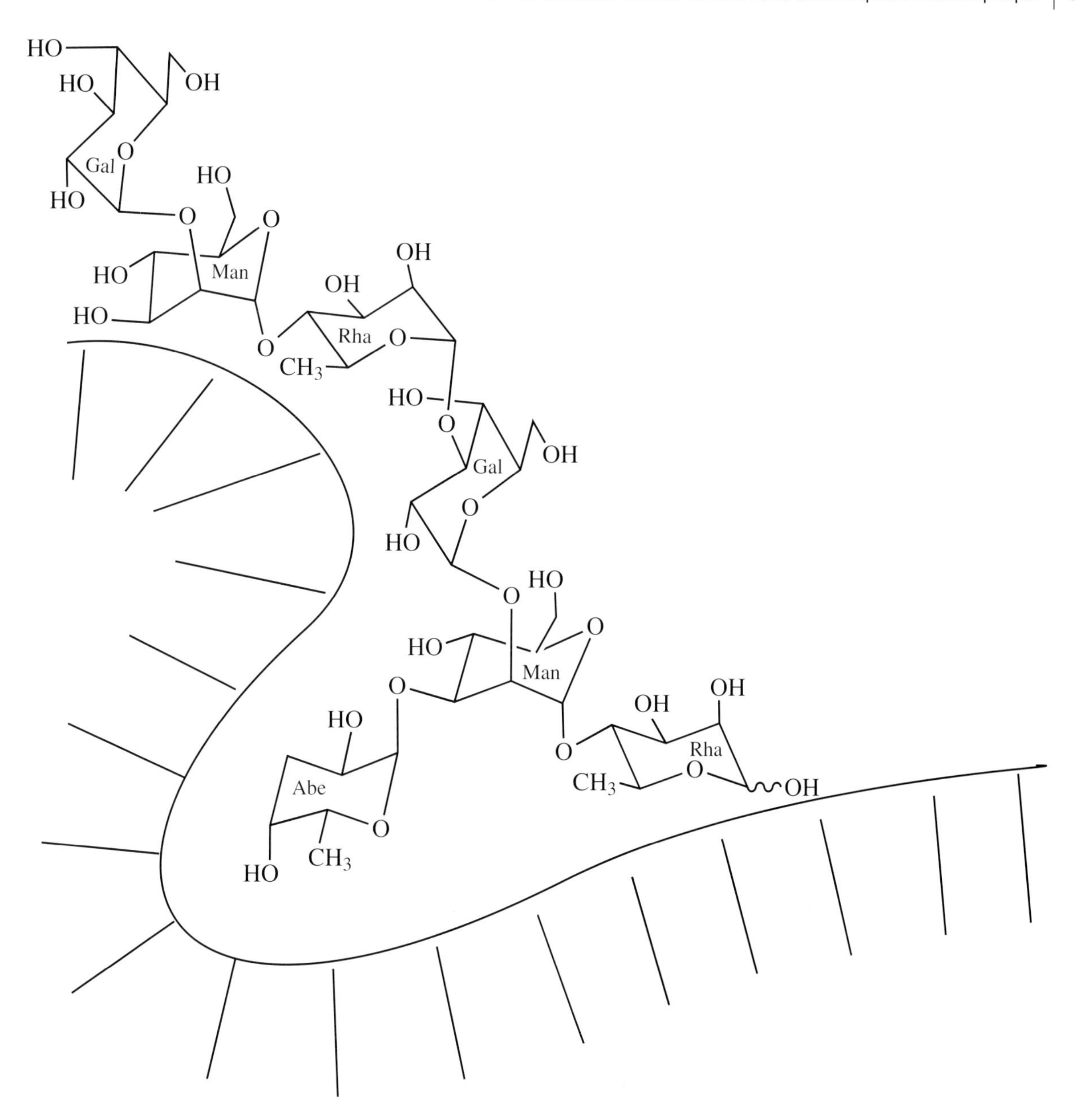

Figure 5a. Heptasaccharide B GalManRhaGal[Abe]ManRha can be accepted in the binding site and places the segment GalManRha in an orientation that does not clash with protein but instead makes favorable contacts (not shown).

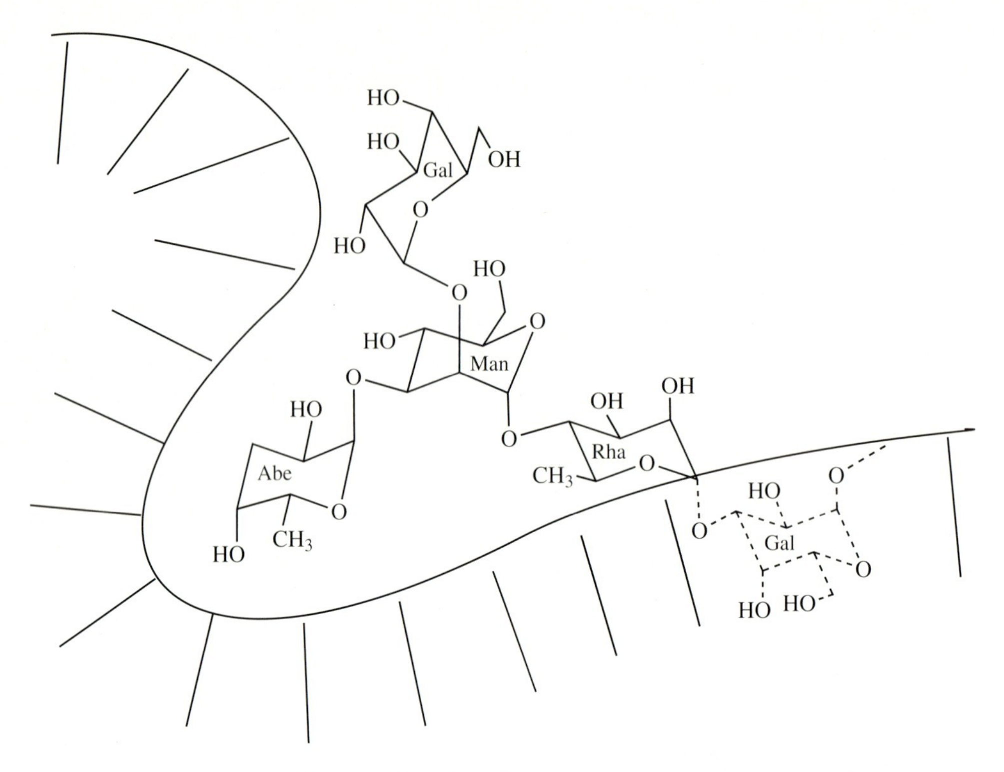

Figure 5b. The heptasaccharide A, Gal[Abe]ManRhaGal-ManRha places its extended chain in the opposite direction to the extended chain of heptasaccharide B. The conformation about either one or both of the internal Man-Rha or the Rha-Gal linkages must adopt Φ, ψ torsional angles that are shifted significantly from those of low energy conformations if the Gal (dotted ring) and subsequent residues, Man-Rha (not shown) are to avoid a steric clash with protein. Heptasaccharide A therefore binds with a more unfavorable entropy than heptasaccharide B (Bundle *et al.*, 1994).

4 Role of Water Molecules

The several solved crystal structures for the Se 155.4 binding site reveal multiple roles for water. The immobilized Wat 1 not only maximizes the number of hydrogen bonds that secure it and abequose to the antibody, it also contours the base of the site to provide improved van der Waals contacts (Figures 2b and 2c). This water molecule and Wat 2 are common features of both bound ligand conformations. The three additional water molecules observed with the bound conformation II in the crystal structure of trisaccharide **2** bound to single chain Fv are particularly interesting. As the Gal-Man torsional angle opens up to give a lower energy conformation, the hydrogen bonds located in the solvent-exposed regions of conformation I can no longer be maintained and Wat 3 and Wat 5 mediate these hydrogen bonds via an extended network of conjugated hydrogen bonds. In the case of Wat 3 it now becomes hydrogen bonded to Wat 4. This rearrangement of ordered water molecules in the immediate vicinity of the galactose residue may be the trigger that accounts for the altered conformation about the Gal-Man glycosidic linkage.

Extended networks of conjugated hydrogen bonds often involving ordered water molecules are common features both in the crystal structure of low molecular weight saccharides (Jeffrey and Sänger, 1991) and in a variety of oligosaccharide-protein complexes (Quiocho, 1989; Bourne *et al.*, 1990, 1992).

Finally, analysis of native Fab structure shows that 5 water molecules occupy the base of the empty binding pocket. Whilst Wat 1 has an occupancy of 1, the four other water molecules have an occupancy of less than 1 per Fab molecule. It seems significant that these water molecules are located at sites that correspond to the position of Abe *O*-2, *O*-4 and *O*-5 and C6. Based on recent estimates of the energy required to position ordered water molecules the displacement of these molecules to bulk solvent can be expected to provide a significant contribution to binding energy (Dunitz, 1994).

5 Solvent-Exposed Hydrogen Bonds

The energetic significance of solvent-exposed hydrogen bonds observed in these crystal structures is of considerable interest. Replacement of galactose by a methyl group results in less than 0.3 kcal mol^{-1} loss in free energy of binding. Further, stereochemical changes Glc for Gal, and functional group substitutions $GalNH_2$ for Gal and replacement of *O*-4 and *O*-6 by chlorine all cause minor losses in ΔG (Bundle *et al.*, 1994). It would therefore seem appropriate to categorize hydrogen bonds to these water molecules as possessing a minor role in complex stability, especially since in solution the competition with water from bulk solvent must provide a dynamic fluctuation. Water molecules such as Wat 3-Wat 5 (Figure 4) do provide a glimpse of the transition that must occur at the interface of the complex with bulk solvent and even though in dynamic equilibrium. The stereochemical requirements for hydrogen bonds oriented toward these water molecules from the first hydration shell highlights the propagation of such ordering on subsequent hydration shells and the possibly dominant role that solvation reorganization seems to play in contributing to the energetics of sugar-protein complexes (Lemieux, 1989, 1993; Lemieux *et al.*, 1991; Beierbeck *et al.*, 1994; Chevenak and Toone, 1994; Siebert *et al.*, this volume).

6 Conformational Change in the Binding Site

In the bound state the conformation of the buried portion of the epitope Abe-Man maintains an almost invariant conformation that lies close to low-energy conformations predicted by force-fields such as HSEA and others that provide a term for the exo-anomeric effect. The situation changes for the solvent-exposed residue and the two observed conformations about the Gal-Man are significantly shifted from low-energy conformations (Meyer, 1990; Stuike-Prill and Meyer, 1990). In conformation I the largest variation is found in the Φ angles while conformation II shows the largest variation from low-energy conformers via a shift in the ψ angle. Formation of inter-saccharide hydrogen bonds and the accommodation of order water molecules appear to be involved in this process but it is not entirely clear what triggers the crystallization of either bound ligand conformation. NMR data and molecular modeling suggest that the galactose residue may adopt a range of conformations in the bound state.

7 Summary

To date three crystal structures of antibody-oligosaccharide complexes have been reported. The general binding motifs seen in different sugar transport proteins and enzymes (Quiocho, 1986, 1989) have also been utilized – with variations – in complexes of sugar ligands with lectins and antibodies. One of the most important themes is the diverse and intriguing role of water. A more detailed understanding of the role of water molecules in carbohydrate-protein complexes is the subject of currently ongoing investigations in several laboratories.

References

Anand NN, Mandal S, MacKenzie CR *et al.* (1991): Bacterial expression and secretion of various single-chain Fv genes encoding proteins specific for a *Salmonella* serotype B *O*-antigen. In *J. Biol. Chem.* **266**:21874–9.

Baumann H, Altman E, Bundle DR (1993): Controlled acid hydrolysis of an *O*-antigen fragment yields univalent heptasaccharide haptens containing one 3,6-dideoxyhexose epitope. In *Carbohydr. Res.* **247**:347–54.

Beierbeck H, Delbaere LTJ, Vandonselaar M *et al.* (1994): Molecular recognition. XIV. Monte Carlo simulation of the hydration of the combining site of a lectin. In *Can. J. Chem.* **72**:463–70.

Boodhoo A, Mol CD, Lee JS *et al.* (1988): Crystallization of immunoglobulin Fab fragments specific for DNA. In *J. Biol. Chem.* **263**:18578–81.

Bott RR, Navia MA, Smith JL (1982): Improving the quality of protein crystals through purification by isoelectric focusing. In *J. Biol. Chem.* **257**:9883–6.

Bourne Y, Rougé P, Cambillau C *et al.* (1990): X-ray structure of a (α-Man(1–3)β-Man(1–4)GlcNAc)-lectin complex at 2.1 Å resolution. The role of water in sugar-lectin interaction. In *J. Biol. Chem.* **265**:18161–5.

Bourne Y, Roussel A, Frey M *et al.* (1990): Three-dimensional structures of complexes of *Lathyrus ochrus* isolectin I with glucose and mannose: fine specificity of the monosaccharide binding site. In *Struct. Funct. Genet.* **8**:365–76.

Bourne Y, Rougé P, Cambillau C (1992): X-ray structure of a biantennary octasaccharide-lectin complex refined at 2.3Å resolution. In *J. Biol. Chem.* **267**:197–203.

Bourne Y, Ayouba A, Rougé P *et al.* (1994): Interaction of a legume lectin with two components of the bacterial cell wall. A crystallographic study. In *J. Biol. Chem.* **269:**9429–35.

Bourne Y, Bolgiano B, Liao DI *et al.* (1994): Crosslinking of mammalian lectin (galectin-1) by complex biantennary saccharides. In *Nature Struct. Biol.* **1**:863–9.

Brändén C-I, Jones TA (1990): Between objectivity and subjectivity. In *Nature* **343**:687–9.

Bundle DR (1990): Synthesis of oligosaccharides related to bacterial *O*-antigens. In *Top. Curr. Chem.* **154**:1–37.

Bundle DR, Altman E, Auzanneau F-I *et al.* (1994): The structure of oligosaccharide-antibody complexes and improved inhibitors of carbohydrate-protein binding. In *Complex Carbohydrates in Drug Research*, (Alfred Benzon Symposium 36) pp 168–81, Copenhagen: Munksgaard.

Bundle DR, Baumann H, Brisson JR *et al.* (1994): The solution structure of a trisaccharide-antibody complex: comparison of NMR measurements with a crystal structure. In *Biochemistry* **33**:5183–92.

Bundle DR, Eichler E, Gidney MAJ *et al.* (1994): Molecular recognition of a *Salmonella* trisaccharide epitope by monoclonal antibody SE155–4. In *Biochemistry* **33**:5172–82.

Casset F, Hamelryck T, Loris R *et al.* (1995): NMR, molecular modeling, and crystallographic studies of lentil lectin-sucrose interaction. In *J. Biol. Chem.* **270:**25619–28.

Chervenak MC, Toone EJ (1994): A direct measure of the contribution of solvent reorganization to the enthalpy of binding. In *J. Am. Chem. Soc.* **116**:10533–9.

Cygler M, Rose DR, Bundle DR (1991): Recognition of a cell surface oligosaccharide epitope of pathogenic *Salmonella* by an antibody Fab fragment. In *Science* **253**:442–6.

Delbaere LTJ, Vandonselaar M, Prasad L *et al.* (1990): Molecular recognition of a human blood group determinant by a plant lectin. In *Can. J. Chem.* **68**:1116–21.

Delbaere LTJ, Vandonselaar M, Prasad L *et al.* (1993): Structures of the lectin IV of *Griffonia simplicifolia* and its complex with the Lewis b human blood group determinant at 2.0 Å resolution. In *J. Mol. Biol.* **230**:950–65.

Dessen A, Gupta D, Sabesan S *et al.* (1995): X-ray crystal structure of the soybean agglutinin cross-linked with a biantennary analog of the blood group I carbohydrate antigen. In *Biochemistry* **34**:4933–42.

Dunitz JD (1994): The entropic cost of bound water in crystals and biomolecules. In *Science* **264**:670.

Evans SV, Sigurskjold BW, Jennings HJ *et al.* (1995): The 2.8 Å resolution structure, and thermodynamics of ligand binding, of an Fab specific for α(2→8)-polysialic acid. In *Biochemistry* **34**:6737–44.

Hester G, Kaku H, Goldstein IJ *et al.* (1995): Crystal structure of mannose-specific snow drop *(Galanthus nivalis)* lectin representative of a new plant lectin family. In *Nature Struct. Biol.* **2**:472–9.

Imberty A, Perez S (1994): Molecular modeling of protein-carbohydrate interactions. Understanding the specificities of two legume lectins toward oligosaccharides. In *Glycobiology* **4**:351–66.

Jeffrey GA, Sänger W (1991): Hydrogen Bonding in Biological Structures. Berlin: Springer Publ. Co..

Jeffrey PD, Bajorath J, Chang CYY *et al.* (1995): The X-ray structure of an anti-tumour antibody complex with antigen. In *Nature Struct. Biol.* **2**:466–71.

Kabat EA (1966): The nature of an antigenic determinant. In *J. Immunol.* **97**:1–11.

Kabat EA (1976): Structural Concepts in Immunology & Immunochemistry: 2nd edition. New York, NY: Holt, Rinehart & Winston.

Lemieux RU (1989): The origin of the specificity in the recognition of oligosaccharides by proteins. In *Chem. Soc. Rev.* **18**:347–74.

Lemieux RU (1993): How proteins recognize and bind oligosaccharides. In *ACS Symp. Ser.* **519**:5–18.

Lemieux RU, Delbaere LTJ, Beierbeck H *et al.* (1991): Involvement of water in host-guest interactions. In *Host-Guest Molecular Interactions: From Chemistry to Biology* (Ciba Foundation Symposium 158) pp 231–48, Chichester: Wiley.

Liao D-I, Kapadia G, Ahmed H *et al.* (1994): Structure of S-lectin, a developmentally regulated vertebrate β-galactoside-binding protein. In *Proc. Natl. Acad. Sci. USA* **91**:1428–32.

Lobsanov YD, Gitt MA, Leffler H *et al.* (1993): X-ray crystal structure of the human dimeric S-Lac lectin, L-14-II, in complex with lactose at 2.9 Å resolution. In *J. Biol. Chem.* **268**: 27034–8.

Loris R, Casset F, Bouckärt J *et al.* (1994): The monosaccharide binding site of lentil lectin: a X-ray and molecular modelling study. In *Glycoconjugate J.* **11**:507–17.

Meyer B. (1990): Conformational aspects of oligosaccharides. In *Top. Curr. Chem.* **154**:143–208.

Naismith JH, Emmerich C, Habash J *et al.* (1994): Refined structure of concanavalin A complexed with methyl α-D-mannopyranoside at 2.0 Å resolution and comparison with the saccharide-free structure. In *Acta Crystallogr. Sect. D* **50**:847–58.

Nisonoff A (1982): Introduction to Molecular Immunology. Sunderland, Massachusetts: Sinauer Associates.

Oomen R, Young NM, Bundle DR (1991): Molecular modeling of antibody-antigen complexes between the *Brucella abortus O*-chain polysaccharide and a specific monoclonal antibody. In *Protein Engineering* **4**:427–33.

Quiocho FA (1986): Carbohydrate-binding proteins: tertiary structures and protein-sugar interactions. In *Annu. Rev. Biochem.* **55**:287–315.

Quiocho FA (1989): Basic molecular features. In *Pure & Appl. Chem.* **61**:1293–306.

Rhodes G (1993): Crystallography Made Crystal Clear. San Diego: Academic Press.

Rini JM, Hardman KD, Einspahr H *et al.* (1993): X-ray crystal structure of a pea lectin-trimannoside complex at 2.6 Å resolution. In *J. Biol. Chem.* **14**:10126–32.

Rose DR, Przybylska M, To RJ *et al.* (1993): Crystal structure to 2.45 Å resolution of a monoclonal Fab specific for the *Brucella* A cell wall polysaccharide antigen. In *Protein Sci.* **2**:1106–13.

Shaanan B, Lis H, Sharon N (1991): Structure of a legume lectin with an ordered *N*-linked carbohydrate in complex with lactose. In *Science* **254**:862–6.

Shibata S, Goldstein IJ, Baker DA (1982): Isolation and characterization of a Lewis b-active lectin from *Griffonia simplicifolia* seeds. In *J. Biol. Chem.* **257**:9324–9.

Spohr U, Hindsgaul O, Lemieux RU (1985): Molecular recognition II. The binding of the Lewis b and Y human blood group determinants by the lectin IV of *Griffonia simplicifolia*. In *Can. J. Chem.* **63**: 2644–52.

Stein PE, Boodhoo A, Armstrong GD *et al.* (1994): The crystal structure of pertussis toxin. In *Structure* **2**:45–57.

Stuike-Prill R, Meyer B (1990): A new force-field program for the calculation of glycopeptides and its application to a heptacosapeptide-decasaccharide of immunoglobulin G_1. In *Eur. J. Biochem.* **194**: 903–19.

Vyas MN, Vyas NK, Meikle PJ *et al.* (1993): Preliminary crystallographic analysis of a Fab specific for the *O*-antigen of *Shigella flexneri* cell surface lipopolysaccharide with and without bound saccharides. In *J. Mol. Biol.* **231**:133–6.

Weis WI, Drickamer K, Hendrickson WA (1992): Structure of a C-type mannose-binding protein complexed with an oligosaccharide. In *Nature* **360:**127–34.

Wright CS (1990): 2.2 Å Resolution structure analysis of two refined *N*-acetylneuraminyl-lactose-wheat germ agglutinin isolectin complexes. In *J. Mol. Biol.* **215**:635–51.

Wright CS, Jaeger J (1993): Crystallographic refinement and structure analysis of the complex of wheat germ agglutinin with a bivalent sialoglycopeptide from glycophorin A. In *J. Mol. Biol.* **232**:620–38.

Zdanov A, Li Y, Bundle DR *et al.* (1994): Structure of a single-chain Fv fragment complexed with a carbohydrate antigen at 1.7 Å resolution. In *Proc. Natl. Acad. Sci. USA* **91**:6423–7.

18 Thermodynamic Analysis of Protein-Carbohydrate Interaction

DIPTI GUPTA AND C. F. BREWER

1 Introduction

Fundamental to an understanding of many biological processes is the relationship between the structure and binding specificity of macromolecular receptors for specific ligands. This is especially true where macromolecules such as enzymes, antibodies and lectins demonstrate remarkable selectivity in binding specific ligands. Specificity is often defined in terms of the affinity constants (K_a) of ligands for a protein, which, in turn, is determined by their Gibbs free energy of binding (ΔG). However, for a deeper understanding of structure-binding activity correlations, changes in the heat of binding (enthalpy) (ΔH) and entropy of binding (ΔS) for a given complex need to be determined since these two parameters establish the magnitude and sign of ΔG (Hinz, 1983). Thus, differences in binding constants among ligands to a protein can be understood in terms of the relative contributions of ΔH, which is usually associated with direct ligand-protein interactions such as hydrogen bonding, ionic bonding and positive van der Waals interactions, and ΔS which is associated with solvent reorganization and other entropic contributions to binding (Hinz, 1983). Furthermore, changes in the heat capacity of binding (ΔCp), which can be determined from changes in ΔH versus temperature, provide another thermodynamic parameter to characterize complex formation (Sturtevant, 1977).

Titration microcalorimetry provides a direct means of investigating the energetics of binding between molecules. This technique permits the direct determination of K_a, ΔH and n, the number of binding sites of a ligand for a macromolecule (Wiseman *et al.*, 1989). From these data, ΔS and ΔCp can also be calculated. Titration microcalorimetry also eliminates the uncertainty of the so-called van't Hoff method of measuring ΔH (and by extension ΔS) by techniques such as equilibrium dialysis as a function of temperature. This latter method provides only an estimate of ΔH since the experimental value oftens differs from the real value in complex systems (Hinz, 1983).

In studies of lectin-carbohydrate interactions, titration microcalorimetry has become an important method for investigating the carbohydrate binding specificity of lectins (Bains *et al.*, 1992; Mandal *et al.*, 1994a, b). Evidence for extended site binding interactions are revealed by greater $-\Delta H$ values for certain oligosaccharides relative to $-\Delta H$ values for monosaccharide binding. In addition, relatively favorable contributions of ΔS to binding may indicate the presence of enhanced entropic mechanisms for complex formation of particular saccharides. Thus, ΔH and ΔS measurements are necessary for determining the presence or absence of extended binding interactions between a lectin and different carbohydrates, and hence the carbohydrate binding specificity of lectins.

Measurements of ΔH and ΔS are also important in establishing which hydroxyl groups of a saccharide are involved in binding to a lectin where deoxy analogs of the sugar are available. A lower $-\Delta H$ value for a deoxysaccharide analog constitutes direct evidence for binding interactions at the site of substitution in the ligand (Mandal *et al.*, 1994a). Thus, detailed mapping of binding epitopes in a saccharide requires ΔH and ΔS data for the parent carbohydrate as well as its analogs.

H.-J. and S. Gabius (Eds.), Glycosciences

ISBN 3-8261-0073-5

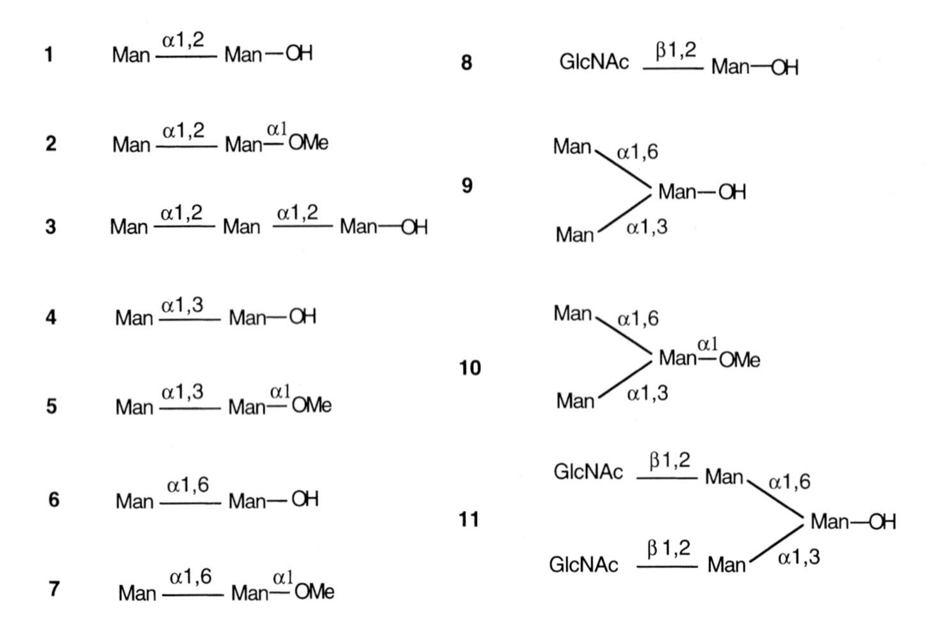

Figure 1. Structures of oligosaccharides **1-12**. Man, GlcNAc and Asn represent mannose, *N*-acetyl glucosamine and asparagine residues, respectively.

Examples of the use of titration microcalorimetry measurements to establish the binding specificity of a lectin and the carbohydrate recognition epitopes involved in binding can be illustrated by recent studies of the lectin concanavalin A (ConA). ConA is a Man/Glc specific lectin isolated from the Jack bean (*Canavalia ensiformis*). Interest in ConA is due to its numerous biological properties which are related to its carbohydrate-binding specificity (Ben-Bassat and Goldblum, 1975; Bittiger and Schnebli, 1976; Brown and Hunt, 1978; Goldstein and Hayes, 1978). Although the x-ray crystal structure of the lectin has been determined with bound monosaccharide (Derewenda *et al.*, 1989), the molecular basis of its binding specificity toward larger asparagine-linked (*N*-linked) carbohydrates has not been determined by similar structural techniques. However, much insight into the extended binding specificity of ConA has recently been gained through the application of titration microcalorimetry measurements (Mandal *et al.*, 1994a, b).

1.1 Background Studies

Early studies by Goldstein and coworkers showed the presence of two classes of linear oligosaccharides which differ in their affinities for the protein. The first class possessed affinities similar to monosaccharide binding and included α(1,3), α(1,4) and α(1,6) oligosaccharides with nonreducing terminal Glc or Man residues (Goldstein *et al.*, 1965) and the second class showed higher affinities and included the α(1,2) oligomannosides (So and Goldstein, 1968). Nuclear magnetic resonance (NMR) measurements led to the suggestion that the enhanced affinities of the α(1,2) oligomannosides (5- to 20-fold relative to monosaccharide binding) were primarily due to their increased probability of binding because of the presence of multiple Man residues with free 3-, 4- and 6-hydroxyl groups in each molecule (Brewer and Brown, 1979). However, subsequent nuclear magnetic relexation dispersion (NMRD) measurements of a series of *N*-linked oligomannose and bisected hybrid-type glycopeptides and complex-type oligosaccharides bound to the lectin showed that they possessed much higher affinities (50- to 3000-fold) and induced different conformational changes in the protein relative to the binding of monosaccharides such as methyl α-mannopyranoside (MeαMan) (Bhattacharyya and Brewer, 1989). Thus, it appeared that ConA possessed enhanced binding specificity for certain *N*-linked carbohydrates. These studies also revealed that the trisaccharide 3,6-di-*O*-(α-*D*-mannopyranosyl)-α-*D*-mannose (**9** in Figure 1) possessed nearly 100-fold higher affinity than MeαMan and yielded NMR data that were similar to that observed with larger *N*-linked glycopeptides (Brewer and Bhattacharyya, 1986; Bhattacharyya and Brewer, 1989). These results suggested that the trimannosyl moiety which is present in all *N*-linked carbohydrates is responsible for the high affinity binding of these naturally carbohydrates to ConA.

Titration microcalorimetry measurements have since been carried out with ConA in the presence of a series of mono- and oligosaccharides, *N*-linked oligosaccharides and glycopeptides, and mono- and dideoxy analogs of 3,6-di-*O*-(α-*D*-mannopyranosyl)-α-*D*-mannose (Mandal *et al.*, 1994a, b). The results of these studies have established a molecular basis for extended binding site interactions of the lectin with *N*-linked carbohydrates, as well as identified hydroxyl group binding epitopes on trimannoside **9** that are responsible for its high-affinity binding.

2 Outline of Experimental Design

2.1 Materials

Native ConA was prepared from jack bean seeds (Sigma) according to the method of Agrawal *et al.* (Agrawal and Goldstein, 1967). Concentration of ConA was determined spectrophotometrically at 280 nm using $A^{1\%,1\,cm} = 13.7$ at pH 7.2 (Goldstein and Poretz, 1986) and 12.4 at pH 5.2 (Derewenda *et al.*, 1968) and expressed in terms of monomer ($M_r = 26\,500$) (Goldstein and Poretz, 1986). The concentrations of carbohydrates were determined by the phenol-sulfuric acid method (Dubois *et al.*, 1956; Saha and Brewer, 1994) using appropriate monosaccharides (Glc, Man or 2-deoxyGlc) as standards.

Methyl α-*D*-glucopyranoside (MeαGlc), methyl α-*D*-mannopyranoside (MeαMan), maltose, maltotriose and carbohydrates **2**, **5**, **7** and **10** in Figure 1 were purchased from Sigma Chemical Co. Carbohydrates **1**, **4**, **6**, **8**, **9** and **11** in Figure 1 were obtained from Dextra Laboratories Ltd, UK. Methyl 2-deoxy α-*D*-glucopyranoside (Meα2-dGlc) and trisaccharide **3** were gifts from Dr. S. Sabesan, Du Pont Company, Wilmington, Delaware and Dr. Fraser Reid of Duke University, respectively). Methyl α-*D*-*N*-acetylglucosamine (MeαGlcNAc) was purchased from Toronto Research Chemicals, Ontario, Canada. Man9 glycopeptide **12** was prepared from the pronase digest of the soybean agglutinin (SBA) (Lis and Sharon, 1978). Methyl 3-*O*-(2-deoxy-α-*D*-man-nopyranosyl)-6-*O*-(α-*D*-mannopyranosyl) -α-*D*-mannopyranoside, **13**; 3-dMan(3,6), methyl 3-*O*- (3-deoxy- α- *D*-mannopyranosyl)-6-*O*-(α-*D*-mannopyranosyl)-α-*D*-mannopyranoside, **14**; 4-dMan(3,6), methyl 3-*O*-(4-deoxy-α-*D*-mannopyranosyl)-6-*O*-(α- *D*-mannopyranosyl)-α-*D*-mannopyranoside, **15**; 6-dMan (3,6), methyl 3-*O*-(6-deoxy-α-*D*-mannopyranosyl)-6-*O*-(α-*D*-mannopyranosyl)-α-*D*-mannopyranoside, **16**; methyl 6-*O*-(6-deoxy-α-*D*-mannopyranosyl)-3-*O*-α-*D*-mannopyranosyl-α-*D*-mannopyranoside, **17**; methyl 6-*O*-(6-deoxy-α-*D*-mannopyranosyl)-3-*O*-α-*D*-mannopyranosyl-α-*D*-mannopyranoside, **18**; methyl 6-*O*-(4-deoxy-α-*D*-mannopyranosyl)-3-*O*-α-*D*-mannopyranosyl-α-*D*-mannopyranoside, **19**; methyl 6-*O*-(6-deoxy-α-*D*-mannopyranosyl)-3-*O*-α-*D*-mannopyranosyl-α-*D*-mannopyranoside, **20**; methyl 6-*O*-(α-*D*-mannopyranosyl)-3-*O*-(α-*D*-mannopyranosyl)-2-deoxy-α-*D*-mannopyranoside, **28**; methyl 6-*O*-(α-*D*-mannopyranosyl)-3-*O*-(α-*D*-mannopyranosyl)-4-deoxy-α-*D*-mannopyranoside, **29**; methyl 3-*O*-(3-deoxy-α-*D*-mannopyranosyl)-6-*O*-(α-*D*-mannopyranosyl)-2-deoxy-α-*D*-mannopyranoside, **30**; and methyl 3-*O*- (3-deoxy- α- *D*-mannopyranosyl) -6-*O*-(α-*D*-mannopyranosyl) -4- deoxy- α- *D*-mannopyranoside. Synthesis of **31** will be described elsewhere.

2.2 Methodology

Isothermal titration microcalorimetry was performed using an OMEGA Microcalorimeter from Microcal, Inc. (Northampton, MA). In individual titrations, injections of 3 or 4 μl of carbohydrate were added from the computer-controlled 100 μl microsyringe at an interval of 4 min into the lectin solution (cell volume = 1.3424 ml) dissolved in the same buffer as the saccharide, while stirring at 350 rpm. Control experiments performed by making identical injections of saccharide into a cell containing buffer with no protein showed insignificant heat of dilution. The experimental data were fitted to a theoretical titration curve using software supplied by Microcal, with ΔH (the enthalpy change in kcal mol^{-1}), K_a (the association constant in M^{-1}) and n (the number of binding sites per monomer), as adjustable parameters. The quantity $c = K_a M_t(0)$, where $M_t(0)$ is the initial macromolecule concentration, is of importance in titration microcalorimetry (Wiseman *et al.*, 1989). All experiments were performed with c values between $1 < c < 200$ in the present study. In the case of maltotriose ($c = 1.2$), the data were fitted with n fixed and K_a and ΔH as floating parameters. The instrument was calibrated by using the calibration kit containing ribonuclease A (RNase A) and cytidine 2'-monophosphate supplied by the manufacturer. The thermodynamic parameters were calculated using the equations,

$$\Delta G = \Delta H - T\Delta S = -RT \ln K_a$$

where ΔG, ΔH and ΔS are the changes in free energy, enthalpy, and entropy upon binding; K_a is the affinity constant; T is the absolute temperature; and R = 1.98 cal md^{-1} K^{-1}.

3 Thermodynamics of Binding of Mono- and Disaccharides

Table 1 shows that MeαGlc, maltose and maltotriose have similar K_a values, and that the oligosaccharides have ΔH values close to that of the monosaccharide (-6.6 kcal mol^{-1}). These values are somewhat less than the ΔH values of -7.3 and -7.2 kcal mol^{-1} reported for MeαGlc and maltose, respectively, by Magnuson and coworkers (Munske *et al.*, 1984). The data, however, are consistent with binding of only the non-reducing terminal glucose residues of α(1,4) glucosyl oligosaccharides to ConA, which agree with studies by Goldstein and coworkers (Goldstein *et al.*, 1965). MeαGlcNAc also has a ΔH similar to that of MeαGlc.

MeαMan has a higher ΔH (-8.2 kcal mol^{-1}) than MeαGlc (Table 1), as well as 4-fold greater affinity which agrees with previous precipitation inhibition measurements (Goldstein *et al.*, 1965). Meα2-dGlc has a ΔH of -7.3 kcal mol^{-1} which is between that of MeαGlc and MeαMan, as is its affinity constant (Goldstein and Poretz, 1986). Thus, the equatorial 2-hydroxyl of MeαGlc destabilizes binding by nearly 1 kcal mol^{-1} relative to Meα2-dGlc, while the axial 2-hydroxyl of MeαMan stabilizes binding by nearly 1 kcal mol^{-1} relative to Meα2-dGlc.

Titration of ConA with α(1,6) dimannoside **7** shows that its K_a and thermodynamic parameters are very similar to those for MeαMan (Table 1). These results demonstrate that the nonreducing terminal residue of **7** binds to the monosaccharide site of ConA, similar to maltose and maltotriose. These findings are consistent with previous NMRD studies (Brewer and Bhattacharyya, 1986, 1988) and a recent microcalorimetric study (Williams *et al.*, 1992) of the binding of **7** to ConA at pH 5.2. Table 1 shows that disaccharide **6** which does not possess an α-anomeric methyl group has a 1 kcal mol^{-1} higher -ΔH and ~2-fold greater affinity than **7**. α(1,3) Mannobiose **4** and its a-methyl derivative **5** bind to ConA with ΔH values of -10.2 and -10.7 kcal mol^{-1}, respectively, which is about -2 kcal mol^{-1} higher than that of MeαMan and **7**. The affinity of **5** is also about 4-fold greater than that of MeαMan and **7**. These

Table 1[a]. Thermodynamic parameters for the binding of ConA with saccharides at 25 °C [b]

Carbohydrate	Carbohydrate conc. (mM)	Lectin conc. (mM)	K_a^c (M^{-1})
MeαGlc	70.0	0.895	1.96 (±0.05)x10^3
MeαGlcNAc	72.0	0.918	1.08 (±0.04)x10^3
Maltose	68.8	0.918	1.31 (±0.07)x10^3
Maltotriose	80.6	0.904	1.34 (±0.03)x10^3
MeαMan	46.0	0.483	0.82 (±0.02)x10^4
Meα2-dGlc	61.7	0.730	2.75 (±0.07)x10^3
1	26.7	0.390	4.17 (±0.08)x10^4
2	26.5	0.380	1.41 (±0.04)x10^5
3	5.15	0.125	3.79 (±0.26)x10^5
4	20.8	0.256	1.41 (±0.02)x10^4
5	14.8	0.247	3.35 (±0.12)x10^4
6	42.9	0.483	1.34 (±0.04)x10^4
7	14.8	0.247	0.81 (±0.03)x10^4
8	35.9	0.420	0.67 (±0.01)x10^4
9	5.93	0.145	3.37 (±0.14)x10^5
10[d]	7.04	0.132	4.90 (±0.15)x10^{5d}
11	8.00	0.145	1.40 (±0.10)x10^6
10[d]	11.0	0.195	5.10 (±0.21)x10^{5d}
12[d]	1.10	0.025	1.10 (±0.10)x10^{6d}

Carbohydrate	$-\Delta H^c$ (kcal mol^{-1})	$-T\Delta S$ (kcal mol^{-1})	n^c(no.of sites/monomer)
MeαGlc	6.6 (±0.1)	2.1	0.98 (±0.02)
MeαGlcNAc	6.2 (±0.2)	2.1	1.03 (±0.03)
Maltose	6.2 (±0.3)	1.9	0.99 (±0.04)
Maltotriose	6.4 (±0.4)	2.1	1.00 (held fixed)
MeαMan	8.2 (±0.1)	2.9	1.04 (±0.01)
Meα2-dGlc	7.3 (±0.1)	2.6	1.05 (±0.01)
1	9.9 (±0.1)	3.6	0.98 (±0.01)
2	10.5 (±0.1)	3.5	1.04 (±0.01)
3	10.7 (±0.1)	3.1	1.05 (±0.01)
4	10.2 (±0.1)	4.5	1.01 (±0.01)
5	10.7 (±0.1)	4.5	1.01 (±0.01)
6	9.4 (±0.1)	3.8	1.00 (±0.01)
7	8.4 (±0.2)	3.1	0.99 (±0.02)
8	5.3 (±0.1)	0.1	0.97 (±0.01)
9	14.1 (±0.1)	6.6	1.03 (±0.01)
10	14.4 (±0.1)	6.6	1.02 (±0.01)
11	10.6 (±0.1)	2.2	0.99 (±0.01)
10[d]	14.3 (±0.1)	6.5	0.96 (±0.01)
12[d]	18.0 (±0.3)	9.8	0.91 (±0.01)

a Values taken from (Mandal *et al.*, 1994b).
b The buffer was 0.1 M HEPES containing 0.9 M NaCl, 1 mM Mn^{2+} and 1 mM Ca^{2+} at pH 7.2.
c Values in parentheses indicate the standard deviation of fit between the experimental binding curve and the calculated curve obtained with the fitted thermodynamic parameters.
d The buffer was 0.05 M dimethyl glutarate containing 0.25 M NaCl, 1 mM Mn^{2+} and 1 mM Ca^{2+} at pH 5.2.

results indicate extended site interactions of ConA with α(1,3) disaccharide **4** and **5**, in contrast to interactions with only the nonreducing terminal residues of the α(1,6) dimannoside(s) and α(1,4) glucose oligosaccharides.

α(1,2) Dimannoside **2** shows a ΔH similar to that of α(1,3) dimannoside **5** (Table 1), however, the affinity of **2** is 4-fold greater. The increase in affinity arises from entropic effects since TΔS is 1 kcal mol^{-1} more positive for **2**. Similar results were obtained for α(1,2) mannobiose **1** relative to α(1,3) mannobiose **4**. These results indicate that the increased affinities of the α(1,2) dimannosides are due to entropic effects which arise from the presence of two Man residues with free 3-, 4- and 6-hydroxyl groups which can independently bind to the lectin and thus increase the number of binding modes. It should be noted that the binding affinity of **1** is about 5-fold higher than MeαMan, while that of **2** is about 17-fold higher than the monosaccharide, which demonstrates the importance of an α-anomeric methyl group at the reducing end of the molecules since α-anomeric pyranosides bind better to the monosaccharide binding site of the lectin than the corresponding free sugars. The greater enthalpy changes for **1** and **2** relative to MeαMan agree with previous work by Goldstein and coworkers (Williams *et al.*, 1981).

Table 1 shows that α(1,2) trimannoside **3** binds with a ΔH of -10.7 kcal mol^{-1}, which is similar to **2**, but with nearly 3-fold higher affinity (Table 1). The enhanced affinity of **3** relative to **2** is primarily due to a positive entropy contribution (0.4 kcal mol^{-1}) because of the presence of a third α(1,2)Man residue with free 3-, 4- and 6-hydroxyl groups.

4 Thermodynamics of Binding of Trimannoside 10

The thermodynamic parameters for binding of methyl 3,6-di-*O*-(α-*D*-mannopyranosyl)-α-*D*-mannopyranoside (**10**) to tetrameric ConA at pH 7.2, 25 °C are shown in Table 1. ΔH is -14.4 kcal mol^{-1}, with $K_a = 4.9 \times 10^5$ M^{-1} and ΔG = -7.8 kcal mol^{-1}. The thermodynamic parameters for **10** are similar to those previously reported at pH 5.4, 25 °C obtained by equilibrium dialysis measurements (ΔH = -14.8 kcal mol^{-1}; $K_a = 6.7 \times 10^5$ M^{-1}) (Mackenzie, 1986). Trisaccharide **9** which lacks the α-anomeric methyl group of **10** yields a ΔH of -14.1 kcal mol^{-1} at pH 7.2, and has about 1.5-fold lower affinity than **10**. Thus, the presence of the α-anomeric methyl group has little effect on the binding of trimannoside to ConA. The ΔH values for **9** and **10** are approximately -6 kcal mol^{-1} greater than that of MeαMan. Furthermore, the K_a and ΔH values for **9** and **10** are also greater than those of disaccharides **4–7** which comprise the two arms of the 3,6-trimannoside. These results provide direct evidence that ConA possesses an extended binding site that recognizes the two non-reducing Man residues of the trimannoside, and that these extended binding interactions are, in part, responsible for its high affinity. These results agree with similar conclusions derived from NMRD studies of the binding of **9** to ConA (Brewer and Bhattacharyya, 1986); Bhattacharyya and Brewer, 1989.

5 Thermodynamics of Binding of Oligosaccharides 8, 11 and 12

Biantennary complex-type oligosaccharide **11** (Figure 1) possesses a terminal β(1,2)GlcNAc residue on each arm of the core trimannoside moiety and is a naturally occurring homolog of **9**. Table 1 shows that **11** possesses nearly 4-fold higher affinity than **9**. However, ΔH for **11** is only -10.6 kcal mol^{-1} as compared to -14.1 kcal mol^{-1} for **9**. On the other hand, TΔS for **11** is -2.2 kcal mol^{-1} whereas that for **9** is -6.6 kcal mol^{-1}. These results indicate that the increase in affinity of **11** is due to favorable entropic effects, and, surprisingly, that it has a lower -ΔH relative to **9**. Thus, even though NMRD studies have suggested similar mechanisms of binding for **9** and **11** to ConA (Bhattacharyya *et al.*, 1987b), the calorimetry data indicate that they have different mechanisms of binding.

Disaccharide **8** which constitutes the two branched chains of **11** also shows a relatively low ΔH but a favorable TΔS contribution (Table 1).

Although the affinity of **8** is only slightly less than MeαMan, ΔH is -5.3 kcal mol^{-1} which is about 3 kcal mol^{-1} less than that of MeαMan, and half of that observed for **11**. The entropy of binding of the disaccharide is relatively more positive (TΔS = –0.1 kcal mol^{-1}), which provides a substantial contribution to the free energy of binding. These results suggest common binding mechanisms for **8** and **11**.

Calorimetric data for binding of Man9 glycopeptide (**12**) with tetrameric ConA at pH 7.2 could not be obtained since the protein is readily cross-linked and precipitated by the glycopeptide under these conditions (Bhattacharyya and Brewer, 1989). ConA also precipitates with **12** in high salt buffer at pH 5.2, suggesting the presence of tetrameric lectin under these conditions. However, binding of **12** to low concentrations of ConA (0.025 mM) in low salt buffer (0.05 M dimethyl glutarate containing 0.25 M NaCl) at pH 5.2 results in no precipitation, indicating that the lectin is essentially a dimer under these conditions. The thermodynamic parameters for binding of the Man9 glycopeptide **12** to ConA at pH 5.2, are shown in Table 1. The value of 0.91 for n suggests some bivalent binding of the glycopeptide via the high affinity pentamannosyl moiety of the α(1,6) arm and the lower affinity α(1,2) trimannosyl moiety on the α(1,3) arm (Bhattacharyya and Brewer, 1989). The ΔH value of -18.0 kcal mol^{-1} for **12** is higher than that for trisaccharide **10**, which may also reflect the formation of a fraction of molecules in solution in cross-linked complexes. The value of K_a is 11×10^5 M^{-1} which is about 2.5-fold higher than that for trisaccharide **10** (Table 1), which suggests that the trimannoside moiety on the α(1,6) arm of **12** is the primary high affinity determinant, as previously suggested from NMRD studies (Bhattacharyya and Brewer, 1989).

6 Binding of Mono- and Dideoxy Derivatives of Trimannoside 10

In order to understand the mechanism of binding of the trimannoside to ConA, we have synthesized and determined the thermodynamics of binding of deoxy and dideoxy derivatives of trimannoside **10** (Figure 2), with ConA. Titration microcalorimetry shows that ΔH is -14.5 kcal mol^{-1} and K_a is 5.7×10^5 M^{-1} for 2-deoxy analog **13**, which are essentially identical to those values for **10** (Table 1). These results demonstrate that the 2-OH of the α(1–3) Man residue of **10** is not involved in binding to ConA. The calorimetry data clearly show that while the 4- and 6-deoxy analogs possess ΔH and K_a values similar to those of **10**, the 3-deoxy derivative, **14**, binds with a 10-fold lower K_a and 3.4 kcal mol^{-1} lesser ΔH. This indicates that the 3-OH of the α(1–3) Man residue of the trimannoside is involved in binding to ConA at a secondary site (by analogy to the requirements for monosaccharide binding (Goldstein and Poretz, 1986)). Because of the magnitude of the loss in the change of the binding enthalpy (3.4 kcal mol^{-1}) of **14** relative to the parent trimannoside (Wells and Fersht, 1986), the extended binding site of ConA appears to include hydrogen bonding of a residue(s) of the lectin with the 3-OH of the α(1–3) Man residue of **10**.

The affinity of the 2-deoxy derivative of the α(1–3) Man residue of the trimannoside (**17**) is 2-fold lower than that of **10**, and **17** possesses similar ΔH and TΔS values as **10**. The affinities of the 3-, 4- and 6-deoxy derivatives (**18–20**, respectively) are about 13- to 19-fold weaker as compared to that of **10**. ΔH for **18–20** are very similar (11.2–11.7 kcal mol^{-1}) and about 3.5 kcal mol^{-1} lower as compared to that of the trimannoside, indicating the involvement of 3-, 4- and 6-OH of the α(1,6) Man residue of **10** in binding. As the magnitude of loss of ΔH is ~ 3.5 kcal mol^{-1} for all three hydroxyl group substititions, all three hydroxyl groups are involved in hydrogen bonds with the ConA combining site. Since the 3-, 4- and 6-OH groups of the monosaccharides Man and Glc are required for binding to ConA (Goldstein and Poretz, 1986), this indicates that the 3-, 4- and 6-OH of the α(1–6) Man residue of **10** are involved in binding to the so-called monosaccharide site of the lectin. The 3-OH groups of the α(1–3) Man residue of **10** therefore binds at an adjacent site in the protein near the monosaccharide site.

The 2-deoxy derivative of the central branching Man residue of **10** (**21**) binds four-fold more

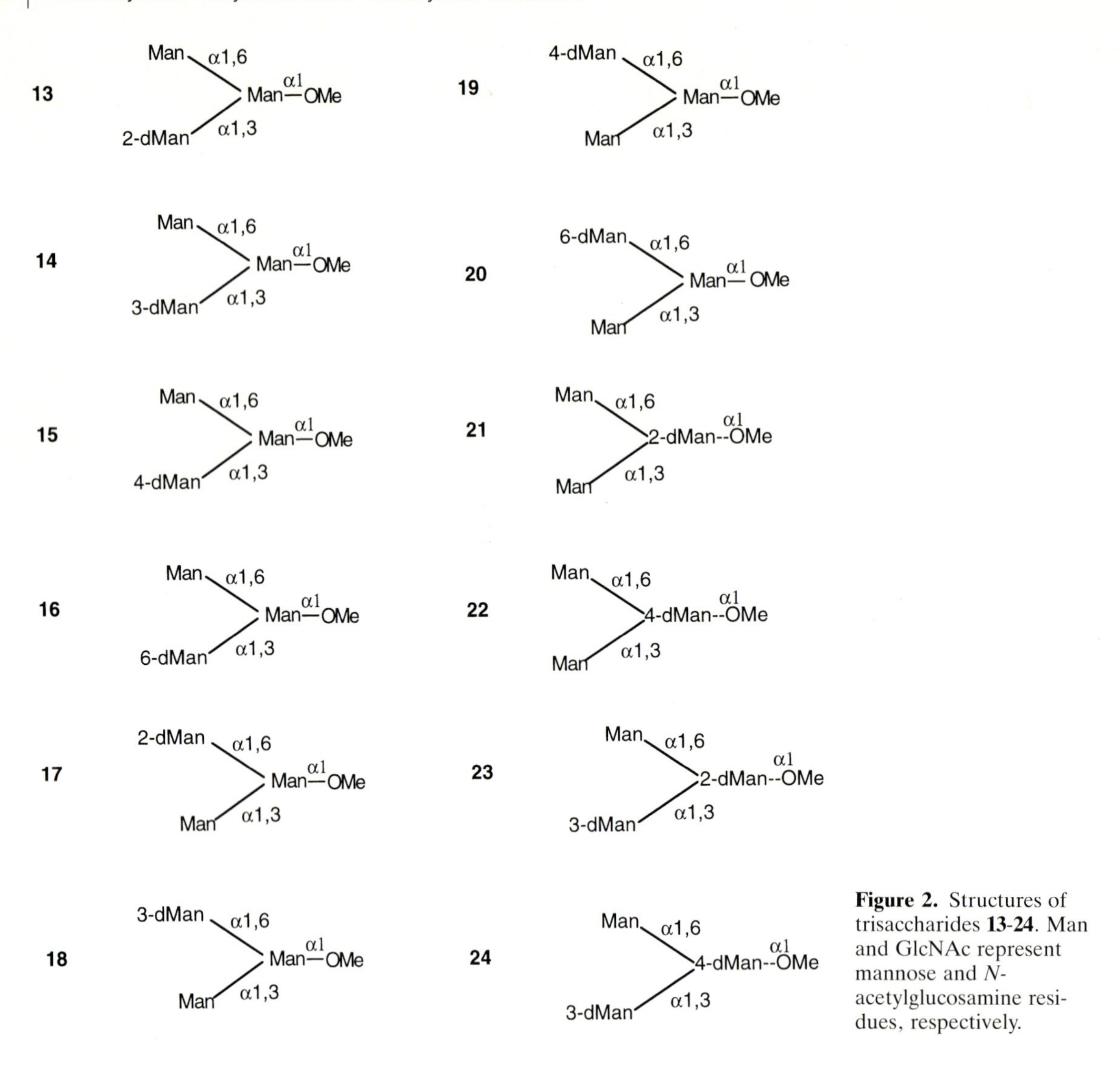

Figure 2. Structures of trisaccharides **13-24**. Man and GlcNAc represent mannose and *N*-acetylglucosamine residues, respectively.

weakly than the trimannoside, whereas the 4-deoxy derivative **22** binds approximately 10-fold weaker than **10**. ΔH for **21** is 1 kcal mol^{-1} lower than that of **10**, while ΔH for **22** is 2.3 kcal mol^{-1} lower. These results indicate the involvement of the 2- and 4-OH of the central Man residue of **10** in binding to ConA, with the central Man residue contributing about -3 kcal mol^{-1} to ΔH, as previously suggested (Mandal *et al.*, 1994a).

Table 2 also shows that the 3,4-dideoxy derivative **24** of the trimannoside binds about 14-fold more weakly than **10**. ΔH for **24** is -9.7 kcal mol^{-1} which is 4.7 kcal mol^{-1} less favorable than that of **10**. This difference is approximately the sum of the ΔH loss of the 3-deoxy α(1,3) derivative **14** and the cental Man 4-deoxy derivative **22** relative to **10**. Similarly, the 2,3-dideoxy derivative **23** binds about 2.5- fold more weakly than **10** and possesses a ΔH of -10.6 kcal mol^{-1} which is a loss of 3.8 kcal mol^{-1} compared to **10**. This is approximately equal to the combined loss in ΔH by the 2-deoxy analog **21** and the 3-deoxy derivative **14** relative to **10**. These results demonstrate that the

Table 2[a]**.** Thermodynamic parameters for the binding of ConA with trisaccharides[b]

Carbohydrate	Carbohydrate conc. (mM)	Lectin conc. (mM)	K_a^c (M^{-1})
13	6.20	0.136	5.68 (±0.29)x10^5
14	9.0	0.140	5.39 (±0.26)x10^4
15	7.20	0.165	5.02 (±0.15)x10^5
16	6.40	0.154	3.96 (±0.12)x10^5
17	6.2	0.230	2.23 (±0.13)x10^5
18	14.5	0.330	3.88 (±0.18)x10^4
19	7.0	0.230	2.54 (±0.14)x10^4
20	12.5	0.330	3.01 (±0.23)x10^4
21	12.5	0.240	1.17 (±0.11)x10^5
22	9.3	0.260	4.63 (±0.13)x10^4
23	42.5	0.540	1.93 (±0.13)x10^5
24	7.0	0.305	3.60 (±0.05)x10^4

Carbohydrate	-ΔH^c (kcal mol^{-1})	-TΔS (kcal mol^{-1})	$n^{c,d}$ (no.of site/monomer)
13[a]	14.5 (±0.1)	6.7	
14[a]	11.0 (±0.2)	4.6	
15[a]	14.3 (±0.1)	6.5	
16[a]	14.0 (±0.1)	6.4	
17	14.9 (±0.3)	7.6	1.02 (±0.02)
18	11.2 (±0.2)	4.9	1.03 (±0.03)
19	11.7 (±0.3)	5.7	1.01 (±0.02)
20	11.6 (±0.1)	5.5	1.03 (±0.03)
21	13.4 (±0.2)	6.5	1.03 (±0.03)
22	12.1 (±0.1)	5.7	1.02 (±0.03)
23	10.6 (±0.2)	3.4	1.02 (±0.02)
24	9.7 (±0.1)	4.8	1.00 (±0.01)

a Values taken from (Mandal *et al.*, 1994a, b).
b The buffer was 0.1 M HEPES containing 0.9 M NaCl, 1 mM $MnCl_2$ and 1 mM $CaCl_2$ at pH 7.2.
c Values in parentheses indicate the standard deviation of fit between the experimental binding curve and the calculated curve obtained with the fitted thermodynamic parameters.
d Values of n were between 0.95 and 1.05 in all cases.

energetics of binding of the two dideoxy analogs **23** and **24** is accountable by the loss in binding at each individual hydroxyl group position in the corresponding monodeoxy analogs of **10**.

7 Summary and Perspectives

Thermodynamic data obtained by titration microcalorimetry provides direct evidence that the 3,6-trimannoside moiety found in all *N*-linked carbohydrates binds to ConA via extended site interactions. The ΔH value for **10** was nearly -6 kcal mol^{-1} greater than that for MeαMan (Table 1). This trimannoside moiety is the primary high affinity epitope on the outer α(1,6) arm of *N*-linked oligomannose and bisected hybrid-type glycopeptides, as well as the inner core region of *N*-linked complex type carbohydrates (Mandal *et al.*, 1994b). Because the α(1,3) arms of many of these *N*-linked carbohydrates also bind to ConA with lower affinity, these naturally occurring oligosaccharides are known to be divalent and to bind and cross-link the lectin (Bhattacharyya *et al.*, 1987a, b; Bhattacharyya and Brewer, 1989).

The thermodynamic data in Tables 1 and 2 for monodeoxy analogs of **10** also show that the 3-, 4- and 6-hydroxyl groups of the α(1,6) Man residue of the trimannoside bind to the lectin, and most likely to the so-called monosaccharide site by analogy to similar binding requirements for monosaccharides. In addition, the 3-hydroxyl of the α(1–3) Man residue and 2- and 4-hydroxyls of the central Man residue of **10** bind to an extended binding site in the lectin. Thus, the combined use of titration microcalorimetry and the use of deoxy oligosaccharide analogs have provided a detailed description of the size of the binding site in the lectin via the binding epitopes in the trimannoside.

In this regard, the titration microcalorimetry data for oligosaccharide **11** serve to make an important point. Although oligosaccharide **11**, which is a naturally occurring complex type carbohydrate, possesses nearly 4-fold higher affinity for ConA relative to **10**, titration microcalorimetry measurements showed that **11** possesses approximately 4 kcal mol^{-1} less favorable -ΔH than the trimannoside. Thus, **11** makes fewer (or weaker) direct binding interactions than **10**, however, **11** binds with a more favorable entropic contribution than does **10** (Table 1). This is also apparent in the thermodyanmic data for disaccharide **8** (Table 1), which represents a portion of the structure of **11**. These results as well as those for disaccharides **1-3**, which also show relatively favorable entropic contributions to binding, demonstrate the importance of obtaining direct measurements of K_a and ΔH (and therefore ΔG and ΔS) for each carbohydrate by titration calorimetry methods. Observation of only the K_a values for these compounds would provide a misleading view of their binding interactions with the lectin.

At the present time, limitations in the use of titration microcalorimetry are the sample volume required for each experiment (~1.6 ml), the concentration of materials (typically greater than 0.05–0.1 mM), and the affinity of the molecules (typically greater 1 mM and less than 1 nM). The amount of heat liberated by the interacting molecules also determines their required concentrations, and cooperative binding interactions requires sophisticated curve-fitting procedures. Future developments in this area include reducing the volume of samples required and increasing the sensitivity of detection of the measurements which would lower the concentration of the molecules involved.

In summary, while techniques such as x-ray crystallography and nuclear magnetic resonance spectroscopy provide important structural information on ligand-macromolecular complexes, thermodynamic data obtained by techniques such as titration microcalorimetry are necessary to understand the binding energetics of such complexes, and ultimately the relationship between the structure of molecules and their binding specificities.

References

Agrawal BBL, Goldstein IJ (1967): Protein-carbohydrate interactions. VI. Isolation of concanavalin A by specific adsorption on cross-linked dextran gels. In *Biochim. Biophys. Acta* **147**:262–71.

Bains G, Lee RT, Lee YC *et al.* (1992): Microcalorimetric study of wheat germ agglutinin binding to *N*-acetylglucosamine and its oligomers. In *Biochemistry* **31**:12624–8.

Ben-Bassat H, Goldblum N (1975): Concanavalin A receptors on the surface membrane of lymphocyte from patients with Hodgkin's disease and other malignant lymphomas. In *Proc. Natl. Acad. Sci. USA* **72**:1046–9.

Bhattacharyya L, Brewer CF (1989): Interactions of concanavalin A with asparagine-linked glycopeptides. Structure-activity relationships of the binding and precipitation of oligomannose and bisected hybrid-type glycopeptides with concanavalin A. In *Eur. J. Biochem.* **178**:721–6.

Bhattacharyya L, Ceccarini C, Lorenzoni P *et al.* (1987a): Concanavalin A interactions with asparagine-linked glycopeptides: bivalency of high mannose and bisected hybrid type glycopeptides. In *J. of Biol. Chem.* **262**:1288–93.

Bhattacharyya L, Haraldsson M, Brewer CF (1987b): Concanavalin A interactions with asparagine-linked glycopeptides. Bivalency of bisected complex type oligosaccharides. In *J. Biol. Chem.* **262**:1294–9.

Bittiger H, Schnebli HP (1976): Concanavalin A as a Tool. New York: John Wiley.

Brewer CF, Brown RD, III (1979): Mechanism of binding of mono- and oligosaccharides to concanvalin A: a solvent proton magnetic relaxation dispersion study. In *Biochemistry* **18**:2555–62.

Brewer CF, Bhattacharyya L (1986): Specificity of concanavalin A binding to asparagine-linked glycopeptides. In *J. Biol. Chem.* **261**:7306–10.

Brewer CF, Bhattacharyya L (1988): Concanavalin A interactions with asparagine-linked glycopeptides. The mechanisms of binding of oligomannose, bisected hybrid, and complex type carbohydrates. In *Glycoconjugate J.* **5**:159–73.

Brown JC, Hunt RC (1978): Lectins. In *Int. Rev. Cytol.* **52**:277–349.

Derewenda Z, Yariv J, Helliwell JR *et al.* (1989): The structure of the saccharide-binding site of concanavalin A. In *EMBO J.* **8**:2189–93.

Dubois M, Gilles KA, Hamilton JK *et al.* (1956): Calorimetric method for determination of sugars and related substances. In *Anal. Chem.* **28**:350–6.

Goldstein IJ, Hayes CE (1978): The lectins: carbohydrate-binding proteins of plants and Animals. In *Adv. Carbohydr. Chem. Biochem* **35**:127–340.

Goldstein IJ, Poretz RD (1986): Isolation, physicochemical characterization, and carbohydrate-binding specificity of lectins, in *The Lectins* (Liener IE, Sharon N *et al.*, eds.) pp 35–244, New York: Academic Press.

Goldstein IJ, Hollerman CE, Smith EE (1965): Protein-carbohydrate interaction. II. Inhibition studies on the interaction of concanavalin A with polysaccharides. In *Biochemistry* **4**:876–83.

Hinz H-J (1983): Thermodynamics of protein-ligand interactions. In *Annu. Rev. of Biophys. Bioeng.* **12**:285–317.

Lis H, Sharon N (1978): Soybean agglutinin – a plant glycoprotein. In *J. Biol. Chem.* **253**:3468–76.

Mackenzie AE (1986): A study of the binding of oligosaccharides and glycopeptides to concanavalin A. Ph.D. Dissertation, University of Toronto, Canada.

Mandal DK, Bhattacharyya L, Koenig SH *et al.* (1994a): Studies of the binding specificity of concanavalin A. Nature of the extended binding site for asparagine-linked carbohydrates. In *Biochemistry* **33**:1157–62.

Mandal DK, Kishore N, Brewer CF (1994b): Thermodynamics of lectin-carbohydrate interactions. Titration microcalorimetry measurements of the binding of *N*-linked carbohydrates and ovalbumin to concanavalin A. In *Biochemistry* **33**:1149–56.

Munske GR, Krakauer H, Magnuson JA (1984): Calorimetric study of carbohydrate binding to concanavalin A. In *Arch. Biochem. Biophys.* **233**:582–7.

Saha SK, Brewer CF (1994): Determination of the concentrations of oligosaccharides, complex type carbohydrates and glycoproteins using the phenol-sulfuric acid method. In *Carbohydr. Res.* **254**:157–67.

So LL, Goldstein IJ (1968): Protein-Carbohydrate Interaction. XIII. The interaction of concanavalin A with α-mannans from a variety of microorganisms. In *J. Biol. Chem.* **243**:2003–7.

Sturtevant JM (1977): Heat capacity and entropy changes in processes involving proteins. In *Proc. Natl. Acad. Sci. USA.* **74**:2236–40.

Wells TNC, Fersht AR (1986): Use of binding energy in catalysis analyzed by mutagenesis of the tyrosyl-tRNA synthetase. In *Biochemistry* **25**:1881–6.

Williams TJ, Homer LD, Shafer JA *et al.* (1981): Characterization of the extended binding site of concanavalin A: specificity for interaction with the nonreducing termini of α(1–2)-linked Disaccharides. In *Arch. Biochem. Biophys.* **209**:555–64.

Williams BA, Chervenak MC, Toone EJ (1992): Energetics of lectin-carbohydrate binding. A microcalorimetric investigation of concanavalin A-oligomannoside Complexation. In *J. Biol. Chem.* **267**:22907–11.

Wiseman T, Williston S, Brandt JF *et al.* (1989): Rapid measurement of binding constants and heats of binding using a new titration calorimeter. In *Anal. Biochem.* **179**:131–5.

19 Analysis of Protein-Carbohydrate Interaction Using Engineered Ligands

Dolores Solís and Teresa Díaz-Mauriño

1 Overview

Recognition of carbohydrates by proteins plays a central role in many intracellular and intercellular processes, such as trafficking of proteins and cells, bacterial and viral infection, normal cell differentiation, development, tumor progression and metastasis (Karlsson, 1991; Sharon and Lis, 1993; Varki, 1993). The ability of carbohydrates to act as recognition determinants derives from their unique potential to generate an enormous variety of complex structures using a relatively limited number of monosaccharides. The information contained in these structures is decoded by complementary sites present on carbohydrate-binding proteins. A high fidelity in the protein-carbohydrate interaction is essential for the normal progression of these various recognition processes.

At the atomic level, the specificity and affinity in these interactions is achieved through hydrogen bonds, with added contributions from van der Waals contacts and stacking interactions of aromatic residues against the hydrophobic regions of the sugar (Vyas, 1991; Bundle and Young, 1992). X-Ray crystallographic studies of several carbohydrate-protein complexes have shown that usually a small number of crucial hydroxyl groups in the carbohydrate molecule are involved in essential hydrogen bonds with the protein, whereas other hydroxyl groups at the periphery of the binding site or exposed to solvent are not essential for the stability of the complex. Chemical mapping studies can help to elucidate this hydrogen bonding scheme and to delineate the topology of the combining site.

Binding studies with synthetic carbohydrate molecules, in which a hydroxyl group has been selectively replaced by hydrogen or by fluorine, can be used to identify hydroxyl groups directly involved in hydrogen bonding to the protein and even to dissect their energetic contribution to the binding. Replacement with hydrogen will result in the abrogation of any hydrogen bond at that position, whereas an electronegative fluorine atom cannot act as a hydrogen-bond donor but it can be a hydrogen bond acceptor, albeit a weak one (Pauling, 1980; Street *et al.*, 1986). Thus, a significant reduction in the binding affinity upon deoxygenation at a given position reveals the involvement of that particular hydroxyl group in hydrogen bonding to the protein. The overall free energy contribution of the hydrogen bond can be estimated from the loss in affinity of the deoxy derivative. In addition, the behavior of the corresponding fluorodeoxy derivative indicates the participation of the hydroxyl group as a hydrogen-bond donor or as an acceptor.

Replacement of a hydroxyl group with an *O*-methyl group provides information not only about the hydrogen-bonding relationships at that position but also on the flexibility of the combining site to steric demands for complex formation (Nikrad *et al.*, 1992). Binding studies using other synthetic derivatives, such as epimeric, amino or *C*-methyl analogs, can provide additional insights into the hydrogen-bonding and steric requirements, as well as information on the distribution of polar and non-polar regions within the combining site. Other synthetic saccharides, including hydrophobic derivatives (e.g. nitrobenzyl monosaccharides), pseudo sugars (compounds in which the ring oxygen of a monosaccharide has been replaced by a different atom such as C, S or

H.-J. and S. Gabius (Eds.), Glycosciences

ISBN 3-8261-0073-5

N) and heteroatom glycosides (*C*-, *S*- or *N*-glycosides, anomerically linked via an atom other than oxygen) have also been used in the analysis of protein-carbohydrate interactions (Loganathan *et al.*, 1992; Magnusson *et al.*, 1994). In addition, these compounds are valuable as inhibitors of enzymes and other carbohydrate-binding proteins and as hydrolytically stable analogs.

The conformational properties of the ligands in solution also play an important role in recognition (Rice *et al.*, 1993). It is commonly assumed that for the binding to a protein a single conformation of the sugar is selected out of the range of possible conformations existing in solution. Therefore, the study of the recognition phenomenon based on engineered ligands must be sustained by the analysis of the possible changes in their conformational preferences. Only when the functional group replacement does not modify substantially the distribution of low energy conformers can the relative affinity of the protein for the engineered ligand be correlated with the presence or the absence of a certain group, and with the size and shape of the hydrophilic and the hydrophobic surfaces of the saccharide molecule. On the other hand, binding studies using conformationally restricted saccharides can provide information on the conformational requirements of the combining site for efficient recognition. Unfortunately, methods for conformational restriction of glycoside bonds are rather limited and rely on the introduction of extra rings or substituents that cause steric hindrance (Magnusson *et al.*, 1994).

The combining sites of several carbohydrate-binding proteins have been extensively mapped using natural and engineered ligands. These include plant and animal lectins, bacterial adhesins, antibodies, glycosyltransferases and other enzymes from various sources. For several of these systems, the conclusions of the chemical mapping studies in solution could be compared with data from crystal structures, allowing the evaluation of the analogs as probes of the molecular and stereochemical requirements of binding, and providing a more detailed appreciation of the atomic features of the interaction. However, for many other carbohydrate-binding proteins no X-ray structures are available, and binding studies with natural and synthetic ligands are the only source of information to identify key interactions stabilizing the complexes and to infer the carbohydrate orientation within the combining site. This information is essential for the rational design of specific inhibitors with enhanced activity.

In the following, we highlight a few examples which illustrate the potential of chemical mapping studies for the analysis of protein-carbohydrate interactions in solution.

2 Analysis of Hydrogen-Bonding and Steric Requirements for Recognition

Chemical mapping studies have proved extremely fruitful for the systematic investigation of the hydrogen-bonding and steric requirements of carbohydrate-binding proteins. Differences in the mode of binding of the basic structure recognized by the protein and/or in the topology of the combining sites will result in a different carbohydrate-binding specificity, and consequently in a different biological activity. For instance, the β-Gal-(1–4/3)-GlcNAc-specific lectins isolated from *Ricinus communis* seeds (ricin and *R. communis* agglutinin) and from bovine heart (galectin-1) display differences in the recognition of naturally occurring oligosaccharides based on this backbone sequence, such as the preferential binding of both ricin and agglutinin to α-NeuAc-(2–6)-substituted rather than α-(2–3)-substituted β-galactoside-terminating oligosaccharides, as opposed to the different preference of bovine lectin. Studies on the molecular recognition of different synthetic monodeoxy, fluorodeoxy, *O*-methyl and other derivatives of methyl β-lactoside by these three lectins (Rivera-Sagredo *et al.*, 1991, 1992b; Solís *et al.*, 1993, 1994b; Fernández *et al.*, 1994) have revealed a different mode of binding of the disaccharide underlying their different oligosaccharide-binding preferences.

As determined by NMR and molecular mechanics calculations (Rivera-Sagredo *et al.*, 1991; Fernández and Jiménez-Barbero, 1993; Fernández *et al.*, 1994), the preferred conformations of the different methyl β-lactoside derivatives were very similar. Therefore, the relative affinity of the lectins for the different structures could be correlated with specific polar and nonpolar interactions. Analysis of the results indicated that the hydroxyl groups at positions 3, 4 and 6 of the galactopyranose unit are key polar groups in the interaction with both ricin and *R. communis* agglutinin (Rivera-Sagredo *et al.*, 1991, 1992b; Solís *et al.*, 1993). *HO*-2' also participates in hydrogen bonding to both lectins and, at the glucopyranose unit, the hydroxyl group at position 6 participates in a minor polar interaction with ricin but not with the agglutinin. In addition, a nonpolar interaction involving position 3 of the glucopyranose moiety plays a significant role in the binding to both lectins (Rivera-Sagredo *et al.*, 1992b; Fernández *et al.*, 1994).

The molecular recognition of methyl β-lactoside by ricin deserves further attention because of the discrepancies between the hydrogen bonding pattern inferred from the chemical mapping studies in solution and the pattern observed in the 0.25 nm refined X-ray crystal structure of the ricin-lactose complex (Rutenber and Robertus, 1991). This highly cytotoxic lectin is composed of two different polypeptide chains: the A chain, which inhibits protein synthesis in eukaryotic cells by inactivating the 60 S ribosomal subunits, and the B chain, which contains two saccharide-binding sites with different affinities. Data derived from the chemical mapping studies mentioned above relate to the high affinity site. According to the X-ray data, only the hydroxyl groups at positions 3 and 4 of the galactopyranose moiety participate in strong hydrogen bonding to the protein in the high affinity site, whereas the C-6' hydroxyl is oriented towards the solvent and is not involved in the binding. In addition, no specific interactions with the C-2' hydroxyl group and with the glucopyranose moiety are seen in the X-ray crystal structure, although these interactions are readily deduced from the chemical mapping studies. A deeper investigation into the origin of these discrepancies revealed a critical influence of the pH required for the crystallization of ricin on the protein-carbohydrate interaction. Binding studies with methyl β-lactoside analogs had been carried out in phosphate-buffered saline at pH 7.2, whereas crystals of ricin were obtained from a solution at pH 4.75 (Villafranca and Robertus, 1977). A study on the recognition of the different monodeoxy methyl β-lactoside derivatives at the galactopyranose unit at the latter pH (Solís *et al.*, 1993) demonstrated a different mode of binding, the involvement of the hydroxyl group at C-6' being strongly dependent on the pH. Thus, at neutral pH the *HO*-6' appears as a key hydroxyl group whereas at acidic pH only a small polar interaction was observed. In addition, the involvement of the glucose moiety in the binding has been very recently confirmed by NMR studies using transferred nuclear Overhauser enhancement experiments and molecular mechanics calculations of the conformation of methyl α-lactoside bound to ricin B chain (Asensio *et al.*, 1995).

Two important conclusions can be drawn from these studies. First, the conditions required for crystallization of the protein may induce local or even global conformational changes, giving rise to a crystalline arrangement that might not reflect exactly the arrangement existing in solution under physiological conditions. And second, binding studies with engineered ligands can be combined to form a versatile tool for the analysis of protein-carbohydrate interactions under various conditions. Therefore, crystal structures should be complemented by chemical mapping studies to obtain a clearer picture of the atomic features of the interaction.

Recapitulating, the studies on the molecular recognition of the synthetic methyl β-lactoside analogs by the two *R. communis* lectins indicate that, although the glucopyranose unit is also involved in the recognition, the key hydroxyl groups all derive from the galactopyranose moiety. On the contrary, a similar study on the recognition of the methyl β-lactoside derivatives by galectin-1 (Solís *et al.*, 1994b) revealed that the essential hydrogen bonds-involved hydroxyl groups derived from both pyranose rings, that is, the hydroxyl groups at positions 4 and 6 of the galactopyranose unit, and at position 3 of the glucopyranose unit. The results also suggest that the hydroxyl group at C-2 in methyl β-lactoside, as well as the *N*-acetyl amino group in *N*-acetyllactosamine, are involved in a polar interaction with the combining site. Thus, hydrogen-bonding interactions are almost exclusively restricted to one side of the molecule whereas the other side of the disaccharide is not involved (H*O*-6 and H*O*-2') or only marginally involved (H*O*-3') in hydrogen bonding. Moreover, *O*-methylation at these positions causes an enhancement of the binding, suggesting favorable interactions of the methyl groups which may come into contact with hydrophobic residues at the periphery of the combining site.

Thus, although the combining sites of the *R. communis* lectins and bovine galectin-1 accommodate the same disaccharide structure, the recognition of different epitopes of this basic sequence generates fine differences in the oligosaccharide binding specificity and hence modulates the biological activity of the lectins. A similar mechanism for the fine tuning of the binding specificity of bacterial adhesins recognizing the α-Gal-(1–4)-Gal sequence has been demonstrated by chemical mapping studies using different synthetic analogs of galabiose (Kihlberg *et al.*, 1989; Haataja *et al.*, 1993, 1994). Also, the combining sites of various monoclonal antibodies and lectins that specifically bind to blood group determinants have been extensively mapped by the use of over 100 modified tri- or tetrasaccharide structures (Lemieux, 1989; Lemieux *et al.*, 1990, 1994; Nikrad *et al.*, 1992; Spohr *et al.*, 1992), demonstrating that the binding geometry of a particular determinant may be different for different receptor proteins. Among these proteins the studies on the molecular recognition of the Lewis b human blood group determinant by the lectin IV of *Griffonia simplicifolia* should be specially mentioned. Using the 0.25 nm resolution crystal structure of the lectin-Le^b-OMe complex as a reference, the effects on binding of deoxygenation and *O*-methylation at the various hydroxyl positions of the Le^b-OMe tetrasaccharide were compared (Nikrad *et al.*, 1992). Hydroxyl groups involved in key polar interactions could neither be deoxygenated nor methylated without virtually complete loss of binding affinity. Hydroxyl groups hydrogen bonded to the protein at the periphery of the combining site could be deoxygenated with some loss in affinity, but their methylation leads to a strong decrease in the stability of the complex due to a loss of complementarity. And finally, hydroxyl groups not involved in hydrogen bonding to the protein and remaining in contact with the aqueous phase may be substituted with little effect on the binding affinity. Furthermore, their methylation may cause an enhancement of the binding as a result of new favorable interactions of the methyl groups, which may come into contact with hydrophobic residues at the periphery of the combining site. Thus, a comparison of the binding activities of the deoxy and *O*-methyl derivatives of the ligand can provide information on the involvement and location of the hydroxyl groups in the complex that otherwise could only be achieved by high-resolution X-ray crystallography.

3 Analysis of Hydrogen-Bonding Energetics and Protein Groups Involved in Recognition

As illustrated by the preceding examples, binding studies with engineered ligands provide information on the contribution of hydrogen bonds to the generation of specificity in protein-carbohydrate interactions. But in addition, they can provide information on the energetic contribution of the hydroxyl groups to the binding, and even on the nature (charged or neutral) of the groups on the protein involved in hydrogen bonding.

Free energy contributions of individual hydrogen bonds can be estimated using the equation

$$\Delta(\Delta G°) = RT \ln \frac{K_{d,2}}{K_{d,1}}$$

where $K_{d,1}$ and $K_{d,2}$ are the dissociation constants for two ligands differing in their ability to form a particular hydrogen bond. Thus, in principle, free energies of hydrogen bonds donated by the hydroxyl groups of the carbohydrate could be calculated from the loss in binding affinity of the different fluoro derivatives ($K_{d,2}$) relative to the parent compound ($K_{d,1}$), while for hydrogen bonds accepted by the hydroxyl groups the values could be calculated from the affinity of the deoxy derivatives ($K_{d,2}$) compared to the corresponding fluoro analogs ($K_{d,1}$). However, the validity of this approach has been questioned in the light of some complications derived from the use of fluorodeoxy carbohydrate derivatives. In several carbohydrate-protein systems (Street *et al.*, 1986; Bhattacharyya and Brewer, 1988; Vermesch *et al.*, 1992) a higher affinity for deoxy sugars than for the fluoro sugars has been demonstrated. To explain this fact, it has been commonly assumed (Street *et al.*, 1986; Bhattacharyya and Brewer, 1988) that a water molecule might bind in the voids created in the complex by deoxygenation of the sugar and thus contribute to the binding affinity, while fluorination would not permit coincident binding of water. However, on the basis of equilibrium binding and high resolution crystallographic studies of the complexes of L-arabinose-binding protein with some deoxy and fluorodeoxy galactose derivatives, Vermesch *et al.* (1992) have recently demonstrated that these assumptions may not necessarily be valid. Thus, no new water molecules moved into the voids created by the replacement of galactose H*O*-1 or H*O*-2 by hydrogen and, contrary to expectations, the binding of a new water molecule was observed on substitution of the sugar H*O*-6 by fluorine. The binding of this water molecule was related to a local conformational change resulting from the repulsion between the electronegative fluoro substituent and the negatively charged carboxylate group of an aspartic acid residue which acts as a hydrogen bond acceptor group of galactose H*O*-6 (Vermesch *et al.*, 1992). This destabilizing effect could account for a lower affinity for the fluorodeoxy analog than for the corresponding deoxy derivative. In that case, the Δ (ΔG^o) values, calculated on the basis of data obtained for the fluorodeoxy analogs for the free energy associated with the participation of the hydroxyl group as a hydrogen bond donor, would be overestimated, and a reasonable measure of the overall energetic contribution of hydrogen bonds at that position would be better obtained from the binding of the deoxy derivatives. However, it still seems plausible that fluoro derivatives might be used as probes for the assessment of hydrogen-bonding energetics when no charged groups are situated in the proximity of the electronegative fluoro substituent, and therefore no destabilizing effect should be expected.

As can be straightforwardly inferred, the free energy contribution of non-essential hydrogen bonds will be much smaller than that of the key hydrogen bonds. These key interactions usually involve charged side chains of the combining site, most frequently the carboxylate groups of aspartate and glutamate residues. Actually, it has been reported (Fersht *et al.*, 1985; Street *et al.*, 1986) that hydrogen bonds involving neutral-charged pairs are stronger than those of the neutral-neutral type. Using site-directed mutagenesis to produce mutants of tyrosyl-tRNA synthetase differing in their abilities to form hydrogen bonds with their substrates, Fersht *et al.* (1985) showed that deletion of a hydrogen bond between an uncharged donor/acceptor pair reduces the affinity by only 2–6 kJ mol^{-1}. However, deletion of a hydrogen bond between a neutral-charged pair weakens binding by 14–19 kJ mol^{-1}, although smaller values (7.5 kJ mol^{-1}) have also been reported (Vermesch *et al.*, 1992). Similar results were found by the complementary approach involving specific modifications of the ligand rather than the protein. Estimates of the net contributions of hydrogen bonds (Street *et al.*, 1986), obtained from systematic measurements of affinity of deoxy and fluorodeoxy analogs compared with refined X-ray crystal structure data on the phosphorylase-glucose complex, were essentially identical to those reported by Fersht *et al.* (1985). Therefore, making allowance for the limitations related to the use of fluorodeoxy analogs, the binding energy losses associated with the replacement of the hydroxyl groups of the ligand may provide an indication of the nature of the groups on the protein involved in hydrogen bonding. For instance, the estimated free energies of hydrogen bonding of the hydroxyl groups of methyl β-lactoside to ricin, calculated on the basis of chemical mapping studies, were in agreement with the involvement of the C-3' and C-4' hydroxyl groups as hydrogen-bond donors to charged residues of the lectin (Solís *et al.*, 1993). And, in fact, according to the X-ray crystal structure of the ricin-lactose complex (Rutenber and Robertus, 1991) an aspartic acid residue accepts hydrogen bonds from the hydroxyl groups at these positions.

4 Mapping of Subsites of Anticarbohydrate Antibodies

Study of the molecular recognition of engineered ligands by anticarbohydrate antibodies provides information on the binding patterns that in turn indicate the arrangement of subsites within the combining site. The mapping of subsites of antigalactan and antidextran antibodies is an excellent example (Glaudemans, 1991; Glaudemans *et al.*, 1994). The combining site of these antibodies can accommodate up to four sequentially linked monosaccharide residues, that is, it contains four subsites each binding a monosaccharide unit of the tetrasaccharide epitope. However, the order of affinity of these subsites does not necessarily correlate with their location. The strategy for the elucidation of the subsite arrangements involved, first, the identification of essential and non-essential hydroxyl groups for the binding of a single monosaccharide, which monosaccharide should inevitably fill the subsite having the highest affinity, and second, the study of the recognition of different di-, tri- and tetrasaccharide ligands in which one or more of the saccharide units had been selectively modified at an essential position, thus forcing such residues away from the highest affinity subsite to optimize binding. This approach was first used for mapping of subsites of two β-(1–6)-D-galactan-binding myeloma antibodies, IgAs J539 and X24 (Glaudemans, 1991). The combining site of these monoclonal antibodies contains two solvent-exposed tryptophanyl residues. From comparison of the affinities of a series of 3-fluorodeoxy and also 3-deoxy derivatives of the methyl β-glycosides of β-(1–6)-linked D-galactosyl oligosaccharides, and the changes in protein fluorescence induced by ligand binding, the arrangement of subsites was deduced to be

Gal – Gal – Gal – MeβGal
C A B D

where the four antibody subsites are labeled A to D in order of decreasing affinity. Similar studies on other monoclonal antigalactan antibodies revealed the same subsite arrangement, while for the antidextran monoclonal immunoglobulin IgA W3129, the subsite affinities diminished in order of their location (Glaudemans, 1991; Glaudemans *et al.*, 1994).

5 Probing the Active Site Requirements of Carbohydrate-Binding Enzymes

Chemical mapping studies on carbohydrate-binding enzymes have proved very useful for the elucidation of the active site requirements not only for substrate recognition and binding but also for efficient catalytic activity. Modifications of substrate hydroxyl groups involved in the recognition by the combining site may produce poorer substrates, if peripheral hydroxyl groups are affected, or even virtually inactive compounds, not bound by the enzyme, when essential hydrogen bonds are deleted. On the other hand, replacement of hydroxyl groups involved in the catalytic mechanism may produce compounds which can be recognized and bound by the combining site but do not show significant activity as substrates, that is, they behave as competitive inhibitors of the enzyme. A study on the role of the hydroxyl groups of methyl β-lactoside in the substrate activity of lactase from the small intestine illustrates this approach. Elucidation of the lactase active site requirements could be very helpful in the design of new lactose analogs for the development of non-invasive diagnostic methods for lactose intolerance (Aragón *et al.*, 1992). Different synthetic monodeoxy, *O*-methyl and halo derivatives of methyl β-lactoside and other lactose analogues were tested as substrates and inhibitors of the enzyme (Rivera-Sagredo *et al.*, 1992a; Fernández *et al.*, 1995). The results confirmed that lactase is not a true β-galactosidase, since it does not show an absolute stereochemical specificity at C-4', and suggested that the galactopyranosyl moiety is deeply embedded in the active site with no space to accommodate molecules with bulky substituents, in agreement with the *exo*-glycosidase character of intestinal lactase. The hydroxyl groups of the glucopyranosyl moiety are involved to various extents in substrate recognition, while H*O*-3' is a key hydroxyl group for the binding since none of the derivatives at C-3' was either a substrate or an inhibitor of lactase. On the contrary, H*O*-2', although not necessary for the initial recognition, was important for the subsequent steps of the catalytic mechanism, since the 2'-deoxy methyl β-lactoside derivative was not a substrate, but behaved as a competitive inhibitor of lactose hydrolysis.

A study on the recognition and hydrolysis of different derivatives of methyl β-lactoside by the β-D-galactosidase from *E. coli* suggested a similar role for the hydroxyl group at C-2' (Bock and Adelhorst, 1990, 1992). It was proposed that H*O*-2' could be involved in the correct positioning of a carboxylate responsible for the stabilization of a galactosyloxocarbonium-ion transition state, or the formation of a covalently linked galactosyl intermediate. A series of deoxy and deoxyfluoro analogs of 2',4'-dinitrophenyl β-D-galacto-pyranoside were also used as mechanistic probes of the *E. coli* galactosidase (McCarter *et al.*, 1992). The contribution of the hydrogen bonding interactions to the stabilization of the transition state was estimated from the kinetic parameters obtained for these derivatives, confirming that the 2-position provides the most important transition-state binding interactions.

Estimates of the contribution of the substrate hydroxyl groups to the stabilization of both ground-state and transition-state complexes may be obtained from the kinetic parameters in the same way that has been used for the evaluation of the hydrogen-bonding energetics in other protein-carbohydrate complexes. Assuming that the replacement of the individual hydroxyl groups does not lead to a change in the mechanism or in a rate-limiting step, the increases in the overall activation free energy $\Delta(\Delta G^{\circ})$ due to these replacements reflect the strengths of interaction at each hydroxyl in the transition state (Street *et al.* 1989). These $\Delta(\Delta G^{\circ})$ values can be calculated using the equation

$$\Delta(\Delta G^{\circ}) = RT\,ln\,\frac{(k_{cat}\,/\,K_M)_2}{(k_{cat}\,/\,K_M)_1}$$

where $(k_{cat}/K_M)_1$ and $(k_{cat}/K_M)_2$ are the kinetic parameters for the two analogs being compared. In addition, the contribution of interactions at the ground state may be estimated from the inhibition constants for a series of inert disubstituted derivatives, obtained by further substitution at the other positions of an analog acting as a competitive inhibitor (Percival and Withers, 1992).

A comparison of the binding energy contributions of the individual hydroxyl groups to the transition-state stabilization for two different enzymes may reveal similarities in the active-site region (Withers and Rupitz, 1990). The transition-state binding interactions can be compared using a linear free energy relationship analysis, by plotting the logarithm of the k_{cat}/K_M values obtained with a series of substrate analogs for one enzyme against the equivalent set of values for the other enzyme. Two enzymes with identical active sites, and therefore identical transition states, will yield a plot with a slope and a correlation coefficient of 1. Any differences in the type or strength of hydrogen-bonding interactions at the transition state will result in scatter of the plot and a poorer correlation coefficient. Therefore, the correlation coefficient and, to a certain extent, the slope of the plot, would provide a measure of the active site similarities.

Using this approach, an extremely high homology between the transition-state complexes of two phylogenetically distant α-glucan phosphorylases was observed (Withers and Rupitz, 1990), in agreement with their high sequence homology. The similarities in the active sites of glucose dehydrogenase, glucose oxidase and glucoamylase were also probed using deoxygenated substrates (Sierks *et al.*, 1992). The high correlation coefficients of their linear free energy relationships indicated significant similarities in transition-state interactions, although in this case the three enzymes lack overall sequence homology.

6 Probing Conformational Requirements for Recognition

Beyond the hydrogen-bonding and steric requirements of the combining site, ligand flexibility may be intimately involved in receptor recognition. Available evidence indicates that some oligosaccharides are flexible molecules which possess relatively well defined conformational preferences in solution. The docking of a flexible oligosaccharide into a protein combining site implies the loss of some degrees of freedom and thus a negative entropic contribution to the binding. When covalently linked to a carrier molecule, the conformational flexibility of the oligosaccharide may be reduced. If one of the permitted conformers satisfies complementarity with the combining site of a possible receptor, the binding to that particular receptor will be favored since a smaller loss in conformational entropy is expected. Presentation on different carrier molecules will then generate a range of affinities for different receptors. This has been proposed as a possible mechanism for the generation of biological specificity, using commonly occurring oligosaccharides in the context of specific carrier proteins (Carver, 1993; Solís *et al.*, 1994a). Similarly, it should be possible to synthesize a conformationally restricted analog of the ligand, matching the conformational requirements of the combining site and thus binding much more strongly than the natural ligand. Therefore, elucidation of the conformational requirements for recognition may be highly relevant for rational drug design.

Very few examples of the use of conformationally restricted saccharides for the analysis of protein-carbohydrate interactions have been reported, and most of them relate to the restriction of conformational flexibility arising from rotation around the C5-C6 linkage in hexoses. This has been accomplished by methyl substitution at the C-6 of the hexopyranose or by formation of a fused ring between *O*-4 and C-6. The first approach was used in a study on the binding of the trisaccharide β-D-Gal-(1–4)-β-D-GlcNAc-(1–6)-β-D-Gal to the monoclonal antibody anti-I Ma (Kabat *et al.*, 1981), and recently it has also been used to investigate the conformational requirements of different viral and bacterial neuraminidases, including influenza A virus neuraminidase, for hydrolysis of, or inhibition by α-D-NeuAc-(2–6)-β-D-Gal sialoside analogs (Sabesan *et al.*, 1994). Elucidation of the neuraminidase conformational requirements is particularly interesting because of the potential of neuraminidase inhibitors as drugs for therapeutic or prophylactic treatment of viral and bacterial infectious processes. The study of the recognition of C-6 C-methylated conformationally biased sialoside analogs indicated that restriction of the rotamer orientation in the trans-gauche (*tg*) mode favors binding to all the neuraminidases investigated.

The second approach, involving the construction of rigid analogs, has been used to explore the conformational preferences of *N*-acetylglucosaminyltransferase V (GlcNAcT-V) (Lindh and Hindsgaul, 1991; Bock *et al.*, 1992), a key enzyme involved in the biosynthesis of highly branched asparagine-linked oligosaccharides. Rigid bicyclic analogs of the trisaccharide β-D-GlcNAc-(1–2)-α-D-Man-(1–6)-β-D-Glc-OR were prepared in which the α-Man-(1–6)-β-Glc linkage was fixed in either a gauche-gauche (*gg*) or gauche-trans (*gt*) conformation. Evaluation of these analogs as substrates for GlcNAcT-V demonstrated that the enzyme preferentially recognizes the carbohydrate chains of asparagine-linked oligosaccharides in their *gg* conformations.

The hydrogen-bonding and steric requirements of several glycosyltransferases have been extensively investigated by chemical mapping studies (Srivastava and Hindsgaul, 1988; Hindsgaul *et al.*, 1991; Lowary and Hindsgaul, 1993; Lowary *et al.*, 1994; Reck *et al.*, 1994), because of their fundamental role in oligosaccharide biosynthesis, and the observed changes in the expression of oligosaccharides and of glycosyltransferases themselves occurring during normal cell development and differentiation, and also associated with the onset of several pathological processes, including cancer. Thus, selective glycosyltransferase inhibitors are of potential therapeutic interest, and in particular they could be useful for the development of new methods of cancer therapy. Studies of the recognition of conformationally restricted analogs by GlcNAcT-V strongly suggested that glycosyltransferases can distinguish discrete conformations of their oligosaccharide acceptors, accounting for the biosynthesis of many unique oligosaccharides. Elucidation of these conformational requirements will be of great value for strategies of drug design.

We thank the Dirección General de Investigación Científica y Técnica (Grant SAF 92–0497) and Fondo de Investigaciones Sanitarias (Grant 93/0317) for financial support.

References

Aragón JJ, Fernández-Mayoralas A, Jiménez-Barbero J *et al.* (1992): Evaluation of rat intestinal lactase in vivo with 4-galactosylxylose. In *Clin. Chim. Acta* **210**:221–6.

Asensio JL, Cañada FJ, Jiménez-Barbero J (1995): Studies of the bound conformations of methyl α-lactoside and methyl β-allolactoside to ricin-B chain using transferred NOE experiments in the laboratory and rotating frames, assisted by molecular mechanics dynamics calculations. In *Eur. J. Biochem.* **233:** 618–30.

Bhattacharyya L, Brewer CF (1988): Lectin-carbohydrate interactions. Studies of the nature of hydrogen bonding between D-galactose and certain D-galactose-specific lectins, and between D-mannose and concanavalin A. In *Eur. J. Biochem.* **176**:207–12.

Bock K, Adelhorst K (1990): Derivatives of methyl β-lactoside as substrates for and inhibitors of β-D-galactosidase from *E. coli*. In *Carbohydr. Res.* **202**:131–49.

Bock K, Adelhorst K (1992): Synthesis of deoxy derivatives of lactose and their hydrolysis by β-galactosidase from *E. coli*. In *Acta Chem. Scand.* **46**:186–93.

Bock K, Duus JO, Hindsgaul O *et al.* (1992): Analysis of conformationally restricted models for the (α 1–6)-branch of asparagine-linked oligosaccharides by n.m.r.-spectroscopy and HSEA calculation. In *Carbohydr. Res.* **228**:1–20.

Bundle DR, Young NM (1992): Carbohydrate-protein interactions in antibodies and lectins. In *Curr. Opinion Struct. Biol.* **2:**666–73.

Carver JP (1993): Oligosaccharides: How can flexible molecules act as signals? In *Pure Appl. Chem.* **65**:763–70.

Fernández P, Jiménez-Barbero J (1993): The conformation of the monomethyl ethers of methyl β-lactoside in D_2O and Me_2SO-d_6 solutions. In *Carbohydr. Res.* **248**:15–36.

Fernández P, Jiménez-Barbero J, Martín-Lomas M *et al.* (1994): Involvement of the glucose moiety in the molecular recognition of methyl β-lactoside by ricin: synthesis, conformational analysis, and binding studies of different derivatives at the C-3 region. In *Carbohydr. Res.* **256**:223–44.

Fernández P, Cañada FJ, Jiménez-Barbero J *et al.* (1995): Substrate-specificity of small-intestinal lactase: study of the steric effects and hydrogen bonds involved in enzyme-substrate interaction. In *Carbohydr. Res.* **271**:31–42.

Fersht AR, Shi JP, Knill-Jones J *et al.* (1985): Hydrogen bonding and biological specificity analysed by protein engineering. In *Nature* **314:**235–8.

Glaudemans CPJ (1991): Mapping of subsites of monoclonal, anti-carbohydrate antibodies using deoxy and deoxyfluoro sugars. In *Chem. Rev.* **91**:25–33.

Glaudemans CPJ, Kovác P, Nashed EM (1994) Mapping of hydrogen bonding between saccharides and proteins in solution. In *Methods Enzymol.* **247**:305–21.

Haataja S, Tikkanen K, Liukkonen J *et al.* (1993): Characterization of a novel bacterial adhesion specificity of *Streptococcus suis* recognizing blood group P receptor oligosaccharides. In *J. Biol. Chem.* **268**:4311–7.

Haataja S, Tikkanen K, Nilsson U *et al.* (1994): Oligosaccharide-receptor interaction of the Galα1–4Gal binding adhesin of *Streptococcus suis*. Combining site architecture and characterization of two variant adhesin specificities. In *J. Biol. Chem.* **269**:27466–72.

Hindsgaul O, Kaur KJ, Srivastava G *et al.* (1991): Evaluation of deoxygenated oligosaccharide acceptor analogs as specific inhibitors of glycosyltransferases. In *J. Biol. Chem.* **266**:17858–62.

Kabat EA, Burzynska MH, Lemieux RU *et al.* (1981): Immunochemical studies on blood-groups. 69. The conformation of the trisaccharide determinant in the combining site of anti-I Ma (Group-1). In *Mol. Immunol.* **18**:873–81.

Karlsson KA (1991): Glycobiology: a growing field for drug design. In *Trends Pharmacol. Sci.* **12**: 265–72.

Kihlberg J, Hultgren SJ, Normark S *et al.* (1989): Probing of the combining site of the PapG adhesin of uropathogenic *Escherichia coli* bacteria by synthetic analogues of galabiose. In *J. Am. Chem. Soc.* **111**:6364–8.

Lemieux RU (1989): The origin of the specificity in the recognition of oligosaccharides by proteins. In *Chem. Soc. Rev.* **18**:347–74.

Lemieux RU, Szweda R, Paszkiewicz-Hnatiw E *et al.* (1990): The effect of substituting key hydroxyl groups by amino groups on the binding of the Lewis b tetrasaccharide by a lectin and a monoclonal antibody. In *Carbohydr. Res.* **205**:c12–7.

Lemieux RU, Du MH, Spohr U (1994): Relative effects of ionic and neutral substituents on the binding of an oligosaccharide by a protein. In *J. Am. Chem. Soc.* **116**:9803–4.

Lindh I, Hindsgaul O (1991): Synthesis and enzymatic evaluation of two conformationally restricted trisaccharide analogues as substrates for *N*-acetylglucosaminyltransferase V. In *J. Am. Chem. Soc.* **113**:216–23.

Loganathan D, Osborne SE, Glick GD *et al.* (1992): Synthesis of high-affinity, hydrophobic monosaccharide derivatives and study of their interaction with concanavalin A, the pea, the lentil, and fava bean lectins. In *Arch. Biochem. Biophys.* **299**:268–74.

Lowary TL, Hindsgaul O (1993): Recognition of synthetic deoxy and deoxyfluoro analogs of the acceptor α-L-Fuc*p*-(1–2)-β-D-Gal*p*-OR by the blood-group A and B gene-specified glycosyltransferases. In *Carbohydr. Res.* **249**:163–95.

Lowary TL, Swiedler SJ, Hindsgaul O (1994): Recognition of synthetic analogues of the acceptor, β-D-Gal*p*-OR, by the blood-group H gene-specified glycosyltransferase. In *Carbohydr. Res.* **256**:257–73.

Magnusson G, Chernyak AYA, Kihlberg J *et al.* (1994): Synthesis of neoglycoconjugates. In *Neoglycoconjugates: Preparation and Applications* (Lee YC, Lee RT, eds) pp 53–143, San Diego: Academic Press.

McCarter JD, Adam MJ, Withers SG (1992): Binding energy and catalysis. Fluorinated and deoxygenated glycosides as mechanistic probes of *Escherichia coli (lac Z)* β-galactosidase. In *Biochem. J.* **286**:721–7.

Nikrad PV, Beierbeck H, Lemieux RU (1992): Molecular recognition. X. A novel procedure for the detection of the intermolecular hydrogen bonds present in a protein-oligosaccharide complex. In *Can. J. Chem.* **70**:241–53.

Pauling L (1980): The Nature of the Chemical Bond, 3rd edition. Ithaca NY: Cornell University Press.

Percival MD, Withers SG (1992): Binding energy and catalysis: deoxyfluoro sugars as probes of hydrogen bonding in phosphoglucomutase. In *Biochemistry* **31**:498–505.

Reck F, Meinjohanns E, Springer M *et al.* (1994): Synthetic substrate analogues for UDP-GlcNAc: Manα1–6R β(1–2)-*N*-acetylglucosaminyltransferase II. Substrate specificity and inhibitors for the enzyme. In *Glycoconjugate J.* **11**: 210–6.

Rice KG, Wu P, Brand L *et al.* (1993): Experimental determination of oligosaccharide three-dimensional structure. In *Curr. Opinion Struct. Biol.* **3**:669–74.

Rivera-Sagredo A, Solís D, Díaz-Mauriño T *et al.* (1991): Studies on the molecular recognition of synthetic methyl β-lactoside analogs by ricin, a cytotoxic plant lectin. In *Eur. J. Biochem.* **197**:217–28.

Rivera-Sagredo A, Cañada FJ, Nieto O *et al.* (1992a): Substrate specificity of small-intestinal lactase. Assessment of the role of the substrate hydroxyl groups. In *Eur. J. Biochem.* **209**:415–42.

Rivera-Sagredo A, Jiménez-Barbero J, Martín-Lomas M *et al.* (1992b): Studies of the molecular recognition of synthetic methyl β-lactoside analogues by *Ricinus communis* agglutinin. In *Carbohydr. Res.* **232:**207–26.

Rutenber E, Robertus JD (1991): Structure of ricin B-chain at 2.5 Å resolution. In PROTEINS: *Struct. Funct. Genet.* **10**: 260–9.

Sabesan S, Neira S, Davidson F *et al.* (1994): Synthesis and enzymatic and NMR studies of novel sialoside probes: unprecedented, selective neuraminidase hydrolysis of and inhibition by C-6-(methyl)-gal sialosides. In *J. Am. Chem. Soc.* **116**:1616–34.

Sharon N, Lis H (1993): Carbohydrates in cell recognition. In *Sci. Amer.* **268**:74–81.

Sierks MR, Bock K, Rehn S *et al.* (1992): Active site similarities of glucose dehydrogenase, glucose oxidase and glucoamylase probed by deoxygenated substrates. In *Biochemistry* **31**:8972–7.

Solís D, Fernández P, Díaz-Mauriño T *et al.* (1993): Hydrogen-bonding pattern of methyl β-lactoside binding to the *Ricinus communis* lectins. In *Eur. J. Biochem.* **214**:677–83.

Solís D, Feizi T, Yuen CT *et al.* (1994a): Differential recognition by conglutinin and mannan-binding protein of *N*-glycans presented on neoglycolipids and glycoproteins with special reference to complement glycoprotein C3 and ribonuclease B. In *J. Biol. Chem.* **269**:11555–62.

Solís D, Jiménez-Barbero J, Martín-Lomas M *et al.* (1994b): Probing hydrogen-bonding interactions of bovine heart galectin-1 and methyl β-lactoside by use of engineered ligands. In *Eur. J. Biochem.* **223**:107–14.

Spohr U, Paszkiewicz-Hnatiw E, Morishima N *et al.* (1992): Molecular recognition. XI. The synthesis of extensively deoxygenated derivatives of the H-type 2 human blood group determinant and their binding by an anti-H-type 2 monoclonal antibody and the lectin 1 of *Ulex europaeus*. In *Can. J. Chem.* **70**:254–71.

Srivastava OP, Hindsgaul O (1988): Recognition of oligosaccharide substrates by *N*-acetyl-glucosaminyltransferase-V. In *Carbohydr. Res.* **179**:137–61.

Street IP, Armstrong CR, Withers SG (1986): Hydrogen bonding and specificity. Fluorodeoxy sugars as probes of hydrogen bonding in the glycogen phosphorylase-glucose complex. In *Biochemistry* **25**:6021–7.

Street IP, Rupitz K, Withers SG (1989): Fluorinated and deoxygenated substrates as probes of transition-state structure in glycogen phosphorylase. In *Biochemistry* **28**:1581–7.

Varki A (1993): Biological roles of oligosaccharides: all of the theories are correct. In *Glycobiology* **3**:97–130.

Vermesch PS, Tesmer JJG, Quiocho FA (1992): Protein-ligand energetics assessed using deoxy and fluorodeoxy sugars in equilibrium binding and high resolution crystallographyic studies. In *J. Mol. Biol.* **226**:923–9.

Villafranca JE, Robertus JD (1977) Crystallographic study of the anti-tumor protein ricin. In *J. Mol. Biol.* **116**:331–5.

Vyas NK (1991): Atomic features of protein-carbohydrate interactions. In *Curr. Opinion Struct. Biol.* **1**:732–40.

Withers SG, Rupitz K (1990): Measurement of active-site homology between potato and rabbit muscle α-glucan phosphorylases through use of a linear free energy relationship. In *Biochemistry* **29**:6405–9.

20 Application of Site-Directed Mutagenesis to Structure-Function Studies of Carbohydrate-Binding Proteins

Jun Hirabayashi

1 General Points on Site-Directed Mutagenesis

Site-directed mutagenesis is a very powerful method to obtain useful information on the structure-function relationship of a protein of special interest. This procedure, combined with X-ray crystallographic studies, now seems to have been established as the standard for this purpose in place of conventional chemical modification studies, though the latter still make an important contribution in some instances. The critical difference between the modern and the conventional procedures is that the modern methods possibly

Table 1. Previous reports on X-ray crystallographic studies and site-directed mutagenesis of lectins

Lectin (Origin)	X-ray (complexed sugar / resolution)	Mutagenesis
Plant (Leguminosae)		
ConA (*Canavalia ensiformis*)	Derewenda et al. (1989) (Me-α-Man / 2.9 Å)	
GS-IV (*Griffonia simplicifolia*)	Delbaere et al. (1990) (Le^b tetrasaccharide / 2.9 Å)	
LOL-1 (Lathyrus ochrus)	Bourne et al. (1990) (trisaccharide / 2.1 Å) Bourne et al. (1992) (biantennary octasaccharide / 2.3 Å)	
EcorL (*Erythrina corallodendron*)	Shaanan et al. (1991) (lactose / 2.0 Å)	
PSA (*Pisum sativum*)	Rini et al. (1993) (trimannoside / 2.6 Å)	van Eijsden et al. (1992) Hoedemaeker et al. (1993) van Eijsden et al. (1994)
SBA (*Glycine max*)	Dessen et al. (1995) (biantenaryc pentasaccharide / 2.6 Å)	
Plant toxin		
Ricin (*Ricinus communis*)	Montfort et al. (1987) (lactose / 2.8 Å)	
Sereal Plant		
WGA		
Animal C-type		
Mannose-binding protein C	Weis et al. (1992) (biantennary pentasaccharide / 1.7 Å)	Drickamer (1992) Iobst et al. (1994) Iobst and Drickamer (1994)
Animal Galectins		
Galectin-1 (bovine)	Liao et al. (1994) (*N*-acetyllactosamine / 1.9 Å)	Abbot and Feizi (1991)
Galectin-1 (bovine)	Bourne et al. (1994) (biantennary nonasaccharide / 2.2 Å)	Hirabayashi and Kasai (1991)
Galectin-2 (human)	Lobsanov et al. (1993) (lactose / 2.9 Å)	Hirabayashi and Kasai (1994)

H.-J. and S. Gabius (Eds.), Glycosciences

ISBN 3-8261-0073-5

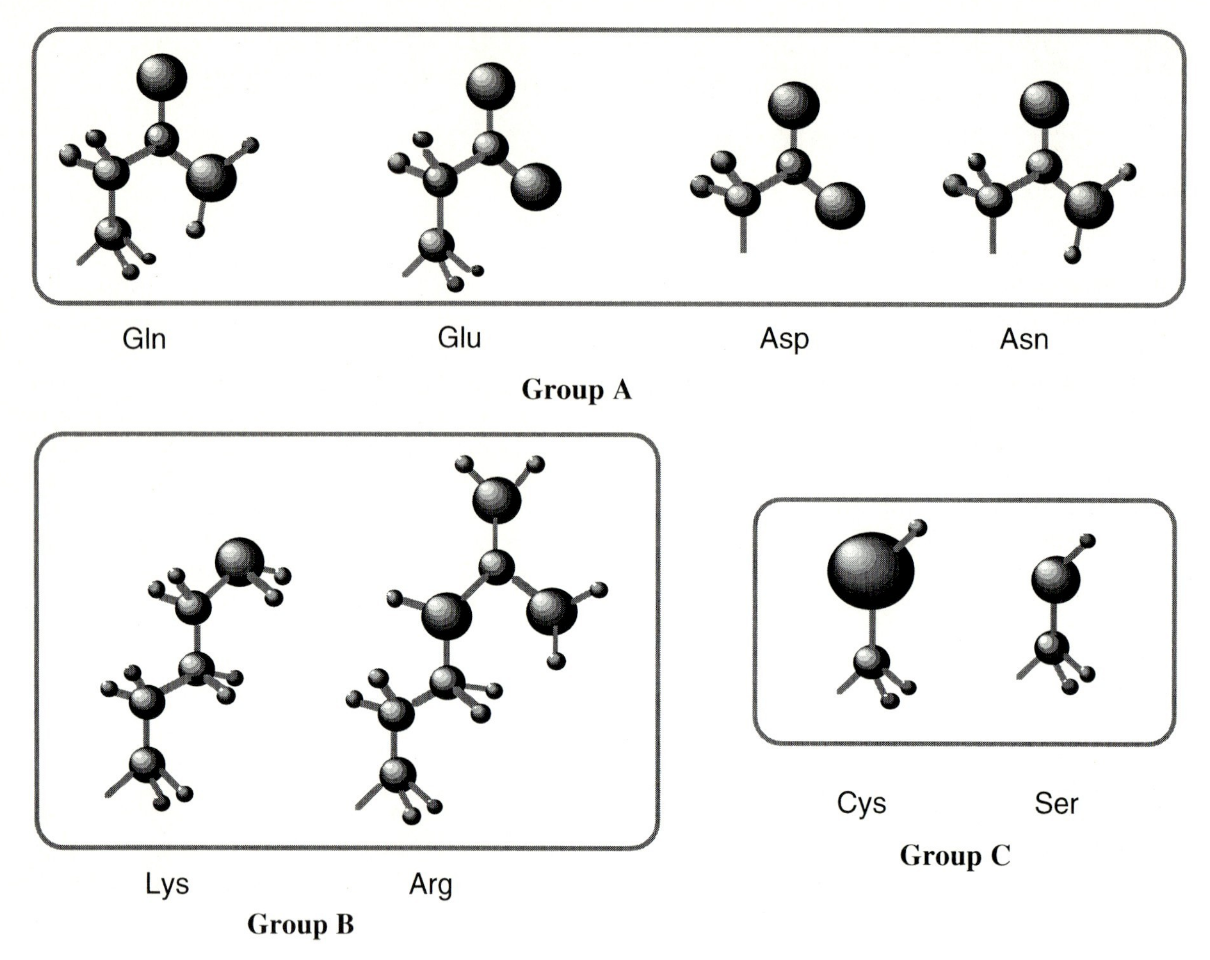

Figure 1. Some groups of homologous amino acids, represented by a ball and stick models. Side chains of homologous amino acids are compared in each group. **Group A**, acidic amino acids (Asp and Glu) and their amides (Asn and Gln); **group B**, basic amino acids (Lys and Arg). As a less basic amino acid, histidine may fall into this group; **group C**, cysteine and serine have rather intermediate properties (neither very hydrophobic nor hydrophilic). Though they differ in only one atom, cysteine is more reactive than serine. A pair of cysteine residues is easily oxidized to make a disulfide bridge.

give much clearer (digital) results and clues for the solution of the structure-function relationship than the older methods, because specific amino acid(s) can be replaced with any desired one using genetic methods. Thus, systematic substitutions are easily achieved, once a gene expression system is established. Such features are difficult to attain by the conventional procedures.

Since the strategy of site-directed mutagenesis is based on molecular genetics, it can be applied to a wide range of proteins including carbohydrate-binding proteins, which are usually called lectins. So far, several groups of lectins have been studied by both X-ray crystallography and site-directed mutagenesis to elucidate the critical amino acid residues involved in the sugar-binding function (see Table 1).

Some information is necessary prior to mutagenesis: amino acid sequences of several homologous proteins are usually compared to choose target amino acids. Information from chemical modification studies is also useful. Here, one can assume that the most important residues are conserved most strongly among homologous proteins, because of "selection pressure" in evolution (Kimura, 1983). Though largely depending on molecular evolutionary rates, at least five sequences are necessary to choose candidate residues. In general, sequences derived from more distant biological species (e.g. mammal and fish) show less sequence similarity than those derived from closer species (e.g. rat and human). Therefore, one can focus on many fewer candidates by comparing sequences of more distant species. Similarly, sequence comparison between different molecular species (e.g. isolectins) in the same biological species can also be useful. However, there is a significant possibility that such different molecular species may have diverged too much to assume a homologous function. For example, estimation of metal-binding sites of animal C-type lectins is meaningful, whereas that of sugar-binding residues, which determine saccharide specificity, would be difficult, because the C-type lectin family is extremely diverse in both structure and sugar-binding specificity, though all of them retain Ca^{2+}-binding properties (Drickamer, 1993; Hirabayashi, 1993).

Though much clearer results may be obtained by mutagenesis than by chemical modification, results obtained by any procedure cannot always be conclusive, because such proteins derived by mutagenesis or chemical modifications are, after all, non-natural proteins. These artificial modifications possibly impair the local conformation of the protein. To avoid such a secondary effect, one should choose amino acids which are as homologous as possible for substitution (see Fig. 1), as the first choice. In this regard, hydrophilic residues are exposed at the protein surface, and thus are expected to permit homologous substitution more freely. On the contrary, substitutions between hydrophobic amino acids are sometimes undesirable for the maintenance of hydrophobic packing, because van der Waals contact, which is responsible for the core packing, depends significantly on both size and shape of the side chains. Nevertheless, site-directed mutagenesis is a most convenient, systematic, and powerful technology to identify essential amino acid residues of proteins.

2 Special Points on Mutagenesis of Carbohydrate-Binding Proteins

The above points on site-directed mutagenesis are applicable to various carbohydrate-binding proteins (lectins). Since lectins interact with carbohydrates mainly by hydrophilic interactions (e.g., hydrogen bonding and electrostatic interaction, Fig. 2), targeting of conserved hydrophilic residues, substitution of which is expected to be less troublesome, should be a main aim in lectin-targeted mutagenesis. Though each hydrogen bond forms a relatively weak interaction, a group of interactions can make a force strong enough for a tight and specific binding. In fact, most of the assumed carbohydrate binding sites have been shown to be formed by an extensive network of hydrogen bonding, to which the predominant contribution is made by hydrophilic side chains, and also backbone carbonyl and amide groups. Water molecules often bridge hydrogen bonds. Such examples include the toxic plant lectin ricin (Montfort *et al.*, 1987), legume lectins such as ConA (Derewenda *et al.*, 1989), LOL (Bourne *et al.*, 1990, 1992) and PSA (Rini *et al.*, 1993), animal C-type lectin like mannose-binding protein (Weis *et al.*, 1992), and animal galectin-1 (Liao *et al.*, 1994; Bourne *et al.*, 1994) and galectin-2 (Lobsanov *et al.*, 1993) (see Table 1).

Though carbohydrates are as a whole hydrophilic, they also have significant hydrophobicity. This can make a second class of interaction between lectins and carbohydrates. Actually, this second class of interaction (hydrophobic or van der Waals interactions) frequently occurs between a few aromatic residues (i.e. Trp, Tyr, Phe) of lectins and a more hydrophobic part (or side, see below) of the ligand saccharides. For example, galectins strongly conserve a single tryptophan in the galactoside-binding site (described later). Lysozyme, hydrolyzing (GlcNAcβ1–4MurNAc) bonds of bacterial peptidoglycans, has relatively many tryptophan residues for its molecular size (6 residues per 14,300 Da polypeptide). Various C-type animal lectins conserve aromatic (Trp, Tyr, Phe and also His)

Hydrogen bonding : Asp, Asn, Glu, His, Arg, etc.

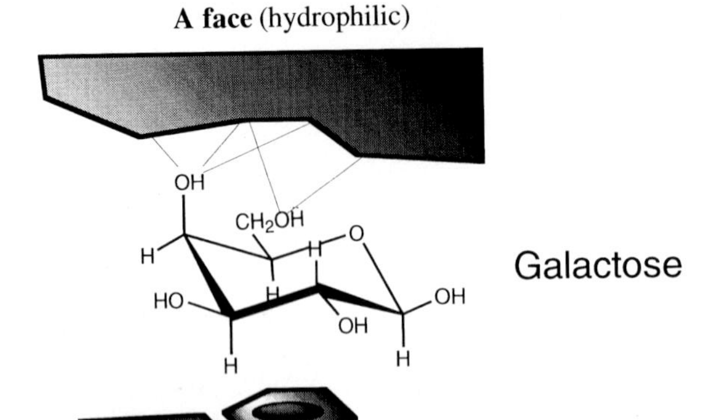

B face (hydrophobic)

Van der Waals interaction : Trp, Tyr, Phe, His etc.

Figure 2. A representative model showing interaction between a carbohydrate-binding protein and a ligand molecule (here galactose). Two kinds of interactions dominate the binding; i.e. hydrogen bonding and van der Waals contact. Exclusive hydrogen bonds are formed between hydroxyl groups of a ligand saccharide and mainly the side chains of hydrophilic residues, but sometimes backbone carbonyl or amide groups. Note that in this case the "A face" (upper side of galactose) is relatively hydrophilic due to the presence of a 4-axial hydroxyl group, while on the other hand the "B face" is relatively hydrophobic, to which an appropriate aromatic residue (most frequently tryptophan) may make contact efficiently.

and aliphatic hydrophobic (Ile, Leu, Val) residues that make contact with the ligand saccharide (Iobst *et al.*, 1994). Ricin has two carbohydrate-binding sites per single polypeptide: one of them (domain 1) has tryptophan in the binding site, and the other (domain 2) has tyrosine instead. It has long been known that tryptophan-containing peptides are significantly retarded by dextran columns, such as Sephadex. In this context, recently investigated "sugar-binding peptides", derived from various legume lectins, usually contain several tryptophan or other aromatic residues. Such affinity of aromatic residues, in particular tryptophan, for carbohydrates may serve as a basis for protein-carbohydrate interaction, though determination of specificity (i.e. discrimination of hydroxyl group configurations) seems to be made mainly by hydrogen bonding, the strength of which depends to a large extent on the length and angles of the bonds.

On the basis of the above understanding, a model for interaction between a lectin and its ligand carbohydrate (here, galactose) is drawn in Fig. 2. The upper side ("A face") of galactose is relatively hydrophilic owing to the presence of a 4-axial hydroxyl group. On the other hand, the "B face" is relatively hydrophobic, because there are no such polar groups on this side. Such a dual character of carbohydrates seems to make an important basis for recognition for many carbohydrate-binding proteins.

3 Actual Procedures for Mutagenesis

Various procedures for site-directed mutagenesis were developed in the middle eighties. These include the methods of Eckstein (Nakamaye and Eckstein, 1986; Sayers *et al.*, 1988), Kunkel (Kunkel, 1985; Kunkel *et al.*, 1987) and Kramer (or gapped-duplex method) (Kramer *et al.*, 1984; Kramer and Fritz, 1987). All of these procedures have been established commercially: e.g. a kit for Eckstein's method is available from Amersham, and kits for both Kunkel's and Cramer's methods can be obtained from Takara. By using these procedures, one can achieve high-mutagenesis efficiency (in general more than 70 %). However, relatively long and complicated steps are required for their accomplishment. A major problem is a need for preparation of a single-stranded template DNA in an appropriate M13-type vector. The use of double-stranded DNA would result in much lower mutagenesis efficiency because of a high non-mutant background. For expression of mutant genes, recloning back into an original expression vector is also necessary. To overcome these problems, some new procedures have now emerged, which are much more convenient and sophisticated (two of them are described in Section 3.2). One is a procedure that has been made more convenient by modifying previous ones by the use of two types of primers, while the other is based on a completely different strategy, i.e. the polymerase-chain reaction (PCR).

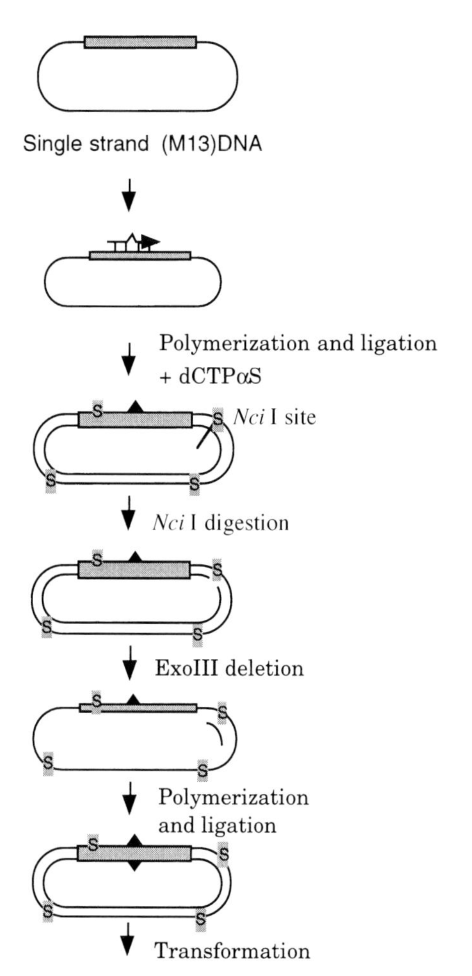

Figure 3. Strategy for site directed-mutagenesis by the method of Eckstein. This is one of the conventional procedures for site-directed mutagenesis, which require cloning in advance of a target gene into an appropriate M13-type vector. In the Eckstein method, both (+) and (-) strands are mutagenized *in vitro* by a strategy using [α-S] dCTP and *Nci* digestion, followed by ExoIII deletion and DNA polymerase I extension. Though the procedures are relatively complicated, basically only mutant plasmids are used for transformation of *E. coli*. On the other hand, other conventional and also more recent strategies (dual-priming methods) select mutant clones by an *in vivo* system after transforming the *MutS E. coli* strain with a hybrid construct, in which only the (-) strand has a mutation. See text and Fig. 4 for more deatils.

In the following sections, a few representative procedures for site-directed mutagenesis are described. Apparently, the more convenient procedures become, the more popular they are. In these procedures, however, confirmation of mutagenesis (i.e. replacement of both target nucleotide bases and translated amino acids) is absolutely necessary, particularly in PCR-aided mutagenesis, because DNA fragments amplified by *Taq* DNA polymerase often introduce errors. It is not necessary to emphasize that the most important point is to ensure that the intended substitutions have been obtained without any possibility of introducing errors. Thus, the so-called "walk, don't run" principle should be applied to site-directed mutagenesis, though the final decision on the choice of method is up to those who carry out this work.

3.1 Conventional Procedures

In conventional mutagenesis procedures, that is, the methods of Kunkel, Kramer and Eckstein, the preparation of a single-stranded form of target gene (cDNA), such as that cloned in filamentous M13 phage viral DNA, is necessary. It is an absolute requirement for efficient hybridization (priming) of mutagenic primers. In any procedure, after mis-matched priming between M13 (+) strand and a mutagenic primer, a complementary (-) strand is synthesized. On the other hand, to replace the non-mutagenic (+) strand, various strategies are available, and this is the major point of difference among these mutagenesis procedures. Here, only the Eckstein method is described in detail.

In the Eckstein method (see Fig. 3), synthesis of the (-) strand is performed with [α-^{35}S] dNTP (e.g., dCTP) to resist digestion by a certain class of restriction endonucleases (e.g., *Nci*I, which recognizes CC(C,G)GG). These enzymes make a nick at the corresponding site only in the original (+) strand. Then a large portion of the (+) strand, which should include a target gene sequence, is removed by the action of *E. coli* 3'-5' exonuclease (ExoIII). A mutagenic site of the (-) strand is regenerated by the action of the appropriate DNA polymerase. Thus, both (+/-) mutagenic strands of recombinant M13 DNA are constructed. This construct is then introduced into an appropriate *E. coli* stain (e.g., TG1) for transformation. Mutant proteins can be expressed after recloning of the mutant gene into some expression vector.

In the above method, and other conventional procedures, a major point is how efficiently a non-mutagenic (+) strand is displaced from a mutagenic strand: in the Eckstein method it is performed by an *in vitro* biochemical strategy, while in both the Kunkel and Kramer methods this is done *in vivo* during the process of plasmid replication by using highly sophisticated devices (i.e. uracylglycosidase and amber/suppressor systems, respectively).

3.2 New Procedures

A few of the new procedures are described in this section. These are well-designed, extremely convenient procedures. However, they are based on completely different strategies, i.e. a dual-priming strategy, and a polymerase chain reaction (PCR) strategy. For the former method, several variations are available, which differ in the selection procedure. Neither of these new procedures requires preparation of a single-stranded target DNA, which is a major problem with the conventional methods. In the PCR-aided method, only a very small amount of target DNA is necessary, because it can be amplified more than 1,000,000-fold. Moreover, PCR does not necessarily require purification of a target DNA, if very specific primers are available.

Whatever procedure one chooses, introduction of an "amber" mutation into a target site is extremely useful for making variable mutant proteins, because from such a single mutant gene one can obtain various mutant proteins by utilizing various *E. coli* suppressor strains that carry different suppressor tRNAs (i.e. amber-Cys, Glu, Phe, His, Lys, Arg, Ser, Tyr, Gln, Leu, Pro and Gly suppressor tRNAs). For this purpose, the Interchange™ *in vivo* Amber Suppressor Mutagenesis System is available from Promega.

3.2.1 Dual priming method

This method was developed by several investigators as a more convenient method directly applicable to any expression system. Variously devised kits for this purpose are commercially available (e.g. from Clontech, Promega, and Takara). In all of these procedures, two kinds of mutagenic oligonucleotide are used: one is a conventional gene-targeting primer and the other is a vector-directed selection primer. The latter primers can be directed either for replacing an amber mutation or for destroying a unique restriction site.

In an amber mutation/suppressor system (Hashimoto-Gotoh *et al.*, 1995; commercial kits supplied from Promega and Takara), a vector-selection primer is designed to replace an amber codon (TAG) with a proper codon, for instance, in the kanamycin-resistant gene (Fig. 4). Here, an important point is that the gene-targeting primer is used in excess relative to the vector-selection primer; e.g. 10 : 1 (mol/mol). In such a biased proportion, most of vector-targeted clones are expected by statistical reasoning to have the gene-specific mutation. After polymerization and ligation by usual procedures, a *SupE*/*MutS* strain (such as BMH71−18*MutS*) is transformed with the hybrid construct. From the *MutS* strain, two types of plasmids should be obtained: one is a doubly mutated (both gene and vector-targeted) plasmid, and the other is a non-mutant (wild-type) plasmid. A mixture of these plasmids is then introduced into a Sup^0 strain (e.g. MV1184) for transformation. In this strain, however, only non-amber clones (i.e. doubly mutated clones) can grow in the presence of kanamycin. On the other hand, the wild-type clone cannot grow, because the amber mutation they carry is not suppressed in the Sup^0 strain. Thus, one can easily choose mutagenized clones by the use of a vector-selection primer that is directed to the amber mutation.

As a variation of the above, a vector-selection primer can be designed to destroy a unique restriction site that is not involved in a target gene. Here again, a gene-targeting primer is used in excess. After transformation of a *MutS* strain with the derived hybrid plasmid, a mixture of doubly mutagenic plasmid and non-mutant plasmid is prepared. Subsequently, these are subjected to digestion with the restriction enzyme, the site of which was first targeted. This treatment results in linearization only of the non-

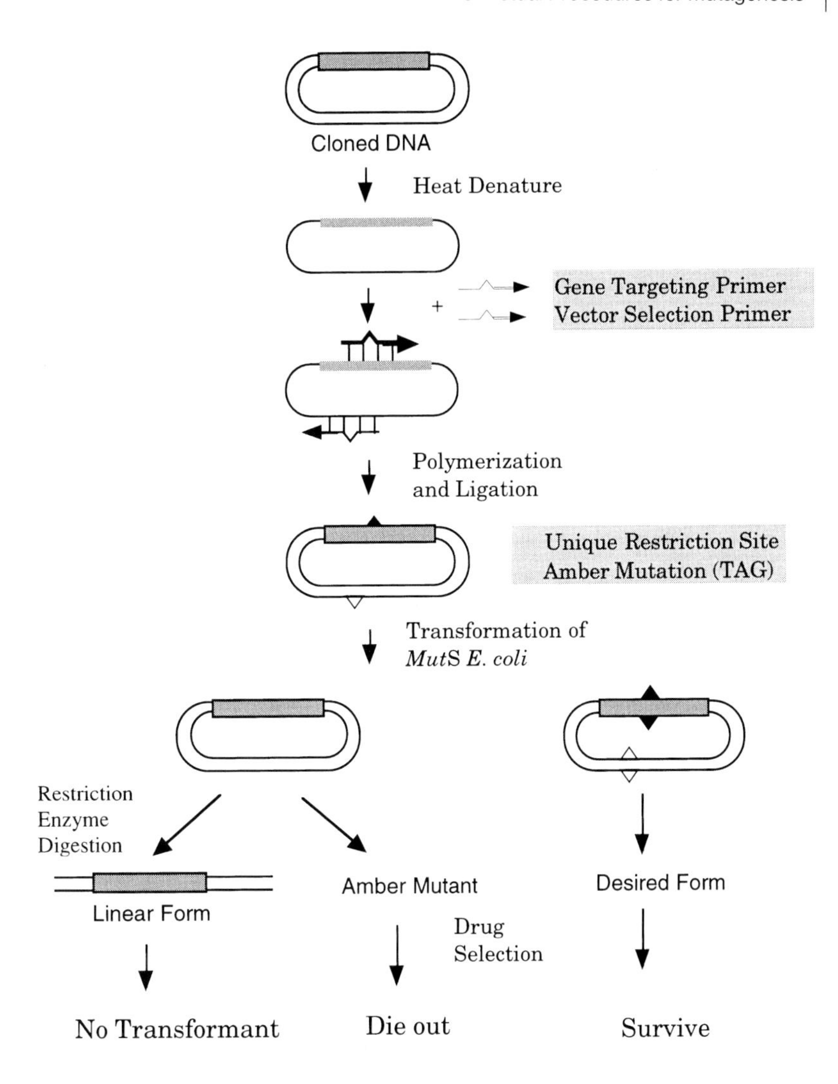

Figure 4. Strategy for site-directed mutagenesis by dual-priming methods. In these procedures, any double-stranded plasmid can be used as a template for priming. Two types of mutagenic primers are hybridized to a heat-denatured plasmid template: one is a gene-targeting primer, and the other is a vector-selection primer. The gene-targeting primer is used in excess relative to the vector-selection primer, e.g. 10:1 (mol/mol). After the priming, the (-) strand is synthesized by polymerase and ligase actions. This results in the construction of a hybrid plasmid, in which only the (-) strand contains mutation(s), whereas the (+) strand remains intact. By statistical reasoning, most of the vector-targeted mutants are expected to bear a gene-targeted mutation as well. Transformation of an appropriate *MutS E. coli* strain (e.g. BMH71 – 18*MutS*) should generate two types of plasmids, i.e. double mutants, single mutant (dominantly gene-targeted) and non-mutant. Among these, only the double mutant clones can grow in the presence of kanamycin after transformation of a Sup^0 strain, As a variant, a vector-target primer can be designed to destroy a unique restriction site. For details, see text.

mutant plasmid, whereas the doubly mutagenic plasmid remains intact. Since the former (linear) form has a much lower transforming ability (less than 1/100) than the latter (circular) form, only clones harboring the mutagenic plasmid will appear. In this procedure, a vector-targeted primer can also be designed to convert one unique restriction site into another so that multiple rounds of mutagenesis can be performed without loss of a unique restriction site, and with no need of recloning (if one desires to introduce double, triple etc. mutations).

Both of the above two methods require no special cloning vector, if either an appropriate amber/suppression or restriction site is available. Thus, the ability to perform direct mutagenesis in extensive expression vectors is a principal advantage of these dual-priming methods.

3.3 PCR-Aided Mutagenesis

A number of recent molecular biological studies have adopted PCR-aided mutagenesis. However, the process for mutagenesis consists of two-step PCR, and thus is somewhat complicated compared with conventional one-step PCR (Fig. 5). In the first PCR, two PCRs are performed separ-

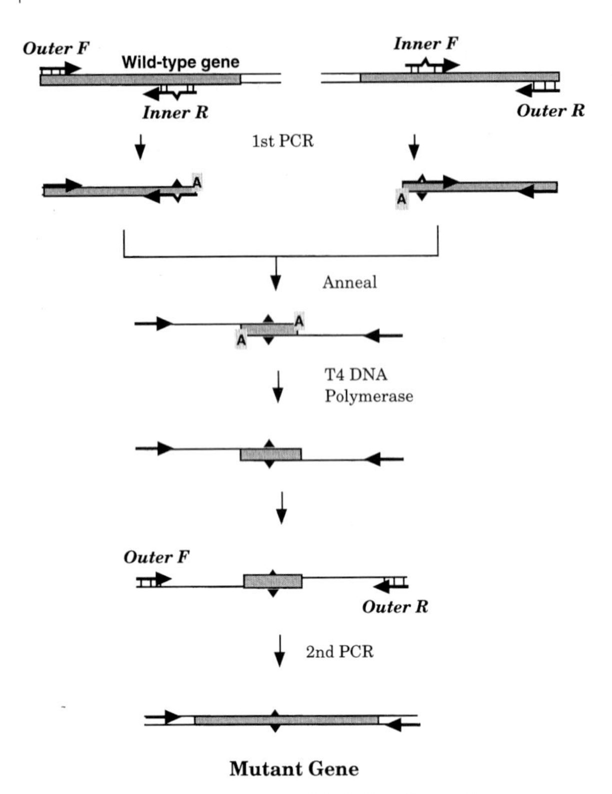

Figure 5. Strategy for PCR-aided site-directed mutagenesis. This is the most convenient strategy for mutagenesis. The procedure consists of two steps of polymerase-chain reactions (PCRs). For the first PCR, two sets of primers are used for amplification of two gene fragments, which overlap each other in the range at least 10 bp. After the first PCR, the two fragments are heat denatured and annealed to make hybrid constructs. After removal of 3'-overhanging adenine residues by T4 DNA polymerase, the second PCR is performed by using a set of outer primers. This results in a mutant gene fragment.

ately: one is done by using a forward outer primer (*Outer F*) and a reverse inner mutagenic primer (*Inner F*), and the other is carried out by using a reverse outer primer (*Outer R*) and a forward inner mutagenic primer (*Inner F*). *Inner F* and *Inner R* must be complementary to each other over a length of at least 10 bp for the subsequent annealing. After the first PCR, 3'-overhanging adenine of both strands, as a usual result of PCR, is best removed, for example, by T4 DNA polymerase having 3'-5' exonuclease activity, if these adenines are mis-matched. Then, complementary strands are synthesized using the same enzyme or *Taq* polymerase. In this process, no primer is necessary. Then, the second PCR is performed by using the derived construct and the two outer primers (*Outer F* and *Outer R*). This gives the full length of the mutant gene fragment. To insert this fragment into an expression vector, an appropriate restriction site is usually introduced at the 5'-site of the outer primers.

In this procedure, both sense and antisense strands are subjected to base replacements in the PCR chemistry. This is a feature distinct from many other mutagenesis procedures (except Eckstein's method).

Confirmation of base replacements is definitely necessary, particularly because *Taq* DNA polymerase has relatively low fidelity. For this reason, a high level of amplification should be avoided. Recently, some thermostable DNA polymerases, such as *PfuI* and *Tli*, which have a proof-reading activity, have become commercially available.

4 Example of Mutagenesis: Human Galectin-1

As an example of site-directed mutagenesis, our previous experiments on human galectin-1 are described. Site-directed mutagenesis was performed by Hirabayashi and Kasai (1991, 1994) according to the Eckstein method. Detailed procedures for the mutagenesis and some important experimental points were described previously (Hirabayashi and Kasai, 1993b).

4.1 What is Galectin?

Galectins are a group of soluble metal-independent-type lectins, found extensively in animal species (for review, see Hirabayashi, 1993, 1994; Hirabayashi and Kasai, 1993a; Barondes *et al.*, 1994a, b). All of them show specificity for β-galactosides and share significant sequence homologies. They were recently renamed based on the proposal of Barondes (Barondes *et al.*, 1994a), because they had been differently named by various investigators, according to the background of the investigators. Mammalian galectins, which show much closer inter-species relationships, are numbered systematically in the order of discovery. Up to the present, galectins-1 to -8 have been sequenced and deposited in appropriate DNA data bases. Among these, however, galectin-1 and galectin-3 are the most common in mammalian species, and thus, have been studied most intensively and extensively.

A recent topic of galectin research is the investigation of invertebrate galectins from the nematode *Caenorhabditis elegans* (Hirabayashi *et al.*, 1992) and from the marine sponge *Geodia cydonium* (Pfeifer *et al.*, 1993). These findings suggest a fundamental role of galectins throuout the animal kingdom. Galectins are supposed to function in various cell-cell and cell-substratum interactions, through which they play important roles essential for multicellular life systems, e.g. development and differentiation. The fact that all galectins so far investigated recognize β-galactosides, e.g., *N*-acetyllactosamine, implies the evolutionary importance of this disaccharide for biological recognition.

Galectins had long been misunderstood to be thiol-sensitive (often called "S-type" lectins) in special contrast to another group of animal lectins, i.e., the Ca-dependent ("C-type") lectins (Drickamer, 1988). This is true for many of the known galectins, including galectin-1, but is not necessarily true for other galectins. However, the fact that galectin-1 loses its sugar-binding activity in the absence of a thiol-reducing reagent prompted us to perform site-directed mutagenesis to examine whether or not any cysteine residue is involved in the function.

4.2 Conserved Amino Acids in the Galectin Family

In 1990, when our mutagenesis studies on human galectin-1 were made, there were nine complete amino acid sequences of galectins, i.e. galectin-1 from human, rat, bovine and mouse, galectin-3 from human, rat and mouse, chicken 14 kDa galectin and chicken 16 kDa galectin (conven tional designations based on apparent subunit molecular weights). Sequence similarities to human galectin-1 are shown schematically in Figure 6, which also includes some other vertebrate galectin sequences reported since 1990. At a glance, one can see that galectins from more distant biological species, such as eel and frog, show much lower sequence similarities to human galectin-1. On the other hand, galectins-3, a human isolectin, shows only 27 % identity to human galectin-1, whereas three galectin-3 from human, rat and mouse show almost 90 % identity to one another. This observation indicates that each galectin member has diverged significantly, though their essential function of recognizing β-galactosides has been strongly conserved.

A number of amino acids are conserved. Most of them are located in the central region, which is relatively hydrophilic. Among these, the most conserved and hydrophilic residues are His44 (conserved in 12 out of 12 sequences), Asn46 (11), Arg48 (12), Asn61 (12), Lys63 (11), Glu71 (12), Arg73 (11), Arg111 (12) and Asp125 (10) (residue numbers are those of human galectin-1). Most of them are encoded by the largest exon (exon 3) of galectin genes as exemplified by human galectin-1 (Gitt and Barondes, 1991) and galectin-2 genes (Gitt *et al.*, 1992) and also by the chick 14 kDa galectin gene (Ohyama and Kasai, 1988).

4.3 Primer Design and Mutagenesis

Oligonucleotide primers used for site-directed mutagenesis must be designed to achieve high mutagenesis efficiencies. They are synthesized in general as approximately 20-mer nucleotides, which correspond to melting temperatures (Tm) of 55–65 °C. Tm is usually simply calculated according to the following empirical formula:

$$Tm = 2 \times N(A+T) + 4 \times N(G+C),$$

where N(A+T) is the number of nucleotides A and T, and N(G+C) is the number of G and C. The length of oligonucleotide should not be very short (< 35-mer), otherwise stable hybridization with a template is not ensured. On the other hand, the use of too long an oligonucleotide (e.g. > 35 mer) will make it difficult to discriminate between mutant clones from non-mutant clones by a conventional hybridization technique that uses the mutagenic primer as a probe (for

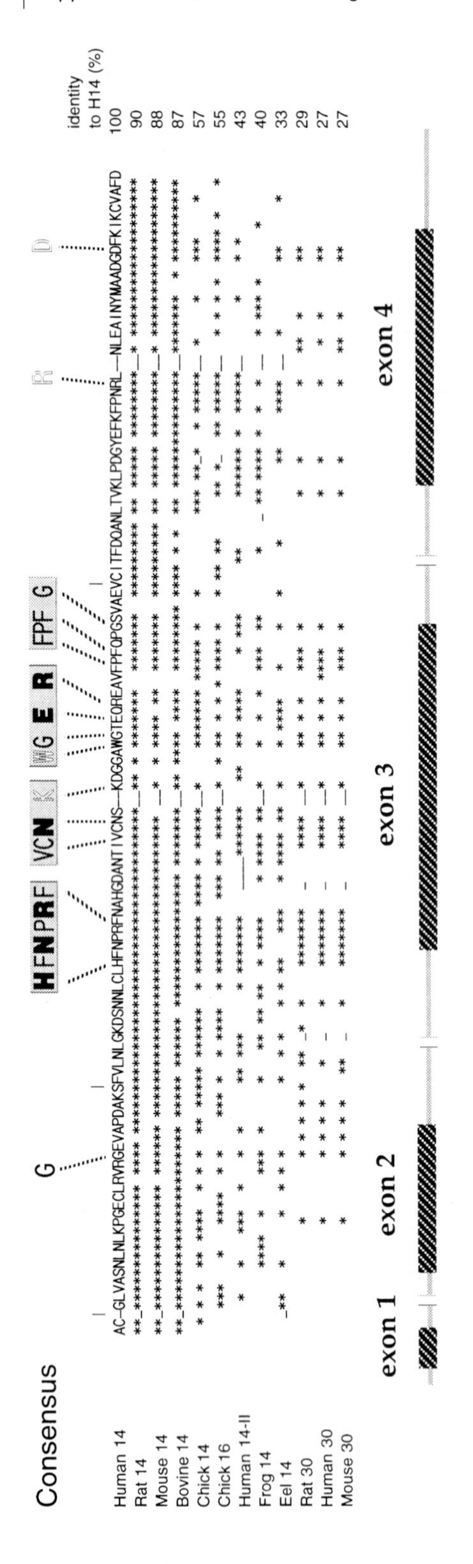

Figure 6. Sequence alignment of representative members of the galectin family. For simplicity, residues identical to those of human galectin-1 (human 14) are shown with asterisks. Strongly conserved residues are also indicated above the top line, and those that make conserved clusters are boxed. Most conserved residues are located in a central region that has been shown to be encoded by the third exon of various 14 kDa-type (proto-type) galectins (Ohyama and Kasai, 1988; Gitt and Barondes, 1991; Gitt *et al.*, 1992). Exon-intron organization is schematically shown below the alignment, and three intron-inserting positions are indicated by short vertical bars above the amino acid sequence of human galectin-1. Target residues, which are described in this chapter, are shown bold. Among them, those proved to be essential for sugar-binding function are drawn in black, whereas those proved not to be essential are drawn in white. For reference to each amino acid sequence, see the appropriate review papers (Hirabayashi, 1993, 1994; Hirabayashi and Kasai, 1993b).

experimental details, see Hirabayashi and Kasai, 1993b).

In general, a mismatched site is located at the center of the mutagenic primer to ensure stable hybridization and to direct proper 5'-3' extension by an appropriate DNA polymerase. For the latter reason, it is better for 3'-sites of primers to be G or C. Primers themselves should neither be self-complementary nor contain inverted or palindromic sequences, which prevent efficient priming. Making mutagenic primers in a repetitive region should also be avoided. Degenerate primers are not usually desirable.

Taking the above points into consideration, a series of mutagenic primers were synthesized for human galectin-1 (see Table 2). These were designed with the following two aims: the focus of the first was the cysteine residues in the context of oxidative inactivation. Whitney *et al.* (1986) reported that hemagglutinating activity of rat 14 kDa β-galactoside-binding lectin (galectin-1) was not affected by modification with a thiol-modifying reagent, monoiodoacetamide. This result seemed to indicate that no cysteine residue was directly involved in the sugar binding function. Moreover, some galectins were found to have no cysteine (e.g. electric eel 14 kDa lectin and nematode 32 kDa lectin). To clarify this problem, each of the six cysteine residues occurring in human galectin-1, i.e. Cys2, Cys16, Cys42, Cys60, Cys88 and Cys130, was substituted with

Table 2. Summary of mutagenesis of human galectin-1

Mutants	Primers used for mutagenesis (amino acid sequence)	TM (°C) (mismatched/ matched)	Yield (mg/l culture)	Binding to asialofetuin	I_{50} of lactose (mM)
Wild type			2.00	Yes	0.54
C2S[a]	CC ATG GCT **TCT** GGT CTG G	54/58	0.93	Yes	0.32
C15S[a]	CT GGA GG **TCC** CTT CGA GTG	60/64	0.42	Yes	0.68
C42S[a]	C AAC CTG **TCC** CTG CAC TTC	56/60	1.92	Yes	0.50
C60S[a]	CC ATC GTG **TCC** AAC AGC AAG	62/66	0.32	Yes	0.32
C88S[a]	CA GAG GTG **TCC** ATC ACC TTC	58/62	0.27	Yes	0.34
C130S[a]	C AAG ATC AAA **TCT** GTG GCC TTT G	64/68	0.15	Yes	0.72
W68Y[a]	C GGG GCC **TAC** GGG ACC G	56/62	1.66	Yes	1.1
H44Q[b]	G TGC CTG **CAG** TTC AAC CC	52/56	–	No	–
N46D[a]	CTG CAC TTG **GAC** CCT CGC	56/60	–	No	–
R48H[b]	C AAC CCT **CAC** TTC AAC GC	54/56	–	No	–
N61D[b]	C GTG TGC **GAC** AGC AAG G	52/56	–	No	–
K63H[b]	GC AC AGC **CAC** GAC GGC GG	58/66	0.69	Yes	0.48
E71Q[a]	G GGG ACC **CAG** CAG GGG G	58/62	–	No	–
R73H[a]	CC GAG CAG **CAC** GAG GCT G	58/66	–	No	–
R111H[b]	C CCC AAC **CAC**CTC AAC C	54/56	2.35	Yes	0.48
D125E[b]	GCT GAC GGT **GAA** TTC AAG ATC	60/62	1.36	Yes	0.36

a Result from Hirabayashi and Kasai (1991);
b from Hirabayashi and Kasai (1994)

serine. These amino acids differ only in one atom ("S" in cysteine and "O" in serine; also see Fig. 1). Nevertheless, the latter amino acid is no longer oxidized to form a disulfide bridge.

The second point involved the conserved hydrophilic residues. This is based on the reason emphasized in section 2. As described in section 4.2, most of the conserved residues encoded by the third exon of human galectin-1 (Gitt and Barondes, 1991) are relatively hydrophilic (Fig. 6). Such conserved hydrophilic residues, i.e. His44, Asn46, Arg48, Asn61, Lys63, Glu71 and Arg73, were the primary targets. Arg111 and Asp125 were also substituted by mutagenesis, although they are located in the C-terminal region encoded by exon 4.

In addition, a single tryptophan is strongly conserved (Trp68 in the case of human galectin-1). It has long been believed that this tryptophan is involved in the sugar-binding site, on the basis of spectrophotometric evidence (Levi and Teichberg, 1981). In our experiment, Trp68 of human galectin-1 was substituted with tyrosine.

4.4 Evaluation of the Results

Using the method by Nakamaye and Eckstein (1986), 16 mutant galectin genes were synthesized, and mutant proteins were produced in *E. coli*. A bacterial expression system for human galectin-1 had already been established (Hirabayashi *et al.*, 1989). In Western blotting experiments using bacterial lysates obtained from the mutant clones, all of the lysates showed an immunostained band at a position of 14 kDa, the size of galectin-1. As the next step, each mutant was expressed on a relatively large scale (e.g. 2 l), and the derived *E. coli* lysate was applied to a column of asialofetuin-agarose to check its binding ability. As a result, all of Cys/Ser mutants (designated C2S, C16S, C42S, C60S, C88S, and C130S) were found to be fully active. They were adsorbed on asialofetuin-agarose, and showed comparable hemagglutinating activity to the wild-type galectin-1. When affinity to lactose was examined in term s of inhibitory power (I_{50}) in a lectin-asialofetuin binding assay (Hirabayashi and

Kasai, 1991), all of the six Cys/Ser mutants showed fairly similar values to that of wild-type galectin (i.e. 0.5 +/- 0.2 mM). This result showed unambiguously that cysteine residues are not necessary for the sugar-binding function of human galectin-1.

On the contrary, most of mutants, in which conservative hydrophilic residues were substituted with homologous ones, failed to bind to the asialofetuin-agarose column. These were H44Q, N46D, R48H, N61D, E71Q and R73H. Even in a sensitive assay utilizing a high-performance liquid chromatography system equipped with a small asialofetuin-Toyopearl column (4.6×50 mm), they showed no sign of retardation (Hirabayashi and Kasai, 1994). This result indicates that substitution of any of these six conserved hydrophilic residues completely abolished the binding activity. On the other hand, substitutions of Lys63, Arg111 and Asp125 had only a small effect on the activity. Actually, Lys63 was later found to be not so highly conserved as the others: two frog (*Xenopus* and *Rana*) lectins, nematode 32 kDa galectin, rat galectin-4 (36 kDa), and two sponge galectins (14 kDa) were shown to lack this residue at the corresponding positions.

When Trp68 was substituted with tyrosine (W68Y), a significant decrease in the activity was noted: in terms of I_{50} for lactose, 1.1 mM. Nevertheless, in contrast to the above six inactive mutants, W68Y mutant still bound to asialofetuin agarose. The result suggests that Trp68 is not essential for the sugar-binding activity of human galectin-1, but possibly contributes to stabilizing the binding through van der Waals contact.

After completion of this work, three independent groups reported X-ray crystallographic structures of complexes; i.e. between human galectin-2 and lactose (Lobsanov *et al.*, 1993), between human galectin-1 and *N*- acetyllactosamine (Liao *et al.*, 1994), and between bovine galectin-1 and biantennary oligosaccharide (Bourne *et al.*, 1994). All of these results agreed completely with our conclusion that no cysteine residue is involved in the saccharide binding, that six conserved hydrophilic residues (His44, Asn46, Arg48, Asn61, Glu71 and Arg73) are essential for the binding, and that Trp68 contributes to the binding to some degree. Actually, extensive hydrogen bonds were observed between the 4-OH group of galactose, and His44, Asn46 and Arg48; between the 6-OH group of galactose, and Arg48 and Asn61; and between the 3-OH group of glucose (or *N*-acetylglucosamine), and Glu71 and Arg73. These results emphasize that tailoring the receptor by site-directed mutagenesis as well as the ligand by chemical synthesis, as discussed by Solís and Díaz-Mauriño in the accompanying chapter, reveals important clues to solve the problem how the sugar molecule interacts with crucial amino acid side chains of the binding site(s) of the lectin.

References

Abbot WM, Feizi T (1991): Soluble 14 kDa β-galactoside-specific bovine lectin. In *J. Biol. Chem.* **266**: 5552–7.

Barondes SH, Castronovo V, Cooper DNW *et al.* (1994a): A family of animal β-galactoside-binding lectins. In *Cell* **76**:597–8.

Barondes SH, Cooper DNW, Gitt MA *et al.* (1994b): Galectins: structure and function of a large family of animal lectins. In *J. Biol. Chem.* **269**:20807–10.

Bourne Y, Rouge P, Cambillau C (1990): X-ray structure of a (α-Man(1–3)β-Man(1–4)GlcNAc)-lectin complex at 2.1-Å resolution. In *J. Biol. Chem.* **265**:18161–5.

Bourne Y, Rouge P, Cambillau C (1992): X-ray structure of a biantennary octasaccharide-lectin complex refined at 2.3-Å resolution. In *J. Biol. Chem.* **267**:97–203.

Bourne Y, Bolgiano B, Liao D-I *et al.* (1994): Cross-linking of mammalian lectin (galectin-1) by complex bianntenary saccharides. In *Nature Struct. Biol.* **1**:863–9.

Delbaere LTJ, Vandonselaar M, Prasad L *et al.* (1990): Molecular recognition of a human blood group determinant by a plant lectin. In *Can. J. Chem.* **68**:1116–21.

Derewenda Z, Yariv J, Helliwel JR *et al.* (1989): The structure of the saccharide-binding site of concanavalin A. In *EMBO J.* **8**:2189–93.

Dessen A, Gupta D, Sabesan S *et al.* (1995): X-ray crystal structure of the soybean agglutinin cross-linked with a biantennary analog of the blood group I carbohydrate antigen. In *Biochemistry* **34**:4933–42.

Drickamer K (1988) Two distinct classes of carbohydrate-recognition domain in animal lectins. In *J. Biol. Chem.* **263**:9557–60.

Drickamer K (1992): Engineering galactose-binding activity into a C-type mannose-binding protein. In *Nature* **360:**183–6.

Drickamer K (1993) Evolution of Ca^{2+}-dependent animal lectins. In *Prog. Nucleic Acid Res. Mol. Biol.* **45**:207–32.

Gitt MA, Barondes SH (1991): Genomic sequence and organization of two members of a human lectin gene family. In *Biochemistry* **30**:82–9.

Gitt MA, Massa SM, Leffler H *et al.* (1992): Isolation and expression of a gene encoding L-14-II, a new human soluble lactose-binding lectin. In *J. Biol. Chem.* **267**:10601–6.

Hashimoto-Gotoh T, Mizuno T, Ogasawara Y *et al.* (1995): An oligodeoxyribonucleotide-directed dual amber method for site-directed mutagenesis. In *Gene* **152**:271–5.

Higuchi R (1990): Recombinant PCR. In *PCR Protocols: A Guide to Methods and Applications* (Innis MA, Gelfand DH, Sninsky JJ, White TJ eds) pp 177–83, San Diego: Academic Press.

Hirabayashi J (1993): A general comparison of two major families of animal lectins. In *Trends Glycosci. Glycotechnol.* **5**:251–70.

Hirabayashi J (1994): Two distinct family of animal lectins: speculations on their raison d'être. In *Lectins: biology, biochemistry, clinical biochemistry.* Vol. 10 (Driessche EV, Fischer J, Beeckmans S, Bøg-Hansen TC eds.), pp. 205–19, Hellerup: Textop.

Hirabayashi J, Kasai K (1991): Effect of amino acid substitution by site-directed mutagenesis on the carbohydrate-recognition and stability of human 14 kDa β-galactoside-binding lectin. In *J. Biol. Chem.* **266**:23648–53.

Hirabayashi J, Kasai K (1993a): The family of metazoan metal-independent β-galactoside-binding lectins: structure, function and molecular evolution. In *Glycobiology* **2**:297–304.

Hirabayashi J, Kasai K (1993b): Construction of mutant lectin genes by oligonucleotide-directed mutragenesis and their expression in *Escherichia coli*. In *Lectins and Glycobiology* (Gabius H-J, Gabius S, eds) pp 482–91, Heidelberg – New York: Springer.

Hirabayashi J, Kasai K (1994): Further evidence by site-directed mutagensis that conserved hydrophilic residues form a carbohydrate-binding site of human galectin-1. In *Glycoconjugate J.* **11**:437–42.

Hirabayashi J, Ayaki H, Soma G *et al.* (1989): Cloning and nucleotidee sequence of a full-length cDNA for human 14 kDa β-galactoside-binding lectin. In *Biochim. Biophys. Acta* **1008**:85–91.

Hirabayashi J, Satoh M, Kasai K (1992): Evidence that *Caenorhabditis elegans* 32 kDa β-galactoside-binding protein is homologous to vertebrate β-galactoside binding lectins: cDNA cloning and deduced amino acid sequence. In *J. Biol. Chem.* **267**:15485–90.

Hoedemaeker FJ, van Eijsden RR, Diaz CL *et al.* (1993): Destabilisation of pea lectin by substitution of a single amino acid in a surface loop. In *Plant Mol. Biol.* **22**:39–46.

Iobst ST, Drickamer K (1994): Binding of sugar ligands to Ca^{2+}-dependent animal lectins: II. Generation of high-affinity galactose binding by site-directed mutagenesis. In *J. Biol. Chem.* **269**:15512–9.

Iobst ST, Wormald MR, Weis WI *et al.* (1994): Binding of sugar ligands to Ca^{2+}-dependent animal lectins: I. Activities of mannose binding by site-directed mutagenesis and NMR. In *J. Biol. Chem.* **269**:15505–11.

Kimura M (1983): The neutral theory of molecular evolution. Cambridge: University Press.

Kramer W, Fritz H-J (1987): Oligonucleotide-directed construction of mutation via gapped duplex DNA. In *Methods Enzymol.* **154**:350–67.

Kramer W, Drutsa V, Jansen H-W *et al.* (1984): The gapped duplex DNA approach to oligonucleotide-directed mutation construction. In *Nucleic Acid Res.* **12**:9441–56.

Kunkel TA (1985): Rapid and efficient site-directed mutagenesis without phenotypic selection. In *Proc. Natl. Acad. Sci. USA* **82**:488–92.

Kunkel TA, Roberts JD, Zakour RA (1987): Rapid and efficient site-specific mutagenesis without phenotypic selection. In *Methods Enzymol.* **154**:367–82.

Levi G, Teichberg VI (1981): Isolation and physicochemical characterization of electrolectin, a β-galactoside-binding lectin from electric organ of *Electrophorus electricus*. In *J. Biol. Chem.* **256**: 5735–40.

Liao D-I, Kapadia G, Ahmed H *et al.* (1994): Structure of S-lectin, a developmentally regulated vertebrate β-galactoside binding protein. In *Proc. Natl. Acad. Sci. USA* **91**:1428–32.

Lobsanov YD, Gitt MA, Leffler H *et al.* (1993): X-ray crystal structure of the human dimeric S-Lac lectin, L-14-II, in complex with lactose at 2.9-Å resolution. In *J. Biol. Chem.* **268**:27034–8.

Montfort W, Villafranca JE, Monzingo AF *et al.* (1987): The three-dimensional structure of ricin at 2.8-Å. In *J. Biol. Chem.* **262**:5398–403.

Nakamaye KL, Eckstein F (1986): Inhibition of restriction endonuclease NciI cleavage by phosphorothioate groups and its application to oligonucleotide-directed mutagenesis. In *Nucleic Acid Res.* **14**:9679–98.

Ohyama Y, Kasai K (1988): Isolation and characterization of the chick 14K β-galactoside-binding lectin gene. In *J. Biochem.* **104**:173–7.

Pfeifer K, Hassemann M, Gamulin V *et al.* (1993): S-type lectins occur also in invertebrates: high conservation of the carbohydrate recognition domain in the lectin genes from the marine sponge *Geodia cydonium*. In *Glycobiology* **3**:179–84.

Quiocho FA (1986): Carbohydrate-binding proteins: tertiary structures and protein-sugar interactions. In *Annu. Rev. Biochem.* **55**:287–315.

Rini JM, Hardman KD, Einspahr H *et al.* (1993): X-ray crystal structure of a pea lectin trimannoside complex at 2.6-Å resolution. In *J. Biol. Chem.* **268**: 10126–32.

Sayers JR, Schmidt W, Eckstein F (1988): 5'-3' Exonucleases in phosphorothioate-based oligonucleotide-directed mutagenesis. In *Nucl. Acids Res.* **16**: 791–802.

Shaanan B, Lis H, Sharon N (1991): Structure of a legume lectin with an ordered *N*-linked carbohydrate in complex with lactose. In *Science* **254**:862–6.

van Eijsden RR, Hoedenmaeker FJ, Diaz CL *et al.* (1992): Mutational analysis of pea lectin: replacement of Asn125 by Asp in the monosaccharide binding site eliminates mannose/glucose binding activity. In *Plant Mol. Biol.* **20**:1049–58.

van Eijsden RR, de Peter BS, Kijne JW (1994): Mutational analysis of the sugar-binding site of pea lectin. In *Glycoconjugate J.* **11**:375–82.

Weis WI, Drickamer K, Hendrickson WA (1992): Structure of a C-type mannose-binding protein complexed with an oligosaccharide. In *Nature* **360**:127–35.

Whitney P, Powell JT, Sanford GL (1986): Oxidation and chemical modification of lung β-galactoside-specific lectin. In *Biochem. J.* **238**:683–9.

Yamamoto K, Konami Y, Kusui K *et al.* (1991): Purification and characterization of a carbohydrate-binding peptide from *Bauhinia purpurea* lectin. In *FEBS Lett.* **281**:258–62.

Yamamoto K, Konami Y, Osawa T *et al.* (1992): Carbohydrate-binding peptides from several anti-H(O) lectins. In *J. Biochem.* **111**:436–9.

21 Bacterial Lectins: Properties, Structure, Effects, Function and Applications

N. GILBOA-GARBER, D. AVICHEZER AND N.C. GARBER

1 Introduction

It is a privilege and a pleasure to write a chapter about bacterial lectins in this book due to its special designation for young scientists. Our very long and rewarding experience with students and young scientists has made it easy for us to accept willingly the editorial policy 'to transmit our opinions and enthusiasm to the readers, rather than giving a tedious description of special details'.

1.1 Prologue

Before starting on the specific subject of the chapter, we wish to introduce our general opinion on the definition and role of lectins, as we have already described in several papers (Gilboa-Garber, 1988; Gilboa-Garber and Garber, 1989, 1992, 1993). This should permit a better understanding of the content and logic of the material, and may also lead to interesting scientific achievements in the future. We regard lectins as "ubiquitous proteins, which, owing to their reversible carbohydrate/receptor-specific binding to macromolecules and cells, function like antibodies, hormones, and 'positioning sites' of enzymes in controlling irreversible, key-enzyme-dependent reactions, in enabling macromolecule and cell *protection, organization, transformation* or *lysis* (in adaptation to environmental, developmental or life cycle drifts) and in supporting *cell nutrition, special functions, contact and fusion with cells, proliferation or death*" (Gilboa-Garber, 1988; Gilboa-Garber and Garber, 1989, 1992, 1993).

The analogy between lectins and antibodies, hormones and binding sites of enzymes is not limited to their role, but is also valid for their applicability in many important analytical-biochemical techniques, industrial biotechniques, scientific or medical research, as well as medical diagnosis and treatment.

1.2 Bacterial Lectinology – Past and Present Status

Those of you who are interested in following the accurate history of bacterial lectinology during the first half of this century may find it in several detailed old manuscripts written by Krüpe (1956), Neter (1956) and Mäkelä (1957), and later on in the comprehensive book of Gold and Balding (1975), in the book of Mirelman (1986) and in more recent reviews (Gilboa-Garber and Garber, 1992; Gilboa-Garber, 1994; Shakhanina *et al.*, 1994). In the present chapter we do not intend to supply such material, but instead to describe selected milestones and examples which may stimulate interest in further exploration of bacterial lectins.

Hemagglutinating activities (on rabbit erythrocytes) in bacterial culture filtrates were first described at the beginning of this century in *Staphylococcus aureus* and *Vibrio cholerae* (Kraus and Ludwig, 1902), *Pseudomonas aeruginosa, Salmonella typhi, Escherichia coli* and *Shigella dysenteriae* (independent studies of Flexner, Kayser, Pearce and Winne cited by Gold and Balding, 1975). A few years later, cell-bound hemagglutinins were found in *E. coli* (Guyot, 1908). During the following years many additional bacterial hemagglutinins were described that interacted selectively with erythrocytes and other cell types.

H.-J. and S. Gabius (Eds.), Glycosciences
© Chapman & Hall, Weinheim, 1997
ISBN 3-8261-0073-5

For example, Rosenthal showed that *E. coli* hemagglutinins also agglutinated leukocytes, thrombocytes, spermatozoa, yeasts, spores of molds and pollen (Rosenthal, 1931, 1943). Moreover, he drew attention to two very important aspects: (a) the effect of culture conditions on the level of hemagglutinating activity, and (b) the possibility that the spermagglutination might be of medical importance (inducing sterility). Several years later, *Haemophilus* hemagglutinins were described in culture filtrates and intact cells (Keogh *et al.*, 1947). Their subsequent purification procedure (Warburton and Fisher, 1951) was probably the first reported purification of a bacterial hemagglutinin (Gold and Balding, 1975). Despite the similarity of bacterial hemagglutinins to phytohemagglutinins in their selective interactions with animal erythrocytes, Boyd and Shapleigh (1954), who coined the term 'lectins' for the latter, did not include the microbial hemagglutinins with them.

In 1955 the first report on inhibition of bacterial hemagglutinin (that of *E. coli*) by mannose (Man) was presented (Collier and Miranda, 1955). Duguid *et al.* (1955), who described in the same year fimbrial hemagglutinins in various enterobacteria, were soon aware of that finding and showed two years later that fimbrial hemagglutinins in *S. dysenteriae* and *E. coli* strains were also inhibited by mannose and methyl-α-mannose (Duguid and Gillies, 1957). Their paper was very important because it already pointed to the linkage of bacterial lectins with adhesion, naming them 'adhesins', and stating that their occurrence in many pathogenic bacteria suggests that they might function in fixing bacteria, or their toxins, to the surface of host cells. Duguid *et al.* (1966) also found that most *Salmonella* strains possess type 1, mannose-sensitive, fimbriae, resembling those of *E. coli*, and that these bacteria also adhere to leukocytes and intestinal epithelial cells. Mannose-inhibitable hemagglutinating activity was also shown in *Pseudomonas echinoides* (Heumann and Marx, 1964).

During the years 1972–1975 the first bacterial galactophilic lectins were discovered in *P. aeruginosa* (Gilboa-Garber, 1972a,b), *Streptomyces* sp. (Fujita *et al.*, 1973, 1975) and *Clostridium botulinum* (Balding *et al.*, 1973). Gilboa-Garber (1972a,b) found the *P. aeruginosa* galactose-binding lectin (PA-IL) in extracts of the bacterial cells. She showed its close similarity in properties and activity to plant lectins – exhibition of sugar-specific selective hemagglutination, relative resistance to heating and proteolysis, and dependence on divalent cations. This lectin was the first microbial lectin purified by means of affinity chromatography on Sepharose (Gilboa-Garber *et al.*, 1972), seven years following the introduction of this method for plant lectins (Agrawal and Goldstein, 1965). The galactophilic lectin of *Streptomyces* sp. was the first bacterial lectin shown to exhibit anti-human blood group (B) specificity comparable with that of plant lectins (Fujita *et al.*, 1973).

According to Mäkelä's system for classification of plant lectins, the mannose-binding bacterial lectins described above are of group III, and the galactophilic lectins belong to group II. L-Fucose-binding bacterial lectins, which are placed in Mäkelä's group I, were found several years later: in *P. aeruginosa*, PA-II lectin (Gilboa-Garber *et al.* 1977a; Gilboa-Garber, 1982), and in *Streptomyces* sp. (Kameyama *et al.* 1979, 1981). The similarity between the *P. aeruginosa* and plant lectins has been further proven by demonstration of the mitogenic effects of PA-I and II lectins on human peripheral lymphocytes (Sharabi and Gilboa-Garber, 1979; Avichezer and Gilboa-Garber, 1987) following Nowell's (1960) observation with a plant lectin (PHA). Despite the reports described above, until 1978 lectins were still described as 'proteins found primarily in plants' (Sharon, 1977) or 'carbohydrate-binding proteins of plant and animals' (Goldstein and Hayes, 1978), but at the same time their status was changing: "the discovery of hemagglutinins in such diverse sources as bacteria, fungi, lichens, fish etc. has resulted in broadening of the term 'lectin' to include carbohydrate-binding proteins without regard to their origin" (Goldstein and Hayes, 1978). Lectin definition (Goldstein *et al.* 1980), which demanded cell-agglutinating activity of saccharide-binding proteins (excluding monomeric proteins) led to a mixed and confused usage of the concepts 'bacterial hemagglutinins', 'agglutinins', 'adhesins', 'lectins' and 'lectin-like' or 'lectinoid' proteins in the literature. Adhesins, which are not saccharide-specific, despite exhibiting lectin-like properties and functions, are not included. This is true for amino acid-binding (Falkow, 1991; Isberg and Tran Van Nhieu, 1994) or lipid-binding (Rostand and Esko, 1993) adhesins.

The scientific and functional potentials and applications of the microbial lectins are only now beginning to be realized (Gilboa-Garber and Garber, 1992). They include: (a) the spectrum of sugars detected by them; (b) their sensitivity to subtle structural differences in glycoconjugates (Gilboa-Garber, 1982; Karlsson, 1989; Lanne *et al.* 1994; Shakhanina *et al.*, 1994); (c) selective specificity to antigenic epitopes (Levene *et al.*, 1994; Garratty, 1995); (d) the molecular (genetic) basis and regulation of their production (Normark *et al.*, 1983, 1986; Avichezer *et al.*, 1992, 1994); (e) the molecular basis of their functions (Gilboa-Garber and Garber, 1992); (f) their involvement in symbiotic and parasitic interactions with higher organisms (enabling adhesion, invasion and expanded infection); (g) the properties of their receptors on host macromolecules and cells; (h) their effects on normal (Gilboa-Garber, 1986) and transformed (Avichezer and Gilboa-Garber, 1991) host cells; (i) their contribution to host defense mechanisms (Hajto *et al.*, 1989; Avichezer and Gilboa-Garber, 1992) as well as (j) their possible applications, including those common to plant lectins as well, and special (mainly medical) ones, such as protection against infectious diseases caused by homologous or heterologous bacteria (Gilboa-Garber and Sudakevitz, 1982; Avichezer and Gilboa-Garber, 1992) and other parasites and viruses as well as against cancer (Gilboa-Garber *et al.*, 1986; Avichezer and Gilboa-Garber, 1991; Beuth *et al.*, 1995). Bacterial lectinology opens new fascinating windows not hitherto known in animal and plant lectinology.

In this chapter, the *P. aeruginosa* PA-I and II lectins, which are multimeric classical lectins (like those of plants and animals), will be used as models for the description of bacterial lectin properties, structure, function and applications. These lectins have been studied more than any other bacterial lectins for their general biological effects and applications.

2 Bacterial Lectin Prevalence, Expression and Detection

Since the beginning of this century, new bacterial lectins have been steadily described in various families of Gram-negative and positive bacteria.

2.1 State of the Art

Lectins were found in the bacterial culture medium, on cell fimbriae/pili [either on their tips or along their structure – e.g. the M, P or S pili of *E. coli* strains (Normark *et al.*, 1986)], on flagella or the cell surface, in the periplasmic space and inside the cells [e.g. the PA-I and II lectins of *P. aeruginosa* (Gilboa-Garber, 1972a, b; Glick and Garber, 1983) or the myxobacterial hemagglutinins (Zusman *et al.*, 1986)]. Most of these lectins were discovered by means of the hemagglutination test and its inhibition by sugars. It seems that lectins could be found in almost all bacteria if the right growth conditions and phase (age) of the cells could be defined and the appropriate analytical techniques used.

Genes coding for some of the bacterial lectins/adhesins have already been described. Examples are those of *E. coli* (Normark *et al.*, 1986; Lindberg *et al.*, 1989), *Myxococcus xanthus* (Romeo *et al.*, 1986), *Salmonella typhimurium*, *Serratia marcescens* (Clegg *et al.*, 1987), *Klebsiella pneumoniae* (Gerlach *et al.*, 1989), *Bordetella pertussis* (Relman *et al.*, 1989), *Streptococcus mutans* (Banas *et al.*, 1990), *Vibrio cholerae* (Häse and Finkelstein, 1991), and *P. aeruginosa* PA-IL (Avichezer *et al.*, 1992, 1994). The expression of the lectin genes is regulated by environmental factors. There is also the possibility of phase variation, shown in *E. coli* P pili, between production and no production (Rhen *et al.*, 1983) and between S and type 1 (M) fimbriae (Nowicki *et al.*, 1985), as well as in *P. aeruginosa* between PA-IL and PA-IIL (Gilboa-Garber, 1982, 1988).

2.2 Dependence of Lectin Production on Growth Conditions

Lectin-enriched bacteria may be obtained by growing them under defined culture conditions. Important factors are basic composition of the medium, pH, temperature, supplements and aeration. Our experience with *P. aeruginosa* has shown that different media are needed for highest production of PA-IL and of PA-IIL (Gilboa-Garber, 1982). Addition of choline to the medium increases the levels of both lectins (Gilboa-Garber, 1972a, 1982). Certain organic and inorganic compounds, cations, anions and antibiotics reduce the levels (Sudakevitz *et al.*, 1979). Highest lectin levels are obtained at neutral pH, 28 °C (with much lower production at 25 °C or 37 °C) with aeration. Highest lectin levels are found in 72-hour-old cells, but considerable lectin activity may appear in a certain medium much earlier than in another.

Cell density and age are important factors in lectin production. Highest levels of PA-IL and PA-IIL are obtained at the stationary stage, linked with some virulence secondary metabolites, including proteolytic activities (e.g. those of elastase and alkaline protease), hemolysin, pyocyanin (Gilboa-Garber, 1983), cyanide, chitinase and staphylolytic activity (Cámera *et al.*, 1995), probably due to coregulation by cell density-sensing regulators or autoinducers (Garber *et al.*, 1995; Cámera *et al.*, 1995).

2.3 Lectin Screening Technology

Lectins may be looked for in the growth medium (those released from the cells), on intact cells (on cell appendages and on the cell surface), on isolated fimbriae/pili, separated from the cells by gentle mechanical treatment, and in cell extracts, obtained by mechanical cell disruption (internal lectins). Some lectins may be found both on the cell surface and inside it, as in the case of PA-I and -II lectins, which are mainly internal, but of which there are also small fractions in the periplasmic space and on the cell surface (Glick and Garber, 1983). Some treatments of the cells (e.g. heating and addition of certain cations or cell damaging compounds) induce higher external lectin distribution (Sheffi *et al.*, 1989; Wentworth *et al.*, 1991).

Most lectins in general, and also the bacterial ones, were discovered by means of the hemagglutination test (Neter, 1956). Some lectins, especially those reacting with sialylated compounds, agglutinate only untreated red blood cells (RBCs). Other lectins (those which are either specific for the subterminal sugars, unmasked by desialylation, or disturbed by the negatively charged sialic acid) agglutinate only sialidase-treated RBCs. However, many lectins agglutinate sialidase/papain-treated cells much more strongly than untreated cells, because the glycolipids of these cells are more exposed following the removal of sialic acid residues and glycoprotein/protein branches.

The use of hemagglutination or other cell agglutination assays is very convenient. RBCs bear a wide spectrum of glycosylated residues which attract lectins of known composition, and the hemagglutination test is very simple, rapid, reliable and cheap. There are also other means for lectin detection, based on the use of native or neo-glycosylated macromolecules, either free or bound to polymeric particles, such as Sepharose, agarose or latex beads (Ascencio *et al.*, 1990).

A positive reaction in any of these tests has to be verified by an inhibition test in the presence of sugar. Examination of sugar specificity of intact bacterial lectins in the hemagglutination-inhibition test may be complicated due to the coexistence of multiple types of adhesive molecules. In the case of *P. aeruginosa*, many strains contain, in addition to PA-IL and PA-IIL, hydrophobic surface hemagglutinins (SHA) which are sugar-insensitive (Garber *et al.*, 1985; Glick *et al.*, 1987), or other lectins/adhesins, which bind sialic acids (Ramphal and Pyle, 1983; Hazlett *et al.*, 1986) and galactose/*N*-acetylgalactosamine (Gal/GalNAc)-bearing cell surface components (Krivan *et al.*, 1988a,b; Baker *et al.*, 1990; Ramphal *et al.*, 1991a,b; Sheth *et al.*, 1994), as well as non-protein compounds, such as alginate (McEachran and Irvin, 1985; Ceri *et al.*, 1986; Doig *et al.*, 1987, 1989; Baker and Svanborg-Edén, 1989; Ramphal *et al.*, 1991b) or lipids (Rostand and Esko, 1993). If adhesin activity masks the lectin sugar, special treatments must be used for selective inactivation or removal of the non-lectin adhesins. This may be achieved by mild denaturing treatments. For example, the SHA of *P. aeruginosa* may be inactivated selectively by mild heating or mechanical disruption, which do not damage the PA-I and II lectins (Glick *et al.*, 1987). Interaction of the bacterial lectins with macromolecules and indissociable cells may be shown indirectly by the hemagglutination-inhibition (adsorption) test, following exposure of the lectin to them.

3 Bacterial Lectin Properties

The classical bacterial lectins are almost indistinguishable from the plant lectins in their multimeric structure, divalent cation-binding, physicochemical properties, sugar specificities and purification procedure, as well as in their interactions with macromolecules and cells (leading to their agglutination) and their effects on them.

3.1 Divalent Cation-Binding

Like most plant and animal C-type lectins (Drickamer, 1988, 1992), *P. aeruginosa* PA-I and II lectins are inhibited by EDTA, since the association with divalent cations is crucial for their sugar binding. Both bind Ca^{2+} (PA-IIL>>PA-IL) and Mg^{2+}, but not Mn^{2+}. PA-IIL also contains bound Zn^{2+} (Avichezer *et al.*, 1994). The PA-IIL activity is reversibly inhibited by heavy metal cations: Co, Pb, Zn and Cd (Sudakevitz *et al.*, 1992), like the soybean lectin (Liener, 1986), but unlike PA-IL and most other lectins which are not sensitive to inhibition by these cations.

3.2 Sugar Specificities of the Bacterial Lectins

The range of sugar specificities of bacterial lectins is similar to that of plant and animal lectins. Like the eukaryotic lectins, each bacterial lectin also interacts with a variety of simple sugars and their derivatives. Mannose-binding bacterial hemagglutinins have been described in various Gram-negative (e.g. *E. coli, S. dysenteriae, P. echinoides* and *Salmonella* strains) and Gram-positive bacteria (e.g. *Corynebacterium parvum*) and in viruses (e.g. Japanese encephalitis virus). Examples of L-fucose (Fuc)-binding lectins are those of *Streptomyces* sp. (Kameyama *et al.*, 1979), *P. aeruginosa* (PA-IIL) and *V. cholerae* (Booth *et al.*, 1986). Galactophilic lectins have been found in *P. aeruginosa* (Gilboa-Garber, 1972a; Krivan *et al.*, 1988a,b; Ramphal *et al.*, 1991a, b), *S. dysenteriae* (Shiga) toxin, *E. coli* P fimbriae (Bock *et al.*, 1985), *Streptomyces* sp. (anti-B), *Actinomyces viscosus* and *Actinomyces naeslundii* (Brennan *et al.*, 1987), *Fusobacterium nucleatum*, *Streptococcus mutans* and *Streptococcus suis* (Haataja *et al.*, 1993, 1994). Lectins which bind glucose and its derivatives, such as *N*-acetyl-D-glucosamine (GlcNAc), were described in *Bartonella bacilliformis, S. mutans, Bordetella bronchiseptica, Chlamydia trachomatis* and *Pasteurella multocida*. Sialic acid-binding lectins, which are especially abundant in viruses (e.g. influenza, polyoma and Sendai viruses), are also present in bacteria. Examples are: *Mycoplasma pneumoniae* and *M. gallisepticum, E. coli* strains, *Streptococcus sanguis* and *P. aeruginosa* (Ramphal and Pyle, 1983).

It has to be emphasized that the classification of bacterial lectins (as is also true for plant and animal lectins) according to simple sugar specificities is an oversimplication. Lectin specificity to a certain sugar exhibits a broad variation in several aspects: (a) the lectin affinity constant for the specific sugar, which may vary between $3x10^{-1}$ M and $1x10^{-6}$ M; (b) the lectin spectrum of interactions with diverse derivatives of this sugar – some lectins prefer hydrophilic derivatives and others prefer hydrophobic ones (probably due to hydrophobic clefts near the sugar-binding site); (c) the lectin interaction with additional monosaccharides. For example, *E. coli* fimbrial type-1 lectin is inhibited by mannose, but not by glucose, while Con A is inhibited by both, and PA-IIL is inhibited by L-fucose in additon to mannose; (d) the lectin profile of relative affinities for a series of diverse sugar derivatives. For example, examination of the relative affinities of PA-IL for the respective sugar derivatives (using equilibrium dialysis and hemagglutination inhibition tests) has revealed the following order: phenyl-β-thioGal>phenyl-β-Gal> isopropyl-β-thioGal>p-nitro-phenyl-β-Gal>methyl-β-thioGal>methyl-α-Gal>melibiose>Gal/p-nitrophenyl-α-Gal>methyl-β-Gal>GalNAc>L-rhamnose (6-deoxy-L-mannose) (Gilboa-Garber, 1982; Pal *et al.*, 1987; Garber *et al.*, 1992). The galactophilic lectin of *Streptomyces* sp. (Fujita *et al.*, 1973), however, shows a different order of affinity: L-rhamnose>methyl-β-Gal>methyl-α-Gal/Gal>L-arabinose, which may be related to its higher anti-B specificity. Finally PA-IIL shows the following order of affinity: p-nitrophenyl-α-Fuc>Fuc> L-fucosylamine/L-Gal >>Man>>>D-fructose (Gilboa-Garber, 1982; Garber *et al.*, 1987). Details of the specificity of the surface lectins of *E. coli, K. pneumoniae* and *S. typhimurium* for mannose derivatives may be found in the paper of Firon *et al.* (1983); (e) lectin sensitivity to the position of the sugar – either terminal or internal; (f) the lectin sensitivity to adjacent sugars surrounding the target sugar. For example, the binding of *Ulex*-I lectin to L-fucose of blood group H antigen is significantly reduced if GalNAc is also bound to the same subterminal galactose in an adjacent location (A antigen), while PA-IIL binding to the same fucose is almost unaffected by the presence of GalNAc at the same site (Gilboa-Garber *et al.*, 1994). Additional examples include: P pili specificity for the digalactoside Gal-α-4Gal in the terminal or internal position (Leffler and Svanborg-Edén, 1986); the *A. viscosus* and *A. naeslundii* lectin specificity for terminal Gal-β-3GalNAc; PA-I high affinity for both Gal-α-3/4Gal (of the B/P^k and P_1 antigens) (Gilboa-Garber *et al.*, 1994) and the biantennary fork (Gal-β-4GlcNAc)$_3$ structure of I blood group antigen (Sudakevitz *et al.*, 1995); and *P. aeruginosa* pili adhesin binding to Gal-β-3/4GlcNAc (Ramphal *et al.*, 1991a); (g) the lectin preference for the carrier molecule, whether proteins or sphingolipids. Glycoproteins are regarded as the targets for type 1 (Sharon, 1986) and *E. coli* K88 fimbriae (involved in adherence of this bacterium to adult brush border enterocytes) (Seignole *et al.*, 1994), while glycolipids are the receptors for the P-fimbriae of *E. coli* strains (Leffler and Svanborg-Edén, 1980; Källenius *et al.*, 1980; Väisänen *et al.*, 1981; Bock *et al.*, 1985; Leffler and Svanborg-Edén, 1986), which interact with P system glycosphingolipid antigens. Likewise, the PA-I and II lectins of *P. aeruginosa* interact with the glycosphingolipid globosides and paraglobosides of human blood group antigens P^k, P_1, B and I (Gilboa-Garber *et al.*, 1994; Sudakevitz *et al.*, 1995). The factors described above determine the receptor/antigen and cell selectivity of the bacterial lectins and may determine their host specificity (Garratty, 1995). Galactose and its derivatives are major components of the glycosylated common histo-blood group antigens. Galactophilic specificity is most frequent among lectins in general, including bacterial lectins.

3.3 Physico-Chemical Properties of the Lectins and their Purification

The microbial lectins, as represented by PA-IL and PA-IIL, are very similar to plant lectins in relative resistance to heating (up to 70°C), extreme pH, proteolysis and other denaturing treatments (Gilboa-Garber, 1982, 1986). These properties are valuable for their purification, achieved by classical procedures of plant/animal lectin purification, including sugar-affinity chromatography (Gilboa-Garber, 1982). Sepharose is used for capturing PA-IL and Sepharose-mannose for PA-IIL. The lectins are eluted from the column, after thorough washing, by galactose or mannose, respectively. The eluates are dialyzed for sugar removal.

4 Bacterial Lectin Size and Structure

Since studies of bacterial lectin structure are far behind those of plant lectins, we will concentrate on PA-IL as an example. The relative molecular mass of microbial lectins ranges from 11 kDa for the monomeric polypeptide of the galactophilic lectin of *Streptomyces* sp. (Fujita *et al.*, 1975) up to 800 kDa for *Streptomyces* SFL 100–2 Fuc/Man-binding particle, which consists of 12 identical subunits (each of 68 kDa) (Matsui *et al.*, 1985;

Matsui and Oishi, 1986). The PA-IL subunit is around 13 kDa (Gilboa-Garber, 1982, 1986; Avichezer *et al.*, 1994) and presumed to be monovalent with respect to galactose-binding, according to Scatchard analysis of the native tetrameric lectin (Garber *et al.*, 1992). The native lectin is considered to be tetrameric, based on results of gel filtration using various biogels (Gilboa-Garber, 1986), and scanning tunneling microscopic observations (Roberts *et al.*, see acknowledgments) that show PA-IL as an oval molecule with size measurement fitting a molecular weight of 52 kDa) (Fig. 1A). The size of PA-IIL subunit is almost the same as that of PA-IL. Both lectins exhibit a profound density-dependent self-aggregation (Gilboa-Garber, 1982), probably due to their high content of glycine, alanine, isoleucine, proline and valine (Gilboa-Garber *et al.*, 1972; Avichezer *et al.*, 1994). This property is shown in Fig. 1B.

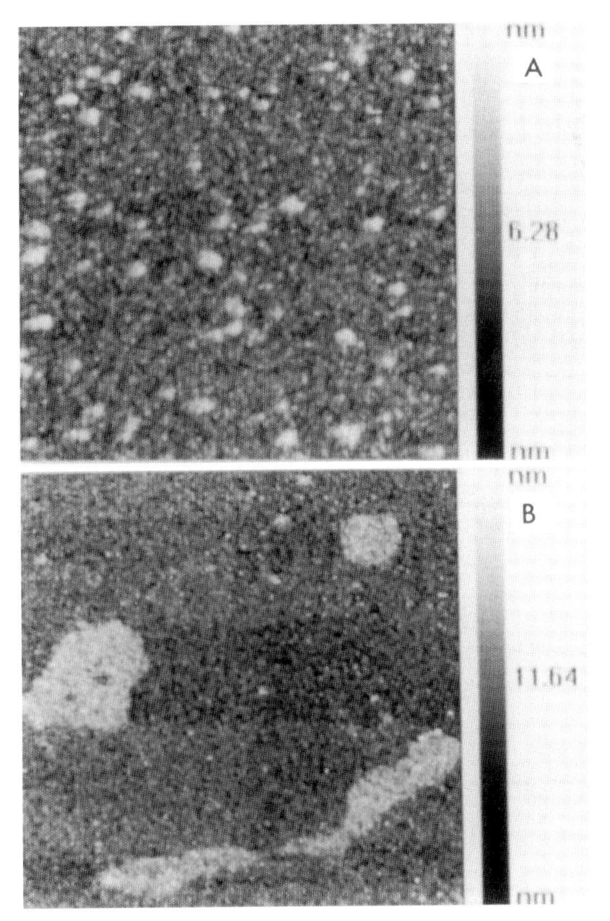

Figure 1. Scanning tunneling microscopic examination of PA-I lectin at concentrations of 10 (A) and 50 (B) μg/ml. X,Y=504 nm in A and 529 nm in B. Performed and provided by CJ Roberts *et al.* (see Acknowledgments).

The 121 amino acid sequence of PA-IL (Avichezer *et al.*, 1992, 1994) has been deduced from its nucleotide sequence following isolation of the structural gene pa-IL, which codes for this lectin (Fig. 2).

The entire amino acid sequence of PA-IIL is not yet known. Its quantitative amino acid composition is similar to that of PA-IL (Gilboa-Garber, 1994), but their *N*-terminal amino acid sequences are different. PA-IIL has less glycine, isoleucine, proline and tyrosine and more leucine, serine, threonine and valine than PA-IL. The amino acid composition of PA-IL is similar to that of most other lectins in its high abundance of aspartic acid/asparagine, alanine and glycine, paucity of arginine and histidine (low pI), and the presence of alanine at the first position (Avichezer *et al.*, 1992). The *N*-terminal region of PA-IL does not contain a typical signal peptide (Fig. 2). Its hydropathic profile (Avichezer *et al.*, 1992) indicates that it contains two predominant hydrophobic stretches one on each side of a major hydrophilic one (Fig. 3A) – a structure which may be associated with membrane surface exposure. Secondary structure and hydrophobic cluster analyses (Fig. 3A,B) indicate that PA-IL is predominantly built up from β sheet strands separated by extended loops and two major helical regions (in the initial *N*-terminal and central domains).

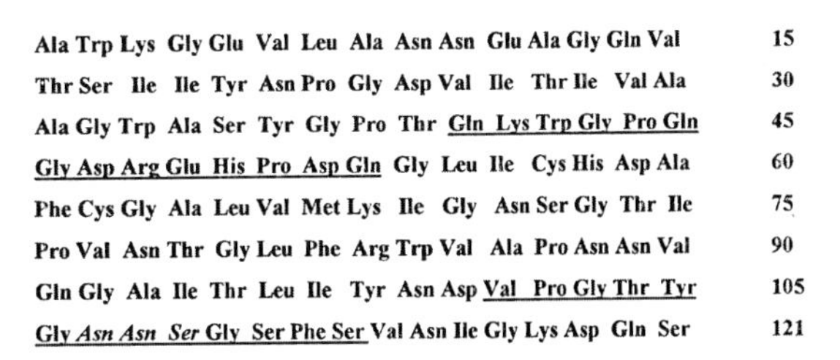

Ala Trp Lys Gly Glu Val Leu Ala Asn Asn Glu Ala Gly Gln Val	15
Thr Ser Ile Ile Tyr Asn Pro Gly Asp Val Ile Thr Ile Val Ala	30
Ala Gly Trp Ala Ser Tyr Gly Pro Thr Gln Lys Trp Gly Pro Gln	45
Gly Asp Arg Glu His Pro Asp Gln Gly Leu Ile Cys His Asp Ala	60
Phe Cys Gly Ala Leu Val Met Lys Ile Gly Asn Ser Gly Thr Ile	75
Pro Val Asn Thr Gly Leu Phe Arg Trp Val Ala Pro Asn Asn Val	90
Gln Gly Ala Ile Thr Leu Ile Tyr Asn Asp Val Pro Gly Thr Tyr	105
Gly *Asn Asn Ser* Gly Ser Phe Ser Val Asn Ile Gly Lys Asp Gln Ser	121

Figure 2. Amino acid sequence of PA-I lectin (Avichezer *et al.*, 1992, 1994). The two underlined sequences represent the composition of synthetic peptides exhibiting the strongest antigenicity, and the italics indicate the potential glycosylation site.

The most extended loop is located from residue 38 up to residue 50. The molecule exhibits several highly antigenic sites (Avichezer *et al.*, 1992). Residues 45–47 from the *N*-terminus are Gln-Gly-Asp (Q-G-D), which are also found, as a conserved sequence, in legume lectins (Young and Oomen, 1992). A similar sequence, Q-P-D, has been suggested (Drickamer, 1992) to contribute to carbohydrate binding of C-type galactose-binding animal lectins (Avichezer *et al.*, 1994). Since a 13 amino acid synthetic peptide analogous to amino acids 101–113 (Fig. 2) exhibited a slight competition with the native lectin in hemagglutination test, it might be that the sugar-binding site is situated near to the C-terminus of the lectin (Avichezer *et al.*, 1994).

The extended loop located at the C-terminal end of the PA-IL sequence, between residues 100–110 (Fig. 3) contains a Asn-Asn-Ser (*N*-*N*-S) sequence (amino acids 107–109) – a putative *N*-glycosylation site (Avichezer *et al.*, 1992) which is likely to be glycosylated due to its exposed position (Rougé and Amar, see acknowledgments). The glycosylation may depend on environmental factors (Goochee and Monica, 1990). The finding of the glycosylation site fits our previous observation, which indicates that the native PA-IL and probably also PA-IIL are glycosylated (Gilboa-Garber *et al.*, 1977a). The mechanism of their glycosylation could be related to that of the eubacterial glycoproteins described by Messner and Sleytr (1988), exhibiting asparagine-linked glycans, such as rhamnose trisaccharide. Virji *et al.* (1993) also presented strong evidence showing that *Neisseria meningitidis* pilin is glycosylated and that it also contains the characteristic consensus sequence *N*-X-S (or T), suggesting *N*-linkage of the carbohydrate. Recently, Castric (1995) has described a glycosylated pilin in *P. aeruginosa*. The post-translational glycosylation of this 15–16 kDa pilin depends on the product of pilO gene, which is involved in attaching an acidic carbohydrate to the pilin protein.

Comparative analyses of the PA-IL amino acid sequence using EMBL/Genbank/DDBJ data banks, revealed only a very slight homology with few other proteins (Avichezer *et al.*, 1992). However, recently a new protein sequence was found to exhibit a high analogy with PA-IL. This protein, which was isolated from the extracellular space of the insect *Manduca sexta* (tobacco hawkmoth/hornworm) midgut (Peterson *et al.*, 1994), was described, to our great satisfaction, to be a trypsin (alkaline) A precursor, catalyzing preferential cleavage between Arg and Lys and exhibiting similarity to other trypsin-like serine proteases. The similarity between this protein and PA-IL ranges from almost complete PA-IL identity to the insect enzyme (amino acids 1–116 out of 121) to about 50 % similarity from amino acids 118–236. There is 46 % similarity including 25.66 % identity (with 4 gaps) over the entire length of the PA-IL protein.

The gene which codes for the PA-I structure (pa-IL), which was isolated several years ago (Avichezer *et al.*, 1992), was expressed in *E. coli* cells, and the recombinant PA-IL, which was purified by gel filtration and affinity chromatography on Sepharose 4B (Avichezer *et al.*, 1994), agglutinated papain-treated human erythrocytes, showing galactose-specificity identical to that of the native PA-I lectin. It migrated in SDS-gel electrophoresis (15 % acrylamide) slightly slower than the purified native PA-I lectin, compatible with the deduced amino acid sequence, and it was specifically precipitated in agar by rabbit anti-native PA-I lectin serum (Avichezer *et al.*, 1994).

5 Bacterial Lectin Target Molecules

The glycosylated molecules and cells that bind the bacterial lectins are many and various. Binding of isolated lectins may be demonstrated by macromolecule precipitation, cell agglutination, adsorbtion of labeled lectins, and inhibition of hemagglutinating activity of the lectin preparations. Alternatively, adsorption of the macromolecules, cells, or isolated cell receptors to lectin-bearing sorbent particles may be also used for showing their interactions.

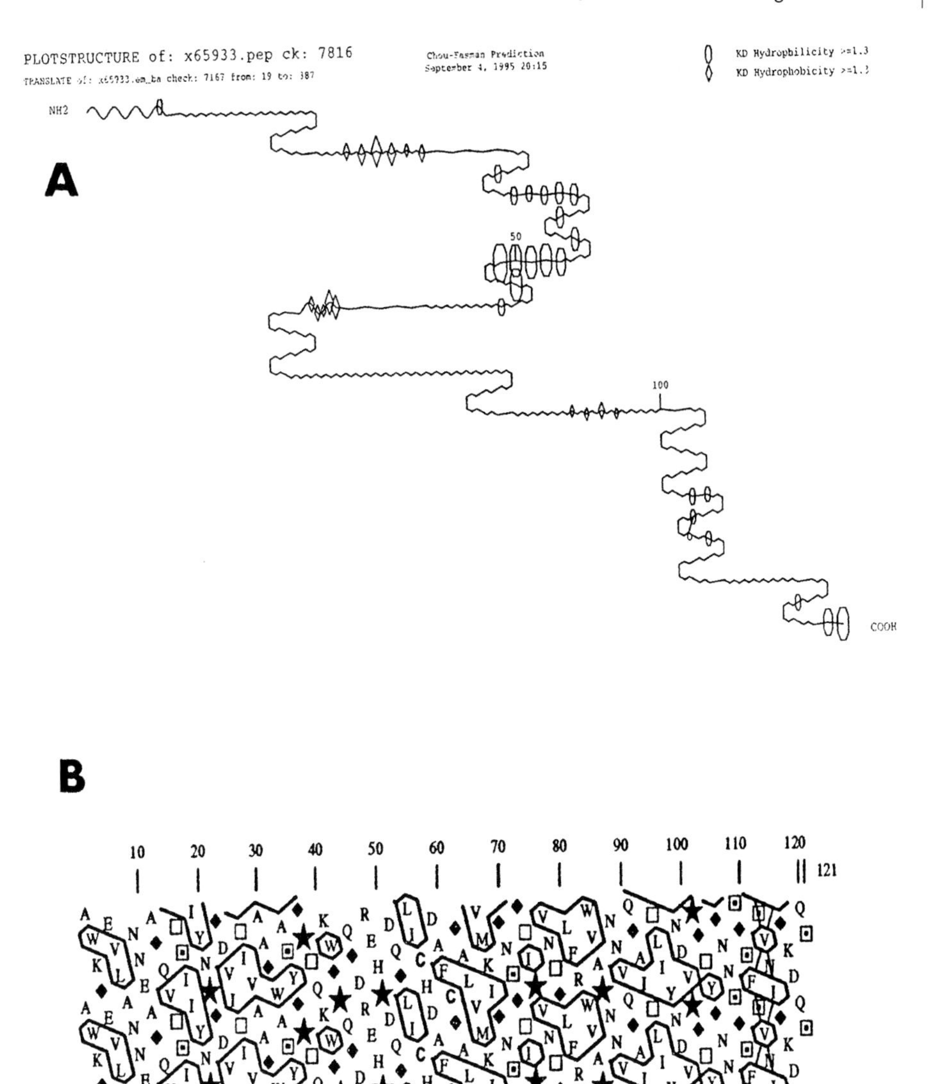

Figure 3. Secondary structure analyses of PA-I lectin. (A) performed by Avichezer *et al.* (1992) according to Chou and Fasman (1978). The α helices are marked by a sine wave, β sheets by a saw-tooth wave, turns by a 180¡ turn, hydrophobicity by pointed rhombi, and hydrophilicity by blunt ellipses. (B) hydrophobic cluster analysis (HCA) performed by Rouge and Amar (see acknowledgments) according to Gaboriaud *et al.* (1987) using the HCA-plot program (Doriane, Paris). Arrows indicate the predicted strands of β sheet. Three main loops are indicated by horizontal lines and an asterisk indicates the putative N-glycosylation site *N-N*-S.

5.1 Interactions of Bacterial Lectins with Free Glycosylated Macromolecules

The interactions of lectins from viruses, bacteria and fungi with glycoconjugates from microbial, plant or animal origins were recently reviewed by Shakhanina *et al.* (1994). We will describe few examples: a) *A. naeslundii* and *A. viscosus* fimbrial lectin reaction with Gal-β-3GalNAc/GalNAc-β-3Gal-containing glycolipids or polysaccharides (McIntire *et al.*, 1983; Brennan *et al.*, 1987; Kolenbrander, 1989); (b) preferential binding of *E. coli* type 1 fimbrial adhesin to $Man_5GlcNAc_2$-peptide (Firon *et al.*, 1983); (c) *E. coli* K88ac fimbrial binding to glycoproteins via Gal-β-3GalNAc and Fuc-α-2Gal-β-3/4GlcNAc (Seignole *et al.*, 1994); (d) P-fimbriae of uropathogenic *E. coli* strains (Bock *et al.*, 1985), *S. suis* (Haataja *et al.*, 1993, 1994) and PA-IL (Lanne *et al.*, 1994), which bind Gal-α-4Gal-containing glycosphingolipids, and (e) PA-IL binding to the glycosylated enzyme peroxidase (Gilboa-Garber and Mizrahi, 1973) and PA-IIL binding to peroxidase and invertase (Gilboa-Garber, 1986).

5.2 Interactions of Bacterial Lectins/Adhesins with Cell Receptors Including Blood Group Antigens

Isolated and bound bacterial lectins have been shown to interact specifically with animal, plant, or microbial cells via their membrane surface polysaccharides, glycoproteins and glycolipids. The binding, which is sugar-specific, depends on the type of molecule that bears the sugar. Some of these glycosylated molecules are well-known blood group antigens, as well as familiar membrane receptors. Examples are: the interaction of K99 fimbriae of enterotoxigenic *E. coli* strains with the sialylated erythrocyte glycophorin A (Parkkinen *et al.*, 1986), and of hemagglutinins of pyelonephritogenic *E. coli* with blood group M (Väisänen *et al.*, 1982), and N (Grünberg *et al.*, 1988) antigens, which are confined to glycophorins; the interactions of lectins of uropathogenic *E. coli* strains (Källenius *et al.*, 1980; Leffler and Svanborg-Edén, 1980), and *S. suis* (Haataja *et al.*, 1993, 1994) with the sphingolipid blood group antigens P^k and P_1; the interactions of PA-IL with blood group antigens B, P^k, P_1 and I (Gilboa-Garber *et al.*, 1991, 1994; Sudakevitz *et al.*, 1995); and the interactions of PA-IIL with the fucosylated glycolipids of H and Le blood group antigens (Gilboa-Garber *et al.*, 1994) and the lectin of *Helicobacter pylori*, which is involved in gastric ulcer (Blaser, 1992; Cover and Blaser, 1995), and with gastric tissue Le^b antigen (containing a second fucosyl on type 1 H antigen chain) (Borén *et al.*, 1993; Falk *et al.*, 1995).

The mannose-specific type 1 fimbriae of *E. coli* interact with the leukocyte CD11 and CD18 antigens, which are themselves adhesion molecules of the integrin type (Gbarah *et al.*, 1991). Integrins were shown to be receptors for microbial adhesion either by binding the microorganisms or by being targets for the microbial adhesins (Isberg and Tran Van Nhieu, 1994). *Yersinia pseudotuberculosis* invasin protein mediates the bacterial entry into mammalian cells by sugar-independent interaction with multiple β1 chain integrins: aspartate at the Arg-Gly-Asp (R-G-D) site of this protein is critical for integrin binding (Leong *et al.*, 1995). There are also sugar-independent bacterial hemagglutinins, which bind specifically to blood group antigens, and which are therefore lectins according to Boyd and Shapleigh (1954), but not according to the definition of Goldstein *et al.* (1980). The best example is a family of *E. coli* hemagglutinins specific for Dr antigen (which is present on erythrocytes, epithelial cells and various tissues), demonstrated in certain entero- and uropathogenic *E. coli* strains (Nowicki *et al.*, 1988a,b, 1990; Johnson *et al.*, 1995). Among the anti-Dr hemagglutinins are the afimbrial adhesins AFA-I and AFA-III, and F-1845 fimbriae of these *E. coli* strains. The structure of Dr antigen is considered to be "chloramphenicol-like", including tyrosine (Nowicki *et al.*, 1988b). The high frequency AnWj antigen was also shown to be the erythrocyte receptor for certain *E. coli* strains and *Haemophilus influenzae* fimbriae (van Alphen *et al.*, 1986, 1990; Garratty, 1995). We agree with Gold and Balding (1975) that the definition of lectins could be "selective and reversible receptor-binding proteins", by analogy with the definition of antibodies, which do not restrict their specificities to sugar-binding, but also include lipid- or peptide-binding immune proteins.

5.3 The Receptor-Specificity of Bacterial Lectins Determines Host Cell Selectivity

The specificity of the bacterial lectins is an important factor in determination of host cell selectivity. Lectins, which interact with widespread receptors, enable bacterial adherence to many tissues in diverse hosts. The presence of several adhesin types on the same bacterium may ensure its adhesion to almost any cell. However, the microenvironment in the host and in each tissue determines production, dominance and efficiency of the bacterial lectin/adhesin. Therefore, although *in vitro* experiments performed on artificial bacterial cultures enrich our information, the *in vivo* conditions, where bacteria are in biofilms, embedded in extracellular matrix, or attached to cells, are different (Costerton *et al.*, 1987; Cacalano *et al.*, 1992; Davies *et al.*, 1993; Brözel *et al.*, 1995), favoring production of certain components. Among the host factors which affect bacterial lectin production and function concomitantly are: (a) the physiological state of the host; even minor temperature changes were shown to reduce lectin production significantly in *P. aeruginosa* (Gilboa-Garber, 1982), and fimbrial expression in the periodontitis-inducing bacterium *Porphyromonas gingivalis* (Amano *et al.*, 1994); (b) the host immune status. It was shown that antibodies against PA-I and -II lectins protected mice completely against otherwise lethal *P. aeruginosa* infection (Gilboa-Garber and Sudakevitz, 1982); antibodies against neisserial pili protected against infection by these bacteria (Brinton *et al.*, 1982); and immunization of germ-free rats with *P. gingivalis* fimbriae protected them against periodontitis induced by that bacterium (Evans *et al.*, 1992); (c) the tissue metabolic state and age. For example, sialophilic adhesins of viruses would attract the viruses to younger cells, which exhibit higher sialylation, and which may support a better viral proliferation; (d) the tissue extracellular matrix (ECM) composition. Bacterial adhesion to host cells may arise from interaction of its adhesins with the extracellular matrix (Woods, 1987; Ramphal *et al.*, 1991b) but ECM components may also inhibit cell infections; (e) competition between adjacent tissues and cells at different growth stages. Differences in cell sialylation and ciliation or cell outgrowth affect bacterial attraction by the tissue or even part of it (Plotkowski *et al.*, 1989, 1991); (f) presence of extracellular proteases or sialidases changing the host cell surface structure (Woods *et al.*, 1981b; Plotkowski *et al.*, 1989); (g) the tissue contact with blood cells, plasma or lymph circulation. Components of the blood/lymph circulation may compete with the target host cell receptors (which may also be present on the blood cells or in the body fluids) for the bacterial adhesins. This, in addition to their ability to lead to complement/phagocyte dependent bacteriolysis; (h) native microflora. Native bacteria may either reduce or increase adhesion of the invading ones. Reduced adhesion may be due to competition either at the host cell receptors [when both populations have adhesins with similar specificities (Imundo *et al.*, 1995)] or at the pathogen adhesins, if the first bacteria bear host-like receptors. For example, PA-I and II lectins adsorb on to cells of *E. coli* strains bearing human blood group B- and H-like antigens, respectively (Gilboa-Garber *et al.* 1977b; Garber *et al.*, 1981). Increased adhesion in the presence of a native bacterium may result if the latter exposes host cell receptors for the pathogen, or if it serves as a bridge between the host cell receptors and the pathogenic bacteria. An example is the aggravation of pyelonephritis caused by *P. aeruginosa* in the presence of *Enterococcus faecalis,* which is part of the "normal" intestinal flora in mice (Tsuchimori *et al.*, 1994); (i) mixed pathogen infections. The effects of such infections may be bilateral: there may be competence on the host cell receptors, if both organisms have similar lectin specificities (Krivan *et al.*, 1988a,b; Yu *et al.*, 1994; Imundo *et al.*, 1995). However, there are solid indications for the reverse condition, where primary infections lead to such changes in the host cells, which enable better adhesion of the secondary pathogen. Pneumococcal adhesion to respiratory epihelium and *Pseudomonas* infections are significantly aggravated following viral or bacterial infections (Ramphal *et al.*, 1980; Plotkowski *et al.*, 1986, 1991), and this may be attributed to removal of sialic acid and glycopeptides which otherwise interfere with their adhesin/lectin binding to the cells (Gilboa-Garber *et al.*, 1972a,b) (Fig. 4). Another case of mixed bacteria cooperation in infection is a chain infection, where one bacterium supports the adhesion of a second one by linking it to the host cells. A very good example of such a case is the microflora associated with oral infections (Cisar *et al.*, 1979); (j) presence of transformed cells. Tumor cells differ from normal ones in their surface composition. They may exhibit neoglycoconjugates and altered levels or position of normal components, exposing more receptors, enabling higher bacterial/viral adherence (and probably supplying a better machinery); (k) host medical treatment. Patient therapy by anticancer or antimicrobial drugs may affect glycosylation of the patient's cells as well as lectin/adhesin production in the infectious agent. It was observed that antibiotics at subinhibitory concentrations can modify bacterial lectin/adhesin production and the ability to adhere. This subject was recently reviewed by Piatti (1994).

In order to anticipate tissue infectability based on receptor affinity, the expression of the receptors on the different tissues must be explored. However, the overall balance would determine the final settlement of the bacterium in the host, since the first binding is reversible. Thus, P-pili, which bind P blood group antigens, are present on most *E. coli* isolates from pyelonephritis patients (Korhonen *et al.*, 1982). *P. aeruginosa* PA-IL, which binds to the same receptors, may also be regarded as contributing to the frequent involvement of *P. aeruginosa* in that disease. Several pyelonephritis-associated *E. coli* isolates, which revealed neither mannose-sensitive- nor P-pili, were found to exhibit Dr specificity. Studies on the Dr antigen expression in different tissues showed it to be present in different parts of the digestive, respiratory, urinary and genital tracts, and in skin. Accordingly, Nowicki *et al.* (1988a,b) suggested that the high density of the Dr-rich cells in colon and urinary tract (especially Bowman's capsule) may permit colonization of these

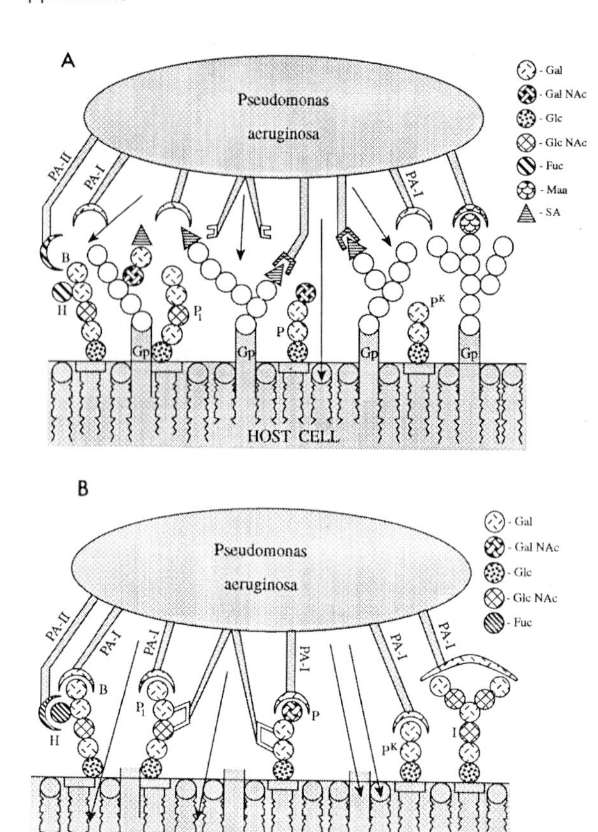

Figure 4. A schematic model of *P. aeruginosa* adherence to host cells without treatment (A) and following their exposure to proteolysis (B). PA-IL and IIL as well as sialophilic, hydrophobic and other adhesins are noted, and the arrows represent toxins/enzymes which are released from the bacterium on to the host cell membrane.

tissues by *E. coli*. The same wide distribution may apply to *P. aeruginosa* lectins, exhibiting ABH (PA-IIL), BI and P system blood group-binding (PA-IL) reactivity. Examination of the distribution of *i. v.*-injected ^{125}I-labeled PA-I and PA-II lectins in mice showed the highest radioactivity in lungs, spleen, kidney and liver (Avichezer and Gilboa-Garber, 1995; Fig. 5). The relative concentrations order of the two lectins in various tissues was different, PA-IL exhibiting highest affinity for lungs and kidneys, while PA-IIL exhibiting outstandingly high binding to spleen (lymphoid tissue) and kidney, followed by lung and liver. Very interesting reports on the association of viral infections with blood group antigens are also

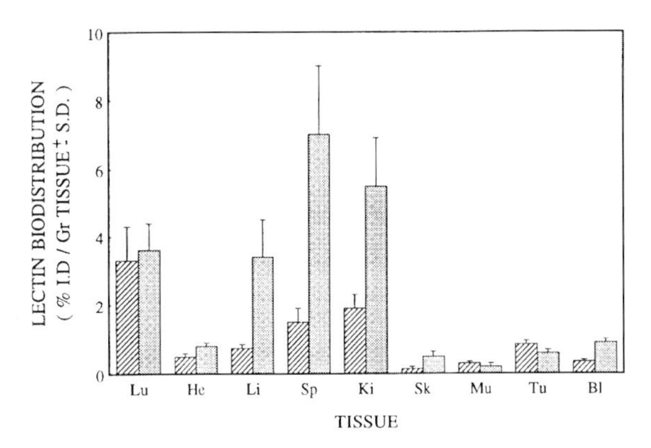

Figure 5. Distribution in tissues of ^{125}I-labeled PA-I and II lectins injected intravenously (*i. v.*) to mice, 24 h before their measurement by a Gamma Counter. The data are expressed as percentage of the injected dose (10^6 cpm per mouse) per gram wet tissue. The mice used were both normal ICR and CD_1 nude female mice bearing subcutaneous human ovarian IGROV-1 tumors. The tissues examined were lung (Lu), heart (He), liver (Li), spleen (Sp), kidney (Ki), skin (Sk), muscle (Mu), tumor (Tu) and blood (Bl) (Avichezer and Gilboa-Garber, 1995).

emerging (Garratty, 1995). A very relevant example is the demonstration by Brown *et al.* (1993, 1994) of the association of P blood group antigens with sensitivity to parvovirus B19 infection, to which the rare negative phenotype individuals are resistant.

6 Bacterial Lectin Effects on Macromolecules and Cells

Bacterial lectin binding to many macromolecules and cell types, which initially is reversible, leads to diverse irreversible alterations in them (Gilboa-Garber and Garber, 1989, 1992, 1993). The effects of bacterial lectins resemble those of plant or animal lectins in their pattern and probably also in their final destination in certain situations.

6.1 Effects on Macromolecules

Lectin binding to free macromolecules may affect their conformation and state of aggregation, as well as their binding to cells, their function and their fate. PA-IL was shown to precipitate glycosylated lectins and enzymes, and PA-IIL was shown to bind to many glycosylated proteins and to link peroxidase to cells (Gilboa-Garber *et al.*, 1977a, Fig. 6). There is reason to expect that bacterial lectins would also function like plant and animal lectins (Komano and Natori, 1985) or antibodies in promotion or inhibition of the activities of enzymes which are linked to them, which could also lead to sequestration of various molecules [e.g. glycoprotein by liver parenchymal enzymes (Weigel, 1992) and various macromolecules by the complement system and phagocyte enzymes] (Gilboa-Garber and Garber, 1989, 1993).

6.2 Effects on Cells

Bacterial lectin binding may affect cell membrane conformation, and may lead to cell agglutination, cell fusion and binding of macromolecules and cells to the bacteria, resembling effects of antibo-

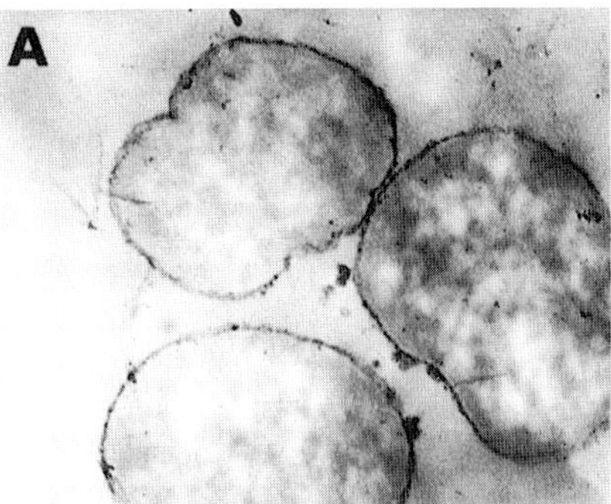

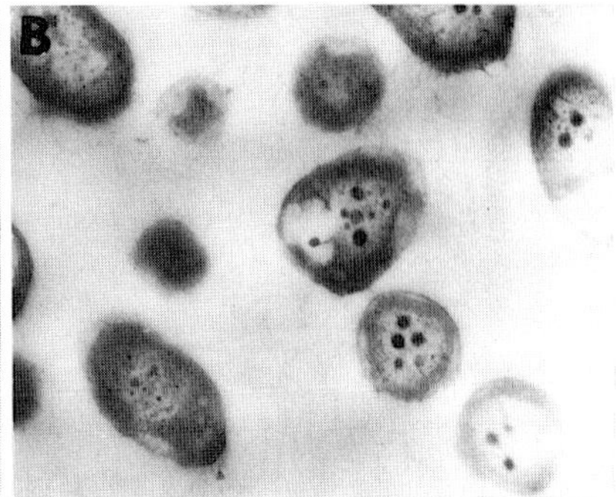

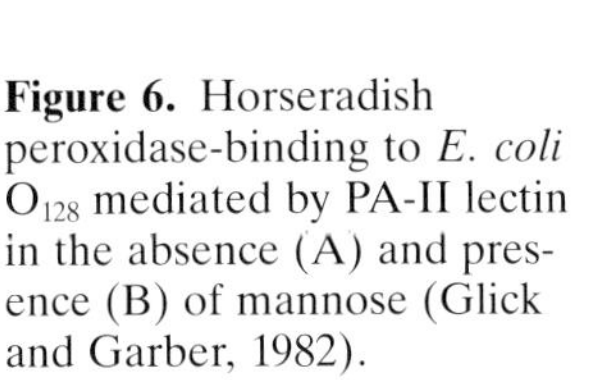

Figure 6. Horseradish peroxidase-binding to *E. coli* O_{128} mediated by PA-II lectin in the absence (A) and presence (B) of mannose (Glick and Garber, 1982).

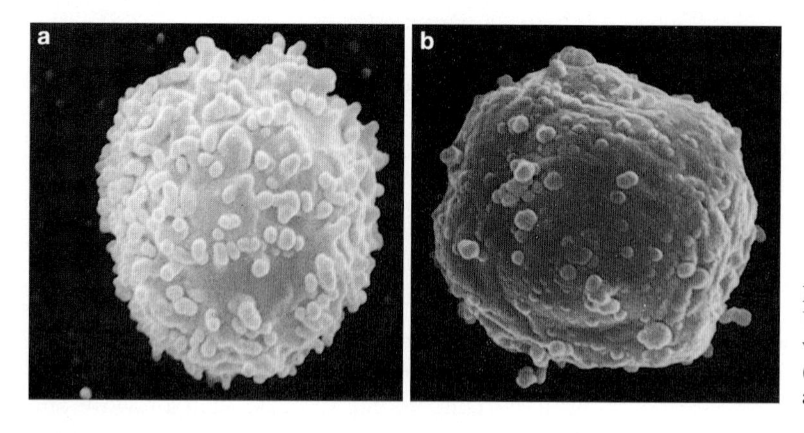

Figure 7. Human lymphocytes. (A) Untreated, showing normal microvilli, and (B) exposed to PA-IIL (0.5 mg/ml) for 60 min, showing disappearance of microvilli.

dies. This may activate key enzymes in the cells, inducing, as hormones do, a cascade of metabolic (hydrolytic or synthetic) reactions, leading to metabolic changes, which may terminate in either cell transformation and proliferation (mitogenic stimulation) or apoptotic death.

Pseudomonas PA-I and -II lectins were shown to resemble plant lectins (Yahara and Edelman, 1973) with respect to induction of receptor clustering and capping in lymphocytes, as well as agglutination of these cells and their metabolic and mitogenic stimulation (Sharabi and Gilboa-Garber, 1979; Avichezer and Gilboa-Garber, 1987). These lectins were also shown to stimulate growth of free living unicellular organisms, and increase phagocytosis in *Tetrahymena pyriformis* (Gilboa-Garber and Sharabi, 1980). On the other hand, at higher concentrations, PA-IL was shown to reduce erythrocyte membrane stability (Gilboa-Garber and Blonder, 1979), induce vacuolization in lymphocytes (Gilboa-Garber, 1972b) and the disappearance of their microvilli (Glick *et al.*, 1981, Fig. 7), decrease phagocytic activity of neutrophils (Sudakevitz and Gilboa-Garber, 1982), and finally to reduce the viability of tumor cells (Avichezer and Gilboa-Garber, 1991) by causing vacuolization which leads to their death (Fig. 8). PA-IIL, which had no direct cytotoxic effect, led to tumor cell death in the presence of splenocytes (Avichezer and Gilboa-Garber, 1991). It has recently been shown that PA-I lectin is also cytotoxic for respiratory epithelial cells in primary cultures of nasal polyps *in vitro;* the percentage of active ciliated cell surface decreased and the epithelial cells showed intracytoplasmic vacuoles (Bajolet-Laudinat *et al.*, 1994). Furthermore, it has been shown that orally ingested PA-I and II lectins induced gut growth and interfered with the polyamine metabolism of rat intestinal cells, just as plant lectins do (Grant

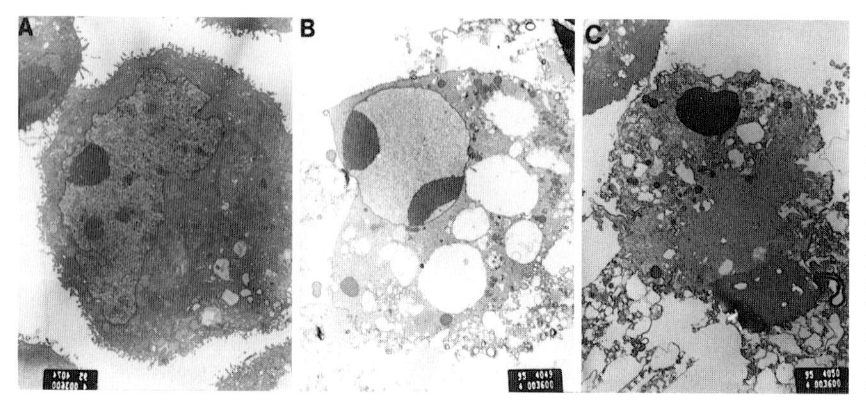

Figure 8. Transmission electron micrographs of ovarian IGROV-1 cancer cells cultured for 24 h in the absence (A) or presence (B, C) of PA-I lectin (10 μg/ml). In the presence of the lectin there is an increase in cytoplasmic vacuolization and chromatin condensation (B) ending in necrotic cell morphology with organelle decomposition (C).

et al., 1995a,b; Pusztai *et al.*, 1995) and that binding of *E. coli* P-fimbriae triggered IL-6 secretion from uroepithelial cell lines *in vitro* (Svensson *et al.*, 1994). Similarly, mannose-specific adhesins, which stimulated B lymphocytes, triggered immunoglobulin secretion from them (Ponniah *et al.*, 1989).

7 Bacterial Lectin Functions

The definition of the exact functions of bacterial lectins is very complicated, as it is with lectins in general. This situation may be viewed following the general lectinological literature in many fields (Komano and Natori, 1985; Barondes, 1986), e.g. the disputes about legume lectin involvement in *Rhizobium* binding (Kijne *et al.*, 1983; Smit *et al.*, 1992) and the debates about liver lectin function, as recently described by Ashwell (1994). Therefore, although our view might not please everybody, we will put it forward to stimulate thought.

We will develop this subject in accordance with our views on the definition of lectins that we summarized in the Introduction (Section 1.1). We shall consider the function of bacterial lectins in the context of irreversible enzyme activities controlled by the lectins, although in many cases it is hard to supply definite proof, because there are always alternative compensating means (e.g. other adhesins) to ensure crucial function, and the connection with an enzyme is not always easy to find. For example, a function such as bacterial adherence to their hosts (plants or animals) for symbiotic or parasitic purposes may be mediated by a series of lectins, other receptor binding proteins, and hydrophobic interactions. Addition of sugar will not prevent this binding, because the lectins are not the sole factors, but this fact already undermines the recognition of their contribution to the system. The insistence on inhibition by sugars as a decisive criterion or proof for lectin involvement must be re-evaluated. Fortunately, there are several systems where the lectin contribution is clear, since it is not obscured by backing factors (e.g. in viral hemagglutinin-dependent adhesion). These have to be used as models.

7.1 Bacterial Self-Protection

Purified lectins (but also other adhesins), at high concentrations, tend to aggregate (Fig. 1B). Bacteria which bear them also aggregate in self aggregation and in co(hetero)aggregation. The individual cells in the center of clusters are protected from adverse physical effects (heating, drying, radiation etc.) and from antibacterial compounds produced in the medium or added to it (e.g. antibiotics, bacteriolysins and antibodies). Pellicle formation on liquid media by floating bacteria (in contrast to sinking ones) may also serve for their exposure to oxygen.

Bacterial lectins involved in adhesion to the extracellular matrix may also protect the bacteria which bear them from environmental damage by covering them or by embedding them in the matrix. Thus, in the concentrated secretions of cystic fibrosis patients, *P. aeruginosa* cells are less susceptible to antibiotics, antibodies and phagocytic cells.

Protection from sweeping by host cell cilia or fluids may be obtained by inactivation of the cilia and adhesion to the host cells (Korhonen *et al.*, 1982). Bajolet-Laudinat *et al.* (1994) showed that PA-IL was cytotoxic to respiratory epithelial cells, destroying their ciliation. There is also bacterial protection inside host cells, even in phagocytes (e.g. *Neisseria*), as a shelter from the host immune system. Bacterial lectins mediate adhesion, followed by invasion of such cells. Plotkowski *et al.* (1994) have shown that *P. aeruginosa* invades endothelial cells by a pilus-dependent selective process and that, following progressive lysis of the cells, the bacteria were released in increasing numbers, indicating that the intracellular residence might afford protection from the host defense mechanism and from antibiotics. Binding and damaging host leukocytes or intestinal cells (see section 6.2) may also be regarded as bacterial protection from extinction by the host immune or digestive systems.

Lectins may also protect bacteria from phagocytosis by repulsion. Tewari *et al.* (1994) reported that *E. coli* bearing PapG (Gal-α-4Gal-binding adhesin) at the distal end of P-fimbriae, in contrast to a PapG⁻mutant, exhibited poor binding to neutrophils. This property was attributed to electrostatic repulsion by the adhesin's negative charge (pI=5.2). Antibodies against PapG overcame this protective effect. The pI of *P. aeruginosa* lectins (Avichezer *et al.*, 1992) is even lower (pI=4.94). Addition of its lectins to neutrophils blocked their phagocytic activity (Sudakevitz and Gilboa-Garber, 1982) and would probably also interfere with leukocyte rolling on endothelial cells.

7.2 Bacterial Cell Organization

The surface and periplasmic lectins of bacteria may partly participate in the binding of glycosylated structural macromolecules such as polysaccharides (alginate) to the cell surface, and of lipopolysaccharide (LPS) to the cell membrane. Some of the lectin-devoid *P. aeruginosa* cells (due to genetic or environmental factors) exhibit lower rigidity or compactness and lower stability to lyophilization (Gilboa-Garber and Kilfin, 1994), probably due to weaker linkage between the components. Such cells are also more leaky. It is possible that bacterial lectins would also be found to interact with the nucleic acids, as has been shown in eukaryotes.

7.3 Cell Contacts Supplying Nutrition and Physiological Functions

Among the special bacterial functions which, it has been established, are due to their lectins are commensal and symbiotic interrelationships for the benefit of both host cells and microbes. Examples are the gut commensal microflora in animals, and the legume-*Rhizobium* symbiosis for nitrogen fixation (Smit *et al.*, 1992). The attached bacteria enjoy nutritional and protective supplies from the host, while the host is provided by a supply of essential compounds from the bacteria.

7.4 Interactions for Proliferation

The purposes of most contacts between bacteria or viruses and host cells are not merely for their survival, but also for their proliferation. As a model for this aspect (because of the reasons stated in section 7), we will use the proliferation of viral entities, which is dependent on cell invasion. Viral hemagglutinins mediate viral adhesion to the host cells (Markwell, 1986; Wiley and Skehel, 1987).

During the last decade, some very elaborate works demonstrated the function of influenza and other viral lectins in fusion of viral and host cell membranes (Wiley and Skehel, 1987; Bullough *et al.*, 1994; Gaudin *et al.*, 1995). The influenza virus hemagglutinin (HA) is synthesized as a single polypeptide chain, and upon maturation it is post-translationally modified by a host cell-related trypsin-like enzyme, which attacks intersubunit region of the molecule, giving rise to covalently linked HA1 and HA2 subunits. Low pH induces a conformational change in HA. The *N*-terminal segment of HA2 undergoes a major refolding of the secondary and tertiary structures of the molecule (Bullough *et al.*, 1994). The apolar fusion peptide moves at least 100 Å to one tip of the molecule. At the other end a helical segment unfolds, a subdomain relocates reversing the chain direction, and part of the structure becomes disordered, resulting in the exposure of hydrophobic fusion peptide which is inserted into and induces conformational changes in the host cell membrane, mediating fusion of the hetero membranes and pore formation in the latter membrane. This model provides us, by deduction, with some evidence for a possible bacterial lectin function for their proliferation in the host.

7.5 Cell Contacts Leading to Bacterial Death

In contrast to the protective (repulsive) effect of P-pili against phagocytosis, type 1 pili (Silverblatt *et al.*, 1979; Mangan and Snyder, 1979), other pili, hydrophobicity (Speert *et al.*, 1986, 1989), and S fimbriae, as well as Dr antigen-specific adhesins (Johnson *et al.*, 1995) were shown to mediate bacterial adherence to human neutrophils. However, as opposed to type 1 fimbriae, adherence mediated by the Dr hemagglutinin did not lead to significantly increased bacterial killing (Johnson *et al.*, 1995). Analogous biological systems, where microbial lectins lead to bacterial death for the sake of the survival of the eukaryotic cell, are ingestion and digestion of bacteria by protozoa and other eukaryotic parasites. For example, *Trichomonas vaginalis* ingests *E. coli* strains bearing mannose-binding lectin-like molecules (Benchimol and Desouza, 1995).

8 Involvement of Bacterial Lectins in Pathogenesis

Bacterial lectins and adhesins participate in pathogenesis and contribute to bacterial virulence. They are involved in several stages of the infection, affecting its final outcome. As already described in section 6.2, even free bacterial lectins induce direct membranal and metabolic changes in host cells that probably facilitate later damage to them. They also target enzymes and toxins to the cells, allowing their penetration. Cell-bound lectins target bacteria to the host cells, enabling their enzymes and toxins to attack the cell membrane (Fig. 4), leading to cell penetration and destruction by the bacteria.

8.1 Enzyme/Toxin-Targeting and Trafficking to Host Macromolecules and Cells

Bacteria may induce remote cell-killing by secretion of toxins/enzymes linked to lectins or to lectin-like proteins. These bacterial toxic proteins express their toxic activities in different ways, such as: disruption of the target membrane integrity, inhibition of protein synthesis, blocking of neurotransmitter release (Menestrina *et al.*, 1991), and elevation of cAMP. Despite this diversity, they have several common properties, such as binding of the toxin to the host membrane and insertion (and translocation) of the toxin into the membrane, frequently with formation of transmembrane pores. Another feature common to different bacterial toxins, which resembles that of influenza hemagglutinin, is the conformational transition of the proteins, induced by acidic pH, exposing hidden hydrophobic domains, which facilitate both the incorporation of the protein into the target membrane and the translocation of the entire protein (or part of it) into the cytoplasm. Many of the bacterial toxins (Middlebrook and Dorland, 1984; Keush *et al.*, 1986), like plant toxins (Olsnes and Sandvig, 1988), are composed of two or three domains. One of them behaves like lectin or lectinoid adhesin (B domain), which binds to the host cell surface, and may function in enabling penetration of the catalytic A domain into the cell. The latter domain mostly is an enzyme, which blocks protein synthesis or changes the metabolism in the target cell. The bacterial A domain function, in many cases, is catalyzing ADP-ribosylation of target key protein (Iglewski *et al.*, 1978; Katada and Ui, 1982; Moss and Vaugham, 1990; Locht and Antoine, 1995). The bacterial lectin effect in these systems is analogous to those of animal antibodies or lectins which function with complement (Yokota *et al.*, 1995). While the aim of the latter animal systems is to kill invading cells for host defense, the bacterial lectinoid toxins defend bacteria by killing the host cells. The toxins of *Corynebacterium diphteriae, S. dysenteriae* (Shiga), *P. aeruginosa* (exotoxin A), *C. tetani, C. botulinum* and of other pathogenic microbes are all composed of A and B domains. However, their cell receptor-binding B domain specificity is not for simple sugars. The B domain of the exoenzyme S of *P. aeruginosa*, which is indistinguishable in function from the B domains described above, was shown to behave as galactophilic adhesin (Baker *et al.*, 1991; Lingwood *et al.*, 1991), exhibiting sugar specificity resembling PA-I lectin. Moreover, there are also a hemagglutinin/protease (Häse and Finkelstein, 1990) and a lectin neuraminidase in *V. cholerae*, which participate in degrading the mucin layer of the gastrointestinal tract. The latter degrades sialylated gangliosides to GM_1, which is the receptor for cholera toxin. Crennell *et al.* (1994) showed that this protein folds into three distinct domains. The central catalytic domain was shown by X-ray crystal structure analysis to have the canonical neuraminidase β-propeller fold, and is flanked by two domains possessing identical legume lectin-like topologies (without the usual metal-binding loops). Crennell *et al.* (1994) who reported this finding, have pointed to the fact that although originally they did not expect lectin-like domains, they suppose that these probably mediate the neuraminidase attachment to the small intestine cells. They have also claimed that "these bacterial lectin folds represent additional members of a growing lectin superfamily". We personally are very pleased to see these findings, because several years ago we claimed that "the general basic lectin role is cofunction (promotion and regulation) of key lytic activities" (Gilboa-Garber and Garber, 1989, 1993). This is also the reason for our satisfaction in finding the high degree of PA-IL analogy with the amino acid sequence comprising half the structure of the insect *Manduca sexta* (tobacco hornworm) midgut protease (Peterson *et al.*, 1994) described in section 4. We propose that this half might well be the binding domain which mediates the attachment of this protease to its target substrate, as in the case of *V. cholerae* hemagglutinin/protease (Booth *et al.*, 1983) or regulates its activity. It would be interesting to find whether this domain is indeed a lectin, which would, of course, further strengthen our claim, above, concerning the role of PA-IL. Closely related to the PA-IL analogy with the animal enzyme domain is the finding of lectin domains in the toxin of *B. pertussis* exhibiting selectin-like specificity (Sandros *et al.*, 1994). The pertussis toxin mediates intoxication of eukaryotic cells by elevation of cAMP, and serves as an adhesin for binding the bacteria to ciliated cells and respiratory macrophages (Sandros *et al.*, 1994). Sandros *et al.* (1994) have commented "that a comparison of pertussis toxin and the animal selectins, involved in leukocyte trafficking, indicates that these prokaryotic and eukaryotic C-type lectins share some elements of primary sequence similarity, three dimensional structure, and biological activities, and that such mimicry suggests a link between eukaryotic cell-cell adhesion motifs and microbial pathogenesis".

8.2 The Role of Bacterial Lectins in Adherence of Individual and Coaggregated Bacteria to Host Cells

Lectins of either host or parasite origin mediate bacterial pathogenicity (Uhlenbruck, 1987) by targeting the bacteria to sensitive host cells, and enabling their adherence to them. This adherence is crucial for initiation of the infection, since it enables the bacterial enzyme/toxin to act on the host cells/macromolecules at optimal efficiency (Fig. 4), allowing invasion of the host cells by the bacteria. In addition, they facilitate coaggregation with other bacteria which function together in damaging the host. An example of the latter is periodontal disease, which is initiated by adherence of bacteria (*Fusobacterium nucleatum* with streptococci) to the subgingival pockets and their coaggregation, which is ensured by multimodal adhesins, including L-arginine-sensitive, L-lysine-sensitive, GalNAc-sensitive and "resistant" factors (Takemoto *et al.*, 1995).

As an example of lectin-dependent bacterial adhesion to host cells that causes very serious complications, we will describe the case of *P. aeruginosa* infection in cystic fibrosis (CF) patients (Widdicombe and Wine, 1991). Chronic colonization and infection of lungs with *P. aeruginosa* is the major cause of morbidity and mortality in CF patients. Buccal, bronchial and pancreatic epithelial cells of these patients exhibit much higher binding of *P. aeruginosa* than normal matched controls (Woods *et al.*, 1980a,b; Imundo *et al.*, 1995). The receptor for this binding was shown to be asialo-GM_1 (aGM_1). *S. aureus*, which was found to bind to the same receptor, was effective in displacing *P. aeruginosa* from the apical membrane. The binding of both bacteria was reduced by free aGM1 and by its tetrasaccharide, Gal-β-3GalNAc-β-4Galβ4Glc-β, or by antibodies to aGM_1. Hence, the tetrasaccharide of aGM_1, which is found in increased abundance in the apical membrane of CF epithelium, is the receptor for *P. aeruginosa* and *S. aureus* and contributes to the pathogenesis of their infections in CF lungs (Imundo *et al.*, 1995). Ramphal and Pyle (1983) have presented evidence that mucins and sialic acids act as receptors for *P. aeruginosa* in the lower respiratory tract. But they found similar adhesion in CF and chronic bronchitis (Ramphal *et al.*, 1989). Imundo *et al.* (1995) have suggested that absence of the CF gene product (CF transmembrane conductance regulator, CFTR), which is an apical membrane Cl^- channel, from intracellular glycosylation compartments impairs sialylation of gangliosides. This defect results in an increase in the aGM_1 on the apical surfaces of the CF epithelia. Woods *et al.* (1983) have suggested that a high salivary protease level in these patients exposes these aGM_1 receptors.

Lectin-mediated reversible binding of bacteria to host cells is followed by irreversible cellular changes due to the lectin effect on the cell membrane and the directing of enzymes and toxins to their targets (Gilboa-Garber and Garber, 1989, Fig. 4). *P. aeruginosa* produces a battery of virulence factors, including exotoxin A (Liu, 1973), hemolysins, phospholipases, exoenzyme S, proteases (including elastase), cytotoxins (including leukocidin, pyocyanin and other pigments), hydrogen cyanide, mucopolysaccharides etc., most of which are toxic to the host cells. If these factors (excluding the exotoxin A and exoenzyme S, which contain private binding domains in their structure) circulate in the blood or body fluids, they may be inhibited by host soluble components (e.g. specific inhibitors of protease and of other enzymes and antibodies against them). But, when released from the *Pseudomonas* cells close to the host cell surface, the toxins and enzymes may function at considerably higher efficiency (Fig. 4). The solid basis to our claim that PA-I and II lectins enable the action of the virulence factors of *P. aeruginosa*, was the finding of co-regulated production of both these lectins and of several of the *P. aeruginosa* potent secondary metabolism virulence factors (SMVFs) mentioned above, including protease (elastase) and hemolysin (Gilboa-Garber, 1983; Gilboa-Garber and Garber, 1989). In *E. coli* there is also an association between hemolysin and adhesin (Hacker *et al.*, 1986). *P. aeruginosa* chitinase and staphylolytic activities, as well as cyanide production, were recently also correlated with the lectin production (Cámera *et al.*, 1995). Moreover, it has been shown that production of all of them is co-regulated. The regulatory mechanism involves *N*-acyl-*L*-homoserine-lactone as a signalling "autoinducer" molecule, represented in Fig. 9 (Garber *et al.*, 1995), which combines with the positive transcriptional activators LasR/VsmR to stimulate the expression of SMVFs. Mutants lacking the lasI/vsmI genes, also required for autoinducer synthesis, do not produce the lectins (Cámera *et al.*, 1995, Table 1). Another preliminary observa-

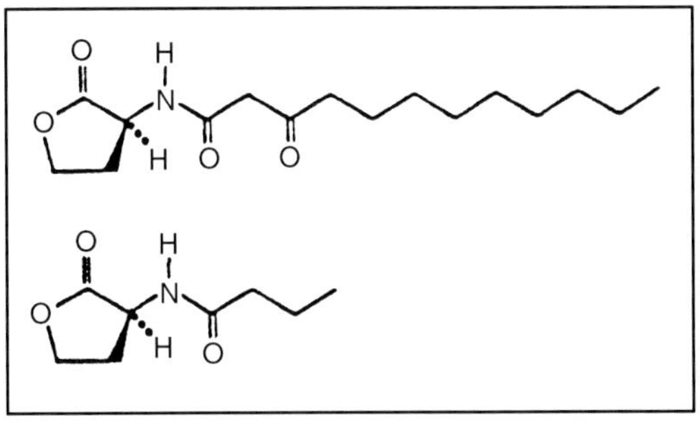

Figure 9. Structure of two *P. aeruginosa* "autoinducers": *N*-butyrylhomoserine lactone (factor 2) and *N*-(3-oxododecanoylhomoserine) lactone (PAI) (Pearson *et al.*, 1994, 1995).

Table 1 Production of lectins and virulence factors by various strains and mutants of *Pseudomonas aeruginosa*.

	S1Wt	S1M	PAO1	S2	S103	PAN067*
PA-I L	H	vL	M	L	-	-
PA-II L	H	vL	M	L	-	-
Protease (elastase)	H	vL	M	L	-	-
Hemolysin	H	vL	M	L	vL	vL
Pyocyanin	H	vL	M	L	-	-
Cyanide	H	ND	M*	ND	ND	-
Chitinase*	ND	ND	M*	ND	ND	-
Staph. lytic activity*	ND	ND	M	ND	ND	-
Phosphatase	M	M	M	M	M	M
Acetylcholinesterase	M	M	M	M	M	M
Exotoxin A	ND	ND	M*	ND	H	M

This table summarizes results described by Gilboa-Garber (1983, 1986) and by *Cámara *et al.* (1995).
S1Wt = strain 1 wild type (*P. aeruginosa* ATCC 33347).
S1M = mutant of S1Wt.
S2 and S103 = strains 2 and 103 (Liu's strain).
PAN067 = mutant of PAO1 in VsmR/LasR and VsmI/LasI.
H, M, L, vL, - = high, medium, low, very low, and no detectable activity (respectively).
ND = not determined.

tion supporting the validity of this SMVF-encompassing multimolecular system is that the pa-1L gene was localized close to virulence-associated genes in the *P. aeruginosa* chromosome (Avichezer *et al.*, 1994).

9 Preventive Strategies against Lectin-Mediated Adherence of Pathogenic Bacteria to Host Cells

The fact that lectins are involved in bacterial pathogenicity caused by the bacterial toxins and enzymes is important for understanding the infection process. A very practical aspect of this understanding is the possibility of devising preventive strategies (Beuth *et al.*, 1995). The treatment must be a combination of several components: anti-adhesins, anti-toxins and antibacterials. There are several ways to prevent the interaction between the bacterial adhesins/lectins and cell receptors, including (a) inhibition of their binding; (b) elimination of the host cell receptors, and (c) repression of lectin/adhesin production, considering the possible existence of multiple adhesive factors, as already discussed.

9.1 Prevention of Lectin-Receptor Interactions

The means of prevention of lectin-receptor interactions may be divided into five groups, according to the type of compounds introduced into the system. The *in vitro* conditions are not identical to the *in vivo* situation, where active compounds may be altered by the host metabolic system, or captured by diverse cells. In order to overcome these two obstacles, the compounds added must be inert to the host metabolic systems and exhibit highest affinity for the target system. The five groups of compounds that can be used are: (a) glycosylated compounds for neutralization of the lectins/adhesins involved in the bacterial adhesion (Williams and Gibbons, 1975; Woods *et al.*, 1981a; Gilboa-Garber and Mizrahi, 1985; Beuth and Pulverer, 1994); (b) monovalent lectin subunits or lectin-like molecules of the infecting bacteria, or their segments (peptides), which will compete with the intact bacterial lectins for their binding. Analogous products of heterologous bacteria may also be used (Rozdzinski *et al.*, 1993; Imundo *et al.*, 1995); (c) antibodies (or their Fab fragments) produced specifically against the bacterial lectins (Gilboa-Garber and Sudakevitz, 1982; Sexton and Reed, 1992); (d) antibodies (or Fab fragments) produced specifically against the host cell receptors (Imundo *et al.*, 1995) and (e) non-specific inhibitors which will inhibit the lectin-receptor interaction by a different mechanism.

9.2 Host Cell Receptor Elimination

Elimination of the host cell receptors is easy *in vitro* but complicated *in vivo*. Ways to attain it include: (a) repression of *de novo* production of the receptor by genetic manipulations, or by inhibition of production or activity of enzymes involved in their synthetic pathways. An example for the latter is the use of a synthetic analog of glucosylceramide, D-threo-1-phenyl-2-decanoyl-amino-3-morpholino-1-propanol (PDMP) which interferes competitively with glycosphingolipid synthesis in intact epithelial cells (Svensson *et al.*, 1994); (b) enzymatic degradation of the terminal dominant epitope. Examples are viral sialidase-catalyzed degradation of sialylated receptors, preventing adhesion of sialophilic lectins; (c) prevention of the activity of enzymes which unmask cell receptors, such as sialidase or protease, which degrade their respective substrate molecules and unmask the glycolipid receptors (Woods *et al.*, 1981b); (d) sialylation or other means leading to receptor masking with perturbation of its original specificity.

9.3 Prevention of the Bacterial Lectin/Adhesin Production

Prevention of bacterial lectin/adhesin production may be obtained by addition of certain extracellular factors which would inhibit lectin production without affecting essential host systems. The most elegant way would be: (a) knocking out a bacterial-specific reaction which is absent in the host, such as the reaction involved in the production of the "autoinducer", *N*-acyl-L-homoserine lactone that regulates the lectin (PA-I and II), together with other SMVF production (Cámera *et al.*, 1995; Garber *et al.*, 1995); (b) antibiotic treatment, even at concentrations below those required to inhibit growth, which may totally inhibit the production of the bacterial lectins and adhesins (Piatti, 1994), as well as their toxic factors (Shibl and Al-Sowaygh, 1980); (c) nutrition, which might affect the expression of the adhesins or the virulence factors (Seddon *et al.*, 1991).

10 Applications of Bacterial Lectins

The applications of bacterial lectins and lectinoids also include those common to plant and animal lectins, in various scientific and medical fields. In the therapeutic area, bacterial lectins may be highly valuable, especially in the vaccine industry, and in cancer diagnosis and therapy. Since this subject was dealt with in detail in the reviews of Gilboa-Garber and Garber (1992) and Shakhanina *et al.* (1994), we will summarize it very briefly.

The bacterial lectins are useful for the following purposes:

(a) identification of carbohydrates in solutions and body fluids;
(b) identification of glycosylation in macromolecules;
(c) detection of defined cell receptors;
(d) blood group typing or differentiation. PA-IL may be useful in differentiating between P positive ($P^k > P_1 > P_2$) and negative (p) erythrocytes. PA-IIL may be useful in differentiating between H-positive (O, A, B or AB) and H-negative (Bombay type) cells (Gilboa-Garber *et al.*, 1994);
(e) bacterial identification;
(f) separation of molecules in mixtures, and purification of glycosylated macromolecules (enzymes and other glycoproteins), cell receptors and other cell components (Lakhtin, 1994);
(g) separation between different normal cell populations, and between normal and transformed (e.g. neoplastic) cells;
(h) labeling glycosylated cellular components, for their localization;

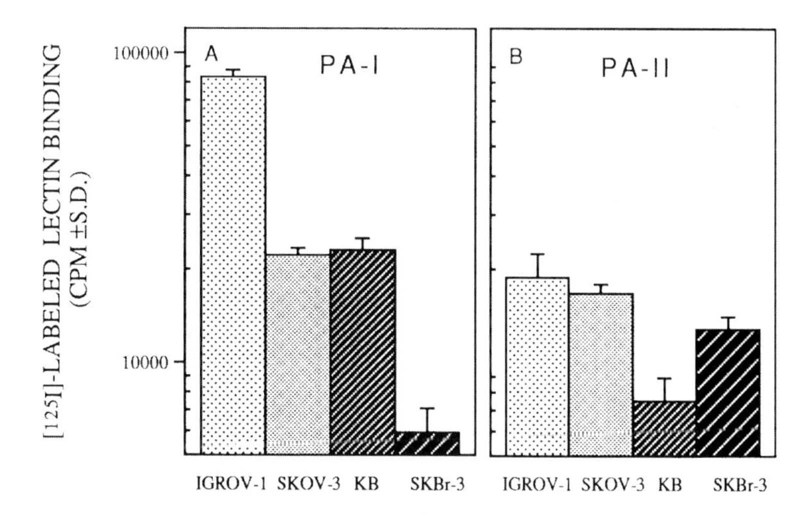

Figure 10. Binding of ^{125}I-labeled PA-I (A) and II (B) lectins to human ovarian (IGROV-1 and SKOV-3), oral epidermoid (KB) and breast (SKBr-3) carcinoma cells grown *in vitro* in monolayers at 37 °C for 24 h.

(i) augmentation or suppression of intercellular interactions, including bacterial phagocytosis by human neutrophils (Sudakevitz and Gilboa-Garber, 1982);
(j) detection of cell membrane alterations;
(k) cancer diagnosis by mitogenic index, skin test, or histochemical tests;
(l) selective *in vitro* anti-tumor cell cytotoxicity and anti-cancer cell activity (Gilboa-Garber *et al.*, 1986; Avichezer and Gilboa-Garber, 1991, 1995; Gilboa-Garber and Avichezer, 1993; and Figs. 5, 8 and 10);
(m) *in vivo* reduction of cancer cell tumorigenicity with antigenic preservation as a vaccine (Avichezer and Gilboa-Garber, 1991);
(n) activation of cell metabolism and functions;
(o) mitogenic stimulation of peripheral lymphocytes (e.g. for genetic assays) and other eukaryotic cells;
(p) trafficking of drugs and other active components to deficient target cells, or cancer cells for their selective damage (Chaudry *et al.*, 1993);
(q) efficient vaccines for active and passive immune protection against *P. aeruginosa* (Gilboa-Garber and Sudakevitz, 1982) and other lectin-bearing bacterial infections;
(r) immunomodulation, increasing resistance against heterologous bacterial infections (Avichezer and Gilboa-Garber, 1992) and probably also against cancer, like plant lectins (Hajto *et al.*, 1989; Gabius *et al.*, 1992; Gabius, 1994).

Acknowledgments

We thank Mrs. Sharon Victor for her great help in preparation of the manuscript and Mrs. Ella Gindi for help in the graphic work. Our research is supported by Bar-Ilan University Research Fund, Health Science Research Center and Mitzi-Dobrin Cancer Research funds. We also thank Drs. CJ Roberts, RCA Hirst, SJB Tendler, MC Davies and P Williams, from the Department of Pharmaceutical Sciences, University of Nottingham U.K., for the scanning tunneling microscopic examination of PA-I lectin (Fig. 1). Thanks to Drs. Pierre Rougé and Michael Amar, Centre National de la Recherche Scientifique, Laboratoire de Pharmacologie et de Toxicologie Fondamentales, France, for the hydrophobic cluster analysis of PA-I lectin of *P. aeruginosa* (Fig. 3B).

References

Agrawal BBL, Goldstein IJ (1965): Specific binding of concanavalin A to cross-linked dexran gels. In *Biochem J.* **96**:23c-5c.

Amano A, Sharma A, Sojar HT *et al.* (1994): Effects of temperature stress on expression of fimbriae and superoxide dismutase by *Porphyromonas gingivalis*. In *Infect. Immun.* **62**:4682–5.

Ascencio F, Aleljung P, Wadström T (1990): Particle agglutination assays to identify fibronectin and collagen cell surface receptors and lectins in *Aeromonas* and *Vibrio* species. In *Appl. Environ. Microbiol.* **56**:1926–31.

Ashwell G (1994): The physiological role of the hepatic lectin for asialoglycoproteins: an evaluation of current hypotheses. In *Lectin Blocking: New Strategies for the Prevention and Therapy of Tumor Metastasis and Infectious Diseases* (Beuth J, Pulverer G, eds) pp 26–36, Stuttgart: Gustav Fischer Verlag.

Avichezer D, Gilboa-Garber N (1987): PA-II, the L-fucose and D-mannose binding lectin of *Pseudomonas aeruginosa* stimulates human peripheral lymphocytes and murine splenocytes. In *FEBS Lett.* **216**:62–6.

Avichezer D, Gilboa-Garber N (1991): Anti-tumoral effects of *Pseudomonas aeruginosa* lectins on Lewis lung carcinoma cells cultured *in vitro* without and with murine splenocytes. In *Toxicon* **29**:1305–13.

Avichezer D, Gilboa-Garber N (1992): Use of *Pseudomonas aeruginosa* lectin preparations for mice protection against bacterial infections. In *Isr. J. Med. Sci.* **28**:74–5.

Avichezer D, Gilboa-Garber N (1995): Effects of *Pseudomonas aeruginosa* lectins on human ovarian, breast and oral epidermoid carcinoma cells. In *Proc. 1st FISEB Meeting* Eilat (Israel) p. 192.

Avichezer D, Katcoff DJ, Garber NC *et al.* (1992): Analysis of the amino acid sequence of the *Pseudomonas aeruginosa* galactophilic PA-I lectin. In *J. Biol. Chem.* **267**:23023–7.

Avichezer D, Gilboa-Garber N, Garber NC *et al.* (1994): *Pseudomonas aeruginosa* PA-I lectin gene molecular analysis and expression in *Escherichia coli*. In *Biochim. Biophys. Acta* **1218**:11–20.

Bajolet-Laudinat O, Girod-de Bentzmann S, Tournier JM *et al.* (1994): Cytotoxicity of *Pseudomonas aeruginosa* internal lectin PA-I to respiratory epithelial cells in primary culture. In *Infect. Immun.* **62**:4481–7.

Baker NR, Svanborg-Edén C (1989): Role of alginate in the adherence of *Pseudomonas aeruginosa*. In *Pseudomonas aeruginosa Infection* (Hoiby N, Pedersen SS, Sland GH *et al.*, eds) pp 72–9, Basel: S. Karger.

Baker N, Hansson GC, Leffler H *et al.* (1990): Glycosphingolipid receptors for *Pseudomonas aeruginosa*. In *Infect. Immun.* **58**:2361–6.

Baker NR, Minor V, Deal C *et al.* (1991): *Pseudomonas aeruginosa* exoenzyme S is an adhesin. In *Infect. Immun.* **59**:2859–63.

Balding P, Gold ER, Boroff DE *et al.* (1973): Observations on receptor specific proteins II. Haemagglutination and haemagglutination-inhibition reactions of *Clostridium botulinum* types A, C, D and E haemagglutinins. In *Immunology* **25**:773–82.

Banas JA, Russel RRB, Ferretti JJ (1990): Sequence analysis of the gene for the glucan-binding protein of *Streptococcus mutans* Ingbritt. In *Infect. Immun.* **58**:667–79.

Barondes SH (1986): Vertebrate lectins: Properties and functions. In *The Lectins: Properties, Functions, and Applications in Biology and Medicine* (Liener IE, Sharon N, Goldstein IJ, eds) pp 437–66, New York: Academic Press.

Benchimol M, Desouza W (1995): Carbohydrate involvement in the association of a prokaryotic cell with *Trichomonas vaginalis* and *Trichomonas foetus*. In *Parasitol. Res.* **81**: 459–64.

Beuth J, Pulverer G, eds (1994): Lectin Blocking: New Strategies for the Prevention and Therapy of Tumor Metastasis and Infectious Diseases. Stuttgart: Gustav Fischer Verlag.

Beuth J, Uhlenbruck G (1994): Adhesive properties of bacteria. In *Principles of Cell Adhesion* (Richardson PD, Steiner M, eds) pp 87–105, Boca Raton (Florida): CRC Press.

Beuth J, Ko HL, Pulverer G *et al.* (1995): Importance of lectins for the prevention of bacterial infections and cancer metastases. In *Glycoconjugate J.* **12**:1–6.

Blaser MJ (1992): Hypotheses on the pathogenesis and natural history of *Helicobacter pylori*-induced inflammation. In *Gastroenterology* **102**:720–7.

Bock K, Breimer ME, Brignole A *et al.* (1985): Specificity of binding of a strain of uropathogenic *Escherichia coli* to Galα1–4Gal – containing glycosphingolipids. In *J. Biol. Chem.* **260**:8545–51.

Booth BA, Boesman-Finkelstein M, Finkelstein RA (1983): *Vibrio cholerae* soluble hemagglutinin/protease is a metalloenzyme. In *Infect. Immun.* **42**:639–44.

Booth BA, Sciortino CV, Finkelstein RA (1986): Adhesins of *Vibrio cholerae*. In *Microbial Lectins and Agglutinins: Properties and Biological Activity* (Mirelman D, ed) pp 169–82, New York: John Wiley & Sons.

Borén T, Falk P, Roth KA *et al.* (1993): Attachment of *Helicobacter pylori* to human gastric epithelium mediated by blood group antigens. In *Science* **262**:1892–5.

Boyd WC, Shapleigh E (1954): Separation of individuals of any blood group into secretors and non-secretors by use of a plant agglutinin (lectin). In *Blood* **9**:1195–8.

Brennan MJ, Joralmon RA, Cisar JO *et al.* (1987): Binding of *Actinomyces naeslundii* to glycosphingolipids. In *Infect. Immun.* **55**: 487–9.

Brinton CC, Woods SW, Brown A *et al.* (1982): The development of a neisserial pilus vaccine for gonorrhea and meningococcal meningitis. In *Bacterial Vaccines; Seminars in Infectious Diseases*, Vol 4 (Weinstein L, Fields BN, eds) pp 140–59, New York: Thieme Stratton.

Brown KE, Anderson SM, Young NS (1993): Erythrocyte P antigen: cellular receptor for B19 parvovirus. In *Science* **262**:114–7.

Brown KE, Hibbs JR, Gallinella G *et al.* (1994): Resistance of parvovirus B19 infection due to lack of virus receptor (erythrocyte P antigen). In *New Engl. J. Med.* **330**:1192–6.

Brözel VS, Strydom GM, Cloete TE (1995): A method for the study of *de novo* protein synthesis in *Pseudomonas aeruginosa* after attachment. In *Biofouling* **8**:195–201.

Bullough PA, Hughson FM, Skehel JJ *et al.* (1994): Structure of influenza haemagglutinin at the pH of membrane fusion. In *Nature* **371**:37–43.

Cacalano G, Kays M, Saiman L *et al.* (1992): Production of the *Pseudomonas aeruginosa* neuraminidase is increased under hyperosmolar conditions and is regulated by genes involved in alginate expression. In *J. Clin. Invest.* **89**:1866–74.

Cámera M, Winson MK, Latifi A *et al.* (1995): Multiple quorum sensing modulons interactively regulate production of virulence determinants and secondary metabolites in *Pseudomonas aeruginosa*. In *Proc. "Pseudomonas 1995" Meeting*, Tsukuba (Japan) p 159.

Castric P (1995): pilO, a gene required for glycosylation of *Pseudomonas aeruginosa* 1244 pilin. In *Microbiol.* **141**:1247–54.

Ceri H, McArthur HAI, Whitfield C (1986): Association of alginate from *Pseudomonas aeruginosa* with two forms of heparin-binding lectin isolated from rat lung. In *Infect. Immun.* **51**:1–5.

Chaudry GJ, Fulton RJ, Draper RK (1993): Variant of exotoxin A that forms potent and specific chemically conjugated immunotoxins. In *J. Biol. Chem.* **268**:9437–41.

Chou PY, Fasman GD (1978): Prediction of the secondary structure of proteins from their amino acid sequence. In *Adv. Enzymol.* **47**:45–148.

Cisar JO, Kolenbrander PE, McIntire FC (1979): Specificity of coaggregation reactions between human oral streptococci and strains of *Actinomyces viscosus* and *Actinomyces naeslundii*. In *Infect. Immun.* **24**:742–52.

Clegg S, Purcell BK, Pruckler J (1987): Characterization of genes encoding type 1 fimbriae of *Klebsiella pneumoniae*, *Salmonella typhimurium* and *Serratia marcescens*. In *Infect. Immun.* **55**:281–7.

Collier WA, Miranda JC (1955): Bakterienhämagglutination. III. Die Hemmung der Colihämagglutination durch Mannose. In *Antonie van Leeuwenhoek J. Microbiol. Serol.* **21**:133–40.

Costerton JW, Cheng K-J, Geesey G *et al.* (1987): Bacterial biofilms in nature and disease. In *Annu. Rev. Microbiol.* **41**:435–64.

Cover TL, Blaser MJ (1995): *Helicobacter pylori*: a bacterial cause of gastritis, peptic ulcer disease, and gastric cancer. In *ASM News* **61**:21–6.

Crennell S, Garman E, Laver G (1994): Crystal structure of *Vibrio cholerae* neuraminidase reveals dual lectin-like domains in addition to the catalytic domain. In *Structure* **2**:535–44.

Davies DG, Chakrabarty AM, Greesey G (1993): Exopolysaccharide production in biofilms: substratum activation of alginate gene expression by *Pseudomonas aeruginosa*. In *Appl. Environ. Microbiol.* **59**:1181–6.

Doig P, Smith NR, Todd T *et al.* (1987): Characterization of the binding of *Pseudomonas aeruginosa* alginate to human epithelial cells. In *Infect. Immun.* **55**:1517–22.

Doig P, Paranchych W, Sastry PA *et al.* (1989): Human buccal epithelial cell receptors of *Pseudomonas aeruginosa*: identification of glycoproteins with pilus binding activity. In *Can. J. Microbiol.* **35**:1141–5.

Drickamer K (1988): Two distinct classes of carbohydrate-recognition domains in animal lectins. In *J. Biol. Chem.* **263**:9557–60.

Drickamer K (1992): Engineering galactose-binding activity into a C-type mannose-binding protein. In *Nature* **360**:183–6.

Duguid JP, Gillies RR (1957): Fimbriae and adhesive properties in dysentery bacilli. In *J. Pathol. Bacteriol.* **74**:397–411.

Duguid JP, Smith IW, Dempster G *et al.* (1955): Nonflagellar filamentous appendages ('fimbriae') and haemagglutinating activity in *Bacterium coli*. In *J. Pathol. Bacteriol.* **70**:335–48.

Duguid JP, Anderson ES, Campbell I (1966): Fimbriae and adhesive properties in *Salmonellae*. In *J. Pathol. Bacteriol.* **92**:107–38.

Evans RT, Klausen B, Sojar HT *et al.* (1992): Immunization with *Porphyromonas (Bacteroides) gingivalis* fimbriae protects against periodontal destruction. In *Infect. Immun.* **60**:2926–35.

Falk PG, Bry L, Holgersson J *et al.* (1995): Expression of human α3/4-fucosyltransferase in the pit cell lineage of FVB/N mouse stomach results in production of Le b-containing glycoconjugates: A potential transgenic mouse model for studying *Helicobacter pylori* infection. In *Proc. Natl. Acad. Sci. USA* **92**:1515–9.

Falkow S (1991): Bacterial entry into eukaryotic cells. In *Cell* **65**: 1099–102.

Firon N, Ofek I, Sharon N (1983): Carbohydrate specificity of the surface lectins of *Escherichia coli*, *Klebsiella pneumoniae* and *Salmonella typhimurium*. In *Carbohydr. Res.* **120**: 235–49.

Fujita Y, Oishi K, Aida K (1973): Sugar specificity of anti-B hemagglutinin produced by *Streptomyces* sp. In *Biochem. Biophys. Res. Commun.* **53**:495–501.

Fujita Y, Oishi K, Suzuki K *et al.* (1975): Purification and properties of an anti-B hemagglutinin produced by *Streptomyces* sp. In *Biochemistry* **14**: 4465–70.

Gabius H-J (1994): Lectinology meets mythology: oncological future for the mistletoe lectin? In *Trends Glycosci. Glycotechnol.* **6**:229–38.

Gabius S, Joshi SS, Kayser K *et al.* (1992): The galactoside-specific lectin from mistletoe as biological response modifier. In *Int. J. Oncol.* **1**:705–8.

Gaboriaud C, Bissery V, Benchetrit T *et al.* (1987): Hydrophobic cluster analysis: an efficient new way to compare and analyse amino acid sequences. In *FEBS Lett.* **224**:149–55.

Garber N, Glick J, Gilboa-Garber N *et al.* (1981): Interactions of *Pseudomonas aeruginosa* lectins with *Escherichia coli* strains bearing blood group determinants. In *J. Gen. Microbiol.* **123**:359–63.

Garber N, Sharon N, Shohet D *et al.* (1985): Contribution of hydrophobicity to the hemagglutination reactions of *Pseudomonas aeruginosa*. In *Infect. Immun.* **50**:336–7.

Garber N, Guempel U, Gilboa-Garber N *et al.* (1987): Specificity of the fucose-binding lectin of *Pseudomonas aeruginosa*. In *FEMS Microbiol. Lett.* **48**:331–4.

Garber N, Guempel U, Belz A *et al.* (1992): On the specificity of the D-galactose-binding lectin (PA-I) of *Pseudomonas aeruginosa* and its strong binding to hydrophobic derivatives of D-galactose and thiogalactose. In *Biochim. Biophys. Acta* **1116**:331–3.

Garber NC, Hammer-Müntz O, Belz A *et al.* (1995): The lux autoinducer stimulates the production of the lectins of *Pseudomonas aeruginosa*. In *ISM Lett.* **15**:164.

Garratty G (1995): Blood group antigens as tumor markers, parasitic/bacterial/viral receptors, and their association with immunologically important proteins. In *Immunol. Invest.* **24**:213–32.

Gaudin Y, Ruigrok RWH, Brunner J (1995): Low pH induced conformational changes in viral fusion proteins: implications for the fusion mechanism. In *J. Gen. Virol.* **76**:1541–56.

Gbarah A, Gahmberg CG, Ofek I *et al.* (1991): Identification of the leukocyte adhesion molecules CD11 and CD18 as receptors for type 1-fimbriated (mannose-specific) *Escherichia coli*. In *Infect. Immun.* **59**:4524–30.

Gerlach G-F, Clegg S, Allen BL (1989): Identification and characterization of the genes encoding the type 3 and type 1 fimbrial adhesins of *Klebsiella pneumoniae*. In *J. Bacteriol.* **171**:1262–70.

Gilboa-Garber N (1972a): Inhibition of broad spectrum hemagglutinin from *Pseudomonas aeruginosa* by D-galactose and its derivatives. In *FEBS Lett.* **20**:242–4.

Gilboa-Garber N (1972b) Purification and properties of hemagglutinin from *Pseudomonas aeruginosa* and its reaction with human blood cells. In *Biochim. Biophys. Acta* **273**:165–73.

Gilboa-Garber N (1982): *Pseudomonas aeruginosa* lectins. In *Methods Enzymol.* **83**:378–85.

Gilboa-Garber N (1983): The biological functions of *Pseudomonas aeruginosa* lectins. In *Lectins: Biology, Biochemistry, Clinical Biochemistry*, Vol 3 (Bøg-Hansen TC, Spengler GA, eds) pp 495–502, Berlin: Walter de Gruyter.

Gilboa-Garber N (1986): Lectins of *Pseudomonas aeruginosa*: properties, biological effects and applications. In *Microbial Lectins and Agglutinins: Properties and Biological Activity* (Mirelman D, ed) pp 255–69, New York: John Wiley & Sons.

Gilboa-Garber N (1988): *Pseudomonas aeruginosa* lectins as a model for lectin production, properties, applications and functions. In *Zentralbl. Bakteriol. Hyg.* [A] **270**:3–15.

Gilboa-Garber N (1994): *Pseudomonas aeruginosa* PA-I and PA-II lectins. In *Lectin Blocking: New Strategies of the Prevention and Therapy of Tumor Metastasis and Infectious Diseases* (Beuth J, Pulverer G, eds) pp 44–58, Stuttgart: Gustav Fischer Verlag.

Gilboa-Garber N, Mizrahi L (1973): Peroxidase attachment to Sepharose mediated by bacterial hemagglutinin of *Pseudomonas aeruginosa*. In *Biochim. Biophys. Acta* **317**:106–13.

Gilboa-Garber N, Blonder E (1979): Augmented osmotic hemolysis of human erythrocytes exposed to the galactophilic lectin of *Pseudomonas aeruginosa*. In *Isr. J. Med. Sci.* **15**:537–9.

Gilboa-Garber N, Sharabi Y (1980): Increase of growth-rate and phagocytic activity of *Tetrahymena* induced by *Pseudomonas* lectin. In *J. Protozool.* **27**:207–11.

Gilboa-Garber N, Sudakevitz D (1982): The use of *Pseudomonas aeruginosa* lectin preparations as a vaccine. In *Advances in Pathology*, Vol 1 (Levy E, ed) pp 31–3, Oxford: Pergamon Press.

Gilboa-Garber N, Mizrahi L (1985): *Pseudomonas* lectin PA-I detects hybrid product of blood group AB genes in saliva. In *Experientia* **41**:681–2.

Gilboa-Garber N, Garber N (1989): Microbial lectin cofunction with lytic activities as a model for a general basic lectin role. In *FEMS Microbiol. Rev.* **63**:211–22.

Gilboa-Garber N, Garber N (1992): Microbial lectins. In *Glycoconjugates: Composition, Structure and Function* (Allen HJ, Kisailus EC, eds) pp 541–91, New York: Marcel Dekker, Inc.

Gilboa-Garber N, Avichezer D (1993): Effects of *Pseudomonas aeruginosa* PA-I and PA-II lectins on tumoral cells. In *Lectins and Glycobiology* (Gabius H-J, Gabius S, eds) pp 380–95, Berlin: Springer Verlag.

Gilboa-Garber N, Garber N (1993): Lectins promote and regulate key lytic reactions: a hypothesis. In *Lectins: Biology-Biochemistry Clinical Biochemistry*, Vol 8 (van Driessche E, Franz H, Beeckmans S, Pfuller U, Kallikorm A, Bøg-Hansen TC, eds) pp 276–87, Hellerup (Denmark): Textop.

Gilboa-Garber N, Kilfin G (1994): Lectin/protease negative *Pseudomonas aeruginosa* are significantly less viable than the positive strains following lyophilization. In *Proc. Annual Meeting of the Israel Society for Microbiology*, Rehovot, Israel, p. 94.

Gilboa-Garber N, Mizrahi L, Garber N (1972): Purification of the galactose-binding hemagglutinin of *Pseudomonas aeruginosa* by affinity column chromatography using Sepharose. In *FEBS Lett.* **28**:93–5.

Gilboa-Garber N, Mizrahi L, Garber N (1977a): Mannose-binding hemagglutinins in extracts of *Pseudomonas aeruginosa*. In *Can. J. Biochem.* **55**:975–81.

Gilboa-Garber N, Nir-Mizrahi I, Mizrahi L (1977b): Specific agglutination of *Escherichia coli* $O_{128}B_{12}$ by the mannose-binding protein of *Pseudomonas aeruginosa*. In *Microbios* **18**:99–109.

Gilboa-Garber N, Avichezer D, Leibovici J (1986): Application of *Pseudomonas aeruginosa* galactophilic lectin (PA-I) for cancer research. In *Lectins: Biology, Biochemistry, Clinical Biochemistry*, Vol. 5 (Bøg-Hansen TC, van Driessche E, eds) pp 329–38, Berlin: Walter de Gruyter.

Gilboa-Garber N, Sudakevitz D, Levene C *et al.* (1991): Preferential interactions of PA-I lectin of *Pseudomonas aeruginosa* with terminal galactosyl residues of blood group antigens B, P^k, Bombay P_1, I(i) and T. In *Proc. 15th Intern. Congr. Biochem.* Jerusalem, p 328.

Gilboa-Garber N, Sudakevitz D, Sheffi M *et al.* (1994): PA-I and PA-II lectin interactions with the ABO(H) and P blood group glycosphingolipid antigens may contribute to the broad spectrum adherence of *Pseu-*

domonas aeruginosa to human tissues in secondary infections. In *Glycoconjugate J.* **11**:414–7.

Glick J, Garber NC (1982): The hemagglutinins (lectins) of *Pseudomonas aeruginosa*: localization, specification and interaction with various cells. In Ph.D. Thesis, (Glick J) pp 1–220, Ramat Gan: Bar-Ilan University.

Glick J, Garber N (1983): The intracellular localization of *Pseudomonas aeruginosa* lectins. In *J. Gen. Microbiol.* **129**:3085–90.

Glick J, Malik Z, Garber N (1981): Lectin-bearing protoplasts of *Pseudomonas aeruginosa* induce capping in human peripheral blood lymphocytes. In *Microbios* **32**:181–8.

Glick J, Garber N, Shohet D (1987): Surface haemagglutinating activity of *Pseudomonas aeruginosa*. In *Microbios* **50**:69–80.

Gold ER, Balding P (1975): *Receptor Specific Proteins. Plant and Animal Lectins*. Amsterdam: Excerpta Medica.

Goldstein IJ, Hayes CE (1978): The lectins: carbohydrate-binding proteins of plants and animals. In *Adv. Carbohydr. Chem. Biochem.* **35**:127–340.

Goldstein IJ, Hughes RC, Monsigny M *et al.* (1980): What should be called a lectin? In *Nature* **285**:66.

Goochee CF, Monica T (1990): Environmental effects on protein glycosylation. In *Bio Technology* **8**:421–7.

Grant G, Bardocz S, Ewen SWB *et al.* (1995a): Purified *Pseudomonas aeruginosa* PA-I lectin induces gut growth when orally ingested by rats. In *FEMS Immun. Med. Microbiol.* **11**:191–6.

Grant G, Buchan WC, Duguid TJ *et al.* (1995b): Orally administered *Pseudomonas aeruginosa* PA-II lectin alters gut metabolism in rats. In *Proc. Biochem. Soc. Meeting, Aberdeen* (In press).

Grünberg J, Perry R, Hoschutzky H *et al.* (1988): Nonfimbrial blood group *N*-specific adhesin (NFA-3) from *Escherichia coli* 020:KX104:H4, causing systemic infection. In *FEMS Microbiol. Lett.* **56**:241–6.

Guyot G (1908) Über die Bakterielle Hämagglutination. In *Zentralbl. Bakteriol. Abt. I. Orig.* **47**:640–53.

Haataja S, Tikkanen K, Liukkonen J *et al.* (1993): Characterization of a novel bacterial adhesion specificity of *Streptococcus suis* recognizing blood group P receptor oligosaccharides. In *J. Biol. Chem.* **268**:4311–7.

Haataja S, Tikkanen K, Nilsson U *et al.* (1994): Oligosaccharide-receptor interaction of the Gal1α–1–4 Gal binding adhesin of *Streptococcus suis*. In *J. Biol. Chem.* **269**:27466–72.

Hacker J, Hof H, Emody L *et al.* (1986): Influence of cloned *Escherichia coli* hemolysin gene, S-fimbriae and serum resistance on pathogenicity in different animal models. In *Microb. Pathogen.* **1**:533–47.

Hajto T, Hostanska K, Gabius H-J (1989): Modulation potency of the β-galactoside-specific lectin from mistletoe extract (Iscador) on the host defense system *in vivo* in rabbits and patients. In *Cancer Res.* **49**:4803–8.

Häse CC, Finkelstein RA (1990): Comparison of the *Vibrio cholerae* hemagglutinin/protease and the *Pseudomonas aeruginosa* elastase. In *Infect. Immun.* **58**:4011–5.

Häse CC, Finkelstein RA (1991): Cloning and nucleotide sequence of the *Vibrio cholerae* hemagglutinin/protease (HA/Protease) gene and construction of an HA/Protease–negative strain. In *J. Bacteriol.* **173**:3311–7.

Hazlett LD, Moon M, Berk RS (1986): *In vivo* identification of sialic acid as the ocular receptor of *Pseudomonas aeruginosa*. In *Infect. Immun.* **51**:687–9.

Heumann W, Marx R (1964): Feinstruktur und Funktion der Fimbrien bei dem stern-bildenden Bakterium *Pseudomonas echinoides*. In *Arch. Mikrobiol.* **47**:325–37.

Iglewski BH, Sadoff JC, Bjorn MJ *et al.* (1978): *Pseudomonas aeruginosa* exoenzyme S: an adenosine diphosphate ribosyl-transferase distinct from toxin A. In *Proc. Natl. Acad. Sci. USA* **75**:3211–5.

Imundo L, Barasch J, Prince A *et al.* (1995): Cystic fibrosis epithelial cells have a receptor for pathogenic bacteria on their apical surface. In *Proc. Natl. Acad. Sci. USA* **92**:3019–23.

Isberg RR, Tran Van Nhieu G (1994): Binding and internalization of microorganisms by integrin receptors. In *Trends Microbiol.* **2**:10–4.

Johnson JR, Skubitz KM, Nowicki BJ *et al.* (1995): Nonlethal adherence to human neutrophils mediated by Dr antigen-specific adhesins of *Escherichia coli*. In *Infect. Immun.* **63**:309–16.

Källenius G, Mollby R, Svenson SB *et al.* (1980): The P^k antigen as receptor for the hemagglutinin of pyelonephritic *Escherichia coli*. In *FEMS Microbiol. Lett.* **7**:297–302.

Kameyama T, Oishi K, Aida K (1979): Stereochemical structure recognized by the L-fucose-specific hemagglutinin produced by *Streptomyces* sp. In *Biochim. Biophys. Acta* **587**:407–14.

Kameyama T, Oishi K, Aida K (1981): Human erythrocyte receptor of L-fucose-specific lectin produced by *Streptomyces* sp. In *Agric. Biol. Chem.* **45**:975–80.

Karlsson KA (1989): Animal glycosphingolipids as membrane attachment sites for bacteria. In *Annu. Rev. Biochem.* **58**:309–50.

Katada T, Ui M (1982): Direct modification of the membrane adenylate cyclase system by islet-activating protein due to ADP-ribosylation of a membrane protein. In *Proc. Natl. Acad. Sci. USA* **79**:3129–33.

Keogh EV, North EA, Warburton MF (1947): Hemagglutinins of the *Haemophilus* group. In *Nature* **160**:63.

Keusch GT, Donohue-Rolfe A, Jacewicz M (1986): Sugar binding bacterial toxins. In *Microbial Lectins and Agglutinins: Properties and Biological Activity* (Mirelman D, ed) pp 271–95, New York: John Wiley & Sons.

Kijne JW, van der Schaal IAM, Diaz CL *et al.* (1983): Mannose-specific lectin and the recognition of pea roots by *Rhizobium leguminosarum*. In *Lectins: Biology, Biochemistry, Clinical Biochemistry*, Vol. 3 (Bøg-Hansen TC, Spengler GA, eds) pp 521–38, Berlin: Walter de Gruyter.

Ko HL, Beuth J, Soelter J *et al.* (1987): *In vitro* and *in vivo* inhibition of lectin mediated adhesion of *Pseudomonas aeruginosa* by receptor blocking carbohydrates. In *Infection* **15**:237–40.

Kolenbrander PE (1989): Surface recognition among oral bacteria: multigenetic coaggregation and their mediators. In *CRC Crit. Rev. Microbiol.* **17**:137–59.

Komano H, Natori S (1985): Participation of *Sarcophaga peregrina* lectin in the lysis of sheep red blood cells injected into the abdominal cavity of larvae. In *Dev. Comp. Immunol.* **9**:31.

Korhonen TK, Väisänen V, Kallio P (1982): The role of pili in adhesion of *Escherichia coli* to human urinary tract epithelial cells. In *Scand. J. Infect. Dis.* **33** (Suppl.) 26–31.

Kraus R, Ludwig S (1902): Über Bakterienhaemagglutinine und Antihaemagglutinine. In *Wiener. Klin. Wochenschr.* **15**:120–1.

Krivan HC, Ginsburg V, Roberts DD (1988a): *Pseudomonas aeruginosa* and *Pseudomonas cepacia* isolated from cystic fibrosis patients bind specifically to gangliotetraosylceramide (asialo GM_1) and gangliotriaosylceramide (asialo GM_2). In *Arch. Biochem. Biophys.* **260**:-6.

Krivan HC, Roberts DD, Ginsburg V (1988b): Many pulmonary pathogenic bacteria bind specifically to the carbohydrate sequence GalNAcβ1–4Gal found in some glycolipids. In *Proc. Natl. Acad. Sci. USA* **85**:6157–61.

Krüpe M (1956): *Blutgruppenspezifische pflanzliche Eiweißkörper* (Phytagglutinine), Stuttgart: Ferdinand Enke Verlag.

Lakhtin VM (1994): Lectin sorbents in microbiology. In *Lectin-Microorganism Interactions* (Doyle RJ, Slifkin M, eds) pp 249–98, New York: Marcel Dekker, Inc.

Lanne B, Cioprag J, Bergström J *et al.* (1994): Binding of the galactose-specific *Pseudomonas aeruginosa* lectin, PA-I, to glycosphingolipids and other glycoconjugates. In *Glycoconjugate J.* **11**:292–8.

Leffler H, Svanborg-Edén C (1980): Chemical identification of a glycosphingolipid receptor for *Escherichia coli* attaching to human urinary tract epithelial cells and agglutinating human erythrocytes. In *FEMS Microbiol. Lett.* **8**:127–34.

Leffler H, Svanborg-Edén C (1986): Glycolipids as receptors for *Escherichia coli* lectins or adhesins. In *Microbial Lectins and Agglutinins: Properties and Biological Activity* (Mirelman D, ed) pp 83–111, New York: John Wiley & Sons.

Leong JM, Morrissey PE, Marra A *et al.* (1995): An aspartate residue of the *Yersinia pseudotuberculosis* invasin protein that is critical for integrin binding. In *EMBO J.* **14**:422–31.

Levene C, Gilboa-Garber N, Garber NC (1994): Lectin-blood group interactions. In *Lectin – Microorganism Interactions* (Doyle RJ, Slifkin M, eds) pp 327–93, New York: Marcel Dekker, Inc.

Liener IE (1986): Nutritional significance of lectins in the diet. In *The Lectins: Properties, Functions and Applications in Biology and Medicine* (Liener IE, Sharon N, Goldstein IJ, eds) pp 527–52, New York: Academic Press.

Lindberg F, Tennert JM, Hultgren SJ *et al.* (1989): Pap D, a periplasmic transport protein in P-pilus biogenesis. In *J. Bacteriol.* **171**:6052–8.

Lingwood CA, Cheng M, Krivan HC *et al.* (1991): Glycolipid receptor binding specificity of exoenzyme S from *Pseudomonas aeruginosa*. In *Biochem. Biophys. Res. Commun.* **175**:1076–81.

Liu PV (1973): Exotoxins of *Pseudomonas aeruginosa*. I. Factors that influence the production of exotoxin A. In *J. Infect. Dis.* **128**:506–13.

Locht C, Antoine R (1995): A proposed mechanism of ADP-ribosylation catalyzed by the pertussis toxin S1 subunit. In *Biochimie* **77**:333–40.

Mäkelä O (1957): Studies in haemagglutinins of Leguminosae seeds. In *Ann. Med. Exp. Biol. Fenn.* **35** (Suppl 11):1–133.

Mangan DF, Snyder IS (1979): Mannose-sensitive interaction of *Escherichia coli* with human peripheral leukocytes *in vitro*. In *Infect. Immun.* **26**:520–7.

Markwell MAK (1986): Viruses as hemagglutinins and lectins. In *Microbial Lectins and Agglutinins: Properties and Biological Activity* (Mirelman D, ed) pp 21–53, New York: John Wiley & Sons.

Matsui I, Oishi K (1986): Carbohydrate-binding properties of L-fucose, D-mannose-specific lectin (SFL 100–2) particles produced by *Streptomyces* sp. In *J. Biochem.* **100**:115–21.

Matsui I, Oishi K, Kanaya K *et al.* (1985): The morphology of L-fucose, D-mannose specific lectin (SFL 100–2) produced by *Streptomyces* no. 100–2. In *J. Biochem.* **97**:399–408.

McEachran DW, Irvin RT (1985): Adhesion of *Pseudomonas aeruginosa* to human buccal epithelial cells: evidence for two classes of receptors. In *Can. J. Microbiol.* **31**:563–9.

McIntire FC, Crosby LK, Barlow JJ *et al.* (1983): Structural preferences of β-galactoside reactive lectins on *Actinomyces viscosus* T14V and *Actinomyces naeslundii* WV445. In *Infect. Immun.* **41**:848–50.

Menestrina G, Pederzolli C, Forti S *et al.* (1991): Lipid interaction of *Pseudomonas aeruginosa* exotoxin A. In *Biophys. J.* **60**:1388–400.

Messner P, Sleytr UB (1988): Asparaginyl-rhamnose: a novel type of protein-carbohydrate linkage in a eubacterial surface-layer glycoprotein. In *FEBS Lett.* **228**:317–20.

Middlebrook JL, Dorland RB (1984): Bacterial toxins: cellular mechanisms of action. In *Microbiol. Rev.* **48**:199–221.

Mirelman D (ed) (1986): Microbial Lectins and Agglutinins: Properties and Biological Activity. New York: John Wiley & Sons.

Moss J, Vaughan M (1990): ADP-ribosylating Toxins and G Proteins. Washington D.C.: American Society for Microbiology Press.

Neter E (1956): Bacterial hemagglutination and hemolysis. In *Bacteriol. Rev.* **20**:166–88.

Normark S, Lark D, Hull R *et al.* (1983): Genetics of digalactoside-binding adhesin from a uropathogenic *Escherichia coli* strain. In *Infect. Immun.* **41**:942–9.

Normark S, Baga M, Goransson M *et al.* (1986): Genetics and biogenesis of *Escherichia coli* adhesins. In *Microbial Lectins and Agglutinins: Properties and Biological Activity* (Mirelman D, ed) pp 113–43, New York: John Wiley & Sons.

Nowell PC (1960): Phytohemagglutinin: An initiator of mitosis in cultures of normal human leukocytes. In *Cancer Res.* **20**:462–6.

Nowicki B, Rhen M, Väisänen-Rhen *et al.* (1985): Kinetics of phase variation between S and type-1 fimbriae of *Escherichia coli*. In *FEMS Microbiol. let.* **28**:237–42.

Nowicki B, Moulds J, Hull R *et al.* (1988a): A hemagglutinin of uropathogenic *Escherichia coli* recognizes the Dr blood group antigen. In *Infect. Immun.* **56**:1057–60.

Nowicki B, Truong L, Moulds J *et al.* (1988b): Presence of the Dr receptor in normal human tissues and its possible role in the pathogenesis of ascending urinary tract infection. In *Am. J. Pathol.* **133**:1–4.

Nowicki B, Labigne A, Moseley S *et al.* (1990): The Dr hemagglutinin, afimbrial adhesins AFA-I and AFA-III, and F1845 fimbriae of uropathogenic and diarrhea-associated *Escherichia coli* belong to a family of hemagglutinins with Dr receptor recognition. In *Infect. Immun.* **58**:279–81.

Olsnes S, Sandvig K (1988): How protein toxins enter and kill cells. In *Immunotoxins* (Fraenkel AE, ed) pp 39–73, New York: Kluwer Academic Publishing.

Pal R, Ahmed H, Chatterjee BP (1987): *Pseudomonas aeruginosa* bacteria Habs type 2a contains two lectins of different specificity. In *Biochem. Arch.* **3**:399–412.

Parkkinen J, Rogers GN, Korhonen T *et al.* (1986): Identification of the *O*-linked sialyloligosaccharides of glycophorin A as the erythrocyte receptors for S-fimbriated *Escherichia coli*. In *Infect. Immun.* **54**:37–42.

Pearson JP, Gray KM, Passador L *et al.* (1994): Structure of autoinducer required for expression of *Pseudomonas aeruginosa* virulence genes. In *Proc. Natl. Acad. Sci. USA* **91**:197–201.

Pearson JP, Passador L, Iglewski BH *et al.* (1995): A second *N*-acylhomoserine lactone signal produced by *Pseudomonas aeruginosa*. In *Proc. Natl. Acad. Sci. USA* **92**:1490–4.

Peterson AM, Barillas-Mury CV, Wells MA (1994): Sequence of three cDNAs encoding an alkaline midgut trypsin from *Manduca sexta*. In *Insect. Biochem. Mol. Biol.* **24**:463–71.

Piatti G (1994): Bacterial adhesion to respiratory mucosa and its modulation by antibiotics at sub-inhibitory concentrations. In *Pharmacol. Res.* **30**:289–99.

Plotkowski MC, Puchelle E, Beck G *et al.* (1986): Adherence of type 1 *Streptococcus pneumoniae* to tracheal epithelium of mice infected with influenza A/PR8 virus. In *Am. Rev. Respir. Dis.* **134**:1040–4.

Plotkowski MC, Beck G, Tournier JM *et al.* (1989): Adherence of *Pseudomonas aeruginosa* to respiratory epithelium and the effect of leucocyte elastase. In *J. Med. Microbiol.* **30**:285–93.

Plotkowski MC, Chevillard M, Pierrot D *et al.* (1991): Differential adhesion of *Pseudomonas aeruginosa* to human respiratory epithelial cells in primary culture. In *J. Clin. Invest.* **87**:2018–28.

Plotkowski MC, Saliba AM, Pereira SHM *et al.* (1994): *Pseudomonas aeruginosa* selective adherence to and entry into human endothelial cells. In *Infect. Immun.* **62**:5456–63.

Ponniah S, Abraham SN, Dockter ME *et al.* (1989): Mitogenic stimulation of human B lymphocytes by the mannose-specific adhesin on *Escherichia coli* type 1 fimbriae. In *J. Immunol.* **142**:992–8.

Ponniah S, Abraham SN, Endres RO (1992): T-cell-independent stimulation of immunoglobulin secretion in resting human B lymphocytes by the mannose-specific adhesin of *Escherichia coli* type 1 fimbriae. In *Infect. Immun.* **60**:5197–203.

Pusztai A, Ewen SWB, Grant G *et al.* (1995): Lectins and also bacteria modify the glycosylation of gut surface receptors in the rat. In *Glycoconjugate J.* **12**:22–35.

Ramphal R, Pyle M (1983): Evidence for mucins and sialic acid as receptors for *Pseudomonas aeruginosa* in the lower respiratory tract. In *Infect. Immun.* **41**:339–44.

Ramphal R, Pier GB (1985): Role of *Pseudomonas aeruginosa* mucoidexopolysaccharide in adherence to tracheal cells. In *Infect. Immun.* **47**:1–4.

Ramphal R, Small PA, Shands JW Jr *et al.* (1980): Adherence of *Pseudomonas aeruginosa* to tracheal cells injured by influenza virus or by endotracheal intubation. In *Infect. Immun.* **27**:614–9.
Ramphal R, Houdret N, Koo L *et al.* (1989): Differences in adhesion of *Pseudomonas aeruginosa* to mucin glycopeptides from sputa of patients with cystic fibrosis and chronic bronchitis. In *Infect. Immun.* **57**:3066–71.
Ramphal R, Carnoy C, Fievre S *et al.* (1991a): *Pseudomonas aeruginosa* recognizes carbohydrate chains containing type 1 (Galβ1–3GlcNAc) or type 2 (Galβ1–4GlcNAc) disaccharide units. In *Infect. Immun.* **59**:700–4.
Ramphal R, Koo L, Ishimoto KS *et al.* (1991b): Adhesion of *Pseudomonas aeruginosa* pilin-deficient mutants to mucin. In *Infect. Immun.* **59**:1307–11.
Relman DA, Domenighini M, Tuomanen E *et al.* (1989): Filamentous hemagglutinin of *Bordetella pertussis*: nucleotide sequence and crucial role in adherence. In *Proc. Natl. Acad. Sci. USA* **86**:2637–41.
Rhen M, Mäkelä PH, Korhonen TK (1983): P-fimbriae of *Escherichia coli* are subject to phase variation. In *FEMS Microbiol. Lett.* **19**:267–71.
Romeo JM, Esmon B, Zusman DR (1986): Nucleotide sequence of the mycobacterial hemagglutinin gene contains four homologous domains. In *Proc. Natl. Acad. Sci. USA* **83**:6332–6.
Rosenthal L (1931): Spermagglutination by bacteria. In *Proc. Soc. Exp. Biol. Med.* **28**:827–8.
Rosenthal L (1943): Agglutinating properties of *Escherichia coli*: agglutination of erythrocytes, leucocytes, thrombocytes, spermatozoa, spores of molds and pollen by strains of *E. coli*. In *J. Bacteriol.* **45**:545–50.
Rostand KS, Esko JD (1993): Cholesterol and cholesterol esters: host receptors for *Pseudomonas aeruginosa* adherence. In *J. Biol. Chem.* **268**:24053–9.
Rozdzinski E, Burnette WN, Jones T (1993): Prokaryotic peptides that block leukocyte adherence to selectins. In *J. Exp. Med.* **178**:917–24.
Sandros J, Rozdzinski E, Zheng J *et al.* (1994): Lectin domains in the toxin of *Bordetella pertussis*: selectin mimicry linked to microbial pathogenesis. In *Glycoconjugate J.* **11**:501–6.
Seddon SV, Krishna M, Davies HA *et al.* (1991): Effect of nutrition on the expression of known and putative virulence factors of *Clostridium difficile*. In *Microb. Ecol. Health Dis.* **4**:303–9.
Seignole D, Grange P, Duval-Iflah Y *et al.* (1994): Characterization of D-glycan moieties of the 210 and 240 kDa pig intestinal receptors for *Escherichia coli* K88ac fimbriae. In *Microbiology* **140**:2467–73.
Sexton N, Reen DJ (1992): Characterization of antibody-mediated inhibition of *Pseudomonas aeruginosa* adhesion to epithelial cells. In *Infect. Immun.* **60**:3332–8.
Shakhanina KL, Kalinin NL, Lakhtin VM (1994): Microbial lectins for the investigation of glycoconjugates. In *Lectin – Microorganism Interactions* (Doyle RJ, Slikfin M, eds) pp 299–326, New York: Marcel Dekker, Inc.
Sharabi Y, Gilboa-Garber N (1979): Mitogenic stimulation of human lymphocytes by *Pseudomonas aeruginosa* galactophilic lectins. In *FEMS Microbiol. Lett.* **5**:273–6.
Sharon N (1977): Lectins. In *Sci. Am.* **236**:108–19.
Sharon N (1986): Bacterial lectins. In *The Lectins: Properties, Functions and Applications in Biology and Medicine* (Liener IE, Sharon N, Goldstein IJ, eds) pp 493–526, New York: Academic Press.
Sheffi M, Shalom S, Garber N *et al.* (1989): Heating of *Pseudomonas aeruginosa* cells leads to lectin externalization (a salvage “in-out” phase variation). In *Isr. J. Med. Sci.* **25**:595.
Sheth HB, Lee KK, Wong WY *et al.* (1994): The pili of *Pseudomonas aeruginosa* strains PAK and PAO bind specifically to the carbohydrate sequence βGalNAc (1–4)βGal found in glycosphingolipids asialo-GM_1 and asialo-GM2. In *Mol. Microbiol.* **11**:715–23.
Shibl AM, Al-Sowaygh IA (1980): Antibiotic inhibition of protease production by *Pseudomonas aeruginosa*. In *J. Med. Microbiol.* **13**:345–9.
Silverblatt FJ, Dreyer JS, Schauer S (1979): Effect of pili on susceptibility of *Escherichia coli* to phagocytosis. In *Infect. Immun.* **24**:218–23.
Smit G, Swart S, Lugtenberg BJJ *et al.* (1992): Molecular mechanisms of attachment of *Rhizobium* bacteria to plant roots. In *Mol. Microbiol.* **6**:2897–903.
Speert DP, Loh BA, Cabral DA (1986): Nonopsonic phagocytosis of nonmucoid *Pseudomonas aeruginosa* by human neutrophils and monocyte-derived macrophages is correlated with bacterial piliation and hydrophobicity. In *Infect. Immun.* **53**:207–12.
Speert DP, Eftekhar F, Puterman ML (1989): Nonopsonic phagocytosis of strains of *Pseudomonas aeruginosa* from cystic fibrosis patients. In *Infect. Immun.* **43**:1006–11.
Sudakevitz D, Gilboa-Garber N (1982): Effect of *Pseudomonas aeruginosa* lectins on phagocytosis of *Escherichia coli* strains by human polymorphonuclear leukocytes. In *Microbios* **34**:159–66.
Sudakevitz D, Gilboa-Garber N, Mizrahi L (1979): Regulation of lectin production in *Pseudomonas aeruginosa* by culture medium composition. In *Isr. J. Med. Sci.* **15**:97.
Sudakevitz D, Gilboa-Garber N, Sofer D (1992): Effect of cations on *Pseudomonas aeruginosa* lectin activity and growth. In *Proc. 21st FEBS Meeting*, Dublin, p 165.
Sudakevitz D, Levene C, Gilboa-Garber N (1995): The galactophilic lectins of *Aplysia* and *Pseudomonas aeruginosa* differentiate between I-positive and I-

negative human erythrocytes. *In Lectins: Biology, Biochemistry, Clinical Biochemistry*, Vol. 11 (In Press).

Svensson M, Lindstedt R, Radin NS *et al.* (1994): Epithelial glucosphingolipid expression as a determinant of bacterial adherence and cytokine production. In *Infect. Immun.* **62**:4404–10.

Takemoto T, Hino T, Yoshida M *et al.* (1995): Characteristics of multimodal co-aggregation between *Fusobacterium nucleatum* and *Streptococci*. In *J. Periodont. Res.* **30**:252–7.

Tewari R, Ikeda T, Malaviya R *et al.* (1994): The PapG tip adhesin of P fimbriae protects *Escherichia coli* from neutrophil bactericidal activity. In *Infect. Immun.* **62**:5296–304.

Tsuchimori N, Hayashi R, Shino A *et al.* (1994): *Enterococcus faecalis* aggravates pyelonephritis caused by *Pseudomonas aeruginosa* in experimental ascending mixed urinary tract infection. In *Infect. Immun.* **62**:4534–41.

Uhlenbruck G (1987): Bacterial lectins: mediators of adhesion. In *Zentralbl. Bakteriol. Hyg.* [A] **263**:497–508.

van Alphen L, Poole J, Overbeeke M (1986): Combined inheritance of epithelial and erythrocyte receptors for *Haemophilus influenzae*. In *FEMS Microbiol. Lett.* **37**:69–71.

van Alphen L, Levene C, Geelen-van den Broek L *et al.* (1990): The Anton blood group antigen is the erythrocyte receptor for *Haemophilus influenzae*. In *Infect. Immun.* **58**:3807–9.

Väisänen V, Elo J, Tallgren LG *et al.* (1981): Mannose-resistant haemagglutination and P antigen recognition are characteristic of *Escherichia coli* causing primary pyelonephritis. In *Lancet* **2**:1366–9.

Väisänen V, Korhonen TK, Jokinen M *et al.* (1982): Blood group M specific haemagglutinin in pyelonephritogenic *Escherichia coli*. In *Lancet* **2**:1192.

Virji M, Saunders JR, Sims G *et al.* (1993): Pilus facilitated adherence of *Neisseria meningitidis* to human epithelial and endothelial cells: modulation of adherence phenotype occurs concurrently with changes in primary amino acid sequence and the glycosylation status of pilin. In *Mol. Microbiol.* **10**:1013–28.

Warburton MF, Fisher S (1951): Haemagglutinin of *Haemophilus pertussis*; extraction of antigen from bacteria and its stabilization by adsorption. In *Aust. J. Exper. Biol. Med. Sci.* **29**:265–72.

Weigel PH (1992): Mechanisms and control of glycoconjugate turnover. In *Glycoconjugates: Composition, Structure and Function* (Allen HJ, Kisailus EC, eds) pp 421–97, New York: Marcel Dekker.

Wentworth JS, Austin FE, Garber N *et al.* (1991): Cytoplasmic lectins contribute to the adhesion of *Pseudomonas aeruginosa*. In *Biofouling* **4**:99–104.

Widdicombe JH, Wine JJ (1991): The basic defect in cystic fibrosis. In *Trends Biochem. Sci.* **16**:474–7.

Wiley DC, Skehel JJ (1987): The structure and function of the hemagglutinin membrane glycoprotein of influenza virus. In *Annu. Rev. Biochem.* **56**:365–74.

Williams RC, Gibbons RJ (1975): Inhibition of streptococcal attachment to receptors on human buccal epithelial cells by antigenically similar salivary glycoproteins. In *Infect. Immun.* **11**:711–8.

Woods DE (1987): Role of fibronectin in the pathogenesis of Gram negative bacillary pneumonia. In *Rev. Infect. Dis.* **9** (Suppl):386–90.

Woods DE, Bass JA, Johanson WG Jr *et al.* (1980a): Role of adherenin the pathogenesis of *Pseudomonas aeruginosa* lung infection in cystic fibrosis patients. In *Infect. Immun.* **30**:694–9.

Woods DE, Straus DC, Johanson WG Jr *et al.* (1980b): Role of pili in adherence of *Pseudomonas aeruginosa* to mammalian buccal epithelial cells. In *Infect. Immun.* **29**:1146–51.

Woods DE, Straus DC, Johanson WG Jr *et al.* (1981a): Role of fibronectin in the prevention of adherence of *Pseudomonas aeruginosa* to buccal cells. In *J. Infect. Dis.* **143**:784–90.

Woods DE, Straus DC, Johanson WG Jr *et al.* (1981b): Role of salivary protease activity in adherence of gram-negative bacilli to mammalian buccal epithelial cells *in vivo*. In *J. Clin. Invest.* **68**:1435–40.

Woods DE, Straus DC, Johanson WG Jr *et al.* (1983): Factors influencing the adherence of *Pseudomonas aeruginosa* to mammalian buccal epithelial cells. In *Rev. Infect. Dis.* **5** (Suppl. 5):S847–51.

Yahara I, Edelman GM (1973): The effects of concanavalin A on the mobility of lymphocyte surface receptors. In *Exp. Cell Res.* **81**:143–55.

Yokota Y, Arai T, Kawasaki T (1995): Oligomeric structures required for complement activation of serum mannan-binding proteins. In *J. Biochem.* **117**:414–19.

Young NM, Oomen RP (1992): Analysis of sequence variation among legume lectins. A ring of hypervariable residues forms the perimeter of the carbohydrate-binding site. In *J. Mol. Biol.* **228**:924–34.

Yu L, Lee KK, Hodges RS *et al.* (1994): Adherence of *Pseudomonas aeruginosa* and *Candida albicans* to glycosphingolipid (asialo-GM_1) receptors. In *Infect. Immun.* **62**:5213–9.

Zhao Z, Panjwani N (1995): *Pseudomonas aeruginosa* infection of the cornea and asialo GM_1. In *Infect. Immun.* **63**:353–5.

Zusman DR, Cumsky MG, Nelson DR *et al.* (1986): Myxobacterial hemagglutinin: a developmentally induced lectin from *Myxococcus xanthus*. In *Microbial Lectins and Agglutinins: Properties and Biological Activity* (Mirelman D, ed) pp 197–216, New York: John Wiley & Sons.

22 Glycobiology of Parasites: Role of Carbohydrate-Binding Proteins and Their Ligands in the Host-Parasite Interaction

HONORINE D. WARD

1 Introduction

Recent advances in the rapidly expanding field of glycobiology have extended to specific interactions between parasite and host and have greatly contributed to our understanding of underlying pathophysiological mechanisms. One of the hallmarks of the host-parasite interaction is specificity, which may be manifested at the level of host species, tissue site, cell type or receptor (Pereira, 1986). Host specificity is exemplified by *Plasmodium falciparum*, a causative agent of malaria, which infects only humans. *Giardia lamblia*, which causes diarrheal disease, in humans exhibits tissue specificity by selectively colonizing the proximal small intestine. *Leishmania* promastigotes, the causative agent of kala-azar and oriental sore, display specificity for cell type by exclusively infecting macrophages. While infecting a number of different cell types *Trypanosoma cruzi*, which causes Chagas' disease is believed to interact with specific receptors common to these cells.

This selectivity exhibited by parasites is to a large extent mediated by specific parasite molecules and complementary ligands on host cells. The specificity of carbohydrates, as well as the presence of proteins that bind to (lectins) and that modulate (glycosyl transferases and glycosidases) them, confers these molecules with recognition and discrimination properties which enable them to contribute to the selectivity of the host cell-parasite interaction.

As in mammalian cell-cell interactions, the host cell-parasite interaction is a complex, stepwise process, involving multiple receptors and complementary ligands on both cell types. A primary event is attachment of the organism to host

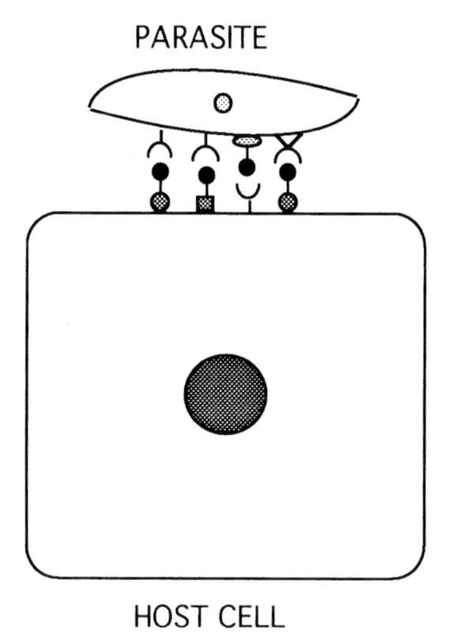

Figure 1. Attachment of parasites to host cells is mediated by carbohydrate-binding proteins and complementary ligands on both cell types.

cells. This may then lead to colonization as occurs with *Giardia lamblia*-enterocyte interaction, lysis of the host cell which is characteristic of *Entameba histolytica* infection or invasion of the host cell as is the case with *Plasmodium falciparum*. In addition, parasites may encounter and interact

H.-J. and S. Gabius (Eds.), Glycosciences

ISBN 3-8261-0073-5

Table 1. Carbohydrate-binding proteins of pathogenic protozoa

Parasite	Stage	Carbohydrate-binding protein	Carbohydrate specificity	Role in host-parasite interaction
Plasmodium falciparum	merozoite	EBA-175	Neu5Ac, α2.3	merozoite-erythrocyte adhesion and invasion
		Pf200/MSA-1	Neu5Ac	merozoite-erythrocyte adhesion and invasion
	sporozoite	circumsporozoite protein	heparan sulphate	sporozoite-hepatocyte adhesion and invasion
	parasitized erythrocyte	?	sulphated glycoconjugates, blood group specific oligosaccharides	rosetting of parasitized and uninfected erythrocytes
		?	chondroitin sulphate	parasitized erythrocyte-endothelial cell adhesion
Trypanosoma cruzi	trypo-mastigote	transialidase	Neu5Ac, α2.3	trypomastigote-host cell adhesion, invasion
		penetrin	heparin, heparan sulphate	trypomastigote-host cell adhesion, invasion trypomastigote-matrix interaction
Leishmania	pro-mastigote	Gene B protein	lipophosphoglycan	promastigote-macrophage adhesion
		?	heparin	promastigote-macrophage adhesion
	amastigote	?	heparin	amastigote-macrophage adhesion
Entamoeba histolytica	trophozoite	Gal/GalNAc-specific lectin	Gal/GalNAc	trophozoite-colonocyte adhesion, contact mediated lysis
		220 kDa lectin	chitotriose	trophozoite-colonocyte adhesion
Giardia lamblia	trophozoite	taglin	Man-6-P	trophozoite-enterocyte adhesion
Cryptospori-dium parvum	sporozoite	Gal/GalNAc-specific lectin	Gal/GalNAc	sporozoite-enterocyte adhesion

with carbohydrate components of the host extracellular matrix as occurs in infection with *Trypanosoma cruzi*. A number of these cell-cell and cell-matrix interactions are believed to be mediated by carbohydrate-binding proteins (CBPs) on the parasite and their specific carbohydrate ligands on the host cell (Fig. 1). These CBPs may be lectins or carbohydrate-binding enzymes such as the transialidase of *T. cruzi*. Host cell ligands may include glycoproteins, glycolipids or proteoglycans. In addition, parasites themselves may express glycoconjugates which mediate adhesion by binding to host lectins (Fig. 1). Examples of CBPs of pathogenic protozoa are shown in Table 1. This chapter will focus on the role of CBPs and their ligands in mediating intercellular recognition and adhesion functions between parasite and host. Although many of the CBPs of these parasites may not be lectins in the classical sense they are included since their role in attachment is mediated via their carbohydrate binding properties.

2 Malaria

Malaria, the disease caused by *Plasmodium* species is estimated to affect up to 300 million people worldwide, and to result in the deaths of up to 2 million of these, mainly children, annually (Greenwood *et al.*, 1987; Sturchler, 1989). Four species of the parasite infect humans, *Plasmodium falciparum*, *Plasmodium vivax, Plasmodium ovale* and *Plasmodium malariae*. Of these

Plasmodium falciparum is the most virulent, causing cerebral malaria which is responsible for much of the mortality. The parasite has a complex life cycle and exists in a number of asexual and sexual developmental stages in the human host as well as in the mosquito vector. Attachment to and invasion of hepatocytes by the infective sporozoite form of the parasite, a primary event in initiating infection occurs within minutes after the bite of infected mosquitoes. Another crucial event in the pathogenesis of *Plasmodium falciparum* malaria is invasion of the red cell membrane by the merozoite form of the parasite. Parasitized erythrocytes adhere to the endothelium of post-capillary venules or to uninfected erythrocytes forming rosettes and both of these events are believed to contribute to the pathogenesis of cerebral malaria (Pasloske and Howard, 1994). CBPs have been implicated in mediating interactions between sporozoites and hepatocytes as well as between merozoite and erythrocytes, parasitized erythrocytes and endothelium and parasitized and uninfected erythrocytes.

2.1 Sialic Acid and Sialic Acid Binding Proteins

A number of lines of evidence suggest that sialic acid on erythrocytes is involved in the adherence and invasion of erythrocytes by *Plasmodium falciparum* (Hadley *et al.*, 1986). Thus, sialidase treatment of human erythrocytes leads to a marked reduction of invasion. In addition, human erythrocytes lacking the major sialoglycoproteins, glycophorin A (Ena-) and glycophorin A and B (MkMk) as well as Tn erythrocytes which lack sialic acid on the *O*-linked tetrasaccharides of glycophorins are resistant to invasion (Camus and Hadley, 1985; Hadley *et al.*, 1987).

One of the parasite molecules believed to mediate merozoite invasion of erythrocytes is a sialic acid-binding protein named EBA-175 (Camus and Hadley, 1985). This protein is released into the culture supernatant of schizont-infected erythrocytes and binds to susceptible uninfected erythrocytes. Binding of EBA-175 also occurs to merozoites, leading to the hypothesis that it might function as a "bridge" between merozoite and erythrocytes. Sialidase treatment of human erythrocytes reduced invasion and abolished binding of EBA-175, suggesting that sialic acid residues were involved. In addition, binding of EBA-175 did not occur to invasion-resistant erythrocytes. Sialic acid did not inhibit binding whereas soluble glycophorin A did (Orlandi *et al.*, 1992). Binding was specific for α2–3 linkage of Neu5Ac since it could be inhibited by oligosaccharides containing Neu5Ac (α2–3)Gal, but not Neu5Ac(α2–6)Gal (Orlandi *et al.*, 1992). Selective cleavage of *O*-linked tetrasaccharides of glycophorin A which contain these Neu5Ac(α2–3)Gal residues resulted in greatly decreased binding of EBA-175 (Orlandi *et al.*, 1992). Binding of EBA-175 to mouse erythrocytes was shown to require *N*-acetylneuraminic acid but not its 9-*O*-acetylated form (Klotz *et al.*, 1992).

EBA-175 appears to be synthesized initially as a 190 kDa protein in schizonts which is processed to a 175 kDa form (Orlandi *et al.*, 1990). Elegant studies using COS cells expressing truncated portions of the protein have identified the erythrocyte binding domain of EBA-175 as being on region II of the molecule (Sim *et al.*, 1994). The role that this sialic acid binding protein, EBA-175 plays in mediating invasion of red cells by merozoites, together with the finding that it is highly conserved have led to the proposal that it may serve as a useful vaccine candidate (Sim *et al.*, 1990, 1994).

A major merozoite surface antigen of *Plasmodium falciparum*, namely Pf200 or MSA-1 has also been shown to mediate sialic acid dependent adhesion (Perkins and Rocco, 1988). Binding of this protein to erythrocytes was also abrogated by neuraminidase treatment, soluble glycophorin as well as a monoclonal antibody directed against the carbohydrate portion of the molecule (Perkins and Rocco, 1988). Pf200/MSA-1 does not appear to be immunologically related to EBA-175 or to share sequence homology (Sim *et al.*, 1990; Barnwell and Galinski, 1991).

2.2 Glycosaminoglycan-Binding proteins

A number of studies have suggested a role for cell surface glycosaminoglycans such as heparin and chondroitin sulphate in mediating various *Plasmodium*-host cell interactions. Perhaps the most striking of these is the interaction of the sporozoite form with hepatocytes, which has been shown to be mediated by binding of a major surface protein, the circumsporozoite protein to heparan sulphate proteoglycans on the basolateral membranes of these cells (Cerami *et al.*, 1992; Frevert *et al.*, 1993). Malarial sporozoites as well as circumsporozoite protein have been shown to bind to sulfated glycoconjugates such as sulfatide, heparin, fucoidan and dextran sulphate (Pancake *et al.*, 1992). In addition, Chinese hamster ovary (CHO) cell mutants defective in glycosaminoglycan (GAG) synthesis bind sporozoites as well as circumsporozoite protein to a lesser degree than the parental cells (Pancake *et al.*, 1992). The circumsporozoite protein was shown to bind to the basolateral membrane of hepatocytes via region II plus, a highly conserved motif of the molecule (Cerami *et al.*, 1992). Binding of this motif of the protein was shown to occur to hepatocyte surface membrane heparan sulphate proteoglycans (Frevert *et al.*, 1993). Thus, heparitinase treatment of human liver sections abrogated binding of the protein to hepatocytes. Circumsporozoite protein also bound specifically to HepG2 cells and this binding was inhibited by heparin, heparan sulphate, dextran sulphate and fucoidan as well as by heparitinase treatment. In addition, recombinant circumsporozoite protein containing region II plus bound specifically to biosynthetically labeled proteoglycans of rat liver and binding was abolished by soluble heparin as well as by heparitinase treatment.

Rosette formation between parasitized and uninfected erythrocytes is believed to occur via a lectin-type of interaction with blood group specific oligosaccharides and/or GAGs on red cells (Carlson and Wahlgren, 1992; Rogerson *et al.*, 1994; Rowe *et al.*, 1994). Rosetting has been shown to be inhibited by sulphated glycoconjugates including sulfatide, heparin, fucoidan and dextran sulphate (Rowe *et al.*, 1994).

Recently, cell surface chondroitin sulphate A has been implicated as being involved in adherence of parasitized erythrocytes to host cells (Rogerson *et al.*, 1995). This was shown by the finding that binding of a selected line of parasitized erythrocytes to CHO cell mutants expressing excess chondroitin sulphate A was increased when compared to that of the parent cells, whereas they failed to bind to mutants deficient in this GAG. Binding to parental CHO cells could be inhibited by low concentrations of chondroitin sulphate A, but not by other GAGs. Treatment of the CHO cells with chondroitinase markedly reduced binding. Binding of parasitized erythrocytes also occurred to immobilized chondroitin sulphate A but not to other GAGS, and this binding could be inhibited by the soluble GAG. These findings suggested that a parasite-derived protein on the surface of infected erythrocytes is involved in binding to host cell chondroitin sulphate A. However the putative protein has not been identified.

2.3 GlcNAc-Binding Proteins

GlcNAc-binding proteins have also been implicated in mediating red cell adhesion and invasion. Earlier studies showed inhibition of invasion by GlcNAc, but subsequent studies suggested that this may have been due to toxicity of the sugar for the parasitized erythrocyte (Hadley *et al.*, 1986). More recently, three merozoite surface GlcNAc-binding proteins have been identified, and have been implicated in binding of merozoites to red cells (El Moudni *et al.*, 1993).

3 Chagas' Disease

Trypanosoma cruzi infects millions of people in Latin America, causing Chagas' disease which is characterized by multisystem involvement, predominantly of the cardiovascular and gastrointestinal systems (Brener, 1973). The epimastigote stage of the parasite replicates extracellularly in the midgut of the insect vector and is transformed to the non-multiplying trypomastigote which transmits the disease. In the human host trypomastigotes migrate from the skin through the extracellular matrix, via the bloodstream to target organs where they invade host cells and are transformed to the intracellular amastigote form. CBPs of the parasite have been implicated in the interaction of trypomastigotes with host cells as well as with the extracellular matrix.

3.1 Sialic Acid and Sialic Acid-Binding Proteins

As in the case of *Plasmodium falciparum*, sialic acid residues on the host cell are believed to function in mediating trypomastigote attachment to and invasion of host cells. The *T. cruzi* molecule which is implicated in sialic acid-dependent adhesion to host cells is transialidase, a unique enzyme which catalyses the transfer of α2,3-linked host-derived sialic to parasite acceptor molecules (Schenkman *et al.*, 1994). This enzyme was initially described as a developmentally regulated sialidase activity with low levels in the epimastigote stage, higher levels in the trypomastigote and none in amastigotes (Pereira, 1983). Subsequently, in addition to sialidase activity the enzyme was also shown to catalyze the transfer of sialic acid to terminal β Gal residues (Schenkman *et al.*, 1991, 1992 a, b) Since its discovery this enzyme has been extensively studied and reviewed (Colli, 1993; Cross and Takle, 1993; Schenkman *et al.*, 1994). It is localized on the parasite surface from where it is shed into the external medium (Prioli *et al.*, 1991). The protein is polymorphic, ranging from 120–220 kDa in molecular weight (Prioli *et al.*, 1990) and is anchored to the parasite surface by a glycophosphatidylinositol anchor (Rosenberg *et al.*, 1991; Schenkman *et al.*, 1992 a, b).

Transialidase is believed to mediate trypomastigote attachment and invasion by binding to host cell sialic acid in a lectin-type of interaction (Ming *et al.*, 1993). This was shown using a series of CHO cell glycosylation mutants defective in the synthesis of sialic acid (Ciavaglia *et al.*, 1993; Ming *et al.*, 1993; Schenkman *et al.*, 1993). Thus, attachment to and infection of these mutants was reduced compared to the parental K1 cell (Ciavaglia *et al.*, 1993; Ming *et al.*, 1993; Schenkman *et al.*, 1993). Resialylation of the Lec-2 mutant by the transialidase restored attachment and invasion to levels comparable to that of the parent, whereas desialylation of the parental cells with exogenous sialidase reduced adhesion and infection (Ciavaglia *et al.*, 1993; Ming *et al.*, 1993). In addition, affinity-purified enzyme inhibited attachment and invasion of sialylated cells, but not of the sialic acid deficient mutant (Ming *et al.*, 1993). The specificity of α2,3-linkage of the sialic acid was shown by the finding that adhesion and invasion were blocked by α2,3-linked sialyllactose but not its α2,6-linked counterpart (Vandekerckhove *et al.*, 1992; Ferrero-Garcia *et al.*, 1993; Scudder *et al.*, 1993). Transialidase is also believed to mediate attachment and invasion by sialylating parasite glycoproteins expressing the Ssp-3 epitope which has been shown to be involved in these processes (Schenkman *et al.*, 1991, 1992 a, b). In addition, since antibodies to the enzyme enhance infectivity it is believed to modulate infection by exerting a negative control (Cavallesco and Pereira, 1988; Prioli *et al.*, 1990). The enzyme is also postulated to be involved in exiting of trypomastigotes from phagosomes as well as from infected cells (Schenkman *et al.*, 1994). Recently soluble transialidase has been shown to enhance virulence in a murine model of Chagas' disease (Chuenkova and Pereira, 1995).

3.2 Heparin-Binding Protein

T. cruzi trypomastigotes express a novel heparin-binding protein named penetrin which mediates attachment of the parasite to GAGs on the host cell as well as to extracellular matrix components (Ortega-Barria and Pereira, 1991). This was initially shown by specific binding of the parasite to immobilized heparin, heparan sulphate and collagen, but not to other GAGs or glycoproteins. Low concentrations of soluble heparin and heparan sulphate inhibited trypomastigote adherence and prevented infection of host cells. The protein purifed by heparin-affinity chromatography has a molecular weight of 60 kDa, is located on the surface of the parasite and promotes adhesion and spreading of cells to the substratum. Binding of purified native as well as recombinant penetrin to host cells could be inhibited by heparin and heparan sulphate. The role of penetrin in mediating invasion of host cells was elegantly demonstrated by studies showing that expression of recombinant penetrin in *E. coli* conferred these normally non-invasive bacteria with the ability to invade non-phagocytic cells. Mediation of trypomastigote invasion of host cells by penetrin was further shown by studies utilizing a series of proteoglycan-deficient CHO cell mutants (Herrera *et al.*, 1994). Thus, adhesion to and invasion of the mutant cell lines was considerably decreased compared to that in the parental cells. Inhibition of GAG synthesis in the parent cells by p-nitrophenyl β-D-glycosides also resulted in abrogation of attachment and invasion. In addition, depolymerization of heparin and heparan sulphate on the parent cells by the lyases, heparinase and heparitinase also inhibited interaction of trypomastigotes with these cells. Lectin activity of purified penetrin was shown by hemagglutination which could be inhibited by extremely low concentrations of heparin and heparan sulphate.

4 Leishmaniasis

Leishmaniasis, a group of diseases in humans, caused by parasites of the genus *Leishmania* affect millions of people worldwide each year (Peters and Killick-Kendrick, 1987). *Leishmania donovani* causes visceral Leishmaniasis or kala-azar and mucocutaneous Leishmaniasis is caused by *L. major*, *L. tropica*, and *L. braziliensis*. The promastigote, the extracellular flagellated form of the parasite is primarily found in the sandfly vector and is responsible for transmitting the disease to man. While in the sandfly gut, the non-infective procyclic promastigotes undergo metacyclogenesis and are transformed into the metacyclic infective form. In the human host promastigotes attach to and invade macrophages where they differentiate into amastigotes, the intracellular stage. In *Leishmania*, in addition to CBPs, glycoconjugates of the parasite are believed to play a major role in mediating attachment by binding to host (macrophage) receptors (Turco and Descoteaux, 1992; McConville and Ferguson, 1993).

4.1 Lipophosphoglycan (LPG) and LPG-Binding Proteins

The surface of *Leishmania* is covered by a complex glycocalyx which plays a key role in enabling the parasite to survive in a hostile environment. The major glycoconjugate of the promastigote is lipophosphoglycan (LPG), a heterogeneous lipid-containing polysaccharide. The structural and functional characteristics of LPG have been extensively studied and reviewed (Turco and Descoteaux, 1992; McConville and Ferguson, 1993). Briefly, LPG consists of 4 domains, a phosphatidylinositol lipid anchor, a phosphosaccharide core, a repeating phosphorylated saccharide region and an oligosaccharide cap structure. The lipid anchor and the phosphosaccharide core are highly conserved whereas the other two regions show considerable variation between species (Turco and Descoteaux, 1992; McConville and Ferguson, 1993). LPG is also developmentally regulated in that transformation of the promastigote during metacyclogenesis is associated with marked structural changes in the molecule (Turco and Descoteaux, 1992). Functionally, LPG has been implicated as playing a major role in mediating attachment of promastigotes to host macrophages (Turco and Descoteaux, 1992). Binding of LPG is thought to occur to macrophages indirectly following opsonization via the complement receptors CR1 and CR3 and directly via the mannose-fucose receptor (Blackwell *et al.*, 1985; Turco and Descoteaux, 1992). Macrophage binding of *L. major* LPG was shown to occur to the phosphorylated oligosaccharide region and to be calcium dependent suggesting the involvement of a macrophage lectin receptor (Kelleher *et al.*, 1992). LPG is also believed to have multiple other functions including protection from proteolysis in the sandfly gut, resistance to complement mediated lysis, as well as several intracellular functions promoting survival in the host phagolysosome (Turco and Descoteaux, 1992). A structurally distinct LPG is also present on the surface of *L. major* amastigotes and has recently been shown to also mediate binding of this form of the parasite to macrophages (Kelleher *et al.*, 1995).

Recently a novel *Leishmania* surface protein which binds to LPG has been described (Smith and Rangarajan, 1995). This protein is encoded by gene B, one of a family of 5 related genes located on a single chromosome. These were identified by differential screening of a cDNA library for genes expressed in the infective stage of the parasite (Coulson and Smith, 1990; Flinn *et al.*, 1994). Antibodies to the recombinant protein were used to demonstrate that the protein was localized to the surface of the parasite. However, analysis of the deduced amino acid sequence predicted a hydrophilic molecule with no signal peptide, no membrane-spanning domain or GPI consensus (Flinn *et al.*, 1994). A region of the molecule containing a repetitive amino acid motif was found to be homologous to the cell wall peptidoglycan-binding domain of *S. aureus* protein A and led to speculation that the gene B protein was attached to LPG in an analagous fashion via the related amino acid repeat (Smith and Rangarajan, 1995). This possibility was strengthened by studies in which antibodies to the recombinant protein and to LPG were used to show co-localization and redistribution following capping on the parasite surface (Pimenta *et al.*, 1994). However, the determinant of LPG that is involved in the binding of the gene B protein has not been identified. The surface localization of the gene B protein, its association with LPG and the finding that it is expressed only at low levels in avirulent stocks of the parasite has led to the hypothesis that the protein may be involved in the host-parasite interaction (Smith and Rangarajan, 1995).

4.2 Heparin-Binding Proteins

Heparin-binding proteins have been described in both developmental stages of the parasite and are postulated to mediate attachment of the parasite to host cell proteoglycans.

Heparin-binding activity on the surface of promastigotes was demonstrated using a radioligand assay, electron microscopy with gold-labeled heparin and flow cytometry using FITC-heparin (Mukhopadhyay *et al.*, 1989; Butcher *et al.*, 1990, 1992). Expression of heparin-binding activity was found to correlate with differentiation of the non-infective promastigote to the infective metacyclic form (Butcher *et al.*, 1992). In addition, adhesion of promastigotes to macrophages was enhanced in the presence of heparin (Butcher *et al.*, 1992).

In the amastigote form of the parasite heparin-binding activity was demonstrated using binding of radiolabeled heparin (Love *et al.*, 1993). Amastigotes attached to a number of cell types expressing cell surface proteoglycans including wild type CHO cells. However CHO cell mutants deficient in proteoglycan synthesis or cells treated with heparitinase failed to bind heparin. Attachment of amastigotes to cells was inhibited by low doses of soluble heparin but not by chondroitin sulphate or hyaluronic acid.

4.3 GlcNAc-Binding Proteins

A number of studies have suggested the presence of GlcNAc-binding activity in *Leishmania* promastigotes. Binding of *Leishmania brasiliensis* promastigotes to macrophage-like cells was inhibited by GlcNAc and chitin and the presence of specific GlcNAc binding on promastigotes was demonstrated using FITC-labeled GlcNAc-BSA (Hernandez *et al.*, 1986). Neoglycoproteins as well as neoglycoenzymes have also been used as tools to identify carbohydrate binding activity on the surface of *Leishmania* parasites. In these studies, the neoglycoproteins GlcNAc-BSA and GalNAc-BSA bound to promastigotes but not amastigotes as determined by agglutination and fluorescence assays (Schottelius, 1992). Binding was calcium-dependent since it could be blocked by EDTA and restored by calcium. Neoglycoenzyme binding studies confirmed these findings and also showed evidence for specific binding of other mono- and disaccharides including sialic acid (Schottelius and Gabius, 1992).

5 Amebiasis

Entameba histolytica infects up to 10 % of the world's population, causing colonic or liver disease in 50 million people and resulting in up to 100,000 deaths annually (Walsh, 1986). The parasite exists in two developmental stages, trophozoite and cyst. Attachment of the trophozoite stage of the parasite to colonic cells is followed by invasion leading to lysis of host tissue and dissemination of the parasite to extraintestinal sites. Lectins have been identified in *E. histolytica* trophozoites and are implicated in mediating attachment and invasion of the trophozoite to host cells.

5.1 Gal/GalNAc-Binding Lectin

One of the best characterized parasite lectins is the Gal/GalNAc-specific lectin of *Entameba histolytica* (Petri and Mann, 1993; McCoy *et al.*, 1994 a, b). The presence of this lectin was initially identified using an adherence assay in which rosetting of trophozoites and CHO cells could be inhibited specifically by Gal and GalNAc (Ravdin and Guerran, 1981). Terminal Gal-containing glycoproteins such as asialofetuin and asialoorosomucoid were 1000 times more potent than Gal monomers in inhibiting adherence (Petri *et al.*, 1987). The role of Gal residues in mediating adherence was studied using a series of CHO cell mutants defective in the production of various *N*- and *O*-linked oligosaccharides (Li *et al.*, 1988, 1989; Ravdin, 1989; Ravdin and Murphy, 1992). Adherence of trophozoites to mutant CHO cells which express increased terminal Gal residues was almost twice that to the wild type CHO cells. Exposure of terminal Gal residues on the parental cells by sialidase treatment resulted in enhanced binding of parasites. Binding was minimal to CHO mutants lacking lactosamine units on β1–6 branched *N*-linked carbohydrates. Binding of purified lectin to CHO cells could be inhibited by α as well as β-linked Gal-containing saccharides (Saffer and Petri, 1991). Binding studies using synthetic multivalent ligands revealed that the purified lectin has the highest affinity for polyvalent *N*-acetylgalactosaminides which were 140,000 and 500,000 times more potent than monovalent GalNAc and Gal, respectively (Adler *et al.*, 1995). Binding to these ligands was maximal in the presence of Ca^{2+} and occurred within a broad pH range of 6 to 9.

The lectin, purified by galactose affinity or immunoaffinity chromatography, has been identified as a 260 kDa heterodimeric protein with disulfide-liked subunits of 170 kDa and 31–35 kDa (Petri *et al.*, 1987, 1989). Three genes encoding the 170 kDa heavy subunit and two encoding the 31–35 kDa light subunit have been cloned and sequenced (Mann *et al.*, 1991; Tannich *et al.*, 1991, 1992; McCoy *et al.*, 1993; Purdy *et al.*, 1993). The 170 kDa subunit has a transmembrane domain whereas the 31 kDa subunit isoform but not the 35 kDa isoform is glycophosphatidylinositol (GPI)-linked (McCoy *et al.*, 1993). Monoclonal antibodies to the heavy subunit have been shown to inhibit as well as enhance adherence *in vitro* (Petri *et al.*, 1990 a, b; Petri and Mann, 1993). While antibodies to the heavy subunit inhibit adherence, those to the light subunit do not, suggesting that the former contains the carbohydrate recognition domain (Petri *et al.*, 1989; McCoy *et al.*, 1994). The native lectin exists as multimeric aggregates of molecular mass 440–660 kDa (Petri *et al.*, 1989). It is present on the surface of amebic trophozoites in membrane-bound form and is also secreted into the extracellular medium (Petri *et al.*, 1987).

The major role of the Gal/GalNAc lectin is to mediate adherence of trophozoites to host cells. This has been demonstrated by a number of studies showing inhibition of adherence by lectin-specific carbohydrates and antibodies as well as by the purified lectin itself (Petri *et al.*, 1987; 1989). In addition to CHO cells, the lectin has also been shown to mediate attachment to colonic epithelial cells, neutrophils and macrophages as well as to colonic mucins which are rich in Gal and GalNAc residues (Ravdin *et al.*, 1985; Chadee *et al.*, 1987; Burchard *et al.*, 1993).

The lectin is also involved in contact dependent cytolysis of host cells which occurs following adherence of trophozoites. This was suggested by the finding that amebic cytolysis of CHO cells could be blocked by Gal and GalNAc as well as by anti-lectin antibodies (Ravdin, 1981; Saffer and Petri, 1991). Furthermore, CHO cell mutants deficient in terminal Gal residues were found to be completely resistant to cytolysis, whereas mutants with increased Gal residues were more susceptible to amebic killing (Ravdin *et al.*, 1989). Lectin mediated target cell cytolysis is associated with a rise in intracellular calcium (Ravdin *et al.*, 1988) and has been shown to occur via a Bcl-2 independent apoptotic mechanism (Ragland *et al.*, 1994). In addition to these functions the Gal/GalNAc lectin also confers resistance to lysis of the parasite by complement (Braga *et al.*, 1992) and may also be involved in cell signaling (Bailey *et al.*, 1990 a, b). The role of this lectin in the pathogenesis of amebiasis has lead to studies investigating its use as a potential vaccine candidate (Petri and Ravdin, 1991; Soong *et al.*, 1995).

5.2 Chitotriose-Binding Proteins

The presence of another lectin in *E. histolytica* trophozoites was initially demonstrated by hemagglutination of erythrocytes which was inhibitable by GlcNAc oligomers such as chitotriose (Kobiler and Mirelman, 1981). A lectin with similar sugar specificity was purified by size exclusion chromatography and shown to have a molecular weight of 220 kDa (Rosales-Encina *et al.*, 1987). The purified protein as well as antibodies to it were shown to inhibit adhesion of trophozoites to host cells (Meza *et al.*, 1987). Recently, antibodies to this lectin were used to obtain a recombinant cDNA clone expressing the protein (Talamas-Rohana *et al.*, 1995).

6 Giardiasis

The intestinal protozoan *Giardia lamblia* is a significant cause of diarrheal disease worldwide (Meyer and Radulescu, 1979). This parasite exists in two developmental forms, trophozoite and cyst. The trophozoite form selectively colonizes the small intestine where it adheres to epithelial cells and exerts its pathogenic effects. The selective tropism of *Giardia lamblia* for intestinal cells and the abundance of glycoconjugates in the glycocalyx coat and microvillus membrane of intestinal cells supports the involvement of lectin-carbohydrate interactions in the recognition and adhesion trophozoites to host cells. Lectins with specificity for mannosyl and phosphomannosyl residues have been described in *Giardia* (Farthing *et al.*, 1986; Lev *et al.*, 1986; Ward *et al.*, 1990).

6.1 Man-6-P-Binding Protein

Trophozoites express a surface membrane associated lectin, taglin, which is specifically activated by limited proteolysis with trypsin, a protease that is present in abundance at the site of infection (Lev *et al.*, 1986; Ward *et al.*, 1987). When activated, taglin agglutinated enterocytes which are the cells to which the parasite adheres *in vivo*, and in addition, bound to isolated brush border membranes of these cells (Ward *et al.*, 1987). The sugar specificity of taglin was determined by a hemagglutination inhibition assay using a wide variety of simple and substituted sugars (Ward *et al.*, 1987). The best monosaccharide inhibitor was mannose-6-phosphate which was seven times as potent an inhibitor as mannose. Mannose-1-phosphate had no inhibitory activity at a concentration five fold higher than mannose-6-phosphate, indicating the importance of phosphorylation at C-6 of mannose. The specificity of taglin for phosphomannosyl residues was confirmed by the finding that phosphomannose-containing compounds, such as yeast phosphomannans and bovine testicular β-galactosidase inhibited taglin induced hemagglutination. Although taglin-induced hemagglutinating activity was most specific for mannose-6-phosphate it is not known whether attachment to enterocytes is mediated by this sugar. Taglin-induced hemagglutination was most active at pH 6.5 and was dependent on divalent cations (Ward *et al.*, 1987). A monoclonal antibody to taglin reacted with the surface membrane of live trophozoites and recognized a protein of 28/30 kDa in lysates of *Giardia* trophozoites by immunoblotting (Ward *et al.*, 1987). This finding was confirmed by direct demonstration of lectin activity by erythrocyte binding to proteins electroblotted to nitrocellulose, which revealed specific red cell binding to giardial protein bands in the same molecular weight range as those recognized by the monoclonal antibody. The anti-taglin monoclonal antibody also inhibited attachment of trophozoites to an intestinal cell line, IEC-6, in an in vitro adhesion assay (Keusch *et al.*, 1991). In addition, clones of a particular strain of the parasite which displayed a greater attachment capacity in this assay also expressed greater amounts of taglin activity as determined by a hemagglutination assay, as well as by scanning densitometry of immunoblots of lysates of the different clones probed with the monoclonal antibody (Keusch *et al.*, 1991). These findings suggest that taglin may play a role in attachment of trophozoites to host cells.

7 Cryptosporidosis

The coccidian parasite *Cryptosporidium parvum* is currently recognized as a significant cause of gastrointestinal disease worldwide (Current and Garcia, 1991). Infection with this parasite in immunocompetent individuals is frequently asymptomatic or self-limiting. However in immunocompromised hosts such as patients with the acquired immunodeficiency syndrome (AIDS), infection with *C. parvum* may cause severe, unrelenting and often fatal diarrhea and wasting. Infection is initiated by the sporozoite form of the parasite which attaches to and invades intestinal epithelial cells where the parasite undergoes further intracellular development via sexual as well as asexual cycles. As is the case with *Giardia*, the abundance of glycoconjugates at the site of infection supports the involvement of lectin-carbohydrate interactions

7.1 Gal/GalNAc-Binding Protein

A surface-associated Gal/GalNAc lectin which may mediate attachment of sporozoites to host cells has been identified in *C. parvum* sporozoites (Thea *et al.*, 1992; Joe *et al.*, 1994; Keusch *et al.*, 1995). Lectin activity in intact and lysed sporozoites was identified using a hemagglutination assay and found to be optimal at a temperature of 4 °C, a pH of 7.5, and in the presence of the divalent cations Ca^{2+} and Mn^{2+}. The sugar specificity of this lectin was determined by a hemagglutination inhibition assay using a wide range of simple and substituted mono- and disaccharides as well as glycoproteins. The results indicated that of 12 monosaccharides tested the lectin was most specific for the monosaccharides galactose (Gal) and *N*-acetylgalactosamine (GalNAc) which inhibited lectin activity at MIC's of 4 and 14 mM, respectively (Joe *et al.*, 1994). Both α- as well as β-linked methyl and paranitrophenyl galactosides bound to the lectin suggesting that there was no specific preference for α or β linkage of the Gal residues. Gal- and GalNAc-containing disaccharides were shown to be potent inhibitors of hemagglutination, the best being Gal(β1–3)GalNAc and Gal(α1–4)Gal (Joe *et al.*, 1994). The disaccharide Gal(α1–4)Gal is present in the carbohydrate moiety of P1 glycoprotein, a mucin-like glycoprotein found in hydatid cyst fluid which is related to the P1 blood group antigen. This glycoprotein was also found to inhibit lectin activity at very low concentrations. In addition the lectin bound avidly to mucins such as bovine submaxillary mucin and hog gastric mucin which is of interest since the parasite exists in a mucin-rich environment. The role of this lectin in mediating attachment was studied using lectin-specific saccharides and glycoproteins to inhibit adherence of sporozoites to intestinal epithelial cells. The monosaccharides Gal and GalNAc, the disaccharides, lactose and melibiose as well as the mucins bovine submaxillary mucin, hog gastric mucin and hydatid cyst P1 glycoprotein inhibited attachment in a dose dependent fashion, further suggesting the possibility that the lectin is involved in these events (Keusch *et al.*, 1995). The role of Gal residues in mediating attachment was further studied using attachment of sporozoites to CHO cell glycosylation mutants. Attachment of sporozoites to the Lec-2 mutant which expresses increased terminal Gal residues was increased compared to the parent strain. In addition, sialidase treatment of intestinal epithelial cells, leading to exposure of terminal Gal residues, resulted in increased attachment of sporozoites to these cells. The lectin has been partially purified by galactose-affinity chromatography (Keusch *et al.*, 1995).

8 Conclusion

In conclusion, a number of significant advances have been made in the field of parasite glycobiology, related in particular to the identification of specific proteins and their carbohydrate ligands involved in recognition and adhesion mechanisms. These have greatly increased our understanding of the molecular mechanisms underlying the host-parasite interaction. However, with most host-parasite interactions the story is far from complete and a number of gaps remain to be filled. In many instances specific carbohydrate binding proteins and/or their complementary ligands have not been identified or fully characterized as yet nor has their role in mediating adhesion been unequivocally proved. To date none of the protozoal CBPs have been characterized with respect to crystal structure nor has their role been conclusively shown by gene knockout experiments. These studies should however, be greatly facilitated by the increasingly rapid technological advances in molecular biology and glycobiology. Identification of specific molecules involved in and elucidation of molecular mechanisms underlying the host-parasite interaction are of vital importance in developing strategies to combat the diseases caused by these organisms, by targeted drug and vaccine development.

Acknowledgements

I thank Miercio E. A. Pereira for critical reading of this chapter and helpful suggestions.

References

Adler P, Wood SJ, Lee YC *et al.* (1995): High affinity binding of the *Entamoeba histolytica* lectin to polyvalent *N*-acetylgalactosaminides. In *J. Biol. Chem.* **270**:5164–71.

Bailey GB, Gilmour JR, McCoomer NE (1990a): Roles of target cell membrane carbohydrate and lipid in *Entamoeba histolytica* interaction with mammalian cells. In *Infect. Immun.* **58**:2389–91.

Bailey GB, Nudelman ED, Day DB *et al.* (1990b): Specificity of glycosphingolipid recognition by *Entamoeba histolytica* trophozoites. In *Infect. Immun.* **58**:43–7.

Barnwell JW, Galinski MR (1991): The adhesion of malaria merozoite proteins to erythrocytes: a reflection of function? In *Res. Immunol.* **142**:666–72.

Blackwell JM, Ezekowitz RA, Roberts MB *et al.* (1985): Macrophage complement and lectin-like receptors bind *Leishmania* in the absence of serum. In *J. Exp. Med.* **162**:324–31.

Braga LL, Ninomiya H, McCoy JJ *et al.* (1992): Inhibition of the complement membrane attack complex by the galactose-specific adhesion of *Entamoeba histolytica*. In *J. Clin. Invest.* **90**:1131–7.

Brener Z (1973): Biology of *Trypanosoma cruzi*. In *Annu. Rev. Microbiol.* **27**:349–81.

Burchard GD, Prange G, Mirelman D (1993): Interaction between trophozoites of *Entamoeba histolytica* and the human intestinal cell line HT-29 in the presence or absence of leukocytes. In *Parasitol. Res.* **79**:140–5.

Butcher BA, Shome K, Estes LW *et al.* (1990): *Leishmania donovani* cell surface heparin receptors of promastigotes are recruited from an internal pool after trypsinization. In *Exp. Parasitol.* **71**:49–56.

Butcher BA, Sklar LA, Seamer LC *et al.* (1992): Heparin enhances the interaction of infective *Leishmania donovani* promastigotes with mouse peritoneal macrophages. A fluorescence flow cytometric analysis. In *J. Immunol.* **148**:2879–86.

Camus D, Hadley TJ (1985): A *Plasmodium falciparum* antigen that binds to host erythrocytes and merozoites. In *Science* **230**:553–6.

Carlson J, Wahlgren M (1992): *Plasmodium falciparum* erythrocyte rosetting is mediated by promiscuous lectin-like interactions. In *J. Exp. Med.* **176**:1311–7.

Cavallesco R, Pereira ME (1988): Antibody to *Trypanosoma cruzi* neuraminidase enhances infection *in vitro* and identifies a subpopulation of trypomastigotes. In *J. Immunol.* **140**:617–25.

Cerami C, Frevert U, Sinnis P *et al.* (1992): The basolateral domain of the hepatocyte plasma membrane bears receptors for the circumsporozoite protein of *Plasmodium falciparum* sporozoites. In *Cell* **70**:14021–33.

Chadee K, Petri W Jr., Innes DJ *et al.* (1987): Rat and human colonic mucins bind to and inhibit adherence lectin of *Entamoeba histolytica*. In *J. Clin. Invest.* **80**:1245–54.

Chuenkova M, Pereira ME (1995): *Trypanosoma cruzi* trans-sialidase: enhancement of virulence in a murine model of Chagas' disease. In *J. Exp. Med.* **181**:1693–703.

Ciavaglia M, de Carvalho TU, de Souza W (1993): Interaction of *Trypanosoma cruzi* with cells with altered glycosylation patterns. In *Biochem. Biophys. Res. Commun.* **193**:718–21.

Colli W (1993): Trans-sialidase: a unique enzyme activity discovered in the protozoan *Trypanosoma cruzi*. In *FASEB J.* **7**:1257–64.

Coulson RM, Smith DF (1990): Isolation of genes showing increased or unique expression in the infective promastigotes of *Leishmania major.* In *Mol. Biochem. Parasitol.* **40**:63–75.

Cross GA, Takle GB (1993): The surface trans-sialidase family of *Trypanosoma cruzi*. In *Annu. Rev. Microbiol.* **47**:385–411.

Current WL, Garcia LS (1991): Cryptosporidiosis. In *Clin. Lab. Med.* **11**:873–97.

El Moudni B, Philippe M, Monsigny M *et al.* (1993): *N*-acetylglucosamine-binding proteins on *Plasmodium falciparum* merozoite surface. In *Glycobiology* **3**:305–12.

Farthing MJ, Pereira ME, Keusch GT (1986): Description and characterization of a surface lectin from *Giardia lamblia*. In *Infect. Immun.* **51**:661–7.

Ferrero-Garcia MA, Trombetta SE, Sanchez DO *et al.* (1993): The action of *Trypanosoma cruzi* trans-sialidase on glycolipids and glycoproteins. In *Eur. J. Biochem.* **213**:765–71.

Flinn HM, Rangarajan D, Smith DF (1994): Expression of a hydrophilic surface protein in infective stages of *Leishmania major.* In *Mol. Biochem. Parasitol.* **65**:259–70.

Frevert U, Sinnis P, Cerami C *et al.* (1993): Malaria circumsporozoite protein binds to heparan sulphate proteoglycans associated with the surface membrane of hepatocytes. In *J. Exp. Med.* **177**:1287–98.

Greenwood BM, Bradley AK, Greenwood AM *et al.* (1987): Mortality and morbidity from malaria among children in a rural area of The Gambia. In *Trans. Royal Soc. Trop. Med. Hyg.* **81**:478–86.

Hadley TJ, Klotz FW, Miller LH (1986): Invasion of erythrocytes by malaria parasites: a cellular and molecular overview. In *Annu. Rev. Microbiol.* **40**:451–77.

Hadley TJ, Klotz FW, Pasvol G *et al.* (1987): Falciparum malaria parasites invade erythrocytes that lack glycophorin A and B (MkMk). Strain differences indicate receptor heterogeneity and two pathways for invasion. In *J. Clin. Invest.* **80**:1190–3.

Hernandez AG, Rodriguez N, Stojanovic D *et al.* (1986): The localization of a lectin-like component on the *Leishmania* cell surface. In *Mol. Biol. Rep.* **11**:149–53.

Herrera EM, Ming M, Ortega-Barria E *et al.* (1994): Mediation of *Trypanosoma cruzi* invasion by heparan sulfate receptors on host cells and penetrin counter-receptors on the trypanosomes. In *Mol. Biochem. Parasitol.* **65**:73–83.

Joe A, Hamer DH, Kelley MA *et al.* (1994): Role of a Gal/GalNAc specific sporozoite surface lectin in *C. parvum*-host cell interaction. In *J. Euk. Microbiol.* 44S.

Kelleher M, Bacic A, Handman E (1992): Identification of a macrophage-binding determinant on lipophosphoglycan from *Leishmania major* promastigotes. In *Proc. Natl. Acad. Sci. USA* **89**:6–10.

Kelleher M, Moody SF, Mirabile P *et al.* (1995): Lipophosphoglycan blocks attachment of *Leishmania major* amastigotes to macrophages. In *Infect. Immun.* **63**:43–50.

Keusch GT, Ward HD, Ortega-Barria E *et al.* (1991): Molecular pathogenesis of *Giardia lamblia*: adherence and encystation. In *Molecular Pathogenesis of Gastrointestinal Infections*, pp 237–246, New York: Plenum Press.

Keusch GT, Hamer D, Joe A *et al.* (1995): Cryptosporidia – who is at risk? In *J. Suisse Med.* **125**:899–908.

Klotz FW, Orlandi PA, Reuter G *et al.* (1992): Binding of *Plasmodium falciparum* 175-kilodalton erythrocyte binding antigen and invasion of murine erythrocytes requires *N*-acetylneuraminic acid but not its *O*-acetylated form. In *Mol. Biochem. Parasitol.* **51**:49–54.

Kobiler D, Mirelman D (1981): Adhesion of *Entamoeba histolytica* trophozoites to monolayers of human cells. In *J. Inf. Dis.* **144**:539–46.

Lev B, Ward H, Keusch GT *et al.* (1986): Lectin activation in *Giardia lamblia* by host protease: a novel host-parasite interaction. In *Science* **232**:71–3.

Li E, Becker A, Stanley S Jr. (1988): Use of Chinese hamster ovary cells with altered glycosylation patterns to define the carbohydrate specificity of *Entamoeba histolytica* adhesion. In *J. Exp. Med.* **167**:1725–30.

Li E, Becker A, Stanley S Jr. (1989): Chinese hamster ovary cells deficient in *N*-acetylglucosaminyltransferase I activity are resistant to *Entamoeba histolytica*-mediated cytotoxicity. In *Infect. Immun.* **57**:8–12.

Love DC, Esko JD, Mosser DM (1993): A heparin-binding activity on *Leishmania* amastigotes which mediates adhesion to cellular proteoglycans. In *J. Cell Biol.* **123**:759–66.

Mann BJ, Torian BE, Vedvick TS *et al.* (1991): Sequence of a cysteine-rich galactose-specific lectin of *Entamoeba histolytica*. In *Proc. Natl. Acad. Sci. USA* **88**:3248–52.

McConville MJ, Ferguson AJ (1993): The structure, biosynthesis and function of glycosylated phosphatidylinositols in the parasitic protozoa and higher eukaryotes. In *Biochem. J.* **294**:305–324.

McCoy JJ, Mann BJ, Vedvick TS *et al.* (1993): Structural analysis of the light subunit of the *Entamoeba histolytica* galactose-specific adherence lectin. In *J. Biol. Chem.* **268**:24223–31.

McCoy JJ, Mann BJ, Petri W Jr. (1994 a): Adherence and cytotoxicity of *Entamoeba histolytica* or how lectins let parasites stick around. In *Infect. Immun.* **62**:3045–50.

McCoy JJ, Weaver AM, Petri W Jr. (1994 b): Use of monoclonal anti-light subunit antibodies to study the structure and function of the *Entamoeba histolytica* Gal/GalNAc adherence lectin. In *Glycoconjugate J.* **11**:432–6.

Meyer EA, Radulescu S (1979): *Giardia* and giardiasis. In *Adv. Parasitol.* 1–47, New York: Academic Press.

Meza I, Cazares F, Rosales-Encina JL *et al.* (1987): Use of antibodies to characterize a 220-kilodalton surface protein from *Entamoeba histolytica*. In *J. Infect. Dis.* **156**:798–805.

Ming M, Chuenkova M, Ortega-Barria E *et al.* (1993): Mediation of *Trypanosoma cruzi* invasion by sialic acid on the host cell and trans-sialidase on the trypanosome. In *Mol. Biochem. Parasitol.* **59**:243–52.

Mukhopadhyay NK, Shome K, Saha AK *et al.* (1989): Heparin binds to *Leishmania donovani* promastigotes and inhibits protein phosphorylation. In *Biochem. J.* **264**:517–25.

Orlandi PA, Sim BK, Chulay JD *et al.* (1990): Characterization of the 175-kilodalton erythrocyte binding antigen of *Plasmodium falciparum*. In *Mol. Biochem. Parasitol.* **40**:285–94.

Orlandi PA, Klotz FW, Haynes JD (1992): A malaria invasion receptor, the 175-kilodalton erythrocyte binding antigen of *Plasmodium falciparum* recognizes the terminal Neu5Ac (α2–3) Gal- sequences of glycophorin A. In *J. Cell Biol.* **116**:901–9.

Ortega-Barria E, Pereira ME (1991): A novel *T. cruzi* heparin-binding protein promotes fibroblast adhesion and penetration of engineered bacteria and trypanosomes into mammalian cells. In *Cell* **67**: 411–21.

Pancake SJ, Holt GD, Mellouk S *et al.* (1992): Malaria sporozoites and circumsporozoite proteins bind specifically to sulfated glycoconjugates. In *J. Cell Biol.* **117**:1351–7.

Pasloske BL, Howard RJ (1994): Malaria, the red cell, and the endothelium. In *Annu. Rev. Med.* **45**:283–95.

Pereira ME (1983): A developmentally regulated neuraminidase activity in *Trypanosoma cruzi*. In *Science* **219**:1444–6.

Pereira ME (1986): Lectins and agglutinins in protozoa. In *Microbial Lectins and Agglutinins: Properties and Biological Activity*, (Mitchell R, ed), pp 297–300, New York: John Wiley.

Perkins ME, Rocco LJ (1988): Sialic acid-dependent binding of *Plasmodium falciparum* merozoite surface antigen, Pf200, to human erythrocytes. In *J. Immunol.* **141**:3190–6.

Peters W, Killick-Kendrick R (1987): The Leishmaniases in Biology and Medicine, London: Academic Press.

Petri W Jr., Ravdin JI (1991): Protection of gerbils from amebic liver abscess by immunization with the galactose-specific adherence lectin of *Entamoeba histolytica*. In *Infect. Immun.* **59**:97–101.

Petri W Jr., Mann BJ (1993): Molecular mechanisms of invasion by *Entamoeba histolytica*. In *Semin. Cell Biol.* **4**:305–13.

Petri W Jr., Smith RD, Schlesinger PH *et al.* (1987): Isolation of the galactose-binding lectin that mediates the *in vitro* adherence of *Entamoeba histolytica*. In *J. Clin. Invest.* **80**:1238–44.

Petri W Jr., Chapman MD, Snodgrass T *et al.* (1989): Subunit structure of the galactose and *N*-acetyl-D-galactosamine-inhibitable adherence lectin of *Entamoeba histolytica*. In *J. Biol. Chem.* **264**:3007–12.

Petri W Jr., Jackson TF, Gathiram V *et al.* (1990 a): Pathogenic and nonpathogenic strains of *Entamoeba histolytica* can be differentiated by monoclonal antibodies to the galactose-specific adherence lectin. In *Infect. Immun.* **58**:1802–6.

Petri W Jr., Snodgrass TL, Jackson TF *et al.* (1990 b): Monoclonal antibodies directed against the galactose-binding lectin of *Entamoeba histolytica* enhance adherence. In *J. Immunol.* **144**:4803–9.

Pimenta PF, Pinto DSP, Rangarajan D *et al.* (1994): *Leishmania major*: association of the differentially expressed gene B protein and the surface lipophosphoglycan as revealed by membrane capping. In *Exp. Parasitol.* **79**:468–79.

Prioli RP, Mejia JS, Pereira ME (1990): Monoclonal antibodies against *Trypanosoma cruzi* neuraminidase reveal enzyme polymorphism, recognize a subset of trypomastigotes, and enhance infection in vitro. In *J Immunol* **144**:4384–91.

Prioli RP, Mejia JS, Aji T *et al.* (1991): *Trypanosoma cruzi*: localization of neuraminidase on the surface of trypomastigotes. In *Trop. Med. Parasitol.* **42**:146–50.

Purdy JE, Mann BJ, Shugart EC *et al.* (1993): Analysis of the gene family encoding the *Entamoeba histolytica* galactose-specific adhesin 170-kDa subunit. In *Mol. Biochem. Parasitol.* **62**:53–60.

Ragland BD, Ashley L , Vaux DL *et al.* (1994): *Entamoeba histolytica*: target cells killed by trophozoites undergo DNA fragmentation which is not blocked by Bcl-2. In *Exp. Parasitol.* **79**:460–7.

Ravdin JI (1989): *Entamoeba histolytica*: from adherence to enteropathy. In *J. Inf. Dis.* **159**: 420–9.

Ravdin JI, Guerrant RL (1981): Role of adherence in cytopathic mechanisms of *Entamoeba histolytica*. Study with mammalian tissue culture cells and human erythrocytes. In *J. Clin. Invest.* **68**: 1305–13.

Ravdin JI, Murphy CF (1992): Characterization of the galactose-specific binding activity of a purified soluble *Entamoeba histolytica* adherence lectin. In *J. Protozool.* **39**:319–23.

Ravdin JI, John JE, Johnston LI *et al.* (1985): Adherence of *Entamoeba histolytica* trophozoites to rat and human colonic mucosa. In *Infect. Immun.* **48**:292–7.

Ravdin JI, Moreau F, Sullivan JA *et al.* (1988): Relationship of free intracellular calcium to the cytolytic activity of *Entamoeba histolytica*. In *Infect. Immun.* **56**:1505–12.

Ravdin JI, Stanley P, Murphy CF *et al.* (1989): Characterization of cell surface carbohydrate receptors for *Entamoeba histolytica* adherence lectin. In *Infect. Immun.* **57**:2179–86.

Rogerson SJ, Reeder JC, Al-Yaman F *et al.* (1994): Sulfated glycoconjugates as disrupters of *Plasmodium falciparum* erythrocyte rosettes. In *Am. J. Trop. Med. Hyg.* **51**:198–203.

Rogerson SJ, Chaiyaroj SC, Ng K *et al.* (1995): Chondroitin sulfate A is a cell surface receptor for *Plasmodium falciparum*-infected erythrocytes. In *J. Exp. Med.* **182**:15–20.

Rosales-Encina JL, Meza I, Lopez-De-Leon A *et al.* (1987): Isolation of a 220-kilodalton protein with lectin properties from a virulent strain of *Entamoeba histolytica*. In *J. Inf. Dis.* **156**:790–7.

Rosenberg I, Prioli RP, Ortega-Barria E *et al.* (1991): Stage-specific phospholipase C-mediated release of *Trypanosoma cruzi* neuraminidase. In *Mol Biochem Parasitol* **46**:303–5.

Rowe A, Berendt AR, Marsh K *et al.* (1994): *Plasmodium falciparum*: a family of sulphated glycoconjugates disrupts erythrocyte rosettes. In *Exp. Parasitol.* **79**:506–16.

Saffer LD, Petri W Jr. (1991): Role of the galactose lectin of *Entamoeba histolytica* in adherence-dependent killing of mammalian cells. In *Infect. Immun.* **59**:4681–3.

Schenkman S, Jiang MS, Hart GW *et al.* (1991): A novel cell surface trans-sialidase of *Trypanosoma cruzi* generates a stage-specific epitope required for invasion of mammalian cells. In *Cell* **65**: 1117–25.

Schenkman S, Kurosaki T, Ravetch JV *et al.* (1992 a): Evidence for the participation of the Ssp-3 antigen in the invasion of non-phagocytic mammalian cells by *Trypanosoma cruzi*. In *J. Exp. Med.* **175**:1635–41.

Schenkman S, Pontes-de-Carvalho L, Nussenzweig V (1992 b): *Trypanosoma cruzi* trans-sialidase and neuraminidase activities can be mediated by the same enzymes. In *J. Exp. Med.* **175**:567–75.

Schenkman RP, Vandekerckhove F, Schenkman S (1993): Mammalian cell sialic acid enhances invasion by *Trypanosoma cruzi*. In *Infect Immun* **61**:898–902.

Schenkman S, Eichenger D, Pereira MEA *et al.* (1994): Structural and functional properties of *Trypanosoma* trans-sialidase. In *Annu. Rev. Microbiol.* **48**:499–523.

Schottelius J (1992): Neoglycoproteins as tools for the detection of carbohydrate-specific receptors on the cell surface of *Leishmania*. In *Parasitol. Res.* **78**:309–15.

Schottelius J, Gabius HJ (1992): Detection and quantitation of cell-surface sugar receptor(s) of *Leishmania donovani* by application of neoglycoenzymes. In *Parasitol. Res.* **78**:529–33.

Scudder P, Doom JP, Chuenkova M *et al.* (1993): Enzymatic characterization of β-D-galactoside α2,3-trans-sialidase from *Trypanosoma cruzi*. In *J. Biol. Chem.* **268**:9886–91.

Sim BK, Orlandi PA, Haynes JD *et al.* (1990): Primary structure of the 175K *Plasmodium falciparum* erythrocyte binding antigen and identification of a peptide which elicits antibodies that inhibit malaria merozoite invasion. In *J. Cell Biol.* **111**:1877–84.

Sim BK, Chitnis CE, Wasniowska K *et al.* (1994): Receptor and ligand domains for invasion of erythrocytes by *Plasmodium falciparum*. In *Science* **264**: 1941–4.

Smith DF, Rangarajan D (1995): Cell surface components of *Leishmania*: identification of a novel parasite lectin? In *Glycobiology* **5**:161–6.

Soong CJ, Kain KC, Abd-Alla M *et al.* (1995): A recombinant cysteine-rich section of the *Entamoeba histolytica* galactose-inhibitable lectin is efficacious as a subunit vaccine in the gerbil model of amebic liver abscess. In *J. Inf. Dis.* **171**:645–51.

Sturchler D (1989): How much malaria is there worldwide? In *Parasitol. Today* **5**:39–40.

Talamas-Rohana P, Schlie-Guzman MA, Hernandez-Ramirez VI *et al.* (1995): T-cell suppression and selective in vivo activation of TH2 subpopulation by the *Entamoeba histolytica* 220-kilodalton lectin. In *Infect. Immun.* **63**:3953–8.

Tannich E, Ebert F, Horstmann RD (1991): Primary structure of the 170-kDa surface lectin of pathogenic *Entamoeba histolytica*. In *Proc. Natl. Acad. Sci. USA* **88**:1849–53.

Tannich E, Ebert F, Horstmann RD (1992): Molecular cloning of cDNA and genomic sequences coding for the 35-kilodalton subunit of the galactose-inhibitable lectin of pathogenic *Entamoeba histolytica*. In *Mol. Biochem. Parasitol.* **55**:225–7.

Thea DM, Pereira ME, Kotler D *et al.* (1992): Identification and partial purification of a lectin on the surface of the sporozoite of *Cryptosporidium parvum*. In *J. Parasitol.* **78**:886–93.

Turco SJ, Descoteaux A (1992): The lipophosphoglycan of *Leishmania* parasites. In *Annu. Rev. Microbiol.* **46**:65–94.

Vandekerckhove F, Schenkman S, Pontes-de-Carvalho L *et al.* (1992): Substrate specificity of the *Trypano-*

soma cruzi trans-sialidase. In *Glycobiology* **2**: 541–8.

Walsh JA (1986): Problems in recognition and diagnosis of amebiasis: estimation of the global magnitude of morbidity and mortality. In *Rev. Infect. Dis.* **8**: 228–38.

Ward HD, Lev BI, Kane AV *et al.* (1987): Identification and characterization of taglin, a mannose 6-phosphate binding, trypsin-activated lectin from *Giardia lamblia*. In *Biochemistry* **26**:8669–75.

Ward HD, Lev BI, Keusch GT *et al.* (1987): Induction of Lectin Activity in *Giardia*. In: *Molecular Strategies of Parasite Invasion*, pp 521–30, New York: Alan R. Liss.

Ward HD, Keusch GT, Pereira MEA (1990): Induction of a phosphomannosyl binding activity in *Giardia*. In *Bio Essays* **12**:211–5.

23 Structure and Function of Plant Lectins

HAROLD RÜDIGER

1 Introduction

About a century ago, a medical student of Dorpat in Estonia described, for the first time, plant proteins that can agglutinate erythrocytes and other cells (Franz, 1988). Later on, based on what we now know was a prematurely optimistic expectation, these proteins were named lectins (Boyd and Shapleigh, 1954). This indicated their ability to distinguish between red cells of different human blood groups. Meanwhile it became necessary to define what should be called a lectin. It is now generally agreed: (a) that lectins are proteins or glycoproteins that bind carbohydrates, and in consequence agglutinate red or other cells if they possess more than one sugar binding site, (b) that lectins merely bind but do not process carbohydrates, and (c) that lectins are produced constitutively, i.e. not as a result of an external stimulus, as are immunoglobulins of higher animals (Goldstein *et al.*, 1980; Kocourek and Hořejší, 1981).

Though these definitions are now widely accepted, it should be pointed out that some borderline cases and exceptions have been described. A lectin known for many decades, and characterized most thoroughly, concanavalin A, has been shown to bind selected peptides from a peptide library at its sugar binding site (Oldenburg *et al.*, 1992; Scott *et al.*, 1992). Thus, certain peptides are apparently able to simulate carbohydrates in shape and in those physicochemical parameters that are essential for binding. Generally, inhibition of the interaction between a lectin and a glycoconjugate by the haptenic sugar is taken as a proof that the interaction is mediated by the lectin's carbohydrate-binding site. This conclusion should also be looked upon with caution. There are cases where an interaction is prevented or inhibited solely because the lectin changes its conformation after having bound the haptenic carbohydrate (Glew and Doyle, 1979; Lafont and Favero, 1995) or where even low salt concentrations prevent sugar binding (Etzler and Borrebaeck, 1980). Even the third definition has become questionable since there are cases where plants respond to an infection or a lesion by the production of lectin (Bolwell, 1987; Scheggia *et al.*, 1988; Millar *et al.*, 1992). Nevertheless, for most lectins the above-mentioned definitions are still valid.

Lectins are used extensively in many branches of science. An abbreviation system has therefore developed which is as illogical as natural languages are. Generally used abbreviations have been listed recently (van Driessche *et al.*, 1994). In the same volume, Peumans and van Damme (1994a) propose a rational system of lectin nomenclature. A comprehensive review of the properties of lectins has been given by Goldstein and Poretz (1986). Purification methods have been reviewed by Rüdiger (1988).

By definition, lectins are proteins. Only a few lectins, however, are pure proteins, whereas most are glycoproteins. Since typical plant lectins are synthesized at the endoplasmatic reticulum, then secreted into its lumen, and finally accumulated in the protein bodies (van Driessche, 1988), glycosylation is very likely to occur. Even lectins that are not glycosylated in their mature form are generated from glycosylated precursors, as for example concanavalin A (Carrington *et al.*, 1985; Herman *et al.*, 1985; Sheldon and Bowles, 1992) or wheat germ agglutinin (Mansfield *et al.*, 1988).

H.-J. and S. Gabius (Eds.), Glycosciences

ISBN 3-8261-0073-5

Nevertheless, glycosylation does not seem to be important for the action of lectins (Desai *et al.*, 1983). Many lectins need bivalent ions such as calcium, magnesium or manganese for full activity. Lectins in one plant often occur as families of so-called isolectins that are identical or nearly identical in carbohydrate specificity. Usually ion exchange chromatography is used to resolve such mixtures. If affinity chromatography is used, isolectins are not separated unless continuous concentration gradients of the desorbants are applied (Cunningham *et al.*, 1972; Rüdiger, 1977). There may be several reasons for this heterogeneity. One reason is incomplete post-translational processing. Typical cases are the isolectins from pea seeds (Hoedemaker *et al.*, 1994) and from soybeans (Mandal *et al.*, 1994) which differ in the lengths of their C-termini. Irregular *N*-terminal processing has been observed in the mistletoe lectin (Gabius *et al.*, 1992). A further reason may be glycosylation (Nikrad *et al.*, 1990). Lectins, like other proteins that follow the endoplasmatic reticulum pathway, generally become glycosylated on their way from the endoplasmatic reticulum to the protein bodies or other vacuolar compartments. The glycosylation is complex enough to allow many variations (Fuhrmann *et al.*, 1985; Elbein, 1988; Faye *et al.*, 1989). Mechanisms that could correct possible mistakes are known in nucleic acid or protein synthesis do not exist in glycosylation. Finally, plants may also be genetically inhomogeneous and contain more than one lectin gene. Since the genetic background of most cultivated plants is not known – not to speak of wild plants – the lectin contents and the isolectin patterns may differ considerably from one cultivar to the other. Even genetically homogeneous plants may contain different closely related gene families.

2 Structure

2.1 Leguminosae

The overwhelming majority of plant lectins that have been isolated and characterized are the lectins of the Leguminosae. This family differs from others because their lectins cover nearly all carbohydrate specificities, in spite of their close structural relatedness. The main groups of specificities found in Leguminosae lectins are for: (a) mannose, glucose and their α-glycosides (all lectins of this group bind mannose as well as glucose), (b) galactose and *N*-acetylgalactosamine (specificities for these sugar often occur together but in many cases one of the two sugars is bound more strongly), (c) *N*-acetylglucosamine (mostly bound more efficiently in its β-1,4-oligomeric form), (d) α-L-fucose, and (e) complex oligosaccharides.

There are also cases in which the same plant contains lectins that are totally different in carbohydrate specificity. Such an example is the common vetch, *Vicia cracca*, which contains a series of typical isolectins that bind the monosaccharide *N*-acetylgalactosamine, its α-glycosides and hence human blood group substance A (Rüdiger, 1977). At the same time, the seeds contain a second lectin which binds mannose/glucose. It is unspecific towards human blood groups and resembles the lectins from pea and lentil (Baumann *et al.*, 1982). Indeed, on inspection of the *N*-terminal sequence it was found that both lectins from *Vicia cracca* are similar. However, the similarity between its mannose/glucose-binding lectin and the lectins of the same specificity from pea or lentil is closer than between both *Vicia cracca* lectins (Baumann *et al.*, 1979). Thus, lectins from different plants may be more similar to each other than different lectins occurring in the same plant. Similar conclusions have also been drawn for other plants (Konami *et al.*, 1991). Further examples of plants harboring lectins of different specificities are *Ulex europaeus*, which in addition to the blood group 0 (H) specific α-L-Fuc-binding lectin, contains a further one that binds to *N*-acetylglucosamine, and the African shrub *Griffonia simplicifolia* which contains a series of isolectins with specificities for α-galactose and α-*N*-acetylgalactosamine, a second lectin that recognizes *N*-acetylglucosamine, and a further one with a specificity for complex oligosaccharides such as the human blood group Le^b (Goldstein and Poretz, 1986). It is not known why some plants contain only one lectin whereas others contain a series of isolectins and yet others completely different lectins.

Amino acid compositions do not tell us very much about the lectins' structure. Usually lectins resemble other proteins of the same plant in amino acid compositions. Thus, the lectins of the Leguminosae are generally rich in glutamic acid, glutamine, aspartic acid, asparagine, serine and threonine, but poor in sulfur amino acids.

Much more can be learnt from sequences than from amino acid compositions. The most detailed knowledge exists about Leguminosae lectin sequences. They all have in common that they are synthesized as peptides of about 30 kDa. A leader or signal sequence of about 20 amino acid residues precedes the chain of the mature lectin and guides its transport into the lumen of the endoplasmatic reticulum. From here the mature lectins make their way to the protein body *via* the Golgi compartment for further processing (van Driessche, 1988). The 30 kDa chains may combine to form 60 kDa dimers or 120 kDa tetramers. These alternatives apply to the majority of the Leguminosae lectins.

In most species of the Vicieae subtribe, the original 30 kDa peptide chain is further processed by cutting it into two dissimilar subunits, called α-(the smaller) and β-(the larger) subunits. Usually some further amino acid residues forming the linker peptide are lost in this process. Therefore, most Vicieae lectins belong to the "two-chain lectins" whereas the majority of the other leguminous lectins are "single-chain lectins" (Foriers *et al.*, 1979). In the precursor peptides of the "two-chain lectins", the β-subunit precedes the α-subunit. The Vicieae are only a small group but contain some economically very important plants such as pea (*Pisum sativum*), lentil (*Lens culinaris*), and broad bean (*Vicia faba*).

The difference between the single-chain and the two-chain lectins is as great as originally believed. Extensive investigations in the genus *Lathyrus* revealed that as expected most lectins of this genus belong to the two-chain type. There are, however, exceptions, namely the lectins from *Lathyrus sphaericus* and *Lathyrus nissolia* (Rougé *et al.*, 1990). These lectins belong to the single-chain type, though there is no notable difference in their amino acid sequences compared with other *Lathyrus* lectins. The main difference is that these lectins remain in their original precursor form. If the *Lathyrus nissolia* lectin is treated with a crude lectin-free extract from immature seeds of the related species *Lathyrus ochrus*, which produces a two-chain lectin, the *Lathyrus nissolia* lectin is also split into two dissimilar chains (Rougé *et al.*, 1989). Thus, the decision whether a two-chain or single-chain lectin is produced is rather a question of the specificity of the processing enzymes than of the lectins' amino acid sequence.

Complete amino acid sequences have been determined for several leguminous lectins, in part by the classical Edman method, in part by deducing the sequence from cDNA. All these lectins are closely related, provided that some insertions or deletions are allowed. In several reviews, lectin sequences have been compiled (Strosberg *et al.*, 1986; van Driessche, 1988; Rougé *et al.*, 1992). Lack of space does not allow discussion of identities, similarities or differences in detail here.

A great surprise came when the sequence of concanavalin A from the legume *Canavalia ensiformis* (Jack bean) was compared with those of several other Leguminosae, such as the soybean. A striking homology was observed but only if residue 1 of the soybean lectin was aligned to residue 122 of concanavalin A. The homology then proceeds up to residue 112 of the soybean lectin which corresponds to residue 237 of concanavalin A, i.e. its C-terminus. The homology then starts again from residue 114/115 of the soybean lectin which corresponds to the *N*-terminus of concanavalin A, and then proceeds to the position 120 of ConA which corresponds to the C terminus of the soybean lectin. So, if both sequences are aligned as circles, a gap between the *N*- and the C-terminus of the soybean lectin is bridged by concanavalin and *vice versa*. Two-chain lectins correspond to the soybean lectin, except that their chains have to be arranged successively in the order β-α.

For some years it remained an enigma how this strange relationship between concanavalin A and the other lectins might have evolved. Some researchers discussed rearrangement of genes whereas others favored rearrangement at the level of the RNA. Finally, it was shown that the concanavalin A precursor has a "normal" sequence comparable with those of soybean and others, but that after translation the precursor chain is split at one site which then forms the *N*- and the C-termini of the mature chains, and what was a sensation, is sealed, presumably by transpeptidation, at another site which then becomes the connection between positions 121 and 122 (Carrington *et al.*, 1985; Herman *et al.*, 1985; Sheldon and Bowles, 1992). In this respect it resembles the well known transpeptidation in the final crosslinking step of the bacterial murein which, however, is not a protein but a peptidoglycan. Thus the well-known and intensely studied concanavalin A represents an exception among the lectins rather than the normal situation. A similar circular permutation as compared to the "regular" lectins is found in other species of the subtribe Diocleae to which *Canavalia* belongs (Richardson *et al.*, 1984; Perez *et al.*, 1990; Fujimura *et al.*, 1993; Moreira *et al.*, 1993).

For several lectins, the three-dimensional structures have been worked out, in particular for Leguminosae lectins. Rougé *et al.* (1991) and Sharon (1993) reviewed the structures of concanavalin A and the lectins from pea, *Vicia faba*, *Griffonia simplicifolia* and *Lathyrus ochrus*. In spite of their differences in the primary structures as one-chain lectins (concanavalin A and the *Griffonia* lectin), two-chain lectins (pea and *Vicia faba* lectins) and as a lectin with circularly permuted sequence (concanavalin A) and of the differences in carbohydrate binding specificities, their three-dimensional structures are surprisingly similar. The subunits look like bell-shaped flattened domes that are basically made up of two antiparallel β sheets. Almost no α helical elements occur. The binding sites for hydrophobic residues, metal ions, carbohydrates and even of particular water molecules (Loris *et al.*, 1994) are highly conserved. The carbohydrate binding sites are formed by shallow depressions at the top of the dome. It is astonishing that even the carbohydrate binding amino acid residues are the same although the lectins differ in their carbohydrate specificities. The amino acid residues, however, bind to the carbohydrate in different orientations, depending on the lectin. Apparently, not only the amino acid residues but also the framework of the peptide chain is important. More recently, Loris *et al.* (1993) determined the structure of the lentil lectin, and Banerjee *et al.* (1994) that of the peanut lectin. Both resemble other Leguminosae lectins regarding the subunit structure, with small alterations in the loops. Apparently, however, the small differences in the peanut lectin are sufficient to make the subunits associate differently from other Leguminosae lectins, in a so-called "open" structure. Recently, the structures of demetallized concanavalin A (Bouckaert *et al.*, 1995), *Phaseolus vulgaris* isolectin L4 (Hamelryck *et al.*, 1995) and *Vicia villosa* lectin (Osinage *et al.*, 1995) have been published.

Independently of X-ray analyses, the group of T. Osawa has added a new aspect to the question of the carbohydrate-binding sites. They isolated from Leguminosae lectins relatively small peptides that are still able to bind or at least to interact with carbohydrate, though with lower efficiency than the intact lectin (Yamamoto *et al.*, 1991, 1992a). From the *Bauhinia purpurea* lectin, which binds galactose and lactose, the short octapeptide stretching from residues Asp-135 to Ser-143 (Asp-Thr-Trp-Pro-Asn-Thr-Glu-Trp-Ser) is still retarded on a column of lactose-agarose. The striking feature of this peptide is its high content of tryptophan, which otherwise is a rare amino acid. The ability of this peptide to bind carbohydrate could also be confirmed by synthesis. In a further elegant study, these workers constructed a hybrid lectin in which the framework of the protein was taken from the *Bauhinia* lectin, but the carbohydrate-binding octapeptide was replaced by the homologous sequence of the lentil (or pea) lectin (Asp-Thr-Phe-Tyr-Asn-Ala-Ala-Trp), which binds mannose and glucose. After expression of the hybrid protein in *Escherichia coli*, it displayed a specificity for mannose (Yamamoto *et al.*, 1992b). This shows that certain peptide stretches may determine the carbohydrate specificity. It has, however, to be kept in mind that these peptides interact weakly with carbohydrates and are only retarded but not bound by affinity columns. Thus, the role of the peptide backbone is to improve the binding strength.

Recently, researchers have begun to study the influence of single amino acid residues by site-directed mutagenesis. Thus, it was found that if

Asn-125 in the pea lectin is substituted by Asp (N125D), this completely eliminates the carbohydrate binding activity (van Eijsden *et al.*, 1992). If residues at neighboring positions are exchanged (Y124R, A126P), the carbohydrate specificity is altered (Hoedemaker *et al.*, 1995). Changes in the ability of the *Erythrina corallodendron* lectin to bind galactose derivatives with bulky substituents has also been achieved by site directed mutagenesis (Arango *et al.*, 1993). Apart from their value in deepening our knowledge, such studies may also lead to practical consequences: Hoedemaker *et al.* (1993) found that in pea lectin the substitution V103A impairs its thermal stability significantly. Since many lectins are toxic to man and animals, a final goal of these experiments would be to construct plants with genetically changed lectins that could fulfill their natural role during the lifetime of the plant, but would then be inactivated in the digestive tract of warm-blooded animals.

2.2 Non-Leguminous Plants

Other plant families have not been studied as thoroughly as the Leguminosae. Solanaceae lectins have been isolated and characterized mostly from potato tubers (Allen and Neuberger, 1978; Matsumoto *et al.*, 1983; van Driessche *et al.*, 1983; McCurrach and Kilpatrick, 1986), tomato fruits (Nachbar and Oppenheim, 1982; Kilpatrick *et al.*, 1983) and thorn-apple seeds (Crowley *et al.*, 1982; Petrescu *et al.*, 1993). These lectins are quite similar to each other but very different from the Leguminosae lectins. They mostly consist of dimeric proteins of 80–100 kDa with identical or nearly identical subunits. These M_r values are not very reliable since Solanaceae lectins are highly glycosylated and contain 40 to 50 % carbohydrate, which is much more than in any of the Leguminosae lectins. Their carbohydrate composition is quite unique: most of it consists of L-arabinose, which otherwise is not very common in glycoproteins. The rest of the carbohydrate moiety consists of D-galactose (Allen, 1983). The amino acid composition of Solanaceae lectins is also unusual because they contain considerable amounts of 4-hydroxyproline, an amino acid that is formed from proline by post-translational hydroxylation. The oligosaccharide chains are attached to hydroxyproline residues *via* the L-arabinose residues (Desai *et al.*, 1981). Immunochemical studies show that Solanaceae lectins are closely related (Kilpatrick *et al.*, 1980).

Generally, the carbohydrate specificities of Solanaceae lectins are similar and primarily directed against β-1–4-linked oligomers of *N*-acetylglucosamine, i.e. against chitin fragments (Goldstein and Poretz, 1986). This is in contrast to the Leguminosae lectins, which differ greatly in their carbohydrate specificities (see above). In detail, however, some differences exist. For instance, the potato lectin interacts weakly with monomeric GlcNAc, in contrast to the others, which only accept the dimer and higher oligomers. The *Datura* lectin also reacts with *N*-acetyllactosamine (Gal β1-4GlcNAc), i.e. it makes no great distinction between *N*-acetylglucosamine or galactose at the nonreducing end. This is in contrast to the potato lectin. Of the potato lectin, an *N*-terminal sequence (Millar *et al.*, 1992) and a few internal sequences (Kieslis-zewski *et al.*, 1994) have been published. The latter are remarkable in that on the one hand they are similar to extensin, a hydroxyproline-rich plant protein that participates in cell extension, and on the other hand to hevein, a chitin-binding protein from the rubber tree *Hevea brasiliensis*, a member of the Euphorbiaceae.

Of other dicotyledonous plants, only some selected lectins have been studied extensively. Therefore comparisons within a particular family are hardly possible. *Ricinus communis* (castor bean), a member of the Euphorbiaceae, contains in its seeds a highly toxic galactose-binding lectin named ricin, which is composed of two dissimilar subunits held together by a disulfide bridge. This toxin belongs to the class of ribosome-inactivating proteins (RIP, see below) and has attracted much attention as a potential candidate for target-directed drug design (Lambert *et al.*, 1991; Brinkmann and Pastan, 1994). In addition, the seeds contain a nontoxic tetrameric lectin usually designated as *R. communis* agglutinin (RCA). The sequences of both have been determined (Lamb *et al.*, 1985; Roberts *et al.*, 1985; Araki *et al.*, 1986; Araki and Funatsu, 1987; Rutenber *et al.*, 1987; Funatsu *et al.*, 1988; Morris and Wool, 1992) and were found to be highly homologous to each other. Interestingly, the ricin sequence is also similar to other RIPs as for example that from *Abrus precatorius*, a member of the Leguminosae.

From the latex of the rubber tree *Hevea brasiliensis* (Euphorbiaceae), a small protein, hevein, has been isolated. It interacts with oligomers of *N*-acetylglucosamine and resembles chitin-binding proteins from distant plant families rather than other Euphorbiaceae lectins (Broekaert *et al.*, 1990).

Lectins have been studied from two genera of Moraceae, namely from *Artocarpus integrifolia* (Jack fruit) and other *Artocarpus* species, and from *Maclura pomifera* (Osage orange). Their sequences are up to 60 % identical (Young *et al.*, 1989) but no homologies could be found to other lectins (Mahanta *et al.*, 1992). A special feature in the architecture of both lectins is an unusually small (2.09 kDa) β-subunit besides the larger (16.2 kDa) α-subunit. Since the β-subunit contains tryptophan, it is believed to be essential for sugar binding. Both subunits combine to $\alpha_4\beta_4$ complexes in solution (Young *et al.*, 1990).

Of most other dicotyledonous species, only lectins from one single representative are known in detail, as for example from Cucurbitaceae (Bostwick *et al.*, 1993, 1994), Caprifoliaceae (Shibuya *et al.*, 1989) and Urticaceae (Beintema and Peumans, 1992; Lerner and Raikhel, 1992).

A monocotyledonous plant family from which several lectins can be compared are the *Gramineae*. The lectin known longest and studied most thoroughly is the wheat (*Triticum vulgare*) germ lectin or agglutinin (WGA). Other lectins from this group are from barley (*Hordeum vulgare*), rye (*Secale cereale*) and rice (*Oryza sativa*) (van Damme and Peumans, 1991). They all are built from 18 kDa subunits which form noncovalent dimers of 36 kDa. The rice lectin is an exception because its monomer is partially cleaved into 8 kDa and 10 kDa subunits. All Gramineae lectins are rich in glycine and cystine, resemble each other immunologically (Stinissen *et al.*, 1983), and are able to exchange subunits with each other (Peumans *et al.*, 1982). The stability of WGA and its relatives is due to the high degree of internal crosslinking by disulfide groups.

Cultivated wheat varieties generally are polyploid. This may be why several isolectins occur. In each case, the two 18 kDa subunits consist of four internally homologous domains. The amino acid sequences (Wright and Olafsdottir, 1986; Raikhel and Lerner, 1991) allow classification of all WGA variants into the group of chitin-binding proteins such as are found in many plant families (Chapot *et al.*, 1986; Beintema, 1994). In its three-dimensional structure, WGA is also completely different from Leguminosae lectins (van Damme and Peumans, 1991; Sharon, 1993). There is only little order in the secondary structure in terms of α-helices or β-structures. Nevertheless, the subunits are not arranged randomly because they are stabilized by disulfide bonds. Similar cystine-stabilized structures have been found in many proteins (Beintema, 1994).

Gramineae lectins are specific for *N*-acetylglucosamine, preferably in oligomeric forms as chito-oligosaccharides. As a specialty, they also recognize *N*-acetylneuraminic acid (Peters *et al.*, 1979; Monsigny *et al.*, 1980). At a first glance, this may be surprising. It can, however, be rationalized by a similar spatial arrangement if both sugars are turned into appropriate positions. The dual specificity against both *N*-acetylglucosamine and *N*-acetylneuraminic acid is not shared by the Solanaceae lectins, which otherwise have nearly the same carbohydrate specificity in spite of differences in composition and structure.

A very interesting group of lectins occurs in the bulbs of Amaryllidaceae (van Damme *et al.*, 1991a, 1992). They specifically bind α-mannose and mannose oligosaccharides, but in contrast to mannose-binding Leguminosae lectins they do not recognize glucose. The first lectin to be isolated was from *Galanthus nivalis* (snow drop) bulbs (van Damme *et al.*, 1991b, 1992, 1994a). It occurs in multiple isoforms that are genetically determined (van Damme *et al.*, 1991c). Other Amaryllidaceae such as *Leukojum, Narcissus, Hippeastrum* and *Clivia*, also contain lectins that are similar to the *Galanthus* lectin in carbohydrate specificity and in amino acid sequence (van Damme and Peumans, 1994). The related Alliaceae family (*Allium cepa* (onion), *A. sativum* (garlic), *A. porrum* (leek) and others) is also a rich source of lectins (van Damme *et al.*, 1991a; Kaku *et al.*, 1992; van Damme and Peumans, 1994). Though the lectins are similar to the Amaryllidaceae lectins in sequence and sugar specificity, they form a more heterogeneous group than the Amaryllidaceae lectins do. Lectins also occur in the leaves of the Orchidaceae *Listera ovata*, *Epipactis helleborine*, *Cymbidium hybr.* (van Damme *et al.*, 1991, 1994b). They are similar to Amaryllidaceae and Alliaceae lectins in carbohydrate specificity and amino acid sequence, but as expected this homology is not as close as between the Orchidaceae lectins themselves.

Remarkably, *Tulipa* (tulip, a Liliacea) bulbs contain a similar lectin which, however, not only binds mannose but also *N*-acetylgalactosamine (van Damme *et al.*, 1995). This is the first example of a lectin that binds two profoundly different carbohydrates. In a certain way it is reminiscent of the recent finding that the correct assembly and folding of the soybean lectin depends on an interaction between a mannose-binding site on the one subunit and a high-mannose oligosaccharide chain on another subunit (Nagai *et al.*, 1993; Nagai and Yamaguchi, 1993), and of a lectin from the legume *Vicia faba* that binds mannose and at the same time acts as an α-galactosidase (Dey and Pridham, 1986; Dey *et al.*, 1990). Certainly lectin sequences can be used to study phylogenetic relationships (Rougé *et al.*, 1992a). On the other hand, it has to be kept in mind that homologous structures, domains or motifs may occur in completely different plants, as in ricin from *Ricinus communis* (Euphorbiaceae) which is related to similarly toxic proteins from *Abrus precatorius* (Leguminosae), *Viscum album* (Viscaceae) and probably other toxins with the same mechanism of action. The chitin-binding domains, as found in WGA from *Triticum vulgare* (Gramineae) are also found in proteins from *Hevea brasiliensis* (Euphorbiaceae), *Urtica dioica* (Urticaceae) and in several plant chitinases (Beintema, 1994).

3 Location

Before discussing the roles that lectins may play in nature, it is important to look at which site and at which time in the life cycle of the plant they occur. The "classical" lectins occur in the storage organs of plants. Mostly these are the seeds, but considerable amounts of lectins have also been found in tubers, bulbs or even in the bark. In some leguminous seeds such as those of *Phaseolus vulgaris* or *Canavalia ensiformis*, the lectins are major constituents among the seed proteins and make up several percent of the seed weight. Within seeds of the Leguminosae, lectins occur predominantly in the cotyledons. Only a little is found in the embryo and in the seed coat. The minute amounts occurring in organs other than seeds can only be determined reliably by immunochemical methods, but not by red cell agglutination. Nevertheless, even immunochemical determinations may leave some doubts about the chemical identity. In particular, there is the danger of false-positive reactions if the antibody used is not directed against the protein but against a carbohydrate moiety. In the plant kingdom, protein glycosylations may be very similar in quite different species (Laurière *et al.*, 1989). Borrebaeck (1984) determined the lectin distribution in *Phaseolus vulgaris* plants. He found that the roots, stems or leaves contain only about 50–100 ng lectin/100 g fresh weight which is negligible compared with the seed lectin, of which there is 1.2 g/100 g (Freier *et al.*, 1985).

The synthesis of lectins during seed development has been studied by several workers but has only rarely been compared with the synthesis of other seed proteins such as the typical storage proteins. Wenzel *et al.* (1993) studied the time course of protein synthesis for the lectin and three storage proteins in developing pea seeds. They found that the first proteins to appear are storage proteins of the vicilin group, closely followed by the lectin. After that convicilin, and much later legumin, two further storage proteins, are produced.

A further question is the intracellular localization of lectins. It is now generally agreed that seed lectins reside in the protein bodies (van

Driessche, 1988), i.e. organelles that are derived from the vacuole. In the later stages of seed development, the single vacuole of leguminous and other dicotyledonous seed cells is gradually replaced by smaller vesicles, the protein bodies, and the content of the vacuole transferred to them. This process not only includes a massive effort by the cell in protein synthesis but also a dramatic increase in membrane material.

The abundance of lectins in seeds does not exclude the possibility that small amounts at other sites have important functions. Lectins have also been found in leaves, tubers, rhizomes and roots (see below). The search for lectins in the vegetative parts of plants has led to interesting results in the legume *Dolichos biflorus*. The seeds of this plant contain a lectin with an extremely high specificity for α-glycosidically bound *N*-acetylgalactosamine residues. In fact, it is the best lectin for tracing red cells of group A_1. When other parts of the *Dolichos* plant were screened for the presence of lectin by an anti-seed lectin antibody, a material was found in stems, leaves and pods, but not in roots that displayed partial immunochemical identity with the seed lectin (Talbot and Etzler, 1978; Roberts and Etzler, 1984; Etzler *et al.*, 1984; Etzler, 1989). The most surprising fact was that this material did not agglutinate erythrocytes and that it did not bind carbohydrate under the conditions usually employed. At low ionic strength, however, the material was able to bind carbohydrate (Etzler and Borrebaeck, 1980). This finding is important because it shows that lectin-carbohydrate interactions may occasionally be abolished even at moderately high ionic strengths, which is in contrast to the popular opinion that lectin-carbohydrate interactions are generally stable even at high ionic strengths. This also emphasizes that it is not justified to confuse the terms "specific" and "unspecific" that characterize the quality of an interaction, with the terms "strong" and "weak" which are thermodynamic parameters. Interestingly, the carbohydrate specificity of the material from *Dolichos* stems and leaves is not quite the same as of the seed lectin. In particular, it not only binds *N*-acetylgalactosamine but also *N*-acetylglucosamine (Etzler and Borrebaeck, 1980). The seed lectin does not bind the latter monosaccharide. In contrast to the seed lectin, the cross-reactive material is not localized in vacuoles or protein bodies but in the cytoplasm and the cell wall (Etzler *et al.*, 1984). It does not occur in the seeds during maturation but appears in the first week of germination exclusively in the leaves and stems above the ground, and remains present during the whole life of the plant. Attempts to find seed lectin or its cross-reactive material also in *Dolichos* roots were unsuccessful. By affinity chromatography of extracts from young roots, however, a further lectin was detected which neither cross-reacts with the seed lectin nor with its cross-reactive material from leaves and stems (Quinn and Etzler, 1987). No other plant has been studied in such detail as *Dolichos biflorus*, so that generalizations cannot be made at present.

In several other plants, counterparts to seed lectins have also been detected in other parts, be it leaves, stems or roots. A particularly interesting case is the bark lectins from *Robinia pseudoacacia* (Hořejší *et al.*, 1978; Yoshida *et al.*, 1994), *Sophora japonica* (Hankins *et al.*, 1988; Ueno *et al.*, 1991), *Sambucus nigra* (Broekaert *et al.*, 1984; Nsimba-Lubaki *et al.*, 1986; Shibuya *et al.*, 1989; Kaku *et al.*, 1990) and some other trees or shrubs (Antonyuk and Lutsik, 1986), in which they occur in considerable amounts. As far as seed lectins occur in these plants, their bark lectins are similar to but not identical with them. Since the bark persists over the whole year, seasonal variations were studied. Most lectin is found during the resting phase but its content drops sharply during the vegetative phase (Greenwood *et al.*, 1986; Nsimba-Lubaki and Peumans, 1986; Baba *et al.*, 1991; Yoshida *et al.*, 1994). The intracellular localization is similar to that of the seed lectins, i.e. they occur in storage organelles comparable to the protein bodies of the seeds (Greenwood *et al.*, 1986; Herman *et al.*, 1988; Baba *et al.*, 1991).

Quite different conditions have been found in cereal lectins. Whereas the leguminous lectins are localized in the storage tissue which are the cotyledons, the wheat germ lectin is confined to the embryo, and nothing is found in the endosperm, which functions as storage tissue in these plants. As electron micrographs show, the lectin is concentrated in outer layers of the root cells, the cells (coleorhiza) that enclose them, and in other exposed cells. At the subcellular level, the lectin resides at the periphery of protein bodies, at the periphery of electron-translucent regions and at the interface between cell wall and cell membrane (Mishkind *et al.*, 1982; Triplett and Quatrano, 1982; Quatrano *et al.*, 1983; Mishkind *et al.*, 1983; Raikhel and Lerner, 1991).

From the Solanaceae, the potato lectin which has been studied most thoroughly occurs in the tubers (see above). A similar lectin has also been found in potato fruits (Kilpatrick, 1980; McCurrach and Kilpatrick, 1986), but this has not been characterized as closely as the tuber lectin. The thorn-apple (*Datura*) contains a lectin in the seeds, which presumably has no related counterparts in other tissues, and in the tomato plant the richest source of the lectin is the fruit juice. Thus, in each case Solanaceae lectins occur predominantly in the storage organs. Reports about the intracellular localization of Solanaceae are contradictory. Most researchers report localization in the cell walls (Kilpatrick *et al.*, 1979; Leach *et al.*, 1982; Casalongué and Pont-Lezica, 1985; Broekaert *et al.*, 1988; Pont-Lezica *et al.*, 1991) but some find lectins also associated with the plasmalemma and the membranes of cell organelles (Jeffree and Yeoman, 1981).

In contrast to the plants discussed so far, the localization of the *Cucurbita* lectin is quite different. In this plant, lectins occur in the phloem exudate and can be collected after the stalks of the developing fruits are cut and the sap that is running out collected. In the intact plant, the lectin is found in sieve elements and companion cells, predominantly in the extrafascicular phloem (Smith *et al.*, 1987; Bostwick *et al.*, 1992). The *Cucurbita* lectin binds to β-1,4-linked GlcNAc oligomers, i.e. to chitin fragments. As a special feature, this lectin is easily oxidized on exposure to air leading to gelation of the sap. This can be prevented by low molecular weight thiols (Sabnis and Hart, 1978; Allen, 1979). Autoxidation of the *Cucurbita* lectin may be responsible for the clotting phenomenon that is observed when *Cucurbita* stalks are cut. The phloem lectin of *Cucurbita* is probably an example of a more general phenomenon which in this case was detected so easily because the plants readily produce phloem exudates in large amounts. Phloem exudates from other plants, among them some trees, also contain lectins with a specificity for *N*-acetylglucosamine oligomers (Gietl and Ziegler, 1980). Isolation and more detailed studies, however, have not yet been performed.

4 Functions

The classical way in biochemical research is to detect a biological phenomenon, such as an enzymatic reaction or a chain of reactions, and then to look for enzymes, activators and inhibitors that regulate it. In the same way that advanced techniques were established and became routine, the reverse situation has often arisen: a substance can be isolated and characterized but it is difficult or even impossible to assign a biological function to it. Because of the rapid progress made in gene technology, we experience this situation quite frequently. There are ample gene libraries, that is DNA fragments with known sequences but unknown functions.

In lectinology the situation is similar. Lectins can be identified by agglutination and by inhibition of agglutination by carbohydrates, and mostly they can also be isolated by affinity chromatography. Nevertheless it is not known why nature produces them. Therefore all that can be said about their biological role has to be regarded as a hypothesis. Of the hypotheses, many of them without experimental evidence, only a few can be presented here. Possible roles of lectins can be divided into two categories: those which are exerted inside the plant and those which are directed externally.

4.1 Internally Directed Activities

Since the bulk of lectins occurs in storage organs of plants such as seeds, tubers, bulbs and bark, it has been proposed that lectins are mainly storage proteins (van Damme *et al.*, 1993; Peumans and van Damme, 1993). Obviously this is true but at the same time self-evident. The arguments in favor of this proposal, for example amino acid composition and spatial and temporal occurrence similar to those of storage proteins, are only indirect and do not take into account the differences between lectins and storage proteins. In seeds, the amounts of lectins are mostly relatively small compared with the amounts of typical storage proteins. Lectins possess one or more sugar-binding sites which are lacking in storage proteins. Lectins are relatively stable *in vitro*. There are also reports that they survive the germination process much longer than storage proteins (Gifford *et al.*, 1983; Moreira and Cavada, 1984; Boisseau *et al.*, 1985). This clearly distinguishes them from the highly labile storage proteins, which in part are already proteolytically cleaved in the maturating seed. Moreover, the assumption that lectins and storage proteins are synthesized simultaneously is not quite correct (Wenzel *et al.*, 1993). Nevertheless, lectins can be regarded as storage proteins in the sense that all available protein material will be utilized by the germinating seed.

A non-specific interaction of lectins with hydrophobic residues has been observed quite frequently (Hardman and Ainsworth, 1973; Ochoa *et al.*, 1979). The hydrophobic binding sites or cavities may be localized in proximity to the sugar binding site, since glycosides with hydrophobic aglycons often bind more tightly than the unsubstituted sugars (Loganathan *et al.*, 1992). The role of these sites, however, is unknown. In ricin, the lectin/toxin from *Ricinus communis* seeds, the hydrophobic region of the A-chain presumably participates in keeping A- and B-chains together (Yamasaki *et al.*, 1988). There are also, however, hydrophobic sites with more specialized affinities. Several workers found low affinity but specific binding to the plant hormone indole acetic acid and to the related amino acid tryptophan (Edelman and Wang, 1978; Chatelain *et al.*, 1982; Umekawa *et al.*, 1990), and high affinity binding to some fluorescent naphthalene derivatives, and to adenine and some of its derivatives, but not to nucleosides nor nucleotides (Roberts and Goldstein, 1982, 1983a, b; Roberts *et al.*, 1986; Maliarik *et al.*, 1989; Gegg *et al.*, 1992; Loganathan *et al.*, 1992; Gegg and Etzler, 1994). The most extensive studies of the adenine binding site have been performed with the lectins from *Phaseolus lunatus limensis* (lima bean), *Phaseolus vulgaris* (kidney bean) and *Dolichos biflorus*. Photoaffinity labeling and tryptic digestion showed that in lima and kidney beans the binding site is represented by a particular peptide in the C-terminal half (Maliarik and Goldstein, 1988), far away from the sugar-binding site (Maliarik *et al.*, 1989). Sugar- and adenine-binding sites do not influence each other. The fact that indole acetic acid as well as adenine derivatives such as 6-benzylaminopurine are phytohormones supports the proposal that lectins may participate in hormonal regulation. Other lectins such as those from *Griffonia simplicifolia*, *Glycine max*, *Amphicarpaea bracteata* (Maliarik *et al.*, 1989), *Canavalia ensiformis*, *Vicia faba* or *Pisum sativum* (Maliarik and Goldstein, 1988) do not bind adenine.

In several leguminous seeds an interaction could be demonstrated between the lectin and the storage proteins from the same seeds (Gansera *et al.*, 1979; Bowles and Marcus, 1981; Einhoff *et al.*, 1986a; Freier and Rüdiger, 1987; Kummer and Rüdiger, 1988). The conditions in pea and lentil seeds were studied more thoroughly. The lectin-binding proteins could be assigned to the legumin and vicilin families. Legumins and unglycosylated vicilins interact by ionic forces whereas glycosylated vicilin binds to the lectin by its sugar binding site. Though electrophoresis did not show significant differences from the bulk of storage proteins, the lectin-binding and non-binding materials differ in several other properties, for example, a strong interaction also in solution, as visualized by laser nephelometry (Freier and Rüdiger, 1987) and calorimetry (Einhoff *et al.*, 1986b), and a completely different activation pattern towards murine lymphocytes (Gebauer *et al.*, 1982; Freier and Rüdiger, 1990).

A further binding partner of lectins is the protein body membrane. The membranes from pea protein bodies react with the pea lectin in a manner that depends mainly on pH and ionic strength (Gers-Barlag *et al.*, 1993), whereas the membranes from soybean protein bodies bind the soybean lectin by its carbohydrate binding site (Rüdiger and Schecher, 1993; Schecher and Rüdiger, 1994). The identity of the membrane components responsible for these interactions is unknown. It would be of particular interest to identify the partner in the membrane of soybean protein bodies because the optimal haptenic monosaccharide of the soybean lectin, *N*-acetylgalactosamine, probably does not occur in the plant kingdom (Lamport, 1980). It has, however, to be considered that it might not necessarily be a carbohydrate component of the membrane that acts as a binding partner. Lectins may be dissociated from their binding partners solely by the conformational change that accompanies sugar binding (Glew and Doyle, 1979; Lafont and Favero, 1995).

In this connection it is noteworthy that protein body membranes apparently contain a tightly bound protein that cross-reacts immunologically with the lectin (Mäder and Chrispeels, 1984; Wenzel *et al.*, 1995). The double binding ability, both to storage proteins and to the membranes, may be interpreted as showing that lectins form a kind of glue intracellularly in much the same way as they do outside the cell.

In Leguminosae, interactions between lectins and enzymes are quite common (Einhoff *et al.*, 1986a). Most of the lectin-reactive enzymes are glycosidases. Some of these interactions add a more dynamic aspect to the function of lectins. Thus, Lorenc-Kubis *et al.* (1981) and Wierzba-Arabska and Morawiecka (1987) found that the lectins from wheat germ and from potato tubers can activate an endogenous phosphatase. The same phenomenon was observed with the edible mushroom *Pleurotus ostreatus* (Conrad and Rüdiger, 1994). An increase in enzymatic activity of up to 50 % could be achieved by addition of the lectin. It is only the maximal velocity that is affected, whereas the Michaelis constant remains virtually unchanged. The activation is dependent upon sugar binding; when the enzyme is tested in the presence of lectin and the haptenic sugar galactose, no activation occurs. Possibly lectins inside the plant or fungus may act as modulators of enzymatic activities.

An α-galactosidase was isolated from *Vicia faba* seeds which is also a lectin with a binding specificity towards glucose and mannose, just like the well-known "classical" lectin from this plant (Dey *et al.*, 1986; Dey and Pridham, 1986). However, the α-galactosidase is not identical with the classical lectin and does not even resemble it. The authors propose that the carbohydrate binding moiety may serve to keep the subunits of the enzyme together, or to fix the enzyme to the cell wall.

4.2 Externally Directed Activities

Externally directed activities of lectins have attracted considerable attention mainly because practical applications are expected from them.

The latex of the economically important rubber tree *Hevea brasiliensis* (Euphorbiaceae) contains a small protein (hevein) which resembles *N*-acetylglucosamine-binding lectins from other sources (van Parijs *et al.*, 1991). Gidrol *et al.* (1994) showed that hevein participates in the coagulation of the latex by forming bridges between the rubber particles via their surface glycoproteins. In addition, due to the chitin-binding ability of hevein, an antifungal effect could also be demonstrated (van Parijs *et al.*, 1991). The dual ability of hevein to support latex coagulation and to inhibit fungal growth may protect the plant from infection after having been wounded.

In vitro interactions between lectins and microorganisms are quite common. Thus numerous papers have appeared that demonstrate an interaction between lectins and fungi, either as spores or during germination and growth. The idea of lectins as protectants against fungi has gained indirect support from the finding that a soybean line that is resistant to infection by *Phytophthora* contains more lectin and releases it faster to the surrounding medium than controls (Gibson *et al.*, 1982). Often, however, when such ideas about protection from pathogens have been advanced, it has not been the well-known "classical" seed lectins that participate in such interactions, but postulated lectin-like factors that are believed to occur at the site of the infection, e.g. in the leaves (Singh and Schroth, 1977; Wydra and Rudolph, 1989). Lectins may also be connected with the action of elicitors (Darvill and Albersheim, 1984), either by interacting with them directly (Garas and Kuc, 1981), or by being induced by them (Bolwell, 1987). In any case, it is hard to decide whether an interaction or a growth inhibition observed *in vitro* is specific enough to be a candidate for an *in vivo* role. Moreover, in such studies it is absolutely essential to work with pure preparations. Thus it had been believed for a long time that the lectin from wheat germ (WGA) inhibits growth of fungi due to its specificity for chitin, a cell wall constituent of fungi. Later it became apparent that the inhibiting effect was due to contamination of the lectin preparation by chitinase (Schlumbaum *et al.*, 1986). This negative result, however, does not contradict a role of lectins in protection against fungal attack. Thus, a lectin isolated from the rhizome of the stinging nettle (*Urtica dioica*), which is an unusually small lectin (8.5 kDa), does indeed inhibit fungal growth by binding to the incomplete chitin polymer (Broekaert *et al.*, 1989). This underlines the importance of investigating each case separately, and of avoiding premature generalizations. The defense capabilities of chitin-binding proteins, including lectins, have been reviewed by Chrispeels and Raikhel (1991).

Recently, proteinaceous factors were found in seeds of pea and tubers of *Cyperus* that agglutinate erythrocytes and can be inhibited by carbohydrate (Torres, Heid-Jehn and Rüdiger, unpublished). These properties would justify classifying them as lectins. In contrast to typical lectins, however, they are extremely thermostable, have low molecular weights and are best inhibited by free glucosamine but not by the more common *N*-acetylated form. Since fungal cell walls contain not only chitin (the polymer of *N*-acetylglucosamine) but also some chitosan (the deacetylated form), these factors were tested for inhibition of fungal growth. At low concentrations of the factors, *Fusarium* spores are inhibited from germinating and forming hyphae, while at high concentrations the spores are even lysed. It is very likely that these factors protect the seeds from fungal infections. Lectin-carbohydrate interactions are mostly tested with monosaccharides or with oligosaccharides from animal or human sources. The work of Rougé and his colleagues adds a new aspect to this. In extensive screening studies they tested the interaction of Leguminosae lectins with muramic acid, *N*-acetylmuramic acid (MurNAc) and *N*-acetylmuraminyl dipeptides (Ayouba *et al.*, 1991, 1992) which are constituents of bacterial cell walls. Due to the similarity of muramic acid derivatives to glucose or mannose, lectins of the *Vicieae* (subtribe of the *Leguminosae*) generally react as strongly or even more strongly than the haptenic monosaccharides. Surprisingly, however, some legume lectins such as those from soybean, *Dolichos biflorus*, and *Butea frondosa*, that do not bind glucose or mannose but do bind galactose and its derivatives, interact with muramic acid derivatives. Surprisingly, of the *N*-acetylmuraminyl dipeptides, the unnatural isomer with two D-amino acids attached to the lactyl residue binds to the lectins better than the natural one with L-Ala-D-isoGln. The repeating disaccharide unit GlcNAc-β-4MurNAc connected to the dipeptide was less reactive than the MurNAc dipeptide, probably due to steric hindrance (Ayouba *et al.*, 1991; Rougé *et al.*, 1992b). To take an example, mixed crystals of the isolectin I from *Lathyrus ochrus* (Vicieae) with muramic acid derivatives were grown, X-ray diffraction patterns taken (Bourne *et al.*, 1994) and the contact amino acids determined. Possibly the interaction between lectins and bacterial cell wall constituents may promote the symbiotic, saprophytic or pathogenic propensities of bacteria. Since these interactions are not restricted to a particular bacterial species, they can only be a prelude to subsequent more specific interactions.

A hypothesis that takes into account a more specific interaction claims that lectins may serve as auxiliary agents in establishing the symbiosis between Leguminosae and rhizobia, the root nodule-forming bacteria which enable Leguminosae to fix atmospheric nitrogen. Such functions may have evolved as a side product of the more general defense role against bacteria. This idea was originally put forward merely because it is typical for Leguminosae both to enter a symbiosis with root nodulating bacteria and to contain considerable amounts of lectins in their seeds. Since nodulation mostly occurs quite specifically it was investigated whether seed lectins would recognize their "own" species of *Rhizobium*. The result was that there are indeed some positive correlations, but also many false-positive and false-negative results. Nevertheless, the idea was attractive enough to be taken up again. The present state of knowledge is comprehensively reviewed by Ho and Kijne (1991) and Kijne *et al.* (1992). Those lectins whose participation in nodulation is well documented are synthesized *de novo* by the roots and do not originate from the seeds. Nevertheless, the possibility cannot be excluded that lectins that are found at the root surface may actually come from the cotyledons. Some seeds are known to release lectin into the surrounding medium (Fountain *et al.*, 1977; Gietl and Ziegler, 1979; Causse and Rougé, 1983; Broekaert *et al.*, 1988) which in turn could be adsorbed to the roots. In *Pisum* roots, however, it was demonstrated convincingly that the root lectin is responsible for specific recognition by the bacteria, since transgenic *Trifolium* roots, after having received the *Pisum* lectin gene, are nodulated by the *Pisum*-specific *Rhizobium* strain (Díaz *et al.*, 1989, 1995). The mechanism, however, is not simple selective adhesion but must be more sophisticated.

The chemical nature of the lectin receptors on the Rhizobia is unknown. Possibly they belong to or are derived from exopolysaccharides that are produced and excreted by the bacteria. This, however, is hard to investigate because exopolysaccharides are polymorphic and change their properties depending on the age of the bacterial culture and on the growth medium. In addition, the ability of the plant to be nodulated is transient, since root cells are only susceptible to nodulation for a short period.

The findings in the *Pisum-Rhizobium* system cannot be generalized since there are additional factors playing important roles in nodulation, such as bacterial lectins (Loh *et al.*, 1993), calcium-binding proteins called rhicadhesins (Smit *et al.*, 1991; Swart *et al.*, 1994), and low molecular substances produced by the plants (flavonoids) (Loh *et al.*, 1994) or by the bacteria (nodulation factors). In recent years some low molecular weight nodulation factors have been isolated that specifically establish the *Rhizobium*-plant symbiotic relationship (Lerouge *et al.*, 1990; Roche *et al.*, 1991a, b; Spaink *et al.*, 1991; Truchet *et al.*, 1991; Sanjuan *et al.*, 1992). Chemically most of them are oligomers of glucosamine linked β-1–4 and *N*-acylated with unsaturated fatty acid residues at their nonreducing end, whereas the other sugar residues are *N*-acetylated, sometimes *O*-sulfated or fucosylated. Since it was suspected that lectins might be receptors or at least part of a receptor for these factors, Rougé and his colleagues studied a possible interaction between the nodulation factors and some lectins whose structural co-ordinates are known. Since attempts to solve this question experimentally failed because of the strong adherence of the lipophilic nodulation factors to plastics, they used computer assisted modeling. It seems likely that interactions may take place with the sugar residues at the nonreducing end provided that they are not substituted at their primary hydroxy groups (Fabre *et al.*, 1994, 1995). A final answer to this question will have to await experimental studies.

Root lectins are not identical or at least similar to the seed lectins in all cases. Law and his colleagues found that the legume *Lotononis spinosa* expresses in its roots a lectin that is completely different from the seed lectin both in molecular architecture and in carbohydrate specificity. This lectin binds to the species of *Rhizobium* that nodulates *Lotononis* roots whereas the seed lectin fails to do so (Law and Strijdom, 1982, 1984a, b). The same group (Kishinevsky *et al.*, 1988; Law *et al.*, 1988; 1991a) isolated from *Arachis hypogaea* (peanut) root nodules two lectins, one resembling the seed lectin in carbohydrate specificity but differing in architecture. From its *N*-terminal sequence it is evident that it is coded by a gene different from the seed lectin gene (Law *et al.*, 1991b). The other nodule lectin is quite different from the others both in architecture and in carbohydrate specificity. Both their subcellular location in root nodules (Law and van Tonder, 1992) and the lack of interaction with peanut nodulation bacteria (*Bradyrhizobium*) (Law *et al.*, 1990) rules out a possible participation in determining the plant-bacteria specificity. The authors propose that they may play a role in packaging or mobilization of storage proteins. Also in other plants root lectins have been detected that differ from seed lectins in molecular architecture and/or carbohydrate specificity (Quinn and Etzler, 1987; Kalsi *et al.*, 1992; Das *et al.*, 1993; Chatelain *et al.*, 1994).

Some lectins are toxic to man and animals. As mentioned in the Introduction, this observation formed the start of lectinology. Toxicity is mostly interpreted as a protection mechanism against animals. This, however, cannot be generalized because animals can differ widely in their sensitivity to a particular lectin.

Two different kinds of lectin toxicity are known. One depends primarily on carbohydrate binding, whereas in the other carbohydrate binding is just an auxiliary reaction and the toxicity is due to an enzymatic attack on the rRNA. The toxicity by carbohydrate binding has been comprehensively reviewed (Pusztai, 1991). It was observed many years ago that animals did not increase in weight at the expected rate if they were fed with flour of certain kinds of seeds. Utilization, however, improved when the moist flour was heated in advance. This suggested that proteins might be responsible. Indeed, in some though not in all cases it could be shown that a lectin inhibited the utilization of food. Though this phenomenon had been observed initially with soybeans, most subsequent experiments have been performed with *Phaseolus vulgaris*. In rats, as little as 1 % lectin in the diet leads to severe loss of weight and will finally kill the animal. Since 10 % of the protein in *Phaseolus* beans consists of lectin, not very much bean flour is necessary to poison an animal. Even for man this lectin is toxic, and there have been many reports about poisoning by improperly cooked beans.

Rats fed with *Phaseolus* lectin show massive lectin binding to the intestinal mucosa, in particular on the microvilli of the proximal region. The lectin is not only bound externally but also enters the epithelial cells by endocytosis. Surprisingly, endocytosis is observed primarily in those animals that harbor a normal intestinal flora, in contrast to germ-free animals. The immediate consequence of lectin binding is that nutrients can no longer be taken up as effectively as before. This is probably due to an impairment of glycosidase and peptidase activities in the brush border. On the other hand, the lectin is a potent growth factor for the intestine and for other organs such as the pancreas. Intestinal cells are induced to produce vast amounts of mucus. Since on the one hand nutrients are required for the production of mucus and for the growth of the stimulated organs and on the other hand the uptake of nutrients is impaired, the situation becomes worse for the rest of the body. Moreover, hormone production is adversely affected: the insulin level decreases, the glucagon level increases, fat mobilization is induced, ketone bodies are formed, and gluconeogenesis is fed solely from amino acids that are generated from degradation of skeletal muscle. This is a metabolic situation very similar to diabetes mellitus. Above a certain concentration of lectin or with chronical exposure, bacteria accumulate at the damaged microvilli that ultimately are disrupted. Bacteria can now penetrate the mucosa and become distributed throughout the body. Eventually, the animals die from bacterial infections in the inner organs.

Other lectins have not been investigated as thoroughly as the severely toxic *Phaseolus* lectin. Not all lectins are as toxic as that from *Phaseolus vulgaris*. There are even lectins that are not toxic at all such as those from pea, lentil, faba beans, wheat and tomatoes. The latter have been consumed by man for a long time without any detrimental effect. Though the toxicity of *Phaseolus* lectin towards several mammals is quite evident, it may not necessarily be toxic for all animals. Initially, a protective function of this lectin due to its toxicity toward insects was also assumed (Janzen *et al.*, 1976), but later the toxic constituent appeared to be a contaminating amylase inhibitor (Huesing *et al.*, 1991a). In addition to the lectin, *Phaseolus* beans contain other anti-nutritional factors such as protease inhibitors, amylase inhibitors and arcelins. Sequence homologies have been detected between the *Phaseolus* lectin on the one hand, and amylase inhibitor and arcelin (Chrispeels and Raikhel, 1991) on the other hand. It is interesting to note that nature has created a set of related proteins with similar protective functions, but with different mechanisms by which they reach their goal.

At present, many lectins are screened for their toxicity against insects (Murdock *et al.*, 1990; Huesing *et al.*, 1991b; Habibi *et al.*, 1993; Powell *et al.*, 1993) with the intention of constructing transgenic, lectin-expressing plants that are poisonous to insects but useful for human or animal nutrition (Peumans *et al.*, 1994b; Hilder *et al.*, 1995; Sauvion *et al.*, 1995).

The lectins that cause toxicity by attacking RNA enzymatically are known under the name of (eukaryotic) ribosome-inactivating proteins (RIPs). The best studied example is ricin from castor beans (*Ricinus communis*), a member of the Euphorbiaceae (Olsnes, 1978a). Lectins acting in the same way occur scattered throughout the plant kingdom, as in jequirity bean (*Abrus precatorius*, Leguminosae) (Olsnes, 1978b), in mistletoe (*Viscum album*, Viscaceae) (Franz, 1989), in *Adenia* (*Modecca*) *digitata* (Passifloracea) (Olsnes *et al.*, 1982), and in *Trichosanthes kirilowii* (Cucurbitaceae) (Maraganor *et al.*, 1987). The toxicity of castor beans has long been known and for many decades it has been recognized that ricin is the toxic principle, and that it inactivates ribosomes. The mechanism of ricin action was elucidated in 1989 (Endo, 1987). It acts highly specifically as an *N*-glycosidase by cleaving off a particular adenine residue (A-4324) from the 28S RNA of eukaryotes. It is surprising that ricin picks out just this particular residue but this probably depends more on the conformation of the RNA than on the specificity of the enzyme. The position 4324 belongs to a loop that is highly exposed and probably binds the aminoacyl-tRNA. After the removal of A-4324, the ribosome is unable to support protein synthesis any more. The fact that ricin acts as an enzyme explains its very low LD_{50}, which is about 1–10 μg/kg (mouse). In contrast, the LD_{50} of the *Phaseolus* lectin, which acts stoichiometrically, is higher by a factor of about 1000. The action of ricin and similar toxins is only indirectly related to their sugar binding site. They are composed of two dissimilar peptide chains linked by a disulfide bridge. The enzymatic activity is localized on one of them (effectomer), the other subunit (haptomer) containing a carbohydrate binding site for galactose and/or *N*-acetylgalactosamine. The latter is responsible for adhesion to and endocytosis by the cells. Since galactose and its derivatives occur on essentially all cell types, RIPs are generally toxic. For many years researchers have tried to construct anti-tumor toxins by substituting the natural haptomer with a more specific one, i.e. an antibody specifically directed against tumor cells (Lambert *et al.*, 1991; Brinkmann and Pastan, 1994).

Toxic proteins that resemble the A subunit of ricin are widely distributed in plants (Stirpe *et al.*, 1986; Bolognesi *et al.*, 1990). All these proteins can inhibit protein synthesis, but do so only in cell-free systems. Since they lack the haptomer, they are not able to enter intact cells.

It is very likely that both types of toxicity, the intestinal mucosa binding and destroying activity as found in *Phaseolus* beans, and the ribosome inactivating activity as exemplified by castor beans, may serve as protection from predators. There remain, however, many open questions as to the species specificities.

5 Conclusion

In view of the fact that many biological roles for lectins have been proposed and for some of them experimental evidence has even been presented, the opinion that lectins are nothing but "flotsam of evolution" (Hymowitz, 1980) is certainly no longer valid. Indirect evidence against this opinion is that the amino acid sequences of lectins are conserved. Structures that fulfill a vital function have usually been optimized during evolution and are therefore relatively stable towards spontaneous mutation. Nevertheless, there is no doubt that one cannot ask for a single role for all lectins. Classification of a protein as a lectin has been made on an artificial and purely operational basis, namely agglutination of cells that a lectin will never meet in nature. So we have to answer the question about a possible biological function for every lectin separately, and even one and the same lectin may assume different roles depending on the stage in the life cycle of a plant.

References

Allen AK (1979): A lectin from the exudate of the fruit of the vegetable marrow (*Cucurbita pepo*) that has a specificity for β-1,4-linked GlcNAc oligomers. In *Biochem. J.* **183**: 133–7.

Allen AK (1983): Potato lectin – a glycoprotein with two domains. In: *Chemical Taxonomy, Molecular Biology and Function of Plant Lectins* (Goldstein IJ, Etzler ME, eds) pp 71–85, New York: Alan R. Liss.

Allen AK, Neuberger A (1978): Potato lectin. In *Methods-Enzymol.* **50**: 340–5.

Antonyuk VA, Lutzik MD (1986): The annual cycle of lectin activity specific to carbohydrates of the D-galactose group in perennials. In *Ukr. Bot. Zh.* **43**: 21–5.

Araki T, Funatsu G (1987) The complete amino acid sequence of the B-chain of ricin E isolated from small castor bean seeds. Ricin E is a recombination product of ricin D and *Ricinus communis* agglutinin. In: *Biochim. Biophys. Acta* **911**: 191–200.

Araki T, Yoshioka Y, Funatsu G (1986): The complete amino acid sequence of the B-chain of the *Ricinus communis* agglutinin isolated from large-grain castor bean seeds. In *Biochim. Biophys. Acta* **872**: 277–85.

Arango R, Rodriguez-Arango E, Adar R *et al.* (1993): Modification by site-directed mutagenesis of the specificity of *Erythrina corallodendron* lectin for galactose derivatives with bulky substituents at C-2. In *FEBS Lett.* **330**: 133–6.

Ayouba A, Chatelain C, Rougé P (1991): Legume lectins interact with muramic acid and *N*-acetylmuramic acid. In *FEBS Lett.* **289**: 102–4

Ayouba A, Martin D, Rougé P (1992): Recognition of muramic acid and *N*-acetylmuramic acid by Leguminosae lectins: possible role in plant-bacteria interaction. In *FEMS Microbiol. Lett.* **92**: 41–6.

Baba K, Ogawa M, Nagano A *et al.* (1991): Developmental changes in the bark lectins of *Sophora japonica*. In *Planta* **181**: 462–70.

Banerjee R, Mande SC, Ganesh V *et al.* (1994): Crystal structure of peanut lectin, a protein with an unusual quaternary structure. In *Proc. Natl. Acad. Sci. USA* **91**: 227–31.

Baumann C, Rüdiger H, Strosberg AD (1979): A comparison of the two lectins from *Vicia cracca*. In *FEBS Lett.* **102**: 216–8.

Baumann CM, Strosberg AD, Rüdiger H (1982): Purification and characterization of a mannose/glucose-specific lectin from *Vicia cracca*. In *Eur. J. Biochem.* **122**: 105–10.

Beintema JJ (1994): Structural features of plant chitinase and chitin-binding proteins. In *FEBS Lett.* **350**: 159–63.

Beintema JJ, Peumans WJ (1992): The primary structure of stinging nettle (*Urtica dioica*) agglutinin. In *FEBS Lett.* **299**: 131–4.

Boisseau C, Moisand A, Père D *et al.* (1985): Immunocytochemical localization of the *Lathyrus ochrus* seed lectin in seeds and seedlings. In *Plant Sci.* **42**: 25–34.

Bolognesi A, Barbieri L, Abbondanza A *et al.* (1990): Purification and properties of new ribosome-inactivating proteins with RNA *N*-glycosidase activity. In *Biochim. Biophys. Acta* **1087**: 293–302.

Bolwell GP (1987): Elicitor induction of the synthesis of a novel lectin-like arabinosylated hydroxyprolin-rich glycoprotein in suspension cultures of Phaseolus vulgaris. In *Planta* **172**: 184–91.

Borrebaeck CAK (1984): Detection and characterization of a lectin from non-seed tissue of *Phaseolus vulgaris*. In *Planta* **161**: 223–8.

Bostwick DE, Thompson GA (1993): Nucleotide sequence of a pumpkin phloem lectin. In *Plant Physiol.* **102**: 693–4.

Bostwick DE, Dannenhofer JM, Skaggs MI *et al.* (1992): Pumpkin phloem lectin genes are specifically expressed in companion cells. In *Plant Cell* **4**: 1539–48.

Bostwick DE, Skaggs MI, Thompson GA. (1994): Organization and characterization of Cucurbita phloem lectin genes. In *Plant Mol. Biol.* **26**: 887–97.

Bouckaert J, Loris R, Poortmans F *et al.* (1995): Concanavalin A: new tales from an old protein. In *16th Interlec Conference (Toulouse)*: 1.

Bourne Y, Ayouba A, Rougé P *et al.* (1994): Interaction of a legume lectin with two components of the bacterial cell wall. A crystallographic study. In *J. Biol. Chem.* **269**: 9429–35.

Bowles DJ, Marcus S (1981): Characterization of receptors for the endogenous lectins of soybean and jackbean seeds. In *FEBS Lett.* **129**: 135–8.

Boyd WC, Shapleigh E (1954): Specific precipitating activity of plant agglutinins (lectins). In *Science* **119**: 419.

Brinkmann U, Pastan I (1994): Immunotoxins against cancer. In *Biochim. Biophys. Acta* **1198**: 27–45.

Broekaert WF, Nsimba-Lubaki M, Peeters B *et al.* (1984): A lectin from elder (*Sambucus nigra*) bark. In *Biochem. J.* **221**: 163–9.

Broekaert WF, Lambrechts D, Verbelen JP *et al.* (1988): *Datura stramonium* agglutinin. Location in the seed and release upon imbibition. In *Plant Physiol.* **86**: 569–74.

Broekaert WF, van Parijs J, Leyns F *et al.* (1989): A chitin-binding lectin from stinging nettle rhizomes with antifungal properties. In *Science* **245**: 1100–2.

Broekaert W, Lee HI, Kush A *et al.* (1990): Wound-induced accumulation of mRNA containing a hevein sequence in laticifers of rubber tree (*Hevea brasiliensis*). In *Proc. Natl. Acad. Sci. USA* **87**: 7633–7.

Carrington DM, Auffret A, Hanke DE (1985): Polypeptide ligation occurs during post-translational modification of concanavalin A. In *Nature* **313**: 64–7.

Casalongué C, Pont Lezica R (1985): Potato lectin: a cell-wall glycoprotein. In *Plant Cell Physiol.* **26**: 1533–9.

Causse H, Rougé P (1983): Lectin release from imbibed soybean seeds and its possible function. In *Lectins: Biology, Biochemistry, Clinical Biochemistry* Vol. 3 (Bøg-Hansen TC, Spengler GA, eds) pp 559–72, Berlin: de Gruyter.

Chapot MP, Peumans WJ, Strosberg AD (1986): Extensive homology between lectins from non-leguminous plants. In *FEBS Lett.* **195**: 231–4.

Chatelin C, Oustrin J, Rougé P (1982): *In vitro* interaction between lectin and the plant auxin β-indole acetic acid. In *Ann. pharm. franç.* **40**: 473–9.

Chatelain C, Roques D, Rougé P (1994): Seeds and tubers of *Lathyrus tuberosus* L. contain different lectins of similar monosaccharide-binding specificities. In *Lectins: Biology, Biochemistry, Clinical Biochemistry* Vol. 10 (van Driessche E, Fischer J, Beeckmans S, Bøg-Hansen TC, eds): 149–55, Hellerup: Textop.

Chrispeels MJ, Raikhel NV (1991): Lectins, lectin genes, and their role in plant defense. In *The Plant Cell* **3**:1–9.

Conrad F, Rüdiger H (1994): The lectin from *Pleurotus ostreatus*: purification, characterization and interaction with a phosphatase. In *Phytochemistry* **36**: 277–83.

Crowley JF, Goldstein, IJ (1982): *Datura stramonium* lectin. In *Methods Enzymol.* **83**: 368–73.

Cunningham BA, Lang JL, Pflumm MN *et al.* (1972): Isolation and proteolytic cleavage of the intact subunit of concanavalin A. In *Biochemistry* **11**: 3233–9.

Darvill AG, Albersheim P (1984): Phytoalexins and their elicitors – a defense against microbial infections in plants. In *Annu. Rev. Plant Physiol.* **35**: 243–75.

Das RH, Kalsi G, Babu CR *et al.*(1993): Lectin activities in developing roots of peanuts (*Arachis hypogaea*). In *Lectins: Biology, Biochemistry, Clinical Biochemistry* Vol. 9 (Basu J, Kundu M, Chakrabarti P, Bøg-Hansen, TC, eds) pp 35–43, New Delhi: Wiley Eastern Ltd.

Desai NN, Allen AK, Neuberger, A (1981): Some properties of the lectin from *Datura stramonium* (thorn-apple) and the nature of its glycoprotein linkage. In *Biochem. J.* **197**: 345–53.

Desai NN, Allen AK, Neuberger A (1983): The properties of potato (*Solanum tuberosum*) lectin after deglycosylation by trifluoromethanesulphonic acid. In *Biochem. J.* **211**: 273–6.

Dey PM, Pridham JB (1986): *Vicia faba* α-galactosidase with lectin activities: an overview. In *Lectins: Biology, Biochemistry, Clinical Biochemistry* Vol. 5 (Bøg-Hansen TC, van Driessche E, eds) pp 161–70, Berlin: de Gruyter.

Dey PM, Naik S, Pridham JB (1986): *Vicia faba* α-galactosidase with lectin activity. In *Phytochemistry* **25**: 1059–61.

Dey PM, Herath NL, Pridham JB (1990): An investigation of the lectin domain of *Vicia faba* α-galactosidase I. In *Lectins: Biology, Biochemistry, Clinical Biochemistry* Vol. 7 (Kocourek J, Freed D, eds) pp 113–9, St. Louis: Sigma Chem. Comp.

Díaz CL, Melchers LS, Hooykaas PJ *et al.* (1989): Root lectin as a determinant of host-plant specificity in the *Rhizobium*-legume symbiosis. In *Nature* **338**: 579–81.

Díaz CL, Spaink HP, Wijffelman CA *et al.* (1995): Genomic requirement of *Rhizobium* for nodulation of white clover hairy roots transformed with the pea lectin gene. In *Mol. Plant-Microbe Interact.* **8**: 348–56.

Edelman GM, Wang JL (1978): Binding and functional properties of concanavalin A and its derivatives. III. Interactions with indoleacetic acid and other hydrophobic ligands. In *J. Biol. Chem.* **253**: 3016–22.

Einhoff W, Fleischmann G, Freier T *et al.* (1986a): Interactions between lectins and other components of leguminous protein bodies. In *Biol. Chem. Hoppe-Seyler* **367**: 15–25.

Einhoff W, Freier T, Kummer H *et al.* (1986b): Microcalorimetric effects of lectin interactions. In *Lectins: Biology, Biochemistry , Clinical Biochemistry* Vol. 5 (Bøg-Hansen TC, van Driessche E, eds) pp 53–6, Berlin: de Gruyter.

Elbein, AD (1988): Glycoprotein processing and glycoprotein processing inhibitors. In *Plant Physiol.* **87**: 291–5.

Endo Y (1989): Mechanism of action of ricin and related toxic lectins on the inactivation of eukaryotic ribosomes. In *Advances in Lectin Research* Vol. 2 (Franz H, Kasai KI, Kocourek J *et al.*, eds) pp 60–73, Berlin: VEB Verlag Volk und Gesundheit.

Etzler ME (1989): Vegetative tissue lectins from *Dolichos biflorus*. In *Methods Enzymol.* **179**: 341–7.

Etzler ME, Borrebaeck C (1980): Carbohydrate binding activity of a lectin-like glycoprotein from stem and leaves of *Dolichos biflorus*. In *Biochem. Biophys. Res. Commun.* **96**: 92–7.

Etzler ME, MacMillan S, Scates S *et al.* (1984): Subcellular localization of two *Dolichos biflorus* lectins. In *Plant Physiol.* **76**: 871–8.

Fabre C, Barre A, Demont N *et al.* (1994): Do Leguminosae lectins interact with Nod factors? In *Lectins: Biology, Biochemistry, Clinical Biochemistry* Vol. 10 (van Driessche E, Fischer J, Beeckmans S *et al.*, eds) pp 142–8, Hellerup: Textop.

Fabre C, Barre A, Esnauld J *et al.* (1995): Modelling of the molecular interaction of *Leguminosae* lectins with a diglycosyl diacylglyceride glycolipid from *Rhizobium* biovar *trifolii* cell membrane. In *16th Interlec Conference (Toulouse)*: 116.

Faye L, Johnson KD, Sturm A *et al.* (1989): Structure, biosynthesis, and function of asparagine-linked glycans on plant glycoproteins. In *Physiol. Plant.* **75**: 309–14.

Foriers A, de Neve R, Strosberg AD (1979): Lectin sequences as a tool for chemotaxonomic classification. In *Physiol. Vég.* **17**: 597–606.

Fountain DW, Foard DE, Repogle WD *et al.* (1977): Lectin release by soybean seeds. In *Science* **197**: 1185–7.

Franz H (1988): The ricin story. In *Advances in Lectin Research* Vol. 1 (Franz H, Kasai KI, Kocourek J *et al.*, eds) pp 10–25, Berlin: VEB Verlag Volk und Gesundheit.

Franz H (1989): Viscaceae lectins. In *Advances in Lectin Research* Vol. 2 (Franz H, Kasai KI; Kocourek J *et al.*, eds.) pp 28–59, Berlin: VEB Verlag Volk und Gesundheit.

Freier T, Rüdiger H (1987): *In vivo* binding partners of the *Lens culinaris* lectin. In *Biol. Chem. Hoppe-Seyler* **368**: 1215–23.

Freier T, Rüdiger H (1990): Lectin-binding proteins from lentil seeds as mitogens for murine B lymphocytes. In *Phytochemistry* **29**: 1459–61.

Freier T, Fleischmann G, Rüdiger H (1985): Affinity chromatography on immobilized hog gastric mucin and ovomucoid. A general method for the isolation of lectins. In *Biol. Chem. Hoppe-Seyler* **366**: 1023–8.

Fuhrmann U, Bause E, Ploegh H (1985): Inhibitors of oligosaccharide processing. In *Biochim. Biophys. Acta* **825**: 95–110.

Fujimura S, Terada S, Jayavardhanan KK *et al.* (1993): Primary structure of concanavalin A-like lectins from seeds of two species of *Canavalia*. In *Phytochemistry* **33**: 985–7.

Funatsu G, Taguchi Y, Kamenosono M *et al.* (1988): The complete amino acid sequence of the A-chain of abrin-a, a toxic protein from *Abrus precatorius*. In *Agric. Biol. Chem.* **52**: 1095–7.

Gabius HJ, Walzel H, Joshi SS *et al.* (1992): The immunomodulatory β-galactoside-specific lectin from mistletoe: partial sequence analysis, cell and tissue binding and impact on intracellular biosignalling of monocytic leukemia cells. In *Anticancer Res.* **12**: 669–76.

Gansera R, Schurz H, Rüdiger H. (1979): Lectin-associated proteins from the seeds of Leguminosae. In *Hoppe-Seyler's Z. Physiol. Chem.* **360**: 1579–85.

Garas NA, Kuc J (1981): Potato lectin lyses zoospores of *Phytophthora infestans* and precipitates elicitors of terpenoid accumulation produced by the fungus. In *Physiol. Plant Pathol.* **18**: 227–37.

Gebauer G, Schimpl A, Rüdiger H (1982): Lectin-binding proteins as potent mitogens for B-lymphocytes from nu/nu mice. In *Eur. J. Immunol.* **12**: 491–5.

Gegg CV, Etzler ME (1994): Photoaffinity labeling of the adenine-binding site of two *Dolichos biflorus* lectins. In *J. Biol. Chem.* **269**: 5687–92.

Gegg CV, Roberts DD, Segel IH *et al.* (1992): Characterization of the adenine binding sites of two *Dolichos biflorus* lectins. In *Biochemistry* **31**: 6938–42.

Gers-Barlag H, Schecher G, Siva Kumar N *et al.* (1993): Protein body membranes as binding partners of lectins. In *Lectins: Biology, Biochemistry, Clinical Biochemistry* Vol. 8 (van Driessche E, Franz H, Beeckmans S *et al.*, eds) pp 97–100, Hellerup: Textop.

Gibson DM, Stack S, Krell K *et al.* (1982): A comparison of soybean agglutinin in cultivars resistant and susceptible to *Phytophthora megasperma* var. *soja* (race 1). In *Plant Physiol.* **70**: 560–6.

Gidrol X, Chrestin H, Tan L *et al.*(1994): Hevein, a lectin-like protein from *Hevea brasiliensis* (rubber tree) is involved in the coagulation of latex. In *J. Biol. Chem.* **269**: 9278–83.

Gietl C, Ziegler H (1979): Lectins in the excretion of intact roots. In *Naturwissenschaften* **66**: 161–2.

Gietl C, Ziegler H (1980): Affinity chromatography of carbohydrate-binding proteins in the phloem exudate from several tree species. In *Biochem. Physiol. Pflanzen* **175**: 50–7.

Gifford DJ, Greenwood JS, Bewley JD (1983): Vacuolation and storage protein breakdown in the castor bean endosperm followig imbibition. In *J. Exp. Botany* **34**: 1433–43.

Glew RH, Doyle RJ (1979): Protection of sodium dodecyl sulfate-induced aggregation of concanavalin A by saccharide ligands. In *Carbohydr. Res.* **73**: 219–26.

Goldstein IJ, Poretz RD (1986): Isolation, physicochemical characterization, and carbohydrate-binding specificity of lectins. In *The Lectins. Properties, Functions and Applications in Biology and Medicine* (Liener IE, Sharon N, Goldstein, eds) pp 33–247, Orlando: Academic Press.

Goldstein IJ, Hughes RC, Monsigny M *et al.* (1980): What should be called a lectin? *Nature* **285**: 66.

Greenwood JS, Stinissen HM, Peumans WJ *et al.* (1986): *Sambucus nigra* agglutinin is located in protein bodies in the phloem parenchyma of the bark. In *Planta* **167**: 275–8.

Habibi J, Bacjus EA, Czapla TH (1993): Plant lectins survival of the potato leafhopper (Homoptera: Cicadellidae). In *J. Econ. Entomol.* **86**: 945–51.

Hamelryck T, Dao-Thi MH, Poortmans F *et al.* (1995): The crystal structure of PHA-L. In *16th Interlec Conference (Toulouse)*: 92.

Hankins CN, Kindinger JI, Shannon LM (1988): The lectins of *Sophora japonica*. II. Purification, properties and *N*-terminal amino acid sequence of five lectins from bark. In *Plant Physiol.* **86**: 67–70.

Hardman KD, Ainsworth CI (1973): Binding of nonpolar molecules by crystalline concanavalin A. In *Biochemistry* **12**: 4442–8.

Herman EM, Shannon LM, Chrispeels MJ (1985): Concanavalin A is synthesized as a glycoprotein precursor. In *Planta* **165**: 23–9.

Herman EM, Hankins CN, Shannon LM (1988): Bark and leaf lectins of *Sophora japonica* are sequestered in protein-storage vacuoles. In *Plant Physiol.* **86**: 1027–31.

Hilder VA, Powell KS, Gatehouse AMR *et al.* (1995): Expression of snowdrop lectin in transgenic tobacco plants results in added protection against aphids. In *Transgenic Res.* **4**: 18–25.

Ho SC, Kijne JW (1991): Lectins in *Rhizobium*-legume symbiosis. In *Lectin Reviews* Vol. 1 (Kilpatrick DC, van Driessche E, Bøg-Hansen TC, eds) pp 171–81, St. Louis: Sigma Chemical Corporation.

Hoedemaeker FJ, van Eijsden RR, Díaz CL *et al.* (1993): Destabilization of pea lectin by substitution of a single amino acid in a surface loop. In *Plant Mol. Biol.* **22**: 1039–46.

Hoedemaker FJ, Richardson M, Díaz CL *et al.* (1994): Pea (*Pisum sativum* L.) seed isolectins 1 and 2 and pea root lectin result from carboxypeptidase-like processing of a single gene product. In *Plant Mol. Biol.* **24**: 75–81.

Hoedemaker F, van Eijsden R, Díaz C *et al.* (1995): Mutational analysis of legume lectins. In *16th Interlec Conference (Toulouse)*: 7.

Hořejší V, Haskovec C, Kocourek J (1978): Studies on lectins. XXXVIII. Isolation and characterization of the lectin from black locust. *Biochim. Biophys. Acta* **532**: 98–104.

Huesing JE, Murdock LL, Shade RE (1991a): Rice and stinging nettle lectins: insecticidal activity similar to wheat germ agglutinin. In *Phytochemistry* **30**: 3565–8.

Huesing JE, Shade RE, Chrispeels MJ *et al.* (1991b): α-Amylase inhibitor, not phytohemagglutinin, explains resistance of common bean to cowpea weevil. *Plant Physiol.* **96**: 993–6.

Hymowitz T (1980): Chemical germplasm investigations in soybeans: the flotsam hypothesis. In *Recent Adv. Phytochem.* **14**: 157–79.

Janzen DH, Juster HB, Liener IE (1976): Insecticidal action of the phytohemagglutinin in black beans on a bruchid beetle. In *Science* **192**: 795–6.

Jeffree CE, Yeoman MM (1981): A study of the intracellular and intercellular distribution of the *Datura stramonium* lectin using an immunofluorescent technique. In *New Phytol.* **87**: 463–71.

Kaku H, Peumans WJ, Goldstein IJ (1990): Isolation and characterization of a second lectin (SNA-II) present in elderberry (*Sambucus nigra*) bark. In *Arch. Biochem. Biophys.* **277**: 255–62.

Kaku H, Goldstein IJ, van Damme EJM *et al.* (1992): New mannose-specific lectins from garlic (*Allium sativum*) and ramsons (*Allium ursinum*) bulbs. In *Carbohydr. Res.* **229**: 347–53.

Kalsi G, Das HR, Babu CR *et al.* (1992): Isolation and characterization of a lectin from peanut roots. In *Biochim. Biophys. Acta* **1117**: 114–9.

Kiesliszewski MJ, Showalter AM, Leykam JF (1994): Potato lectin. A modular protein sharing sequence similarities with the extensin family, the hevein lectin family and snake venom disintegrins (platelet aggregation inhibitors). In *Plant J.* **5**: 849–61.

Kijne J, Díaz C, de Pater S *et al.* (1992): Lectins in the symbiosis between *Rhizobia* and leguminous plants. In *Advances in Lectin Research* Vol. 5 (Franz H, van Driessche E, Kasai KI *et al.*, eds) pp 15–50, Berlin, Ullstein Mosby.

Kilpatrick DC (1980): Isolation of a lectin from the pericarp of potato fruits. In *Biochem. J.* **191**: 273–5.

Kilpatrick DC, Yeoman, MM (1978): Purification of the lectin from *Datura stramonium*. In *Biochem. J.* **175**: 1151–3.

Kilpatrick DC, Yeoman MM, Kilpatrick SP (1978): A lectin from seed extracts of *Datura stramonium*. In *Plant Sci. Lett.* **13**: 35–40.

Kilpatrick DC, Yeoman MM, Gould AR (1979): Tissue and subcellular distribution of the lectin from *Datura stramonium* (thorn apple). In *Biochem. J.* **184**. 215–9.

Kilpatrick DC, Jeffree CE, Lockhardt CM *et al.* (1980): Immunochemical evidence for structural similarities among lectins from species of the Solanaceae. In *FEBS Lett.* **113**: 129–3.

Kilpatrick DC, Weston J, Urbaniak SJ (1983): Purification and separation of tomato isolectins by chromatofocusing. In *Anal. Biochem.* **134**: 205–9.

Kishinevsky BD, Law IJ, Strijdom BW (1988): Detection of lectin in nodulated peanut and soybean plants. In *Planta* **176**: 10–8.

Kocourek J, Hořejší V (1981): Defining a lectin. In *Nature* **290**: 188.

Konami Y, Yamamoto K, Toyoshima S *et al.* (1991): The primary structure of the *Laburnum alpinum* seed lectin. In *FEBS Letters* **286**: 33–8.

Kummer H, Rüdiger H (1988): Characterization of a lectin-binding storage protein from pea (*Pisum sativum*). In *Biol. Chem. Hoppe-Seyler* **369**: 639–46.

Lafont V, Favero J (1995): Jacalin, a lectin which displays in vitro anti-HIV properties binds CD4 independently of its oligosaccharide moiety. In *16th Interlec Conference (Toulouse)*: 84.

Lamb FI, Roberts LM, Lord JM (1985): Nucleotide sequence of cloned cDNA coding for preproricin. In *Eur. J. Biochem.* **148**: 265–70.

Lambert JM, Goldmacher VS, Collinson AR *et al.* (1991): An immunotoxin prepared with blocked ricin: a natural plant toxin adapted for therapeutic use. In *Cancer Res.* **51**: 6236–42.

Lamport DTA (1980): Structure and function of plant glycoproteins. In *The Biochemistry of Plants* Vol. 3 (Preiss J, ed) pp 501–41, New York: Academic Press.

Laurière M, Laurière C, Chrispeels MJ *et al.* (1989): Characterization of a xylose-specific antiserum that reacts with the complex asparagine-linked glycans of extracellular and vacuolar glycoproteins. In *Plant Physiol.* **90**: 1182–8.

Law IJ, Strijdom BW (1982): *Lotononis bainesii* seed and root lectins and their interaction with *Rhizobium*. In *South Afr. J. Sci.* **78**: 375.

Law IJ, Strijdom BW (1984a): Properties of lectin in the root and seed of *Lotononis bainesii*. In *Plant Physiol.* **74**: 773–8.

Law IJ, Strijdom BW (1984b): Role of lectins in the specific recognition of *Rhizobium* by *Lotononis bainesii*. In *Plant Physiol.* **74**: 779–85.

Law IJ, van Tonder HJ (1992): Localization of mannose- and galactose-binding lectins in an effective peanut nodule. *Protoplasma* **167**: 10–8.

Law IJ, Haylett T, Strijdom BW (1988): Differences in properties of peanut seed lectin and purified galactose- and mannose-binding lectin from nodules of peanut. In *Planta* **176**: 19–27.

Law IJ, Kriel MM, Strijdom BW (1990): Differences in the distribution of mannose- and galactose-binding lectins in peanut tissues. In *Plant Sci.* **71**: 129–35.

Law IJ, Brandt WF, Mort AJ (1991a): Evidence of differences between mannose-binding lectins from nodules and cotyledons of peanut. In *Plant Sci.* **79**: 127–33.

Law IJ, Haylett T, Mort AJ *et al.* (1991b): Evidence of differences between related galactose-specific lectins from nodules and seeds of peanut. In *Plant Sci.* **75**: 123–7.

Leach JE, Cantrell MA, Sequeira L (1982): A hydroxyproline-rich bacterial agglutinin from potato: its localization by immunofluorescence. In *Physiol. Plant. Pathol.* **21**: 319–25.

Lerner DR, Raikhel NN (1992): The gene for stinging nettle lectin (*Urtica dioica* agglutinin) encodes both a lectin and a chitinase. In *J. Biol. Chem.* **267**: 11085–91.

Lerouge P, Roche P, Faucher C *et al.* (1990): Symbiontic host-specificity of *Rhizobium meliloti* is determined by a sulphated and acylated glucosamine oligosaccharide signal. In *Nature* **344**: 781–4.

Loganathan D, Osborne SE, Glick GD *et al.* (1992): Synthesis of high-affinity, hydrophobic monosaccharide derivatives and study of their interaction with concanavalin A, the pea, the lentil and fava bean lectins. In *Arch. Biochem. Biophys.* **299**: 268–74.

Loh JT, Ho SC, de Feijter AW *et al.* (1993): Carbohydrate binding activities of *Bradyrhizobium japonicum*: unipolar localization of the lectin BJ38 on the bacterial surface. In *Proc. Natl. Acad. Sci. USA* **90**: 3033–37.

Loh JT, Ho SC, Wang JL *et al.* (1994): Carbohydrate binding activities of *Bradyrhizobium japonicum*. IV. Effect of lactose and flavones on the expression of the lectin, BJ38. In *Glycoconjugate J.* **11**: 363–70.

Lorenc-Kubis I, Morawiecka B, Wieczorek E *et al.* (1981): Effects of lectins on enzymatic properties. Plant acid phosphatases and ribonucleases. In *Lectins: Biology, Biochemistry, Clinical Biochemistry* Vol. 1 (Bøg-Hansen TC, ed) pp 168–78, Berlin: de Gruyter.

Loris R, Steyart J, Maes D *et al.* (1993): Crystal structure determination and refinement at 2.3-Å resolution of the lentil lectin. In *Biochemistry* **32**: 8712–8.

Loris R, Stas PPG, Wyne L (1994): Conserved waters in legume lectin crystal structures. The importance of bound water for the sequence-structure relationship within the legume lectin family. In *J. Biol. Chem.* **269**: 26722–33.

Mäder M, Chrispeels MJ (1984): Synthesis of an integral protein of the protein-body membrane in *Phaseolus vulgaris*. In *Planta* **160**: 330–40.

Mahanta SK, Sanker S, Prasad Rao NVSA *et al.* (1992): Primary structure of a Thomsen-Friedenreich antigen-specific lectin, jacalin [*Artocarpus integrifolia* (jack fruit) agglutinin]. In *Biochem. J.* **284**: 95–101.

Maliarik MJ, Goldstein IJ (1988): Photoaffinity labeling of the adenine binding site of the lectin from Lima bean, *Phaselous lunatus*, and the kidney bean, *Phaseolus vulgaris*. In *J. Biol. Chem.* **263**: 11274–9.

Maliarik M, Plessas NR, Goldstein IJ *et al.* (1989): ESR and fluorescence studies on the adenine binding site of lectins using a spin-labeled analog. In *Biochemistry* **28**: 912–7.

Mandal DK, Nieves E, Bhattacharyya L *et al.* (1994): Purification and characterization of three isolectins of soybean agglutinin. Evidence for C-terminal truncation by electrospray ionization mass spectrometry. *Eur. J. Biochem.* **221**: 547–53.

Mansfield MA, Peumans WJ, Raikhel NV (1988): Wheat-germ agglutinin is synthesized as a glycosylated precursor. In *Planta* **173**: 482–9.

Maraganore JM, Joseph M, Baley MC (1987): Purification and characterization of trichosanthin. Homology to the ricin A chain and implication as to mechamism of abortefacient activity. In *J. Biol. Chem.* **262**: 11628–33.

Matsumoto I, Jimbo A, Mizuno Y *et al.* (1983): Purification and characterization of potato lectin. *J. Biol. Chem.* **258**: 2886–91.

McCurrach PM, Kilpatrick DC (1986): Purification of potato lectin (*Solanum tuberosum* agglutinin) from tubers or fruits using chromatofocusing. In *Anal. Biochem.* **154**: 492–6.

Millar DJ, Allen AK, Smith CG *et al.* (1992): Chitin-binding proteins in potato (*Solanum tuberosum*) tuber. Characterization, immunolocalization and effects of wounding. In *Biochem. J.* **283**: 813–21.

Mishkind M, Raikhel NV, Palevitz BA *et al.* (1982): Immunocytochemical localization of wheat germ agglutinin in wheat. In *J. Cell Biol.* **92**: 753–64.

Mishkind ML, Raikhel NV, Palevitz BA *et al.* (1983): The cell biology of wheat germ agglutinin and related lectins. In *Chemical Taxonomy, Molecular Biology and Function of Plant Lectins* (Goldstein IJ, Etzler ME, eds) pp 163–176, New York: Alan R. Liss.

Monsigny M, Roche AC, Sene C *et al.* (1980): Sugar-lectin interactions: how does wheat-germ agglutinin bind sialoglycoconjugates? In *Eur. J. Biochem.* **104**: 147–53.

Moreira RA, Cavada BS (1984): Lectin from *Canavalia brasiliensis*. Isolation, characterization and behaviour during germination. In *Biol. Plant.* **26**: 113–20.

Moreira RDA, Cordeiro EDF, Cavada SBS *et al.* (1993): Plant seed lectins. A possible marker for chemotaxonomy of the genus *Canavalia*. In *Rev. Bras. Fisiol. Veg.* **5**: 127–32.

Morris KN, Wool JG (1992): Determination by systematic deletion of the amino acids essential for catalysis by ricin A chain. In *Proc. Natl. Acad. Sci. USA* **89**: 4869–73.

Murdock LL, Huesing JE, Nielsen SS *et al.* (1990): Biological effects of plant lectins on the cowpea weevil. In *Phytochemistry* **29**: 85–9.

Nachbar MS, Oppenheim JD (1982): Tomato (*Lycopersicon esculentum*) lectin. In *Methods Enzymol.* **83**: 363–8.

Nagai K, Yamaguchi H (1993): Direct demonstration of the essential role of the intramolecular high-mannose oligosaccharide chains in the folding and assembly of soybean (*Glycine max*) lectin polypeptides. In *J. Biochem.* **113**: 123–5.

Nagai K, Shibata K, Yamaguchi H. (1993): Role of intramolecular high-mannose chains in the folding and assembly of soybean (*Glycine max*) lectin polypeptides: studies by the combined use ion spectroscopy and gel filtration size analysis. In *J. Biochem.* **114**: 830–4.

Nikrad PV, Pearlstone JR, Carpenter MR *et al.* (1990): Molecular mass heterogeneity of *Griffonia simplicifolia* lectin IV subunits. In *Biochem. J.* **272**: 343–50.

Nsimba-Lubaki M, Peumans WJ (1986): Seasonal fluctuations of lectins in barks of elderberry (*Sambucus nigra*) and black locust (*Robinia pseudoacacia*). In *Plant Physiol.* **80**: 747–51.

Nsimba-Lubaki M, Peumans WJ, Allen AK (1986): Isolation and characterization of glycoprotein lectins from the bark of three species of elder. In *Planta* **168**: 113–8.

Ochoa JL, Kristiansen T, Påhlman S (1979): Hydrophobicity of lectins. I. The hydrophobic character of concanavalin A. In *Biochim. Biophys. Acta* **577**: 102–9.

Oldenburg KR, Loganathan D, Goldstein IJ *et al.* (1992): Peptide ligands for a sugar-binding protein isolated from a random peptide library. In *Proc. Natl. Acad. Sci. USA* **89**: 5393–7.

Olsnes S (1978a): Ricin and *Ricinus* agglutinin from castor beans. In *Methods Enzymol.* **50**: 330–5.

Olsnes S (1978b): Toxic and nontoxic lectins from *Abrus precatorius*. In *Methods Enzymol.* **50**: 323–30.

Olsnes S, Haylett T, Sandvig K (1982): The toxic lectin modeccin. In *Methods Enzymol.* **83**: 357–62.

Osinaga E, Tello D, Batthyany C *et al.* (1995): Preliminary crystallographic analysis and amino acid sequence of isolectin B_4 from *Vicia villosa* . In *16th Interlec (Toulouse)*, 16.

Perez G, Hernandez M, Mora E (1990): Isolation and characterization of a lectin from the seeds of *Dioclea lehmanni*. In *Phytochemistry* **29**: 1745–49.

Peters BP, Ebisu S, Goldstein IJ *et al.* (1979): Interaction of wheat germ agglutinin with sialic acid. In *Biochemistry* **18**: 5505–11.

Petrescu SM, Petrescu A, Rüdiger HEF (1993): Purification and partial characterization of a lectin from *Datura innoxia* seeds. In *Phytochemistry* **34**: 343–8.

Peumans WJ, van Damme EJM (1993): Plant lectins: storage proteins with a defensive role In *Lectins: Biology, Biochemistry, Clinical Biochemistry* Vol. 9 (Basu J, Kundu M, Chakrabarti P *et al.*, eds) pp 27–34, New Delhi: Wiley Eastern Ltd.

Peumans WJ, van Damme EJM (1994a): Proposal for a novel system of nomenclature of plant lectins. In *Lectins: Biology, Biochemistry, Clinical Biochemistry* Vol. 10 (van Driessche E, Fischer J, Beeckmans S *et al.*, eds) pp 105–17, Hellerup: Textop.

Peumans WJ, van Damme, EJM (1994b): The role of lectins in the plant's defense against insects. In *Lectins Biology, Biochemistry, Clinical Biochemistry* Vol. 10 (van Driessche E, Fischer J, Beeckmans S *et al.*, eds) pp 128–41, Hellerup: Textop.

Peumans WJ, Stinissen HM, Carlier AR (1982): Subunit exchange between lectins from different cereal species. In *Planta* **154**: 568–72.

Pont-Lezica RF, Taylor R, Varner JE (1991): *Solanum tuberosum* agglutinin accumulation during tuber development. In *J. Plant Physiol.* **137**: 453–8.

Powell KS, Gatehouse AMR, Hilder VA *et al.* (1993): Antimetabolic effects of plant lectins and plant and fungal enzymes on the nymphal stages of two important rice pests, *Nilaparvata lugens* and *Nephotettix cincitepa*. In *Entomol. Exp. Appl.* **66**: 119–26.

Pusztai A (1991) Plant lectins, pp 105–205, Cambridge: University Press.

Quatrano RS, Hopkins R, Raikhel NV (1983): Control of the synthesis and localization of wheat germ agglutinin during embryogenesis. In: *Chemical Taxonomy, Molecular Biology and Function of Plant Lectins* (Goldstein IJ, Etzler ME, eds) pp 117–130, New York: Alan R. Liss.

Quinn JM, Etzler ME (1987): Isolation and characterization of a lectin from the roots of *Dolichos biflorus*. In *Arch. Biochem. Biophys.* **258**: 535–44.

Raikhel NV, Lerner DR (1991): Expression and regulation of lectin genes in cereals and rice. In *Develop. Gen.* **12**: 255–60.

Richardson M, Campos FDA, Moreira RA *et al.* (1984): The complete amino acid sequence of the major α-subunit of the lectin from the seeds of *Dioclea grandiflora*. In *Eur. J. Biochem.* **144**: 101–10.

Roberts DD, Goldstein IJ (1982): Hydrophobic binding properties of the lectin from Lima beans (*Phaseolus lunatus*). In *J. Biol. Chem.* **257**: 11274–7.

Roberts DD, Goldstein IJ (1983a): Adenine binding sites of the lectin from Lima beans (*Phaseolus lunatus limensis*). In *J. Biol. Chem.* **258**: 13820–4.

Roberts DD, Goldstein IJ (1983b): Binding of hydrophobic ligands to plant lectins: titration with arylaminonaphthalenesulfonates. In *Arch. Biochem. Biophys.* **224**: 479–84.

Roberts DD, Arjunan P, Townsend LB *et al.* (1986): Specificity of adenine binding to lima bean lectin. In *Phytochemistry* **25**: 589–93.

Roberts DM, Etzler ME (1984): Development and distribution of a lectin from the stems and leaves of *Dolichos biflorus*. In *Plant Physiol.* **76**: 879–84.

Roberts LM, Lamb FI, Pappin DJC *et al.* (1985): The primary sequence of *Ricinus communis* agglutinin in comparison with ricin. In *J. Biol. Chem.* **260**: 15682–6.

Roche P, Debellé F, Maillet F *et al.* (1991a): Molecular basis of symbiontic host specificity in *Rhizobium meliloti*: nodH and nodPQ genes encode the sulfation of lipooligosaccharide signals. In *Cell* **67**: 1131–43.

Roche P, Lerouge P, Ponthus C *et al.* (1991b): Structural determination of bacterial nodulation factors involved in the *Rhizobium meliloti*-alfalfa symbiosis. In *J. Biol. Chem.* **266**: 10933–40.

Rougé P, Père D, Bourne Y *et al.* (1989): *In vitro* cleavage of the *Lathyrus nissolia* isolectin. In *Plant Sci.* **62**: 181–9.

Rougé P, Père D, Bourne Y *et al.* (1990): Single- and two-chain legume lectins: a revisited question. In *Lectins: Biology, Biochemistry, Clinical Biochemistry* Vol. 7 (Kocourek J, Freed D, eds) pp 105–12, St. Louis: Sigma Chem. Comp.

Rougé P, Cambillau C, Bourne Y (1991): The three-dimensional structure of legume lectins. In *Lectin Reviews* Vol. 1 (Kilpatrick DC, van Driessche E, Bøg-Hansen TC, eds) pp 143–59, St. Louis: Sigma Chem. Comp.

Rougé P, Barre A, Chatelain C *et al.* (1992a): Leguminous lectins as phenetic and phylogenetic tools. In *Advances in Lectin Research* Vol. 5 (Franz H, van Driessche E, Kasai KI *et al.*, eds) pp 95–122, Berlin: Ullstein Mosby.

Rougé P, Ayouba A, Cambillau C *et al.* (1992b): Interaction of legume lectins with components of the bacterial cell-wall. In *14th Interlec Conference (Darjeeling)*: 22.

Rüdiger H (1977): Purification and properties of blood group specific lectins from *Vicia cracca*. In *Eur. J. Biochem.* **72**: 317–22.

Rüdiger H (1988): Preparation of plant lectins. In *Advances in Lectin Research* Vol. 1 (Franz H, Kasai KI, Kocourek, J *et al.*, eds) pp 26–72, Berlin: VEB Verlag Volk und Gesundheit.

Rüdiger H, Schecher G (1993): The protein body membrane of soybean seeds as a possible lectin-binding seed component. In *Lectins: Biology, Biochemistry, Clinical Biochemistry* Vol. 8 (van Driessche E, Franz H, Beeckmans S *et al.*, eds) pp 101–4, Hellerup: Textop.

Rutenber E, Ready M, Robertus JD (1987): Structure and evolution of ricin B-chain. In *Nature* **326**: 624–6.

Sabnis DD, Hart JW (1978):The isolation and some properties of a lectin (haemagglutinin) from *Cucurbita* phloem extract. In *Planta* **142**: 97–101.

Sanjuan J, Carlson RW, Spaink HP *et al.* (1992): A 2-*O*-methylfucose moiety is present in the lipo-oligosaccharide nodulation signal of *Bradyrhizobium japonicum*. In *Proc. Natl. Acad. Sci USA* **89**: 8789–93.

Sauvion N, Rahbé Y, Gatehouse AMR (1995): Effects and mechanisms of toxicity of convanavalin A and the snowdrop lectin towards the pea aphid, *Acyrthosiphon pisum* (Harris). In *16th Interlec Conference (Toulouse)*: 120.

Schecher G, Rüdiger H (1994): Interaction of the soybean (*Glycine max*) seed lectin with components of the soybean protein body membrane. In *Biol. Chem. Hoppe-Seyler* **375**: 829–32.

Scheggia C, Prisco AE, Dey PM *et al.* (1988): Alteration of lectin pattern in potato tuber by virus X. In *Plant Sci.* **58**: 9–14.

Schlumbaum A, Mauch F, Vögeli U *et al.* (1986): Plant chitinases are potent inhibitors of fungal growth. In *Nature* **324**: 365–7.

Scott JK, Loganathan D, Easley RB *et al.* (1992): A family of concanavalin A-binding peptides from a hexapeptide epitope library. In *Proc. Natl. Acad. Sci. USA* **89**: 5398–402.

Sharon N (1993): Lectin-carbohydrate complexes of plants and animals: an atomic review. In *TIBS* **18**: 221–7.

Sheldon PS, Bowles DJ (1992): The glycoprotein precursor of concanavalin A is converted to an active lectin by deglycosylation. In *EMBO J.* **11**: 1297–301.

Shibuya N, Tazaki K, Song Z *et al.* (1989): A comparative study of bark lectins from three elderberry (*Sambucus*) species. In *J. Biochem.* **106**: 1098–103.

Sing VO, Schroth MN (1977): Bacteria-plant cell surface interactions: active immobilization of saprophytic bacteria in plant leaves. In *Science* **197**: 759–61.

Smit G, Tubbing DMJ, Kijne JW *et al.* (1991): Role of Ca^{2+} in the activity of rhicadhesin from *Rhizobium leguminosarum* biovar *vicia*e which mediates the first step in attachment of Rhizobiaceae cells to plant root hair tips. In *Arch. Microbiol.* **155**: 278–83.

Smith LM, Sabnis DD, Johnson RPC (1987): Immunocytochemical localisation of phloem lectin from *Cucurbita maxima* using peroxidase and colloidal-gold labels. In *Planta* **170**: 461–70.

Spaink HP, Sheeley DM, van Brussel AAN *et al.* (1991): A novel highly unsaturated fatty acid moiety of lipo-oligosaccharide signals determines host specificity of *Rhizobium*. In *Nature* **354**: 125–30.

Stinissen HM, Peumans WJ, Carlier AR (1983): Occurrence and immunological relationship of lectins in Gramineae species. In *Planta* **159**: 105–11.

Stirpe F, Barbieri L (1986): Ribosome inactivating proteins up-to-date. In *FEBS Lett.* **195**: 1–8.

Strosberg AD, Buffard D, Lauwereys M *et al.* (1986): Legume lectins, a large family of homologous proteins. In *The Lectins, Properties, Functions and Applications in Biology and Medicine* (Liener IE, Sharon N, Goldstein IJ, eds) pp 249–64, Orlando: Academic Press.

Swart S, Logman TJJ, Smit G *et al.* (1994): Purification and partial characterization of a glycoprotein from pea (*Pisum sativum*) with receptor activity for rhicadhesin, an attachment protein of Rhizobiaceae. In *Plant Mol. Biol.* **24**: 171–83.

Talbot CF, Etzler ME (1978): Isolation and characterization of a protein from leaves and stems of *Dolichos biflorus* that cross reacts with antibodies to the seed lectin. In *Biochemistry* **17**: 1474–9.

Triplett BA, Quatrano RS (1982): Timing, localization, and control of wheat germ agglutinin synthesis in developing wheat embryos. In *Dev. Biol.* **91**: 491–6.

Truchet G, Roche P, Lerouge P *et al.* (1991): Sulphated lipo-oligosaccharide signals of *Rhizobium meliloti* elicit root nodule organogenesis in alfalfa. In *Nature* **351**: 670–3.

Ueno M, Ogawa H, Matsumoto I *et al.* (1991): A novel mannose-specific and sugar specifically aggregated lectin from the bark of the Japanese pagoda tree (*Sophora japonica*). In *J. Biol. Chem.* **266**: 3146–53.

Umekawa H, Takao K, Fujihara M *et al.* (1990): Interaction of Tora-mame (*Phaseolus vulgaris*) lectin with indole derivatives. In *Agric. Biol. Chem.* **54**: 3295–9.

van Damme EJM, Peumans WJ (1991): Lectins from monocotyledonae. In *Lectin Reviews* Vol. 1 (Kilpatrick, DC, van Driessche, E., Bøg-Hansen, TC, eds) pp 161–70, St. Louis: Sigma Chem. Corp.

van Damme EJM, Peumans WJ (1994): Molecular cloning of the mannose-binding lectins from Amaryllidaceae and Alliaceae species. In *Lectins: Biology, Biochemistry, Clinical Biochemistry* Vol. 10 (van Driessche E, Fischer J, Beeckmans S *et al.*, eds) 166–77, Hellerup: Textop.

van Damme EJM, Goldstein IJ, Peumans WJ (1991a): A comparative study of mannose-binding lectins from the Amaryllidaceae and Alliaceae. In *Phytochemistry* **30**: 509–14.

van Damme EJM, Kaku H, Perini F *et al.* (1991b): Biosynthesis, primary structure and molecular cloning of snowdrop (*Galanthus nivalis*) lectin. In *Eur. J. Biochem.* **202**: 23–30.

van Damme EJM, De Clercq N, Claessen F *et al.* (1991c): Molecular cloning and characterization of multiple isoforms of the snowdrop (*Galanthus nivalis* L.) lectin. In *Planta* **186**: 35–43.

van Damme EJM, Goldstein IJ, Vercammen G *et al.* (1992): Lectins of members of the Amaryllidaceae are encoded by multigene families which show extensive homology. In *Physiol. Plant.* **86**: 245–52.

van Damme EJM, Smeets K, van Driessche E *et al.* (1993): Plant lectins: reflections on recent developments and future prospects. In *Lectins: Biology, Biochemistry, Clinical Biochemistry* Vol. 8 (van Driessche E, Franz H, Beeckmans S *et al.* eds) pp 73–81, Hellerup, Textop.

van Damme EJM, Smeets K, van Leuven F *et al.* (1994a): Molecular cloning of mannose-binding lectin from *Clivia miniata*. In *Plant Mol. Biol.* **24**: 825–30.

van Damme EJM, Smeets K, Torrekens S *et al.* (1994b): Characterization and molecular cloning of mannose-binding lectins from the Orchidaceae species *Listera ovata*, *Epipactis helleborine* and *Cymbidium hybrid*. In *Eur. J. Biochem.* **221**, 769–77.

van Damme EJM, van Leuven F, Peumans WJ (1995): Tulip (*Tulipa* cv. Apeldoorn) bulbs contain a lectin with two distinct binding sites for N-acetylgalactosamine and mannose. In *16th Interlec Conference (Toulouse)*: 126.

van Driessche E (1988): Structure and function of *Leguminosae lectins*. In *Advances in Lectin Research* Vol. 1 (Franz H, Kasai KI, Kocourek J *et al.*, eds) pp 73–134, Berlin: VEB Verlag Volk und Gesundheit.

van Driessche E, Beeckmans S, Dejaegere R *et al.* (1983): A critical study on the purification of potato lectin (*Solanum tuberosum* L). In *Lectins: Biology, Biochemistry, Clinical Biochemistry* Vol. 3 (Bøg-Hansen TC, Spengler GA, eds) pp 629–38, Berlin: de Gruyter.

van Driessche E, Fischer J, Beeckmans *et al.*, eds (1994): Abbreviations for lectins. In *Lectins: Biology, Biochemistry, Clinical Biochemistry* **10**, pp xxxi-xxxiv.

van Eijsden RR, Hoedemaeker FJ, Díaz CL *et al.* (1992): Mutational analysis of pea lectin. Substitution of Asn125 for Asp in the monosaccharide binding site eliminates mannose/glucose-binding activity. In *Mol. Biol.* **20**: 1049–58.

van Parijs J, Broekaert WF, Goldstein IJ *et al.* (1991): Hevein: an antifungal protein from rubber tree (*Hevea brasiliensis*) latex. In *Planta* **183**: 258–64.

Wenzel M, Gers-Barlag H, Schimpl A *et al.* (1993): Time course of lectin and storage protein biosynthesis in developing pea (*Pisum sativum*) seeds. In *Biol. Chem. Hoppe-Seyler* **374**: 887–94.

Wenzel M, Gers-Barlag H, Rüdiger H (1995): Cross-reaction with pea lectin of a protein from protein body membranes. In *Phytochemistry* **38**: 825–9.

Wierzba-Arabska E, Morawiecka B (1987): Purification and properties of lectin from potato tubers and leaves; interaction with acid phosphatase from potato tubers. In *Acta Biochim. Polon.* **34**: 407–20.

Wright CS, Olafsdottir S (1986): Structural differences in the two major wheat germ agglutinin isolectins. In *J. Biol. Chem.* **261**: 7191–5.

Wydra K, Rudolph K (1989): Interactions between cell surface polymers of *Phaseolus vulgaris* and bacterial cell surface polymers of *Pseudomonas syringae* pv. *Phaseolicola.* In *Proc. 7th Int. Conf. Plant Path. Budapest 1989*: 63–8.

Yamamoto K, Konami Y, Kusui K *et al.* (1991): Purification and characterization of a carbohydrate-binding peptide from *Bauhinia purpurea* lectin. In *FEBS Lett.* **281**: 258–62.

Yamamoto K, Konami Y, Osawa T *et al.* (1992a): Carbohydrate-binding peptides from several anti-H(0) lectins. In *J. Biochem.* **111**: 436–9.

Yamamoto K, Konami Y, Osawa T *et al.* (1992b): Alteration of the carbohydrate-binding specificity of the *Bauhinia purpurea* lectin through the preparation of a chimeric lectin. In *J. Biochem.* **111**: 87–90.

Yamasaki N, Nagase Y, Funatsu G (1988): Hydrophobicities of ricin D and its constituent polypeptide chains. In *Agric. Biol. Chem.* **52**: 1021–6.

Yoshida K, Baba K, Yamamoto N *et al.* (1994): Cloning of a lectin cDNA and seasonal changes in levels of the lectin and its mRNA in the inner bark of *Robinia pseudoacacia*. In *Plant Mol. Biol.* **25**: 845–53.

Young NM, Johnston RAZ, Szabo AG *et al.* (1989): Homology of the D-galactose specific lectins from *Artocarpus integrifolia* and *Maclura pomifera* and the role of an unusual small polypeptide subunit. In *Arch. Biochem. Biophys.* **270**: 596–603.

Young NM, Johnston RAZ, Watson DC (1990): The primary and quaternary structure of jacalin and *Maclura pomifera* agglutinin. In *12th Interlec Conference (Davis)*: 49.

24 Lectins and Carbohydrates in Animal Cell Adhesion and Control of Proliferation

JEAN-PIERRE ZANETTA

1 Introduction

For many years, lectins were considered as molecules restricted to lower organisms and representative of archaic biological functions lost during evolution. With increasing data, the concept of the lost function appears old-fashioned, since some lectins present in higher and lower organisms share common structural features (Saukkonen *et al.*, 1992; Fiedler and Simons, 1994; Fiedler *et al.*, 1994). There is increasing evidence that glycobiological interactions involving glycoconjugates and carbohydrate-binding proteins have fundamental roles in cell proliferation, cell adhesion, organogenesis and human pathology. There is now no doubt that lactose-binding proteins constitute a large family of soluble molecules, endowed with a highly conserved carbohydrate recognition domain (CRD) during evolution, termed galectins in mammals (Barondes *et al.*, 1994a). The C-type lectins (Drickamer, 1988) constitute a very large family of calcium-dependent lectins, characterized by a CRD in which calcium is directly involved in the binding of monosaccharide. They include lectins with various physico-chemical and carbohydrate-binding properties, such as the membrane-bound asialoglycoprotein receptor (Masayuki *et al.*, 1988; Krebs *et al.*, 1994), the soluble mannose-binding proteins (Drickamer *et al.*, 1986; Taylor *et al.*, 1989), the selectins (Bevilacqua *et al.*, 1989; Springer, 1991; Foxall *et al.*, 1992; Lasky, 1992), the chaperone calnexin (Ware *et al.*, 1995) and the lectins on the surface of natural killer cells (Bezouska *et al.*, 1994; Lanier *et al.*, 1994). A large number of calcium-independent lectins have been isolated, including the mannose-binding lectins R1 and CSL (Zanetta *et al.*, 1985, 1987a) and some heparin-binding proteins including the fibroblast growth factor (FGF) family. Carbohydrate-binding properties have also been detected in cytokines such as interleukin 1 (Hession *et al.*, 1987; Muchmore and Decker, 1987; Brody and Durum, 1989), interleukin 2 (Sherblom *et al.*, 1989) and tumor necrosis factor alpha (Sherblom *et al.*, 1988; Lucas *et al.*, 1994).

This review will summarize the role of mammalian lectins in the mechanisms of cell adhesion and cell proliferation, with an emphasis on newer concepts. The role of galectins will not be discussed extensively since the accompanying chapter by D. W. Ohannesian and R. Lotan is devoted to these molecules, with a focus on their role in tumor biology.

2 The Galectins

Galectins are the mammalian members (Barondes *et al.*, 1994a) of a larger family of lactose-binding proteins highly conserved during evolution. The three-dimensional structure and the amino acid residues involved in the lactose-binding site are known (Lobsanov *et al.*, 1993; Barondes *et al.*, 1994b). Although the galectins display galactose-dependent agglutinating activities *in vitro* for a variety of cells (Joubert *et al.*, 1987), the endogenous involvement of these molecules as cell adhesion molecules is not clear. The reason is that in many cells the high affinity ligand is laminin (Cooper *et al.*, 1991; Ochieng *et al.*, 1992), a large glycoprotein of the extracellular matrix consisting of more than 50 % carbo-

H.-J. and S. Gabius (Eds.), Glycosciences

ISBN 3-8261-0073-5

hydrates. Surprisingly, the L-14 galectin shows an inhibition of binding of cells to laminin substrata (Cooper *et al.*, 1991). This property may explain the higher metastatic potential of cells expressing the galectins extracellularly (for review see Ohannesian and Lotan, this volume), since these lectins may inhibit the adhesion of cells to laminin-containing structures. A role in neurite fasciculation was suggested (Joubert *et al.*, 1989; Kuchler *et al.*, 1989a) due to the expression of galectins at the surface of growing axons (axons do grow individually, but as fascicles following a pioneer axon). This role was recently substantiated (Mahanthappa *et al.*, 1994). In contrast, their putative role in interneuronal recognition and in synapse formation between neurons was not compatible with the ultrastructural localization of the lectin, since, although the lectin is transiently over-expressed close to the sites of synapse formation, it is never externalized (Joubert *et al.*, 1989; Kuchler *et al.*, 1989a). The finding that L-14 binds actin (Joubert *et al.*, 1992) may rather suggest a role in the intracellular traffic of ligands to post-synaptic areas (Kuchler *et al.*, 1989a).

There is no evidence that these lectins play an extracellular role as inducers of cell proliferation. When this specific problem was studied, the exogenously added lectins showed an inhibitory effect (Caron *et al.*, 1986). The reason for this negative effect is unknown, but one interpretation is that their major ligands at the cell surface are glycolipids. Bivalent lectins may produce clustering, followed by subsequent internalization and degradation of glycolipids. Degradation products of glycosphingolipids modulate the activity of enzymes involved in signal transduction pathways, triggering cell proliferation (Olivera and Spiegel, 1992; Jayadev *et al.*, 1995; Nagai, 1995). Thus, it may be hypothesized that the inhibitory effect on cell proliferation results from this regulation by degradation products of glycosphingolipids. However, another specific intracellular role is suggested for CBP-35 (localized in the cytoplasm and in the nucleus; Jia and Wang, 1988) since it regulates pre-mRNA splicing (Dagher *et al.*, 1995). The over-expression of this molecule may serve as a modulator of the expression of molecules involved in cell proliferation.

3 The C-Type Lectins

3.1 The Selectins

The selectins are well-known C-type lectins involved in the homing of lymphocytes, or in leukocyte-recruitment. The mechanism of adhesion is well-documented although some ambiguities remain concerning the actual endogenous ligands of the E (endothelial)-selectin and P (platelet)-selectin. Sialyl-Le^x is an important structure (Foxall *et al.*, 1992), but a substantial affinity is directed to sialyl-Le^a (Larkin *et al.*, 1992). L(leukocyte)- and P-selectins also bind (Green *et al.*, 1992; Needham and Schnaar, 1993) to sulfated oligosaccharides containing the HNK-1 epitope (glucuronic acid-3-sulfate). Some endogenous ligands bearing these carbohydrate chains have been identified. For example, the NCA (one of the carcinoembryonic antigens, glycoproteins of the superfamily of immunoglobulins) and the ESL-1 antigen (structurally related to the FGF receptor) constitute major ligands of E-selectins on circulating cells (Stocks and Kerr, 1993; Steegmaier *et al.*, 1995). An endogenous ligand of L-selectin on activated endothelial cells has been identified as the GlyCAM-1 adressin (Imai *et al.*, 1993), a mucin-type glycoprotein bearing a 6'-sulfated sialyl-Le^x oligosaccharide (Hemmerich and Rosen, 1994; Hemmerich *et al.*, 1994). Besides their important role in adhesion, these membrane-bound molecules are potentially involved in signal transduction, since they can be phosphorylated on their intracytoplasmic domain (Crovello *et al.*, 1993). Furthermore, expression of P-selectin regulates the production of pro-inflammatory cytokines by stimulated $CD4^+$ T cells (Damle *et al.*, 1992). The observation that ligands are molecules directly involved in important signal transduction pathways (like the FGF receptor) is certainly of fundamental importance (Steegmaier *et al.*, 1995). Their properties as specific adhesion molecules for homing and recruitment of circulating cells are of fundamental importance in the fields of cancer and autoimmune diseases.

3.2 The NK Sub-Family of C-Type Lectins

The CD69 "activation inducer molecule" is a C-type glycosylated membrane-bound lectin (Hamann *et al.*, 1993; Santis *et al.*, 1994; Ziegler *et al.*, 1994) induced during lymphocyte activation. It plays an important role in lymphocyte activation, in particular inducing the differentiation of T cells. Sequence homologies defined other surface antigens of human and murine leukocytes (especially NK cells) as lectins (Yabe *et al.*, 1993; Adamkiewicz *et al.*, 1994; Bezouska *et al.*, 1994). Their biological function is still poorly documented, but it can be hypothesized that these surface lectins are involved in the association of specific molecular complexes containing their ligands. The evidence for a role in B cell activation was obtained for the C-type lectins CD23 and CD72 (Gordon, 1994).

3.3 The Soluble C-Type Lectins

Soluble calcium-dependent mannose-binding proteins have been isolated from liver and from blood (Drickamer *et al.*, 1986; Oka *et al.*, 1987, 1988). These lectins are members of a larger family known as MBPs (mannose-binding proteins), surfactant proteins, collectins, or conglutinins (Benson *et al.*, 1985; Andersen *et al.*, 1992; Haagsman, 1994; Kuroki and Voelker, 1994), which have a CRD associated with a collagenous tail. These molecules have been used for crystal studies (Weis *et al.*, 1992) as well as for site-directed mutagenesis (Quesenberry and Drickamer, 1992). The CRD has a complex structure since calcium ions are important both for the stabilization of the three dimensional structure of the molecule and for the binding of the monosaccharide. Pioneer works demonstrated that site-directed mutagenesis of two residues changes a mannose-binding protein to a galactose-binding protein (Drickamer, 1992; Iobst and Drickamer, 1994).

The MBP and surfactant proteins participate to essential immunological processes, since they interact with complement constituents (Kawasaki *et al.*, 1989; Ohta *et al.*, 1990; Ohta and Kawasaki, 1994). Evidence was provided that MBP and surfactant proteins neutralize (in a complement-dependent mechanism) the pathogenicity of influenza type A virus (Anders *et al.*, 1990, 1994), favoring its opsonization (Benne *et al.*, 1994; Hartshorn *et al.*, 1994; McNeely and Coonrod, 1994). They also facilitate the opsonization of specific bacteria (Summerfield, 1993). Surfactant protein A binds through its CRD to the lipopolysaccharide of the bacterial surface (Lim *et al.*, 1994; Van Iwaarden *et al.*, 1994). The possible role of these molecules in the processes of leukocyte activation has not been studied yet, although such a role may be expected as for the other mannose-binding proteins.

4 Soluble Calcium-Independent Mannose-Binding Lectins

The CSL (cerebellar soluble lectins) were isolated from the cerebellum of young rats (Zanetta *et al.*, 1987a). Immunological as well agglutinating properties indicate that a family of CSL molecules is present with subunits of different Mr depending on the cell type. CSL subunits of 31.5 and 33 kDa were found in neurons and astrocytes, whereas subunits of 43 kDa are found in myelinating cells and activated neutrophils. The CSL recognizes preferentially $Man_6GlcNAc_2$ *N*-glycans, hybrid type *N*-glycans sulfated and fucosylated on the core GlcNAc residues, and heparin (with a lower affinity). Preliminary structural data indicate that the 31.5 kDa form has a significant homology with a galactose-binding lectin of a sponge (Buck *et al.*, 1992) and a lower homology (14.5 % for two interrupted domains) with the L-14-II human galectin. Shorter domains present homologies with FGF_β. Interestingly, L-14-II and FGF_β domains homologous to CSL correspond to domains presenting similar three-dimensional organization in the crystal structure, found also in interleukin 1 and TNF- (Eriksson *et al.*, 1991; Zhang *et al.*, 1991). Studies of the biological function of CSL molecules indicate that this molecule is important both for cell adhesion (by forming bridges between surface glycans present on different cells) and for signal generation (by clustering ligands at the surface of the same cell).

4.1 Glycoprotein Ligands of Lectin CSL

In non-malignant cells, the endogenous ligands of CSL are a few cell-specific glycoproteins. For example, CSL recognizes the early form of the myelin-associated glycoprotein (MAG; Lai *et al.*, 1987) in early stages of myelination (Badache *et al.*, 1992, 1995). The late form of MAG, differing from the former by its *N*-glycans and its intracytoplasmic domain, has no affinity for CSL. In compact myelin of the peripheral nervous system (Kuchler *et al.*, 1989b), a major CSL ligand is the P0 glycoprotein (Uyemura and Kitamura, 1991). On the neuronal surface one of the major ligand of CSL is the P31/CD24/nectadrin/heat stable antigen (Kay *et al.*, 1991; Kadmon *et al.*, 1992; Nédelec *et al.*, 1992), a 31 amino-acid peptide containing four *N*-glycans and a glycosyl-phosphatidylinositol (GPI) anchor. In human T cells, glycosylated forms of CD3 are ligands of CSL (very low Mr constituents are suggested to be GPI-anchored glycoproteins of the B7 family) whereas in B cells, the major ligands are the same low Mr glycoproteins and CD24. This cell-specific profile of CSL ligands is no longer observed in cancer cells, where the profiles appear very complex and the quantity of CSL ligands is dramatically increased (Zanetta *et al.*, 1991; Maschke *et al.*, 1993).

4.2 Lectin CSL as an Adhesion Molecule

One interesting aspect of the CSL ligands is that some of them have been identified as glycoproteins involved in cell adhesion mechanisms. Some of them belong to the superfamily of immunoglobulins (CAMs) such as MAG and P0 (Sakamoto *et al.*, 1987), or to the GPI-anchored glycoproteins: P31/CD24/nectadrin/heat stable antigen, Thy-1 and possibly members of the B7 family (VanGool *et al.*, 1995). For the MAG, only the early form, which has CSL-binding *N*-glycans, has an adhesive role, whereas the late form, which does not bind to CSL, is a poorly adhesive molecule (Zanetta *et al.*, 1994a). This suggests that homophilic mechanisms (interaction between peptide domains of the same molecule at the surface of different cells) suggested previously (Edelman, 1986) are not predominant. This view was confirmed by the minimal effect of the inactivation of the gene for the neural cell adhesion molecule (*N*-CAM) on the development of the brain (Cremer *et al.*, 1994). In contrast, the role of specific *N*-glycans in adhesion mechanisms, induced by glycoproteins supposed to act by homophilic mechanisms, has been documented in transfected cells. For example, the increased adhesion of CHO cells transfected with the gene for P0 (Filbin *et al.*, 1990) is no longer observed (Filbin and Tennekoon, 1991) in mutant CHO cells lacking the *N*-acetyl-glucosaminyl-transferase I (GNTase I), which produce Man_5 $GlcNAc_2$ *N*-glycans (not recognized by CSL). In fact, hybrid- and complex-type *N*-glycans seem to play an essential role in embryonic development, since inactivation of the gene for GNTase I provokes the death of the embryo at mid-gestation (Ioffe and Stanley, 1994; Metzler *et al.*, 1994). The hypothesis that *N*-glycans are necessary for the stabilization of the true conformation of the hypothetical homophilic binding site (Filbin and Tennekoon, 1993) is contradicted by the finding that the isolated *N*-glycans of the P0 glycoprotein are strong inhibitors of P0-dependent adhesion (Yazaki *et al.*, 1992). The hypothesis that *N*-glycans are important because CAMs have lectin activity is not proven. In fact, considering soluble polyvalent extracellularly expressed lectins, there is no need for homophilic domains or for a lectin activity for these glycoproteins. The observation that the interactions, presumably attributable to homophilic mechanisms, are inhibited by antibodies against carbohydrate structures identical to those recognized by CSL, strongly suggests that CSL itself could be responsible for this adhesion. Indeed, a strong inhibitor of adhesion is the L3 antibody, which recognizes $Man_6GlcNAc_2$, and another is the HNK-1 antibody (Bollensen and Schachner, 1987; Kucherer *et al.*, 1987; Fahrig *et al.*, 1990). The former epitope is relatively rare in mammals but has been identified as a high affinity CSL ligand (Marschal *et al.*, 1989). The latter is present on several molecules including glycolipids. In our hands, the HNK-1 type antibody inhibits the binding of CSL to its glycoprotein ligands. But glycolipids or glycoproteins that have this epitope (such as the late form of MAG) are not ligands of CSL. Thus, it is suggested that only specific hybrid type *N*-glycans having the HNK-1 epitope are CSL ligands and are actually involved in adhesion. The inhibitory effect of HNK-1 antibodies (Griffith *et al.*, 1992; Hall *et al.*, 1993) on adhesion (homotypic aggregation or adhesion of neural cells on CAMs and laminin) may be due to steric hindrance of the antibody to reactivity with a lectin like CSL. These two oligosaccharide determinants also appeared to be essential (Kadmon *et al.*, 1990b) for the "assisted homophilic interaction between Ng/CAM and *N*-CAM" (Kadmon *et al.*, 1990a), a mechanism which proposes that the adhesive properties of *N*-CAM are increased when the later associates with Ng-CAM.

4.2.1 CSL and neuron migration

One well documented cell adhesion mechanism is contact guidance of neuron migration along astrocytic processes. In nervous tissue, neurons proliferate in specific areas and have to migrate along pre-existing astrocytic processes in order to reach their adult localization. In cultured cerebellar explants, the migration of neurons along astrocytic processes is completely inhibited by low concentrations of monovalent antibodies against CSL and against its major ligand at the neuronal surface (Lehmann *et al.*, 1990), the P31/CD24/nectadrin/heat stable antigen (Kuchler *et al.*, 1989c). The interpretation of these data, based on the ultrastructural immunolocalization of CSL and P31, is that CSL (produced and externalized by astrocytes) forms bridges between ligands at the surface of neurons and ligands at the surface of astrocytes. This model is reinforced by the observation that the homotypic adhesion between astrocytes (which have surface CSL ligands) is inhibited by the same anti-CSL Fab fragments (Kuchler *et al.*, 1989d).

4.2.2 CSL and myelination

Anti-CSL antibodies also inhibit the formation of compact myelin *in vitro* (Kuchler *et al.*, 1988). Furthermore, when a monoclonal anti-CSL antibody that inhibits the lectin activity of CSL was injected into rat brain ventricles, it opened the ependymal barrier lining the cerebral ventricles (Kuchler-Bopp *et al.*, submitted) and provoked demyelination *in vivo*. This specific effect can be explained because (1) CSL bridges between its ligands at the surface of ependymal cells are responsible for the tightening of the ependymal barrier (which can be also opened by CSL ligands such as HRP (Broadwell *et al.*, 1983) or mannose-containing neoglycoproteins (Kuchler *et al.*, 1994)) and (2) CSL bridges between ligands at the surface of myelin are involved in myelin compaction (Kuchler *et al.*, 1988). These effects of anti-CSL antibodies mimic periventricular anomalies found in patients with multiple sclerosis. These patients have systematic autoantibodies to CSL in their cerebrospinal fluid (Zanetta *et al.*, 1990, 1994b) and in their blood (Zanetta *et al.*, 1990b), suggesting that CSL is an important immunological target in multiple sclerosis.

4.2.3 CSL and early aggregation of activated lymphocytes

The early calcium-independent aggregation of human lymphocytes observed after activation with phorbol esters appears to be largely CSL-dependent (CSL is an extremely rapidly expressed antigen at the surface of activated cells), and this adhesion mechanism can be reproduced in completely artificial systems using fixed resting human lymphocytes and addition of exogenous CSL (Zanetta *et al.*, 1995). The aggregation can be inhibited by anti-CSL Fab fragments and monoclonal antibody and by CSL ligands. The role of CSL in the aggregation of lymphocytes is confirmed by the observation that CD24/nectadrin, a major glycoprotein ligand of CSL in B cells, plays an important role in the aggregation of activated B cells (Kadmon *et al.*, 1994).

4.2.4 CSL and homotypic adhesion of cancer cells

The calcium-independent homotypic adhesion between C6 glioblastoma cells is completely inhibited by low concentrations of the same Fab fragments (Maschke *et al.*, 1996). This homotypic adhesion is similarly inhibited when C6 cells are cultured in the presence of *N*-glycosylation inhibitors (castanospermine, and deoxynojirimycin which produce $Glc_{3-2}Man_9GlcNAc_2$ *N*-glycans which are not recognized by CSL). In contrast, the homotypic adhesion was fully preserved when cells were cultured in the presence of deoxymannojirimycin and swainsonine (which produce $Man_{9-6}GlcNAc_2$ and short hybrid-type *N*-glycans respectively, which are CSL ligands). In contrast to C6 cells, the homotypic adhesion of CHO cells is less sensitive to anti-CSL antibodies (Lehmann *et al.*, 1991), suggesting that other mechanisms are also involved.

4.3 Lectin CSL as a Mitogen

As with most mannose-binding lectins, the CSL molecules are very active promoters of cell proliferation. A negative example concerns astrocytes. Although these cells have ligands of CSL involved with CSL in homotypic adhesion, exogenous CSL is not mitogenic. Another example concerns rat fibroblasts which do not produce CSL and CSL ligands. On the other hand, in several cell types (lymphocytes, myelinating cells (oligodendrocytes and Schwann cells) and some cancer cells), CSL is an endogenous mitogen.

4.2.1 CSL and human lymphocyte activation

Extensive studies have been made on the mechanisms of lymphocyte activation (for reviews, see Altman *et al.*, 1990; Bierer and Burakoff, 1991; Izquierdo and Cantrell, 1992; Schlessinger and Ulrich, 1992). It is generally assumed that the clonal activation (activation of cells specific for one antigen) of T lymphocytes follows the recognition by the T cell receptor (TCR) of the antigen presented with the MHC by antigen presenting cells. This recognition induces the expression of cytokine receptors, cytokines, oncogene products, increased aggregation properties and, 2–3 days later, cell proliferation. This specific activation is followed by a general activation of the immune system, attributed to cytokines (produced by activated cells) which can bind to those cells that have the proper receptor system. However, it has been shown that activation can be also produced by membranes of activated cells, suggesting that *in vivo* non-specific activation also depends on adhesion. The role of *N*-glycans (Hart, 1982) and, more precisely, mannose-rich *N*-glycans (Savage *et al.*, 1993) has been documented. The activation by specific plant lectins is compatible with such a view, but a similar effect can be obtained using polyvalent antibodies to specific surface molecules, or cross-linking agents. Thus, besides a specific role of surface glycans, the concept is emerging that the specific clustering of surface molecules may induce activation. But activation can be also obtained using substances acting directly on the signal transduction pathway. This is the case with phorbol esters such as PMA, an activator of protein kinase C, which induces a rapid aggregation of the cells, suggesting that an adhesion molecule is expressed early at the surface.

One of these adhesion molecules is the 31.5 kDa form of CSL. CSL is an antigen expressed at a very low level in resting human lymphocytes and is apparently not expressed at the lymphocyte surface. When cells are activated, its level increases by a factor ten within two hours using PMA, and within 24 h using mitogenic plant lectins (Zanetta *et al.*, 1995). Part of the lectin is externalized, binding to surface ligands including CD3 on T cells and CD24 on B cells. The switch-over of major tyrosine-phosphorylated proteins characteristic of the early stage of activation, from $p56^{lck}$ to $p59^{fyn}$ (two tyrosine kinases of the *src* family), occurs with the same kinetics as the expression of CSL (within two hours using PMA and 24 h using mitogenic plant lectins). This switch-over is complete 20 min after addition of exogenous CSL. Furthermore, anti-CSL Fab fragments inhibit the switch-over of tyrosine phosphorylation associated with activation, but are ineffective against the phosphorylation of resting cells. These data suggest that CSL is an "endogenous amplifier of activation signals".

The mass of data accumulated in the field of lymphocyte activation allows one to go further into the explanation of these data. There is general agreement that clustering of TCR complexes is one of the key events in T cell activation (Beyers *et al.*, 1992). The TCR complex is formed from the TCR itself, strongly associated with CD3 molecules (two of which are *N*-glycosylated). Other molecules, including kinases and phosphatases, are also tightly associated. A classical way to activate T lymphocytes is the use of a divalent anti-CD3 antibody, but the CD3-dependent activation process needs additional molecules. For example, B7 molecules are co-stimulatory with anti-CD3 for T cell activation (Van Gool *et al.*, 1995). Since CSL is highly polyvalent, it might be expected that it would cause its ligands to associate at the T cell surface (CD3 and the low Mr glycoprotein ligands, presumably B7 molecules) to produce higher order cell surface organization (Fig. 1). Such an organization, due to the endogenous CSL, could constitute a unique persistent assembly of surface complexes to generate a constant signal for committing lymphocytes to express interleukin 2 receptors, oncogene products and cytokines necessary for proliferation.

The identification of the ligands of plant lectins which induce activation (Concanavalin A and *Robinia pseudoacacia* lectin) indicates that minor glycoproteins with the same Mr as the CSL ligands were involved, but that the major ligands were different. Thus, plant lectins cause TCR complexes to associate with other surface complexes which are not the proper partners. Consequently, they are not able to induce the switch-over of major tyrosine phosphorylation, but instead produce minor signals which induce the slow synthesis of CSL. In contrast, PMA, by activating protein kinase C, supports directly the expression of CSL. The common observation that phorbol esters and plant lectins have an antagonistic effect on lymphocyte activation could be explained by an initial competition between plant lectins and CSL for the same glycoprotein ligands.

Besides the important role of CSL in lymphocyte activation, the demonstration of a large variety of lectins at the leukocyte surface suggests a very complex glycobiological mechanism for the regulation of the activation processes and the association of surface glycoconjugates (and consequently of surface complexes). Some of these lectins are constitutively expressed at the leukocyte surface and are unchanged or reduced (L-selectin) during activation. Many other molecules (including glycoproteins) appear later at the leukocyte surface during the activation process. Furthermore, the initial ligands may be internalized rapidly. Thus, it may be hypothesized that the expression of a lectin and its ligands at the cell surface may modify pre-existing glycobiological interactions, establishing a different organization that persists for a period and which is finally succeeded by other mechanisms which are responsible for cell proliferation and cell differentiation.

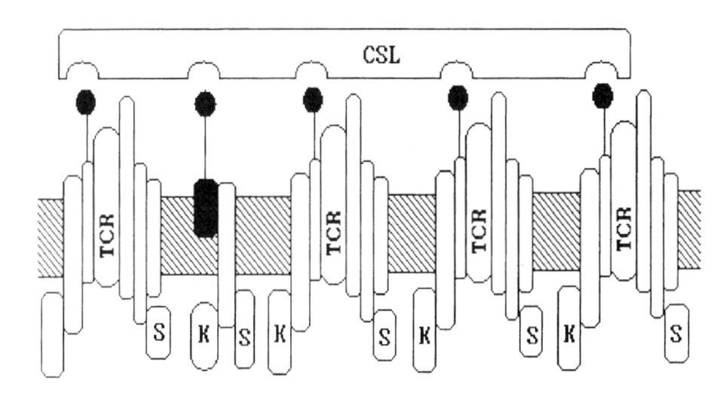

Figure 1. Schematic drawing of the proposed association of molecular complexes through the polyvalent lectin CSL during lymphocyte activation. Surface complexes other than the TCR complex can be associated. The clustering of the glycoprotein ligands induces intracellular interactions, especially phosphorylation-dephosphorylation mechanisms. K = kinases; S = substrate of kinases.

4.2.2 CSL and Schwann cell proliferation

The 43 kDa form of CSL appeared to be an autocrine mitogen for Schwann cells (Badache *et al.*, 1995), the myelinating cells of the peripheral nervous system. The externalized CSL bound to a few ligands at the Schwann cell surface (the major being the early form of MAG), probably clustering its ligands. This clustering provokes specific intracytoplasmic phosphorylation, triggering a proliferative response. Using inhibitors of *N*-glycosylation, a correlation between the presence of CSL-binding *N*-glycans, proliferation and phosphorylation on tyrosine and serine/threonine residues was observed. Schwann cell proliferation can be inhibited by anti-CSL Fab fragments, or stimulated by the addition of exogenous CSL (Badache *et al.*, 1995) or of plasma membranes of different origins (Sobue and Pleasure, 1985), including myelin and axonal membranes. Based on the solubilizing effect of heparin, the mitogenic activity of axonal membranes was initially attributed to a heparin-binding protein attached to an axonal heparan sulfate proteoglycan. In fact, this mitogenic activity is completely inhibited by monovalent anti-CSL antibodies, suggesting that CSL bound to specific glycoprotein ligands of these membranes is the mitogen. The solubilizing effect of heparin can be explained, since heparin is a ligand of CSL which inhibits *in vitro* the binding of CSL to glycoproteins (such as P0, MAG, P31) that contain only *N*-glycans. The evidence that CSL is the axonal mitogen was obtained by the specific solubilization of CSL (and of the mitogenic activity) from axonal membranes using $Man_6GlcNAc_2Asn$ *N*-glycans (Badache *et al.*, 1995), the ligands of CSL at the axonal surface being the P31/CD24/nectadrin glycoprotein and higher Mr glycoproteins, which migrate as the L1/Ng-CAM antigen.

4.2.3 CSL and C6 glioblastoma cell proliferation

Like many malignant cells (and in contrast to normal cells), C6 glioblastoma cells showed very abundant CSL ligands, presenting a complex "cancer-specific" profile similar to malignant cells from different tissues and species. The proliferation of C6 glioblastoma cell was strongly inhibited by anti-CSL Fab fragments, and was also inhibited when cells were cultured in the presence of inhibitors which produce *N*-glycans not recognized by CSL. This pattern of inhibition was identical to that for adhesion of C6 cells (Maschke *et al.*, 1996). This essential role of CSL and its ligands in both adhesive and proliferative properties of C6 cells suggested that these compounds play an important role in the loss of contact inhibition of malignant cells. In normal cells, the quantity of CSL ligands is low. When CSL is externalized, these ligands are rapidly saturated with CSL so that the induced adhesion and proliferation cannot exceed a certain level. In cancer cells, the large quantity of CSL ligands never becomes saturated, so that when CSL is externalized, it produces increased adhesion and proliferation. Consequently, proliferating cells are able to built three dimensional structures *in vitro* as a result of the loss of contact inhibition.

The modulation of the proliferation of C6 cells using inhibitors of *N*-glycosylation allowed the examination of their effects on signal transduction pathways considered to be important for cell proliferation. In contrast to Schwann cells, tyrosine phosphorylation does not seem to be involved in the generation of proliferative signals, but instead, phosphorylation on serine and threonine residues is required. Furthermore, the levels of the *abl/bcr* (tyrosine kinases) and of *erbB2* oncogene products (tyrosine-phosphorylated proteins) are modified whether C6 cells are proliferating or not. In contrast, the expression of the *ras* oncogene product was strongly decreased using inhibitors which produces *N*-glycans that are not recognized by CSL.

5 Heparin-Binding Growth Factors

Several growth factors isolated by their affinity to heparin have a biological function which depends almost entirely on the presence of exogenous heparin or of endogenous cell surface heparan sulfate proteoglycans. This is the case for the fibroblast growth factors, FGFα and FGFβ. It was initially assumed that these factors bound to the highly negatively charged heparin through their cationic domains. However, site directed mutagenesis of these putative cationic sites to hydrophilic or acidic amino acids (Presta *et al.*, 1992) produced FGF molecules with similar affinities for heparin and which induced only subtle differences in their biological properties (i.e. receptor binding and induction of cell proliferation). Studies of the carbohydrate moieties of heparin interacting with FGFβ defined a minimal tetrasaccharide structure [L-IduUraβ*1–3*(or D-GlcUraβ*1–3*)GlcN(2-SO_4)β*1–4*]$_2$ (Maccarana *et al.*, 1993; Rustani *et al.*, 1994). This suggested a CRD rather than ionic interactions. The similarity of the three-dimensional structure of FGFβ with L-14, TNFα and interleukin 1, which have calcium-independent lectin activities, suggests that FGFβ is actually a lectin.

Heparin needs repetitive carbohydrate sequences to increase the activity of FGF, since shorter oligomers are inhibitory (Maccarana *et al.*, 1993). This suggests that the binding of heparin does not primarily induce a conformational change from a poorly active to an active form of the FGF receptor. Recent data (Spivak-Kroizman *et al.*, 1994; Obrien, 1995) resolved this critical problem. The binding of FGF to specific repetitive domains of heparin induces a clustering of FGF molecules which, consequently, induces an oligomerization of FGF receptor molecules (Fig. 2). This results in trans-phosphorylation of the FGF receptor complexes and therefore the generation of an intracellular signal. The question why FGF needs heparin to induce the proliferation of some cells, and does not need heparin for others, may be answered by considering the previous mechanism. When cells do not have an endogenous surface proteoglycan recognized by FGF, heparin is needed for clustering FGF receptor molecules. When endogenous proteoglycan ligands of FGFβ at the cell surface are present, the binding of FGFβ to its high affinity receptor provokes the association of the latter with the proteoglycan-containing complex (Fig. 2). Therefore, the resulting signal may be also interpreted in terms of clustering of surface molecules, as in other models concerned with lectins. The structural sequence homologies with FGFβ suggest that other heparin-binding molecules including other FGF molecules, ECGF and INT-2, also have a lectin activity. Unfortunately, it is not known if these molecules, defined as heparin-binding recognize the same oligosaccharide moiety of heparin. The problem remains for the other heparin-binding proteins which have been less extensively studied, and particularly the very basic proteins of the midkine family. Site-directed mutagenesis of individual cationic amino acids may help to understand whether the heparin binding is due to strong ionic interactions or to specific glycobiological interactions.

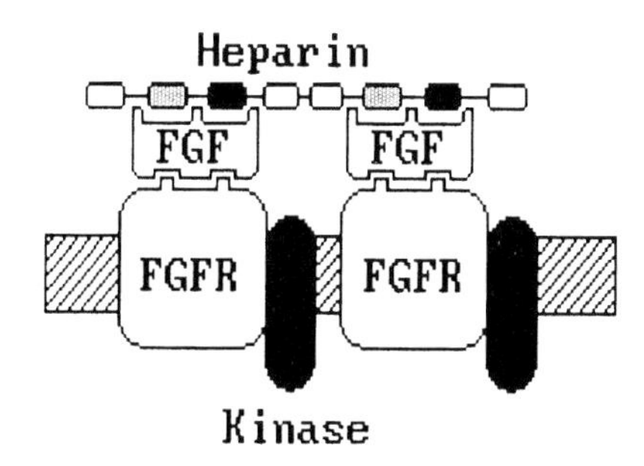

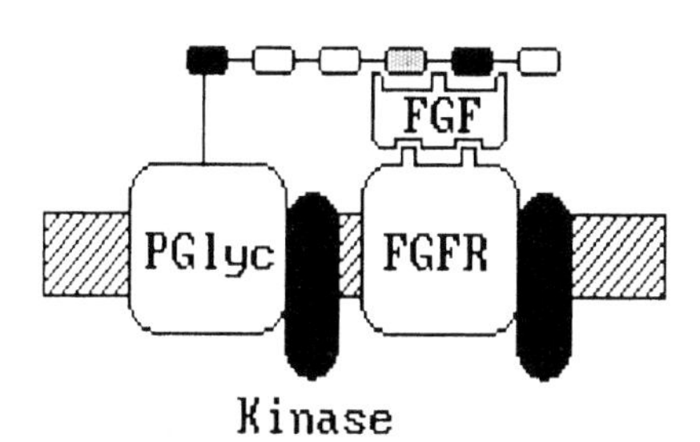

Figure 2. Schematic drawing of the interactions between FGF and heparin or heparan sulfate proteoglycans. In (a), the binding of FGF molecules on the template heparin provokes the oligomerization of complexes containing the FGF receptor, resulting in intracellular phosphorylation of the receptor molecules. In (b), the binding of FGF molecules to an endogenous heparan sulfate proteoglycan associates the FGF receptor with the former.

6 Cytokines

6.1 Interleukin 2 (IL-2)

IL-2 is a 14 kDa molecule, produced essentially by activated T cells, which can stimulate the proliferation of T cells, B cells and other leukocytes having IL-2 receptors at their surface. Three IL-2 receptors (IL-2R) have been identified (α, β, γ), the IL-2Rβ being constitutively expressed on resting cells (Zola *et al.*, 1991). Site-directed mutagenesis of IL-2 and domain-specific antibodies (Cohen *et al.*, 1986; Ju *et al.*, 1987) demonstrated that IL-2 needs two domains for expressing its full biological activity: one is involved in the binding to its receptors and a second is necessary for the expression of the biological activity measured by cell proliferation. But no indication of the nature and mechanism of action of this second site was proposed. Several years ago, IL-2 was shown to be a lectin that bound to Man_5 and Man_6 but not to Man_9 oligomannosidic *N*-glycans in the absence of calcium at neutral pH (Sherblom *et al.*, 1989). In contrast, the binding to Man_5 and Man_6 oligomannosidic *N*-glycans was calcium-dependent at pH 5.0. Due to low sequence homologies, it was suggested that IL-2 was a C-type lectin (Sherblom *et al.*, 1989). In fact, the high affinity binding of IL-2 for Man_5 and Man_6 *N*-glycans is observed in the presence of 5 mM EDTA (Zanetta *et al.*, 1996), demonstrating that IL-2 is a calcium-independent lectin.

Studies on the signal transduction pathway of IL-2 binding (for reviews, see Taniguchi and Minami, 1993; Waldman, 1991) showed that, although IL-2Rβ has no kinase activities, its intracytoplasmic domain is phosphorylated at tyrosine after fixation of IL-2 (Farrar *et al.*, 1990). This suggests that the IL-2 fixation on IL-2Rβ induces its association with other surface molecular complexes. Since this phosphorylation is due to the kinase p56^{lck} (Shibuya *et al.*, 1994), considered to be associated with the TCR complex, it was suggested that IL-2 bound to its receptor, associating IL-2Rβ with the TCR complex. One possibility to explain this association was that IL-2 could bind to glycans of molecules of the TCR complex through its CRD. The similarities of the carbohydrate-binding properties of CSL and IL-2 and the demonstration that glycosylated forms of CD3 are CSL ligands gave credence to this hypothesis (Fig. 3a). Using a complex experimental design (Fig. 3b), it appeared that this mechanism did occur. Indeed, in the presence of IL-2 (and only in the presence of IL-2), IL-2Rβ was co-immunoprecipitated with the TCR com-

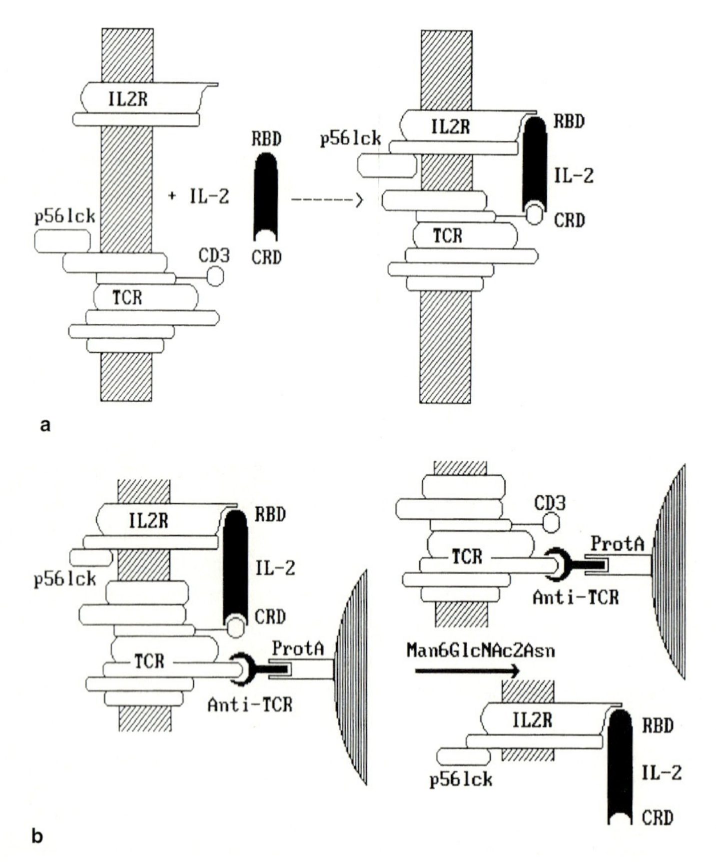

Figure 3. (a) Schematic drawing of the molecular mechanism proposed for the association of the TCR complex with IL2-Rβ (IL2R) through the bi-functional IL-2 (having a receptor binding domain (RBD) and a carbohydrate-binding domain (CRD). (b) Schematic representation of the experimental design used for demonstrating the IL-2 and Man_6 dependence of the interaction between IL-2Rβ and the TCR complex.

plex using an anti-TCR antibody. However, IL-2Rβ can be detached from the immunoprecipitate by a Man_9-independent, Man_6-dependent mechanism. This specifically detached IL-2Rβ co-immunoprecipitated with the kinase $p56^{lck}$, indicating a strong association between IL-2Rβ and $p56^{lck}$ (Zanetta *et al.*, 1996). Thus, the lectin activity of IL-2 provokes the association between IL-2Rβ and glycoprotein constituents of the TCR complex. Using CSL to detect the putative IL-2 ligands in the TCR complex, two glycoprotein subunits were identified, the one having the same Mr as a glycosylated form of CD3, while the other was an unidentified glycoprotein of 55 kDa. This suggested that after fixation of IL-2 to its receptor, two molecules of IL-2Rβ could be associated with the TCR complex (Fig. 3a). Due to the similarities of the carbohydrate-binding properties of CSL and IL-2, it might be suggested that IL-2 bound to its receptor can also associate IL-2Rβ with other complexes containing other glycoprotein ligands of IL-2. Consequently, although this point is not documented, the lectin activity of IL-2 may associate IL-2Rβ with several molecular complexes at the surface of human lymphocytes. However, IL-2 is not able to associate TCR and B7 as did CSL (Fig. 1).

This mechanism (very similar to the one proposed for FGF on cells having an endogenous heparan sulfate ligand) is particularly stimulating since a cytokine, endowed with a monovalent lectin activity, can provoke the surface association between different types of surface receptor complexes, because it is actually bifunctional (having a receptor-binding site and a lectin site). It is able to associate surface molecules and this provokes intracytoplasmic associations, through the relatively well known mechanism of intracellular recognition. The $p56^{lck}$ possesses the SH2 domain of the kinase of the *src* family, which recognizes short intracytoplasmic sequences including a phosphotyrosine residue. The jump of $p56^{lck}$ from one receptor complex to another is representative of these changes in intracellular association (Garnett *et al.*, 1993). These experiments are also stimulating because they demonstrate for the first time that an extracellular glycobiological interaction which occurs *in vivo* can modify the intracellular organization of molecules.

The experiments described above were short term experiments on resting cells (IL-2Rβ is rapidly internalized and degraded). In activated cells, the situation is more complex since CSL, endowed with a similar carbohydrate-binding specificity to that of IL-2, is also expressed at the cell surface. Thus, the possibility of competition and/or cooperative effect could be postulated. Since over-expression of IL-2 and IL-2R (α, β and γ) is a relatively late mechanism, it may be hypothesized that, at a certain stage of the activation process, an IL-2 dependent clustering remodels the initial CSL-dependent clustering of surface complexes, by competition with the same ligands. The example of these two lectins, endowed with similar carbohydrate-recognition properties, may provide a very efficient mechanism of regulating signaling systems, which cannot be understood without considering glycobiological interactions.

6.2 Interleukin 1 (IL-1) and Tumor Necrosis Factor (TNF)

IL-1α and IL-1β are cytokines of 17 and 14 kDa which are produced by macrophages, epithelial cells and T and B lymphocytes. These molecules have multiple activities but one considered to be the most important is the induction of the proliferation of thymocytes co-stimulated with the plant lectin PHA and B cell activation. TNFα is a 17 kDa protein produced by macrophages whereas the 18 kDa TNFβ (lymphotoxin) is produced by T cells. This component is directly cytotoxic for malignant cells but also stimulates the production of other cytokines, activates endothelial cells and macrophages and plays essential roles in inflammation.

Evidence that IL-1 is a lectin was obtained from its interactions with the *N*-glycans of uromodulin (Hession *et al.*, 1987; Muchmore and Decker, 1987; Brody and Durum, 1989). This interaction appears to be dependent on complex polyantennary *N*-glycans. As with other cytokines, the fixation of IL-1 to its receptor also induces its association with a kinase (Martin *et al.*, 1994). Similar association of kinases with the interleukin receptors appears as a general effect of the fixation of the cytokine to its receptor (Venkitaraman and Cowling, 1994).

The first evidence that TNFα could have a lectin-like activity was obtained through its interaction with the uromodulin glycoprotein (Sherblom *et al.*, 1988). However, TNF could bind to *Shigella flexneri* bacteria (Lucas *et al.*, 1994) by a N,N'-di-*N*-acetyl-chitobiose-dependent mechanism which does not involve the TNF receptor. Site directed mutagenesis and inhibition by synthetic peptides against the putative CRD allowed the definition of an area comprising the CRD of TNFα (Lucas *et al.*, 1994). It should be emphasized that the TNF receptors, like the cytokine receptors, do not have a kinase activity. The three dimensional structure of TNF resembles that of IL-1 (Eriksson *et al.*, 1991; Zhang *et al.*, 1991) suggesting that they belong to the same family. Recent data indicate that molecules produced by leukocytes have strong sequence homologies with TNFα including TNFβ, CD70/CD27 ligand and the CD40 ligand (Hintzen *et al.*, 1994). Thus, it may be speculated that these molecules also have lectin activities.

7 Conclusions and Perspectives

Recent data indicate that complex glycobiological interactions of fundamental importance take place at the cell surface. Such a possibility was underlined several years ago (Feizi and Childs, 1987). It is now evident that a large variety of lectins is found at the surface of cells, which may play a role in intercellular contacts or in the organization of the cell surface glycoconjugate ligands. These lectins are either integral membrane components (many C-type lectins), or externalized soluble compounds (calcium-independent lectins), or bifunctional molecules having a CRD and a receptor-binding site (some cytokines and growth factors).

One prominent function attributed to lectins is the specific association of cell surface complexes. The evidence for the involvement of extracellular lectin-glycan interactions in the stabilization of these complexes is lacking as yet, although such interactions are expected if a lectin and its ligands are constitutively present at the cell surface. In contrast, lectins appear to be essential for the functional associations between surface complexes. Externalized soluble polyvalent lectins (like CSL), bifunctional soluble lectins (like cytokines), and membrane-bound lectins can induce a clustering of surface complexes containing their ligands. This remodeling of the organization of surface molecules provokes new intracellular interactions, including phosphorylation-dephosphosphorylation mechanisms (or potentially addition or deletion of *O*-GlcNAc residues) specific to the clustered complexes. This results in the true intracellular signal which is transduced to the nucleus and, consequently, in the induction of the synthesis of molecules necessary for the true cellular response (expression of new surface receptors, production of growth factors, expression of specific oncogene products). The concept that oncogene expression may be regulated by specific extracellular signals is largely accepted, but its glycobiological nature is frequently denied. The finding that some cytokines or growth factors have lectin activities helps to overcome this reluctance, since everybody accepts that binding of a cytokine or a growth factor to its specific receptor regulates the expression of oncogene products (Sato *et al.*, 1991).

The observation that some cytokines and growth factors are bifunctional (one receptor binding site and one carbohydrate binding site) requires a new explanation for the observation that the binding to a receptor is not sufficient for the expression of the biological activity. The carbohydrate-binding site is necessary as well. Based on structural similarities (including three-dimensional ones) between cytokines and growth factors, it is expected that at least some of them have a carbohydrate-binding site. Structure-activity studies emphasize the presence of a domain, opposite the receptor-binding site, which is necessary for the activity of these molecules. Phosphorylation and "jumping" of kinases to the receptor are frequently observed after binding to a receptor which does not itself have a kinase activity (other receptors, like the EGF receptors, are kinases). Thus, a systematic search for lectin activities for cytokines (interleukins, members of the TNF family, interferons) and growth factors (heparin-binding growth factors), as well as the identification of their glycoprotein ligands, is a promising area for the future.

The prevalence of extracellular glycobiological interactions for generation of signals raised a fundamental question in the field of glycobiology, concerned with the actual effect of *N*-glycosylation inhibitors (for a review, see Elbein, 1987) on cell proliferation. The alternatives may be summarized as follows: (1) are these compounds directly inhibiting the expression and/or the traffic of molecules (such as oncogene products) that are necessary for the proliferation of the cells?; (2) are these inhibitors affecting primarily the surface *N*-glycans which are necessary for maintaining a surface signal which governs the expression of molecules necessary for cell proliferation (such as oncogene products)? The absence of *N*-glycans (using tunicamycin) and the production of immature *N*-glycans (using deoxynojirimycin or castanospermine) inhibit both basal cell proliferation and the cell proliferation induced by lectins or cytokines. Several cell surface receptors are unable to transduce a signal when their *N*-glycans are absent or incorrect (use of glycosylation inhibitors or site directed mutagenesis (Leconte *et al.*, 1992)), although the receptor molecules are normally expressed at the cell surface and are able to bind their ligands. The concept of glycobiological clustering of receptor complexes provides an explanation for this inability to generate a signal. This hypothesis is compatible with the sequence of expression of molecules in the process of lymphocyte activation. The clustering lectin CSL is expressed very early at the surface, much earlier than the *ras* oncogene product and other molecules (such as IL-2 and IL-2Rβ). If the correct *N*-glycans are not present, the true surface signal cannot occur and, consequently, neither can the expression of later expressed molecules (Bowlin *et al.*, 1989, 1991).

A second important aspect is the role of lectins in cell adhesion mechanisms. Depending on the nature of the ligand, the same lectin may have adhesive or anti-adhesive properties. The interaction of galectins with laminin is inhibitory (probably due to steric hindrance) for another type of interaction which remains to be discovered. Other lectins may be involved in recognition between two cells, but not in adhesion mechanisms. For example, the neuronal membrane-bound lectin R1 is transiently externalized by target neurons to phagocyte glycoproteins at the surface of incoming axons (Zanetta *et al.*, 1987b). This transient contact seems to be necessary for committing the two neurons to synthesize the molecules involved in the formation of the synapse (a specialized contact), but is not concerned with the stabilization of this contact.

A large number of glycoproteins have been involved in adhesion mechanisms, but most of the studies have been concerned with the polypeptide chains and, with a few exceptions, not with the carbohydrate moieties. Polysialosyl α2–8 sequences, specifically found on *N*-CAM, show anti-adhesive properties *in vitro*, but experiments on inactivation of the gene for *N*-CAM indicate that these sequences do not play a major role during embryogenesis or development of the brain. *N*-glycans of the P0 glycoprotein are involved in adhesion, but it is not clear which of the 5 variants is effective. The GPI-anchored brain Thy-1 contains about 20 different glycan variants at the same glycosylation site. Such a complexity is puzzling and contributes to the difficulty of these studies. The fact that multiple alternative splicing variants (192 possibilities for *N*-CAM resulting from a single gene) can be produced (including GPI-anchored forms) makes it difficult to know if one particular *N*-glycan on a particular polypeptide chain is involved in adhesion mechanisms or not. The evidence that splicing variants have completely different functions is provided by the studies on MAG. The early form of MAG (but not the late form) plays a role in both adhesion and signal transduction (its intracytoplasmic domain contains specific sites for phosphorylation). Since the *N*-glycans and the localization of the two forms are different, it is tempting to hypothesize that the intracytoplasmic domain orients the intracellular traffic of the glycoprotein to special Golgi compartments where different *N*-glycosylations can occur. Although this finding has yet to be confirmed, a second mRNA for P0 glycoprotein has been described (Poduslo, 1990), which was found at early stages of myelination and during remyelination of the lesioned sciatic nerve. Thus, a possibility is emerging that there is a coupling between an intracytoplasmic sequence and the synthesis of specific *N*-glycans involved in cell adhesion mechanisms.

A new concept that has emerged from the glycobiological approach in the field of adhesion and signal transduction is the importance of glycoproteins having a small polypeptide chain and a very high level of glycosylation, especially *N*-glycosylation. Such glycoproteins can exist as major components of the cell surface (in the developing rat cerebellum, they represent more than 10 % of the total protein). Examples are the Thy-1 antigen, the CD24 and B7 glycoproteins. Although they are involved in signal transduction in adult tissues or leukocytes (the surface complexes containing these molecules have associated kinases), the GPI-anchoring of these "carbohydrate-carriers" is a unique feature for glycobiological interactions. They are not associated directly with the cytoskeleton, so that, when bound to another membrane through a polyvalent lectin, the two membranes can move relative to one another. One problem in the field of neurobiology is to understand how an oligodendrocyte or a Schwann cell can make several tens of rolls of membranes around an axon without twisting the neurotubules and neurofilaments. A GPI-anchored "carbohydrate carrier" may function as a "ball bearing" in the areas of contact between axons and myelin. The relatively recent discovery of the GPI anchors and the difficulties in handling (wrong apparent Mr on SDS gels, amphipathic properties) and sequencing these molecules may explain why they have not been studied extensively. The study of the carbohydrate moieties of these glycoproteins will be a major breakthrough in the field of cell adhesion.

8 Appendix

Terminology: Some confusion exists in the literature concerning the terminology "homophilic", "homotypic", "heterophilic" and "heterotypic". The terminology "homotypic" and "heterotypic" refers only to the nature of the cell types in contact and not to a molecular mechanism. A cell contact is "homotypic" if cells have the same phenotype. It is "heterotypic" if cells have different phenotypes. For example, if cells transfected with the gene of a specific molecule are put into contact with the wild type cells, the contact is necessarily "heterotypic". When considering the mechanism of interaction itself, it may be "homophilic" or "heterophilic". The mechanism is "homophilic" if one domain of a molecule at the surface of one cell is interacting with the same domain of the molecule at the surface of the other cell. The involvement of molecules of the superfamily of immunoglobulins (CAMs) or cadherins (in the latter case, the homophilic interaction is considered as calcium-dependent) has been considered as "homophilic". If the interaction involves two complementary molecules at the surface of the two cells (Fig. 4), the mechanism is "heterophilic" whether or not the cells are the same. This includes, for example, lectin-carbohydrate, and integrin-ligand recognition, these interactions being calcium-dependent or not. In the case of the interaction of a polyvalent soluble lectin with its surface ligands, the interaction has to be considered as "heterophilic", because it results from at least two "heterophilic interactions".

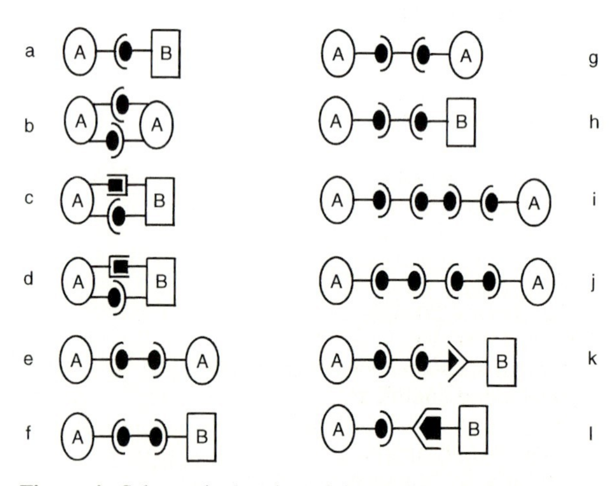

Figure 4. Schematic drawing of theoretical possibilities for glycobiological interactions involved in cell homotypic (b, e, g, i, j) and heterotypic (a, c, d, f, h, k, l) adhesion. (a) correspond to lymphocyte homing. (g) and (h) are situations encountered for CSL (Kuchler *et al.*, 1988; Lehmann *et al.*, 1990). Evidence for situation (i) has been obtained in CHO cells (Lehmann *et al.*, 1991).

References

Adamkiewicz TV, Mcsherry C, Bach FH *et al.* (1994): Natural killer cell lectin-like receptors have divergent carboxy-termini, distinct from C-type lectins. In *Immunogenetics* **40**:318.

Altman A, Mustelin T, Coggeshall KM (1990): T lymphocyte activation: a biological model of signal transduction. In *Crit. Rev. Immunol.* **10**:347–91.

Anders EM, Hartley CA, Jackson DC (1990): Bovine and mouse serum β inhibitors of influenza A viruses are mannose-binding lectins. In *Proc. Natl. Acad. Sci. USA* **87**:4485–9.

Anders EM, Hartley CA, Reading PC *et al.* (1994): Complement-dependent neutralization of influenza virus by a serum mannose-binding lectin. In *J. Gen. Virol.* **75**:615–22.

Andersen O, Friis P, Nielsen EH *et al.* (1992): Purification subunit characterization and ultrastructure of 3 soluble bovine lectins. Conglutinin, mannose-binding protein and the pentraxin serum amyloid P-component. In *Scand. J. Immunol.* **36**:131–41.

Asou H, Miura M, Kobayashi M *et al.* (1992): Cell adhesion molecule L1 guides cell migration in primary reaggregation cultures of mouse cerebellar cells. In *Neurosci. Lett.* **144**:221–4.

Badache A, Burger D, Villarroya H *et al.* (1992): Carbohydrate moieties of MAG, P0 and other myelin glycoproteins potentially involved in cell adhesion. In *Dev. Neurosci.* **14**:342–50.

Badache A, Lehmann S, Kuchler-Bopp S *et al.* (1995): An endogenous lectin and its glycoprotein ligands are triggering basal and axon-induced Schwann cell proliferation. In *Glycobiology* **5**: 371–83.

Barondes SH, Castronovo V, Cooper DNW *et al.* (1994a): Galectins. A family of animal beta-galactoside-binding lectins. In *Cell* **76**:597–8.

Barondes SH, Cooper DNW, Gitt MA *et al.* (1994b): Galectins. Structure and function of a large family of animal lectins. In *J. Biol. Chem.* **269**:20807–10.

Benne CA, Kraaijeveld CA, Vanstrijp JAG *et al.* (1995): Interactions of surfactant protein A with influenza A viruses: Binding and neutralization. In *J. Infect. Dis.* **171**:335–41.

Benson B, Hawgood S, Shilling J *et al.* (1985): Structure of a canine pulmonary surfactant apoprotein: cDNA and complete amino acid sequence. In *Proc. Natl. Acad. Sci. USA* **82**:6379–83.

Bevilacqua MP, Stengeli S, Gimbrone MA *et al.* (1989): Endothelial leukocyte adhesion molecule I: an inducible receptor for neurophils related to complement regulatory proteins and lectins. In *Science* **243**:1160–4.

Beyers AD, Spruyt LL, Williams AF (1992): Multimolecular associations of the T-cell antigen receptor. In *Trends Cell Biol.* **2**:253–5.

Bezouska K, Vlahas G, Horvath O *et al.* (1994): Rat natural killer cell antigen, NKR-P1, related to C-type animal lectins is a carbohydrate-binding protein. In *J. Biol. Chem.* **269**:16945–52.

Bierer BE, Burakoff SJ (1991): T-cell receptors. Adhesion and signaling. In *Advances in Cancer Research* **56**:49–76.

Bollensen E, Schachner M (1987): The peripheral myelin glycoprotein P0 expresses the L2/HNK-1 and L3 carbohydrate structures shared by neural adhesion molecules. In *Neurosci. Lett.* **82**:77–82.

Bowlin TL, McKown BJ, Kang MS *et al.* (1989): Potentiation of human lymphokine-activated killer cell activity by swainsonine an inhibitor of glycoprotein processing. In *Cancer Res.* **49**:4109–13.

Bowlin TL, Schroeder KK, Fanger BO (1991): Swainsonine, an inhibitor of mannosidase II during glycoprotein processing, enhances concanavalin A-induced T cell proliferation and interleukin 2 receptor expression exclusively via the T cell receptor complex. In *Cell. Immunol.* **137**:111–7.

Broadwell RD, Balin BJ, Salcman M *et al.* (1983): Blood-brain barrier? Yes or no. In *Proc. Natl. Acad. Sci. USA* **80**:7352–6.

Brody DT, Durum SK (1989): Membrane IL-1. IL-1-alpha precursor binds to the plasma membrane via a lectin-like interaction. In *J. Immunol.* **143**:1183–7.

Buck F, Luth C, Strupat K *et al.* (1992): Comparative investigations on the amino-acid sequences of different isolectins from the sponge *Axinella polypoides* (Schmidt). In *Biochim. Biophys. Acta* **1159**:1–8.

Caron M, Joubert R, Bladier D (1986): Cell growth factors and soluble lectins. In *Trends Biochem. Sci.* **11**:319.

Cinek T, Horejsi V (1992): The nature of large non-covalent complexes containing glycosyl-phosphatidylinositol-anchored membrane glycoproteins and protein tyrosine kinases. In *J. Immunol.* **149**:2262–70.

Cohen FE, Kosen PA, Kuntz ID *et al.* (1986): Structure-activity studies of interleukin-2. In *Science* **234**:349–52.

Cooper DNW, Massa SM, Barondes SH (1991): Endogenous muscle lectin inhibits myoblast adhesion to laminin. In *J. Cell Biol.* **115**:1437–48.

Cremer H, Lange R, Christoph A *et al.* (1994): Inactivation of the *N*-CAM gene in mice results in size reduction of the olfactory bulb and deficits in spatial learning. In *Nature* **367**:455–9.

Crovello CS, Furie BC, Furie B (1993): Rapid phosphorylation and selective dephosphorylation of P-selectin accompanies platelet activation. In *J. Biol. Chem.* **268**:14590–3.

Dagher SF, Wang JL, Patterson RJ (1995): Identification of galectin-3 as a factor in pre-mRNA splicing. In *Proc. Natl. Acad. Sci. USA* **92**:1213–7.

Damle NK, Klussman K, Dietsch MT *et al.* (1992): GMP-140 (P-selectin/CD62) binds to chronically stimulated but not resting CD4$^+$ lymphocyte-T and regulates their production of proinflammatory cytokines. In *Eur. J. Immunol.* **22**:1789–93.
Drickamer K (1988): Two distinct classes of carbohydrate-recognition domains in animal lectins. In *J. Biol. Chem.* **263**:9557–60.
Drickamer K (1992): Engineering galactose-binding activity into a C-type mannose-binding protein. In *Nature* **360**:183–6.
Drickamer K, Dordal MS, Reynold L (1986): Mannose-binding proteins isolated from rat liver contain carbohydrate-recognition domains linked to collagenous tails. In *J. Biol. Chem.* **261**:6878–87.
Edelman GM (1986): Cell adhesion molecules in the regulation of animal form and tissue pattern. In *Annu. Rev. Cell Biol.* **2**:81–116.
Elbein AD (1987): Inhibitors of the biosynthesis and processing of *N*-linked oligosaccharide chains. In *Annu. Rev. Biochem.* **56**:497–534.
Eriksson AE, Cousens LS, Weaver LH *et al.* (1991): Three-dimensional structure of human basic fibroblast growth factor. In *Proc. Natl. Acad. Sci. USA* **88**:3441–5.
Fahrig T, Schmitz B, Weber D *et al.* (1990): Two monoclonal antibodies recognizing carbohydrate epitopes on neural adhesion molecules interfere with cell interactions. In *Eur. J. Neurosci.* **2**:153–61.
Farrar WL, Linnekin D, Evans G *et al.* (1990): Interleukin-2 regulation of a tyrosine kinase associated with the P70–75 beta-chain of the receptor complex. In *Mol. Cell. Biol. Cytokines* **10**:265–9.
Feizi T, Childs RA (1987): Carbohydrates as antigenic determinants of glycoproteins. In *Biochem. J.* **245**:1–11.
Fiedler K, Simons K (1994): A putative novel class of animal lectins in the secretory pathway homologous to leguminous lectins. In *Cell* **77**:625–6.
Fiedler K, Parton RG, Kellner R *et al.* (1994): VIP36, a novel component of glycolipid rafts and exocytic carrier vesicles in epithelial cells. In *EMBO J.* **13**:1729–40.
Filbin MT, Tennekoon GI (1991): The role of complex carbohydrates in adhesion of the myelin protein P0. In *Neuron* **7**:845–55.
Filbin MT, Tennekoon GI (1993): Homophilic adhesion of the myelin P0 protein requires glycosylation of both molecules in the homophilic pair. In *J. Cell Biol.* **122**:451–9.
Filbin MT, Walsh FS, Trapp BD *et al.* (1990): Role of myelin P0 protein as a homophilic adhesion molecule. In *Nature* **344**:871–2.
Foxall C, Watson SR, Dowbnko D *et al.* (1992): The three members of the Selectin receptor family recognize a common carbohydrate epitope, the sialyl Lewis^x oligosaccharide. In *J. Cell Biol.* **117**:895–902.
Garnett D, Barclay AN, Carmo AM *et al.* (1993): The association of the protein tyrosine kinases p56^{lck} and p60^{fyn} with the glycosyl phosphatidylinositol-anchored proteins Thy-1 and CD48 in rat thymocytes is dependent on the state of cellular activation. In *Eur. J. Immunol.* **23**:2540–4.
Gordon J (1994): B-cell signalling via the C-type lectins CD23 and CD72. In *Immunol. Today* **15**:411–7.
Green PJ, Tamatani T, Watanabe T *et al.* (1992): High affinity binding of the leucocyte adhesion molecule L-selectin to 3'-sulphated-Le^a and 3'-sulphated-Le^x oligosaccharides and the predominance of sulphate in this interaction demonstrated by binding studies with a series of lipid-linked oligosaccharides. In *Biochem. Biophys. Res. Commun.* **188**:244–51.
Griffith LS, Schmitz B, Schachner M (1992): L2/HNK-1 carbohydrate and protein-protein interactions mediate the homophilic binding of the neural adhesion molecule P0. In *J. Neurosci. Res.* **33**:639–48.
Haagsman HP (1994): Surfactant proteins A and D. In *Biochem. Soc. Trans.* **22**:100–6.
Hall H, Liu L, Schachner M *et al.* (1993): The L2/HNK-1 carbohydrate mediates adhesion of neural cells to laminin. In *Eur. J. Neurosci.* **5**:34–42.
Hamann J, Fiebig H, Strauss M (1993): Expression cloning of the early activation antigen CD69, a type-II integral membrane protein with a C-type lectin domain. In *J. Immunol.* **150**:4920–7.
Hart GW (1982): The role of asparagine-linked oligosaccharides in cellular recognition by thymic lymphocytes. Effects of tunicamycin on the mixed lymphocyte reaction. In *J. Biol. Chem.* **257**:151–8.
Hartshorn KL, Sastry K, White MR *et al.* (1993): Human mannose-binding protein functions as an opsonin for influenza-A viruses. In *J. Clin. Invest.* **91**:1414–20.
Hartshorn KL, Crouch EC, White MR *et al.* (1994): Evidence for a protective role of pulmonary surfactant protein D (SP-D) against influenza A viruses. In *J. Clin. Invest.* **94**:311–9.
Hemmerich S, Rosen SD (1994): 6'-Sulfated sialyl Lewis^x is a major capping group of Glycam-1. In *Biochemistry* **33**:4830–5.
Hemmerich S, Bertozzi CR, Leffler H *et al.* (1994): Identification of the sulfated monosaccharides of Glycam-1, an endothelial-derived ligand for L-selectin. In *Biochemistry* **33**:4820–9.
Hession C, Decker JM, Sherblom AP *et al.* (1987): Uromodulin (Tamm-Horsfall glycoprotein): a renal ligand for lymphokines. In *Science* **237**:1479–84.
Hintzen RQ, Lens MS, Beckmann MP *et al.* (1994): Characterization of the human CD27 ligand, a novel

member of the TNF gene family. In *J. Immunol.* **152**:1762–73.

Imai Y, Rosen SD (1993): Direct demonstration of heterogeneous, sulfated *O*-linked carbohydrate chains on an endothelial ligand for L-selectin. In *Glycoconjugate J.* **10**:34–9.

Iobst ST, Drickamer K (1994): Binding of sugar ligands to $Ca2^+$-dependent animal lectins. 2. Generation of high-affinity galactose binding by site-directed mutagenesis. In *J. Biol. Chem.* **269**:15512–9.

Ioffe E, Stanley P (1994): Mice lacking *N*-acetylglucosaminyltransferase I activity die at mid-gestation, revealing an essential role for complex or hybrid *N*-linked carbohydrates. In *Proc. Natl. Acad. Sci. USA* **91**: 728–32.

Izquierdo M, Cantrell DA (1992): T-cell activation. In *Trends Cell Biol.* **2**:268–71.

Jayadev S, Liu B, Bielawska AE *et al.* (1995): Role for ceramide in cell cycle arrest. In *J. Biol. Chem.* **270**:2047–52.

Jia S, Wang JL (1988): Carbohydrate binding protein 35. In *J. Biol. Chem.* **263**:6009–11.

Joubert R, Caron M, Bladier D (1987): Brain lectin mediated agglutinability of dissociated cells from embryonic and postnatal mouse brain. In *Develop. Brain Res.* **36**:146–50.

Joubert R, Kuchler S, Zanetta J-P *et al.* (1989): Immunohistochemical localization of a β-galactoside-binding lectin in rat central nervous system. I. Light and electron-microscopical studies on developing cerebral cortex and corpus callosum. In *Dev. Neurosci.* **11**:397–413.

Joubert R, Caron M, Avellana-Adalid V *et al.* (1992): Human brain lectin. A soluble lectin that binds actin. In *J. Neurochem.* **58**:200–3.

Ju G, Collins L, Kaffka K *et al.* (1987): Structure-function analysis of human interleukin-2. In *J. Biol. Chem.* **262**:5723–31.

Kadmon G, Kowitz A, Altevogt P *et al.* (1990 a): The neural cell adhesion molecule *N*-CAM enhances L1-dependent cell-cell interactions. In *J. Cell Biol.* **110**:193–208.

Kadmon G, Kowitz A, Altevogt P *et al.* (1990 b): Functional cooperation between the neural adhesion molecules L1 and *N*-CAM is carbohydrate dependent. In *J. Cell Biol.* **110**:209–18.

Kadmon G, Eckert M, Sammar M *et al.* (1992): Nectadrin, the heat-stable antigen is a cell adhesion molecule. In *J. Cell Biol.* **118**:1245–58.

Kadmon G, Halbach FVU, Schachner M *et al.* (1994): Differential, LFA-1-sensitive effects of antibodies to nectadrin, the heat-stable antigen, on B lymphoblast aggregation and signal transduction. In *Biochem. Biophys. Res. Commun.* **198**:1209–15.

Kawasaki N, Kawasaki T, Yamashina I (1989): A serum lectin (mannan-binding protein) has complement-dependent bactericidal activity. In *J. Biochem.* **106**:483–9.

Kay R, Rosten PM, Humphries RK (1991): CD24 a signal transducer modulating B-cell activation responses is a very short peptide with a glycosyl phosphatidylinositol membrane anchor. In *J. Immunol.* **147**:1412–6.

Krebs A, Depew WT, Szarek WA *et al.* (1994): Binding of D-galactose-terminated ligands to rabbit asialoglycoprotein receptor. In *Carbohydr. Res.* **254**: 223–44.

Kucherer A, Faissner A, Schachner M (1987): The novel carbohydrate epitope L3 is shared by some neural cell adhesion molecules. In *J. Cell Biol.* **104**:1597–602.

Kuchler S, Fressinaud C, Sarlième LL *et al.* (1988): Cerebellar soluble lectin is responsible for cell adhesion and participates in myelin compaction in cultured rat oligodendrocytes. In *Dev. Neurosci.* **10**: 199–212.

Kuchler S, Joubert R, Avellana-Adalid V *et al.* (1989a): Immunohistochemical localization of α β-galactoside-binding lectin in rat central nervous system. II. Light- and electron-microscopical studies in developing cerebellum. In *Dev. Neurosci.* **11**: 414–27.

Kuchler S, Rougon G, Marschal P *et al.* (1989b): Localization of a transiently expressed glycoprotein in developing cerebellum delineating its possible ontogenetic roles. In *Neuroscience* **33**:111–24.

Kuchler S, Perraud F, Sensenbrenner M *et al.* (1989c): An endogenous lectin found in rat astrocyte cultures has a role in cell adhesion but not in cell proliferation. In *Glia* **2**:437–45.

Kuchler S, Herbein G, Sarlième LL *et al.* (1989d): An endogenous lectin CSL interacts with major glycoprotein components in peripheral nervous system myelin. In *Cell. Mol. Biol.* **35**:581–96.

Kuchler S, Graff M-N, Gobaille S *et al.* (1994): Mannose-dependent tightening of the ependymal cell barrier. In vivo and in vitro study using neoglycoproteins. In *Neurochem. Int.* **24**: 43–55.

Kuchler S, Badache A, Linington C *et al.* (1995): Specific invasion of white matter and demyelination induced by intraventricular injection of antibodies against the cerebellar soluble lectin. *Submitted*.

Kuroki Y, Voelker DR (1994): Pulmonary surfactant proteins. In *J. Biol. Chem.* **269**:25943–6.

Lai C, Brow MA, Nave K-A *et al.* (1987): Two forms of 1B236 / myelin-associated glycoprotein, a cell adhesion molecule for postnatal development, are produced by alternative splicing. In *Proc. Natl. Acad. Sci. USA* **84**:4337–41.

Lanier LL, Chang CW, Phillips JH (1994): Human NKR-P1A. A disulfide-linked homodimer of the C-type lectin superfamily expressed by a subset of NK and T lymphocytes. In *J. Immunol.* **153**:2417–28.

Larkin M, Ahern TJ, Stoll MS *et al.* (1992): Spectrum of sialylated and nonsialylated fuco-oligosaccharides bound by the endothelial-leukocyte adhesion molecule E-selectin. Dependence of the carbohydrate binding activity on E-selectin density. In *J. Biol. Chem.* **267**:13661–8.

Lasky LA (1992): Selectins. Interpreters of cell-specific carbohydrate information during inflammation. In *Science* **258**:964–9.

Leconte I, Auzan C, Debant A *et al.* (1992): *N*-Linked oligosaccharide chains of the insulin receptor-beta subunit are essential for transmembrane signaling. In *J. Biol. Chem.* **267**:17415–23.

Lehmann S, Kuchler S, Théveniau M *et al.* (1990): An endogenous lectin and one of its neuronal glycoprotein ligands are involved in contact guidance of neuron migration. In *Proc. Natl. Acad. Sci. USA* **87**: 6455–9.

Lehmann S, Kuchler S, Badache A *et al.* (1991): Involvement of the endogenous lectin CSL in adhesion of Chinese hamster ovary cells. In *Eur. J. Cell Biol.* **56**:433–42.

Lim BL, Wang JY, Holmskov U *et al.* (1994): Expression of the carbohydrate recognition domain of lung surfactant protein D and demonstration of its binding to lipopolysaccharides of gram-negative bacteria. In *Biochem. Biophys. Res. Commun.* **202**:1674–80.

Lobsanov YD, Gitt MA, Leffler H *et al.* (1993): Crystallization and preliminary X-ray diffraction analysis of the human dimeric S-lac lectin (L-14-II). In *J. Mol. Biol.* **233**:553–5.

Lucas R, Magez S, Deleys R *et al.* (1994): Mapping the lectin-like activity of tumor necrosis factor. In *Science* **263**:814–7.

Maccarana M, Casu B, Lindahl U (1993): Minimal sequence in heparin/heparan sulfate required for binding of basic fibroblast growth factor. In *J. Biol. Chem.* **268**:23898–905.

Mahanthappa NK, Cooper DNW, Barondes SH *et al.* (1994): Rat olfactory neurons can utilize the endogenous lectin, L-14, in a novel adhesion mechanism. In *Development* **120**:1373–84.

Marschal P, Reeber A, Neeser J-R *et al.* (1989): Carbohydrate and glycoprotein specificity of two endogenous cerebellar lectins. In *Biochimie* **71**:645–53.

Martin M, Bol GF, Eriksson A *et al.* (1994): Interleukin-1-induced activation of a protein kinase co-precipitating with the type I interleukin-1 receptor in T cells. In *Eur. J. Immunol.* **24**:1566–71.

Masayuki I, Kawasaki T, Yamashina I (1988): Structural similarity between the macrophage lectin specific for galactose/*N*-acetylgalactosamine and the hepatic asialoglycoprotein binding protein. In *Biochem. Biophys. Res. Comm.* **155**:720–5.

Maschke S, Robert J, Coindre J-M *et al.* (1993): Malignant cells have increased levels of common glycoprotein ligands of the endogenous cerebellar soluble lectin. In *Eur. J. Cell Biol.* **62**:163–72.

Maschke S, Kuchler-Bopp S, Zanetta J-P (1996) Role of an endogenous lectin and its glycoprotein ligands in the proliferation and adhesion of C6 glioblastoma cells. In *Glycobiology* submitted.

Matsushita M, Fujita T (1992): Activation of the classical complement pathway by mannose-binding protein in association with a novel C1s-like serine protease. In *J. Exp. Med.* **176**:1497–502.

McNeely TB, Coonrod JD (1994): Aggregation and opsonization of type-A but not type-B hemophilus-influenzae by surfactant protein-A. In *Amer. J. Respir. Cell. Mol. Biol.* **11**:114–22.

Metzler M, Gertz A, Sarkar M *et al.* (1994): Complex asparagine-linked oligosaccharides are required for morphogenic events during post-implantation development. In *EMBO J.* **13**:2056–65.

Muchmore AU, Decker J-M (1987): Evidence that recombinant IL1 exhibits lectin-like specificity and binds to homogeneous uromodulin via *N*-linked oligosaccharides. In *J. Immunol.* **138**:2541–52.

Nagai Y (1995): Functional roles of gangliosides in biosignaling. In *Behav. Brain Res.* **66**:99–104.

Nédelec J, Pierres M, Moreau H *et al.* (1992): Isolation and characterization of a novel glycosylphosphatidylinositol-anchored glycoconjugate expressed by developing neurons. In *Eur. J. Biochem.* **203**:433–42.

Needham LK, Schnaar RL (1993): The HNK-1 reactive sulfoglucuronyl glycolipids are ligands for L-selectin and P-selectin but not E-selectin. In *Proc. Natl. Acad. Sci. USA* **90**:1359–63.

Obrien C (1995): Cell biology. Neuronal adhesion molecules signal through FGF receptor. In *Science* **267**:1263–4.

Ochieng J, Gerold M, Raz A (1992): Dichotomy in the laminin-binding properties of soluble and membrane-bound human galactoside-binding protein. In *Biochem. Biophys. Res. Commun.* **186**:1674–80.

Ohta M, Kawasaki T (1994): Complement-dependent cytotoxic activity of serum mannan-binding protein towards mammalian cells with surface-exposed high-mannose type glycans. In *Glycoconjugate J.* **11**: 304–8.

Ohta M, Okada M, Yamashina I *et al.* (1990): The mechanism of carbohydrate-mediated complement activation by the serum mannan-binding protein. In *J. Biol. Chem.* **265**:1980–4.

Oka S, Itoh N, Kawasaki T *et al.* (1987): Primary structure of rat liver mannan-binding protein deduced from its cDNA sequence. In *J. Biochem.* **101**: 135–44.

Oka S, Ikeda K, Kawasaki T *et al.* (1988): Isolation and characterization of two distinct mannan-binding proteins from rat serum. In *Arch. Biochem. Biophys.* **260**:257–66.

Olivera A, Spiegel S (1992): Ganglioside-G(M1) and sphingolipid breakdown products in cellular proliferation and signal transduction pathways. In *Glycoconjugate J.* **9**:110–7.

Poduslo JF (1990): Golgi sulfation of the oligosaccharide chain of P0 occurs in the presence of myelin assembly but not in its absence. In *J. Biol. Chem.* **265**: 3719–3725.

Presta M, Statuto M, Isacchi A *et al.* (1992): Structure-function relationship of basic fibroblast growth factor. Site-directed mutagenesis of a putative heparin-binding and receptor-binding region. In *Biochem. Biophys. Res. Commun.* **185**:1098–107.

Quesenberry MS, Drickamer K (1992): Role of conserved and nonconserved residues in the $Ca2^+$-dependent carbohydrate-recognition domain of a rat mannose-binding protein. Analysis by random cassette mutagenesis. In *J. Biol. Chem.* **267**: 10831–41.

Rusnati M, Coltrini D, Caccia P *et al.* (1994): Distinct role of 2-*O*-, *N*-, and 6-*O*-sulfate groups of heparin in the formation of the ternary complex with basic fibroblast growth factor and soluble FGF receptor-1. In *Biochem. Biophys. Res. Commun.* **203**:450–8.

Sakamoto Y, Kitamura K, Yoshimura K *et al.* (1987): Complete amino acid sequence of PO protein in bovine peripheral nerve myelin. In *J. Biol. Chem.* **262**:4208–14.

Santis AG, Lopez-Cabrera M, Hamann J *et al.* (1994): Structure of the gene coding for the human early lymphocyte activation antigen CD69: A C-type lectin receptor evolutionarily related with the gene families of natural killer cell-specific receptors. In *Eur. J. Immunol.* **24**:1692–7.

Sato T, Nakafuku M, Miyajima A *et al.* (1991): Involvement of ras p21 protein in signal-transduction pathways from interleukin-2, interleukin-3 and granulocyte macrophage colony-stimulating factor but not from interleukin-4. In *Proc. Natl. Acad. Sci. USA* **88**:3314–8.

Saukkonen K, Burnette WN, Mar VL *et al.* (1992): Pertussis toxin has eukaryotic-like carbohydrate recognition domains. In *Proc. Natl. Acad. Sci. USA* **89**: 118–22.

Savage SM, Donaldson LA, Sopori ML (1993): T-cell-B cell interaction. Autoreactive T-cells recognize B-cells through a terminal mannose-containing superantigen-like glycoprotein. In *Cell. Immunol.* **146**:11–27.

Schlessinger J, Ullrich I (1992): Growth factor signaling by receptor tyrosine kinases. In *Neuron* **9**, 383–391.

Sherblom AP, Decker JM, Muchmore AV (1988): The lectin-like interaction between recombinant tumor necrosis factor and uromodulin. In *J. Biol. Chem.* **263**:5418–24.

Sherblom AP, Sathyamoorthy N, Decker JM *et al.* (1989): IL-2, a lectin with specificity for high mannose glycopeptides. In *J. Immunol.* **143**:939–44.

Shibuya H, Kohu K, Yamada K *et al.* (1994): Functional dissection of $p56^{lck}$, a protein tyrosine kinase which mediates interleukin-2-induced activation of the c-fos gene. In *Mol. Cell. Biol.* **14**:5812–9.

Sobue G, Pleasure D (1985): Adhesion of axolemmal fragments to Schwann cells: a signal- and target-specific process closely linked to axolemmal induction of Schwann cell mitosis. In *J. Neurosci.* **5**:379–87.

Spivak-Kroizman T, Lemmon MA, Dikic I *et al.* (1994): Heparin-induced oligomerization of FGF molecules is responsible for FGF receptor dimerization, activation, and cell proliferation. In *Cell* **79**:1015–24.

Springer TA (1991): Cell adhesion. Sticky sugars for selectins. In *Nature* **349**:196–7.

Steegmaier M, Levinovitz A, Isenmann S *et al.* (1995): The E-selectin-ligand ESL-1 is a variant of a receptor for fibroblast growth factor. In *Nature* **373**:615–20.

Stocks SC, Kerr MA (1993): Neutrophil NCA-160 (CD66) is the major protein carrier of selectin binding carbohydrate groups Lewis(x) and sialyl Lewis(x). In *Biochem. Biophys. Res. Commun.* **195**:478–83.

Summerfield JA (1993): The role of mannose-binding protein in host defence. In *Biochem. Soc. Transact.* **21**:473–7.

Taniguchi T, Minami Y (1993): The IL-2/IL-2 receptor system: a current overview. In *Cell* **73**:5–8.

Taylor ME, Brickell PM, Craig RK *et al.* (1989): Structure and evolutionary origin of the gene encoding a human serum mannose-binding protein. In *Biochem. J.* **262**:763–71.

Uyemura K, Kitamura K (1991): Comparative studies on myelin proteins in mammalian peripheral nerve. In *Comp. Biochem. Physiol.* **98**:63–72.

Van Gool SW, Kasran A, Wallays G *et al.* (1995): Accessory signalling by B7–1 for T cell activation induced by anti-CD2: Evidence for IL-2-independent CTL generation and CsA-resistant cytokine production. In *Scand. J. Immunol.* **41**:23–30.

Van Iwaarden JF, Pikaar JC, Storm J *et al.* (1994): Binding of surfactant protein A to the lipid. A moiety of bacterial lipopolysaccharides. In *Biochem. J.* **303**:407–11.

Venkitaraman AR, Cowling RJ (1994): Interleukin-7 induces the association of phosphatidylinositol 3-kinase with the a chain of the interleukin-7 receptor. In *Eur. J. Immunol.* **24**:2168–74.

Waldmann TA (1991): The interleukin-2 receptor. In *J. Biol. Chem.* **266**:2681–4.

Ware FE, Vassilakos A, Peterson PA *et al.* (1995): The molecular chaperone calnexin binds Glc_1Man_9 $GlcNAc_2$ oligosaccharide as an initial step in recognizing unfolded glycoproteins. In *J. Biol. Chem.* **270**:4697–704.

Weis WI, Drickamer K, Hendrickson WA (1992): Structure of a C-type mannose-binding protein complexed with an oligosaccharide. In *Nature* **360**:127–34.

Yabe T, McSherry C, Bach FH *et al.* (1993): A multigene family on human chromosome-12 encodes natural killer-cell lectins. In *Immunogenetics* **37**:455–60.

Yazaki T, Miura M, Asou H *et al.* (1992): Glycopeptide of P0 protein inhibits homophilic cell adhesion. Competition assay with transformants and peptides. In *FEBS Lett.* **307**:361–6.

Zanetta J-P, Dontenwill M, Meyer A *et al.* (1985): Isolation and immunohistochemical localization of a lectin like molecule from the rat cerebellum. In *Dev. Brain Res.* **17**:233–43.

Zanetta J-P, Meyer A, Kuchler S *et al.* (1987a): Isolation and immunochemical study of a soluble cerebellar lectin delineating its structure and function. In *J. Neurochem.* **49**:1250–7.

Zanetta J-P, Dontenwill M, Reeber A *et al.* (1987b): Expression of recognition molecules in the cerebellum of young and adult rats. In *NATO ASI Series H* **2**:92–104.

Zanetta J-P, Warter J-M, Kuchler S *et al.* (1990a): Antibodies to cerebellar soluble lectin CSL in multiple sclerosis. In: *Lancet* **335**: 1482–4.

Zanetta J-P, Warter J-M, Lehmann S *et al.* (1990b): Presence of antibodies to lectin CSL in the blood of multiple sclerosis patients. In *C. R. Acad. Sci. Paris* **311**:327–31.

Zanetta J-P, Staedel C, Kuchler S *et al.* (1991): Malignant transformation in hepatocytes is associated with the general increase of glycoprotein ligands specifically binding to the endogenous lectin CSL. In *Carbohydr. Res.* **213**:117–26.

Zanetta J-P, Badache A, Maschke S *et al.* (1994a): Carbohydrates and soluble lectins in the regulation of cell adhesion and proliferation. In *Histol. Histopathol.* **9**:385–412.

Zanetta J-P, Tranchant C, Kuchler-Bopp S *et al.* (1994 b): Presence of anti-CSL antibodies in the cerebrospinal fluid of patients. A sensitive and specific test in the diagnosis of multiple sclerosis. In *J. Neuroimmunol.* **52**:175–82.

Zanetta J-P, Vantyghem J, Kuchler-Bopp S *et al.* (1995): Human lymphocyte activation is associated with the early and high level expression of the endogenous lectin CSL. In *Biochem. J.* **311**:629–36.

Zanetta J-P, Alonso C, Michalski J-C (1996): Interleukin-2 associates its receptor to the T cell receptor through its lectin activity. *Biochem. J.* submitted.

Zhang J, Cousens LS, Barr P *et al.* (1991): Three-dimensional structure of human basic fibroblast growth factor, a structural homolog of interleukin 1β. In *Proc. Natl. Acad. Sci. USA* **88**: 3446–50.

Ziegler SF, Levin SD, Johnson L *et al.* (1994): The mouse CD69 gene. Structure, expression and mapping to the NK gene complex. In *J. Immunol.* **152**:1228–36.

Zola H, Weedon H, Thompson GR *et al.* (1991): Expression of IL-2 receptor p55 and p75 chains by human lymphocyte. Effects of activation and differentiation. In *Immunology* **72**:167–73.

25 Galectins in Tumor Cells

DAVID W. OHANNESIAN AND REUBEN LOTAN

1 Introduction

The ability of various monosaccharides to be linked in a number of anomeric configurations confers on oligosaccharides the potential to encode a large amount of positional and spatial information (Sharon, 1975). The utilization of this "code" requires complementary molecules to recognize ("decode") this information. Therefore, the discovery of proteins endowed with carbohydrate-binding activity in plants has received considerable attention (Sharon and Lis, 1990). Although their functions still remain an active area of inquiry, knowledge obtained about plant lectins and their utilization as carbohydrate-specific reagents have contributed to much of our knowledge of the structure of the oligosaccharide side chains of cellular glycoconjugates in both plants and animals. The identification of endogenous lectins in animal cells provided a new incentive for further investigations into the functions and physiological role(s) of these proteins in vertebrates. A lectin has been defined as a carbohydrate-binding protein of nonimmune origin that agglutinates cells or precipitates polysaccharides or glycoconjugates. More recently, the less restrictive definition of a carbohydrate-binding protein other than an enzyme or an antibody has been proposed (Barondes, 1988). Carbohydrate-recognition mechanisms have been implicated in specific cellular interactions that occur during embryonic development (Thorpe *et al.*, 1988; Kimber, 1990; Feizi, 1991), during cell growth and differentiation, and during cancer metastasis (Raz and Lotan, 1987; Harrison, 1991; Hughes, 1992; Lotan, 1992; Zhou and Cummings, 1992). These interactions may be mediated in part by lectins produced by vertebrate cells and by complementary glycoconjugates that are expressed on the surface of adjacent cells or are present in the extracellular matrix (ECM). This chapter is focused on galactoside-binding proteins and their cognate glycoconjugates that may be responsible for cellular recognition and adhesion processes.

2 Galectins in Normal Cells

The most widely expressed vertebrate lectins bind β-galactosides (Harrison, 1991; Lotan, 1992; Zhou and Cummings, 1992; Barondes *et al.*, 1994b). These lectins are soluble in aqueous buffers and the activity of some of the members of this group depends on the presence of thiol reducing agents (Drickamer and Taylor, 1993). Now called galectins, these lectins constitute a family of proteins with related amino acid sequences (Barondes *et al.*, 1994a, b). The best studied among the galectins are galectin-1 (13–14.5 kDa) that has been referred to as galaptin (Harrison, 1991) and L-14.5 (Raz *et al.*, 1987) and galectin-3 (29–35 kDa) that has been referred to as CBP35, RL-29, L-34, L-31, low affinity IgE-binding protein, and Mac-2 (Leffler *et al.*, 1986; Albrandt *et al.*, 1987; Raz *et al.*, 1987; Cherayil *et al.*, 1989; Lotan, 1992; Wang *et al.*, 1992; Hughes, 1994).

2.1 Galectin-1

Galectin-1 was discovered in the electric organ of the eel (*Electrophorus electricus*) by Teichberg *et al.* (1975) and has been isolated from a number of

H.-J. and S. Gabius (Eds.), Glycosciences
© Chapman & Hall, Weinheim, 1997
ISBN 3-8261-0073-5

different tissues and a variety of cell types. The expression of galectin-1 is developmentally regulated in several vertebrate tissues including intestine, skin, lung, muscle and nervous system (Powell and Whitney, 1980; Oda and Kasai, 1984; Regan *et al.*, 1986; Akimoto *et al.*, 1992; Barondes *et al.*, 1994b). The developmental regulation of galectin-1 expression has led to speculation that this protein may be involved in the control of cellular proliferation and differentiation. The cDNA for galectin-1 has been cloned from various species and it has been found to exhibit considerable sequence homology (more than 85 %) among mouse, rat, bovine, and human at the amino acid level that also indicated an important function for this protein (Raz *et al.*, 1987; Paroutaud *et al.*, 1989; Lotan, 1992; Zhou and Cummings, 1992; Barondes *et al.*, 1994b). Although a clear demonstration of such a function has proven difficult, there are indications for a role of this lectin in cell growth and adhesion.

The subcellular location of galectin-1 has provided few clues to its function. In most cell types this lectin has a rather broad distribution throughout the cytoplasm with a lower but detectable level of expression on the cell surface. The requirement for a reducing agent for galectin-1 activity (Whitney et *al.*, 1986) suggested an intracellular function. However, the presence of a carbohydrate-binding activity at the cell surface (Gabius 1991; Ohannesian *et al.*, 1994) and the secretion of galectin-1 (Cooper and Barondes, 1990) argues for an extracellular function for this lectin in addition to the putative intracellular function. Binding of the secreted galectin to the cell surface or to ECM stabilizes the protein and protects it from inactivation (Cho and Cummings, 1995).

The ability of galectin-1 to stimulate DNA synthesis in a lactose-inhibitable process was demonstrated in aortic smooth muscle cells and pulmonary endothelial cells (Sanford and Harris-Hooker, 1990). In contrast, in mouse embryo fibroblasts galectin-1 was found to act as an autocrine negative growth inhibitor endogenously produced and secreted by these cells (Wells and Malluci, 1991). Because the latter activity was not inhibited by galactosides, it may be unrelated to the carbohydrate-binding function of the galectin.

Recently, galectin-1 expression was analyzed during mouse embryogenesis. It was found to first appear at 4.5 days post coitum in trophectodermal cells of the hatched blastocyst (Poirier *et al.*, 1992). This led to the proposal that galectin-1 may play a role in the implantation of the embryo. In addition, galectin-1 was detected in most of the motor neurons of the brain and spinal cord and in the sensory neurons of the dorsal root ganglia suggesting that the lectin may be involved in the specification of axonal pathways. Galectin-1 expression was also abundant in the myotomes which give rise to muscle tissue. Galectin-1 null mice obtained through gene targeting by homologous recombination demonstrated no detectable phenotype and were viable and fertile (Poirier and Robertson, 1993). It was suggested that the galectin-3 gene, which is expressed in many of the same tissues as galectin-1 may compensate for galectin-1 loss in these null mice. Redundancy in function and expression of several of the numerous members of the expanding galectin family (Gitt and Barondes, 1986; Gitt *et al.*, 1992, 1995) could also explain the lack of a phenotype in the galectin-1 deficient mice.

2.2 Galectin-3

The 35 kDa lactose-binding lectin (galectin-3) initially purified from mouse 3T3 fibroblasts (Wang *et al.*, 1992; Hughes, 1994) has also been cloned, sequenced, and found to have 85 % homology at the amino acid level to a low affinity IgE-binding protein first isolated from rat basophilic leukemia cells. (Albrandt *et al.*, 1987). Galectin-1 and galectin-3 are encoded by two distinct genes, however, there are regions of complete homology between the two lectins including a 39 residue sequence in which there are 15 invariant amino acids (Lotan, 1992; Wang *et al.*, 1992; Barondes, 1994a, b). These residues lie in the carbohydrate recognition domains of the S-type lectins. Site-directed mutagenesis of galectin-1 revealed that sequences conserved in this region between galectin-1 and galectin-3 are critical for the carbohydrate-binding activity of galectin-1 (Hirabayashi and Kasai, 1991).

More recently, galectin-3 has been shown to be identical to the murine macrophage cell surface antigen, Mac-2 (Cherayil *et al.*, 1989). Mouse galectin-3 also exhibits amino acid sequence homology at its amino terminus with heterogeneous nuclear ribonucleoproteins (hnRNPs) and can be localized in cell nuclei (Wang *et al.*, 1992). The level of galectin-3 protein and mRNA is low in quiescent mouse fibroblasts and increases rapidly after mitogenic stimulation (Agrwal *et al.*, 1989). It has been proposed that galectin-3 may function in the nucleus in the transport of mRNA to the cytosol (Wang *et al.*, 1992). Although carbohydrate has been found on proteins in the nucleus it is usually not found in the form of extended oligosaccharides as it is in membrane glycoproteins but rather consists of single *N*-acetylglucosamine residues in an *O*-glycosidic linkage to serine and threonine residues in the polypeptide (Hart *et al.*, 1989). This form of glycosylation has been observed on proteins that make up the nuclear pore complex as well as in a number of transcription factors including Sp1. Because *O*-GlcNAc residues can function as substrates for polylactosamine biosynthesis, the possibility exists that some of these proteins could be converted to a form that would be recognized by nuclear galectin-3.

3 Putative Functions of Galectins

Various functions have been proposed for the galectins. These include a role in cell-cell interactions, cell adhesion to ECM, organization of the ECM, and tissue remodeling (Hughes, 1992; Lotan, 1992; Zhou and Cummings, 1992; Barondes *et al.*, 1994b). However, evidence implicating the lectins in these events has often been indirect and the subject of controversy. One of the most likely functions of these lectins is to interact with complementary carbohydrate-containing glycoconjugates. Indeed, galectin-1 from chicken embryonic skin was found to bind to a polylactosaminoglycan present in the same tissue using photoaffinity labeling techniques (Oda and Kasai, 1984). A thorough study of the oligosaccharide binding profiles of galectin-1 and -3 purified from lung showed that both lectins bound to polylactosaminoglycans with highest affinity (Leffler *et al.*, 1986; Sparrow *et al.*, 1987).

Although galectins lack a signal sequence or hydrophobic sequence that could serve as a transmembrane domain they can be secreted to the extracellular milieu. In mouse myoblasts, galectin-1 is found intracellularly but is secreted by myoblasts prior to their fusion into myotubes (Cooper and Barondes, 1990). The extracellular galectin-1 has been demonstrated to inhibit myoblast fusion into myotubes through the binding of oligosaccharides on laminin present in the ECM (Cooper *et al.*, 1991). Galectin-3 secretion from baby hamster kidney (BHK) cells also occurred through a non-classical secretory pathway (Sato *et al.*, 1993; Hughes, 1994).

There is accumulating evidence that one role of the galectins on the cell surface may be to function as accessory adhesion molecules. In this way, cell surface lectin could modulate adhesive interactions of more established cell adhesion molecules (CAMs) such as integrins, selectins, or members of members of the immunoglobulin superfamily. This could occur through the masking of important recognition determinants of the adhesion molecules by galectin binding to oligosaccharides on the CAMs. Alternatively, the binding of galectin directly to a CAM may modulate its ability to recognize its normal ligand. As mentioned above galectin-1 was shown to modulate the interaction of myogenic cells with laminin. Recently, it was demonstrated that galectin-1 binds to the $\alpha_7\beta_1$ integrin, which is the major laminin binding integrin in these cells, in differentiating rat skeletal muscle (Gu *et al.*, 1994). Galectin-1 was shown to selectively interfere with $\alpha_7\beta_1$ binding to laminin but not fibronectin. The inhibitory effect of galectin-1 was largely removed in the presence of the competitive sugar, thiodigalactoside (Gu *et al.*, 1994). A positive role of galectin-1 in binding of cells to laminin has also been demonstrated in other cell type (Mahantahappa *et al.*, 1994). Galectin-3 is also present on the surface of certain cells, however, previous work had demonstrated that galectin-3 inhibited integrin-mediated attachment and spreading of BHK cells on EHS-laminin (Sato and Hughes, 1992) and the adhesion of breast carcinoma cells to laminin could not be inhibited by galactosides (Ochieng *et al.*, 1992).

4 Galectins in Tumor Cells

Cell-cell and ECM interactions are critical events in elaboration of the tumorigenic and metastatic phenotypes (Liotta *et al.*, 1986). A host of cellular adhesion molecules (CAMs) have been demonstrated to be involved in adhesion in normal cells as well as in tumor cells including integrins, selectins, and members of the immunoglobulin superfamily such as carcinoembryonic antigen (CEA). The role of protein-carbohydrate interactions in tumor cell adhesion has focused increasing attention on their role in events related to metastasis such as their interaction with the components of the ECM (Gabius, 1987, 1991; Raz and Lotan, 1987). The ECM, is composed mainly of laminin, fibronectin, type IV collagen, and various proteoglycans many of which are heavily glycosylated and whose oligosaccharide side chains can provide recognition determinants for CAMs.

Some of the first indications that lectin activity was present in tumor cells came from observations in mouse neuroblastoma cells however the properties of the endogenous lectin detected was not well characterized (Teichberg *et al.*, 1975). In the ensuing years, lectins have been identified in a wide variety of different tumors and tumor cell lines (Allen *et al.*, 1987; Gabius, 1987; Raz and Lotan, 1987; Lotan, 1992).

Much of the evidence for lectin activity in tumor cells was initially based on the indirect observations that single cell suspensions from several human and mouse melanoma cell lines, a human cervical adenocarcinoma, and a murine fibrosarcoma aggregated when incubated with the serum glycoprotein fetuin and its desialylated derivative asialofetuin (Raz and Lotan, 1981). In addition, aqueous extracts from these cells could cause hemagglutination of trypsinized, glutaraldehyde-fixed rabbit erythrocytes. These galactoside-binding lectins were subsequently identified as galectin-1 (L-14.5) and galectin-3 (L-31) (Raz *et al.*, 1987). Lactose-binding lectins have now been isolated from a variety of tumors and tumor cell lines where they have been proposed to play a role in the expression of the transformed phenotype and in metastasis (Allen *et al.*, 1987; Gabius, 1987; Raz and Lotan, 1987; Gabius *et al.*, 1989). Flow cytometric analysis of galectin expression on the cell surface or by indirect immunofluorescence indicated that several sublines and related variants of the B16, K1735 melanoma, or the UV2237 fibrosarcoma expressed higher levels of lectin on the surface of the more metastatic cell lines (Raz and Lotan, 1987). Anti-lectin antibodies have been shown to inhibit the growth of several tumor cell lines and to suppress their ability to grow in semi-solid medium (Lotan *et al.*, 1985). Further, preincubation of metastatic cells with an anti-lectin monoclonal antibody abolished their ability to form lung tumor colonies after injection into the tail vein of syngeneic mice (Meromsky *et al.*, 1986). Most of the tumors and tumor cell lines analyzed contained primarily galectin-1 although several of these also expressed galectin-3. The level of galectin-3 is higher in virally transformed 3T3 fibroblasts than in their untransformed counterparts (Raz *et al.*, 1987). This observation has also been reported for normal and transformed lymphocytes (Carding *et al.*, 1985). Similarly, variants of several tumor cell lines that express a higher metastatic potential in vivo than the parental line have been shown to have qualitative and quantitative differences in lectin expression (Gabius *et al.*, 1987).

More recent results have implicated galectin-1 in malignant transformation following the finding that the stable overexpression of galectin-1 in BALB3T3 fibroblasts results in the acquisition of a transformed phenotype in these cells (Yamaoka *et al.*, 1991). These authors isolated and cloned the cDNA for galectin-1 in search of a growth promoting activity they referred to as TGFg that was secreted by murine Molony sarcoma virus-transformed cells. The galectin-1 overexpressing cells they produced demonstrated a loss of contact inhibition of growth in monolayer culture, grew well in soft agar and were also tumorigenic when injected into nude mice. Similarly, the transfection of galectin-3 into immortalized fibroblastic cells increased their ability to form colonies in agar, tumorigenicity, and lung colonization propensity (Raz *et al.*, 1990).

There have been several reports on the presence of lactose-binding lectins in human colon carcinoma (Allen *et al.*, 1987; Gabius *et al.*, 1987, 1989, 1990; Lee *et al.*, 1990; Irimura *et al.*, 1991; Rosenberg *et al.*, 1991; Castronovo *et al.*, 1992a). Most of these studies detected the presence of galectin-1 in primary colon tissue and tumor samples after affinity chromatography on immobilized lactose. Gabius *et al.* (1989) have demonstrated differences in the lectin profiles in metastatic lesions to either lung or liver from similar primary colon carcinomas. Rosenberg *et al.* (1991) obtained evidence for galectin-3 expression in 6 out of 7 HCC cell lines they examined. We found that galectin-3 was present in 20 of 21 colon carcinoma cell lines (Ohannesian *et al.*, 1994) as well as in surgical specimens of colorectal (Irimura *et al.*, 1991) and gastric carcinoma (Lotan *et al.*, 1994a). This expression was related to progression to metastatic stages in cancer patients (Irimura *et al.*, 1991). Specifically, the levels of galectin-3 in primary tumors of patients having distant metastases (Dukes' stage D) were significantly higher than in those from patients without detectable metastases (Dukes' stages B1 and B2). Likewise, the level of galectin-3 correlated significantly with the serum CEA level in the same patients. In contrast, the variation in the level of galectin-1 among the different specimens was smaller and there was no correlation between the amount of galectin-1 and cancer stage or CEA levels. The results indicated that the relative amount of galectin-3 increases as the colorectal cancer progresses to the more malignant stage. However, galectin-3 has been reported to be found primarily in the cytoplasm of human colon carcinoma in tissue sections where staining of adjacent normal colonic epithelium shows the majority of the lectin localized in the nucleus (Lotz *et al.*, 1993). The intensity of staining for galectin-3 was also more pronounced in normal colonic epithelium than it was in colon carcinomas. These authors have also found that the expression of galectin-3 increases along the crypt-to-surface axis in normal colonic epithelium. The reasons for the differences between results obtained in different laboratories are not clear. The small sample size of the study by Lotz *et al.* may be the reason because in an independent study, Bresalier *et al.* (1995) have analyzed a large number of specimens from benign, premalignant, and malignant colon tissues and have concluded that the expression of galectin-3 increased with tumor progression thus confirming with immunohistochemical data what was suggested based on immunoblotting data by Irimura *et al.* (1991). An increase in the expression of galectin-1 was found in rat transformed thyroid cancer cell lines relative to untransformed cells and also in human thyroid cancer specimens at the mRNA level (Chiarotti *et al.*, 1992). We have extended this observation to a large collection of 30 specimens from human thyroid tumors and adjacent normal tissue and found that the normal and benign adenoma tissue did not express either galectin-1 or galectin-3, however all the papillary and follicular tumors expressed both galectins (Xu *et al.*, 1995). The mechanism of this upregulation of galectin expression after malignant transformation and the role of galectin in thyroid cancer merit further investigation.

5 Carbohydrate Specificity of Galectins

The highest affinity carbohydrate structures that are bound by the galectins are -galactosides (Leffler and Barondes, 1986; Sparrow *et al.*, 1987; Barondes *et al.*, 1994b). Recently, the interactions of galectin-2 with lactose has been examined in detail by X-ray crystallography (Lobsanov *et al.*, 1993). While the major interaction of the galectins occurs with portions of the galactose residue in lactose, important contacts are also made with equatorial positions in glucose and comparable sites in more complex oligosaccharides. Studies of galectin-1 and -3 isolated from human lung indicate that while these lectins have very similar carbohydrate-binding profiles, differences in the carbohydrate specificity of the two lectins for complex oligosaccharides can be observed (Sparrow *et al.*, 1987). Both galectin-1 and 3 have increased affinities for oligosaccharides possessing polylactosamines which also reflects on the extended nature of their carbohydrate recognition domains (Sparrow *et al.*,

1987; Knibbs *et al.*, 1993). Galectin-3 from human lung was found to have a much higher affinity for 1–6-branched poly-*N*-acetyl lactosaminoglycans than for lactose (Sparrow *et al.*, 1987). Galectin-1 also binds these oligosaccharides with a greater affinity than lactose but does so less well than galectin-3. Oligosaccharides containing polylactosamine are developmentally regulated in normal tissues and many tumor cells as are the galectins (Fukuda, 1985; Thorpe *et al.*, 1988; Hakomori, 1989; Feizi *et al.*, 1991). The malignant transformation of certain immortalized cell lines results in the upregulation of GlcNac transferase V which is responsible for producing the GlcNac-β1–6Man-α1–6Man-β- moiety at the trimannosyl core (Dennis *et al.*, 1989). This structure is readily extended with polylactosamine and increases in tetraantennary oligosaccharides have been detected by leukoagglutinating phytohemaglutinin (L-PHA) staining in sections of human breast and colon carcinomas (Fernandes *et al.*, 1991). Such observations have heightened speculation that biologically relevant interactions between the galectins and polylactosaminoglycans may be important *in vivo*.

6 Galectin Ligands

Because the major questions involving the galectins revolve around their function(s) both within cells and extracellularly, it is presumed that the identification of their cognate ligands or associated binding proteins will provide some answers to these questions. The ability of endogenously produced proteins to interact with galectin-1 has been examined primarily in rodent tissues or cell lines. Powell and Whitney (1984) demonstrated that a group of four glycoproteins with three components of M_r 160,000 to 200,000 and a smaller component of 75,000 could be affinity purified from extracts of rat lung on a column of immobilized galectin-1. Galectin-3 has been demonstrated to bind to two glycoproteins that are present in several cell types including human colon carcinoma (Rosenberg *et al.*, 1991) and breast carcinoma (Koth *et al.*, 1993) These proteins referred to as Mac-2 binding proteins have molecular weights of ~92kDa and ~70 kDa. Preliminary sequencing data revealed that the lower molecular protein is likely a degradative product of the larger. Recently, the Mac-2 binding protein has been identified as a secreted glycoprotein present in breast milk and in a number of other tissues and was first identified as the lung tumor antigen L3 (Koths *et al.*, 1993). Other complementary molecules for both galectins are laminin, lysosomal-associated membrane proteins 1 and 2 (lamp-1 and lamp-2), and CEA, as described below.

Because the process of blood born organ colonization in metastasis is thought to require interactions between circulating tumor cells and endothelial cells underlying blood vessels, it is possible that galectins that were found to be expressed on the surface of endothelial cells (Lotan *et al.*, 1994b) participate in tumor cell-endothelial cell adhesion relevant for metastasis.

6.1 Laminin

Laminin is a major component of the basement membrane and it mediates interactions of normal and tumor cells with the ECM primarily by means of integrins but other adhesion molecules have also been implicated in this adhesion (Mecham, 1991). Zhou and Cummings (1993) found that laminin was one of the major glycoproteins purified on immobilized porcine galectin-1 from F9 cell extracts. They have shown that exogenously added galectin-1 was capable of mediating the adhesion of F9 and CHO cells to laminin *in vitro*. Castronovo *et al.* (1992b) have also identified galectin-1 as a laminin-binding protein in a human melanoma cell line.

A role for galectin-1 in preventing myoblast adhesion to laminin and the subsequent fusion of these cells into myotubes has been shown previously (Cooper *et al.*, 1991). In contrast, galectin-1 was found to mediate adhesion of rat olfactory cells to laminin (Mahanthappa *et al.*, 1994). Apparently, galectin-1 may function both positively and negatively in mediating cell interactions with laminin.

Galectin-3 was found to be the major non-integrin laminin-binding protein in activated mouse macrophages where it was identified as the macrophage cell surface differentiation antigen, Mac-2 (Woo *et al.*, 1990). This galectin was also found to be present on the surface of colon carcinoma cells (Lee *et al.*, 1991; Ohannesian *et al.*, 1994). The binding of recombinant human galectin-3 to mouse EHS laminin has also been reported (Ochieng *et al.*, 1992; Massa *et al.*, 1993). We found that both galectin-1 and galectin-3 bind laminin (Ohannesian *et al.*, 1994, 1995). Therefore, this lectin may play a role in tumor cell binding to ECM, an important event in cancer cell invasion and metastasis (Liotta *et al.*, 1986). However, it should be remembered that the mere expression of galectin-3 on the surface of a cell may not be sufficient to mediate carbohydrate-specific adhesion to laminin (Ochieng *et al.*, 1992).

6.2 Lysosome-Associated Membrane Glycoproteins (LAMPs)

The lysosomal-associated membrane glycoproteins lamp-1 and lamp-2 are heavily glycosylated with over 50 % of their molecular weights contributed by carbohydrate side chains and are the major carriers of poly-*N*-acetyllactosamine on the cell surface (Fukuda, 1991). The major function of the lamps is thought to be the protection of the lysosomal membranes from the acid hydrolases which they contain (Fukuda, 1991). Previous studies have shown that lamp-1 and lamp-2, although predominantly intracellular, are present on the surface of various tumor cells.

The presence of β1–6-branched oligosaccharides on cellular glycoproteins such as that found on the lamp molecules and others has been strongly implicated in the expression of the metastatic phenotype, including that of colon carcinoma (Dennis *et al.*, 1989; Fernandes *et al.*, 1991). This branching produces a high affinity substrate for the addition of repeating units of Galβ1–4GlcNAc or polylactosamine. It is not yet clear why this alteration in oligosaccharide side chains should confer a more metastatic phenotype on cells. One possibility is that proteins capable of recognizing these polylactosaminoglycans such as the galectins or selectins may be involved in mediating this effect.

In this context, it is interesting to note that the major galectin-1-binding protein in CHO cells was found to be lamp-1 (Do *et al.*, 1990). Further, polymerized human galectin-1 has also been found to bind to lamp-1 and/or lamp-2 on the surface of human ovarian carcinoma A121 cells and mediate cell attachment (Scrincosky *et al.*, 1993). We have also found that galectin-1 purified from human placenta and recombinant galectin-3 bind lamp-1 and that lamp-1 was a major complementary molecule for both galectins in colon carcinoma cells (Ohannesian *et al.*, 1995).

6.3 Carcinoembryonic Antigen

CEA, a 180–200 kDa glycoprotein, is a member of the immunoglobulin superfamily of proteins and is over 50 % of its weight being contributed by carbohydrate (Yamashita *et al.*, 1987). Increased CEA expression is a characteristic of metastatic colon carcinoma cells and this glycoprotein is used clinically as a marker to monitor recurrence in HCC. We were interested in exploring the interaction of galectin-3 with this glycoprotein for two reasons: a) some of its carbohydrate side chains contain polylactosamine sequences favored by the lectin (Yamashita *et al.*, 1987); and b) CEA has been found to be involved in calcium-independent homotypic and heterotypic adhesion in colon carcinoma cells (Zhou *et al.*, 1990). Indeed, we discovered that both galectin-1 and galectin-3 bind to CEA and that galectin-3 is co-distributed on the surface of colon carcinoma cells with CEA (Ohannesian *et al.*, 1994). The observation of a loss of polarized expression of CEA in colon carcinoma has led to the suggestion that changes in cell-cell interactions relevant to invasion and metastasis may be mediated in part by the overproduction and abnormal distribution of CEA in these cells. Interactions between galectins and CEA at the cell surface could modulate CEA's distribution in the membrane and its efficiency as an adhesive molecule.

7 Conclusions

Major progress has been made over the last few years in the understanding of the structure of the galectins. However, important questions remain to be answered. Foremost among these questions is the function(s) of the lectins in the nucleus, cytoplasm, cell surface and the extracellular matrix. There are also questions concerning the mechanism by which lectins are secreted, the mechanism by which they are associated with the cell membrane, the reasons why many cells contain two or more galactoside-specific lectins, and the identity of the relevant endogenous glycoconjugates with which the galectins interact to execute their functions. The elucidation of the mechanisms by which the galectin genes are regulated during differentiation and development and the interplay between the carbohydrate-binding domain and other functional domain within the galectin molecule are also challenges for future research in this area.

References

Agrwal N, Wang JL, Voss P (1989): Carbohydrate-binding protein 35. Levels of transcription and mRNA accumulation in quiescent and proliferating cells. In *J. Biol. Chem.* **264**:17236–42.

Akimoto Y, Kawakami H, Oda Y *et al.* (1992): Changes in expression of the endogenous β-galactoside-binding 14-kDa lectin of chick embryonic skin during epidermal differentiation. In *Exp.Cell. Res.* **199**:297–304.

Albrandt K, Orida NK, Liu F-T (1987): An IgE-binding protein with a distinctive repetitive sequence and homology with an IgG receptor. In *Proc. Nat. Acad. Sci. USA* **84**:6859–63.

Allen HJ, Karakousis C, Piver MS *et al.* (1987): Galactoside-binding lectin in human tissues. In *Tumor Biol.* **8**:218–29.

Barondes SH (1988): Bifunctional properties of lectins: lectins redefined. In *Trends Biochem. Sci.* **13**:480–2.

Barondes SH, Castronovo V, Cooper DNW *et al.* (1994): Galectins: a family of animal β-galactoside-binding lectins. In *Cell* **76**:597–8.

Barondes SH, Cooper DNW, Gitt MA *et al.* (1994): Galectins: structure and function of a large family of animal lectins. In *J. Biol. Chem.* **269**:20807–10.

Carding SR, Thorpe SJ, Thorpe R *et al.* (1985): Transformation and growth related changes in levels of nuclear and cytoplasmic proteins antigenically related to mammalian galactoside-binding lectin. In *Biochem. Biophys. Res. Commun.* **127**:680–6.

Castronovo V, Campo E, van den Brule FA *et al.* (1992a): Inverse modulation of steady-state messenger RNA levels of two non-integrin laminin-binding proteins in human colon carcinoma. In *J. Natl. Cancer Inst.* **84**:1161–9.

Castronovo V, Luyten F, van den Brule F *et al.* (1992b): Identification of a 14 kD laminin binding protein (HLBP14) in human melanoma cells that is identical to the 14 kD galactoside binding lectin. In *Arch. Biochem. Biophys.* **297**:132–8.

Cherayil BJ, Weiner SJ, Pillai S (1989): The Mac-2 antigen is a galactose-specific lectin that binds IgE. In *J. Exp. Med.* **170**:1959–72.

Chiarotti L, Berlingieri MT, De Rosa P *et al.* (1992): Increased expression of the negative growth factor, galactoside-binding protein, gene in transformed thryoid cells and in human thyroid carcinomas. In *Oncogene* **7**:2507–11.

Cho M, Cummings RD (1995): Galectin-1, a β-galactoside-binding lectin in Chinese hamster ovary cells. II. Localization and biosynthesis. In *J. Biol. Chem.* **270**:5207–12.

Cooper DNW, Barondes SH (1990): Evidence for export of a muscle lectin from cytosol to extracellular matrix and for a novel secretory mechanisms. In *J. Cell Biol.* **110**:1681–91.

Cooper DNW, Massa SM, Barondes SH (1991): Endogenous muscle lectin inhibits myoblast adhesion to laminin. In *J. Cell Biol.* **115**:1437–8.

Dennis JW, Laferte S, Yagel S *et al.* (1989): Asparagine-linked oligosaccharides associated with metastatic cancer. In *Cancer Cells* **1**:87–92.

Do K-Y, Smith DF, Cummings RD (1990): LAMP-1 in CHO cells is a primary carrier of poly-*N*-acetyllactosamine chains and is bound preferentially by a mammalian S-type lectin. In *Biochem. Biophys. Res. Commun.* **173**:1123–8.

Drickamer K, Taylor ME (1993): Biology of animal lectins. In *Annu. Rev. Cell Biol.* **9**:237–64.

Feizi T (1991): Carbohydrate differentiation antigens: probable ligands for cell adhesion molecules. In *Trends Biochem. Sci.* **16**:84–6.

Fernandes B, Sagman U, Auger M *et al.* (1991): β 1–6 branched oligosaccharides as a marker of tumor progression in human breast and colon neoplasia. In *Cancer Res.* **51**:718–23.

Fukuda M (1985): Cell surface glycoconjugates as onco-differentiation markers in hematopoietic cells. In *Biochim. Biophys. Acta* **780**:119–50.

Fukuda M (1991): Lysosomal membrane glycoproteins: structure, biosynthesis, and intracellular trafficking. In *J. Biol. Chem.* **266**:21327–30.

Gabius H-J (1987): Endogenous lectins in tumors and the immune system. In *Cancer Investig.* **5**:39–46.

Gabius H-J (1991): Detection and functions of mammalian lectins – with emphasis on membrane lectins. In *Biochim. Biophys. Acta* **1071**:1–18.

Gabius H-J, Bandlow G, Schirrmacher V *et al.* (1987): Differential expression of endogenous sugar-binding proteins (lectins) in murine tumor model systems with metastatic capacity. In *Int. J. Cancer* **39**:643–8.

Gabius H-J, Ciesiolka T, Kunze E *et al.* (1989): Detection of metastasis-associated differences for receptors of glycoconjugates (lectins) in histomorphologically unchanged xenotransplants from primary and metastatic lesions of human colon adenocarcinomas. In *Clin. Expl. Metastasis* **7**:571–84.

Gitt MA, Barondes SH (1986): Evidence that a human soluble β-galactoside-binding lectin is encoded by a family of genes. In *Proc. Natl. Acad. Sci. USA* **83**:7603–7.

Gitt MA, Massa SM, Leffler H *et al.* (1992): Isolation and expression of a gene encoding L-14-II, a new human soluble lactose-binding lectin. In *J. Biol. Chem.* **267**:10601–6.

Gitt MA, Wiser MF, Leffler H *et al.* (1995): Sequence and mapping of galectin-5, a β-galactoside-binding lectin, found in rat erythrocytes. In *J. Biol. Chem.* **270**:5032–8.

Gu M, Wang W, Song WK *et al.* (1994): Selective modulation of the interaction of α7β1 integrin with fibronectin and laminin by L-14 lectin during skeletal muscle differentiation. In *J. Cell Sci.* **107**:175–81.

Hakomori S (1989): Aberrant glycosylation in tumors and tumor-associated carbohydrate antigens. In *Adv. Cancer Res.* **52**:257–331.

Harrison FL (1991): Soluble vertebrate lectins: ubiquitous but inscrutable proteins. In *J. Cell Sci.* **100**:9–14.

Hart GW, Haltiwanger RS, Holt GD *et al.* (1989): Glycosylation in the nucleus and cytoplasm. In *Annu. Rev. Biochem.* **58**:841–74.

Hirabayashi J, Kasai K-I (1991): Effect of amino acid substitution by site-directed mutagenesis of the carbohydrate recognition and stability of human 14 kDa β-galactoside-binding lectin. In *J. Biol. Chem.* **266**: 23648–55.

Hughes RC (1992): Role of glycosylation in cell interactions with extracellular matrix. In *Biochem. Soc. Transact.* **20**:279–84.

Hughes RC (1994): Mac-2: a versatile galactose-binding protein of mammalian tissues. In *Glycobiology* **4**:5–12.

Irimura T, Matsushita Y, Sutton RC *et al.* (1991): Increased content of an endogenous lactose-binding lectin in human colorectal carcinoma progressed to metastatic stages. In *Cancer Res.* **51**:387–93.

Kimber SJ (1990): Glycoconjugates and cell surface interactions in pre- and peri-implantation mammalian embryonic development. In *Int. Rev. Cytol.* **120**:53–167.

Knibbs RN, Agrwal N, Wang JL *et al.* (1993): Carbohydrate-binding protein 35: analysis of the interaction of the recombinant polypeptide with saccharides. In *J. Biol. Chem.* **268**:14940–7.

Koths K, Taylor E, Halenbeck R *et al.* (1993): Cloning and characterization of a human Mac-2-binding protein, a new member of the superfamily defined by the macrophage scavenger receptor cysteine-rich domain. In *J. Biol. Chem.* **268**:14245–9.

Lee EC, Woo H-J, Korzelius CA *et al.* (1991): Carbohydrate-binding protein 35 is the major cell-surface laminin-binding protein in colon carcinoma. In *Arch. Surg.* **126**:1498–502.

Leffler H, Barondes SH (1986): Specificity of binding of three soluble rat lung lectins to substituted and unsubstituted mammalian β-galactosides. In *J. Biol. Chem.* **261**:10119–26.

Liotta LA, Rao CN, Wewer UM (1986): Biochemical interactions of tumor cells with the basement membrane. In *Annu. Rev. Cell Biol.* **55**:1037–57.

Lobsanov YD, Gitt MA, Leffler H *et al.* (1993): X-ray crystal structure of the human dimeric S-lac lectin, L-14-II, in complex with lactose at 2.9 angstrom resolution. In *J. Biol. Chem.* **268**:27034–8.

Lotan R (1992): β-Galactoside-binding vertebrate lectins: synthesis, molecular biology, function. In *Glycoconjugates: Composition, Structure, and Function* (Allen H, Kisailus E, eds) pp 635–71, New York: Marcel Dekker, Inc.

Lotan R, Ito H, Yasui W *et al.* (1994a): Expression of a 31-kDa lactoside-binding lectin in normal human gastric mucosa and in primary and metastatic gastric carcinomas. In *Int. J. Cancer* **56**:474–80.

Lotan R, Belloni PN, Tressler RJ *et al.* (1994b): Expression of galectins on microvessel endothelial cells and their involvement in tumor cell adhesion. In *Glycoconjugate J.* **11**:462–8.

Lotan R, Lotan D, Raz A (1985): Inhibition of tumor cell colony formation in culture by a monoclonal antibody to endogenous lectins. In *Cancer Res.* **45**:4349–53.

Lotz MM, Andrews CW, Korzelius C *et al.* (1993): Decreased expression of Mac-2 (carbohydrate-binding protein 35) and loss of its nuclear localization are associated with the neoplastic progression of colon carcinoma. In *Proc. Natl. Acad. Sci. USA* **90**: 3466–70.

Mahanthappa NK, Cooper DNW, Barondes SH *et al.* (1994): Rat olfactory neurons can utilize the endogenous lectin, L-14, in a novel adhesion mechanism. In *Development* **120**:1373–84.

Mecham RP (1991): Laminin receptors. In *Annu. Rev. Cell Biol.* **7**:71–91.

Meromsky L, Lotan R, Raz A (1986): Implications of endogenous tumor cell surface lectins as mediators of cellular interactions and lung colonization. In *Cancer Res*. **46**:5270–2.

Ochieng J, Gerold M, Raz A (1992): Dichotomy in the laminin-binding properties of soluble and membrane-bound human galactoside-binding protein. In *Biochem. Biophys. Res. Commun.* **186**: 1674–80.

Oda Y, Kasai K-I (1984): Photochemical cross-linking of β-galactoside-binding lectin to polylactosaminoproteoglycan of chick embryonic skin. In *Biochem. Biophys. Res. Commun.* **123**:1215–20.

Ohannesian DW, Lotan D, Lotan R (1994): Concomitant increases in galectin-1 and its glyconjugate ligands (carcinoembryonic antigen, lamp-1, and lamp-2) in cultured human colon carcinoma cells by sodium butyrate. In *Cancer Res.* **54**:5992–6000.

Ohannesian DW, Lotan D, Thomas P *et al.* (1995): Carcinoembryonic antigen and other glycoconjugates act as ligands for galectin-3 in human colon carcinoma cells. In *Cancer Res.* **55**:2191–9.

Paroutaud P, Levi G, Teichberg VI *et al.* (1989): Extensive amino acid sequence homology between animal lectins. In *Proc. Natl. Acad. Sci. USA* **84**:6345–50.

Poireir F, Robertson EJ (1993): Normal development of mice carrying a null mutation in the gene encoding the L14 S-type lectin. In *Development* **119**: 1229–36.

Poirier F, Timmons PM, Chan C-TJ *et al.* (1992): Expression of the L-14 lectin during mouse embryogenesis suggests multiple roles during pre- and post-implantation development. In *Development* **115**: 143–55.

Powell JT, Whitney PL (1980): Postnatal development of rat lung. In *Biochem. J.* **188**:1–8.

Powell JT, Whitney PL (1984): Endogenous ligands of rat lung β-galactoside-binding protein (galap tin) isolated by affinity chromatography on carboxyamidomethylated-galaptin-Sepharose. In *Biochem. J.* **223**: 769–74.

Raz A, Lotan R (1981): Lectin-like activities associated with human and murine neoplastic cells. In *Cancer Res*. **41**:3642–7.

Raz A, Lotan R (1987): Endogenous galactoside-binding lectins: a new class of functional tumor cell surface molecules related to metastasis. In *Cancer Metastasis Rev.* **6**:433–52.

Raz A, Avivi A, Pazerini G *et al.* (1987): Cloning and expression of cDNA for two endogenous UV-2237 fibrosarcoma lectin genes. In *Exp. Cell Res.* **173**: 109–16.

Raz A, Carmi P, Pazerini G (1988): Expression of two different endogenous galactoside-binding lectins sharing sequence homology. In *Cancer Res.* **48**: 645–9.

Raz A, Zhu D, Hogan V *et al.* (1990): Evidence for the role of 34-kDa galactoside-binding lectin in transformation and metastasis. In *Int. J. Cancer* **46**: 871–7.

Regan LJ, Dodd J, Barondes SH *et al.* (1986): Selective expression of endogenous lactose-binding lectins and lactoseries glycoconjugates in subsets of rat sensory neurons. In *Proc. Natl. Acad. Sci. USA* **83**:2248–52.

Rosenberg I, Cherayil BJ, Isselbacher KJ *et al.* (1991): Mac-2 binding glycoproteins: putative ligands for a cytosolic β-galactoside lectin. In *J. Biol. Chem.* **266**: 18731–6.

Sanford GL, Harris-Hooker S (1990): Stimulation of vascular cell proliferation by β-galactoside specific lectins. In *FASEB J.* **4**:2912–8.

Sato S, Burdett I, Hughes RC (1993): Secretion of the baby hamster kidney 30-kDa galactose-binding lectin from polarized and nonpolarized cells: a pathway independent of the endoplasmic reticulum-Golgi complex. In *Exp. Cell Res.* **207**:8–18.

Sharon N (1975): Complex Carbohydrates: Their Chemistry, Biosynthesis, and Functions. Reading, MA: Addison-Wesley.

Sharon N, Lis H (1990): Legume lectins– a large family of homologous proteins. In *FASEB J.* **4**:3198–208.

Skrincosky DM, Allen HJ, Bernacki RJ (1993): Galaptin-mediated adhesion of human ovarian carcinoma A121 cells and detection of cellular galaptin-binding glycoproteins. In *Cancer Res.* **53**:2667–75.

Sparrow CP, Leffler H, Barondes SH (1987): Multiple soluble β-galactoside-binding lectins from human lung. In *J. Biol. Chem.* **262**:7383–90.

Teichberg VI, Silman I, Beitsch DD *et al.* (1975): A β-D-galactoside binding protein from electric organ tissue of *Electrophorus electricus*. In *Proc. Natl. Acad. Sci. USA* **72**:1383–7.

Thorpe SJ, Bellairs R, Feizi T (1988): Developmental patterning of carbohydrate antigens during early embryogenesis of the chick: expression of antigens of the poly-*N*-acetyllactosamine series. In *Development* **102**:193–210.

Wang JL, Werner EA, Laing JG *et al.* (1992) Nuclear and cytoplasmic localization of a lectin-ribonucleoprotein complex. In *Biochem. Soc. Transact.* **20**:269–78.

Wells V, Mallucci L (1991): Identification of an autocrine negative growth factor: mouse β-galactoside-binding protein is a cytostatic factor and cell growth regulator. In *Cell* **64**:91–7.

Whitney PL, Powell JT, Sanford GL (1986): Oxidation and chemical modification of lung β-galactoside-specific lectin. In *Biochem. J.* **238**:683–9.

Woo H-J, Shaw LM, Messier JM *et al.* (1990): The major non-integrin laminin binding protein of macrophages is identical to carbohydrate binding protein 35 (Mac-2). In *J. Biol. Chem.* **265**:7097–9.

Xu X-C, El-Naggar AK, Lotan R (1995) Differential expression of galectin-1 and galectin-3 in thyroid tumors: potential diagnostic implications. In *Am. J. Pathol.* **147**:815–22.

Yamaoka K, Ohno S, Kawasaki H *et al.* (1991): Overexpression of a β-galactoside binding protein causes transformation of 3T3 fibroblast cells. In *Biochem. Biophys. Res. Commun.* **179**:272–9.

Yamashita K, Totani K, Kuroki M *et al.* (1987): Structural studies of the carbohydrate moieties of carcinoembryonic antigens. In *Cancer Res.* 47: 3451–9.

Zhou H, Fuks A, Stanners CP (1990): Specificity of intercellular adhesion mediated by various members of the immunoglobulin supergene family. In *Cell Growth Differ.* **1**:209–15.

Zhou Q, Cummings RD (1992): Animal lectins: a distinct group of carbohydrate-binding proteins involved in cell adhesion, molecular recognition, and development. In *Cell Surface Carbohydrates and Cell Development* (Fukuda M, ed) pp 99–126, Boca Raton: CRC Press.

Zhou Q, Cummings RD (1993): L-14 lectin recognition of laminin and its promotion of in vitro cell adhesion. *Arch. Biochem. Biophys.* **300**: 6–17.

26 Glycoconjugate-Mediated Drug Targeting

KEVIN G. RICE

1 Introduction

Targeted drug delivery involves the design and synthesis of carriers displaying ligands that mediate binding of a drug/carrier complex to a receptor. Subsequent internalization of the carrier/drug complex leads to accumulation of the drug in the target cells and exclusion from non-target cells which lack the requisite receptor.

Biomolecules that have been employed for active drug targeting include monoclonal antibodies (MoAbs), transferrin, insulin, epidermal growth factor, melanocyte stimulating hormone, folate and biotin (Gregoriadis, 1977, 1983; Venter, 1982). To operate successfully *in vivo*, targeted drug delivery systems must evade phagocytosis, identify their target site, and internalize into cells (Ghose and Blair, 1987; Pimm, 1988; Koppel, 1990; Vaickus and Foon, 1991; Kosmas *et al.*, 1993). The expectation that MoAbs would achieve these functions and serve as programmable "magic bullets" (Ehrlich, 1906) has been largely unfulfilled. MoAb targeting often fails because of the heterogeneity of cell surface antigens and/or the down regulation of cell surface receptors, resulting in populations of target cells that escape selection.

Growing knowledge of the function of carbohydrates has lead to their increased consideration as ligands for drug targeting. Carbohydrates are attractive drug targeting candidates because of their low molecular weight, their extraordinary structural diversity with potential to select numerous target receptors, and their frequently encountered entry into target cells via receptor mediated endocytosis (Shen and Ryser, 1981; Monsigny *et al.* 1988a; Karlsson, 1991).

Carbohydrates are bound and internalized into mammalian cells by membrane spanning lectins. The binding specificity of carbohydrate/lectin interactions is dictated by the identity and valency of terminal sugar residues. Multiple non-reducing residues of oligo-antennary chains are recognized as a clustered determinant by multi-subunit lectins. Biologically active oligosaccharides are seldom found unconjugated to macromolecules but instead are typically bound to proteins or lipids and function to control the biodistribution and serum half-life of glycoproteins and to mediate cell/cell interactions. Optimal presentation occurs when two or more carbohydrate determinants are properly spaced on the antenna of *N*-linked oligosaccharides or directly linked to the protein surface in the form of *O*-linked oligosaccharides.

Several cell surface lectins have been discovered that endocytose and sort their ligands to the lysosomes (Ashwell and Hardford, 1982; Wileman *et al.*, 1985). Because lectins recognize and bind to the terminal sugar residues of oligosaccharides, endocytosis is relatively uninfluenced by the size or composition of the aglycone. Thus, carbohydrate ligands may be derivatized with drugs to provide an efficient route to shuttle foreign molecules into cells. Since mammalian lectins are usually narrowly distributed to a single cell type and have a well-conserved ligand specificity across many species, the potential exits for developing and testing carbohydrate mediated drug delivery systems in animal models that may also operate when tested in humans. The following sections document the utility of carbohydrates in drug targeting.

H.-J. and S. Gabius (Eds.), Glycosciences

ISBN 3-8261-0073-5

2 Mammalian Lectins that Mediate Drug Delivery

Ashwell and co-workers discovered the first mammalian lectin when they observed clearance of ceruloplasmin after removal of sialic acid from its *N*-linked oligosaccharides (Van Den Hamer *et al.*, 1970). A hepatic asialoglycoprotein receptor (ASGP-R) was later isolated, characterized and found to be responsible for this activity. It functions by selectively removing glycoproteins that have been de-sialylated, and thereby expose terminal galactose residues (Ashwell and Hardford, 1982; Schwartz, 1984).

In addition to asialo-glycoproteins, the ASGP-R binds a broad range of molecules that possess terminal galactose or *N*-acetylgalactosamine residues presented in a variety of geometries. This unusual feature of the ASGP-R allowed the design and testing of structurally diverse glycoconjugates for drug delivery. The relative ease at which ASGP-R-mediated targeting has been achieved has raised expectations for the general utility of mammalian lectins as targets for site specific drug delivery. Even though the discovery of carbohydrate receptors has progressed rapidly in the past decade, many of these mammalian lectins may have less immediate utility in drug targeting applications because of their strict ligand requirements and because their natural function does not involve the internalization of their ligands. Thus, one approach to discover new lectins which are applicable to the design of drug delivery systems is to probe for lectin activity by analyzing ligand uptake *in vivo* using radiolabeled glycoconjugates (Kojima *et al.* 1990; Gupta and Surolia, 1994; Chiu *et al.* 1994, 1995).

Ligand-binding affinity is probably the most important consideration when designing carbohydrate mediated drug delivery systems. This is because drug-loaded carrier molecules must ultimately compete for receptor binding with other endogenous ligands *in vivo*. Although it is true that the ASGP-R binds and internalizes most galactose terminated ligands, it shows markedly different affinities for them depending on structure (Lee *et al.*, 1983). Monovalent ligands such as galactose, lactose, and monoantennary galactosides bind with a millimolar dissociation constant, which is far below that needed to achieve drug targeting under physiological conditions. In comparison, divalent galactose-terminated oligosaccharides bind with affinities that are three orders of magnitude higher but still fail to target the receptor appreciably *in vivo* due to competition with higher affinity glycoproteins and elimination by renal filtration (Chiu *et al.*, 1994). Ligands containing a cluster of three galactose residues bind to the receptor with a nanomolar dissociation constant which appears to be the affinity threshold needed to achieve appreciable liver targeting under physiological conditions (Chiu *et al.*, 1994, 1995). This relationship between sugar clustering and enhanced binding affinity is related to the structure of the ASGP-R receptor which is composed of three subunits, each of which contains a galactose-binding site (Rice *et al.*, 1990).

Among the mammalian lectins discovered so far, the binding specificity of the ASGP-R has been studied in greatest detail. It is the prototype of a family of "C-type" lectins, named for their calcium dependent ligand binding. The receptors of this family possess similar carbohydrate recognition domains (CRD) of 115–134 amino acids which contain 18 highly conserved and 14 invariant residues (Drickamer, 1988, 1993). To date, over thirty different mammalian proteins have been classified as members of this family because they possess the "lectin motif" and have demonstrated sugar binding activity (Hirabayashi, 1993).

C-type lectins have been sub-classified as either membrane integrated or soluble receptors. Primarily the membrane-spanning receptors are most relevant to carbohydrate-mediated targeting, since these lectins present their CRD to the extracellular space and thereby can bind and internalize ligands that are extracellular (Drickamer, 1993).

Type II and Type VI membrane-spanning lectins that possess specificity for galactose and mannose have been used to target drug internalization into hepatocytes or macrophages of various origins (Table 1). At present, the ligand binding specificity for most of the C-type lectins has only been explored using monosaccharide or neoglycoprotein ligands. However, it is likely, as

Table 1. Localization and specificity of membrane spanning C-type lectins

Name/Tissue Location	Sugar Specificity	Reference
Type II[a]		
Asialo-glycoprotein Receptor / Liver Hepatocytes	Gal & GalNAc	Spiess *et al.* (1985)
Kupffer Cell Receptor / Liver Kupffer Cells	Gal & Fuc	Hoyle *et al.* (1988)
Macrophage Receptor / Liver Kupffer Cells	Gal & GalNAc	Ii *et al.* (1990)
IgE Fc Receptor / B cells	Gal	Kikutani *et al.* (1986)
Placental Receptor / Placenta	Fuc & Man	Curtis *et al.* (1992)
Type IV		
L-Selectin / Leukocytes	Fuc & Sialic Acid	Lasky *et al.* (1989)
E-Selectin / Endothelial Cells	Fuc & Sialic Acid	Bevilacqua *et al.* (1989)
P-Selectin / Platelets	Fuc & Sialic Acid	Johnston *et al.* (1989)
Type VI		
Mannose Receptor / Lung and Liver Macrophages	Man & Fuc	Taylor *et al.* (1992)

a Lectin types have been defined by Drickamer (1993)

in the case of the ASGP-R, that branched oligosaccharides or multivalent *O*-linked oligosaccharide bind with higher affinity and selectivity than simple glycosides. Nevertheless, although information on the identity of their natural ligands is being accumulated, both the ASGP-R of hepatocytes and the mannose receptor of macrophages have been successfully targeted for the purpose of drug delivery.

Other membrane spanning lectins, such as those categorized as Type IV are presently of enormous interest but not because of their potential use in drug targeting but because of their established function in mediating leukocyte trafficking (Varki, 1992). E-selectin is constitutively expressed on endothelial cells at sites of inflammation leading to rolling and infiltration of leukocytes into these tissues (Bevilacqua *et al.*, 1991). The carbohydrate ligand present on leukocytes is believed to be a glycoconjugate which displays a cluster of sialyl Lewis x oligosaccharide determinants (Lowe *et al.*, 1990). Although E-selectin has not been shown to endocytose its ligands, its inducible expression in various tissues may be an attractive feature for future glycoconjugate mediated target drug delivery systems.

Recently, a new subclass of the lectins has emerged which are known as I-type lectins (Powell and Varki, 1995). The members of this family, which include CD22 and CD33, are all also members of the IgG super family. These adhesion molecules bind to glycoprotein ligands possessing clusters of α2–6-linked sialic acid on their oligosaccharides. Although at present it is uncertain how many members of this lectin family exist, their restricted cell distribution and selective ligand requirements may allow their selective targeting for the purpose of drug delivery.

In addition to membrane-spanning lectins, the galectin family of carbohydrate receptors (see review of Lotan in this volume) represents a new and promising target for drug delivery. There are now eight members of the galectin family all of which recognize *N*-acetyllactosamine containing oligosaccharides and may function in mediating cell/cell interactions. Since individual members of this family are expressed in unique tissues, it may be possible to discover carbohydrate ligands that selectively bind to one galectin for the purpose of developing targeting agents (Barondes *et al.*, 1994).

3 Design of Glycoconjugate Carriers for Drug Delivery

Ligands used in the design of glycoconjugate carriers can be classified as either glycoproteins, glycopeptides, neoglycoproteins, neoglycopeptides or glycosylated polymers. These differ in their size, shape, and degree of heterogeneity. All possess in common a carbohydrate of varying complexity as the ligand which is attached to an anchor or scaffolding molecule which is usually proteinaceous.

Glycoproteins are frequently used during the initial stages of developing a drug targeting agent since they are readily available and possess a sufficient cluster of terminal sugar residues that serve as ligands for mammalian lectins. The amine and carboxyl groups on the protein surface can be modified with drug molecules without disrupting the affinity between a glycoprotein and a lectin. The desialylated form of bovine fetuin and human orosomucoid are usually selected as suitable ligands for the ASGP-R on hepatocytes (Meijer *et al.*, 1992).

Even though asialoglycoproteins are easy to prepare and possess high affinity for the ASGP-R, they lack some qualities desirable of refined drug targeting agent for human use. Most glycoproteins contain an array of oligosaccharide structures at multiple glycosylation sites which results in the formation of numerous glycoforms (Spellman, 1990). Since it is presently not possible to separate glycoforms, oligosaccharide heterogeneity represents a significant obstacle to developing well-characterized drug target agents. Further complications arise from the protein portion which contains extraneous bulk, resulting in drug delivery agents that are high molecular weight and possess complexity beyond the capability of current structure proofing technology.

Neoglycoproteins have been used in place of glycoproteins to minimize oligosaccharide heterogeneity. They are prepared by covalently attaching monosaccharides, disaccharides, or oligosaccharides to a protein such as bovine serum albumin (BSA) (Stowell and Lee, 1980, Lee and Lee, 1994). The conjugation of sugars can be controlled to systematically vary the loading density and hence the degree of clustering, resulting in glycoconjugates having predictable affinities for lectins (see the review by Lee and Lee in this volume).

Although originally developed as lectin probes, neoglycoproteins have also found applications in drug targeting (Gabius, 1988; Monsigny *et al.*, 1988; Gabius and Bardosi, 1991). BSA modified with either mannose, galactose, or lactose is used most frequently, however; BSA conjugates possessing mannose-6-phosphate have also been developed (Roche *et al.*, 1990).

There are two potential drawbacks to the use of neoglycoproteins for human drug delivery studies. First, the targeting specificity in certain cases is influenced by the chemical linkers used to attach sugars to the protein. This can result in artifactual recognition by certain endogenous lectins (Lee and Lee, 1994) leading to re-direction of the carrier system and potentially a greater disparity between results in animals versus humans. Secondly, neoglycoproteins prepared from BSA are immunogenic when dosed repeatedly in other species. However, this drawback may be minimized using a homologous albumin from the species in which the drug delivery system is being tested (Fiume *et al.*, 1987)

Direct glycosylation of polylysine or copolymers of polyhydroxypropyl-methacrylamide create glycoconjugate carriers that have a dramatically different shape. (Fiume *et al.*, 1986a; Kopecek, 1991; Magnusson *et al.* 1994). These polymeric glycoconjugates possess lower molecular weights and higher sugar densities relative to neoglycoproteins. In addition, the polymer backbones may be selected for a desired property such as biodegradation, metabolic stability, or charge character (Duncan *et al.*, 1983; Rejmanova *et al.*, 1985; Ohsumi *et al*, 1988). Glycoconjugates of this type have found their greatest utility as components of non-viral gene delivery systems (Midoux *et al.*, 1993).

Even simpler glycoconjugate ligands are accessible by attaching two or more sugar residues to a di- or tri-peptide resulting in neoglycopeptides. The branching afforded by the peptide is designed to mimic multivalent *N*-linked oligosaccharides (Lee and Lee, 1987; Plank *et al.*, 1992; Haensler and Szoka, 1993). Glycosylated Tris is another example of a neoglycoconjugate designed to create tight clusters of terminal sugar residues (Kempen *et al.*, 1984). These low molecular weight ligands are attractive because they retain high affinity for the ASGP-R and are available in large quantities through synthesis.

A further refinement in glycoconjugate design is the incorporation of natural *N*-linked oligosaccharides into a targeted drug delivery system (Wadhwa *et al.*, 1995). Since these are high affinity ligands, it is conceivable that incorporation of just a few oligosaccharides may be sufficient to mediate efficient targeting of macromolecules, thereby alleviating the need for more bulky and heterogenous ligands. In certain cases, the natural conformation of an *N*-linked oligosaccharide may be critical for selecting target receptors with high affinity. In addition, *N*-linked oligosaccharides are non-antigenic since they are found ubiquitously in animals.

N-Linked oligosaccharides are most conveniently prepared by purification from glycoproteins, although this approach is time consuming and expensive (Rice and Silva, 1996). Nonetheless, in certain applications such as gene therapy, the expense of the carrier molecule may be outweighed by the high affinity and selectivity afforded by glycoconjugates composed of well-characterized *N*-linked oligosaccharides (Wadhwa *et al.*, 1995).

The site-specific conjugation of drug molecules to the carrier is a critical component toward developing targeted drug delivery systems. A variety of schemes have been devised for conjugating drugs with protein carriers as discussed in detail elsewhere (Ghose *et al.*, 1983; Pietersz, 1990; Brinkley, 1992; Meijer *et al.*, 1992). However, a special design consideration when attaching drugs to glycoconjugates is to avoid sterically blocking the non-reducing end of the sugar moiety which is critical for receptor recognition. Likewise, drug labeling schemes which disrupt the clustering of sugar residues may reduce the overall receptor affinity and negatively influence the targeting efficiency *in vivo*.

Targeted drug delivery systems also need to be built in mechanisms that cause intracellular liberation of the drug from the carrier. To achieve this, drugs have been linked to carriers *via* acid-labile linkers that are cleaved following endocytosis (Shen and Ryser, 1981; Greenfield *et al.*, 1990). Linkers that are stable in the circulation, but are degraded by proteases present in lysosomes have also been developed (Trouet *et al.*, 1982; Duncan *et al.*, 1983; Rejmanova *et al.*, 1985; Franssen *et al.*, 1991). Additionally, an important consideration towards achieving greater efficiency may be the need for including subcellular targeting mechanisms into targeted drug delivery systems (Kaneda *et al.*, 1989). This may be of less immediate concern when targeting low molecular weight drugs which readily diffuse through membranes, but may be essential for macromolecular proteins or nucleic acids which may not operate unless delivered to the appropriate organelle.

4 Applications of Carbohydrate-Mediated Drug Targeting

Many examples of glycoconjugate-mediated drug targeting aimed at either hepatocytes or macrophages have been reported during the last decade. From these reports one can conclude that different classes of drug molecules which range in size from a few hundred molecular weight up to tens of millions are amendable to glycoconjugate mediated targeting. This concept has been further extended to target liposome and lipoprotein particles that are 100 nm in diameter.

A further observation is that glycoconjugates of varying complexity are used and that no single type of glycoconjugate is universally accepted to target drugs of a given class. Thus, a particular drug targeting application may benefit by exploring glycoconjugate carriers with multiple designs before deciding on how to refine the system.

Also, it is evident that despite their promise in both *in vitro* and *in vivo* applications, very few glycoconjugate targeting systems have been refined into homogenous low molecular weight delivery systems. This may be a necessary requirement to create targeted drug delivery systems that operate safely in humans but also represents an economical barrier towards their further development and commercialization. The following describes some of the many applications reported for glycoconjugate targeted drug delivery systems which are grouped according to the relative size of the drug molecule attached to the glycoconjugate carrier.

4.1 Targeting Low Molecular Weight Drug Molecules to Hepatocytes and Macrophages

Fiume and colleagues have firmly established the concept of glycoconjugate targeting to hepatocytes *via* the ASGP-R by improving the activity of antiviral agents following their attachment to a carrier molecule. Trifluorothymidine, adenine-9-β-D-arabinofuranoside (Ara-A) and adenine-9-β-D-aribinofuranoside 5'-monophosphase (Ara-AMP) were 2–4 times more potent at inhibiting viral synthesis in mouse hepatocytes following conjugation to asialo-fetuin (Balboni *et al.*, 1976; Fiume *et al.*, 1980). Neoglycoproteins and glycosylated polylysine were also effective in targeting the antiviral Ara-AMP to hepatocytes (Fiume *et al.*, 1986 a, b). However, further studies revealed that certain carriers were antigenic while others were toxic (Fiume *et al.*, 1987, 1994). Nonetheless, a clinical trial with Ara-AMP bound to lactosylated serum albumin was found promising for treating chronic hepatitis B (Fiume *et al.*, 1988).

The dose of primaquine needed to treat hepatitis was reduced when a lactosylated serum albumin was used as carrier to deliver the drug to hepatocytes in mice. This allowed the administration of a single curative dose of conjugated primaquine at a concentration below the mean lethal dose of the free drug (Hofsteenge *et al.*, 1986). Likewise, a conjugate consisting of uridine monophosphate covalently linked to asialo-orosomucoid was targeted to the ASGP-R on rat hepatoma cells and resulted in complete inhibition of galactosamine toxicity (Wu *et al.*, 1988).

Alternatively, Midoux and colleagues (1990) have demonstrated that the mannose receptor on macrophages can be efficiently targeted using mannosylated polylysine to transport the antiviral, phosphonylmethoxyethyl-adenine, into human macrophages *in vitro*. Neoglycoproteins containing mannose increased targeting of 3'-axido-3-deoxythymidine monophosphate to human T_4 lymphocytes *in vitro* and enhanced the inhibition of HIV activity over that of free drug (Molema *et al.*, 1990).

Mannosylated serum albumin-muramyl dipeptide (MDP) conjugates have been used to activate alveolar macrophages both in mice and rats for up to 72 hours after *i. v.* or *i.p.* administration (Monsigny *et al.*, 1984). Most importantly, the MDP-neoglycoprotein conjugates provided greater protection to tumor metastasis in mice (Roche *et al.*, 1985). Glycosylated serum albumin and glycosylated polylysine were compared as carriers for MDP *in vitro* (Petit *et al.*, 1990) Both were effective as measured by the cytotoxicity resulting from the secreted tumor necrosis factor.

Methotrexate and doxorubicin have been conjugated to mannosylated human serum albumin and targeted to macrophages for the treatment of leishmaniasis (Sett *et al.*, 1993a, b). Similarly, allopurinol riboside was conjugated to mannosylated polylysine and found to be 50 times more effective than the free drug in an *in vitro* model (Negre *et al.*, 1992). The anti-inflammatory agent naproxen was improved by targeting to macrophages following its conjugation to mannosylated human serum albumin (HSA), and to hepatocytes by lactosylated HSA (Franssen *et al.*, 1993).

4.2 Targeting Antisense Oligonucleotides

Antisense oligonucleotides composed of fifteen or more nucleotides are medium sized molecules for glycoconjugate mediated targeting. Targeting antisense oligonucleotides involves either their covalent or non-covalent attachment to neoglycopeptide, neoglycoprotein, or asialo-glycoprotein carriers. Mannose-6-phosphate labelled bovine serum albumin was covalently coupled to the 3' end of an antisense oligonucleotide which enhanced internalization of the conjugate 20-fold in mouse macrophages compared to the free oligonucleotide (Bonfils *et al.*, 1992a). Alternatively, a biotinylated oligonucleotide was non-covalently complexed with mannosylated streptavidin in order to increase its uptake into macrophages (Bonfils *et al.*, 1992b). Oligonucleotides have also been non-covalently complexed with asialoglycoprotein-polylysine conjugates resulting in increased uptake into hepatocytes (Bunnell *et al.*, 1992; Reinis *et al.*, 1993) and then in specific inhibition of hepatitis B viral gene expression in infected HepG2 cells (Wu and Wu, 1992). Recently, the neoglycopeptide (YEE) bearing three terminal *N*-acetylgalactosamine residues was covalently bound to an antisense oligonucleotide to improve its internalization into hepatocytes (Hangeland *et al.*, 1995).

4.3 Glycoconjugate Targeting of Enzymes

The glycoconjugate targeting concept has also been successful at improving the *in vivo* activity of medium sized molecules, such as enzymes. Replacement of glucocerebrosidase in patients that suffer a deficiency of this enzyme is one approach to treat Gaucher's disease. However, direct infusion of purified glucocerebrosidase failed to provide clear clinical benefits due to the preferential uptake of the enzyme by hepatocytes instead of Kupffer cells (Furbish *et al.*, 1978; Brady, 1984).

To alter the biodistribution of the enzyme, exoglycosidases were used to trim the *N*-linked oligosaccharides on glucocerebrosidase in order to reveal terminal mannose residues on the *N*-linked oligosaccharides. Since Kupffer cells express mannose receptors, glucocerebrosidase could be re-directed to internalize into these cells (Stahl *et al.*, 1978; Furbish *et al.*, 1981). To improve the targeting efficiency, glucocerebrosidase was also chemically mannosylating or entrapping it in mannosylated liposomes (Doebber *et al.*, 1982; Das *et al.*, 1985). However, the best results were obtained by oligosaccharide remodeling (Murray, 1987). The remodelled mannose-terminated glucocerebrosidase (Alglucerase) has been clinically used in patients and has shown excellent therapeutic effects in treating type I Gaucher's disease (Barton *et al.*, 1991).

The principles learned in targeting glucocerebrosidase may also be applied to other enzyme deficiency related disorders. Superoxide dismutase (SOD) has been used to treat rheumatoid arthritis in humans (Menander-Huber and Huber, 1977). The enzyme functions in defending against tissue injury by eliminating superoxide free radicals released into the bloodstream (Fridovich, 1983). However, exogenously administered SOD is rapidly excreted by renal filtration. Galactosylation of the enzyme redirects it to liver parenchymal cells in mice while modification with mannose allows non-parenchymal cells to be targeted (Fujita *et al.*, 1992). *In vitro* studies using peritoneal macrophages established that mannosylated SOD was superior at inhibiting superoxide radical compared to unmodified SOD (Takakura *et al.*, 1994).

4.4 Glycoconjugate-Mediated Targeting of DNA

Perhaps one of the most challenging areas of drug targeting is the development of carrier molecules capable of targeting macromolecular (10 million Da) plasmid DNA. Gene therapy involves the replacement or augmentation of missing or defective genes for therapeutic benefit. An important advantage in gene therapy is that DNA does not need to be delivered to cells at a high concentration since each gene can potentially express multiple copies of its polypeptide product. However, to optimize these as potential therapeutic agents, DNA should not only be targeted to gain cell entry but should also be localized to the nucleus (Mulligan, 1993).

Efficient gene delivery requires carrier systems which protect the DNA from nucleases and compact it into a small size. Wu and Wu (1988a, b) demonstrated receptor-mediated gene delivery in rat hepatocytes both *in vitro* and *in vivo* using asialo-orosomucoid polylysine carriers that target the ASGP-R. This system was used to correct genetic analbuminemia in rats and to decrease hypercholesterolemia in rabbits (Wu *et al.*, 1991; Wilson *et al.*, 1992a). However, in the absence of partial hepatectomy of the liver, transgene expression levels were low (Wilson *et al.*, 1992b).

Wagner and colleagues also targeted the ASGP-R with a tetragalactosyl neoglycopeptide and found that reporter gene expression in hepatoma cells could be enhanced by co-administration of fusogenic peptides designed to provide endosomal escape of internalized DNA complexes (Plank *et al.*, 1992, 1994). This was consistent with earlier results using carriers specific for the transferrin receptor and also with results using adenovirus (Wagner *et al.*, 1992). Cristiano *et al.* (1993) demonstrated expression of a phenylalanine hydroxylase (PAH) gene in PAH deficient primary hepatocytes using an asialo-orosomucoid polylysine carrier co-transfected with adenovirus. Targeted gene delivery has also been achieved using glycosylated polylysine both *in vitro* (Midoux *et al.*, 1993) and *in vivo* (Perales *et al.*, 1994). With respect to glycosylated polylysine, the degree of glycosylation strongly influences the transfection efficiency in HepG2. Greater substitution of lysine side chains with lactose creates more clustered sugar-binding sites, which improves the binding affinity to the ASGP-R, but this is offset by a reduction in the ionic binding of the glycosylated peptide for the anionic DNA (Erbacher *et al.*, 1995).

Neoglycopeptides, such as GalNAc terminated YEE, have also been used for targeted gene delivery (Merwin *et al.*, 1994). Despite the small size and homogeneity of these ligands, their linkage to HSA prior to coupling to polylysine, diminishes the advantage in using these as ASGP-R ligand mimetics.

4.5 Glycoconjugate Targeting of Liposomes and Lipoproteins

Liposomes are bilayered phospholipid vesicles that have been used as carriers for many types of drug molecules (Ryman and Tyrrell, 1980; Gregoriadis and Florence, 1993). Liposomes offer several advantages as drug carriers including high loading capacity, biodegradability, low toxicity, good solubilization, and stability of incorporated drugs. However, liposomes do not mediate active targeting, but instead are passively phagocytosed by reticulo-endothelial cells (macrophages) of the liver and spleen. Accordingly, surface bound asialo-fetuin enhanced the targeting of liposomes to hepatocytes in rats while ordinary liposomes were mainly distributed to the non-parenchymal cells (Hara *et al.*, 1987). Similarly, liposomes modified with glycoconjugates were shown to enhance the uptake of human interferon-γ into isolated rat hepatocytes (Ishihara *et al.*, 1990). Liposomes have also been labelled with mannose to augment targeting to macrophages, taking advantage of the mannose receptor on these cells (Ghosh and Bachhawat, 1980; Barratt *et al.*, 1986). Mannosylated liposomes improved the therapeutic benefit of the antifungal drug hamycin used for the treatment of aspergillosis in mice (Moonis *et al.*, 1993). Interestingly, liposomes labelled with mannose were reportedly able to cross the blood brain barrier and accumulated preferentially in glial cells (Umezawa and Eto, 1988). Liposomes containing surface-bound neoglycoproteins have also been described (Yamazaki *et al.*, 1992).

Low density lipoprotein (LDL) particles have been used as drug carriers (De Smidt and Van Berkel, 1990) by covalently attaching lactose in order to target Kupffer cells (Bijsterbosch *et al.*, 1989). Later it was determined that the degree of lactosylation influenced targeting specificity. Low level lactosylation resulted in targeting to hepatocytes whereas high levels of lactosylation afforded preferential accumulation of LDL in Kupffer cells despite their much fewer numbers compared to hepatocytes (Bijsterbosch and Van Berkel, 1990).

Both LDL and HDL (high density lipoprotein) particles have been targeted by incorporating a Tris-galactoside-terminated cholesterol (TGC) derivative (Van Berkel *et al.*, 1985a). TGC-lipoprotein targeted galactose receptors in liver. In the case of HDL, targeting was predominantly to the ASGP-R on hepatocytes while LDL targeted mainly Kupffer cells (Van Berkel *et al.*, 1985b). This difference in targeting was ascribed to the size difference between LDL and HDL: the larger LDL particles were recognized by Kupffer cells but failed to be internalized by the ASGP-R probably due to size restrictions imposed for endocytosis through coated pits.

5 Conclusions

The feasibility of glycoconjugate-mediated targeting has been firmly established by demonstrated applications in targeting small drug molecules such as nucleosides, intermediate sized antisence oligonucleotides and enzymes, and very large plasmid DNA molecules. The adaptability of this approach to target drug molecules of such diverse size illustrates the advantage of targeting receptors that internalize through coated pits. In addition, the ability of mammalian lectins to selectively recognize their saccharide ligands when presented in multiple geometries, and when bound to numerous aglycones, provides the opportunity to use glycoconjugates to target virtually any type of drug molecule. The ubiquitous nature of carbohydrates in biological organisms and their established function in mediating binding and uptake of proteins makes glycoconjugate-mediated drug targeting an attractive approach for future exploration.

Abbreviations

Ara-A : adenine-9-β-D-arabinofuranoside; Ara-AMP : adenine-9-β-D-arabinofuranoside 5'-monophosphate; ASGP-R : asialo-glycoprotein receptor; AZT : 3'-azido-3'-dexoythymidine; BSA : bovine serum albumin; CRD : carbohydrate recognition domain; Fuc : fucose; Gal : galactose; Gal NAc : *N*-acetylgalactosamine; GlcNAc : *N*-acetyl glucosamine; HDL : high density lipoprotein; HIV : human immunodeficiency virus; HSA : human serum albumin; LDL : low density lipoprotein; Man : mannose; MDP : muramyl dipeptide; MoAb : monoclonal antibody; PAH : phenylalanine hydroxylase; SOD: superoxide dismutase; TGC : Tris-galactoside cholesterol derivative.

References

Ashwell G, Harford J (1982): Carbohydrate specific receptors of the liver. In *Annu. Rev. Biochem.*, **51**:531–54.

Balboni PG, Minia A, Grossi MP *et al.* (1976): Activity of albumin conjugates of 5-fluorodeoxyuridine and cytosine arabinoside on poxviruses as a lysosomotropic approach to antiviral chemotherapy. In *Nature* **264**:181–3.

Barondes SH, Cooper DNW, Gitt MA *et al.* (1994): Galectins. In *J. Biol. Chem.* **269**:20807–10.

Barratt G, Tenu J, Yapo A *et al.* (1986): Preparation and characterization of liposomes containing mannosylated phospholipids capable of targeting drugs to macrophages. In *Biochim. Biophys. Acta* **862**: 153–64.

Barton NW, Brady RO, Dambrosia JM *et al.* (1991): Replacement therapy for inherited enzyme deficiency – macrophage targeted glucocerebrosidase for Gaucher's disease. In *New Engl. J. Med.* **324**: 1464–70.

Bevilacqua MP, Stengelin S, Gimbrowne MA Jr *et al.* (1989): Endothelial leukocyte adhesion molecule 1: An inducible receptor for neutrophils related to complement regulatory proteins and lectins. In *Science* **243**:1160–4.

Bevilacqua M, Butcher E, Furie B *et al.* (1991): Selectins: a family of adhesion receptors. In Cell **67**:233.

Bijsterbosch MK, Ziere GJ, Van Berkel TJC (1989): Lactosylated low density lipoprotein: a potential site-specific delivery of drugs to Kupffer cells. In *Mol. Pharmacol.* **36**:484–6.

Bijsterbosch MK, Van Berkel TJC (1990): Uptake of lactosylated low-density lipoprotein by galactose-specific receptors in rat liver. In *Biochem. J.* **270**: 233–9.

Bonfils E, Depierreux C, Midoux P *et al.* (1992a): Drug targeting: synthesis and endocytosis of oligonucleotide-neoglycoprotein conjugates. In *Nucl. Acids Res.* **20**:4621–9.

Bonfils E, Mendes C, Roche AC *et al.* (1992b): Uptake by macrophages of a biotinylated oligo-α-deoxythymidylate by using mannosylated streptavidin. In *Bioconjugate Chem.* **3**:277–84.

Brady RO (1984): In *The Molecular Basis of Lysosomal Storage Disorders* (Barranger JA, Brady RO, eds) pp 461–78, Academic Press, Orlando, FL.

Brinkley M (1992): A brief survey of methods for preparing protein conjugates with dyes, haptens, and cross-linking reagents. In *Bioconjugate Chem.* **3**:2–13.

Bunnell BA, Askari FA, Wilson JM (1992): Targeted delivery of antisense oligonucleotides by molecular conjugates. In *Somatic Cell Mol. Genet.* **18**:559–69.

Chiu M, Tamura T, Wadhwa M *et al.* (1994): *In vivo* targeting function of *N*-linked oligosaccharides with terminating galactose and *N*-acetylgalactosamine residues. In *J. Biol. Chem.* **269**:16195–202.

Chiu M, Thomas HV, Stubbs HJ *et al.* (1995): Tissue targeting of multi-valent Le^X terminated *N*-linked oligosaccharides. In *J. Biol. Chem.* **270**:24024–31.

Cristiano RJ, Smith LC, Woo SLC (1993): Hepatic gene therapy: adenovirus enhancement of receptor-mediated gene delivery and expression in primary hepatocytes. In *Proc. Natl. Acad. Sci. USA* **90**:2122–6.

Curtis BM, Scharnowske S, Watson AJ (1992): Sequence and expression of a membrane-associated C-type lectin that exhibits CD4-independent binding of human immunodeficiency virus envelop glycoprotein gp120. In *Proc. Natl. Acad. Sci. USA* **89**: 8356–60.

Das PK, Murray GJ, Zirzow GC *et al.* (1985): Lectin-specific targeting of beta-glucocerebrosidase to different liver cells via glycosylated liposomes. In *Biochem. Med.* **33**:124–31.

De Smidt PC, Van Berkel TJ (1990): LDL-mediated drug targeting. In *Crit. Rev. Ther. Drug Carrier Sys.* **7**:99–120.

Doebber TW, Wu MS, Bugianesi RL *et al.* (1982): Enhanced macrophage uptake of synthetically glycosylated human placental beta-glucocerebrosidase. In *J. Biol. Chem.* **257**:2193–9.

Drickamer K (1988): Two distinct classes of carbohydrate-recognition domains in animal lectins. In *J. Biol. Chem.* **263**:9557–60.

Drickamer K (1993): Ca^{2+}-dependent carbohydrate-recognition domains in animal proteins. In *Curr. Opinion Struct. Biol.* **3**:393–400.

Duncan R, Cable HC, Lloyd JB *et al.* (1983): Polymers containing enzymatically degradable bonds: 7. Design of oligopeptide side chains in poly *N*-(2-hydroxypropyl)methacrylamide copolymers to promote efficient degradation by lysosomal enzymes. In *Macromol. Chem.* **184**:1997- 2008.

Ehrlich P (1906): In *Collected Studies on Immunity*, pp 442–7, Wiley, New York.

Erbacher P, Roche AC, Monsigny M *et al.* (1995): Glycosylated polyllysine/DNA complexes: gene transfer efficiency in relation with the size and the sugar substitution level of glycosylated polylysines and with the plasmid size. In *Bioconjugate Chem.* **6**:401–10.

Fiume L, Mattioli A, Busi C *et al.* (1980): Selective inhibition of ectromelia virus DNA synthesis in hepatocytes by adenine-9-β-D-arabinofuranoside (ARA-A) and adenine-9-β-D-arabinofuranoside 5'-monophosphate (ARA-AMP) conjugated to asialofetuin. In *FEBS Lett.* **116**:185–8.

Fiume L, Bassi B, Busi C *et al.* (1986a): Galactosylated poly(L-lysine) as a hepatotropic carrier of 9-β-D-arabinofuranosyladenine 5'-monophosphate. In *FEBS Lett.* **203**:203–6.

Fiume L, Bassi B, Busi C *et al.* (1986b): Drug targeting in antiviral chemotherapy: a chemically stable conjugate of 9-β-D-arabinofuranosyl-adenine 5'-monophosphate with lactosaminated albumin accomplishes a selective delivery of the drug to liver cells. In *Biochem. Pharmacol.* **35**:967–72.

Fiume L, Busi C, Preti P *et al.* (1987): Conjugates of ara-AMP with lactosaminated albumin: a study on their immunogenicity in mouse and rat. In *Cancer Drug Deliv.* **4**:145–50.

Fiume L, Cerenzia MRT, Bonino F *et al.* (1988): Inhibition of hepatitis B virus replication by vidarabine monophosphate conjugated with lactosaminated serum albumin. In *Lancet* **2**:13–5.

Fiume L, Stefano GD, Busi C *et al.* (1994): A conjugate of lactosaminated poly-L-lysine with adenine arabinoside monophosphate, administered to mice by intramuscular route, accomplishes a selective delivery of the drug to the liver. In *Biochem. Pharmacol.* **47**:643–50.

Franssen EJF, Van Amsterdam RGM, Visser J *et al.* (1991): Low molecular weight proteins as carriers for renal drug targeting: naproxen-lysozyme. In *Pharm. Res.* **8**:1223–30.

Franssen EJF, Jansen, RW, Vaalburg M *et al.* (1993): Hepatic and intrahepatic targeting of an anti-inflammatory agent with human serum albumin and neoglycoproteins as carrier molecules. In *Biochem. Pharmacol.* **45**:1215–26.

Fridovich I (1983) Superoxide radical: an endogenous toxicant. In *Annu. Rev. Pharmacol.* **23**: 239–57.

Fujita T, Nishikawa M, Tamaki C *et al.* (1992): Targeted delivery of human recombinant superoxide dismutase by chemical modification with mono- and polysaccharide derivatives. In *J. Pharmacol. Exp. Ther.* **263**:971–8.

Furbish FS, Steer CJ, Barranger JA *et al.* (1978): The uptake of native and desialylated glucocerebrosidase by rat hepatocytes and Kupffer cells. In *Biochem. Biophys. Res. Commun.* **81**:1047–53.

Furbish FS, Steer CJ, Krett NL *et al.* (1981): Uptake and distribution of placental glucocerebrosidase in rat hepatic cells and effects of sequential deglycosylation. In *Biochim. Biophys. Acta* **673**:425–34.

Gabius HJ (1988): Tumor lectinology: at the intersection of carbohydrate chemistry, biochemistry, cell biology and oncology. In *Angew. Chem. Int. Ed. Eng.* **27**:1267–76.

Gabius HJ, Bardosi A (1991): Neoglycoproteins as tools in glycohistochemistry. In *Prog. Histochem. Cytochem.* **22**:1–66.

Ghose TI, Blair AH (1987) The design of cytotoxic-agent-antibody conjugates. In *CRC Crit. Rev. Therap. Drug Carrier Sys.* **3**:263–359.

Ghose TI, Blair AH, Kulkarni PN (1983): Preparation of antibody-linked cytotoxic agents. *Meth. Enzymol.* **93**:280–333.

Ghosh P, Bachhawat BK (1980): Grafting of different glycosides on the surface of liposomes and its effect on the tissue distribution of ^{125}I-labelled globulin encapsulated in liposomes. In *Biochim. Biophys. Acta* **632**:562–72.

Greenfield RS, Kaneko T, Daues A *et al.* (1990): Evaluation *in vitro* of adriamycin immunoconjugates synthesized using an acid-sensitive hydrazone linker. In *Cancer Res.* **50**:6600–7.

Gregoriadis G (1977): Targeting of drugs. In *Nature* **265**:407–11.

Gregoriadis G (1983): Targeting of drugs with molecules, cells and liposomes. In *Trends Pharmacol. Sci.* **4**:304–7.

Gregoriadis G, Florence AT (1993): Liposomes in drug delivery. Clinical, diagnostic and ophthalmic potential. In *Drugs* **45**:15–28.

Gupta D, Surolia A (1994): Synthesis of neoglycopeptides and analysis of their biodistribution in vivo to identify tissue specific uptake of novel putative membrane lectins. In *Glycoconjugate J.* **11**:43–55.

Haensler J, Szoka FC Jr. (1993): Synthesis and characterization of a trigalactosylated bisacridine compound to target DNA to hepatocytes. In *Bioconjugate Chem.* **4**:85–93.

Hangeland JJ, Levis JT, Lee YC *et al.* (1995): Cell-type specific and ligand-specific enhancement of cellular association of oligodeoxynucleoside methylphosphonates covalently linked with a neoglycopeptide, YEE $(ah\text{-}GalNAc)_3$. In *Bioconjugate Chem.* **6**:695–701.

Hara T, Aramaki Y, Tsuchiya S *et al.* (1987): Specific incorporation of asialofetuin-labeled liposomes into hepatocytes through the action of galactose-binding protein. In *Biopharmaceut. Drug Disposition* **8**: 327–39.

Hirabayashi J (1993): A general comparison of two major families of animal lectins. *Trends Glycosci. Glycotechnol.* **5**:251–70.

Hofsteenge J, Capuano A, Altszuler R *et al.* (1986): Carrier-linked primaquine in the chemotherapy of malaria. In *J. Med. Chem.* **29**:1765–9.

Hoyle GW, Hill RL (1988): Molecular cloning and sequencing of a cDNA for a carbohydrate binding receptor unique to rat Kupffer cells. In *J. Biol.-Chem.* **263**:7487–92.

Ii M, Kurata H, Itoh N *et al.* (1990): Molecular cloning and sequence analysis of cDNA encoding the macrophage lectin specific for galactose and *N*-acetylgalactosamine. In *J. Biol. Chem.* **265**:11295–8.

Ishihara H, Hara T, Aramaki Y *et al.* (1990): Preparation of asialofetuin-labeled liposomes with encapsulated human interferon-γ and their uptake by isolated rat hepatocytes. In *Pharm. Res.* **7**:542–6.

Johnston GI, Cook RG, McEver RP (1989): Cloning of GMP-140, a granule membrane protein of platelets and endothelium: sequence similarity to proteins involved in cell adhesion and inflammation. In *Cell* **56**:1033–44.

Kaneda Y, Iwai K, Uchida T (1989): Increased expression of DNA cointroduced with nuclear protein in adult rat liver. In *Science* **243**:375–8.

Karlsson KA (1991): Glycobiology: a growing field for drug design. In *Trends Pharmacol. Sci.* **12**:265–72.

Kempen HJM, Hoes C, Van Boom JH *et al.* (1984): A water-soluble cholesteryl-containing trisgalactoside: synthesis, properties, and use in directing lipid-containing particles to the liver. In *J. Med. Chem.* **27**:1306–12.

Kojima S, Ishido M, Kubota K *et al.* (1990): Tissue distribution of radioiodinated neoglycoproteins and mammalian lectins. In *Biol. Chem. Hoppe Seyler* **371**:331–8.

Kikutani H, Inui S, Sata R *et al.* (1986): Molecular structure of human lymphocyte receptor for immunoglobulin E. In *Cell* **47**:657–65.

Kopecek J (1991): Targetable polymeric anticancer drugs. Temporal control of drug activity. In *Ann. NY Acad. Sci.* **618**:335–44.

Koppel GA (1990): Recent advances with monoclonal antibody drug targeting for the treatment of human cancer. In *Bioconjugate Chem.* **1**:13–23.

Kosmas C, Linardou H, Epenetos A (1993): Review: advances in monoclonal antibody tumour targeting. In *J. Drug Targeting* **1**:81–91.

Lasky LA, Singer MS, Yednock TA *et al.* (1989): Cloning of a lymphocyte homing receptor reveals a lectin. In *Cell* **56**:1045–55.

Lee RT, Lee YC (1987): Preparation of cluster glycosides of *N*-acetylgalactosamine that have subnanomolar binding constants towards the mammalian hepatic Gal/GalNAc-specific receptor. In *Glycoconjugate J.* **4**:317–28.

Lee RT, Lee YC (1994): Enhanced biochemical affinities of multivalent neoglycoconjugates. In *Neoglycoconjugates: Preparation and Applications* (Lee YC, Lee RT, eds) pp 23–50, San Diego: Academic Press.

Lee YC, Townsend RR, Hardy MR *et al.* (1983): Binding of synthetic oligosaccharides to the hepatic gal/galNac lectin. Dependence on fine structural features. In *J. Biol. Chem.* **258**:199–202.

Lowe JB, Stoolman LM, Nair RP *et al.* (1990): ELAM-1-dependent cell adhesion to vascular endothelial determined by a transfected human fucosyltransferase cDNA. In *Cell* **63**:475–84.

Magnusson G, Chernyak AY, Kihlberg J *et al.* (1994): Synthesis of neoglycoconjugates. In *Neoglycoconjugates: Preparation and Applications* (Lee YC, Lee RT, eds) pp 53–143, San Diego: Academic Press.

Meijer DKF, Jansen RW, Molema G (1992): Drug targeting for antiviral agents: options and limitations. In *Antiviral Res.* **18**:215–58.

Menander-Huber KB, Huber W (1977): Orgotein, the drug version of bovine Cu-Zn superoxide dismutase. II. A summary of clinical trials in man and animals. In *Superoxide and Superoxide Dismutases* (Michelson AM, McCord JM, Fridovich I, eds) pp 537–56, London: Academic Press.

Mervin JR, Noell GS, Thomas WL *et al.* (1994): Targeted delivery of DNA using YEE$(GalNAcAH)_3$, a synthetic glycopeptide ligand for the asialoglycoprotein receptor. In *Bioconjugate Chem.* **5**:612–20.

Midoux P, Negre E, Roche AC *et al.* (1990): Drug targeting: anti-HSV-1 activity of mannosylated polymer-bound 9-(2-phosphonylmethoxyethyl)-adenine. In *Biochem. Biophys. Res. Comm.* **167**: 1044–9.

Midoux P, Mendes C, Legrand A *et al.* (1993): Specific gene transfer mediated by lactosylated poly-L-lysine into hepatoma cells. In *Nucleic Acids Res.* **21**:871–8.

Molema G, Jansen RW, Pauwels R *et al.* (1990): Targeting of antiviral drugs to T_4-lymphocytes: anti-HIV activity of neoglycoprotein-AZTMP conjugates *in vitro*. In *Biochem. Pharmacol.* **40**:2603–10.

Monsigny M, Roche AC, Bailly P (1984): Tumoricidal activation of murine alveolar macrophages by muramyldipeptide substituted mannosylated serum albumin. In *Biochem. Biophys. Res. Comm.* **121**:579–84.

Monsigny M, Roche AC, Midoux P (1988): Endogenous lectins and drug targeting. In *Ann. NY Acad. Sci.* **551**:399–414.

Moonis M, Ahmad I, Bachhawat BK (1993): Mannosylated liposomes as carriers for hamycin in the treatment of experimental aspergillosis in Balb/C mice. In *J. Drug Targeting* **1**:147–55.

Mulligan RC (1993): The basic science of gene therapy. *Science* **260**:926–32.

Murray GJ (1987): Lectin-specific targeting of lysosomal enzymes to reticuloendothelial cells. *Meth. Enzymol.* **149**:25–42.

Negre E, Chance ML, Hanboula SY *et al.* (1992): Antileishmanial drug targeting through glycosylated polymers specifically internalized by macrophage membrane lectins. In *Antimicrobial Agents Chemotherapy* **36**:2228–32.

Ohsumi Y, Chen VJ, Wold F *et al.* (1988): Interaction between new neoglycoproteins and the D-Man/L-Fuc receptor of rabbit alveolar macrophages. In *Glycoconjugate J.* **5**:99–106.

Perales JC, Ferkol T, Beegen H *et al.* (1994): Gene transfer *in vivo*: sustained expression and regulation of genes introduced into the liver by receptor-targeted uptake. In *Proc. Natl. Acad. Sci.* USA **91**:4086–90.

Petit C, Monsigny M, Roche AC (1990): Macrophage activation by muramyl dipeptide bound to neoglycoproteins and glycosylated polymers: cytotoxic factor production. In *J. Biol. Response Modif.* **9**:33–43.

Pietersz GA (1990):The linkage of cytotoxic drugs to monoclonal antibodies for the treatment of cancer. In *Bioconjugate Chem.* **1**:89–95.

Pimm MV (1988): Drug-monoclonal antibody conjugates for cancer therapy: potentials and limitations. In *CRC Crit. Rev. Therap. Drug Carrier Sys.* **5**:189–227.

Plank C, Zatloukal K, Cotten M *et al.* (1992): Gene transfer into hepatocytes using asialoglycoprotein receptor mediated endocytosis of DNA complexed with an artificial tetra-antennary galactose ligand. In *Bioconjugate Chem.* **3**:533–9.

Plank C, Oberhauser B, Mechtler K *et al.* (1994): The influence of endosome-disruptive peptides on gene transfer using synthetic virus-like gene transfer systems. In *J. Biol. Chem.* **269**:12918–24.

Powell L, Varki A (1995): I-type lectins. In *J. Biol. Chem.* **270**:14243–6.

Reinis M, Damkova M, Korec E (1993): Receptor-mediated transport of oligodeoxynucleotides into hepatic cells. *J. Virol. Meth.* **42**:99–106.

Rejmanova P, Kopecek J, Duncan R *et al.* (1985): Stability in rat plasma and serum of lysosomally degradable oligopeptide sequences in *N*-(2-hydroxypropyl)methacrylamide copolymers. In *Biomaterials* **6**:45–8.

Rice KG, Silva M (1996): Preparative purification of Tyrosinamide *N*-linked oligosaccharides. In *J. Chromat.***720**: 235–49.

Rice KG, Weisz OA, Barthel T *et al.* (1990): Defined geometry of binding between triantennary glycopeptide and the asialoglycoprotein receptor of rat hepatocytes. In *J. Biol. Chem.* **265**:18429–34.

Roche AC, Bailly P, Monsigny M (1985): Macrophage activation by MDP bound to neoglycoproteins: metastasis eradication in mice. In *Invasion Metastasis* **5**:218–32.

Roche AC, Midoux P, Pimpaneau V *et al.* (1990): Endocytosis mediated by monocyte and macrophage membrane lectins – application to antiviral drug targeting. In *Res. Virol.* **141**:243–9.

Ryman BE, Tyrrell DA (1980): Liposomes – bags of potential. In *Essays Biochem.* **16**:49–98.

Schwartz AL (1984): The hepatic asialoglycoprotein receptor. In *CRC Crit. Rev. Biochem.* **16**: 207–33.

Sett R, Sarkar K, Das PK (1993a): Pharmacokinetics and biodistribution of methotrexate conjugated to mannosyl human serum albumin. In *J. Antimicr. Chemother.* **31**:151–9.

Sett R, Sarkar K, Das PK (1993b): Macrophage-directed delivery of doxorubicin conjugated to neoglycoprotein using leishmaniasis as the model disease. In *J. Infect. Dis.* **168**:994–9.

Shen WC, Ryser JP (1981): *Cis*-aconityl spacer between daunomycin and macromolecular carriers: a model of pH-sensitive linkage releasing drug from a lysosomotropic conjugate. In *Biochem. Biophys. Res. Commun.* **102**:1048–54.

Shen TY (1987): Saccharide determinants in selective drug delivery. In *Ann. NY Acad. Sci.* **507**: 272–80.

Spellman MW (1990): Carbohydrate characterization of recombinant glycoproteins of pharmaceutical interest. In *Anal.Chem.* **62**:1714–22.

Spiess M, Schwartz AL, Lodish HF (1985): Sequence of human asialoglycoprotein receptor cDNA. In *J. Biol. Chem.* **260**:1979–82.

Stahl PD, Rodman JS, Miller MJ *et al.* (1978): Evidence for receptor-mediated binding of glycoproteins, glycoconjugates, and lysosomal glycosidases by alveolar macrophages. In *Proc. Natl. Acad. Sci. USA* **75**:1399–403.

Stowell CP, Lee YC (1980): Neoglycoproteins. The preparation and application of synthetic glycoproteins. In *Adv. Carbohydr. Chem. Biochem.* **37**: 225–81.

Takakura Y, Masuda S, Tokuda H *et al.* (1994): Targeted delivery of superoxide dismutase to macrophages via mannose receptor-mediated mechanism. In *Biochem. Pharmacol.* **47**:853–8.

Taylor ME, Bezouska K, Drickamer K (1992): Contribution to ligand binding by multiple carbohydrate-recognition domains in the macrophage mannose receptor. In *J. Biol. Chem.* **267**: 1719–26.

Trouet A, Masqueller M, Baurain R *et al.* (1982): A covalent linkage between daunorubicin and proteins that is stable in serum and reversible by lysosomal hydrolases, as required for a lysosomotropic drug-carrier conjugate: *in vitro* and *in vivo* studies. In *Proc. Natl. Acad. Sci. USA* **79**:626–9.

Umezawa F, Eto Y (1988): Liposome targeting to mouse brain: mannose as a recognition marker. In *Biochem. Biophys. Res. Comm.* **153**:1038–44.

Varki A (1992): Selectins and other mammalian sialic acid-binding lectins. In *Curr. Opinion Cell Biol.* **4**:257–66.

Van Den Hamer CJ, Morell AG, Scheinberg IH *et al.* (1970): Physical and chemical studies on ceruloplasmin. IX. The role of galactosyl residues in the clearance of ceruloplasmin from the circulation. In *J. Biol. Chem.* **245**:4397–402.

Van Berkel TJC, Kar Kruijt J, Spanjer HH *et al.* (1985a): The effect of a water-soluble tris-galactoside-terminated cholesterol derivative on the fate of low density lipoproteins and liposomes. In *J. Biol. Chem.* **260**:2694–9.

Van Berkel TJC, Kar Kruijt J, Kempen HJ (1985b): Specific targeting of high density lipoproteins to liver hepatocytes by incorporation of a tris-galactoside-terminated cholesterol derivative. In *J. Biol. Chem.* **260**:12203–7.

Vaickus L, Foon KA (1991): Overview of monoclonal antibodies in the diagnosis and therapy of cancer. In *Cancer Invest.* **9**:195–209.

Venter JC (1982): Immobilized and insolubilized drugs, hormones, and neurotransmitters: properties, mechanisms of action and applications. In *Pharmacol. Rev.* **34**:153–87.

Wadhwa MS, Knoell D, Young T *et al.* (1995): Targeted gene delivery with a low molecular weight glycopeptide. In *Bioconjugate Chem.* **6**:283–91.

Wagner E, Zatloukal K, Cotten M *et al.* (1992): Coupling of adenovirus to transferrin-polylysine/DNA complexes greatly enhances receptor-mediated gene delivery and expression of transfected genes. In *Proc. Natl. Acad. Sci. USA* **89**:6099–103.

Wileman T, Harding C, Stahl P (1985): Receptor mediated endocytosis. In *Biochem. J.* **232**:1–14.

Wilson JM, Grossman M, Wu CH *et al.* (1992a): Hepatocyte-directed gene transfer *in vivo* leads to transient improvement of hypercholesterolemia in low density lipoprotein receptor-deficient rabbits. In *J. Biol. Chem.* **267**:963–7.

Wilson JM, Grossman M, Cabrera JA *et al.* (1992b): A novel mechanism for achieving transgene persistence *in vivo* after somatic gene transfer into hepatocytes. In *J. Biol. Chem.* **267**:11483–9.

Wu GY, Wu CH (1988a): Evidence for targeted gene delivery to HepG2 hepatoma cells *in vitro*. In *Biochemistry* **27**:887–92.

Wu GY, Wu CH (1988b): Receptor-mediated gene delivery and expression *in vivo*. *J. Biol. Chem.* **263**: 14621–4.

Wu GY, Wu CH (1992): Specific inhibition of hepatitis B viral gene expression *in vitro* by oligonucleotides. In *J. Biol. Chem.* **267**:12436–9.

Wu GY, Keegan-Rogers V, Franklin S *et al.* (1988): Targeted antagonism of galactosamine toxicity in normal rat hepatocytes *in vitro*. In *J. Biol. Chem.* **263**:4719–23.

Wu GY, Wilson JM, Shalaby F *et al.* (1991): Receptor-mediated gene delivery *in vivo*. Partial correction of genetic analbuminemia in nagase rats. In *J. Biol. Chem.* **266**:14338–42.

Yamazaki N, Kojima S, Gabius S *et al.* (1992): Studies on carbohydrate-binding proteins using liposome-based systems – I. Preparation of neoglycoprotein-conjugated liposomes and the feasibility of their use as drug-targeting devices. In *Int. J. Biochem.* **24**: 99–104.

27 Glycobiology of Signal Transduction

Antonio Villalobo, José Antonio Horcajadas, Sabine André, and Hans-J. Gabius

1 Introduction

The homeostatic control of cellular metabolism, cell growth and differentiation, and the completion of specialized cellular functions encompass a series of processes by which the information arriving at a cell from the extracellular environment and/or neighboring cells is transduced into appropriate cellular responses elicited by well defined signal pathways. The arriving chemical or electrophysiological signals are amplified by the cell using the transient increase in the concentration of some ions and molecules called second messengers such as calcium ions, cyclic nucleotides or polyphosphoinositides. The different components of a signal transduction system will be referred to as effectors, receptors, transducers, operators and terminators, and are depicted in Fig. 1 within a simplified view of the vectorial informational flow occurring in a classical signal transduction pathway. The effectors could be freely diffusible hormones, growth factors, differentiation factors, cytokines, antigens, mating factors, and neurotransmitters, or both plasma membrane-bound and extracellular matrix-associated molecules. They bind to specific receptors that are typically located in the plasma membrane or in the cell nucleus which initiate the transduction of the message to attain a given cellular response. These receptors could have different activities or functions which are used within the signal transduction events. Among these are protein kinase and protein phosphatase activities, channel activity for specific ions, or the capacity to bind to response elements in the DNA, promoting the transcription of specific genes.

Participants in the signal transduction systems include accessory molecules denoted transducers such as G proteins, adaptor proteins and nucleotide exchanger proteins. They interconnect the activated receptors with other elements of the system named operators, which are capable of performing distinct biochemical tasks. Among them are enzymes some of which synthesize second messenger molecules such as nucleotide cyclases and phospholipases that supply cyclic nucleotides, polyphosphoinositides, diacylglycerol or arachidonic acid, and others such as protein kina-

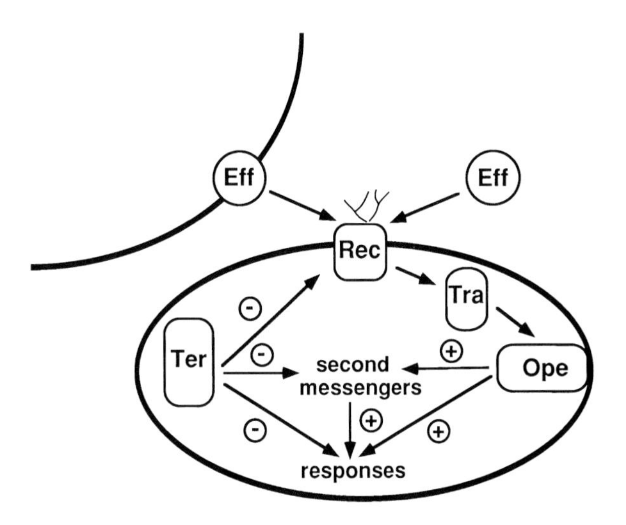

Figure 1. A typical signal transduction pathway. Binding of a soluble, membrane-bound, or intercellular matrix-associated effector (Eff) to a receptor (Rec) located in the plasma membrane initiates a process which involves generation of activated transducers (Tra) and finally operators (Ope) relaying positive signals (+). The signaling processes are abolished (-) by terminator (Ter) systems acting at different levels.

H.-J. and S. Gabius (Eds.), Glycosciences

ISBN 3-8261-0073-5

ses and protein phosphatases that phosphorylate and dephosphorylate, respectively, target molecules; channels that mobilize second messenger ions such as Ca^{2+}; and ion exchangers such as the Na^+/H^+-exchanger that modify the ionic composition and pH of the cytosol.

Finally, to terminate the signal transduction events certain systems sense changes in second messenger levels and/or other cellular parameters relaying a feed-back signal to shut off effector-mediated signals by acting either on the receptors, on the systems controlling the concentration of second messengers, and/or the operator systems. Among these terminators are phosphodiesterases that degrade cAMP and cGMP, protein phosphatases that dephosphorylate target proteins, and Ca^{2+}-transport systems such as Ca^{2+}-ATPases and the Na^+/Ca^{2+} exchanger that decrease the concentration of free Ca^{2+} in the cytosol to resting levels.

However, some complexity is introduced in this simplified classification of the components involved in signal transduction pathways when the function of any of these elements are shared by other elements upstream or downstream of the signal transduction cascade. As an example, Ca^{2+}-mobilization could be attained by both the activation of receptors having itself channel properties, such as the glutamate receptors, or by distinct Ca^{2+}-channels that can be activated by electrophysiological or chemical signals. We have referred Ca^{2+}-ions earlier as operator molecules. Additionally, a tyrosine kinase receptor could mediate, by a series of sequential reactions, both the synthesis or mobilization of second messengers and the direct phosphorylation of target proteins required to trigger cellular responses. Furthermore, the direct communication between cells mediated by gap junction channels could bypass some of the components of a typical signal transduction pathway, since second messengers such as Ca^{2+}, cAMP, cGMP and inositol 1,4,5-triphosphate are able to pass directly from the cytosol of one cell to the cytosol of an adjacent cell via these intercellular channels (Loewenstein, 1987).

It is interesting to notice that the three major second messenger systems, namely Ca^{2+}, polyphosphoinositides and cyclic nucleotides, are interconnected. Polyphosphoinositides play an important role in cellular regulation and its synthesis is mediated by agonist-coupled receptors (Monaco and Gershengorn, 1992; Pike, 1992). The synthesis of phosphatidylinositol-4,5-biphosphate (PtdIns-4,5-P_2) first requires the phosphorylation of phosphatidylinositol (PtdIns) by phosphatidylinositol-4-kinase yielding phosphatidylinositol-4-phosphate (PtdIns-4-P), and secondly the phosphorylation of the later by phosphatidylinositol-4-phosphate 5-kinase.

PtdIns-4,5-P_2 is an important precursor molecule since its hydrolysis by phospholipase $C\gamma$ forms inositol 1,4,5-triphosphate (Ins-1,4,5-P_3) and 1,2-diacylglycerol (DAG), two major intracellular second messengers. The former is an activator of Ca^{2+}-channels from the endoplasmic reticulum, and the later is an activator of certain types of protein kinase C. In addition to Ca^{2+} mobilization mediated by Ca^{2+}-channels located in the endoplasmic reticulum and regulated by Ins-1,4,5-P_3, the activation of Ca^{2+}-channels by leukotriene-C_4 (LT-C_4) and the activation of G proteins-regulated voltage-activated Ca^{2+}-channels located in the plasma membrane contribute to the increase of the cytosolic concentration of Ca^{2+}during cell activation by different agonist receptors. The two major pathways for the mobilization of Ca^{2+} by the epidermal growth factor receptor (EGFR) are depicted in Figs. 2a and 2b.

Synthesis of cAMP by adenylyl cyclases is controlled by stimulatory and inhibitory receptor-regulated trimeric G_s and G_i proteins, respectively (Gilman, 1987) (see Fig. 3). The activated G protein α-subunit containing bound GTP dissociates from its $\beta\gamma$-dimer and plays an essential role in this process. However, the dissociated $\beta\gamma$-dimer also modulates the activity of some types of adenylyl cyclases. Furthermore, the $\beta\gamma$-dimer stimulates phospholipase A_2, and phospholipase $C\beta$ (Clapham and Neer, 1993) which synthesize two additional major second messengers, arachidonic acid (AA) and DAG, respectively.

The other cyclic nucleotide of special relevance in cell physiology is cGMP. Interestingly, one of the most important control mechanisms on the synthesis of cGMP phosphorylation of target proteins by PKC relay a signal (Panel B). Activation

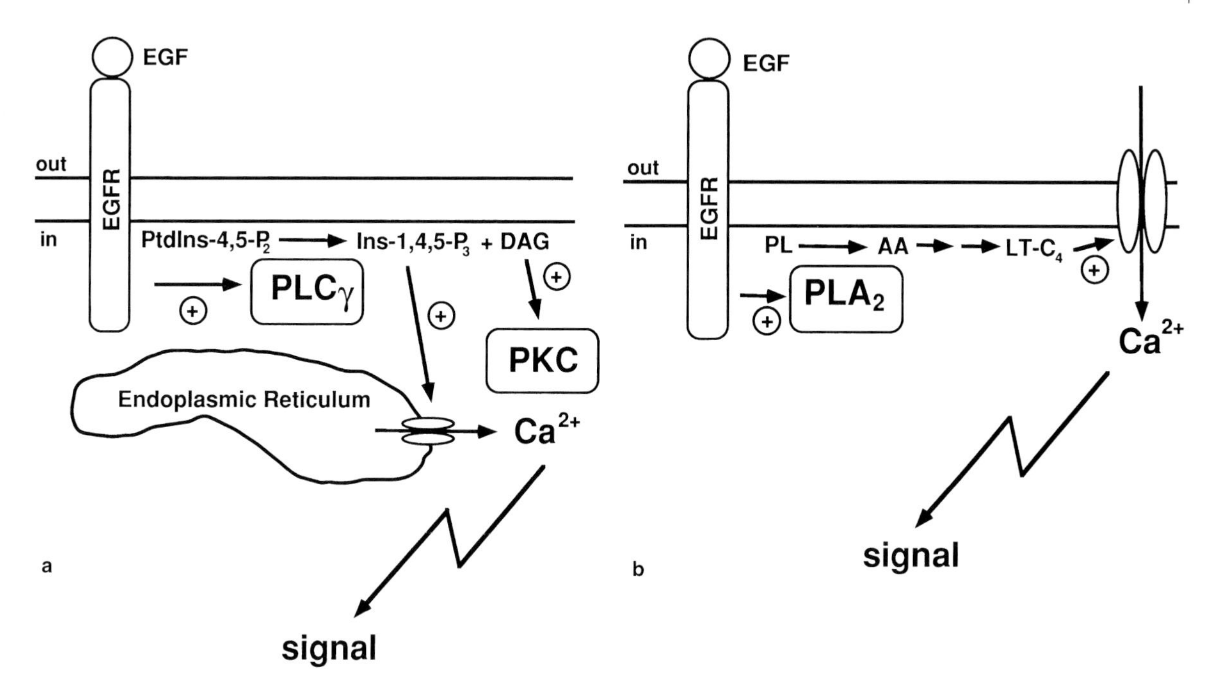

Figure 2. Signaling by the epidermal growth factor receptor mediated by inositol 1,4,5-triphosphate, arachidonic acid and Ca^{2+}(Panel A). Activation of the epidermal growth factor receptor (EGFR) by transphosphorylation upon ligand binding activates phospholipase Cγ (PLCγ) by phosphorylation leading to the production of inositol 1,4,5-triphosphate (Ins-1,4,5-P_3) and diacylglycerol (DAG). The former activates Ca^{2+}-channels in the endoplasmic reticulum and the later activates protein kinase C (PKC). Both the increase in concentration of cytosolic Ca^{2+} and the modification of the epidermal growth factor receptor (EGFR) by transphosphorylation upon ligand binding activate phospholipase A_2 (PLA_2) by phosphorylation which in turn hydrolyzes certain phospholipids yielding arachidonic acid (AA). This compound is the precursor of leukotriene C_4 (LT-C_4) synthesized by the 5-lipoxygenase pathway which activates Ca^{2+}-channels located in the plasma membrane. The increase in concentration of cytosolic Ca^{2+} relays a signal.

by guanylyl cyclase is excerted by nitric oxide. This very reactive free radical gas acts as a second messenger and is synthesized from L-arginine in mammalian cells by different forms of nitric oxide synthases, some of which are regulated in turn by Ca^{2+} and calmodulin (Ignarro, 1990; Marletta, 1994).

To add even more complexity to the signal transduction mechanisms, it is important to realize that two distinct effectors could share common signaling pathways. Conversely, a given effector could operate by conveying a message through more than one transduction pathway, working simultaneously or in an alternative mode. This provides the cell with a network of redundant systems for the efficient simultaneous and/or sequential transduction of a variety of incoming signals. Inherent redundancy ensures the capacity of a cell to express pleiotropicity in response to a given signal, as well as temporal plasticity, what facilitates its adaptive response to complex and rapidly variable conditions.

The study of signal transduction mechanisms in the field of glycobiology is an active area of research that can be examined under a dual point of view. On the one hand, exogenous lectins which are used as tools and do not play a physiological role in a given organism are able to mimic or to prevent the action of physiological effectors by interacting with the glycoside part of receptors or other glycoproteins involved in their signal transduction pathways. Thus, plant or invertebrate agglutinins are widely used as useful tools to dissect intricate signal transduction pathways

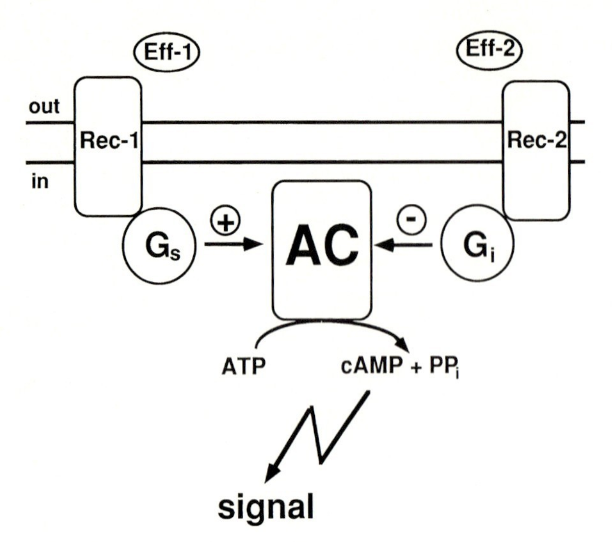

Figure 3. Signaling by cAMP. Two different effectors (Eff-1 and Eff-2) control the activity of adenylyl cyclase by distinct receptors (Rec-1 and Rec-2) associated to either stimulatory (G_s) or inhibitory (G_i) G proteins. cAMP relays signals by activating protein kinase A and by binding to proteins associated to genomic cAMP response elements.

(Lis and Sharon, 1986). As an example, it has been shown that the ability of the 60 kDa glycosylated receptor for the vasoactive intestinal peptide to communicate with the adenylyl cyclase system can be impaired upon its binding of wheat germ lectin (Chochola *et al.*, 1993).

On the other hand, endogenous lectins which display functional significance *in situ* have defined signal transduction pathways by which their interaction with glycoconjugates leads to specific cellular responses. Therefore, the recognitive interplay and the physiological signals induced by cellular lectins via their interaction with specific plasma membrane-bound or soluble receptors are studied to understand the details of their mode of operation (Ashwell and Harford, 1982; Barondes, 1984; Gabius, 1991; Drickamer and Taylor, 1993). Interestingly, some lectins can apparently associate with non-carbohydrate ligands in addition to their capacity to interact with glycoside residues in their target molecules (Barondes, 1988; Gabius, 1994a). The presence of different molecular modules in one type of protein calls for adequate control reactions and cautious interpretations of the results.

Besides tissue lectins, potential glycoligands, too, can be assayed. For example, glycopeptides isolated from chorionic gonadotropin cause inhibition of adenylyl cyclase activity in rat corpora luteal tissue, suggesting the presence of a membrane-bound lectin and its involvement in the regulation of the adenylyl cyclase by the glycopart of the hormone (Calvo and Ryan, 1985; Hartree and Renwick, 1992).

One of the early events required for transmembrane signaling appears to be ligand-induced dimerization of membrane receptors (see for review Metzger, 1992). In principle, this reaction is reminiscent of cell aggregation, when lectins

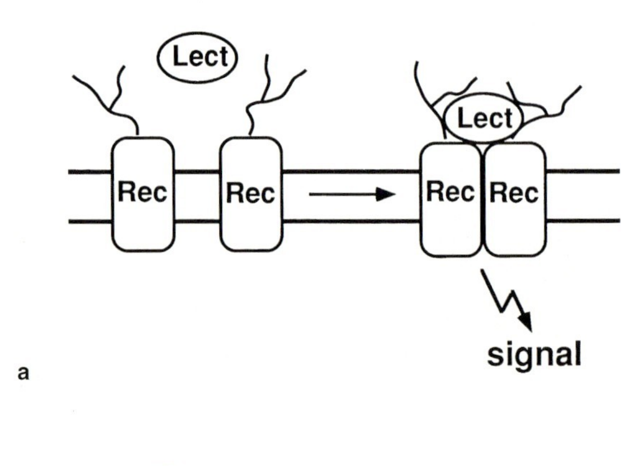

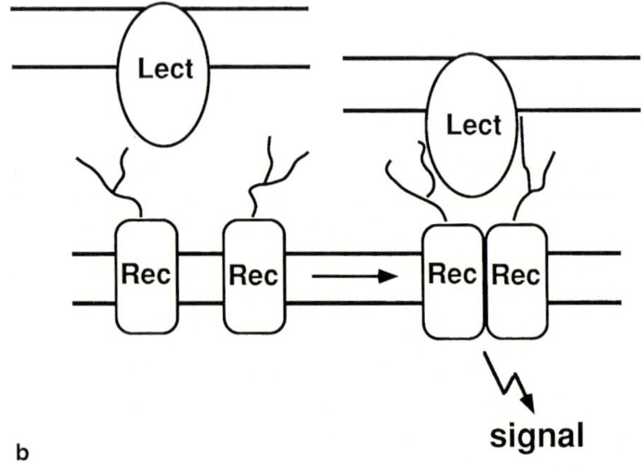

Figure 4. Signaling by lectins. Soluble (panel a) and membrane-bound (panel b) lectins (Lect) interact with the glycopart of membrane-bound receptors (Rec) inducing lateral diffusion within the plane of the membrane and receptor dimerization. This process activates the receptors which relay a signal.

exert their activity to mediate epitope association. Therefore, the underlying molecular mechanism by which exogenous lectins elicit signal transduction events could well lead to homologue dimerization of receptors and/or the heterologue clustering of receptors and other molecules required for signal transduction. Likewise, an endogenous lectin could induce dimerization of an appropriate binding partner like a receptor as prerequisite for ensuing processes like transphosphorylation. These circumstances are depicted in a simplified manner in Fig. 4.

Besides signaling that is dependent on receptor transphosphorylation other main routes of information transfer based on second messenger molecules are well described and investigated. Commercially available fluorescent probes for convenient monitoring some of the events associated to these signaling process, such as changes in the cytosolic concentration of free Ca^{2+} or changes in the intracellular pH, made these pathways accessible to investigation.

In the next paragraph we shall discuss some examples of how exogenous and endogenous lectins appear to employ distinct pathways of signal transduction to elicit measurable cellular responses such as mitogenesis, immunomodulation or apoptosis following their binding to glycoligands.

2 Lectin-Induced Mitogenesis and Immunomodulation

The mitogenic activation of quiescent cells by growth factors is accompanied by a series of early events among which the activation of both Ca^{2+}-channels and the Na^+/H^+ exchanger are prominent features, leading to the transient increase in the concentration of free Ca^{2+} and the alkalization of the cytosol, respectively (Rozengurt, 1986). Lectins are known to mimic the action of growth factors in this regard. Explicitly, Ca^{2+}-mobilization (Wheeler *et al.*, 1987) and the stimulation of the Na^+/H^+ exchanger (Green and Muallem, 1989) have been shown to be induced by a mannose-binding plant lectin, concanavalin A. Moreover, concanavalin A also stimulates the tyrosine kinase activity of the EGF receptor from rat liver by a mechanism independent of EGF binding (Zeng *et al.*, 1995). This transphosphorylation will then stimulate SH2-domain-containing proteins. Phospholipase C is one of the SH2-domain-containing enzymes activated by this mechanism, leading to the formation of DAG, an activator of protein kinase C, and Ins-1,4,5-P_3, an activator of Ca^{2+}-channels in the endoplasmic reticulum (see Fig. 2a). Moreover, the SH2-SH3-domain-containing adaptor protein denoted growth factor receptor bound protein 2 (Grb2) interacts with both the autophosphorylated EGF receptor and the nucleotide-exchanger protein denoted Son of sevenless (Sos). The activation of Sos in turn activates the low molecular mass monomeric G protein Ras by replacing its bound GDP for GTP. Activated Ras interacts

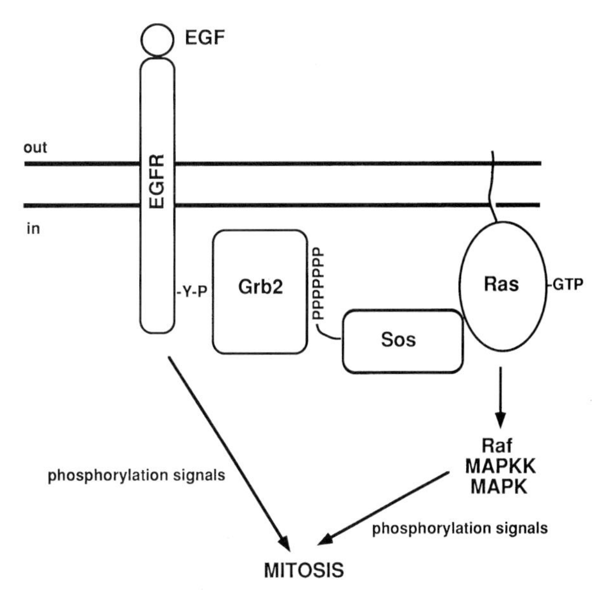

Figure 5. Signaling of the epidermal growth factor receptor mediated by Ras. Activation of the epidermal growth factor receptor (EGFR) by transphosphorylation upon ligand binding induces the association of the SH2-domain of the Grb2 adaptor protein to its phosphorylated carboxyl-terminal tail. The SH3-domain of Grb2 then binds to a proline-rich sequence in the carboxyl-terminal of the nucleotide exchanger protein Sos which in turn exchanges GDP for GTP in Ras. Activated Ras initiates the activation of the mitogen-activated protein kinase cascade formed by Raf, MAPKK and MAPK. Phosphorylation signals from the MAP-kinase cascade and from other EGFR-controlled protein kinases are essential for cell proliferation.

with Raf-1, the first step of the mitogen-activated protein (MAP)-kinase cascade which end up activating transcription factors such as c-Jun and c-Myc leading to the synthesis of DNA (Davis, 1993; Lemmon *et al.*, 1994) (see Fig. 5).

The mitogenic activity of lectins is not at all restricted to concanavalin A and/or to lymphocytes (Borrebaeck and Carlsson, 1989). For example, a mannose receptor of macrophages induces mitogenesis in bovine airway smooth muscle cells (Lew *et al.*, 1994), and several β-galactoside-specific lectins from plant and mammalian origin have mitogenic action on vascular cells including smooth muscle cells (Sanford and Harris-Hooker, 1990).

Growth factors such as EGF, NGF and βFGF induce a decrease in the phosphorylation of a protein first denoted as Nsp100 and later identified as elongation factor-2 which is involved in protein synthesis. Phosphorylation of Nsp100 by Ca^{2+}/calmodulin-dependent protein kinase III inhibits its activity and its dephosphorylation abrogates this inhibition. This growth factor-mediated dephosphorylation appears to require a Ca^{2+}-dependent mechanism since it was prevented by the pretreatment of cells with EGTA (Hashimoto *et al.*, 1990). Lectins appear to be able to interfere with this system, since the dephosphorylation of Nsp100 is prevented by the binding of wheat germ lectin to the receptor for bFGF (Hashimoto *et al.*, 1990).

Mitogenesis and immunomodulation are complex and interdependent processes which are regulated by multiple factors. Therefore, the current state of investigation does not allow for a straightforward and brief description in this short chapter. However, the analysis clearly reveals main lines of influence of lectins on these processes to be pursued.

When antigenic molecules enter the blood stream they are processed by lymphocytes inducing their proliferation and maturation. In these series of events lymphokines play a key role. Activation of T-lymphocytes is initiated by interaction of its antigen receptors with specific antigens, which are processed and presented with a major histocompatibility complex (MHC) molecule by antigen-presenting cells. Cytotoxic T-lymphocytes use class I MHC molecules and inducer T-lymphocytes use class II MHC molecules (Schwartz, 1985). Remarkably, mitogenic lectins induce proliferation of lymphocytes by interacting with the glycoside part of several cell surface glycoproteins like the α/β and perhaps the γ T-lymphocyte receptor molecules (Chilson and Kelly-Chilson, 1989). Besides cell activation, the pattern of cytokines which are secreted can be changed quantitatively, enhancing the impact of lectin binding. This response can thus modulate immune parameters *in vitro* and *in vivo*. Due to current considerations to test a plant lectin with respect to antitumoral activity *in vivo* on the basis of its *in vitro* activity (Gabius, 1994b), it is noteworthy that the galactoside-specific lectin from mistletoe induces the phosphorylation of a 28 kDa protein, the phosphorylation of PtdIns to PtdIns-4,5-P_2, and the mobilization of Ca^{2+} in leukemia cells besides enhanced secretion of interleukin-1 (IL-1), interleukin-6 (IL-6) and tumor necrosis factor-α (Gabius *et al.*, 1992).

Although without remarkable clinical perspective, concanavalin A, and phytohaemagglutinin have long been recognized as mitogenic lectins inducing a variety of cellular signal transduction events in T-lymphocytes from polyphosphoinositides turnover, release of arachidonic acid, activation of protein kinase C by DAG, Ca^{2+}-mobilization, activation of K^+-channels, and increase of the intracellular concentrations of cAMP and cGMP (for reviews see Hadden, 1988; Borrebaeck and Carlsson, 1989).

Recently, the application of such a lectin enabled to add a further member to the family of molecular mediators. Thus, the human tumor lymphocyte Jurkat cell line stimulated by phytohemagglutinin has been used to identify a new small messenger molecule denoted Ca^{2+}-influx factor (CIF). It stimulates Ca^{2+}-influx from the extracellular medium into the cytosol when intracellular Ca^{2+}-stores are emptied (Randriamampita and Tsien, 1993).

The mitogenic activation of T-lymphocytes by antigens or mitogenic lectins in the presence of IL-1 produced by macrophages is mediated by the synthesis of interleukin-2 (IL-2), which potentiates T-lymphocyte proliferation and thereby increases the duration and magnitude of the immune response. During this process activation of phospholipase C takes place with the subsequent production of diacylglycerols, activators of some types of protein kinase C, and Ins-1,4,5-P_3, involved in the activation of Ca^{2+}-channels that transiently increases the cytosolic concentration of this cation. However, the activation of protein kinase C does not appear to be an obligatory step in the production of IL-2 by T-lymphocytes (Mills *et al.*, 1989). The concanavalin A-induced turnover of PtdIns-4,5-P_2, IL-2 production, and DNA synthesis in lymphocytes is potentiated by macrophages. However, the addition of IL-1 or the protein kinase C activator 12-*O*-tetradecanoylphorbol-13-acetate, an analog of DAG, in combination with concanavalin A replaces the need of macrophages for the production of IL-2 and DNA synthesis in lymphocytes, although it does not affect the hydrolysis of polyphosphoinositides (Grier and Mastro, 1988).

In line with deciphering lectin-dependent intracellular signal cascades, the nature of the ligand molecules for mitogenic lectins is being unraveled. An illustrative example for thymocytes is selected to illustrate this aspect. Accordingly, concanavalin A binds to both glycoprotein(s) associated with a 25 kDa G protein that probably is involved in the hydrolysis of polyphosphoinositides by phosphoinositide-specific phospholipase C, and to ligand(s) associated with an inhibitory G protein controlling the adenylyl cyclase (Wang *et al.*, 1989). Protein phosphorylation is another event accompanying lectin-mediated activation of lymphocytes. Thus, binding of concanavalin A to T-lymphocytes activates protein kinase C that in turn phosphorylates target proteins (Hadden, 1988). Among the concanavalin A-induced serine/threonine-phosphorylated substrates is a 50 kDa Ca^{2+}-binding protein related to the lymphocyte specific protein-1 (LSP-1). This protein appears to have a function downstream of the polyphosphoinositide signaling pathway (Matsumoto *et al.*, 1993). Moreover, phosphorylation of several proteins in tyrosine residues is also induced by concanavalin A. Concomitant with these events, the non-receptor protein-tyrosine kinase $p72^{syk}$ of lymphocytes has been described to be activated by this lectin (Takeuchi *et al.*, 1993).

Binding of mitogenic lectins to lymphocyte ligands leads to important metabolic consequences for the cell. The activation of Na^+-dependent and Na^+-independent amino acid transport systems and the stimulation of glucose transport has been described (see Hadden, 1988). Moreover, lectin binding can activate enzymes of the intermediate metabolism, such as the mitochondrial pyruvate dehydrogenase responsible for the oxidation of pyruvate to CO_2. As noticed, such effects strikingly resemble events within the metabolic activation of cells by insulin upon binding to its receptor (Beachy *et al.*, 1981).

The inflammatory response of tissues is an important step in the development of an effective immunological defense. Within this complex process branched sugars generated from cellular components play a prominent role. Heparan sulfate chains are cleaved from proteoglycans located on the surface of endothelial or parenchymal cells and activate macrophages which produce prostaglandin E2, IL-1 and IL-6 that in turn control the proliferation of T-lymphocytes. Heparan sulfate signaling includes activation of tyrosine kinases and protein kinase C, the increase of the cytosolic concentrations of both inositolpolyphosphates and Ca^{2+} and the nuclear translocation of NF-κB (Wrenshall *et al.*, 1995). Moreover, some of these processes appear to be initiated by interaction with glycosylphosphatidylinositol-linked receptors (Wrenshall *et al.*, 1995). This class of molecules shows that cell surface-derived compounds can be part of the information-bearing system.

Similarly, endogenous lectins are beginning to be appreciated as an integral part of this context. In this respect, the contact of adhesion molecules is known to also affect intracellular parameters. Circulating neutrophils adhere to the vascular endothelium and are extravasated into tissues at the point of injury producing an inflammatory reaction. In the adhesion and extravasation processes L-selectin and β2-integrins play an important and sequential role. After establishing contact, activated neutrophils then produce the highly reactive superoxide anion (O_2^{*-}) which contributes to kill invading microorganisms. It has been shown that the cross-linking of L-selectin and Mac-1, a β2-integrin, induces O_2^{*-}

production and increases the concentration of Ca^{2+} in the cytosol of neutrophils (Crockett-Torabi *et al.*, 1995). Moreover, increased tyrosine phosphorylation and activation of MAP-kinase in neutrophils is linked to ligation of L-selectin (Waddell *et al.*, 1995). Furthermore, the glycophospholipid-linked plasma membrane receptor for IgGs denoted FcγRIII also mediates Ca^{2+}-mobilization and the burst in O_2^{*-} release. It has been proposed that the oligosaccharide chain of this receptor binds to a lectin-like domain of the adherence-promoting receptor molecule CR3 to elicit these responses (Sehgal *et al.*, 1993). Like the immunomodulatory mistletoe lectin, galectin-3, an endogenous human β-galactoside-binding lectin, too, with capacity to bind to IgE and to the high-affinity IgE receptor FcεRI, stimulates the production of O_2^{*-} in neutrophils (Timoshenko and Gabius, 1993; Yamaoka *et al.*, 1995). Remarkably, the triggering of signal production is not confined to these cellular lectins.

The low-affinity IgE receptor FcεRII from B-cells, later denoted CD23, has been shown to recognize galactose residues (Kijimoto-Ochini and Uede, 1995). Furthermore, CD23 and the B-cell differentiation antigen Lyb-2, later denoted CD72, are membrane-bound lectins that signal to the cell interior e.g. mobilization of Ca^{2+}. This takes place upon activation of phospholipase C, which hydrolyses PtdIns-4,5-P_2 into Ins-1,4,5-P_3 and DAG (Gordon, 1994). In addition, CD23 also mediates an increase in the intracellular concentration of cAMP and the activation of both CD23 and CD72 enhances the activity of tyrosine kinases (Gordon, 1994).

CD22 is a B-cell-specific receptor that recognizes sialoglycoproteins, among which the T-lymphocyte receptor-linked phosphotyrosine phosphatase CD45 is its major ligand. CD45 in turn is involved in coupling the antigen-receptor/CD3 complex to intracellular signaling pathways in T-lymphocytes (Sgroi *et al.*, 1995). Thus, CD22-CD45-CD3 interactions result in the activation of phosphatidylinositol-specific phospholipase Cγ1 by phosphorylation of tyrosine residues. The activation of this phospholipase renders available DAG and Ins-1,4,5-P_3 and they in turn activate protein kinase C and mobilize Ca^{2+} from intracellular stores, respectively (Sgroi *et al.*, 1995).

3 Lectin-Induced Apoptosis

Apoptosis, the programmed cell death, is a unique type of cellular response to external and internal signals. This process is an essential part of the normal development of multicellular organisms and also plays a prominent role in host defense. Hence, cytotoxic T-lymphocytes induce disruption in target cells by both osmotic lysis and apoptosis.

Lectins have already been implicated in the induction of apoptosis in different cell types. The isolectin of *Griffonia simplicifolia* is internalized by several murine tumor cell types and can trigger programmed cell death (Kim *et al.*, 1993). Moreover, ricin, a galactose-specific lectin that inhibits protein synthesis by removal of an adenine molecule from the 28S RNA of the eukaryotic 60S ribosomal subunit, has likewise been shown to induce apoptosis in macrophages and prevents the adhesion of these cells to a solid substrate (Khan and Waring, 1993). However, macrophages that are preadhered to a solid substrate became more resistant to ricin-induced apoptosis. Among the processes characteristically linked to apoptosis, a signal mediated by Ca^{2+} appears to be responsible for the activation of endonucleases that fragments the chromosomal DNA at the linker regions between the nucleosomes, yielding a characteristic pattern consisting of regular DNA fragments (Gerschenson and Rotello, 1992). Ricin-induced apoptosis, however, appears to be mediated by a Ca^{2+}-independent mechanism and its action appears to be independent of its effect on protein synthesis (Khan and Waring, 1993).

4 Signaling by Membrane-Bound Mammalian Lectins

A number of receptors for glycoconjugates has been identified biochemically in mammalian cells including among others the asialoglycoprotein receptor, the mannose/*N*-acetylglucosamine receptor, the fucose-specific receptor and the phosphomannosyl receptor (Ashwell and Harford, 1982; Gabius, 1991; Weigel, 1994; Stockert,

1995). Presence at the cell surface is not a constant feature, but is regulated by external parameters like substances which induce cell differentiation or by biotin which activates a membrane-associated guanylyl cyclase (Gabius *et al.*, 1990; Stockert, 1995). Receptor recycling can be accompanied by changes in its phosphorylation. The asialoglycoprotein (galactosyl) receptor undergoes internalization upon ligand binding and protein kinase C appears to be implicated in this process since phorbol esters cause its hyperphosphorylation and internalization. However, internalization does not appear to be coupled to the generation of second messengers from the metabolism of polyphosphoinositides (Medh *et al.*, 1989).

Another relevant receptor for certain glycoconjugates in this context is the cation-independent mannose-6-phosphate/insulin-like growth factor-II receptor. It is involved in the translocation of proteins containing a mannose-6-phosphate recognition marker to lysosomes, the endocytosis of glycoproteins containing this residue and transmembrane signaling. This receptor binds IGF-II and mannose-6-phosphate at different sites (Kornfeld, 1992). The transduction mechanism of the signal mediated by IGF-II involves the activation of phospholipase C with the subsequent production of Ins-1,4,5-P_3 and this signal is potentiated by mannose-6-phosphate (Rogers and Hammerman, 1989). A pertussis toxin-sensitive calcium mobilization process has also been shown to be mediated by IGF-II in cells sequentially pretreated with PDGF and EGF, and direct interaction of a G_i protein α-subunit with the mannose-6-phosphate/insulin-like growth factor-II receptor reconstituted into phospholipid vesicles has been demonstrated (see Kornfeld, 1992).

5 Conclusions

The application of plant lectins as tools for cell activation especially in immunology has established a familiarity with the fact that glycoligand recognition by lectins can trigger physiological consequences. The increasing extent of effort to characterize mammalian lectins and to describe their functions reflects their assumed relevance for multiple physiological processes. As a consequence of measuring a response to lectin binding the question of how this effect is triggered needs to be addressed. The detailed analysis of signaling pathways and the availability of suitable routine assay systems enable to conveniently check defined determinants for lectin-dependent alteration. This strategy has already solidified the validity of the notion that lectin-carbohydrate interaction can be referred to in line with other signaling cascade-initiating modes of recognition of hormones, neurotransmitters or growth factors. Coordination and integration of the incoming signals and cross-talk among messengers and target constituents of activated pathways present another level of mysteries to be solved in this area of research.

Acknowledgments

This article was prepared during a sabbatical leave by A.V. at the Brigham & Women's Hospital, Harvard Medical School, Boston MA, USA. Work at the authors' laboratories was funded by grants (to A. V.) from the Dirección General de Investigación Científica y Técnica (PR94–343), from the Comisión Interministerial de Ciencia y Tecnología (SAF392/93), and from the Consejería de Educación de la Comunidad Autónoma de Madrid (AE16/94) Spain, by a grant (to H.-J. G.) from the Dr.-M.-Scheel-Stiftung für Krebsforschung Germany, and a grant (to A. V. and H.-J. G.) from the program Acciones Integradas (H-A 106) between Spain and Germany. The support from the scientific program of the North Atlantic Treaty Organization is also acknowledged.

Abbreviations

AA, arachidonic acid; bFGF, basic fibroblast growth factor; cAMP, cyclic AMP; cGMP, cyclic GMP; CIF, Ca^{2+}-influx factor; DAG, 1,2-diacylglycerol; EGF, epidermal growth factor; EGFR, epidermal growth factor receptor;

EGTA, ethylene glycol bis(β-aminoethyl)-N,N'-tetraacetic acid; Fc«RI, IgE receptor I; Fc«RII, IgE receptor II; FcγRIII, IgG receptor III; G_i, inhibitory G protein; G_s, stimulatory G protein; Grb2, growth factor receptor bound protein 2; IgG, immunoglobulin G; IgE, immunoglobulin E; IGF-II, insulin-like growth factor II; IL-1, interleukin-1; IL-2, interleukin-2; IL-6, interleukin-6; Ins-1,4,5-P_3, inositol 1,4,5-triphosphate; LSP-1, lymphocyte specific protein-1; LT-C_4, leukotriene C_4; MAPK, mitogen-activated protein kinase; MAPKK, mitogen-activated protein kinase; MHC, major histocompatibility complex; NGF, neural growth factor; PDGF, platelet-derived growth factor; PKC, protein kinase C; PLA_2, phospholipase A_2; PLC, phospholipase C; PL, phospholipid; PtdIns, phosphatidylinositol; PtdIns-4-P, phosphatidylinositol-4-phosphate; PtdIns-4,5-P_2, phosphatidylinositol-4,5-biphosphate; SH2, src homology region 2; SH3, src homology region 3; Sos, son of sevenless nucleotide exchanger protein.

References

Ashwell G, Harford J (1982): Carbohydrate-specific receptors of the liver. In *Annu. Rev. Biochem.* **51**:531–54.

Barondes SH (1984): Soluble lectins: a new class of extracellular proteins. In *Science* **223**:1259–64.

Barondes SH (1988): Bifunctional properties of lectins: lectins redefined. In *Trends Biochem. Sci.* **13**:480–2.

Beachy JC, Goldman D, Czech MP (1981): Lectins activate lymphocyte pyruvate dehydrogenase by a mechanism sensitive to protease inhibitors. In *Proc. Natl. Acad. Sci.* USA **78**:6256–60.

Borrebaeck CAK, Carlsson R (1989): Lectins as mitogens. In *Adv. Lectin Res.* **2**:10–27.

Calvo FO, Ryan RJ (1985): Inhibition of adenylyl cyclase activity in rat corpora luteal tissue by glycopeptides of human chorionic gonadotropin and the α-subunit of human chorionic gonadotropin. In *Biochemistry* **24**:1953–9.

Chilson OP, Kelly-Chilson AE (1989): Mitogenic lectins bind to the antigen receptor on human lymphocytes. In *Eur. J. Immunol.* **19**:389–96.

Chochola J, Fabre C, Bellan C *et al.* (1993): Structural and functional analysis of the human vasoactive intestinal peptide receptor glycosylation. Alteration of receptor function by wheat germ agglutinin. In *J. Biol. Chem.* **268**:2312–8.

Clapham DE, Neer EJ (1993): New role for G-protein βγ-dimers in transmembrane signalling. In *Nature* **365**:403–6.

Crockett-Torabi E, Sulenbarger B, Smith CW *et al.* (1995): Activation of human neutrophils through L-selectin and Mac-1 molecules. In *J. Immunol.* **154**: 2291–302.

Davis RJ (1993): The mitogen-activated protein kinase signal transduction pathway. In *J. Biol. Chem.* **268**: 14553–6.

Drickamer K, Taylor ME (1993): Biology of animal lectins. In *Annu. Rev. Cell Biol.* **9**:237–64.

Gabius H-J (1991): Detection and functions of mammalian lectins – with emphasis on membrane lectins. In *Biochim. Biophys. Acta* **1071**:1–18.

Gabius H-J (1994a): Non-carbohydrate binding partners/domains of animal lectins. In *Int. J. Biochem.* **26**:469–77.

Gabius H-J (1994b): Lectinology meets mythology: oncological future for the mistletoe lectin? In *Trends Glycosci. Glycotechnol.* **6**:229–38.

Gabius S, Yamazaki N, Hanewacker W *et al.* (1990): Regulation of distribution, amount and ligand affinity of sugar receptors in human colon carcinoma cells by treatment with sodium butyrate, retinoic acid and phorbol ester. In *Anticancer Res.* **10**:1005–12.

Gabius H-J, Walzel H, Joshi SS *et al.* (1992): The immunomodulatory β-galactoside-specific lectin from mistletoe: partial sequence analysis, cell and tissue binding, and impact on intracellular biosignaling of monocytic leukemia cells. In *Anticancer Res.* **12**:669–76.

Gerschenson LE, Rotello RJ (1992): Apoptosis: a different type of cell death. In *FASEB J.* **6**:2450–5.

Gilman AG (1987): G proteins: transducers of receptor-generated signals. In *Annu. Rev. Biochem.* **56**: 615–49.

Gordon J (1994): B-cell signalling via the C-type lectins CD23 and CD72. In *Immunol. Today* **15**:411–7.

Green J, Muallem S (1989): A common mechanism for activation of the Na^+/H^+ exchanger by different types of stimuli. In *FASEB J.* **3**:2408–14.

Grier CE III, Mastro AM (1988): Lectin-induced phosphatidylinositol metabolism in lymphocytes is potentiated by macrophages. In *J. Immunol.* **141**: 2585–92.

Hadden JW (1988): Transmembrane signals in the activation of T-lymphocytes by lectin mitogens. In *Mol. Immunol.* **25**:1105–12.

Hartree AS, Renwick AGC (1992): Molecular structure of glycoprotein hormones and functions of their carbohydrate structures. In *Biochem. J.* **287**: 665–79.

Hashimoto S, Hagino A, Amagai Y (1990): Fibroblast growth factor-induced decrease in the phosphorylation of Nsp100 mediated through a calcium-dependent mechanism and blocked by lectins. In *Cell Struct. Funct.* **15**:181–9.

Ignarro LJ (1990): Haem-dependent activation of guanylate cyclase and cyclic GMP formation by endogenous nitric oxide: a unique transduction mechanism for transcellular signaling. In *Pharmacol. Toxicol.* **67**:1–7.

Khan T, Waring P (1993): Macrophage adherence prevents apoptosis induced by ricin. In *Eur. J. Cell Biol.* **62**:406–14.

Kijimoto-Ochini S, Uede T (1995): CD23 molecule acts as a galactose-binding lectin in the cell aggregation of EBV-transformed human B-cell lines. In *Glycobiology* **5**:443–8.

Kim M, Rao MV, Tweardy DJ *et al.* (1993): Lectin-induced apoptosis of tumor cells. In *Glycobiology* **3**:447–53.

Kornfeld S (1992): Structure and function of the mannose-6-phosphate/insulin-like growth factor II receptors. In *Annu. Rev. Biochem.* **61**:307–30.

Lemmon MA, Ladbury JE, Mandiyan V *et al.* (1994): Independent binding of peptide ligands to the SH2 and SH3 domains of Grb2. In *J. Biol. Chem.* **269**:31653–8.

Lew DB, Songu-Mize E, Pontow SE *et al.* (1994): A mannose receptor mediates mannosyl-rich glycoprotein-induced mitogenesis in bovine airway smooth muscle cells. In *J. Clin. Invest.* **94**:1855–63.

Lis H, Sharon N (1986): Lectins as molecules and as tools. In *Annu. Rev. Biochem.* **55**:35–67.

Loewenstein WR (1987): The cell-to-cell channel of gap-junctions. In *Cell* **48**:725–6.

Marletta MA (1994): Nitric oxide synthase: aspects concerning structure and catalysis. In *Cell* **78:**927–30.

Matsumoto N, Toyoshima S, Osawa T (1993): Characterization of the 50 kDa protein phosphorylated in concanavalin A-stimulated mouse T cells. In *J. Biochem.* **113**:630–6.

Medh JD, Haynes PA, Weigel PH *et al.* (1989): Ligand binding and internalization by the rat hepatic asialoglycoprotein receptor does not generate polyphosphoinositide derived second messengers. In *Life Sci.* **45**:2285–94.

Metzger H (1992): Transmembrane signaling: the joy of aggregation. In *J. Immunol.* **149**:1477–87.

Mills GB, May C, Hill M *et al.* (1989): Role of protein kinase C in interleukin 1, anti T3, and mitogenic lectin-induced interleukin 2 secretion. In *J. Cell. Physiol.* **141**:310–7.

Monaco ME, Gershengorn MC (1992): Subcellular organization of receptor-mediated phosphoinositide turnover. In *Endocrine Rev.* **13**:707–18.

Pike LJ (1992): Phosphatidylinositol 4-kinases and the role of polyphosphoinositides in cellular regulation. In *Endocrine Rev.* **13**:692–706.

Randriamampita C, Tsien RY (1993): Emptying of intracellular Ca^{2+} stores releases a novel small messenger that stimulates Ca^{2+} influx. In *Nature* **364**: 809–14.

Rogers SA, Hammerman MR (1989): Mannose-6-phosphate potentiates insulin-like growth factor-II-stimulated inositol triphosphate production in proximal tubular basolateral membranes. In *J. Biol. Chem.* **264**:4273–6.

Rozengurt E (1986): Early signals in the mitogenic response. In *Science* **234**:161–6.

Sanford GL, Harris-Hooker S (1990): Stimulation of vascular cell proliferation by β-galactoside-specific lectins. In *FASEB J.* **4**:2912–8.

Sehgal G, Zhang K, Todd RF III *et al.* (1993): Lectin-like inhibition of immune complex receptor-mediated stimulation of neutrophils. Effects on cytosolic calcium release and superoxide production. In *J. Immunol* **150**:4571–80.

Schwartz RH (1985): T-lymphocyte recognition of antigen in association with gene products of the major histocompatibility complex. In *Annu. Rev. Immunol.* **3**:237–61.

Sgroi D, Koretzky GA, Stamenkovic I (1995): Regulation of CD45 engagement by the B-cell receptor CD22. In *Proc. Natl. Acad. Sci. USA* **92**:4026–30.

Stockert RJ (1995): The asialoglycoprotein receptor: relationship between structure, function, and expression. In *Physiol. Rev.* **75**:591–609.

Takeuchi F, Taniguchi T, Maeda H *et al.* (1993): The lectin concanavalin A stimulates a protein-tyrosine kinase p72syk in peripheral blood lymphocytes. In *Biochem. Biophys. Res. Commun.* **194**:91–6.

Timoshenko AV, Gabius H-J (1993): Efficient induction of superoxide release from human neutrophils by the galactoside-specific lectin from mistletoe. In *Biol. Chem. Hoppe-Seyler* **374**:237–43.

Waddell TK, Fialkow L, Chan CK *et al.* (1995): Signaling functions of L-selectin. Enhancement of tyrosine phosphorylation and activation of MAP kinase. In *J. Biol. Chem.* **270**:15403–11.

Wang P, Matsumoto N, Toyoshima S *et al.* (1989): Concanavalin A receptor(s) possibly interacts with at least two kinds of GTP-binding proteins in murine thymocytes. In *J. Biochem.* **105**:4–9.

Weigel PH (1994): Galactosyl and *N*-acetylgalactosamimyl homeostasis: a function for mammalian asialoglycoprotein receptor. In *BioEssays* **16**: 519–24.

Wheeler LA, Sachs G, De Vries G, *et al.* (1987): Manoalide, a natural sesterterpenoid that inhibits calcium channels. In *J. Biol. Chem.* **262**:6531–8.

Wrenshall LE, Cerra FB, Singh RK et al (1995): Heparan sulfate initiates signals in murine macrophages leading to divergent biologic outcomes. In *J. Immunol.* **154**:871–80.

Yamaoka A, Kuwabara I, Frigeri LG *et al.* (1995): A human lectin, galectin-3 (ε/Mac-2), stimulates superoxide production by neutrophils. In *J. Immunol.* **154**:3479–87.

Zanetta JP (1996): Lectins and carbohydrates in animal cells adhesion and control of proliferation. *In Glycosciences: Status and Perspectives (Gabius H-J, Gabius S, eds)* pp 438–58, Weinheim: Chapman & Hall.

Zeng F-Y, Benguría A, Kafert S *et al.* (1995): Differential response of the epidermal growth factor receptor tyrosine kinase activity to several plant and mammalian lectins. In *Mol. Cell. Biochem.* **142**:117–24.

28 Glycobiology of Host Defense Mechanisms

Hans-J. Gabius, Klaus Kayser, Sabine André, and Sigrun Gabius

1 Introduction

The adaptive immune response of mammals is based upon sophisticated gene rearrangement processes which yield an enormous recombinatorial diversity of antibodies and cellular receptors. However, this part of the immunological system is not sufficient to protect a mammalian organism at any stage of its development. Prior to its full efficiency at an age of around 18 months in humans, non-adaptive host defense mechanisms are instrumental in keeping infections in check. Similarly, the innate immunity-conferring molecules may also serve as a first-line of defense at locations with low abundance of antibody presentation. These effector molecules would be directed against common constituents of the pathogenic organism, reducing the probability of rapid epitope alteration to evade pattern recognition. Moreover, the non-clonal response should discriminate infectious non-self from harmless self (Janeway, 1992). Thus, in addition to the clonally expanded immune reactions, contemporary vertebrates and also members of other phyla, deal with the problem of self/non-self recognition by expression of non-immunoglobulin recognition proteins. Lectins form an important group within this class of animal defense molecules.

The participation of lectins in host defense includes among others their activity as agglutinins to enhance phagocytosis, opsonins, complement activators, cell recognition molecules, and as elicitors of mediator release. It is thus not surprising that the detection of homologous mechanisms in invertebrates and vertebrates is possible (Marchalonis and Schluter, 1990; Vasta *et al.*, 1994). For example, the mammalian acute-phase reactants (pentraxins) and certain tunicate or horseshoe crab lectins are considered to belong to an ancient group of recognition molecules in both protostomes and deuterostomes. Remarkably, lectins appear to serve as defense mechanisms not only in the animal kingdom but also in plants (Etzler, 1985; Chrispeels and Raikhel, 1991; Peumans and van Damme, 1995; Rüdiger, 1996). This widespread utilization of carbohydrate epitope binding of pathogens as a product of diverse evolutionary histories underscores the importance of lectins for coping with infections.

2 The Family of Lectins in Host Defense

2.1 Acute-Phase Reactants (Pentraxins)

C-reactive protein and serum amyloid P component form discoid cyclic pentameric structures which are composed of non-covalently associated monomers, hence the term 'pentraxin' (Steel and Whitehead, 1994). They share significant sequence homology, indicating a common reactivity during the acute-phase response, in which often only one of the pentraxins behaves as a reactant for the analyzed species (Gitlin and Colten, 1987; Ballou and Kushner, 1992). As might be expected from the origin of its name, from the binding of phosphocholine of the C-polysaccharide of *Pneumococcus*, C-reactive protein acts as an opsonin for bacteria. Its several calcium-dependent binding capacities include galactosyl residues when immobilized on to a surface and kept in a mildly acidic environment

H.-J. and S. Gabius (Eds.), Glycosciences

ISBN 3-8261-0073-5

(Köttgen *et al.*, 1992). At sites of inflammation this lectin activity, for example towards foreign galactans, may render the galactans susceptible to phagocytosis. In the cascade of steps to fight infectious microorganisms, this feature of the pentraxin may therefore provide enhanced protection against pathogens, limit tissue damage and promote a rapid return to homeostasis. It has been suggested that the recently reported expression of a membrane-associated form of this pentraxin in monocytes permits a selectin-like adhesion process, which is assumed to be involved in monocyte infiltration and/or homing (Kolb-Bachofen *et al.*, 1995).

Like C-reactive protein, serum amyloid P component binds to various ligands in the presence of physiological levels of Ca^{2+}, including chromatin, histones, and also heparan sulfate, mannose-6-phosphate, and 3'-sulfated derivatives of glucuronic acid, galactose and *N*-acetylgalactosamine (Loveless *et al.*, 1992; Tennent and Pepys, 1994). Clearance of nuclear and lysosomal material from necrotic areas during inflammation offers a reasonable explanation for the respective binding activities. Interestingly, its sole biantennary glycosylation at Asn-32 lacks any microheterogeneity, and enzymatic desialylation leads to rapid endocytosis into hepatocytes and to changes in the surface accessibilities of certain amino acid residues (Pepys *et al.*, 1994; Siebert *et al.*, 1995). In addition to its presence in the circulation, this pentraxin is a constituent of amyloid deposits, primarily associating with glycosaminoglycans and amyloid fibrils. Although its role in amyloidogenesis is currently not precisely understood, it is obvious that several distinct functional motifs establish the complete molecule with roles in defense and structural assembly of amyloid plaques.

2.2 Collectins

The term "collectin" is used to denote a group of Ca^{2+}-dependent lectins containing collagen-like domains (Malhotra *et al.*, 1992). As already noted in the last paragraph for pentraxins, the collectins display at least one further distinct structural moiety besides the lectin domain. Such an integration of functionally indispensable building blocks is often seen in animal lectins (Gabius, 1994a). In serum collectins, the recognition domain (lectin) is linked to a potential effector mechanism resembling the organization of the first complement component C1q (Sastry and Ezekowitz, 1993; Holmskov *et al.*, 1994; Hoppe and Reid, 1994; Miyamura *et al.*, 1994).

2.2.1 Serum collectins

To fulfil their assumed role of binding avidly to sugar epitopes on the surface of pathogens, it would be advantageous to cluster the carbohydrate-recognition domains. Indeed, mannan-binding proteins from serum form a triple helix with their collagenous tails, which is stabilized by disulfide bridges in the *N*-terminal cysteine-rich region (Thiel, 1992; Reid, 1993; Summerfield, 1993). Further aggregation of three to six trimers is similarly promoted. The arrangement of the two types of structural domain resembles a bouquet of flowers, the collagen tails representing the stalks for the lectin-like flower heads. Each carbohydrate-binding domain in a trimer is separated from the others by 53 Å, rendering feasible self/non-self discrimination by high-affinity binding only to widely spaced, repetitive sugar arrays on the surface of bacteria, yeasts and other types of parasites (Sheriff *et al.*, 1994; Weis and Drickamer, 1994). Notably, aberrant glycosylation of tumor cells may also be a target for this serum lectin (Mann *et al.*, 1994; Ohta and Kawasaki, 1994; Fujita *et al.*, 1995).

Binding of the oligomeric structure to clustered ligands on a surface is the prerequisite for C1q-similar activation of the classical pathway of complement. It involves a C1r,s-like serine protease (Matsushita and Fujita, 1992). The structural integrity of both domains and an unique conformational feature in the middle oligomers that confers resistance to bacterial collagenase is critically important for the sugar-dependent complement activation (Schweinle *et al.*, 1993; Yokota *et al.*, 1995). Target glycoligands for the lectin are exposed, as for example on Gram-negative enteric bacteria, yeasts and viruses, as substantiated by the identification of the lectin as the β-inhibitor of influenza A virus infections and as the Ra-reactive factor which binds to Ra-chemotype strains of bacteria and yeasts (for reviews, see Sastry and Ezekowitz, 1993; Hoppe and Reid, 1994; Reid and Turner, 1994). Its action as agglutinin, complement activator and opsonin can elicit an enhanced oxidative burst by neutrophils, target cell lysis and phagocytosis. It is thus consistent that a mutation at functionally crucial points of the lectin, such as codon 54 which disrupts the fifth gly-Xaa-Yaa repeat in the collagen-like part of exon 1, will predispose the individual to a lifelong increased risk of infections (Summerfield, 1993; Lipscombe *et al.*, 1995; Summerfield *et al.*, 1995). This point mutation, and also an alteration in codon 57, are responsible for the occurrence of recurrent infections in infants, linking lectin functionality to a clinical syndrome (Summerfield, 1993).

Since the allele frequencies of these defects are between 0.11 and 0.29, it is reasonable to assume that low lectin concentrations may have beneficial effects, balancing the reduced anti-pathogenic activity (Summerfield, 1993). The causative agents of infectious diseases themselves are the principal driving force in maintaining polymorphism of genes. Heterosis by selective advantage for carriers of the sickle cell hemoglobin allele in Africa is a classic example. Consequently, the interaction with the lectin of intracellularly destined pathogens like mycobacteria or herpes simplex viruses type 2 will not eventually lead to their destruction, but can provide them with a lectin-dependent route of cell entry (Fischer *et al.*, 1994; Garred *et al.*, 1994). Attenuated complement activation at the site of inflammation in carriers of the mutant allele may additionally limit the local extent of this response. To emphasize deliberately the possible advantages, tissue damage and endotoxin release would be reduced under these circumstances. Heterozygotic carriers of variant alleles (especially those with decreased amounts of complement regulators or lipopolysaccharide neutralizers) may thus profit from low but sufficient amounts of the lectin to reduce a risk of severe infections.

Another member of the serum collectin group has been detected by its ability to mediate agglutination of serum-reacted erythrocytes (conglutination) *via* interaction with iC3b, derived from surface-deposited complement component C3b. It is present in bovidae and is considered to be an antibacterial-antiviral homologue of the mannan-binding protein, despite differences in structural organization with a cruciform arrangement of 38 nm-long arms which end in globular heads (Jensenius *et al.*, 1994; Reid, 1994). It does not, however, activate complement. To participate in first-line host defense to eradicate pathogens, conglutinin appears to enhance interactions of iC3b-coated pathogens and C1q receptor-exposing effector cells. Interestingly, the protein context in the vicinity of the glycosylation site at Asn-917 of the complement glycoprotein C3 or at Asn-34 of RNase B markedly influences the presentation of the glycoligand structure, which explains the selectivity of sugar-specific binding of conglutinin to a commonly occurring high mannose-type oligosaccharide (Solis *et al.*, 1994).

Bovine serum contains another independent but closely related collectin in addition to conglutinin, named CL-43 (Jensenius *et al.*, 1994). Its triple helix apparently does not aggregate to establish a mannan-binding protein-like structural organization. The carbohydrate specificity shares common features with other collectins, the lack of binding to *N*-acetylglucosamine-terminating *N*-linked sugar chains being a major difference (Loveless *et al.*, 1995). Despite this difference its presently known functions enable one to assign to this lectin a role in innate antibacterial immunity of ruminants (Jensenius *et al.*, 1994; Reid, 1994). Such a role is not confined to carbohydrate-binding proteins in serum. The continuous challenge by inhaled pathogens similarly necessitates efficient defense mechanisms in the respiratory tract. The large size of circulating oligomeric collectins alone will preclude their appearance by diffusion on the alveolar surface. Further requirements, such as a role in the organization of the surfactant into ordered tubular myelin, add to the demand in alveoli for locally produced proteins like collectins which are at least bifunctional. Indeed, two surfactant proteins are referred to as collectins (Haagsman, 1994; Kuroki and Voelker, 1994).

2.2.2 Surfactant proteins A and D

These two collectins are considered to be the alveolar counterparts of the serum lectins. Their structures resemble either the mannan-binding protein in the case of surfactant protein A with the flower bouquet-like arrangement, or conglutinin in the case of surfactant protein D with cruciform spacing of the globular heads connected to 46 nm long rods (Holmskov *et al.*, 1994). Both surfactant proteins bind to lipids as well as carbohydrate, that is, the major surfactant phospholipid phosphatidylcholine and also phosphatidylinositol (Miyamura *et al.*, 1994). In addition to participation in the structural organization of the lipid film, and the inherent modulation of surfactant secretion and turnover, their carbohydrate-binding and other domains react with bacterial lipopolysaccharides and with glycostructures of influenza A viruses (Haagsman, 1994; Miyamura *et al.*, 1994). Enhanced phagocytosis of bound targets and stimulation of O_2^{-*} production or chemotaxis of alveolar macrophages contribute to keeping such pathogens confined to the upper respiratory tract. Appraising the value of the documented *in vitro* activities, these two collectins appear to behave as multi-functional "doormen" to restrict the entry of infectious organisms into the lower respiratory tract, reducing the incidence of persistent infections.

It is intriguing that the concept derived from alveolar surfactant organization may be applicable to other tissues. From our point-of-view it is especially notable that expression of pulmonary surfactant protein A has recently been detected in epithelial cells of small and large intestine (Rubio *et al.*, 1995). This result is in accordance with the idea of including surfactant collectins among the active participants in the front line of host defense.

2.3 Selectins

As with collectins, the structure of the members of the group of selectins displays a modular organization. As reviewed by Zanetta in a preceding chapter of this volume, the amino-terminal and extracellular lectin domain is associated with an epidermal growth factor-like motif, consensus repeats found in complement-binding proteins, a transmembrane portion and a cytoplasmic tail. The lectin activity of each selectin mediates cellular adhesion such as lymphocyte homing and leukocyte attachment at inflammatory sites by recognition of glycoligands. Consequently, cellular failure to synthesize these epitopes causes an inherited disease, the leukocyte adhesion deficiency type 2 (Etzioni *et al.*, 1992). Like the situation in patients with mutant alleles of the mannan-binding protein, this syndrome is characterized by recurrent severe infections. On the other hand, our understanding of the molecular basis of leukocyte migration may prove therapeutically beneficial. Symptoms of typical inflammatory disorders such as autoimmune diseases or graft rejection may be reduced by substances that mimic selectin ligands which are designed to impair effectively leukocyte attachment to activated endothelium (Adams and Shaw, 1994; Parekh and Edge, 1994; Welply *et al.*, 1994).

2.4 NK-Cell Lectins

Natural killer cells are a subgroup of lymphocytes which are functionally defined by their cytolytic activity against certain tumor and virally infected cells (Trinchieri, 1989). Recent studies involving cloning of surface antigens have started to shed light on the molecular details that govern target recognition (Yokoyama, 1993; Chambers and Brisette-Storkus, 1995). Due to the presence of a sequence with a strong resemblance to the features of a Ca^{2+}-dependent carbohydrate-recognizing domain, three groups of NK cell surface determinants assumed to be lectins have been described, namely Ly-49, NKR-P1 and NKG2 (Chambers *et al.*, 1993). Ly-49A, Ly-49C and NKR-P1 have been shown either to bind to sulfated polysaccharides with a dependence on the presence of fucose moieties, or to sulfated tetrasaccharides of heparin derived from heparinase I digestion (Bezouska *et al.*, 1994; Daniels *et al.*, 1994; Brennan *et al.*, 1995). Detailed delineation of the ligand structures for these lectins on NK cells which discriminate normal from aberrant self may increase our glycobiological arsenal to make malignant cells susceptible to destruction.

2.5 Miscellaneous Lectins

Tumoricidal effects can also be exerted by macrophages. Activated macrophages employ a galactose/*N*-acetylgalactosamine-binding lectin to adhere to tumor cells prior to lysis (Kawakami *et al.*, 1994). This lectin facilitates glycoprotein uptake in inflammatory macrophages, thus serving independent functional roles in different macrophage populations (Kawakami *et al.*, 1994). Its upregulation in a subset of mononuclear cells in cardiac allografts phenomenologically connects this lectin to the chronic cardiac rejection process (Russell *et al.*, 1994). In addition to this galactose-specific lectin, macrophages possess a mannose-binding protein (MW 180 kDa) with repetitive recognition domains, which carries out phagocytosis of pathogens and acts as a scavenger for lysosomal enzymes (Stahl, 1992). The substantial sequence homology of this lectin to the human 180 kDa receptor for secretory phospholipase A_2 has been suggested to underscore its participation in clearance (Ancian *et al.*, 1995).

Facilitation of clearance and/or of reduction of cytotoxicity of tumor necrosis factor-α has been proposed as a functional consequence of its lectin-like behavior (Muchmore *et al.*, 1990). Its trypanolytic activity against salivary trypanosomes is also linked to its capacity to accept *N,N'*-diacetylchitobiose as a partner for interactions, as well as the known protein-type receptors (Lucas *et al.*, 1994). Binding to an immune mediator is another mechanism by which a lectin can act in host defense. The major ligand of the sialic acid-binding sarcolectin in human placenta is the lymphokine macrophage migration inhibitory factor (Zeng *et al.*, 1993). Its intratumoral presence is positively correlated with prognosis in lung cancer patients and it appears to play a central role in the response to endotoxemia (Bernhagen *et al.*, 1993; Kayser *et al.*, 1994). Regulation of accessibility by *in situ* association of the two proteins may influence factor-dependent modulatory responses. The elicitation of such cellular activities by an effector represents another level of action in the glycobiology of host defense mechanisms.

3 Lectin-Dependent Activation of Defense Mechanisms

An important humoral effector pathway is established by the availability of cytokines, as already indicated in the case of sarcolectin's interaction with a lymphokine. Instead of applying recombinant cytokines clinically in non-physiologically high concentrations, the deliberate alteration of cytokine synthesis and secretion *in vivo* affords the opportunity to enhance purposefully potential anti-tumoral mechanisms. Rigorous testing of this concept requires solid evidence that lectin-cell surface glycoligand binding can trigger this desired response. Indeed, a series of *in vitro* and *in vivo* experiments has substantiated that a plant lectin, namely the galactoside-binding agglutinin from *Viscum album* (mistletoe), elicits enhanced cytokine secretion, including interleukin-1, interleukin-6 and tumor necrosis factor-α, and antitumoral/antimetastatic responses in animal tumor models at a dose of 1 ng/kg body weight with biweekly subcutaneous injections (Gabius *et al.*, 1992, 1994; Gabius, 1994b). However, a premature judgement on the clinical efficiency of this modality should be strictly avoided, until convincing clinical data are presented. It is also unmistakably fundamental to emphasize that increased cytokine levels, such as in the case of interleukin-6, can promote tumor growth *in vivo* (Ravoet *et al.*, 1994; Hilbert *et al.*, 1995).

Since the lectin recognizes galactose moieties irrespective of their anomeric linkage to the subterminal position (Lee *et al.*, 1992, 1994), the sugar-dependent response to binding of this rather abundant epitope allows one to infer that binding to common, not just rare, epitopes initiates cell type-specific reactions. Interestingly, this activity as a biological response modifier is not restricted to plant lectins. For example, a loach and a flesh fly lectin cause cytotoxin release (Itoh *et al.*, 1984; Okutomi *et al.*, 1987). This activity of the galactose-specific flesh fly lectin, which is synthesized in response to injury, is considered to be essential for elimination of foreign cells introduced into the body cavity of larvae. Similarly, its documented role in the development of imaginal discs provides evidence that a defense molecule can play an important role in an ontogenetic process (Kawaguchi *et al.*, 1991). The list of lectins with this immunomodulatory potency extends to mammalian proteins, remarkably with specificity to galactosides (Kajikawa *et al.*, 1986; Jeng *et al.*, 1994). Just as with the mistletoe lectin, the exposure of human neutrophils to the mammalian galectin-3 stimulates further activities of the host defense system such as O_2^{-*}-production, the clinical relevance of which is presently unclear (Timoshenko and Gabius, 1992; Timoshenko *et al.*, 1993; Yamaoka *et al.*, 1995).

These examples describe the clinically exploitable reactivities of lectins as biological response modifiers. In this context, the other side of the coin, namely the cell surface lectin-reactive glycoligand, also warrants consideration. The enhanced secretion of tumor necrosis factor-α by alveolar macrophages in the presence of *Candida albicans* mannan or of interleukin-10 by spleen cells in the presence of a lacto-*N*-fucopentaose III-exposing neoglycoconjugate illustrates the capacity of carbohydrate elements to act as immune modulators (Garner *et al.*, 1994; Velupillai and Horn, 1994). It is a safe bet that further examples in this area will be published in due course.

4 Perspectives

Cellular and humoral effector pathways cooperate to account for the exquisite efficiency of the host defense system. Whereas during recent decades glycoscientific aspects in this area have mainly been attributed to the antigenic properties of glycoconjugates and the production of carbohydrate-binding antibodies, it is now firmly appreciated that the lectinological defense line against pathogens by non-recombinatorial sugar receptors operates successfully in organisms from different branches of the evolutionary tree. In metaphorical terms, these lectins are reliable interpreters of deviations from the self-code usage of carbohydrate sequences. Beyond this self/non-self-discrimination, we are currently becoming aware that lectin-dependent responses in the immune system should not merely be considered as instrumental aspects in basic research such as plant lectin-mediated mitogenicity of lymphocytes. The thorough analysis of responses which are provoked by lectin binding to immune cells has already revealed consequential reactions with a clinical perspective, defined and reproducible alterations in cytokine availability *in vitro* and *in vivo* illustrating a currently investigated topic in oncology. As outlined in the preceding paragraph, certain lectins thus rightfully deserve to be referred to as biological response modifiers. It is thus reasonable to encourage further efforts along this line of research to understand the function of lectins in immunological communication and to try and take clinical advantage of this progress.

References

Adams DH, Shaw S (1994): Leukocyte-endothelial interactions and regulation of leukocyte migration. In *Lancet* **343**:831–6.

Ancian P, Lambeau G, Mattei M-G *et al.* (1995): The human 180 kDa receptor for secretory phospholipase A_2. In *J. Biol. Chem.* **270**:8963–70.

Ballou SP, Kushner I (1992): C-reactive protein and the acute phase response. In *Adv. Intern. Med.* **37**:313–36.

Bernhagen J, Calandra T, Mitchell RA *et al.* (1993): MIF is a pituitary-derived cytokine that potentiates lethal endotoxaemia. In *Nature* **365**:756–9.

Bezouska K, Yuen C-T, O'Brien J *et al.* (1994): Oligosaccharide ligands for NKR-P1 protein activate NK cells and cytotoxicity. In *Nature* **372**:150–7 and 380, 559 (correction).

Brennan J, Takei F, Wong S *et al.* (1995): Carbohydrate recognition by a natural killer cell receptor, Ly-49C. In *J. Biol. Chem.* **270**:9691–4.

Chambers WH, Adamkiewcz T, Houchins JO (1993): Type II integral membrane proteins with characteristics of C-type animal lectins expressed by natural killer (NK) cells. In *Glycobiology* **3**:9–14.

Chambers WH, Brisette-Storkus CS (1995): Hanging in the balance: natural killer cell recognition of target cells. In *Chemistry & Biology* **2**:429–35.

Chrispeels MJ, Raikhel NV (1991): Lectins, lectin genes, and their role in plant defence. In *Plant Cell* **3**:1–9.

Daniels BF, Nakamura MC, Rosen SD *et al.* (1994): Ly-49A, a receptor for H-2D^d, has a functional carbohydrate recognition domain. In *Immunity* **1**:785–92.

Etzioni A, Frydman M, Pollack S *et al.* (1992): Recurrent severe infections caused by a novel leukocyte adhesion deficiency. In *N. Engl. J. Med.* **327**: 1789–92.

Etzler ME (1985): Plant lectins: molecular and biological aspects. In *Annu. Rev. Plant Physiol.* **36**: 209–34.

Fischer PB, Ellermann-Eriksen S, Thiel S *et al.* (1994): Mannan-binding protein and bovine conglutinin mediate enhancement of herpes simplex virus type 2 infections in mice. In *Scand. J. Immunol.* **39**: 439–45.

Fujita T, Taira S, Kodama N *et al.* (1995): Mannose-binding protein recognizes glioma cells: *in vitro* analysis of complement activation on glioma cells via the lectin pathway. In *Jpn. J. Cancer Res.* **86**:187–92.

Gabius H-J (1994a): Non-carbohydrate binding partners/domains of animal lectins. In *Int. J. Biochem.* **26**:469–77.

Gabius H-J (1994b): Lectinology meets mythology: oncological future for the mistletoe lectin? In *Trends Glycosci. Glycotechnol.* **6**:229–38.

Gabius S, Joshi SS, Kayser K *et al.* (1992): The galactoside-specific lectin from mistletoe as biological response modifier. In *Int. J. Oncol.* **1**:705–8.

Gabius H-J, Gabius S, Joshi SS *et al.* (1994): From ill-defined extracts to the immunomodulatory lectin: will there be a reason for oncological application of mistletoe? In *Planta Med.* **60**:2–7.

Garner RE, Rubanowice K Sawyer RT *et al.* (1994): Secretion of TNF-α by alveolar macrophages in response to *Candida albicans mannans.* In *J. Leukocyte Biol.* **55**:161–8.

Garred P, Harboe M, Oettinger T *et al.* (1994): Dual role of mannan-binding protein in infections: another case of heterosis? In *Eur. J. Immunogenet.* **21**:125–31.

Gitlin JD, Colten HR (1987): Molecular biology of the acute phase plasma proteins. In *Lymphokines* **14**: 123–53.

Haagsmann HP (1994): Surfactant proteins A and D. In *Biochem. Soc. Transact.* **22**:100–6.

Hilbert DM, Kopf M, Mock BA *et al.* (1995): Interleukin-6 is essential for *in vivo* development of B lineage neoplasms. In *J. Exp. Med.* **182**:243–8.

Holmskov U, Malhotra R, Sim RB *et al.* (1994): Collectins: collagenous C-type lectins of the innate immune system. In *Immunol. Today* **15**:67–74.

Hoppe H-J, Reid KBM (1994): Collectins: soluble proteins containing collagenous regions and lectin domains and their roles in innate immunity. In *Protein Science* **3**:1143–58.

Itoh A, Iizuka K, Natori S (1984): Induction of TNF-like factor by murine macrophage-like cell line J774.1 on treatment with Sarcophaga lectin. In *FEBS Lett.* **175**:59–62.

Janeway CA (1992): The immune system evolved to discriminate infectious non-self from non-infectious self. In *Immunol. Today* **13**:1–6.

Jeng KCG, Frigeri LG, Liu F-T (1994): An endogenous lectin, galectin-3, potentiates Il-1 production by human monocytes. In *Immunol. Lett.* **42**:113–6.

Jensenius JC, Laursen SB, Zheng Y *et al.* (1994): Conglutinin and CL-43, two collagenous C-type lectins (collectins) in bovine serum. In *Biochem. Soc. Transact.* **22**:95–100.

Kajikawa T, Nakajima Y, Hirabayashi J *et al.* (1986): Release of cytotoxin by macrophages on treatment with human placenta lectin. In *Life Sci.* **39**:1177–81.

Kawaguchi N, Komano H, Natori S (1991): Involvement of Sarcophaga lectin in the development of imaginal discs of *Sarcophaga peregrina* in an autocrine manner. In *Dev. Biol.* **144**:86–93.

Kawakami K, Yamamoto K, Toyoshima S *et al.* (1994): Dual function of macrophage galactose/*N*-acetylgalactosamine-specific lectins: glycoprotein uptake and tumoricidal cellular recognition. In *Jpn. J. Cancer Res.* **85**:744–9.

Kayser K, Bovin NV, Korchagina EY *et al.* (1994): Correlation of expression of binding sites for synthetic blood group A-, B- and H-trisaccharides and for sarcolectin with survival of patients with bronchial carcinoma. In *Eur. J. Cancer* **30A**:653–7.

Kolb-Bachofen V, Puchta-Teudt N, Egenhofer C (1995): Expression of membrane-associated C-reactive protein by human monocytes: indications for a selectin-like activity participating in adhesion. In *Glycoconjugate J.* **12**:122–7.

Köttgen E, Hell B, Kage A *et al.* (1992): Lectin specificity and binding characteristics of human C-reactive protein. In *J. Immunol.* **149**:445–53.

Kuroki Y, Voelker DR (1994): Pulmonary surfactant proteins. In *J. Biol. Chem.* **269**:25943–6.

Lee RT, Gabius H-J, Lee YC (1992): Ligand-binding characteristics of the major mistletoe lectin. In *J. Biol. Chem.* **267**:23722–7.

Lee RT, Gabius H-J, Lee YC (1994): The sugar-combining area of galactose-specific toxic lectin of mistletoe extends beyond the terminal sugar residue: comparison with a homologous toxic lectin, ricin. In *Carbohydr. Res.* **254**:265–76.

Lipscombe RJ, Sumiya M, Summerfield JA *et al.* (1995): Distinct physiochemical characteristics of human mannose-binding protein expressed by individuals of differing genotype. In *Immunology* **85**:660–7.

Loveless RW, Floyd-O'Sullivan G, Raynes JG *et al.* (1992): Human serum amyloid P is a multispecific adhesive protein whose ligands include 6-phosphorylated mannose and the 3-sulfated saccharides galactose, *N*-acetylgalactosamine and glucuronic acid. In *EMBO J.* **11**:813–9.

Loveless RW, Holmskov U, Feizi T (1995): Collectin-43 is a serum lectin with a distinct pattern of carbohydrate recognition. In *Immunology* **85:**651–9.

Lucas R, Magez S, DeLeys R *et al.* (1994): Mapping the lectin-like activity of tumor necrosis factor. In *Science* **263**:814–7.

Malhotra R, Haurum J, Thiel S *et al.* (1992): Interaction of C1q receptor with lung surfactant protein A. In *Eur. J. Immunol.* **22**:1437–45.

Mann KK, André S, Gabius H-J *et al.* (1994): Phenotype-associated lectin-binding profiles of normal and transformed blood cells: a comparative analysis of mannose- and galactose-binding lectins from plants and human serum/placenta. In *Eur. J. Cell Biol.* **65**:145–51.

Marchalonis JJ, Schluter SF (1990): On the relevance of invertebrate recognition and defence mechanisms to the emergence of the immune response of vertebrates. In *Scand. J. Immunol.* **32**:13–20.

Matsushita M, Fujita T (1992): Activation of the classical complement pathway by mannose-binding protein in association with a novel C1s-like serine protease. In *J. Exp. Med.* **176**:1497–502.

Miyamura K, Reid KBM, Holmskov U (1994): The collectins: mammalian lectins containing collagen-like regions. In *Trends Glycosci. Glycotechnol.* **5**:286–309.

Muchmore A, Decker J, Shaw A *et al.* (1990): Evidence that high mannose glycopeptides are able to functionally interact with recombinant tumor necrosis factor and recombinant interleukin 1. In *Cancer Res.* **50**:6285–90.

Ohta M, Kawasaki T (1994): Complement-dependent cytotoxic activity of serum mannan-binding protein towards mammalian cells with surface-exposed high-mannose type glycans. In *Glycoconjugate J.* **11**: 304–8.

Okutomi T, Nakajima Y, Sakakibara F *et al.* (1987): Induction of release of cytotoxin from murine bone marrow cells by an animal lectin. In *Cancer Res.* **47**:47–50.

Parekh RB, Edge CJ (1994): Selectins – glycoprotein targets for therapeutic intervention in inflammation. In *Trends Biotechnol.* **12**:339–45.

Pepys MB, Rademacher TW, Amatayakul-Chantler S *et al.* (1994): Human serum amyloid P component is an invariant constituent of amyloid deposits and has a uniquely homogeneous glycostructure. In *Proc. Natl. Acad. Sci. USA* **91**:5602–6.

Peumans WJ, van Damme EJM (1995): The role of lectins in plant defence. In *Histochem. J.* **27**:253–71.

Ravoet C, De Greve J, Vandewoude K *et al.* (1994): Tumor-stimulating effects of recombinant human interleukin-6. In *Lancet* **344**:1576–7.

Reid KBM (1993): Structure/function relationships in the collectins. In *Biochem. Soc. Transact.* **21**: 464–8.

Reid KBM, Turner MW (1994): Mammalian lectins in activation and clearance mechanisms involving the complement system. In *Springer Sem. Immunopathol.* **15**:307–25.

Rubio S, Lacaze-Masmonteil T, Chailley-Heu B *et al.* (1995): Pulmonary surfactant protein A is expressed by epithelial cells of small and large intestine. In *J. Biol. Chem.* **270**:12162–9.

Rüdiger H (1996): Structure and functions of plant lectins. In this volume.

Russel ME, Utans U, Wallace AF *et al.* (1994): Identification and upregulation of galactose/*N*-acetylgalactosamine macrophage lectins in rat cardiac allografts with arteriosclerosis. In *J. Clin Invest.* **94**:722–30.

Sastry K, Ezekowitz RA (1993): Collectins: pattern recognition molecules involved in first line host defence. In *Curr. Opinion Immunol.* **5**:58–66.

Schweinle JE, Nishiyasu M, Ding TQ *et al.* (1993): Truncated forms of mannose-binding protein multimerize and bind to mannose-rich *Salmonella montevideo* but fail to activate complement *in vitro*. In *J. Biol. Chem.* **268**:364–70.

Sheriff S, Chang CY, Ezekowitz RAB (1994): Human mannose-binding protein carbohydrate recognition domain trimerizes through a triple α-helical coiled-coil. In *Nature Struct. Biol.* **1**:789–94.

Siebert H-C, André S, Reuter G *et al.* (1995): Effect of enzymatic desialylation of human serum amyloid P component on surface exposure of laser photo CIDNP (chemically induced dynamic nuclear polarization)-reactive histidine, tryptophan and tyrosine residues. In *FEBS Lett.* **371**:13–6.

Solis D, Feizi T, Yuen C-T *et al.* (1994): Differential recognition by conglutinin and mannan-binding protein of *N*-glycans presented on neoglycolipids and glycoproteins with special reference to complement glycoprotein C3 and ribonuclease B. In *J. Biol. Chem.* **269**:11555–62.

Stahl PD (1992): The mannose receptor and other macrophage lectins. In *Curr. Opinion Immunol.* **4**: 49–52.

Steel DM, Whitehead AS (1994): The major acute phase reactants C-reactive protein, serum amyloid P component and serum amyloid A protein. In *Immunol. Today* **15**:81–8.

Summerfield JA (1993): The role of the mannose-binding protein in host defence. In *Biochem. Soc. Transact.* **21**:473–7.

Summerfield JA, Ryder S, Sumiya M *et al.* (1995): Mannose-binding protein gene mutations associated with unusual and severe infections in adults. In *Lancet* **345**:886–9.

Tennent GA, Pepys MB (1994): Glycobiology of pentraxins. In *Biochem. Soc. Transact.* **22**:74–9.

Thiel S (1992): Mannan-binding protein, a complement-activating animal lectin. In *Immunopharmacology* **24**:91–9.

Timoshenko AV, Gabius H-J (1993): Efficient induction of superoxide release from human neutrophils by the galactoside-specific lectin from *Viscum album*. In *Biol. Chem. Hoppe-Seyler* **374**:237–43.

Timoshenko AV, Kayser K, Drings P *et al.* (1993): Modulation of lectin-triggered superoxide release from neutrophils of tumor patients with and without chemotherapy. In *Anticancer Res.* **13**:1789–92.

Trinchieri G (1989): Biology of natural killer cells. In *Adv. Immunol.* **47**:187–376.

Vasta GR, Ahmed H, Fink NE *et al.* (1994): Animal lectins as self/non-self recognition molecules. In *Annu. NY Acad. Sci.* **712**:55–73.

Velupillai P, Harn DA (1994): Oligosaccharide-specific induction of interleukin-10 production by $B220^+$ cells from schistosome-infected mice: a mechanism for regulation of $CD4^+$ T-cell subsets. In *Proc. Natl. Acad. Sci.* USA **91**:18–22.

Weis WI, Drickamer K (1994): Trimeric structure of a C-type mannose-binding protein. In *Structure* **2**: 1227–40.

Welply JK, Keene JL, Schmuke JJ *et al.* (1994): Selectins as potential targets of therapeutic intervention in inflammatory diseases. In *Biochim. Biophys. Acta* **1197**:215–26.

Yamaoka A, Kuwabara I, Frigeri LG *et al.* (1995): A human lectin, galectin-3, stimulates superoxide production by neutrophils. In *J. Immunol.* **154**:3479–87.

Yokota Y, Arai T, Kawasaki T (1995): Oligomeric structures required for complement activation of serum mannan-binding proteins. In *J. Biochem.* **117**:414–9.

Yokoyama WM (1993): Recognition structures on natural killer cells. In *Curr. Opinion Immunol.* **5**:67–73.

Zeng F-Y, Weiser WY, Kratzin H *et al.* (1993): The major binding protein of the interferon antagonist sarcolectin in human placenta is the macrophage migration inhibitory factor. In *Arch. Biochem. Biophys.* **303**:74–80.

29 Transgenic Approaches to Glycobiology

HELEN J. HATHAWAY AND BARRY D. SHUR

1 Introduction

Glycobiologists have been faced for years with a dilemma not shared by their colleagues in other areas of biology and biochemistry. Despite the fact that glycoconjugates are ubiquitous in plants and animals, the function of many of these molecules is still largely unknown. In contrast, the roles of nucleic acids and polypeptides were elucidated decades ago. The difficulties in studying glycoconjugates stem in large part from their extreme complexity and heterogeneity, but it is these properties that make this class of macromolecule so intriguing.

For several years it has been supposed that carbohydrate heterogeneity might endow glycoconjugates with specific information, such as positional cues required for subcellular localization or for interaction with specific receptor-like molecules (Varki, 1993). In support of this hypothesis, many oligosaccharide sequences are expressed in a cell-, tissue-, and stage-specific manner. Notably, glycoconjugate profiles change during embryonic development and during the acquisition of the malignant phenotype (Feizi, 1985; Dennis *et al.*, 1987; Muramatsu, 1988; Hakamori, 1989; see chapter by Mann *et al.*, this volume).

Traditionally, the function of glycoconjugates has been addressed using exogenously applied reagents (antibodies, lectins, competitors, etc.) to perturb their function *in vitro* (Surani, 1979; Bayna *et al.*, 1988). Such studies have provided the basis for our current understanding of glycoconjugate function, namely, that carbohydrate residues endow stability, targeting, and/or recognition functions to their polypeptide backbones. However, it is wise to interpret the results cautiously, since inhibitory reagents are unavailable for many classes of carbohydrates, and the specificity of others is unclear in the cell culture environment. Finally, inhibitory reagents with short half-lives are inadequate to address glycoconjugate function in the more complex *in vivo* environment. To overcome these difficulties, glycobiologists have increasingly turned to transgenic technology to manipulate the expression of specific classes of glycoconjugates. By genetically altering gene expression or function, mutations can be permanently introduced into an organism, and the consequences can be monitored throughout development or during the progression of a disease state.

The field of glycobiology has, and will continue to see, a rapid expansion of progress with the advent of transgenic technology in mammals. With the ability to genetically manipulate the expression of molecules in whole animals comes the ability to address specific questions about the function of those components in new ways. Such studies have confirmed the importance of glycosylation in a variety of critical events, including embryonic development, glycoprotein stability, and cellular interactions (Varki *et al.*, 1991; Mayadas *et al.*, 1993; Arbones *et al.*, 1994; Ioffe and Stanley, 1994; Metzler *et al.*, 1994; Youakim *et al.*, 1994; see chapters by Keszler-Moll *et al.* and Sinowatz *et al.*, this volume). However, the results have often yielded surprises, and in many instances appear to contradict what has been learned about the role of glycoconjugates using traditional *in vitro* inhibition approaches (Ioffe and Stanley, 1994; Metzler *et al.*, 1994; Shur, 1994; Thall *et al.*, 1995). These apparent paradoxes will likely be clarified by further analysis of genetically altered animals and as our under-

H.-J. and S. Gabius (Eds.), Glycosciences

ISBN 3-8261-0073-5

standing improves of the complex interplay of intermolecular pathways. Nevertheless, what we have already learned thus far from genetically manipulating glycoconjugate expression in the intact animal has given new insight into the function of this intriguing, and poorly understood class of macromolecules.

2 Protein Glycosylation

Glycosylation of proteins occurs in a highly regulated fashion in the endoplasmic reticulum and Golgi apparatus through the action of glycosyltransferases and glycosidases (Schachter, 1986; see chapters by Brockhausen and Schachter; Pavelka; Sharon and Lis, this volume). Oligosaccharide chains can be attached to the protein backbone in either an *N*-glycosidic linkage to asparagine residues, or in an *O*-glycosidic linkage to serine or threonine residues. *N*-linked oligosaccharides occur as either high mannose-type structures, or are further processed to more complex oligosaccharides whose composition can vary dramatically, but normally includes *N*-acetylglucosamine, galactose, sialic acid, and fucose (Kornfeld and Kornfeld, 1985). Hybrid structures with features of both the high mannose- and complex-type oligosaccharides have also been described. The other predominant class of oligosaccharide chains, those *O*-linked to serine and threonine, are normally composed of *N*-acetylglucosamine, *N*-acetylgalactosamine, galactose, fucose, and sialic acid. Terminal modifications such as sulfation and phosphorylation of specific monosaccharides are common in both *N*- and *O*-linked oligosaccharides. Adding to the complexity of glycoconjugates is the fact that a variety of oligosaccharide structures can be found on a single polypeptide backbone, each of which can show tissue- and stage-specific alterations in composition.

Genetic manipulation of glycosides and the molecules that interact with them would not be possible without the considerable recent progress in the cloning of glycosyltransferases, glycosidases, and carbohydrate-binding proteins. Genes for most of the steps of *N*-linked oligosaccharide processing have been elucidated, and rapid progress is also being reported in isolating genes for many additional glycosyltransferases, glycosidases, and other molecules that modify oligosaccharide structure (Lowe, 1991).

3 Approaches to Genetic Manipulation of Oligosaccharide Function

3.1 Overexpression of Gene Products

Gene manipulation in mice was first described in 1982 with the introduction of exogenous copies of the human growth hormone gene into the murine genome (Palmiter *et al.*, 1982). When microinjected into mouse zygotes, exogenous DNA is stably integrated into the genome making the "transgene" inheritable. By itself, this is an extremely powerful technique, but it has been further refined and modified by the incorporation of constitutive, inducible, and tissue-specific regulatory elements, making it possible, at least in theory, to activate transgene expression at will (Palmiter *et al.*, 1993; Furth *et al.*, 1994). In this way, expression systems can be designed that are best suited to the question to be addressed.

More recently, improvements have been made that allow greater control of inducible transgenes. Two new approaches employ a transgene that can be either activated or repressed via tetracycline or its derivatives (Furth *et al.*, 1994; Gossen *et al.*, 1995). Induction of the transgene in one study was as high as one thousand-fold (Gossen *et al.*, 1995). These approaches will be especially useful to alter the expression of gene products that are otherwise lethal or detrimental at early stages of development, but which have interesting functions later on as well.

3.2 Elimination of Genes

In addition to manipulating gene products by overexpression, important biological questions can be addressed by removing a particular gene of interest. In 1987, the first report appeared of the complete and specific elimination of a target gene from the murine genome (Thomas and Capecchi, 1987). Since that time this technique, known as homologous recombination, gene targeting, or "knockout", has become the method of choice to study the function of hundreds of gene products. Although homologous recombination is a powerful tool to study gene function, millions of years of evolution has resulted in the duplication of critical genes, which form the basis for gene families in vertebrates. Members of gene families frequently have similar or overlapping functions, complicating the analysis of null mutations in vertebrates. Furthermore, convergent evolution has resulted in a variety of structurally unrelated gene products with similar functions. What this means to the researcher hoping to gain insights into the function of a particular gene product is the need to pay critical attention to subtle differences in phenotype that exemplify the true function of a gene.

3.3 Tissue- and Stage-Specific Alteration of Gene Products

Ideally, the biologist desires a system in which the gene of interest can be manipulated at a precise developmental stage or in a specific tissue. This is particularly important when the gene product is expressed at multiple times or places and is thought to have critical functions in each of these instances. Altering the expression of a particular gene that is required early in development prevents the investigator from examining its function during subsequent developmental events. One way to circumvent this problem is the use of inducible or tissue-specific transgenes, as discussed above; however, other approaches must be employed if one desires to eliminate the gene in a tissue- or stage-specific manner.

Tissue- and stage-specific gene "knockouts" can be created using the *cre/lox* recombinase system (Lakso *et al.*, 1992; Orban *et al.*, 1992) (see Figure 1). This approach takes advantage of the bacterial gene product recombinase (*cre*, an acronym for "causes recombination") to excise DNA that is flanked by the *loxP* sequences (*lox* is an acronym for "locus of crossing over"; Sauer and Henderson, 1989). In one scenario, homologous recombination technology is used to introduce *loxP* sites that flank the gene of interest, but which do not by themselves alter gene expression. A second line of mice is created that contains a *cre* transgene under the control of tissue- or stage-specific regulatory elements. Since the expression of *cre* by itself in transgenic mice has no effect, these mice are also phenotypically normal. By crossing the two groups of mice, the expression of *cre* causes excision of the *loxP*-flanked target gene in a tissue- or stage-specific manner (Gu *et al.*, 1994; Kuhn *et al.*, 1995). Alternatively, the *cre/lox* recombinase system can be used to activate a transgene of interest by introducing a transgenic construct containing a "stop" sequence between *loxP* sites. The stop sequences prevent expression of the transgene until crossed with the mice bearing *cre*, at which time tissue- or stage-specific activation is accomplished (Sauer, 1993).

The study of glycobiology using molecular biology and transgenic technology has one additional complication. Genetic manipulation by its very nature targets the gene product, or polypeptide, and therefore, studies of glycoconjugate function are directed at the enzymes that synthesize, metabolize, or otherwise alter carbohydrates, or at the carbohydrate-binding proteins that interact with them. Consequently, multiple glycoconjugates may be affected by a single transgenic event. Misexpressing one specific enzyme involved in glycoconjugate metabolism may alter the structure of an entire class of oligosaccharides. Similarly, carbohydrate binding proteins normally have an invariant or preferred specificity for a unique carbohydrate epitope, but these epitopes frequently occur on a variety of glycoconjugate structures. This complication makes the use of tissue- and stage-specific, or inducible systems, all the more useful to look at specific, rather than global events.

All of the previously described technologies have recently been applied to the study of glycoconjugate function in mammals, and will doubtless continue (Shur, 1994). While this field is still in its infancy, the results from these pioneering studies have already provided tantalizing puzzles, a few answers, and more questions.

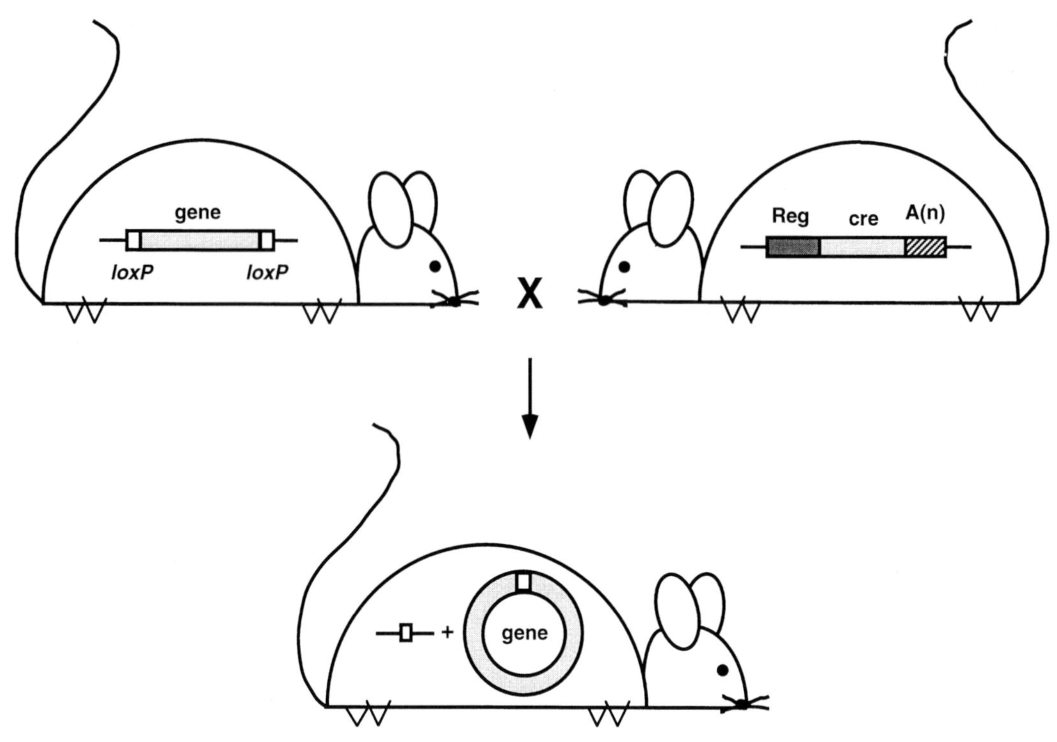

Figure 1. The *cre/lox* recombinase system can be used to create stage- or tissue-specific transgenic and "knockout" mice. The first line of transgenic mouse is created containing *loxP* sites flanking a gene of interest. The second line of transgenic mouse harbors a transgene consisting of *cre* under the control of tissue- or stage-specific regulatory elements. When the two mice are crossed, the gene flanked by *loxP* sites is excised only where or when *cre* is expressed, resulting in a tissue- or stage-specific "knockout". Reg, regulatory sequences; A(n), polyadenylation signal.

4 Genetic Manipulation of Glycosylation Pathways

4.1 Altering Terminal Glycosylation

Several groups have altered the expression of specific enzymes involved in oligosaccharide biosynthesis in transgenic mice. In one study, α1,3-galactosyltransferase, the enzyme responsible for transferring terminal galactose in an α1,3 linkage to a second galactosyl residue (Galα1–3Gal), was constitutively overexpressed in mice (Ikematsu *et al.*, 1993). The mice developed to term and to adulthood normally, although some evidence of early death was noted. Furthermore, the transgenic embryos demonstrated slightly reduced preimplantation development when analyzed *in vitro*. Since most of these transgenic mice survive to adulthood and are fertile, they present an opportunity to determine if overexpression of α1,3-galactosyltransferase affects tumor susceptibility, immune function, or disease resistance, all of which are thought to depend upon glycosylated components.

An alternative terminal modification to many *N*- and *O*-linked carbohydrate moieties is the addition of sialic acids (see chapter by Reutter *et al.*, this volume). Some of these residues become *O*-acetylated in a stage- and tissue-specific manner during development. *O*-Acetylation changes the size and charge of the sialic acid residue, suggesting that the regulation of *O*-Acetylation may be an important factor in the function of carbohydrates bearing these moieties. In order to address this question, transgenic mice were created that express a 9-*O*-acetylesterase, which selectively destroys the *O*-acetyl groups on sialic acid residues (Varki *et al.*, 1991). Expression of acetylesterase activity was lethal during preimplantation stages of embryogenesis. However, through the use of a tissue-specific promoter, or regulatory element, acetyltransferase was expressed in the adrenal gland and retina leading to developmental defects in these tissues. These results implicate *O*-acetylated sialic acids in normal morphogenesis and development. Since such moieties are found on a variety of carbohydrate structures, including *N*- and *O*-linked oligosaccharides of glycoproteins, as well as on glycolipids, more investigation will be required to determine if a specific subset of glycoconjugates containing *O*-acetylated sialic acids is responsible for the developmental defects observed in these mice.

4.2 Altering Complex *N*-linked Glycosylation

In vitro analyses using glycosylation inhibitors have demonstrated that the absence of *N*-linked glycosylation leads to preimplantation lethality (Surani, 1979). However, these inhibitors eliminate all *N*-linked carbohydrates, so a distinction between the relative importance of the high mannose-type, the complex-type, and the hybrid-type oligosaccharide modifications was not possible. As a first step toward resolving this problem, two labs have recently described the elimination of the gene encoding *N*-acetylglucosaminyltransferase I (GlcNAc-TI; Ioffe and Stanley, 1994; Metzler *et al.*, 1994). This enzyme is responsible for the first step in the biosynthesis of all complex-type *N*-linked oligosaccharides. Since virtually all mammalian cells and tissues express these structures, and since they appear to be required for some biological functions *in vitro* (Varki, 1993), it was surprising to find that embryos completely lacking GlcNAc-TI survived and appeared normal for the first nine days of development. At this stage, abnormalities such as growth retardation, neural tube deformities, abnormal vasculature, and left-right asymmetry inversion were apparent (Ioffe and Stanley, 1994; Metzler *et al.*, 1994). By the tenth day, the GlcNAc-TI-null mutation was lethal in homozygous embryos.

While it is comforting to have definitive proof that complex-type *N*-linked oligosaccharides are required for viability, the fact that development appeared normal during the first nine days was a surprise considering the complex regulation of these structures throughout embryogenesis (Muramatsu, 1988). Several explanations should be considered. First, it is possible that the *in vitro* systems employed in the earlier studies do not adequately represent the *in vivo* sequence of events, and that complex-type *N*-linked oligosaccharides are in fact not required for pre- and early postimplantation development. Second, it is possible that the gene encoding GlcNAc-TI was not completely inactivated, although in both reports, several lines of evidence argue against this (Ioffe and Stanley, 1994; Metzler *et al.*, 1994). Third, another gene product may be capable of partially compensating for the function of the inactivated GlcNAc-TI activity during early development. Elimination of GlcNAc-TI deletes complex-type *N*-linked glycans, but it does not affect other classes of oligosaccharides such as high mannose glycans, *O*-linked glycans and glycolipids, all of which play key roles in various biological processes (Varki, 1993). Finally, it is possible that maternal sources of complex-type *N*-linked oligosaccharides are available to the early embryo, since the GlcNAc-TI-null embryos are derived from crosses of heterozygous adults. Recent evidence suggests that the latter scenario is in fact the case, and thus explains the viability of GlcNAc-TI-deficient embryos for the first ten days of development (Campbell *et al.*, 1995).

4.3 Altering Glycosidases

In addition to altering the enzymes responsible for the synthesis of glycoconjugates, it is also possible to perturb the expression of glycosidases, the enzymes involved in glycan processing and turnover. In humans, a number of diseases collectively referred to as lysosomal storage diseases result from mutations in glycosidase genes leading to an accumulation of glycoconjugates in the lysosomes. Lysosomal storage diseases are characterized by mental retardation, skeletal abnormalities, and connective tissue lesions (Mononen *et al.*, 1993). As with many other human disease states, targeted deletion of the homologous genes in mice provides a potentially powerful tool with which to model and study the human condition. To that end, investigators have recently deleted the acid β-galactosidase gene in mice to produce an animal model for the lysosomal storage disorder G_{M1}-gangliosidosis (Matsuda *et al.*, 1995). At two months of age, the homozygous mutant animals appear to be healthy even though they do not express enzyme activity. Preliminary evidence suggests a vacuolar accumulation in some cells resembling the human condition. It remains to be determined if these mice will eventually develop symptoms characteristic of the human disease.

5 Altering Carbohydrate Receptors

5.1 Selectins

Selectins are a family of lectin-like carbohydrate receptors that are expressed on cells of the immune system and preferentially recognize and bind to ligands expressing sialic acid, fucose and in some cases sulfate residues (Imai *et al.*, 1991; Larsen *et al.*, 1992). (see chapters by Zanetta; Kannan and Nair, this volume). Three selectins have been described, E-, P-, and L-selectin, each of which has a distinct function reflected in its temporal and spatial pattern of expression on blood cells and/or endothelial cells. In general, the selectins appear to mediate leukocyte interactions with the endothelium by facilitating the initial binding of the blood cell to the endothelial wall, an interaction known as leukocyte "rolling." Rolling is followed by a tighter, higher affinity attachment that is mediated through the leukocyte integrins and unrelated glycoprotein receptors (Springer, 1990). All three selectins have recently been the subject of intensive investigation, since an understanding of their biological properties has important clinical and therapeutic implications.

Null mutations in the genes for both L-selectin and P-selectin have been reported (Mayadas *et al.*, 1993; Arbones *et al.*, 1994). The resulting phenotypes have essentially confirmed previous *in vitro* findings, in that mice lacking L-selectin or P-selectin have impaired leukocyte rolling, extravasation, and reduced leukocyte response to inflammation. These transgenic models should prove extremely useful in the study of carbohydrate-mediated immune responses.

Of equal interest to the glycobiology community is the transgenic manipulation of the selectin ligands. Recently, it has been shown that ligands for L-selectin have elevated expression in mice transgenic for an oncoprotein (Onrust *et al.*, 1995). Elevated ligand expression correlates with lymphocytic infiltration of transgene-induced tumors suggesting a role for these ligands in immune-mediated tumor destruction. This study also exemplifies the utility of transgenic animal models to probe complex molecular pathways that become altered by the perturbation of a single gene product.

5.2 Galectin: L14

In contrast to the results obtained with selectin "knockouts", the galactose-binding lectin, L14, appears to be completely dispensable, since L14-null mice appear normal (Poirier and Robertson, 1993). Since L14 is one of a family of galactose-binding lectins, called galectins, the lack of a phenotype in L14-deficient mice suggests that other family members share similar or overlapping functions, or at least can compensate for the lack of L14 in these mice (see chapter by Ohannesian and Lotan, this volume).

5.3 β1,4-Galactosyltransferase

β1,4-Galactosyltransferase (GalTase) is unusual among the glycosyltransferases in that it is found in two subcellular compartments; the Golgi apparatus where it participates in glycoprotein biosynthesis, and on the plasma membrane where it mediates cellular interactions (Lopez *et al.*, 1991). The dual subcellular distribution of GalTase results, at least in part, from the fact that the GalTase gene encodes two polypeptides (Shur,

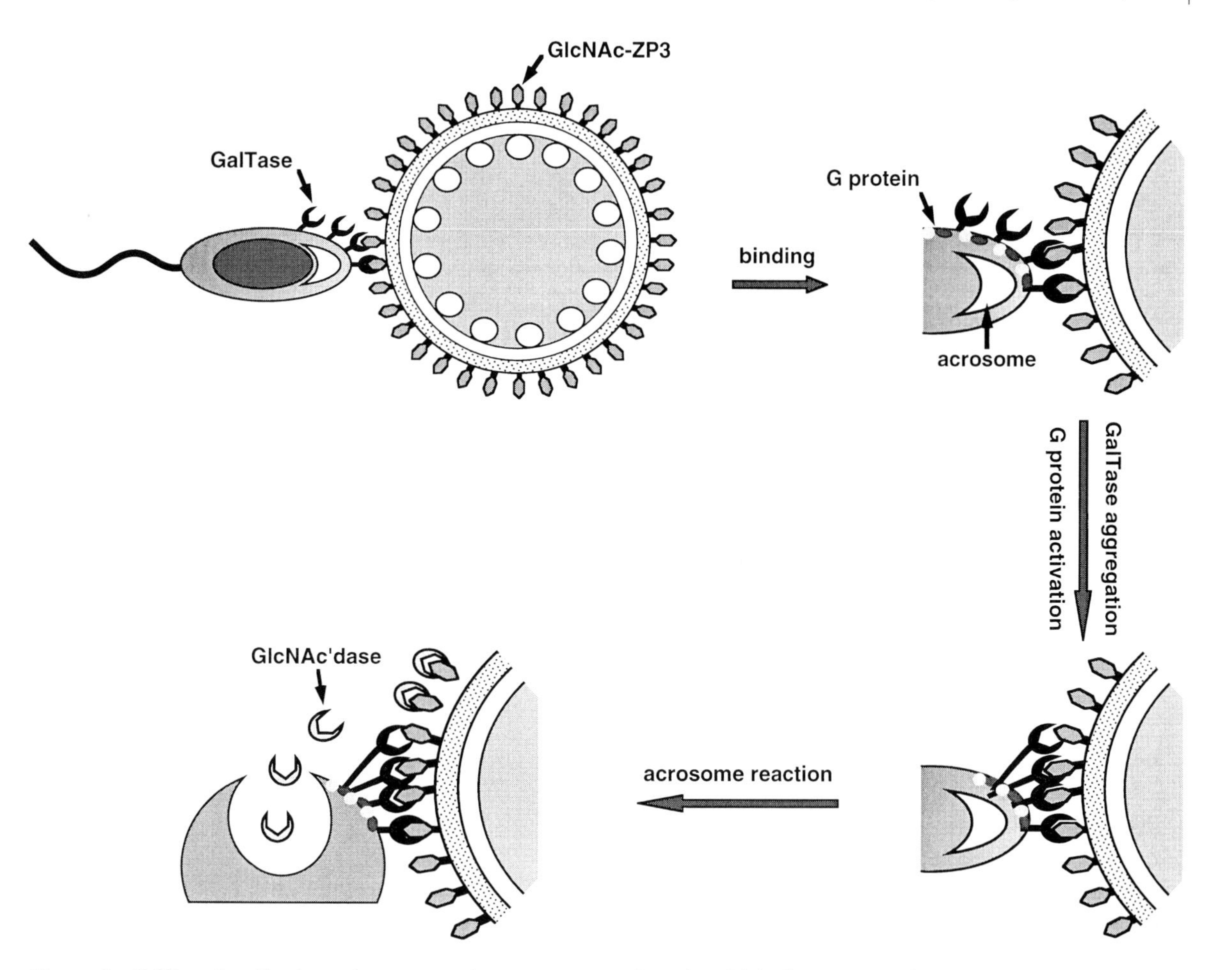

Figure 2. GalTase localized to the sperm plasma membrane binds to complementary *N*-acetylglucosamine (GlcNAc) residues on the egg coat glycoprotein ZP3. Multivalent oligosaccharides on ZP3 induces GalTase aggregation and activation of a heterotrimeric G protein cascade. G protein activation triggers the acrosome reaction, in which the acrosomal contents are released. The acrosome contains several enzymes, including *N*-acetylglucosaminidase, that cleave GalTase substrates from ZP3 to facilitate sperm penetration through the egg coat.

1993). The two GalTase proteins differ from one another in their cytoplasmic domains; the longer protein having an additional 13 amino acids not found in the shorter polypeptide. The 13 amino acid extension appears to override the Golgi retention signal within the transmembrane domain enabling this protein to be targeted to the plasma membrane (Evans *et al.*, 1995).

On the cell surface, GalTase functions non-catalytically in a variety of cellular interactions by binding its *N*-acetylglucosamine-terminating oligosaccharide substrate on adjacent cells or in the extracellular matrix (Shur, 1993). Considerable *in vitro* evidence has accumulated in support of this hypothesis. For example, antibody and hapten inhibition studies have shown that GalTase participates in sperm-egg binding, morula compaction, and migration of mesenchymal, neural, and neural crest cells on laminin-containing extracellular matrices (Shur, 1993). The application of transgenic technology now allows an *in vivo* analysis of GalTase function on the cell surface, and towards this end, transgenic mice have been created that overexpress the long, plasma membrane-as-

associated form of GalTase (Youakim *et al.*, 1994). As expected, these mice show defects in sperm:egg binding, mammary gland morphogenesis, and neonatal development, but do so in unexpected ways.

During fertilization, GalTase facilitates gamete interactions by serving as a sperm surface receptor for oligosaccharide ligands in the egg coat (Miller *et al.*, 1992) (see Figure 2). GalTase is localized on the plasma membrane overlying the acrosome, in which location it can bind to its *O*-linked oligosaccharide ligand on the egg coat glycoprotein ZP3. Multivalent *O*-linked oligosaccharides on the ZP3 polypeptide induce aggregation of GalTase on the sperm surface, activating a heterotrimeric G-protein-coupled signal cascade, culminating in the acrosome reaction (Gong *et al.*, 1995). The acrosome reaction releases hydrolytic enzymes from the fertilizing sperm that digest a penetration path through the egg coat (Miller *et al.*, 1993).

Most transgenic mice that overexpress GalTase on the cell surface are lethal in homozygous form, i.e. mice that carry two copies of the transgene (Hathaway and Shur, unpublished). However, since developing spermatids share messenger RNA and protein between them, heterozygous males produce a biochemically homogeneous population of sperm even though two different genotypes (transgenic and wild-type) coexist within the testis. This enables one to analyze the effects of elevated GalTase expression on sperm fertilizing ability.

As expected, sperm from transgenic mice that overexpress GalTase have elevated levels of GalTase on their surface and bind more ZP3 ligand than do wild-type sperm (Youakim *et al.*, 1994). This simple result confirms the ability of sperm surface GalTase to function as a receptor for the egg coat ligand. This result also illustrates how transgenic analyses can be used to resolve conflicting *in vitro* observations. Some workers have postulated that gamete recognition depends upon a sperm protein unrelated to GalTase that binds terminal α-galactosyl residues on ZP3 (Bleil and Wassarman, 1988). Recently, oocytes from mice homozygous null for α1,3-galactosyltransferase were shown to be fertilized normally despite the absence of terminal α-galactosyl residues (Thall *et al.*, 1995). Whether or not elimination of the appropriate *N*-acetylglucosaminyltransferase in oocytes prevents binding by sperm GalTase and thereby inhibits fertilization awaits testing.

Nevertheless, the fact that overexpressing surface GalTase leads to increased ZP3 binding is consistent with GalTase serving as a ZP3 receptor. It is therefore surprising that transgenic sperm that overexpress GalTase have a significantly reduced ability to bind to the intact egg coat (Youakim *et al.*, 1994). This apparent paradox is due, in part, to the fact that the elevated expression of GalTase renders the transgenic sperm hypersensitive to their ZP3 ligand, resulting in accelerated G-protein activation leading to higher rates of acrosome reactions (Youakim *et al.*, 1994; Gong *et al.*, 1995). Since acrosome-reacted sperm bind the egg coat with lower affinity than do acrosome-intact sperm, the transgenic sperm display a reduced adhesion to the egg coat as compared to wild-type sperm.

As mentioned above, the majority of transgenic mice that are homozygous for the increased GalTase expression die within the first three weeks of life. Preliminary evidence suggests morphogenetic defects in kidney, gut, liver, and lung, including abnormal vasculature of these and other tissues (Hathaway and Shur, unpublished). Of the few homozygous mice that survive this critical neonatal period, the resulting females have severely impaired mammary gland morphogenesis. This results in the complete inability to lactate at parturition. Biochemical, histological, and immunohistochemical analysis of transgenic mammary glands demonstrates a reduced ability to interact with the basement membrane at all stages of mammary gland development (Hathaway and Shur, submitted). As a result, the expression of milk proteins is severely inhibited during pregnancy, and milk is not secreted at parturition. This defective mammary gland morphogenesis extends earlier *in vitro* studies suggesting that the mammary epithelium uses surface GalTase to interact with the underlying basal lamina (Barcellos-Hoff, 1992). These data also support the conclusion that cell surface GalTase functions more than simply as a receptor for extracellular carbohydrates. By interacting with appropriate ligands in the basal lamina, GalTase may also trigger intracellular signals that are critical for cellular differentiation. Whether or not surface GalTase on mammary epithelial cells is coupled to a heterotrimeric G-protein cascade as it is on sperm awaits testing.

6 Future Directions

There is no question that the use of transgenic and gene targeting techniques will continue to provide valuable information regarding the function of glycosylation and carbohydrate recognition events. Rather than simply confirming previously held hypotheses based on *in vitro* data, in most cases, altering gene expression in the context of the whole organism has revealed novel functions. Furthermore, in many cases this technology allows us to address biological systems that have no *in vitro* counterpart. The few reports of gene knockout experiments in glycobiology have also reinforced the notion that redundancy of function is a common feature of complex biological systems. What this means to the researcher is not that the molecule of interest is trivial, but that its actual function is probably narrower than originally supposed and partially overlaps with the function of other components. It will therefore be necessary to delve deeper into the biology of the phenotypically "normal" transgenic mouse to dissect the more subtle phenotypes that are a consequence of altered gene expression.

Exciting times are ahead for glycobiologists, as the ability to manipulate the synthesis, expression, and recognition of carbohydrates *in vivo* presents almost limitless opportunities to understand glycoconjugate function. Of primary importance will be the development of animal models for human diseases known to be mediated or affected by altered glycosylation, such as the lysosomal storage diseases and other metabolic defects such as HEMPAS disease, characterized by global underglycosylation (Fukuda *et al.*, 1987). In addition, it will now be possible to test rigorously the hypothesis that a variety of normal and abnormal biological events, such as immune recognition and cancer cell metastasis, are mediated through carbohydrate recognition as has been suggested from *in vitro* analyses. Finally, developmental biologists are now in a position to ask how the developmental regulation of carbohydrate structures, realized for over a decade, is relevant to morphogenesis, growth, and differentiation.

References

Arbones ML, Ord DC, Ley K *et al.* (1994): Lymphocyte homing and leukocyte rolling and migration are impaired in L-selectin-deficient mice. In *Immunity* **1**:247–60.

Barcellos-Hoff MH (1992): Mammary epithelial reorganization on extracellular matrix is mediated by cell surface galactosyltransferase. In *Exp. Cell Res.* **201**: 225–34.

Bayna EM, Shaper JH, Shur BD (1988): Temporally specific involvement of cell surface β1,4-galactosyltransferase during mouse embryo morula compaction. In *Cell* **53**:145–57.

Bleil JD, Wassarman PM (1988): Galactose at the nonreducing terminus of *O*-linked oligosaccharides of mouse egg zona pellucida glycoprotein ZP3 is essential for the glycoprotein's sperm receptor activity. In *Proc. Natl. Acad. Sci. USA* **85**:6778–82.

Butcher EC (1991): Leukocyte-endothelial cell recognition: three (or more) steps to specificity and diversity. In *Cell* **67**:1033–6.

Campbell RM, Metzler M, Granovsky M *et al.* (1995): Complex asparagine-linked oligosaccharides in *Mgat1*-null embryos. In *Glycobiology* **5**:535–43.

Dennis JW, Laferte S, Waghorne C *et al.* (1987): β1–6 branching of Asn-linked oligosaccharides is directly associated with metastasis. In *Science* **236**:582–5.

Evans SC, Youakim A, Shur BD (1995): Biological consequences of targeting β1,4-galactosyltransferase to two different subcellular compartments. In *BioEssays* **17**:261–8.

Feizi T (1985): Demonstration by monoclonal antibodies that carbohydrate structures of glycoproteins and glycolipids are onco-developmental antigens. In *Nature* **314**:53–7.

Fukuda MN, Dell A, Scartezzini P (1987): Primary defect of congenital dyserythropoietic anemia type II. In *J. Biol. Chem.* **262**:7195–206.

Furth PA, One LS, Boger H *et al.* (1994): Temporal control of gene expression in transgenic mice by a tetracycline-responsive promoter. In *Proc. Natl. Acad. Sci. USA* **91**:9302–6.

Gong X, Dubois DH, Miller DJ *et al.* (1995): Activation of a G protein complex by aggregation of β-1,4-galactosyltransferase on the surface of sperm. In *Science* **269**:1718–21.

Gossen M, Freundleib S, Bender G *et al.* (1995): Transcriptional activation by tetracyclines in mammalian cells. In *Science* **268**:1766–9.

Gu H, Marth JD, Orban PC *et al.* (1994): Deletion of a DNA polymerase β gene segment in T cells using cell type-specific gene targeting. In *Science* **265**:103–6.

Hakomori S-I (1989): Aberrant glycosylation in tumors and tumor-associated carbohydrate antigens. In *Adv. Cancer Res.* **52**:257–331.

Ikematsu S, Kaname T, Ozawa M *et al.* (1993): Transgenic mouse lines with ectopic expression of α1,3-galactosyltransferase: production and characteristics. In *Glycobiology* **3**:575–80.

Imai Y, Singer MS, Fennie C *et al.* (1991): Identification of a carbohydrate-based endothelial ligand for a lymphocyte homing receptor. In *J. Cell Biol* **113**: 1213–22.

Ioffe E, Stanley P (1994): Mice lacking *N*-acetylglucosaminyltransferase I activity die at mid-gestation, revealing an essential role for complex or hybrid *N*-linked carbohydrates. In *Proc. Natl. Acad. Sci. USA* **91**:728–32.

Kornfeld, R, Kornfeld S (1985): Assembly of asparagine-linked oligosaccharides. In *Annu. Rev. Biochem.* **54**:631–64.

Kuhn R, Schwenk F, Aguet M *et al.* (1995): Inducible gene targeting in mice. In *Science* **269**:1427–9.

Lakso M, Sauer B, Mosinger B *et al.* (1992): Targeted oncogene activation by site-specific recombination in transgenic mice. In *Proc. Natl. Acad. Sci. USA* **89**:6232–6.

Larsen GR, Sako D, Ahern TJ *et al.* (1992): P-selectin and E-selectin. Distinct but overlapping leukocyte ligand specificities. In *J. Biol. Chem.* **267**:11104–10.

Lasky LA (1992): Selectins: interpreters of cell-specific carbohydrate information during inflammation. In *Science* **258**:964–9.

Lopez LC, Youakim A, Evans SC *et al.* (1991): Evidence for a molecular distinction between Golgi and cell surface forms of β1,4-galactosyltransferase. In *J. Biol. Chem.* **266**:15894–991.

Lowe JB (1991): Molecular cloning, expression, and uses of mammalian glycosyltransferases. In *Sem. Cell Biol.* **2**:289–307.

Matsuda J, Suzuki O, Oshima A *et al.* (1995): Targeted disruption of the mouse acid β-galactosidase gene: an animal model for G_{M1}-gangliodosis. In *Glycoconjugate J.* **12**:46a.

Mayadas TN, Johnson RC, Rayburn H *et al.* (1993): Leukocyte rolling and extravasation are severely compromised in P-selectin-deficient mice. In *Cell* **74**:541–54.

Metzler M, Gertz A, Sarkar M *et al.* (1994): Complex asparagine-linked oligosaccharides are required for morphogenic events during post-implantation development. In *EMBO J.* **13**:2056–65.

Miller DJ, Macek MB, Shur BD (1992): Complementarity between sperm surface β1,4-galactosyltransferase and egg coat ZP3 mediates sperm-egg binding. In *Nature* **357**:589–93.

Miller DJ, Gong X, Shur BD (1993): Sperm require β-*N*-acetylglucosaminidase to penetrate through the egg zona pellucida. In *Development* **118**:1279–89.

Mononen I, Fisher KJ, Kaartinen V *et al.* (1993): Aspartylglycosaminuria: protein chemistry and molecular biology of the most common lysosomal storage disorder of glycoprotein degradation. In *FASEB J.* **7**:1247–56.

Muramatsu T (1988): Developmentally regulated expression of cell surface carbohydrates during mouse embryogenesis. In *J. Cell. Biochem.* **36**:1–14.

Onrust SV, Hartl PM, Rosen SD *et al.* (1996): Modulation of L-selectin ligand expression during an immune response accompanying tumorigenesis in transgenic mice. In *J. Clin. Invest.* **97**:54–64.

Orban PC, Chui D, Marth JD (1992): Tissue- and site-specific DNA recombination in transgenic mice. In *Proc. Natl. Acad. Sci. USA* **89**:6861–5.

Passaniti A, Hart GW (1990): Metastasis-associated murine melanoma cell surface galactosyltransferase: characterization of enzyme activity and identification of the major surface substrates. In *Cancer Res.* **50**:7261–71.

Palmiter RD, Brinster RL, Hammer RE *et al.* (1982): Dramatic growth of mice that develop from eggs microinjected with metallothionein-growth hormone fusion genes. In *Nature* **300**:611–5.

Palmiter RD, Sandgren EP, Koeller DM *et al.* (1993): Distal regulatory elements from the mouse metallothionein locus stimulate gene expression in transgenic mice. In *Mol. Cell. Biol.* **13**:5266–75.

Poirier F, Robertson EJ (1993): Normal development of mice carrying a null mutation in the gene encoding the L14 S-type lectin. In *Development* **119:** 1229–36.

Sauer B, Henderson N (1989): Cre-stimulated recombination at *loxP*-containing DNA sequences placed into the mammalian genome. In *Nucleic Acids Res.* **17**:147–61.

Sauer B (1993): Manipulation of transgenes by site-specific recombination: use of cre recombinase. In *Methods Enzymol.* **225**: 890–900.

Schachter H (1986): Biosynthetic controls that determine the branching and microheterogeneity of protein-bound oligosaccharides. In *Biochem. Cell Biol.* **64**:163–81.

Shur BD (1993): Glycosyltransferases as cell adhesion molecules. In *Curr. Opinion Cell Biol.* **5**:854–63.

Shur BD (1994): The beginning of a sweet tale. In *Current Biology* **4**:996–9.

Springer TA (1990): Adhesion receptors of the immune system. In *Nature* **346**:425–34.

Surani MAH (1979): Glycoprotein synthesis and inhibition of glycosylation by tunicamycin in preimplantation mouse embryos: compaction and trophoblast adhesion. In *Cell* **18**:217–27.

Thall AD, Maly P, Lowe JB (1995): Oocyte Galα1,3Gal epitopes implicated in sperm adhesion to the zona pellucida glycoprotein ZP3 are not required for fertilization in the mouse. In *J. Biol. Chem.* **270**: 21437–40.

Thomas KR, Capecchi MR (1987): Site-directed mutagenesis by gene targeting in mouse embryo-derived stem cells. In *Cell* **51**:503–12.

Varki A (1993): Biological roles of oligosaccharides: all of the theories are correct. In *Glycobiology* **3**: 97–130.

Varki A, Hooshmand F, Diaz S *et al.* (1991): Developmental abnormalities in transgenic mice expressing a sialic acid-specific 9-*O*-acetylesterase. In *Cell* **65**: 65–74.

Youakim A, Hathaway HJ, Miller DJ *et al.* (1994): Overexpressing sperm surface β1,4-galactosyltransferase in transgenic mice affects multiple aspects of sperm-egg interactions. In *J. Cell Biol.* **119**:1229–36.

30 Biomodulation, the Development of a Process-Oriented Approach to Cancer Treatment

PAUL L. MANN, REBECCA WENK, AND MARY A. RAYMOND-STINTZ

1 Introduction

The focus for current, traditional treatment modalities is the end result, the physical manifestation of the disease process and not the process itself. This apparent fixation with technology's ability by itself to supply the "magic bullet" for human disease represents a strange dichotomy for the science community. When the research aim is abstract a very problem-oriented [rational, systematic, integrated with the problem] approach is pursued. However, when it comes to human disease management and especially those conditions which catch the general public's attention, a completely different style of approach emerges. Instead of the time-tested method of problem identification and hypothesis generation followed by testing and re-definition, the scientific community seems to pursue outcome [the answer] without regard for process. This chapter will attempt to place the search for answers in the general area of disease management and specifically cancer management in a more appropriate light.

Instead of using the tools and assets of technology as components of an overall strategy to define the disease process and then develop a "menu" of intervention approaches, technology has itself become the single focus of our attention. As individuals, communities of scientists and clinicians, and as society as a whole we have become almost fanatical in our devotion to the imagined power of technology. By allowing the "machines" of science to take on the total responsibility for both problem and solution, we have abdicated our responsibility in defining the problem and developing the process by which many solutions are tested and only those that can be integrated back to the hypotheses in some meaningful way are maintained. The reason for this change in a time honored process in the specific case of human disease is the elusive "magic bullet", the cure. This is the single dominant reason for technology's abysmal track record in terms of cost:benefit analyses. Technology without process can only improve the accuracy of the result, not its validity.

Where did this concept that everything in biology can be reduced to an engineering process equation start? According to an excellent essay by McKeown (McKeown, 1990) the concept that human health is under external, mechanism-based, engineering intervention is almost 300 years old. This 17th century concept dissected the organismic entity into its component parts and then made the leap in logic that, because it could be "described" at the microscopic level, it could be completely understood and therefore "fixed". The philosophical outgrowth of this [or its cause if you prefer] is LaPlacian determinism (LaPlace, 1749) which has resulted in the conceptual linearization of process to match these mechanical processes.

Much of the "data" which was used to justify this mechanistic/engineering approach is dubious (McKeown, 1990). Streptomycin, discovered in the late 1940's, was heralded by the scientific community as the "cure" for infectious disease. As evidence for this contention they quoted "data" that indicated a 50% decrease in death rate due to tuberculosis in Great Britain. These data did not reflect the appropriate context for the death rate due to tuberculosis. When compared to data from the earlier part of the 19th century, the decrease in death rate attributable to streptomycin was only 3%. Apparently, the

H.-J. and S. Gabius (Eds.), Glycosciences

ISBN 3-8261-0073-5

researchers of the day were so excited and focused on their own discovery they failed to observe that by far the most significant decrease in disease frequency was due to other factors, namely personal and public health awareness. Similar analyses can be found for other major causes of death during this period (McKeown, 1990).

It is this combination of linear thinking, inappropriate data presentation [experimental design], and human arrogance which has led us to this conundrum whereby society has placed all its faith and resources in technology-based systems which, by definition, are incapable of matching expectations. Basic scientists pursue microscopic "pieces" of the overall puzzle, which after years of hard work, start to take on the spectre of absolute truth. Clinical scientists deal with the sheer volume of conflicting information necessarily generated by this infinite, non-integrated, microscopic database, by reassuring their patients ["cancer can be beaten"] and generating highly restrictive and linear regulatory mechanisms.

What has been lost with this outcome driven search for the "magic bullet" is the definition of the problem. As in the streptomycin example discussed above, without a clear context-based definition of the problem, the conclusions drawn from any study will be, at best, very difficult to interpret. Problem identification is an art which is making a comeback in our educational systems (Kaufman, 1985; Baca *et al.*, 1990; Mennin *et al.*, 1992; Obenshain and Galey, 1992). Student-centered problem-based learning [SCPBL] is being used in a variety of formats in health-rated professional schools. These same institutions had used a traditional strictly didactic approach to education for generations. In the didactic format, information [content] is presented in a linear configuration with application of the content and integration into the problem left until much later in the curriculum. This didactic learning leads to the misperception that linear knowledge-based arguments could be used to attain any desired outcome. In the problem-based learning environment the student discovers that learning begins with the clear definition of the problem [not knowing something is just as important as knowing, in this process]. This problem identification is then followed by hypothesis generation exercises designed to discover everything known about the problem. A prioritization of those data, with respect to quality, structural level, and ability to integrate with other data is followed by a specific definition of the unanswered important, and relevant questions. The interface between knowledge and "new knowledge" is thereby established and the process of acquiring new information that bears on the problem is initiated. This information is then used to refine or re-define the hypotheses and eventually, after a number of iterations, this process results in a conclusion, and more importantly, a further course of action. The quality of that conclusion is only related to the quality of the input information and open to further refinement when more information is available. The analogy here is that SCPBL is three-dimensional in form and process, whereas didactic learning is very linear.

If the current changes in professional education can be taken as an indication of what is to come in the research and clinical application arena, then this 300-year tradition of simplistic linear thinking and outcome-oriented strategies will be replaced with problem focused integrated approaches.

2 Problem Definition

Of course this is where the "new" controversies will be generated, but for the sake of the present argument we will assume a simple model. The problem will be identified as a behavioral one in the case of cancer – cancer, in its multiple forms, is an interruption in the natural [i.e. inherent] organismic recognition system, which in turn is a fundamental communication among components of the host. One of the many outcomes of this interruption in recognition is the imbalance between differentiative/proliferative control at the cellular level. The problem is the mismanagement of information, a problem which requires modulation, not eradication. This is the fundamental difference between the Biomodulator and the traditional approach. Evidence will be presented which shows there is an active interface between host and aberrance [the process], and that both systems are amenable to manipulation [Biomodulation]. The hypothesis will then be generated that the control level of this problem rests with the arrangement of cell surface oligosaccharides.

3 A Brief Summary of Other Approaches

The only thing that is absolutely clear from all the literature on the treatment and management of cancer is that it is anything but a single disease or a even family of diseases. Virtually every cell type in the human body has a malignant counterpart. There are as many "styles" of induction, growth, and prognosis as there are variants within one sub-classification. This pleiotropic array of diseases should only be simplified at its induction point and not when it manifests itself. Most cytotoxic chemotherapeutic regimens are founded on the concept that cancers represent aberrant growth. The identified problem in this case is the outcome of a process, i.e. the tumor, and not the process which produced it. The chemotherapeutic agent then is used to preferentially destroy the cancer by virtue of this aberrant growth. But cancer, and especially the advanced cancer, has developed an organ status in terms of its complexity, and therefore, any single outcome-based approach [aberrant growth] misperceives the overall problem of the organ-based and organ/host interrelationships.

Surgery is one of the oldest treatments for cancer and in certain well defined circumstances probably is still the optimal outcome-based treatment. The underlying assumption for this treatment is that the excised growth is a single aberrance, and not the result of a process upon which the surgery can have no effect. A less invasive ablation modality is radiation therapy. In the old parlance, the radiation is used like a high energy, invisible scalpel, "to cut out and destroy simultaneously" the malignancy. Here the underlying linear assumption was that somehow the treatment was specific for the aberrance, and therefore would not affect the host. As the frequency of unrelated malignancies in radiation "cured" patients increases, this linear analysis of cause and effect is coming under increased scrutiny.

Feld (1994) analyzed the "early and late toxicities" of current treatment regimens for small cell lung cancer. In this cancer, chemotherapy, radiation, and surgery are all used with very limited success. The thrust is, however, directed at making the treatment a little less noxious. This would be defensible and rational if the original treatment were highly successful, and therefore the emphasis would shift towards making the treatment a few percentage points better or the patient more comfortable. Clearly this is not the case. In an extensive review by Faguet (1994), the completely ablative characteristics of some high dose chemotherapeutic regimens have been used to destroy the chronic lymphocytic leukemia [CLL] patient's bone marrow. After bone marrow transplantation and use of other "rescue" regimens, a subset of CLL patients is being identified with high potential for complete remission from their disease. It is striking that in this case the value of chemotherapy is diametrically opposed to its hypothesized selectivity. The other disturbing feature of the chemotherapy/radiation dyad of treatment is the increasing frequency of new cancers [especially leukemia and myelodysplasia] after successful treatment of solid tumors (Kantarjian *et al.*, 1993).

4 The Biological Response Modifier [BRM] Approach

The general area of biological response modifiers [BRMs] as it is practiced today is another example of technology without context, and particularly appropriate to our discussion of the importance of problem identification. In the 1970's it was clear to basic and clinical investigators that the host's immune system was central in all surveillance and recognition phenomena, and therefore, a suppressed or inappropriately functioning immune system could be the problem in cancer patient.

The development of the modern day BRM (Biological Response Modifiers, 1983) is illustrative of how intense the search for specificity in biological interactions has become. In this context, specificity is a negative attribute because it is focused too tightly on outcome and not on the process. Again, in oversimplified terms, the analogy would be that chemotherapy treatment "A" is destroying some of the tumor but in the process the patient is succumbing to massive infection; therefore, we will now give BRM treatment "B", designed to expand the peripheral lymphocyte count to ameliorate the problem induced by the first treatment. The basis for this must be a linear faith in the connection between cause and effect, it seems more rational to question the use of the original treatment.

Originally, BRM was coined by a special subcommittee of the Division of Cancer Treatment, at the National Institutes of Health (Biological Response Modifiers, 1983). Their original recommendations were broad in scope and designed to encompass the "problem" of host/aberrance interactions. Since that time there has been a progressive narrowing of the interpretation of the "problem" to the point that BRM is now synonymous with immune activation, clearly a linear, outcome-based interpretation. The pioneers in this field, such as George Mathe and his group (Mathe *et al.*, 1969), saw BRMs as generically useful in re-balancing the host/immune/aberrance triad. These broad-based BRMs were offered as an alternative to the therapies of the day. The hypothesis might have been that the presence of a neoplasia, either of itself or through some secondary process, created an imbalance in the host system which permitted the tumor to become established and to thrive. The "addition" of the BRM facilitated the host in re-establishing original balance. The mechanism(s) for this involve *recognition* of the host/aberrance interface, and then the *activation* of a specific immune function. These host-dependent functions must be integrated. It was this host-dependent integrated rebalancing function that facilitated the resolution of the problem. In this analysis the problem has shifted from the tumor itself to the host/tumor relationship, and as an extension of that hypothesis to a non-functioning recognition system [the host's immune system].

The Hellstrom group (Hellstrom and Hellstrom, 1969; Hellstrom and Brown, 1979) at about the same time described this "imbalance" as an excess of immune product and/or immune complexes [blocking antibody]. The investigators identified an appropriate problem, in this case a malfunctioning immune system, and hypothesized that removal of the inappropriate complexes might permit the host to reestablish its appropriate recognition function. Thus, in the first instance the idea was to add a non-specific agent, and in the other to subtract a specific one, all for the same result. To LaPlacian thought, these concepts are mutually contradictory; however, in the real non-linear world these are simply shades of the same answer. Both of these pioneering groups [and others as well], although restricting themselves to the immune system, were discussing the complex control over growth/differentiation patterns within the host, as well as how access to imbalanced systems might be achieved.

Over the last decade, this very broad and comprehensive BRM strategy seems to have transformed itself into a search for more and more specific biological molecules for the purpose of inducing specific effector mechanisms [immune activation]. This search seems to correlate with technology's ability to produce better quality, higher purity reagents.

5 Recent Uses of BRM/Biomodulator Terminology

In the past several years the use of the terms BRM and Biomodulator in the literature has been primarily from within the context of immunostimulators. As discussed above, this application is aimed at the amelioration of a treatment-induced problem rather than focusing on the original problem. Mitchell (1992 a, b, c) presents an overview of the clinical use of Biomodulators in which he attempts to reconcile the diametrically opposed objectives of the use of cytotoxic drugs and Biomodulators to treat malignant disease.

Herbal extracts and other natural products are starting to make a stronger appearance in the literature under the broad descriptor of biomodulator (Kao *et al.*, 1994). These natural products range from plant extracts with B-cell mitogenic properties (Coppi *et al.* 1994) to lipopolysaccharide preparations administered in a liposomal form (Petrov *et al.*, 1994) which cause significant increases in humoral, mitogenic and lymphokine production. It is interesting that these natural products are apparently being developed with a single purpose and point of intersection with the physical manifestations of the disease, as is the case with synthetic drugs, instead of using these materials in the context of their folk medicine origins. Polysaccharide [alpha-glucan] has been shown to decrease the growth of implanted tumors in immune competent animals and not in their T-cell deficient counterparts, indicating that individual compartments of the immune system are accessible (Takeda *et al.*, 1994). Fish oil at 6 grams/day over the first year post renal transplantation has been shown to increase renal hemodynamics in cyclosporine treated recipients (van der Heide *et al.*, 1993). Natural biological products derived from the biological system itself such as RNA [or nucleotides] (Jyonouchi *et al.*, 1993), melatonin (Caroleo *et al.*, 1994), granulocyte-colony stimulating factor [G-CSF], or T-cell helper factor [γ-2] (Barak *et al.*, 1992) have all been shown to have an effect on immunostimulation in a variety of models.

The only tested and implemented application of biomodulation, which encompasses the major precepts of the original work on BRMs, is the Prosorba column produced by IMRÉ Corporation [Seattle, USA]. This protein-A-based immunoabsorbant device removes specific aberrant immune complexes [see discussion of the Hellstrom studies above] from the patient's circulation periodically during a short treatment period [for a more detailed discussion see Mann, 1993]. This modulation [interestingly, in this case, through extraction – all other examples are via addition of something] in the concentration of a specific aberrant component results in reversal of symptomatology, and in a significant proportion of treated patients an actual remission of the disease. It appears to be pleiotropic in its application, having shown success in idiopathic thrombocytopenia [ITP], thrombotic thrombocytopenia purpura [TTP], feline leukemia virus in cats, and most recently in rheumatoid arthritis, with clinical trials in kidney transplantation and human brain cancer in the advanced planning stage.

What is the Problem? Cancer, in its multiple forms, is an interruption in the natural [i.e., inherent] organismic recognition system, which in

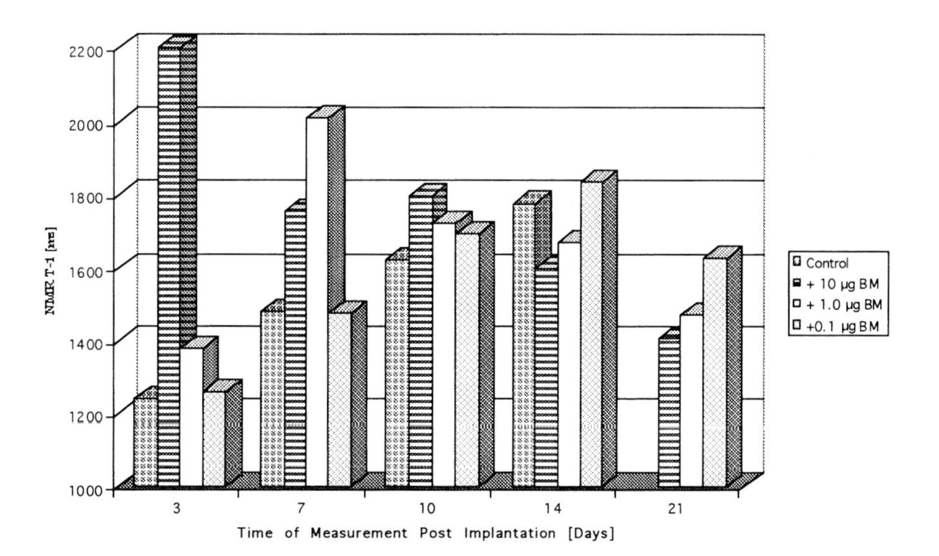

Figure 1. The effect of Biomodulator Treatment on host/tumor T-1

Table 1. T-1 measurements [rNU/CG]

Time Post Implantation [Days]	3	7	10	14	21
Control	1241 [n=3]	1483 [n=6]	1624 [n=6]	1778 [n=6]	all controls were sacrificed due to large tumor size
+ 10 μg BM	2202 [n=3]	1758 [n=6]	1798 [n=6]	1598 [n=6]	1409 [n=6]
[P to Control]	[P=0.007]	[P=0.001]	[P=0.019]	[P=0.008]	
+ 1.0 μg BM	1379 [n=3]	2012 [n=6]	1727 [n=6]	1674 [n=6]	1474 [n=6]
[P to Control]	[P=0.104]	[P=0.007]	[P=0.070]	[P=0.060]	
+ 0.1 μg BM	1262 [n=3]	1478 [n=6]	1697 [n=6]	1838 [n=6]	1631 [n=6]
[P to Control]	[P=0.612]	[P=0.94]	[P=0.98]	[P=0.15]	

turn is a fundamental communication among components of the host. Our hypothesis places the mechanism for aberrant growth development at the host/aberrance interface.

Initial Hypothesis. Biomodulation [a subset of BRMs] is the natural, inherent mechanism(s) by which the complex biological system manages information including recognition, coordination of effort, and integration of purpose. We have described several classes of Biomodulator [dissection of the whole for understanding only] which range from natural products [plant and microbial products] to their synthetic analogs (Mann *et al.*, 1991). This multi-faceted approach is considered important as it connects the definition of detailed mechanism(s), i.e. the development of the synthetics, to the inherent, endogenous control and integration process.

In Figure 1 and Table 1, we present data from a series of experiments designed to investigate the host/tumor interface. In this experiment canine glioma cells were implanted into the flank of rNU [T-cell deficient rats] and treatment groups were treated with intraperitoneal injections of Biomodulator [in this experiment pokeweed mitogen, a plant mitogen with differentiative as well as the traditional proliferative properties] at a number of concentrations [10, 1.0 and 0.1 ug/injection shown here]. The animals are then imaged on a 1.5 Tesla NMR imaging unit and the T-1 [T-1 is the spin-lattice relaxation time and relates to the exchange of nuclear magnetization in a direction parallel to the magnetic field] of the tumor was assessed. The opposite flank was used for a control [sham injected]. Normal control muscle tissue in this model yields approximately a 1200 ms T-1 value. The control animals with tumor growing in their flank show a slow increase in T-1 which mirrors the growth curve of the tumor itself [1241 ms on day four and 1778 ms on day 14]. The animals are sacrificed before the tumor becomes a health problem, and therefore, the day 21 data is absent. In other experiments however, the T-1 continues to increase as the tumor grows. The physiological meaning of increased T-1 is related to increased tumor growth with its vascularization and accumulation of excess water.

The 10 μg dose of Biomodulator shows a marked increase in the T-1 [2202 vs 1241 P=0.007] on day four. This is tumor-specific and independent of the injection process. The effect is also dose-dependent, time-dependent, and duration of treatment-dependent. The 1.0 μg dose shows a slightly lower T-1 [2014 vs 2202 for the 10 μg dose] response and a peak activity delay until day seven. The 0.1 μg dose is delayed even further until day 14. This Biomodulator related T-1 modulation is directly correlated with the eventual tumor regression. The changes in T-1 are also predictive of eventual tumor regression [day 21 to day 25].

From the perspective of our problem this data indicates that the Biomodulator has specificity for the tumor (Mann *et al.*, 1991). Once it traffics to the tumor interface it changes the physiological interaction between the host and aberrant tissue [increased in T-1]. This is even more striking when it is realized that on day four the actual tumor size is still microscopic. The T-1 measurements could signify an increased water concentration [water is assumed because it is the most prevalent proton source – but this is just an assumption] or an increased ordering of the proton sources already present. At one level this is recognition. It should be noted that if this is host/aberrance recognition then it is chemical in nature. In a series of histological examinations of the tumor/tissue interactions there is no indication of lytic activity at this stage. On the contrary, this purported recognition is just that – an induced heightening of an inherent difference which subsequently leads to host responsiveness [the outcome].

There appear to be at least three events which occur as a result of this purported recognition [note the use of a linear argument structure here]. The first event class has a short time frame [probably minutes] and encompasses fundamental, chemistry-based changes (Busse *et al.*, 1989). The Biomodulator traffics preferentially to the site of the host/aberrance interface, where it modulates the interface. This modulation is then the basis for the host recognition and subsequent responsiveness. What is the basis for this modulation? It is clearly not lytic or anti-metabolic like the traditional approaches. It appears to involve the arrangement of the cell surface oligosaccharides (Mann *et al.*, 1988). This conformation of cell surface oligosaccharides allows for the hypothesis that it is the arrangement of existing sugars which has changed and therefore caused the interruption of the recognition process rather than any overt change in the content of these cell surface structures which would require very significant changes in biochemistry. This hypothesis is supported by the NMR modulation of proton exchange rates and the very short time span for the biomodulation of this effect (Busse *et al.*, 1989). This purported structural change in cell surface oligosaccharide arrangement is also found as a component of "normal" regulatory cellular process(es) (Mann *et al.*, 1991, 1992). There is also an induced change at the physiology level between the normal and aberrant tissue. In other experiments (Bitner *et al.*, 1993) where radiolabeled injections of Biomodulator are given to animals with mature tumor loads, the data shows that a differential change in the rate of washout occurs between the aberrant and host tissues. This differential has been used to image aberrant tissue without the need for a tissue-specific labeled ligand.

In these systems the cell surface oligosaccharide display of cell lines [both normal and aberrant] was characterized and shown to have several novel features directly related to extended structure. In these models, using lectins and monoclonal antibodies as binding probes for individual carbohydrate specificities, it was shown that there are "super structures" of individual carbohydrate specificities into apparent "affinity classes". Using the equilibrium binding data from the lectin studies, it is possible to calculate an overall "barrier energy" for this super structure. When this is done normal cells can be shown to down-regulate this surface feature as a function of plating density and stage of differentiation (Mann *et al.*, 1992). Treatment with the Biomodulator modifies this oligosaccharide display [not the number but the arrangement of the carbohydrates] and this can be related to physical changes in exchange frequency of water at the cell surface (Busse *et al.*, 1989). These data suggest that the integrating factor in all of the Biomodulator treatment effects is the three-dimensional structural arrangement of the cell surface oligosaccharides. This is reminiscent of the Roseman hypothesis (Roseman, 1970), in which Saul Roseman hypothesized that basic communication [of which recognition is just a subset] is conducted between cells at the sugar level. There is an array of possible effector mechanisms; ligand:receptor, carbohydrate:carbohydrate interactions, intercalation of divalent cations within extended oligosaccharide structures and many others.

The second event class [hours to days post treatment] involves the host histological level. Studies of tissue taken from control and treatment groups support the concept developed above whereby the Biomodulator traffics to the host/aberrance interface [?specificity?], induces a physical change [T-1], which in turn results in a

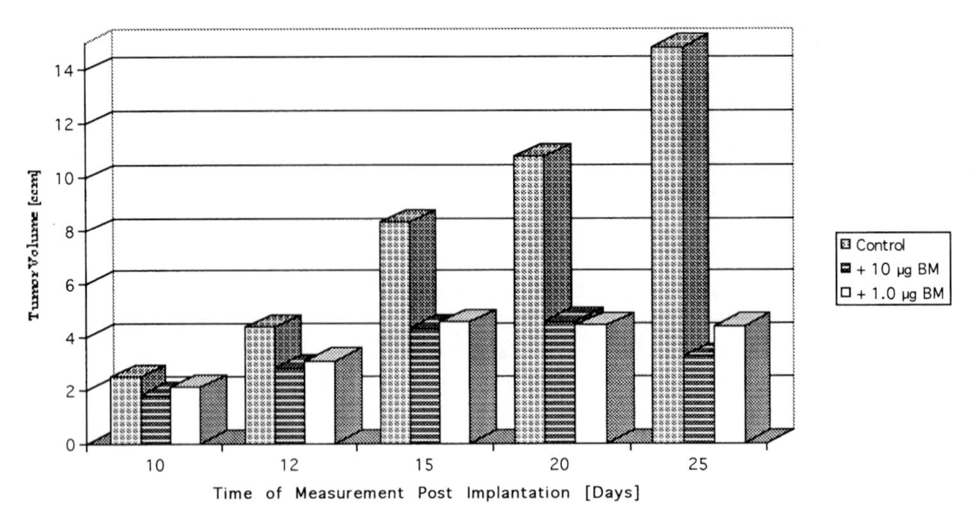

Figure 2. The effect of Biomodulators on tumor volume

tissue response. By 24 to 72 hours after the initiation of treatment the tissue interface has increased numbers of mast cells. Towards the end of this period there also appears to be a recruitment and/or differentiation stage at the tissue whereby there is a further increase of mast cells to the area. Again, this is specific to the host/aberrance interface and the effect is dose-dependent, time-dependent and duration of treatment-dependent. It should also be noted that during these first two phases, the tumors are growing [see Figure 2 and Table 2]. When the various growth curves are subjected to mathematical modeling it is clear that in both control and treatment there is a rapid growth phase which begins between five and seven days post tumor implantation. In the case of the treatment groups the slope of the curve is not as steep as in the case of the control, however, there is no significant difference in the volume of the tumor in any case until approximately day 15 post-implantation. This suggests that perhaps the effect of the Biomodulator is on tumor or host/tumor during the early stages of the initiation of tumor growth. Because this tumor is so aggressive in this rNU model it was not technically feasible to assess the Biomodulator effect after the tumor was well established [day 10 to 12]. The observations that there is a strict dose dependency and that there are titrable effects on both T-1 peak heights and tumor regression potential for the same sub-optimal dose of Biomodulator indicate that either there is a threshold to the host induction or the recognition phenomenon or both. It appears from the correlation between T-1 data and the tumor regression that the earliest event so far described must attain a specific level for maximal tumor regression to occur. If this is not attained, or delayed kinetically beyond a specific point, then the outcome [tumor regression] is proportionalized. This is a function of the model system [in terms of aggression in growth pattern] and therefore the emphasis of this approach to the problem should be the early events, and our understanding of the recognition process itself.

In other experiments with the same tumor implanted into canine brain, where the Biomodulator treatment could be delayed until NMR evidence of the implantation-induced neo-vascularization was obtained, similar results to those in the rNU model were observed. This again suggests that it is the recognition phenomenon which is critical and not the subsequent host effector mechanism(s). In the case of the canine implantation model, there is even less time for induction of the full-blown immune response before there is evidence of growth control by the host over the tumor.

Table 2. Growth of canine glioma tumor in rNU rats

Time Post Implantation [Days]	10	12	15	20
Control	2.474 [n=47]	4.361 [n=49]	8.294 [n=52]	10.787 [n=55]
+10 µg BM	1.808 [n=8]	2.835 [n=8]	4.336 [n=9]	4.58 [n=8]
[P to Control]	[P=0.512]	[P=0.324]	[P=0.049]	[P=0.007]
+1.0 µg BM	2.091 [n=8]	3.090 [n=8]	4.549 [n=8]	4.408 [n=8]
[P to Control]	[P=0.717]	[P=0.420]	[P=0.065]	[P=0.006]

The third event class occurs between day seven and day 19, and involves activation of host immune systems and the physical regression of the tumor in the treatment groups. During this phase of the Biomodulator effect, host immunocytes become activated in a number of ways. Splenocytes from treated rNU rats show increased T-cell-specific mitogen responses, increased humoral response [both specific and nonspecific], and increased cytotoxic T-cell lymphoctoxicity [CTL] and natural killer [NK] activity. Again, this effect appears to be both host and tumor specific. Treated tumor cells are more effective targets in NK assays, and treated leukocytes are more effective killers. These alterations in immune effector-function are correlated with changes in the tumor cell surface oligosaccharide display [the affinity of the surface increases in a dose-dependent manner with Biomodulator treatment and this is related to increased effector-function either in terms of target or effector-function]. All these traditional effects are observed concurrently with the tumor regression, and therefore, it is tempting to conclude that the Biomodulator effector mechanism is simply immune activation. This appears to be too simplistic. First, the real significance of the apparent immune activation is questionable, as all the histological studies performed on these *in vivo* models during the regression process never show massive cell death, or tissue necrosis, or any other signs of massive effector effects. Rather, the histology shows that the tumors regress from the host vascular bed in a manner more reminiscent of a mild delayed hypersensitivity response. There are accumulations of host cells in the area [typical of any mildly inflammatory condition], but not what would be expected if a 20 to 30 ccm tumor was treated with a chemotherapy agent. Second, the actual numerical analyses of the overall effect of an induced level of cytotoxicity from 30 % kill at an effector:target ratio of 100:1 to a 50 % kill at ratios of 10:1 is at best dubious. With tumor loads of 10–30 gms it is difficult to reconcile the ability of CTL to kill targets and the actual kinetics of tumor regression.

We hypothesize that the Biomodulator is inducing differentiation at both the host and tumor cell levels, and this is its major effect in terms of tumor regression. This may be a significant finding, not from the perspective of the tumor regression but more from the perspective that this is a host/aberrance recognition-based modulation.

Thus, even this rudimentary attempt at problem identification [multiple iterations of this process are required] has resulted in some interesting data, and, more importantly, more interesting questions. Is the host/aberrance interface the appropriate point of investigation? Are physical reactions [water/proton concentration and/or structure] the primary basis for biological recognition phenomena? Is the major integrator for the organismic system the cell surface diffusional barrier? Is this purported barrier under oligosaccharide conformational control? Can this approach, which apparently modulates biological integration, be used as a preventive and specific treatment modality?

References

Baca E, Mennin S, Kaufman A *et al.* (1990): Comparison between a problem-based, community-oriented track and a traditional track within one medical school in New Mexico. In *Innovation in Medical Education: An Evaluation of its Present Status.* (Khattab T, Schmidt H, eds) pp 9–26, New York: Springer Publ. Co..

Barak Y, Hahn T, Pecht M *et al.* (1992): Thymic humoral factor gamma-2, an immunoregulatory peptide, enhances human hematopoietic progenitor cell growth. In *Exp. Hematol.* **20**:173–7.

Biological Response Modifiers (1983): Sub-committee report. National Cancer Institute Monograph **63**. In *NIH Publication 83–2606*, Bethesda Maryland.

Bitner DM, Mann PL, D'Souza P *et al.* (1993). Enhanced tumor imaging with pokeweed mitogen. In *Nucl. Med. Biol.* **20**:203–6.

Busse, SC, Mann PL, Griffey RH (1989): Senescence-induced alterations in cell surface carbohydrates correlated using proton NMR spectroscopy and a lectin-based affinity binding assay. In *Biochim. Biophys. Acta* **984**:183–6.

Caroleo MC, Doria G, Nistico G (1994): Melatonin restores immunodepression in aged and cyclophosphamide treated mice. In *Ann. NY Acad. Sci.* **719**:343–52.

Coppi G, Falcone A, Manzardo S (1994): The protective effects of pidotimod against experimental bacterial infections in mice. In *Drug Res.* **44**:1417–21.

Faguet GB (1994): Chronic lymphocytic leukemia: an updated review. In *J. Clin. Oncol.* **12**:1974–90.

Feld R (1994): Complications associated with the treatment of small cell lung disease. In *Lung Cancer* **10**:307–17.

Hellstrom I, Hellstrom KE (1969): Studies in cellular immunity and serum mediated inhibition in moloney virus induced mouse sarcoma. In *Int. J. Cancer* **4**:587–92.

Hellstrom KE, Brown JP (1979): Tumor antigens. In *The Antigens* (Sela M, ed) pp 1–82, New York: Academic Press.

Jyonouchi H, Zhang L, Tomita Y (1993): Studies of immunomodulating actions of RNA/nucleotides. In *Pediatr. Res.* **33**:548–65.

Kantarjian HM, Estey EH, Keating MJ (1993): Treatment of therapy-related leukemia and myelodysplasia. In *Hematol. Oncol. Clin. North Am.* **7**:81–107.

Kaufman A. (1985): Implementing Problem Based Medical Education: Lessons from Successful Innovations. New York: Springer Publ. Co.

Kao SF, Kuo HL, Lee YC *et al.* (1994): Immunostimulation by *Alsophila spinulosa* extract fraction VII of both humoral and cellular immune responses. In *Anticancer Res.* **14**:2439–43.

LaPlace P-S (1749–1827): French mathematician and astronomer " We may regard the present state of the universe as the effect of its past and the cause of its future. An intellect which at any given moment knew all of the forces that animate nature and the mutual positions of the beings that compose it, if this intellect were vast enough to submit the data to analysis, could condense into a single formula the movement of the greatest bodies of the universe and that of the lightest atom; for such an intellect nothing could be uncertain and the future just like the past would be present before its eyes".

Mann PL (1993): Biomodulation: an alternate approach to disease management. In *Curr. Opinion Investig. Drugs* **2**:1–8.

Mann PL, Swartz CM, Holmes DT (1988): Cell surface oligosaccharide modulation during differentiation. IV. Normal and transformed cell growth control. In *Mech. Ageing Devel.* **44**: 17–30.

Mann PL, Eshima D, Bitner DM *et al.* (1991): Biomodulation: An integrated approach to access and manipulate biological information. In *Lectins and Cancer* (Gabius H-J, Gabius S, eds) pp 179–206, Heidelberg/New York: Springer Publ. Co..

Mann PL, Busse SC, Griffey RH *et al.* (1992): Cell surface oligosaccharide modulation during differentiation. VI. The effect of biomodulation on the senescent and the neoplastic cell phenotype. In *Mech. Ageing Devel.* **62**:79–90.

Mathe G, Amiel TL, Schwartzenberg L *et al.* (1969): Active immunotherapy for acute lymphoblastic leukaemia. In *Lancet* **1**:697–9.

McKeown T (1990): Determinants of health. In *The Nation's Health*, 3rd edition (Lee PR, Estes CL, eds) pp 6–13, Boston: Jones and Bartlett.

Mennin S, Friedman M, Woodward C (1992): Evaluating innovative medical education programs. In *Ann. Community-Oriented Ed.* **5**:123–33.

Mitchell MS (1992a): Principles of combining biomodulators with cytotoxic agents in vivo. In *Semin. Oncol.* **19**:51–6.

Mitchell MS (1992b): Biomodulators in cancer treatment. In *J. Clin. Pharmacol.* **32**:2–9.

Mitchell MS (1992c): Chemotherapy in combimation with biomodulation: a 5-year experience with cyclophosphamide and interleukin-2. In *Sem. Oncol.* **19**:80–7.

Obenshain S, Galey W (1992): In *The New Mexico Experience: Lessons Learned*. LCME, Chicago, IL, July meeting.

Petrov AB, Kolenko VM, Koshkina NV *et al.* (1994): Non-specific modulation of the immune response with liposomal meningococcal lipopolysaccharide: role of different cells and cytokines. In *Vaccine* **12**:1064–70.

Roseman S (1970): The synthesis of complex carbohydrates by multiglycosyltransferase systems and their potential function in intracellular adhesion. In *Chem. Phys. Lipids* **5**:270–97.

Takeda Y, Tanaka M, Miyazaki H *et al.* (1994): Analysis of effector T cells against the murine syngeneic tumor MethA in mice orally administered antitumor polysaccharide SPR-901. In *Cancer Immunol. Immunother.* **38**:143–8.

van der Heide JJ, Bilo HJ, Danker JM *et al.* (1993): Effect of dietary fish oil on renal function and rejection in cyclosporine treated recipients of renal transplants. In *N. Engl. J. Med.* **329**:769–73.

31 Glycobiology in Xenotransplantation Research

David K.C. Cooper and Raffael Oriol

1 Introduction

Organ allotransplantation (transplantation between members of the same species, e.g. human-to-human) is one of the success stories of the latter part of the twentieth century. It's role as a form of therapy in patients with end-stage disease of any vital organ is limited, however, by the number of donor organs that become available worldwide each year.

Data from the United Network for Organ Sharing (UNOS) in the USA indicate that some 35,000 patients awaited a donor organ at the end of 1994, and yet only 7000 organs became available during that year. The total number of patients on the UNOS waiting list is currently increasing by approximately 15 % per year, yet the number of donors available is increasing only marginally. Furthermore, the median waiting period for several types of donor organs has increased significantly during recent years. In particular, patients in need of livers or hearts are waiting approximately twice as long today as they did 5 years ago.

It has been estimated that the current need for donor organs worldwide is at least 50–60,000 organs per year (Cooper, 1993). However, this estimate does not take into account a number of important factors. The first of these is whether the number of patients on the waiting list truly reflects the number that might benefit from an organ transplant. There is considerable evidence that borderline candidates are not being added to the waiting list as surgeons are anxious to use donors only in those patients with an optimum chance of long-term survival. Secondly, there are a number of countries in which, for religious or cultural reasons, cadaveric organ donation is rare or nonexistent. Japan is the prime example of a country with advanced medical technological skills and yet where, for cultural reasons, cadaveric allotransplantation does not take place.

This worldwide donor shortage could be overcome if we were able to transplant animal organs into humans (xenotransplantation). The advantages are obvious. An unlimited number of donor organs would be available electively, and therefore patients would not have to wait in a debilitated state for months or even years for a suitable organ to become available. The operative procedures could be carried out on routine operating lists and not as emergency procedures in the middle of the night, as is frequently the case today. It could be ensured that the organs were totally viable as they would be excised from healthy anesthetized animals (rather than from brain-dead human donors on life-sustaining drugs), the ischemic time would be minimized (as organs would not need to be transported long distances from donor to recipient centers), and the donor animal could be screened for possible infection, which could be excluded (as is not always the case in human donors today). The ability to carry out a transplant electively (rather than maintain a transplant candidate in hospital for several weeks as is frequently the case today) would be associated with considerable reduction in costs.

2 Xenotransplantation – Basic Immunobiology

Transplantation across species barriers can be divided into that between closely related species

H.-J. and S. Gabius (Eds.), Glycosciences
© Chapman & Hall, Weinheim, 1997
ISBN 3-8261-0073-5

(concordant) (e.g. baboon-to-human) and that between widely-disparate species (discordant) (e.g. pig-to-human). After concordant xenografting, the rejection that occurs may be antibody-mediated or cellular (or mixed) in nature, and generally takes a few days to cause complete graft failure (but is more rapid than when allografting is carried out). After discordant xenotransplantation, rejection is uniformly hyperacute; it is antibody-mediated and occurs within a few minutes or hours of the transplant procedure.

To some extent, rejection following concordant xenografting can be controlled by the presently available pharmacologic drugs that are effective in patients with allografts. However, these drugs have no effect in preventing the hyperacute rejection that occurs between discordant species.

Although the immunologic barrier is less formidable in concordant xenografting, the relatively limited number of nonhuman primates that are closely related to humans, as well as several other factors, make it unlikely that these species will prove suitable for organ donation to humans on a large scale. Current research is therefore being directed towards the use of more widely disparate animals, such as the pig, as donors for humans. The pig has many advantages in this respect, which have been discussed elsewhere (Cooper *et al.*, 1991).

The major problem is, of course, the extremely rapid antibody-mediated rejection that occurs following the transplantation of a pig organ into a human (or into a nonhuman primate). We know that hyperacute rejection is the result of the presence of preformed "natural" antibodies in human serum directed against epitopes on the surface of pig vascular endothelium. The antibody-antigen reaction activates complement, which leads to endothelial cell destruction with massive interstitial hemorrhage and edema, resulting in rapid graft failure (Platt and Bach, 1991). Depletion or inhibition of the anti-pig antibodies prolongs graft survival (Cooper *et al.* 1988a; Alexandre *et al.*, 1989), by preventing activation of complement and possibly by prevention of other mechanisms that follow (Inverardi *et al.*, 1992). Inhibition of complement activation alone, however, prolongs graft survival for several days, but does not prevent the delayed vascular rejection that subsequently takes place (Leventhal *et al.*, 1993; Kobayashi *et al.*, 1995).

At the present time, it seems unlikely that antibodies can be depleted or inhibited permanently and therefore the long-term outcome of discordant xenografting remains unknown. However, there are considerable data available to us from allografting across the ABO histo-blood group barrier (in which the mechanism of rejection is similar to that in discordant xenografting) to suggest that even when the antibodies return to normal levels, rejection is not inevitable (Bach *et al.*, 1991). In order to understand the problems relating to discordant xenotransplantation and the potential role of glycobiology in research in this field, a brief description of our knowledge of allotransplantation across the ABO barrier is helpful.

3 Allotransplantation Across the ABO Histo-Blood Group Barrier

The histo-blood group A and B epitopes, against which anti-A and anti-B antibodies are directed, are carbohydrates, the structures of which have been accurately defined (Lloyd *et al.*, 1968; Watkins, 1972, 1974, 1980; Yamamoto *et al.*, 1990). The anti-ABH antibodies, like anti-pig antibodies, are not present at birth, but develop within a few weeks or months as soon as the neonate's gastrointestinal tract becomes colonized by micro-organisms (Wiener, 1951; Springer and Horton, 1969).

In approximately two-thirds of patients receiving organ transplants from an AB*O*-incompatible donor, hyperacute rejection occurs within a few hours (Cook *et al.*, 1987; Cooper, 1990). Long-term graft survival has been documented in the remaining one-third of patients, although the reasons and mechanisms whereby antibody-mediated rejection does not occur in this group remain unknown. A similar situation has been demonstrated after AB*O*-incompatible heart transplantation in the baboon, but in this species spontaneous survival of the organ occurs in approximately two-thirds of recipients (Cooper *et al.*, 1988b), as opposed to only one-third in humans.

There is evidence from both clinical (Alexandre *et al.*, 1985, 1987; Bannett *et al.*, 1987) and experimental (Cooper, 1992; Cooper *et al.*, 1992, 1993c) studies of allografting across the ABO blood group barrier that, if anti-A or anti-B natural antibodies can be depleted (or otherwise inhibited) for a relatively short, but hitherto uncertain, period of time (but days rather than weeks), then an AB*O*-incompatible organ transplanted during this "window" will not be hyperacutely rejected despite the subsequent return of normal levels of antibody and complement. This concept, first introduced by Alexandre *et al.* (1985,1987), has variously been termed anergy, adaptation, or accommodation (Bach *et al.*, 1991). The mechanism by which accommodation occurs remains unclear, but there is some evidence that the target antigens are masked or capped, possibly by blocking antibodies (Cooper, 1992).

The terminal oligosaccharides of the A and B epitopes have been synthesized (Lemieux, 1978; Pinto and Bundle, 1983) and, when attached to a solid support, can act as highly efficient immunoaffinity columns. Plasma or blood repeatedly passed through such a column is rapidly depleted of the respective antibody, as first demonstrated by Bensinger *et al.* (1981a,b) and Bannett *et al.* (1987).

Following successful preliminary studies by Romano *et al.* (1987a,b,c), in 1987 we began an exploration of the possibility of preventing hyperacute rejection of AB*O*-incompatible baboon cardiac allografts by the intravenous infusion of the specific A or B trisaccharide against which the anti-A or anti-B antibodies were directed. If the trisaccharide were infused continuously into the host circulation, the specific antibodies bound to the trisaccharide and were thus "neutralized" – that is, they were no longer freely available to "attack" the AB*O*-incompatible transplanted organ. In the anti-A or anti-B hypersensitized baboon receiving both trisaccharide infusion and conventional pharmacologic immunosuppressive therapy (cyclosporine, azathioprine or cyclophosphamide, and corticosteroids), survival of AB*O*-incompatible cardiac allografts was extended from a mean of 19 **minutes** to more than 28 **days** (Cooper, 1992; Cooper *et al.*, 1992, 1993c). Accommodation was clearly achieved in several cases. In one baboon, the graft functioned for >30 days after cessation of trisaccharide infusion.

As the trisaccharide is a small molecule, a hapten-like effect results, leading neither to immunogenicity nor complement activation. The trisaccharide is rapidly excreted in the urine and bile, immune complexes are not formed, and therapy in baboons for up to 19 days proved totally innocuous.

4 Identification of Human Anti-Pig Antibodies as Anti-α-Galactosyl Antibodies

In 1990, it occurred to us that similar techniques of antibody adsorption or "neutralization" would be possible if the antibody (or antibodies) that initiated discordant xenograft destruction also proved to be directed against a carbohydrate epitope. By repeatedly infusing human plasma through pig hearts or kidneys and eluting out the anti-pig antibodies, we were able to demonstrate that these belonged to three immunoglobulin classes – IgG, IgA and IgM (Koren *et al.*, 1992, 1993). These anti-pig antibodies were then tested against a panel of synthetic carbohydrate hapten-BSA conjugates and the specificity of the antibodies was determined. By far the strongest binding was to oligosaccharides with an α-galactosyl (αGal) terminal non-reducing residue, linked in 1–3 (Figure 1). This original finding was presented at the First Congress on Xenotransplantation held in Minneapolis in August, 1991 (Cooper, 1992; Good *et al.*, 1992), and has subsequently been reported in more detail (Cooper *et al.*, 1993a).

Some of the individual eluted antibody preparations also bound to other carbohydrate antigens, but none of these antigens was bound by significant levels of antibody in all preparations (Good *et al.*, 1992; Cooper *et al.*, 1993a).

Initial studies using the chromium (^{51}Cr) release assay with labeled target cells of pig kidney cell line LLCPK1, pig erythrocytes, and pig lymphocytes, demonstrated that cytotoxicity could be significantly reduced after passage of

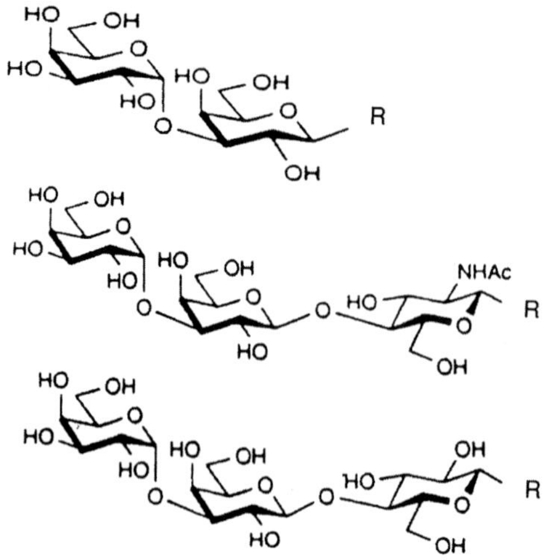

αGal(1→3)βGal-R

B disaccharide

αGal(1→3)βGal(1→4)βGlcNAc-R

linear B type 2

αGal(1→3)βGal(1→4)βGlc-R

linear B type 6

Figure 1. Chemical structures of B-disaccharide and linear B type 2 and linear B type 6 trisaccharides. R = $(CH_2)_8COOCH_3$.

human serum of any ABO group through an immunoaffinity column of either the αGal disaccharide or αGal trisaccharides illustrated in Figure 1 (Good *et al.*, 1992). Antibody-mediated lysis of sheep and bovine erythrocytes was also reduced by pre-adsorption of human plasma with matrix-bound αGal1–3Gal oligosaccharides, suggesting that human anti-αGal antibodies may be directed against the same αGal epitopes in several non-primate mammals. Lysis of pig kidney cells was also virtually abolished when human plasma was incubated with any one of the three soluble αGal1–3Gal oligosaccharides.

These initial experiments demonstrated some heterogeneity in the population, since the cytotoxic antibodies found in some plasma apparently required adsorption by the linear B type 2 trisaccharide (αGal1–3βGal1–4βGlcNAc-R) while the B αGal disaccharide (αGal1–3βGal-R) was sufficient for adsorption of the cytotoxic antibodies in other plasma. Similarly, the cytotoxicity of some sera was more markedly reduced by immunoadsorption with linear B type 2 trisaccharide than by linear B type 6 trisaccharide (αGal1–3βGal1–4βGlc-R), and vice versa.

A recent study by Cairns *et al.* (1995a) has confirmed this heterogeneity. They confirmed by ELISA that all 20 human sera tested had equivalent binding of IgM to three αGal antigens, namely the disaccharide (Galα1–3Galβ1-HSA), the trisaccharide (Galα1–3Galβ1–4GlcNAcβ1-HSA), and the pentasaccharide (Galα1–3Gal β1–4GlcNAcβ1–3Galβ1–4Glcβ1-HSA). However, with regard to IgG there were three distinct patterns of binding. Fifty-five percent of sera bound to the di-, tri- and pentasaccharides, with 35 % binding to the di- and pentasaccharides only, and 10 % binding to the disaccharide only. They therefore concluded that there are distinct subgroups of IgG anti-αGal antibodies. Further work in our own laboratory in the development of anti-idiotypic antibodies against anti-αGal antibody has also shown that at least two (and probably more) anti-αGal idiotypes can be demonstrated (Koren *et al.*, 1995).

Extensive subsequent studies utilizing a live/dead staining technique (Koren *et al.*, 1994) and pig kidney PK15 cells as well as pig vascular endothelial cells as the antigenic targets have confirmed that cytotoxicity of both human and baboon sera can only be significantly reduced or abolished by αGal structures, particularly those with an α1–3 linkage, and not by a large selection of other carbohydrates (Figure 2) (Neethling *et al.*, 1994). Recent *in vitro* studies have demonstrated considerable variation in the degree of cytotoxicity of individual human (Kujundzic *et al.*, 1994) and baboon (Neethling *et al.*, 1995c) sera towards cultured pig PK15 cells. Samples of human serum from 75 individuals showed widely

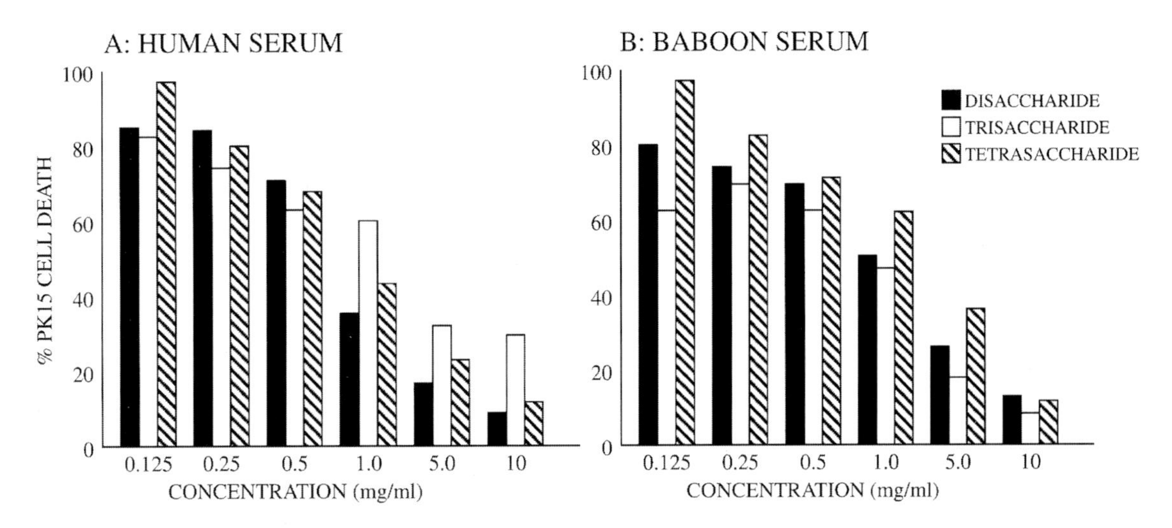

Figure 2. Reduction in cytotoxicity of (A) human and (B) baboon serum on PK15 cells after incubation of serum with increasing concentrations of three different αGal oligosaccharides (Dextra). B disaccharide = αGal1–3Gal; B trisaccharide = αGal1–3βGal1–4Gal; B tetrasaccharide = αGal1–3βGal1–4αGal1–3Gal. (From Neethling *et al.*, 1994). The tri- and tetrasaccharides studied have reducing ends different from the pig major αGal glycolipid αGal1–3βGal1–4βGlcNAc1–3βGal1–4βGlc. They reduce the cytotoxicity of the serum, but they are not as effective as the trisaccharide αGal1–3βGal1–4βGlcNAc with a structure similar to the terminal trisaccharide of the pig glycolipid (Table 2).

different cytotoxicity to pig cells, with PK15 cell death ranging from 1–100 % under standard conditions. The degree of cytotoxicity was associated with the titer of anti-αGal antibody, which also varied considerably among human sera. Approximately 2/3 of individuals tested, however, demonstrated >50 % cytotoxicity of their sera against PK15 cells.

The importance of anti-αGal antibodies in the hyperacute rejection of pig organs by human serum has subsequently been confirmed by several other groups (Fournier *et al.*, 1993; Galili, 1993; Karlsson *et al.*, 1993; Sandrin *et al.*, 1993; Cairns *et al.*, 1994a,b).

Anti-αGal antibodies were first described by Galili *et al.* (1984). Like the anti-ABH antibodies, it is suggested that they develop in neonates in response to the presence of certain microorganisms (Galili *et al.*, 1988a), and therefore probably play a role in the body's defense mechanisms against infection. Several strains of bacteria, virus and parasites are known to express αGal epitopes (discussed by Cooper *et al.*, 1994). Anti-αGal antibodies can only be detected in humans, apes and Old World monkeys and are not present in lower nonhuman primates or other mammals. The αGal antigen, however, is known to be present in lower mammals (Galili *et al.*, 1987, 1988b). The gene encoding for α1–3-galactosyl-transferase, which is essential for the production of αgalactose, is present in humans as a processed pseudogene (Joziasse *et al.*, 1991) or has undergone a frameshift mutation and is no longer functional (Larsen *et al.*, 1990).

Galili and his colleagues focused their studies primarily on autoimmune disease (Galili, 1991), but in 1993 they also presented evidence confirming that anti-αGal antibodies play a significant role in the hyperacute rejection of transplanted pig organs (Galili, 1993).

The presence of anti-αGal antibodies in Old World monkeys, including baboons, makes these ideal experimental animals for the study of discordant xenotransplantation using pig organs. We have found similar variations in serum cytotoxicity and anti-αGal antibody level in baboons as in humans (Neethling *et al.*, 1995c). We have confirmed a low level of anti-αGal antibodies in

Table 1. Concentrations (in uM) of various oligosaccharides needed to obtain 50 % inhibition of cytotoxicity of unmodified human or baboon serum on pig kidney (PK15) cells

OLIGOSACCHARIDE	SERUM	
	Human	Baboon
1. ***Fucα1–2**Gal*β1-R	>10000	>10000
2. Gal*β**1-R**	>10000	>10000
3. Galα1-***2Galβ1-R'**	7000	>10000
4. Galα1–3Gal	386(±149)	301(±44)
5. Galα1–3Galβ1–4***Gal**	163(±73)	141(±60)
6. Galα1–3Galβ1–4***Galα1–3Gal**	54(±31)	119(±30)
7. Galα1–3Galβ1–4GlcNAc	27(±11)	31(±4)

* Bold type indicates structural differences of the oligosaccharide with themajor pig vascular endothelium glycolipid Galα1–3Galβ1–4GlcNAcβ1–3Galβ1–4Glc β1-Cer; R represents 1–3 or 1–4 linkages to Gal or to GlcNAc; R' is -$O(CH_2)_3NHCOCF_3$. The results of the strong inhibitors (4–7) are expressed as mean ± SD (n=3). (From Neethling *et al.*, 1995b).

baboons 6 weeks old and this correlates with reduced cytotoxicity of their sera to PK15 cells and a delay in rejection of pig cardiac xenografts in these animals (Neethling *et al.*, 1995a).

Studies using a modified PK15 cell cytotoxicity assay, that is both rapid and inexpensive (Neethling *et al.*, 1995d), have demonstrated that the synthetic monosaccharide, methyl-α-Gal, gives only weak inhibition of the cytotoxic reaction of human and baboon natural antibodies on pig kidney PK15 cells in culture. The disaccharide, Galα1–3Gal, substantially inhibits the cytotoxic reaction, with 50 % inhibition at a concentration of 400 μM, and the trisaccharide, Galα1–3Galβ1–4GlcNAc, is 10 times more efficient (50 % inhibition at a concentration of 30 μM) (Table 1) (Neethling *et al.*, 1995b). Modification from α to β anomeric configuration of the non-reducing end of the oligosaccharide results in a complete loss of inhibitory activity, while substitutions at the reducing end of the oligosaccharide induce only a partial loss of activity. These observations suggest that natural anti-αGal antibodies recognize the oligosaccharide epitope from its non-reducing end, but that substitutions at the reducing terminus can also modify the antibody-binding capacity. The reducing end of the oligosaccharide therefore contributes some energy to the reaction with the antibodies, indicating that certain oligosaccharides will be of more potential clinical use than others.

5 Identification of Oligosaccharide Epitopes on Pig Vascular Endothelium

In 1992 we turned our attention towards confirming the nature of the carbohydrates that are present on the vascular endothelium of the pig. Other pig tissues were also studied in detail, but it is clearly the vascular endothelium which is the first structure that comes into contact with the host xenoreactive antibodies. Staining of pig vascular endothelium utilizing a panel of carbohydrate-specific lectins and immunoaffinity antibodies clearly demonstrated the presence of 3 main car-

Table 2. Structure of the main carbohydrate epitopes exposed at the surface of human and porcine vascular endothelium (only the epitopes shown in bold type and underlined are different between the two species)

Human	Pig
βGal1-4βGlcNAcl-R*	βGal1-4βGlcNAc1-R
ABH-βGal1-4βGlc NAc1-R	**αGal1-3**βGlcNAc1-R
αNeuAc2-3βGal1-4βGalNAc1-R	αNeuAc2-3βGal1-4βGlcNAc1-R

* R are glycolipid or glycoprotein carrier molecules anchored in the cell membrane. (From Cooper *et al.*, 1994)

bohydrate epitopes (Table 2) (Oriol *et al.*, 1993). As we have found no evidence that human or baboon plasma contain large amounts of antibodies directed against sialic acid or lactosamine, and since human tissues, including the vascular endothelium (Table 2), contain both of these carbohydrates, it seems unlikely that either of these last two epitopes plays a role in the vascular rejection that takes place when pig organs are transplanted into primates.

The unfucosylated monomorph linear B (αGal) antigen was found at the surface of all porcine vascular endothelial cells. This pig αGal antigen reacts strongly with the anti-αGal isolectin B4 from *Griffonia simplicifolia 1* (GS_1B_4) and with human natural anti-αGal antibody specifically purified by affinity chromatography on synthetic oligosaccharides containing the terminal non-reducing αGal1–3βGal-R disaccharide. These antigenic determinants are destroyed by the treatment of pig tissues with α-galactosidase, which removes the terminal non-reducing αGal. This completely abolished the positive reactions obtained with the GS_1B_4 isolectin, which is specific for the αGal1–3βGal-R structure. The localization of this epitope on vascular endothelium and its reactivity with natural human anti-αGal antibodies suggest that it plays a major role in hyperacute vascular rejection of pig-to-human organ xenografts.

Anthropoid apes have the same epitopes as humans in vascular endothelium (Oriol *et al.*, 1984). Baboons, therefore, are very suitable as recipients in discordant xenotransplantation animal models because they have ABH antigens on vascular endothelium and have natural circulating anti-αGal antibodies as do humans.

We have investigated the vascular endothelial cells from organs from several breeds of domestic pigs and wild boars, and they were all found to be positive when tested with affinity-purified human anti-αGal and the GS_1B_4 lectin (Oriol *et al.*, 1994). We therefore conclude that the αGal epitope of pig vascular endothelium is either monomorphic or at least has a high incidence in the porcine species, since we have not found any pig negative for this antigen. However, it may be worth looking for αGal-negative pigs, because such animals could exist at a very low frequency. Such pigs might be compared with humans with the Bombay phenotype, the incidence of which is estimated to be 1:10000. The breeding of such αGal-negative pigs, if they are ever found, would be easier than producing αGal-negative pigs by genetic manipulation.

Samuelsson and his colleagues in Goteborg have confirmed the structure of the epitope on glycolipids extracted from porcine aorta (Karlsson *et al.*, 1993), and have determined the complete pentasaccharide structure of the major kidney αGal neutral glycosphingolipid to be αGal1–3βGal1–4βGlcNAc1–3βGal1–4βGlc1–-1Ceramide (Backer *et al.*, 1993; Samuelsson *et al.*, 1994). The pig αGal epitope has been independently identified in both glycolipids (Backer *et al.*, 1993; Cairns *et al.*, 1994b) and glycoproteins (Vaughan *et al.*, 1993).

In a search for possible αGal-negative animals, we have studied oligosaccharide expression on the vascular endothelium (and other tissues) of more than 20 different animal species, ranging from primates to nonprimate mammals to birds and reptiles (Oriol *et al.*, 1995). The vascular endothelia of most mammals express a monomorphic αGal epitope reacting only with GS_1B_4, while that of birds, including large ratites (e.g. the ostrich) express a βGal epitope (reacting only with *Arachis hypogaea* (PNA)). In some species, these oligosaccharide epitopes are predominantly expressed on vascular endothelial cells, while in others they are better expressed on red blood cells, or are similarly expressed on both. Epithelial cells producing exocrine secretions also express different oligosaccharide epitope patterns, each species showing a specific pattern of reactivity, suggesting that different oligosaccharide epitopes have been selected through evolution by each species.

Although the ostrich or emu, or even the alligator, can be "farmed" in large numbers in captivity, and therefore might be considered as potential donors of some organs for humans, the macroscopic and microscopic structure of their organs show significant differences from those of humans, and it is unlikely that they will prove to be of clinical use.

6 Experimental Studies in Baboons

Having identified both the host antibody and the donor epitope, and shown *in vitro* that the antibody could be readily adsorbed from human or baboon plasma by an immunoaffinity column of αGal trisaccharide and, furthermore, that the adsorbed plasma was not cytotoxic to pig PK15 cells but that the anti-α-Gal antibodies eluted from the column were highly cytotoxic (Neethling *et al.*, 1994), an effort to perform *in vivo* studies was undertaken.

Plasma taken from baboons that were (i) unmodified, (ii) splenectomized, (iii) receiving conventional pharmacologic immunosuppressive therapy, or (iv) splenectomized and receiving immunosuppression, was highly cytotoxic to pig kidney PK15 cells. The continuous intravenous infusion of carbohydrates devoid of terminal αGal, such as dextrose, dextran or mannitol, even in very high concentrations (of up to 40 g/h) had no effect on reducing cytotoxicity (Ye *et al.*, 1994a,b).

Initially, neither the αGal disaccharide nor trisaccharides were available to us in the large quantities required for extracorporeal immunoabsorption or continuous intravenous infusion in adult baboons. As our *in vitro* studies had demonstrated that melibiose (αGal1–6Glc) and arabinogalactan (an impure polysaccharide of plant origin with an αGal terminal residue) had some protective effect from the cytotoxic activity of both human and baboon sera (Neethling *et al.*, 1994), we chose to infuse these, even though they were effective *in vitro* only at 50–100 times the concentration of the αGal disaccharide.

Infusion of melibiose and/or arabinogalactan at extremely high concentrations (up to 50g/h) completely eliminated the cytotoxicity of the serum to PK15 cells in 4 of 15 baboons, and greatly reduced it in all other cases. However, if continued for more than a few days, it led to toxic effects manifest by hemorrhage in the kidneys and lungs. Neither melibiose nor arabinogalactan, therefore, were suitable to allow experimental pig organ transplantation to be successful in the non-human primate.

Pig hearts were transplanted into 4 baboons, 3 of whom received melibiose +/– arabinogalactan infusions. There was increased graft survival in one baboon where cytotoxicity was demonstrated to be 5 % of control, but this baboon died of the toxic effect of the infusion within 18 hours.

In an *ex vivo* study from another center, in which human blood was perfused through pig kidneys, the addition of αGal1–3Gal disaccharide (at 70 mM) completely inhibited red cell occlusion of glomerular, but not of intertubular, capillaries, although there was residual platelet thrombus in the glomeruli (Cairns *et al.*, 1995b). The authors of this study concluded that the αGal disaccharide and trisaccharide could be used (in concentrations shown for other carbohydrate inhibitors to be non-toxic) for inhibition of hyperacute pig-to-human xenograft rejection. The inhibition was incomplete, however, and other antigen specificities and other rejection mechanisms were thought likely to be involved.

Extracorporeal immunoadsorption of baboon plasma, utilizing immunoaffinity columns of melibiose on 4 consecutive days, reduced cytotoxicity to PK15 cells to 20 % (Ye *et al.*, 1994a,b). After the fourth immunoadsorption, the addition of an arabinogalactan or melibiose intravenous infusion reduced serum cytotoxicity further to 2–12 %.

Based on the *in vitro* studies of Rieben *et al.* (1995), continuing *in vivo* studies in our laboratory of the daily extracorporeal immunoadsorption of baboon plasma through a column of αGal disaccharide has demonstrated complete loss of serum cytotoxicity towards PK15 cells after 2 days therapy in a baboon that was also receiving pharmacologic immunosuppression (in the form of cyclosporine, cyclophosphamide and methylprednisolone) (Taniguchi *et al.*, unpublished). Minimal cytotoxicity was noted for a subsequent period of approximately 7 days after which antibody levels began to rise. Although no pig heart transplant has been carried out in any of these experiments to-date, the good correlation we have found previously between the PK15 cell assay and pig graft survival suggests that graft survival would be significantly prolonged.

7 Clinical Relevance of Anti-α-Gal Antibody in Xenotransplantation

It would appear, therefore, that hyperacute rejection is primarily initiated by preformed anti-α-Gal antibodies present in human (and ape and Old World monkey) serum directed against αGal epitopes on the surface of the vascular endothelium of the pig. Present evidence is that no other antibodies play a major role, although this cannot be totally excluded in all cases. Activation of the complement system follows, leading to rapid destruction of the transplanted organ. There may also be other mechanisms by which antibody leads to early graft destruction (Inverardi *et al.*, 1992; Leventhal *et al.*, 1993; Kobayashi *et al.*, 1995). For example, when complement activity is completely negated by the administration of cobra venom factor to the host, features suggestive of vascular or mixed vascular/cellular rejection, although not hyperacute, still develop within a few days.

It should therefore prove possible to develop immunoaffinity columns of one or more αGal oligosaccharides capable of being used in an extracorporeal immunoadsorption system to deplete anti-α-Gal antibodies from the host, during which period the pig organ would be transplanted into the human recipient. In addition (or alternatively), the continuous intravenous infusion of one or more αGal-terminated oligosaccharides into the host to "neutralize" any remaining low levels of antibody (or new antibody that is produced) would be beneficial until the organ is "accommodated" and is no longer susceptible to hyperacute rejection.

Pharmacologic immunosuppressive therapy will clearly be required to prevent the subsequent development of cellular rejection and perhaps also to reduce the production of new antibody by B lymphocytes, particularly during the early post-transplant period while accommodation is taking place. If accommodation proves not to be a mechanism by which discordant xenotransplantation can be achieved, then acceptance of the discordant xenograft might still be possible if, following removal of the anti-pig antibodies, their titer could be maintained at very low levels by one or a combination of immunosuppressive drugs (discussed by Cooper *et al.*, 1994).

The major remaining problem before these theories can be tested fully in the baboon, and before a clinical trial can be considered, relate to the availability of relatively large amounts of the αGal oligosaccharides required. Clearly, relatively little oligosaccharide would be required to produce an immunoaffinity column. However, if the oligosaccharide were to be infused continuously to "neutralize" the anti-α-Gal antibodies, it is estimated that something in the region of 500 mg/hr might be required to bring about the desired effect. Based on our similar studies in the AB*O*-incompatible baboon, it is likely that the infusion would need to be continued for perhaps 10–14 days.

Synthetic methods of producing oligosaccharides clearly have significant limitations in the production of large quantities of the oligosaccharide, and therefore another source needs to be sought. The most likely source will be from the enzymatic production of oligosaccharides, which is a rapidly developing methodology. Alternatively, as pig gastric mucin is readily available in large quantities at minimal cost, this may prove a major source of such oligosaccharides. In a collaborative study at our own center, in the laboratory of Dr. Richard Cummings, we have identified a subfraction of pig gastric mucin that is highly potent in inhibiting the cytotoxicity of human and baboon sera towards pig PK15 cells (Li *et al.*, 1995).

We have purified neutral oligosaccharides from porcine stomach mucin and tested their efficiency at neutralizing human anti-αGal antibody. In ELISA assay, these oligosaccharides were able to inhibit the binding of purified anti-αGal antibody to mouse laminin, which contains α-galactosyl determinants. When the oligosaccharide was size-fractionated, the larger oligosaccharides were more potent in inhibiting anti-α-Gal antibodies than the smaller. The inhibitory activity was abolished by treating the oligosaccharide with α-galactosidase. β-Galactosidase had no effect. To further isolate the oligosaccharides that specifically bind anti-αGal antibody, purified anti-αGal IgG was immobilized. Oligosaccharides that bound the anti-αGal IgG column, representing

1% of the total, were nearly 1000-fold more potent in inhibiting anti-α-Gal antibodies than the unbound oligosaccharides. When added to baboon or human sera, this subfraction completely protected PK15 cells from lysis in concentrations as low as 5 and 250 μg/ml, respectively. These porcine stomach mucin-derived oligosaccharides are the most potent inhibitors of anti-α-Gal antibodies that have to-date been identified.

Finally, it may be possible to produce the required oligosaccharide by the genetic engineering of certain plants that are known to produce sugars in large quantities.

8 The Genetically Engineered α-Gal-Negative Pig

8.1 Deletion of the Gene Encoding α1,3-Galactosyltransferase

Three groups have independently suggested that the ultimate solution to the problem of discordant xenotransplantation might be to genetically engineer a pig that does not express αGal epitopes on its vascular endothelium (Cooper *et al.*, 1993b; Galili, 1993; Sandrin *et al.*, 1993). The expression of terminal α-galactose is based on the proper function of a single gene encoding for the enzyme α1,3-galactosyltransferase (α1,3GT) (Spiro and Bhoyroo, 1984; Galili *et al.*, 1988b). If this could be rendered nonfunctional or dysfunctional, then there would be no target for human anti-αGal antibodies. To prevent expression of the α1,3GT, the gene would need to be deleted, interrupted, or replaced, either within the coding region or within the regulatory sequences, so that enzyme is not produced. This might be achieved either by homologous recombination of the gene in embryonic stem cells to eliminate expression of the gene, or by microinjection of antisense cDNA constructs into embryos to inactivate or decrease expression of α1,3GT.

Genes encoding α1,3GT from various species have been identified (Joziasse *et al.*, 1989; Larsen *et al.*, 1989; Lowe *et al.*, 1989), including a pig genomic DNA clone (Dabkowski *et al.*, 1993; Sandrin *et al.*, 1993). Although techniques to inactivate or "knock out" the expression of the galactosyltransferase gene in pigs have not yet been demonstrated, such techniques are established in the mouse and, with the current rate of advancement in the field of genetic engineering, it is likely that they will prove feasible in swine in the near future. The use of pigs devoid of αGal epitopes would certainly be expected to help in preventing hyperacute vascular rejection, and may also allow for the detection of other targets which may be masked by the violence of the reaction of the anti-αGal antibodies.

Mice that do not express αGal, however, have been reported to have a propensity to certain anatomical defects, such as cataracts that form at a very early age (d'Apice AJF, 1995, personal communication), and it therefore remains uncertain whether similarly genetically engineered pigs will be fully viable. By analogy with humans with the Bombay phenotype, who express no ABH epitopes (the equivalent to αGal-negative pigs) and yet who remain fully healthy, however, such αGal-negative pigs may be fully viable (see also chapter of Hathaway and Shur, this volume).

8.2 Increased Expression of Alternate Epitopes

Alternative methodologies to produce animals with altered expression of α1,3GT have been suggested (discussed by Cooper *et al.*, 1994). For example, the introduction of a gene for an enzyme that could mask the αGal epitope might prove feasible. The DNA encoding for another enzyme (such as fucosyltransferase or sialyltransferase) that would modify the sugar structures could be inserted into the embryo where it would be incorporated into the chromosomes and expressed to modify or mask the immunoreactivity of the αGal structure on the cell surfaces (Table 3).

Good progress in this respect is being made in mice by McKenzie's group in Australia, who have inserted the gene encoding the H blood group antigen, which encodes a fucosyltransferase and uses the same substrate as does α1,3GT (namely βGal1–4GlcNAc) (McKenzie *et al.*, personal communication).

Table 3. Fucosylation or sialylation of N-acetyl-lactosamine (βGal1–4βGlcNAc-R) can impair the formation of terminal non-reducing αGal epitopes because the presence of fucose or sialic acid represents a steric hindrance for the enzyme acitvity of the pig α(1,3)galactosyltransferase. (Added fucose or sialic acid are shown in bold type and underlined)

Chemical structure	Epitope name
αFuc1–2βGal1–4 βGlcNAc-R*	H type 2
βGal1–4βGlcNAc-R 1–3**αFuc**	Lewis^x (Le^x)
αNeuAc2–3βGal1–4 βGlcNAc-R 1–3**αFuc**	Sialyl-Lewis^x
αNeuAc2–3βGal1–4 βGlcNAc-R	Sialyl-3-lactosamine
αNeuAc2–6βGal1–4 βGlcNAc-R	Sialyl-6-lactosamine

* R are glycolipid or glycoprotein carrier molecules anchored in the cell membrane. (From Cooper *et al.*, 1994)

Even a very slight change of carbohydrate epitope could successfully reduce antibody binding. For example, in our own studies, the oligosaccharide αGal1–4βGal-R (the P1 antigen) had virtually no effect in reducing the cytotoxicity of human plasma on pig kidney cells (Neethling *et al.*, 1994), and therefore a change of epitope from αGal1–3βGal-R to αGal1–4βGal-R on pig vascular endothelium may be sufficient to prevent hyperacute rejection of a transplanted pig organ. A well-known illustration of this point is the A and B histo-blood group epitopes that differ by only one *N*-acetyl group on a single carbohydrate unit, yet show a completely different immunogenicity.

9 Other Therapeutic Approaches – Gene Therapy

The possibility of removing αGal from the surface of the cells by treating the organ with α-galactosidase after excision from the pig, but before transplantation into the human recipient, has been considered. Extensive removal of a specific type of sugar from a complex cell-surface "glycocalyx" is, however, considered virtually impossible (Varki, 1992), even using large quantities of enzyme and/or extended incubation periods. Furthermore, after transplantation the natural glycosyltransferase would soon replace the removed sugar.

Recent developments in gene therapy techniques, however, suggest possibly simpler and less expensive methods for depleting or masking the αGal epitope than are required by genetic engineering. The introduction by transfection of a gene encoding an α-galactosidase (Den Herder *et al.*, 1992) into the vascular endothelium of the donor organ would theoretically prevent the production and expression of αGal in those tissues. Alternatively, transfer of the H gene would lead to expression of the H blood group oligosaccharide (αFuc1–2βGal1–4βGlcNAc) and reduce (or completely prevent) expression of the αGal epitope. A third possibility might be the introduction of the α1,4GT gene which would compete with the α1,3GT gene, as discussed above.

Relatively simple techniques of introducing a gene directly into the endothelium of an organ, either *in vivo* (Lim *et al.*, 1991) or *ex vivo* (Ardehali *et al.*, 1995) by organ perfusion of the arterial system (e.g. of the heart by the coronary circulation) have recently been described. The selected gene could therefore be introduced into the pig organ's arterial system either by the use of a balloon catheter before or immediately after excision. Such gene function in the vascular endothelium would be anticipated to persist for some weeks, which should be sufficient time to allow accommodation to develop.

10 Comment

The work by our group has clearly been directed towards inhibiting or blocking the binding of anti-αGal antibodies to the αGal epitopes on the surface of the pig vascular endothelium. Oligosaccharides could, however, be involved in a second, non-specific anti-inflammatory therapeutic approach, for example, by blocking the selectin-mediated leukocyte adhesion or by modifying complement.

There are other problems relating to xenotransplantation that are receiving attention. These include the possibility of transferring infection with the donor organ to the human recipient, infections that have been termed "xenozoonoses". These would appear likely to be less of a problem with regard to pig organs than to nonhuman primate organs (Ye *et al.*, 1994c).

Finally, it remains unknown if pig organs will function adequately in the human "milieu interieur". We already have evidence that organs such as pig kidneys (Alexandre *et al.*, 1989) and heart (Kobayashi *et al.*, 1995) will function in nonhuman primates for periods in excess of three weeks, and therefore there is some hope that these organs will support humans for prolonged periods of time. However, it seems inconceivable that the pig liver will carry out all of the functions required of the human liver, and this may clearly limit the value of liver xenotransplantation, although genetic engineering techniques may also help to produce pig livers that will manufacture enzymes and other products that will function in humans.

Glycobiology is therefore playing a major role in our research into the antibody-mediated rejection that takes place after discordant xenotransplantation. This field of research has been, and will continue to be, an exciting area in which to participate for the foreseeable future. The benefits of a successful breakthrough are clearly substantial. Our ability to perform clinically successful discordant xenotransplantation of organs, using the pig or other suitable non-primate mammal as a donor, will result in an enormous increase in the number of organ transplants performed worldwide each year. Patients will be referred earlier in their disease process, will undergo elective transplantation, and retransplantation when indicated. Patients with relative contraindications or in an advanced state or organ failure will not be so readily denied transplantation. Clinical transplantation will expand rapidly in Japan and several other countries. Furthermore, if every diabetic patient currently requiring insulin could receive a xeno-islet cell graft instead (Rydberg *et al.*, 1994), the demand for xenografts will clearly expand significantly.

At this relatively early stage of its development, it is difficult for us to envisage the role that xenotransplantation may play in clinical medicine in 40–50 years' time. The ready availability of an organ to replace a diseased one will prove too great a temptation to the average patient or physician to allow either to persevere with inadequate medical therapy, including dialysis, that maintains the patient in a suboptimal quality of life.

References

Alexandre GPJ, Squifflet JP, De Bruyere M *et al.* (1985): Splenectomy as a prerequisite for successful human AB*O*-incompatible renal transplantation. In *Transplant. Proc.* **17**:138–43.

Alexandre GPJ, Squifflet JP, De Bruyere EM *et al.* (1987): Present experiences in a series of 26 AB*O*-incompatible living donor renal allografts. In *Transplant. Proc.* **19**:4538–42.

Alexandre GPJ, Gianello P, Latinne D *et al.* (1989): Plasmapheresis and splenectomy in experimental renal xenotransplantation. In *Xenograft* **25** (Hardy MA, ed) pp 259–66, New York: Elsevier.

Ardehali A, Fyfe A, Laks H *et al.* (1995): Direct gene transfer into donor hearts at the time of harvest. In *J. Thorac. Cardiovasc. Surg.* **109**:716–9.

Bach FH, Platt JL, Cooper DKC (1991): Accommodation – the role of natural antibody and complement in discordant xenograft rejection. In: *Xenotransplantation* (Cooper DKC *et al.*, eds) pp 81–99, Heidelberg: Springer.

Backer AE, Holgersson J, Karlsson H *et al.* (1993): Structural characterization of glycosphingolipids in different organs of a semi-inbred pig strain using GC/MS of ceramide glycanase-released oligosaccharides. Presented to the *Second International Congress on Xenotransplantation*, p 52 (abstract 24), Cambridge, UK.

Bannett AD, McAlack RF, Raja R *et al.* (1987): Experiences with known AB*O*-mismatched renal transplants. In *Transplant. Proc.* **19**:4543–6.

Bensinger WI, Baker DA, Buckner CD *et al.* (1981a): Immunoadsorption for removal of A and B blood group antibodies. In *N. Engl. J. Med.* **304**:160–2.

Bensinger WI, Baker DA, Buckner CD *et al.* (1981b): *In vitro* and *in vivo* removal of anti-A erythrocyte antibody by adsorption to a synthetic immunoadsorbent. In *Transfusion* **21**:335–42.

Cairns T, Hammelmann W, Gray D *et al.* (1994a): Enzymatic removal from various tissues of the galactose-α1,3-galactose target antigens of human anti-species antibodies. In *Transplant. Proc.* **26**:1279–80.

Cairns T, Karlsson E, Holgersson J *et al.* (1994b): Confirmation of a major target epitope of human natural IgG and IgM anti-pig antibodies: terminal galactose-α1,3-galactose. In *Transplant. Proc.* **26**:1384.

Cairns T, Lee J, Goldberg L *et al.* (1995a): Polymorphism within the human IgG anti-pig repertoire. Presented to the British Transplantation Society, Manchester, UK.

Cairns T, Lee J, Goldberg L *et al.* (1995b): Inhibition of the pig-to-human xenograft reaction, using soluble Galα1–3Gal and Galα1–3Galβ1–4GlcNAc. submitted.

Cook DJ, Graver B, Terasaki PI (1987): AB*O*-incompatibility in cadaver donor kidney allografts. In *Transplant. Proc.* **19**:4549–52.

Cooper DKC (1990): Clinical survey of heart transplantation between AB*O*-blood group incompatible recipients and donors. In *J. Heart Transplant.* **9**:376–81.

Cooper DKC (1992): Depletion of natural antibodies in non-human primates – a step towards successful discordant xenografting in humans. In *Clin. Trans.* **6**:178–83.

Cooper DKC (1993): Xenografting: How great is the clinical need? In *Xeno* **1**:25–6.

Cooper DKC, Human PA, Lexer G *et al.* (1988a): Effects of cyclosporine and antibody adsorption on pig cardiac xenograft survival in the baboon. In *J. Heart Transplant.* **7**:238–46.

Cooper DKC, Lexer G, Rose AG *et al.* (1988b): Cardiac allotransplantation across major blood group barriers in the baboon. In *J. Med. Primatol.* **17**:333–46.

Cooper DKC, Ye Y, Rolf LL Jr. *et al.* (1991): The pig as potential organ donor for man. In *Xenotransplantation* (Cooper DKC *et al.*, eds) pp 481–500, Heidelberg: Springer.

Cooper DKC, Ye Y, Kehoe M *et al.* (1992): A novel approach to "neutralization" of preformed antibodies: cardiac allotransplantation across the AB*O*-blood group barrier as a paradigm of discordant xenotransplantation. In *Transplant. Proc.* **24**:566–71.

Cooper DKC, Good AH, Koren E *et al.* (1993a): Identification of αgalactosyl and other carbohydrates that are bound by human anti-pig antibodies: relevance to discordant xenografting in man. In *Transplant. Immunol.* **1**:198–205.

Cooper DKC, Koren E, Oriol R (1993b): Genetically engineered pigs. In *Lancet* **342**:682–3.

Cooper DKC, Ye Y, Niekrasz M *et al.* (1993c): Specific intravenous carbohydrate therapy – a new concept in inhibiting antibody-mediated rejection: experience with AB*O*-incompatible cardiac allografting in the baboon. In *Transplantation* **56**:769–77.

Cooper DKC, Koren E, Oriol R (1994): Oligosaccharides and discordant xenotransplantation. In *Immunol. Rev.* **141**:31–58.

Dabkowski PL, Vaughan HA, McKenzie IFC *et al.* (1993): Characterization of a cDNA clone encoding the pig α1,3galactosyltransferase: implications for xenotransplantation. In *Transplant. Proc.* **25**:2921.

Den Herder IF, Mateo Rosell AM, Van Zuilen CM *et al.* (1992): Cloning and expression of a member of the *Aspergillus niger* gene family encoding αgalactosidase. In *Mol. Gen. Genet.* **233**:404–10.

Fournier AM, Birchall IE, Kyriazis AG *et al.* (1993): A human naturally occurring antibody, anti-Gal, recognizes epitopes in pig kidney, heart and liver and is cytotoxic to endothelial cells in the presence of rabbit complement. Presented to the *Second International Congress on Xenotransplantation,* p 142 (abstract 113), Cambridge, UK.

Galili U (1991): The natural anti-Gal antibody: evolution and autoimmunity in man. In *Immunol. Series* **55**:355–73.

Galili U (1993): Interaction of the natural anti-Gal antibody with αgalactosyl epitopes: a major obstacle for xenotransplantation in humans. In *Immunol. Today* **14**:480–2.

Galili U, Rachmilewitz EA, Peleg A *et al.* (1984): A unique natural human IgG antibody with anti-αgalactosyl specificity. In *J. Exp. Med.* **160**: 1519–31.

Galili U, Clark MR, Shohet SB *et al.* (1987): Evolutionary relationship between the natural anti-Gal antibody and the Galα-1–3Gal epitope in primates. In *Proc. Natl. Acad. Sci. USA* **84**:1369–73.

Galili U, Mandrell RE, Hamadeh RM *et al.* (1988a): The interaction between the human natural anti-αgalactosyl IgG (anti-Gal) and bacteria of the human flora. In *Infect. Immun.* **56**:1730–7.

Galili U, Shohet SB, Kobrin E *et al.* (1988b): Man, apes and Old World monkeys differ from other mammals in the expression of αgalactosyl epitopes on nucleated cells. In *J. Biol. Chem.* **263**:17755–62.

Good H, Cooper DKC, Malcolm AJ *et al.* (1992): Identification of carbohydrate structures which bind human anti-porcine antibodies: implications for discordant xenografting in man. In *Transplant. Proc.* **24**:559–62.

Inverardi L, Samaja M, Marelli F *et al.* (1992): Cellular immune recognition of xenogeneic vascular endothelium. In *Transplant. Proc.* **24**:459–61.

Joziasse DH, Shaper JH, Van Den Eijnden DH *et al.* (1989): Bovine α1–3 galactosyltransferase: isolation and characterization of a cDNA clone. Identification of homologous sequences in human genomic DNA. In *J. Biol. Chem.* **264**:14290–7.

Joziasse DH, Shaper JH, Jabs EW *et al.* (1991): Characterization of an αgal1–3-galactosyltransferase homologue on human chromosome 12 that is organized as a processed pseudogene. In *J. Biol. Chem.* **266**:6991–8.

Karlsson E, Cairns T, Holgersson J *et al.* (1993): Pig-to-human xenotransplantation. Confirmation of a major target epitope of preformed human natural IgG and IgM anti-pig-terminal galactose α1,3galactose. In *Glycoconjugate J.* **10**:297.

Kobayashi T, Taniguchi S, Ye Y *et al.* (1995): Prolongation of graft survival following pig-to-baboon heart transplantation by cobra venom factor (CVF): without natural antibody depletion. Presented to the *International Society for Heart and Lung Transplantation*, San Francisco.

Koren E, Neethling FA, Ye Y *et al.* (1992): Heterogeneity of preformed human anti-pig xenogeneic antibodies. In *Transplant. Proc.* **24**:598–601.

Koren E, Neethling FA, Richards S *et al.* (1993): Binding and specificity of major immunoglobulin classes of preformed human anti-pig heart antibodies. In *Transplant. Int.* **6**:351–3.

Koren E, Neethling FA, Koscec M *et al.* (1994): *In vitro* model for hyperacute rejection of xenogeneic cells. In *Transplant. Proc.* **26**:1166.

Koren E, Milotec F, Neethling FA *et al.* (1996): Mouse monoclonal anti-idiotypic antibodies directed against human anti-αGal antibodies reduce cytotoxicity of human and baboon sera to pig cells in culture. In *Transplantation* Proc. **28**:559.

Kujundzic M, Koren E, Neethling FA *et al.* (1994): Variability of anti-αGal antibodies in human serum and their relation to serum cytotoxicity against pig cells. In *Xenotransplantation* **1**:58–65.

Larsen RD, Rajan VP, Ruff M *et al.* (1989): Isolation of a cDNA encoding murine UDP galactose: β-D-galactosyl-1,4-*N*-acetyl-D-glucosaminide α-1,3-galactosyltransferase: Expression cloning by gene transfer. In *Proc. Natl. Acad. Sci. USA* **86**:8227–31.

Larsen RD, Rivera-Marrero CA, Ernst LK *et al.* (1990): Frameshift and non-sense mutations in a human genomic sequence homologous to a murine UDP-Gal:β-D-Gal(1–4)-D-GlcNAcα(1–3)-galactosyl transferase cDNA. In *J. Biol. Chem.* **265**: 7055–61.

Lemieux RU (1978): Human blood groups and carbohydrate structures. In *Chem. Soc. Rev.* **7**:423–52.

Leventhal JR, Dalmasso AP, Cromwell JW *et al.* (1993): Prolongation of cardiac xenograft survival by depletion of complement. In *Transplantation* **55**: 857–65 (Discussion 865–6).

Li S, Yeh J-C, Cooper DKC *et al.* (1995): Inhibition of human anti-αGal IgG by oligosaccharides derived from porcine stomach mucin. In *Xenotransplantation,* in press.

Lim CS, Chapman GD, Gammon RS *et al.* (1991): Direct *in vivo* gene transfer into the coronary and peripheral vasculatures of the intact dog. In *Circulation* **83**:2007–11.

Lloyd KO, Kabat EA, Licerio E (1968): Immunochemical studies on blood groups. Structures and activities of oligosaccharides produced by alkaline degradation of blood-group Lewis[a] substance. Proposed structure of the carbohydrate chains of human blood-group A, B, H, Le[a], and Le[b] substances. In *Biochemistry* **7**:2976–90.

Lowe JB, Larsen RD, Ruff MM *et al.* (1989): Isolation of cloned cDNAs that direct expression of oligosaccharides recognized by *Griffonia simplicifolia* isolectin B_4. In *J. Cell. Biochem.* **13**A:129.

Neethling FA, Koren E, Ye Y *et al.* (1994): Protection of pig kidney (PK15): cells from the cytotoxic effect of anti-pig antibodies by αgalactosyl oligosaccharides. In *Transplantation* **57**:959–63.

Neethling F, Cooper DKC, Xu H *et al.* (1995a): Newborn baboon serum anti-α-galactosyl antibody levels and cytotoxicity to cultured pig kidney (PK15) cells. (Letter). In *Transplantation* **60**:520–1.

Neethling F, Joziasse D, Bovin N *et al.* (1995b): The reducing end of αGal oligosaccharides contributes to their efficiency in blocking natural antibodies of human and baboon sera. In *Transplant. Int.*, in press.

Neethling FA, Koren E, Cooper DKC (1995c): Variability of anti-αGal antibodies in human and baboon sera. Presented to the *International Society for Heart and Lung Transplantation*, San Francisco.

Neethling FA, Koren E, Cooper DKC *et al.* (1995d): A rapid and inexpensive microcytotoxicity assay to monitor potential xenograft hyperacute rejection following organ xenotransplantation, submitted.

Oriol R, Cooper JE, Davies DR *et al.* (1984): ABH antigens in vascular endothelium and some epithelial tissues of baboons. In *Lab. Invest.* **50**:514–8.

Oriol R, Ye Y, Koren E *et al.* (1993): Carbohydrate antigens of pig tissues reacting with human natural antibodies as potential targets for hyperacute vascular rejection in pig-to-man organ xenotransplantation. In *Transplantation* **56**:1433–42.

Oriol R, Barthod F, Bergemer A-M *et al.* (1994): Monomorphic and polymorphic carbohydrate antigens on pig tissues: implications for organ xenotransplantation in the pig-to-man model. In *Transplant. Int.* **7**:405–13.

Oriol R, Candelier JJ, Taniguchi S *et al.* (1996): Major oligosaccharide epitopes found in tissues of 20 animal species: implications for experimental and clinical organ xenotransplantation. In *Transplant. Proc.*, **28**:794.

Pinto BM, Bundle DR (1983): Preparation of glycoconjugates for use as artificial antigens: a simplified procedure. In *Carbohydr. Res.* **124**:313–8.

Platt JL, Bach FH (1991): Mechanism of tissue injury in hyperacute xenograft rejection. In: *Xenotransplantation* (Cooper DKC *et al.*, eds) pp 69–79, Heidelberg: Springer.

Rieben R, Von Allmen E, Korchagina EY *et al.* (1995):

Detection, immunoadsorption, and inhibition of cytotoxic activity of anti-αGal antibodies using newly developed substances with synthetic Galα1–3Gal disaccharide epitopes. In *Xenotransplantation* **2**:98–106.

Romano EL, Bhardwaj J, Kelly S *et al.* (1987a): Interaction of IgG and IgM anti-A with synthetic oligosaccharides. In *Transplant. Proc.* **19**:4479–83.

Romano EL, Soyano A, Linares J (1987b): Preliminary human study of synthetic trisaccharide representing blood group substance A. In *Transplant. Proc.* **19**: 4475–8.

Romano EL, Soyano A, Linares J *et al.* (1987c): Neutralization of ABO blood group antibodies by specific oligosaccharides. In *Transplant. Proc.* **19**: 4426–30.

Rydberg L, Cairns TDH, Groth CG *et al.* (1994): Specificities of human IgM and IgG anti-carbohydrate xenoantibodies found in the sera of diabetic patients grafted with fetal pig islets. In *Xenotransplantation* **1**:69–79.

Samuelsson BE, Rydberg L, Breimer ME *et al.* (1994): Natural antibodies and human xenotransplantation. In *Immunol. Rev.* **141**:151–68.

Sandrin MS, Vaughan HA, Dabkowski PL *et al.* (1993): Anti-pig IgM antibodies in human serum react predominantly with Gal(α1–3)Gal epitopes. In *Proc. Natl. Acad. Sci. USA* **90**:11391–5.

Spiro RG, Bhoyroo VD (1984): Occurrence of α-D-galactosyl residues in the thyroglobulins from several species. In *J. Biol. Chem.* **259**:9858–66.

Springer GF, Horton RF (1969): Blood group isoantibody stimulation in man by feeding blood group-active bacteria. In *J. Clin. Invest.* **48**:1280–91.

Varki A (1992): Selectins and other mammalian sialic acid-binding lectins. In *Curr. Opinion Cell Biol.* **4**: 257–66.

Vaughan HA, Dabkowski PL, McKenzie IFC *et al.* (1993): Biochemical analysis of pig xenoantigens detected by human antibodies. In *Transplant. Proc.* **25**:2919–20.

Watkins WM (1972): Blood group specific substances. In *Glycoproteins: Their Composition, Structure and Function* (Gottschalk A, ed) pp 830–91, Amsterdam: Elsevier.

Watkins WM (1974): Genetic regulation of the structure of blood group-specific glycoproteins. In *Biochem. Soc. Symp.* **40**:125–46.

Watkins WM (1980): Biochemistry and genetics of the ABO, Lewis, and P blood group systems. In *Advances in Human Genetics,* Vol. **10** (Harris H, Hirschhorn K, eds) pp 1–136, New York: Plenum.

Wiener AS (1951): Origin of naturally occurring hemagglutinins and hemolysins: a review. In *J. Immunol.* **66**:287–95.

Yamamoto F, Clausen H, White T *et al.* (1990): Molecular genetic basis of the histo-blood group ABO system. In *Nature* **345**:229–33.

Ye Y, Neethling FA, Niekrasz M *et al.* (1994a): Intravenous administration of α-galactosyl carbohydrates reduces *in vivo* baboon serum cytotoxicity to pig kidney cells and transplanted pig hearts. In *Transplant. Proc.* **26**:1399.

Ye Y, Neethling FA, Niekrasz M *et al.* (1994b): Evidence that intravenously administered α-galactosyl carbohydrates reduce baboon serum cytotoxicity to pig kidney (PK15): cells and transplanted pig hearts. In *Transplantation* **58**:330–7.

Ye Y, Niekrasz M, Kosanke S *et al.* (1994c): The pig as a potential organ donor for man: a study of potentially transferable disease from donor pig to recipient man. In *Transplantation* **57**:694–703.

32 Modern Glycohistochemistry: A Major Contribution to Morphological Investigations

André Danguy, Isabelle Camby, Isabelle Salmon, and Robert Kiss

1 Introduction

Many proteins and lipids produced by unicellular and multicellular eukaryotic organisms contain covalently linked carbohydrate chains, and are called glycoproteins and glycolipids respectively. The addition of glycans to proteins and lipids is highly regulated by the structure of the molecule, the cell environment, and extracellular effectors such as hormones. The sugar part of a glycoprotein confers an additional set of potential functional properties on the protein backbone (Blithe, 1993; Lis and Sharon, 1993; Varki, 1993). For these reasons, it is clear that glycosylation represents an important biological event for the exchange of molecular information within and between cells (Rademacher *et al.*, 1988; Drickamer and Carver, 1992; Varki, 1993; Dwek, 1995). Furthermore, due to the extensive heterogeneity of oligosaccharide moieties, glycoconjugates are capable of encoding a potentially enormous amount of biologically critical information. This characteristic separates glycans from the rest of the biological macromolecules (Fukuda, 1994).

While they can demonstrate the presence of glycans in glycoproteins, neither "conventional" carbohydrate histochemical procedures nor procedures using labeled lectins can detect the sugar moieties of the glycolipids. The failure to detect sugar residues in glycolipids is due to : (1) each glycolipid carrying a single oligosaccharide side chain, whereas the number of such moieties can be enormous in the case of glycoproteins; and (2) the use of organic solvents prior to the embedding of tissues in either paraffin or resins, which extracts the majority of glycolipids.

2 Why Lectins and Neoglycoproteins are Attractive Reagents in Histology

Recent advances in the field of glycobiology and the ensuing start to the unravelling of their functional significance have stimulated the introduction of new probes and techniques. The identification of glycans in tissues has developed significantly in the last decade and constitutes one of the most fascinating chapters of histochemistry (Damjanov, 1987; Alroy *et al.*, 1988; Spicer and Schulte, 1992; Roth, 1993; Spicer, 1993; Danguy *et al.*, 1994 for reviews).

Before the introduction of lectin histochemistry, the identification of carbohydrates was carried out by conventional methods. Amongst these methods, the most popular procedures were: the use of alcian blue (AB); the periodic acid-Schiff (PAS) method; iron diamines, and the observation of metachromasia by means of toluidine blue. Readers are referred to the following English language books for complete and fully detailed technical procedures and original publications. We also recommend these books for a further understanding of the theoretical background of the histochemical methods developed for carbohydrate location: Pearse (1980–1985); Clark (1981); Kiernan (1990); Lyon (1991).

Despite their ability to differentiate between the main categories of glycoproteins, these conventional procedures are still rather limited in their specificity for particular sugar moieties, and yield incomplete information on the structural details of the glycans. The introduction of lectins in histochemistry has revolutionized our knowledge.

H.-J. and S. Gabius (Eds.), Glycosciences

ISBN 3-8261-0073-5

Table 1. Lectin characteristics

Latin name	Source	Acronym	Nominal specificity
α/β-D-Galactosyl-specific lectins			
Agaricus bisporus	Common mushroom	ABA	Galβ3GalNAc-*O*-Ser/Thr
Erythrina cristagalli	Coral tree	ECL	Galβ4GlcNAcβ
Griffonia simplicifolia I	unknown	GSI	α-Gal/α-GalNAc
Griffonia simplicifolia I-B_4	unknown	GSI-B_4	Galα3Gal
Artocarpus integrifolia	Jackfruit	Jacalin	Galβ3GalNAc
Maclura pomifera	Osage orange	MPA	Gal-β3GalNAc
Arachis hypogaea	Peanut	PNA	Gal-β3GalNAc
N-Acetyl-D-galactosaminyl-specific lectins			
Cytisus scoparius	Scotch broom	CSA	GalNAc
Dolichos biflorus	Horse gram	DBA	GalNAcα3GalNAc/GalNAcα3Gal
Helix pomatia	Garden snail	HPA	α-GalNAc
Glycine max	Soybean	SBA	GalNAcα3GalNAc/GalNacα/β3/4Gal
Sophora japonica	Pagoda tree	SJA	Galβ3GalNAc
N-Acetyl-D-glucosaminyl-specific lectins			
Datura stramonium	Jimson weed	DSA	(GlcNAcβ4)2–4
Griffonia simplicifolia II	unknown	GSII	(GlcNAcβ4)2–3
Lycopersicon esculentum	Tomato	LEL	(GlcNAcβ4)1–4
Solanum tuberosum	Potato	STL	(GlcNAcβ4)2–3
Triticum vulgare (succincylated form)	Wheat germ	s-WGA	(GlcNAcβ4)2–4
D-Mannosyl/D-glucosyl-specific lectins			
Canavalia ensiformis	Jack bean	Con-A	branched *N*-linked high mannose-type oligosaccharides
Galanthus nivalis	Snowdrop	GNA	Manα3Man
Lens culinaris	Lentil	LCA	Fucα6GlcNAc-*N*-Asn containing *N*-linked oligosaccharides
Pisum sativum	Pea	PSA	idem
α-L-Fucosyl-specific lectins			
Anguilla anguilla	Eel	AAA	α-L-Fuc
Aleuria aurantia	Orange peel fungus	OPA	Fucα1,6/(3)GlcNAc
Ulex europaeus	Gorse	UEA-I	Fucα2Galβ
Lotus tetragonolobus	Asparagus	LTA	Fucα2Galβ4(Fucα3)GlcNAcβ
Sialic acid binding lectins			
Limax flavus	Slug	LFA	Neu5Ac/Neu5Gc
Maackia amurensis	unknown	MAA	Neu5Ac2,3Galβ4Glc/GlcNAc
Sambucus nigra	Elderberry bark	SNA	Neu5Ac2,6Gal/GalNAc
Triticum vulgare	Wheat germ	WGA	Neu5Ac/GlcNAc
Lectins with complex specificities			
Phaseolus vulgaris E	Kidney bean	PHA-E	bisected bi- and tri-antennary N-linked sequences
Phaseolus vulgaris L	Kidney bean	PHA-L	highly branched (in tri-antennary or more) non-bisected sequences

Carbohydrate sequences are presented according to the short form described in the 1985 IUPAC-IUB JCBN recommendations. Man: mannose; Glc: glucose; Gal: galactose; GlcNAc: *N*-acetylglucosamine; GalNAc: *N*-acetylgalactosamine; Fuc: fucose; Neu5Ac: neuraminic acid; Ser: serine; Thr: threonine.

Lectins obtained from plants and animals are defined as a group of non-enzymatic glycan-binding proteins of non-immune origin. They react specifically with carbohydrates and the oligosaccharide sequences of glycoconjugates (Table 1). Lectins have found widespread use as specific markers for locating and characterizing the glycan profiles of cells and tissues. They are classified according to their affinity with D-pyranose monosaccharides (Goldstein and Poretz, 1986; Wu *et al.*, 1988). However, whereas some lectins bind to glycans irrespective of the configuration of adjacent molecules, others react only if additional conditions are fulfilled (Lis and Sharon, 1986) as, for example, linear or branched carbohydrate arrangements or differences in anomeric specificity (α-β-forms). Some lectins require extended binding sites such as di- or trisaccharides. While most lectins only detect terminal sugar residues, a few (e.g. WGA) also bind to internal components of the saccharide chain. Moreover, lectins enable a degree of differentiation to be established between *N*- and *O*-linked oligosaccharides. Lectins with fewer additional requirements usually identify significantly more sites than those with high specificities. Staining with batteries of lectins representing the same nominal monosaccharide affinity, but with different binding requirements, is therefore potentially valuable for determining complex oligosaccharide molecules in histochemistry (Alroy *et al.*, 1988; Spicer and Schulte, 1992; Danguy *et al.*, 1994; Menghi and Materazzi, 1994). Furthermore, it is accepted that negative lectin staining does not always indicate the absence of a given carbohydrate residue. Some lectin binding sites located in the internal core of oligosaccharide chains may have their lectin affinity inhibited by the presence of adjacent sugar residues.

Although it is not widely appreciated that lectins are at least as abundant in animal as in plant tissues, it is now clear that this group of bivalent non-enzymatic carbohydrate-binding proteins, unrelated to antibodies, fulfills the necessary hypotheses for a physiologically meaningful glycobiological interaction. Their ubiquitous occurrence, which is definitely not restricted to organisms from only a few branches of the evolutionary tree, emphasizes their key role in a wide variety of cellular activities. Animal lectins can be grouped into two main families based on a number of properties, e.g. the nature of their glycan ligands, their subcellular location, and their dependence on metal co-factors.

Excellent reviews and comments have recently been published on the biological structure and importance of these macromolecules (Drickamer, 1988, 1994; Caron *et al.*, 1990; Gabius, 1991; Harrison, 1991; Barondes *et al.*, 1994; Hirabayashi, 1994; Müller, 1994; Varki, 1994). The heightened interest in the location of endolectins in tissues has stimulated the synthesis of new, artificially "tailor made" probes which have been named neoglycoproteins. These have already proved their practical value, enabling investigators to monitor the other side of recognitive protein-carbohydrate interactions (Danguy *et al.*, 1991; Gabius and Bardosi, 1991; Danguy and Gabius, 1993; Gabius *et al.*, 1993; Lee and Lee, 1994). The neoglycoproteins used and their specificities are listed in Table 2. The foremost reason for constructing any neoglycoprotein is to obtain glycoconjugates containing glycan groups of known structure and assured purity.

Table 2. Neoglycoproteins and their specificities

Receptor specificity	Type of neoglycoprotein
β-Galactoside	Lac-BSA Gal-β-(1,3)-glcNAc-BSA Asialofetuin
α-Galactoside	Melibiose-BSA
N-acetylated sugars	*N*-acetylglucosam.-BSA N-acetylgalactosam.-BSA
α-Glucoside	Maltose-BSA
β-Glucoside	Cellobiose-BSA
α-Fucoside	Fucose-BSA
α-Mannoside	Mannose-BSA
Phosphorylated sugars	Mannose-6-P-BSA Galactose-6-P-BSA
Sugars with a carboxylgroup	Sialic acid-BSA Glucuronic acid-BSA
β-Xyloside	Xylose-BSA
L-Rhamnose	Rhamnose-BSA

3 Methods

For glycohistochemical studies, sections 5 μm thick were prepared from paraffin-embedded tissues. These sections were subsequently dewaxed in toluene, rehydrated through an alcohol series of decreasing concentrations, and H_2O, and placed in phosphate-buffered saline (PBS), pH 7.3 ± 0.1.

3.1 Lectin Cytolabeling

The sequence comprised the following steps : (1) incubation in methanol containing 0.3 % H_2O_2 for 30 min to block endogenous peroxidase activity; (2) incubation with avidin D followed by biotin (blocking kit, Vector Laboratories). This procedure ensures that all the endogenous biotin, biotin receptors, or avidin binding sites present in the tissues are blocked prior to the addition of biotin-avidin system reagents; (3) incubation with biotinylated lectins (10 μg/ml PBS containing 0.1 mM Ca^{2+}, Mg^{2+}, Mn^{2+}) for 15 min in a moist chamber; (4) application of the avidin-biotinylated horseradish peroxidase complex (Vectastain Elite ABC kit) solution for 30 min. Between steps, the sections were rinsed thoroughly for 10 min in PBS. After the last rinse, staining was performed for 10 min in H_2O containing 16 mM 3,3'-diaminobenzidine tetrahydrochloride (DAB; Sigma Chemical Co., St. Louis, MO, USA) and 0.015 % hydrogen peroxide, to visualize the peroxidase activity. Finally, the sections were washed in running tap water, lightly counterstained with hematoxylin, dehydrated, cleared and mounted in DPX (BDH Chemicals, Poole, England) according to standard procedures.

All the steps were performed at room temperature. The sections were examined with a light microscope and the staining intensity of the different structures was graded semiquantitatively on an estimated scale from zero (unreactive) to three units (strongly reactive). Controls included: (1) omission of the respective lectin; (2) omission of the ABC kit reagents; (3) incubation of the section with a solution that contained, in addition to the biotinylated lectin, 0.2–0.3 M of the specific inhibitory sugar.

3.2 Reverse Lectin Histochemistry

Over the past few years increasing effort has been devoted to elucidating the presence and function of endogenous lectins (endolectins) (Gabius, 1991; Gabius and Gabius, 1993; Gabius *et al.*, 1993; Hirabayashi, 1993; Barondes *et al.*, 1994; Drickamer, 1994). Neoglycoconjugates are powerful tools for the detection and location of accessible endolectins *in situ*. Along with lectins, these synthetic reagents offer the opportunity to monitor both sides of protein-carbohydrate interactions and to assign fundamental relevance to endolectins in glycobiological interactions (Gabius *et al.*, 1993, 1994; Lee and Lee, 1994).

Generally speaking, the various steps employed in our standardized staining protocol are as follows :

(1) After rinsing in phosphate-buffered saline (PBS, pH 7.3 ± 0.1), the sections were preincubated with 0.1 % carbohydrate-free bovine serum albumin (BSA)-0.1 M PBS solution for 15 min to block the non-specific binding of the BSA-biotin derivatives used in the following steps.
(2) Excess solution was blotted from the slides and the sections were incubated for 45 min with 25–100 μg/ml of the different biotinylated carbohydrate-BSA conjugates (neoglycoproteins) dissolved in PBS containing 0.1 % carbohydrate-free BSA.
(3) As step (4) described above (Lectin cytolabeling). The staining intensity of the different structures was graded, as reported for lectin cytolabeling. Controls to ascertain specificity of neoglycoprotein binding included:
(a) Omission of any biotinylated probe so as to be able to exclude any non-specific binding of the kit reagents, e.g. the glycoproteins avidin and peroxidase.
(b) Performance of the complete protocol with non-glycosylated, biotinylated BSA at the same concentration as the labeled neoglycoprotein, so as to be able to exclude any binding by protein-protein interaction.

(c) Competitive inhibition experiments using the homologous unlabeled neoglycoproteins, which had been preincubated on the sections in order to saturate specific binding sites, and then applied in a mixture accompanied by the labeled neoglycoprotein. This experimental series is complemented by heterologous inhibition with a neoglycoprotein carrying a carbohydrate residue with little or no affinity to the detected binding sites.

3.3 Special Considerations

Several procedural and technical details deserve additional comment. Differences in lectin staining have been reported to exist between unfixed, fixed frozen, and paraffin sections (Alroy *et al.*, 1988; Danguy and Gabius, 1993). Ideally, tissue lectin and neoglycoprotein acceptors should be studied in both frozen and paraffin sections. Bouin-fixed, paraffin-embedded tissue was used throughout our studies. There were no clear differences observed in lectin and neoglycoprotein binding patterns between frozen and Bouin-fixed paraffin-embedded specimens. Formalin is a fixative used most commonly in routine laboratory practice; although glycosubstances are well preserved with this fixative, it is not the optimal choice for glycohistochemistry. Although glutaraldehyde fixes by a mechanism similar to formaldehyde or paraformaldehyde (Hoyer *et al.*, 1991), the glycosubstances did not stain well in sections from glutaraldehyde-fixed tissue. We believe that this difference is related to the cross-linking capacity of aldehyde fixatives, which is greatest with glutaraldehyde, a dialdehyde. Apparently the powerful cross-linking of proteins by glutaraldehyde results in the masking of the oligosaccharide side chains. While cross-linking of glycoproteins also occurs with formaldehyde and paraformaldehyde (monoaldehyde fixatives), this appears to be less potent.

Carnoy's fixative causes no cross-linking between protein chains. This may leave more glycoconjugate sites exposed to accommodate binding. The results with Bouin's fixative are due to the picric acid component, which is a powerful protein precipitant that forms protein-picrate in tissues, but glycans are probably retained by their association with proteins. We selected Bouin's solution because it produces good protein antigenicity and excellent morphological preservation. However, Bouin's fluid does not preserve intracellular membranes well, and therefore the definition of the intracellular compartmentalization of some molecules is lost. Paraffin embedding and subsequent deparaffinization of 5 μm thick tissue sections before glycocytochemical staining does not appear to alter lectin or neoglycoprotein binding. Processing tissue in this fashion allows the lectins and neoglycoproteins access to free intracellular binding sites that may be more difficult to label if thicker frozen sections are employed.

The choice of marker lies between a fluorescent compound (fluorescein or rhodamine), or an enzyme such as horseradish peroxidase. Colloidal gold has been used as a third type of label (Skutelsky *et al.*, 1987). However, this latter methodology is mainly employed as an ultrastructural marker (Beesley, 1989). The advantage of using fluorescent dyes as labels is their simple technical aspects, but a major drawback is background fluorescence, which interferes with the interpretation of specific staining. Moreover, an additional disadvantage is the impermanence of preparations labeled in this way. Our laboratory prefers 3,3'-diaminobenzidinetetrahydrochloride (DAB) as the chromogen, because the brown granules of DAB polymers are stable and insoluble in organic solvents, and so provide a permanent record when slides are stored. The glycohistochemical method preferred in our laboratory employs both biotinylated lectins and neoglycoproteins as probes, and the avidin-biotin-peroxidase complex (ABC). This choice is based largely on experience in light microscopy.

Avidin (MW 70 000) is a glycoprotein of egg-white. It has a very high affinity for the small molecule biotin (MW 244) (Green, 1975) that can be covalently linked to several types of amino acid side chains. Since a broad spectrum of molecular probes, including lectins, neoglycoproteins, antibodies and enzymes can be efficiently biotinylated, avidin-based systems are widely employed technique for the detection and visualization of various binding reactions (Hsu *et al.*,

1981; Danguy and Genten, 1990; Kiernan, 1990). It is notable that the binding of the glycoprotein avidin as well as horseradish peroxidase via endogenous sugar receptors may occur in the presence of abundant mannose-binding sites (Kuchler *et al.*, 1990, 1992).

In contrast to egg-white avidin, streptavidin isolated from *Streptomyces avidinii* has an isoelectric point close to neutral pH and contains no carbohydrate residues. Consequently, it has been reported that detection methods based on the biotin-streptavidin system do not suffer from the problems of non-specific binding associated with egg-white avidin (Chaiet and Wolf, 1964; Jones *et al.*, 1987). The pitfall of endogenous biotin in tissue can be circumvented by the pre-exposure of the section to unlabeled avidin followed by biotin (Vector blocking kit) before the addition of the biotinylated probe.

In our laboratory, the use of the Vectastain Elite ABC kit and the blocking kit gave excellent specific and reproducible results. Very recently, some authors (Ellis and Halliday, 1992; Grumbach and Veh, 1995) reported that the Vectastain Elite ABC system, a modified ABC version, generated a considerably higher signal intensity. Grumbach and Veh (1995) found that the Elite ABC complex consisted of avidin D and biotinylated peroxidase in a molar ratio of about 1:1 whereas in the conventional ABC complex avidin and biotinylated enzyme are in a molar ratio of about 2:1. These authors concluded that the roughly equimolar ratio improved the performance of the Elite system and that an ABC complex intended to maximize signal intensity must consist of an excess of peroxidase over avidin molecules. They therefore developed a new ABC procedure.

The binding of a lectin molecule to a carbohydrate does not involve the formation of covalent bonds, but is similar in nature to the attachment of an antigen to its specific antibody. As is the case with carbohydrate-specific antibodies, the affinity of a lectin is often influenced by the neighboring one to three units in the chain (Goldstein and Poretz, 1986); consequently, a particular glycoconjugate may be able to bind different amounts of different lectins of the same group. This may also explain why, in some cases, the inhibition of lectin binding may not be complete even using 0.3 M of hapten sugar; this is particularly seen with lectins which recognize complex oligosaccharide sequences such as PHA-E and PHA-L.

A lot of papers exist on the contributions of modern carbohydrate histochemistry, and many reports have emphasized how work in this field has proved fruitful. However, the present chapter is not intended to be exhaustive and it is proposed to discuss selected topics in the following paragraphs in order to exemplify the direction taken by glycohistochemical investigation.

4 Functional Morphological Data

4.1 The Kidney

The histological investigation of the vertebrate kidney generally reveals the presence of nephrons. The nephron, consisting of the Malpighian corpuscle and convoluted tubules (which is typically differentiated into proximal and distal portions), is considered to be the functional unit for the process of urine formation (Bulger, 1988). There is a clear-cut correlation between the architecture of the vertebrate kidney and its ability to manufacture a urine that is hypertonic in relation to body fluids. Kidneys capable of producing a hyperosmotic urine (i.e. those of birds and mammals) all have nephrons featuring the loop of Henle (Schmidt-Nielsen, 1991). In general, the longer the loop, the greater the concentrating power of the nephron.

In mammals, the kidney is an organ in which the glycoconjugates detected commonly differ between cell types (Truong *et al.*, 1988). The lectin acceptors in proximal nephron tubules are, as a rule, unlike those in the distal tubule and the loop of Henle (Coppée *et al.*, 1993). A similar observation has recently been made concerning endolectins (Coppée *et al.*, 1993). To substantiate the fact that environmental conditions can affect this physiological characteristic in animal species, a comparative analysis has been completed between the laboratory mouse and the golden spiny mouse (*Acomys russatus*), which lives in deserts (Coppée *et al.*, 1993). Since the kidney undoubtedly has a role in the adaptation of vertebrates to dry climates (Schmidt-Nielsen, 1991), the expression of defined determinants in different segments of the nephron of these species may be correlated with distinct functional aspects contributing to an explanation of the adaptive response.

In different groups of fish, the adaptation to a fresh- or salt-water environment may lead to changes in the nephron structure (Dantzler, 1989; Schmidt-Nielsen, 1991; Gambaryan, 1994). This great morphological and functional variety has provided several important animal models for the study of basic renal mechanisms (Hentschel and Elger, 1987). Data obtained in our laboratory have emphasized the glycohistochemical differences between the trout (*Oncorhynchus mykiss*), a freshwater species, and a marine teleost (*Gadus luscus*). In the trout, renal glomeruli expressed glycan acceptors with specificity for negatively charged sugars that contain a carboxyl group, and for xylose. Neoglycoprotein binding was observed at comparatively high levels in the cells of the second portion of proximal tubules of both teleosts. Sugar receptors with glucuronic acid, α-mannose and xylose specificities were densely distributed in the brush borders of the proximal tubules; however, some differences in staining intensity can be seen across species, and topologically in relation to the segment of the proximal tubule. Accessible sugar receptors with glucuronic acid, mannose, xylose and β-galactose specificities were expressed in the trout distal tubule. Some receptor expression was also located by fucose and *N*-acetylgalactosamine as a ligand probe, whereas α-glucoside-specific receptors were expressed in phagosomes and lysosomes of distal tubule cells, as shown by the maltosylated marker. In addition, these glucoside-specific receptors were detected in the second segment of the trout proximal tubules. Marked specific differences were also observed in connection with the collecting duct system.

The differences between the nephron segments in lectin and neoglycoprotein binding strongly suggest various glycobiological interactions. It is tempting to speculate that there might be different transport systems, thus indicating a functional heterogeneity in line with structural peculiarities along the nephron of both teleosts (Hentschel and Walther, 1993). However, the present lack of comprehensive biochemical data prevents any further attempts to ascribe functional significance to these results. Consequently, our study, referred to above, forms a basis for subsequent glycobiological investigations which may provide an insight into the nature of the detectable binding sites in the fish nephron.

4.2 Endocrine Status and the Glycohistochemical Expression

Similarly, parallel investigations into the mouse submandibular glands (Akif *et al.*, 1993), prostate (Akif *et al.*, 1994) and endometrium (Akif *et al.*, 1995) illustrate the potential of this approach. We have reported that in these organs the hormonal status changes the expression of lectin-binding sites and endolectins, thereby modifying both sides of a presumably physiologically significant protein-glycan interaction.

4.3 The Integument

The skin forms the continuous external surface or integument of the body.

Various epidermal appendages (glands, hairs, scales, claws, horns etc.) can be observed. In mammals, the skin consists of the epidermis, a superficial epithelial component, and underlying connective tissue components, which are the dermis and hypodermis (Matoltsy, 1986). The epidermal component of the skin consists of a stratified squamous keratinizing epithelium. The epidermis is a constantly renewed tissue. Proliferation takes place in the basal layer of the keratinocytes, which is an inexhaustible source of new keratinocytes as a result of mitosis. Cells undergo terminal differentiation as they move through the suprabasal layers to the tissue surface. Mammalian skin thus provides an excellent model for the study of cell differentiation and maturation.

Distinct patterns of carbohydrate expression in human epidermis (Ookusa *et al.*, 1983; Schaumburg-Lever *et al.*, 1984; Virtanen *et al.*, 1986; Kariniemi and Virtanen, 1989), in eccrine glands, sebaceous glands (Ookusa *et al.*, 1983; Schaumburg-Lever, 1984) and hair follicles (Ohno *et al.*, 1990) have been observed using a wide range of commercially available lectins. Some lectins such as ConA, WGA and RCA bind to all the living cell layers, while others, such as LFA and DBA, prefer keratinocytes in the basal layer. Suprabasal cell layers are reactive to PNA and SBA, whereas the onset of UEA-I binding is mid-way through the spinous layers (Watt and Jones, 1992). Glycoproteins originally detected with lectins have now been identified as desmosomal glycoproteins and integrins. Thus, lectin studies may provide new insights into the changes in the adhesive properties of keratinocytes which occur during terminal differentiation. Although glycoproteins play an important role in cutaneous biology, as is exemplified by lectin histochemistry, only one study has been performed to our knowledge on the distributions of glycans during the development of epidermis in hairless animals (Miyazawa *et al.*, 1994). These authors reported differences in lectin staining in the epidermis of young dogs between hairless and haired animals. GSL-I, BPA, MPA, and especially UEA-I, exhibit a restricted distribution of lectin staining. In hairless animals, there is complete aplasia of the hair follicles but the epidermis has epidermal ingrowths into the dermis, similar to rudimentary follicular remnants. The lectin histochemical approach in hairless dogs showed that there were different developmental steps in the keratinocytes between infants and adults.

Fish epidermis is known to display a great range of variation in its structural complexities. In most species, it is mucogenic in nature and provided with several cell types such as filament, sensory and chloride cells (Whitear, 1986; Zaccone *et al.*, 1986; Fasulo *et al.*, 1993). Mucins are one of the major constituents of fish epidermis; they are highly *O*-glycosylated proteins involved in the protection of cells from extracellular agents. These macromolecules were, and still are located by means of conventional histochemical techniques. As reported above, these methods largely fail to identify different classes of glycoproteins, and it has been postulated that their sensitivity is limited (Spicer and Schulte, 1992; Danguy *et al.*, 1994). The application of a battery of lectins to the examination of fish skin disclosed marked differences in staining among the species investigated (Genten and Danguy, 1990b). Ostariophysi and Gonorhynchiformes share a physiological feature which is absent in all other groups of teleostean fish, namely, the fright reaction elicited by an alarm substance (a pheromone-like substance) produced by giant ovoid cells located in the mid-epidermal layers (Pfeiffer, 1977; Whitear and Zaccone, 1984; Genten and Danguy, 1990a,b; Iger *et al.*, 1994a). Moderate to strong binding with some lectins, and a lack of staining by others, demonstrate the presence of only some oligosaccharide sequences attached to the polypeptide backbone (Witt and Reutter, 1988; Genten and Danguy, 1990b). The alarm substance cells have been reported to incorporate several enzymatic activities and polysaccharides (Pfeiffer, 1974). Zaccone *et al.* (1990) demonstrated the presence of serotonin in the club cells of two species of Teleosts. As the nature of the active agent has been reported to be a pterin (Pfeiffer, 1977), we suggest that the plentiful glycoproteins demonstrated by lectin histochemistry may act as a carrier or diluent of the active agent(s).

Among all abiotic factors affecting toxicity in aquatic organisms, anthropogenic acidification has the most adverse effects (Fromm, 1980; Wendelaar Bonga and Dederen, 1986). Histopathological and cytochemical reports on fish epidermis exposed to water with a low pH are uncommon (Genten and Danguy, 1990a; Iger and Wendelaar Bonga, 1994; Iger *et al.*, 1994a,b; Balm *et al.*, 1995). Moreover, very few studies using lectins as probes have been conducted on fish epidermis following limited and controlled exposure to acid and alkaline environments (Genten and Danguy, 1990a). By lectin histochemistry, modifications in the intensity of reaction of common mucous epithelial cells and club cells have been observed following exposure to either low pH or alkaline water. Furthermore, marked differences were shown among the species examined (Genten and Danguy, 1990a). Our observations indicate firstly, that several cell types respond to stressors with a change in production of glycoproteins, and secondly, that Club cells are also involved in stress response, in addition to the production of the pheromones reported in the literature.

The neoglycoprotein staining of fish skin demonstrated endogenous sugar acceptors. Our data clearly emphasize the specific and extensive binding of these probes to tissue components in different species, indicating the presence of accessible endogenous lectin. Moreover, semi-quantitative evaluation disclosed both quantitative and qualitative site-associated differences in endolectin expression (Danguy *et al.*, 1991). Convinced by this glycohistochemical approach, we investigated the influence on trout skin of experimental short- (24 h) and long-term (three weeks) exposure to copper. We observed drastic changes in both lectin and neoglycoprotein binding patterns during this exposure to metal pollution. This study thus emphasizes the feasibility and versatility of glycohistochemistry in the field of aquatic toxicology (Genten *et al.*, 1995). Lectins have been shown to be good markers for monitoring the differentiation of amphibian epidermis (Navas *et al.*, 1987) and its development (Villalba and Navas, 1989). Recently, flask cells, a specific cell type, were selectively demonstrated in adult frogs by UEA-I only (Villalba *et al.*, 1993). Moreover, in amphibia the epidermis carries out a unique function in the regulation of gas exchange and osmoregulation (Fox, 1986). On the other hand, amphibians inhabit a wide spectrum of habitats including arid regions. It has been thought that the varied function performed by amphibian skin may be reflected in the composition of its whole stock of mucosubstances (glycoproteins) (Danguy and Genten, 1989, 1990). Differences in epidermis and glandular lectin binding patterns between toads (*Bufo bufo*) and African clawed frogs (*Xenopus laevis*) have been observed (Danguy and Genten, 1989, 1990). Whether these differences are correlated with habitat selection or phylogenesis deserves further research on a large number of species. Since *Xenopus laevis* is widely used as a experimental animal, studies have been made of the mucous glands in the skin of this frog (Schumacher *et al.*, 1989). An enormous variety of different mucosubstances has been demonstrated in *X. laevis* skin secretions. The strong to moderate binding of galactosamine- and galactose-specific lectins to the mucus indicates a high level of *O*-glycosylation, whereas a low concentration of *N*-linked glycans is disclosed (Schumacher *et al.*, 1989). Lectins have proved their suitability for the characterization and comparison of saccharide sequences in tissue sections of fish and anuran extracellular matrices of the skin (Genten and Danguy, 1990c). A non-uniform binding pattern has been discovered between *B. bufo* and *X. laevis* connective skin tissue.

Prolactin is involved in an extraordinary range of functions varying in major ways from one animal group to another (Bentley, 1982). Examples include the promotion of the growth of the mammary glands and milk synthesis in mammals; the stimulation of the production of crop milk in pigeons; the stimulation of nest-building and incubation behavior in some mammals; the promotion of survival of certain fresh water fish by exerting favorable effects on ion and water exchange in the gills, the kidneys, and other organs; the stimulation of salt-gland secretion in birds; and the promotion of the "water-drive" effect in some newts (which causes them to seek water in preparation for breeding) (Bern, 1975; Bentley, 1982). The ability of prolactin to promote changes in the integument of the red-spotted newt (*Notophthalmus viridescens*) has been widely reported (Dent, 1975). This adenohypophyseal hormone may be involved in seasonal changes in skin texture and cutaneous secretion that are observed in both sexes. Both the integumental and behavioral effects of prolactin in the red-spotted newt may derive from the osmoregulatory action of this hormone (Bern, 1975). Lectin histochemistry has been employed to stress possible relations between prolactin-induced changes in texture and secretory activity, and changes in the distribution and expression of glycans in the skin of the adult red-spotted newt (Singhas and Ward, 1993). These authors report that prolactin increases levels of sialic acid and *N*-acetylglucosamine in the stratum corneum. In contrast, glycoproteins containing fucose, galactose, *N*-acetylgalactosamine and galactose-(1,3)-*N*-acetylgalactosamine are reduced by prolactin in the two glands and the epidermis (Singhas and Ward, 1993). The authors suggest that the integumental effects associated with prolactin in this newt species are mediated, at least in part, through the alteration of epidermal and glandular glycoproteins.

4.4 Glycan Expression in Skeletal Muscle

Nearly all vertebrate skeletal muscles consist of different types of myocytes (also called fibers), and the pattern often reflects the type of activity of the muscle. Three different functional types have been described. These are usually termed : (a) fast-contracting, fast-fatiguing (type IIb); (b) fast-contracting, fatigue-resistant (type IIa); (c) slow contracting (type I). Slow twitch fibers are adapted to regenerate ATP by oxidative processes. They contain many mitochondria and a dense network of capillaries to supply oxygen and substrates. Moreover, they contain appreciable quantities of myoglobin, a protein related to hemoglobin, which facilitates the entry of oxygen into the muscle fibers. Fast twitch fibers are predominantly glycolytic. These cells contain fewer mitochondria, a less extensive capillary network, and less myoglobin than slow twitch fibers. The third muscle fiber type, fast-contracting and fatigue-resistant, is generally intermediate in its properties between the other two.

A number of lectins have been shown to bind to both the sarcoplasm and sarcolemma of normal skeletal myoctyes, the muscle capillaries and connective tissue (Pena *et al.*, 1981; Capaldi *et al.*, 1985; Christie and Thomson, 1989). In the normal skeletal rat muscle Con-A, WGA and RCA-I bind strongly to the endomysial region of the fibers but not to the sarcoplasm. Other lectins (DBA, GSA, UEA-I, SBA and PNA) bind poorly or not at all to normal muscle (Gulati and Zalewski, 1985). Pena *et al.* (1981) had previously demonstrated in human skeletal muscle that Con-A, RCA and WGA permitted a clear visualization of the plasmalemma-basement membrane unit, tubular profiles in the interior of myocytes, blood vessels, and connective tissue. PNA gave no intracellular staining while SBA and UEA-I selectively stained blood vessels. Finally DBA was unique in providing visualization of nuclei. Some of these results were later confirmed by Capaldi *et al.* (1985). In the mouse, the myocytes were stained by DBA and PNA; the latter was also bound to the fasciae (Watanabe *et al.*, 1981). Recent work conducted at the ultrastructural level on skeletal rat fibers (Bonilla and Moggio, 1987) demonstrated the binding of WGA and LPA, which recognize sialic acid, to the glycocalyx, caveolae and the basal lamina of the muscle fibers. Having observed the specific binding of WGA in the myogenic zone of regenerating muscle in the rat, Gulati and Zalewski (1985) hypothesized that an environment rich in GlcNAc or NeuAc may be favorable to myoblast proliferation, migration, and fusion.

Very recently, Danish researchers have directed their attention to skeletal myocytes in an attempt to understand better the glycobiological processes involved in muscle ageing and disease. Kirkeby *et al.* (1991) point to strong cytoplasmic staining in some fibers treated with GSII and ConA, while other myocytes reacted weakly. They demonstrated that amylase treatment strongly inhibited the sarcoplasmic GSII reaction while the Con-A staining was still detectable. The sarcoplasmic WGA reaction was not influenced by the amylase digestion of the myocytes. These authors suggest that GSII displays mainly muscle glycogen or some glycoproteins linked to glycogen. WGA detected no muscle glycogen, only sarcoplasmic glycoproteins. They also suggest that if the glycoconjugates are membrane-bound, the most likely sites may be mitochondria, the tubule system or the sarcoplasmic reticulum. Kirkeby *et al.* (1992a) suggest that in rat skeletal muscle, the lectin-staining pattern might be related to development, the specialization, and the function of the individual muscles, and that there is no relationship between the mosaic lectin binding pattern and that observed after incubation for myosin ATPase or succinic dehydrogenase.

More recently, the same authors applied three different fucose-specific lectins (AAA, UEA-I, LTA) to normal human and rat muscle, to patients with Duchenne muscular dystrophy, and to rat muscle degenerating as a result of denervation and devascularization (Kirkeby *et al.*, 1993). They demonstrated that both the sarcoplasm and the sarcolemma of diseased myocytes display an altered fucose expression. In a mouse experimental model of muscular dystrophy, Kirkeby *et al.* (1992b) demonstrated a modified histochemical lectin binding pattern in comparison with normal mice. These authors suggested that the changes of muscle glycan components might be involved in the expression of dystrophic muscle. Modification in the expression of sarcoplasmic glycoconjugates is also involved in atrophic skeletal myocytes from ageing rats (Kirkeby, 1994).

Since capillary density depends on the fiber type distribution of the muscle, it is interesting for anybody studying normal physiology and pathology to display fiber types and the capillary system on the same slide. As UEA-I is a sensitive marker for endothelium in the human, and NADH tetrazolium reductase is a useful enzyme for fiber typing, Paljärvi and Naukkarinen (1990) have developed a quick and highly reproducible method to demonstrate capillaries and fibers simultaneously.

4.5 Lectin Histochemistry of the Teleost Intestine

Lectin cytochemical techniques have yielded data indicating that the mucus-secreting cells (goblet cells) are positive to a large array of lectins and that species differences can be observed, a factor which suggests the existence of different functional sub-populations of goblet cells (Burkhardt *et al.*, 1987). The goblet cells and the enterocytes exhibit in part different lectin staining patterns. No major staining differences were noted in animals given different diets (Burkhardt *et al.*, 1987).

The distal intestinal mucosa of the rainbow trout was investigated in our laboratory because its function is still debated, even though its involvement in antigen transport from the intestinal lumen and the absorption of protein in macromolecular form have been claimed (Pajak and Danguy, 1993). A rich supply of glycans was observed. At the intestinal level, cell-cell recognition finds an expression in parasite-host interplay. Evidence has been adduced that various parasites such as viruses, bacteria, protozoa and worms possess saccharide-binding proteins which can interact with carbohydrates on the surfaces of cells with which the parasites associate (Rademacher *et al.*, 1988). The lectin-binding pattern observed may be related to the specificity of colonization by the parasites. This binding may also reflect the broad range of glycans attached to the protein cores of enzymes involved in the transport and digestion of macromolecules. Furthermore, the glycan part of glycoproteins as demonstrated by the histochemical lectin approach may have some relevance in the future, i.e. for anal vaccination (Pajak and Danguy, 1993).

4.6 Sugar-Binding Sites and Lectin Acceptors in Prokaryotes and Eukaryotic Parasites

An important step in microbial pathogenesis is the interaction of microbes with host cells and substances. Several papers have demonstrated that lectins are instrumental in interactions between the host and intestinal parasites (bacteria, protozoans) as well as diverse non-intestinal species (Mandrell *et al.*, 1994; Strout *et al.*, 1994).

The presence of lectins has been shown on the sporozoïtes in three species of avian coccidia; in addition, each species has a different sugar specificity (Strout *et al.*, 1994). This diversity in lectin expression allowed the authors to suggest that lectins aid in the site selection of parasite infection. Similarly, the cell-surface glycoconjugates of the various developmental stages of *Trypanosoma cruzi*, the causative agent of Chagas' disease in the New World, have been characterized (Previato *et al.*, 1994). Even a decade ago, seven lectins were shown to discriminate between pathogenic and non-pathogenic trypanosomes. Tomato lectin (LEL), *Bauhinia pupurea* agglutinin and SJA were specific to *T. cruzi*, VVA reacted with *T. rangeli*, and PNA, LTA, and UEA-I were specific to *T. conorhini* (de Miranda Santos and Pereira, 1984). Schottelius (1990) confirmed that lectins can be used for the identification of the epimastigotes of these three trypanosome species. Furthermore, this investigator reported that in certain areas of South America there is some correlation between both lectin types and the transmission of *T. cruzi*.

Malaria is an endemic fever caused by blood protozoa (*Plasmodium* spp.). Conditions that favor the presence of *Anopheles* mosquitoes, and the binding of the parasite to its tissues tend, to cause an increase of the disease. By means of fluoresceinated lectins, interspecific variations of salivary gland glycan residues have been demonstrated, together with intraspecific differences between anopheline strains (Mohamed *et al.*, 1991).

Attention has also been paid to the cuticle of pathogenic nematodes, and various lectin acceptors have been identified (Rudin, 1990; Peixoto and De Souza, 1992).

Certain diseases in tropical and subtropical regions are produced by nematode worms, or filariae, of which the adults of both sexes live in lymphatics, connective tissues or mesenteries. Recent investigations emphasize the role played by glycans of the microfilariae in the host-parasite interaction, and the precise distribution of carbohydrate residues in sections of Lowicryl-embedded mature microfilariae of *Wucheria bancrofti* and *Brugia malayi* have been demonstrated using gold-labelled lectins (Araujo *et al.*, 1993).

Using a neoglycoprotein-gold method at the ultrastructural level, Morioka *et al.* (1994) studied the sugar-binding sites on sections of *Staphylococcus aureus* (bacteria), one of the most common human pathogens. Their data suggest that molecules that bind to *N*-acetylglucosamine occur in the cell wall and cytoplasm of the bacterium. These authors show convincingly that at least some of the sugar-binding sites are lectin-like substances. They propose that staphylococcal toxins may be responsible for the sugar-binding sites revealed in the study.

Dynamic variations in the galactomannan coat of insect-pathogenic fungi have been investigated by means of fluorescence and electron microscopy (Pendland and Boucias, 1993). These researchers demonstrated the constant rearrangement of the carbohydrate components of surface entities on the fungus cells and suggest that these modifications may be relevant to host-insect cell reaction to the pathogen.

5 Conclusions and Perspectives

In the past two decades, our knowledge of carbohydrate biology has expanded enormously and "glycobiology", a novel scientific field, has been born. Currently, this field is growing due to the joint efforts of scientists from various disciplines, e.g. biochemistry, cell biology, histochemistry and pathology. Histochemistry/cytochemistry is a biological discipline that permits the chemical characterization of cells and tissues in relation to structural organization. It is thus complementary to the biochemical approach. Indeed, biochemical analysis alone is inadequate because the homogenization of an organ obscures its structural heterogeneity, which extends to the molecular level. In histochemistry, morphological relationships are maintained by means of tissue sections. The availability of lectins and neoglycoproteins with a high degree of specificity for different glycoforms has enabled a number of exciting observations to be made. It is expected that glycohistochemistry will advance our knowledge in this area and serve as a guideline for further study that will attempt to unravel the significance of protein-glycan interactions.

Acknowledgments

This work is partly funded by an F.N.R.S. (Fonds National de la Recherche Scientifique, Belgique) grant to A.D. The authors are grateful to P. Miroir for her outstanding secretarial assistance.

References

Akif F, Gabius H-J, Danguy A (1993): Effects of castration and thyroidectomy on expression of lectin-binding sugar moieties and endolectins in mouse submandibular glands. A glycohistochemical study. In *In vivo* **7**:37–44.

Akif F, Gabius H-J, Danguy A (1994): Androgen status and expression of glycoconjugates and lectins in the epithelial cells of the mouse ventral prostate. A glycohistochemical approach. In *Histol. Histopath.* **9**:705–13.

Akif F, Gabius H-J, Danguy A (1995): Estrous cycle-related alterations in the expression of glycoconjugates and lectins in the mouse endometrium shown histochemically. In *Tissue and Cell* **27**:197–206.

Alroy J, Ucci A, Pereira MEA (1988): Lectin histochemistry : An update. In *Advances in Immunohistochemistry* (De Lellis RA, ed) pp 93–131, New York: Raven Press.

Araujo ACG, Souto Padrón T, de Souza W (1993): Cytochemical localization of carbohydrate residues in microfilariae of *Wuchereria bancrofit* and *Brugia malayi*. In *J. Histochem. Cytochem.* **41**:571–8.

Balm PHM, Iger Y, Prunet P *et al.* (1995): Skin ultrastructure in relation to prolactin and MSH function in rainbow trout (*Oncorhynchus mykiss*) exposed to environmental acidification. In *Cell Tissue Res.* **279**: 351–8.

Barondes SH, Cooper DNW, Gitt MA *et al.* (1994): Galectins. Structure and function of a large family of animal lectins. In *J. Biol. Chem.* **269**:20807–10.

Beesley JE (1989): Colloidal Gold: A New Perspective for Cytochemical Marking. Oxford Rochal Microscopical Society, Handbook Series, vol. 17, Oxford: Oxford University Press.

Bentley PJ (1982): Comparative Vertebrate Endocrinology, 2nd ed. Cambridge: Cambridge University Press.

Bereither-Hahn J, Matoltsy AG, Richards KS (1986): Biology of the Integument. 2. Vertebrates. Berlin, Heidelberg: Springer Verlag.

Bern HA (1975): Prolactin and osmoregulation. In *Am. Zool.* **15**:937–948.

Blithe DL (1993): Biological functions of oligosaccharides on glycoproteins. *Trends Glycosci. Glycotechnol.* **5**:81–98.

Bonilla E, Moggio M (1987): Electron cytochemical study of the muscle cell surface. In *Histochemistry* **86**: 503–7.

Bulger RE (1988): The urinary system. In *Cell and Tissue Biology. A Textbook of Histology*, 6th edition (Weiss L, ed) pp 815–50, Baltimore: Urban & Schwarzenberg.

Burkhardt P, Schumaker U, Welsch U *et al.* (1987): Lectin histochemistry of the teleost intestine and liver under normal and starved conditions. In *Z. Mikrosk.-Anat. Forsch.* **101**:301–17.

Capaldi MJ, Dunn MJ, Sewry CA *et al.* (1985): Lectin binding in human skeletal muscle : a comparison of 15 different lectins. In *Histochem. J.* **17**: 81–92.

Caron M, Bladier D, Joubert R (1990): Soluble galactoside-binding vertebrate lectins: A protein family with common properties. In *Int. J. Biochem.* **22**: 1379–85.

Chaiet L, Wolf FJ (1964): The properties of streptavidin, a biotin binding protein produced by *Streptomyces*. In *Arch. Biochem. Biophys.* **106**:1–5.

Christie KN, Thomson C (1989): *Bandeiraea simplicifolia* lectin demonstrates significantly more capillaries in rat skeletal muscle than enzyme methods. In *J. Histochem. Cytochem.* **37**:1303–4.

Clark G (1981): Staining procedures, 4th edition. Baltimore: Williams and Wilkins.

Coppée I, Gabius H-J, Danguy A (1993): Histochemical analysis of carbohydrate moieties and sugar-specific acceptors in the kidneys of the laboratory mouse and the golden spiny mouse (*Acomys russatus*). In *Histol. Histopath.* **8**:673–83.

Damjanov I (1987): Lectin cytochemistry and histochemistry. In *Lab. Invest.* **57**:5–20.

Danguy A, Genten F (1989): Comparative lectin-binding patterns in the epidermis and dermal glands of *Bufo bufo* (L) and *Xenopus laevis* (Daudin). In *Biol. Struct. Morphol.* **2**:94–101.

Danguy A, Genten F (1990): Lectin histochemistry on glucoconjugates of the epidermis and dermal glands of *Xenopus laevis* (Daudin 1802). In *Acta. Zool.* **71**:17–24.

Danguy A, Gabius H-J (1993): Lectins and neoglycoproteins: attractive molecules to study glycoconjugates and accessible sugar-binding sites of lower vertebrates histochemically. In *Lectins and Glycobiology* (Gabius H-J, Gabius S, eds) pp 241–50, Berlin, New York: Springer Verlag.

Danguy A, Kiss R, Pasteels JL (1988): Lectins in histochemistry : A survey. In *Biol. Struct. Morphol.* **1**:93–106.

Danguy A, Genten F, Gabius H-J (1991): Histochemical evaluation of application of biotinylated neoglycoproteins for the detection of endogenous sugar receptors in fish skin. In *Eur. J. Appl. Histochem.* **35**:341–57.

Danguy A, Akif F, Pajak B *et al.* (1994): Contribution of carbohydrate histochemistry to glycobiology. In *Histol. Histopath.* **9**:155–71.

Dantzler WH (1989): Comparative Physiology of the Vertebrate Kidney. Berlin, New York: Springer Verlag.

de Miranda Santos IFK, Pereira MEA (1984): Lectin discriminates between pathogenic and nonpathogenic South American trypanosomes. In *Am. J. Trop. Med. Hyg.* **33**:839–44.

Dent JN (1975): Integumentory effects of prolactin in the lower vertebrate. In *Am. Zool.* **15**:923–35.

Drickamer K (1988): Two distinct classes of carbohydrate-recognition domains in animal lectins. In *J. Biol. Chem.* **263**:9557–60.

Drickamer K (1994): Molecular structure of animal lectins. In *Molecular Glycobiology* (Fukuda M, Hindsgaul O, eds) pp 53–87, Oxford: IRL Press, Oxford University Press.

Drickamer K, Carver J, eds (1992): Carbohydrates and glycoconjugates. In *Curr. Opinion Struct. Biol.* **2**: 653–710.

Dwek RA (1995): Glycobiology: towards understanding the function of sugars. In *Biochem. Soc. Transact.* **23**:1–25.

Egea G, Marsal J (1992): Carbohydrate patterns of the pure cholinergic synapse of torpedo electric organ : A cytochemical electron microscopic approach. In *J. Histochem. Cytochem.* **40**:513–21.

Ellis J, Haliday G (1992): A comparative study of avidin-biotin peroxidase complexes for the immunohistochemical detection of antigens in neural tissue. In *Biotech. Histochem.* **67**:367–71.

Fasulo S, Tagliafierro G, Contini A *et al.* (1993): Ectopic expression of bioactive peptides and serotonin in the sacciform gland cells of teleost skin. In *Arch. Histol. Cytol.* **56**:117–25.

Fox H (1986): The skin of amphibia. In *Biology of the Integument. Vertebrates* (Bereiter-Hahn J, Matoltsy AG, Richards KS, eds) pp 78–135, Berlin, Heidelberg: Springer Verlag.

Fromm PO (1980): A review of some physiological and toxicological responses of freshwater fish to acid stress. In *Environ. Biol. Fish* **5**:79–93.

Fukuda M (1994): Cell surface carbohydrates : Cell-type specific expression. In *Molecular glycobiology* (Fukuda M, Hindsgaul O, eds) pp 1–52, Oxford: IRL Press, Oxford University Press.

Gabius H-J (1991): Detection and functions of mammalian lectins with emphasis on membrane lectins. In *Biochim. Biophys. Acta* **1071**:1–18.

Gabius H-J, Bardosi A. (1991): Neoglycoproteins as tools in glycohistochemistry. In *Prog. Histochem. Cytochem.* **22**:1–66.

Gabius H-J, Gabius S, eds (1993): *Lectins and Glycobiology* Berlin, New York: Springer Verlag.

Gabius H-J, Gabius S, Zemlyanukhina T.V *et al.* (1993): Reverse lectin histochemistry: design and application of glycoligands for detection of cell and tissue lectins. In *Histol. Histopath.* **8**:369–83.

Gabius H-J, André S, Danguy A *et al.* (1994): Detection and quantification of carbohydrate-binding sites on cell surfaces and in tissue sections by neoglycoproteins. In *Meth. Enzymol.* **242**:37–46.

Gambaryan SP (1994): Microdissectional investigation of the nephrons in some fishes, amphibians, and reptiles inhabiting different environments. In *J. Morphol.* **219**:319–39.

Genten F, Danguy A (1990a): Light microscopic characterization of glycoconjugates and cell proliferation in teleost epidermis exposed to acute levels of pH. In *Z. Mikrosk.-Anat. Forsch.* **104**:119–39.

Genten F, Danguy A (1990b): A comparative histochemical analysis of glycoconjugates in secretory cells of fish epidermis by use of biotinylated lectins. In *Z. Mikrosk.-Anat. Forsch.* **104**:835–55.

Genten F, Danguy A (1990c): Comparative lectin histochemical characterization of fish and anuran skin extracellular matrices. In *Eur. Arch. Biol.* **101**: 133–52.

Goldstein IJ, Poretz RD (1986): Isolation, physiochemical characterization and carbohydrate-binding specificity of lectins *The Lectins: Properties, Functions and Applications in Biology and Medicine* (Liener IE, Sharon N, Goldstein IJ, eds) pp 33–247, New York: Academic Press Inc.

Green NM (1975): Avidin. In *Adv. Prot. Chem.* **29**:85–133.

Grumbach IM, Veh RW (1995): The SA/rABC technique: A new ABC procedure for detection of antigens at increased sensitivity. In J. *Histochem. Cytochem.* **43**:31–7.

Gulati AK, Zalewski AA (1985): An immunofluorescent analysis of lectin binding to normal and regenerating skeletal muscle of rat. In *Anat. Rec.* **212**:113–7.

Harrison FL (1991): Soluble vertebrate lectins: Ubiquitous but inscrutable proteins. In *J. Cell Sci.* **100**:9–14.

Hentschel H, Elger M (1987): The distal nephron in the kidney of fishes. In *Adv. Anat. Cell Biol.* **108**:1–51.

Hentschel H, Walther P (1993): Heterogeneous distribution of glycoconjugates in the kidney of dogfish *Scyliorhinus caniculus* (L) with reference to changes in the glycosylation pattern during ontogenetic development of the nephron. In *Anat. Rec.* **235**:21–32.

Hirabayashi J (1994): Two distinct families of animal lectins : speculations on their Raison d'Etre. In *Lectins: Biology, Biochemistry, Clinical Biochemistry*, vol. 10 (Van Driessche E, Fisher J, Beeckmans S, Bøg-Hansen TC, eds) pp 205–19, Hellerup, Denmark: Textop.

Hoyer PE, Lyon H, Moller M *et al.* (1991): Tissue processing: III. Fixation, general aspects. In *Theory and Strategy in Histochemistry* (Lyon H, ed) pp 171–86, Berlin: Springer Verlag.

Hsu SM, Raine L, Fanger H (1981): Use of avidin-biotin-peroxidase complex (ABC) in immunoperoxidase techniques : A comparison between ABC and unlabelled antibody (PAP) procedures. In *J. Histochem. Cytochem.* **29**:577–80.

Iger I, Wendelaar Bonga SE (1994): Cellular responses of the skin and change in plasma cortisol levels of carp (*Cyprinus carpio*) exposed to acidified water. In *Cell Tissue Res.* **275**:481–92.

Iger I, Balm PHM, Wendelaar Bonga SE (1994a): Cellular responses of the skin of trout (*Oncorhynchus mykiss*) exposed to acidified water. In *Cell Tissue Res.* **278**:535–42.

Iger I, Abraham M, Wendelaar Bonga SE (1994b): Response to club cells in the skin of the carp *Cyprinus carpio* to exogenous stressors. In *Cell Tissue Res.* **277**:485–91.

Iglesias M, Ribera J, Esquerda JE (1992): Treatment with digestive agents reveals several glycoconjugates specifically associated with rat neuromuscular junction. In *Histochemistry* **97**:125–31.

Kariniemi AL, Virtanen I (1989): *Dolichos biflorus* agglutinin (DBA) reveals a similar basal cell differentiation in normal and psoriatic epidermis. In *Histochemistry* **93**:129–32.

Kiernan J.A. (1990): Histological and Histochemical methods. Theory and Practice, 2nd edition. Oxford: Pergamon Press.

Kirkeby S (1994): Glycosylation pattern and enzyme activities in atrophic, angulated skeletal muscle fibres from aging rats. In *Virchows Arch.* **424**:279–85.

Kirkeby S, Bøg-Hansen TL, Moe D *et al.* (1991): Lectin binding in skeletal muscles. Evaluation of alkaline phosphatase conjugated avidin staining procedures. In *Histochem. J.* **23**:345–54.

Kirkeby S, Garbarsch C, Matthiessen ME *et al.* (1992): Changes of soluble glycoproteins in dystrophic (dy/dy) mouse muscle shown by lectin binding. In *Pathobiology* **60**:297–302.

Kirkeby S, Bøg-Hansen TL, Moe D (1992): Mosaic lectin and enzyme staining patterns in rat skeletal muscle. In *J. Histochem. Cytochem.* **40**:1511–6.

Kirkeby S, Moe D, Bøg-Hansen TL (1993): Fucose expression in skeletal muscle : a lectin histochemical study. In *Histochem. J.* **25**:619–27.

Kobata A (1992): Structures and functions of the sugar chains of glycoproteins. In *Eur. J. Biochem.* **209**: 483–501.

Kuchler S, Zanetta J-P, Vincendon G *et al.* (1990): Detection of binding sites for biotinylated neoglycoproteins and heparin (endogenous lectins) during cerebellar ontogenesis in the rat. In *Eur. J. Cell Biol.* **52**:87–97.

Kuchler S, Zanetta J-P, Vincendon G *et al.* (1992): Detection of binding sites for biotinylated neoglycoproteins and heparin (endogenous lectins) during cerebellar ontogenesis in the rat: an ultrastructural study. In *Eur. J. Cell Biol.* **59**:373–81.

Lee YC (1992): Biochemistry of carbohydrate-protein interaction. In *FASEB J.* **6**:3193–200.

Lee YC, Lee RT, eds (1994): *Neoglycoconjugates: Preparation and Application.* San Diego: Academic Press.

Liener I.E, Sharon N, Goldstein IT, eds (1986): The Lectins. Properties, Functions and Applications in Biology and Medicine. Orlando, Fl: Academic Press.

Lis H, Sharon N (1986): Lectins as molecules and as tools. In *Annu. Rev. Biochem.* **55**:35–67

Lis H, Sharon N (1993): Protein glycosylation. Structural and functional aspects. In *Eur. J. Biochem.* **218**:1–27.

Lyon H, ed (1991): Theory and Strategy in Histochemistry. Berlin: Springer Verlag, **35**:1836–90.

Mandrell RE, Apicella MA, Lindstedt R *et al.* (1994): Possible interaction between animal lectins and bacterial carbohydrates. In *Methods Enzymol.* **236**:231–54.

Matoltsy EG (1986): Structure and function of the mammalian epidermis. In *Biology of the Integument. 2. Vertebrates* (Bereiter-Hahn J, Matoltsy AG, Richards KS, eds) pp 255–71, Berlin, Heidelberg: Springer Verlag.

Menghi G, Materazzi G (1994): Exoglycosidases and lectins as sequencing approaches of salivary gland oligosaccharides. In *Histol. Histopath.* **9**:173–83.

Miyazawa M, Kimura T, Itagaki S (1994): Lectin histochemistry on the skin of hairless descendants of Mexican hairless dogs. In *Tissue and Cell* **26**:19–27.

Mohamed HA, Ingram GA, Molyneux DH *et al.* (1991): Use of fluorescein-labelled binding of salivary glands to distinguished between *Anopheles stephensi* and *A. albimanus* species and strains. In *Insect Biochem.* **21**:767–73.

Morioka H, Suganuma A, Tachibana M (1994): Localization of sugar-binding sites in *Staphylococcus aureus* using gold-labeled neoglycoprotein. In *J. Histochem. Cytochem.* **42**:1609–13.

Müller K (1994): Animal lectins: Recent progress. In *Lectins: Biology, Biochemistry, Clinical Biochemistry*, vol. 10 (Van Driessche E, Fischer J, Beeckmans S, Bøg-Hansen TC, eds) pp 234–43, Hellerup, Denmark: Textop.

Navas P, Villalba JM, Burón MI *et al.* (1987): Lectins as markers for plasma membrane differentiation of amphibian keratinocytes. In *Biol. Cell* **60**:225–34.

Ohno J, Fukuyama K, Epstein WL (1990): Glycoconjugate expression of cells of human anagen hair follicles during keratinization. In *Anat. Rec.* **228**:1–6.

Ookusa Y, Takata K, Nagashima M *et al.* (1983): Distribution of glycoconjugates in normal human skin using biotinyl lectins and avidin-horseradish peroxidase. In *Histochemistry* **79**:1–7.

Pajak B, Danguy A (1993): Characterization of sugar moieties and oligosaccharide sequences in the distal intestinal epithelium of the rainbow trout by means of lectin histochemistry. In *J. Fish Biol.* **43**:709–22.

Paljärvi L, Naukkarinen A (1990): Histochemical method for simultaneous fiber typing and demonstration of capillaries in skeletal muscle. In *Histochemistry* **93**:385–7.

Pearse AGE (1980–1985): Histochemistry. Theoretical and Applied, 4th edition, vol. 1 Preparative and Optical Technology, vol. 2 Analytical Technique. Edinburgh: Churchill Livingstone.

Peixoto CA, De Souza W (1992): Cytochemical characterization of the cuticle of *Caenorhabditis elegans* (Nematoda: Rhabditoidea). In *J. Submicrosc. Cytol. Pathol.* **24**:425–35.

Pena SDJ, Gordon BB, Karpati G *et al.* (1981): Lectin histochemistry of human skeletal muscle. In *J. Histochem. Cytochem.* **29**:542–6.

Pendland JC, Boucias DG (1993): Variations in the ability of galactose- and mannose-specific lectins to bind to cell wall surfaces during growth of the insect pathogenic fungus *Paecilomyces farinosus.* In *Eur. J. Cell Biol.* **60**:322–30.

Pfeiffer W (1974): Pheromones in fish and amphibia. In *Pheromones. Frontiers of Biology* (Birch MC, ed) pp 269–96, Amsterdam: North Holland Publ. Comp.

Pfeiffer W (1977): The distribution of fright reaction and alarm substance cells in fishes. In *Copeia* **1977**:653–65.

Previato JO, Jones C, Gonçalves LPB *et al.* (1994): *O*-glycosidically linked *N*-acetylglucosamine-bound oligosaccharides from glycoproteins of *Trypanosoma cruzi*. In *Biochem. J.* **301**:151–9.

Rademacher TW, Parekh RB, Dwek RA (1988): Glycobiology. In *Annu. Rev. Biochem.* 57:785–838.

Ribera J, Esquerda JE, Comella JX (1987): Phylogenetic polymorphism on lectin binding to junctional and non-junctional basal lamina at the vertebrate neuromuscular junction. In *Histochemistry* **87**: 301–307.

Roth J (1993): Cellular sialoglycoconjugates: a histochemical perspective. In *Histochem. J.* **25**:687–710.

Rudin W (1990): Comparison of the cuticular structure of parasitic nematodes recognized by immunocytochemical and lectin binding studies. In *Acta Trop.* 47:255–68.

Schaumburg-Lever G, Alroy J, Ucci A *et al.* (1984): Distribution of carbohydrate residues in normal skin. In *Arch. Dermatol. Res.* **276**:216–23.

Schmidt-Nielsen K (1991): Animal Physiology: Adaptation and Environment. 4th edition, Cambridge: Cambridge University Press.

Schottelius J (1990): New World trypanosomatidae: The importance of lectins for their differentiation with special emphasis on *Trypanosoma cruzi*, the agent of Charga's disease. In *Lectins: Biology, Biochemistry, Clinical Biochemistry.*, vol. 7 (Kocourek J, Freed DLJ, eds) pp 321–7, St. Louis, Missouri: Sigma Chemical Company.

Schumacher U, Adam E, Hauser F *et al.* (1989): Molecular anatomy of a skin gland: Histochemical and biochemical investigations on the mucous glands of *Xenopus laevis*. In *J. Histochem. Cytochem.* **42**: 57–65.

Sharon N, Lis H (1989): Lectins as cell recognition molecules. In *Science* **246**:227–34.

Sharon N, Lis H (1993): Carbohydrates in cell recognition. In *Scient. Am.* **268**:74–81.

Singhas CA, Ward DL (1993): Prolactin alters the expression of integumental glycoconjugates in the Red-stopped newt, *Notophthalmus viridesceus*. In *Anat. Rec.* **236**:537–46.

Skutelsky E, Goyal W, Alroy J (1987): The use of avidin-gold complex for light microscopic localization of lectin receptors. In *Histochemistry* **86**: 291–5.

Spicer SS (1993): Advantages of histochemistry for the study of cell biology. In *Histochem. J.* **25**:531–47.

Spicer SS, Schulte BA (1992): Diversity of cell glycoconjugates shown histochemically : A perspective. In *J. Histochem. Cytochem.* **40**:1–38.

Strout RG, Alroy J, Lukacs NW *et al.* (1994): Developmentally regulated lectins in *Eimeria* species and their role in avian coccidiosis. In *J. Parasitol.* **80**: 946–51.

Truong LD, Phung VT, Yoshikawa Y *et al.* (1988): Glycoconjugates in normal human kidney: A histochemical study using 13 biotinylated lectins. In *Histochemistry* **90**:51–60.

Varki A (1993): Biological roles of oligosaccharides: all of the theories are correct. In *Glycobiology* **3**: 97–130.

Varki A (1994): Selectin ligands. In *Proc. Natl. Acad. Sci. USA* **91**:7390–7.

Villalba JM, Navas P (1989): Polarization of plasma membrane glycoconjugates in amphibian epidermis during metamorphosis. In *Histochemistry* **90**: 453–8.

Villalba JM, Roldán JM, Navas P (1993): Flask cells and flask-shaped glandular cells of amphibian skin specifically produce fucose-rich glycoproteins. In *Histochemistry* **99**:363–7.

Virtanen I, Kariniemi AL, Holthofer H *et al.* (1986): Fluorochrome-coupled lectins reveal distinct cellular domains in human epidermis. In *J. Histochem. Cytochem.* **34**:307–15.

Watanabe M, Muramatsu T, Shirane H *et al.* (1981): Discrete distribution of binding sites for *Dolichos biflorus* agglutinin (DBA) and for peanut agglutinin (PNA) in mouse organ tissue. In *J. Histochem. Cytochem.* **29**:779–90.

Watt FM, Jones PH (1992): Changes in cell-surface carbohydrate during terminal differentiation of human epidermal keratinocytes. In *Biochem. Soc. Transact.* **20**:285–8.

Wendelaar Bonga SE, Dederen LHT (1986): Effects of acidified water on fish. In *Endeavour* **10**:198–202.

Whitear M (1986): The skin of fishes including cyclostomes. In *Biology of the Integument* (Bereiter-Hahn J, Matoltsy AG, Richards KS, eds) pp 8–64, Berlin: Springer Verlag.

Whitear M, Zaccone G (1984): Fine structure and histochemistry of club cells in the skin of three species of eel. In *Z. Mikrosk.-Anat. Forsch.* **98**:481–501.

Witt M, Reuter K (1988): Lectin histochemistry on mucous substances of the taste buds and adjacent epithelia of different vertebrates. In *Histochemistry* **88**:453–61.

Wu AM, Sugii S, Herp A (1988): A table of lectin carbohydrate specificities. In *Lectins. Biology, Biochemistry, Clinical Biochemistry*, vol. 6 (Bøg-Hansen TC, Freed DLJ, eds) pp 723–40, St. Louis, Missouri: Sigma Chemical Company.

Zaccone G, Fasulo S, Lo Cascio P *et al.* (1986): 5-Hydroxytryptamine immunoreactivity in the epidermal sacciform gland cells of the clingfish *Lepadogaster candollei* Risso. In *Cell Tissue Res.* **246**:679–82.

Zaccone G, Tagliafierro G, Fasulo S *et al.* (1990): Serotonin-like immunoreactivity in the epidermal club cells of teleost fishes. In *Histochemistry* **93**: 355–7.

33 Lectins and Neoglycoproteins in Histopathology

S. Kannan and M. Krishnan Nair

1 Introduction

1.1 Relevance of Histopathology

Health and disease are complementary to each other, and all living organisms experience these states during their lives. Health is the condition in which the organism is in complete accord with its surroundings, with that exquisite coordination of the different functions which characterizes the living animal or plant. The diseased state is a change in the above condition as a result of which the organism suffers from discomfort (disease) (Boyd, 1970). The healthy state is the normal condition in an organism, and once in a while the organism becomes diseased, due either to physiological dysfunction or to pathogenic infection. In the management of disease, pathology has a key role. Pathology is the study of disease, its nature and its causes, and is basic to both the clinical and the laboratory phases of medicine (Boyd, 1970). In recent years pathology has been developed very extensively and divided into many branches. Of these, surgical pathology or histopathology is very important and is the most developed branch (Rosai, 1989). It includes examination of the diseased tissues or cells under the microscope to detect the exact causes for the present condition, and also the identification and classification of the disease (diagnosis) for proper treatment. The characteristics of cells or tissues in various pathological conditions have been documented vividly (Rosai, 1989). Hence histopathologists compare the cellular characteristics of the sample with documented cases and make the diagnosis.

In classical histopathology, diagnosis is mostly based on morphological and structural features of cells and tissues, and to a lesser extent on the chemistry of the cell (histochemistry) (Rosai, 1989). This method of diagnosis is efficient and is still sufficient for the majority of cases even these days. Only in a few cases does classical histopathology meet with some limitations in diagnosis. Various methods have recently been devised to overcome these limitations. The development of immunohistochemisry has led to a renaissance in the field of histopathology (Seifert and Caselitz, 1989). Immunohistochemistry is utilized for diagnostic purposes by exploiting the differential expression and the quantitative changes in various proteins in healthy as well as in pathological conditions of the cells (Seifert and Caselitz, 1989). This has been achieved using specific antibodies against these proteins. In pathological conditions, as well as proteins, the sugar part of cellular glycoconjugates often shows alterations in expression (Rademacher *et al.*, 1988). It should be remembered in this context that in the mammalian system many proteins require this post-translational modification (glycosylation) to become functionally active (West, 1986; Hughes, 1992). Aberrant glycosylation is a common feature of many pathologic cells, especially neoplastic cells (Hakomori, 1989). Although first described many years ago, the role of aberrant patterns of glycosylation is not fully understood (Chammas *et al.*, 1994). Protein-carbohydrate interactions constitute a system of molecular interaction with relevance to pathological conditions (Lee, 1992). The variations in carbohydrate composition can also be used for histopathological diagnosis.

H.-J. and S. Gabius (Eds.), Glycosciences

ISBN 3-8261-0073-5

1.2 Glycoconjugates and Pathology

Much interest has been devoted to cell surface carbohydrates, as they are related to differentiation and maturation phenomena and show marked changes in various disease conditions (Jones and Smith, 1983). This emphasis stems from the observations that the cell surface membrane may determine several characteristics of neoplastic cells, such as decreased adhesion, loss of contact inhibition, increased growth rate, prolonged survival, increased invasiveness and motility, expression of "new" antigens, and escape from immunological surveillance. Membrane glycoconjugates are oligosaccharides linked either to lipids or to proteins, and can apparently play an important role in the development of these neoplastic features (Jones and Smith, 1983; Yogeeswaran, 1983; Sell, 1990). The structural complexity in carbohydrate structures, which reflects the large amount of information available for their interactions, makes these carbohydrate antigens ideal markers for various cellular processes (Gabius, 1988; Mandel *et al.*, 1988; Glinsky, 1992). The most widely studied surface glycoproteins are blood group antigens, and the expression of these antigens shows profound alterations during various disease states, especially in malignancy (Caselitz, 1987; Dabelsteen *et al.*, 1988, 1992). An important and convenient class of probes for the analysis of alterations in the expression of carbohydrates is the family of lectins (Lis and Sharon, 1986; Caselitz, 1987; Alroy *et al.*, 1988; Walker, 1989). Until recently, only lectins of plant origin have been used extensively to study the sugar expression patterns in various types of cells and tissues. Recently, as well as these plant lectins, lectins from animals have also been isolated and characterized (Gabius, 1991; Lotan, 1992; Zhou and Cummings, 1992). For studying the physiological and pathological relevance of the vertebrate endogenous lectins, a new variety of tailor-made probes, called neoglycoproteins (NGPs) has been designed (Gabius and Gabius, 1992; Gabius *et al.*, 1993; Lee and Lee, 1994). The present chapter discusses briefly the current status and perspectives on the application of lectins and NGPs in histopathology.

2 Lectins

Various aspects of the structure, properties, functions and uses of lectins are reviewed in detail elsewhere in this book (Cummings, 1996; Danguy *et al.*, 1996; Gilboa-Garber *et al.*, 1996; Ohannesian and Lotan, 1996; Rice, 1996; Rüdiger, 1996; Zanetta, 1996). Of the many possible monosaccharides, seven are particularly important in the construction of oligosaccharides found as glycolipids or glycoproteins on the surface of human cells as well as in secreted mucins or glycoproteins (Yogeeswaran, 1980). These include D-glucose, *N*-acetyl-D-glucosamine, D-galactose, *N*-acetyl-D-galactosamine, D-mannose, *N*-acetyl-D-neuraminic acid (sialic acid) and L-fucose (Sell, 1990). Lectins are now known and are commercially available with specificities to these monosaccharides, and can be grouped according to their sugar specificity. The majority of the lectins used in histochemistry are of plant origin, except for a few of animal origin. For example, the lectin isolated from a snail (*Helix pomatia*) has been used extensively in histopathology. In glycobiology, lectins can be classified broadly as exogenous or endogenous, based on their presence in the human tissues. The most extensively characterized exogenous lectins in histopathology are ConA, UEA, PNA, WGA, DBA, RCA-I, JFL, PHA, LCA, SBA, BSA, PSA, GSA-I, VVA, LTA and HPA (Walker, 1989; Danguy *et al.*, 1994; Vijayakumar *et al.*, 1994).

2.1 Methods in Lectin Histochemistry

Principles and procedures in lectin histochemical techniques are almost identical to those for immunohistochemical methods. Here, instead of primary antibodies, the respective lectins are used. Immunohistochemical methods such as direct, indirect, peroxidase anti-peroxidase (PAP), avidin-biotin and avidin-biotin complex can also be employed in lectin histochemistry (Denk, 1987). Tissue fixation, processing and staining procedures for using lectins in histopathology have been reviewed extensively by Mason and Matthews (1988), Walker (1989) and

Danguy *et al.* (1994). For the detection of lectin binding, especially exogenous lectins, the direct method is generally preferred. In the direct method the lectins are conjugated to labels such as enzymes or fluorochromes, making the method simple so that it can be performed in a single step. The avidin-biotin-peroxidase complex (ABC) method is also not uncommon in lectin histochemical studies (Hsu and Raine, 1982). This is a three-step procedure and more sensitive than the direct method. Consequently, a reduced amount of lectin is sufficient, and this method is commonly used in studies with animal lectins. As in immunohistochemistry, lectin binding also can be studied at the electron microscopical level (Horisberger, 1984). Lectins conjugated to electron dense labels such as ferritin or colloidal gold are used in lectin-electron microscopical studies. The colloidal gold-labeling method can also be used for light microscopical studies (Roth *et al.*, 1992). Horseradish peroxidase, using DAB as a substrate, can also be used in electron microscopy (Jones and Stoddart, 1986). There are two methods for lectin incubation. One is the pre-embedding procedure, in which a small piece of the tissue or a thick, frozen section is incubated with the lectin and subsequently processed for electron microscopy. The other method is post-embedding staining, in which an ultrathin section is incubated with a labelled lectin (Walker, 1989). With the advent of new embedding polymers for electron microscopy such as LR White the second method has become more efficient. In all these methods, rigorous controls have to be included to assess the specificity of lectin binding. For this, the lectin should be preincubated with an excess of monosaccharides specific for the lectin to be assessed, and then the mixture is placed on the section, which is processed in the same way as all other sections. If the control section shows staining, then either the purity of the lectin must be suspected and re-examined, or binding can take place via other non-carbohydrate-binding domains of the lectin.

2.2 Exogenous Lectins in Histopathology

Comparatively few reports are available about the potential use of lectins in histopathology. Most of the studies in lectin biology have been on their agglutination, immunomodulatory and cytotoxic properties (Liener *et al.*, 1986; Gabius and Gabius, 1993; Vijayakumar *et al.*, 1994) and are mostly confined to cancer biology, because significant alterations in the composition of cellular glycoconjugates have been shown more often in malignant conditions than in other disease states (Gabius and Gabius, 1991). Reviews and monographs on the use of lectins in histopathology are available and are recommended reading to get wider perspectives on lectin histochemistry (Liener *et al.*, 1986; Caselitz, 1987; Damjanov, 1987; Alroy *et al.*, 1988; Danguy *et al.*, 1988; Gabius, 1988; Sharon and Lis, 1989; Walker, 1989; Gabius and Gabius, 1991; Vierbuchen, 1991; Spicer and Schulte, 1992; Gabius and Gabius, 1993; Danguy *et al.*, 1994; Vijayakumar *et al.*, 1994). In the present chapter, it is intended to discuss only recent reports on lectin histochemistry.

A few studies have reported the diagnostic potential of lectins in oral premalignant and malignant lesions (Saku and Okabe, 1989; Vigneswaran *et al.*, 1990; May and Sloan, 1991). Studies with JFL in oral mucosal lesions have demonstrated its usefulness in diagnosis, in early detection, and in predicting the radioresponsiveness of the tumors (Vijayan *et al.*, 1987; Bhattathiri *et al.*, 1991). A recent study using eight lectins in normal tissue, and in premalignant and malignant lesions in oral mucosa, showed that lectins, including PNA and ConA, can be used as histochemical probes to detect the dysplastic and malignant status of the oral mucosa (Mazumdar *et al.*, 1993). However, the results of our study, using fresh frozen sections of normal tissue, and premalignant and malignant lesions of the oral mucosa, showed that the lectins PNA and ConA have only limited value in early diagnosis or prediction of the biological behavior of potentially malignant oral lesions (Kannan *et al.*, 1993). A marked difference in the binding pattern of lectins such as WGA, ECA, PNA and UEA was observed between ameloblastoma and odontogenic keratocyst, and this difference can be used in diagnosis (Aguirre *et al.*, 1989). In esophageal carcinomas, HPA binding has been shown to correlate significantly with increasing depth of cancer invasion, venous invasion, tumor stage and poor prognosis. Thus, the HPA staining method is apparently useful for the prognosis of patients with carcinoma of the esophagus (Yoshida *et al.*, 1993, 1994).

In lectin histochemistry, extensive work has been done in breast tissues, but few studies showed remarkable results. The studies on tissue binding properties JFL and DSA in benign and malignant lesions of the breast showed that neoplastic cells have more affinity towards these lectins than the normal or benign cells. This result suggests the potential of these lectins as histochemical markers in detection of breast neoplastic cells (Remani *et al.*, 1989; Hiraizumi *et al.*, 1992). However, a study with a panel of eight lectins in fibroadenoma, cystosarcoma phyllodes, fibrocystic disease and infiltrating duct carcinoma showed only a limited value for lectins in diagnosis (Karuna *et al.*, 1992). The binding pattern of HPA has been studied extensively in breast lesions for its prognostic significance. Initial studies have reported the prognostic significance of *Helix pomatia* lectin (HPA) staining on disease-free and overall survival in primary breast carcinomas. The studies revealed that for patients in whom axillary lymph node dissection has not been performed, HPA staining status alone or in combination with other markers such as DNA ploidy and c-erbB-2 expression seems to be a powerful tool to discriminate subpopulations with a high recurrence risk and shorter survival who should undergo more aggressive therapy (Alam *et al.*, 1990; Brooks and Leathem, 1991; Brooks *et al.*, 1993; Noguchi *et al.*, 1993 a, b; Thomas *et al.*, 1993). However, a large multicentric study on the prognostic value of HPA binding for the survival of 684 primary breast cancers by the International (Ludwig) Breast Cancer Study Group showed that HPA is of no clinical predictive value (International (Ludwig) Breast Cancer Study Group, 1993).

In pancreatic cancer, lectins appear to have some potential for diagnosis because a few tumor-associated glycoproteins have been well characterized. These glycoproteins from pancreatic cancer showed intense and specific reactivity for BPA, VVA, WGA, PNA lectins and thus should provide a useful approach to the diagnosis of pancreatic cancer (Kawa *et al.*, 1991, 1992; Parker *et al.*, 1992; Shue *et al.*, 1993). Lectins also proved to be useful in diagnosis of thyroid malignancies. The binding properties of lectins JFL, WBL, and PNA in benign and malignant lesions of the thyroid showed that PNA and JFL conjugates exhibited differential positive binding to the cells in different lesions, while WBL, despite having a common inhibitory sugar, did not bind even after neuraminidase treatment (Vijayakumar *et al.*, 1992). UEA lectin has been useful in the identification of a case of malignant hemangioendothelioma of the thyroid (Beer, 1992). In a study on microacinar medullary carcinomas of the thyroid, the pattern of binding of lectins did not disclose any significant differences from that of follicular medullary carcinomas of the thyroid (Sobrinho-Simoes *et al.*, 1990).

Lectin-binding studies in human bronchial tissue are comparatively rare. The binding pattern of lectins RCA-I and UEA-I showed significant differences in staining pattern between the primary adenocarcinomas of the lungs and their regional and distant metastatic lesions (Sugiyama *et al.*, 1992). A recent study with BPA, PHA and MPA on normal and neoplastic lung tissue revealed only limited usefulness of these lectins in diagnosis of pulmonary neoplasms (Sarker *et al.*, 1994). JFL proved to be a good marker in differentiating various bronchopulmonary neoplasms both at the histological and the cytological levels (Pillai *et al.*, 1992). Lectins such as PNA, HPL, SBA, LAL, and LCL have demonstrated their usefulness in the identification of malignant cells in pleural effusions (Abramenko *et al.*, 1991). However, these lectins were unsuccessful in determining reliably the tumor cell origin. A few studies in experimental models of lung metastasis showed that HPA and WGA display a significant association with metastatic potential of lung carcinomas (Saito *et al.*, 1992; Kjnniksen *et al.*, 1994). These experimental models apparently illustrate the potential of lectins in prognosis and diagnosis of the human bronchial neoplasms.

Lectin-binding studies on endometrial carcinomas showed that UEA-I and PNA binding can serve as prognostic and diagnostic markers (Aoki *et al.*, 1990; Ookuma *et al.*, 1994). Perchloric acid-soluble glycoprotein fractions from the cyst fluids of human ovarian cystadenomas, as well as borderline and ovarian clear cell carcinomas, exhibited a distinctive pattern of reaction with RCA-I, PNA and VGA (or VUA) which can permit biochemical diagnosis of the degree of malignancy of ovarian cystomas (Yanagi *et al.*, 1990). Lectins can also be used to discriminate between endocervical mucosa and endometrium and between benign, premalignant and malignant conditions of the cervical epithelium (Bychkov and Toto, 1986, 1988; Byrne, 1989a,b). Moreover, one of these studies showed differences in the lectin-binding pattern in endometrium within the course of the menstrual cycle, between normal and gestational endometrium, and in transition to malignancy (Bychkov and Toto, 1988). In this respect, JFL has been demonstrated to be a good diagnostic marker in differentiating potentially premalignant lesions in cervical epithelium both at the histological and the cytological levels (Remani *et al.*, 1990; Pillai *et al.*, 1994).

In the intracranial germ cell tumors, HPA may be useful in predicting the radiotherapy responses (Niikawa *et al.*, 1993a). SBA and VVA can be helpful in the diagnosis of subacute motor neuronopathy (Nagao *et al.*, 1994). Examination of normal and pathological conditions of peripheral nerves permitted the conclusion that lectins have a certain degree of value as markers in diagnosis of various neural diseases (Matsumura *et al.*, 1993 a,b). To distinguish gliomas from higher and lower grades, the binding pattern of RCA-I lectin can be considered as a histochemical marker (Niikawa *et al.*, 1993 b). Studies on oligodendrogliomas and on neoplastic human choroid plexus cells with a panel of lectins suggest that the pattern of lectin expression can undergo substantial changes in the course of differentiation and malignant transformation (Cruz-Sanchez *et al.*, 1991; Kaneko *et al.*, 1991; Niikawa *et al.*, 1993c).

In end stage renal diseases, the identification of nephron segments which are involved in pathological changes is usually impossible by routine histologic examination alone. In this regard the potential of lectins such as PNA, PHA and *Tetragonolobus purpuras* agglutinin as nephron-segment-specific renal epithelial markers in identifying the type of tubular atrophy in end stage kidney diseases has been reported (Ulrich *et al.*, 1985; Ito *et al.*, 1993; Silva *et al.*, 1993; Nadasdy *et al.*, 1994). By analysis of the lectin-binding pattern and survival in transitional cell carcinoma the degree of expression of LTA was found to correlate with increased metastatic potential of bladder cancer and with poor patient survival times (Shirahama *et al.*, 1993). Out of a panel of 16 lectins employed to detect the changes in the expression of glycoconjugate structures in nonmalignant and malignant lesions of urothelium, only the two lectins TKA and SBA demonstrated positive staining reactions in all tumor grades without binding to normal epithelium (Recker *et al.*, 1992). Another study with PNA and WGA lectins on normal and malignant urothelium showed that the binding patterns have a good correlation with DNA content, histopathological grade, and the presence or absence of invasion (Langkilde *et al.*, 1989). A recent investigation reported increases in binding of DBA, PHA and WFA, and a decrease in PNA binding, after neuraminidase treatment, during tumor progression in urinary bladder (Knox *et al.*, 1993). In this field an independent study documented an increase in binding of PNA without neuraminidase treatment in invasive carcinomas of urinary bladder in relation to carcinoma-in-situ, and proposed its usefulness in differential diagnosis (Nakanishi *et al.*, 1993).

UEA has been used as a marker to quantify angiogenesis in cutaneous malignant melanoma (Barnhill *et al.*, 1992). Changes in glycoconjugate expression defined by lectins are observed in disorders of hair growth (Ohno *et al.*, 1990) and in psoriatic epithelium (Bell and Skerrow, 1985). The application of lectins in the diagnosis of a Merkel cell tumor has also been reported (Waibel *et al.*, 1991). The lectin UEA-I has been shown to be useful in differentiating basal cell carcinoma from squamous cell carcinomas of the skin (Heng *et al.*, 1992).

In the identification of Reed-Sternberg cells and variants in Hodgkin's disease, BPA can be regarded as a potential marker due to its high detection rate, reproducible staining pattern, and resistance of its tissue ligands to fixatives (Sarker *et al.*, 1992). In an evaluation of lectins in the diagnosis of myelodysplastic-myeloproliferative disorders, none of the 23 lectins employed can be recommended unequivocally (Schumacher *et al.*, 1991). The utility of different lectins in identifying various forms and types of malignant lymphoproliferative diseases should yet not be excluded (Skliarenko *et al.*, 1990). The most common and widely used lectin in studying angiogenesis and the circulation pattern in normal as well as in various diseased tissues is UEA-I (Miettinen *et al.*, 1983; Kuruvilla and Madhavan, 1992). The potential of UEA-I in cytological diagnosis of cardiac myxomas has been demonstrated (Iwa and Yutani, 1993). Application of UEA-I lectin confirmed the diagnosis of angiosarcoma and its demarcation from other poorly differentiated sarcomas (Kuratsu *et al.*, 1991).

Lectins have been evaluated for diagnosis of malignant lesions of gastrointestinal tract and liver. Studies with 12 lectins clearly demonstrated differences in the glycoconjugate compositions in specimens from human hepatocellular carcinomas (HCC), cirrhotic patients, and normal subjects. This difference can be utilized for diagnostic purposes (Zhang *et al.*, 1989). The potential of the endothelial cell marker UEA-I in the diagnosis of hepatocellular carcinoma has also been underscored (Hattori *et al.*, 1991). The α-fetoprotein (AFP) fraction which binds to the lectins LCA, PHA, and PSA is shown to be a better marker than total AFP in diagnosis and early detection of HCC (Wu, 1990; Ono *et al.*, 1991; Tu *et al.*, 1992; van Staden *et al.*, 1992; Yin *et al.*, 1993).

It has been suggested that lectins such as HPA, PHA, PNA, DBA, and WGA can serve as prognostic and diagnostic markers in gastric cancer (Li, 1992; Li and Lei, 1993; Kakeji *et al.*, 1994). In colorectal and gastric carcinomas the potential of lectins such as GSI-A4 and VVA in diagnosis has been demonstrated (Shue *et al.*, 1993, Chen *et al.*, 1994). Histochemical expression of *Helix pomatia* agglutinin-binding sites is suggested to aid the prognosis of patients with advanced colorectal cancer, which encourages further histopathological and biochemical studies (Ikeda *et al.*, 1994; Schumacher *et al.*, 1994). The potential of UEA-I in identification of high-risk colorectal cancer has also been reported (Kuroki *et al.*, 1991). Not all aspects of glycoconjugate expression are linked to prognosis. A retrospective study with the mistletoe lectins I and III on patients with colorectal carcinoma gave no evidence for a correlation between survival and the presence of lectin binding (Dixon *et al.*, 1994). In colorectal adenomas as well lectins apparently have the capacity to differentiate hyperplastic and dysplastic cases (Kolar and Altavilla, 1989; Orntoft *et al.*, 1991). A recent study with PNA could not distinguish between pseudo and invasive colonic adenomas (Fucci *et al.*, 1993). The changes in vascularization and the formation of sinusoids during the pathogenesis of various pathological conditions in liver and colon can be studied with lectin UEA-I (Ohtani and Sasano, 1989; Dhillon *et al.*, 1992).

2.3 Endogenous Lectins in Histopathology

Compared with plant lectins, animal lectins have been investigated only recently. The study of endogenous animal lectins and glycoconjugates has become an important area of research in biomedical sciences, as these molecules are believed to play important roles in a variety of biological processes (Gabius, 1991; Harrison, 1991; Lotan, 1992; Zhou and Cummings, 1992). The discovery of endogenous lectins having specific and high affinity for the carbohydrate portions of glycoproteins has opened up new directions in the field of cell adhesion and cell recognition (Gabius, 1987; Zanetta *et al.*, 1992). This aspect has been discussed in detail by Zanetta in the one of the preceding chapters (Zanetta, 1996). Protein-carbohydrate recognition may be involved in an array of molecular interactions at the cellular and subcellular levels (Gabius and Nagel, 1988; Lee, 1992; Hebert and Monsigny, 1993). One of the growing points in the study of endogenous lectin is its role in organotropic metastasis in malignancy. Specific lectins have already been implicated in certain functional steps of the metastatic cascade (Raz and Lotan,

1987; Glaves *et al.*, 1990; Raz *et al.*, 1990). Biochemical details regarding endogenous lectins with special reference to galactose-specific lectins (galectins) in tumors are discussed by Ohannesian and Lotan in this monograph.

The first reported endogenous lectin in mammals was the receptor for asialoglycoproteins of the liver (Ashwell and Harford, 1982). An extensively studied group of endogenous lectins are β-galactoside-specific receptors with molecular weights of 14 and 29 to 35 kDa which are found to be modulated during cell proliferation and progression to metastasis (Raz and Lotan, 1987; Ohannesian and Lotan, 1996). A member of this group is present in the extracellular matrix in mammals. Studies on a human ovarian cell line showed that galaptin (galectin-1) plays an important role in tumor cell adhesion, and perhaps metastasis, and may serve as suitable target for therapy (Woynarowska *et al.*, 1994). ε-BP is an IgE-binding protein also known as Mac-2, CBP35 or galectin-3 (carbohydrate-binding protein, 35 kDa). The expression of ε-BP depends on the degree of cellular differentiation in human keratinocytes (Konstantinov *et al.*, 1994). Like galectins the class of selectins are involved in recognitive interactions. P-Selectin is a major C-type lectin, also known as GMP-140, mediating the adhesion of human leucocytes to activated platelets and endothelium and also participating in inflammatory processes (Lasky, 1991). The sialic acid-binding sarcolectin, an interferon-α/β antagonist and growth regulator, had been shown to bind the lymphokine macrophage migration inhibitory factor (MIF) (Zeng *et al.*, 1993). A study on lung cancers showed that the binding pattern of sarcolectin correlated with the survival of the patients (Kayser *et al.*, 1994). CSL and R1 are two endogenous mannose-specific lectins, initially isolated from the rat cerebellum, which have a wide distribution in mammalian tissues, and are known to participate in mechanisms of cell adhesion. The over-expression of the glycoprotein ligands of these lectins in most transformed cells can provide new tools for understanding the underlying mechanism of malignant transformation, as well as the generation of signals through cell adhesion (Zanetta *et al.*, 1992). The set of specificities of endogenous lectins is not restricted to the examples given. Our present lack of precise knowledge of the actual functions notwithstanding, purification is undoubtedly an important step for further investigations. A heparin-specific endogenous lectin (AHL) with subunits of molecular weights of 14–18 kDa has, for example, been isolated from human placenta (Gabius *et al.*, 1991a). A galactose- and *N*-acetylgalactosamine-specific Ca^{2+}-dependent lectin is a family member of membrane-bound C-type animal lectins commonly seen in inflammatory and tumoricidal macrophages with a supposed role in antitumoral defence (Yamamoto *et al.*, 1994). As noted for galectin-3, various names can be used for the same protein (Ackerman *et al.*, 1993).

The description of the state of knowledge in this field illustrates the exquisite specificity displayed by animal lectins for their carbohydrate ligands, and suggests possible mechanisms of promoting or interfering with lectin binding for the purpose of therapeutic treatment. The alterations in expression of these endogenous lectins in various pathological conditions may also aid in diagnosis of the diseases, as reviewed in the preceding section for plant lectins.

3 Neoglycoproteins

Over the last few years increasing interest in endogenous lectins has been generated because it has been shown that they are involved in a number of biological processes such as endocytosis, intercellular contacts, cell growth, fertilization and biosignaling (Gabius, 1991; Harrison, 1991; Lotan, 1992; Zhou and Cummings, 1992; Lee and Lee, 1994). The search for suitable probes for endogenous lectins, to study their expression pattern and involvement in various physiological processes, has resulted in the design of NGPs (Stowell and Lee, 1980; Gabius and Gabius 1992; Gabius *et al.*, 1993). The synthesis, the usefulness in various types of techniques, and the contribution of these tools in morphological investigations have been dealt with in detail by Lee and Lee (1996) and by Danguy *et al.* (1996) in this book. Hence for more details regarding biochemical features of NGPs and their application in histology these chapters are recommended.

3.1 Neoglycoprotein Staining Protocols

A neoglycoprotein comprises a carrier, a label, a spacer between the carrier and the carbohydrate group, and a histochemically crucial carbohydrate structure (Lee and Lee, 1994). Usually, the label is biotin. However, other labels commonly used in immunohistochemistry, such as fluorochromes and colloidal gold particles, can also be employed. The protocols for staining are identical to those for lectins with the exception that NGPs in appropriate concentrations are used instead of lectins. The standard avidin-biotin or avidin-biotin complex method can be employed for the detection of the binding pattern of biotinylated NGPs in tissue sections (Denk, 1987). To ensure the specificity of the staining, rigorous controls must be employed by preincubating the sections with an excess of the appropriate ligand, as in the case of lectins. Moreover, the use of a battery of NGPs with various sugar specificities is recommended initially to identify the endogenous sugar receptors such as lectins, which show prominent alterations during pathological conditions in a particular tissue (Gabius *et al.*, 1993).

3.2 Neoglycoproteins in Histopathology

As with lectins, the majority of studies with NGPs are also on tumor tissues. Bardosi and colleagues analyzed the binding patterns of various NGPs in normal and pathological conditions of muscles, nerves, brain and heart, and found some alterations in binding of certain NGPs during specific diseases (Bardosi *et al.*, 1988, 1989, 1990 a,b; Bardosi and Gabius, 1991). In heart, differences in lectin expression were observed between layers of endocardial tissue, myocardial cell constituents, connective-tissue elements, and vascular structures. Overall, the observed patterns of lectin expression may serve as a guideline to elucidate the precise physiological relevance of lectins and to compare different pathological conditions (Bardosi and Gabius, 1991). Studies using a battery of NGPs revealed pronounced quantitative differences in expression of fucose-, mannose- and lactose-specific proteins between normal and atherosclerotic arterial walls (Kayser *et al.*, 1993). Proliferation of vascular cells was shown to be increased by β-galactoside-specific endogenous lectins (Sanford and Harris-Hooker, 1990). The study on different types of synovitis showed that NGPs appear to be reliable markers for phenotypic characterization of synovial cells (Zschäbitz *et al.*, 1991).

NGPs also showed some potential use in the identification and treatment of pathogens. *Leishmania* express calcium-dependent sugar receptors on the cell surface, promastigote and amastigote forms of human *Leishmania enrietti* being different in this respect (Schottelius, 1992). These results also suggest a potential general approach to intracellular targeting of clinical agents against *Leishmania* (Chakraborty *et al.*, 1990). Similarly, various NGPs bind to pathogens such as *Toxoplasma gondii*, *Trypanosoma cruzi*, *Listeria monocytogenes* and also in HIV (Cottin *et al.*, 1990; Degett *et al.*, 1990; Robert *et al.*, 1991; Haurum *et al.*, 1993). Our study on oral squamous cell carcinoma revealed that 90 % of the lesions express accessible binding sites for lactosylated NGP and 40 % of the lesions expressed *N*-acetyl-D-galactosamine-specific binding sites. These NGP-binding sites were mostly concentrated in immature basaloid cells, indicating a possible association with cell proliferation (Kannan *et al.*, 1994). It has also been found that the expression of the receptors for the lactosylated probe can have a significant correlation with tumor progression in oral mucosa (Kannan *et al.*, 1994). Hence, lactosylated NGPs may be useful in early diagnosis of oral premalignant lesions. When using a panel of NGPs, non-uniform binding patterns are commonly detected. For example, in a similar study in head and neck cancers, pronounced cytoplasmic staining was seen with the NGPs with reactive sugars such as sialic acid, glucuronic acid, *N*-acetylglucosamine (glcNAc), *N*-acetylgalactosamine (β-galNAc), lactose, maltose, mannose, and mannose-6-phosphate, and no specific staining was seen for fucoidan, heparin, and the α-anomeric form of *N*-acetylgalactosamine sugars (Steuer *et al.*, 1991).

Studies on the tumors of central and peripheral nervous tissue revealed significant differences in binding of various NGPs, and correlations with the differentiation of lesions (Bardosi *et al.*,

1991). On the basis of the distinct binding spectrum of NGPs, the presence of an additional subtype of meningiomas called submalignant meningioma has been proposed (Bardosi *et al.*, 1988). A study of cutaneous lesions, basal cell epithelioma, squamous cell carcinoma and malignant melanoma showed a mixed pattern of staining for the different types of NGPs such as α- or β-galactosides, α-mannosides, α-fucosides and α-glucosides (Gabius *et al.*, 1990a). NGPs carrying β-galactosides or α- and β-anomers of *N*-acetyl-D-galactosamine showed that staining of tumor cells was more intense than staining of normal cell types in breast cancer (Gabius *et al.*, 1990b). In comparing the efficiency of NGPs in detecting the endogenous lectins with that of antibodies and natural ligands for the endogenous lectins in breast cancer, it was found that the staining pattern was almost the same with all these probes, serving as a further control reaction (Gabius *et al.*, 1991b). In lung cancers NGPs carrying maltose, fucose, and mannose sugars are useful for separating small cell anaplastic carcinomas from non-small cell anaplastic carcinomas, and also for separation of mesothelioma from pleuritis carcinomatosa, a NGP carrying *N*-acetylglucosamine being as useful as monoclonal antibodies (Kayser *et al.*, 1989).

Another important area in which the NGPs may prove helpful is the biology of metastasis. Because cell surface characteristics mainly constituted by glycoconjugates and their receptors are believed to affect recognition processes within the cascade of interactions, they are involved in the pathway to the clinically often fatal establishment of metastasis (Nicolson, 1988; Ponta *et al.*, 1994). The direct participation of endogenous lectins in metastasis has been made likely by transfection of the mouse homologue of the L-31 endogenous lectin gene into non-metastatic cells, resulting in expression of the metastatic phenotypes (Raz *et al.*, 1990). In colorectal carcinomas a higher level of expression of lectins is observed in advanced stages of disease than in early stages (Irimura *et al.*, 1991). Biochemical and histopathological studies on colorectal and gastric carcinomas emphasize that a clear-cut difference between primary and secondary lesions has not been unequivocally defined so far (Gabius and Engelhardt, 1988; Gabius *et al.*, 1991; Lotan *et al.*, 1994).

Histopathological evaluation of a battery of NGPs in malignant diseases may offer a new approach for uncovering properties of organ-related structural growth, and may additionally aid in refining histopathological diagnosis and tumor classification (Gabius *et al.*, 1994). In any case, histochemical application of labeled NGPs is valuable to discern the presence, localization and developmental pattern of binding sites for the carbohydrate part of glycoconjugates, on which further biochemical and cell biological studies can consequently be based (Gabius and Gabius, 1992; Gabius *et al.*, 1991c, 1993; Lee and Lee, 1994).

4 Precautions to be Taken in Lectin Histochemistry

From the preceding discussion about the application of lectins in histopathology, it is clear that identical results for the same lectin in the same tissue type may not automatically be obtained. Notable variations in results are seen among comparable studies, for example PNA and ConA binding results in oral cancer (Kannan *et al.*, 1993; Mazumdar *et al.*, 1993). Except for UEA as an endothelial marker, no other lectins displayed strict uniformity in the results. HPA has been advocated as a reliable prognostic marker in breast cancer by several studies (Alam *et al.*, 1990; Brooks and Leathem, 1991; Brooks *et al.*, 1993; Noguchi *et al.*, 1993a,b; Thomas *et al.*, 1993), but a multicentric study on large group of breast cancer patients opposed its clinical usefulness (International (Ludwig) Breast Cancer Study Group, 1993). Thus, we should be more cautious in interpreting the results with lectins. It has already been mentioned that two lectins with identical sugar specificities to monosaccharides may bind differently in the same tissues, as shown in thyroid carcinoma (Vijayakumar *et al.*, 1992). Moreover, it has been documented that exogenous and endogenous lectins having similar sugar specificities bind differently in same tissue, as reported in oral and breast tumours (Gabius *et al.*, 1991; Kannan *et al.*, 1994). Hence, a few precautions have to be taken in using lectins in histopathology. One important point to be remembered in lectin histochemistry is the sugar specificity of each lectin. This has been well-reviewed by Walker (1989). Some lectins show strong anomeric specificity, whereas others will react equally with both α and β forms. Many lectins will react only with terminal non-reducing sugars, whereas others bind to monosaccharides which are internal or external constituents of glycan chains of glycoproteins or glycolipids. Some lectins can bind to several different disaccharide structures, but with differing affinities. Hence the binding properties of a lectin, such as intensity, number of lectin molecules per polysaccharide molecule, and affinity may vary between oligo- and polysaccharides having similar monosaccharide constituents. Thus, to arrive at a firm conclusion about the presence of glycoconjugates with certain sugar groups, a battery of lectins should be employed initially, and interpretation of the results in relation to the sugar specificity characteristics of each lectins should follow.

Another major reason for inconsistency in results is due to the variations in tissue preparation and lectin staining protocols (Mason and Matthews, 1988). It is well-established that formalin fixation, paraffin processing and prolonged storage results in the loss and alteration of antigens (Miettinen, 1990). Loss of glycoconjugates during the chemical processing of the tissue such as fixation, dehydration, clearing and embedding is a problem in lectin histochemistry as in immunohistochemisry (Walker, 1989). Rittman and MacKenzie (1983) have clearly shown that the binding pattern of the same lectin in oral mucosa and skin varied with type of fixative and processing protocol used. This may also be the main reason for the variations in results between two studies with same lectin. Methods with minimum loss of the materials of interest should be standardized by trial and error before starting the actual study. Otherwise, the results may not be authentic and may sometimes be erroneous. The recommended material for lectin-histochemical studies is frozen sections. Frozen sections are essential for studies on glycolipids since lipids are readily soluble in most of the organic solvents used in tissue processing. The optimal fixative for a particular tissue can also be identified by comparing the staining pattern of the lectin on samples fixed in a variety of fixatives with that of frozen sections (Walker, 1989). It should also be remembered that the optimum fixative for each type of tissue, as well as for each lectin, will vary. Most of the lectins showed satisfactory results with formalin-fixed, routinely processed paraffin sections. As in immunohistochemistry, epitope demasking is required in lectin studies. For example PNA binding can be enhanced by pre-treating the paraffin sections with neuraminidase (Mason and Matthews, 1988). Thus, before starting any lectin histochemistry studies, standardization in tissue fixing, processing, embedding and staining protocol is essential. To ensure uniformity in the results, the quality and quantity of lectins, incubation time and temperature in each step, and maintenance of controls to assess the speci-

▶

Figure 1. Normal keratinizing epithelium of oral mucosa showing intense membrane binding in basal and parabasal cells with PNA (100x).

Figure 2. Dysplastic epithelium of oral mucosa showing intense membrane and cytoplasmic staining in basal and spinal cells with PNA (100x).

Figure 3. Dysplastic epithelium of oral mucosa showing mild cytoplasmic staining in lower layer cells and intense binding in subepithelial tissues with ConA (100x).

Figure 4. Poorly differentiated squamous cell carcinoma of the oral mucosa showing intense cytoplasmic and membrane staining in malignant cells with PNA (100x).

Figure 5. Well differentiated squamous cell carcinoma of the oral mucosa showing mild cytoplasmic staining in malignant cells and intense binding in basement membrane with ConA (100x).

Figure 6. Well differentiated squamous cell carcinoma of the oral mucosa showing intense membrane and moderate cytoplasmic staining in malignant cells with JFL (100x).

Figure 7. Well differentiated squamous cell carcinoma of the oral mucosa showing intense cytoplasmic and membrane staining in basaloid cells with lactosylated neoglycoprotein. The differentiated cells have intense membrane binding only (250x).

Figure 8. Adjacent epithelium of well differentiated squamous cell carcinoma showing intense staining in lower layer cells of hyperplastic area (arrow) and most of the intraepithelial carcinoma cells (arrow head) with lactosylated neoglycoprotein (40x).

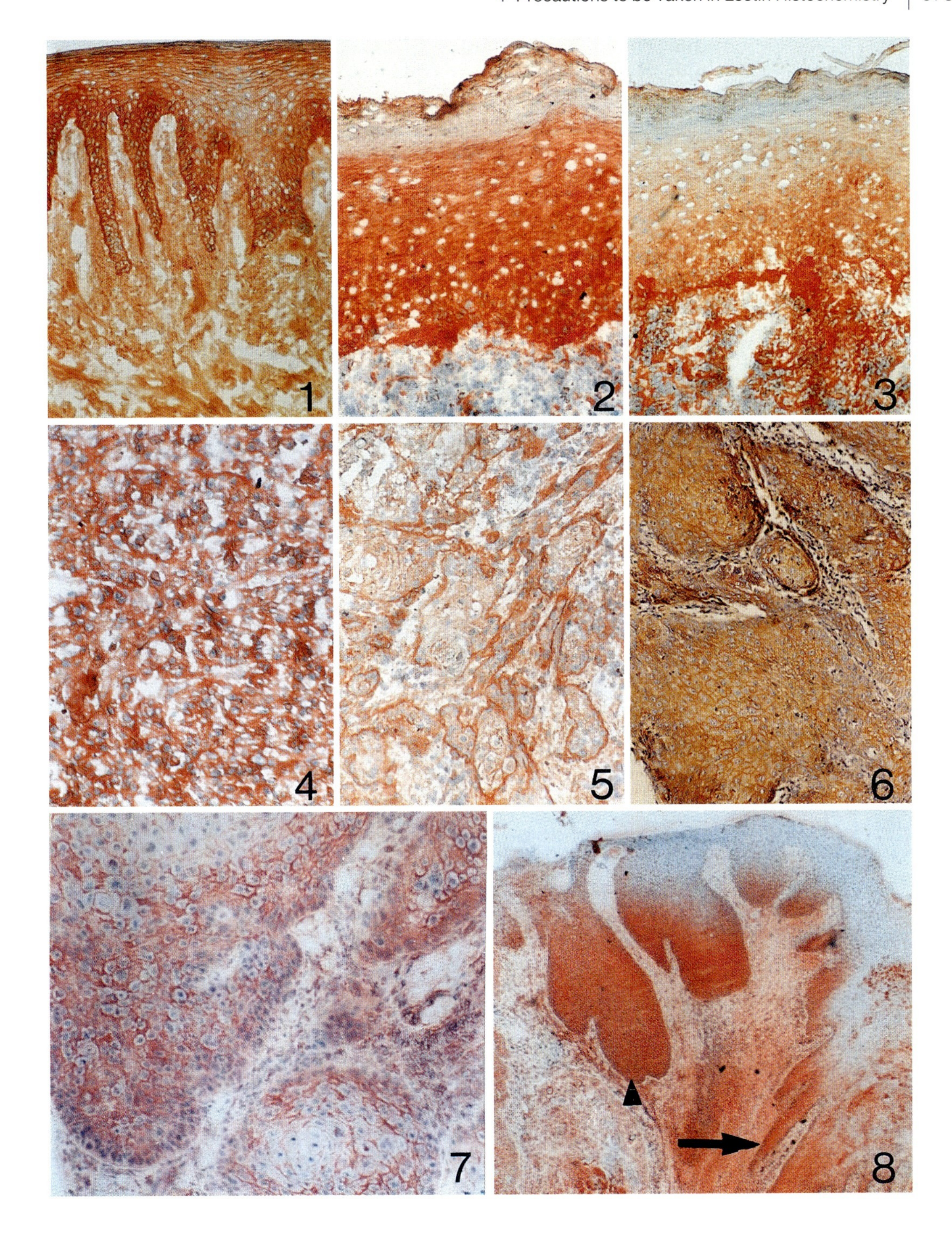
1
2
3
4
5
6
7
8

ficity, are required, and should be done in a standard protocol acceptable to every institute. With respect to NGPs, the precautions discussed can be obeyed easily because they are custom-made synthetic probes (Gabius *et al.*, 1993). Similar problems are also encountered in immunohistochemistry, but by adopting strict controls and quality assurance procedures as suggested herein, and by the introduction of epitope-defined monoclonal antibodies, problems can be greatly reduced. The potential problems occurring in a new immunohistochemistry laboratory have been outlined in detail by us elsewhere (Pillai *et al.*, 1993). If similar precautions are adopted, the problems in a lectin histochemistry laboratory can likewise be addressed, leading to results such as those illustrated in Figs. 1–17.

5 Perspectives

5.1 Modern Roles Played by Histopathology

In recent years, histopathology has gained new dimensions in addition to classical diagnostic histopathology. These are early diagnosis, prognosis and disease staging (Preisler and Raza, 1992). One of the most important area in modern histopathology is prognostic histopathology (Hermanek *et al.*, 1990). Prognosis of a lesion is dependent on the biological composition of the cells which determines the clinical characteristics, such as response to therapy, recurrence, metastasis, etc. It is found that the response of malignant cells to any treatment protocol is directly related to various morphological as well as biochemical features (Fenoglio-Preiser, 1992). Research has revealed new markers that may improve our ability to estimate disease outcome (Preisler and Raza, 1992). They include amplified oncogenes, oncoproteins, paracrine and autocrine growth factors, extent of aneuploidy, proliferation markers, and cell receptors, which include the glycobiological determinants reviewed here. In addition, prognostic markers will also perform new roles in early cancer detection and throw new light on the process of carcinogenesis (Fenoglio-Preiser, 1992).

Staging of the disease is an essential requirement for effective therapy (Preisler and Raza, 1992). The optimal staging system of malignant neoplasms describes the extent of spread of the particular cancer and relates to its natural course (Henson, 1988). Though the "T" staging and lymph node involvement are of prognostic significance in malignancies, they often fail in the accurate prediction of disease course and outcome. It is with these limitations in mind that the American Joint Committee on Cancer has stated in

▶

Figure 9. Well differentiated squamous cell carcinoma of the oral mucosa showing intense staining in connective tissue with galactoside-specific VAA (100x).

Figure 10. Well differentiated squamous cell carcinoma of the oral mucosa showing intense cytoplasmic-binding in basaloid cells with β-galactoside-specific human 14 kDa lectin (100x).

Figure 11. Large cell keratinizing squamous cell carcinoma of the uterine cervix showing intense cytoplasmic binding in basaloid cells with β-galactoside-specific human 14 kDa lectin (250x).

Figure 12. Benign reactive mesothelial cells in ascitic fluid showing mild staining with JFL (1000x).

Figure 13. Metastatic adenocarcinoma cells in ascitic fluid showing intense staining with JFL (1000x).

Figure 14. Squamous cell carcinoma of the bronchial epithelium showing intense cytoplasmic staining in differentiated cells with JFL (250x).

Figure 15. Small cell anaplastic carcinoma of the bronchial epithelium showing mild staining in a few malignant cells (arrow). The dysplastic cells in superficial epithelium show moderate staining with JFL (arrow head) (250x).

Figure 16. Normal squamous cells from the uterine cervix showing no staining with JFL (400x).

Figure 17. Small cell non-keratinizing squamous cell carcinoma of the uterine cervix showing intense staining in malignant cells with JFL (400x).

[Figures 1 to 11 except 6 were acetone-fixed fresh frozen sections and figures 6,14 and 15 were formalin-fixed paraffin-embedded sections. Figures 12, 13, 16 and 17 were cytological preparations fixed in alcohol-ether mixture. Figures 1 to 5 were stained with peroxidase-conjugated PNA or ConA and aminoethyl carbazole (AEC) as chromogen. Figures 6 and 12 to 17 were stained with peroxidase-conjugated JFL and diaminobenzidine (DAB) as chromogen. Figures 7 to 11 were stained with respective biotinylated probes, the binding being detected with the avidin-biotin method and AEC as chromogen.]

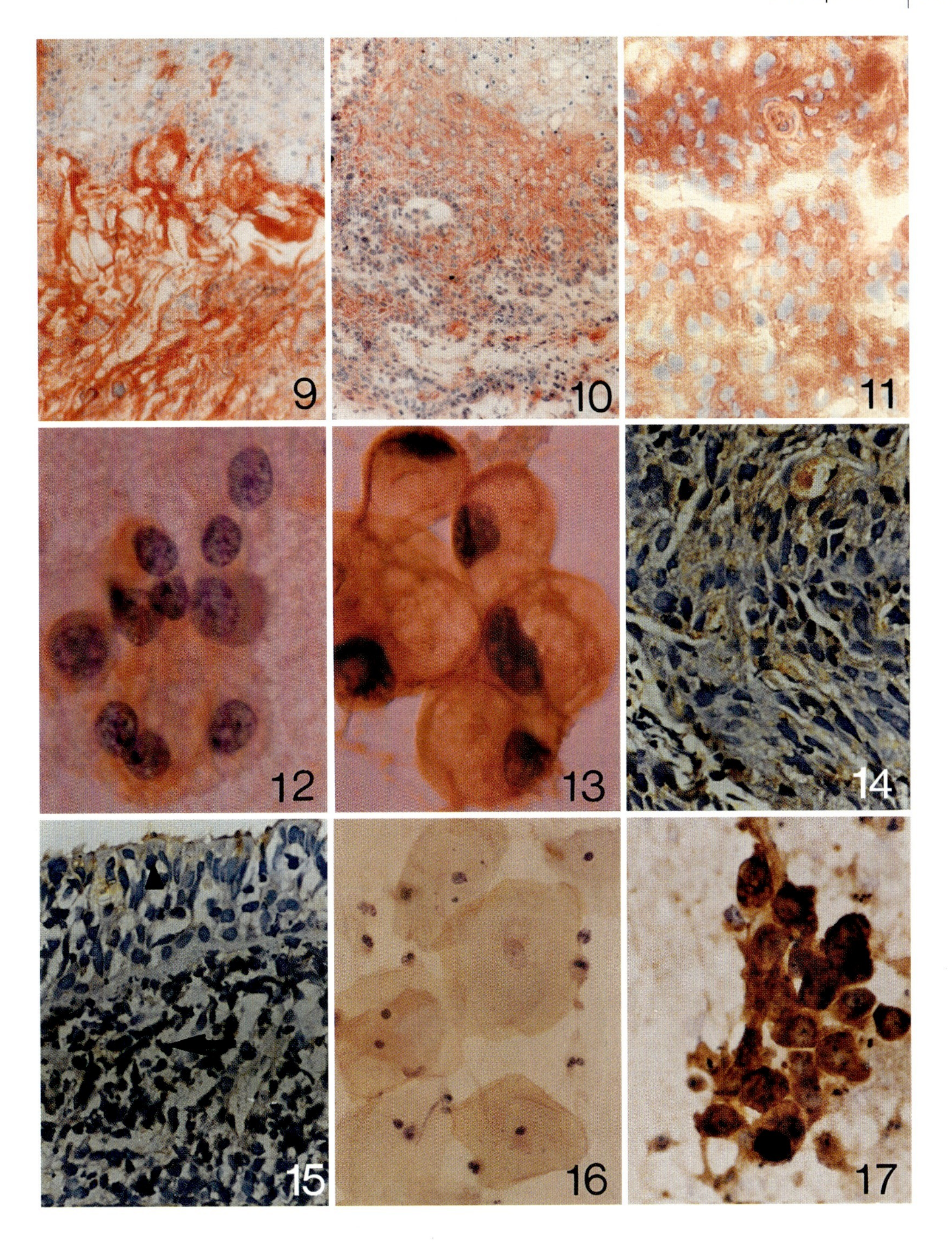
9
10
11
12
13
14
15
16
17

their recent publication that "in the future biologic markers and other parameters may have to be added to those of anatomic extent in classifying cancer" (Henson, 1992). In this context lectins and NGPs have the potential to perform an important role as biological markers in classifying and staging the diseases.

5.2 Perspectives of Lectins and Neoglycoproteins in Histopathology

As mentioned repeatedly, the functional roles played by glycoconjugates are numerous both in intracellular and intercellular processes. The involvement of glycoconjugates and their endogenous receptors in various diseases has already been described above. The necessary combination of methods to define NGP-binding capacities in normal and pathological states is the prerequisite for meaningful monitoring to trace clear-cut correlations between probe binding and clinically and functionally relevant processes. From the point of view of a histochemist, it is challenging that the wealth of information on glycoconjugate expression that is being accumulated by investigations with plant lectins can be supplemented by complementary analysis with defined carbohydrate structures as ligands (Gabius *et al.*, 1994). Studies in this direction are encouraging, and, especially with respect to the participation of endogenous lectins in various crucial biochemical processes, the studies with NGPs will definitely provide a significant contribution. Both classical and modern histopathology will benefit from this glycobiological development. Besides their usefulness in histopathology, lectins and NGPs have another potential role in disease management by permitting drug targeting. This aspect has been discussed in detail by Rice in this monograph (Rice, 1996). Advances in separate fields are certain to converge in glycohistochemistry, to accelerate the pace of acquisition of knowledge about the physiological relevance of lectin-carbohydrate recognition. Organic chemistry will supply tailor-made ligands and carriers available for histochemical use as well as for molecular characterization of endogenous lectins (Gabius *et al.*, 1993; Lee and Lee, 1994). In conclusion, lectins and NGPs appear to be suitable probes or markers in modern histopathology, which are at the stage of thorough research to delineate precisely the functional significance of the results and to prepare – if warranted – their application in routine histopathological investigation.

References

Abramenko IV, Gluzman DF, Sklyarenko LM *et al.* (1991): Immunocytochemical staining of cells in 153 pleural effusions with a panel of monoclonal antibodies and lectins. In *Anticancer Res.* **11**:629–34.

Ackerman SJ, Corrette SE, Rosenberg HF *et al.* (1993): Molecular cloning and characterization of human eosinophil Charcot-Leyden crystal protein (lysophospholipase): similarities to IgE binding proteins and the S-type animal lectin superfamily. In *J. Immunol.* **150**:456–68.

Aguirre A, Takai Y, Meenaghan M *et al.* (1989): Lectin histochemistry of ameloblastomas and odontogenic keratocysts. In *J. Oral. Pathol. Med.* **18**:68–73.

Alam SM, Whitford P, Cushley W *et al.* (1990): Flow cytometric analysis of cell surface carbohydrates in metastatic human breast cancer. In *Br. J. Cancer* **62**:238–42.

Alroy J, Ucci AA, Pereira MEA (1988): Lectin histochemistry: an update. In *Advances in Immunohistochemistry* (De Lellis RA, ed) pp 93–131, New York: Raven Press.

Aoki D, Nozawa S, Iizuka R *et al.* (1990): Differences in lectin binding patterns of normal endometrium and endometrial adenocarcinoma, with special reference to staining with *Ulex europaeus* agglutinin 1 and peanut agglutinin. In *Gynecol. Oncol.* **37**:338–45.

Ashwell G, Harford J (1982): Carbohydrate-specific receptors of the liver. In *Annu. Rev. Biochem.* **51**:531–44.

Bardosi A, Gabius H-J (1991): Neoglycoproteins in glycohistochemistry. In *Progr. Histochem. Cytochem.* **22**:1–66.

Bardosi A, Dimitri T, Gabius H-J (1988): (Neo)glycoproteins as tools in neuropathology: histochemical patterns of the extent of expression of endogenous carbohydrate-binding receptors like lectins in meningiomas. In *Virch. Arch. B [Cell. Pathol.]* **56**:35–43.

Bardosi A, Dimitri T, Wosgien B *et al.* (1989): Expression of endogenous receptors for neoglycoproteins, especially lectins, which allow fiber typing on formaldehyde-fixed, paraffin-embedded muscle biopsy specimens. A glycohistochemical, immunohistochemical, and glycobiochemical study. In *J. Histochem. Cytochem.* **37**:989–98.

Bardosi A, Bardosi L, Hendrys M *et al.* (1990a): Spatial differences of endogenous lectin expression within cellular organization of human heart: a glycohistochemical, immunohistochemical, and glycobiochemical study. In *Am. J. Anat.* **188**:409–18.

Bardosi A, Bardosi L, Lindenblatt R *et al.* (1990b): Detection and mapping of endogenous receptors for carrier-immobilized constituents of glycoconjugates (lectins) by labelled neoglycoproteins and by affinity chromatography in human adult mesencephalon, pons, medulla oblangata and cerebellum. In *Histochemistry* **94**:285–91.

Bardosi A, Brkovic D, Gabius H-J (1991): Localization of endogenous sugar-binding proteins (lectins) in tumors of the central and peripheral nervous system by biotinylated neoglycoproteins. In *Anticancer Res.* **11**:1183–8.

Barnhill RL, Fandrey K, Levy MA *et al.* (1992): Angiogenesis and tumor progression of melanoma. Quantification of vascularity in melanocytic nevi and cutaneous malignant melanoma. In *Lab. Invest.* **67**:331–7.

Beer TW (1992): Malignant thyroid haemangioendothelioma in a non-endemic goitrous region, with immunohistochemical evidence of a vascular origin. In *Histopathol.* **20**:539–41.

Bell CM, Skerrow CJ (1985): Lectin binding to psoriatic epidermis. In *Br. J. Dermatol.* **113**:205–12.

Bhattathiri VN, Remani P, Pillai KR *et al.* (1991) Radiation dose-related changes in lectin-staining pattern in oral squamous cell carcinoma. In *Br. J. Rad.* **64**:124.

Boyd W (1970): A textbook of pathology. In *Structure and Function in Disease*, 8th edition, pp 3–12, London: Henry Kimpton.

Brooks SA, Leathem AJ (1991): Prediction of lymph node involvement in breast cancer by detection of altered glycosylation in the primary tumour. In *Lancet* **338**:71–4.

Brooks SA, Leathem AJ, Camplejohn RS *et al.* (1993): Markers of prognosis in breast cancer – the relationship between binding of the lectin HPA and histological grade, SPF, and ploidy. In *Breast Cancer Res. Treat.* **25**:247–56.

Bychkov V, Toto PD (1986): Lectin binding to normal, displastic and neoplastic cervical epithelium. In *Am. J. Clin. Pathol.* **85**:542–7.

Bychkov V, Toto PD (1988): Application of lectin binding in gynecologic pathology. In *Lectins. Biology, Biochemistry, Clinical Biochemistry*, Vol.6 (Bog-Hansen TC, Freed DLJ, eds) pp 677–82, St. Louis: Sigma Chemical Co..

Byrne P, Williams A, Rollason T (1989a): Studies of lectin binding to the human cervix uteri. I. Normal cervix. In *Histochem. J.* **21**:311–22.

Byrne P, Williams A, Rollason T (1989b): Studies of lectin binding to the human cervix uteri. II. Cervical intraepithelial neoplasia and invasive squamous carcinoma. In *Histochem. J.* **21**:323–36.

Caselitz J (1987): Lectins and blood group substances as 'tumor markers'. In *Morphological Tumor Markers. General Aspects and Diagnostic Relevances* (Seifert G, ed) pp 245–77, New York: Springer Publ. Co.

Chakraborty P, Bhaduri AN, Das PK (1990): Neoglycoproteins as carriers for receptor-mediated drug targeting in the treatment of experimental visceral leishmaniasis. In *J. Protozool.* **37**:358–64.

Chammas R, Jasiulinois MG, Cury PM *et al.* (1994): Functional hypotheses for aberrant glycosylation in tumor cells. In *Braz. J. Med. Biol. Res.* **27**:505–7.

Chen YF, Boland CR, Kraus ER *et al.* (1994): The lectin *Griffonia simplicifolia* I-A4 (GS I-A4) specifically recognizes terminal α-linked *N*-acetylgalactosaminyl groups and is cytotoxic to the human colon cancer cell lines LS174T and SW1116. In *Int. J. Cancer* **57**:561–7.

Cottin J, Loiseau O, Robert R *et al.* (1990): Surface *Listeria monocytogenes* carbohydrate-binding components revealed by agglutination with neoglycoproteins. In *FEMS Microbiol.* Lett. **56**:301–5.

Cruz-Sanchez FF, Rossi ML, Buller JR *et al.* (1991): Oligodendrogliomas: a clinical, histological, immunocytochemical and lectin-binding study. In *Histopathol.* **19**:361–7.

Cummings RD (1996): Glycoconjugate purification and characterization by lectins as tools. In *Glycosciences: Status and Perspectives* (Gabius H-J, Gabius S, eds) this volume, Weinheim: Chapman & Hall.

Dabelsteen E, Clausen H, Holmstrup S *et al.* (1988): Premalignant and malignant oral lesions are associated with changes in the glycosylation pattern of carbohydrates related to ABH blood group antigens. In *APMIS* **96**:813–8.

Dabelsteen E, Clausen H, Mandel U (1992): Carbohydrate changes in squamous cell carcinomas. In *APMIS* **100** (Suppl.27):130–8.

Damjanov I (1987): Lectin cytochemistry and histochemistry. In *Lab. Invest.* **57**:5–20.

Danguy A *et al.* (1996): Modern glycohistochemistry: a major contribution to morphological investigations. In *Glycosciences: Status and Perspectives* (Gabius H-J, Gabius S, eds) this volume; Weinheim: Chapman & Hall.

Danguy A, Kiss R, Pasteels JL (1988): Lectins in histochemistry: a survey. In *Biol. Struct. Morphol.* **1**: 93–106.

Danguy A, Akif F, Pajak B *et al.* (1994): Contribution of carbohydrate histochemistry to glycobiology. In *Histol. Histopath.* **9**:155–71.

Degett J, Souto-Padron T, De Souza W (1990): Localization of sugar-binding proteins in *Trypanosoma cruzi* using gold-labeled neoglycoproteins. In *Microsc. Electron. Biol. Celular* **14**:11–7.

Denk H (1987): Immunohistochemical methods for the demonstration of tumor markers. In *Morphological Tumor Markers. General Aspects and Diagnostic Relevances* (Seifert G, ed) pp 47–65, New York – Springer Publ. Co.

Dhillon AP, Colombari R, Savage K *et al.* (1992): An immunohistochemical study of the blood vessels within primary hepatocellular tumours. In *Liver* **12**:311–8.

Dixon A, Schumacher U, Pfuller U *et al.* (1994): Is the binding of mistletoe lectins I and III a useful prognostic indicator in colorectal carcinoma? In *Eur. J. Surg. Oncol.* **20**:648–52.

Fenoglio-Preiser CM (1992): Selection of appropriate cellular and molecular biologic diagnostic tests in the evaluation of cancer. In *Cancer* **69**:1607–32.

Fucci L, Valentini AM, Caruso ML (1993): Can peanut agglutinin distinguish between pseudo and true invasion in colonic adenomas? In *Eur. J. Histochem.* **37**:335–44.

Gabius H-J (1987): Endogenous lectins in tumors and the immune system. In *Cancer Investig.* **5**:39–46.

Gabius H-J (1988): Tumor lectinology: at the intersection of carbohydrate chemistry, biochemistry, cell biology, and oncology. In *Angew. Chem. Int. Ed.* Engl. **27**:1267–76.

Gabius H-J (1991): Detection and functions of mammalian lectins – with emphasis on membrane lectins. In *Biochim. Biophys. Acta* **1071**:1–18.

Gabius H-J, Engelhardt R (1988): Sugar receptors of different types in human metastases to lung and liver. In *Tumor Biol.* **9**:21–36.

Gabius H-J, Nagel GA (eds) (1988) *Lectins and Glycoconjugates in Oncology.* Heidelberg: Springer Publ. Co.

Gabius H-J, Gabius S (eds) (1991): *Lectins and Cancer.* Heidelberg: Springer Publ. Co.

Gabius H-J, Gabius S (1992): Chemical and biochemical strategies for the preparation of glycohistochemical probes and their application in lectinology. In *Adv. Lectin Res.* **5**:123–57.

Gabius H-J, Gabius S (eds) (1993): *Lectins and Glycobiology.* Heidelberg: Springer Publ. Co.

Gabius H-J, Gabius S, Brinck U *et al.* (1990a): Endogenous lectins with specificity to β-galactosides and α- or β-*N*-acetyl-galactosaminides in human breast cancer. Their glycohistochemical detection in tissue sections by synthetically different types of neoglycoproteins, their quantitation on cultured cells by neoglycoenzymes and their usefulness as targets in lectin-mediated phototherapy in vitro. In *Path. Res. Pract.* **186**:597–607.

Gabius H-J, Heil MS, Berger H (1990b): Glycohistochemistry of endogenous lectins in cutaneous cancer. In *Anticancer Res.* **10**:1627–31.

Gabius H-J, Kohnke-Godt B, Leichsenring M *et al.* (1991a) Heparin-binding lectin of human placenta as a tool for histochemical ligand localization and isolation. In *J. Histochem. Cytochem.* **39**:1249–56.

Gabius H-J, Wosgien B, Brinck U *et al.* (1991b): Localization of endogenous β-galactoside-specific lectins by neoglycoproteins, lectin-binding tissue glycoproteins and antibodies and of accessible lectin-specific ligands by mammalian lectin in human breast carcinomas. In *Path. Res. Pract.* **187**:839–47.

Gabius H-J, Grote T, Gabius S *et al.* (1991c): Neoglycoprotein binding to colorectal tumour cells: comparison between primary and secondary lesions. In *Virch. Arch. A [Pathol. Anat.]* **419**:217–22.

Gabius H-J, Gabius S, Zemlyanukhina TV *et al.* (1993): Reverse lectin histochemistry: design and application of glycoligands for detection of cell and tissue lectins. In *Histol. Histopath.* **8**:369–83.

Gabius H-J, Brinck U, Kayser K *et al.* (1994): Neoglycoproteins: use in tumor diagnosis. In *Neoglycoconjugates: Preparation and Applications* (Lee YC, Lee RT, eds) pp 403–24, San Diego: Academic Press.

Gilboa-Garber N *et al.* (1996): Structure and functions of bacterial lectins. In *Glycosciences: Status and Perspectives* (Gabius H-J, Gabius S, eds) Weinheim: Chapman & Hall.

Glaves D, Weiss L, Vidal-Vanaclocha F (1991): Site-associated differences in endogenous lectin expression by mouse colon carcinoma cells. In *Lectins and Cancer* (Gabius H-J, Gabius S, eds) pp 137–51, Heidelberg: Springer Publ. Co.

Glinsky GV (1992): The blood group antigens-related glycoepitopes: a key structural determinant in immunogenesis and cancer pathogenesis. In *Crit. Rev. Oncol./Hematol.* **12**:151–66.

Hakomori SI (1989): Aberrant glycosylation in tumors and tumor-associated antigens. In *Adv. Cancer Res.* **52**:257–331.

Harrison FL (1991): Soluble vertebrate lectins: ubiquitous but inscrutable proteins. In J. Cell Sci. **100**: 9–14.

Hattori M, Fukuda Y, Imoto M *et al.* (1991): Histochemical properties of vascular and sinusoidal endothelial cells in liver diseases. In *Gastroenterol. Jpn.* **26**:336–43.

Haurum JS, Thiel S, Jones IM *et al.* (1993): Complement activation upon binding of mannan-binding protein to HIV envelope glycoproteins. In *AIDS* **7**:1307–13.

Hebert E, Monsigny M (1993): Oncogenes and expression of endogenous lectins and glycoconjugates. In *Biol. Cell* **79**:97–109.

Heng MC, Fallon-Friedlander S, Bennett R (1992): Expression of *Ulex europaeus* agglutinin I lectin-binding sites in squamous cell carcinomas and their absence in basal cell carcinomas. Indicator of tumor type and differentiation. In *Am. J. Dermatopathol.* **14**:216–9.

Henson DE (1988): The histological grading of neoplasms. In *Arch. Pathol. Lab. Med.* **112**:1091–6.

Henson DE (1992): Future directions for the American joint committee on cancer. In *Cancer* **69**:1639–44.

Hermanek P, Hutter RVP, Sobin LH (1990): Prognostic grouping: the next step in tumor classification. In *J. Cancer Res. Clin. Oncol.* **116**:513–6.

Hiraizumi S, Takasaki S, Ohuchi N *et al.* (1992): Altered glycosylation of membrane glycoproteins associated with human mammary carcinoma. In *Jpn. J. Cancer Res.* **83**:1063–72.

Horisberger M (1984): Lectin cytochemistry. In *Immunolabelling for Electron Microscopy* (Polak JM, Varndell TM, eds) pp 249–58, Amsterdam: Elsevier Sci. Publ.

Hsu S-M, Raine L (1982): Versatility of biotin-labelled lectins and avidin-biotin-peroxidase complex for localization of carbohydrate in tissue sections. In *J. Histochem. Cytochem.* **30**:157–61.

Hughes RC (1992): Role of glycosylation in cell interactions with extracellular matrix. In *Biochem. Soc. Transact.* **20**:279–84.

Ikeda Y, Mori M, Adachi Y *et al.* (1994): Prognostic value of the histochemical expression of *Helix pomatia* agglutinin in advanced colorectal cancer. A univariate and multivariate analysis. In *Dis. Colon Rectum* **37**:181–4.

International (LUDWIG) Breast Cancer Study Group (1993): Prognostic value of *Helix pomatia* in breast cancer. In *Br. J. Cancer* **68**:146–50.

Irimura T, Matsushita Y, Sutton RC *et al.* (1991): Increased content of an endogenous lactose-binding lectin in human colorectal carcinoma progressed to metastatic stages. In *Cancer Res.* **51**:387–93.

Ito F, Horita S, Yanagisawa H *et al.* (1993): Bellini's duct tumor associated with end stage renal disease: a case diagnosed by lectin immunohistochemistry. In *Hinyokika Kiyo* **39**:735–8.

Iwa N, Yutani C (1993): Cytology of cardiac myxomas: presence of *Ulex europaeus* agglutinin-I (UEA-I) lectin by immunoperoxidase staining. In *Diagn. Cytopathol.* **9**:661–4.

Jones CD, Smith AB (1983): Glycoconjugates in Human Disease, 2nd edition. London: Novo Press.

Jones CJP, Stoddart RW (1986): A post embedding avidin-biotin peroxidase system to demonstrate the light and electron microscopic localization of lectin-binding sites in rat kidney tubules. In *Histochem.* J. **18**:371–9.

Kakeji Y, Maehara Y, Tsujitani S *et al.* (1994): *Helix pomatia* agglutinin-binding activity and lymph node metastasis in patients with gastric cancer. In *Semin. Surg. Oncol.* **10**:130–4.

Kaneko Y, Iwaki T, Matsushima T *et al.* (1991): Comprehensive lectin histochemistry of normal and neoplastic human choroid plexus cells: alteration of lectin-binding patterns through neoplastic transformation. In *Acta Neuropathol.* **82**:127–33.

Kannan S, Balaram P, Chandran GJ *et al.* (1993): Expression of lectin-specific cellular glycoconjugates during oral carcinogenesis. In *J. Cancer Res. Clin. Oncol.* **119**:689–94.

Kannan S, Gabius HJ, Chandran GJ *et al.* (1994): Expression of galactoside-specific endogenous lectins and their ligands in human oral squamous cell carcinoma. In *Cancer Lett.* **85**:1–7.

Karuna V, Shanthi P, Madhavan M (1992): Lectin binding patterns in benign and malignant lesions of the breast. In *Indian J. Pathol. Microbiol.* **35**:289–97.

Kawa S, Kato M, Oguchi H *et al.* (1991): Preparation of pancreatic cancer-associated mucin expressing CA19–9, CA50, SPA*N*-1, SIALYL SSEA-1, and DUPA*N*-2. In *Scand. J. Gastroenterol.* **26**:981–92.

Kawa S, Kato M, Oguchi H *et al.* (1992): Clinical evaluation of pancreatic cancer-associated mucin expressing CA19–9, CA50, SPA*N*-1, SIALYL SSEA-1, and DUPA*N*-2. In *Scand. J. Gastroenterol.* **27**: 635–43.

Kayser K, Heil M, Gabius H-J (1989): Is the profile of binding of a panel of neoglycoproteins useful as a diagnostic marker in human lung cancer? In *Path. Res. Pract.* **184**:621–9.

Kayser K, Bartels S, Yoshida Y *et al.* (1993): Atherosclerosis-associated changes in the carbohydrate-binding capacities of smooth muscle cells of various human arteries. In *Zentralbl. Pathol.* **139**:307–12.

Kayser K, Bovin NV, Korchagina EY *et al.* (1994): Correlation of expression of binding sites for synthetic blood group A-, B-, H-trisaccharides and for sarcolectin with survival of patients with bronchial carcinoma. In *Eur. J. Cancer* **30**A:653–7.

Kjnniksen I, Rye PD, Fodstad O (1994): *Helix pomatia* agglutinin binding in human tumour cell lines: correlation with pulmonary metastases in nude mice. In *Br. J. Cancer* **69**:1021–4.

Knox WF, Mcmahon RF, Mosley SM (1993): Human urinary bladder glycoconjugates. In *J. Pathol.* **170** (Suppl.): 370A.

Kolar Z, Altavilla G (1989): Lectin histochemistry of colonic adenomas. In *Appl. Pathol.* **7**:42–53.

Konstantinov KN, Shames B, Izuno G *et al.* (1994): Expression of ε-BP, a β-galactoside-binding soluble lectin, in normal and neoplastic epidermis. In *Exp. Dermatol.* **3**:9–16.

Kuratsu J, Seto H, Kochi M *et al.* (1991): Metastatic angiosarcoma of the brain. In *Surg. Neurol.* **35**: 305–9.

Kuroki T, Kubota A, Miki Y *et al.* (1991): Lectin staining of neoplastic and normal background colorectal mucosa in nonpolyposis and polyposis patients. In *Dis. Colon Rectum* **34**:679–84.

Kuruvilla S, Madhavan M (1992): Diagnostic parameters of vascular neoplasms by immunohistochemistry. In *Indian J. Pathol. Microbiol.* **35**:180–7.

Langkilde NC, Wolf H, Orntoft TF (1989): Binding of wheat and peanut lectins to human transitional cell carcinomas and correlation with histopathologic grade, invasion, and DNA ploidy. In *Cancer* **64**: 849–53.

Lasky LA (1991): Lectin cell adhesion molecules (LEC-CAMs): a new family of cell adhesion proteins involved with inflammation. In *J. Cell. Biochem.* **45**:139–46.

Lee YC (1992): Biochemistry of carbohydrate-protein interaction. In *FASEB J.* **6**:3193–200.

Lee YC, Lee RT (eds) (1994): Neoglycoconjugates: Preparation and Applications. San Diego: Academic Press.

Lee YC, Lee RT (1996): Neoglycoconjugates. In *Glycosciences: Status and Perspectives* (Gabius H-J, Gabius S, eds) this volume; Weinheim: Chapman & Hall.

Li CM, Lei DN (1993): Lectins in gastric carcinoma and precancerous lesions (Chinese). In *Chung Hua I Hsueh Tsa Chih* **73**:283–5.

Li N (1992): Research of lectin receptors in early gastric carcinoma (Chinese). In *Chung-hua Ping Li Hsueh Tsa Chih* **21**:215–7.

Liener IE, Sharon N, Goldstein IJ (eds) (1986): The Lectins. Properties, Functions and Applications in Biology and Medicine. Orlando: Academic Press.

Lis H, Sharon N (1986): Lectins as molecules and as tools. In *Annu. Rev. Biochem.* **55**:35–67.

Lotan R (1992): β-Galactoside-binding vertebrate lectins: synthesis, molecular biology, function. In *Glycoconjugates: Composition, Structure, and Function* (Allen H, Kisailus E, eds) pp 635–71, New York: Marcel Dekker.

Lotan R, Ito H, Yasui W *et al.* (1994): Expression of a 31-kDa lactoside-binding lectin in normal human gastric mucosa and in primary and metastatic gastric carcinomas. In *Int. J. Cancer* **56**:474–80.

Mandel U, Clausen H, Vedtofte P *et al.* (1988): Sequential expression of carbohydrate antigens with precursor-product relation characterizes cellular maturation in stratified squamous epithelium. In *J. Oral Pathol.* **17**:506–11.

Mason GI, Matthews JB (1988): A comparison of lectin binding to frozen, formalin-fixed paraffin-embedded, and acid-decalcified paraffin-embedded tissues. In *J. Histotechnol.* **11**:223–9.

Matsumura K, Nakasu S, Nioka H *et al.* (1993a): Lectin histochemistry of normal and neoplastic peripheral nerve sheath. 1. Lectin binding pattern of normal peripheral nerve in man. In *Acta Neuropathol.* **86**:554–8.

Matsumura K, Nakasu S, Nioka H *et al.* (1993b): Lectin histochemistry of normal and neoplastic peripheral nerve sheath. 2. Lectin binding patterns of schwannoma and neurofibroma. In *Acta Neuropathol.* **86**:559–66.

May DP, Sloan P (1991): Lectin binding to normal mucosa and squamous cell carcinoma of the oral cavity. In *Med. Lab. Sci.* **48**:6–18.

Mazumdar S, Sengupta SK, Param R *et al.* (1993): Binding pattern of eight different lectins in healthy subjects and patients with dysplastic and malignant lesions of the oral cavity. In *Int. J. Oral. Maxillofac. Surg.* **22**:301–5.

Miettinen M (1990): Immunohistochemistry of solid tumors. Brief review of selected problems. In *APMIS* **98**:191–9.

Miettinen M, Holthofer H, Lehto V-P *et al.* (1983): *Ulex europaeus* I lectin as a marker for tumors derived from endothelial cells. In *Am. J. Clin. Pathol.* **79**:32–6.

Nadasdy T, Laszik Z, Blick KE *et al.* (1994): Tubular atrophy in the end-stage kidney: a lectin and immunohistochemical study. In *Hum. Pathol.* **25**:22–8.

Nagao M, Nakamura M, Oka N *et al.* (1994): Abnormal glycosylation of motor neurons with *N*-acetyl-D-galactosamine in a case of subacute motor neuronopathy associated with lymphoma. In *J. Neurol.* **241**:372–5.

Nakanishi K, Kawai T, Suzuki M (1993): Lectin binding and expression of blood group-related antigens in carcinoma in situ and invasive carcinoma of urinary bladder. In *Histopathol.* **23**:153–8.

Nicolson GL (1988): Cancer metastasis: tumor cell and host organ properties important in metastasis to specific secondary sites. In *Biochim. Biophys. Acta* **948**:175–224.

Niikawa S, Sakai N, Yamada H *et al.* (1993a): Histochemistry with *Helix pomatia* agglutinin in human germ cell tumors: detection of non-germinomatous components and correlation between HPA reactivity and radiosensitivity in germinomas. In *Childs Nerv. Syst.* **9**:266–70.

Niikawa S, Hara A, Shirakami S *et al.* (1993b): Relationship between *Ricinus communis* agglutinin-1 binding and nucleolar organizer regions in human gliomas. In *Neurol. Med. Chir.* **33**:345–9.

Niikawa S, Ito T, Murakawa T (1993c): Recurrence of choroid plexus papilloma with malignant transformation – case report and lectin histochemistry study. In *Neurol. Med. Chir.* **33**:32–5.

Noguchi M, Thomas M, Kitagawa H *et al.* (1993a): DNA ploidy and *Helix pomatia* lectin binding as predictors of regional lymph node metastases and prognostic factors in breast cancer. In *Breast Cancer Res. Treat.* **26**:67–75

Noguchi M, Thomas M, Kitagawa H *et al.* (1993b):

Further analysis of predictive value of *Helix pomatia* lectin binding to primary breast cancer for axillary and internal mammary lymph node metastases. In *Br. J. Cancer* **67**:1368–71.

Ohannesian DW, Lotan R (1996): Lectins in tumors – with focus on galectins. In *Glycosciences: Status and Perspectives* (Gabius H-J, Gabius S, eds) this volume; Weinheim: Chapman & Hall.

Ohno J, Fukuyama K, Epstein WL (1990): Glycoconjugate expression of cells of human anagen hair follicles during keratinization. In *Anat. Rec.* **258**:403–8.

Ohtani H, Sasano N (1989): Microvascular changes in the stroma of human colorectal carcinomas: an ultrastructural histochemical study. In *Jpn. J. Cancer Res.* **80**:360–5.

Ono M, Ishikawa Y, Sekiya C *et al.* (1991): Demonstration of a yolk sac tumor type sugar chain of α-fetoprotein produced by cholangiocellular carcinoma. In *J. Tumor Marker Oncol.* **6**:73–83.

Ookuma Y, Hachisuga T, Iwasaka T *et al.* (1994): Assessment of lectin binding for prognosis in endometrial carcinoma. In *Pathology* **26**:225–9.

Orntoft TF, Langkilde NC, Wiener H *et al.* (1991): Cellular localization of PNA binding in colorectal adenomas: comparison with differentiation, nuclear:cell height ratio and effect of desialylation. In *APMIS* **99**:275–81.

Parker N, Makin CA, Ching CK *et al.* (1992): A new enzyme-linked lectin/mucin antibody sandwich assay (CAM 17.1/WGA) assessed in combination with CA 19–9 and peanut lectin binding assay for the diagnosis of pancreatic cancer. In *Cancer* **70**:1062–8.

Pillai KR, Remani P, Augustine J *et al.* (1992): Jack fruit lectin binding pattern in the exfoliative cytology of bronchopulmonary neoplasia. In *In Vivo* **6**: 117–22.

Pillai KR, Kannan S, Chandran GJ (1993): The immunohistochemistry of solid tumours: potential problems for new laboratories. In *Nat. Med. J. India* **6**:71–5.

Pillai KR, Remani P, Kannan S *et al.* (1994): Jack fruit lectin specific glycoconjugates expression during the progression of cervical intraepithelial neoplasia. A study on exfoliated cells. In *Diagn. Cytopathol.* **10**:342–6.

Ponta H, Sleeman J, Herrlich P (1994): Tumor metastasis formation: cell-surface proteins confer metastasis-promoting or -suppressing properties. In *Biochim. Biophys. Acta* **1198**:1–10.

Preisler HD, Raza A (1992): The role of emerging technologies in the diagnosis and staging of neoplastic diseases. In *Cancer* **69**:1520–6.

Rademacher TW, Parekh RB, Dwek RA (1988): Glycobiology. In *Annu. Rev. Biochem.* **57**:785–838.

Raz A, Lotan R (1987): Endogenous galactoside-binding lectins: a new class of functional tumor cell surface molecules related to metastasis. In *Cancer Metast. Rev.* **6**:433–52.

Raz A, Zhu D, Hogan V *et al.* (1990): Evidence for the role of 34-kDa galactose-binding lectin in transformation and metastasis. In *Int. J. Cancer* **46**: 871–7.

Recker F, Otto T, Rubben H (1992): Lectins in diagnosis of bladder carcinoma. In *Urol. Int.* **48**:149–53.

Remani P, Augustine J, Vijayan KK *et al.* (1989): Jack fruit lectin-binding pattern in benign and malignant lesions of the breast. In *In Vivo* **3**:275–8.

Remani P, Augustine J, Vijayan KK *et al.* (1990): Jack fruit lectin binding pattern in carcinoma of uterine cervix. In *J. Exp. Pathol.* **5**:89–96.

Rice K (1996): Glycobiology of endocytosis and lectin mediated drug targeting. In *Glycosciences: Status and Perspectives* (Gabius H-J, Gabius S, eds) Weinheim: Chapman & Hall.

Rittman BR, MacKenzie IC (1983): Effect of histological processing on lectin binding patterns in oral mucosa and skin. In *Histochem. J.* **15**:467–74.

Robert R, de la Jarrige PL, Mahaza C *et al.* (1991): Specific binding of neoglycoproteins to Toxoplasma gondii tachyzoites. In *Infect. Immun.* **59**:4670–3.

Rosai J (1989): *Ackerman's Surgical Pathology*, Vol 1&2, 7th Edition. St. Louis: The C. V. Mosby Company.

Roth J, Saremaslani P, Warhol J *et al.* (1992): Improved accuracy in diagnostic immunohistochemistry, lectin histochemistry and in situ hybridization using a gold-labelled horseradish peroxidase antibody and silver intensification. In *Lab. Invest.* **67**:263–9.

Rüdiger H (1996): Structure and functions of plant lectins. In *Glycosciences: Status and Perspectives* (Gabius H-J, Gabius S, eds) Weinheim: Chapman & Hall.

Saito K, Uda H, Tanaka S *et al.* (1992): A light and electron microscopic histochemical study on lectin binding to cells with high metastatic potential in Lewis lung carcinoma. In *J. Exp. Pathol.* **6**:123–32.

Saku T, Okabe H (1989): Differential lectin-binding in normal and precancerous epithelium and squamous cell carcinoma of the oral mucosa. In *J. Oral. Pathol.* Med. **18**:438–45.

Sanford GL, Harris-Hooker S (1990): Stimulation of vascular cell proliferation by β-galactoside-specific lectins. In *FASEB J.* **4**:2912–8.

Sarker AB, Akagi T, Jeon HJ *et al.* (1992): *Bauhinia purpurea*: a new paraffin section marker for Reed-Sernberg cells of Hodgkin's disease. A comparison with LEU-M1 (CD15), LN2 (CD74), peanut agglutinin, and BER-H2 (CD30). In *Am. J. Pathol.* **141**: 19–23.

Sarker AB, Koirala TR, Aftabuddin et al (1994): Lectin histochemistry of normal lung and pulmonary carcinoma. In *Indian J. Pathol. Microbiol.* **37**:29–38.

Schottelius J (1992): Neoglycoproteins as tools for the detection of carbohydrate-specific receptors on the cell surface of Leishmania. In *Parasitol. Res.* **78**: 309–15.

Schumacher U, Horny HP, Welsch U *et al.* (1991): Lectin histochemistry of human bone marrow: investigation of biopsy specimens in normal and reactive states and neoplastic disorders. In *Histochem. J.* **23**:215–20.

Schumacher U, Higgs D, Loizidou M *et al.* (1994): *Helix pomatia* agglutinin binding is a useful prognostic indicator in colorectal carcinoma. In *Cancer* **74**:3104–7.

Seifert G, Caselitz J (1989): General aspects and diagnostic relevance of morphological tumor markers. In *J. Tumor Marker Oncol.* **4**:1–22.

Sell S (1990): Cancer-associated carbohydrates identified by monoclonal antibodies. In *Hum. Pathol.* **21**:1003–19.

Sharon N, Lis H (1989): Lectins as cell recognition molecules. In *Science* **246**:227–46.

Shirahama T, Ikoma M, Muramatsu T *et al.* (1993): The binding site for fucose-binding proteins of Lotus tetragonolobus is a prognostic marker for transitional cell carcinoma of the human urinary bladder. In *Cancer* **72**:1329–34.

Shue GL, Kawa S, Kato M *et al.* (1993): Expression of glycoconjugates in pancreatic, gastric, and colonic tissue by *Bauhinia purpurea, Vicia villosa,* and peanut lectins. In *Scand. J. Gastroenterol.* **28**:599–604.

Silva FG, Nadasdy T, Laszik Z (1993): Immunohistochemical and lectin dissection of the human nephron in health and disease. In *Arch. Pathol. Lab. Med.* **117**:1233–9.

Skliarenko LM, Gluzman DF, Lutsik MD (1990): Opredelenie retseptorov lektinov na poverkhnostnykh membranakh limfoidnykh kletok. [Determination of lectin receptors on the surface membranes of lymphoid cells]. In *Lab. Delo.* **17**:18–20.

Sobrinho-Simoes M, Sambade C, Nesland JM (1990): Lectin histochemistry and ultrastructure of medullary carcinoma of the thyroid gland. In *Arch. Pathol. Lab. Med.* **114**:369–75.

Spicer SS, Schulte BA (1992): Diversity of cell glycoconjugates shown histochemically: a perspective. In *J. Histochem. Cytochem.* **40**:1–38.

Steuer MK, Gabius HJ, Bardosi A *et al.* (1991): Histochemical identification of endogenous lectins using labelled neoglycoproteins in human head and neck squamous cell carcinoma. In *Laryngorhinootologie* **70**:243–9.

Stowell CP, Lee YC (1980): Neoglycoproteins. The preparation and application of synthetic glycoproteins. In *Adv. Carbohydr. Chem. Biochem.* **37**:225–81.

Sugiyama K, Kawai T, Nakanishi K *et al.* (1992): Histochemical reactivities of lectins and surfactant apoprotein in pulmonary adenocarcinomas and their metastases. In *Mod. Pathol.* **5**:273–6.

Thomas M, Noguchi M, Fonseca L *et al.* (1993): Prognostic significance of *Helix pomatia* lectin and c-erbB-2 oncoprotein in human breast cancer. In *Br. J. Cancer* **68**:621–6.

Tu Z, Yin Z, Wu M (1992): Prospective study on the diagnosis of hepatocellular carcinoma by using α-fetoprotein reactive to lentil lectin. In *Chin. Med. Sci.* J. **7**:191–6.

Ulrich W, Horvat R, Krisch K (1985): Lectin histochemistry of kidney tumours and its pathomorphological relevance. In *Histopathol.* **9**:1037–50.

van Staden L, Bukofzer S, Kew MC *et al.* (1992): Differential lectin reactivities of α-fetoprotein in hepatocellular carcinoma: diagnostic value when serum α-fetoprotein levels are slightly raised. In *J. Gastroenterol. Hepatol.* **7**:260–5.

Vierbuchen M (1991): Lectin receptors. In *Cell Receptors. Morphological Characterization and Pathological Aspects* (Seifert G, ed) pp 271–361, Berlin: Springer Publ. Co.

Vigneswaran N, Peters KP, Hornstein OP *et al.* (1990): Alteration of cell surface carbohydrates associated with ordered and disordered proliferation of oral epithelia: a lectin histochemical study in oral leukoplakias, papillomas and carcinomas. In *Cell Tissue Kinet.* **23**:41–55.

Vijayakumar T, Augustine J, Mathew L *et al.* (1992): Tissue binding pattern of plant lectins in benign and malignant lesions of thyroid. In *J. Exp. Pathol.* **6**:11–23.

Vijayakumar T, Remani P, Ankathil R *et al.* (1994): Plant lectins in immunology, cytology and hematology: a bibliography, and projects for development. In *Hematol.* Rev. **8**:151–68.

Vijayan KK, Remani P, Beevi VMH *et al.* (1987): Tissue binding pattern of lectins in premalignant and malignant lesions of the oral cavity. In *J. Exp. Pathol.* **3**:295–304.

Waibel M, Richter K, von Lengerken W *et al.* (1991): Neuroendokrines Karzinom der Haut: Immunohistochemische und lektinhistochemische Befunde. [Merkel cell tumor (neuroendocrine carcinoma of the skin) in an unusual location. Immunohistochemical and lectin histochemical findings]. In *Zentralbl. Pathol.* **137**:140–50.

Walker RA (1989): The use of lectins in histopathology. In *Path. Res. Pract.* **185**:826–35.

West CM (1986): Current ideas on the significance of protein glycosylation. In *Mol. Cell. Biochem.* **72**: 3–20.

Woynarowska B, Skrincosky DM, Hagg A *et al.* (1994): Inhibition of lectin-mediated ovarian tumor cell adhesion by sugar analogs. In *J. Biol. Chem.* **269**: 22797–803.

Wu JT (1990): Serum α-fetoprotein and its lectin reactivity in liver diseases: a review. In *Annu. Clin. Lab. Sci.* **20**:98–105.

Yamamoto K, Ishida C, Shinohara Y *et al.* (1994): Interaction of immobilized recombinant mouse C-type macrophage lectin with glycopeptides and oligosaccharides. In *Biochemistry* **33**:8159–66.

Yanagi K, Ohyama K, Yamakawa T *et al.* (1990): Biochemical characterization of glycoprotein components in human ovarian cyst fluids by lectins. In *Int. J. Biochem.* **22**:659–63.

Yin ZF, Tu ZX, Cui ZF *et al.* (1993): α-Fetoprotein reaction to *Pisum sativum* agglutinin in differentiation of benign liver diseases from hepatocellular carcinoma. In *Chin. Med. J.* **106**:615–8.

Yogeeswaran G (1980): Surface glycolipid and glycoprotein antigen. In *Cancer Markers: Diagnostic and Developmental Significance* (Sell S, ed) pp 371–401, New Jersey: Humana Press.

Yogeeswaran G (1983): Cell surface glycolipids and glycoproteins in malignant transformation. In *Adv. Cancer Res.* **38**:289–350.

Yoshida Y, Okamura T, Shirakusa T (1993): An immunohistochemical study of *Helix pomatia* agglutinin-binding on carcinomas of the esophagus. In *Surg. Gynecol. Obstet.* **177**:299–302.

Yoshida Y, Okamura T, Yano K *et al.* (1994): Silver stained nucleolar organizer region proteins and *Helix pomatia* agglutinin immunostaining in esophageal carcinoma: correlated prognostic factors. In *J. Surg. Oncol.* **56**:116–21.

Zanetta J-P (1996): Lectins and carbohydrates in animal cell adhesion and control of proliferation. In *Glycosciences: Status and Perspectives* (Gabius H-J, Gabius S, eds) this volume; Weinheim: Chapman & Hall.

Zanetta J-P, Kuchler S, Lehmann S *et al.* (1992): Glycoproteins and lectins in cell adhesion and cell recognition processes. In *Histochem. J.* **24**:791–804.

Zeng F-Y, Weiser WY, Kratzin H *et al.* (1993): The major binding protein of the interferom antagonist sarcolectin in human placenta is a macrophage migration inhibitory factor. In *Arch. Biochem. Biophys.* **303**:74–80.

Zhang SM, Wu M, Chen H *et al.* (1989): Changes of glycoconjugates in human hepatocellular carcinoma. In *Histochemistry* **92**:171–5.

Zhou Q, Cummings RD (1992): Animal lectins: a distinct group of carbohydrate-binding proteins involved in cell adhesion, molecular recognition, and development. In *Cell Surface Carbohydrates and Cell Development* (Fukuda M, ed) pp 99–126, Florida: CRC.

Zschäbitz A, Gabius H-J, Stofft E *et al.* (1991): Demonstration of endogenous lectins in synovial tissue. In *Scand. J. Rheumatol.* **20**:242–51.

34 Glycobiology of Development: Spinal Dysmorphogenesis in Rat Embryos Cultured in a Hyperglycemic Environment

Lori Keszler-Moll, Amy Garcia, Marisa Braun, Mary A. Raymond-Stintz, Paul L. Mann, Jim Hanosh, and Robert O. Kelley

1 Introduction

Animal models have been used as experimental tools to delineate temporal relationships and potential mechanism(s) for analogous processes observed in human disease since the beginning of experimental science. Often these models come to represent such interesting basic science issues that the connection between the model and its true application is lost. This is especially true of neural tube defects. This general descriptor encompasses a complex series of anatomically based disorders which have a diverse set of physical manifestations and yet share common embryonic bases (Copp and Bernfield, 1994; Dias and Pang, 1995; Urui and Oi, 1995). In this chapter, we will use the rat embryo model to investigate the relationship between cell surface changes within the neural tube during the closure process under hyperglycemic conditions. The rationale for this approach is that although cellular biochemical, temporal sequencing and gross anatomical changes during neural tube closure may be species specific the cellular and molecular events should be more uniformly related to process itself and therefore share basic mechanisms. The relevant questions which represent the focus for this study are: (1) Why is the frequency of neural tube defects higher in infants of diabetic mothers? and, (2) Can an intervention be designed to prevent this?

After an early flurry of activity in the 1930s whole embryo culture did not receive much attention until the work of New and his co-workers (New and Stein, 1964; New and Daniel, 1969). This systematic study first explored the use of a plasma clot technique (New and Stein, 1964), and followed by the development of an easier homologous serum culture technique (New, 1966; New and Daniel, 1969). The results of both techniques were comparable. Cockroft (1973) extended the technique to 15.5 days of gestation. New (New *et al.*, 1973) further modified the original static culture techniques by introducing the rotating culture technique.

Studies into the contribution of the serum culture medium (Steele and New, 1974) showed that the use of a rapid centrifugation of the serum provided for better growth; whereas, improperly prepared serum was teratogenic. These studies have stimulated the investigation of the serum for essential developmental factors (Greenberg, 1976; New and Cockroft, 1978; Morriss and New, 1979) and the development of a systematic model for the study of teratogenesis (Sadler *et al.*, 1982; Sanyal and Naftolin, 1983).

This rat embryo culture model was crucial for the investigation of the hyperglycemic environment on neural tube closure. The risk of having offspring with malformations is greater [2 to 4 times] in diabetic women with elevated levels of glycosylated hemoglobins [first trimester] compared with women whose metabolic control is maintained, and to the general population. Sadler (1989) hypothesized that the origin of the diabetic embryopathy was multifactorial, with glucose and hydroxybutyrate parameters accounting for only a small component of the total teratogenic effect (Buchanan *et al.*, 1994). D-Glucose at 12 to 15 mg/ml was observed to have teratogenic [and hyperosmolar] effects in the whole rat embryo culture model (Cockroft and Coppola, 1977). The hyperosmolarity was itself also shown to be teratogenic to embryos grown in culture,

H.-J. and S. Gabius (Eds.), Glycosciences

ISBN 3-8261-0073-5

establishing a potential dichotomy between physical and biochemical mechanism(s). Although the levels of D-glucose used in these studies are significantly higher than found in circulation of most diabetics, the potential still exists for locally high tissue concentrations. Aside from high D-glucose, streptozotocin-induced diabetic models have been shown to produce higher levels of developmental malformations, especially neural tube defects, heart and caudal regression (Deuchar, 1977). Lumbosacral defects were later shown to be reversible if the induced diabetes was carefully managed (Baker *et al.*, 1981; Eriksson *et al.*, 1982).

Sadler and his co-workers (Sadler, 1980; Garnham *et al.*, 1983) showed that hyperglycemia was teratogenic to both rat and mouse whole embryos during the early head fold stage of development. There was a differential dose-dependent effect between axial rotation and neural tube closure defects. Reece (Reece *et al.*, 1985) studied ultrastructural malformations of the embryonic neural axis under hyperglycemic culture conditions and showed significant cytoarchitectural alterations in the neuroepithelium. Goldman (Goldman *et al.*, 1985) presented data suggesting that the mechanisms of the glucose-dependent teratogenic effects were mediated by a functional deficiency of arachidonic acid at a critical period of organ differentiation. This group pointed out the similarity between palatal shelf fusion and neural tube closure. In both processes there are specific cell movements, apposition of advancing cellular layers, and local cell effects at the point of fusion.

Baker (Baker *et al.*, 1990) demonstrated that the failure of rostral neural tube fusion in mouse embryos incubated with excess D-glucose could be modulated by the addition of myo-inositol. Taken together, these data suggest that modulations in both the arachidonic acid and myo-inositol pathways, with subsequent perturbations of prostaglandin production, may be the critical biochemical events implicated in neural tube defects. Are these biochemical events involved in cellular movement, layer apposition, and/or local cellular communication during the critical phase of local neural fusion? Sadler (Hunter and Sadler, 1992) indicated that hyperglycemia can alter histotrophic function of the visceral yolk sac after prolonged exposure. Free radical mediation has also been suggested as a biochemical mediator of the process (Erickson and Borg, 1993). More recently other mechanisms have been proposed for neural tube defects (Metzler *et al.*, 1994; Chambers *et al.*, 1995). The Chambers group found that anti-laminin antibodies reduce the nutrient exchange kinetics across the yolk sac endoderm, inducing neural tube defects. The Metzler group produced a model whereby the genetic mutant lacked both alleles for the *Mgat*-1 complex [encoding for *N*-linked complex oligosaccharides]. In this model, development was relatively normal through day 9 of gestation, showing that complex oligosaccharides play little or no role during this period of development. By Day 10.5 embryonic death showed that these same complex *N*-linked oligosaccharides were essential for development.

2 Problem Identification

The problem seems to be to integrate the various proposed mechanisms for the hyperglycemic-dependent neural tube defect animal model into a multi-level approach which can then be related to the human problem. The biochemical level, and the multi-cell levels are the most developed in terms of proposed mechanism. Our study is aimed at the single-cell or cell-cluster level and hopefully can be used as an integrator. By using lectin staining on sections from various levels and stages of neural tube closure, it is possible to construct a pattern of oligosaccharide changes topographically associated with specific cells or small clusters of cells involved in the closure process. The temporal changes in oligosaccharide pattern can then be used to differentiate a very basic question in the integrated mechanism(s): is the problem one of temporal displacement, where the appropriate biochemistry occurs within and on the cell surfaces but not at the appropriate time for closure; or is/are the cellular molecular mechanism(s) so inappropriate that the "wrong" surface components are presented?

2.1 The Model, Embryo Culture

The method of New, as modified by Sadler (New *et al.*, 1973; Sadler and New, 1981; Sadler, 1989) was used throughout these studies. It should be noted that reproducible results are strongly dependent on following this methodology. The culture medium was 3 ml of specially prepared rat serum with added Tyrodes solution or graded amounts of exogenous D-glucose. In some cases serum from diabetic was used.

Immediately after the embryo transfer, the culture bottle was gassed (New *et al.*, 1976a).

2.2 The Model, Serum Preparation

Male, retired Sprague-Dawley rats were exsanguinated under anesthesia. Blood was removed from the abdominal aorta and immediately centrifuged at 1200 g for 5' in B-D Vacutainer Evacuated Serum Separator tubes. The serum was separated, heat inactivated at 56 °C for 30'. Penicillin G at 200 U/ml and streptomycin at 200 μg/ml were added to the serum. The serum was then aliquotted and frozen at –70 °C.

In order to expose all the embryos to the same overall culture conditions, the control serum was diluted with Tyrodes (0.5 ml Tyrodes + 2.5 ml of serum). Hyperglycemic culture medium was prepared by adding varying amounts of D-glucose at 3.5 mg/ml to 15 mg/ml to the control serum. This resulted in increases in glucose concentration from 2.5 to 12 times the normal control glucose values (1 to 2 mg/ml). Sterile double distilled water was added to the hyperglycemic sera in order to counteract the hyper-osmolar effects of the glucose. The osmolarity of the various media was assessed with a Vapor Pressure Osmometer (5130, Wescor, Inc.).

Serum from diabetic donors was taken and processed as described above from Sprague-Dawley rats treated with streptozotocin according to Baker (Baker *et al.*, 1981). One week post drug treatment the rats were bled and the glucose levels assayed using the Sigma glucose oxidase kit. Lipemic sera were centrifuged at 12,000 g for 10′ to remove excess lipids.

2.3 Embryo Morphology

Figure 1 shows a schematic of rat embryo development during the culture period of these studies. The scale across the top of the figure shows the somite count and the corresponding gestational age of the embryos. The 48 hour culture period [9 day to 11 day gestational age] covers the development of the embryo from 0 to 25 somites and includes the development and closure of the neural tube. The three arrows on the left side of the figure show the range of developmental stages at which the addition of exogenous D-glucose was made, with the middle arrow representing the median time of addition. On the right side of the figure the three major classes of anatomical defects are shown. The least serious was termed the *side-tail* [bottom], and was observed occasionally under the control culture conditions. The *side-tail* appears to be the result of incomplete rotation of the most distal caudal region (Deuchar, 1971). This defect appears to be temporally isolated, as all other developmental aspects of the embryo's development appear normal.

The second morphological aberration is the *squirrel-tail* [middle] (Sadler *et al.*, 1982). In this case, the rotation appeared to be defective in its entirety, resulting in the juxtaposition of the cranium and the tail region of the neural tube. The neural tube is closed with the exception of a small posterior neuropore in some cases. The cranial neural tube appears fully fused in all cases. The relative locations of some anatomical features, such as pharyngeal arches in relation to the diencephalon, appeared displaced. This is probably more related to the maintenance of the aberrant reverse curvature of the body rather than to any additional anomaly.

The third, and most serve defect was termed *fused-tail* [top] (Cockroft and Coppola, 1977; Sadler *et al.*, 1982). This morphological aberration appeared similar to the *squirrel-tail* but with the addition of a fusion between the tail and cranial surfaces at the point of juxtaposition. This fusion resulted in a much tighter kink in the tail. The brain region diencephalon to hindbrain always appeared to fuse properly.

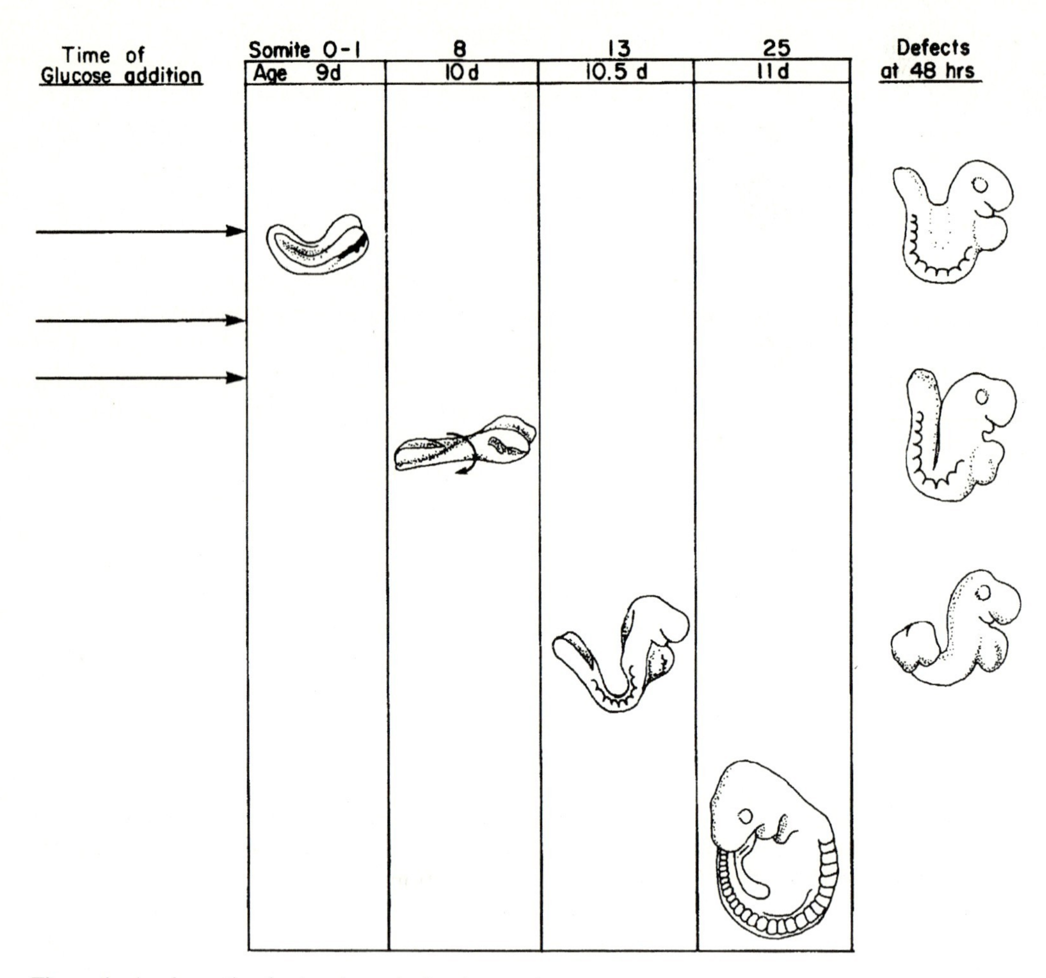

Figure 1. A schematic of rat embryonic development

Table 1 shows the general morphological description of the embryos at the end of the 48 h culture period. Some embryos were removed at 24 hours of culture to permit examination of the intermediate stages in the development of the embryos and any morphological defects.

Embryos grown under the *in vitro* culture conditions are similar in all respects to those left *in vivo* for the same period of time (New *et al.*, 1976b). Comparisons of somite count, circulation, yolk sac dimensions are shown in Table 1. The neural tube was completely closed, the pharyngeal arches, otic and optic vesicles were well developed. The heart beat was strong and the circulation was well developed throughout the yolk sac. Rotation of the embryo from the inverted to the concave aspect was complete (Deuchar, 1971).

Table 1 also shows a comparison of morphological data between the controls and those exposed to growth in hyperglycemic sera. The yolk sac measurement shows that high glucose causes a smaller and less developed sac when compared to the normals. The subjective assessment of the circulation indicated a slight retardation in the moderate to high glucose exposure. Somite counts were comparable but it was difficult to make accurate counts on very deformed embryos. The total protein content of the embryos were comparable, although a trend to

Table 1. General growth characteristics in the rat model

Characteristics	Normal	7 TO 10 mg/ml	11 TO 15 mg/ml	Diabetic Serum	Diabetic + 7 mg/ml
Yolk Sac Diameter [mm]	3.03 ± 0.50 [N=35]	2.99 ± 0.95 [N=16]	2.64 ± 0.65 [N=39]	3.28 ± 1.11 [N=7]	3.36 ± 0.41 [N=11]
Circulation	2.65 ± 0.06 [N=38]	1.89 ± 0.53 [N=19]	2.35 ± 0.35 [N=44]	2.60 ± 0.56 [N=7]	2.37 ± 0.18 [N=10]
Somite Count	20 ± 3.6 [N=12]	20 ± 1.5 [N=10]	17 ± 4.3 [N=14]	22 ± 1.4 [N=4]	23 ± 1.9 [N=6]
Total Protein [mg]	0.33 ± 0.07 [N=24]	0.34 ± 0.05 [N=12]	0.32 ± 0.1 [N=12]		
Percentage no growth	11 [N=5]	25 [N=8]	27 [N=18]		
Osmolarity [mOs]	324 ± 4 [N=5]	321 ± 4 [N=5]	322 ± 9 [N=4]		

lower growth rates with increasing hyperglycemia was apparent. The number of "no-growth" was significantly higher in the hyperglycemic cultures.

Table 2 shows data collected from experiments with 129 embryos exposed to different culture conditions. It is clear that the D-glucose concentration has a profound effect on the overall development of the morphological defects [5/40 from the control series versus 30/50 from the 11 to 15 mg/ml series]. D-Glucose concentration also effects the severity of the defect observed [0/40 from the control series versus 21/50 from the test series showed *squirrel-* or *fused*-tail defects]. There is also a dramatic shift to higher frequencies of more severe defects observed as the stage of development at the beginning of culture is lowered from 4 to 6 somite to the base-line of 0 to 1 somite. The diabetic serum does not appear to be any more effective at inducing the morphological defects than the simple addition of D-glucose to normal rat serum.

2.4 The Kinetics of Neural Tube Defect Formation

The kinetics of defect formation are important in the development of mechanism(s) for the morphological changes observed above. Some embryos were removed from culture after 24 hours of culture. In this more limited study only normal rat serum controls and hyperglycemic sera at 15 mg/ml were compared.

Normal embryos cultured in normal medium for 24 hours developed to the 9 to 11 somite stage, and the neural tube was partially fused [see

Table 2. Summary of defect frequency

Somite number	Normal culture conditions		7 to 10 mg/ml D-glucose		11 to 15 mg/ml D-glucose		Diabetic serum culture		Diabetic + 7 mg/ml D-glucose	
	side-tail	squirrel + fused	side-tail	squirrel + fused	side-tail	squirrel + fused	side-tail	squirrel + fused	side-tail	squirrel + fused
0 TO 1	3/25	0/25	1/8	4/8	4/19	11/19	1/5	1/5	1/5	1/5
2 TO 3	0/3	0/3	1/5	1/5	3/14	6/14	ND	ND	ND	ND
4 TO 6	2/12	0/12	1/6	0/6	2/17	4/17	0/3	0/3	0/7	0/7
TOTAL	5/40	0/40	3/19	5/19	9/50	21/50	1/8	1/8	1/12	1/12

Fig. 1]. The posterior neuropore was open. The cranial neural tube was open, the second fusion point (cranial region) was just beginning to fuse at this time point. Axial rotation was also just starting. Standard histological staining demonstrated the side-ways displacement of the distal tail region. It was hypothesized that this was related to increased hydrostatic pressure exerted on the amnion. When it was removed the embryo "snapped" to its concave orientation, confirming the hypothesis. These embryos appeared grossly normal.

The majority of the embryos exposed to hyperglycemic media showed no fusion of the neural folds whatsoever below the first somite. Cranial fusion was partial in some cases and complete in others [defined as from the diencephalon to the hind brain]. Rotation had not begun and the neural folds from the brain and tail regions were closely apposed.

Standard histological staining through fore-brain (optic region), mid and tail region of embryos after 48 hours in culture revealed positional displacement of various surfaces. The *squirrel-tail* defect showed a very distorted, closed neural tube throughout its length because of the lack of mobility of the tail associated with the fusion of the cranial and tail surfaces [see Fig. 1].

The fore-brain region of the *fused-tail* embryos show that the neural tube associated with the head has fused and the distal tail region has a ridge of flattened neural tissue, whereas, the neural tissue ventral to it is morphologically normal. A slightly lower section through the hind brain shows the roof of the neural tube has thinned significantly, and an ectodermal fusion between the tail and hind brain is apparent, although the neural tubes are still distinct. A more caudal section, just below the pharyngeal arches, show the complete dissolution of the roof of both the hind brain and tail region neural tubes. This results in a single large malformed neural cavity. Sections distal to this show a very large irregular, but fully closed neural tube.

3 Lectin Staining Patterns

Groups of embryos, control and tests, were used for these assays. Embryos were washed several times in Tyrodes and then fixed in 2% formaldehyde (Tousimis) supplemented with 0.5% cetylpyridinium chloride (CPC) in 0.1 M phosphate buffer (pH 7.2) for 1 to 2 hours at room temperature. The embryos were then processed for paraffin embedment with Tissue Prep (56 °C to 62 °C, Fisher). Tissue sections (8 micron) were prepared on an American Optical microtome. A standard light microscope was used to position the cut sections on Carlson 3-spot slides (Carlson, Inc) and deparaffinized using standard techniques.

FITC-conjugated lectins (Vector Labs) were used to stain specific carbohydrate residues as follows. The sections were blocked with 0.1% bovine serum albumin in phosphate buffered saline [BSA/PBS] for 16 hours at 4 °C to reduce non-specific lectin uptake. Sections were stained with lectins at 5 μg/ml in BSA/PBS for 90' at 37 °C in a humidified atmosphere. After staining the sections were incubated in three changes of PBS before mounting in Immunomount (Shandon). The fluorescence staining pattern was viewed and photographed under epi-illumination with a Leitz photomicroscope. Specificity of staining was confirmed by using appropriate and inappropriate soluble carbohydrate competitors for the lectin binding.

Table 3 shows summary data for these lectin staining experiments. Various anatomical regions were assessed for fluorescence intensity on a relative scale of 0 to +4. Staining patterns were also assessed and scaled at various time points during the culture period [8.9 days to 11 days of gestation].

ConA [mannose] staining was found to be ubiquitous on virtually all cells in all sections. The RCA-120 [galactose] staining pattern was more selective, showing specificity for both endodermal and ectodermal derived tissues. This is in basic agreement with the findings in the mouse model (Wilson and Wyatt, 1995). In this study ConA was found to stain basal and intercellular surfaces but not luminal surfaces in early mouse embryos [9 to 10 somites]. By 26 to 30 somites the luminal staining was evident.

Table 3. The lectin staining pattern of rat embryonic tissues during development

	Day	Aer	Ectoderm [non-ridge]	Basal Lamina	Mesoderm [sub-ridge]	Core [mesoderm]
WGA	8.9	++	++	−	++	++
	9.5	++++	++++	+	+++	++
	10.0	++++	++++	+	+++	++
	11.0	+++	+++	−	++	+
PNA	8.9	++	++	−	−	−
	9.5	+++	++	−	++	++
	11.0	+++	+++	+++	+	−
	11.0	+++	+++	+++	+	−
SBA	8.9	+	−	−	−	−
	9.5	+++	+	−	−	−
	10.0	+	+	+	+	+
	11.0	−	−	−	−	−
RCA	8.9	+	+	−	−	−
	9.5	++++	+++	+	++	+
	10.0	+++	+	+++	+	+
	11.0	−	+	−	−	−

The RCA-120 staining specificity for the "lining" structures appears to be unaffected by the neural tube defects.

Figure 2. shows a section through the head region of the *fused-tail* defect stained with PNA which shows high specificity for the luminal surface of the neural tube. The CPC addition to the fixation mixture provides a better preservation of the extra-cellular matrix; if it is omitted from the procedure the specificity of the PNA disappears leading to the conclusion that the galactose/*N*-acetyl galactose residues are part of the matrix. At this section level the tail region is observed in its aberrant dorsal position, and the neural tube is complete and separate [tail region not shown in Figure 2].

The preferential PNA staining of the luminal neural tube surface, and the absence of staining on the roof region of the neural tube is a defect-specific characteristic. This is clearly visible on the extended roof area of the hind brain section, but can also be seen on the dorsal fusion point of the tail section. In the *fused-tail* defect the PNA preferentially stains the luminal surface. Figure 3

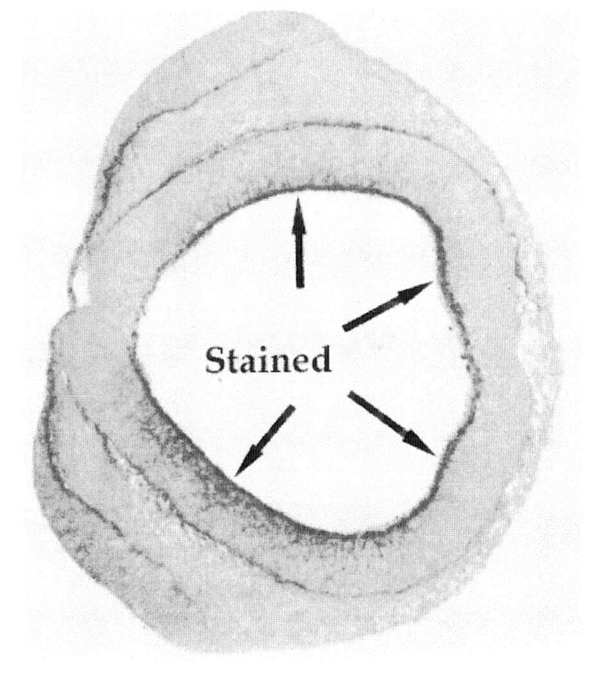

Figure 2. PNA staining pattern of a fused-tail defect

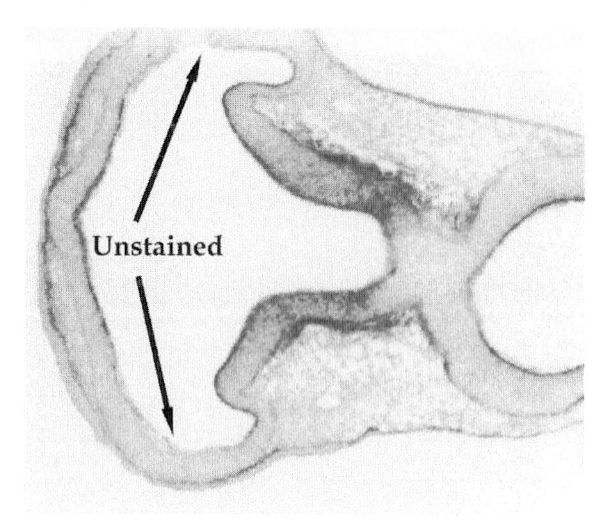

Figure 3. PNA staining pattern of a fused-tail defect

shows a PNA stained section where there are two non-staining regions located at the top and bottom of the Y-shaped thickened section of the left hand side of the neural tube complex. It appears that these two unstained regions represent abnormal fusion points of the opposed neural tubes.

4 Surface Oligosaccharide and Fusion Mechansim

During the developmental period [9 to 11 days gestational] axial rotation occurs and the closure of the neural tube is accomplished. When embryos are exposed to high concentrations of D-glucose [12 to 15 mg/ml] a series of three dose-dependent defects can be documented. The *side-tail* defect is the least severe and appears to result from an interruption in either the signal or the stimulus which initiates axial rotation. As described by Deuchar (1971), similar partial rotations could be achieved by surgical intervention at different levels along the trunk of the embryo. This, therefore, suggest that either chemical or mechanical trauma can induce morphological defects at this stage of development.

The *squirrel-tail* defect (Cockroft and Coppola, 1977; Sadler *et al.*, 1982) is more severe and presumably results from the interruption of the axial rotation at a very early stage. Neural tubes are completely closed indicating that closure precedes the rotational event or that other physical factors prevented the close apposition of the anterior and posterior neural folds.

In the third and most severe defect, *fused-tail*, the tail is again non-rotated as seen in the *squirrel-tail* lesion but there is also an aberrant fusion of the neural folds (Cockroft and Coppola, 1977). These results indicate that in all defects the fusion mechanism was operative and that by 48 hours of culture [11.5 days gestation] was complete. One of the causal events in these defects is the inappropriate or delayed rotation of the embryo which in turn presents the inappropriate neural folds to each other for fusion. This type of defect can also be caused by hyperosmolarity (Cockroft and Coppola, 1977), suggesting that overall tonicity can play a role in maintaining appropriate shape, and therefore the appropriate tissue/surface juxtaposition. This may provide added significance to the observation of the differential movement of the cultured embryos [normal and those exposed to hyperglycemic levels of glucose] when the visceral yolk sac was removed. These membranes have been shown to have a mediation role in the absorption of macromolecules (Freeman *et al.*, 1981) and on general growth characteristics (Payne and Deuchar, 1972).

The dose-dependency of the defects is related to severity, incidence, D-glucose concentration and the stage of development at the point of explanation. Lectin-staining patterns did not indicate any gross changes in surface carbohydrate composition regardless of exposure. This is in spite of the observations (Chandramotili *et al.*, 1977) that long term exposure of diabetic red cells in vivo caused changes in the lectin binding properties of cell surface glycoproteins. The differences in model are obvious, long term versus shorter exposure times, with the long term exposures favoring non-enzymatic glycosylation. This does, however, point to the importance of integrating proposed biochemical mechanisms with real biological processes.

The specific absence of PNA-dependent staining pattern of the neural ridge, either before fusion or after the overgrowth of the roof tissue, indicates a cell [or cell-cluster] specific mechanism. PNA shows a high degree of specificity for β1,3-Gal-GalNAc. This residue as a terminal specificity is generally confined to fetal and oncogenic expression models (Rosenstraus *et al.*, 1982). The fusion area/point at the top of the neural folds has been shown to undergo structural differentiation just prior to the fusion event in the mouse (Waterman, 1976), and later in the rat embryo (Morriss-Kay, 1981), and has been associated with increased proteoglycan production and orientation of the apical cytoskeletal elements. The lack of surface staining with PNA at this particular neural fold/ridge region may indicate a localized proteoglycan production/secretion. If these patterns are indicative of the fusion process itself, it is interesting to note that these fusion points remain "active" for at least 24 to 30 hours after the fusion event has occurred. This suggests that there is considerable temporal flexibility in the biochemical events leading [?initiating?] to fusion, and little or no flexibility in the positional parameters required for fusion. This conclusion, that temporospatial relationships are primary in determining the extent and severity of the neural tube defect, is supported by the work of Wilson (Wilson and Wyatt, 1995). In this study, the mouse mutation loop-tail was shown to have lectin staining patterns consistent with the appropriate cell surface pattern appearing at the appropriate time in development, compared with normal development. Furthermore, the staining pattern indicated that the staining pattern was stable for some hours after the fusion event normally occurred. The defect was related to faulty spatial apposition of the tissues.

The challenge for the future will be to develop these animal models by integrating the various classes of effects [biochemical, anatomical, and physical] discussed herein, into a coherent and relavent process-oriented picture. With these models in place a series of prioritized intervention protocols can then be developed.

References

Baker E, Egler JM, Klein SH *et al.* (1981): Meticulous control of diabetes during organogenesis prevents congenital lumbosacral defects in rats. In *Diabetes* **30**:955–60.

Baker L, Piddington R, Goldman A *et al.* (1990): Myo-inositol and prostaglandins reverse the glucose inhibition of neural tube fusion in cultured mouse embryos. In *Diabetologia* **33**:593–6.

Buchanan TA, Denno KM, Sipos GF *et al.* (1994): *In vitro* evidence for a multifactorial etiology with little contribution from glucose per se. In *Diabetes* **43**: 656–60.

Chambers BJ, Klein WW, Nosel PG *et al.* (1995): Methionine overcomes neural tube defects in rat embryos cultured on sera from laminin immunized monkeys. In *J. Nutr.* **125**:1587–99.

Chandramorili V, Williams S, Marshal JS *et al.* (1977): Cell surface changes in diabetic rats: studies of lectin binding to liver cell plasma membranes. In *Biochim. Biophys. Acta* **465**:19–26.

Cockroft DL (1973): Development in culture of rat fetuses explanted at 12.5 and 13.5 days of gestation. In *J. Embryol. Exp. Morph.* **29**:473–80.

Cockroft DL, Coppola PT (1977): Teratogenic effects of excess glucose on head-fold rat embryos in culture. In *Teratology* **16**:141–4.

Copp AJ, Bernfield M (1994): Etiology and pathogenesis of human neural tube defects: insights from mouse models. In *Curr. Opinion Pediatr.* **6**:624–31.

Deuchar, EM (1971): The mechanism of axial rotation in the rat embryo: an experimental study *in vivo*. In *J. Embryol. Exp. Morphol.* **25**:189–93.

Deuchar EM (1977): Embryonic malformations in rats, resulting from maternal diabetes. In *J. Embryol. Exp. Morphol.* **42**:93–101.

Dias MS, Pang D (1995): Split cord malformations. In *Neurosurg. Clin. N. Am.* **6**:339–58.

Eriksson UJ, Dahlstrom E, Larson KG *et al.* (1982): Increased incidence of congenital malformations in the offspring of diabetic rats and their prevention by maternal insulin therapy. In *Diabetes* **31**:1–11.

Eriksson UJ, Borg LAH (1993): Diabetes and embryonic malformations: the role of substrate induced free oxygen radical production for dysmorphogenesis in cultured rat embryos. In *Diabetes* **42**:411–8.

Freeman SJ, Beck F, Lloyd JB (1981): The role of the visceral yolk sac in mediating protein utilization by rat embryos cultured *in vitro*. In *J. Embryol. Exp. Morphol.* **66**:223–6.

Garnham EA, Beck F, Clark CA *et al.* (1983): Effects of glucose in rat embryos in culture. In *Diabetologia* **25**:291–4.

Goldman AS, Baker L, Piddington R *et al.* (1985): Hyperglycemia-induced teratogenesis is mediated by functional deficiency of arachidonic acid. In *Proc. Natl. Acad. Sci. USA* **82**:8227–31.

Greenberg DL (1976): *In vitro* development of post-implantation rat embryos cultured on dialyzed rat serum. In *Teratology* **14**:65–9.

Hunter ES, Sadler TW (1992): The role of the visceral yolk sac in hyperglycemia-induced embryopathies in mouse embryos in vitro. In *Teratology* **45**:195–203.

Metzler M, Getz A, Sarkar M *et al.* (1994): Complex asparagine linked oligosaccharides are required for morphogenic events during post implantation development. In *EMBO J.* **13**:2056–65.

Morriss-Kay GM (1981): Growth and development of patterns in the cranial neural epithelium of rat embryos during neurulation. In *J. Embryol. Exp. Morphol.* **65**:225–31.

Morriss-Kay GM, New DAT (1979): Effect of oxygen concentration on morphogenesis of cranial neural folds and neural crest in cultured rat embryos. In *J. Embryol. Exp. Morphol.* **54**:17–22.

New DAT (1966): Development of rat embryos cultured in blood sera. In *J. Reprod. Fert.* **12**:509–14.

New DAT, Stein KF (1964): Cultivation of post-implantation mouse and rat embryos in plasma clots. In *J. Embryol. Exp. Morphol.* **12**:101–6.

New DAT, Daniel JC (1969): Cultivation of rat embryos explanted at 7.5 to 8.5 days of gestation. In *Nature* **223**:515–8.

New DAT, Cockroft DL (1978): A rotating bottle culture method with continuous replacement of the gas phase. In *Experientia* **33**:138–42.

New DAT, Coppola PT, Terry S (1973): Culture of explanted rat embryos in rotating bottles. In *J. Reprod. Fert.* **35**:135–41.

New DAT, Coppola PT, Cockroft DL (1976a): Improved development of head fold rat embryos in culture resulting from new oxygen and modifications of the culture serum. In *J. Reprod. Fert.* **48**: 219–22.

New DAT, Coppola PT, Cockroft DL (1976b): Comparison of growth *in vitro* and *in vivo* of post-implantation rat embryos. In *J. Embryol. Exp. Morphol.* **36**:133–41.

Payne GS, Deuchar EM (1972): An *in vitro* study of the functions of embryonic membranes in the rat. In *J. Embryol. Exp. Morphol.* **27**:533–9.

Reece EA, Pinter E, Lebranth CZ *et al.* (1985): Ultrastructural analysis of malformations of the embryonic neural axis induced by in vitro hyperglycemic conditions. In *Teratology* **32**:363–73.

Rosenstraus MJ, Hannis M, Kupatt LJ (1982): Isolation and characterization of peanut agglutinin resistant embryonic carcinoma cell surface variants. In *J. Cell. Physiol.* **112**:162–6.

Sadler TW (1980): Effects of maternal diabetes on early embryogenesis. In *Teratology* **21**:349–53.

Sadler TW (1989): Culture of early somite mouse embryos during organogenesis. In *J. Embryol. Exp. Morphol.* **49**:17–22.

Sadler TW, New DAT (1981): Culture of mouse embryos during neurulation. In *J. Embryol. Exp. Morphol.* **66**:109–13.

Sadler TW, Horton WE, Warner CW (1982): Whole embryo culture: a screening technique for teratogens. In *Teratog. Carcinog. Mutag.* **2**:243–6.

Sanyal MK, Naftolin F (1983): In vitro development of mammalian embryos. In *J. Exp. Zool.* **228**:235–43.

Steele CE, New DAT (1974): Serum variants causing the formation of double hearts and other abnormalities in explanted rat embryos. In *J. Embryol. Exp. Morphol.* **31**:707–11.

Urui S, Oi S (1995): Experimental study of embryogenesis of open spinal dysraphism. In *Neurosurg.* **6**:195–202.

Waterman RE (1976): Topographical changes along the neural fold associated with neurulation in the hamster and mouse. In *Am. J. Anat.* **146**:151–6.

Wilson DB, Wyatt DP (1995): Patterns of lectin binding during mammalian neurogenesis. In *J. Anat.* **186**:209–16.

35 Glycobiology of Fertilization

Fred Sinowatz, Edda Töpfer-Petersen, and Juan J. Calvete

1 A Short Review of the Fertilization Pathway

Fertilization is a fundamentally significant event, which has been intensely studied. Many of the early studies concentrated on fertilization of marine organisms and lower vertebrates, where large numbers of female as well as male gametes can be easily obtained (Dunbar *et al.*, 1994). Due to technical advances in morphology, biochemistry and molecular biology (reviewed by Dunbar and O'Rand, 1991; Wassarman, 1991) a better understanding of mammalian fertilization has been achieved. In particular, the notion that glycobiological interactions play a central role in this process, as suggested a decade ago (Gabius, 1987), has become consolidated.

In mammals, fertilization can be viewed as a sequence of specific cellular interactions between the sperm and the egg. At insemination, several hundred million spermatozoa are released into the female genital tract. During their transport through the uterus and oviduct, spermatozoa are capacitated and subsequently undergo a change in motility called hyperactivation that is associated with the sperm's ability to fertilize the oocyte. Fully capacitated spermatozoa pass through several layers of somatic cells of the former cumulus oophorus before reaching the zona pellucida (zona pellucida), an acellular glycoprotein coat which surrounds the oocyte.

The first contact between male and female gametes appears to be a disconcertingly random affair (Aitken, 1995). But, once made, contact is established by an intimately tuned recognition mechanism between sperm and the zona pellucida. Once bound, sperm undergo the acrosome reaction, by which hydrolytic enzymes are released from the acrosome of the sperm head. This permits a sperm to degrade the matrix of the zona pellucida and to fuse with the plasma membrane of the egg (Burks *et al.*, 1995). Immediately after the first spermatozoon fuses with the oocyte, exocytosis of a set of secretory granules, the cortical granules of the oocyte, is triggered. The contents of these granules modify the egg membrane and the zona pellucida so that any further penetration is prevented, the so-called block to polyspermia.

Sperm-zona interaction involves the binding of sperm to species-specific receptor molecules present within the zona (Yanagimachi, 1981; Wassarman, 1987). A number of recent studies have shown that carbohydrate moieties present within the zona pellucida and corresponding carbohydrate-binding proteins on the sperm surface (Sinowatz *et al.*, 1988, 1989) regulate this interaction (Boldt *et al.*, 1989). Competitive inhibition studies using various monosaccharides, polysaccharides or glycoproteins have indicated a role for specific carbohydrate moieties in mediating sperm-zona interaction in hamsters (Ahuja, 1982, 1985; Huang *et al.*, 1982), rats (Shalgi *et al.*, 1986) and mice (Shur and Hall, 1982; Lambert, 1984). Treatment of either sperm or zona pellucida coated eggs with glycosidases also inhibits sperm binding to the zona in a variety of species (Shur and Hall, 1982; Ahuja, 1985; Florman and Wassarman, 1985; Shalgi *et al.*, 1986; Boldt *et al.*, 1989).

The nomenclature used throughout this review is that proposed by Wassarman (1990) and follows that used in other biological interactions which are analogous to fertilization in several respects,

H.-J. and S. Gabius (Eds.), Glycosciences

ISBN 3-8261-0073-5

e.g. the initial interaction between bacteriophages and bacteria, and between viruses and animal cells. The egg surface components (either from the zona pellucida or the plasma membrane) to which sperm bind, are called sperm receptors, while the sperm surface components, to which oocytes bind, are termed egg-binding proteins (Wassarman, 1992).

2 Structure and Function of the Zona Pellucida

2.1 The Glycoprotein Constituents of the Zona Pellucida

All mammalian eggs are enveloped by the zona pellucida, a unique extracellular coat that protects the egg and the early embryo from physical damage. Before completion of the fertilization process, spermatozoa must penetrate this highly organized matrix, which, in consequence, regulates species-specific recognition and sperm binding, induces the acrosome reaction, and is involved in the establishment of the block to polyspermy following gamete fusion (for review see Yanagimachi, 1994). The most obvious interspecies difference is the relative size of the zona pellucida with a thickness varying from 1–2 μm in the opossum to 27 μm in the cow (Dunbar *et al.*, 1994). The mammalian zona pellucida is constructed from only three glycoproteins, which build up the typical complex fibrillogranular structure by noncovalent interactions (Wassarman and Mortillo, 1991).

Studies in the mouse provided insight into the ultrastructural organization of the matrix of the zona pellucida. Heterodimeric units, formed by the proteins ZP2 and ZP3, are periodically arranged in filaments. These filaments appear to be interconnected by ZP, the largest mouse zona pellucida protein, resulting in the well-organized zona pellucida network (Wassarman and Mortillo, 1991). At present, however, it can only be speculated whether or to what extent the information on the structure obtained for the zona pellucida of the mouse is also relevant for other species.

It has been recently shown that the glycoproteins of the zona pellucida, commonly denoted as ZP1, ZP2 and ZP3 (according to their electrophoretic mobility), are in most, if not all species, the products of three major gene families referred to as ZPA, ZPB and ZPC, according to the size of their cDNAs (Harris *et al.*, 1994). The DNA and deduced amino acid sequences within one family are highly conserved among species, including the number and position of most cysteine residues and potential *N*-glycosylation sites. Although the processing that yields mature proteins differs among species, as can be concluded from their molecular masses and the known partial *N*-terminal sequences, the high degree of sequence homology, and in particular the conserved position of cysteine residues, suggests that members of the same protein family may possess similar tertiary structures (Dunbar *et al.*, 1994; Harris *et al.*, 1994). Surprisingly, this evident structural homology and the same genetic origin do not necessarily imply in every case that homologous proteins have comparable biological functions. Studies in rodents (mouse, hamster and rabbit) and other species indicate that the identity of individual zona proteins that exhibit sperm receptor activity may differ among species (reviewed by Dunbar *et al.*, 1994).

Recognition and sperm binding takes place in a relatively species-specific manner and appears to be mediated by the carbohydrate portion of the zona glycoproteins. In the mouse, the sperm receptor activity has been mapped to ZP3, a member of the zona pellucida protein family (Florman and Wassarman, 1985). In the pig, however, the major biologically active component appears to be pZP3α, which belongs to the ZPB protein familiy (Sacco *et al.*, 1989; Yurewicz *et al.*, 1993), although other zona proteins may contribute to the binding process (Hedrick, 1993). Experiments with transgenic mice bearing the hamster ZPC gene (coding for the hZP3-protein) lead to the expression of a fully functional hamster sperm receptor (Kinloch *et al.*, 1992). In other rodents such as the rabbit, another member of the ZPB protein family, the 55RC protein seems to be involved in the binding of sperm to the zona pellucida (Dunbar *et al.*, 1994). Hence, it is tempting to speculate that acquisition of sperm receptor activity may be the result of a species-specific glycosylation process of the distinct zona protein backbones. Thus, though the enzymes responsible for carbohydrate processing are present in all mammalian cells, the expression of glycosyltransferases related to the maturation of the sugar chains is quite different between animal species, leading to a different glycosylation pattern in the individual species (Kobata, 1992). In this respect, the knowledge of the carbohydrate structures and of the extent of their interspecies variations are prerequisites for understanding the structural requirements of gamete interaction and the expression of species-specificity during the early events of fertilization.

2.2 Characterization of the Zona Pellucida Oligosaccharides by Lectins

One approach to characterize the oligosaccharide chains of the zona proteins is to employ lectins as tools. The wide panel of lectins with narrow sugar specificity allows the identification of distinct carbohydrate structures and particularly their *in situ* localization (Shalgi *et al.*, 1991; Aviles *et al.*, 1994; Maymon *et al.*, 1994; Skutelsky *et al.*, 1994). Comparative cytochemical studies demonstrate the species-dependent variations in the expression and distribution of sugar moieties throughout the thickness of the zona pellucida. Thus, the organization of the zona into distinct layers, which have been demonstrated by microscopic studies in various species (Shalgi *et al.*, 1991), appears to be accompanied by a spatial distribution of defined sugar structures throughout the thickness of the zona pellucida. Different species are characterized by a highly specific lectin-binding pattern. In rodents, this pattern varies only in the expression of ligand sites for α-galactose and/or β-*N*-acetylgalactosamine-recognizing lectins. Interestingly, binding of these lectins is not detectable in the dog, cat, pig and human. In general, the variation of the lectin-binding patterns increases with the evolutionary distance of species. Some carbohydrate structures are found in all species examined. These are preferentially sugar residues such as mannose and *N*-acetylglucosamine, which are usually part of the core region of *N*-linked oligosaccharides. Although *Ulex europaeus* agglutinin-I (UEA I), specific for fucose, did not bind to the zona pellucida in any species examined (Maymon *et al.*, 1994; Skutelsky *et al.*, 1994), the *Aleuria aurantia* agglutinin (AAA) detected the presence of fucose residues, α1,6-linked to the inner core region of *N*-linked zona pellucida oligosaccharides in rats (Aviles *et al.*, 1994), cows and pigs (Blase and Töpfer-Petersen, unpublished observations), indicating the occurrence of this structure in rodents and other species. Most variations appear to affect the terminal non-reducing region of the zona pellucida sugar chains, supporting the idea that even small structural differences may contribute to the establishment of the species-specific nature of the gamete interaction.

2.3 Oligosaccharides of the Porcine Zona Pellucida

Due to the development of methods for the large scale isolation of zona pellucida from pig ovaries, the structural elucidation of the zona oligosaccharides has been thoroughly investigated in this species. Porcine zona glycoproteins can be resolved electrophoretically into two components, with molecular masses of about 55 kDa and 90 kDa (Hedrick and Wardrip, 1987). The 55 kDa component, representing about 80 % of the total protein content of the porcine zona pellucida, consists of two distinct polypeptides, referred to as ZP3α and ZP3β by analogy to the mouse. Sperm receptor activity has been correlated to pZP3α (of the ZPB family), whereas pZP3β is the homologue of the mouse ZP3 protein, a member of the ZPC-familiy. Both proteins can be separated only after partial deglycosylation with endo-β-galactosidase (Sacco *et al.*, 1989). To avoid changes of the sugar portion and modification of potential binding sites, structural elucidation of the carbohydrate chains was done using the whole 55 kDa component fraction containing both zona pellucida proteins (see below). According to the deduced cDNA encoded amino acid sequence, each of the proteins may possess up to five potential *N*-glycosylation sites (Harris *et al.*, 1994). Carbohydrate composition analysis, however, suggested that the pZP3α and pZP3β proteins contain three or four *N*-linked oligosaccharide chains and additionally three (pZP3α) and six (pZP3β) *O*-linked glycan chains (Yurewicz *et al.*, 1992).

2.3.1 *N*-linked carbohydrates

The *N*-linked carbohydrate portion of the porcine 55 kDa component, released by hydrazinolysis, belongs to the complex type of glycoproteins with a fucose moiety in (α1,6-linkage to the proximal *N*-acetylglucosamine residue (Mori *et al.*, 1989; Noguchi and Nakano, 1992). They can be separated into neutral (28 %) and acidic (72 %) fractions. More than 30 different structures of *N*-linked oligosaccharides have been identified in the neutral fraction. The most abundant neutral structure is a very common biantennary oligosaccharide which is also found in other glycoproteins. Some structures lack a terminal non-reducing galactose residue at the manα1,3- or manα1,6-arms of the trimannosyl core, or contain linear *N*-acetyllactosamine repeats in their outer chains (Mori *et al.*, 1989; Noguchi and Nakano, 1992). The acidic *N*-linked sugar chains of the 55 kDa components occur as di-, tri-, and tetra-antennary structures. They can be sialylated and/or, as in most cases, sulfated at the poly-*N*-acetyllactosamine units. The major acidic non-sulfated components are differently sialylated (with Neu5Gc/Ac residues) biantennary and 4,2 branched tri-antennary acidic chains. The sulfate groups are exclusively 1,6-linked to the *N*-acetylglucosamine residues of the *N*-acetyllactosamine repeats (Noguchi and Nakano, 1992). In spite of their general similarity to keratan sulfate structures, the acidic sulfated *N*-linked carbohydrate chains of the porcine zona pellucida are unique in the position of the sulfate group and the unusual extension of poly-*N*-acetyllactosamine chains from a complex-type carbohydrate core.

Recently, the first attempts have been made to investigate the *N*-linked carbohydrate chains of the mouse zona glycoproteins (Noguchi *et al.*, 1992; Nagada *et al.*, 1994). Although the structure of the murine zona oligosaccharides is basically the same as that described for the porcine zona pellucida, both species differ in their content of sialic acid, sulfate and the distribution of di-, tri-, and tetra-antennary structures. The most significant difference, however, appears to be the occurrence of terminal non-reducing α-galactose and β-*N*-acetylgalactosamine residues in the mouse glycoproteins which have not been found in the porcine *N*-linked oligosaccharides (Noguchi *et al.*, 1992).

2.3.2 *O*-linked oligosaccharides

The acidic poly-*N*-acetyllactosamine backbone of the *O*-linked oligosaccharides exhibits the same structure as that in the *N*-linked glycans extending from Galβ1,3GalNAc disaccharide core. Interestingly, the longer chains terminate predominantly with Neu5Gc/Ac residues, which are then α2,3 linked to a galactose residue of the outermost lactosamine unit, whereas shorter chains with no or only one *N*-acetyllactosamine unit can also be α2,6-sialylated at the proximal *N*-acetylgalactosamine residue. Neutral structures with α-galactose and β-*N*-acetylglucosamine residues, and minor components with α-galactose and β-*N*-acetylgalactosamine residues at their non-reducing termini have also been identified (Hirano *et al.*, 1993; Hokke *et al.*, 1993, 1994). The amount of sulfated lactosamine repeats and the degree of sialylation contribute to the enormous heterogeneity of the carbohydrate portion of the zona matrix. More than 30 different *O*-linked structures have been identified in porcine zona glycoproteins and the diversity of the acidic *N*-linked structures appears to be even larger.

2.4 Carbohydrates Involved in Sperm-Zona Pellucida Interaction

Species-specific recognition and binding between spermatozoa and the oocyte take place at the surface of the complex three-dimensional zona pellucida, a structure consisting of specifically arranged polypeptide chains and carbohydrates. In mouse and pig, the sperm receptor activity has been attributed mainly to a certain class of *O*-

linked oligosaccharides (Florman and Wassarman, 1985; Yurewicz *et al.*, 1991). At present, this interpretation is not easily reconcilable with a contribution of neutral *N*-linked carbohydrates in the binding event (Noguchi *et al.*, 1992). In both species, the clustering of the biologically active carbohydrate moieties appears to be necessary to achieve high affinity binding. In the mouse, the active sperm receptor region has been mapped to the C-terminal 28 kDa peptide of mZP3 (Rosiere and Wassarman, 1992). In the pig, trypsin digestion of isolated pZPα and pZPβ glycoproteins produced a single *O*-glycosylated domain from each glycoprotein containing three to five *O*-linked glycosylation sites, respectively. Only the *O*-glycosylated domain of the pZPα-protein retained the sperm receptor activity, indicating both zona pellucida glycoproteins may have a different *O*-glycosylation pattern. The poly-*N*-acetlyllactosamine units, however, do not appear to play a role in sperm binding (Yurewicz *et al.*, 1992). The amino acid sequences of the *O*-glycosylated regions are highly conserved within the ZPB- and ZPC-protein families. However, it is interesting to note that the *O*-glycosylated domain of porcine ZPC-protein (pZP3β), that contains the receptor activity does not reside on a different protein from the active site (Rosiere and Wassarman, 1992; Yurewicz *et al.*, 1992).

Carbohydrates are mandatory for the receptor binding, however, in order to achieve high affinity interaction; the biologically active oligosaccharides must be presented as a multivalent ligand within the supramolecular architecture of the zona pellucida (Yurewicz *et al.*, 1993). Despite increasing information on zona oligosaccharide structures, neither the distribution of sugars along the different glycoprotein layers of the zona pellucida nor the molecular identity of the oligosaccharides with sperm-binding ability have been established in any mammalian species.

3 Sperm-Associated Zona Pellucida- and Carbohydrate-Binding Proteins

Sperm proteins with affinity for zona pellucida glycoproteins have been identified in a number of mammalian species by (a) incubating Western blots of sperm proteins with labeled zona pellucida proteins; (b) with the help of specific polyclonal or monoclonal antisera, which inhibit sperm-zona binding; or (c) by using crosslinking reagents. However, only some of them have been characterized biochemically or structurally (Table 1). Furthermore, to complicate the matter, carbohydrate-binding ability has been demonstrated only in few instances. Therefore, considering the wide variety of putative zona pellucida binding molecules identified even in a single species, it is a reasonable assumption that fertilization is most likely mediated by multiple complementary receptor-ligand systems. A classification of the actual sperm lectins that act as primary gamete adhesion molecules is still in its infancy. Moreover, little is known about the type of binding (carbohydrate-protein or protein-protein), the actual hierachy of interactions, and cross-talking mechanisms between different zona-binding proteins, which may collectively govern the species-specificity of gamete recognition.

The most thoroughly characterized sperm surface proteins known to have specific zona pellucida- and carbohydrate-binding sites are: the mouse proteins β1,4-galactosyltransferase (GalTase) (Miller *et al.*, 1992) and sp56 (Bookbinder *et al.*, 1995); the rabbit protein Sp17 (Richardson *et al.*, 1994; Yamasaki *et al.*, 1995); a human mannose-binding protein (Benoff, 1993); and the members of the porcine spermadhesin protein family known as AQ*N*-1, AQ*N*-3, and AWN (Calvete *et al.*, 1995; Töpfer-Petersen *et al.*, 1995). Mouse GalTase, rabbit Sp17, and human mannose-binding protein are sperm type-I integral membrane proteins (Miller *et al.*, 1992; Benoff, 1993), whereas mouse sp56 and the porcine spermadhesins are peripherally associated proteins (Cheng *et al.*, 1994; Dostàlovà *et al.*, 1994).

Table 1. Mammalian sperm proteins identified by their zona pellucida-binding capability. Species code: M, mouse; Ha, hamster; Hu, human; R, rat; Ra, rabbit; P, pig; GP, guinea pig; Ho, horse; D, dog; W, wide distribution; [1]GalTase, β-1,4-*N*-acetylglucosamine:galactose transferase; [2]integral plasma membrane protein; [3]peripheral protein; [4]SPI-B, serine proteinase inhibitor-binding protein; [5]RTK, receptor tyrosine kinase; [6]MPB, mannose-binding protein; [7]possibly GPI-anchored in the membrane; PH-20 harbors hyaluronidase activity; [8]Sp17 mRNA is present in these species; [9]this carbohydrate-binding activity is deduced from the structure of the sperm protein; [10]SGR, sperm galactosyl receptor; [11]acrosomal protein: may only play a role as secondary adhesion molecule to bind acrosome-reacted sperm to the zona pellucida.

Protein	Species	Carbohydrate-binding activity	Reference
GalTase[1,2]	M	terminal GlcNAc residues	Miller *et al.* (1992)
α-D-mannosidase[2]	M,Ha,Hu, R	α1–3/α1–2-Man	Cornwall *et al.* (1991)
15–20 kDa SPI-B[3,4]	M,Hu		Boetger-Tong *et al.* (1993)
sp56[3]	M,Ha	terminal α-Gal	Bookbinder *et al.* (1995)
APz (52 kDa)	P		Peterson *et al.* (1991)
RTK[5] 95 kDa	M,Hu		Burks *et al.* (1995)
C-type MBP[6]	Hu	D-mannose	Benoff (1993)
SLIP1[3] (68 kDa)	W	sulfogalactolipids	Anphaichitr *et al.* (1993)
PH-20[7]	W		Overstreet *et al.* (1995)
p105/45[2]	P		Hardy and Garbers (1994)
Sp17[2] (17 kDa)	Ra (M,Hu)[8]	(galactose)[9]; heparin	Richardson *et al.* (1994)
54 kDa SGR[10]	Ra,Hu	(galactose)[9]	Mertz *et al.* (1995)
Spermadhesins[3]	P,Ho,D,Hu	β-Gal in oligosaccharides	Calvete *et al.* (1995)
(Pro)acrosin[11]	W	sulfated polysaccharides	Jones (1991) Urch and Patel (1991)

3.1 Sperm Membrane C-Type Lectins

Human sperm mannose-binding protein and rabbit Sp17 belong to the expanding group of Ca^{2+}-dependent (C-type) lectins (Drickamer, 1993). C-Type carbohydrate-recognition domains (CRD) share a common 120 amino acid sequence motif characterized by 14 invariant and 18 other residues which are conserved (Fig. 1). The overall fold of the domain, as revealed by X-ray crystallography, consists of an structure with two conserved disulfide bonds and two bound Ca^{2+} ions, one of which participates in ligand binding (Weis, 1994; Weis and Drickamer, 1994). Two types of C-type CRDs have been identified: those that bind carbohydrates with 3- and 4-OH groups in an equatorial-axial position (i.e. galactose and *N*-acetylgalactosamine), and those with affinity for mannose, *N*-acetylglucosamine, and glucose which share bi-equatorially positioned hydroxyl groups at positions 3 and 4 (Drickamer, 1993).

```
χ (8-12) ϕ (2) θ (3) C (6) θ (2) θ (1) Ω (2) E (2) χ θ (6-14) ϕ θ G θ (1) Ω (1-

                                                  185 187          193
3) Ω (2) ϕ (1) χ (1-2) G (2-4) χ (1-6) χ (2) W (3) Z P Ω Ω (5-9) E Ω C θ (1)
                                               31 Q P D  N  (9)  E N (8)

      205 206
χ (3-8) G (1) W N   D (2) C (5) χ (1) C
                N F D56
```

Figure 1. Alignment of the consensus amino acid sequence motif and residue spacing in C-type CRDs (21) and the putative galactose-binding motif of rabbit Sp17 (residues 31–56) (11). Invariant residues are indicated in the one-letter code: E, glutamic acid; G, glycine; W, tryptophan; P, proline; C, cysteine; N, asparagine; D, aspartic acid; Q, glutamine; F, phenylalanine; χ, either aromatic or aliphatic; ϕ, aromatic; θ, aliphatic; Ω, oxygen-containing; cysteine residues 1–4 and 2–3 form two invariant disulfide bridges; Z^{185} and W at position 187 are E and N, respectively, in those CRDs with affinity for carbohydrates with similarly positioned equatorial 3- and 4-OH groups (mannose, glucose, and *N*-acetylglucosamine), but are Q and D, respectively, in lectins that bind galactose and *N*-acetylgalactosamine with shared equatorial-axial arrangement of hydroxyl groups in the 3 and 4 positions.

Invariably, while the galactose-like-binding CRDs contain Q and D at positions 185 and 187 (Fig. 1), these residues are replaced by D and Q, respectively, in mannose-like-binding CRDs (Drickamer, 1992). From these considerations, it has been proposed that rabbit Sp17, whose amino acid sequence contains some of the invariant amino acids of a C-type CRD, may belong to the subfamily of proteins with affinity for galactose (Richardson *et al.*, 1994). On the other hand, the sequence of a complete cDNA of the human sperm mannose-binding lectin encodes a protein with seven putative mannose-like-binding CRDs (Benoff *et al.*, 1993).

3.1.1 Rabbit Sp17

The amino acid sequence of rabbit sp17 indicates that the protein may contain a single CRD which includes essential residues found in galactose-binding Ca^{2+}-dependent CRDs (Richardson *et al.*, 1994) (Fig. 1). The carbohydrate-binding site of C-type CRD is relatively small, located in a shallow groove on the protein surface, and the affinity of a single domain for its carbohydrate ligand is rather weak (K_i in the mM range). A common strategy utilized by C-type lectins for enhancing both the affinity and specificity of carbohydrate recognition consists of multivalent interactions with complex oligosaccharides. Interestingly, it has been reported that an aggregated form of Sp17 promotes high affinity binding and anchorage of rabbit spermatozoa to their homologous zona pellucida (O'Rand *et al.*, 1988). Recombinant Sp17 has binding affinity for dextran sulphate and the zona pellucida; using disuccinimidyl suberate, Sp17 was found crosslinked to both rabbit zona glycoproteins R45 and R55 (Yamasaki *et al.*, 1995). Rabbit R55 and R45 belong to the zona pellucida (ZP) B- and C-type gene families (Harris *et al.*, 1994). Analysis of Sp17 mRNA by Northern blots has shown that a similar mRNA is present in mouse, baboon, and human testes, but not in any somatic tissue analyzed (Yamasaki *et al.*, 1995). Immunofluorescence localization of Sp17 demonstrated that the protein is inaccessible to antisera before the acrosome reaction begins (Richardson *et al.*, 1994). Since acrosome-reacted rabbit spermatozoa can bind to the zona pellucida, the authors hypothesized that fertilizing spermatozoa which might have begun the acrosome reaction within the cumulus oophorus before reaching the oocyte's extracellular glycoprotein mesh would be available for initial binding to the zona pellucida.

3.1.2 Human sperm mannose-binding lectin

Mannose ligands on the human zona pellucida have been suggested to be involved in sperm-egg recognition (Mori *et al.*, 1989). Recent studies (Bennoff, 1993) have provided direct evidence for the presence of receptors for mannose-containing glycoligands on the surface of human spermatozoa subjected to capacitating procedures, and their reduced expression in specimens from males failing to bind to human zona pellucida and fertilize oocytes during in vitro fertilization (Benoff, 1993). This author has proposed a model for the action of the human sperm mannose-binding protein: freshly isolated motile spermatozoa contain a store of mannose-specific receptors in the subplasmalemmal space underlying the rigid, cholesterol-stabilized plasma membrane. Following capacitation, the cholesterol content of the plasma membrane falls to 0.001 $\mu mol/10^9$ cells and mannose-binding sites are translocated to the sperm surface. The receptors concentrate in the plasma membrane overlying the acrosome. Cooperative interaction of multiple carbohydrate recognition domains may lead to high-affinity mannose ligand binding. The receptors subsequently aggregate and move to the equatorial/post-acrosomal segment of the sperm head, activating a mechanism culminating in acrosomal exocytosis.

3.2 Mouse Sperm β1,4-Galactosyltransferase

The enzyme β-1,4-galactosyltransferase (GalTase) has been viewed traditionally as a biosynthetic component of the Golgi complex, although it is also found on the surface of many cells. The GalTase has been found on all mammalian sperm tested so far and appears to be confined to the plasma membrane in those species whose sperm has been analyzed by subcellular fractionation (Miller *et al.*, 1992). The GalTase gene is expressed in murine testicular germ cells and the enzyme is localized on the surface of spermatogonia, spermatocytes, and spermatids (Pratt and Shur, 1993). In spermatids, GalTase is localized in the sperm membrane overlying the acrosome. The β-*N*-acetylglucosaminidase isozyme found in sperm is the β-hexosaminidase B, a b/b homodimer (Miller *et al.*, 1993).

GalTase has been reported to mediate sperm-egg recognition in mice by binding to *N*-acetylglucosaminyl residues present on zona pellucida glycoconjugates because (a) sperm GalTase transfers galactose selectively from UDP-Gal to mouse zona pellucida glycoprotein3 (ZP3), and blocking or removing the binding sites for GalTase on ZP3 inhibits its ability to bind sperm (Miller *et al.*, 1992); and (b) aggregation of β1,4-galactosyltransferase on mouse sperm induced the acrosome reaction (Macek *et al.*, 1991). However, its participation in sperm binding to the zona pellucida is a matter of debate, since a set of oligosaccharides terminating in galactose, either in α- or β-linkage, has been shown to inhibit binding of mouse sperm to eggs in the concentration range 1–10 μM, whereas oligosaccharides with β-*N*-acetylglucosamine at the non-reducing end did not significantly affect the number of sperm bound to eggs at equivalent concentrations (Litscher *et al.*, 1995). Although this result does not rule out a biological role for GlcNAc residues or other sugar residues as well, it provides further support for the proposal that galactose at the non-reducing end of zona pellucida *O*-linked oligosaccharide(s) may play an essential role in sperm binding (Bleil and Wassarman, 1988).

Coincident with the acrosome reaction, GalTase is redistributed to the sides of the sperm head where it can participate in secondary events of sperm adhesion to the zona pellucida.

3.3 Sperm Membrane-Associated Zona Pellucida-Binding Proteins

3.3.1 Mouse sp56

The 56 kDa protein, termed sp56, was localized at the head of acrosome-intact but not acrosome-reacted mouse sperm. It was radiolabeled preferentially by the photoactivatable heterobifunctional Denny-Jafee crosslinker covalently linked to purified mouse ZP3. Sp56 has been proposed as a candidate for the role of a mouse ZP3-binding protein (Bleil and Wassarman, 1990). Purified sp56 eluted from size exclusion columns as a homodimer, and immunohistochemical studies demonstrated that it is a peripheral membrane protein located on the sperm head plasma membrane where sperm initiates binding to the zona pellucida (Cheng *et al.*, 1994).

A full-length sp56 cDNA has been sequenced (Bookbinder *et al.*, 1995). It encodes a presumptive 32-amino acid signal peptide followed by a 547-amino acid open reading frame. The sp56 amino acid sequence indicates that it contains multiple consensus repeats of approximately 60 amino acids in length, termed Sushi domains (Bookbinder *et al.*, 1995). The cDNA-derived polypeptide sequence does not indicate the presence of regions having similarity with CRDs of known lectins, suggesting that sp56 may recognize galactose by a novel mechanism.

A mouse 2 kb mRNA was detected by Northern blot analysis in testes, but not in other tissues, and by immunoblot analysis sp56 protein was found in round and elongated spermatids and in testicular spermatozoa, but was not detected in earlier stages of spermatogenesis (Bookbinder *et al.*, 1995). Mouse and hamster sperm, which bind to the zona pellucida of mouse eggs, contained sp56. However, sp56 was not detected (by Western or Northern blot analysis) in guinea pig or human sperm, which do not bind to mouse egg zona pellucida (Bookbinder *et al.*, 1995). These results suggest that sp56 could account for species-specific sperm-egg interaction.

3.3.2 Porcine spermadhesins

Spermadhesins are a group of 12–16 kDa polypeptides found in the seminal plasma and/or associated with the spermatozoal surface of several mammalian species, i.e. pig, bull, horse, dog, and human (Calvete *et al.*, 1995; Töpfer-Petersen *et al.*, 1995). Accumulating evidence (discussed below) indicates that members of the boar spermadhesin family with affinity for heparin and zona pellucida glycoconjugates might play a role in at least two aspects of porcine fertilization, namely sperm capacitation and initial gamete interaction (Dostàlovà *et al.*, 1995b).

Boar spermadhesins AQ*N*-1, AQ*N*-3, PSP-I, PSP-II, AW*N*-1, and AW*N*-2 (Fig. 2) are major secretory products of the seminal vesicle epithelium; their concentrations in seminal fluid range from 0.6 to 7 mg/ml (Dostàlovà *et al.*, 1994). AW*N*-1 is also synthesized by rete testis and

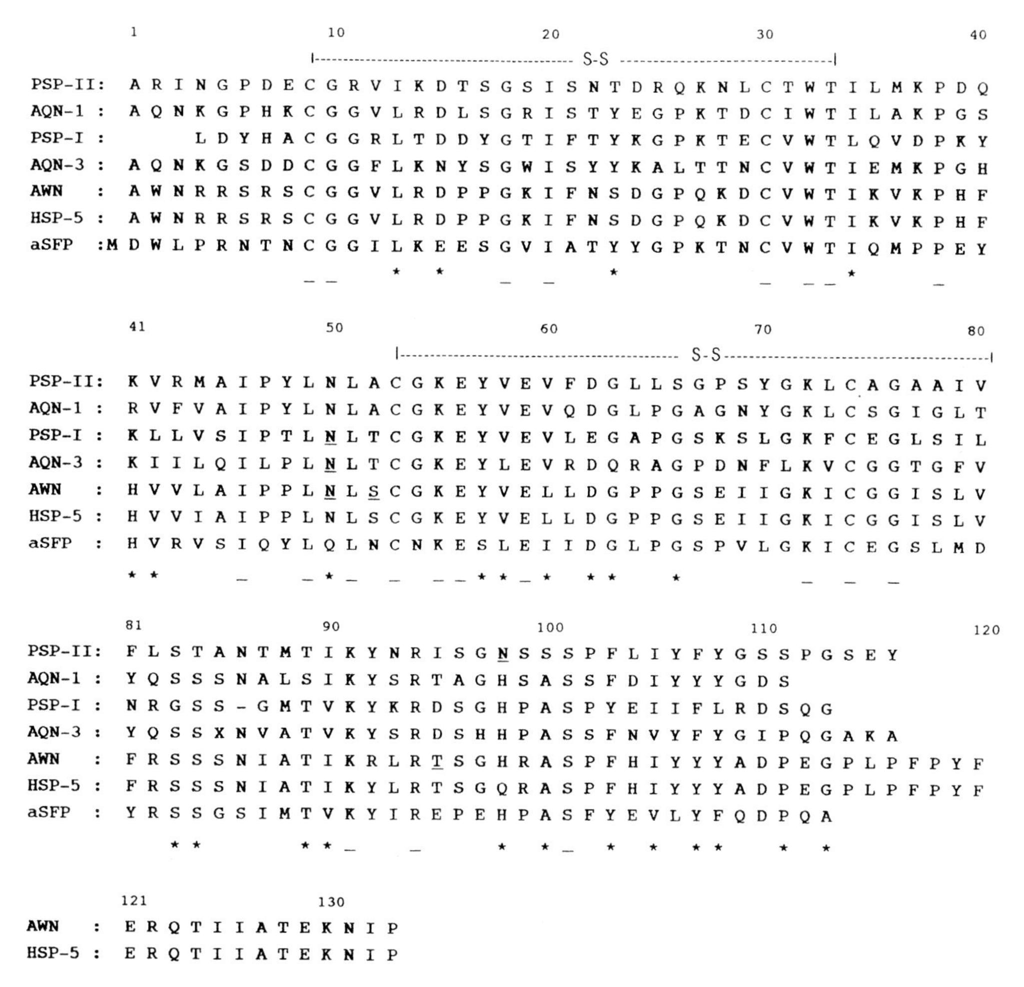

Figure 2. Alignment of the amino acid sequences of boar spermadhesins PSP-II, AQ*N*-1, PSP-I, AQ*N*-3, and AWN, equine HSP-5 (homologue of AWN), and bovine aSFP. AW*N*-1 and AW*N*-2 differ only in that the latter possesses an acetylated *N*-terminal residue. Positions with absolutely conserved residues (-) and conservative substitutions (*) in all polypeptides are shown. N^{50}, asparagine residues which are constitutively glycosylated in PSP-I and in glycoforms of AQ*N*-3 and AWN; N^{98}, glycosylated asparagine in PSP-II; S^{52} and T^{95}, serine and threonine residues which are *O*-glycosylated in glycoforms of AWN.

tubuli recti (Sinowatz *et al.*, 1995) and is the only member of its family present on epididymal sperm, where there are 6–7 million molecules/cell (Dostàlovà *et al.*, 1994). Following ejaculation 12–60 million molecules of each AQ*N*-1, PSP-1, AQ*N*-2, AQ*N*-3, AW*N*-1 and AW*N*-2 become coated on the apical third of the acrosomal cap. However, most of this coating material is released during *in vitro* capacitation when the level of each spermadhesin falls to 5–6 million molecules/spermatozoon (Dostàlovà *et al.*, 1994). A specific interaction between non-aggregated AW*N*-1 and AQ*N*-3 molecules and *o*-phosphorylethanolamine has been reported (Dostàlovà *et al.*, 1994), suggesting that these molecules could form the first layer of the coating material by interacting with the lipid bilayer. Aggregated spermadhesins may serve as acrosome-stabilizing factors and might constitute the subpopulation that is released during capacitation; the remaining spermadhesin molecules on capacitated sperm are in a good location to act as primary zona pellucida-binding proteins.

Isolated spermadhesin AW*N*-1 bound zona pellucida glycoproteins, with a dissociation constant (Kd) in the low micromolar range, and displayed similar binding affinity for glycoproteins containing a Galβ1,3-GalNAc sequence in *O*-linked structures (Dostàlovà *et al.*, 1995a). α2,6-Sialylated forms of this oligosaccharide appeared to lower the affinity constant approximately five-fold, suggesting that the oligosaccharide binding surface may include a cationic subsite. AW*N*-1 also displayed affinity for Neu5Acα2,3/6-Galβ1,4-GlcNAc oligosaccharides in *N*-linked triantennary structures of fetuin. However, peptides containing the *N*-linked oligosaccharide sequence inhibited the binding of the parent glycoprotein to immobilized AW*N*-1 5–45 times less effectively than those carrying the *O*-linked trisaccharide. Remarkably, most of the neutral *N*- and *O*-linked carbohydrate structures of porcine zona pellucida glycoproteins contain a terminal Gal-β1,4-GlcNAc sequence; neutral *O*-linked oligosaccharides have an internal Gal-β1,3-GalNAc sequence. Both disaccharide structures are also present in the anionic *O*-linked sugar chains (Mori *et al.*, 1991; Noguchi and Nakano, 1992; Noguchi *et al.*, 1992; Hirano *et al.*, 1993; Hokke *et al.*, 1993, 1994). On the other hand, the sperm receptor activity of porcine zona pellucida appears to be mediated by *O*-linked saccharide chains (Sacco *et al.*, 1989), although neutral *N*-linked sugar chains have also been reported to possess sperm-binding activity (Sacco *et al.*, 1989; Noguchi *et al.*, 1992; Hedrick, 1993). Thus, most of the carbohydrate structures that have been involved in sperm attachment to porcine oocytes contain information for AWN binding.

Using indirect immunofluorescence microscopy, AW*N*-1 molecules were observed on *in vitro* capacitated spermatozoa bound to isolated porcine zona pellucida as well as to isolated zona-encased porcine oocytes (Dostàlovà *et al.*, 1995b). Upon incubation with the oocyte for short periods of time (1–3 min), these *in vitro* capacitated spermatozoa were able to perform the acrosome reaction and start penetration of the zona pellucida. These results indicate that AW*N*-1 is present on the surface of fertile porcine spermatozoa. Furthermore, AW*N*-1 has been immunologically demonstrated on spermatozoa recovered from the uterotubal junction of sexual mature gilts inseminated between 8–12 hours before surgery. In the pig, the uterotubal junction is the major barrier for sperm ascent to the site of fertilization, the ampulla, and the place where spermatozoa seem to complete capacitation (Yanagimachi, 1994). Circumstantial evidence has suggested that these spermatozoa represent the subpopulation which eventually is capable of fertilizing the egg (Yanagimachi, 1994). Hence, spermadhesin AW*N*-1 seems to be at the place and time expected for a physiologically relevant zona pellucida-binding molecule. Whether AW*N*-1 functions merely to anchor spermatozoa to the zona pellucida as a prerequisite to enable other sperm (membrane) proteins to find their ligands, or whether AW*N*-1 also participates in coupling sperm binding to the acrosome reaction-inducing mechanism, remains to be established.

In dog and stallion, AW*N*-like molecules are already detectable on the equatorial segment of some, but not all, testicular spermatozoa. On the other hand, all dog and stallion sperm cells taken from the rete testis as well as ejaculated spermatozoa, displayed anti-AWN immunoreactive material, whose topography was restricted to the equatorial segment (Töpfer-Petersen *et al.*, 1995).

Purified stallion seminal plasma AWN, like its porcine homologue, had the capability to bind to isolated equine zonae pellucidae. This result, together with the observation that frozen stallion epididymal spermatozoa possess fertilizing capacity (Barler and Gandier, 1957), suggests that spermadhesin AWN may be one of the factors contributing to the reproductive ability of horse epididymal sperm.

Stallion and boar AWN molecules differ only in three amino acids (T49) (Fig. 2). Since perissodactyls (i.e. horse) and artiodactyls (i.e. pig) had a common ancestry over 50 million years ago, the unusually low mutational rate might suggest that the whole AWN structure is under a strong selective pressure, pointing to a highly conserved conformation and most probably of a biological function of spermadhesin AWN in both vertebrate species.

The carbohydrate-binding site of spermadhesin AW*N*-1 and AQ*N*-3 has been located around a conserved asparagine residue at position 50 (Fig. 2) (Calvete *et al.*, 1993a, b). The amino acid sequence of spermadhesins, however, does not show any discernible similarity to known carbohydrate-recognition domains, indicating that spermadhesins may belong to a novel group of animal lectins. A sequence pattern-search analysis has revealed that spermadhesins belong to a family of 16 functionally diverse proteins, many of which are known to be involved in developmental processes (Bork and Beckmann, 1993). The structure of these proteins consists of a combination of several modules, but all of them share the *CUB* domain, first identified in *C*omplement subcomponents C1r/C1s, *U*egf, and *B*mp1 (Macek *et al.*, 1991). Spermadhesins, 110–133 amino acid spanning polypeptides (Fig. 1), form a subgroup of the CUB domain family, as expected by their single domain architechture for which a 9 anti-parallel β-strand topology like that observed in immunoglobulin V regions has been proposed (Bork and Beckmann, 1993). X-ray crystallographic studies are underway to elucidate the three-dimensional structure of spermadhesins and the structural basis of their carbohydrate-binding capability.

4 Inhibition of Sperm-Egg Interactions by Saccharides and Glycosidases

4.1 Sperm-Binding Assays

Sperm-binding assays have been introduced as an experimental method to study the mechanisms of sperm-egg interaction as well as to establish a diagnostic test for predicting the fertilization potential (Burkman *et al.* 1988; Oehninger *et al.*, 1990). The number of spermatozoa bound to the zona pellucida is positively correlated with fertilization rates *in vitro*. Therefore, assessment of sperm-zona interaction provides a useful tool for evaluating the fertilization potential of spermatozoa. Two important methods have been described to assess the zona-binding ability of spermatozoa: the hemizona assay (Burkman *et al.* 1988; Oehninger *et al.*, 1990) and the competitive zona-binding assay. Various parameters are known to influence significantly the *in vitro* binding of spermatozoa to the zona. Among them, purified zona pellucida proteins and small glycopeptides isolated from zona proteins have been shown to inhibit binding of sperm to eggs *in vitro* (Wassarman, 1992; Litscher *et al.*, 1995). During the last few years the hypothesis that carbohydrates are involved in mammalian gamete adhesion has been supported by a variety of experimental evidence, including the ability of various saccharides to inhibit sperm-egg interactions.

4.2 Inhibition of Sperm-Zona Binding by Saccharides

Early competitive inhibition studies using various monosaccharides, polysaccharides or glycoproteins indicated a role for specific carbohydrate moieties in mediating sperm-zona interaction in several mammalian species. This evidence includes reports that various monosacharides, disaccharides, and fucoidan, a polysaccharide consisting primarily of sulfated fucose, inhibited binding of sperm to eggs *in vitro,* for example, *N*-acetyl-D-galactosamine, *N*-acetyl-D-glucosamine, α-D-mannose, and α-methylmannoside at 1 mM, and fucoidan at 1 mg/ml inhibited binding of rat sperm to eggs (Shalgi *et al.*, 1991). Similarly, *N*-acetyl-D-glucosamine, *N*-acetyl-D-galactosamine, and *N*-acetylmannosamine, at concentrations between 250 mM and 1 mM, and fucoidan at 50 μg/ml inhibited binding of hamster sperm to eggs (Ahuja, 1982). Fucoidan at 500 μg/ml also prevents binding of guinea pig sperm to eggs (Jones, 1991) and at 1 mg/ml inhibited binding of human sperm to eggs (Oehninger *et al.*, 1990). In each of these cases, inhibition of gamete adhesion was attributable to binding of the monosaccharide or polysaccharide to the sperm.

In a recent study, Litscher *et al.* (1995) tested 15 *O*-linked-related oligosaccharide constructs with defined structures for their ability to inhibit binding of mouse sperm to ovulated eggs, and to induce sperm to undergo the acrosome reaction *in vitro*. The results suggest that several factors can influence the effectiveness of oligosaccharides as inhibitors of gamete adhesion in mice. These include the nature of the sugar moiety at the non-reducing end of the oligosaccharides, the number of sugar residues in the oligosaccharides, and the branching pattern of the oligosaccharides. Among the oligosaccharides tested, a biantennary dodecasaccharide and an octadecasaccharide with galactose in α-linkage at their non-reducing ends were very active inhibitors of sperm binding to eggs. Similarly, a tetraantennary tetradecasaccharide with galactose in β-linkage at its non-reducing end was a very good inhibitor, although a closely related biantennary decasaccharide lacked this capacity. Also, analogous oligosaccharide constructs with *N*-acetylglucosamine at their non-reducing end did not inhibit binding of sperm to eggs at equivalent concentrations. These findings emphasize the great importance of the nature of the sugar at the non-reducing end of oligosaccharides in determining their affinity for sperm (Litscher *et al.*, 1995). They also support the proposal that galactose at the non-reducing end of oligosaccharides present at the mouse ZP3 combining site may play an essential role in sperm binding in this species (Bleil and Wassarman, 1988).

However, the finding that oligosaccharides with galactose in β-linkage are also effective inhibitors of sperm-egg binding was puzzling. Earlier investigations had provided evidence that *O*-linked oligosaccharides serve as sperm ligands when galactose is in α-linkage with the penultimate sugar (Bleil and Wassarman, 1988; 1990). Removal of the galactose by digestion with α-galactosidase or conversion of its C-6 hydroxyl to an aldehyde by oxidation with galactose oxidase inactivates the oligosaccharides of the sperm receptor. Nevertheless, the inhibitiory effect of oligosaccharides with galactose in β-linkage could be due to the fact that the oligosaccharides are in solution (i.e. not bound to polypeptide). They can under these conditions retain enough flexibility to adopt common topographical features (Litscher *et al.*, 1995).

The monosaccharides galactose and *N*-acetylglucosamine, as well as the disaccharide melibiose, did not inhibit binding of mouse sperm to eggs, even at a concentration of 5 mM. Similarly, an unbranched oligosaccharide consisting of five sugars and a biantennary oligosaccharide consisting of either six or eight sugars, all with galactose at their non-reducing end, did not affect sperm binding at a concentration of 10 mM. Biantennary oligosaccharides consisting of 12 sugars and tetraantennary oligosaccharides consisting of either 14 or 18 sugars, all with galactose at their non-reducing end (in α- or β-linkages), however, significantly inhibited sperm binding at concentrations of 4 and 10 mM, respectively. These findings suggest that the number of sugars in an oligosaccharide is important for its ability to bind to sperm and, consequently, to interfere with binding of sperm to eggs (Litscher *et al.*, 1995).

4.3 Do Carbohydrates also Play a Role in the Fusion of the Oocyte Plasma Membrane with the Spermatozoon?

Whereas the role of carbohydrates for sperm-zona binding appears well-established in several species, comparatively little is known about the molecular events of the interplay between the sperm cell and the egg plasma membrane. Recently, the putative involvement of sperm membrane proteins containing a disintegrin domain and integrin counter-receptors of the egg plasma membrane has received much attention (Blobel *et al.*, 1992; Myles *et al.*, 1994; Almeida *et al.*, 1995; Wolfsberg *et al.*, 1995). In addition, the possibility that sperm-egg fusion involves carbohydrate components of the sperm and/ or egg plasma membrane is suggested by observations in other membrane fusion systems, where glycoproteins play a critical role in the fusion process, for example fusion of paramyxoviruses with the plasma membrane of the host cells (Poste and Pasternak, 1978), or the fusion of myoblasts to form multinucleate myotubes (Gilfix and Sanwal, 1980; Knudsen, 1985). However, Ahuja (1982) demonstrated that certain saccharides and glycoproteins competively inhibit sperm-zona binding in zona-intact hamster eggs but not in zona-free eggs. These findings do not support a role for carbohydates in sperm-egg fusion. Futhermore, experiments, in which either sperm or zona-free hamster eggs were treated with exoglycosidic enzymes prior to *in vitro* fertilization also failed to reveal an inhibition of sperm-egg fusion (Hirao and Yanagimachi, 1978). Explicitly, Boldt *et al.* (1988) showed for mice that treatment of zona-free eggs with several exoglycosidic enzymes did not interfere with the ability of the egg's plasma membrane to fuse with sperm. They also studied the potential involvement of cell-surface carbohydrates in sperm-egg fusion in the mouse employing various saccharides and glycoproteins as inhibitors. Of the sugars tested, L-fucose appeared to be the most potent inhibitor but *N*-acetylglucosamine and galactose were also effective. Each of these sugars caused significant inhibition of penetration relative to control only when tested at high concentrations, i.e. 5 mg/ml and 10 mg/ml. The effect of L-fucose appeared to be stereospecific, since the percentage of inhibition observed with L-fucose at 10 mg/ml was approximately twice that obtained with the D-isomer. Fucoidan and ascophyllan, which are L-fucose-containing sulfated heteropolysaccharides, also inhibited sperm egg fusion *in vitro*. In contrast, none of the other glycocompounds studied, including fetuin, asialofetuin, ovalbumin, ovomucoid and the sulfated glycosaminoglycan chondroitin sulfate, impaired sperm-egg fusion when present during incubation at concentrations up to 5 mg/ml. Treatment of spermatozoa, but not zona-free eggs, with fucosidase prior to incubation caused significant reductions in sperm penetration levels. Other glycosidic enzymes, including glucosidase, galactosidase, and *N*-acetylglucosaminidase, had no inhibitory effect on the sperm. These data suggest that an L-fucose-containing component of the sperm surface is involved in sperm-egg fusion in the mouse. However, the fact that sperm-egg fusion in hamster could not be inhibited by a variety of saccharides and glycoproteins poses a note of caution, arguing against the idea of a rather species-independent and fixed set of recognition mechanisms. This conclusion sets the stage for continued efforts to decipher the glycobiology of fertilization.

Acknowledgements

The work of the authors' laboratories has been financed by grants 01KY9103 from Bundesministerium für Forschung und Technologie and the DFG grant Si 279/4–3 and DFG grant Tö 114/ 1–3.

References

Ahuja KK (1982): Fertilization in the hamster. The role of cell surface carbohydrates. In *Exp. Cell Res.* **140**:353–6.

Ahuja KK (1985): Carbohydrate determinants involved in mammalian fertilization. In *Am. J. Anat.* **174**: 207–33.

Aitken J (1995): The complexities of conception. In *Science* **269:**39–40.

Almeida EAC, Huovilla APJ, Sutherland AE *et al.* (1995): Mouse egg integrin $\alpha_6\beta_1$ functions as a sperm receptor. In *Cell* **81**:1095–104.

Aviles M, Martinez-Menarguez JA, Castells MT *et al.* (1994): Cytochemical characterization of oligosaccharide side chains of the glycoproteins of rat zona pellucida: an ultrastructural study. In *Anat. Rec.* **239**: 127–49.

Barler CAV, Gandier JCC (1957): Pregnancy in a mare resulting from frozen epididymal spermatozoa. In *Can. J. Comp. Med.* **XXI**:47–51.

Benoff S (1993): Preliminaries to fertilization. The role of cholesterol during capacitation of human spermatozoa. In *Hum. Reprod.* **8**:2001–8.

Benoff S, Hurley I, Cooper GW *et al.* (1993): Head-specific mannose-ligand receptor expression in human spermatozoa is dependent on capacitation-associated membrane cholesterol loss. In *Hum. Reprod.* **8**:2141–54.

Bleil JD, Wassarman PM (1988): Galactose at the nonreducing terminus of *O*-linked oligosaccharides of mouse egg zona pellucida glycoprotein ZP3 is essential for the glycoprotein's sperm receptor activity. In *Proc. Natl. Acad. Sci.* USA **85**:6778–82.

Bleil JD, Wassarman PM (1990): Identification of a ZP3-binding protein on acrosome-intact mouse sperm by photoaffinity crosslinking. In *Proc. Natl. Acad. Sci.* USA **87**:5563–67.

Blobel CP, Wolfsberg TG, Turck CW *et al.* (1992): A potential fusion peptide and an integrin ligand domain in a protein active in sperm egg fusion. In *Nature* **356**:248–52

Boettger-Tong H, Aarons DJ, Biegler BE *et al.* (1993) Binding of a murine proteinase inhibitor to the acrosome region of the human sperm head. In *Mol. Reprod. Dev.* **36**:346–53.

Boldt J, Howe AM, Preble J (1988): Enzymatic alteration of the ability of mouse egg plasma membrane to interact with sperm. In *Biol. Reprod.* **39**: 19–27.

Boldt J, Howe AM, Parkerson JB *et al.* (1989): Carbohydrate involvement in sperm-egg fusion in mice. In *Biol. Reprod.* **40**:887–96.

Bookbinder LH, Cheng A, Bleil JD (1995): Tissue- and species-specific expression of sp56, a mouse sperm fertilization protein. In *Science* **269**:86–9.

Bork P, Beckmann G (1993): The CUB domain. A widespread module in developmentally regulated proteins. In *J. Mol. Biol.* **231**:539–45.

Burkman LJ, Coddington CC, Franken DR *et al.* (1988): The hemizona assay (HZA): development of a diagnostic test for the binding of human spermatozoa to the human zona pellucida to predict fertilization potential. In *Fert. Steril.* **49**:668–97.

Burks DJ, Carballada R, Moore HDM *et al.* (1995): Interaction of a tyrosine kinase from human sperm with the zona pellucida at fertilization. In *Science* **269**:83–6.

Calvete JJ, Sanz L, Dostàlovà Z *et al.* (1993a): Characterization of AW*N*-1 glycosylated isoforms helps to define the zona pellucida and serine proteinase inhibitor-binding region on boar spermadhesins. In *FEBS Lett.* **334**:37–40.

Calvete JJ, Solís D, Sanz L *et al.* (1993b): Characterization of two glycosylated boar spermadhesins. In *Eur. J. Biochem.* **218**:719–25.

Calvete JJ, Sanz L, Dostàlovà Z *et al.* (1995): Spermadhesins: sperm-coating proteins involved in capacitation and zona pellucida binding. In *Fertilität* **11**: 35–40.

Cornwall GA, Tulsiani DRP, Orgebin-Crist MC (1991): Inhibition of the mouse sperm surface α-D-mannosidase inhibits sperm-egg binding in vitro. In *Biol. Reprod.* **44**:913–21.

Dostàlovà Z, Calvete JJ, Sanz L *et al.* (1994): Quantitation of boar spermadhesins in accessory sex gland fluids and on the surface of epididymal, ejaculated and capacitated spermatozoa. In *Biochim. Biophys. Acta* **1200**:48–54.

Dostàlovà Z, Calvete JJ, Sanz L *et al.* (1995a): Boar spermadhesin AW*N*-1. Oligosaccharide- and zona pellucida-binding characteristics. In *Eur. J. Biochem.* **230**:329–36.

Dostàlovà Z, Calvete JJ, Töpfer-Petersen E (1995b): Interaction of non-aggregated boar AW*N*-1 and AQ*N*-3 with phospholipid matrices. A model for coating of spermadhesins to the sperm surface. In *Biol. Chem. Hoppe-Seyler* **376**:237–42.

Drickamer K (1992): Engineering galactose-binding activity into a C-type mannose-binding protein. In *Nature* **360**:183–6.

Drickamer K (1993): Ca^{2+}-dependent carbohydrate-recognition domains in animal proteins. In *Curr. Opinion Struct. Biol.* **3**:393–400.

Dunbar BS, O'Rand M (1991): A Comparative Overview of Mammalian Fertilization. New York: Plenum Press.

Dunbar BS, Avery S, Lee V *et al.* (1994): The mammalian zona pellucida: its biochemistry, immunochemistry, molecular biology and development expression. In *Reprod. Fert. Develop.* **6**:331–47.

Florman HM, Wassarman PM (1985): *O*-Linked oligosaccharides of mouse egg ZP3 account for its sperm receptor activity. In *Cell* **41**:313–24.

Gabius HJ (1987): Vertebrate lectins and their possible role in fertilization, development and tumor biology. In *In Vivo* **1**:75–84.

Gilfix BM, Sanwal BD (1980): Inhibition of myoblast fusion by tunicamycin and pantomycin. In *Biochem. Biophys. Res. Commun.* **96**:1184–91.

Hardy DM, Garbers DL (1994): Species-specific binding of sperm proteins to the extracellular matrix (zona pellucida) of the egg. In *J. Biol. Chem.* **269**:19000–4.

Harris JD, Hibler DW, Fonteno GK *et al.* (1994): Cloning and characterization of zona pellucida genes and cDNAs from a variety of mammalian species: the ZPA, ZPB and ZPC gene families. In *DNA Sequence – J. Sequenc. Map.* **4**:361–93.

Hedrick JL (1993): The pig zona pellucida: sperm binding ligands, antigens, and sequence homologies. In *Reproductive Immunology* (Dondero F, Johnson PM, eds) Serono Publication 97, pp 59–65, New York: Elsevier Science Publisher.

Hedrick JL, Wardrip NJ (1987): Isolation of the zona pellucida and purification of its glycoprotein families from pig oocytes. In *Anal. Biochem.* **157**:63–70.

Hirano T, Takasaki S, Hedrick JL *et al.* (1993): *O*-Linked neutral sugar chains of porcine zona pellucida glycoproteins. In *Eur. J. Biochem.* **214**:763–9.
Hirao Y, Yanagimachi R (1978): Effects of various enzymes on the ability of hamster egg plasma membranes to fuse with spermatozoa. *In Gamete Res.* **1**:3–12.
Hokke CH, Damm JBL, Kamerling JP *et al.* (1993): Structure of three acidic *O*-linked carbohydrate chains of porcine zona pellucida glycoproteins. In *FEBS Lett.* **329**:29–34.
Hokke CH, Damm JBL, Penninkhof B *et al.* (1994): Structure of the *O*-linked carbohydrate chains of porcine zona pellucida glycoproteins. In *Eur. J. Biochem.* **221**:491–512.
Huang TTF, Ohzu E, Yanagimachi R (1982): Evidence suggesting that *L*-fucose is part of a recognition signal for sperm-zona pellucida attachment in mammals. In *Gamete Res.* **5**:355–61.
Jones R (1991): Interaction of zona glycoproteins, sulphated carbohydrates and synthetic polymers with proacrosin, the putative egg-binding protein from mammalian spermatozoa. In *Development* **111**:1155–63.
Kinloch RA, Mortillo S, Wassarman PM (1992): Transgenic mouse eggs with functional hamster sperm receptor in their zona pellucida. In *Development* **115**:937–46.
Knudsen KA (1985): The calcium-dependent myoblast adhesion that precedes cell fusion is mediated by glycoproteins. In *J. Cell Biol.* **101**:891–7.
Kobata A (1992): Structures and functions of the sugar chains of glycoproteins. In *Eur. J. Biochem.* **209**: 483–501.
Lambert H (1984): Role of sperm-surface glycoproteins in gamete recognition in two mouse species. In *J. Reprod. Fert.* **70**:281–4.
Litscher ES, Juntunen K, Seppo A *et al.* (1995): Oligosaccharide constructs with defined structures that inhibit binding of mouse sperm to unfertilized eggs in vitro. In *Biochemistry* **34**:4662–9.
Macek MB, Lopez LC, Shur BD (1991): Aggregation of β1,4-galactosyltransferase on mouse sperm induces the acrosome reaction. In *Dev. Biol.* **147**:440–4.
Maymon BB, Maymon R, Ben-Nun I *et al.* (1994): Distribution of carbohydrates in the zona pellucida of human oocytes. In *J. Reprod. Fert.* **102**:81–6.
Mertz JR, Banda PW, Kierszenbaum AL (1995): Rat sperm galactosyl receptor: purification and identification by polyclonal antibodies raised against multiple antigen peptides. In *Mol. Reprod. Dev.* **41**:374–83.
Miller DJ, Macek MB, Shur BD (1992): Complementarity between sperm surface β1,4-galactosyltransferase and egg-coat ZP3 mediates sperm-egg binding. In *Nature* **357**:589–93.
Miller DJ, Gong X, Shur BD (1993): Sperm require β-*N*-acetylglucosaminidase to penetrate through the egg zona pellucida. In *Development* **118**: 1279–89.
Mori K, Daitoh T, Irahara M *et al.* (1989): Significance of D-mannose as a sperm receptor site on the zona pellucida in human fertilization. In *Am. J. Obstet. Gynecol.* **161**:207–21.
Mori E, Takasaki S, Hedrick JL *et al.* (1991): Neutral oligosaccharide structures linked to asparagines of porcine zona pellucida glycoproteins. In *Biochemistry* **30**:2078–87.
Myles DG, Kimmel LH, Blobel CP *et al.* (1994): Identification of a binding site in the disintegrin domain of fertilin required for sperm-egg fusion. In *Proc. Natl. Acad. Sci.* **91**:4195–8.
Nagada SK, Araki Y, Chayko CA *et al.* (1994): *O*-Linked trisaccharide and *N*-linked poly-*N*-acetyllactosaminyl glycans present on mouse ZP2 and ZP3. In *Biol. Reprod.* **51**:262–72.
Noguchi S, Nakano M (1992): Structure of the acidic *N*-linked carbohydrate chains of the 55-kDa glycoprotein family (pZP3) from porcine zona pellucida. In *Eur. J. Biochem.* **209**:883–94.
Noguchi S, Hatanaka Y, Tobita T *et al.* (1992): Structural analysis of the *N*-linked carbohydrate chains of the 55-kDa glycoprotein family (PZP3) from porcine zona pellucida. In *Eur. J.Biochem.* **209**:1089–100.
O'Rand MG, Widgren EE, Fisher SJ (1988): Characterization of the rabbit sperm membrane autoantigen, RSA, as a lectin-like zona-binding protein. In *Dev. Biol.* **129**:231–40.
Oehninger S, Acosta AA, Hodgen GD (1990): Antagonistic and agonistic properties of saccharide moieties in the hemizona assay. In *Fertil. Steril.* **53**:143–9.
Overstreet JW, Lin Y, Yudin AI *et al.* (1995): Location of the PH-20 protein on acrosome-intact and acrosome-reacted spermatozoa of Cynomolgus macaques. In *Biol. Reprod.* **52**:105–14.
Peterson RN, Campell P, Hunt WP *et al.* (1991): Two-dimensional polyacrylamide gel electrophoresis characterization of APz, a sperm protein involved in zona binding in the pig and evidence for its binding to specific zona glycoproteins. In *Mol. Reprod. Dev.* **28**:260–71.
Poste G, Pasternak CA (1978): Virus induced cell fusion. In *Cell Surface Reviews* (Poste G, Nicholson GC, eds) pp 305–67, Amsterdam: Elsevier.
Pratt SA, Shur BD (1993): β1,4-Galactosyltransferase expression during spermatogenesis: stage-specific regulation by t-alleles and uniform distribution in ±spermatids and t-spermatids. In *Dev. Biol.* **156**: 80–93.
Richardson RT, Yamasaki N, O'Rand MG (1994): Sequence of a rabbit sperm zona pellucida binding

protein and localization during the acrosome reaction. In *Dev. Biol.* **65**:688–701.
Rosiere T, Wassarman PM (1992): Identification of a region of mouse zona pellucida glycoprotein mZP3 that possesses sperm receptor activity. In *Dev. Biol.* **154**:309–17.
Sacco AG, Yurewicz EC, Subramanian MG *et al.* (1989): Porcine zona pellucida: association of sperm receptor activity with a glycoprotein component of the Mr= 55,000 family. In *Biol. Reprod.* **41**:523–32.
Shalgi R, Mitityahu A, Nebel L (1986): The role of carbohydrates in sperm-egg interaction in rats. In *Biol. Reprod.* **34**:446–52.
Shalgi R, Maymon R, Bar-Shira B *et al.* (1991): Distribution of lectin receptor sites in the zona pellucida of follicular and ovulated oocytes. In *Mol. Reprod. Dev.* **29**:365–72.
Shur BD, Hall NG (1982): A role for mouse sperm surface galactosyltransferase in sperm binding to the egg zona pellucida. In *J. Cell Biol.* **95**:674–9.
Sinowatz F, Gabius HJ, Amselgruber W (1988): Surface sugar binding components of bovine spermatozoa as evidenced by fluorescent neoglycoproteins. In *Histochemistry* **88**:395–9.
Sinowatz F, Voglmayr JK, Gabius HJ *et al.* (1989): Cytochemical analysis of mammalian sperm membranes. Progr. Histochem. Cytochem. **19**:1–74.
Sinowatz F, Amselgruber W, Töpfer-Petersen E *et al.* (1995): Immunocytochemical localization of spermadhesin AWN in the porcine genital tract. In *Cell Tissue Res.* **282**:175–9.
Skutelsky E, Ranen E, Shalgi R (1994): Variations in the distribution of sugar residues in the zona pellucida as possible species-specific determinants of mammalian oocytes. In *J. Reprod. Fert.* **100**:35–41.
Tanaphaichitr N, Smith J, Mongkolsirikieart S *et al.* (1993): Role of gamete specific sulfoglycolipid immobilizing protein on mouse sperm-egg binding. In *Dev. Biol.* **156**:164–75.
Töpfer-Petersen E, Calvete JJ, Dostàlovà Z *et al.* (1995): One year in the life of the spermadhesin family. In *Fertilität* **11**:233–41.
Urch UA, Patel H (1991) The interaction of boar sperm acrosin with its natural substrate, the zona pellucida, and with polysulfated polysaccharides. In *Development* **111**:1165–72.
Wassarman PM (1987): The biology and chemistry of fertilization. In *Science* **235**:553–60.
Wassarman PM (1990): Profile of a mammalin sperm receptor. In *Development* **108**:1–17.
Wassarman PM (1991): Cell surface carbohydrates and mammalian fertilization. In *Cell Surface Carbohydrates and Cell Development* (Fukuda M, ed) pp 140–64, Boca Raton: CRC Press.
Wassarman PM (1992): Regulation of mammalian fertilization by gamete adhesion molecules. In *Spermatogenesis, Fertilization, Contraception. Molecular, Cellular and Endocrine Events in Male Reproduction* (Nieschlag E, Habenicht UF, eds) pp 345–66, Berlin: Springer Publ. Co.
Wassarman PM, Mortillo S (1991): Structure of the mouse egg extracellular coat, the zona pellucida. In *Int. Rev. Cytol.* **130**:85–110.
Weis WI (1994): Lectins on a roll: the structure of E-selectin. In *Structure* **2**: 47–50.
Weis WI, Drickamer K (1994): Trimeric structure of a C-type mannose-binding protein. In *Structure* **2**: 1227–40.
Wolfsberg TG, Straight PD, Gerena RL *et al.* (1995): ADAM, a widely distributed and developmentally regulated gene family encoding membrane proteins with a disintegrin and metalloprotease domain. In *Dev. Biol.* **169**:378–83.
Yamasaki N, Richardson RT, O'Rand MG (1995): Expression of the rabbit sperm protein Sp17 in COS cells and interaction of recombinant Sp17 with the rabbit zona pellucida. In *Mol. Reprod. Dev.* **40**: 48–55.
Yanagimachi R (1981): Mechanism of fertilization in mammals. In *Fertilization and Embryonic Development In Vitro* (Mastroianni L, Biggers JD, eds) pp 81–82, New York: Plenum Press.
Yanagimachi R (1994): Mammalian fertilization. In *Physiology of Reproduction*, 2nd edition (Knobil E, Neill JD, eds) pp 189–317, New York: Raven Press.
Yurewicz EC, Pack BA, Sacco AG (1991): Isolation, composition, and biological activity of sugar chains of porcine oocyte zona pellucida 55K glycoprotein. In *Mol. Reprod. Dev.* **30**:126–34.
Yurewicz EC, Pack BA, Sacco AG (1992): Porcine oocyte zona pellucida Mr 55,000 glycoproteins: identification of *O*-glycosylated domains. In *Mol. Reprod. Dev.* **33**:182–8.
Yurewicz EC, Pack BA, Armant R *et al.* (1993): Porcine zona pellucida ZP3 glycoprotein mediates binding of the biotin-labelled Mr 55,000 familiy (ZP3) to boar sperm membrane vesicles. In *Mol. Reprod. Dev.* **36**:382–9.

36 Glycobiology of Consciousness

RAYMONDE JOUBERT-CARON, DIDIER LUTOMSKI, DOMINIQUE BLADIER, AND MICHEL CARON

1 Introduction

Neuroscientists are still far from agreeing on how consciousness should be studied or even defined (Edelman, 1987; Crick, 1989). In 1987, Gerald M. Edelman, who shared a Nobel prize in 1972 for research on antibodies, contended that our sense of awareness stems from a process he called Neural Darwinism, in which groups of neurons compete with one another to create an effective representation of the world. A few years later, Francis Crick, who shared a Nobel prize for the discovery of DNA's structure in 1953, and Christof Koch proclaimed that the time was right for an assault on consciousness. Only by examining neurons and the interactions between them could scientists accumulate the kind of empirical knowledge that is required to create truly scientific models of consciousness.

More generally, understanding how the brain works from a biochemical perspective means understanding how the neurobiochemical organization and dynamics of the brain translate into behavioral functions. There is general agreement concerning the extent to which mechanisms involved in developmental neural plasticity may be regarded as an analogue, or at least a model, of mechanisms related to consciousness, i.e. memory, psychic disorders etc. Among the functional molecules which are involved in neural plasticity, important for cognitive functions of the brain, endogenous lectins, along with glycoproteins, are assumed to play significant roles (Nicolini and Zatta, 1994).

2 Brain Lectins

The brain is not an object – a sort of impervious box – whose internal structure is unknown and even irrelevant for ever and ever. Many aspects of human behavior and mental functioning cannot be fully understood without some knowledge of the underlying biochemical processes. Because a given neuron is able to communicate with other cells and eventually with itself, a neuronal network can be created, even linking memory and emotion. Which molecules combine to initiate neurite outgrowth, elongation, branching and cessation, to program normal cell death, and to generate axonal or dendritic geometry? *In vitro* these phenomena can be modified by a variety of soluble and substrate-bound factors, suggesting that, *in vivo*, control over morphological differentiation is multifactorial. The role of lectins in the nervous system was initially investigated at the same time as that of the cell adhesion molecules (CAMs) (Edelman, 1985, 1988), but did not attract much attention for many years. This was due to the difficulty of inferring their precise biological functions and the lack of identification of their specific ligands. Moreover, in the 1970s and 1980s neuroscientists considered lectins as exogenous proteins extracted from plants, used for decades for analytical purposes such as detection of glycoconjugates on nerve cell surfaces (Caron *et al.*, 1981; Raedler and Raedler, 1985). In addition, some plant lectins such as concanavalin A (ConA) were described as modulating neuronal membrane properties, promoting neurite regeneration in cultured neurons, modulating the responses of some neurons to the glutamate neurotransmitter and modifying the synaptic

H.-J. and S. Gabius (Eds.), Glycosciences

ISBN 3-8261-0073-5

specificity and synaptic efficiency (Levitan *et al.*, 1990).

The important finding that animal nervous tissue contains endogenous lectins stimulated the researchers to pay more attention to lectin molecules. In recent years, a growing number of lectins with various sugar-binding specificities has been characterized in brain and cerebellum (Caron *et al.*, 1990; Gabius and Bardosi, 1990; Zanetta *et al.*, 1992). If we look back on the past, at the turn of 80's the investigation of the environmental factors which might influence brain development *in vivo* led to the study of the effect of brain extract components on nerve cells *in vitro*. Brain extracts of different species were found to promote proliferation and differentiation of nervous cells. Thus, adult rat brain extract added to the culture medium increased the reassociation and clustering of mouse embryonic nerve cells (Joubert *et al.*, 1985). From these observations, our study began when we asked a simple question: what is the factor responsible for this reaggregation? In 1987, we described a rat brain lectin specific for lactosyl residues (Caron *et al.*, 1987) which was able to aggregate either cultured nerve cells or dissociated brain cells maintained in suspension. A protein showing similar properties was then characterized in human brain (Bladier *et al.*, 1989). Our findings agreed with the discovery of the expression of glycoconjugates of the lactoseries on subsets of central and peripheral neurons by Jessell and collaborators from Columbia University (Jessel *et al.*, 1990). We could finally take up the task of purifying and sequencing the lactosyl-specific human brain lectin, which was found to be similar to a protein isolated from other tissues by different groups, galectin-1 (Gal-1). This name was adopted for galactoside-specific lectins with molecular mass of about 14,500 previously known under a number of different designations, including human brain lectin (HBL). Galectins are gradually revealing their structural secrets, but thus far, their physiological functions remain unknown. However, various hypotheses have been proposed concerning the biological functions of lectins – not only galectins – found in nervous tissue extracts. These speculations are generally based on the localization of the lectin molecules and the kind of ligands facing them. The ontogenic regulation of Gal-1, as well as its localization in dendritic spines, encouraged our group to propose multiple distinct roles for this lectin, especially in neurite fasciculation and in the intracellular traffic in relationship to the formation of synapses (Joubert *et al.*, 1989).

The saccharidic specificity of galectins for lactosamines is interesting, particularly in nervous tissue where glycoconjugates of the lactoseries are present in high density at definite localizations on neurons and glial cells. Glycosylated and non-glycosylated ligands have been determined, leading us to consider that the reversible association of brain Gal-1 with both actin and glycoconjugates may be one of the complex systems which regulate synaptic plasticity.

Two lectins with mannose specificity have been described in cerebellum. The first one, R1, located exclusively intracellularly, was proposed to be involved in normal cerebellar synaptogenesis. The nature of the endogenous glycoligands of R1 is particularly interesting. These transiently expressed glycans also bind to ConA, which is a mannose-specific plant lectin, and to a monoclonal antibody of the HNK-1 family. The same group reported a possible role of the R1 lectin in repair after cerebellar lesions (Zanetta *et al.*, 1992). This report constitutes a demonstration that re-expression of recognition molecules, as a basis for a repair mechanism in adult, can occur, analogous to that occurring during normal construction of the nervous system. A similar observation was made by our group concerning the re-expression of galectin-1 after adult nerve injury. The second mannose-binding lectin described by this group was named CSL (Zanetta *et al.*, 1987). Its role in contact guidance of neurons migrating during ontogenesis was deduced from the localization of CSL in the cerebellar premigratory zone, its accumulation in Bergmann fibers, and its externalization. Studies of the role of CSL in the adhesion mechanism in astrocyte cell cultures showed that CSL and its glycoprotein ligands were responsible for the formation of contacts between the astrocytes. The involvement of CSL in myelin compaction was considered, and this idea was reinforced by the presence of two major CSL-binding ligands in oligodendrocytes. The postulated role of CSL in adhesion mechanisms occurring during myelination conflicts with the homophilic mechanism involving CAMs (Zanetta, 1994, this volume).

Several glycosaminoglycan-binding lectins were isolated from soluble brain extracts. One of them binds to hyaluronic acid (Delpech and Halavent, 1981). The involvement of this lectin in cell migration in white matter is still speculative. The amount of hyaluronic acid is considerably higher in younger mammalian brain than in adult tissue, and when tissue remodelling and regeneration take place. Consequently, the presence in brain of appropriate ligands is of some importance. Multiple heparin-binding proteins were also purified from brain tissue. Among them two classes of low-molecular weight soluble proteins have been studied: fibroblast growth factors (FGFs) and heparin-inhibitable lectins. Our group has uncovered a relationship between these two classes of molecules which may function as parts of control mechanisms that modulate autocrine regulation of epithelial cells and paracrine stimulations (Eloumami *et al.*, 1990). Other heparin-binding molecules have been isolated after solubilization in detergent (Rauvala and Pihlaskari, 1987). They display strong adhesive properties for neuroblasts. The role of heparin in inhibiting cell adhesion or glial cell proliferation indicates that a heparin-binding mechanism in the nervous system could be important.

All the lectins expressed in brain and cerebellum seem to be involved in major processes such as the construction and the repair of nervous tissue, and the establishment of neuronal connections through synapses. For example, the cognitive processes of perception and memory are mediated via interactions between neuronal pathways. Thus far, the molecular basis of emotional events remains unknown because most of our knowledge about how the brain links information, emotions and memory has been perceived more through behaviorism than biochemistry.

3 Neuroimmunomodulation

Just like the central nervous system (CNS), the immune system (IS) is a complex evolutionary unit. Both these systems have perhaps evolved beyond the apparent evolutionary state of the species in which they function. They have been modulated by different elements linked not only to the internal evolution of their elementary genes but also by coevolution with elements in the internal environment and behavior. The comparative study of IS and CNS shows a dynamic and diverse world. In some circumstances the CNS seems to maintain an ambivalent relation to the IS. On the one hand more and more evidence accumulates that the CNS upholds an important regulatory influence on aspects of the IS such as fever or stress; on the other hand the CNS is well known for its tendency not to elicit an immune response within its territory. One of the key points where the CNS could mount an access control on the IS is antigen presentation. In contrast to cells from other organs, cells of the CNS do not normally express major histocompatibility complex (MHC)-class II molecules. This peculiarity of the CNS limits and prevents the immune response within this organ. However, in CNS some cells such as microglial cells or astrocytes start to express MHC-class II molecules when they have been sufficiently stimulated by interferon γ. In this case, what appears well designed for resistance to microbial or viral agents in an otherwise immune-privileged site might have its negative counterpart, when it comes to unwanted defensive reactions within the nervous tissue in some diseases such as multiple sclerosis or *myasthenia gravis*. As suggested by Cleveland (1986) in a metaphorical article entitled "Crime and punishment in the society of lymphocytes", the immunological network can be compared to a human society. It can be divided into two categories: one category that consists of an establishment composed of law-abiding citizens who do not commit offenses or crimes against each other, and a second category, referred to as a fringe element, whose members assail and destroy the establishment. In the autoimmune process a network consisting of a fringe element faces the establishment. Sometimes a fringe element is destroyed by the establishment.

Gal-1 has been reported to have interesting immunoregulatory properties, playing a part in the release of tumor necrosis factor from macrophages (Kajikawa *et al.*, 1986) and suppressing experimental autoimmune encephalomyelitis (Offner *et al.*, 1990). As brain cells express a significant amount of this lectin, it is not out of the question that Gal-1 might play a role as a neuroimmunomodulator in the interaction between CNS and IS. Gal-1-immunoreactive material has been detected and measured in human cerebrospinal fluid and serum (Eloumami *et al.*, 1991). We then proceeded out to characterize this immunoreactive material. It was identified as a population of oligoclonal β-galactoside-binding IgG (lectin-like IgG), which was found to react with cellular antigenic determinants expressed in mouse brain after surgical injury (Joubert-Caron *et al.*, 1994, 1995).

This discovery suggested that these lectin-like immunoglobulins could be natural anti-idiotype antibodies implicated in the liquidation of the consequences of tissue injury. Moreover, the existence of these anti-idiotype immunoglobulins raised questions about the presence in the sera of natural anti-galectin-1 autoantibodies belonging to the same network. The finding in the serum of healthy individuals of antibodies reacting both toward Gal-1 and lectin-like IgG (Lutomski *et al.*, 1995) was consistent with the view that Gal-1 participates in an immunoregulatory network. Anti-Gal-1 antibody levels measured in patients with neurological diseases suggested that changes in their concentration may, in certain situations, reflect an aberrant regulation of the immune system during the development of clinical symptoms, including troubles of consciousness and psychic disorders. Furthermore, the highest levels of lectin-like IgG were in patients with low anti-Gal-1 titers, while patients with high anti-Gal-1 titers often have very little detectable anti-idiotypic immunoglobulin. The coincidental association of high levels of anti-idiotype with low levels of anti-Gal-1 antibodies suggests that they may exert a suppressive effect on anti-Gal-1 antibody production.

In this example we identified indications for a molecular basis of overlapping of immunity and neurological disorders. In the field of stress study, it is not so easy to appreciate the consequence of stress on IS. It is known intuitively that stress decreases the host defences and favors the emergence of microbial and viral infections or even of malignant proliferation. It was suggested that an increased secretion of corticosteroids would be responsible for this situation (Mormède, 1991). However, recently clinical and epidemiological studies contradicted this concept. To understand the eventual results of stress on IS, it is important to consider that stress constitutes a psychobiological response to physical stimulations with a non-negligible implication for the individual characteristics, and the individual's own way of perceiving a precise situation. Moreover, other neuroendocrine systems such as the autonomous nervous system – not only the corticotrope axis – can play a role in the individual response to stress, and prolactin is a molecule that may be implicated in the individual's reactivity to stress. The IS is a complex functional entity whose elements react differently to external modulations, and the same neuroendocrine factors or environmental elements may influence the response in an opposite direction. In this context, glucocorticoids can regulate the inflammatory reactions, for example, during the autoimmune diseases, but they can also increase some immune functions.

In the field of self and non-self recognition, the CNS and IS serve similar functions on different levels. The CNS assumes the cognition of many aspects of human behavior and mental functioning, particularly the differentiation between self and non-self. The IS assumes similar functions in non-cognitive relations between the tissues and their environment. This bipolarity of the human organism, separated in cognitive and non-cognitive domains, is directed on one side by the CNS and on the other side by the IS. This consideration has prompted some neuroscientists to evaluate the immunological consequences of the use of psychotropic drugs. These pharmacological approaches suggested a way to demonstrate a new loop in neuroimmunomodulation, and also suggest a molecular basis for immunoregulation by anxiety (Zavala, 1991).

4 Brain Glycoproteins

Memory formation can be regarded as a special case of neuronal plasticity. It involves the making of new synapses or the modification of those already in existence in the brain. As glycoproteins are major components of synaptic membranes where they play a role in the specificity of intercellular recognition properties, the possibility that they would be important in regulating plastic changes is attractive. In the chick (Rose, 1989), learning-induced changes in the glycoprotein fucosylation pattern has been demonstrated. There is an increased uptake of fucose, a precursor for glycoproteins, after memory formation, and this increased incorporation is concentrated in synaptic membrane fractions. This synthesis may be involved in the post-synaptic remodelling of dendritic spines, thereby modulating the connectivity of synapses so as to form the neural representation of memory.

Given that synaptic connectivity change is a consistent anatomical feature of memory storage, it is possible to suggest that cognitive functions may be mediated by molecules involved in neurodevelopmental events. An important class of glycoproteins of the Ig-superfamily, neural cell adhesion molecules (NCAM), are therefore prime candidates for participation in such events (Edelman, 1985, 1988; Fazelli and Walsh, 1994).

NCAM is present in multiple isoforms which have been identified on the basis of differing molecular weights. The 140 and 180 kDa polypeptides (NCAM 140 and NCAM 180) contain transmembrane domains, whereas the 120 kDa polypeptide (NCAM 120) is attached to the membrane by a glycosyl-phosphatidyl-inositol anchor. The strength of NCAM-mediated cell adhesion is developmentally altered through regulating the extent to which its *N*-linked glycans are sialylated with homopolymers of α2–8-linked sialic acid (Hoffman and Edelman, 1983). A particular form of cell adhesion molecule, telencephalin, has its expression restricted to the telencephalon (Yosihara and Mori, 1994). The telencephalon occupies the largest portion of the mammalian brain and presumably takes charge of functions such as memory, learning, and sensory perception.

In the developing nervous system, most of the NCAM is highly polysialylated as 140 and 120 isoforms. A few days after birth, when cellular migration is completed and synaptogenesis occurs, the NCAM 180 isoform reaches a detectable level and the majority of the polysialic acid is lost. Continued expression of NCAM in the adult brain indicates that, in addition to its roles in development, NCAM may also be involved in the maintenance of adult neuronal connectivity and therefore is in a position to act as a modulator of plasticity. Changes in the NCAM sialylation state induced by memory formation were observed in the hippocampus (Doyle *et al.*, 1992). A significant increase in NCAM sialylation state becomes apparent at 12 h after training. Thus, these changes were supposed to be involved in memory consolidation rather than acquisition. The learning-induced change in NCAM sialylation state is reminiscent of that which occurs during development. It is associated with the NCAM 180 isoform which has the potential to elaborate synapse structure. A learning-specific polysialylated 210 kDa isoform of NCAM was detected that was believed to arise from NCAM 180.

In the adult rat, the expression of polysialylated NCAM was studied in the hypothalamo-neurohypophysial system, which undergoes important neuronal-glial and synaptic rearrangement in response to physiological stimuli (Theodosis *et al.*, 1991). This study suggested that polysialylation may confer on nervous cells the ability to reversibly change their morphological plasticity in adulthood, by rendering them less adhesive. This raises the possibility that the degree of glycosylation of NCAM may serve as a regulator of synaptic plasticity under physiological or pathological stimulations.

Clearly, studies on the functional role of lectins and glycoproteins provide not only insight into important mechanisms defining cell proliferation and cell-cell adhesion. The view that endogenous lectins and glycoproteins act simply in a "glue-like" fashion is no longer tenable. Their interactions may be involved in synaptic plasticity in the adult. In this way it is possible for these molecules to play an active role in cognitive phenomena through changes in molecular connections. Consequently, one can confidently predict that the research on glycobiology in the field of neurobiology and neuropathology may give rise to interesting applications in therapy.

More generally, as is evident from the diversity of themes reviewed in this book, glycosciences will certainly play a growing role in the understanding of animal and human biology in the near future. Independent lines of investigation are pointing to carbohydrate structures of glycoproteins and glycolipids as being important during cellular adhesion and development, infection and host defence mechanisms, ontogenesis and oncogenesis and so on. Undoubtedly, the interactions of specific carbohydrate sequences with complementary binding proteins, namely endogenous lectins, are important topics of current research. As a result, the information supplied by the glycobiological approach will appear more and more clearly as absolutely necessary for our comprehension of physiology and pathology at the level of cells, organs and organisms.

References

Bladier D, Joubert R, Avellana-Adalid V *et al.* (1989): Purification and characterization of a galactoside-binding lectin from human brain. In *Arch. Biochem. Biophys.* **269**:433–9.

Caron M, Deugnier MA, Albe X *et al.* (1981): Rhodamine isothiocyanate coupled peanut lectin for quantitative studies of D-galactosyl receptors of neuroblastoma cells. In *Experientia* **37**:1154–6.

Caron M, Joubert R, Bladier D (1987): Purification and characterization of a β-galactoside-binding soluble lectin from rat and bovine brain. In *Biochim. Biophys. Acta* **925**:290–6.

Caron M, Bladier D, Joubert R (1990): Soluble galactoside-binding vertebrate lectins: a protein family with common properties. In *Int. J. Biochem.* **22**:1379–85.

Cleveland WL (1986): Crime and punishment in the society of lymphocytes: a speculation on the structure of the putative idiotype network. In *Antiidiotypes, Receptors, and Molecular Mimicry* (Linthicum DS, Farid NR eds) pp 199–229, Heidelberg: Springer Publ. Co.

Crick F (1989): Neural Edelmanism. In *Trends Neurosci.* **12**:240–8.

Delpech B, Halavent C (1981): Characterization and purification from human brain of a hyaluronic acid-binding glycoprotein, hyaluronectin. In *J. Neurochem.* **36**:855–9.

Doyle C, Nolan PM, Bell R *et al.* (1992): Hippocampal NCAM180 transiently increases sialylation during the acquisition and consolidation of a passive avoidance response in the adult rat. In *J. Neurosci. Res.* **31**:513–23.

Edelman GM (1985): Cell adhesion and the molecular processes of morphogenesis. In *Annu. Rev. Biochem.* **54**:135–69.

Edelman GM (1987): *Neural Darwinism*, Basic Books.

Edelman GM (1988): Morphoregulatory molecules. In *Biochemistry* **27**:3533–43.

Eloumami H, Bladier D, Caruelle D *et al.* (1990): Soluble heparin-binding lectins from human brain: purification, specificity and relationship to an heparin-binding growth factor. In *Int. J. Biochem.* **22**:539–44.

Eloumami H, Caron M, Joubert R *et al.* (1991): Human brain lectin immunoreactive material in cerebrospinal fluids determined by enzyme immunoassay (EIA). In *J. Neurol. Sci.* **105**:6–11.

Fazelli MS and Walsh FS (1994): *N*-CAM and L1: role of glycoproteins of the immunoglobulin superfamily in nervous system development. In *Glycobiology and the Brain* (Nicolini M, Zatta PF, eds) pp 63–81, Oxford: Pergamon Press.

Gabius H-J, Bardosi A (1990): Regional differences in the distribution of endogenous receptors for carbohydrate constituents of cellular glycoconjugates, especially lectins, in cortex, hippocampus, basal ganglia and thalamus of adult human brain. In *Histochemistry* **93**:581–92.

Hoffman S, Edelman GM (1983): Kinetics of homophilic binding by embryonic and adult forms of the neural cell adhesion molecule. In *Proc. Natl. Acad. Sci. USA* **80**:5762–6.

Jessel TM, Hynes MA, Dodd J (1990): Carbohydrates and carbohydrate-binding proteins in the central nervous system. In *Annu. Rev. Neurosci.* **13**:227–55.

Joubert R, Caron M, Deugnier MA *et al.* (1985): Effect of adult rat brain extracts on cultures of mouse brain cells: consequence of the depletion in a carbohydrate-binding fraction. In *Cell. Mol. Biol.* **31**:131–8.

Joubert R, Kuchler S, Zanetta JP *et al.* (1989): Immunochemical localization of a β-galactoside-binding lectin in central nervous system. Light- and electron-microscopical studies on developing cerebral cortex and corpus callosum. In *Dev. Neurosci.* **11**:397–413.

Joubert-Caron R, Caron M, Bocher P *et al.* (1994): Oligoclonal β-galactoside binding imunoglobulins antigenically related to 14 kDa lectin in human serum and cerebrospinal fluid: purification and characterization. In *Int. J. Biochem.* **26**:813–23.

Joubert-Caron R, Lutomski D, Le Saux F *et al.* (1995): Oligoclonal β-galactoside-specific γ-immunoglobulins allow the immunocytochemical detection of cellular antigenic determinants expressed in mouse brain after surgical injury. In *Neurochem. Int.* **26**:607–13.

Kajikawa T, Nakajima Y, Hirabayashi J *et al.* (1986): Release of cytotoxin by macrophages on treatment with human placenta lectin. In *Life Sci.* **39**:1177–81.

Levitan IB, Carrow GM, Dagan D *et al.* (1990): Neuronal plasticity: lectin-induced changes in the membrane properties of cultured *Aplysia* neurons. In *Progress in Cell Res.* **1**:251–61.

Lutomski D, Caron M, Bourin P *et al.* (1995): Purification and characterization of natural antibodies that recognize a human brain lectin. In *J. Neuroimmunol.* **57**:9–15.

Mormède P (1991): Stress et immunité. In *Proceedings of INTERMED, 3° Forum International sur le médicament*, Bordeaux 3–5 october 1991, pp 46–7.

Nicolini M, Zatta PF (eds) (1994): *Glycobiology and the Brain*, Oxford: Pergamon Press.

Offner M, Celnik B, Bringman TS *et al.* (1990): Recombinant human β-galactoside-binding lectin suppressed clinical and histological signs of experimental autoimmune encephalomyelitis. In *J. Neuroimmunol.* **28**:177–84.

Raedler A, Raedler E (1985): The use of lectins to study normal differentiation and malignant transformation. In *J. Cancer Res. Clin. Oncol.* **109**: 245–51.

Rauvala H, Pihlaskari R (1987): Isolation and some characteristics of an adhesive factor of brain that enhances neurite outgrowth in central neurons. In *J. Biol. Chem.* **262**:16625–35.

Rose SPR (1989): Glycoprotein synthesis and postsynaptic remodelling in long-term memory. In *Neurochem. Int.* **14**:299–307.

Theodosis DT, Rougon G, Poulain DA (1991): Retention of embryonic features by an adult neuronal system capable of plasticity: polysialylated neural cell adhesion molecule in the hypothalamo-neurohypophysial system. In *Proc. Natl. Acad. Sci. USA* **88**:5494–8.

Yoshihara Y, Mori K (1994): Telencephalin: a neuronal area code molecule? In *Neurosci. Res.* **21**:119–24.

Zanetta JP (1994): A re-evaluation of the concept of homophilic interactions? In *Glycobiology* **4**:243–9.

Zanetta JP, Meyer A, Kuchler S *et al.* (1987): Isolation and immunochemical study of a soluble cerebellar lectin delineating its structure and function. In *J. Neurochem.* **49**:1250–7.

Zanetta JP, Kuchler S, Lehmann S *et al.* (1992): Cerebellar lectins. In *Int. Rev. Cytol.* **135**:123–54.

Zavala F (1991): Psychotropes et système immunitaire. In *Proceedings of INTERMED, 3° Forum International sur le médicament*, Bordeaux 3–5 October 1991, pp 52–5.

Subject Index

A

abequose 294, 319, 320–327
Abrus precatorius lectin 420, 421, 429
acetylcholinesterase 223, 224, 386
Acomys russatus 552
acrosome reaction 514, 600–602
acryloylation 68
actin 151
Actinomyces naeslundii 373, 374, 378
- *viscosus* 373, 374, 378

activator protein 171, 172
acute-phase reactants (see pentraxins; C-reactive protein; serum amyloid P component)
Adenia digitata lectin 429
adenylate cyclase 153, 486–488, 491
adhesin 5, 369, 372, 378, 379, 383, 385, 386, 388 (see also lectin)
adipocyte 236
ADP-ribosylation 138, 385
affinity enhancement 56
Agaricus bisporus lectin 20, 548
aggrecan 208
aggregometer 280
aglycon 68, 98, 205, 479
AIDS 157, 287, 408
alcian blue 547
Aleuria aurantia lectin 20, 548, 597
alginate 372
algorithm 306
alkaline phosphatase 197, 223, 224
Allium cepa lectin 420
- *sativum* lectin 420

Allomyrina dichotoma lectin 117
allotransplantation 531, 532
almond 123
Alzheimer's disease 181, 215
Amadori rearrangement 22
amber/suppressor system 360
amebiasis 407
ameloblastoma 565
amidination 59, 60
2-aminobenzamide 17
2-aminopyridine 16
amperometry 15, 229
Amphicarpaea bracteata lectin 424
amphomycin 230
ampulla 604
α-amylase 58
amyloidogenesis 448
amyloidosis 134
analbuminemia 478
angiosarcoma 568
Anguilla anguilla lectin 20, 548, 556
annexin 127
anomericity 1, 5, 6, 22, 31, 44, 45, 168, 549
anthranilic acid 16
antibodies (see also immunoglobulin)
- carbohydrate-binding 5, 84, 251, 293, 316, 319–329, 348, 350, 532
- CD2 148, 150
- CD3 442–444, 452
- CD4 157, 440
- CD14 7
- CD22 19, 249, 473, 492
- CD23 441, 492
- CD31 213
- CD33 248, 473
- CD34 270
- CD40 450
- CD44 203, 2313, 215
- CD45 492
- CD48 236
- CD69 441
- CD70 450
- CD72 442, 444, 445, 492
- CDW60 246
- α-gal 533–541
- glycosylation 134, 149
- targeting 471

antigen
- galectin 462, 463
- HNK-1 272, 440, 442, 612
- I 92, 93, 98, 142
- i 92, 98, 142
- MAC-1 491
- MAC-2 459, 569

H.-J. and S. Gabius (Eds.), Glycosciences

ISBN 3-8261-0073-5

- sialyl T_n 99
- stage-specific embryonic 173, 174
- T_n 144, 401

antithrombin III 6, 206, 215, 298
aorta 537
apoptosis 407, 492
appican 215
arabinofuranose 135
arabinogalactan 538
arabinose 1, 374, 419
arachidonic acid 485–487
Arachis hypogaea 427, 537 (see also peanut)
arcelin 428
archaebacteria 138, 164, 172
arteriosclerosis 206, 570
Arthrobacter 63, 64
arthrosides 168
Artocarpus integrifolia lectin 195, 420
arylsulfatase 171, 172, 178
ascophyllan 607
asialofetuin 365, 366, 406, 462, 476, 607
- glycoprotein 154, 192, 439, 472–479
- orosomucoid 406, 476, 477

asparagine amidase 121
aspartate aminotransferase 63
aspergillosis 478
Aspergillus saitoi 229
Asterias rubens 142
astrocyte 175, 215, 441, 443, 612, 613
autophagy 128
autophosphorylation 175
avidin 63, 550–552, 564, 565
Avogadro's number 2

B

baboon 534–538, 601
Bacillus stearothermophilus 136
Bacteroides cellulosolvens 138
Bartonella bacilliformis 373
batroxobin 141
benzylamine 70
betaglycan 205
biglycan 210
biomodulation 519–525
biotin 63, 197, 281, 471, 550, 564, 565
- 2,6-diaminopyridine derivative 63

bis(sulfosuccinimidyl)suberate 65
bladder 150
blastocyst 460
blood clotting 133, 138, 145
blood group 81, 84, 93, 96, 99, 142, 143, 150, 158, 166, 197, 375, 400, 532, 536 (see also Lewis)
- A 93, 100, 176, 416
- B 79, 93, 100, 534
- H(O) 93, 94, 282, 294, 374, 540
- P_1 408

blood transfusion 93, 176
Bombay phenotype 93, 158, 540
bone marrow 150
Bordetella bronchiseptica 373
- *pertussis* 139, 371

borohydride degradation 18
botulinum toxin 250
Bouin's fixative 551
Bowman's capsule 380
brain 82, 85, 86, 89, 125, 166, 180, 213, 215, 523, 611, 612
brefeldin A 118
bronchiectasis 145, 146
Brønsted acid 43
Brugia malayi 558
brush border 407
Bufo bufo 555

C

CAAT box 83
cadherin 452
Caenorhabditis elegans 89, 211, 363
calcium channel 175
calmodulin 487
calnexin 7, 116, 127, 147, 439
calreticulin 116, 117, 125
Candida albicans 502
capacitation 601, 603, 604
capillary electrophoresis 25, 26, 151
carbonic anhydrase 262, 263, 269
carcinoembryonic antigen 440, 462, 463, 465
carcinogenesis 139
carcinoma
- associated antigen 99, 166, 180
- breast 98, 100, 389, 571
- cervical 462
- colon 97, 98, 100, 211, 463, 465, 568, 571
- embryonal 145
- esophageal 566
- gastric 463, 568, 571
- kidney 143
- lung 95, 566

Carnoy's fixative 551
β-casein 63
castanospermine 451
CD spectroscopy 21
cell adhesion molecule
- GlyCAM-1 262, 263, 270, 430
- ICAM 203
- L1 152, 446
- neural (N-CAM) 89, 143, 152, 209, 248, 442, 451, 465, 615

central nervous system 57, 180, 613, 614

ceramide 18, 46, 165
– glycanase 65, 128
– signaling 175
cerebroside sulfotransferase 170
cerebrospinal fluid 156, 443, 614
ceruloplasmin 472
Chagas' disease 399, 402, 403, 557
chaperone 7, 127, 147, 439
chick embryo 85, 86
chick testis 86
Chinese hamster ovary (CHO) cells 87, 142, 146, 149, 153, 206, 211, 214, 271, 402, 403, 405, 406, 442
chitin 7, 405, 420, 423, 425, 426
chitinase 7, 372, 386, 387
chitotriose 400, 407
Chlamydia trachomatis 373
chloroplast 164, 172, 173
cholera 139
– toxin 7, 65, 179, 250, 385
cholesterol 478, 601
chondrocyte 203, 212
chondroitin sulfate 65, 201, 202, 204, 205, 208, 215, 278, 400, 402, 405, 607
chondroitinase 24
chorionic gonadotropin 148, 151, 264, 488
chromatin condensation 382
chromatography
– affinity 56, 193, 312, 366, 370, 408, 463
– gas-liquid 16, 168
– high pH anion exchange 15, 20, 25, 229
– hydrophobic 125, 227, 228
– HPLC 16, 25, 193, 197, 225, 229
– iatrobead 228
– ion exchange 197
– lectin-affinity 56, 193–197
– paper 197
– porous graphitized carbon 21, 24
– reversed phase 24–26, 63, 70
– thin layer 57, 64, 65, 225, 228
– weak-affinity 281
chromogen 551
circular dichroism (see CD)
Clostridium botulinum 370, 385
– *difficile* 151
– *symbiosum* 136
– *tetani* 385
– *thermocellum* 138
Club cells 554
cluster effect 56, 66, 71
coagulation 26
collagen 135, 204, 210, 211, 215, 403, 462
collagenase 498
collectin 18, 414, 498, 499
colloid chemistry 282
colominic acid 7
colon epithelium 19
colony-stimulating factor 133, 150, 215, 523
colorimetry 19, 333, 336–342
complement 385, 441, 499
– C1q 149, 203, 498, 499
– C1r,s-like protease 498
– CR1, CR3 404
– iC3b 499
– regulatory domain 208, 209, 500
concanavalin A 6, 19, 73, 116, 117, 125, 127, 148, 192, 194, 196, 318, 334–342, 355, 415–418, 445, 489, 491, 548, 554, 564, 571, 590, 611
conglutinin 441, 499
convicilin 421
cooperativity 18
Corynebacterium diphtheriae 385
– *parvum* 373
COS cells 233, 401
C-reactive protein 497, 498
Crithidia fasciculata 246
Cryptosporidium parvum 400, 408
crystallization 306, 312, 313
– heavy atom derivative 313
CUB-domain 605
cyclodextrins 3
cyclophosphamide 538
cyclosporine 538
cystic fibrosis 137, 145, 146, 386
Cytisus scoparius lectin 548
cytokine 444, 450, 501, 502
(see also interleukin)
cytotoxin 387, 501

D

Datura stramonium lectin 192, 419
decay-accelerating factor 233
decorin 210, 211
deglucosylation/reglucosylation 115, 147
demarcation 286
denervation 556
deoxynojirimycin 451
dermatan sulfate 6, 201, 202, 204, 205, 207, 210, 211
dermis 553
deuterostomes 497
diabetes 22, 428
diacylglycerol 485–487
diarrhea 407
diastereoselectivity 32, 42
Dictyostelium discoideum 137, 138, 262
(see also slime mold)
diencephalon 587
differentiation 93, 97, 98, 100, 133, 173–175, 177, 352, 515, 569, 612
digoxigenin 100
Dioclea grandiflora lectin 20
distearylphosphatidyl ethanolamine 64

DNA polymerase
- I 359
- *PfuI* 362
- T4 362
- *Taq* 359, 362
dolichol phosphate-mannose synthase 238, 239
- pyrophosphate 79, 115, 122, 230
Dolichos biflorus lectin 20, 422, 426, 548, 554, 556, 564, 568
doxorubicin 476
Drosophila melanogaster 135, 142
drug targeting 155, 419, 471–479
Duchenne muscular dystrophy 556
Dukes' stage 463

E

ectoderm 93
Edman method 417
eel 364
effectomer 429
elastase 272, 387
electron density map 313, 314
Electrophorus electricus lectin 460
β-elimination 18, 21, 59, 64
embryo 16, 143, 442
embryogenesis 91, 122, 124, 128, 174, 460, 511
embryoglycan 143, 283
encephalomyelitis 614
endo-α-N-acetylgalactosaminidase 24, 128
- β-galactosidase 24
- β-hexosaminidase 58
- β-N-acetylglucosaminidase 63, 128
endoglycosidase 18, 24
- H 24, 82
endoplasmic reticulum 79, 80, 115–118, 121, 122, 127, 152, 166, 169, 231, 416, 508
- ERGIC 116
- ERGIC53 117
endotoxemia 501
endotoxin 499 (see also lipopolysaccharide)
Entameba histolytica 399, 400, 406, 407
Enterococcus faecalis 378
enterocyte 374, 407
enthalpy 292, 293, 295, 299, 300, 333, 336–339, 341, 342
entropy 292, 293, 299, 300, 302, 333, 336–339, 341, 342, 352
epidermal growth factor 145, 150, 203, 208, 213, 271, 471, 486, 487, 500
epimer 6
equilibrium dialysis 281, 333, 373
Erythrina corallodendron lectin 148, 152, 317, 355, 419, 565
- *cristagalli* lectin 117, 548
erythrocyte 92, 97, 370, 378, 382, 400–402, 407, 415, 422, 426, 462, 533
erythropoietin 250
Escherichia coli 7, 148, 150, 152, 156, 157, 351, 359, 365, 373, 374, 378, 381, 387, 404, 418
- suppressor strain 360, 361
E-selectin 20, 23, 65, 94, 95, 157, 158, 249, 269, 271, 284, 440, 473
estradiol 205
estrogen 268
exoglycosidase 16, 17, 19, 151, 170
exon shuffling 100
exonuclease 359, 362
expression system 83

F

Faber disease 178
Fabry disease 178
fasciculation 440
fertilization 84, 123, 156, 514, 569, 595–607
fetuin 124, 141, 462, 474, 607
fibrin 151, 250
fibrinogen 250
fibroadenoma 566
fibroblast 86, 87, 91, 124, 174, 175, 211, 212, 444, 460, 461
fibroblast growth factor 6, 152
fibroglycan 212
fibromodulin 210, 211
fibronectin 152, 173, 213, 461, 462
fibrosarcoma 462
Fischer-Helferich procedure 34
fish 84, 128, 143, 268, 357, 554
Flavobacterium meningosepticum 22, 122–124
flavonoid 427
flipase 231
flow cytometry 462
fluorescence 16, 19, 405, 551, 558
fluorescence energy transfer 21
fluorimeter 281
folate 471
follitropin 264, 266
Forssmann glycolipid 277
Fourier transformation 313
frog 89, 90
fructose 1
fucoidan 402, 606, 607
fucose 1, 15, 32, 36, 138, 150, 157, 280, 373, 374, 540, 570, 597, 606, 607
- binding protein 61, 192, 404, 416, 492, 556, 570
fucosidase 607
fucosyltransferase 61, 79, 540
- α1,2 93, 94
- α1,3/4 93, 94, 197
- α1,6 80, 81, 94–96
Fusobacterium nucleatum 386

G

Gadus lyscus 553
galabiose 348
galactan 498
galactocerebroside 166, 170
α-galactosidase 178, 421, 425, 540, 607
β-galactosidase 73, 171, 178, 351, 407, 512, 607
galactoside-binding protein (see also galectin; lectin, hepatic)
– detection 73
– hepatocyte 67, 71
galactosylceraminidase 178
galactosyltransferase
– α1,3- 79, 84, 156, 510, 514, 535, 540
– β1,3- 84, 97
– β1,4- 81–84, 116, 117, 156, 197, 266, 512–515, 599–602
– mutation 205
galacturonic acid 1
Galanthus nivalis lectin 20, 318, 420
galaptin (see galectin)
galectin 357, 362, 439, 440, 459–466, 473, 569
– 1 192, 198, 296–298, 316, 317, 346–348, 355, 362–366, 459, 460, 512, 569, 613, 614
– 2 318, 355, 463
– 3 363, 460, 461, 493, 502, 569
– null mutation 460
ganglioside (see also glycolipids)
– degradation 170, 172
– GM_1 7, 65, 167, 174, 175, 179, 385
– GM_2 171, 282
– GM_3 174
– molecular interaction 282–286
– nomenclature 166–168
– receptor 57
– structure 167–169
– synthesis 37, 88, 169, 170
gapped-duplex method 358
gastrulation 124, 125
Gaucher's disease 178, 477
GC box 83
gene duplication 84
Geodia cydonium lectin 363
germ cell 84
germination 128
gestation 91
Giardia lamblia 399, 400, 407, 408
Gibbs free energy 292, 293, 299, 300, 329, 333, 336, 342, 348, 349, 351
gland
– lacrimal 269
– mammary 83
– salivary 93, 150
– submandibular 553
– submaxillary 85, 86, 97, 99, 118, 269
glioblastoma 443, 446
glucoamylase 351
β-glucocerebrosidase 133, 155, 156, 171, 178, 477
glucocorticoid 210
glucosamine 6-sulfotransferase 206
glucose dehydrogenase 351
– oxidase 351, 587
glucosidase I, II 80, 116
glucosyltransferase 116, 127
glucuronate 5-epimerase 207
glucuronic acid 1, 36, 135, 164, 202, 206, 207, 272, 549
glucuronyl-3-sulfotransferase 272
glucuronyltransferase 206, 272
glycamine 70
Glycine max lectin (see soybean)
glycoamidase 58
glycoasparagine 121
glycodelin 141
glycogen 133, 153, 556
glycogenin 138
glycolipid
– animal 165–168, 173–181
– bacterial 163, 165, 172, 173
– carbohydrate-carbohydrate interaction 278–286
– functions 172–181
– heterogeneity 56
– interaction 174
– lectin binding 374
– metabolism 169–172
– nomenclature 166–168
– physicochemical properties 169
– plant 163–165, 172, 173
– structure 163–169
glycophorin 145, 153, 283, 378, 401
glycoprotein
– bacteria 143, 144
– cytoskeletal 19
– function 148–158
– glycoform analysis 24, 25, 55, 134, 474, 603
– information content 6–8
– isolation 192–197
– microheterogeneity 134, 507
– nuclear 19
– plant 135–138, 154
– structure 134–148, 192–197
glycosaminoglycan (see also proteoglycan)
– biological activity 6
– linkage 18
– role in protozoan infection 402–404
– structure 201, 202
glycosciences
– enthusiasm of authors 1–617
glycosequencer 16, 17
glycosidase 4, 62, 79, 121, 127, 170, 350, 399, 508, 511, 512, 595, 605
glycosides
– ω-aldehydo/carboxyl 68

- allyl 68, 69
- p-aminophenyl 60
- deoxy 73
- dibromoalkyl 68
- fluorodeoxy 345, 348–350
- n-pentenyl 68
- polyethylene glycol 69

glycosphingolipids 43
glycosulfatase 248
glycosyl carboxylate 51
glycosyl halide 35–37
glycosylamine 70
glycosylation
- aberrant (histopathological detection) 563–576
- chemical 43–51, 58–65, 70
- drug targeting 472–479
- enzymatic 80–99
- genetic manipulation 508–515
- topology 115–118

glycosylphosphatidylinositol anchor
- function 234–237, 403, 406
- mutants 237–239
- structure 137, 225–233, 442
- transamidase 232

glycosyltransferases
- binding-site mapping 346, 352
- evolution 5
- glycan biosynthesis 79–100
- glycolipid synthesis 169, 170
- in chemoenzymatic synthesis 32, 33, 61, 62
- localization 115–118
- promiscuity 79
- substrate sequence 18, 19

glypican 213
goblet cells 87, 117, 557
Golgi apparatus 79, 83, 88, 115–118, 122, 166, 169, 205, 207, 213, 235, 508
- cisternae 87, 89, 116, 117, 248
- compartment 417, 451, 601
- network 87, 116
- stacks 116, 117, 170
- targeting 87, 117

gonadotroph 268
gonadotropin-releasing factor 263
G protein 153, 485, 491, 493, 513, 515
granulocyte 242
Griffonia simplicifolia lectin 20, 294, 319, 348, 355, 416, 424, 492, 537, 548, 556, 564, 568
growth factor 58
(see also receptor; epidermal or fibroblast growth factor)
guanylyl cyclase 487, 493

H

halobacteria 143
haptomer 429
heart 89
heat capacity of binding 333
HeLa cells 117
Helicobacter pylori 7, 378
Helix pomatia lectin 118, 195, 196, 548, 564, 566, 568, 571
hemagglutination 100
hemangioendothelioma 566
hematopoiesis 176, 270
hematoxylin 550
hemocyanin 141, 142
hemolysin 372, 386
hemostasis 157
Henle's loop 552
heparan sulfate 6, 154, 201, 205, 212–214, 400, 402–404, 447, 491, 498
heparin 6, 154, 201, 204, 206, 400, 402–404, 446, 569, 600, 603, 613
- labeled 405

heparinase 24, 402, 404, 500
hepatitis 477
hepatocyte 67, 68, 87, 154, 192, 400–402, 473, 477–479, 498
hepatoma 89, 91, 248
heterosis 499
Hevea brasiliensis lectin 420, 421, 425
(see also hevein)
hevein 425
β-hexosaminidase A 171, 178
- B 602

hippocampus 615
histocompatibility 214
histone 498
HL60 cells 98, 271
Hodgkin's disease 568
homology modeling 295–297
Hordeum vulgare lectin 420
hyaluronate synthase 202
hyaluronectin 208
hyaluronic acid 26, 202, 203, 278, 405, 613
hyaluronidase 142, 600
hybridoma technology 18, 180
hydrazinolysis 18
hyosophorin 123, 124
hypercholesterinemia 478
hypersialylation 100
hypodermis 553

I

I-cell disease 155
iduronate 2-sulfotransferase 207
iduronic acid 1, 135, 202, 206
iduronidase 155
immunoelectron microscopy 83
immunofluorescence 83, 255, 604

immunoglobulin 58, 148, 303, 312, 452, 465, 473, 605 (see also antibody)
immunogold labeling 116
immunomodulation 390, 489, 502
immunotoxin 180
induced fit 292
inflammation 8, 15, 95, 149, 154, 157, 158, 203, 476, 527
influenza virus 150, 156, 250–255, 352, 373, 384, 441, 499, 500
inositol-1,4,5-triphosphate 486, 487, 490, 492, 493
insect 94, 98
insulin 471, 491
integrin 6, 152, 174, 212, 284, 378, 452, 461, 491, 554
integument 553, 554
interferon-γ 133, 213, 450, 569, 613
interleukin
- 1 210, 237, 439, 441, 449, 491, 501
- 2 98, 158, 439, 445, 447–449, 491
- 6 383, 491, 501
- 8 213
- 10 213, 502
intestinum 92, 93, 156
iodoacetamide 364
ion transport 172
ionisation
- fast atom bombardment 22, 228
- liquid secondary ion 22
- matrix-assisted laser desorption 22, 228
ischemia 158, 181
isomer permutation 1, 8–11
isomorphous replacement 313

J

J-splitting 23
Jurkat cells 97

K

K562 cells 153
kala-azar 399
kallidinogenase 141
kanamycin 360
keratan sulfate 26, 201, 204, 207
keratinocyte 553, 569
kidney 85, 100, 125, 150, 212, 380, 461, 514, 534, 536, 538, 540, 552, 553
Klebsiella pneumoniae 157, 371, 374
Koenigs-Knorr procedure 34–40
Krabbe disease 178
Kupffer cells 154, 156, 264, 473, 477, 478

L

lacdiNAc 140
α-lactalbumin 81
lactase 350
lactosylceramide 165
lactotransferrin 141
laminin 137, 212, 213, 283, 439, 451, 461, 462, 586
Langmuir-Blodgett technique 282
laser nephelometry 424
laser photo CIDNP 306
latex coagulation 425
Lathyrus nissolia lectin 417
- *ochrus lectin* 299, 317, 355, 417, 418, 426
lectin 5, 6, 95, 148, 154, 157, 158, 173, 270, 370, 383, 386, 441, 445, 473, 500, 512, 540, 569
- animal 439–452 (see also galectin; selectin)
- bacterial 369–390 (see also adhesin)
- chemical mapping 345–352
- crystallography 296–298, 316–318, 345, 355, 463
- cytotoxic 88
- definition 369, 370, 415
- families 192
- frequency of citation 6
- glycosylation 152
- hepatic 56, 71, 154, 192, 439, 472–479, 493
- histochemistry 116–118, 547–558, 564–569
- labeling 565
- mosaic-like structure 125
- mutagenesis 355–366, 418, 419, 441, 460
- plant 416–429
- protozoan infection 400–409
- purification 67, 72, 365, 366, 370, 406, 422
- purification of glycoproteins 192–198
- recognition 6, 7
- resistant cells 142, 143
- role of water 300, 328, 329
- signaling 487–493 (see also adhesin; galectin)
- specificity determination 72
- splicing 440
- thermodynamics of ligand binding 291–295, 333–342
lectinophagocytosis 156, 157
leech 65
legumin 421, 424
Leishmania braziliensis 404, 405
- *enritii* 570
- *major* 404
- *spp.* 223, 234, 400
- *tropica* 404
lemurs 84
Lens culinaris lectin 20, 100, 194, 316, 416, 417, 548, 564, 566
Lepromonas samueli 141
leukemia 91
- acute lymphocytic 98
- acute myelogenous 98, 100

- chronic lymphocytic 98
- chronic myelogenous 98, 99, 521
- promyelocytic 214
leukocidin 387
leukodystrophy 178
leukosialin 98
leukotriene 486, 487
Lewis acid 43, 48
- blood groups 99, 176, 294
- - Lewis a 93–95
- - Lewis b 93, 94, 294, 319, 348, 378, 416
- - Lewis x 20, 23, 93, 150, 277, 278, 280–284, 540
- - Lewis x, sialyl 23, 46, 47, 50, 84, 93–95, 150, 269, 271, 440, 473, 540
- - Lewis y 93, 94, 284
Limax flavus lectin 117, 548
linkage analysis 17
lipid A 164
lipophosphoglycan 234, 400, 404, 405
lipopolysaccharide 164, 441, 499, 500
(see also endotoxin)
lipoprotein 475, 478
- lipase 213
liposome 279, 280, 284, 475, 477, 478
lipoxygenase 487
Listeria monocytogenes 570
- *ovata* lectin 421
liver 85, 86, 88, 97, 99, 125, 156, 264, 380, 406, 514
locus of crossing-over (*lox*) 508
Locusta migratoria 141
Lotonosis spinosa 427
Lotus tetragonolobus lectin 20, 192, 195, 548, 556, 564, 567
L-selectin 157, 158, 173, 213, 269, 270, 473, 491, 492
lumican 210
lung 86, 89, 150, 156, 380, 514, 538
luteinizing hormone 262–269, 272
Lycopersicon esculentum lectin 192, 195, 548, 557
Lymnaea stagnalis 141
lymphocyte 98, 154, 174–176, 263, 270, 382, 390, 424, 441, 443–445, 452, 476, 613
- B lymphocyte 383, 447
- T lymphocyte 148, 153, 180, 447, 462, 491, 500
lymphokine 501
lymphoma 254, 256, 279
lymphotoxin 449
lysoganglioside 65
lysosomal protein 7
lysosome 116, 121, 124, 155, 166
lysosome-associated membrane glycoprotein 465
lysozyme 357

M

Maackia amurensis lectin 195, 196, 548
Maclura pomifera lectin 420, 548, 554
macrophages 149, 154, 180, 404, 406
Madin-Darby canine cells 147, 156
magic bullet 519
Maillard reaction 22
malaria 237, 400, 401
malfolding 127
Malphigian corpuscle 552
maltose 336, 337
maltotriose 336, 337
Manduca sexta 376, 386
mannan 124, 125
(see also mannose(mannan)-binding protein)
mannitol 538
mannose(mannan)-binding protein 5, 20, 65, 157, 317, 355, 421, 439, 441, 473, 476, 498–500, 569, 599–601, 612
mannosidase I 80, 153, 229
- II 80, 81, 117
mannuronic acid 1
mass spectrometry 4, 16, 18, 21, 22, 25, 65, 278
- electrospray 22, 228
- plasma desorption 22
mast cell 214
melanoma 85, 86, 92, 141, 279, 282, 462, 567
meningioma 571
mercury salt 35
mesoderm 93
mesothelioma 571
metacyclogenesis 404
metastasis 87, 92, 93, 203, 285, 462, 465, 566, 569, 571
Methanothermus fervidus 136
methotrexate 476
methylprednisolone 538
Michael addition 68
microcalorimetry 19, 333–342
microembolus 286
microflora 379, 380
mistletoe (see *Viscum album*)
Mitsunobu reagent 41
molecular dynamics 300, 314
- mechanics 300, 346, 347
- modeling 6
monkey 84
Monte Carlo simulation 300
mucin 92, 97, 99, 137, 142, 144–148, 203, 270, 406, 408, 440
mucolipidosis II 155
Mucor 63
mucosa 92, 98
multiple sclerosis 443, 613
muraminic acid 36, 426
muramyl dipeptide 476
myasthenia gravis 613
Mycobacterium leprae 57
Mycoplasma gallisepticum 373

Mycoplasma pneumoniae 373
myelin 65, 173, 174, 272, 441, 443, 613
 – tubular 499
myelin-associated glycoprotein 249, 442, 445, 451
myocytes 556
myoglobin 556
myotube 461
Myxococcus xanthus 371

N

α-N-acetylglucosaminidase 155
β-N-acetylglucosaminidase 602, 607
N-acetylgalactosamine-4-sulfotransferase 206, 266, 267
N-acetylgalactosaminyltransferase 84, 97, 118, 266–268
N-acetylglucosamine phosphotransferase 155
N-acetylglucosaminyltransferase
 – β1,2- 80, 81, 89–91, 116, 117, 442, 511
 – β1,3- 97–99
 – β1,4- 79, 81, 90–92
 – β1,6- 81, 90–93, 98
 – mutations 82, 91, 92, 146
N-acetylneuraminic acid 15, 48, 51, 135, 142, 166, 598
 – synthesis 247, 248 (see also sialic acid)
N-bromosuccinimide 37
N-butyldeoxynojirimycin 157
N-deacetylase/N-sulfotransferase 206
N-glycanases 121–128
N-glycolylneuraminic acid 15, 32, 62, 135, 142, 146, 166, 246, 598
N-iodosuccinimide 37
Na^+/Ca^{2+}-exchanger 486
Na^+/H^+-exchanger 486, 489
nasal polyps 382
nectadrin 442, 443, 446
Neisseria gonorrhoeae 146
 – *meningitidis* 23, 138, 376
neoganglioprotein 65, 66, 174
neoglycoconjugate
 – application 71–73, 281, 502
 – definition 55
 – lectin characterization 197
 – ligand incorporation 68–70
neoglycoenzyme 73, 405
neoglycolipids 55, 56, 64, 65, 72
neoglycopeptide 476–478
neoglycopolymers 55, 66
 – polyacrylamide 66
 – polystyrene 47
neoglycoprotein
 – application 372, 443, 474, 476, 547, 550–558, 569–571
 – definition 55, 549
 – labeling 71, 405, 558
 – synthesis 58–64
neoproteoglycans 65
nephritogenoside 137
neural tube 91
neuraminidase (see sialidase)
neurite outgrowth 175
neuroblastoma 174, 175, 462
neurocan 209
neurofilament 175, 452
neuroimmunomodulator 614
neurotransmitter 485, 493, 611
neurotubule 452
neutrophil 6, 15, 156, 158, 382–384, 406, 441, 491, 492, 499, 502
nitric oxide synthase 487
nitrile effect 45
nitrogen fixation 139, 262, 384
NK cells 153, 214, 439, 441, 500, 527
NMR
 – chemical shift library 3
 – comparison to X-ray crystallography 302, 305, 312
 – conformational analysis 15, 19, 168
 – COSY 23
 – HPLC 25
 – imaging 524, 525
 – ligand affinity 334, 335
 – ROESY 23, 303
 – TOCSY 23
nodulation 427
NOE 3, 6, 23, 303–305, 311, 347
 – derived distances 303, 304
 – transferred NOE 204, 305, 324
Northern blot 602
Notophthalmus viridescens 555
nuclear magnetic relaxation dispersion 335, 338
nuclear magnetic resonance (see NMR)
nuclear Overhauser effect (see NOE)
nucleosome 492

O

octulosonic acid 48
oligodendrocytes 173, 444, 452
oligodendroglioma 567
oligonucleotides
 – anti-sense 476, 477
oligosaccharide
 – analysis 25, 301–304, 306
 – branching 7–11, 31, 56
 – cleavage 18
 – derivatives 4, 5, 25
 – library 9, 10
 – N-linked 24, 25, 70, 81–95, 139–143
 – O-acetylated 22
 – O-linked 19, 22, 96–100, 144–146
 – separation 24–26, 64, 65

– sulfated 261–272
– synthesis 4, 31–48
– transferase 122
oncogene 87, 100, 444, 446, 450, 451
– abl/bcr 446
– erbB2 446, 566
oncogenesis 136
Oncorhynchus mykiss 553
oozyte 122, 173, 514, 595
opsonin 497
opsonization 441
orosomucoid 474
Oryza sativa lectin 420
Oryzias latipes 122
ovalbumin 63, 127, 134, 136, 137, 262, 607
ovary 265, 268
oviduct 91
oxidative burst 499
oxygen radical 203
ozonolysis 18

P

papain 372, 376
papovavirus 254, 256
Paralichthys olivaceus 123
Paramecium primaurelia 235
paramyxovirus 607
Parkinson's disease 181
paroxysmal nocturnal hemoglobinuria 225, 238
parvovirus 381
Pasteurella multocida 373
peanut agglutinin 20, 100, 118, 548, 554, 564, 565, 567, 568, 571, 591–593
penetrin 403, 404
pentraxin 491, 498
peptide N-glycanase F 24
peptidoglycan 418
periodate oxidation 18, 64, 72, 153, 168
periodic acid-Schiff method 547
periodontitis 379
perlecan 211, 212
pertussis toxin 386, 493
phagocytes 154, 156, 451
phagocytosis 384
Phaseolus vulgaris lectin 20, 192, 194, 195, 421, 428, 548, 552, 568
phenol-sulfuric acid assay 15, 335
phorbol ester 444, 445, 491, 493
phosphate donor 32
phosphatidylinositol 137, 486, 500
phosphatidylinositol-4-kinase 486
phosphite method 48–50
phosphocholine 497, 499
phosphoenolpyruvate 247, 248
phosphoglucomutase 33, 145
phospholipase 142, 486, 487
– A_2 487, 501
– C 223, 226, 236, 489, 491, 492
phosphonate 11
phosphorylase 35
photoaffinity labeling 461
photosynthesis 172
phytohormone 424
pistil 7
Pisum sativum lectin 20, 100, 192, 194, 316, 355, 416, 417, 424, 548, 564
placenta 85, 88, 473, 501, 569
plasminogen 151
Plasmodium falciparum 223, 224, 230, 231, 233, 236, 399–403
– *malariae* 401
– *ovale* 401
– *vivax* 401
platelets 6, 98, 214, 269, 473
Plecoglossus altivelis 123
Pleurotus ostreatus 425
Pneumocystis carinii 157
P_o glycoprotein 262, 272, 442, 451
pokeweed mitogen 20
pollen tube 7
poly(N-acetyllactosamine) 24, 81, 96, 142, 146, 153, 461, 463, 465
polyagglutinability 97
polymerase chain reaction 358–362
polyoma virus 254, 373
polyphosphoinositides 485, 486, 491
polysialic acid 86, 89, 143, 248, 451, 615
polyspermia 595
Porphyromonas gingivalis 379
pregnancy 149
probes
– digoxigenylated 100
– fluorescent 57
– HRP-labeled 118
promoter 87, 100,
proopiomelanocortin 262, 263, 269
prosaposin 172
prostaglandin 491
protein A 312, 523
– body 421, 422, 424, 425
– kinase 174, 175, 485
– – mitogen-activated 489, 492
– – A 488
– – C 175, 444, 445, 486, 490, 491, 493
proteinuria 212
proteoglycan (see also glycosaminoglycan)
– definition 201
– families 208–214
– sugar composition 135
– synthesis 204–208
protostomes 497

P-selectin 19, 157, 158, 173, 269, 284, 440, 473, 569
– ligand 271, 373
pseudogene 84
Pseudomonas aeruginosa 157, 369
– lectin 370–389
Pseudomonas echinoides 370
Psophocarpus tetragonolobus lectin 294
pyelonephritis 380
pyocyanin 372, 386
pyrophosphatase 33
pyruvate dehydrogenase 491
pyruvate kinase 33

Q

quality control 127, 148

R

rainbow trout 146
Ramachandran plot 315
Raphanus sativus 128
reagent array analysis method 21, 25
receptor
– β-adrenergic 152
– FGF 152, 213, 440 446, 447
– IGF 153, 493
– interleukin-2 448, 449
– LDL 212
– mannose-6-phosphate 153, 154, 198, 407, 408, 476, 492, 493
– muscarinic acetylcholine 152
– somatostatin 250
recombinase *(cre/lox)* 509, 510
reductive amination 58, 59, 61, 64, 70
Reed-Sternberg cell 567
regioselectivity 31, 32
Reichert's membrane 212
renal medulla 156
reptiles 84
restriction endonuclease 359, 360
R-factor 314
rhamnose 15, 36, 136, 137, 143, 164, 320–325, 327, 328, 374, 376, 549
rheumatoid arthritis 149, 203, 523
Rhizobium 1, 154, 383, 426, 427
ribonuclease A 151, 336
– B 25, 151, 499
– U 138
ribose 1
ribosome-inactivating protein 419, 429
ricin 346, 347, 349, 355, 357, 429, 492
Ricinus communis lectin 20, 117, 152, 192, 195, 196, 346–348, 419, 421, 424, 429, 554, 564, 567, 590, 591
ring size 1
Robinia pseudoacacia lectin 422, 445
root ganglia 173
rRNA N-glycosidase 429

S

Saccharomyces cerevisiae 148, 235
Salmonella typhimurium 371, 374
salvage compartment 116
Sambucus nigra lectin 20, 195, 196, 422, 548
Sandhoff disease 178
β-sandwich structure 296
Saphora japonica lectin 422
saposin 171, 172
sarcolectin 501, 569
sarcolemma 556
sarcoplasma 556
scanning tunneling microscopy 375
Scatchard analysis 124, 375
scatter factor 213
scavenger 501
Schiff base 22, 58, 153
Schistosoma mansoni 145
schizont 401
Schwann cell 444–446, 452
Se locus 93
Secale cereale lectin 420
selectins 5, 11, 18, 84, 158, 269, 284, 386, 440, 465, 500, 512, 540
(see also E-, L-, and P-selectin)
Sendai virus 100, 373
senescence 128
serglycin 204
serum amyloid P component 134, 250, 497, 498
SH2-domain 449, 489
SH3-domain 489
Shigella dysenteriae 369, 373
– *flexnerii* 157, 450
sialate O-acetylesterase 248, 511
sialic acid 1, 15, 84, 88, 94, 99, 117, 135, 146, 153, 157, 196, 204, 245–255, 280, 304, 372, 401, 403, 405, 420, 473, 498, 508, 510, 512, 549
sialidase 171, 248, 251, 253, 352, 372, 385, 403, 408, 566, 572
sialoadhesin 18, 20
sialylmotif 87–89, 99
sialyltransferase 62, 79, 96, 248, 267, 540
– α2,3 61, 63, 84–86, 88, 99–100
– α2,6 61, 84–88, 99, 116, 117, 197
– α2,8 86, 88
sickle cell 499
signal peptidase 232
– sequence 232
Silene alba 128
silylation 41
simian virus 40 255

site-directed mutagenesis 355–366, 418, 419, 441, 460
slime mold 136
smooth muscle cell 6
Solanum tuberosum lectin 192, 419, 423
somatostatin 250
son of sevenless (sos) 489
sorting signal 147
– di-lysine 117
– KDEL 117
soybean lectin 318, 335, 373, 421, 424, 554, 556, 591
sperm 155, 166, 173, 379, 514, 595, 596, 601–607
spermadhesin 599, 600, 603–605
spermatid 83
sphingolipid 66
sphingomyelinase 228
sphingosine 65, 165, 171, 175
spinal cord 180
spiro-diazirine 41
spleen 89, 125, 297, 298, 380
splice variant 208
Staphylococcus aureus 369, 386, 405, 558
starfish 136, 141
stereocenter 32
stomach 93
storage protein 421, 423, 424, 427
streptavidin 63, 197, 281, 552
Streptococcus mutans 371
– *sanguis* 156
– *avidinii* 552
streptomycin 519, 520
streptozotocin 586, 587
substantia nigra 181
sulfatide 166, 402
sulfonate 11
sulfonylation 41
sulfoquinovosyldiacylglycerol 164, 172
superoxide anion 491, 500, 502
– dismutase 213, 477
surface plasmon resonance 19
surfactant proteins 441, 500
Sushi domain 602
synapse formation 440
syndecan 154, 204, 205, 212, 213
syndrome
– carbohydrate-deficient glycoprotein 91
– Ehlers-Danlos 205
– Guillain-Barré 179
– leukocyte adhesion deficiency type II 500
– Wiscott-Aldrich immunodeficiency 98

T

taglin 407
Tamm-Horsfall glycoprotein 262
TATA box 83
Tay-Sachs disease 178
T-cell activation 177, 445
– receptor 444, 445, 448, 449
telencephalin 615
teratocarcinoma 93, 98, 279
tetanus toxin 250
Tetrahymena pyriformis 382
thiocarbamylation 60
thioglycosides 39, 45, 51, 59, 68
– cyanomethyl derivative 59
thrombocytes 370
thrombomodulin 215
thrombospondin 213
thylakoid sacs 172
thymocytes 175
thymus 125
thyroglobulin 136, 137, 141, 262, 272
thyroid-stimulating hormone 264, 266, 269
thyrotropin 264
titration microcalorimetry 333–342
Toxoplasma gondii 223, 230, 234
trachea 98
trans-sialidase 62, 403
transcription factor 83, 87, 461
transfection 92, 100
transferrin 149, 471
transforming growth factor
– β 210, 211, 213
– γ 462
transglutaminase 63
transglycosylase 62
transglycosylation 39, 62–64, 65
transition-state stabilization 351
transmission electron microscopy 382
transpeptidation 418
travellers' diarrhea 139
trehalose 8, 164
Tribolodon hakomensis 123
Trichinella spiralis 137, 141
trichloroacetimidate activation 34, 41–48
Trichomonas vaginalis 385
trimethylsilyl ethers 16
Triticum vulgaris lectin (see wheat germ agglutinin)
tRNA 202
trophoblast 265
Trypanosoma brucei 142, 223, 224, 229, 230, 233, 239
– *congolense* 231
– *cruzi* 62, 136, 250, 399, 400, 402, 403, 557, 570
trypsin inhibitor 133
tumor necrosis factor 158, 237, 439, 449, 450, 501, 502, 614
tumor suppressor 211
tunicamycin 149, 150, 153, 451
tyrosine kinase 174, 236, 237, 444, 446, 489, 491, 492, 600
tyvelose 137

U

ubiquitin 125
Ulex europaeus lectin 20, 195, 374, 416, 548, 554–556, 565–567, 571, 597
uracil glycosidase 360
urokinase 262, 263
uromodulin 449, 450
Urtica dioica lectin 421, 426

V

vaccine 61, 390, 401, 557
vacuolization 382
van der Waals interaction 148, 294, 297, 299, 300, 322, 333, 345, 357, 358, 366
van't Hoff method 333
vapour diffusion 313
vector 361
- M13 358, 359
- plasmid 359, 360

venom 141, 150
Vero cells 254
versican 208
Vibrio cholerae 139, 369, 371, 385
Vicia cracca 416
- *faba* 417, 418, 421
- *villosa* lectin 118, 564, 568

vicilin 421, 424
virulence factor 151, 179, 387
visceral yolk 592
Viscum album lectin 304, 305, 420, 429, 492, 501, 502
vitellogenin 123

W

wallaby 84
wheat germ agglutinin 20, 152, 196, 318, 355, 420, 426, 548, 554, 556, 564–566, 591
whooping cough 139
Wilms' tumor 143
Wistaria floribunda lectin 20, 195
Wucheria bancrofti 558

X

X-ray crystallography
- antibody-sugar complexes 316, 319–329
- comparison to NMR 302, 305, 312
- data bank 295, 296, 315
- diffraction data 313
- general 19, 21, 23, 148
- immunoglobulin 149
- lectin-sugar complexes 296–298, 316–318, 345, 355, 418
- precision of model 315

X-ray diffraction 169
Xenopus laevis 365, 555
xenotransplantation 15, 84, 531–542
xenozoonoses 542
xylonic acid 65
xylose 1, 15, 65, 128, 135, 139, 142, 150, 204, 205, 549, 553
xylosyltransferase 205, 207

Y

Yersinia pseudotuberculosis 378

Z

zona pellucida 83, 595–607
- ZP1 596
- ZP2 596
- ZP3 83, 84, 156, 513, 514, 596–599, 602, 606

zygote 508
zymosan 281, 283, 286, 287